Toxins and Hemostasis

R. Manjunatha Kini · Kenneth J. Clemetson ·
Francis S. Markland · Mary Ann McLane ·
Takashi Morita
Editors

Toxins and Hemostasis

From Bench to Bedside

Springer

Editors
Dr. R. Manjunatha Kini
National University of Singapore
Department of Biological Sciences
14, Science Drive 4
Singapore 117543
Singapore
dbskinim@nus.edu.sg

Dr. Kenneth J Clemetson
University of Berne
Department of Haematology
Inselspital
CH-3010 Bern
Switzerland
clemetson@tki.unibe.ch

Dr. Francis S. Markland
University of Southern California
Biochemistry and Molecular Biology
Keck School of Medicine
Health Sciences Campus/9270
Los Angeles, California 90089
USA
markland@usc.edu

Dr. Mary Ann McLane
University of Delaware
Medical Technology
305G Willard Hall Educ Bld
Newark Delaware 19716
USA
mclane@udel.edu

Takashi Morita
Meiji Pharmaceutical University
Biochemistry
Nishio Kiyose 2-522-1
Tokyo 204-8588
Japan
tmorita@my-pharm.ac.jp

ISBN 978-90-481-9294-6 e-ISBN 978-90-481-9295-3
DOI 10.1007/978-90-481-9295-3
Springer Dordrecht Heidelberg London New York

Library of Congress Control Number: 2010933098

Printed on acid-free paper

Springer is part of Springer Science+Business Media (www.springer.com)

Preface

Those who are inspired by a model other than Nature, a mistress above all masters, are laboring in vain.

– Leonardo da Vinci

Nature is wonderful; organisms, whether plants, animals or microbes, have evolved through competition (including predator-prey and host-parasite competitions) and natural selection. In this long drawn out process, they have developed, been fine-tuned and improvised a number of strategies to stay ahead of the competition. Human beings have been fascinated how living organisms have solved very complicated problems by simple innovative methods (For details, see http://asknature.org/). These clever strategies have inspired a number of new technologies, such as Velcro fasteners, Gecko adhesive tapes without glue, photosynthesis-inspired fuel cells, and insect-inspired autonomous robots (For details, see http://www.biomimicryinstitute.org/).

A number of pharmaceutical drugs were designed, based on natural compounds. A classical example is the development of antimalarial drugs from the alkaloid quinine isolated from the barks of *Cinchona* trees. Anecdotal evidence for the antimalarial properties of *Cinchona* dates back to seventeenth century although the indigenous populations used it even earlier. Since then several hundreds of drugs were designed and developed based on plant secondary metabolites and substances isolated from animals, microbes and fungi. Despite our computational and synthetic organic chemistry capabilities, nature has inspired the design of new chemical entities. Epibatidine – an alkaloid isolated from a frog (*Epipedobates tricolor*), is a potent analgesic (200 times as effective as morphine). Based on this nicotinic acetylcholine receptor antagonist, a potent pain-killer ABT-594 (Tebanicline) was designed and developed. Thus nature provided a large number of templates for the development of therapeutic agents.

Over the last two decades, protein biotherapeutics have gained popularity and are fast becoming the major part of the pharmaceutical market. The high affinity, high specificity binding to target receptors/ion channels and the consequent low toxicity and side-effects profile of protein drugs are extremely attractive as lead molecules in drug development. Endogenous proteins (such as factor VIIa, erythropoietin and insulin) as well as humanized or chimeric monoclonal antibodies to targeted specific antigens (such as tumor necrosis factor α, vascular endothelial growth factor, and cytokines) are some of the successful protein therapeutics. In addition, we also use exogenous proteins as potential prototypes in the development of protein biotherapeutics. These exogenous factors have evolved for millions of years as a part of predator-prey and host-parasite competitions. They target specific and critical physiological processes that play crucial roles in the survival of the organism. The aim in editing this book is to bring out the complexities of structure-function relationships and mechanisms of exogenous proteins with crucial roles in cardiovascular and hematological disorders and recent progress in understanding them. Such exogenous factors are found in venoms of snakes, scorpions, spiders and other venomous animals, saliva of hematophagous animals as well as in many microorganisms. The 43 review chapters are written by leading experts in their field from 22 different countries. The book focuses on various aspects ranging from modern approaches to the analysis of venom/saliva contents and identification for new active components to the wealth and diversity of structures and mechanisms of these proteins. It also highlights the development of these proteins as (a) novel therapeutic agents for the treatment/prevention of cardiovascular and cerebrovascular diseases as well as cancer; and (b) diagnostic agents for the identification of a number of hereditary defects and other hematological disorders.

This book is a product of concerted team effort of all members of the Registry of Exogenous Factors Affecting Thrombosis and Haemostasis, a subcommittee of the Scientific Standardization Committee of the International Society of Thrombosis and Haemostasis as well as some of members of the International Society of Toxinology. The primary intent in producing this book is to elevate awareness and enthusiasm in the field of exogenous factors. We hope that it will provide greater impetus to the search for novel proteins based on naturally occurring exogenous factors. The book should also remind us of the importance of preservation of diversity of species, even those that are often considered dangerous and nuisance organisms, to allow access in the future to venoms, salivary gland extracts and other tissues, which could permit new discoveries and development of novel therapeutics for a number of life-threatening diseases. It celebrates the role of biotechnology and advances in newer technologies that are faster and more sensitive thus the way to many new discoveries. As the first book dealing extensively with exogenous factors in the last 25 years it should provide a modern, easily accessible reference to the different approaches being used to solve complex, contemporary problems in protein chemistry and pharmacology of exogenous factors. It also updates our understanding of the structure-function relationships and mechanisms of action of exogenous factors and provides insights into future directions for solving many remaining challenges. Thus this book helps to foster a wider interest in isolation and characterization of novel proteins, to entice new, talented researchers into this field, and to generate more enthusiasm in the field.

Singapore, Singapore — R. Manjunatha Kini
Berne, Switzerland — Kenneth J. Clemetson
Los Angeles, California — Francis S. Markland
Wilmongton, Delaware — Mary Ann McLane
Tokyo, Japan — Takashi Morita

Acknowledgments

In 1985 the first meeting of scientists working on snake venom proteins which affected thrombosis and hemostasis was organized as the "Symposium on Animal Venoms and Hemostasis (July 20–21, 1985)" in conjunction with the Xth Congress of the International Society of Thrombosis and Haemostasis. The organizing committee included Hubert Pirkle, Francis S. Markland, Patrick Gaffney, Zbigniew Latallo and Neville Marsh. Marcel Dekker, Inc. published the proceedings of this meeting as a book "Hemostasis and Animal venoms" edited by Hubert Pirkle and Francis Markland. No meeting was organized for the next 15 years. However, several years ago, as a Co-Chairman of the Registry of Exogenous Hemostatic Factors, a subcommittee of the International Society on Thrombosis and Haemostasis (ISTH), I proposed that the committee should consider organizing an international conference focused on the exogenous factors. The then chairman, Neville Marsh, and other co-chairmen, Francis Markland and Jan Rosing recognized the need for such international conferences. The committee unanimously agreed to organize international conferences every 4 years after the Congress of the ISTH. When I was the Chairman of the subcommittee, I discussed this issue with Ian Peake and Gilbert White, the President and Executive Director of ISTH, respectively, and they strongly supported the idea and encouraged us. As a result, the Second International Conference on Exogenous Factors Affecting Thrombosis and Hemostasis was held at the Pasteur Institute, Paris, France in 2001 under the leadership of Cassian Bon. The Third Conference was held at the University of Technology Sydney, Australia in 2005 under the leaderships of Kevin Broady and Neville Marsh. The proceedings of both these conferences were published as a special issue of Haemostasis, and Pathophysiology of Haemostasis and Thrombosis, respectively. As the Fourth Conference coincided with the 25th year of the first meeting, we decided to publish a book highlighting the progress made in the last 25 years. The Fourth Conference was held at the University of Massachusetts, Boston, USA under the leadership of MaryAnn McLane and the main content of this book is the proceedings of the meeting. To have a complete coverage, we also approached some of the leading scientists in the field. They all readily accepted our invitation to contribute to this book.

Firstly, I would like to thank all the Chairmen, Co-Chairmen and members of our subcommittee, past and present, for their continuous and unconditional support and encouragement. It was this true spirit that made it possible to organize the

International conferences every 4 years. I am especially indebted to the strong support by Neville Marsh, Francis Markland, Jan Rosing, Kenneth Clemetson, Takashi Morita and MaryAnn McLane. Secondly, I thank Gilbert White and Ian Peake along with Cathy Cole, ISTH Executive Secretary, for their support and encouragement during the initial stages. I thank Dr. Ryan McCleary for his assistance during proof reading of the book. I also thank all the subsequent Presidents of ISTH Congresses (Dominique Meyer, Colin Chesterman, Barbara Furie and Bruce Furie) for their support. Along with the other editors I thank all the contributors to this book for their timely and very professional submission of the chapters. Without their efforts, this book would not have been possible. I appreciate all their time and efforts.

I had approached several publishers to ask if they were interested in publishing this book. All the publishers I talked to were very supportive of the project and wanted to publish the book. But Thijs van Vlijmen, Springer Science (Biomedical Unit | Life Sciences) approached this project very efficiently and was the first one to come up with all the necessary paperwork. Throughout the project, he was extremely supportive. I appreciate his energy and enthusiasm, and his confidence in us. I also thank the Springer staff, Andre Tournois (Production) and Nina Wilhelm (Marketing) for their efforts. I appreciate the efforts of the Integra staff, Leah Georgina for timely typesetting and the fantastic layout of the book. I thank Anne Nee Yong for selecting the exciting picture and eStudio Calamar S.L. for the beautiful cover design.

I am especially grateful to my wife, Nobue, who is ever so supportive, tolerant and considerate. Finally, I thank my family, friends and teachers for their confidence and constant moral support.

Contents

Contributors

John F. Andersen Section of Vector Biology, Laboratory of Malaria and Vector Research, National Institute of Allergy and Infectious Diseases, National Institutes of Health, Rockville, MD 20852-8132, USA

Avner Bdolah Department of Zoology, George S. Wise Faculty of Life Sciences, Tel Aviv University, Ramat Aviv, Tel Aviv, 69978 Israel

Paul E. Bock Department of Pathology, Vanderbilt University School of Medicine, Nashville, TN 37232-2561, USA

Bernard F. Le Bonniec INSERM U765, Université Paris Descartes, 4 Av. de l'Observatoire, 75006 Paris Cedex, France

Juan J. Calvete Laboratorio de Proteinómica Estructural, Instituto de Biomedicina de Valencia, CSIC, Jaime Roig 11, 46010 Valencia, Spain

Esteban Cantu Natural Toxins Research Center, College of Arts and Sciences, Texas A&M University-Kingsville, Kingsville, TX, USA

D.E. Champagne Department of Entomology, and Center for Tropical and Emerging Global Diseases, University of Georgia, Athens, GA 30602-2603, USA

Linda Christian Carrijo-Carvalho Laboratory of Biochemistry and Biophysics, Instituto Butantan, São Paulo, SP, Brazil

Jindřich Chmelař Laboratory of Genomics and Proteomics of Disease Vectors, Institute of Parasitology, Biology Centre, ASCR v.v.i., 37005 Ceske Budejovice, Czech Republic

Ana Marisa Chudzinski-Tavassi Laboratory of Biochemistry and Biophysics, Instituto Butantan, São Paulo, SP, Brazil

Ching-Hu Chung Institute of Pharmacology and Toxicology, Tzu Chi University, Hualien, Taiwan

Kenneth J. Clemetson Department of Haematology, University of Berne, Inselspital, CH-3010 Berne, Switzerland

R. Conde Departamento de Medicina Molecular y Bioprocesos, Instituto de Biotecnología, Universidad Nacional Autónoma de México, Avenida Universidad 2001, Col. Chamilpa, Cuernavaca Morelos, 62210, México

Fritz Costa USC/Norris Comprehensive Cancer Center, University of Southern California, Keck School of Medicine, Cancer Research Laboratory #106, 1303 N. Mission Road, Los Angeles, CA 90033, USA

Dermot Cox Molecular and Cellular Therapeutics, Royal College of Surgeons in Ireland, Dublin, Ireland

Grazyna Faure Institut Pasteur, Unité d'Immunologie Structurale; CNRS, URA 2185, Département de Biologie Structurale et Chimie, F-75015, Paris, France

Sérgio Henrique Ferreira Department of Pharmacology, University of São Paulo, Av. dos Bandeirantes, 3900, Campus USP, 14049-900, Ribeirão Preto, SP, Brazil

Jay W. Fox Department of Microbiology, University of Virginia School of Medicine, Charlottesville, VA 22908, USA

I.M.B. Francischetti Section of Vector Biology, Laboratory of Malaria and Vector Research, National Institute of Allergy and Infectious Diseases, National Institutes of Health, Rockville, MD 20852-8132, USA

Pablo Fuentes-Prior Institut de Recerca, Hospital de la Santa Creu i Sant Pau, 08025 Barcelona, Spain

Belsy Guerrero Laboratorio de Fisiopatología, Centro de Medicina Experimental, Instituto Venezolano de Investigaciones Científicas (IVIC), Caracas, República Bolivariana de Venezuela

Corey Helchowski USC/Norris Comprehensive Cancer Center, University of Southern California, Keck School of Medicine, Cancer Research Laboratory #106, 1303 N. Mission Road, Los Angeles, CA 90033, USA

Chun-Chieh Hsu Institute of Pharmacology, College of Medicine, National Taiwan University, Taipei, Taiwan

Tur-Fu Huang Institute of Pharmacology, College of Medicine, National Taiwan University, Taipei, Taiwan

Mária Kazimírová Institute of Zoology, Slovak Academy of Sciences, Dúbravská Cesta 9, SK-84506, Bratislava, Slovakia

Steven Kerrigan Molecular and Cellular Therapeutics, Royal College of Surgeons in Ireland, Dublin, Ireland

R. Manjunatha Kini Protein Science Laboratory, Department of Biological Sciences, National University of Singapore, Singapore, 117543, Singapore; Department of Biochemistry and Molecular Biophysics, Medical College of Virginia, Virginia Commonwealth University, Richmond, VA 13298, USA

Karsten Knobloch Plastic, Hand and Reconstructive Surgery, Hannover Medical School, Hannover, Germany

Cho Yeow Koh Protein Science Laboratory, Department of Biological Sciences, National University of Singapore, Singapore 117543, Singapore

Michalis Kotsyfakis Laboratory of Genomics and Proteomics of Disease Vectors, Institute of Parasitology, Biology Centre, ASCR v.v.i., 37005 Ceske Budejovice, Czech Republic

Heather K. Kroh Department of Pathology, Vanderbilt University School of Medicine, Nashville, TN 37232-2561, USA

Ren Lai Kunming Institute of Zoology, Chinese Academy of Sciences, Kunming 650023, Yunnan, China

Julie Libraire INSERM U765, Université Paris Descartes, 4 Av. de l'Observatoire, 75006 Paris Cedex, France

Cunbao Liu Kunming Institute of Zoology, Chinese Academy of Sciences, Kunming 650023, Yunnan, China

Stephen P. Mackessy School of Biological Sciences, University of Northern Colorado, 501 20th St., CB 92, Greeley, CO 80639-0017, USA

B.J. Mans Parasites, Vectors and Vector-Borne Diseases, Onderstepoort Veterinary Institute, Pretoria, South Africa and the Department of Veterinary Tropical Diseases, University of Pretoria, Pretoria, South Africa

Francis S. Markland Department of Biochemistry and Molecular Biology, University of Southern California, Keck School of Medicine, Cancer Research Laboratory #106, 1303 N. Mission Road, Los Angeles, CA 90033, USA

Radu Minea USC/Norris Comprehensive Cancer Center, University of Southern California, Keck School of Medicine, Cancer Research Laboratory #106, 1303 N. Mission Road, Los Angeles, CA 90033, USA

Robson Q. Monteiro Instituto de Bioquímica Médica, Universidade Federal do Rio de Janeiro, Rio de Janeiro, RJ, Brazil

Takashi Morita Department of host defense and biological Research, Juntendo University School of Medicine, Tokyo 113-8421, Japan

Ana Cristina Ferraz Nogueira Instituto de Bioquímica Medica, Centro de Ciências da Saúde, UFRJ, Rio de Janeiro, Brazil

Carlo José Freire de Oliveira Section of Vector Biology, Laboratory of Malaria and Vector Research, National Institute of Allergy and Infectious Diseases, National Institutes of Health, Rockville, MD 20852-8132, USA; Department of Biochemistry and Immunology, School of Medicine of Ribeirão Preto, University of São Paulo, Ribeirão Preto, SP 14049-900, Brazil

Peter Panizzi Center for Systems Biology, Massachusetts General Hospital, Boston, MA 02114, USA

Anna Maria Perchuc August Lenz Foundation at the Institute for Prevention of Cardiovascular Diseases, University of Munich, Munich, Germany; Swiss Tropical Institute, MPI, Basel, Switzerland

Karl-Uwe Petersen PAION Deutschland GmbH, Martinstrasse 10–12, 52062 Aachen, Germany

Lourival D. Possani Departamento de Medicina Molecular y Bioprocesos, Instituto de Biotecnología, Universidad Nacional Autónoma de México, Avenida Universidad 2001, Col. Chamilpa, Cuernavaca Morelos, 62210, México

R. Rajesh Department of Studies in Biochemistry, University of Mysore, Manasagangothri, Mysore 570 006, Karnataka, India; Department of Microbiology and Immunology, University of Maryland, Baltimore, MD 21201, USA

Md. Abu Reza Department of Biological Sciences, National University of Singapore, Singapore 117 543 Singapore; Department of Genetic Engineering and Biotechnology, University of Rajshahi, Rajshahi 6205, Bangladesh

Alexis Rodríguez-Acosta Immunochemistry Section, Tropical Medicine Institute of the Universidad Central de Venezuela, Caracas, República Bolivariana de Venezuela

Ponlapat Rojnuckarin Division of Hematology, Department of Medicine, King Chulalongkorn Memorial Hospital, Rama IV Rd, Patumwan, Bangkok 10330, Thailand

Jan Rosing Department of Biochemistry, Cardiovascular Research Institute Maastricht (CARIM), Maastricht University, P.O. Box 616, Maastricht, 6200 MD, The Netherlands

Elda E. Sánchez Natural Toxins Research Center, College of Arts and Sciences, Texas A&M University-Kingsville, Kingsville, TX, USA

Anderson Sá-Nunes Laboratory of Experimental Immunology, Department of Immunology, Institute of Biomedical Sciences, University of São Paulo, São Paulo, SP 05508 900, Brazil

Frederick Saul Institut Pasteur, Unité d'Immunologie Structurale; CNRS, URA 2185, Département de Biologie Structurale et Chimie, F-75015, Paris, France

H. V. Shivaprasad Department of Studies in Biochemistry, University of Mysore, Manasagangothri, Mysore 570 006, Karnataka, India

Ene Siigur National Institute of Chemical Physics and Biophysics, Tallinn 12618, Estonia

Jüri Siigur Laboratory of Bioorganic Chemistry, National Institute of Chemical Physics and Biophysics, Akadeemia tee 23, Tallinn 12618, Estonia

Mariola Söhngen PAION Deutschland GmbH, Martinstrasse 10–12, 52062 Aachen, Germany

Wolfgang Söhngen PAION Deutschland GmbH, Martinstrasse 10–12, 52062 Aachen, Germany

Stephen Swenson USC/Norris Comprehensive Cancer Center, University of Southern California, Keck School of Medicine, Cancer Research Laboratory #106, 1303 N. Mission Road, Los Angeles, CA 90033, USA

P. Takáč Institute of Zoology, Slovak Academy of Sciences, Bratislava, Slovakia

Soichi Takeda National Cardiovascular Center Research Institute, Suita, Osaka 565-8565, Japan

Guido Tans Department of Biochemistry, Cardiovascular Research Institute Maastricht (CARIM), Maastricht University, P.O. Box 616, Maastricht, 6200 MD, The Netherlands

H. Tsujimoto Department of Entomology, and Center for Tropical and Emerging Global Diseases, University of Georgia, Athens, GA 30602-2603, USA

N. A. Valdez-Cruz Departamento de Medicina Molecular y Bioprocesos, Instituto de Biotecnología, Universidad Nacional Autónoma de México, Avenida Universidad 2001, Col. Chamilpa, Cuernavaca Morelos, 62210, México

B.S. Vishwanath Department of Studies in Biochemistry, University of Mysore, Manasagangothri, Mysore 570 006, Karnataka, India

Marianne Wilmer Department DXSC, Roche Diagnostics Ltd., Rotkreuz, Switzerland

Haijin Xu Institut Pasteur, Unité d'Immunologie Structurale; CNRS, URA 2185, Département de Biologie Structurale et Chimie, F-75015, Paris, France

M. Yariswamy Department of Studies in Biochemistry, University of Mysore, Manasagangothri, Mysore 570 006, Karnataka, India

F. Z. Zamudio Departamento de Medicina Molecular y Bioprocesos, Instituto de Biotecnología, Universidad Nacional Autónoma de México, Avenida Universidad 2001, Col. Chamilpa, Cuernavaca Morelos, 62210, México

Samuel Zidovetzki USC/Norris Comprehensive Cancer Center, University of Southern California, Keck School of Medicine, Cancer Research Laboratory #106, 1303 N. Mission Road, Los Angeles, CA 90033, USA

Russolina B. Zingali Instituto de Bioquímica Medica, Centro de Ciências da Saúde, UFRJ, Rio de Janeiro, Brazil

About the Editors

R. Manjunatha Kini Dr. Kini graduated from University of Mysore, India in 1983 and did his postdoc in Kyushu University, Japan and Virginia Commonwealth University, Richmond, U.S.A. Then he joined National University of Singapore, Singapore in 1994. Currently, he is a Professor in the Department of Biological Sciences, National University of Singapore. He also holds an Affiliate Professor position in the Department of Biochemistry at Medical College of Virginia, Virginia Commonwealth University, Richmond. He is the founder and Chief Scientific Officer of a small start-up company ProTherapeutics Private Limited, Singapore, which engineers orally and sublingually active therapeutic peptides. He has been the Chairman or Co-Chairman of the Registry of Exogenous Hemostatic Factors, a sub-committee of the International Society on Thrombosis and Haemostasis since 1998. He was also a member of the Council of the International Society on Toxinology. Currently, he is a nominated Member of the Nomenclature Committee of the International Society on Toxinology. Dr. Kini is on the editorial board of several international scientific journals. He has published 148 original articles, 29 reviews, 15 book chapters and 31 patent applications. He edited a monograph on Snake venom phospholipases and was guest editor for three special issues. His research interests include structure-function relationships and mechanism of action of proteins, particularly snake venom toxins. The functional sites are used in designing proteins with novel biological activities and prototypes of therapeutic agents.

Kenneth J. Clemetson Dr. Clemetson graduated from University of Cambridge 1968 and completed postdocs at the University of Alberta, Canada and the University of California at Santa Barbara. He was awarded Sc.D. by University of Cambridge in 2007. He was elected as a Fellow of the Royal Society of Chemistry. He is Professor Emeritus of biochemistry in the Department of Medicine at the University of Berne, Switzerland. Currently, he is working in the Department of Haematology. His research interests include structure-function relationships and mechanism of action and signalling of blood platelet receptors. Snake venom toxins are interesting and novel tools to investigate platelet receptor function. They also provide lead structures for the development of pharmacologically active drugs for treatment of disorders involving platelets such as infarcts and stroke. He was a member of Council of the International Society on Thrombosis and Haemostasis and President of the European Thrombosis Research Organization. He has also been chairman and co-chairman of the Platelet Physiology subcommittee and am currently co-chairman of the Registry of Exogenous Hemostatic Factors subcommittee of the International Society on Thrombosis and Haemostasis. Now, he is on the editorial board of several international scientific journals in the platelet field. He has published over 230 original articles, reviews, and book chapters and has several patents.

Francis S. Markland, Jr. Dr. Markland He received his Ph.D. in Physiological Chemistry from Johns Hopkins University in 1964, and completed a two year NIH postdoctoral fellowship in the Biological Chemistry Department, UCLA, School of Medicine. He was appointed assistant professor at UCLA in 1966 and received an NIH Career Development Award from 1968 to 1973. In 1974 he moved to the USC School of Medicine as Associate Professor of Biochemistry and a member of the USC Cancer Center. In 1983 he was appointed Professor of Biochemistry. Currently, he is Professor of Biochemistry and Molecular Biology, University of Southern California (USC), Keck School of Medicine (KSOM). He served as Acting Chair of the Biochemistry Department (1986–1988) and Associate Dean for Scientific Affairs at KSOM (2004–2008). He has served on numerous NIH study sections. Dr. Markland is an internationally recognized expert on proteins from snake venoms with potential clinical application; a metalloproteinase he purified from southern copperhead venom recently advanced to Phase III clinical trials as therapy for peripheral arterial occlusive disease. He is presently working with novel adhesive peptides with antiangiogenic activity. His laboratory recently succeeded in using an engineered bacterial strain to produce a recombinant version of the venom peptide. He has numerous patents related to his technologies.

Mary Ann McLane Dr. McLane pursued a doctorate in Physiology from Temple University, Philadelphia PA in 1993 and completed a molecular biology postdoc in the Thrombosis Research Center at Temple University in 1996. She is a tenured full Professor in the Department of Medical Technology, University of Delaware (Newark DE, USA) with a joint appointment in the Department of Biological Science. She is a certified medical laboratory scientist specializing in clinical biochemistry, and is the 2009–2010 President of the American Society for Clinical Laboratory Science. She was named national Laboratorian of the Year in 1999, and received a Lindback Teaching Award in 2002. Dr. McLane has been the Chairman of the Registry of Exogenous Hemostatic Factors, a sub-committee of the International Society on Thrombosis and Haemostasis, and organized the 4th International Conference in Boston MA, USA in July 2009 from which came the manuscripts published in this collection. Her research focuses on the anti-cancer activity of the disintegrin eristostatin from the viper *Eristocophis macmahoni*.

Takashi Morita Dr. Morita was born in Osaka, Japan in 1943, and graduated from Osaka University in 1967. After four years of working at pharmaceutical company, he finished his Ph. D. in 1977 at Osaka University. After two years of postdoctoral working in Washington University School of Medicine, St Louis, USA, he returned to Japan to be an assistant professor of Kyushu University, Faculty of Science in 1979. He has been a Professor of Biochemistry, Meiji Pharmaceutical University, Tokyo, between 1986 and 2009. He is now Professor Emeritus at Meiji Pharmaceutical University, and is a Visiting Professor of Host Defense and Biological Research, Juntendo University School of Medicine, Tokyo. His research interests include structure and function of Vitamin K-dependent coagulation factors and snake venom proteins such as VEGF and disintegrins.

Part I
From Bench to Bedside

Chapter 1
Introduction

Kenneth J. Clemetson and R. Manjunatha Kini

Abstract Haemostasis is critical to normal health by preventing blood loss following injury and contributing to the maintenance of the vasculatory system, which supplies oxygen and nutrients to all parts of the body as well as performing other essential tasks. Thrombosis is the pathological variant leading to blockage of vessels, cutting off the blood supply to vital organs. Because of these important functions there is great interest in the development of new drugs that can prevent or cure thrombosis. Since platelet malfunction is a major contributor to thrombus formation and platelet inhibition reduces risk it is a prime target for new drug development. Many animals have developed naturally proteins or peptides that affect mammalian haemostasis. They use these mechanisms either to immobilise or kill prey in the case of snakes or to enhance blood feeding in the case of haematophagous insects or vampire bats. Often these proteins are effective in extremely small amounts. This book provides descriptions of a number of different classes of such proteins from snakes and insects and their target molecules on platelets, as a way of developing new approaches to anti-thrombotic drugs.

Introduction

Blood circulation has a vital role in vertebrate survival, including man. It carries nutrients and oxygen to every tissue and organ throughout the body, including the brain, and transports the waste products and carbon dioxide to be removed via the liver, kidneys and lungs. It also has a critical role in defense against a wide range of pathogens by conveying leukocytes and other immune cells to all parts of the body and recruiting them to the appropriate site in the case of a pathogen attack. Any major problem with the hemostatic and cardiovascular systems can lead to

R.M. Kini (✉)
Protein Science Laboratory, Department of Biological Sciences, National University of Singapore, Singapore, 117543, Singapore; Department of Biochemistry and Molecular Biophysics, Medical College of Virginia, Virginia Commonwealth University, Richmond, VA 13298, USA
e-mail: dbskinim@nus.edu.sg

R.M. Kini et al. (eds.), *Toxins and Hemostasis*, DOI 10.1007/978-90-481-9295-3_1,

death or severe debility therefore they are normally tightly controlled by a complex system of proteins and cells that prevent blood loss from injuries or major blockages by pathological thrombi and emboli. Hemostasis is a complex process and its pathological variant thrombosis reflects disequilibrium, often due to poor lifestyle choices but, in some cases, also thought to have possible genetic origins. Hemostasis involves ongoing repair mechanisms that maintain the vascular system by covering subendothelium exposed by loss of endothelium cells. This is accentuated under high shear conditions, particularly at sites like vascular bifurcations where turbulence occurs, as a side effect of high blood pressure. Of course, stopping bleeding after menstruation and repair of injuries whether minor or major are also important functions. Populations living on Western diets, smoking and lacking regular exercise tend to acquire vascular plaque and other defects that favor heart disease and stroke. The major mechanisms regulating hemostasis and thrombosis involve the factors of the coagulation cascade as well as platelets and there is also increasing evidence that changes in endothelial cells have a role.

Various animals have developed mechanisms to target these systems and exploit this vulnerability for their own ends. In some species (for example, snakes), such mechanisms are used to immobilize and kill the prey/victim, and these are also used as defensive strategies against attacks by larger animals. In others (for example leeches, or insects such as mosquitoes and ticks), they are used to provide a continuous supply of blood as food. These mechanisms include, but are not limited to, procoagulant and anticoagulant agents that affect the coagulation cascade and platelet aggregation, as well as altering vasodilatory responses. In all these various animals, the mechanisms have evolved to perfection over millions of years, in parallel with the development of their hosts, to allow their survival. Since many of the components present in venom and saliva target critical hemostatic pathways they have been and are of great interest in basic research into the mechanisms of these pathways, as well as in possible applications to modify deviant behavior. Over the last 3–4 decades, due to the efforts of scientists from various backgrounds including biology, protein chemistry, molecular biology, pharmacology, hematology, and structural biology, significant progress has been made in understanding the structure-function relationships, as well as the mechanism of action of a number of exogenous factors from various animals, that affect blood coagulation, platelet aggregation and vasodilation. These exogenous factors have contributed significantly to the development of research tools as well as providing new therapeutic agents.

The increasing average age of world populations, coupled with recent changes in life style, has lead to significant increases in cardiovascular and hematological disorders. Thus, both academic as well as pharmaceutical industry scientists are working to develop better therapeutic agents to improve both quality of life and life expectancy. This impetus has lead to the search for novel agents from various sources that interfere with cardiovascular and hematological processes and could be used to regulate these. Thus, exogenous factors are excellent sources of such novel therapeutic lead molecules. In addition, some of them affect angiogenesis of blood vessels or immunological responses that are only briefly touched upon here. There are also antimicrobial peptides that may help in design of new antibiotics.

Animal venoms are complex mixtures of many components, mainly but not exclusively, proteins and peptides. Within venom these various components often have synergistic roles aimed at the rapid immobilization or inactivation of their prey. Most efficient and quick immobilization and death of the prey is achieved by targeting its main critical systems, namely hemostatic and neuromuscular systems. Since the venomous animals prey on many different species as well as having an effective defense against still more, they produce both proteins and peptides with specific molecular targets as well as those that are active across this wide range of prey species. Further, the same species from widely differing environments produce venoms with significant variation in their components. They often show different patterns of expression, qualitative and/or quantitative, of the various components. Thus, venom may represent a form of accelerated evolution in responding more rapidly to varying prey populations. Similarly, salivary glands in hematophagous animals also produce a wide range of closely related proteins, which have specific targets in the haemostatic system. These proteins rarely exhibit neurotoxic effects, as these animals do not intend to kill their host, but depend on them in the longer term for continuous blood feeds. Despite the key difference, protein components in both venoms and salivary gland secretions have evolved through gene duplication events and acquiring new functions through accelerated evolution of protein coding regions. Particularly in snake venoms, it is becoming clear that the venom proteins have been adapted during evolution from various tissues, often by gene duplication. A comparison of the sequences of the "original" and adapted genes can cast light on evolutionary processes involved in the predator-prey competitions. But the origin of saliva proteins is not yet clear.

Previously, identification and characterization of individual components relied primarily on various methods in protein chemistry and subsequent cloning of specific genes. Individual components were isolated with narrowly defined and highly specific functions. Such fully defined proteins are of greater interest to the pharmacological scientist trying to tackle a specific problem. In recent years, transcriptomics and proteomics techniques have been used in deciphering the structural and compositional aspects of the venome/sialome. These approaches have lead to a rapid increase in the amount of sequence data available for various classes of venom proteins. While it has enabled more profound insight into relations between families of venomous/ hematophagous animals and the enormous diversity in toxin sequence, it has the disadvantage of not providing any direct input concerning the targets (or even biological properties) of these proteins. This requires individual expression of DNA sequences as proteins and careful checking that they have refolded correctly followed by testing against possible targets in blood or the vascular system. Despite much research, these approaches do not always lead efficiently to the required knowledge. Nevertheless, these approaches are the only feasible ones for dealing with the tiny size of the saliva glands of hematophagous animals. Earlier methods required the dissection of thousands of insects to provide enough material for classical protein chemistry approaches. In contrast, proteomic and transcriptomic approaches require a small number of salivary glands. Expression studies allow the preparation of antibodies that may be used to investigate the protein in the original saliva since here only micrograms are necessary. These approaches have also allowed investigation of the composition of salivary gland proteins as well as the venom of colubrid snakes, which generally produce only very small amounts of venom and also that of a range of other reptiles, such as Gila monsters and Komodo dragons (Fry et al., 2006, 2009) that were also inaccessible earlier.

Platelet aggregation inhibitors have been isolated from various animal sources. They are proteins or glycoproteins with molecular weight ranging from 5000 to several tens of thousands. These factors inhibit platelet aggregation by different mechanisms. A large number of these inhibitors have no enzymatic activity. In contrast, some of them do exhibit enzymatic activities, such as phospholipase A_2s (PLA_2 – but not all are active, lacking essential catalytic residues), proteinases and nucleotidases. In general, the mechanism of inhibition of platelet aggregation is well understood for several groups of non-enzymatic proteins. However, further research is required to delineate the mechanism of inhibition by some of the enzymes. A number of components from animal venom/saliva, which affect blood coagulation, platelet aggregation, and cardiovascular system have been purified and characterized. They fall into a wide range of molecular categories ranging from several categories of enzymes to non-enzymatic proteins and polypeptides. Enzymatic proteins exert their effects on hemostatically critical components either by an inactivating cleavage or in many cases an activating cleavage of a precursor molecule. They also can exert their effects by cleaving agonists such as ADP in platelet aggregation. Non-enzymatic proteins exert their effects by binding to various receptors or ligands and act either as inhibitors or activators of the physiological process. Factors inducing platelet aggregation remove platelets from the circulation with high efficiency and prevent hemostasis. The biologically-active molecules tend to form into separate classes with a number of representatives of each class, often with specific functions, present in each venom/saliva. Snake venoms in particular contain a wide range of different classes.

Although at first glance the above exogenous factors appear to function as "villains", causing undesirable effects, several life-saving drugs have been developed based upon these factors. Such drugs or drug leads include those that inhibit the angiotensin converting enzyme (ACE) (Captopril (Capoten®) and Enalapril (Vasotec®)), that block platelet receptors (Eptifibatide (Integrilin®) and Tirofiban (Aggrastat®)), or that inhibit thrombin (bivalirudin (Angiomax®)) to name but a few. Briefly, we describe below the development of these drugs starting from exogenous factors.

Captopril was the first ACE inhibitor which helps in controlling hypertension designed in 1975 based on the structure of bradykinin potentiating peptides (BPPs) found in Brazilian pit viper *Bothrops jararaca* venom (Brunner et al., 1978). It was among the first successes of structure-based drug design. Ferreira and Rocha e Silva purified BPPs from *B. jararaca* venom in 1965 (Ferreira and Rocha e Silva, 1965). They are proline-rich oligopeptides of 5–14 amino acid residues which have a pyroglutamic acid at their N-terminal end. They are naturally occurring ACE inhibitors. Enalapril (Sweet et al., 1981) and Quinapril (Kaplan et al., 1984) were developed based on these to reduce the undesirable side effects (see Chapter 2).

Eptifibatide (Integrilin®) (Tcheng et al., 1995) is based on the structure of a functional and structural loop present in disintegrins, the snake venom components, which inhibit integrins present in a range of cells, including platelets, endothelial and smooth muscle cells. During development the idea was to find a disintegrin that was specific for platelet αIIbβ3 and that did not inhibit other integrins. A wide range of disintegrins from different snakes were tested until barbourin from the venom of *Sistrurus m. barbouri* was identified as an αIIbβ3-specific integrin antagonist (Scarborough et al., 1991). Eptifibatide is based on the KGD containing conformationally-constrained loop, which is the active region of this disintegrin. It has become a very effective drug for preventing thrombus formation leading to occlusion of vessels (or stents) after surgery by preventing fibrinogen and fibrin from binding to activated αIIbβ3 on platelets.

Tirofiban (Aggrastat®) is a non-peptidic αIIbβ3 inhibitor that was designed based on the RGD-containing, the conformationally-constrained loop of another disintegrin, echistatin, isolated from *Echis carinatus* (Hartman et al., 1992). Tirofiban is also used extensively in surgery for the same purposes as eptifibatide. Roxifiban is an oral version of tirofiban (Mousa et al., 1999). Originally, there were great hopes of using orally-available αIIbβ3 inhibitors like roxifiban and its competitors to prevent thrombus formation in patients at risk. However, most of this class of cyclic peptides and non-peptidic drugs bind preferentially to the active state of the integrin and thus tip an equilibrated situation in the direction of activation. It might be thought that if the activated integrin is in any case blocked this would not be a problem. However, when the αIIbβ3 integrin is activated it signals to the platelet interior and causes more platelet activation. In patients this may lead to increased thrombosis. Thus, although this route works well in acute situations it is less applicable to chronic ones (Hantgan and Stahle, 2009).

Bivalirudin (Angiomax®) (Ofosu et al., 1992) and lepirudin (Römisch et al., 1994) are both direct thrombin inhibitors based on the structure of the leech protein hirudin that block both the active site and exosites on thrombin. In the leech this salivary protein is largely responsible for inhibition of coagulation at the bite site. Even after the leech has been removed the bite continues to bleed for a long time showing how efficient this protein is. Thrombin inhibitors of this class are particularly useful where heparin or low molecular weight heparin cannot be used because of evidence of heparin-induced thrombocytopenia however careful dosing is necessary to avoid a bleeding risk caused by these powerful inhibitors (Tschudi et al., 2009).

All of these drugs were the result of long research programs involving investigation of a wide range of similar molecules from different snakes or other hematophagous animals to select appropriate starting proteins. This was particularly important in the development of eptifibatide. After this stage the molecule was then gradually adapted to yield the desired pharmacology.

In other cases, the entire exogenous protein is used in the treatment of certain diseases. For example, Batroxobin (Reptilase) from *Bothrops atrox* and Ancrod from *Callosellasma rhodostomata* are thrombin-like proteases used to digest and remove fibrinogen from the blood to reduce risk of, or treat stroke or venous thrombosis in PTCA (Bell, 1997).

Although the use of macromolecules may be acceptable for a single use, multiple chronic uses raise the risk of eliciting immune responses in the patients. In these cases, smaller and/or less immunological molecules (with tighter folding or exposed hydrophobic surfaces) need to be developed.

Venom components are extensively used as diagnostic tools in assays of haemostatic function (Schoni, 2005; see also Chapter 43). Reptilase is used as a rapid fibrinogen assay in samples containing heparin and also in assay of antithrombin in plasma where thrombin would interfere with the assay. Snake venom prothrombin-activating enzymes are widespread, particularly in Australian elapids and are used in prothrombin assays.

Venom disintegrins as well as snaclecs (snake C-type lectin-like) have been used extensively to investigate the function of platelet and other cell receptors and have played important roles in the discovery and characterization of new receptors.

A look into the distant future might see applications for venom/saliva components in treating complex diseases such as cancer and neurological disorders. However, this book is mainly aimed at the exogenous proteins regulating hemostasis and thrombosis and at using such components to develop assays for physiological and pathological conditions.

Several new and exciting success stories are currently unfolding. In this book, recent studies on some of the exogenous factors that play crucial roles in cardiovascular and hematological disorders are reviewed in order to consolidate efforts in this area of research and to entice new, talented researchers into this field by demonstrating the interesting options available. The review chapters, each written by experts in their field, are devoted to exogenous factors affecting platelet aggregation, anticoagulant and procoagulant proteins, fibrinolytic proteins and hypotensive agents.

The first section starts with several reviews of modern approaches to the analysis of overall venom/saliva contents, covering applications of transcriptomics and snake venomics as well as a detailed investigation of wasp venom and then passes on to a consideration of differences in hemostatic activity between venoms from different species. Finally in this section, the identification and characterization of new inhibitors from ticks is described. The second section deals with anticoagulant proteins from snakes, hematophagous animals, ticks and, surprisingly, scorpions, which are generally thought of as having neurotoxic venoms only. The third section deals with proteins that inhibit platelet function. Many of these affect platelet integrins, the receptors linking extracellular matrix to the platelet cytoskeleton or aggregating platelets via fibrinogen/fibrin and involved in clot retraction. From their specificity for integrins and their ability to inhibit their function they have been named disintegrins. A major target of snake disintegrins is the platelet integrin αIIbβ3 preventing platelet aggregation to avoid clot formation. This contributes towards prey blood losses on the one hand and facilitates swallowing the prey and its digestion on the other. However, other integrins including αvβ3, as well as β1 and β2 families are also targeted and the function of inhibiting these is only starting to be uncovered. Inhibition of angiogenesis by disintegrins is a major topic of discussion. Anti-platelet agents from haematophagous animals are then discussed. These mainly fall into categories different from snake venom components. However, one class of inhibitor found in both sources is apyrases which destroy ADP and ATP, both important platelet activators. A fascinating aspect of haematophagous salivas is the range of molecules adapted to tight binding to small molecule platelet activators such as ADP and which therefore provide a mopping up function keeping concentrations below a critical platelet activating concentration in the vicinity of the feeding wound. Finally, in this section, snaclecs (snake C-type lectins) that inhibit platelet function are discussed. These fall mainly into those proposed to block GPIbα function based on the assay system used but which, however, may not be their real function as well as snaclecs that inhibit specific integrin functions.

The fourth section covers fibrinolytic proteins from various organisms. Although there was interest in direct fibrinogen/fibrin cleaving enzymes at one stage, in recent years this has shifted to enzymes which activate the physiological plasminogen to plasmin pathway since this is thought less likely to lead to bleeding complications in patients. The fifth section deals with a wide range of procoagulant proteins from various sources including bacteria, caterpillars and plants as well as snakes.

The sixth section covers some of the toxins that activate platelets starting with a detailed look at the snaclec, aggretin/rhodocytin that has been recently of great interest because of its use to identify and characterize CLEC-2 as an important activatory receptor in platelets and an increasing range of other cells. This is followed by a general review of the various types of snaclec that activate platelets and their different targets. Lastly, proteins derived from bacteria that activate platelets are described and the effects that these have on patient health.

The seventh section deals with proteins from venoms that reduce the blood pressure in their prey – hypotensive factors. Various classes from different sources are described. The eighth section deals with more generalized effects of envenomation on the overall haemostatic and coagulation system of the prey and the use of antivenoms to reduce major and side-effects. This is an area with a lot of open questions remaining, in particular about long term effects.

Finally, the last section deals with some of the effective applications of toxin research. These include the use of leeches in microsurgery, reducing scar tissue formation and maintaining microcirculation. Toxins, particularly from snake venoms, have also been used extensively to develop assays for both pathological haemostatic conditions and for levels of coagulation factors. This is an active area with new assays being continually developed. Vascular endothelial growth factor-like proteins are common components of snake venoms and may function by inhibiting the physiological receptors. This is an area of considerable interest because of the possibility of designing drugs based on these structures that are effective in treating particular types of tumour or diseases such as anomalous retinal vessel development.

There remain many highly interesting areas, such as the small phospholipase A_2s, where non-enzymatic aspects of their mechanisms of cell activation or inhibition are poorly understood, as well as their possible receptors and which may be addressed in the future. As well as affecting haemostasis, venom proteins in particular have effects that go way beyond control of bleeding. These involve the access of the toxins to areas outside the vasculature, critical for transporting the venom. Permeabilization of the vessel wall, including removal of endothelial cells and digestion of extracellular matrix, allow toxins to reach critical muscles, such as those of the heart, and attack the nervous system including the brain. Many aspects of heart disease and stroke are touched upon in this way. Most of the venom components involved in these effects have been relatively little investigated and constitute a new area for future research.

References

Bell, W.R.J., 1997. Defibrinogenating enzymes. Drugs 54 (Suppl. 3), 18–30.

Brunner, H.R., Gavras, H., Turini, G.A., Waeber, B., Cappuis, P., McKinstry, D.N., 1978. Long-term treatment of hypertension in man by an orally active angiotensin-converting enzyme inhibitor. Clin. Sci. Mol. Med. Suppl. 4, 293s–295s.

Ferreira, S.H., Rocha e Silva, M., 1965. Potentiation of bradykinin and eledoisin by BPF (bradykinin potentiating factor) from *Bothrops jararaca* venom. Experientia 21, 347–349.

Fry, B.G., Vidal, N., Norman, J.A., Vonk, F.J., Scheib, H., Ramjan, S.F., Kuruppu, S., Fung, K., Hedges, S.B., Richardson, M.K., Hodgson, W.C., Ignjatovic, V., Summerhayes, R., Kochva, E., 2006. Early evolution of the venom system in lizards and snakes. Nature 439, 584–588.

Fry, B.G., Wroe, S., Teeuwisse, W., van Osch, M.J., Moreno, K., Ingle, J., McHenry, C., Ferrara, T., Clausen, P., Scheib, H., Winter, K.L., Greisman, L., Roelants, K., van der Weerd, L., Clemente, C.J., Giannakis, E., Hodgson, W.C., Luz, S., Martelli, P., Krishnasamy, K., Kochva, E., Kwok, H.F., Scanlon, D., Karas, J., Citron, D.M., Goldstein, E.J., McNaughtan, J.E., Norman, J.A., 2009. A central role for venom in predation by *Varanus komodoensis* (Komodo Dragon) and the extinct giant *Varanus (Megalania) priscus*. Proc. Natl. Acad. Sci., U.S.A. 106, 8969–8974.

Hantgan, R.R., Stahle, M.C., 2009. Integrin priming dynamics: mechanisms of integrin antagonist-promoted αIIbβ3:PAC-1 molecular recognition. Biochemistry 48, 8355–8365.

Hartman, G.D., Egbertson, M.S., Halczenko, W., Laswell, W.L., Duggan, M.E., Smith, R.L., Naylor, A.M., Manno, P.D., Lynch, R.J., Zhang, G.X., Chang, C.T.C., Gould, R.J., 1992. Non-peptide fibrinogen receptor antagonists. 1. Discovery and design of exosite inhibitors. J. Med. Chem. 35, 4640–4642.

Kaplan, H.R., Cohen, D.M., Essenburg, A.D., Major, T.C., Mertz, T.E., Ryan, M.J., 1984. CI-906 and CI-907: new orally active nonsulfhydryl angiotensin-converting enzyme inhibitors. Fed. Proc. 43, 1326–1329.

Mousa, S.A., Kapil, R., Mu, D.X., 1999. Intravenous and oral antithrombotic efficacy of the novel platelet GPIIb/IIIa antagonist roxifiban (DMP754) and its free acid form, XV459. Arterioscler. Thromb. Vasc. Biol. 19, 2535–2541.

Ofosu, F.A., Fenton, J.W., II., Maraganore, J., Blajchman, M.A., Yang, X., Smith, L., Anvari, N., Buchanan, M.R., Hirsh, J., 1992. Inhibition of the amplification reactions of blood coagulation by site-specific inhibitors of alpha-thrombin. Biochem. J. 283, 893–897.

Römisch, J., Stöhr, H.A., Stauss, H., Koschinsky, R., Stüber, W., Pâques, E.P., 1994. Inhibition of in vitro clot growth by r-hirudin is more effective and longer sustained than by an analogous peptide. Thromb. Haemost. 71, 320–324.

Scarborough, R.M., Rose, J.W., Hsu, M.A., Phillips, D.R., Fried, V.A., Campbell, A.M., Nannizzi, L., Charo, I.F., 1991. Barbourin. A GPIIb-IIIa-specific integrin antagonist from the venom of *Sistrurus m. barbouri*. J. Biol. Chem. 266, 9359–9362.

Schoni, R., 2005. The use of snake venom-derived compounds for new functional diagnostic test kits in the field of haemostasis. Pathophysiol. Haemost. Thromb. 34, 234–240.

Sweet, C.S., Gross, D.M., Arbegast, P.T., Gaul, S.L., Britt, P.M., Ludden, C.T., Weitz, D., Stone, C.A., 1981. Antihypertensive activity of N-[(S)-1-(ethoxycarbonyl)-3-phenylpropyl]-L-Ala-L-Pro (MK-421), an orally active converting enzyme inhibitor. J. Pharmacol. Exp. Ther. 216, 558–566.

Tcheng, J.E., Harrington, R.A., Kottke-Marchant, K., Kleiman, N.S., Ellis, S.G., Kereiakes, D.J., Mick, M.J., Navetta, F.I., Smith, J.E., Worley, S.J., Miller, J.A., Joseph, D.M., Sigmon, K.N., Kitt, M.M., Dumee, C.P., Califf, R.M., Topol, E.J., 1995. Multicenter, randomized, double-blind, placebo-controlled trial of the platelet integrin glycoprotein IIb/IIIa blocker Integrelin in elective coronary intervention. IMPACT Investigators. Circulation 91, 2151–2157.

Tschudi, M., Lämmle, B., Alberio, L., 2009. Dosing lepirudin in patients with heparin-induced thrombocytopenia and normal or impaired renal function: a single-center experience with 68 patients. Blood 113, 2402–2409.

Part II
The Omics – The Search Is On

Chapter 2
From the *Bothrops Jararaca* Bradykinin Potentiating Peptides to Angiotensin Converting Enzyme Inhibitors

Sérgio Henrique Ferreira

Abstract When Rocha e Silva received the first samples of synthetic bradykinin from Boissonnas et al. (1960), was found that it was much less potent than "natural" bradykinin made by incubation of plasma with the Bothrops jararaca venom. Erdös and Sloane (1962) demonstrated that the major inactivating enzyme in plasma had the characteristics of pancreatic carboxypeptidase B. This enzyme was a zinc metal-enzyme (Vallee, 1961). This knowledge prompted us to investigate the bradykinin potentiating activity of several metal chelating agents. BAL (dimercaptopropanol) inhibited bradykinin inactivation by plasma, and potentiated its actions "in vitro" and "in vivo" (Ferreira and Rocha e Silva, 1962, 1963; Ferreira et al., 1962; Rocha e Silva, 1963). This venom when added to an isolated guinea pig ileum preparation causes tachyphylactic contractions. Investigating if BAL could potentiate those contractions, we observed that the venom itself had a powerful peptidic potentiating substance, which we named bradykinin potentiating peptide, BPF. As the enzyme that inactivates bradykinin also converts angiotensin in the hypertensive active peptide the structure of BPF was instrumental for the development of the new synthetic family of antihypertensive drugs, named Angiotensin Converting Enzyme Inhibitors.

We started studying the pharmacological potentiation of bradykinin by investigating agents that are able to inhibit its inactivation. Following the observations that the major inactivating enzyme in plasma had the characteristics of a carboxypeptidase (Erdos and Sloane, 1962) and that pancreatic carboxypeptidase B was a metal-enzyme (Vallee, 1961), we then demonstrated that several metal chelating agents (in particular, dimercaptopropanol) inhibited bradykinin inactivation by plasma, and potentiated its actions "in vitro" and "in vivo" (Ferreira and Rocha e Silva, 1962, 1963; Ferreira et al., 1962; Rocha e Silva, 1963). This line of research was initiated when Rocha e Silva received the first samples of synthetic bradykinin from

S.H. Ferreira (✉)
Department of Pharmacology, University of São Paulo, Av. dos Bandeirantes, 3900, Campus USP, 14049-900, Ribeirão Preto, SP, Brazil
e-mail: shferrei@fmrp.usp.br

R.M. Kini et al. (eds.), *Toxins and Hemostasis*, DOI 10.1007/978-90-481-9295-3_2,

Fig. 2.1 A historic moment: Diana Bartelt (*left*) showing the deduced structure of BPF5a to Sérgio Ferreira (*middle*) and Lewis Greene (*right*). The N-terminal of BPF5a was blocked by pyroglutamic acid, which at that time was a major problem for determining the peptide sequence. Greene had a classical strategy. However, Bartelt decided, quietly, to deduce its structure on her own. The peptide synthesis of BPF5a by John Stewart showed that the structure proposed by Bartelt was correct

Boissonnas et al. (1960), which were found to be much less potent than "natural" bradykinin.

Bradykinin was discovered by incubating plasma with the venom of the Brazilian snake *Bothrops jararaca* (Rocha e Silva et al., 1949). In the same venom I found a strong bradykinin potentiating factor, which was named BPF (Ferreira, 1965). This factor explained why "natural" bradykinin was more potent than the synthetic one. BPF potentiated bradykinin in several pharmacological tests, possibly by inhibiting its destruction, since it increased bradykinin half-life both in plasma and in the circulation (Ferreira, 1966; Ferreira and Vane, 1967a; Erdos, 1966).

It was assumed that bradykinin inactivation and angiotensin conversion were accomplished by the same tissue/plasma enzyme. This hypothesis was supported by the observation made at Prof. J. R. Vane's laboratory, in London, where we demonstrated that the major site of inactivation of circulating bradykinin was in the pulmonary vascular bed (Ferreira and Vane, 1967b). At that time, synthetic angiotensin-I became available and Ng and Vane (1968) found that, during the passage through the pulmonary vascular bed, it was converted to angiotensin-II. This conversion appeared to be similar to bradykinin inactivation, since the carboxy-terminal amino acids of both peptides could be hydrolyzed by the same enzymatic activity. However, it was Dr Bakhle, at the same Department, who using our preparation of BPF first demonstrated its inhibitory action on in vitro enzymatic conversion of angiotensin-I (Bakhle, 1968; Bakhle et al., 1969). Thus, BPF provided the first clear demonstration of the association of the conversion of angiotensin-I with kininase activity. This was later confirmed with purified kininase-II by Yang et al. (1971).

Our association with Dr L.T. Greene of Brookhaven National Laboratory resulted in the isolation and characterization of nine low molecular weight pharmacologically active peptides (Ferreira et al., 1970a; Greene et al., 1970). A general association between the bradykinin potentiation and angiotensin-I conversion activities was described for those peptides fractions (Ferreira et al., 1970b). The smallest peptide (PCA-Lys-Tryp-Ala-Pro) was sequenced, synthesized and named BPP5a. It showed identical activities as BPF and was used to make the first demonstration of the participation of angiotensin system on a renal experimental model of hypertension (Krieger et al., 1971). Our work clearly demonstrated that among the *Bothrops* peptides the nonapeptide (BPP_{9a}) was the most potent one (Ferreira et al., 1970a, b). BPP_{9a} structure and synthesis was performed at Squibb by Ondetti et al. (1971), being rebaptized as converting enzyme inhibitor teprotide, SQ 20,88 1. Around 1973, the Squibb group realized the impossibility to develop classical peptides for oral antihypertensive drugs and allowed the publication showing the effectiveness of SQ 20,88 1 in hypertension (Gavras et al., 1974) and in human hypertension angiotensin dependent (Rosenthal et al., 1979). The late awareness of the work of Byers and Wolfenden (1972), showing the strategy to develop non-peptidic inhibitors from peptides, directed the synthesis of captopril by Cushman et al. (1977). Byers and Wolfenden proposed that the carboxy-terminal amino acid of a carboxypeptidase-A substrate would interact with the active site of the enzyme, and that the carbonyl of the peptide bond interacted with a zinc ion tightly bound to the enzyme. Based upon this model, inhibitors of carboxypeptidase-A, such as D-2 benzylsuccinic acid, were designed. Cushman et al. assumed that the active site of angiotensin converting enzyme inhibitor (ACEi) was analogous to that of carboxypeptidase-A although the distance between the carboxyl-binding group and the zinc ion was greater than that in carboxypeptidase-A. The SQ 20,88 1 and BBP_{5a} had a proline as carboxy terminal amino acid. When a succinyl radical was added to proline, a weak specific ACEi, but oral active resulted, leading to the invention of captopril and creating a new class of antihypertensive, ACEi (Cushman et al., 1977).

Although ACEi was marketed as a specific drug for the blockade of conversion of Angiotensin-II, there were several papers calling attention that they obviously had a potent action on bradykinin effect. Martorana et al. (1991) showed a bradykinin dependent antiischemic effect of ACEi. More recently, it was described that the bradykinin antagonist (icatibant) reduces the hypotensive effect of captopril oral administration in hypertensive subjects (Gainer et al., 1998).

ACEi has several side effects that frequently limit its use (allergic reaction, chest pain cough, loss of taste sensation, etc). This fact, associated with the introduction of angiotensin receptor antagonists and renin inhibitors in the market, may explain why ACEi international market is slowly declining. However, if bradykinin is relevant for the therapeutic of some special physiopathologic processes (due to bradykinin influence) some reflection should be made regarding its substitution by angiotensin receptor antagonists and renin inhibitors.

The history of the development of the angiotensin converting enzyme inhibitors reflects the combined efforts of basic and drug-orientated research, a classical example of a relationship between university and industry research.

As in the case of Captopril, the major pre-clinical discoveries were made by universities or research institutes, while the clever and practical development of useful compounds was accomplished by industry scientists

Last but not least, I would like to draw attention to the central role of the bioassay in the development of this area. The discovery of angiotensin, the demonstration of the conversion of angiotensin-1 by lungs or tissue homogenates and its inhibition by BPF, clearly illustrate this point. It is important for medical students and young scientists to understand that it is not only with the use of sophisticated methodology that relevant discoveries can be made. Indeed, as shown by captopril history, the contractions of a guinea-pig ileum led to the discovery of a new class of antihypertensive drugs.

References

Bakhle, Y.S., 1968. Conversion of angiotensin I to angiotensin II by cell-free extracts of dog lung. Nature 220, 919–921.

Bakhle, Y.S., Reynard, A.M., Vane, J.R., 1969. Metabolism of the angiotensins in isolated perfused tissues. Nature 222, 956–959.

Boissonnas, R.A., Guttmann, S., Jaquenoud, P.A., Konzett, H., Stuermer, E., 1960. Synthesis and biological activity of peptides related to bradykinin. Experientia 16, 326.

Byers, L.D., Wolfenden, R., 1972. A potent reversible inhibitor of carboxypeptidase A. J. Biol. Chem. 247, 606–608.

Cushman, D.W., Cheung, H.S., Sabo, E.F., Ondetti,M.A., 1977. Design of potent competitive inhibitors of angiotensin-converting enzyme. Carboxyalkanoyl and mercaptoalkanoyl amino acids. Biochemistry 16, 5484–5491.

Erdos, E.G., 1966. Hypotensive peptides: bradykinin, kallidin, and eledoisin. Adv. Pharmacol. 4, 1–90.

Erdos, E.G., Sloane, E.M., 1962. An enzyme in human blood plasma that inactivates bradykinin and kallidins. Biochem. Pharmacol. 11, 585–592.

Ferreira, S.H., 1965. A Bradykinin-potentiation factor (BPF) present in the venom of *Bothrops Jararaca*. Br. J. Pharmacol. 24, 163–169.

Ferreira, S.H., 1966. Bradykinin-potentiating factor, in: Erdos, E.G. (Ed.), Hypotensive Peptides. Springer Verlag, New York, pp. 356–367.

Ferreira, S.H., Bartelt, D.C., Greene, L.J., 1970a. Isolation of bradykinin-potentiating peptides from *Bothrops jararaca* venom. Biochemistry 9, 2583–2593.

Ferreira, S.H., Corrado, A.P., Rocha e Silva, M., 1962. Potenciação do efeito hipotensor da bradicinina por inibição da enzima bradicininolitica plasmática. Ciências e Cultura 14, 238.

Ferreira, S.H., Greene, L.H., Alabaster, V.A., Bakhle, Y.S., Vane, J.R., 1970b. Activity of various fractions of bradykinin potentiating factor against angiotensin I converting enzyme. Nature 225, 379–380.

Ferreira, S.H., Rocha e Silva, M., 1962. Potentiation of bradykinin by dimercaptopropanol (BAL) and other inhibitors of its destroying enzyme in plasma. Biochem. Pharmacol. 11, 1123–1128.

Ferreira, S.H., Rocha e Silva, M., 1963. Potenciação de polipeptideos por um fator presente no veneno de *Bothrops jararaca*. Ciências e Cultura 15, 276.

Ferreira, S.H., Vane, J.R., 1967a. Half-lives of peptides and amines in the circulation. Nature 215, 1237–1240.

Ferreira, S.H., Vane, J.R., 1967b. The detection and estimation of bradykinin in the circulating blood. Br. J. Pharmacol. 29, 367–377.

Gainer, J.V., Morrow, J.D., Loveland, A., King, D.J., Brown, N.J., 1998. Effect of bradykinin-receptor blockade on the response to angiotensin-converting-enzyme inhibitor in normotensive and hypertensive subjects. N. Engl. J. Med. 339, 1285–1292.

Gavras, H., Brunner, H.R., Laragh, J.H., Sealey, J.E., Gavras, I., Vukovich, R.A., 1974. An angiotensin converting-enzyme inhibitor to identify and treat vasoconstrictor and volume factors in hypertensive patients. N. Engl. J. Med. 291, 817–821.

Greene, L.J., Stewart, J.M., Ferreira, S.H., 1970. Bradykinin-potentiating peptides from the venom of *Bothrophs jararaca*, in: Sicuteri, M., Rocha e Silva, M., Back, N. (Eds.), Bradykinin and Related Kinins: Cardiovascular, Biochemical and Neural Actions. Plenum Press, New York, pp. 81–87.

Krieger, E.M., Salgado, H.C., Assan, C.J., Greene, L.J., Ferreira, S.H., 1971. Potential screening test for detection of overactivity of renin-angiotensin system. Lancet 1, 269–271.

Martorana, P.A., Linz, W., Scholkens, B.A., 1991. Does bradykinin play a role in the cardiac antiischemic effect of the ACE-inhibitors? Basic Res. Cardiol. 86, 293–296.

Ng, K.K., Vane, J.R., 1968. Fate of angiotensin I in the circulation. Nature 218, 144–150.

Ondetti, M.A., Williams, N.J., Sabo, E.F., Pluscec, J., Weaver, E.R., Kocy, O., 1971. Angiotensin-converting enzyme inhibitors from the venom of *Bothrops jararaca*. Isolation, elucidation of structure, and synthesis. Biochemistry 10, 4033–4039.

Rocha e Silva, M., 1963. The physiological significance of bradykinin. Ann. N.Y. Acad. Sci. 104, 190–211.

Rocha e Silva, M., Beraldo, W.T., Rosenfeld, G., 1949. Bradykinin, a hypotensive and smooth muscle stimulating factor released from plasma globulin by snake venoms and by trypsin. Am. J. Physiol 156, 261–273.

Rosenthal, T., Kissin, E., Jacob, E.T., Stern, N., Levi, Y., Adar, R., 1979. Angiotensin-converting enzyme inhibitor (SQ 20881) in the diagnosis of renovascular hypertension. Angiology 30, 123–128.

Vallee, B.L., 1961. The "active catalytic site," an approach through metalloenzymes. Fed. Proc. 20, 71–80.

Yang, H.Y., Erdos, E.G., Levin, Y., 1971. Characterization of a dipeptide hydrolase (kininase II: angiotensin I converting enzyme). J. Pharmacol. Exp. Ther. 177, 291–300.

Sialomic Perspectives on the Evolution of Blood-Feeding Behavior in Arthropods: Future Therapeutics by Natural Design

Chapter 3
Sialomic Perspectives on the Evolution of Blood-Feeding Behavior in Arthropods: Future Therapeutics by Natural Design

B.J. Mans and I.M.B. Francischetti

Abstract Blood-feeding behavior evolved more than 20 times independently in Arthropods. This happened at least 6 times in the Arachnida (Acari) and 15 times in the Hexapoda (Neoptera). This is recapitulated when transcriptomes from the secretory component of salivary glands (Sialomes) are compared. As such, unique protein families are found for the different lineages that adapted to a blood-feeding lifestyle with only a limited number of protein families conserved across all lineages. Closely related lineages might share similar sets of protein families in their sialomes, even if no apparent orthologous or conserved functional relationships exist. This suggests that sialomes of such lineages where already defined before adaptation to a blood-feeding lifestyle, with subsequent innovation. In this regard, the same sets of shared protein families tend to be abundant and prone to lineage specific expansion (gene duplication) with specialized functions associated with various gene duplicates. Perhaps not surprisingly, all sialomes show evidence of convergent evolution in regard modulatory strategies that targets host defenses, even if the molecular mechanisms differ. As such, a checklist of expected functions can be composed for any blood-feeding arthropod not yet characterized. The diversity of mechanisms that counteract vertebrate host immune and hemostatic systems is a veritable pharmacopoeia, optimized by natural evolution, that can be exploited for therapeutic use.

Independent Evolution of Hematophagy in the Insects

Blood-feeding behavior evolved more than 20 times independently in Arthropods. It occurred at least 6 times in the Arachnida (Acari) and 15 times in the Hexapoda (Neoptera). Independent evolution of hematophagy is evident at the Order level in the insects (5 times independent) when phylogenetic position and the relatively

---

B.J. Mans (✉)
Parasites, Vectors and Vector-Borne Diseases, Onderstepoort Veterinary Institute, Pretoria, South Africa; Department of Veterinary Tropical Diseases, University of Pretoria, Pretoria, South Africa
e-mail: mansb@arc.agric.za

R.M. Kini et al. (eds.), *Toxins and Hemostasis*, DOI 10.1007/978-90-481-9295-3_3,

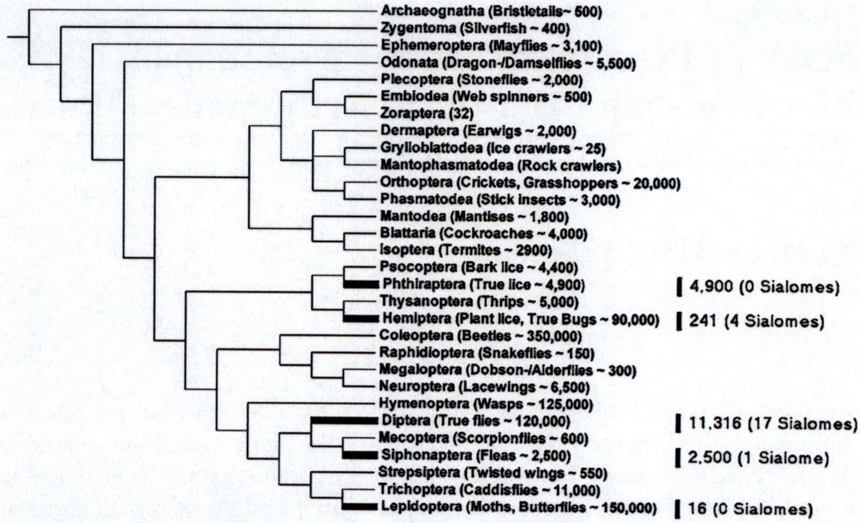

Fig. 3.1 A phylogeny of the insects at order level. *Bold lines* indicate lineages where blood-feeding evolved. Common names and number of species are indicated in *brackets*. *Vertical lines* indicate lineages where blood-feeding lifestyles evolved, followed by the number of blood-feeding species and the number of sialomes constructed thus far in *brackets*. Phylogenetic relationships and species numbers adapted from Grimaldi and Engel (2005)

low number of species that feed on blood is taken into account relative to their non-feeding relatives (Fig. 3.1) (Grimaldi and Engel, 2005). In the case of lice and fleas where all members of the Order are obligate blood-feeders, the monophyletic nature of blood-feeding behavior seems to be relatively strait-forward. The same probably holds for moths where only 16 species have been described that feeds on blood, all members of the same genus, Calyptra.

Within the Hemiptera and Diptera, where the majority of species are non-hematophagous, blood-feeding behavior evolved several times independently. In the Hemiptera, blood-feeding behavior evolved at least twice in the Cimicomorpha, in the Cimicidae (bed bugs) and the reduvids (assassin bugs) (Fig. 3.2). Even so, it is suspected that blood-feeding behavior in reduvid bugs are polypheletic

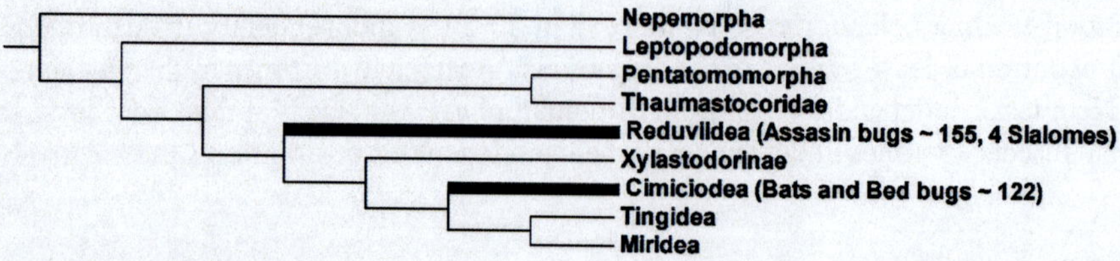

Fig. 3.2 A phylogeny of the Chimicomorpha (Hemiptera) at family level as adapted from Schuh et al. (2009). *Bold lines* indicate lineages where blood-feeding evolved. Common names and number of species are indicated in *brackets* with the number of sialomes described where appropriate

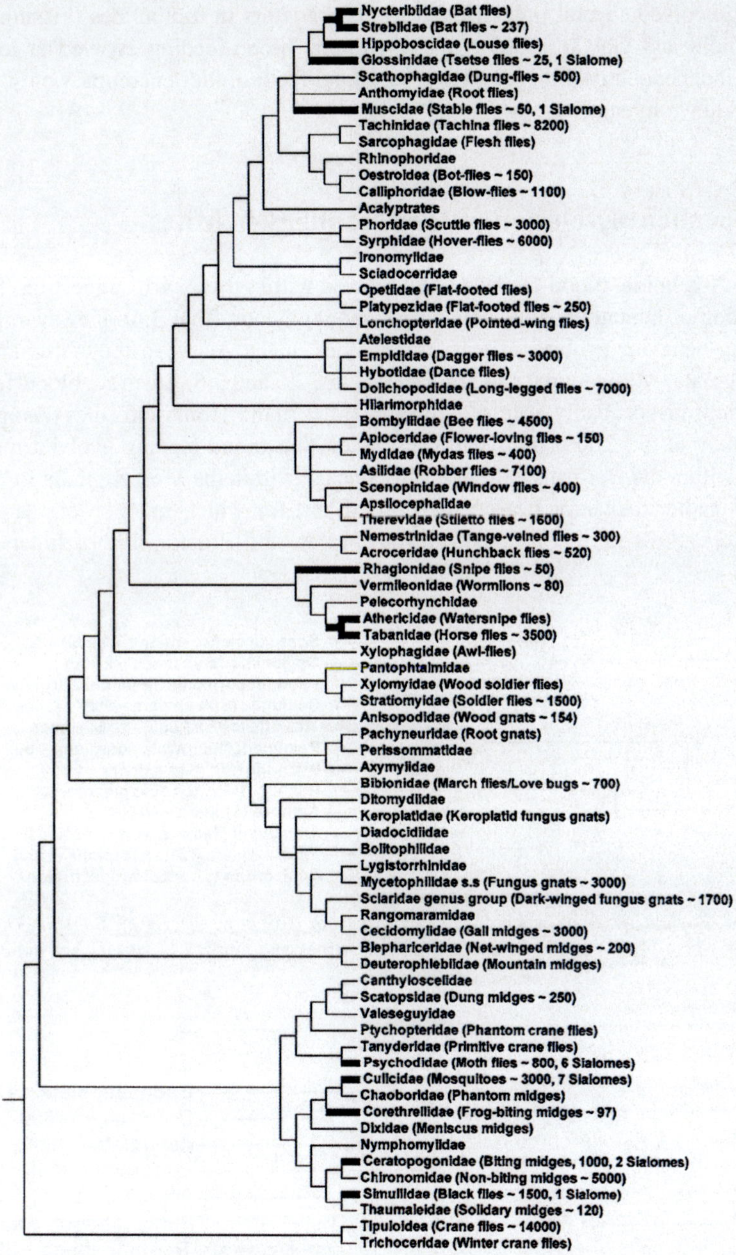

Fig. 3.3 A phylogeny of the Diptera at family level as adapted from Grimaldi and Engel (2005). *Bold lines* indicate lineages where blood-feeding evolved. Common names and number of species are indicated in *brackets* with number of sialomes described where appropriate

having evolved several times from various predators in triatomines (kissing bugs) (Schofield and Galvão, 2009). In the Diptera, blood-feeding evolved at least 11 times independently (Fig. 3.3). Even so, blood-feeding flies encompass only ∼10% of all known fly species.

Independent Evolution of Hematophagy in Mites

In the Arachnida, blood-feeding only evolved within the Acari, suggesting that the evolution of hematophagous behavior is monophyletic (Fig. 3.4). However, blood-feeding behavior evolved at least three times in the Acari, in the Astigmata and Prostigmata (Mesostigmata and Ixodida) (Figs. 3.5 and 3.6). Even so, blood-feeding behavior most probably occurred multiple times in the Trombidiformes (Astigmata) (Bochkov et al., 2008). It is even suspected that blood-feeding evolved multiple times within the Dermanyssina, the only lineage within the Mesostigmata for which blood-feeding behavior has been recorded and for which most species are non-hematophagous mites (Radovsky, 1969; Walter and Proctor, 1999). Independent

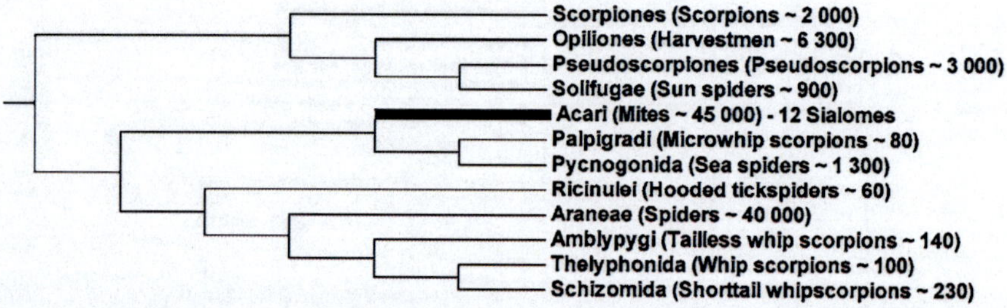

Fig. 3.4 Phylogeny of the arachnids as adapted from Giribet et al. (2002). *Bold lines* indicate lineages where blood-feeding evolved. Common names and number of species are indicated in *brackets*

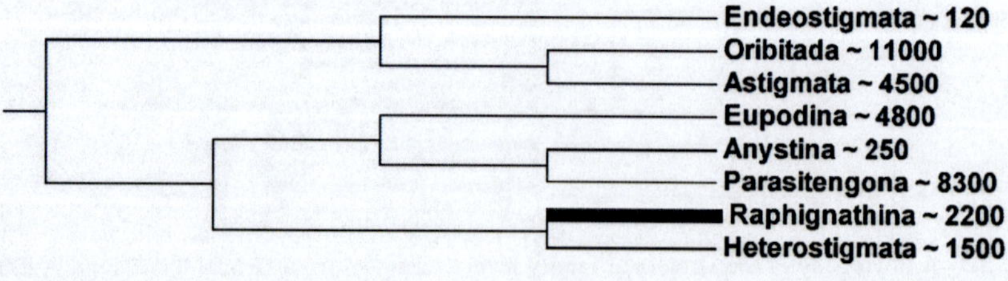

Fig. 3.5 Phylogeny of the Acariformes at suborder level as adapted from Klompen et al. (2007) and Bochkov et al. (2008). *Bold lines* indicate lineages where blood-feeding evolved. Approximate number of species was adapted from Walter and Proctor (1999)

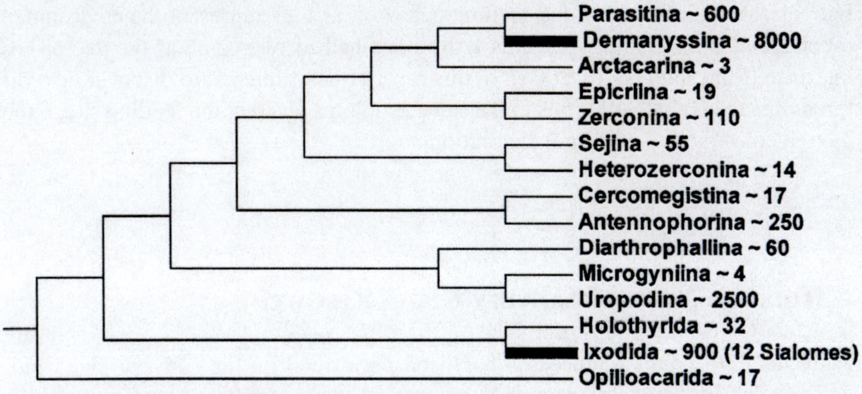

Fig. 3.6 Phylogeny of the Parasitiformes at suborder level as adapted from Klompen et al. (2007). *Bold lines* indicate lineages where blood-feeding evolved. Approximate number of species was adapted from Walter and Proctor (1999)

evolution of blood-feeding behavior was also suggested for hard (Ixodidea) and soft ticks (Argasidae), based on the extensive differences in their biochemical salivary repertoires and blood-feeding biology (Mans and Neitz, 2004; Mans and Ribeiro, 2008a, b; Mans et al., 2002, 2008a, b, c).

The Vector-Host Interface and Salivary Gland Pharmacopoeias

It was probably a given that hematophagous behavior would have arisen once a circulatory system that transported nutrients evolved in multi-cellular organisms. The common denominator that arthropods faced during the adaptation to a blood-feeding lifestyle is the vector-host interface or in simpler terms, the feeding-site (Champagne, 2004; Ribeiro, 1995). Here arthropods encountered the host's defenses against blood-loss and foreign invasion, namely, the hemostatic and immune systems. These defense mechanisms are complex, redundant and inter-connected and in order to become successful feeders, arthropods had to evolve effective counter-mechanisms (Francischetti et al., 2009).

Secretion of bio-active components into the feeding site that specifically recognize and neutralize the molecules involved in hemostasis and immunity were evolved and these derived mostly from the salivary glands (Ribeiro, 1995; Ribeiro and Francischetti, 2003). Blood-feeding arthropods thus became veritable pharmacopeias of bio-active salivary molecules specific for their vertebrate host's defenses. On the one hand the study of these molecules is of interest due to their anti-hemostatic, anti-inflammatory and immuno-modulatory capabilities that may have therapeutic potential (Champagne, 2004). On the other hand, their study should shed light on the evolution of blood-feeding behavior, as they represent the environmental challenges that hematophagous arthropods had to overcome at the vector-host interface (Mans and Neitz, 2004). In this regard it is of interest to discover how different lineages found solutions to the same problems faced at the feeding site, either though convergent or divergent evolution.

A Hundred Years of Salivary Gland Research

More than 100 years of research has been performed on the salivary gland biology of blood-feeding arthropods, starting with the identification of anti-clotting activity in whole tick extracts from *Ixodes ricinus,* after a similar discovery was made in leeches (Sabbatani, 1899). Sabbatani made the generalization that all organisms that feed on blood will possess anti-hemostatics to counter-act the host's defenses. This has been confirmed by numerous biochemical studies on all the major classes of blood-feeding arthropods, starting with the identification of biochemical functions in crude salivary gland extracts or saliva in the first half of the previous century (Cornwall and Patton, 1914; Nuttall and Strickland, 1909). Analysis of salivary components became increasingly complex in the last half of the previous century, especially in the last decade when numerous salivary gland components were purified and characterized at a molecular level (Law et al., 1992; Ribeiro and Francischetti, 2003; Ribeiro, 1995). This yielded a wealth of knowledge on mechanism, function and identity of the dominant functions present in saliva of blood-feeding arthropods. This was obtained by laborious biochemical approaches that advanced our knowledge a step at a time.

The introduction of high through-put methods at the turn of the last millennium changed the field of vector salivary gland biology in significant ways that is still being realized. The first salivary gland transcriptome was described for the sand fly *Lutzomyia longipalpis* with the description of 5 cDNAs implicated in feeding (Charlab et al., 1999). This was an interesting paper that was a harbinger of what was to come. It was a fusion of the old school biochemical approach, with the fractionation of the salivary gland components by chromatography, followed by Edman sequencing, the description of the transcriptome and biochemical confirmation of the functions predicted by the transcriptome in salivary gland extracts. This study was expanded by the high through-put sequencing of salivary gland transcriptomes and their proteomic profiling for the mosquitoes *Aedes aegypti*, *Anopheles gambiae* and the tick *Ixodes scapularis* (Francischetti et al., 2002a; Valenzuela et al., 2002a, b). Since then 34 transcriptomes have been described for various blood-feeding arthropods with extensive confirmation of proteins present in salivary glands by proteomic methods (Table 3.1). These include coverage of most major orders of insects and acarids, except for lice, bedbugs and mites other than ticks (Figs. 3.1, 3.2, 3.3, 3.4, 3.5, and 3.6).

Table 3.1 Major protein families found in the sialomes of blood-feeding arthropods

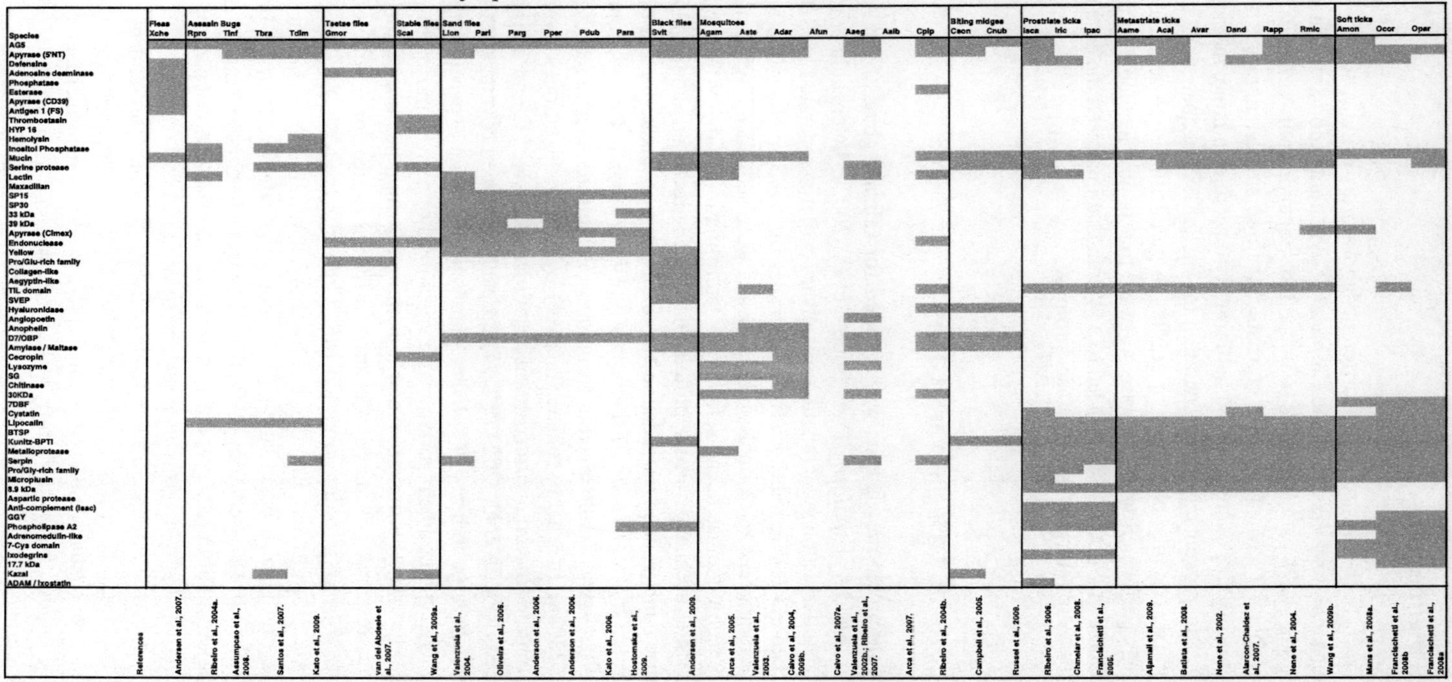

The presence of a protein family is indicated by *shaded areas*. Four letter abbreviations are used for species descriptions. These are *Xenopsylla cheopsis* (Xche), *R. prolixus* (Rpro), *Triatoma infestans* (Tinf), *T. brasiliensis* (Tbra), *T. dimidiate* (Tdim), *Glossina morsitans* (Gmor), *Stomoxys calcitrans* (Scal), *L. longipalpis* (Llon), *Phlebotomus ariasi* (Pari), *P. argentipes* (Parg), *P. perniciosus* (Pper), *P. duboscі* (Pdub), *P. arabicus* (Para), *Simulium vitatum* (Svit), *Anopheles gambia* (Agam), *A. stephensi* (Aste), *A. darlingi* (Adar), *A. funestus* (Afun), *Aedes aegypti* (Aaeg), *A. albopictus* (Aalb), *Culex pipiens quinquefasciatus* (Cpip), *Culicoides sonorensis* (Cson), *C. nubeculosus* (Cnub), *Ixodes scapularis* (Sca), *I. ricinus* (Iric), *I. pacificus* (Ipac), *Amblyomma americanum* (Aame), *A. cajannense* (Acaj), *A. variegatum* (Avar), *D. andersoni* (Dand), *Rhipicephalus appendiculatus* (Rapp), *R. microplus* (Rmic)

Trancriptome Sequencing

Transcriptomes have been constructed in different laboratories using a variety of techniques. In our laboratory the use of the Clontech cDNA library construction kit coupled with lambda phage packaging has been used to sequence approximately 23 transcriptomes (Table 3.1). Most libraries consisted of three smaller libraries that were fractionated using size exclusion chromatography to produce a large, medium and small library. This prevents overrepresentation of smaller EST sequences in the final library. Library assembly, clustering of contigs to yield full-length sequences and annotation was performed with in-house software that is now available in the public domain (Guo et al., 2009). This has led to a system that increased the ease of comparative analyses for salivary gland transcriptomes.

Proteome Studies of Tick Salivary Gland

The first large-scale proteomic analysis was reported for salivary glands from *Ae. aegypti*, *A. gambiae* and *I. scapularis* (Francischetti et al., 2002a; Valenzuela et al., 2002a, b). It consisted of Edman degradation of salivary proteins separated by SDS gel which were then electroblotted to PVDF membranes in CAPS-methanol buffer. After staining the membrane in Coomassie blue, bands were sequenced by Edman degradation. This technique has been useful since determination of the N-terminus of the protein indicates precisely how the proteins have been processed and found in saliva. On the other hand, Edman degradation has relatively low sensitivity and our studies demonstrated that only a small fraction of proteins predicted by sequencing the cDNA library was found by this technique (Francischetti et al., 2002a; Valenzuela et al., 2002a). In order to enhance recovery by Edman degration, salivary gland of the triatomine *Rhodnius prolixus* was fractionated by HPLC and amino-terminal sequenced (Ribeiro et al., 2004a). This approach enhanced coverage at the expenses of a relative large number of salivary glands used for the experiments, a condition not always evident when working with specimens of a given mosquito or tick which availability may be limited. Another technique which has been widely used in proteomics research is separation of protein by a 2-dimensional gel electrophoresis followed by tryptic digestion and MS (Gutierrez et al., 2009; Oleaga et al., 2007). While 2D gels may improve coverage substantially when compared to 1D gel followed by Edman degradation it was not more effective than 1D gel followed by HPLC-MS/MS (Andersen et al., 2007; Francischetti et al., 2008a, b; Mans et al., 2008a). In addition, 2D gels consume relative large amounts of protein – approximately 100 μg of sample proteins is needed to achieve good results.

More recently, we have attempted to separate salivary proteins from soft ticks and other blood-sucking arthropods by 1D gel followed by tryptic digestion and HPLC-MS/MS of the bands (Andersen et al., 2007, 2009; Francischetti et al., 2008a, b). This technique has significantly enhanced coverage and proved more informative than Edman degradation and 2D gel followed by MS/MS. It requires salivary glands to be solubilized and resolved by SDS/PAGE. Then, each gel lane is sliced into approximately 30 gel slices, de-stained, and digested overnight with trypsin. Peptides are subsequently extracted and desalted and resuspended in 0.1% TFA prior to HPLC-MS/MS analysis using standard techniques. The results are blasted against the cDNA library constructed with the salivary gland of the arthropod in focus and the protein identified when it matches the corresponding transcript or classified as "unknown" when it does not. Overall, our results have consistently demonstrated that this approach has been the most informative and has been elected as the preferable technique when studying blood-sucking salivary glands. Nevertheless, Edman degradation and 2D gel followed by MS may provide complementary information to 1D gel HPLC-MS/MS. Figure 3.7 compares the 1D gel followed by Edman degradation (Fig. 3.7a), 2D gel followed by tryptic digestion and MS (Fig. 3.7b) and 1D gel followed by tryptic digestion and HPLC-MS/MS

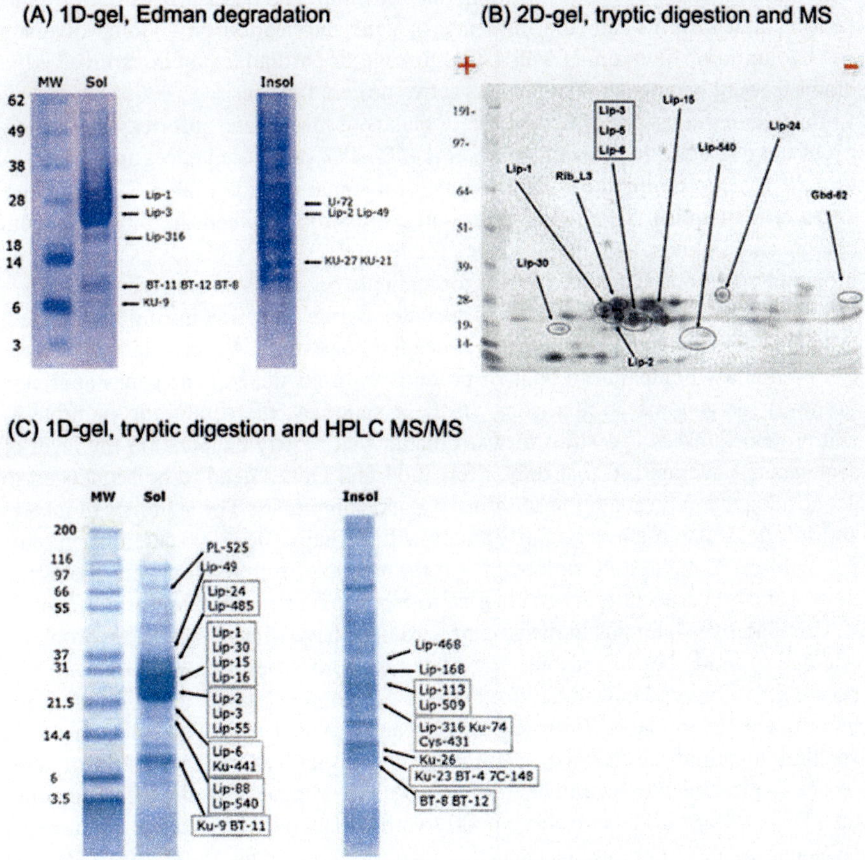

Fig. 3.7 Different proteome approaches to study the salivary gland of the soft tick *O. parkeri.* (**a**) 1D gel followed by Edman degradation, (**b**) 2D gel followed by tryptic digestion and MS, (**c**) 1D gel followed by tryptic digestion and HPLC-MS/MS. For details see (Francischetti et al., 2008)

(Fig. 3.7c) of the salivary gland of *Ornithodoros parkeri* (Francischetti et al., 2008a). In cases where sufficient salivary gland material is available, fractionation by chromatographic methods, followed by MS-MS can give quantitative data on the relative concentrations of various proteins expressed in the sialome, thereby correlating protein and transcript abundance (Mans et al., 2008a).

Sialome Structure and Problems for Comparative Sialomics

The aim of comparative sialomics would be the derivation of insights into the evolution of salivary gland secretory protein function and protein family repertoires. This requires firstly, the comprehensive mapping of protein family members present in a given sialome, and secondly, the accurate annotation of function associated with these proteins. This allows for the identification and description of multi-gene families and insights into the influence of gene duplication on sialome diversity and evolution of function as well as the linkage of particular functions involved in blood-feeding behavior with their respective protein families.

In sialomes constructed from salivary glands of insects and soft ticks, the majority of contigs code for secretory proteins (50–90% of all sequences in the cDNA libraries), as exemplified by the presence of a signal peptide (Table 3.1 and references cited therein). There is also a good correlation between the most abundant contigs and the most abundant proteins found in the salivary glands as analyzed by proteomic methods. This is expected for a simple organ system, where proteins are synthesized and stored in secretory organelles before secretion during fast feeding events, with low metabolic activity between feeding events. As such, salivary glands can be literally compared to bags of proteins vis a vis the bags of genes approach followed for genome comparisons. In these sialomes, the cataloging of proteins and protein families is a strait-forward matter that mostly depends on the level of sequencing coverage, so that only a few thousand clones need to be sequenced to get an adequate representation of salivary gland complexity. The sialomes of insects and soft ticks also seem to be fairly small, with perhaps 100–200 secretory proteins per sialome. Comparative sialomics (vs. for example, full genome sequencing) is thus attainable, especially when coupled to biochemical confirmation of function.

Comparative sialomics in the case of hard ticks have shown to be more problematic due to a number of reasons. Hard ticks feed for several days and during this period the salivary glands undergo extensive development associated with the feeding process (Binnington, 1978; Binnington and Stone, 1981). The salivary glands are thus dynamic entities with significant background expression levels of non-secretory proteins involved in cellular metabolism. The number of contigs found in the cDNA libraries thus correlates poorly with the abundance of secretory proteins (Aljamali et al., 2009; Francischetti et al., 2005; Nene et al., 2002, 2004; Ribeiro et al., 2006; Valenzuela et al., 2002a). Differential expression of genes over the feeding period also dilute expression signal, as the levels of fluctuation occurs within a high background. Higher numbers of potential secretory proteins (>500) has also been detected in the salivary gland of hard ticks (Ribeiro et al., 2006). To gain a representative view of sialome diversity in hard ticks a much larger sequencing effort is thus required.

Confirmation of protein expression in the salivary glands of hard ticks is further hampered by the relatively low levels of secretory protein that can be detected using proteomic methods (Francischetti et al., 2005; Ribeiro et al., 2006; Valenzuela et al., 2002a). This again is due to the more "constitutive-like" release of protein during the prolonged feeding period compared to the bolus released by soft ticks or insects during their shorter feeding periods. A further complication is the secretion of non-salivary gland derived proteins from the host that can mask tick-derived secretory proteins (Madden et al., 2002).

Generalizations Derived from Sialomes for the Evolution of Blood-Feeding Behavior

Several comprehensive reviews of salivary gland transcriptomes have been written recently that comprises catalogues of salivary gland protein families found in different blood-feeding arthropods (Francischetti et al., 2009; Fry et al., 2009). The current study aims at drawing broad generalizations concerning the evolution of blood-feeding behavior as viewed from a sialomic perspective.

Unique Sialomes are Found for Different Lineages

In spite of problems associated with comparative sialomics, the analysis of sialomes has indicated that distinct protein families are expressed in salivary glands from different phylogenetic lineages (Table 3.1). This correlates with the independent adaptation of different arthropods to a blood-feeding environment. It implies that the salivary gland sialomes of ancestral non-hematophagous arthropods from different lineages have already diverged and was adapted to their previous lifestyles before evolution of blood-feeding behavior. Ancestral lifestyles include being plant feeders (mosquitoes, flies and moths), scavengers (ticks, various fly families), predators (reduviid bugs, mesostigmatid mites) and non-hematophagous parasites (mesostigmatid mites) (Grimaldi and Engel, 2005; Lehane, 2005; Radovsky, 1969; Walter and Proctor, 1999). Their previous lifestyles would have predisposed them towards a blood-feeding lifestyle, either via adaptations that allows host finding and feeding, or being in the proximity of the host, either as parasite or as a commensal habitant of the hosts nest or burrow.

This would have predisposed the salivary glands to have certain functional molecules and protein families that were specific for a particular lifestyle. Plant feeders could be predisposed towards sugar metabolism. Scavengers might have a variety of proteases, anti-microbials and protease inhibitors in their repertoire.

Predators would have toxins and digestive enzymes, while non-hematophagous parasites could have the beginnings of immunosuppressive molecules.

In each case, the ancestral lineage would be dependent on the set of protein domains found in their salivary glands at the time of the evolution of blood-feeding behavior for the evolution of new function. This could explain the abundance of some protein families within specific lineages, that expanded by gene duplication and subsequent evolution of novel function.

Closely Related Lineages, Share Similar Sets of Protein Families in Their Sialomes, Even If No Apparent Orthologous or Conserved Functional Relationships Exist

Comparison of closely related lineages indicates that in many instances the same protein families are found in their salivary gland repertoires even if evidence suggest that they evolved blood-feeding behavior independently. As such, hard and soft ticks share a similar protein family repertoire for their abundant protein families that includes the lipocalins, BPTI-Kunitz, BTSP, metalloprotease families (Mans et al., 2008a). The major protein family in *Triatoma* and *Rhodnius* are the lipocalins (>80% of all salivary gland protein) (Assumpcao et al., 2008; Kato et al., 2008; Ribeiro et al., 2004a; Santos et al., 2007). The D7-protein family is found in sand flies, mosquitoes, biting midges and black flies (Arca et al., 2005, 2007; Andersen et al., 2009; Anderson et al., 2006; Calvo et al., 2004, 2006, 2007a, 2009a; Campbell et al., 2005; Francischetti et al., 2002a; Hostomská et al., 2009; Kato et al., 2006; Oliveira et al., 2006; Ribeiro et al., 2004b; Russell et al., 2009; Valenzuela et al., 2002b, 2003, 2004).

However, biochemical analysis of function has indicated that in most cases, those functions considered to be important in vector-host interaction are not orthologous or even belong to the same protein family (Mans and Neitz, 2004). If proteins with similar functions belong to the same protein family, their phylogenetic analysis will indicate no clear orthologous relationships (Mans et al., 2008a). Even if closely related lineages possess the same protein families, members in different lineages will not have similar functions.

For example, while the D7-proteins in mosquitoes possess an extra α-helix involved in the binding of biogenic amines, this helix is absent from the D7-proteins from sand flies and biting midges, implying that they will not possess biogenic amine-binding function (Mans et al., 2007). Even though all triatomines possess lipocalins, all lipocalins from the genus Triatoma lacks the conserved proximal histidine residue involved in binding of the heme group in the nitrophorins from *R. prolixus* and thus cannot function as nitrix oxide carriers (Andersen, 2010; Santos et al., 2007). Within the Cimicomorpha, nitric oxide carrier function has been evolved in the Triatomines (*Rhodnius*) and bed bugs (*Cimex lectularius*). In bed bugs, this function is performed by the inositol polyphosphate 5′ phosphatase family that evolved to bind heme (Weichsel et al., 2005). The inositol polyphosphate 5′

phosphatase family is also present in triatomines but presumably function in a more canonical manner (Andersen and Ribeiro, 2006).

In ticks, lipocalins evolved biogenic amine-binding function in both hard and soft tick lineages, but there are significant differences in their mechanisms of binding, while thromboxane and leukotriene binding also seem to have evolved independently (Beaufays et al., 2008; Mans and Ribeiro, 2008a, b; Mans et al., 2008c; Paesen et al., 1999; Sangamnatdej et al., 2002). Hard and soft ticks all possess Kunitz-BPTI proteins, but in the different tick families these inhibitors target different enzymes and receptors (Corral-Rodriguez et al., 2009; Mans and Neitz, 2004). Even in the case of the BPTI thrombin inhibitors, ornithodorin from the soft tick *Ornithodoros moubata,* and boophilin from the hard tick *Rhipicephalus microplus*, significant differences in mechanism exist that would indicate convergent evolution of function within the same protein family (Corral-Rodriguez et al., 2009; Macedo-Ribeiro et al., 2008).

This would imply that even though closely related lineages shared similar protein repertoires before adaptation to a blood-feeding environment, the evolution of novel functions associated with blood-feeding occurred after speciation, utilizing the same common building blocks (protein family repertoires).

The Same Protein Families are Prone to Expansion in Closely Related Lineages

Lineage specific expansion (gene duplication) of different protein families occurred in almost all of the blood-feeding lineages (Table 3.1). In fleas this occurred in the phosphatase and FS1 antigen families (Andersen et al., 2007). Lipocalins expanded in ticks and assassin bugs (Francischetti et al., 2009; Mans et al., 2008a; Ribeiro et al., 2006). The odorant-binding/D7-family expanded in mosquitoes, sand flies and biting midges (Anderson et al., 2006; Campbell et al., 2005; Francischetti et al., 2002a).

In many cases functions involved in regulation of the hosts hemostatic and immune systems are associated with these families (Francischetti et al., 2009; Fry et al., 2009; Mans and Neitz, 2004). Most of these expanded families are also those which are abundant at protein level. Possible reasons behind these expansions include increasing the effective secretory concentrations of salivary gland proteins involved in functions where high concentrations are important, such as scavenging of bioactive ligands such as ADP, biogenic-amines and eicosanoids (Andersen, 2010). The generation of polymorphism for important functions coupled with differential expression to evade the host's immune responses (Francischetti et al., 2009). Proteins that occur at high levels would also be ideal for the evolution of novel function, as these already exist in abundant concentrations.

The expansion of highly expressed protein families in the salivary glands could also indicate that evolution of novel function in the salivary glands is influenced more by the current salivary gland repertoire available than exaptation and recruitment of other genes from the existing genome for expression in the salivary glands. This might be linked to the salivary glands unique role in blood-feeding as a storage and delivery organ of vector-derived proteins to the external environment.

Specific Host Defense Mechanisms Seem to be Recurring Targets in the Evolution of Modulating Molecules

Similar strategies were evolved in most blood-feeding arthropods that target common host defense mechanisms (Ribeiro, 1995). This include targeting of specific serine proteases involved in the blood coagulation cascade (notably FXa and thrombin), targeting of small molecules involved in platelet aggregation, inflammation and vasoconstriction such as ADP, ATP, histamine, serotonin, nor-epinephrine and epinephrine, leukotrienes and thromboxane A2 (Francischetti et al., 2009). The fact that the same host defense mechanisms are targeted by distinct arthropods indicates that these are central defense mechanisms of the vertebrate host that evolved and were maintained in diverse vertebrates such as amphibians, mammals, birds and reptiles. As such, it might be possible to construct a checklist of things you need to evolve if you want to become a blood-feeding organism. In many cases, the convergence of function correlates with those protein families that show lineage specific expansions.

This leads to the evolution of common functions (as constraint by the hosts defense mechanisms), but with innovative solutions that depends on the protein families present in the sialome and their structural scaffolds. Their structures will determine which becomes scavengers of bio-molecules (such as the D7 and lipocalins), or inhibitors of various enzymes and receptors (smaller proteins are more prone to this), or enzymes that targets host molecules and proteins (enzymes might normally be exapted).

The fact that most blood-feeding lineages evolved their lifestyles independently, at the same vector-host interface, facing similar challenges, make convergent evolution a central feature of blood-feeding behavior in arthropods. Each lineage using different building blocks (sialomes) still found similar solutions to the problems they faced. This highlights the importance that the vertebrate host's hemostatic and immune defenses played in the evolution of blood-feeding behavior. It also provides tremendous opportunities to exploit the diversity of potential therapeutic agents that has been generated during the evolution of blood-feeding behavior in arthropods.

Functional Transcriptomics: Providing New Leads for Drug Discovery

High throughput sequencing of cDNAs has emerged as a powerful technique which has changed the field of exogenous secretions. As the set of transcripts and proteomes from different salivary glands are revealed, our understanding of hematophagy, vector-host interactions, evolutionary aspects and phylogeny of blood sucking arthropods has been modified substantially. Further, massive sequencing has opened our potential to find new biological functions, and has re-emerged as a platform for the discovery of novel drugs with potential therapeutic applications. This database, which has been increasing exponentially, has produced more questions than anticipated, and properties never before hypothesized to exist in hematophagous saliva.

The question now is how to tackle salivary glands in terms of finding novel biological activities, based on information available from functional transcriptomes. Is there a rational, or a particularly informative approach that may produce reliable and consistent results? At present, it appears that advances in methodologies may help us in this endeavor. For example, gene synthesis turned out to be affordable and reliable eliminating time-consuming steps such as cloning, and sequence verification. Expression in bacteria has also been facilitated by new strains available, user-friendly molecular biology techniques, and refolding protocols that take into account complex protein domains. Expression in eukaryotic system such as HEK293 cells has proven to produce significant amount of proteins which are properly folded. Expression libraries, novel purification techniques and highly sensitive mass spectrometry procedures – among other methodologies such as protein arrays, and crystal structure – may also contribute to our task in identifying salivary proteins with novel biological functions. Finally, screening assays are often useful and may serve for fast and reliable function identification. Below we summarize how different approaches have helped investigators to reveal novel aspects of hematophagy which may potentially contribute to drug discovery.

Conventional Purification Techniques

These techniques have been largely used in the first attempts to molecularly characterize salivary proteins with a given property (e.g., an anticoagulant). When the protein is purified to homogeneity, the N-terminus is identified by Edman degradation which is then used to generate probes needed to screen a cDNA library. This technique lead to the discovery of anophelin, a mosquito anti-thrombin (Francischetti et al., 1999; Valenzuela et al., 1999), which displays potential therapeutic applications as reported for hirudin (Greinacher and Warkentin, 2008). While this approach has been widely used (Francischetti et al., 2009; Fry et al., 2009; Koh and Kini, 2009; Maritz-Olivier et al., 2007; Steen et al., 2006), the most important limitation is the large amount of saliva needed to isolate a molecule which may be present in minutes amounts.

Discovery by Sequence Similarity

cDNA library construction generates hundreds of transcripts and has eliminated the need to collect a large and often impractical numbers of salivary glands to isolate a protein. In addition, sequence similarity to a protein previously deposited in the database provides leads that help investigators to confirm the existence of a given activity in the saliva. This was the case of Ixolaris, a two-Kunitz Tissue Factor Pathway Inhibitor (TFPI) from *I. scapularis* which displays sequence similarity to three-Kunitz human TFPI (Francischetti et al., 2002b). However, this approach can be potentially tricky. Even in cases of Ixolaris, several other single Kunitz-containing molecules displayed sequence similarity to TFPI but turned out to be inactive towards tissue factor. Furthermore, two-Kunitz-containing proteins which display homology to Ixolaris have been reported to inhibit contact phase proteins (Decrem et al., 2009), or thrombin (Lai et al., 2004). Accordingly, finding the right sequence to be expressed is critical. Also, evolutionary and molecular plasticity of some protein families (e.g. Kunitz, lipocalins) may produce surprises in terms of how these molecules work mechanistically. For instance, Ixolaris displays a remarkable mechanism of action which – in contrast to human TFPI – was shown to bind to Factor X (besides FXa). This property accounts for the long half-life of Ixolaris *in vivo*, a notable advantage for an anticoagulant with anti-thrombotic properties (Nazareth et al., 2006).

Massive Expression of Proteins from the Same Family – Nature's Version of Directed Evolution

The numerous members from the same protein family found in the sialomes of different arthropods or within the sialome of a particular species provide us with a plethora of genes with similar or different functions that has been optimized or tinkered with during the course of natural evolution. As such, proteins with high sequence similarity may have very specific but related functions that have been optimized to a remarkable extent. Examples are the ADP, prostanoid and biogenic amine-binding proteins found in ticks, assassin bugs and mosquitoes that show specificity, for ADP, nitric oxide, histamine, serotonin, epinephrine, norepineprhine, leukotrienes B4, C4, D4, E4 and thromboxane A2 (Andersen, 2010; Andersen et al., 1997, 2003, 2005; Beaufays et al., 2008; Calvo et al., 2006, 2009a; Francischetti et al., 2002c; Mans and Ribeiro, 2008a, b; Mans et al., 2008c; Montfort et al., 2000). Site-directed mutagenesis linked with crystallography allows for the accurate determination of mechanisms and the identification of specific residues involved in molecular interaction and lays the foundation for the understanding of molecular causes behind functional specificity (Calvo et al., 2009a; Mans and Ribeiro, 2008a; Mans et al., 2007, 2008c). This in turn can allow us to re-design and optimize functions via a variety of methods, such as for example directed evolution or rational engineering, to be even more specific and efficient, as was shown to be the case for mammalian derived lipocalins known as anticalins (Skerra, 2008). Incorporation of binding moieties, such as the interactive site of the lipocalin OMCI with complement C5, can prolong the half-life of lipocalins in the body (Hepburn et al., 2007; Mans and Ribeiro, 2008a). In this regard, tick derived lipocalins have been tested in some allergic conditions (Mans, 2005).

Expression Libraries

The identification of chemokine-binding proteins from *R. sanguineus*, known as evasins, has been carried out by using pools of cDNA clones from a library followed by transfection into HEK293 cells. Then, conditioned media from the transfected cells were tested for chemokine binding activity and shown to display selectivity for chemokines (Frauenschuh et al., 2007) and to inhibit experimental colitis (Vieira et al., 2009). This alternative approach is particularly useful because it potentially screens several activities simultaneously and is also applicable when a given salivary component is expressed at very low levels which preclude purification.

Discovery Based on Screening Assays

More recently, we have successfully identified the function of a protein that was typically characterized as an "unknown". For many years, an abundant protein of 30 kDa, also known as salivary allergen, was subject of study by different groups. We have expressed this protein from the salivary gland of *Ae. aegypti* and named it aegyptin. Its unique sequence was uninformative concerning its biological function. Using initially a number of screening assays, and later SPR experiments, we have discovered that aegyptin is a specific and high-affinity collagen-binding protein (Calvo et al., 2007b). Aegyptin has now been tested in rats and shown to prevent thrombosis (Calvo et al., 2010). This is a typical example of how studying "unknowns" may lead to new drug discoveries. One limitation of screening assays is finding the right bioassay since these molecules often display high specificity. Therefore, success rate in function identification will depend on the number of biological activities tested and novel technologies applied to the discovery process.

Concluding Remarks

We have challenging and exciting times ahead of us in the field of exogenous secretions. This is particularly true in the sense that many sequences are reportedly without databases hits, and accordingly their functions are not easily identifiable. These are the so-called "unknowns", and these account for $\sim 30\%$ of the sequences depending on the salivary gland in focus. To illustrate this scenario, these are the "ice" hidden below the tip of iceberg that, in turn, represents what we know. While challenging, these are the most interesting molecules because they have perhaps evolved to interfere with biological processes that have not yet been identified in human or veterinary physiology. For these cases, screening using experimental procedures designed to study Vascular Biology, Immunology, Pain and Inflammation, Channel Function, Enzymology, Vector Host interaction, Vaccinology etc will hopefully allows us to characterize the activity of "unknowns". In fact, we have most often identified salivary proteins which interact with proteins, channels and receptors that have been described in textbooks as critical components of vertebrate biology (e.g. coagulation factors, adhesion receptors, enzymes etc). This explains why anti-thrombin or specific receptor antagonists, just to mention few examples, have been thoroughly characterized.

However, study of salivary gland proteins may closely accompany a new discovery, or alternatively, may be a source of information that leads to new discoveries. For example, the Nobel Prize in Physiology or Medicine were awarded in 1998 for the identification of NO as a novel vasodilator in the mid-1980s (Howlett, 1998). Soon thereafter it was shown that a vasodilator existed in the salivary secretions of *R. prolixus* and that it was delivered by a NO-carrying molecule (nitrophorins) (Ribeiro et al., 1990, 1993). Another example of the versatility of blood-sucking salivary gland was the discovery of prostaglandins, which were awarded with the Nobel Prize in Physiology or Medicine in 1982 (Oates, 1982). This most potent anti-platelet inhibitor has was found in the saliva of *I. scapularis* shortly after (Ribeiro et al., 1985). Of note, both NO-pathway and prostacyclin display therapeutic application (Kato, 2008; Krug et al., 2009).

Salivary components may also anticipate discovery in human physiology. For example, a novel apyrase was molecularly characterized in the salivary gland of the bedbug *Cimex lectularius* (Valenzuela et al., 1998). A *Cimex*-type apyrase was later found in eukaryotes and to be fully active as a nucleotidase and potentially useful as an antithrombotic (Dai et al., 2004). Therefore, this discovery validates and exemplifies how the study of salivary proteins as sources of (known) biologic activities may provide leads for the identification of the (unknown) function of human counterparts based on sequence similarity. It is possible, and likely, that we have before us molecules in the salivary gland that awaits identification of its human counterparts, or molecules that interact with receptors not yet characterized, or as ligands for orphan receptors. In conclusion, study of blood-sucking salivary gland has unfolded in several directions as reported for snake venoms (Fox and Serrano, 2007), and may be regarded as a fertile field of research for years to come.

Acknowledgements This work was supported by the Intramural Research Program of the Division of Intramural Research, National Institute of Allergy and Infectious Diseases, National Institutes of Health. Because IMBF is a government employee and this is a government work, the work is in the public domain in the United States. Notwithstanding any other agreements, the NIH reserves the right to provide the work to PubMedCentral for display and use by the public, and PubMedCentral may tag or modify the work consistent with its customary practices. You can establish rights outside of the U.S. subject to a government use license.

References

Alarcon-Chaidez, F.J., Sun, J., Wikel, S.K., 2007. Transcriptome analysis of the salivary glands of *Dermacentor andersoni* stiles (Acari: Ixodidae). Insect Biochem. Mol. Biol. 37, 48–71.

Aljamali, M.N., Hern L., Kupfer D., Downard S., So S., Roe B.A., Sauer J.R., Essenberg R.C., 2009. Transcriptome analysis of the salivary glands of the female tick *Amblyomma americanum* (Acari: Ixodidae) Insect Mol. Biol. 18, 129–154.

Andersen, J.F., Champagne, D.E., Weichsel, A., Ribeiro, J.M., Balfour, C.A., Dress, V., Montfort, W.R., 1997. Nitric oxide binding and crystallization of recombinant nitrophorin I, a nitric oxide transport protein from the blood-sucking bug *Rhodnius prolixus*. Biochemistry 36, 4423–4428.

Andersen, J.F., Francischetti, I.M., Valenzuela, J.G., Schuck, P., Ribeiro, J.M., 2003. Inhibition of hemostasis by a high affinity biogenic amine-binding protein from the saliva of a blood-feeding insect. J. Biol. Chem. 278, 4611–4617.

Andersen, J.F., Gudderra, N.P., Francischetti, I.M., Ribeiro, J.M., 2005. The role of salivary lipocalins in blood feeding by *Rhodnius prolixus*. Arch. Insect Biochem. Physiol. 58, 97–105.

Andersen, J.F., Ribeiro, J.M., 2006. A secreted salivary inositol polyphosphate 5-phosphatase from a blood-feeding insect: allosteric activation by soluble phosphoinositides and phosphatidylserine. Biochemistry 45, 5450–5457.

Andersen, J.F., Hinnebusch, B.J., Lucas, D.A., Conrads, T.P., Veenstra, T.D., Pham, V.M., Ribeiro, J.M., 2007. An insight into the sialome of the oriental rat flea, *Xenopsylla cheopis* (Rots). BMC Genomics 8, 102.

Andersen, J.F., Pham, V.M., Meng, Z., Champagne, D.E., Ribeiro, J.M., 2009. Insight into the sialome of the Black Fly, *Simulium vittatum*. J. Proteome Res. 8, 1474–1488.

Andersen, J.F., 2010. Structure and mechanism in salivary proteins from blood-feeding arthropods. Toxicon (In press).

Anderson, J.M., Oliveira, F., Kamhawi, S., Mans, B.J., Reynoso, D., Seitz, A.E., Lawyer, P., Garfield, M., Pham, M., Valenzuela, J.G., 2006. Comparative salivary gland transcriptomics of sandfly vectors of visceral leishmaniasis. BMC Genomics 7, 52.

Arca, B., Lombardo, F., Valenzuela, J.G., Francischetti, I.M., Marinotti, O., Coluzzi, M., Ribeiro, J.M., 2005. An updated catalogue of salivary gland transcripts in the adult female mosquito, *Anopheles gambiae*. J. Exp. Biol. 208, 3971–3986.

Arca, B., Lombardo, F., Francischetti, I.M., Pham, V.M., Mestres-Simon, M., Andersen, J.F., Ribeiro, J.M., 2007. An insight into the sialome of the adult female mosquito *Aedes albopictus*. Insect Biochem. Mol. Biol. 37, 107–127.

Assumpcao, T.C., Francischetti, I.M., Andersen, J.F., Schwarz, A., Santana, J.M., Ribeiro, J.M., 2008. An insight into the sialome of the blood-sucking bug *Triatoma infestans*, a vector of Chagas′ disease. Insect Biochem. Mol. Biol. 38, 213–232.

Batista, I.F.C., Chudzinski-Tavassi, A.M., Faria, F., Simons, S.M., Barros-Batestti, D.M., Labruna, M.B., Leão, L.I., Ho, P.L., Junqueira-de-Azevedo, I.L., 2008. Expressed sequence tags (ESTs) from the salivary glands of the tick *Amblyomma cajannense* (Acari: Ixodidae). Toxicon 51, 823–834.

Beaufays, J., Adam, B, Menten-Dedoyart, C, Fievez, L., Grosjean, A., Decrem, Y., Prevot, P., Santini, S., Brasseur, R., Brossard, M., Vanhaeverbeek, M., Bureau, F., Heinen, E., Lins, L., Vanhamme, L., Godfroid, E., 2008. Ir-LBP, an *Ixodes ricinus* tick salivary LTB4-binding lipocalin, interferes with host neutrophil function. PLoS 3, 1–13.

Binnington, K.C., 1978. Sequenctial changes in salivary gland structure during attachment and feeding of the cattle tick, *Boophilus microplus*. Int. J. Parasitol. 8, 97–115.

Binnington, K.C., Stone, B.F., 1981. Developmental changes in morphology and toxin content of the salivary gland of the australian paralysis tick *Ixodes holocyclus*. Int. J. Parasitol. 11, 343–351.

Bochkov, A.V., Connor, B.M., Wauthy, G., 2008. Phylogenetic position of the mite family Myobiidae within the infraorder Eleutherengona (Acariformes) and origins of parasitism in eleutherengone mites. Zool. Anz. 247, 15–45.

Calvo, E., Andersen, J., Francischetti, I.M., de, L.C.M., deBianchi, A.G., James, A.A., Ribeiro, J.M., Marinotti, O., 2004. The transcriptome of adult female *Anopheles darlingi* salivary glands. Insect Mol. Biol. 13, 73–88.

Calvo, E., Mans, B.J., Andersen, J.F., Ribeiro, J.M., 2006. Function and evolution of a mosquito salivary protein family. J. Biol. Chem. 281, 1935–1942.

Calvo, E., Dao, A., Pham, V.M., Ribeiro, J.M., 2007a. An insight into the sialome of *Anopheles funestus* reveals an emerging pattern in anopheline salivary protein families. Insect Biochem. Mol. Biol. 37, 164–175.

Calvo, E., Tokumasu, F., Marinotti, O., Villeval, J.L., Ribeiro, J.M., Francischetti, I.M., 2007b. Aegyptin, a novel mosquito salivary gland protein, specifically binds to collagen and prevents its interaction with platelet glycoprotein VI, integrin alpha2beta1, and von Willebrand factor. J. Biol. Chem. 282, 26928–26938.

Calvo, E., Mans, B.J., Ribeiro, J.M., Andersen, J.F., 2009a. Multifunctionality and mechanism of ligand binding in a mosquito antiinflammatory protein. Proc. Natl. Acad. Sci., U.S.A. 106, 3728–3733.

Calvo, E., Pham, V.M., Marinotti, O., Andersen, J.F., Ribeiro, J.M., 2009b. The salivary gland transcriptome of the neotropical malaria vector *Anopheles darlingi* reveals accelerated evolution of genes relevant to hematophagy. BMC Genomics 10, 57.

Calvo, E., Tokumasu, F., Mizurini, D.M., McPhie, P., Narum, D.L., Ribeiro, J.M., Monteiro, R.Q., Francischetti, I.M., 2010. Aegyptin displays high-affinity for the von Willebrand factor binding site (RGQOGVMGF) in collagen and inhibits carotid thrombus formation *in vivo*. FEBS J. 277, 413–427.

Campbell, C.L., Vandyke, K.A., Letchworth, G.J., Drolet, B.S., Hanekamp, T., Wilson, W.C., 2005. Midgut and salivary gland transcriptomes of the arbovirus vector *Culicoides sonorensis* (Diptera: Ceratopogonidae). Insect Mol. Biol. 14, 121–136.

Champagne, D.E., 2004. Antihemostatic strategies of blood-feeding arthropods. Curr. Drug Targets – Cardiovasc. Haematol. Dis. 4, 375–396.

Charlab, R., Valenzuela, J.G., Rowton, E.D., Ribeiro, J.M., 1999. Toward an understanding of the biochemical and pharmacological complexity of the saliva of a hematophagous sand fly *Lutzomyia longipalpis*. Proc. Natl. Acad. Sci. U.S.A 96, 15155–15160.

Chmelař, J., Anderson, J.M., Mu, J., Jochim, R.C., Valenzuela, J.G., Kopecký, J., 2008. Insight into the sialome of the castor bean tick, *Ixodes ricinus*. BMC Genomics 9, 233.

Cornwall, J.W., Patton, W.S., 1914. Some observations on the salivary secretion of the commoner blood-cuking insects and ticks. Ind. J. Med. Res. 2, 569–593.

Corral-Rodrıguez, C.A., Macedo-Ribeiro, S., Pereira P.J.B., Fuentes-Prior, P., 2009. Tick-derived Kunitz-type inhibitors as antihemostatic factors. Insect Biochem. Mol. Biol. 39, 579–595.

Dai, J., Liu, J., Deng, Y., Smith, T.M., Lu, M., 2004. Structure and protein design of a human platelet function inhibitor. Cell 116, 649–659.

Decrem, Y., Rath, G., Blasioli, V., Cauchie, P., Robert, S., Beaufays, J., Frere, J.M., Feron, O., Dogne, J.M., Dessy, C., Vanhamme, L., Godfroid, E., 2009. Ir-CPI, a coagulation contact phase inhibitor from the tick *Ixodes ricinus*, inhibits thrombus formation without impairing hemostasis. J. Exp. Med. 206, 2381–2395.

Fox, J.W., Serrano, S.M., 2007. Approaching the golden age of natural product pharmaceuticals from venom libraries: an overview of toxins and toxin-derivatives currently involved in therapeutic or diagnostic applications. Curr. Pharm. Des. 13, 2927–2934.

Francischetti, I.M., Valenzuela, J.G., Ribeiro, J.M., 1999. Anophelin: kinetics and mechanism of thrombin inhibition. Biochemistry 38, 16678–16685.

Francischetti, I.M., Valenzuela, J.G., Pham, V.M., Garfield, M.K., Ribeiro, J.M., 2002a. Toward a catalog for the transcripts and proteins (sialome) from the salivary gland of the malaria vector *Anopheles gambiae*. J. Exp. Biol. 205, 2429–2451.

Francischetti, I.M., Valenzuela, J.G., Andersen, J.F., Mather, T.N., Ribeiro, J.M., 2002b. Ixolaris, a novel recombinant tissue factor pathway inhibitor (TFPI) from the salivary gland of the tick, *Ixodes scapularis*: identification of factor X and factor Xa as scaffolds for the inhibition of factor VIIa/tissue factor complex. Blood 99, 3602–3612.

Francischetti, I.M., Andersen, J.F., Ribeiro, J.M., 2002c. Biochemical and functional characterization of recombinant *Rhodnius prolixus* platelet aggregation inhibitor 1 as a novel lipocalin with high affinity for adenosine diphosphate and other adenine nucleotides. Biochemistry 41, 3810 3818.

Francischetti, I.M., My Pham, V., Mans, B.J., Andersen, J.F., Mather, T.N., Lane, R.S., Ribeiro, J.M., 2005. The transcriptome of the salivary glands of the female western black-legged tick *Ixodes pacificus* (Acari: Ixodidae). Insect Biochem. Mol. Biol. 35, 1142–1161.

Francischetti, I.M., Mans, B.J., Meng, Z., Gudderra, N., Veenstra, T.D., Pham, V.M., Ribeiro, J.M., 2008a. An insight into the sialome of the soft tick, *Ornithodorus parkeri*. Insect Biochem. Mol. Biol. 38, 1–21.

Francischetti, I.M., Meng, Z., Mans, B.J., Gudderra, N., Hall, M., Veenstra, T.D., Pham, V.M., Kotsyfakis, M., Ribeiro, J.M., 2008b. An insight into the salivary transcriptome and proteome of the soft tick and vector of epizootic bovine abortion, *Ornithodoros coriaceus*. J. Proteomics 71, 493–512.

Francischetti, I.M.B., Sa-Nunes, A., Mans, B.J., Santos, I.M., Ribeiro J.M.C., 2009. The role of saliva in tick feeding. Front. Biosci. 14, 2051–2088.

Frauenschuh, A., Power, C.A., Deruaz, M., Ferreira, B.R., Silva, J.S., Teixeira, M.M., Dias, J.M., Martin, T., Wells, T.N., Proudfoot, A.E., 2007. Molecular cloning and characterization of a highly selective chemokine-binding protein from the tick *Rhipicephalus sanguineus*. J. Biol. Chem. 282, 27250–27258.

Fry, B.G., Roelants, K., Champagne, D.E., Scheib, H., Tyndall, J.D.A., King, G.F., Nevalainen, T.J., Norman, J.A., Lewis, R.J., Norton, R.S., Renjifo, C., Rodriguez de la Vega, R.C.R., 2009. The toxicogenomic multiverse: convergent recruitment of proteins into animal venoms. Annu. Rev. Gen. Hum. Genet. 10, 23.1–23.29.

Giribet, G., Edgecombe, G.D., Wheeler, W.C., Babbitt, C., 2002. Phylogeny and systematic position of opiliones: a combined analysis of chelicerate relationships using morphological and molecular data. Cladistics 18, 5–70.

Greinacher, A., Warkentin, T.E., 2008. The direct thrombin inhibitor hirudin. Thromb. Haemost. 99, 819–829.

Grimaldi D., Engel, M.S., 2005. Evolution of the Insects, first ed. Cambridge University Press, New York.

Guo, Y., Ribeiro, J.M., Anderson, J.M., Bour, S., 2009. dCas: a desktop application for cDNA sequence annotation. Bioinformatics 25, 1195–1196.

Gutierrez, J.M., Lomonte, B., Leon, G., Alape-Giron, A., Flores-Diaz, M., Sanz, L., Angulo, Y., Calvete, J.J., 2009. Snake venomics and antivenomics: proteomic tools in the design and control of antivenoms for the treatment of snakebite envenoming. J. Proteomics 72, 165–182.

Hepburn, N.J., Williams, A.S., Nunn, M.A., Chamberlain-Banoub, J.C., Hamer, J., Morgan, B.P., Harris, C.L. 2007. *In vivo* characterization and therapeutic efficacy of a C5-specific inhibitor from the soft tick *Ornithodoros moubata*. J. Biol. Chem. 282, 8292–8299.

Hostomská, J., Volfová, V., Mu, J., Garfield, M., Rohoušová, I., Volf, P., Valenzuela, J.G., Jochim, R.C., 2009. Analysis of salivary transcripts and antigens of the sand fly *Phlebotomus arabicus*. BMC Genomics 10, 282.

Howlett, R., 1998. Nobel award stirs up debate on nitric oxide breakthrough. Nature 395, 625–626.

Kato, G.J., 2008. Novel small molecule therapeutics for sickle cell disease: nitric oxide, carbon monoxide, nitrite, and apolipoprotein A-I. Hematology Am. Soc. Hematol. Educ. Program 186–192.

Kato, H., Anderson, J.M., Kamhawi, S., Oliveira, F., Lawyer, P.G., Pham, V.M., Sangare, C.S., Samake, S., Sissoko, I., Garfield, M., Sigutova, L., Volf, P., Doumbia, S., Valenzuela, J.G., 2006. High degree of conservancy among secreted salivary gland proteins from two geographically distant *Phlebotomus duboscqi* sandflies populations (Mali and Kenya). BMC Genomics 7, 226.

Kato, H., Jochim, R.C., Sakoda, R., Iwata, H., Gomez, E.A., Valenzuela, J.G., Hashiguchi, Y., 2008. A repertoire of the dominant transcripts from the salivary glands of the blood-sucking bug, *Triatoma dimidiata*, a vector of Chagas disease. Inf. Gen. Evol. 10, 184–191.

Klompen, H., Lekveishvili, M., Black., W.C., 2007. Phylogeny of parasitiform mites (Acari) based on rRNA. Mol. Phyl. Evol. 43, 936–951.

Koh, C.Y., Kini, R.M., 2009. Molecular diversity of anticoagulants from haematophagous animals. Thromb. Haemost. 102, 437–453.

Krug, S., Sablotzki, A., Hammerschmidt, S., Wirtz, H., Seyfarth, H.J., 2009. Inhaled iloprost for the control of pulmonary hypertension. Vasc. Health Risk Manag. 5, 465–474.

Lai, R., Takeuchi, H., Jonczy, J., Rees, H.H., Turner, P.C., 2004. A thrombin inhibitor from the ixodid tick, *Amblyomma hebraeum*. Gene 342, 243–249.

Law, J., Ribeiro, J.M.C., Wells, M., 1992. Biochemical insights derived from diversity in insects. Ann. Rev. Biochem. 61, 87–112.

Lehane, M., 2005. The Biology of Blood-Sucking in Insects, first ed. Cambridge University Press, New York.

Macedo-Ribeiro, S., Almeida, C., Calisto, B.M., Friedrich, T., Mentele, R., Sturzebecher, J., Fuentes-Prior, P., Pereira, P.J.B. 2008. Isolation, cloning and structural characterisation of boophilin, a multifunctional Kunitz-type proteinase inhibitor from the cattle tick. PLoS One 3, 1–17.

Madden, R.D., Sauer, J.R., Dillwith, J.W., 2002. A proteomics approach to characterizing tick salivary secretions. Exp. Appl. Acarol. 28, 77–87.

Mans, B.J., 2005. Tick histamine-binding proteins and related lipocalins: potential as therapeutic agents. Curr. Opin. Investig. Drugs 6, 1131–1135.

Mans, B.J., Louw, A.I., Neitz, A.W., 2002. Evolution of hematophagy in ticks: common origins for blood coagulation and platelet aggregation inhibitors from soft ticks of the genus *Ornithodoros*. Mol. Biol. Evol. 19, 1695–1705.

Mans, B.J., Neitz, A.W., 2004. Adaptation of ticks to a blood-feeding environment: evolution from a functional perspective. Insect Biochem. Mol. Biol. 34, 1–17.

Mans, B.J., Calvo, E., Ribeiro, J.M.C., Andersen, J.F. 2007. The crystal structure of D7r4, a salivary biogenic amine-binding protein from the malaria mosquito *Anopheles gambiae*. J. Biol. Chem. 282, 36626–36633.

Mans, B.J., Andersen, J.F., Francischetti, I.M., Valenzuela, J.G., Schwan, T.G., Pham, V.M., Garfield, M.K., Hammer, C.H., Ribeiro, J.M., 2008a. Comparative sialomics between hard and soft ticks: implications for the evolution of blood-feeding behavior. Insect Biochem. Mol. Biol. 38, 42–58.

Mans, B.J., Andersen, J.F., Schwan, T.G., Ribeiro, J.M., 2008b. Characterization of anti-hemostatic factors in the argasid, *Argas monolakensis*: implications for the evolution of bloodfeeding in the soft tick family. Insect Biochem. Mol. Biol. 38, 22–41.

Mans, B.J., Ribeiro, J.M., Andersen, J.F., 2008c. Structure, function, and evolution of biogenic amine-binding proteins in soft ticks. J. Biol. Chem. 283, 18721–18733.

Mans, B.J., Ribeiro, J.M.C., 2008a. Function, mechanism and evolution of the moubatin-clade of soft tick lipocalins. Insect Biochem. Mol. Biol. 38, 841–852.

Mans B.J., Ribeiro, J.M.C. 2008b. A novel clade of cysteinyl leukotriene scavengers in soft ticks. Insect Biochem. Mol. Biol. 38, 862–870.

Maritz-Olivier, C., Stutzer, C., Jongejan, F., Neitz, A.W., Gaspar, A.R., 2007. Tick antihemostatics: targets for future vaccines and therapeutics. Trends Parasitol. 23, 397–407.

Montfort, W.R., Weichsel, A., Andersen, J.F. 2000. Nitrophorins and related antihemostatic lipocalins from *Rhodnius prolixus* and other blood-sucking arthropods. Biochim. Biophys. Acta 1482, 110–118.

Nazareth, R.A., Tomaz, L.S., Ortiz-Costa, S., Atella, G.C., Ribeiro, J.M., Francischetti, I.M., Monteiro, R.Q., 2006. Antithrombotic properties of Ixolaris, a potent inhibitor of the extrinsic pathway of the coagulation cascade. Thromb. Haemost. 96, 7–13.

Nene, V., Lee, D., Quackenbush, J., Skilton, R., Mwaura, S., Gardner, M.J., Bishop, R., 2002. AvGI, an index of genes transcribed in the salivary glands of the ixodid tick *Amblyomma variegatum*. Int. J. Parasitol. 32, 1447–1456.

Nene, V., Lee, D., Kang'a, S., Skilton, R., Shah, T., de Villiers, E., Mwaura, S., Taylor, D., Quackenbush, J., Bishop, R., 2004. Genes transcribed in the salivary glands of female *Rhipicephalus appendiculatus* ticks infected with *Theileria parva*. Insect Biochem. Mol. Biol. 34, 1117–1128.

Nuttall, G.H.F., Strickland, B.A., 1909. On the presence of an anticoagulin in the salivary glands and intestines of *Argas persicus*. Parasitology 1, 302–310.

Oates, J.A., 1982. The 1982 Nobel Prize in Physiology or Medicine. Science 218, 765–768.
Oleaga, A., Escudero-Poblacion, A., Camafeita, E., Perez-Sanchez, R., 2007. A proteomic approach to the identification of salivary proteins from the argasid ticks *Ornithodoros moubata* and *Ornithodoros erraticus*. Insect Biochem. Mol. Biol. 37, 1149–1159.
Oliveira, F., Kamhawi, S., Seitz, A.E., Pham, V.M., Guigal, P.M., Fischer, L., Ward, J., Valenzuela, J.G. 2006. From transcriptome to immunome: identification of DTH inducing proteins from a *Phlebotomus ariasi* salivary gland cDNA library. Vaccine 24, 374–390.
Paesen, G.C., Adams, P.L., Harlos, K., Nuttall, P.A., Stuart, D.I., 1999. Tick histamine-binding proteins: isolation, cloning, and three-dimensional structure. Mol. Cell 3, 661–671.
Radovsky, F.J., 1969. Adaptive radiation in the parasitic mesostigmata. Acarologia 11, 450–483.
Ribeiro, J.M., Makoul, G.T., Levine, J., Robinson, D.R., Spielman, A., 1985. Antihemostatic, anti-inflammatory, and immunosuppressive properties of the saliva of a tick, *Ixodes dammini*. J. Exp. Med. 161, 332–344.
Ribeiro, J.M.C., Marinotti, O., Gonzales, R., 1990. A salivary vasodilator in the blood-sucking bug, *Rhodnius prolixus*. Br. J. Pharmacol. 101, 932–936.
Ribeiro, J.M., Hazzard, J.M., Nussenzveig, R.H., Champagne, D.E., Walker, F.A., 1993. Reversible binding of nitric oxide by a salivary heme protein from a bloodsucking insect. Science 260, 539–541.
Ribeiro, J.M., 1995. Blood-feeding arthropods: live syringes or invertebrate pharmacologists? Infect. Agents Dis. 4, 143–152.
Ribeiro, J.M., Francischetti, I.M., 2003. Role of arthropod saliva in blood feeding: sialome and post-sialome perspectives. Annu. Rev. Entomol. 48, 73–88.
Ribeiro, J.M., Andersen, J.F., Silva-Neto, M.A., Pham, V.M., Garfield, M.K., Valenzuela, J.G., 2004a. Exploring the sialome of the blood-sucking bug *Rhodnius prolixus*. Insect Biochem. Mol. Biol. 34, 61–79.
Ribeiro, J.M., Charlab, R., Pham, V.M., Garfield, M., Valenzuela, J.G., 2004b. An insight into the salivary transcriptome and proteome of the adult female mosquito *Culex pipiens quinquefasciatus*. Insect Biochem. Mol. Biol. 34, 543–563.
Ribeiro, J.M., Alarcon-Chaidez, F., Francischetti, I.M., Mans, B.J., Mather, T.N., Valenzuela, J.G., Wikel, S.K., 2006. An annotated catalog of salivary gland transcripts from *Ixodes scapularis* ticks. Insect Biochem. Mol. Biol. 36, 111–129.
Ribeiro, J.M., Arca, B., Lombardo, F., Calvo, E., Phan, V.M., Chandra, P.K., Wikel, S.K., 2007. An annotated catalogue of salivary gland transcripts in the adult female mosquito, *Aedes aegypti*. BMC Genomics 8, 6.
Russell, C.L., Heesom, K.J., Arthur, C.J., Helps, C.R., Mellor, P.S., Day, M.J., Torsteinsdottir, S., Björnsdóttir, T.S., Wilson, A.D., 2009. Identification and isolation of cDNA clones encoding the abundant secreted proteins in the saliva proteome of *Culicoides nubeculosus*. Insect Mol. Biol. 18, 383–393.
Sabbatani, L., 1899. Fermento anticoagulante de l' "*Ixodes ricinus*". Arch. Ital. Biol. 31, 37–53.
Sangamnatdej, S., Paesen, G.C., Slovak, M., Nuttall, P.A., 2002. A high affinity serotonin- and histamine-binding lipocalin from tick saliva. Insect Mol. Biol. 11, 79–86.
Santos, A., Ribeiro, J.M.C., Lehane, M.J., Gontijo, N.F., Veloso, A.B., Sant′ Anna, M.R.V., Nascimento Araujo, R., Grisard, E.C., Pereira, M.H., 2007. The sialotranscriptome of the blood-sucking bug *Triatoma brasiliensis* (Hemiptera, Triatominae). Insect Biochem. Mol. Biol. 37, 702–712.
Schofield, C.J., Galvão, C., 2009. Classification, evolution, and species groups within the Triatominae. Acta. Trop. 110, 88–100.
Schuh, R.T., Weirauch, C., Wheeler, W.C., 2009. Phylogenetic relationships within the cimicomorpha (Hemiptera: Heteroptera): a total-evidence analysis. Sys. Entomol. 34, 15–48.
Skerra, A., 2008. Alternative binding proteins: anticalins – harnessing the structural plasticity of the lipocalin ligand pocket to engineer novel binding activities. FEBS J. 275, 2677–2683.
Steen, N.A., Barker, S.C., Alewood, P.F., 2006. Proteins in the saliva of the ixodida (ticks): pharmacological features and biological significance. Toxicon 47, 1–20.

Valenzuela, J.G., Charlab, R., Galperin, M.Y., Ribeiro, J.M., 1998. Purification, cloning, and expression of an apyrase from the bed bug *Cimex lectularius*. A new type of nucleotide-binding enzyme. J. Biol. Chem. 273, 30583–30590.

Valenzuela, J.G., Francischetti, I.M., Ribeiro, J.M., 1999. Purification, cloning, and synthesis of a novel salivary anti-thrombin from the mosquito *Anopheles albimanus*. Biochemistry 38, 11209–11215.

Valenzuela, J.G., Francischetti, I.M., Pham, V.M., Garfield, M.K., Mather, T.N., Ribeiro, J.M., 2002a. Exploring the sialome of the tick *Ixodes scapularis*. J. Exp. Biol. 205, 2843–2864.

Valenzuela, J.G., Pham, V.M., Garfield, M.K., Francischetti, I.M., Ribeiro, J.M.C., 2002b. Toward a description of the sialome of the adult female mosquito *Aedes aegypti*. Insect Biochem. Mol. Biol. 32, 1101–1122.

Valenzuela, J.G., Francischetti, I.M., Pham, V.M., Garfield, M.K., Ribeiro, J.M., 2003. Exploring the salivary gland transcriptome and proteome of the *Anopheles stephensi* mosquito. Insect Biochem. Mol. Biol. 33, 717–732.

Valenzuela, J.G., Garfield, M., Rowton, E.D., Pham, V.M., 2004. Identification of the most abundant secreted proteins from the salivary glands of the sand fly *Lutzomyia longipalpis*, vector of *Leishmania chagasi*. J. Exp. Biol. 207, 3717–3729.

Van Den Abbeele, J., Caljon, G., Dierick, J.F., Moens, L., De Ridder, K., Coosemans, M., 2007. The *Glossina morsitans* tsetse fly saliva: general characteristics and identification of novel salivary proteins. Insect Biochem. Mol. Biol. 37, 1075–1085.

Vieira, A.T., Fagundes, C.T., Alessandri, A.L., Castor, M.G., Guabiraba, R., Borges, V.O., Silveira, K.D., Vieira, E.L., Goncalves, J.L., Silva, T.A., Deruaz, M., Proudfoot, A.E., Sousa, L.P., Teixeira, M.M., 2009. Treatment with a novel chemokine-binding protein or eosinophil lineage-ablation protects mice from experimental colitis. Am. J. Pathol. 175, 2382–2391.

Walter, D.E., Proctor, H.C., 1999. Mites: Ecology, Evolution and Behaviour, first ed. CABI Publishing, Wallingford.

Wang X., Ribeiro J.M.C., Broce A.B., Wilkerson M.J., Kanost M.R., 2009a. An insight into the transcriptome and proteome of the salivary gland of the stable fly, *Stomoxys calcitrans*. Insect Biochem. Mol. Biol. 39, 607–614.

Wang M., Guerrero F.D., Pertea G., Nene V.M., 2009b. Global comparative analysis of ESTs from the southern cattle tick, *Rhipicephalus (Boophilus) microplus*. BMC Genomics 8, 368.

Weichsel, A., Maes, E.M., Andersen, J.F., Valenzuela, J.G., Shokhireva, T., Walker, F.A., Montfort, W.R., 2005. Heme-assisted S-nitrosation of a proximal thiolate in a nitric oxide transport protein. Proc. Natl. Acad. Sci., U.S.A. 102, 594–599.

Chapter 4
Snake Venomics, Antivenomics, and Venom Phenotyping: The *Ménage à Trois* of Proteomic Tools Aimed at Understanding the Biodiversity of Venoms

Juan J. Calvete

Abstract This review covers the application of proteomic protocols ("venomics", "antivenomics", and "venom phenotyping") to studying the composition and natural history of snake venoms, and the crossreactivity of antivenoms against homologous and heterologous venoms. Toxins from the same protein family present in venoms from snakes belonging to different genera often share antigenic determinants. This circumstance offers the possibility of defining the minimal set of venoms containing the epitopes necessary to generate therapeutic broad-range polyvalent antisera. Recent work shows how the knowledge of evolutionary trends along with venom phenotyping may be used to replace the traditionally used phylogenetic hypothesis for antivenom production strategies by cladistic clustering of venoms based on proteome phenotype and immunological profile similarities.

Introduction: Aims and Goals of Venomics

Venomous organisms are widely spread throughout the animal kingdom, comprising more than 100,000 species distributed among all major phyla, such as chordates (reptiles, fishes, amphibians, mammals), echinoderms (starfishes, sea urchins), molluscs (cone snails, cephalops), annelids (leeches), nemertines, arthropods (arachnids, insects, myriapods) and cnidarians (sea anemones, jellyfish, corals) (De Lima et al., 2009; Kem and Turk, 2009; Mackessy, 2009). In any habitat there is a competition for resources and every ecosystem on Earth supporting life contains poisonous or venomous organisms. One of the most fascinating techniques of capturing prey or defending oneself is the use of poisons or venoms. Venom represents an adaptive trait and an example of convergent evolution (Fry et al.,

J.J. Calvete (✉)
Laboratorio de Proteinómica Estructural, Instituto de Biomedicina de Valencia, CSIC, Jaime Roig 11, 46010 Valencia, Spain
e-mail: jcalvete@ibv.csic.es

R.M. Kini et al. (eds.), *Toxins and Hemostasis*, DOI 10.1007/978-90-481-9295-3_4,

2009a, b). Venom proteins are the result of the duplication of ordinary body proteins, often involved in key physiological processes, that were recruited through still elusive mechanism(s) into the venom proteome (Fry, 2005). Venoms are deadly cocktails, each comprising unique mixtures of peptides and proteins naturally tailored by Natural Selection to act on vital systems of the prey or victim. Venom toxins disturb the activity of critical enzymes, receptors, or ion channels, thus disarranging the central and peripheral nervous systems, the cardiovascular and the neuromuscular systems, blood coagulation and homeostasis. The more extensively studied, medically important clades of venomous animals are cone snails (700 or so known species) (Norton and Olivera, 2006; Olivera, 2006; Olivera et al., 2009), snakes (about 725 venomous species) (Doley and Kini, 2009; Fox and Serrano, 2005a, 2009; Mackessy, 2009), scorpions (~1,500 species) (Billen et al., 2008; Chippaux and Goyffon, 2008; Espino-Solís et al., 2009; Favreau et al., 2006), and spiders (ca. 37,000 known species) (Escoubas et al., 2008; Espino-Solís et al., 2009; Liang, 2008; Vetter and Isbister, 2008; Wood et al., 2009).

Envenomation is a highly relevant global public health issue, as there are venomous organisms in every continent and almost every country. Venomous animals are particularly abundant in tropical regions, which represent the kitchen of Evolution. Arthropod stings inflicted by bees, wasps, ants, spiders, scorpions, and -to a lesser extent- millipedes and centipedes, constitute the most common cause of envenoming by animals, although around 80% of deaths by envenomation worldwide are caused by snakebites, followed by scorpion stings, which cause 15%. The annual number of scorpion stings exceeds 1.2 million leading to more than 3,250 deaths (Chippaux and Goyffon, 2008; Espino-Solís et al., 2009). The yearly number of envenomation accidents by snakebite and its associated mortality are difficult to estimate, since there are many countries where this pathology is not appropriately reported, and since epidemiological data are often fragmentary. Furthermore, the majority of snakebite victims seek traditional treatment and may die at home unrecorded. Nevertheless, the actual incidence of snakebite envenoming world wide may exceed 2.5 millions resulting in 125,000 deaths; an even large number of victims end up with permanent physical disability (Kasturiratne et al., 2008; Stock et al., 2007). Therefore, if this pathology is analyzed in terms of DALYs ("disability-adjusted life years") lost, its impact is even greater (WHO, 2007). Most envenomation accidents occur in low-income countries of Africa, Asia and Latin America and affect mainly rural workers and children. Although about 2.3 billion of the world population are at risk of this pathology, it has been systematically neglected by health authorities. Envenomation has been recognised by the World Health Organization as a "neglected tropical disease" (Gutiérrez et al., 2006, 2009a; Stock et al., 2007; WHO, 2007).

Adequate treatment of envenoming is critically dependent on the ability of antivenoms to neutralize the lethal toxins reversing thereby the symptoms of envenoming (Espino-Solís et al., 2009; Gutiérrez and León, 2009). Effective treatment of envenoming is based on immunotherapy using animal-derived antivenoms, which is not entirely satisfactory. Current antivenoms consist of purified immunoglobulins, which have reduced the incidence and severity of treatment-induced serum sickness and anaphylactic shock. However, the venom-immunisation protocols have not changed in over a century and make no attempt to direct the immune response to the most pathogenic venom proteins (many venom proteins are not toxic and many low molecular mass venom proteins are highly toxic but weakly immunogenic). Consequently, the dose-efficacy of antivenoms is thought to suffer from the presence of redundant antibodies to non-toxic molecules and a lack of potent neutralising antibodies to small molecular mass toxins. This in turn results in the need for high volumes for effective treatment and a consequent increase in the risk of serum-sickness and adverse anaphylactic effects. Improvements in therapeutic management would reduce the lethality very significantly. In this respect, a thorough knowledge of venom composition and of ontogenetic, individual, and geographic venom variability ("venomics") may have a positive effect on the treatment of bite victims and in the selection of specimens for the generation of improved antidotes (Calvete, 2009a; Gutiérrez et al., 2009). Hence, besides ecological and taxonomical implications, a major aim of venomics is to gain a deeper insight into the spectrum of medically important toxins in venoms and to understand the molecular mechanisms and evolutionary forces that underlie venom variation to uncover clues in order to improve the medical effectiveness of antivenoms.

Venom proteomes may also be employed as taxonomic signatures (species-specific fingerprints) for unambiguous species identification independently of geographic origin and morphological characteristics (Tashima et al., 2008), and to infer intraspecific effects of gene regulation on protein expression (Barlow et al., 2009; Gibbs et al., 2009). In addition to understanding how venoms evolve, the application of venomics to characterize the large molecular variability within toxin families may contribute to a deeper understanding of the biological effects of the venoms. The occurrence of multiple isoforms within each major toxin family provides evidence for the emergence of paralogous groups of multigene families across taxonomic lineages where gene duplication events occurred prior to their divergence, and suggests an important role for balancing selection in maintaining high levels of functional variation in venom proteins within populations. The mechanism leading to this mode of selection is unclear but it has been speculated that it may be related to the unpredictability with which a sit-and-wait predator like a rattlesnake encounters different types of prey, each of which are most efficiently subdued with different venom proteins (Gibbs et al., 2009; Sanz et al., 2006). Thus, to deal with this uncertainty, snakes are required to have a variety of proteins "available" in their venom at all times to deal with different prey. The selection pressure leading to high levels of variation in venom genes may parallel the selection pressures acting by the birth-and-death model of protein evolution (Nei et al., 1997).

Paradoxically, although bites can be deadly, snake venoms also contain components of theraupeutic value. On relatively rare occasions, toxins are potential therapeutic agents that have been used for the treatment of pathophysiological conditions in homeopathy, in folk remedies, and in Western and Chinese traditional medicine (Fox and Serrano, 2007; Harvey et al., 1998; Koh et al., 2007). Understanding how toxin genes are regulated and how toxins mutate in an accelerated fashion may reveal not only the molecular basis for adaptive variations in snake phenotypes (Gibbs et al., 2009), but also how to use deadly toxins as therapeutic agents (Calvete, 2009b).

The field of venomics faces the exciting challenges of delineating structure-function correlations and designing "a la carte" antivenom production strategies. However, the cost of production of high-quality antivenoms is high relative to most other drugs. This is mainly due to the complexity of their elaboration, from husbandry and immunization of large animals to processing of plasma and large-scale antibody purification. Antivenoms have become expensive for fledging health systems of low-income countries where snake bites are a real medical problem. This makes production by biotech industries less attractive, which either increase prices, fueling a vicious circle, or more frequently stop production (Stock et al., 2007; Theakston and Warrell, 2000). A challenge the venomic techniques described below is to help to reduce antivenom prices through the identification of venom mixtures for generating geographically relevant polyvalent antivenoms against the widest possible range of species.

A further challenge of venomics is unravelling the evolutionary mechanisms of recruitment and transformation of ordinary proteins into deadly toxins. Modern venom proteomics have been under investigation since the very earliest biochemical studies (Fox and Serrano, 2009), though the field has began to flourish owing to the recent application of mass spectrometry, coupled with improvements of databases of venom protein sequences provided by transcriptomic analyses of venom glands (Junqueira de Azevedo et al., 2009), for the characterisation of toxin proteomes (Calvete et al., 2007a; Serrano and Fox, 2008). The significance of proteomics methodologies for biological research has been granted by the pace of genome sequencing projects. However, organisms with unsequenced genomes, including venomous animals, still represent the overwhelming majority of species in the biosphere. Thus, despite its exponential progress, venomics is still in its infancy.

To really understand how venoms work, quantitative data on the occurrence of individual toxins in a given venom are required. Abundant venom proteins may perform generic killing and digestive functions that are not prey specific whereas low abundance proteins may be more plastic either in evolutionary or ecological timescales. Mutations provide the ground on which natural selection operates to create functional innovations. A subset of low abundance proteins may serve to "customize" individual venom to feeding on particular prey or may represent orphan molecules evolving under neutral selection "in search of a function" (Juárez et al., 2008). Hence, whereas abundant proteins are the primary targets for immunotherapy, minor components may represent scaffolds for biotechnological developments (Calvete et al., 2009a).

Snake Venomics: Strategy and Applications

To explore putative venom components, several laboratories have carried out transcriptomic analyses of the Duvernoy's (venom) glands of viperid snake species, including *Bothrops insularis* (Junqueira-de-Azevedo and Ho, 2002), *Bitis gabonica* (Francischetti et al., 2004), *Bothrops jararacussu* (Kashima et al., 2004), *Bothrops jararaca* (Cidade et al., 2006), *Bothrops atrox* (Neiva et al., 2009), *Agkistrodon acutus* (Qinghua et al., 2006; Zhang et al., 2006), *Echis ocellatus* (Wagstaff and

Harrison, 2006; Wagstaff et al., 2009), *Lachesis muta* (Junqueira-de-Azevedo et al., 2006), *Crotalus durissus collilineatus* (Boldrini-França et al., 2009), *Agkistrodon piscivorus leucostoma* (Jia et al., 2008), and *Sistrurus catenatus edwardsii* (Pahari et al., 2007). Transcriptomic investigations provide catalogues of partial and full-length transcripts that are synthesized by the venom gland. However, transcriptomes include translated and non-translated mRNAs, transcripts encoding non-secreted proteins, housekeeping, and cellular genes, in addition to toxin precursor genes. Also, toxins may undergo posttranslational processing, which will not be evident from a transcriptomic analysis. Thus, outlining the full map of native toxins that actually constitute the venom requires a combined biochemical and proteomic approach. In addition, the transcriptome does not reflect within-species ontogenetic, individual and geographic heterogeneity of venoms, which may account for differences in the clinical symptoms observed in envenomation.

For a detailed characterization of the toxin content of snake venoms ("venome"), the approach that we have coined "*snake venomics*" (Juárez et al., 2004) starts with the fractionation of the crude venom by reverse-phase HPLC, followed by the initial characterization of each protein fraction by a combination of N-terminal sequencing, SDS-PAGE analysis, and mass spectrometric determination of the molecular masses and the cysteine (SH and S–S) content of the isolated toxins. For the further fractionation of venom proteins we use a reverse-phase HPLC C_{18} column (250 × 4 mm, 5 μm particle size) eluting at 1 ml/min with a linear gradient of 0.1% TFA in water (solution A) and acetonitrile (solution B) (typically, isocratic elution with 5% B for 5 min, followed by linear gradients of 5–15% B over, 20 min, 15–45% B over 120 min, and 45–70% B over, 20 min) (Fig. 4.1a). In our experience, this procedure allows the quantitative recovery of all venom components present in the apparent molecular mass range of 7–150 kDa that can be separated by conventional 2D-SDS-PAGE. On the other hand, the initial part of the acetonitrile gradient of the reverse-phase chromatography resolves peptides and small proteins (0.4–7 kDa), which would not be recovered from a 2D-electrophoretic separation. Moreover, for the accurate determination of toxin-specific features, such as the native molecular mass, the quaternary structural arrangement and the number of sulfhydryl groups and disulfide bonds, toxins need to be available in solution. In addition, given that the wavelength of absorbance of a peptide bond is, 190–230 nm, protein detection at 215 nm allows the estimation of the relative abundances (expressed as percentage of the total venom proteins) of the different protein families from the relation of the sum of the areas of the reverse-phase chromatographic peaks containing proteins from the same family to the total area of venom protein peaks in the reverse-phase chromatogram. In a strict sense, and according to the Lambert-Beer law, the calculated relative amounts correspond to the "% of total peptide bonds in the sample", which is a good estimate of the % by weight (g/100 g) of a particular venom component. The relative contributions of different proteins eluting in the same chromatographic fraction can be estimated by densitometry after SDS-PAGE separation (Fig. 4.1b).

Protein fractions showing a single electrophoretic band and N-terminal sequence (i.e. fractions 7, 8, 11, 12, 23, 29 in panel b of Fig. 4.1) can be straightforwardly assigned by BLAST analysis (http://www.ncbi.nlm.nih.gov/BLAST) to a known

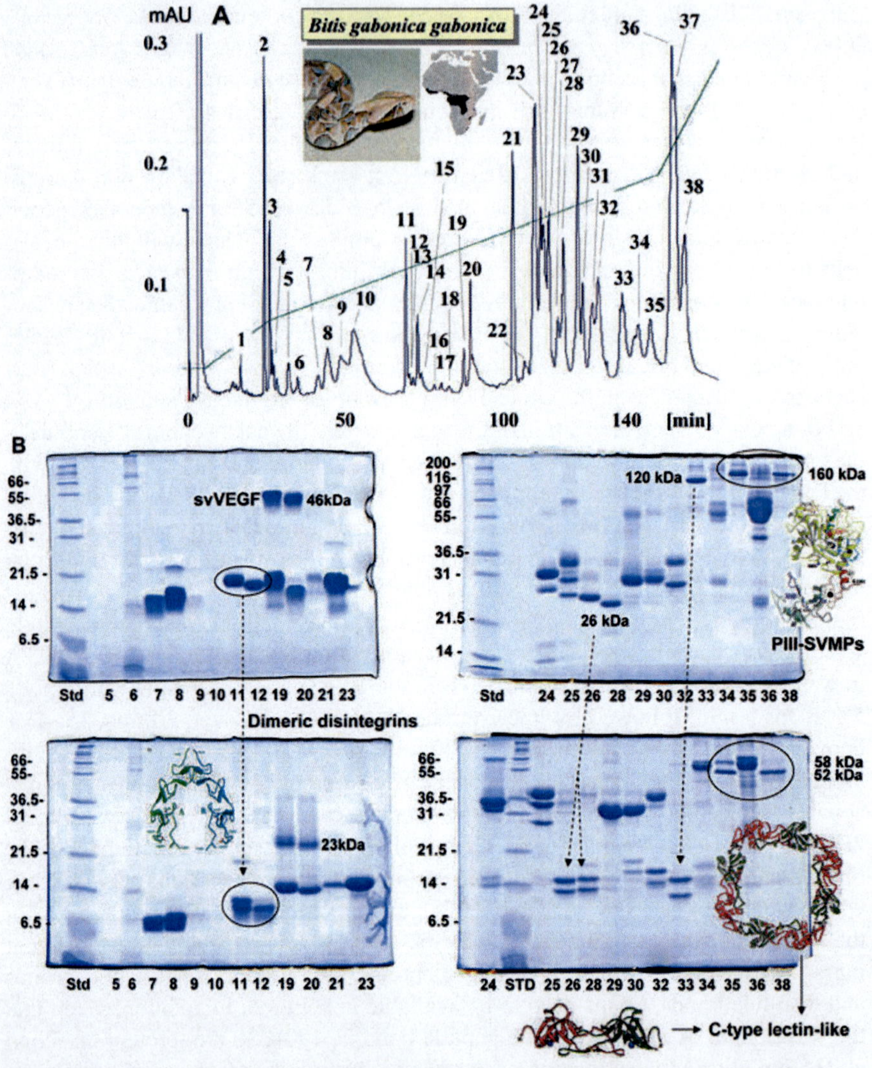

Fig. 4.1 Snake venomics. Schematic representation of some of the steps typically followed in a snake venomics project. (**a**) Reverse-phase chromatographic separation of the venom proteins. Each fraction, from *Bitis gabonica gabonica* in this illustration, was subjected to N-terminal sequencing, molecular mass and disulfide bond content determinations, and SDS-PAGE analysis; (**b**) SDS-PAGE of the RP-HPLC isolated proteins run under non-reduced (*upper panel*) and reduced (*lower panels*) conditions; fractions containing multimeric proteins are highlighted. (**c**) Determination of the molecular masses of the proteins contained in a reverse-phase fraction isolated as in (**a**); this fraction contained a serine proteinase (29 kDa) and a PLA_2molecule (13 kDa); and (**d**) amino acid sequence determination by nanospray-ionization CID-MS/MS of a doubly-and tryptic peptide ion. Sequence-specific "b" and "y" ions are labelled. In this example, the deduced amino acid sequence (231.2)DYEEF(I/L)E(I/L)AK identified an L-amino acid oxidase by a BLAST similarity search

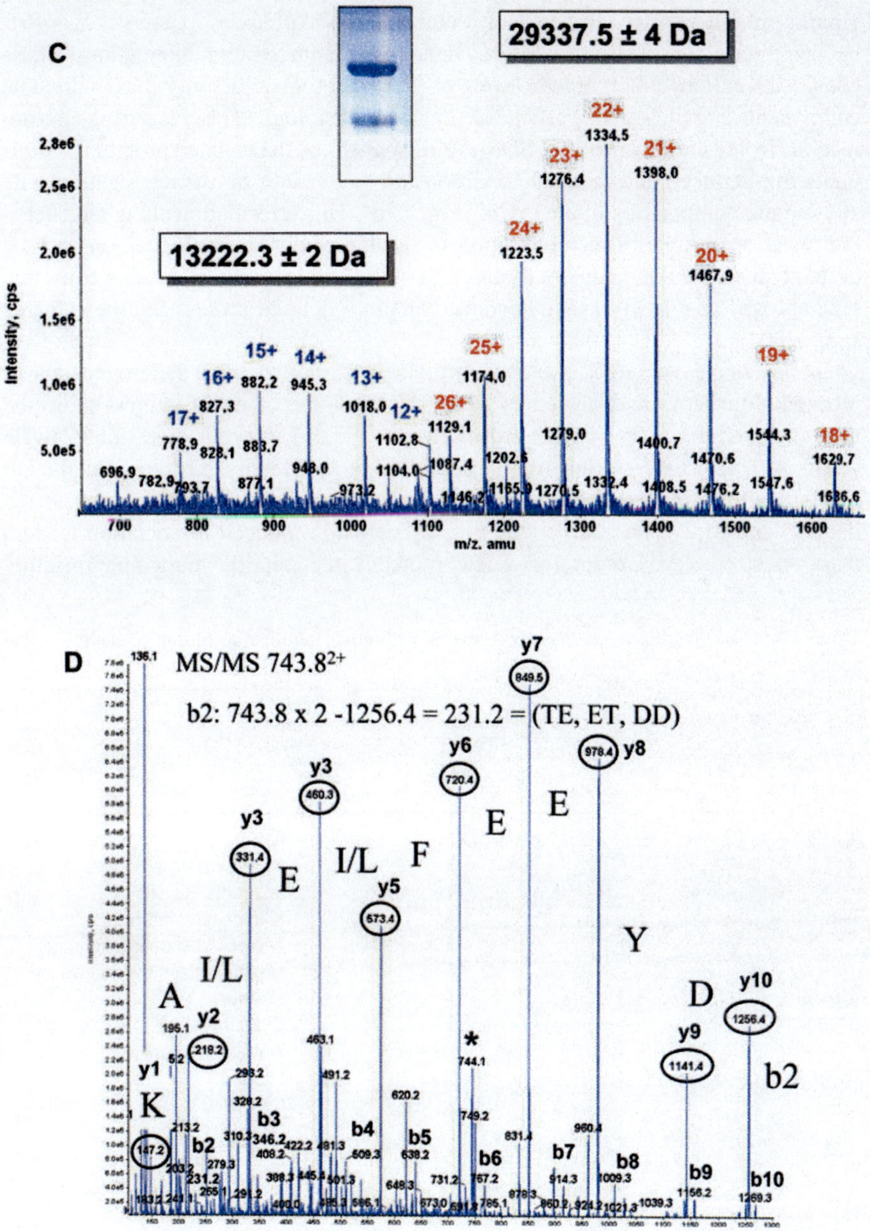

Fig. 4.1 (continued)

viperid protein family. Most venom proteins and polypeptides exist as monomers, but some form complexes through covalent and/or non-covalent interactions. These complexes exhibit much higher levels of pharmacological activity than individual components and play an important role in pathophysiological effects during envenomation (Doley and Kini, 2009). SDS-PAGE analysis of the isolated protein fractions under non-reduced and reduced conditions provides clues on the aggregation state and subunit composition of the toxins (Fig. 4.1b). This information along with accurate mass spectrometric determination of the molecular mass (Fig. 4.1c) and the cysteine/disulfide bond content (Table 4.1) of the native toxins represents a fast and efficient way to classify snake venom toxins into defined protein families (Juárez et al., 2004).

In our snake venomics protocol, protein fractions showing heterogeneous or blocked N-termini are analyzed by SDS-PAGE and the bands of interest subjected to automated reduction, carbamidomethylation, and in-gel tryptic digestion (Calvete et al., 2007a). The resulting tryptic peptides are then analyzed by MALDI-TOF mass fingerprinting followed by amino acid sequence determination of selected doubly- and triply-charged peptide ions by collision-induced dissociation tandem mass spectrometry. Except for a few proteins, the peptide mass fingerprinting

Table 4.1 Classification of snake venom toxins to protein families according to their cysteine content

Molecular mass range (kDa)	Total cysteine residues		Protein family
	–SH	S–S	
1.6–2	–	1	C-NP
4–5	–	3	Myotoxin
	–	4	Short disintegrin
6–8	–	3	Kunitz-type inhibitor
	–	5	Dimeric disintegrin subunit
	–	6	Medium-sized disintegrin
10–12	1	–	Ohanin
13–15	–	2	Cystatin
	–	(2[a]+ 4[b])	Dimeric disintegrin
	–	7	PLA_2
23–33	–	8	CRISP
	1	4	PI-SVMP
	–	6	Serine proteinase
	–	(1[a] + 3[b])	αβ snaclec (C-type lectin)
	–	(1[a] + 4[b])	svVEGF
	–	13	DC-fragment
46–58	–	3	LAO
	1	18	PIII-SVMP

C-NP C-type natriuretic peptide, *CRISP* cysteine-rich secretory protein, *SVMP* snake venom metalloproteinase, *DC* disintegrin-like/cysteine-rich domains of PIII-SVMP, *svVEGF* snake venom vascular endothelial growth factor, *LAO* L-amino acid oxidase
[a]intersubunit disulfide bonds; [b]intrasubunit disulfide linkages

approach alone is unable to identify any protein in the databases. In addition, as expected from the complete absence of sequenced snake genomes, and the rapid amino acid sequence divergence of venom proteins evolving under accelerated evolution (Ohno et al., 1998), with a few exceptions, the product ion spectra do not match any known protein using the ProteinProspector (http://prospector.ucsf.edu) or the MASCOT (http://www.matrixscience.com) search programs against the 1,200 entries of the UniProtKB/Swiss-Prot entries from taxon *Serpentes* (http://www.expasy.ch/sprot/tox-prot). Although the lack of any complete snake genome sequence is a serious drawback for the identification of venom proteins, high-quality MS/MS peptide ion fragmentation spectra usually yield sufficient amino acid sequence information derived from almost complete series of sequence-specific b- and/or y-ions to unambiguously identify a homologuous protein in the current databases by BLAST analysis of de novo sequenced tryptic peptide ions (Fig. 4.1d). Since some snake venom toxin families, such as the hemorrhagic Zn^{2+}-dependent metalloproteinases, comprise multidomain protein subfamilies, whose pharmacological activities are modulated by their different domain structures (Fox and Serrano, 2005b), accurate molecular mass and disulfide linkage content determination along with large sequence coverage are needed to distinguish between the different protein classes (Fig. 4.2 illustrates this point).

m/z	z	Amino acid sequence (X = I or L)	Protein domain
651.8	(2+)	NKYQTXYXTNR	Zn^{2+}-Metalloproteinase
719.5	(2+)	SVGXVEDHSSDHR	Zn^{2+}-Metalloproteinase
691.7	(3+)	NXXVANTMAHEXGHNXGXR	Zn^{2+}-Metalloproteinase
901.3	(3+)	(201.1)FQFSDCSKNKYQTYXTNR	Zn^{2+}-Metalloproteinase
679.3	(2+)	VAXVGXEYQXPR	Zn^{2+}-Metalloproteinase
826.3	(2+)	(FE)ENVVVADX(546.3)	Zn^{2+}-Metalloproteinase
764.9	(2+)	ATVAQGGCFEFNR	Zn^{2+}-Metalloproteinase
752.8	(2+)	GVTPSYQFSDGXPR	Zn^{2+}-Metalloproteinase
893.4	(2+)	TDXVSPPVCGNEXVER	Disintegrin-like
737.9	(3+)	XHSWVQCGNSECCQQCR	Disintegrin-like
627.3	(2+)	NNPCCNAEVCK	Disintegrin-like
910.8	(2+)	GQPSCNNNNGFCYGSK	Disintegrin-like/Cys-rich
964.9	(2+)	XYCSYDSFGYQXPCR	Cysteine-rich
698.3	(2+)	XFCETVPQSCR	Cysteine-rich
957.6	(2+)	ENNXENNTNVXEYVXR	Cysteine-rich
566.3	(2+)	XGNQYGYCR	Cysteine-rich
764.9	(2+)	ATVAQDACHQFR	Cysteine-rich

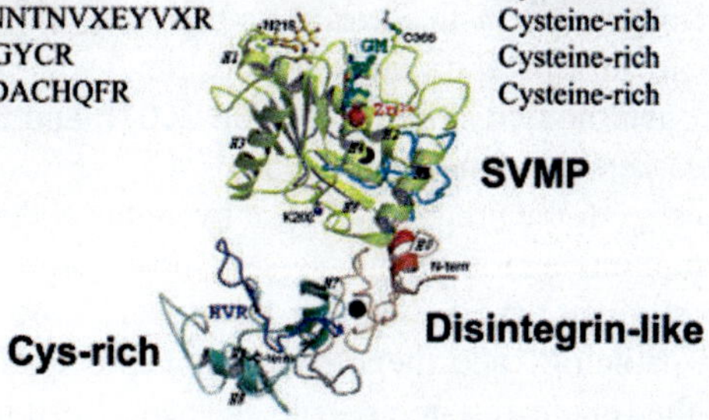

Fig. 4.2 Sequence coverage. List of tryptic peptide ions sequenced by collision-induced dissociation MS/MS from an in-gel tryptic digest of a Zn^{2+}-dependent snake venom metalloproteinase. The simultaneous occurrence of ions corresponding to the metalloproteinase (SVMP), disintegrin-like, and cysteine-rich (Cys-rich) domains identified this particular metalloproteinase as a PIII-SVMP

The outlined venomics strategy allows us to assign unambiguously all the isolated venom toxins representing $\geq$0.05% of the total venom proteins to known protein families. A major conclusion from these studies is that in spite of the fact that viperid venoms may contain well over 100 protein components (Serrano et al., 2005), venom proteins belong to only a few major protein families, including enzymes (serine proteinases, Zn^{2+}-metalloproteases, L-amino acid oxidases, group II PLA_2s) and proteins without enzymatic activity (ohanin, disintegrins, C-type lectins (or snaclecs in the nomenclature proposed by Clemetson et al., 2009), natriuretic peptides, myotoxins, CRISP toxins, nerve and vascular endothelium growth factors, cystatin and Kunitz-type protease inhibitors) (Table 4.2). This situation may reflect the fact that toxins likely evolved from a restricted set of protein families with normal physiological functions that were recruited into the venom proteome before the diversification of the advanced snakes, at the base of the Colubroidea radiation (Fry, 2005; Fry and Wüster, 2004; Fry et al., 2008; Vidal, 2002).

A long-term goal of our snake venomics project is a detailed analysis of all viperid venomes. Understanding how venoms work requires quantitative data on the occurrence of individual toxins in a given venom. To date, we have explored a number of venom proteomes of medically relevant vipers (Table 4.2) from different regions of different continents (Fig. 4.3) (Alape-Girón et al., 2008; Angulo et al., 2008; Bazaa et al., 2005; Calvete et al., 2007b, c, 2009b, c; Gutiérrez et al., 2008; Juárez et al., 2004, 2006; Lomonte et al., 2008; Núñez et al., 2009; Sanz et al., 2006, 2008a, b; Tashima et al., 2008; Wagstaff et al., 2009). A few other laboratories have also reported qualitative proteomic studies on several venoms, including those from the European vipers, *Vipera ammodytes ammodytes* and *Vipera ammodytes meridionalis* (Georgieva et al., 2008), *Vipera aspis* (Ferquel et al., 2007), Asian *Daboia russelli siamensis* (Risch et al., 2009), Amazonian *Bothrops atrox* (Guércio et al., 2006), and Brazilian *Bothrops jararaca* (Fox et al., 2006) and *Bothrops insularis* (Valente et al., 2009). These studies have allowed the inference of phylogenetic alliances within genera Bitis (Calvete et al., 2007c) and *Sistrurus* (Sanz et al., 2006); to rationalize the envenomation profiles of *Atropoides* (Angulo et al., 2008) and *Bothrops* (Gutiérrez et al., 2008) species; to reveal gene regulation effects on venom protein expression in *Sistrurus* rattlesnakes (Gibbs et al., 2009); to correlate geographic intraspecific venom composition variation and reappraisal of *Vipera aspis* venom neurotoxicity (Ferquel et al., 2007); and to define venom-associated taxonomic markers (Tashima et al., 2008).

A critical step in the preparation of antivenoms is the selection of venoms to be used in the immunizing mixture. Assessing the occurrence of qualitative differences in the composition of the venom of the same species are of fundamental importance in snakebite pathology and therapeutics, since, as a rule, ophidian envenomation results from the venom of a single snake. Knowledge of intraspecies venom variability is necessary for the selection of the regions from which snake specimens have to be collected for the preparation of the reference venom pool. We have applied snake venomics to establish the molecular basis of geographic, individual, and ontogenetic

Table 4.2 Overview of the relative occurrence of proteins (in percentage of the total HPLC-separated proteins) of the toxin families in the venoms of *Sistrurus catenatus catenatus* (SCC), *Sistrurus catenatus tergeminus* (SCT), *Sistrurus catenatus edwardsii* (SCE), *Sistrurus miliarius barbouri* (SMB), *Cerastes cerastes cerastes* (CCC), *Cerastes vipera* (CV) and *Macrovipera lebetina transmediterranea* (MLT), *Bitis arietans* (BA), *Bitis gabonica gabonica* (BGG), *Bitis gabonica rhinoceros* (BGR), *Bitis nasicornis* (BN), *Bitis caudalis* (BC), *Echis ocellatus* (EO), *Lachesis muta* (LM), *Crotalus atrox* (CA), *Agkistrodon contortrix contortrix* (ACC), *Macrovipera lebetina obtusa* (Mlo), *Vipera raddei* (Vr), *Atropoides picadoi* (Api), *Atropoides mexicanus* (Amex), *Bothops asper* (Caribbean versant of Costa Rica) (Bas(C)), *Bothops asper* (Pacific versant of Costa Rica) (Bas(P)), *Bothops caribbaeus* (Bcar), *Bothrops lanceolatus* (Blan), *Bothops cotiara* (Bco), *Bothrops fonsecai* (Bfon), *Bothriechis lateralis* (Bolat), *Bothriechis schlegeli* (Bosch), *Lachesis stenophrys* (Lste), *Bothrops colombiensis* (Bcol), *Bothrops atrox* (Colombia) (Bax_Co), *Bothrops atrox* (Brazil) (Bax_Br), *Crotalus simus* (adult) Cs(A), *Crotalus simus* (neonate) CsNeo, *Crotalus durissus durissus* (dryinas) (Cdd), *Crotalus durissus cumanensis* (CdCu), *Crotalus durissus ruruima* (Cdru), *Crotalus durissus terrificus* (Cdt), *Crotalus durissus collilineatus* (Crotamine +) ($Cdco^+$), *Crotalus durissus collilineatus* (Crotamine –) ($Cdco^-$)

Venom																
	SCC	*SCT*	*SCE*	*SMB*	*CCC*	*CV*	*MLT*	*BA*	*BGG*	*BGR*	*BN*	*BC*	*EO*	*LM*	*CA*	*ACC*
Protein family	% of total venom proteins															
Disintegrins																
–Long	–	–	–	–	–	–	–	17.8	–	–	–	–	–	–	–	–
–Medium	2.5	4.2	0.9	7.7	–	–	–	–	–	–	–	–	–	–	6.5	–
–Dimeric	–	–	–	–	8.1	<1	6.0	–	3.4	8.5	3.5	–	4.2	–	–	1.5
–Short	–	–	–	–	–	–	<1	–	–	–	–	–	2.6	–	–	–
Myotoxin	0.4	<0.1	–	–	–	–	–	–	–	–	–	–	–	–	–	–
C-type BPP/NP	–	–	<0.1	<0.1	–	–	<1	–	2.8	0.3	–	–	–	14.7	2.1	<0.1
Kunitz-type inhibitor	–	–	<0.1	<0.1	–	–	–	4.2	3.0	7.5	–	3.2	–	–	–	–
Cystatin	–	–	–	–	–	–	–	1.7	9.8	5.3	4.2	–	–	–	–	–
DC-fragment	<0.1	<0.1	<0.1	1.3	–	–	1.0	–	0.5	0.6	<0.1	–	1.7	–	–	<0.1
NGF/svVEGF	<0.1	<0.1	<0.1	<0.1	–	–	2.1	–	1.0	–	–	–	–	–	–	–
Ohanin-like	–	–	–	–	–	–	–	–	–	–	–	–	–	–	–	<0.1
CRISP	0.8	1.3	10.7	2.9	–	–	–	–	2.0	1.2	1.3	1.2	1.5	1.8	4.2	–

Table 4.2 (continued)

Venom																
	SCC	*SCT*	*SCE*	*SMB*	*CCC*	*CV*	*MLT*	*BA*	*BGG*	*BGR*	*BN*	*BC*	*EO*	*LM*	*CA*	*ACC*
Protein family	% of total venom proteins															
PLA_2	29.9	31.6	13.7	**32.5**	20.0	21.1	4.0	4.3	11.4	4.8	20.1	**59.8**	12.6	8.7	16.3	18.5
Crotamine	–	–	–	–	–		–	–	–	–	–	–	–	–	–	–
Serine proteinase	18.2	20.4	24.4	17.1	9.1	20.0	9.2	19.5	**26.4**	23.9	21.9	15.1	2.0	**31.2**	10.1	13.8
Snaclec (C-type lectin)	< 0.1	< 0.1	< 0.1	< 0.1	24.0	0.9	10.1	13.2	14.3	14.1	4.2	4.9	7.0	8.1	1.6	–
L-amino acid oxidase	4.2	1.6	2.5	2.1	12.0	9.0	–	–	1.3	2.2	3.2	1.7	1.4	2.7	8.0	2.2
Zn^{2+}-metalloproteinase	**43.8**	**40.6**	**48.6**	**36.1**	**37.0**	**48.1**	**67.1**	**38.5**	**22.9**	**30.8**	**40.9**	11.5	**67.0**	**31.9**	**51.1**	**63.6**
Disintegrins																
– Long	–	–	–	–	–	–	1.5	–	–	–	–	–	–			
– Medium	–	–	< 0.1	2.5	2.1	1.4	–	–	1.2	4.4	–	–	–			
– Dimeric	8.5	9.7	–	–	–	–	–	–	–	–	–	–	–			
– Short	2.8	–	–	–	–	–	–	–	–	–	–	–	–			
Myotoxin	–	–	–	–	–	–	–	–	–	–	–	–	–			
C-type BPP/NP	5.3	6.0	1.8	8.6	–	–	–	–	–	–	11.1	13.4	14.7			
Kunitz-type inhibitor	–	0.1	–	–	–	–	–	–	–	–	–	–	–			
Kazal-type inhibitor	–	–	–	–	–	–	–	–	–	–	–	8.3	–			
Cystatin	–	–	–	–	–	–	–	–	–	–	–	–	–			
DC-fragment	1.7	0.1	< 0.1	< 0.1	< 0.1	< 0.1	< 0.2	–	0.5	0.7	–	–	–			
NGF/svVEGF	–	2.4	< 0.1	< 0.1	–	–	–	–	3.3	3.9	0.5	–	0.4			
Ohanin-like	–	–	–	–	–	–	–	–	–	–	–	–	–			

Table 4.2 (continued)

Venom	*Mlo*	*Vr*	*Api*	*Amex*	*Bas(C)*	*Bas(P)*	*Bcar*	*Blan*	*Bco*	*Bfon*	*Bolat*	*Bosch*	*Lste*
Protein family	% of total venom proteins												
3-Finger toxin	–	–	–	< 0.1	–	–	–	–	–	–	–	–	–
CRISP	2.6	7.4	4.8	1.9	0.1	0.1	2.6	–	3.6	2.4	6.5	2.1	–
PLA_2	14.6	23.8	9.5	**36.5**	**28.8**	**45.1**	12.8	8.6	–	**30.1**	8.7	**43.8**	12.3
Crotamine	–	–	–	–	–	–	–	–	–	–	–	–	–
Serine proteinase	14.9	8.4	13.5	22.0	18.2	4.4	4.7	14.4	14.4	4.1	11.3	5.8	25.6
Snaclec (C-type lectin)	14.8	9.6	1.8	1.3	0.5	0.5	–	< 0.1	< 0.1	9.8	0.9	–	3.6
L-amino acid oxidase	1.7	0.2	2.2	9.1	9.2	4.6	8.4	2.8	3.8	1.9	6.1	8.9	5.3
Zn^{2+}-metalloproteinase	**32.1**	**31.6**	**66.4**	18.2	**41.0**	**44.0**	**68.6**	**74.3**	**73.1**	**42.5**	**55.1**	17.7	**38.2**
Disintegrins													
– Long	–	–	–	–	–	–	–	–	–	–	–		
– Medium	5.6	1.7	< 0.1	8.1	2.1	–	–	–	0.5	0.5	0.4		
– Dimeric	–	–	–	–	–	–	–	–	–	–	–		
– Short	–	–	–	–	–	–	–	–	–	–	–		
Myotoxin	–	–	–	–	–	–	–	–	–	–	–		
C-type BPP/NP	0.8	0.3	< 0.1	3.5	4.0	0.9	1.1	< 0.1	2.3	–	–		

Table 4.2 (continued)

Venom													
	Mlo	*Vr*	*Api*	*Amex*	*Bas(C)*	*Bas(P)*	*Bcar*	*Blan*	*Bco*	*Bfon*	*Bolat*	*Bosch*	*Lste*
Protein family	% of total venom proteins												
Kunitz-type inhibitor	–	–	–	–	–	–	–	–	–	–	–		
Kazal-type inhibitor	–	–	–	–	–	–	–	–	–	–	–		
Cystatin	–	–	–	–	–	–	–	–	–	–	–		
DC-fragment	0.5	< 0.1	0.9	–	–	–	–	–	–	–	–		
NGF/svVEGF	–	–	–	–	–	–	–	–	–	3.1	–		
Ohanin-like	–	–	–	–	–	–	–	–	–	–	–		
3-Finger toxin	–	–	–	–	–	–	–	–	–	–	–		
CRISP	0.1	2.6	1.8	–	–	–	–	–	–	–	–		
PLA_2	**44.3**	24.1	14.3	7.6	**55.9**	**70.0**	6.0	**82.7**	**59.5**	**75.9**	**95.1**		
– Crotoxin	–	–	–	4.1	**52.7**	**68.0**	2.6	**82.7**	**59.5**	**71.3**	**83.8**		
Crotamine	–	–	–	–	–	18.0	13.1	1.5	19.0	20.8	–		
Serine proteinase	<1.0	10.9	4.6	5.3	**36.0**	5.1	1.9	8.1	8.2	1.9	0.5		
Snaclec (C-type lectin)	–	7.1	0.7	–	–	<0.2	2.5	4.3	1.7	0.1	0.1		
L-amino acid oxidase	5.7	4.7	2.1	3.8	–	3.6	2.7	<0.5	4.5	0.5	2.6		
Zn^{2+}-metalloproteinase	**42.1**	**48.5**	**72.1**	**71.7**	2.0	2.4	**64.8**	2.9	4.8	0.4	2.3		

Data taken from snake venomics projects carried out in the author's laboratory. Major toxin families in each venom are highlighted in boldface

Fig. 4.3 Geographical distribution of viperid snakes. Phylogeographical distribution of species whose venom has been investigated within the framework of our snake venomics project. The relative occurrence of proteins (in percentage of the total HPLC-separated proteins) of the different toxin families found in the venoms is listed in Table 4.2

venom variations in *B. atrox* (Núñez et al., 2009) and *Bothrops asper* (Alape-Girón et al., 2008); to assess the impact of such regional variation in the design of antivenoms (Gutiérrez et al., 2010); to reconstruct the natural history and cladogenesis of *Bothrops colombiensis* (Calvete et al., 2009b), a medically important pitviper of the *Bothrops atrox-asper* family endemic to Venezuela; and to identify evolutionary trends in venom toxicity evolution among neotropical *Crotalus* species (Calvete et al., 2010), which may affect the selection of specimens for the generation of improved antidotes targeting intra- and intergeneric species (Gutiérrez et al., 2009).

The crisis in the availability of therapeutic antivenoms calls for an international effort to facilitate the transfer of technology to affected countries (WHO, 2007). With this in mind, in addition to snake venomics, we have developed a simple proteomic protocol termed "snake antivenomics" (Gutiérrez et al., 2009) for investigating which venom proteins bear epitopes recognized by an antivenom and which toxins escape the immunological response of the hyperimmunized animal. Antivenomics can aid in assessing the cross-reactivity of a polyvalent antivenom against venoms not included in the immunization mixture, thus expanding its range of clinical application. A further aim of antivenomics is to define the intra- and interspecific complexity of venoms in terms of common and unique antigenic determinants. This information is relevant for defining the minimal set of venoms containing all epitopes necessary to generate therapeutic broad-range polyvalent antisera.

Antivenomics: Expanding the Range of Possible Clinical Use of Antivenoms by Defining Evolutionary and Immunological Trends

The deficit ("crisis") of antivenom supply in some regions of the world can be addressed to a certain extent by optimizing the use of existing antivenoms and through the design of novel immunization mixtures for producing broad-range polyspecific antivenoms. Toxins from the same protein family present in venoms from snakes belonging to different genera often share antigenic determinants. A practical consequence of this circumstance is that it might be possible to formulate a mixture of venoms on an immunologically sound basis for generating antivenoms against a wide range of species, i.e. by including a representative set of cross-reacting epitopes from each toxin family present in the venom pool. A deep insight into inter- and intraspecific variation of the antigenic constituents of venoms from snakes of different geographic origin represents the key for designing novel polyvalent pan-generic antivenoms. Polyspecific antivenoms that cover the range of snakes in a given area are preferred to monospecific ones, because their use simplifies production, distribution, and therapeutic procedures. However, the extent of cross-protection of many monospecific or polyspecific antivenoms against heterologous venoms not included in the immunizing mixtures remains largely unexplored.

Antivenomics is based on the immunodepletion of toxins upon incubation of whole venom with antivenom followed by the addition of a secondary antibody (Gutiérrez et al., 2008, 2009; Lomonte et al., 2008). Antigen-antibody complexes immunodepleted from the reaction mixture contain the toxins against which antibodies in the antivenom are directed. By contrast, venom components that remain in the supernatant are those which failed to raise antibodies in the antivenom, or which triggered the production of low-affinity antibodies. These components can be easily identified by comparison of reverse-phase HPLC separation of the non-precipitated fraction with the HPLC pattern of the whole venom previously characterized by venomics. According to their immunoreactivity towards antivenoms, toxins may be conveniently classified as: C-toxins, **c**ompletely immunodepletable toxins; P-toxins, **p**artly immunodepleted toxins; and N-toxins, **n**on-inmunodepleted proteins (Fig. 4.4). Assuming a link between the in vitro toxin immunodepletion capability of an antivenom and its in vivo neutralizing activity towards the same toxin molecules, improved immunization protocols should make use of mixtures of immunogens to generate high-affinity antibodies against class P and class N toxins. On the other hand, our antivenomics approach is simple and easy to implement in any protein chemistry laboratory, and may thus represent another useful protocol for investigating the immunoreactivity, and thus the potential therapeutic usefulness, of

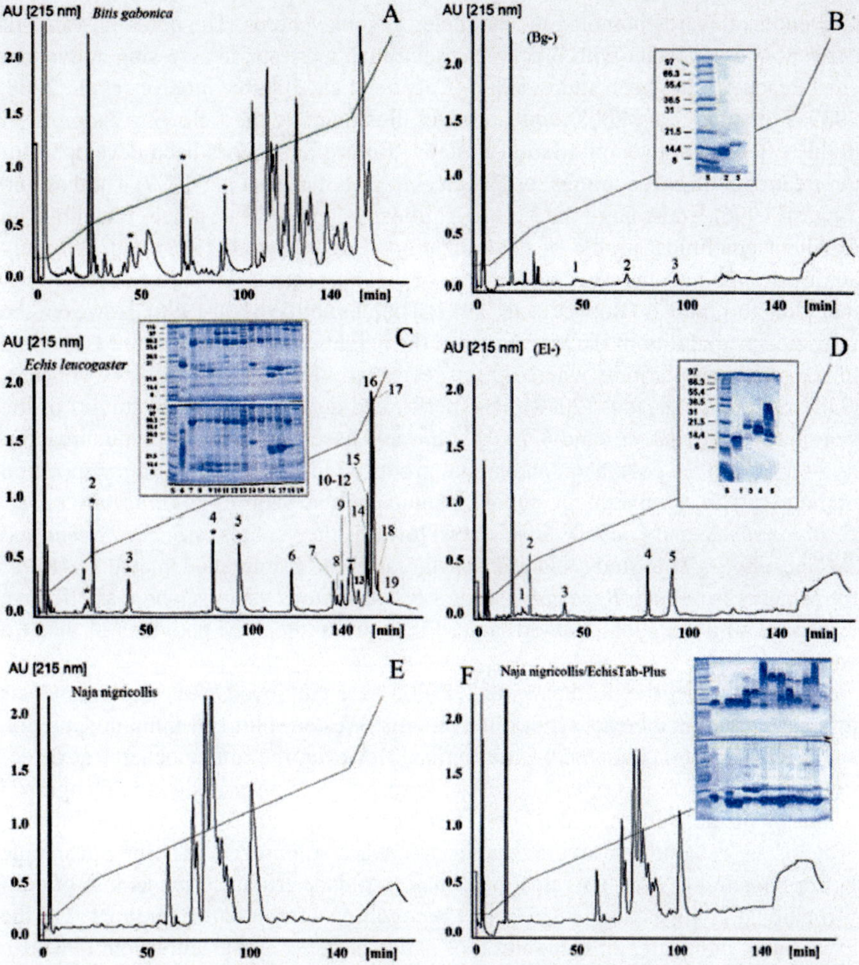

Fig. 4.4 Antivenomics. Panels **b**, **d**, and f show, respectively, reverse-phase separations of the proteins recovered after incubation of the crude venoms of *Bitis gabonica* (**a**), *Echis leucogaster* (**c**), and *Naja nigricollis* (**e**) with the polyspecific EchisTab-Plus-ICP® antivenom (generated against a mixture of the venoms of *Echis ocellatus*, *Bitis arietans* and *Naja nigricollis* (at a weight ratio of 1:1:1.33), as described by Gutiérrez et al., 2005), followed by rabbit anti-horse IgG antiserum and immunoprecipitation. The *inserts* shows SDS-PAGE analyses of β-mercaptoethanol-reduced fractions labelled as in the respective chromatograms. Molecular mass markers (in kDa) are indicated at the side of each gel. Except for the fractions labelled with asteriks, all *B. gabonica* toxins represent C-toxins. Fractions 1–3, and fractions 1–5 from *E. leucogaster*, are examples of P-toxins (disintegrins, PLA_2 molecules). The toxins of *N. nigricollis* venom were essentially not immunorecognized by the antivenom and were thus quantitatively recovered in the supernatant of the antivenomic assay. These venom proteins (neurotoxins and PLA_2 molecules) were classified as N-toxins. The failure of snake antivenoms to recognize specific components of the venom used for hyperimmunization has been previously reported (Fry et al., 2001; Gutiérrez et al., 2008; Judge et al., 2006; Lomonte et al., 2008). In particular, a study on the spectrum of venom proteins from the Australian snakes *Pseudonaja affinis affinis*, *P. textilis* and *P. nuchalis* recognized and neutralized

antivenoms towards homologous and heterologous venoms. The potential value of antivenomics, together with preclinical neutralization tests, in assessing antivenom cross-reactivity has been addressed in (Calvete et al., 2009b; Gutiérrez et al., 2008, 2009; Lomonte et al., 2008) and is clearly illustrated by the following example. A highly effective antivenom (Sanofi-Pasteur "Bothrofav®") has been developed for the treatment of envenomings by *B. lanceolatus* (Bucher et al., 1997; Thomas and Tyburn, 1996), endemic to the Lesser Antillean island of Martinique. It exhibits an excellent preclinical profile of neutralization (Bogarín et al, 1999) and its timely administration prevents the development of the most serious effects of envenoming, including thrombosis (Bucher et al., 1997; Thomas and Tyburn, 1996). However, the restricted availability of the antivenom in the neighboring island of Saint Lucia and in zoos and herpetariums where these species may be kept is a matter of concern. Gutiérrez and colleagues (2008) have performed detailed proteomic studies of the venoms of *B. caribbaeus* and *B. lanceolatus* and have evaluated the immunoreactivity of a Crotalinae polyvalent antivenom produced in Costa Rica (by immunization of horses with a mixture of equal amounts of the venoms of *Bothrops asper*, *Crotalus simus*, and *Lachesis stenophrys*) towards the venoms of *B. caribbaeus* and *B. lanceolatus*. This study showed that the antivenom immunodepleted ~80% of the proteins from both *B. caribbaeus* and *B. lanceolatus* venoms, and was effective in neutralizing the lethal, hemorrhagic, PLA_2 and proteolytic activities of the two venoms.

Several authors (i.e. Kulkeaw et al., 2007) have employed Western blot analysis to assess the immunoreactivity of antivenoms. Western blot and immunodepletion analyses yield complementary information. However, the immunochemical detection of blotted proteins provides a Yes/No response: a given protein is recognized or not by the antivenom, and it is essentially a non-quantitative technique. Further, proteins are denatured to an unknown degree when solubilized by boiling in sample buffer containing SDS. This treatment may introduce artifacts such as loss of conformational epitopes and/or artifactual recognition of non-native epitopes. On the other hand, the degree of recognition of native proteins by the antivenom IgGs can be easily quantitated by measuring the amount of non-immunodepleted proteins.

A thorough understanding of snake speciation and venom diversification during ophidian evolution may influence the selection of specimens for the generation of improved antidotes targeting intra- and intergeneric species (Gutiérrez et al., 2009). This point is illustrated by comparative proteomic studies of the venom composition of Costa Rican *B. asper* and Venezuelan *B. colombiensis* populations pointing at the ancestor of *B. colombiensis* as the founding Middle American *B. asper* ancestor species (Calvete et al., 2009b). The close kinship between *B. asper* and *B. colombiensis* is also indicated by the virtually indistinguishable

Fig. 4.4 (continued) by a brown snake antivenom (Judge et al., 2006), raised in horse against the venom of *P. textilis* by the Australian Commonwealth Serum Laboratories (CSL), highlighted a deficiency in the interaction between this antivenom and protein constituents of the Pseudonaja sp. snakes of 6 kDa to <32 kDa, predominantly PLA_2 proteins and α-neurotoxins immunological cross-reactivity of a divalent Venezuelan antivenom (raised against a mixture of *B. colombiensis* and *Crotalus durissus cumanensis* venoms) and a Costa Rican trivalent antivenom (generated against a mixture of *B. asper*, *Crotalus simus*, and *Lachesis stenophrys* venoms) towards the venoms of *B. colombiensis* and *B. asper* (Calvete et al., 2009b). This antivenom study suggests the possibility of using either the Venezuelan or the Costa Rican antivenoms for the management of snakebites by any of these Bothrops species. On the other hand, the high degree of of cross-reactivity of the Costa Rican antivenom against *B. atrox* venoms (Núñez et al., 2009) also suggests the close evolutionary relationship between *B. asper* and *B. atrox*.

Proteomic analysis of the venoms of *Bothrops atrox* from Colombia, Brazil, Ecuador, and Perú showed the existence of two geographically differentiated venom phenotypes (Núñez et al., 2009). The toxin profiles of the venoms of Brazilian, Ecuadorian, and Peruvian *B. atrox* correlate with the venom phenotype of juvenile *B. asper* from Costa Rica, whereas the venom composition of Colombian *B. atrox* is more similar to that of adult *B. asper*. These observations suggest that paedomorphism (the retention in the adult of juvenile characters first described by Garstang in, 1922) represented a selective trend during the trans-Amazonian southward expansion of *B. atrox* through the Andean Corridor. Strikingly, the Costa Rican polyvalent antivenom is more efficient immunodepleting proteins from the paedomorphic venoms of *B. atrox* from Brazil, Ecuador, and Perú than from Colombia (Núñez et al., 2009). Such behaviour may be rationalized by the lower content of poorly immunogenic toxins, such as PLA_2 molecules and PI-SVMPs in the paedomorphic venoms. The immunological profile of the Costa Rican antivenom strongly suggests the possibility of using this antivenom for the management of snakebites by *B. atrox* in the Amazon regions of Colombia, Ecuador, Perú and Brazil, thus expanding the range of its clinical use.

The recent realization that neurotoxicity represents an adaptive paedomorphic trend along *Crotalus* dispersal in South America (Calvete et al., 2010) may also have implications for improving current antivenoms. The driving force behind paedomorphism is often competition or predation pressure. The increased concentration of neurotoxins crotoxin and crotamine in South American rattlesnake venoms strongly argue that the gain of neurotoxicity and lethal venom activities to mammals may have represented the key axis along which overall venom toxicity has evolved during *Crotalus durissus* invasion of South America. The proteomic and antivenomic characterization of Middle and South American rattlesnake venoms (Calvete et al., 2010) provides a molecular ground for understanding a) the failure of the polyvalent antivenom generated in Costa Rica using the venom of adult specimens of *C. simus* in the immunization mixture to neutralize the neurotoxicity and rhabdomyolysis induced by young Central American rattlesnake venoms (Gutiérrez et al., 1991, 2001; Gutiérrez, 2009), and b) the ineffectiveness of an anti-crotalic antivenom produced at Instituto Butantan (São Paulo, Brazil) against adult *C. d. terrificus* venom to neutralize the hemorrhagic activity of *C. simus* and *C. d. cumanensis* venoms (Saravia et al., 2002). The immunization mixtures used in Costa Rica and Brazil may simply lack, respectively, immunogenic concentrations of neurotoxins (crotamine and crotoxin, which represent the major toxins in the venoms of neonate *C. simus* and adult *C.d. terrificus*) and hemorrhagins (PIII-SVMPs, the more abundant toxins in adult *C. simus venom*), respectively (see Table 4.2). To assess this assumption we have investigated the immunodepleting ability of the Costa Rican polyvalent antivenom towards the toxins of *C. simus* (adult and neonate) and *C. d. terrificus* venoms using an antivenomics approach. The antivenom partially immunodepleted (~80% for PIII-metalloproteinase) or not at all (disintegrins, crotoxin acid and basic subunits, crotamine, and certain serine proteinases) toxins from the venom of adult *C. simus*. The negligible titre of this antivenom towards crotoxin was consistently confirmed in experiments with venoms from newborn *C. simus* and adult *C. durissus terrificus*. In addition, the antivenomic results also explain why the Costa Rican antivenom effectively neutralizes the hemorrhagic activity of adult *C. simus* and *C. cumanensis* venoms but does not protect against the neurotoxicity produced by adult *C. durissus* and newborn *C. simus* envenoming (Calvete et al., 2010; Saravia et al., 2002).

Venom Phenotyping: Clues for Formulating a Pan-Generic Antivenoms

The identification of evolutionary trends may help to define how to prepare the mixture of venoms for immunization to produce effective antivenoms. For example, the evolutionary and immunological trends among *Bothrops* species revealed by our antivenomic investigations (Calvete et al., 2009b; Núñez et al., 2009), and described above, expands the geographic range for clinical use of the Costa Rican antivenom. In addition, the finding that both, the Costa Rican and the Venezuelan antivenoms immunodepleted with higher efficiency the toxins from paedomorphic *B. atrox* venoms than the toxins from venoms exhibiting a "adult-type" phenotype, let us predict that these antivenoms may also neutralize the lethal activity of other PIII-SVMP-rich bothropoid venoms. This hypothesis can be easily tested through venom phenotyping by reverse-phase HPLC, antivenomic analysis, and functional neutralization assays (Gutiérrez et al., 2009), and is being addressed in our laboratory. Similarly, the venoms of neotropical *Crotalus* subspecies belong to one of two distinct phenotypes, which broadly correspond to type I (high levels of SVMPs and low toxicity, LD_{50} >1 μg/g mouse body weight) and type II (low metalloproteinase activity and high toxicity, LD_{50} <1 μg/g mouse body weight) venoms defined by Mackessy (2008), who has demonstrated an inverse relationship between toxicity and metalloprotease activity in the crude venoms of 25 taxa of North American rattlesnakes. We have initiated a venomic project aimed at phenotyping the venoms of neartic crotalid species. Accumulating evidence suggests the occurrence of a high degree of toxin profile overlapping between type I North American venoms and neotropical venoms exhibiting an "adult *C. simus/C. d. cumanensis*" phenotype. On the other hand, neurotoxic (i.e. Mojave toxin positive) neartic type II venoms possess toxin phenotypes closely resembling those of neonate *C. simus* and *C. d. terrificus* (Fig. 4.5). It is relevant to this point that the subunits of the Mojave toxin

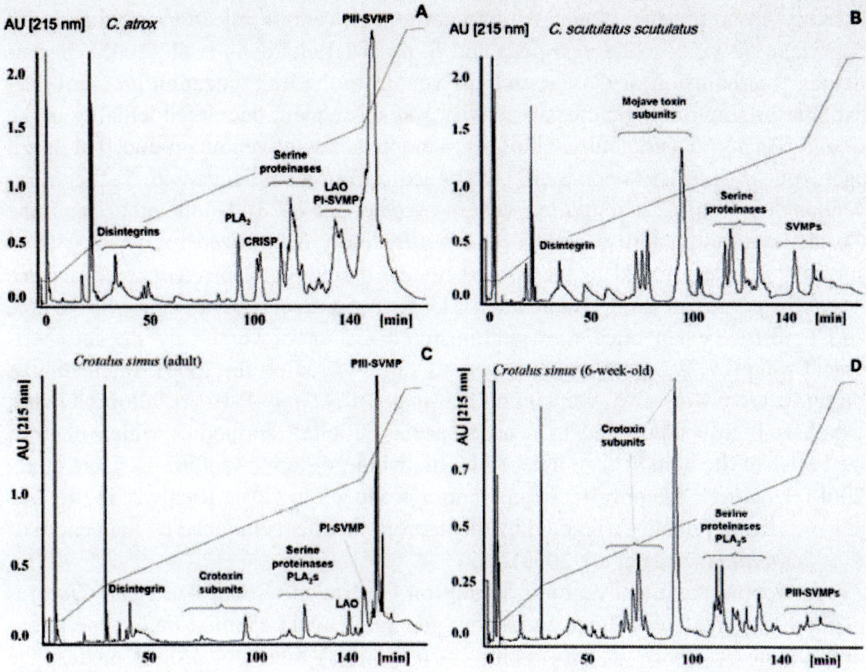

Fig. 4.5 Venom phenotyping. Panels **a** and **b** display, respectively, reverse-phase HPLC separations of the toxins from type-I (western diamondback rattlesnake, *C. atrox*) and type-II (Mojave rattlesnake, *C. scutulatus scutulatus* type A) neartic Crotalus venom phenotypes. Panels **c** and **d** show the reverse-phase HPLC toxin profiles of adult and neonate *Crotalus simus* (Costa Rica), respectively, and illustrate the ontogenetic transition from a neurotoxin (type-I) to a hemorrhagic (type-II) venom phenotype. Peaks belonging to major toxin families are labelled

(SwissProt accession codes P18998 and P62023), a heterodimeric PLA_2 presynaptically acting neurotoxin, share 95 and 100% amino acid sequence identity with the acidic (A, P08878) and the basic (CB1, P62022) subunits of crotoxin. The major PIII-SVMPs of *C. simus* venom (Q2QA02) also displays 96–97% sequence identity with homologuous proteins from *C. atrox* (catrocollastatin, Q90282; vascular apoptosis-inducing protein 2A, BAF56420; and VMP-III, ACV83931) and *C. v. viridis* (VMP-III, ACV83933).

The extensive immunochemical similarities among neotropical and neartic crotalid venoms is further highlighted by the realization that the antivenom Antivipmyn (Fab_2H), produced by Instituto Bioclon in Mexico against a mixture of *C. tzabcan* and *B. asper* venoms, cross-reacted with a number of North American venoms and neutralized their lethal, hemorrhagic, fibrinolytic, and gelatinase proteolytic activities (Dart and McNally, 2001; Sánchez et al., 2003a, b). In addition, a polyvalent antivenom (CroFab, FabAV) produced using venom from 4 North American crotaline snakes, eastern and western diamondback rattlesnake (*C. adamanteus* and *C. atrox*, respectively), Mojave rattlesnake (*C. scutulatus scutulatus*), and the cottonmouth (*Agkistrodon piscivorus*), crossreacted and decreased lethality in a murine model of intraperitoneal venom injection of the South American pit vipers, *C d. terrificus* and *B. atrox* (Richardson et al., 2005). In particular, the inclusion of *C. s. scutulatus* venom in the immunization protocol may explain the immunologic cross-reactivity and subsequent decreased lethality in the *C. d. terrificus* envenomation. Similarly, a monovalent antivenom produced in Brazil against *C. d. terrificus* venom effectively neutralized a challenge with $5LD_{50}$ intravenous dosage of *C. s. scutulatus* venom (Arce et al., 2003). On the other hand, the Costa Rican antivenom described above effectively neutralized the venoms from several pit vipers found in the United States, including *A. piscivorus piscivorus*, *A. contortix laticinctus*, *C. adamanteus*, *C. horridus atricaudatus*, *C. viridis viridis*, and *C. atrox*, when mice were administered $4LD_{50}$ of venom by the intraperitoneal route or $2LD_{50}$ by the intravenous route (Arce et al., 2003). Interestingly, immunization with cDNA encoding the major *C. simus* P-III metalloproteinase, which is highly conserved in neartic species, elicited antibodies which neutralized 69% of the hemorrhage induced by the whole venom (Azofeifa-Cordero et al., 2008). However, the polyvalent antivenom produced in Costa Rica was ineffective at neutralizing lethality triggered by the neurotoxic effects induced by the venom of *C. s, scutulatus* (Arce et al., 2003).

Our venomic and antivenomic studies on Central and South American *Crotalus* explain why the Costa Rican antivenom generated against a mixture of type I venoms (adult *C. simus*, *B. asper*, and *L. stenophrys*) does not protect against the neurotoxicity produced by type II venoms. This conclusion is in line with the suggestion of Arce et al. (2003) that "a venom containing presynaptically-active neurotoxic phospholipases A_2 related to "mojave toxin" needs to be introduced in the immunizing mixture in order to increase the neutralizing scope of this product in North America". Our findings also suggest that an effective pan-American anti-*Crotalus* antivenom should primarily neutralize the toxic actions of four major toxin groups, PIII-SVMPs, crotoxin, crotamine, and thrombin-like serine proteinases. Such antivenom might be achievable by hyperimmunizing with a mixture of neartic and neotropical venoms comprising conserved antigenic determinants for each of the major toxin families of the genus. Work is in progress in our laboratories to define this *magic-bullet* venom mixture.

Acknowledgements The author gratefully thank the many colleagues (particularly from the Instituto Clodomiro Picado, Costa Rica) who, over the years, have provided precious venoms and antivenoms within the framework of collaborative projects and have contributed insight and suggestions. Funding for the projects described in this paper was provided by grant BFU2007-61563 from the Ministerio de Educacíón y Ciencia, Madrid, CRUSA-CSIC (2007CR0004), and CYTED (206AC0281). Travelling between Spain and Costa Rica was financed by Acciones Integradas, 2006CR0010 between CSIC and the University of Costa Rica (UCR).

References

Alape-Girón, A., Sanz, L., Escolano, J., Flores-Díaz, M., Madrigal, M., Sasa, M., Calvete, J.J., 2008. Snake venomics of the lancehead pitviper *Bothrops asper*: geographic, individual, and ontogenetic variations. J. Proteome Res. 7, 3556–3571.

Angulo, Y., Escolano, J., Lomonte, B., Gutiérrez, J.M., Sanz, L., Calvete, J.J., 2008. Snake venomics of Central American pitvipers: clues for rationalizing the distinct envenomation profiles of *Atropoides nummifer* and *Atropoides picadoi*. J. Proteome Res. 7, 708–719.

Arce, V., Rojas, E., Ownby, C.L., Gutiérrez, J.M., 2003. Preclinical assessment of the ability of polyvalent (*Crotalinae*) and anticoral (*Elapidae*) antivenoms produced in Costa Rica to neutralize the venoms of North American snakes. Toxicon 41, 851–860.

Azofeifa-Cordero, G., Arce-Estrada, V., Flores-Diaz, M., Alape-Giron, A., 2008. Immunization with cDNA of a novel P-III type metalloproteinase from the rattlesnake *Crotalus durissus durissus* elicits antibodies which neutralize 69% of the hemorrhage induced by the whole venom. Toxicon 52, 302–308.

Barlow, A., Pook, C.E., Harrison, R.A., Wüster, W., 2009. Coevolution of diet and prey-specific venom activity supports the role of selection in snake venom evolution. Proc. Biol. Sci. 276, 2443–2449.

Bazaa, A., Marrakchi, N., El Ayeb, M., Sanz, L., Calvete, J.J., 2005. Snake venomics: comparative analysis of the venom proteomes of the Tunisian snakes *Cerastes cerastes*, *Cerastes vipera* and *Macrovipera lebetina*. Proteomics 5, 4223–4235.

Billen, B., Bosmans, F., Tytgat, J., 2008. Animal peptides targeting voltage-activated sodium channels. Curr. Pharm. Des. 14, 2492–2502.

Bogarín, G., Romero, M., Rojas, G., Lutsch, C., Casadamont, M., Lang, J., Otero, R., Gutiérrez, J.M., 1999. Neutralization, by a monospecific *Bothrops lanceolatus* antivenom, of toxic activities induced by homologous and heterologous *Bothrops* snake venoms. Toxicon 37, 551–557.

Boldrini-França, J., Rodrigues, R.S., Fonseca, F.P., Menaldo, D.L., Ferreira, F.B., Henrique-Silva, F., Soares, A.M., Hamaguchi, A., Rodrigues, V.M., Otaviano, A.R., Homsi-Brandeburgo, M.I., 2009. *Crotalus durissus collilineatus* venom gland transcriptome: analysis of gene expression profile. Biochimie 91, 586–595.

Bucher, B., Canonge, D., Thomas, L., Tyburn, B., Robbe-Vincent, A., Choumet, V., Bon, C., Ketterlé, J., Lang, J., 1997. Research group of snake bites in Martinique. Clinical indicators of envenoming and serum levels of venom antigens in patients bitten by *Bothrops lanceolatus* in Martinique. Trans. Royal Soc. Trop. Med. Hyg. 91, 186–190.

Calvete, J.J., 2009a. Venomics. J. Proteomics 72, 121–282.

Calvete, J.J., 2009b. Digging into the evolution of venomous systems and learning to twist nature to fight pathology. J. Proteomics 72, 121–126.

Calvete, J.J., Juárez, P., Sanz, L., 2007a. Snake venomics. Strategy and applications. J. Mass Spectrom. 42, 1405–1414.

Calvete, J.J., Marcinkiewicz, C., Sanz, L., 2007b. Snake venomics of *Bitis gabonica gabonica*. Protein family composition, subunit organization of venom toxins, and characterization of dimeric disintegrins bitisgabonin-1 and bitisgabonin-2. J. Proteome Res. 6, 326–336.

Calvete, J.J., Escolano, J., Sanz, L., 2007c. Snake venomics of Bitis species reveals large intragenus venom toxin composition variation: application to taxonomy of congeneric taxa. J. Proteome Res. 6, 2732–2745.

Calvete, J.J., Fasoli, E., Sanz, L., Boschetti, E., Righetti, P.G., 2009a. Exploring the venom proteome of the western diamondback rattlesnake, *Crotalus atrox*, via snake venomics and combinatorial peptide ligand library approaches. J. Proteome Res. 8, 3055–3067.

Calvete, J.J., Borges, A., Segura, A., Flores-Díaz, M., Alape-Girón, A., Gutiérrez, J.M., Diez, N., De Sousa, L., Kiriakos, D., Sánchez, E., Faks, J.G., Escolano, J., Sanz, L., 2009b. Snake venomics and antivenomics of *Bothrops colombiensis*, a medically important pitviper of the *Bothrops atrox-asper* complex endemic to Venezuela: contributing to its taxonomy and snakebite management. J. Proteomics 72, 227–240.

Calvete, J.J., Sanz, L., Cid, P., De La Torre, P., Flores-Diaz, M., Dos Santos, M.C., Borges, A., Bremo, A., Angulo, Y., Lomonte, B., Alape-Giron, A., Gutiérrez, J.M., 2010. Snake venomics of the Central American rattlesnake Crotalus simus and the South American Crotalus durissus complex points to neurotoxicity as an adaptive paedomorphic trend along Crotalus dispersal in South America. J. Proteome Res. 9, 528–544.

Campbell, J.A., Lamar, W.W., 2004. The Venomous Reptiles of the Western Hemisphere (vols. I and II). Cornell University Press.

Chippaux, J.P., Goyffon, M., 2008. Epidemiology of scorpionism: a global appraisal. Acta Trop. 107, 71–79.

Cidade, D.A.P., Simão, T.A., Dávila, A.M.R., Wagner, G., Junqueira-de-Azevedo, I.L.M., Ho, P.L., Bon, C., Zingali, R., Albano, R.M., 2006. *Bothrops jararaca* venom transcriptome: analysis of the gene expression pattern. Toxicon 48, 437–461.

Clemetson, K.J., Morita, T., Kini, R.M., 2009. Classification and nomenclature of snake venom C-type lectins and related proteins. Toxicon 54, 83.

Dart, R.C., McNally, J., 2001. Efficacy, Safety, and Use of Snake Antivenoms in the United States. Ann. Emerg. Med. 37, 181–188.

De Lima, M.E., Pimenta, A.M.C., Martin-Euclaire, M.F., Zingali, R.B., 2009. Animal Toxins: State of the Art. Perspectives in Health and Biotechnology. Editora UFMG, Belo Horizonte.

Doley, R., Kini, R.M., 2009. Protein complexes in snake venom. Cell. Mol. Life Sci. 66, 2851–2871.

Escoubas, P., Quinton, L., Nicholson, G.M., 2008. Venomics: unravelling the complexity of animal venoms with mass spectrometry. J. Mass Spectrom. 43, 279–295.

Espino-Solís, G.P., Riaño-Umbarila, L., Becerril, B., Possani, L.D., 2009. Antidotes against venomous animals: state of the art and prospectives. J. Proteomics 72, 183–199.

Favreau, P., Menin, L., Michalet, S., Perret, F., Cheneval, O., Stöcklin, M., Bulet, P., Stöcklin, R., 2006. Mass spectrometry strategies for venom mapping and peptide sequencing from crude venoms: case applications with single arthropod specimen. Toxicon 47, 676–687.

Ferquel, E., de Haro, L., Jan, V., Guillemin, I., Jourdain, S., Teynié, A., d'Alayer, J., Choumet, V., 2007. Reappraisal of *Vipera aspis* venom neurotoxicity. PLoS One 21, e1194.

Fox, J.W., Serrano, S.M., 2005a. Snake toxins and hemostasis. Toxicon 45, 951–1181.

Fox, J.W., Serrano, S.M.T., 2005b. Structural considerations of the snake venom metalloproteinases, key members of the M12 reprolysin family of metalloproteinases. Toxicon 45, 969–985.

Fox, J.W., Serrano, S.M., 2007. Approaching the golden age of natural product pharmaceuticals from venom libraries: an overview of toxins and toxin-derivatives currently involved in therapeutic or diagnostic applications. Curr. Pharm. Des. 13, 2927–2934.

Fox, J.W., Serrano, S.M., 2009. Timeline of key events in snake venom metalloproteinase research. J. Proteomics 72, 200–209.

Fox, J.W., Ma, L., Nelson, K., Sherman, N.E., Serrano, S.M., 2006. Comparison of indirect and direct approaches using ion-trap and Fourier transform ion cyclotron resonance mass spectrometry for exploring viperid venom proteomes. Toxicon 47, 700–714.

Francischetti, I.M., My-Pham, V., Harrison, J., Garfield, M.K., Ribeiro, J.M.C., 2004. *Bitis gabonica* (Gaboon viper) snake venom gland: towards a catalog of full-length transcripts (cDNA) and proteins. Gene 337, 55–69.

Fry, B.G., Wickramaratna, J.C., Jones, A., Alewood, P.F., Hodgson, W.C., 2001. Species and regional variations in the effectiveness of antivenom against the in vitro neurotoxicity of death adder (*Acanthophis*) venoms. Toxicol. Appl. Pharmacol. 175, 140–148.

Fry, B.G., Wüster, W., 2004. Assembling an arsenal: origin and evolution of the snake venom proteome inferred from phylogenetic analysis of toxin sequences. Mol. Biol. Evol. 21, 870–873.

Fry, B.G., 2005. From genome to "venome": molecular origin and evolution of the snake venom proteome inferred from phylogenetic analysis of toxin sequences and related body proteins. Genome Res. 15, 403–420.

Fry, B.G., Scheib, H., van der Weerd, L., Young, B., McNaughtan, J., Ramjan, S.F., Vidal, N., Poelmann, R.E., Norman, J.A., 2008. Evolution of an arsenal: Structural and functional diversification of the venom system in the advanced snakes (Caenophidia). Mol. Cell Proteomics 7(2), 215–246. Epub 2007 Sep 12.

Fry, B.G., Vidal, N., van der Weerd, L., Kochva, E., Renjifo, C., 2009a. Evolution and diversification of the Toxicofera reptile venom system. J. Proteomics 72, 127–136.

Fry, B.G., Roelants, K., Champagne, D.E., Scheib, H., Tyndall, J.D.A., King, G.F., Nevalainen, T.J., Norman, J.A., Lewis, R.J., Norton, R.S., Renjifo, C., Rodríguez de la Vega, R.C., 2009b. The toxicogenomic multiverse: convergent recruitment of proteins into animal venoms. Annu. Rev. Genomics Hum. Genet. 10, 483–511.

Garstang, W., 1922. The theory of recapitulation: a critical re-statement of the biogenetic law. Linn. Soc. Jour. Zool. XXXV, 81–101.

Guércio, R.A.P., Shevchenko, A., Shevchenko, A., López-Lozano, J.L., Paba, J., Sousa, M.V., Ricart, C.A.O., 2006. Ontogenetic variations in the venom proteome of the Amazonian snake *Bothrops atrox*. Proteome Sci. 4, 11.

Georgieva, D., Risch, M., Kardas, A., Buck, F., von Bergen, M., Betzel, C., 2008. Comparative analysis of the venom proteomes of *Vipera ammodytes ammodytes* and *Vipera ammodytes meridionales*. J. Proteome Res. 7, 866–886.

Gibbs, H.L., Sanz, L., Calvete, J.J., 2009. Snake population venomics: proteomics-based analyses of individual variation reveals significant gene regulation effects on venom protein expression in Sistrurus rattlesnakes. J. Mol. Evol. 68, 113–125.

Gutiérrez, J.M., dos Santos, M.C., Furtado, M.F., Rojas, G., 1991. Biochemical and pharmacological similarities between the venoms of newborn Crotalus durissus durissus and adult Crotalus durissus terrificus rattlesnakes. Toxicon. 29(10), 1273–1277.

Gutiérrez, J.M., Dos Santos, M.C., Furtado, M.F., Rojas, G., 2001. Biochemical and pharmacological similarities between the venoms of newborn *Crotalus durissus durissus* and adult *Crotalus durissus terrificus* rattlesnakes. Toxicon 29, 1273–1277.

Gutiérrez, J.M., Rojas, E., Quesada, L., León, G., Núñez, J., Laing, G.D., Sasa, M., Renjifo, J.M., Nasidi, A., Warrell, D.A., Theakston, R.D.G., Rojas, G., 2005. Pan-African polyspecific antivenom produced by caprylic acid purification of horse IgG: an alternative to the antivenom crisis in Africa. Trans. Roy. Soc. Trop. Med. Hyg. 99, 468–475.

Gutiérrez, J.M., Theakston, R.D.G., Warrell, D.A., 2006. Confronting the neglected problem of snake bite envenoming: the need for a global partnership. PLoS Med. 3, e150.

Gutiérrez, J.M., Sanz, L., Escolano, J., Fernández, J., Lomonte, B., Angulo, Y., Rucavado, A., Warrell, D.A., Calvete, J.J., 2008. Snake venomic of the Lesser Antillean pit vipers *Bothrops caribbaeus* and *Bothrops lanceolatus*: correlation with toxicological activities and immunoreactivity of a heterologous antivenom. J. Proteome Res. 7, 4396–3408.

Gutiérrez, J.M., 2009. *Bothrops asper*: from natural history to public health. Toxicon 54, 901–1028.

Gutiérrez, J.M., León, G., 2009. Snake antivenoms, in: De Lima, M.E., Pimenta, A.M.C., Martin-Euclaire, M.F., Zingali, R.B. (Eds.), Animal Toxins: State of the Art. Perspectives in Health and Biotechnology. Editora UFMG, Belo Horizonte, Brazil, pp. 393–421.

Gutiérrez, J.M., Lomonte, B., León, G., Alape-Girón, A., Flores-Díaz, M., Sanz, L., Angulo, Y., Calvete, J.J., 2009. Snake venomics and antivenomics: proteomic tools in the design and control of antivenoms for the treatment of snakebite envenoming. J. Proteomics 72, 227–240.

Gutiérrez, J.M., Sanz, L., Flores-Diaz, M., Figueroa, L., Madrigal, M., Herrera, M., Villalta, M., León, G., Estrada, R., Borges, A., Alape-Girón, A., Calvete, J.J., 2010. Impact of regional variation in *Bothrops asper* snake venom on the design of antivenoms: integrating antivenomics and neutralization approaches. J. Proteome Res. 9, 564–577.

Harvey, A.L., Bradley, K.N., Cochran, S.A., Rowan, E.G., Pratt, J.A., Quillfeldt, J.A., Jerusalinsky, D.A., 1998. What can toxins tell us for drug discovery? Toxicon 36, 1635–1640.

Jia, Y., Cantu, B.A., Sánchez, E.E., Pérez, J.C., 2008. Complementary DNA sequencing and identification of mRNAs from the venomous gland of *Agkistrodon piscivorus leucostoma*. Toxicon 51, 1457–1466.

Juárez, P., Sanz, L., Calvete, J.J., 2004. Snake venomics: characterization of protein families in *Sistrurus barbouri* venom by cysteine mapping, N-terminal sequencing, and tandem mass spectrometry analysis. Proteomics 4, 327–328.

Juárez, P., Wagstaff, S.C., Oliver, J., Sanz, L., Harrison, R.A., Calvete, J.J., 2006. Molecular cloning of disintegrin-like transcript BA-5A from a *Bitis arietans* venom gland cDNA library: a putative intermediate in the evolution of the long-chain disintegrin bitistatin. J. Mol. Evol. 63, 142–152.

Juárez, P., Comas, I., González-Candelas, F., Calvete, J.J., 2008. Evolution of snake venom disintegrins by positive Darwinian selection. Mol. Biol. Evol. 25, 2391–2407.

Judge, R.K., Henry, P.J., Mirtschin, P., Jelinek, G., Wilce, J.A., 2006. Toxins not neutralized by brown snake antivenom. Toxicol. Appl. Pharmacol. 213, 117–125.

Junqueira-de-Azevedo, I.L., Ho, P.L., 2002. A survey of gene expression and diversity in the venom glands of the pitviper snake *Bothrops insularis* through the generation of expressed sequence tags (ESTs). Gene 299, 279–291.

Junqueira-de-Azevedo, I.L.M., Ching, A.T.C., Carvalho, E., Faria, F., Nishiyama M.Y., Jr., Ho, P.L., Diniz, M.R.V., 2006. *Lachesis muta* (*Viperidae*) cDNAs reveal diverging pitviper molecules and scaffolds typical of cobra (*Elapidae*) venoms: implications in snake toxin repertoire evolution. Genetics 173, 877–889.

Junqueira de Azevedo, I.L.M., Diniz, M.R.V., Ho, P.L., 2009. Venom gland transcriptomic analysis, in: De Lima, M.E., Pimenta, A.M.C., Martin-Euclaire, M.F., Zingali, R.B. (Eds.), Animal Toxins: State of the Art. Perspectives in Health and Biotechnology. Editora UFMG, Belo Horizonte, Brazil, pp. 693–713.

Kashima, S., Roberto, P.G., Soares, A.M., Astolfi-Filho, S., Pereira, J.O., Giuliati, S., Faria M., Jr., Xavier, M.A.S., Fontes, M.R.M., Giglio, J.R., Franca, S.C.., 2004. Analysis of *Bothrops jararacussu* venomous gland transcriptome focusing on structural and functional aspect: I-gene expression profile of highly expressed phospholipases A_2. Biochimie 86, 211–219.

Kasturiratne, A., Wickremasinghe, A.R., de Silva, N., Gunawardena, N.K., Pathmeswaran, A., Premaratna, R., Savioli, L., Lalloo, D.G., de Silva, H.J., 2008. The global burden of snakebite: a literature analysis and modelling based on regional estimates of envenoming and deaths. PLoS Med. 5, e218.

Kem, W.R., Turk T., 2009. Cnidarian toxins and venoms. Toxicon 54, 1029–1214.

Koh, D.C.I., Armugam, A., Jeyaseelan, K., 2007. Snake venom components and their applications in biomedicine. Cell. Mol. Life Sci. 63, 3030–3041.

Kulkeaw K, Chaicump W, Sakolvaree Y, Tongtawe P, Tapchaisri P., 2007. Proteome and immunome of the venom of the Thai cobra, *Naja kaouthia*. Toxicon 49, 1026–1041.

Liang, S., 2008. Proteome and peptidome profiling of spider venoms. Expert Rev. Proteomics 5, 731–746.

Lomonte, B., Escolano, J., Fernández, J., Sanz, L., Angulo, Y., Gutiérrez, J.M., Calvete, J.J., 2008. Snake venomics and antivenomics of the arboreal neotopical pitvipers *Bothriechis lateralis* and *Bothriehis schlegelii*. J. Proteome Res. 7, 2445–2457.

Mackessy, S.P., 2008. Venom composition in rattlesnakes: trends and biological significance, in: Hayes, W.K., Beaman, K.R., Cardwell, M.D., Bush, S.P. (Eds.), The Biology of Rattlesnakes. Loma Linda University Press, Loma Linda, CA, pp. 495–510.

Mackessy, S.P., 2009. Handbook of Venoms and Toxins of Reptiles. CRC Press, Taylor & Francis, Boca Ratón, FL, pp. 1–507.

Nei, M., Gu, X., Sitnikova, T., 1997. Evolution by the birth-and-death process in multigene families of the vertebrate immune system. Proc. Natl. Acad. Sci. U.S.A. 94, 7799–7806.

Neiva, M., Arraes, F.B., de Souza, J.V., Rádis-Baptista, G., Prieto da Silva, A.R., Walter, M.E., Brigido, M. de M., Yamane, T., López-Lozano, J.L., Astolfi-Filho, S., 2009. Transcriptome analysis of the Amazonian viper *Bothrops atrox* venom gland using expressed sequence tags (ESTs). Toxicon 53, 427–436.

Norton, R.S., Olivera, B.M., 2006. Conotoxins down under. Toxicon 48, 789–798.

Núñez, V., Cid, P., Sanz, L., De La Torre, P., Angulo, Y., Lomonte, B., Gutiérrez, J.M., Calvete, J.J., 2009. Snake venomics and antivenomics of *Bothrops atrox* venoms from Colombia and the Amazon regions of Brazil, Perú and Ecuador suggest the occurrence of geographic variation of venom phenotype by a trend towards paedomorphism. J. Proteomics 73, 57–78.

Ohno, M., Menez, R., Ogawa, T., Danse, J.M., Shimohigashi, Y., Fromen, C., Ducancel, F., Zinn-Justin, S., Le Du, M.H., Boulain, J.C., Tamiya, T., Menez, A., 1998. Molecular evolution of snake toxins: is the functional diversity of snake toxins associated with a mechanism of accelerated evolution? Prog. Nucleic Acid Res. Mol. Biol. 59, 307–364.

Olivera, B.M., 2006. Conus peptides: biodiversity-based discovery and exogenomics. J. Biol. Chem. 281, 31173–31177.

Olivera, B.M., Bulaj, G., Garrett, J., Terlau, H., Imperial, J., 2009. Peptide toxins from the venoms of cone snails and other toxoglossan gastropods, in: De Lima, M.E., Pimenta, A.M.C., Martin-Euclaire, M.F., Zingali, R.B. (Eds.), Animal Toxins: State of the Art. Perspectives in Health and Biotechnology. Editora UFMG, Belo Horizonte, Brazil, pp. 25–48.

Pahari, S., Mackessy, S.P., Kini, R.M., 2007. Rattlesnake (*Sistrurus catenatus edwardsii*): towards an understanding of venom composition among advanced snakes (Superfamily Colubroidea). BMC Mol. Biol. 8, 115.

Qinghua, L., Xiaowei, Z., Wei, Y., Chenji, L., Yijun, H., Pengxin, Q., Xingwen, S., Songnian, H., Guangmei, Y., 2006. A catalog for transcripts in the venom gland of the *Agkistrodon acutus*: identification of the toxins potentially involved in coagulopathy. Biochem. Biophys. Res. Commun. 341, 522–531.

Richardson, W.H., Tanen, D.A., Tong, T.C., Betten, D.P., Carstairs, S.D., Williams, S.R., Cantrell, F.L., Clark, R.F., 2005. Crotalidae polyvalent immune Fab (ovine) antivenom is effective in the neutralization of South American Viperidae venoms in a murine model. Ann. Emerg. Med. 45, 595–602.

Risch, M., Georgieva, D., von Bergen, M., Jehmlich, N., Genov, N., Arni, R.K., Betzel, C., 2009. Snake Venomics of the Siamese Russell's Viper (*Daboia russelli siamensis*) – relation to pharmacological activities. J. Proteomics 72, 256–269.

Sánchez, E.E., Ramírez, M.S., Galán, J.A., López, G., Rodríguez-Acosta, A, Pérez, J.C., 2003a. Cross reactivity of three antivenoms against North American snake venoms. Toxicon 41, 315–320.

Sánchez, E.E., Galán, J.A., Rodríguez-Acosta, A, Chase, P., Pérez, J.C., 2003b. Cross reactivity of three antivenoms against North American snake venoms. Toxicon 41, 357–365.

Sanz, L., Gibbs, H.L., Mackessy, S.P., Calvete, J.J., 2006. Venom proteomes of closely related Sistrurus rattlesnakes with divergent diets. J. Proteome Res. 5, 2098–2112.

Sanz, L., Escolano, J., Ferretti, M., Biscoglio, M.J., Rivera, E., Crescenti, E.J., Angulo, Y., Lomonte, B., Gutiérrez, J.M., Calvete, J.J., 2008a. Snake venomics of the South and Central American Bushmasters. Comparison of the toxin composition of *Lachesis muta* gathered from proteomic versus transcriptomic analysis. J. Proteomics 71, 46–60.

Sanz, L., Ayvazyan, N., Calvete, J.J., 2008b. Snake venomics of the Armenian mountain vipers *Macrovipera lebetina obtusa* and *Vipera raddei*. J. Proteomics 71, 198–209.

Saravia, P., Rojas, E., Arce, V., Guevara, C., López, J.C., Chaves, E., Velásquez, R., Rojas, G., Gutiérrez, J.M., 2002. Geographic and ontogenic variability in the venom of the neotropical rattlesnake *Crotalus durissus*: pathophysiological and therapeutic implications. Rev. Biol. Trop. 50, 337–346.

Serrano, S.M.T., Shannon, J.D., Wang, D., Camargo, A.C., Fox, J.W., 2005. A multifaceted analysis of viperid snake venoms by two-dimensional gel electrophoresis: an approach to understanding venom proteomics. Proteomics 5, 501–510.

Serrano, S.M.T., Fox, J.W., 2008. Exploring snake venom proteomes: multifaceted analyses for complex toxin mixtures. Proteomics 8, 909–920.

Stock, R.P., Massougbodji, A., Alagón, A., Chippaux, J-P., 2007. Bringing antivenoms to Sub-Saharan Africa. Nat. Biotech. 2, 173–177.

Tashima, A.K., Sanz, L., Camargo, A.C.M., Serrano, S.M.T., Calvete, J.J., 2008. Snake venomics of the Brazilian pitvipers *Bothrops cotiara* and *Bothrops fonsecai*. Identification of taxonomy markers. J. Proteomics 71, 473–485.

Theakston, R.D.G., Warrell, D.A., 2000. Crisis in snake antivenom supply for Africa. Lancet 356, 2104.

Thomas, L., Tyburn, B., 1996. The research group of snake bite in Martinique. *Bothrops lanceolatus* bites in Martinique: clinical aspects and treatment, in: Bon, C., Goyffon, M. (Eds.), Envenomings and Their Treatments. Fondation Marcel Mérieux, Lyon, pp. 255–265.

Valente, R.,H., Guimarães, P.R., Junqueira, M., Neves-Ferreira, A.G.C., Soares, M.R., Chapeaurouge, A., Trugilho, M., León, I.R., Rocha, S., Oliveira-Carvalho, A.L., Serrão, L.W., Dutra, D., Leão, L.I., Junqueira-de-Azevedo, I., Ho, P.L., Zingali, R.B., Perales, J., Domont, G.B., 2009. *Bothrops insularis* venomics: a proteomic analysis supported by transcriptomic-generated sequence data. J. Proteomics 72, 241–255.

Vetter, R.S., Isbister, G.K., 2008. Medical aspects of spider bites. Annu. Rev. Entomol. 53, 409–429.

Vidal, N., 2002. Colubroid systematics: evidence for an early appearance of the venom apparatus followed by extensive evolutionary tinkering. J. Toxicol. Toxin Rev. 21, 21–41.

Wagstaff, S.C., Harrison, R.A., 2006. Venom gland EST analysis of the saw-scaled viper, *Echis ocellatus*, reveals novel $\alpha_9\beta_1$ integrin-binding motifs in venom metalloproteinases and a new group of putative toxins, renin-like proteases. Gene 377, 21–32.

Wagstaff, S.C., Sanz, L., Juárez, P., Harrison, R.A., Calvete, J.J., 2009. Combined snake venomics and venom gland transcriptomic analysis of the ocellated carpet viper, *Echis ocellatus*. J. Proteomics 71, 609–623.

World Health Organization, 2007. Rabies and Envenomings. A Neglected Public Health Issue: Report of a Consultative Meeting. WHO, Geneva. Available: http://www.who.int/bloodproducts/animal_sera/Rabies.pdf.

Wood, D.L., Miljenović, T., Cai, S., Raven, R.J., Kaas, Q., Escoubas, P., Herzig, V., Wilson, D., King, G.F., 2009. ArachnoServer: a database of protein toxins from spiders. BMC Genomics 10, 375.

Zhang, B., Liu, Q., Yin, W., Zhang, X., Huang, Y., Luo, Y., Qiu, P., Su, X., Yu, J., Hu, S., Yan, G., 2006. Transcriptome analysis of *Deinagkistrodon acutus* venomous gland focusing on cellular structure and functional aspects using expressed sequence tags. BMC Genomics 7, 152.

Chapter 5
Insights in to Venom and Toxin Activities and Pharmacological/Therapeutic Potential Using Gene Expression Profiling

Jay W. Fox

Abstract Traditionally, venoms are assessed, usually following a standard list of potential activities including hypotensive, hemorrhagic, edema and neurotoxicity. This is then followed by the search of toxins in the venoms which contribute to the observed pathophysiological effect. Typically this is done by isolation of the toxin from the venom via biochemical procedures. Alternatively, toxins are cloned and expressed from cDNA libraries generated from venom gland RNA followed by assessment of biological/biochemical function and ultimately pathophysiological effect on the experimental model. These approaches have provided the field with a rich understanding of the potential role of toxins, but may fall short in terms of fully describing the actual function of the toxin in the context of the whole venom and whole animal. This mode of investigation likely does not provide a full understanding of the complete range of biological activities of the venoms and toxins since the suite of assays conducted are generally limited. One data-rich approach that has recently been exploited for further understanding the biological activities and functional potential of venoms and toxins is the study of their ability to alter gene expression in the host. This chapter will discuss how this approach may be of value in terms of drug discovery based on venom produced alteration of gene expression profiles using Connectivity Maps.

Introduction

The study of venoms, animal, insect and others, generally follows a standard experimental approach. The pathophysiological effects of the venoms are assessed, usually following a standard list of potential activities such as hypotensive, hemorrhagic, edema, neurotoxicity etc. This is then followed by the search of toxins in the venoms which contribute to the observed pathophysiological effect. Typically this is done

J.W. Fox (✉)
Department of Microbiology, University of Virginia School of Medicine, Charlottesville, VA 22908, USA
e-mail: jwf8x@virginia.edu

R.M. Kini et al. (eds.), *Toxins and Hemostasis*, DOI 10.1007/978-90-481-9295-3_5,

by isolation of the toxin from the venom via biochemical procedures whereby the biological or biochemical action is assayed over the course of the isolation (Fox and Serrano, 2007). Alternatively, as is now often the case, toxins are cloned and expressed from cDNA libraries generated from venom gland RNA followed by assessment of biological/biochemical function and ultimately pathophysiological effect on the experimental model (Pahari et al., 2007). This approach has provided the field with a rich understanding of the potential role of toxins during an envenomation event, but may fall short in terms of fully describing the actual function of the toxin in the context of the whole venom and whole animal (Gutierrez et al., 2009).

The effects described for the venoms and toxins are likely due to the overt biochemical activities of the toxins in the venoms. For example, the production of hemorrhage by snake venom is very likely due to the presence and activities of snake venom metalloproteinases in the venom and the sort of study described above does provide significant information on the action of the venom and its constituent toxins to afford some logical basis for therapeutic intervention (Calvete, 2009; Fox and Serrano, 2007, 2009). However, this mode of investigation likely does not provide a full understanding of the complete range of biological activities of the venoms and toxins since the suite of assays conducted are generally limited and rather overt and straightforward, i.e. inject hemorrhagic toxin and observe production of hemorrhage.

One could argue that subtle, less overt and heretofore unassayed for activities may also be elicited by venoms and toxins. One data-rich approach that has recently been exploited for further understanding the biological activities and functional potential of venoms and toxins is the study of the ability of venoms and toxins to alter gene expression in the host. In the remainder of this treatise I will discuss the experimental efforts that have been conducted to date and the insights these investigators have provided on the activity of the venoms and toxins with regard to altering gene expression and how this may impact, albeit perhaps only slightly, the overall pathophysiology of the venoms. Finally I will discuss how this approach may be of value in terms of drug discovery based on venom produced alteration of gene expression profiles using Connectivity Maps (Lamb et al., 2006).

Snake Venoms and Toxins

In 2005 Gallagher and colleagues (Gallagher et al., 2003) reported on the use of microarrays (GeneChips™) on the subtoxic effects of two crotalid snake venoms, *Crotalus atrox* and *Bothrops jararaca* on human umbilical vein endothelial cells. The authors dosed the cells at levels which did not produce any overt phenotypic effect on the cells and then examined the changes in gene expression in the treated endothelial cells particularly looking at changes with regard to ontological categories and pathways. Interestingly there were both similarities as well as differences observed between the results from the two venoms underscoring the different animal source of the venoms. The main similarity observed between the results for the two venoms was the up-regulation of genes involved in Fas ligand/TNF-α receptor apoptotic pathway. As this was the first reported such study with animal venoms it most importantly demonstrated the potential power of toxicogenomics to investigate the subtoxic effects of venoms and discover novel activities which had not been previously considered.

These studies were followed by an investigation using a PIIIa SVMPs jararhagin isolated from the venom of *B. jararaca* on its effect upon the gene expression profile of human fibroblasts (Gallagher et al., 2005). The most significant finding from this study was that the metalloproteinase, although delivered at doses below that which causes phenotypic changes in the cells, altered the gene expression pattern of the fibroblasts toward a pro-inflammatory profile with up-regulation of transcripts for IL-1β, IL-6, CXCL1, CXCL2, CXCL8 (IL-8) thus eliciting an inflammatory, pro-apoptotic response (Fig. 5.1). Further investigations using jararhagin were performed in vivo in mice where samples from affected gastrocnemius tissue were subjected to laser capture microdissection followed by quantitative real time-PCR (qRT-PCR) and were observed to be similar to the results of gene expression analysis via Affymetrix GeneChips^{tm} of the toxin on human fibroblasts. Taken together these results suggest that there is a potential that the venom and thus its toxins could play a pathological function via alteration of gene expression profiles in the host tissue.

The studies on the effect of SVMPs on gene expression profiles was complemented by Cominetti and colleagues (Cominetti et al., 2004) when they used a PIIIb disintegrin-like/cystine-rich domain fragment termed Alt-C that was isolated from the venom of *Bothrops alternatus*. This protein displayed the ability to block fibroblast adhesion to collagen I and supported the adhesion of human umbilical vein endothelial cells (HUVECs) suggesting the functionality of integrin ligation. Furthermore Alt-C induced proliferation of HUVECs. In light of these observed activities the authors hypothesized that they were modulated by signal transduction pathways which ultimately result in an altered gene expression. What was observed was that when fibroblasts were grown on either Collagen I or Alt-C, 45 genes were up-regulated including VEGF. This was confirmed by detection of VEGF production by ELISA thus providing a potential explanation for the proliferation activity of the protein modulated by regulation by gene expression.

Alteration of gene expression in human glial cells by a snake venom three-fingered neuroxin, candoxin, isolated from the venom of *Bungarus candidus* was assayed by Pachiappan and colleagues (Pachiappan et al., 2005). A variety of genes and pathways were detected to be affected by treatment with the neurotoxin including signal transduction, ubiquitin-inflammation and others. The conclusion of the authors was that the toxin caused neurodegeneration via toxin induced glial inflammation, DNA-damage and degeneration.

Spider Venom

Brown recluse (*Loxosceles* spp.) spiders produce necrotic arachnidism which can be rather sever and persist over a significant period of time (Swanson and Vetter, 2006). Sphingomyelinase D is thought to be one of the predominant toxins in the venom that causes demonecrotic lesions and is considered to not be a direct inducer

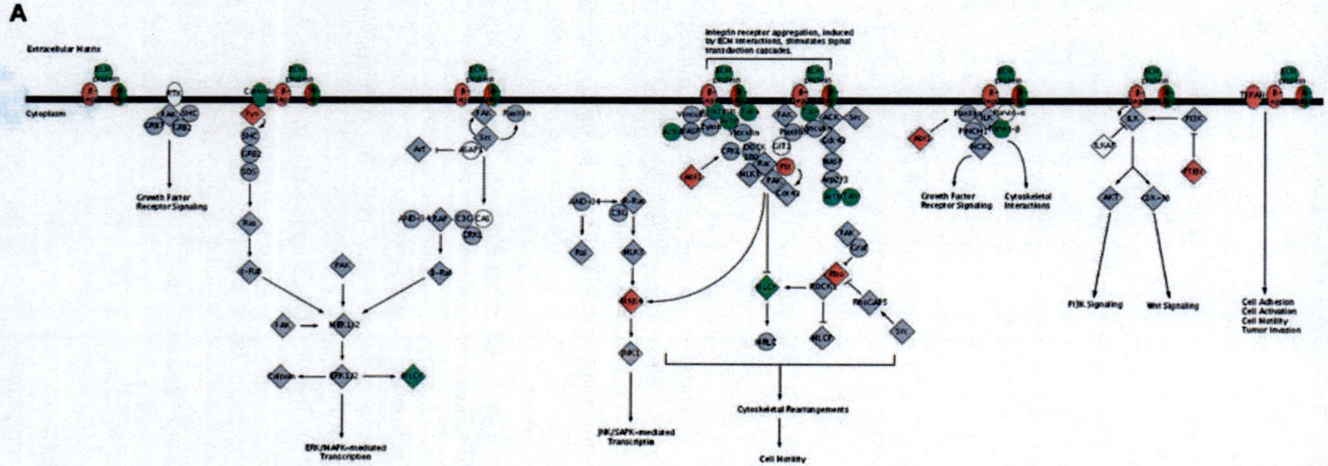

Fig. 5.1 Functional pathways indentified as being significantly altered in jararhagin-treated HS-68 fibroblasts. (**a**) Integrin Signaling pathway; (**b**) p38 MAPK pathway. This pathway is up-regulated in jararhagin-treated fibroblasts and thus contributes to the production of cytokines and promotes apoptosis; (**c**) The Death Receptor pathway. *Boxes* colored in *green* represent genes identified as being present and down-regulated compared to controls; *red colored boxes* represent genes identified a being present and up-regulated compared to controls; *grey colored boxes* were genes indentified as present, but unchanged compared to controls; and *white boxes* represent genes in the pathways that were identified as being absent. (Figure reproduced with permission, Gallagher et al., 2005)

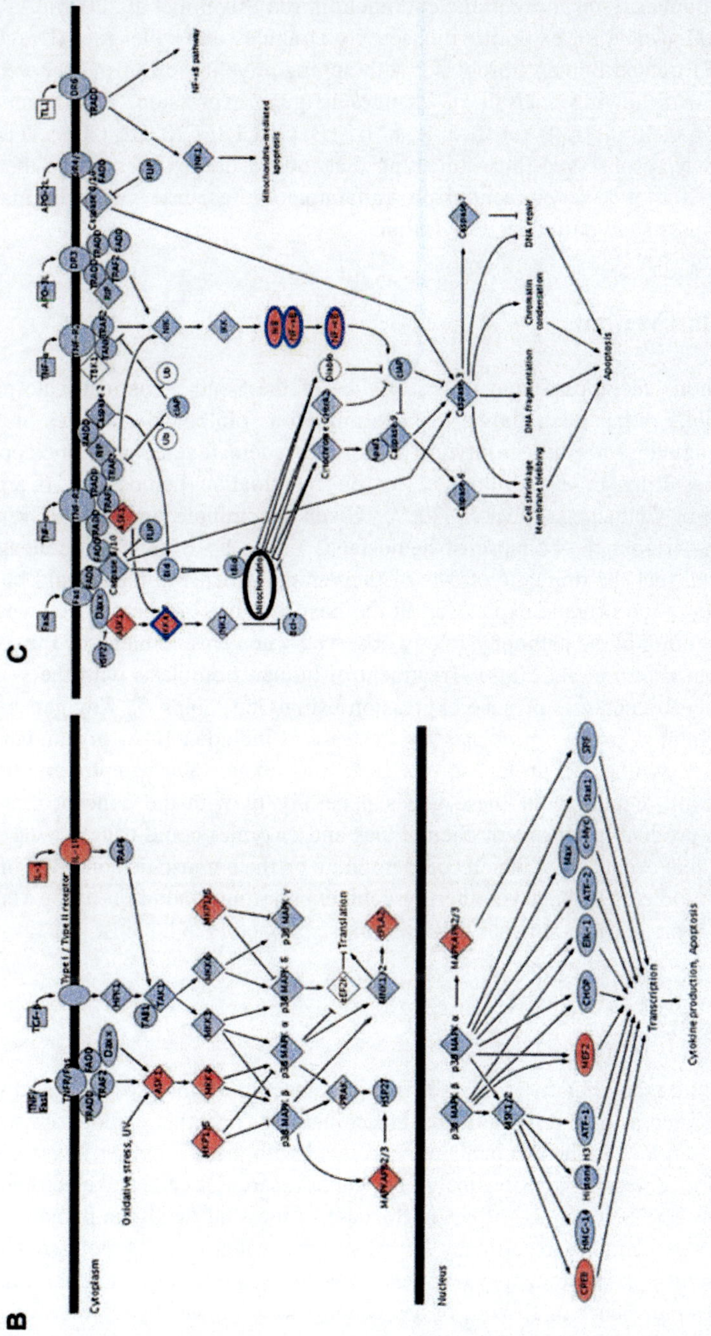

Fig. 5.1 (continued)

of the necrosis but instead functions as an initiator of the pathology by launching an inflammatory response in the endothelium (da Silveira et al., 2006). As one experimental avenue for exploring this activity Dragulev and colleagues (Dragulev et al., 2007) treated human fibroblasts with sphingomyelinase D from the venom of *L. recluse* following analysis of changes in gene expression. Significant upregulation in the transcripts for IL-6, IL-8, IL-1B, CXCL1, CXCL2, CCL5, TBF-α and NF-κB were observed thus indicating that indeed the toxin could be the upstream initiator of a very potential pro-inflammatory response which ultimately results in a persistent dermonecrotic lesion.

Caterpillar Venom

The venomous secretion found in the bristles of the South American caterpillar *Lonomia oblique* has been shown to have numerous biological activities including pro-coagulant, fibrin(ogen)olytic, hemolytic, edematogenic and nociceptive activities resulting in a profound effect on mammalian hemostatsis (Carrijo-Carvalho and Chudzinski-Tavassi, 2007). Given the minute amounts of venom necessary to trigger the collapse of hemostatsis in the host Pinto and colleagues hypothesized that the dramatic effects of the venom, at least in part, could be the result of alteration of gene expression in the host to cause biochemical/enzymatic cascade resulting in the pathophysiology observed when envenomated by the caterpillar venom (Pinto et al., 2008). Treatment of human fibroblasts with the venom was followed by analysis of gene expression using GeneChips^{TM}. Key genes that were upregulated in the fibroblasts by the venom included IL-8, prostaglandin-endoperoxide synthase 2, urokinase-type plasminogen activator receptor and tissue factor. The up-regulation of these genes generally fit with the concept that the increase in production of potent chemokines and enzymes could trigger a cascade of biochemical events all of which could impinge on the hemostatic condition of the host in this indirect manner via altering gene expression in addition to the venom acting in a typical, direct manner on key hemostatic factors in the host.

Bee Venom

In light of the many putative remedies that have been traditionally associated with bee venom such as pain relief and anti-inflammatory. To further explore the activities of bee venom and the mechanisms associated with these activities investigators have performed gene expression analysis of chondrosarcoma cells and macrophages (Jang et al., 2009; Yin et al., 2005). In the case of the study with the human chondrosarcoma cells in one experiment the cells were treated with lipopolysaccharide (LPS) to induce an inflammatory response somewhat equivalent to that observed in arthritis in the presence or absence of bee venom. They showed by microarray analysis that the bee venom was capable of lowering the expression of LPS up-regulated genes such as IL-6 receptor, matrix metalloproteinase 15 (MMP-15), tumor necrosis factor (ligand) superfamily-10, caspase-6 and tissue inhibitor of metalloproteinase-1 (TIMP-1).

In a similar experimental paradigm Jang and colleagues treated LPS-activated RAW 264.7 macrophage cells with bee venom to determine if it could attenuate the effects of LPS as assessed with microarray analysis (Jang et al., 2009). It was seen that bee venom inhibited the LPS-induced expression of several of the downstream inflammatory factors regulated by NF-kB including MAP3K8, TNF, TNF-a-induced proteins 3 suppressor of cytokine signaling 3 (SOCS3), TNF receptor-associated factor 1 (TRAF1), JUN and CREB binding protein. Thus the authors concluded that bee venom via alteration of gene expression of targets in the NF-kB/MAPK pathways attenuated the effects of LPS activated macrophages.

Use of Gene Expression Profiling as a Tool for Venom Drug Lead Discovery

Venoms have long been considered a potentially rich source of pharmaceutically important toxins or drug lead compounds (Fox and Serrano, 2007; Kini, 2006). Traditionally, when querying venom for a particular activity of interest a specific assay is applied to the venom and if a positive resulted is yielded then the toxins is isolated from the venom based on that activity. This is problematic in that in some cases the positive result for the activity in the venom may be due to a synergistic effect of two or more toxins and even if the result of one toxin isolation of the toxin may be difficult. Furthermore it is unlikely novel activities will be found since the discovery is circumscribed by the assays used, which in toxinology are generally of a limited number. We hypothesize that another approach may be valid for identifying novel activities of pharmacological value in the venoms via microarray analysis and an informatics approach. Lamb and colleagues (Lamb et al., 2006) have developed an approach where by numerous compounds, many with know effects, are screened for their impact on the gene expression of various cell lines with the goal of categorizing pathways associated with disease states that may be perturbed by alterations of the gene expression profile. Using this database one can perform a similar assay with the drug or in our case venom or toxin, determine its effect on the gene expression profile of a specific cell line and then correlate those results with the Connectivity Map database and look for novel activities elicited by the venom. One could then isolate the toxin(s) of interest by following the fractions' effect on gene expression via microarray analysis or perhaps more efficiently by q-PCR. Of course this approach also suffers from the potential problem of the readout from the experiment being due to the activity of two or more toxins acting synergistically or against one another in terms of affecting gene expression. Nevertheless, this may be one approach for mining novel activities from venoms for drug lead discovery.

To test this hypothesis we have examined the effect of Gila monster venom (*Heloderma suspectum*) on human breast MCF-7 cells and analyzed the gene chip data using the Connectivity Map database. What we have observed is that several of the pathways upregulated by the venom match that for known diabetes drugs. This is exciting in that a peptide, extendin-4, which was isolated from Gila monster venom. Extendin-4 has a structure similar to glucagon-like peptide 1 (GLP-1) and functions to regulate blood glucose and satiety (Eng et al., 1992). Interestingly, extendin-4 served as a drug-lead compound for the development of the drug exenatide which is currently FDA approved for the treatment of certain forms of diabetes and marketed under the trade name Byetta. Microarray analysis of pancreas from mice treated with exendin-4 showed up-regulation of the mitogenic Reg gene family suggesting that the peptide may function in its role in terms of attenuating the effects of diabetes by stimulating pancreatic growth (De Leon et al., 2006). Similarly, the drug exenatide was used in normal and diabetic islets under lipotoxic conditions (Ghanaat-Pour and Sjoholm, 2009). What was observed following gene expression analysis was that exenatide significantly affect epigenetic processes affecting development and cell proliferation. Further, expression of Bcl2-like 1 and Bcl2 modifying factor transcripts were altered suggesting an effect on moderating apoptosis in the islets.

Thus, although not fully developed, these studies suggest that mining the whole venom for novel functionalities associated with regulation of gene transcription associated with disease states may be an important new approach for identification of novel potential drug leads found in animal venoms.

Conclusion

As highlighted in this review the use of microarray analysis to explore the subtle, indirect effect of venoms and their toxins on the host's gene expression has proven successful. The approach has generated many interesting, novel insights into the mechanism of action of venoms associated with the observed pathophysiological outcome and thus serves to enhance our understanding of envenomation. Further, this approach may be of value in discovering new activities in the venom based on their ability to alter the gene expression of cells in such a manner corresponds to those of other drugs or disease states.

References

Calvete, J.J., 2009. Antivenomics and venom phenotyping: a marriage of convenience to address the performance and range of clinical use of antivenoms. Toxicon. Dec 29 [Epub ahead of print].

Carrijo-Carvalho, L.C., Chudzinski-Tavassi, A.M., 2007. The venom of the *Lonomia* caterpillar: an overview. Toxicon 49, 741–757.

Cominetti, M.R., Terruggi, C.H.B., Ramos, O.H.P., Fox, J.W., Mariano-Oliveria A., De Freitas, M.S., Figueiredo, C.C., Morandi, V., Selistre-De-Araujo, H.S., 2004. Alternagin-C, a disintegrin-like protein, induces vascular endothelial cell growth factor (VEGF) expression and endothelial cell proliferation *in vitro*. J. Biol. Chem. 279, 18247–18255.

da Silveira, R.B., Pigozzo, R.B., Chaim, O.M., Appel, M.H., Dreyfuss, J.L., Toma, L., Mangili, O.C., Gremski, W., Dietrich, C.P., Nader, H.B., Veiga, S.S., 2006. Molecular cloning and functional characterization of two isoforms of dermonecrotic toxin from *Loxosceles intermedia* (brown spider) venom gland. Biochimie 88, 1241–1253.

De Leon, D.D., Farzad, C., Crutchlow, M.F. Brestelli, J., Tobias, J., Kaestner, K.H., Stoffers, D.A., 2006. Identification of transcriptional targets during pancreatic growth after partial pancreatectomy and exendin-4 treatment. Physiol. Genomics 24, 133–143.

Dragulev, B., Bao, Y., Ramos-Cerrillo, B., Vazquez, H., Olvera, A., Stock, R., Algaron, A., Fox, J.W., 2007. Up-regulation of IL-6, IL-8, CXCL1, and CXCL2 dominates gene expression in human fibroblast cells exposed to *Loxosceles reclusa* sphingomyelinase D: insights into spider venom dermonecrosis. J. Invest. Dermatol. 127, 1264–1266.

Eng, J., Kleinman, W.A., Singh, L., Singh, G., Raufman, J.P., 1992. Isolation and characterization of exendin-e, an exendin-3 analogue from *Heloderma suspectum* venom. Further evidence for an exendin receptor o dispersed acini from guinea pig pancreas. J. Biol. Chem. 267, 7402–7405.

Fox, J.W., Serrano, S.M., 2009. Timeline of key events in snake venom metalloproteinase research. J. Proteomics 72, 200–209.

Fox, J.W., Serrano, S.M.T., 2007. Approaching the golden age of natural product pharmaceuticals from venom libraries: an overview of toxins and toxin-derivatives currently involved in therapeutic or diagnostic applications. Curr. Pharm. Design 13, 2977–2934.

Gallagher, P.G., Bao, Y., Serrano, S.M.T., Kamiguti, A.S., Theakston, R.D., Fox, J.W., 2003. Use of microarrays for investigating the subtoxic effects of snake venoms: insights into venom-induced apoptosis in human umbilical vein endothelial cells. Toxicon 41, 429–440.

Gallagher, P.G., Bao, Y., Serrano, S.M.T., Laing, G.D., Theakston, R.D.G., Gutierrez, J.M., Escalante, T., Moura-da-Silva, A.M., Nischt, R., Mauch, C., Moskaluk, C., Fox, J.W., 2005. Role of the snake venom toxin jararhagin in proinflammatory pathogenesis: in vitro and in vivo gene expression analysis of the effects of the toxin. Arch. Biochem. Biophys. 441, 1–15.

Ghanaat-Pour, H., Sjoholm, A., 2009. Gene expression regulated by pioglitazone and exenatide in normal and diabetic rat islets exposed to lipotoxicity. Diabetes/Metab Res. Rev. 25, 163–184.

Gutierrez, J.M., Rucavado, A., Escalante, T., Lomonte, B., Angulo, Y., Fox, J.W. 2009. Tissue pathology induced by snake venoms: how to understand a complex pattern of alterations from a systems biology perspective. Toxicon 55, 166–170.

Jang, H.-S., Chung, H.-S., Ko, E., Shin, J.S., Shin, M.K., Hong, M.C., Kim, Y., Min, B.I., Bae, H., 2009. Microarray analysis of gene expression profiles in response to treatment with bee venom in lipopolysaccharide active RAW 264.7 cells. J. Ethnopharm. 121, 213–220.

Kini, R.M., 2006. Anticoagulant proteins from snake venoms: structure, function and mechanism. Biochem. J. 397, 377–387.

Lamb, J. Crawford, E.D., Peck, D., Modell, J.W., Blat, I.C., Wrobel, M.J., Lerner, J., Brunet, J.P., Subramanian, A., Ross, K.N., Reich, M., Hieronymus, H., Wei, G., Armstrong, S.A., Haggarty, S.J., Clemons, P.A., Wei, R., Carr, S.A., Lander, E.S., Golub, T.R., 2006. The Connectivity Map: using gene-expression signatures to connect small molecules, genes and disease. Science 313, 1929–1935.

Pachiappan, A., Thwin, M.M., Manikandan, J., Gopalakrishnakone, P., 2005. Glial inflammation and neurodegeneration induced by candoxin, a novel neurotoxin from *Bungarus candidus* venom: global gene expression analysis using microarray. Toxicon 46, 883–899.

Pahari, S., Mackessy, S.P., Kini, R.M. 2007. The venom gland transcriptome of the Desert Massasauga rattlesnake (*Sistrusus catenatus edwardsii*) towards an understanding of venom composition among advanced snakes (Superfamily Colubroidea). BMC Mol. Biol. 8, 115–121.

Pinto, A.F., Dragulev, B., Guimaraes, J.A., Fox, J.W., 2008. Novel perspectives on the pathogenesis of Lonomia oblique caterpillar envenomation based on assessment of host response by gene expression analysis. Toxicon 51, 1119–1128.

Swanson, D.L., Vetter, R.S., 2006. Loxoscelism. Clin. Dermatol. 24, 213–221.

Yin, C.S., Lee, H.J., Hong, S.J., Chung, J.H., Koh, H.G., 2005. Microarray analysis of gene expression in chondrosarcoma cells treated with bee venom. Toxicon 45, 81–91.

Chapter 6
Bioactive Peptides and Proteins from Wasp Venoms

Ren Lai and Cunbao Liu

Abstract The members of Vespidae family include hornets (genera *Vespa* and *Dolichovespula*), yellow jackets (genus *Vespula*) and paper wasps (genus *Polistes*). The multi-sting capacity of their stingers together with their highly toxic venoms, makes them more aggressive in the defense of the colony or capture of the pray. Clinical symptoms induced in humans include local reactions (pain, wheal, edema and swelling) caused by biologically active peptides such as bradykinin-like peptides, chemotactic peptides and mastoparans, immunological reactions caused by venom allergens such as phospholipase A (PLA), hyaluronidase, antigen 5 and serine proteases which usually leading to anaphylaxis with subsequent anaphylactic shock, and systemic toxic reactions caused by large doses of venoms, resulting in hemolysis, coagulopathy, rhabdomyolysis, acute renal failure, hepatotoxicity, aortic thrombosis and cerebral infarction. The active components in wasp venoms, especially those acts on the cardiovascular system, nervous system and immunological systems of mammal, including humans, may show a promising perspective for the future discovery and application of potential pharmacological drugs.

Introduction

The members of Vespidae family include hornets (genera *Vespa* and *Dolichovespula*), yellow jackets (genus *Vespula*) and paper wasps (genus *Polistes*). They all possess highly toxic venom, which is a complex mixture of amines, small peptides and high molecular weight proteins such as enzymes, allergens and toxins (de Graaf et al., 2009; Habermann, 1972; Nakajima, 1984). The venoms from these stinging wasps are important weapons both in the defense of the colony or capture of the prey. The wasp inserts and withdraws its stinger much more easily than the bee because of its smooth outer lining, while the bee's stinger has a barbed outer sheath

R. Lai (✉)
Kunming Institute of Zoology, Chinese Academy of Sciences, Kunming 650023, Yunnan, China
e-mail: rlai@mail.kiz.ac.cn

R.M. Kini et al. (eds.), *Toxins and Hemostasis*, DOI 10.1007/978-90-481-9295-3_6,

looking like a fish hook, making it easy to insert but difficult to extract. Thus, the multi-sting capacity of the wasp bestows on this predator a more aggressive nature.

The sting apparatus of the wasp has a venom sac containing the venom and a redundant egg laying tube, and acts rather like the needle on a hypodermic syringe. This needle allows the wasp to curl its abdomen, squeeze the venom from the sac into the sting apparatus and inject the venom into the victim. The amount of venom released during a sting varies in different groups. Generally, the volume of venom delivered by a wasp sting is much less than that delivered by a bee. This phenomenon may be attributed to its repeated stinging abilities. For example, bee sting releases an average of 50–140 μg of venom per sting. Wasps generally inject less venom per sting: 1.7–3.1, 2.4–5.0, and 4.2–17 μg for *Vespula* stings, *Dolichovespula* stings and *Polistes* stings, respectively. In fact, the dry weight of wasp venom storage per sac is up to 260 μg (Biló et al., 2005; Edery et al., 1978; Hoffman and Jacobson, 1984; Schumacher et al., 1994).

A single wasp sting could kill or paralyze its prey insect instantly, and the clinical symptoms induced in humans include local reactions (pain, wheal, edema and swelling), immunological reactions usually leading to anaphylaxis with subsequent anaphylactic shock, and systemic toxic reactions caused by large doses of venom, resulting in hemolysis, coagulopathy, rhabdomyolysis, acute renal failure, hepatotoxicity, aortic thrombosis and cerebral infarction (Chao and Lee, 1999; Chen et al., 2004; Evans and Summers, 1986; Korman et al., 1990; Sakhuja et al., 1988; Watemberg et al., 1995).

Massive envenomation can produce both immediate and delayed toxic reactions in man, sometimes leading to death. Nevertheless, fatality due to wasp venom is more frequently related to allergic responses of the victim than to envenomation. With one sting, up to 17 μg of the wasp venom can be injected into the skin, which is sufficient for sensitization in man (Reisman and Livingston, 1992). Wasp venom allergy is an IgE-mediated allergic hypersensitivity of non-atopic origin (Johansson et al., 2001) and the most frequent clinical patterns are (i) large local reactions exceeding 10 cm in diameter and 24 h in duration, and (ii) rapid-onset (usually within 10 min after the sting) generalized immediate-type hypersensitivity reactions such as pruritus, urticaria, angioedema, nausea, vomiting, diarrhea, rhinoconjunctivitis, bronchiospasm, hypertension, cardiovascular collapse and unconsciousness (Ebo, 2007). Systemic reactions have been reported to occur in 0.8–5.0% of the general population, and they may be severe and even life-threatening with 0.09–0.45 deaths per million within the general population (Charpin et al., 1990, 1992; Mosbech, 1983).

Today venom immunotherapy (VIT) (i.e. subcutaneous injection of increasing doses of allergen preparations) is the only curative treatment against wasp venom allergies, and beneficial effects can persist for many years (Bousquet et al., 1998; Müller, 2001) though the precise mechanisms responsible for the beneficial effects of immunotherapy remain a matter of research and debate. Possible mechanisms are: (i) *Reduction of specific IgE* – Patients usually receive a course of allergen, building up gradually until a plateau or maintenance dose is achieved. Allergen-specific immunoglobulin IgE levels rise temporarily during the initial phase, but fall back to pretreatment levels during maintenance therapy. The late-phase skin test response is virtually abolished after successful immunotherapy. (ii) *Induction of IgG* – Immunotherapy also induces allergen-specific IgG antibodies, an observation that led to suggestions that antibodies might intercept the allergen and block the allergic response. This hypothesis is supported by the fact that, in patients treated for venom anaphylaxis, the development of allergen–specific IgG antibody correlates with clinical efficacy. (iii) *Reduction of effector cell recruitment* – Successful immunotherapy is accompanied by a reduction in T-cell and eosinophil recruitment in response to allergen challenge. (iv) *Altered T-cell cytokine balance* – The shift in the balance of T-helper 1 (Th1) and Th2 cytokine expression has been demonstrated in allergic patients after immunotherapy treatment. Increased number of Th1 cells produce more IL-10 which has a complex series of actions on the immune response, including stimulating production of the IgG4 subclass, which may therefore rise as an indicator of the beneficial effect (Frew, 2003; Levings et al., 2001; Mamessier et al., 2006; Reiman et al., 1984).

Vaccines used for immunotherapy are based on extracts derived from natural allergen sources. For the production of wasp venom vaccines, it is laborious and cost intensive because it is difficult to obtain large amounts of natural allergen preparations with homogeneous or standardized allergen composition (Backman et al., 1991; Littler et al., 1985; Müller, 1998). Moreover, it is unnecessary for patients to encounter the whole panel of allergens within the preparations that may cause adverse effects. Anaphylactic side reactions might be a particular problem during venom immunotherapy because wasp venoms contain a variety of pharmacologically active amines and enzymes that trigger IgE-independent mast cell mediator release that present a fatal risk to wasp venom allergic patients (Dudler et al., 1995; King et al., 1978; Machado et al., 1996). As a result, discovery of wasp venom allergens is necessary. The major wasp venom allergens are diagnosed as phospholipase A (PLA), hyaluronidase, antigen 5 and serine proteases (Asgari et al., 2003; Habermann, 1972; Nakajima, 1984; Yamamoto et al., 2007; Yang et al., 2007).

Bioactive Proteins from Wasp Venoms

Antigen 5

Antigen 5 is the major protein component of wasp venom and is considered the most potent allergen component (Hoffman, 1978, 1985, 1993), but it is not found in honeybee venom and its biological function has not yet been defined (Cascone et al., 1995; Henriksen et al., 2001; King and Guralnick, 2004; King et al., 1987). Antigen 5 is one member of a secreted proteins superfamily, and its partial sequence is similar to other proteins from diverse sources, especially other hymenoptera venom allergens. For example, antigen 5 has 44–46% sequence identity with phospholipases of wasps, and 73–92% sequence identity with hyaluronidases (King and Guralnick, 2004). Antigen 5 from species of the same genus share 98% sequence similarity, while it is 57% when antigen 5 comes from different genera (Hoffman, 1993). These sequence similarities between different allergens may explain why most vespid-allergic patients show multiple reactions to vespid venom (King et al., 1985; Lichtenstein et al., 1979).

Hyaluronidase

Hyaluronidase is a glycoprotein of approximately 43 kDa belonging to family 56 of the glycosyl hydrolases, whose enzymatic specificity is of the endo-*N*-acetylhexosaminidase type. It plays a key role in degrading extracellular matrix and increasing endothelial permeability by catalyzing the removal of *N*-acetyl-D-glucosamine from hyaluronic acid, found in the extracellular matrix of almost all tissue as a component of the substance that connects protein filament, collagen fibers and the connective tissue cell. The hyaluronidase in the wasp venom is thus responsible for the breakdown of the hyaluronic acid after the insect sting (Kasahara et al., 1989; King et al., 1996; Kolarich et al., 2005). Several studies have observed the high sequence similarity of different hyaluronidases contained in bee or wasp venoms and a sequence identity of about 50% between these allergens (Hemmer et al., 2004; King and Valentine, 1987; Wypych et al., 1989). This similarity of the glycan structures of wasp and bee venom hyaluronidases could well explain why hyaluronidase is largely responsible for the cross-reactivity with sera of allergic patients.

Phospholipases (PLA)

PLAs are relatively common in Hymenoptera venoms. PLA_1s comprise 6–14% of the total dry weight of vespid venom and PLA_2s comprise 12–15% of the dry weight of bee venom (de Oliveria and Palma, 1998; Habermann, 1972; King et al., 1983, 1984). PLA_1 and PLA_2 hydrolyze the *sn*-1 and *sn*-2 of the ester bonds of 1,2-diacyl-3-*sn*-glycerophospholipids, respectively, to produce lysophosphatidic acid (LPA) (Kini, 1997). LPA is a lipid mediator with multiple biological functions: inducing platelet aggregation, smooth muscle contraction and stimulating cell proliferation (Aoki, 2004; Baldini et al., 2005; Goetzl and An, 1998; Yang et al., 2007). As a result, despite their allergenic activities (Hoffman, 1986; King et al., 1996; Lu et al., 1995., Sanchez et al., 1994; Soldatova et al., 1993), PLA may be responsible for the uncommon systemic toxic reactions like aortic thrombosis and cerebral infarction following wasp stings. Recently, a PLA_1 named magnifin was isolated from venom of *V. magnifica* wasp. Magnifin strongly induces platelet aggregation at low concentrations and thrombosis in vivo. The hydrolysis product of magnifin, 1-acyl-lysophosphatidic acids, can mediate platelet aggregation and accelerate thrombus formation (Yang et al., 2007). PLAs also are able to disrupt the phospholipid packing of several types of biological membranes, leading to pore-formation and/or cell lysis, which may be responsible for their hemolytic activities (Dotimas et al., 1987; Kuchler et al., 1989).

Serine Protease

Serine proteases belong to trypsin family S1 of clan SA, the largest family of peptidases (Halfon and Craik, 1998; Serrano and Maroun, 2005). Some serine proteases with allergic or melanization-inhibitory activity have been found in wasp venoms (Asgari et al., 2003; Winningham et al., 2004). Many serine proteases have been characterized from animal venoms, especially snakes. They can affect the haemostatic system by acting on a variety of components of the coagulation cascade, on the fibrinolytic and kallikrein-kinin systems (Markland, 1997; Pirkle, 1998; Seegers and Ouyang, 1979). Recently, a serine protease named magnvesin was isolated from wasp venom of *V. magnifica*, which exerts anticoagulant function by hydrolyzing coagulant factors TF, VII, VIII, IX and X (Han et al., 2008).

Bioactive Peptides from Wasp Venoms

Local reactions may come from biologically active peptides such as bradykinin-like peptides, chemotactic peptides and mastoparans (Higashijima et al., 1979; Piek, 1984). These major actions are in vitro antimicrobial effects and inflammation induction, including initial lysis of cell membranes or mast cell degranulation, leading to histamine release and consequent vasodilation, increasing neutrophil and T helper cell chemotaxis (Argiolas and Pisano, 1985; Hancock et al., 1995; Nakajima et al., 1986; Wu and Hancock, 1999).

Chemotactic Peptides

Vespid chemotactic peptides are an important component in wasp venoms. They have potent hemolytic activity. Moreover, they have been observed to kill microbes effectively and induce the cellular chemotactic response (Dohtsu et al., 1993; Higashijima et al., 1979; Konno et al., 2006; Xu et al., 2006a; Yasuhara et al., 1983; Yu et al., 2007). Interestingly, vespid chemotactic peptides share high similarity with temporins (antimicrobial peptides from amphibians). However, the enzyme-cutting sites and the possible processing enzymes for both peptides are different, which for vespid chemotactic peptides are dipeptidyl peptidase IV (DPP-IV) and trypsin-like proteases, while for temporins are only trypsin-like protease. The sequence AXPX, tandem at N-terminus of vespid chemotactic peptides, is the key reason for the difference. DPP-IV, a highly specific dipeptidyl aminopeptidase, is characterized by the ability to release an N-terminal dipeptide, X-Y-↓-Z-, from a polypeptide, preferentially when Y is proline, provided Z is neither proline nor hydroxyproline (Boonacker and Van Noorden, 2003; Kühn-wache et al., 2003; Mentlein et al., 1986; Yu et al., 2007). It seems that DPP-IV is a common enzyme in the processing of peptide components in wasp venoms because the AXPX tandem has also been found in two other families of bioactive peptides including kinins and mastoparans (Xu et al., 2006b; Zhou et al., 2006).

Mastoparan

Mastoparan is the mostly extensively studied bioactive peptide from wasp venoms. It was first reported in the venom of *Vesoula lewisii* and then many homologues of mastoparan were also identified from other wasp species (Cerovský et al, 2007; Hirai et al., 1979a, b; Ho and Hwang, 1991; Konno et al., 2000, 2006; Mendes et al., 2005). They are a family of small peptides composed of 14 amino acid residues with a C-terminal amidated leucine. Mastoparans exert several biological activities, including (i) antimicrobial activity (Konno et al., 2001; Xu et al., 2006b); (ii) platelet activating activity (Ozaki et al., 1990); (iii) activation of phospholipase A_2, C and D, G-proteins and guanylate cyclase, in addition to causing cell lysis (Higashijima et al., 1988, 1990; Mizuno et al., 1995; Nakahata et al., 1990; Perianin and Snyderman, 1989; Song et al., 1993; Todokoro et al., 2006). They also affect phosphoinositide hydrolysis by interacting with lipid rafts (Okano et al., 1985; Sugama et al., 2005), which may be responsible for increasing the permeability of ions and small molecules through the biological membranes by forming pores at high peptide concentrations (Pfeiffer et al., 1995). As an agonist of G proteins, mastoparans compete with G protein receptors (GPRs) for binding to the G proteins (GP). Moreover, such binding is often selective, so as to make mastoparans a great interest in advanced medicine; for example, improved peptide drugs, to target GP (Höller et al., 1999); (iv) mast cell degranulating activity. Mast cells are a group of secretory cells vital to specific and innate immunity, allergy and inflammation. In specific IgE-mediated immune responses, mast cells are activated by antigens to release chemical mediators such as histamine, protease, prostaglandin, and cytokines. In innate immune responses against bacteria, mast cells promote neutrophil phagocytosis and lymph node hyperplasia via the production of tumor necrosis factor. Consistent with their role in innate defense, mast cells can directly kill phagocytes and bacteria (McCafferty et al., 1999; Nakajima and Yasuhara, 1977; Simmaco et al., 1999). Mastoparans are strong mediators of mast cell degranulation and histamine release (Xu et al., 2006b); (v) regulation of Ca^{2+} release. Mastoparan can bind to specific glycoproteins to regulate sarcoplasmic reticular Ca^{2+} release in skeletal muscles (Hirata et al., 2000, 2003), which may be responsible for bronchiospasm or hypertension after wasp sting.

Bradykinin

As an aggressive predator, wasps paralyze their pray (insects or spiders) by injecting venoms into the victim, and the immobilized prey is then used to feed the wasp's larvae. Therefore, wasp venom contains a variety of compounds, especially neurotoxins. In fact, research has been focused on wasp venom neurotoxins and some of them have proven to be highly selective to mammalian nervous systems. Thus, in addition to being used for studying neuronal mechanisms, these neurotoxins also show high prospects in the application of neuropathology (Harvey et al., 1998; Piek, 1991; Piek and Spanjer, 1986).

An important family of bioactive compounds having regulatory or hormonal functions from wasp is the bradykinin-like peptide (vespakinin), a counterpart of mammalian and amphibian bradykinins. They participate in a broad spectrum of biological activities and events in pathophysiological conditions by irreversibly blocking the synaptic transmission of the nicotinic acetylcholine receptors in the insect central nervous system (Bhoola et al., 1992; Kitamura et al., 1987; Nakajima, 1984; Piek et al., 1984, 1987,; Yasuhara et al., 1987). The generation of bradykinin in mammalian blood systems by the action of kallikrein-kinin system has been well documented (Bhoola et al., 1992; Kitamura et al., 1987). Bradykinin is a hydrolytic product from the limited proteolysis of kallikrein on kininogens. There are three types of kininogens in mammals: high molecular weight, low molecular weight kininogens, and T-kininogens (Kitagawa et al., 1987; Nakanishi, 1987; Takano et al., 1997). They are single chain glycoproteins consisting of three domains: a bradykinin moiety, an N-terminal heavy chain, and a C-terminal light chain, bridged by a disulfide linkage. Additionally, the heavy chain of kininogen may act as a cysteine proteinase inhibitor. Different from the precursors encoding mammalian bradykinin, most of the precursors (kininogen) encoding amphibian bradykinin are composed of several copies of a peptide segment unit including mature bradykinin plus a spacer peptide (Lai et al., 2001). Wasp kininogen is composed of a predicted signal peptide, an acidic peptide and a mature bradykinin-like peptide. Furthermore, the kininogens from mammals and amphibians contain a mono- or di-basic site for trypsin-like proteinases while the wasp kininogen lacks the mono- or di-basic site, suggesting they do not share the same processing pathways (Zhou et al., 2006). By blocking the synaptic transmission of the nicotinic acetylcholine receptor or glutamate receptors (Eldefrawi et al., 1988; Piek et al., 1990) or the slowing/blocking of sodium channel inactivation (Piek, 1982; Sahara et al., 2000), these brakykinin-like peptides effectively block the transmission of the vertebrate or invertebrate nervous systems, some of them showing anti-nociceptive effects more potent than that of morphine (Mortari et al., 2007).

In general, the wasp venom gland is a biochemically, pharmacologically and physiologically complex organ which fulfills a wide range of functions necessary for wasp survival. The venoms contain various molecules, especially bioactive proteins and peptides acting on the cardiovascular system, nervous system and immunological system of mammals, including humans, and these biomolecules show a promising perspective for the future discovery and application of potential pharmacological drugs.

References

Aoki, J., 2004. Mechanisms of lysophosphatidic acid production. Semin. Cell. Dev. Biol. 15, 477–489.

Argiolas, A., Pisano, J.J., 1985. Bombolitins, a new class of mast cell degranulating peptides from the venom of the bumblebee *Megabombus pennsylvanicus*. J. Biol. Chem. 260, 1437–1444.

Asgari, S., Zhang, G., Zareie, R., Schmidt, O., 2003. A serine proteinase homolog venom protein from an endoparasitoid wasp inhibits melanization of the host hemolymph. Insect Biochem. Mol. Biol. 33, 1017–1024.

Backman, A., Belin, L., Dreborg, S., Halvorsen, R., Malling, H.J., Weeke, B., 1991. Standardization of allergenic preparations. Comments with reference to the second edition of the common Nordic guidelines for registration of allergenic preparations. Allergy 46, 81–84.

Baldini, P.M., De Vito, P., D'aquilio, F., Vismara, D., Zalfa, F., Bagni, C., Fiaccavento, R., Di Nardo, P., 2005. Role of atrial natriuretic peptide in the suppression of lysophosphatydic acid-induced rat aortic smooth muscle (RASM) cell growth. Mol. Cell Biochem. 272, 19–28.

Bhoola, K.D., Figueroa, C.D., Worthy, K., 1992. Bioregulation of kinins: kallikreins, kininogens, and kininases. Pharmacol. Rev. 44, 1–80.

Biló, B.M., Rueff, F., Mosbech, H., Bonifazi, F., Oude-Elberink, J.N., 2005. EAACI interest group on insect venom hypersensitivity. Diagnosis of hymenoptera venom allergy. Allergy 60, 1339–1349.

Boonacker, E., Van Noorden, C.J., 2003. The multifunctional or moonlighting protein CD26/DPPIV. Eur. J. Cell. Biol. 82, 53–73.

Bousquet, J., Lockey, R., Malling, H.J., 1998. Allergen immunotherapy: therapeutic vaccines for allergic diseases. J. Allergy Clin. Immunol. 102, 558–562.

Cascone, O., Amaral, V., Ferrara, P., Vita, N., Guillemot, J.C., Diaz, L.E., 1995. Purification and characterization of two forms of antigen 5 from *Polybia scutellaris* venom. Toxicon 33, 659–665.

Cerovský, V., Pohl, J., Yang, Z., Alam, N., Attygalle, A.B., 2007. Identification of three novel peptides isolated from the venom of the neotropical social wasp *Polistes major major*. J. Pept. Sci. 13, 445–450.

Chao, S.C., Lee, Y.Y., 1999. Acute rhabdomyolysis and intravascular hemolysis following extensive wasp stings. Int. J. Dermatol. 38, 135–137.

Charpin, D., Birnbaum, J., Lanteaume, A., Vervloet, D., 1992. Prevalence of allergy to hymenoptera stings in different samples of the general population. J. Allergy Clin. Immunol. 90, 331–334.

Charpin, D., Vervloet, D., Haddi, E., Segalen, C., Tafforeau, M., Birnbaum, J, et al., 1990. Prevalence of allergy to Hymenoptera stings. Allergy Proc. 11, 29–32.

Chen, D.M., Lee, P.T., Chou, K.J., Fang, H.C., Chung, H.M., Chen, D.M., Chang, L.K., 2004. Descending aortic thrombosis and cerebral infarction after massive wasp stings. Am. J. Med. 116, 567–569.

De Graaf, D.C., Alerts, M., Danneels, E., Devreese, B., 2009. Bee, wasp and ant venomics pave the way for a component-resolved diagnosis of sting allergy. J. Proteomics 72, 145–154.

De Oliveria, M.R., Palma, M.S., 1998. Polybitoxins: a group of phospholipases A_2 from the venom of the neotropical social wasp paulistinha (*Polybia paulista*). Toxicon 36, 189–199.

Dohtsu, K., Okumura, K., Hagiwara, K., Palma, M.S., Nakajima, T., 1993. Isolation and sequence analysis of peptides from the venom of *Protonectarina sylveirae* (Hymenoptera-Vespidae). Nat. Toxins 1, 271–276.

Dotimas, E.M., Hamid, K.R., Hider. R.C., Ragnarsson, U., 1987. Isolation and structure analysis of bee venom mast cell degranulating peptide. Biochim. Biophys. Acta 911, 285–293.

Dudler, T., Machado, D.C., Kolbe, L., Annand, R.R., Rhodes, N., Gelb, M.H., Koelsch, K., Suter, M., Helm, B.A., 1995. A link between catalytic activity, IgE-independent mast cell activation, and allergenicity of bee venom phospholipase A2. J. Immunol. 155, 2605–2613.

Ebo, D.G., 2007. Hymenoptera venom allergy. Verh. K. Acad. Geneeskd. Belg. 69, 213–230.

Edery, H., Ishay, J., Gitter, S., Joshua, H., 1978. Venom of vespidae, in: Bettini, S. (Ed.), Arthropod Venoms. Sprinter Verlag, Berlin, New York, pp. 691–771.

Eldefrawi, A.T., Eldefrawi, M.E., Konno, K., Mansour, N.A., Nakanishi, K., Oltz, E., Usherwood, P.N., 1988. Structure and synthesis of a potent glutamate receptor antagonist in wasp venom. Proc. Natl. Acad. Sci. U.S.A. 85, 4910–4913.

Evans, R., Summers, S., 1986. Clinical aspects of hymenoptera sensitivity, in: Levine, M.I., Lockey R.F. (Eds.), American Academy of Allergy and Immunology. Monography on Insect Allergy. Lambert Associates, Pittsburgh, pp. 23–28.

Frew, A.J., 2003. Immunotherapy of allergic disease. J. Allergy Clin. Immunol. 111(2 Suppl), S712–S719.

Goetzl, E.J., An, S., 1998. Diversity of cellular receptors and functions for the lysophopholipid growth factors lysophosphatidic acid and sphingosine 1-phosphate. FASEB J. 12, 1589–1598.

Habermann, E., 1972. Bee and wasp venoms. Science 177, 314–322.

Halfon, S., Craik, C.S., 1998. Introduction: serine peptidases and their clan, in: Barrett, A.J., Rawlings, N.D., Woessner, J.F. (Eds.), Handbook of Proteolytic Enzyme. Academic Press, London, pp. 3–4.

Han, J., You, D., Xu, X., Han, W., Lu, Y., Lai, R., Meng, Q., 2008. An anticoagulant serine protease from the venom of *Vespa magnifica*. Toxicon 51, 914–922.

Hancock, R.E., Falla, T., Brown, M., 1995. Cationic bactericidal peptides. Adv. Microb. Physiol. 37, 135–175.

Harvey, A.L., Bradley, K.N., Cochran, S.A., Rowan, E.G., Pratt, J.A., Quillfeldt, J.A., Jerusalinsky, D.A., 1998. What can toxins tell us for drug discovery? Toxicon 38, 745–746.

Hemmer, W., Focke, M., Kolarich, D., Dalik, I., Gotz, M., Jarisch, R., 2004. Identification by immunoblot of venom glycoproteins displaying immunoglobulin E-binding N-glycans as cross-reactive allergens in honeybee and yellow jacket venom. Clin. Exp. Allergy 34, 460–469.

Henriksen, A., King, T.P., Mirza, O., Monsalve, R.I., Meno, K., Ipsen, H., Larsen, J.N., Gajhede, M., Spangfort, M.D., 2001. Major venom allergen of yellow jackets, Ves v 5: structural characterization of a pathogenesis-related protein superfamily. Proteins 4, 438–448.

Higashijima, T., Burnier, J., Ross, E.M., 1990. Regulation of Gi and Go by mastoparan, related amphiphilic peptides, and hydrophobic amines. Mechanism and structural determinants of activity. J. Biol. Chem. 265, 14176–14186.

Higashijima, T., Inubushi, T., Ueno, T., Miyazawa, T., 1979. NMR saturation transfer and line shape analyses of cyclic tetradepsipeptide AM toxin II: conformational equilibrium with very unequal populations. FEBS Lett. 105, 337–340.

Higashijima, T., Uzu, S., Nakajima, T., Ross, E.M., 1988. Mastoparan, a peptide toxin from wasp venom, mimics receptors by activating GTP-binding regulatory proteins (G proteins). J. Biol. Chem. 263, 6491–6494.

Hirai, Y., Kuwada, M., Yasuhara, T., Yoshida, H., Nakajima, T., 1979a. A new mast cell degranulating peptide homologous to mastoparan in the venom of Japanese hornet (*Vespa xanthoptera*). Chem. Pharm. Bull (Tokyo) 27, 1945–1946.

Hirai, Y., Yasuhara, T., Yoshida, H., Nakajima, T., Fujino, M., Kitada, C., 1979b. A new mast cell degranulating peptide "mastoparan" in the venom of *Vespula lewisii*. Chem. Pharm. Bull (Tokyo) 27, 1942–1944.

Hirata, Y., Atsumi, M., Ohizumi, Y., Nakahata, N., 2003. Mastoparan binds to glycogen phosphorylase to regulate sarcoplasmic reticular Ca^{2+} release in skeletal muscle. Biochem. J. 371, 81–88.

Hirata, Y., Nakahata, N., Ohizumi, Y., 2000. Identification of a 97-kDa mastoparan-binding protein involving in Ca^{2+} release from skeletal muscle sarcoplasmic reticulum. Mol. Pharmacol. 57, 1235–1242.

Ho, C.L., Hwang, L.L., 1991. Structure and biological activities of a new mastoparan isolated from the venom of the hornet *Vespa basalis*. Biochem. J. 274, 453–456.

Hoffman, D.R., 1978. Allergens in hymenoptera venom V: identification of some of the enzymes and demonstration of multiple allergens in yellow jacket venom. Ann. Allergy 40, 171–176.

Hoffman, D.R., 1985. Allergens in hymenoptera venom XV: the immunologic basis of vespid venom cross-reactivity. J. Allergy Clin. Immunol. 75, 611–613.

Hoffman, D.R., 1986. Allergens in hymenoptera venom XVI: studies of the structures and cross-reactivities of vespid venom phospholipases. J. Allergy Clin. Immunol. 78, 337–343.

Hoffman, D.R., 1993. Allergens in hymenoptera venom XXV: the amino acid sequences of antigen 5 molecules and the structural basis of antigenic cross-reactivity. J. Allergy Clin. Immunol. 92, 707–716.

Hoffman, D.R., Jacobson, R.S., 1984. Allergens in hymenoptera venom XII: how much protein is in a sting? Ann. Allergy 52, 276–278.

Höller, C., Freissmuth, M., Nanoff, C., 1999. G proteins as drug targets. Cell. Mol. Life Sci. 55, 257–270.

Johansson, S.G., Hourihane, J.O., Bousquet, J., Bruijnzeel-Koommen, C., Dreborg, S., Haahtela. T., Kowalski, M.L., Mygind, N., Ring, J., van Cauwenberge, P., van Hage-Hamsten, M., Wuthrich, B., 2001. A revised nomenclature for allergy. An EAACI position statement from the EAACI nomenclature task force. Allergy 56, 813–824.

Kasahara, M., Gutknecht, J., Brew, K., Spurr, N., Goodfellow, P.N., 1989. Cloning and mapping of a testis-specific gene with sequence similarity to a sperm-coating glycoprotein gene. Genomics 5, 527–534.

King, T.P., Alagon, A.C., Kuan, J., Sobotka, A.F., Lichtestein, L.M., 1983. Immunochemical studies of yellow jacket venom proteins. Mol. Immunol. 20, 297–308.

King, T.P., Guralnick, M., 2004. Hymenoptera allergens. Clin. Allergy Immunol. 18, 339–353.

King, T.P., Joslyn, A., Kochoumian, L., 1985. Antigenic cross-reactivity of venom proteins from hornet, wasps, and yellow jackets. J. Allergy Clin. Immunol. 75, 621–628.

King, T.P., Kochoumian, L., Joslyn, A., 1984. Wasp venom proteins: phospholipase A1 and B. Arch. Biochem. Biophys. 230, 1–12.

King, T.P., Kochoumian, L., Lam, T., 1987. Immunochemical observations of antigen 5, a major venom allergen of hornets, yellow jackets and wasps. Mol. Immunol. 24, 857–864.

King, T.P., Lu, G., Gonzalez, M., Qian, N., Soldatova, L., 1996. Yellow jacket venom allergens, hyaluronidase and phospholipase: sequence similarity and antigenic cross-reactivity with their hornet and wasp homologs and possible implications for clinical allergy. J. Allergy Clin. Immunol. 98, 588–600.

King, T.P., Sobotka, A.K., Alagon, A., Kchoumian, L., Lichtenstein, L.M., 1978. Protein allergens of white-faced hornet, yellow hornet, and yellow jacket venom. Biochemistry 17, 5165–5174.

King, T.P., Valentine, M.D., 1987. Allergens of hymenopteran venoms. Clin. Rev. Allergy 5, 137–148.

Kini, R.M., 1997. Phospholipase A_2 – a complex multifunctional protein puzzle, in: Kini, R.M. (Ed.), Venom Phospholipase A_2 Enzymes: Structure, Function and Mechanism. Wiley, England, pp. 1–28.

Kitagawa, H., Kitamura, N., Hayashida, H., Miyata, T., Nakanishi, S., 1987. Differing expression patterns and evolution of the rat kininogen gene family. J. Biol. Chem. 262, 2190–2198.

Kitamura, N., Ohkubo, H., Nakanishi, S., 1987. Molecular biology of the angiotesinogen and kininogen genes. J. Cardiovasc. Pharmacol. 7(Suppl), 49–53.

Kolarich, D., Léonard, R., Hemmer, W., Altmann, F., 2005. The N-glycans of yellow jacket venom hyaluronidases and the protein sequence of its major isoform in *Vespula vulgaris*. FEBS J. 272, 5182–5190.

Konno, K., Hisada, M., Fontana, R., Lorenzi, C.C., Naoki, H., Itagaki, Y., Miwa, A., Kawai, N., Nakata, Y., Yasuhara, T., Ruggiero Neto, J., de Azevedo, W.F., Jr., Palma, M.S., Nakajima, T., 2001. Anoplin, a novel antimicrobial peptide from the venom of the solitary wasp *Anoplius samariensis*. Biochim. Biophys. Acta 1550, 70–80.

Konno, K., Hisada, M., Naoki, H., Itagaki, Y., Fontana, R., Rangel, M., Oliveira, J.S., Cabrera, M.P., Neto, J.R., Hide, I., Nakata, Y., Yasuhara, T., Nakajima, T., 2006. Eumenitin, a novel antimicrobial peptide from the venom of the solitary eumenine wasp *Eumenes rubronotatus*. Peptides 27, 2624–2631.

Konno, K., Hisada, M., Naoki, H., Itagaki, Y., Kawai, N., Miwa, A., Yasuhara, T., Morimoto, Y., Nakata, Y., 2000. Structure and biological activities of eumenine mastoparan-AF (EMP-AF), a new mast cell degranulating peptide in the venom of the solitary wasp (*Anterhynchium flavomarginatum micado*). Toxicon 38, 1505–1515.

Korman, S.H., Jabbour, S., Harari, M.D., 1990. Multiple hornet (*Vespa orientalis*) stings with fatal outcome in a child. J. Paediatr. Child Health 26, 283–285.

Kuchler, K., Gmachl, M., Sippl, M.J., Kreil, G., 1989. Analysis of the cDNA for phospholipase A_2 from honeybee venom glands. The deduced amino acid sequence reveals homology to the corresponding vertebrate enzymes. Eur. J. Biochem. 184, 249–254.

Kühn-wache, K., Hoffmann, T., Manhart, S., Brandt, W., Demuth, H., 2003. The specificity of DPP-IV for natural substrates is peptide structure determined, in: Hildebrandt, M., Klapp, B., Hoffmamm, T., Demuth, H.U. (Eds.), Dipeptidyl Aminopeptidase in Health and Disease. Kluwer Academic/Plenum Publishers, New York, pp. 57–63.

Lai, R., Liu, H., Lee, W.H., Zhang, Y., 2001. A novel bradykinin-related peptide from skin secretions of toad *Bombina maxima* and its precursor containing six identical copies of the final product. Biochem. Biophys. Res. Commun. 286, 259–263.

Levings, M.K., Sangregorio, R., Galbiati, F., Squadrone, S., de Waal Malefyt, R., Roncarolo, M.G., 2001. IFN-alpha and IL-10 induce the differentiation of human type 1 T regulatory cells. J. Immunol. 166, 5530–5539.

Lichtenstein, L.M., Valentine, M.D., Sobotka, A.K., 1979. Insect allergy: the state of the art. J. Allergy Clin. Immunol. 64, 5–12.

Littler, S., Wypych J.I., Noble, R.W., Abeyounis, C.J., Reisman, R.E., 1985. Allergenic components of bald-faced hornet (*V. maculata*) venom. Int. Arch. Allergy Appl. Immunol. 76, 1–8.

Lu, G., Kochoumian, L., King, T.P., 1995. Sequence identity and antigenic cross-reactivity of white face hornet venom allergen, also a hyaluronidase, with other proteins. J. Biol. Chem. 270, 4457–4465.

Machado, D.C., Horton, D., Harrop, R., Peachell, P.T., Helm, B.A., 1996. Potential allergens stimulate the release of mediators of the allergic response from cells of mast cell lineage in the absence of sensitization with antigen-specific IgE. Eur. J. Immunol. 26, 2972–2980.

Mamessier, E., Birbaum, J., Dupuy, P., Vervloet, D, Magnan, A., 2006. Ultra-rush venom immunotherapy induces differential T cell activation and regulatory patterns according to the severity of allergy. Clin. Exp. Allergy 36, 704–713.

Markland F.S. Jr., 1997. Snake venoms. Drugs 54(Suppl 3), 1–10.

McCafferty, D.G., Cudic, P., Yu, M.K., Behenna, D.D., Kruger, R., 1999. Synergy and duality in peptide antibiotic mechanisms. Curr. Opin. Chem. Biol. 3, 672–680.

Mendes, M.A., de Souza, B.M., Palma, M.S., 2005. Structural and biological characterization of three novel mastoparan peptides from the venom of the neotropical social wasp *Protopolybia exigua* (Saussure). Toxicon 45, 101–106.

Mentlein, R., Rix, H., Feller, A.C., Heymann, E., 1986. Characterization of dipeptidyl peptidase IV from lymphocytes of chronic lymphocytic leukemia of T-type. Biomed. Biochim. Acta 45, 567–574.

Mizuno, K., Nakahata, N., Ohizumi, Y., 1995. Mastoparan-induced phosphatidylcholine hydrolysis by phospholipase D activation in human astrocytoma cells. Br. J. Pharmacol. 116, 2090–2096.

Mortari, M.R., Cunha, A.O., Carolino, R.O., Coutinho-Netto, J., Tomaz, J.C., Lopes, N.P., Coimbra, N.C., Dos Santos, W.F., 2007. Inhibition of acute nociceptive responses in rats after i.c.v. injection of Thr6-braykinin, isolated from the venom of the wasp of the social wasp, *Polybia occidentalis*. Br. J. Pharmacol. 151, 860–869.

Mosbech H., 1983. Deaths resulting from bee and wasp stings in Denmark 1960–1980. Ugeskr. Laeger 145, 1757–1760.

Müller, U.R., 1998. Hymenoptera venom hypersensitivity: an update. Clin. Exp. Allergy 28, 4–6.

Müller, U.R., 2001. New developments in the diagnosis and treatment of hymenoptera venom allergy. Int. Arch. Allergy. Immunol. 124, 447–453.

Nakahata, N., Abe, M.T., Matsuoka, I., Nakanishi, H., 1990. Mastoparan inhibits phosphoinositide hydrolysis via pertussis toxin-insensitive G-protein in human astrocytoma cells. FEBS Lett. 260, 91–94.

Nakajima, T., 1984. Biochemistry of vespid venoms, in: Tu, A.T. (Ed.), Handbook of Natural Toxins. Marcel Dekker, New York, pp. 109–133.

Nakajima, T., Uzu, S., Wakamatsu, K., Saito, K., Miyazawa, T., Yasuhara, T., Tsukamoto, Y., Fujino, M., 1986. Amphiphilic peptides in wasp venom. Biopolymers 25(Suppl), S115–S121.

Nakajima, T., Yasuhara, T., 1977. A new mast cell degranulating peptide, granuliberin-R, in the frog (*Rana rugosa*) skin. Chem. Pharm. Bull (Tokyo) 25, 2464–2465.

Nakanishi, S., 1987. Substance P precursor and kininogen: their structures, gene organizations, and regulation. Physiol. Rev. 67, 1117–1142.

Okano, Y., Takagi, H., Tohmatsu, T., Nakashima, S., Kuroda, Y., Saito, K., Nozawa, Y., 1985. A wasp venom mastoparan-induced polyphosphoinositide breakdown in rat peritoneal mast cells. FEBS Lett. 188, 363–366.

Ozaki, Y., Matsumoto, Y., Yatomi, Y., Higashihara, M., Kariya, T., Kume, S., 1990. Mastoparan, wasp venom, activates platelets via pertussis toxin-sensitive GTP-binding proteins. Biochem. Biophys. Res. Commun. 170, 779–785.

Perianin, A., Snyderman, R., 1989. Mastoparan, a wasp venom peptide, identifies two discrete mechanisms for elevating cytosolic calcium and inositol trisphosphates in human polymorphonuclear leukocytes. J. Immunol. 143, 1669–1673.

Pfeiffer, D.R., Gudz, T.I., Novgorodov, S.A., Erdahl, W.L., 1995. The peptide mastoparan is a potent facilitator of the mitochondrial permeability transition. J. Biol. Chem. 270, 4923–4932.

Piek, T., 1982. Delta-philanthotoxin, a semi-irreversible blocker of ion-channels. Comp. Biochem. Physiol. C. 72, 311–315.

Piek, T., 1984. Pharmacology of hymenoptera venom, in: Tu, A.T. (Ed.), Handbook of Natural Toxins (vol. 2). Marcel Dekker, New York, pp. 135–185.

Piek, T., 1991. Neurotoxic kinins from wasp and ant venoms. Toxicon 29, 139–149.

Piek, T., Hue, B., Mantel, P., Nakajima, T., Pelhate, M., Yasuhara T., 1990. Threonine 6-bradykinin in the venom of the wasp *Colpa interrupta* (F.) presynaptically blocks nicotinic synaptic transmission in the insect CNS. Comp. Biochem. Physiol. C. 96, 157–162.

Piek, T., Hue, B., Mony, L., Nakajima, T., Pelhate, M., Yasuhara, T., 1987. Block of synaptic transmission in insect CNS by toxins from the venom of the wasp *Megascolia flavifrons* (Fab.). Comp. Biochem. Physiol. C. 87, 287–295.

Piek, T., Spanjer, W., 1986. Chemistry and pharmacology of solitary wasp venoms, in: Piek, T. (Ed.), Venoms of the Hymenoptera. Academic Press, London, pp. 161–307.

Piek, T., Mantel, P., Van Ginkel, C.J., 1984. Megascoliakinin, a bradykinin-like compound in the venom of *Megascolia flavifrons* Fab (Hymenoptera: Scoliidae). Comp. Biochem. Physiol. C. 78, 473–474.

Pirkle, H., 1998. Thrombin-like enzymes from snake venoms: an updated inventory. Scientific and standardization committee's resgistry of exogenous hemostatic factors. Thromb. Haemost. 79, 675–683.

Reiman, R.E., Müller U.R., Wypych, J.I., Lazell, M.I., 1984. Studies of coexisting honeybee and vespid-venom sensitivity. J. Allergy Clin. Immunol. 73, 246–252.

Reisman, R.E., Livingston, A., 1992. Venom immunotherapy: 10 years of experience with administration of single venoms and 50 micrograms maintenance doses. J. Allergy Clin. Immunol. 89, 1189–1195.

Sahara, Y., Gotoh, M., Konno, K., Miwa, A., Tsubokawa, H., Robinson, H.P., Kawai, N., 2000. A new class of neurotoxin from wasp venom slows inactivation of sodium current. Eur. J. Neurosci. 12, 1961–1970.

Sakhuja, V., Bhalla, A., Pereira, B.J., Kapoor, M.M., Bhusnurmath, S.R., Chugh, K.S., 1988. Acute renal failure following multiple hornet stings. Nephron 49, 319–321.

Sanchez, F., Blanca, M., Miranda, A., Carmona, M.J., Garcia, J., Fernandez, J., Torres, M.J., Rondon, M.C., Juarez, C., 1994. Comparison of *Vespula germanica* venoms obtained from different sources. Int. Arch. Allergy Immunol. 104, 385–389.

Schumacher, M.J., Tveten, M.S., Egen, N.B., 1994. Rate and quantity of delivery of venom from honeybee stings. J. Allergy Clin. Immunol. 93, 831–835.

Seegers, W.H., Ouyang, C., 1979. Snake venoms and blood coagulation, in: Lee, C.Y. (Ed.), Handbook of Experimental Pharmacology (vol. 52), pp. 684–750.

Serrano, S.M., Maroun, R.C., 2005. Snake venom serine proteases: sequence homology vs. substrate specificity, a paradox to be solved. Toxicon 45, 1115–1132.

Simmaco, M., Mignogna, G., Barra, D., 1999. Antimicrobial peptides from amphibian skin: what do they tell us? Biopolymers 47, 435–450.

Soldatova, L., Kochoumian, L., King, T.P., 1993. Sequence similarity of a hornet (*D. maculata*) venom allergen phospholipase A1 with mammalian lipases. FEBS Lett. 320, 145–149.

Song, D.L., Chang, G.D., Ho, C.L., Chang, C.H., 1993. Structural requirements of mastoparan for activation of membrane-bound guanylate cyclase. Eur. J. Pharmacol. 247, 283–288.

Sugama, J., Ohkubo, S., Atsumi, M., Nakahata, N., 2005. Mastoparan changes the cellular localization of Gαq/11 and Gβ through its binding to ganglioside in lipid rafts. Mol. Pharmacol. 68, 1466–1474.

Takano, M., Kondo, J., Yayama, K., Otani, M., Sano, K., Okamoto, H., 1997. Molecular cloning of cDNAs for mouse low-molecular-weight and high-molecular-weight prekininogens. Biochim. Biophys. Acta 1352, 222–230.

Todokoro, Y., Yumen, L., Fukushima, K., Kang, S.W., Park, J.S., Kohno, T., Wakamatsu, K., Akutsu, H., Fujiwara, T., 2006. Structure of tightly membrane-bound mastoparan-X, a G-protein-activating peptide, determined by solid-state NMR. Biophys. J. 91, 1368–1379.

Watemberg, N., Weizman, Z., Shahak, E., Aviram, M., Maor, E., 1995. Fatal multiple organ failure following massive hornet stings. J. Toxicol. Clin. Toxicol. 33, 471–474.

Winningham, K.M., Fitch, C.D., Schmidt, M., Hoffman, D.R., 2004. Hymenoptera venom protease allergens. J. Allergy Clin. Immunol. 114, 928–933.

Wu, M., Hancock, R.E., 1999. Interaction of the cyclic antimicrobial cationic peptide bectenecin with the outer and cytoplasmic membrane. J. Biol. Chem. 274, 29–35.

Wypych, J.I., Abeyounis, C.J., Reisman, R.E., 1989. Analysis of differing patterns of cross-reactivity of honeybee and yellow jacket venom-specific IgE: use of purified venom fractions. Int. Arch. Allergy Appl. Immunol. 89, 60–66.

Xu, X., Li, J., Lu, Q., Yang, H., Zhang, Y., Lai, R., 2006a. Two families of antimicrobial peptides from wasp (*Vespa magnifica*) venom. Toxicon 47, 249–253.

Xu, X., Yang, H., Yu, H., Li, J., Lai, R., 2006b. The mastoparanogen from wasp. Peptides 27, 3053–3057.

Yamamoto, T., Arimoto, H., Kinumi, T., Oba, Y., Uemura, D., 2007. Identification of proteins from venom of the paralytic spider wasp, *Cyphononyx dorsalis* elastase. Insect Biochem. Mol. Biol. 37, 278–286.

Yang, H.L., Xu, X.Q., Ma, D.Y., Zhang, K.Y., Lai, R., 2007. A phospholipase A1 platelet activator from the wasp venom of *Vespa magnifica* (Smith). Toxicon 51, 289–296.

Yasuhara, T., Mantel, P., Nakajima, T., Piek, T., 1987. Two kinins isolated from an extract of the venom reservoirs of the solitary wasp *Megascolia flavifrons*. Toxicon 25, 527–535.

Yasuhara, T., Nakajima, T., Fukuda, K., Tsukamoto, Y., Mori, M., Kitada, C., Fujino, M., 1983. Structure and activity of chemotactic peptide from the venom sac of Vespinae, in: Munekata, E. (Ed.), Peptide Chemistry. Protein Research Foundation, Osaka, pp. 185–190.

Yu, H., Yang, H., Ma, D., Lv, Y., Liu, T., Zhang, K., Lai, R., 2007. Vespid chemotactic peptide precursor from the wasp *Vespa magnifica* (Smith). Toxicon 50, 377–382.

Zhou, Z., Yang, H., Xu, X., Wang, X., Lai, R., 2006. The first report of kininogen from invertebrates. Biochem. Biophys. Res. Commun. 347, 1099–1102.

Chapter 7
The Theory of Intraspecies Variation is Not the Exception, But Simply the Rule: The Diverse Hemostatic Activities of Snake Venoms

Alexis Rodríguez-Acosta, Belsy Guerrero, and Elda E. Sánchez

Abstract Clinical unpredictability of snake envenomation by the same species has been previously illustrated. These symptoms have been well known in the most common genera, *Bothrops* and *Crotalus*. Variations between venoms among various geographic regions may be due to evolutionary environmental pressure, which continually acts on separated populations. Studies concerning the disparity of snake venom are fundamental for the understanding of snake phylogeny and most significantly for the investigation and production of suitable antivenoms to treat ophidic envenomation. In the Americas, a geographic intraspecific variation in snake venom composition has been described for *Crotalus* and *Bothrops*. Intraspecific venom differences takes place among specific snakes, most likely due to seasonal variation, diet, habitat, age, sexual dimorphism, along with other unidentified factors that could possibly be contributing to the individual variability of venom composition. Furthermore, diverse hemostatic activities of snake venoms differ in their biochemical structure and pharmacological profile, not only between different species, but also within species, and in snakes of diverse ages, sex and geographical localities. Most of these studies scarcely demonstrate the interspecies variation of venoms in specimens from far and near geographical locations, which sustain the need to incorporate pools of venoms of the same species found in different geographical environments that will be employed in immunization protocols for the production of antivenoms.

Introduction

It has been well established that there are intraspecies differences in the snake venoms. Clinical variability of envenomation by snakes within the same species has been described (Aguilar et al., 2007; Sano-Martins et al., 2001). These clinical

A. Rodríguez-Acosta (✉)
Immunochemistry Section, Tropical Medicine Institute of the Universidad Central de Venezuela, Caracas, República Bolivariana de Venezuela
e-mail: rodriguezacosta1946@yahoo.es

R.M. Kini et al. (eds.), *Toxins and Hemostasis*, DOI 10.1007/978-90-481-9295-3_7,

symptoms have been noted in the most common genera, *Bothrops* and *Crotalus*. The diverse symptoms observed were not associated with the severity of envenomation, or to the age or sex of the patients. Differences in venom composition are a significant characteristic of intraspecific variability in the family Viperidae, even though reasons for this variability have not yet been elucidated.

A phylogenetic "advance" to the origin and continuation of species diversity would require the sampling of all species within a clade. In addition, verification that they are evolutionarily distinct entities and knowledge of their geographical distributions would be also required. Compositional variations between venoms amid diverse geographic territories may be due to evolutionary environmental pressure, which constantly acts on separated populations. All species distribution shows restriction by obstacles of unsuitable sedimentary habitats, extreme temperatures, broad regions of rivers and lakes and sea levels above 1,400 km. The geographical limits of similar species show limited or no overlap, demonstrating that the speciation mode is predominantly allopatric. Additionally, scale of expansion following speciation seems to have been restricted, since a high degree of allopatry is preserved through the known domains of the phylogeny. This may be explained by habitat specialization, uncommon long-distance colonization, and possibly by inter- and intra-specific competition (Calsbeek, 2009).

Studies regarding the variability of snake venom are vital for the comprehension of snake phylogeny and most importantly for the research and production of appropriate antivenoms to treat ophidic accidents. In the Americas, a geographic intraspecific variation in snake venom composition has been reported for *Crotalus* (Aguilar et al., 2007; Francischetti et al., 2000; Sánchez et al., 2008) and *Bothrops* (Alape-Girón et al., 2008; Girón et al., 2008; Salazar et al., 2009). Descriptive studies comparing snake venom activities on hemostasis from intraspecific species of different countries have been conducted (Aguilar et al., 2007; Salazar et al., 2009). The hemostatic activities of individual snake venoms from the same species have been quantified in terms of lethality, hemorrhagic, procoagulant and fibrino(geno)lytic activities (Giron et al., 2008; Salazar et al., 2008). Results have demonstrated venom differences amongst similar snake species from close localities in different countries.

This chapter is intended to accumulate a series of studies regarding the comparative hemostatic characterization of snake venoms of the same species of the genus *Bothrops* and *Crotalus* from the family Viperidae, specifically *C. durissus cumanensis*, *B. atrox*, *B. colombiensis* and *B. isabelae* from Venezuela; *B. asper* from Costa Rica; *C. oreganus helleri, C. scutulatus scutulatus*, and *C. viridis lutosus* from the United States of America (Aguilar et al., 2007; Alape-Girón, 2008, 2009; Girón et al., 2008; Salazar et al., 2008; Sánchez, 2004; Sánchez et al., 2005; Yoshida-Kanashiro et al., 2003). These venoms come from geographically isolated populations from several locations in the Americas, which include venoms from the same species captured in the same geographical areas. The crude venoms were analyzed for their lethality, hemorrhagic, proteolytic, coagulant, and fibrino(geno)lytic activities. These results infer patterns of venom diversification within the same species. A large degree of variability was observed in these venoms as defined by several hemostatic assays. These significant variations certainly have an effect on how antivenom neutralizes venoms, thus, complicating the management of snakebites and the production of antivenoms.

Venoms of the Genus *Crotalus*

The genus *Crotalus* are venomous pitvipers found only in the Americas ranging from Canada to Argentina (Campbell and Lamar, 2004). *Crotalus* comes from the Greek word *krotalon* meaning rattler or castanet. These snakes have a rattler at the end of their tail making them a very unique group.

Venezuela has 24 species of medically significant venomous snakes belonging to the family Viperidae in which 20.8% are from the genus *Crotalus* (Rengifo and Rodríguez-Acosta, 2005), and 15% of snakebites are due to this genus (Rodríguez-Acosta et al., 1995). Despite having more than 80% of *Crotalus* accidents in Venezuela being essentially neurotoxic, reports of *Crotalus* envenomations have described unusual hemostatic alterations, such as increase of clotting time, partial thromboplastin time (PTT), prothrombin time (PT), fibrinogen consumption, platelets and hemorrhages in the area of the bite (Yoshida-Kanashiro et al., 2003). Since most *Crotalus* species in South America are not considered hemorrhagic, these reports are an exception in South American *Crotalus* (Sano-Martins et al., 2001) and have been found in identical species from different geographical areas.

In Brazil, Amaral et al. (1988) first reported that spontaneous bleeding had rarely been observed in *C. durissus terrificus* envenomed patients, and some patients did not present detectable fibrinogen and thrombocytopenia after the accident. In addition, Kamiguti and Sano-Martins (1995) showed that Brazilian rattlesnake venoms rendered blood incoagulable due to fibrinogen consumption. Sano-Martins et al. (2001) showed that the envenoming by *C. durissus* was commonly associated with hemostatic disorders, which were possibly attributed principally to the action of the thrombin-like enzymes, with possible further secondary effects due to the potent myotoxic activity of the venom.

Until 2007, fibrinolytic activity in South American *Crotalus* venom had not been reported. For this reason, we started to evaluate this activity in *C. d. cumanensis* venom from different Venezuelan locations. Our findings demonstrated that *C. d. cumanensis* venoms from different locations presented different fibrinolytic activity by amidolytic and fibrin plate methods (Aguilar et al., 2007). The results with chromogenic substrates revealed, in all venoms, a high kallikrein-like activity and a low t-PA-like activity (Table 7.1). On fibrin plasminogen rich plates, the venoms were also active depending on the snake's geographical location. Significant fibrinolytic variations were observed, thus the kallikrein-like amidolytic activity was more elevated in those snake venoms from Santa Teresa (Miranda state) and Margarita Island. In contrast, higher fibrinolytic activity was observed in Lagunetica and Carrizales venoms.

Table 7.1 Variation in venom activities of the genus *Crotalus* from South America and USA

					Activities					
						Amidolytic-like enzymes (mUA/min/μg)		Coagulant		
Species	Country	Geographic locations	MHD	LD_{50}	Fibrinolytic-fibrin plate ($mm^2/10\ \mu g$)	Thrombin (S-2238)	Kallikrein (S-2302)	Plasma	Purified fibrinogen	References
C. durissus cumanensis	Venezuela	Lagunetica	4.1	0.86	51.3 ± 4.0	4.68 ± 1.9	306.7 ± 7.8	3.4	< 1	Aguilar et al. (2007)
C. durissus cumanensis	Venezuela	Santa Teresa	–	0.43	6.4 ± 2.5	9.5 ± 3.6	510.0 ± 6.0	24.0 ± 0.1	31.3 ± 0.2	
C. durissus cumanensis	Venezuela	Guarenas	–	0.66	14.7 ± 2.3	8.42 ± 2.7	442.0 ± 5.6	23.8 ± 0.2	30.2± 0.2	
C. durissus cumanensis	Venezuela	Anzoátegui	16.2	0.60	25.0 ± 1.0	5.70 ± 2.1	390.7 ± 6.0	3.8 ± 0.2	3.4 ±0.2	
C. durissus cumanensis	Venezuela	Aragua	–	0.66	32.4 ± 5.2	9.38 ± 4.6	460.57 ± 7.6	18.6 ± 0.2	30.6 ± 0.2	
C. durissus cumanensis	Venezuela	Carrizales	14.3	0.86	49.2 ± 6.3	1.98 ± 0.2	192.4 ± 5.6	Neg	Neg	
C. durissus cumanensis	Venezuela	Margarita	–	0.18	3.4 ± 1.1	9.10 ± 5.3	524.0 ± 7.2	12.6 ± .4	16.2 ± 0.3	
C. scutulatus scutulatus	USA	Southern California and Southwest Arizona	NT	0.24	NT	NT	NT	NT	NT	Glenn and Straight (1978)
C. scutulatus scutulatus	USA	Northwest and north central Arizona	NT	2.80	NT	NT	NT	NT	NT	

Table 7.1 (continued)

Species	Country	Geographic locations	MHD	LD_{50}	Activities: Fibrinolytic-fibrin plate (mm^2/10 μg)	Amidolytic-like enzymes (mUA/min/μg): Thrombin (S-2238)	Amidolytic-like enzymes (mUA/min/μg): Kallikrein (S-2302)	Coagulant: Plasma	Coagulant: Purified fibrinogen	References
C. scutulatus scutulatus	USA	Southeastern Arizona and 1 Specimen from North Tucson	NT	0.28	NT	NT	NT	NT	NT	Glenn et al. (1983)
C. scutulatus scutulatus	USA	Northern Tucson	NT	3.33	NT	NT	NT	NT	NT	
C. viridis lutosus	USA	Northeast Utah	800[a]	NT	NT	NT	NT	NT	NT	Adame et al. (1990)
C. v. lutosus	USA	Northcentral Utah	400[a]	NT	NT	NT	NT	NT	NT	
C. v. lutosus	USA	Southern Utah	200[a]	NT	NT	NT	NT	NT	NT	
C. scutulatus scutulatus	USA	Culberson Co., TX	–	0.37	–	NT	NT	NT	NT	Sánchez et al. (2005)
C. scutulatus scutulatus	USA	Culberson Co., TX	–	0.48	–	NT	NT	NT	NT	
C. scutulatus scutulatus	USA	Culberson Co., TX	–	0.47	–	NT	NT	NT	NT	
C. scutulatus scutulatus	USA	Jeff Davis Co., TX	–	0.42	+	NT	NT	NT	NT	
C. scutulatus scutulatus	USA	Jeff Davis Co., TX	–	0.42	–	NT	NT	NT	NT	

Table 7.1 (continued)

					Activities					
						Amidolytic-like enzymes (mUA/min/μg)		Coagulant		
Species	Country	Geographic locations	MHD	LD_{50}	Fibrinolytic-fibrin plate ($mm^2/10\ \mu g$)	Thrombin (S-2238)	Kallikrein (S-2302)	Plasma	Purified fibrinogen	References
C. scutulatus scutulatus	USA	Huspeth Co., TX	–	0.35	–	NT	NT	NT	NT	
C. scutulatus scutulatus	USA	Maricopa Co., AZ	–	0.83	–	NT	NT	NT	NT	
C. scutulatus scutulatus	USA	Maricopa Co., AZ	–	0.84	–	NT	NT	NT	NT	
C. scutulatus scutulatus	USA	Pima Co., AZ	–	1.05	+	NT	NT	NT	NT	
C. scutulatus scutulatus	USA	Maricopa Co., AZ	12.5	5.10	+	NT	NT	NT	NT	
C. scutulatus scutulatus	USA	Pinal Co., TX	25	3.54	+	NT	NT	NT	NT	
C. scutulatus scutulatus	USA	Pinal Co., TX	6.25	3.90	+	NT	NT	NT	NT	
C. scutulatus scutulatus	USA	Pinal Co., TX	12.5	2.90	+	NT	NT	NT	NT	
C. scutulatus scutulatus	USA	Pinal Co., TX	12.5	5.5	+	NT	NT	NT	NT	
C. oreganus helleri	USA	California Yellow Venom	0.71±0.14	1.84	+	NT	NT	NT	NT	Galán et al. (2004)
C. oreganus helleri	USA	California White Venom	2.06±0.89	2.95	+	NT	NT	NT	NT	

Table 7.1 (continued)

			Activities							
						Amidolytic-like enzymes (mUA/min/μg)		Coagulant		
Species	Country	Geographic locations	MHD	LD_{50}	Fibrinolytic-fibrin plate (mm^2/10 μg)	Thrombin (S-2238)	Kallikrein (S-2302)	Plasma	Purified fibrinogen	References
C. oreganus helleri	USA	Riverside 1 CA	8.1 ± 1.3	2.5	13.4 ± 0.26	2.23 ± 1.1	611 ± 29	7.0 ± 0.1	10.6 ± 0.6	Salazar et al. (2009)
C. oreganus helleri	USA	Riverside 2 CA	Neg	0.6	0	11.84 ± 7.1	120 ± 23	3.8 ± 0.1	11.0 ± 0.5	
C. oreganus helleri	USA	San Bernardino 1 CA	Neg	0.7	0	2.86 ± 0.9	320 ± 25	12.2 ± 0.2	26.1 ± 0.3	
C. oreganus helleri	USA	San Bernardino 2 CA	2.8 ± 0.5	3.0	15.6 ± 0.10	2.13 ± 1.0	207 ± 53	0	0	
C. oreganus helleri	USA	San Bernardino 3 CA	2.5 ± 0.8	5.7	67.2 ± 0.21	0.93 ± 0.4	142 ± 20	< 1.0	2.9 ± 0.1	

MHD minimal amount of protein in micrograms that produces a 10 mm hemorrhagic spot, LD_{50} = The amount of venom protein that will kill 50% of a mouse population per body weight (mg/kg), − = no activity, + = activity, *NT* not tested, [a] Hemorrhagic activity was reported as specific hemorrhagic activity using the following equation: mean of cross-diameter of hemorrhagic spot (mm)/0.005 mg

Glenn and Straight (1978) were the first to report significant venom variability in the venoms of the Mohave rattlesnakes (*C. scutulatus scutulatus*) from North America. This work classified the venoms of this snake as Type A (neurotoxic), Type B (hemorrhagic), and Type (A and B) (both neurotoxic and hemorrhagic). The hemorrhagic venom of this snake is primarily found in a limited area of Arizona (Glenn and Straight, 1978; Glenn et al., 1983; Sánchez et al., 2005, 2006). Lethality doses revealed that the hemorrhagic venoms were less potent with LD_{50} ranging from 2.9 to 5.5 mg/kg body weight (Table 7.1), while the neurotoxic venom ranged from 0.35 to 0.48 mg/kg (Sánchez et al., 2005). Genomic DNA analysis was done on both types of venoms to determine if the Mojave toxin genes were present. Results revealed that the hemorrhagic venom lacked the genes (Sánchez et al., 2005), while the neurotoxic snakes contained them. In addition, these snakes were also analyzed to determine if a disintegrin gene was present. The disintegrin gene was present only in the Type B snakes and not in the Type A snakes. Furthermore, Type (A and B) snakes contained both disintegrin and Mojave toxin genes and their LD_{50} ranged from 0.83 to 1.05 mg/kg body weight (Sánchez et al., 2005).

Galán et al. (2004) reported differences in the venoms of southern Pacific rattlesnakes (*C. oreganus helleri*). The fact that the color of these two venoms was very different (one white and the other yellow) sparked an interest to further test these venoms for qualitative and quantitative differences. The minimal hemorrhagic activities for these southern Pacific rattlesnakes were 0.7 and 2.1 μg, while the LD_{50} were 1.84 and 2.95 mg/kg body weight, respectively. In addition, the North American Crotalid Antivenom (Wyeth^{TM} Pharmaceuticals) was more efficient in neutralizing the more potent venom ($ED_{50} = 84$ mg/kg) as opposed to the less toxic venom ($ED_{50} = 357$ mg/kg). Furthermore, sera from the North American Virginia opossum (*Didelphis virginiana*) was not able to neutralize the most potent venom, but was able to neutralize the least potent one.

These prior studies done with these two southern Pacific rattlesnake venoms and personal communications with Dr. Sean Bush, MD of Loma Linda University regarding the difficulties of neutralizing the venom of these snakes, initiated another study involving not only the hemorrhagic and lethality activities, but the hemostatic activities as well (Salazar et al., 2009). This study showed that hemostatic activity of venoms from five specimens of *C. oreganus helleri* snakes from close geographical locations varies, both qualitatively and quantitatively. The differences in hemorrhagic, neurotoxic, and fibrinolytic activities pointed to intraspecies variation (Table 7.1), such that some possessed neurotoxic and/or proteolytic actions and others did not (Salazar et al., 2009).

C. oreganus helleri from two counties of southern California (SB: San Bernardino county; RS: Riverside county) presented significant thrombin-like activity in their venoms, which by the coagulant method was considerably higher in SB 1 venom; however, by the amidolytic method, RS 2 venom was more active. The highest amidolytic thrombin-like activity in RS 2 venom was explained by other serine proteases containing trypsin-like activity with actions on S-2238 substrate (Friberger, 1982). SB 2 and 3 venoms were not active or presented an insignificant thrombin-like activity.

C. oreganus helleri venoms also contained fibrinolytic enzymes, which were distinct in the different specimens. SB 3 venom showed the highest activity on fibrin plates, in the presence or absence of plasminogen, which was inhibited by EGTA, EDTA and 1,10 phenantroline (metalloproteinase inhibitors) (Enghild et al., 1989; Salazar et al., 2009). Additionally, the amidolytic method established that all venoms contained kallikrein-like activity (Pućkowska et al., 2008). Nevertheless, RS 1 venom showed the highest kallikrein-like activity, which was neutralized by serine protease inhibitors, indicating the presence of serine proteases that could perhaps in part activate the plasminogen and thus show fibrinolytic activity. In relation to the hemolytic activity, the results revealed that RS 1 and SB 2 venoms were the most active, which correlated with their high phospholipase concentrations (Condrea et al., 1964; Gutiérrez et al., 2005).

RS 2 and SB 1 venoms showed the highest thrombin-like on plasma and/or purified fibrinogen and lethal activities. Nevertheless, they presented neither hemorrhagic nor fibrinolytic activities. SB 2 and 3 venoms, with the modest coagulant and lethal activities, showed the highest fibrinolytic and hemorrhagic activities. These venoms can induce severe hemorrhagic problems in snake bitten patients due to their high hemorrhagic and/or primary fibrinolytic activities (Matsui et al., 2000; Swenson and Markland, 2005). These results support the differential presence of metalloproteinases with activity against fibrin, a criteria that will allow a different therapeutic conduct according with the regional ophidic accident. Venom differences have critical implications for snakebite treatment, since diagnosis may possibly be complicated by intraspecies differences in clinical manifestations, and antivenom made against venom of an assured species may not be efficient against envenomations of specimens of the same species (Chippaux et al., 1991). The hemostatic variations in the venoms of the Southern Pacific rattlesnakes contribute to our understanding of substrate specificities involved in the hemostasis pathway and further reaffirm the differential composition of venoms from the same snake species.

Minton and Weinstein (1986) compared geographic differences in venom from the Western Diamondback Rattlesnake (*C. atrox*) captured from diverse areas of the United States (i.e., Southern Arizona, Southern New Mexico, Texas, Oklahoma, and Western Arkansas). Snakes from the southwest had the highest lethality, while those from the northeast region of the range had the lowest. Geographic differences have also been accounted in the venom of the Great Basin Rattlesnake (*C. lutosus*; Table 7.1) and from several snakes distributed in Central America (Adame et al., 1990).

The majority of the venoms have yet to be characterized. For example, venom of the Grand Canyon Rattlesnake (*C. abyssus*) has been poorly studied. Young et al. (1980) used isoelectric focusing to determine relationships of this species and several other species and subspecies (i.e., *C. lutosus*, *C. concolor,* and *C. viridis nuntius*) within the *C. viridis* complex (see Douglas et al., 2002 for a phylogenetic revision of this complex). Based on venom proteins, Young et al. (1980) noted that *C. abyssus* appeared more closely related to *C. lutosus* than to others in the complex. However, their approach did not specifically characterize the venom of *C. abyssus*, but merely used its nascent properties in a taxonomic fashion.

Venoms of the Genus *Bothrops*

The genus *Bothrops* is found in Central and South America (McDiarmid et al., 1999). The name comes from the Greek words *bothros* and *ops* that mean pit and face, which refer to the loreal pit organs. The *Bothrops* species are responsible for more human morbidity than any other group of venomous snakes (Campbell and Lamar, 2004). In Venezuela and Brazil, *B. colombiensis, B. atrox* and *B. jararaca* are responsible for more than 80% of all recorded snakebites (Kornacker, 1999; Rengifo and Rodríguez-Acosta, 2005). *B. asper* remains the primary medically important pit viper in Central America (Fernández and Gutiérrez, 2008) since it is responsible for approximately 50% of snakebite envenomations in that region. In Costa Rica, Alape-Girón et al. (2008) reported that two *B. asper* populations, divided since the late Miocene or early Pliocene (8–5 $\times 10^6$ years) by the Guanacaste Mountain Range, Central Mountain Range, and Talamanca Mountain Range, have both similar and different (iso)enzymes from the PLA_2, serine proteinase, and metalloproteinase families. The similarity of venom proteins between the two *B. asper* populations was inferred to be 52%.

Hemostatic disorders are the major signs of systemic envenoming in bothropic accidents. Patients from the same geographical location and probably envenomed by the same snake species are described by physicians of having different systemic symptoms such as internal hemorrhage, hypotension, renal failure and shock, which may depend not only on human physiological variations against toxins, but may also depend on varying components in venoms of the same species (Aguilar et al., 2007; Salazar et al., 2008).

Several investigators (Estrade et al., 1989; Kamiguti et al., 1986) have suggested that disseminated intravascular coagulation occurs in *Bothrops* envenomations. Alterations of the coagulation system are a consequence of venom enzymatic activity on fibrinogen (thrombin-like enzyme) or of the activation of prothrombin and/or factor Xa, which in turn induces a decrease of fibrinogen concentration and other clotting factors. Significant fibrinogen degradation is a characteristic of Viperidae bites in South America (Amaral et al., 1988).

A number of studies have established that blood coagulation disorders are frequently reported in patients following systemic envenoming mainly by young *Bothrops* (Furtado et al., 1991; Ribeiro and Jorge, 1990). These authors also reported higher levels of prothrombin and factor X activators in the venom of juvenile *B. jararaca* compared with adult snakes. Snake venoms from different genera or species by age differ in their biochemical constitutions. These discrepancies are also observed in intra-species and in snakes of different sex and age (Furtado et al., 1991; Zelanis et al., 2007). The prerequisite for the venom to immobilize prey and begin digestion may modify with the snake dimension, which is related to the age (Alape-Girón et al., 2008).

Our group has also observed (Salazar et al., 2007) that venoms from the same species, but from diverse geographical locations presented variations in the number of molecular mass electrophoretic bands and chromatographic profiles. Analysis concerning the proportions of the main proteins showed differences among the four studied venoms; Serranía del Cuao (SC), Parguasa (P), Puerto Ayacucho 1 (PA1) and Puerto Ayacucho 2 (PA2). SC venom had the highest proportion of high molecular-mass components, while P and PA1 venoms contained low molecular weight proteins in high proportions. These profiles also evidenced significant differences in the venoms of closed geographical localities (PA1 and PA2). These results were in agreement with previous observations carried out with the venom of *B. alcatraz* from Brazil (Furtado, 2005). Furthermore, the analysis of these *B. atrox* venom samples on fibrin-zymography showed the majority of the activity in bands of approximately 27, 25 and 45–50 kDa in P and PA venoms, respectively (Salazar et al., 2007).

A comparative analysis of the clotting and fibrinolytic activities of these snake venoms was also carried out. The results indicated the presence in PA2 venom of P-III class active metalloproteinases with molecular masses of 50 kDa. On the other hand, in P and PA1 venoms, the fibrinogenolytic activity can be associated with a serine proteinase, trypsin-like or P-I class active metalloproteinases with molecular masses of 25 kDa. Additionally, these components could be in very low concentration in SC venom, since no well-defined active bands in SC venom were detected in the fibrin-zymography.

Fibrinolytic activity was evaluated with S-2251, S-2302, S-2444 and S-2288 chromogenic substrates, which displayed specificity for plasmin, kallikrein, urokinase, and t-PA, respectively. In all venoms, a high t-PA-like activity was observed (Table 7.2). The fibrinolytic activity also was estimated using fibrin plates. All venoms were active on plasminogen-rich fibrin plates. PA1 venom presented the higher fibrinolytic activity by both methods. An appreciable difference in specific activity among venoms of different areas was observed. The activity determined on fibrin plates with 10 μg protein was higher in PA1 venom. At this concentration, venoms from P and PA2 displayed a fibrin area without lysis inside a lysis area (indicating the presence of possible fibrinolytic inhibitors). These results suggested the profound effects that *Bothrops* venoms causes on the hemostatic system of preys and humans, which would not only depend on the quantity of venom, age and sex of the snake, but also on the balance between activators and inhibitors in the venom. These findings could also reflect on the final hemostatic balance of a bitten patient. For instance, PA2 venom could induce a more intense hemorrhagic syndrome than other venoms, via endogen fibrinolysis activation by t-PA-like activity. The amidolytic activities present in these four venoms were also assayed in the presence of protease inhibitors. The results showed that factor Xa-like and t-PA-like activities decreased 70% in the presence of EDTA—Na, EGTA—Na and 1,10-phenantroline, and 10% in the presence of soybean trypsin inhibitor, benzamidine, and aprotinine, suggesting that metalloproteinases are among the most abundant enzymes found in these venoms.

Intraspecific venom disparity takes place among individual snakes, probably due to seasonal variation, diet, habitat, age, and sexual dimorphism and other unknown factors (Daltry et al., 1996a). Venom variation is relevant to research since considerations of the differences in venom constituents are pertinent for antivenom production. Studies of captive bred snakes may suggest that the intraspecific

Table 7.2 Variation in venom activities of the genus *Bothrops* from South America

					Activities						
						Amidolytic-(mUA/min/μg)			Coagulant (IU/mL Th)		
Species	Country	Geographic locations	MHD	LD_{50}	Fibrinolytic-fibrin plate (mm^2/10 μg)	Thrombin (S-2238)	Factor Xa (S-2222)	Kallikrein (S-2302)	Plasma	Purified Fibrinogen	References
B. atrox	Venezuela	Parguasa	2.5	4.0	100	438 ± 2.1	294	25	30.7±2.0	24.3±0.7	Salazar et al. (2008)
B. atrox	Venezuela	Puerto Ayacucho 1	5.6	6.1	257	190	126	48	17.5±1.1	18.6±0.2	
B. atrox	Venezuela	Puerto Ayacucho 2	5.6	7.9	121	473	177	27	35.2±1.5	39.6±1.4	
B. atrox	Venezuela	Serranía del Cuao	4.6	8.3	51	383	235	28	28.0±1.3	25.3±0.3	
B. colombiensis	Venezuela	El Guapo	5.3	11.6	168	568	116	1126	18.0±1.0	31.0±1.6	Giron et al. (2008)
B. colombiensis	Venezuela	Caucagua	13.8	5.8	306	502	56	446	11.0 ±0.8	14.0±0.5	
B. isabelae	Venezuela	Trujillo	9	5.9	500	132	21	125	NT	NT	Not published

MHD minimal amount of protein in micrograms that produces a 10 mm hemorrhagic spot, LD_{50} = The amount of venom protein that will kill 50% of a mouse population per body weight (mg/kg), *NT* not tested disparity in venom is genetically inherited more than environmentally induced (Daltry et al., 1996b). Nevertheless, in agreement with Sasa (2002), microevolutionary influences other than selection for vicinity prey should also be proposed as a basis of the high variation of venom components among populations living in the same habitat.

Pathological symptoms in hemostasis are the main manifestations of systemic bothropic envenomation. The presence of numerous clinical manifestations and hemostatic considerations may be provoked by individual variability of venoms. Snake venoms differ in their molecular composition and pharmacological activity, not only among diverse species, but also within a single species and in snakes of varied ages and different sexes (Aguilar et al., 2007).

Venoms of *B. colombiensis* from El Guapo (EG) and Caucagua (C) localities were assayed for hemorrhagic, lethality, as well as fibrino(geno)lytic and coagulant activities (Table 7.2). Both crude venoms, by amidolytic methods, showed a moderate factor Xa-like activity and an elevated thrombin-like activity. These results were further corroborated by the coagulant method. EG crude venom had higher activity than C crude venom. These results evidenced the presence of prothrombin and/or factor X activators as reported by others (Kamiguti and Sano-Martins, 1995). In the past, the coagulant activity of venom from *B. colombiensis* had not been characterized. Our data suggest intraspecies variations, both quantitative and qualitative, in the procoagulant components in these venoms (Table 7.2).

The presence of varied clinical symptoms and hemostatic parameters is more than likely induced by individual variability of venom composition. Snake venoms differ in their biochemical structure and pharmacological profile, not only between different species, but also within species, and in snakes of different ages, sex and geographical localities (Saravia et al., 2002). These studies strongly illustrate the intraspecies variation of venoms in specimens from far and near geographical locations, which supports the need to include pools of venoms of the same species captured at different geographical regions that will be used in immunization protocols for the production of antivenoms.

References

Adame, B.L., Soto, J.G., Secraw, D.J., Pérez, J.C., 1990. Regional variation of biochemical characteristics and antigenicity in Great basin rattlesnake (*Crotalus viridis lutosus*) venom. Comp. Biochem. Physiol. B. 97, 95–101.

Aguilar, I., Guerrero, B., Salazar, A.M., Girón, M.E., Pérez, J.C., Sánchez, E.E., Rodríguez-Acosta, A., 2007. Individual venom variability in the South American rattlesnake *Crotalus durissus cumanensis*. Toxicon 50, 214–224.

Alape-Girón, A., Flores-Díaz, M., Sanz, L., Madrigal, M., Escolano, J., Sasa, M., Calvete, J.J., 2009. Studies on the venom proteome of *Bothrops asper*: perspectives and applications. Toxicon 54, 938–948.

Alape-Girón, A., Sanz, L., Escolano, J., Flores-Díaz, M., Madrigal, M., Sasa, M., Calvete, J.J., 2008. Snake venomics of the lancehead pitviper *Bothrops asper*: geographic, individual, and ontogenetic variations. J. Proteome Res. 7, 3556–3571.

Amaral, C.F., Rezende, N.A., Pedrosa, T.M., da Silva, O.A., Pedroso, E.R., 1988. Afibrinogenemia secondary to crotalid snake bite (*Crotalus durissus terrificus*). Rev. Inst. Med. Trop. Sao Paulo 30, 288–292.

Calsbeek, R., 2009. Experimental evidence that competition and habitat use shape the individual fitness surface. J. Evol. Biol. 22, 97–108.

Campbell, J.A., Lamar, W.W., 2004. The Venomous Reptiles of the Western Hemisphere. Comstock Publishing Associates, Ithaca and London.

Chippaux, J.P., Williams, V., White, J., 1991. Snake venom variability: methods of study, results and interpretation. Toxicon 29, 1279-1303.

Condrea, E., Fabian, I., De Vries, A., 1964. Action of plastid phospholipase D on free and lipoprotein-bound phospholipids. Experientia 20, 557–558.

Daltry, J.C., Ponnudurai, G., Shin, C.K., Tan, N.H., Thorpe, R.S, Wüster, W., 1996a. Electrophoretic profiles and biological activities: intraspecific variation in the venom of the Malayan pit viper (*Calloselasma rhodostoma*). Toxicon 34, 67–79.

Daltry, J.C., Wuster, W., Thorpe, R.S., 1996b. Diet and snake venom evolution. Nature 379, 537–540.

Douglas, M.E., Douglas, M.R., Schuett, G.W., Porras, L., Holycross, A.T., 2002. Phylogeography of the Western Rattlesnake (*Crotalus viridis*) complex (Reptilia: Viperidae), with emphasis on the Colorado plateau, in: Schuett, G.W., Hoggren, M., Douglas, M.E., Greene, H.W. (Eds.), Biology of the Vipers. Eagle Mountain Press, Salt Lake City, UT, pp. 11–50.

Enghild, J.J., Salvesen, G., Brew, K., Nagase, H., 1989. Interaction of human rheumatoid synovial collagenase (matrix metalloproteinase 1) and stromelysin (matrix metalloproteinase 3) with human alpha 2-macroglobulin and chicken ovostatin. Binding kinetics and identification of matrix metalloproteinase cleavage sites. J. Biol. Chem 264, 8779–8785.

Estrade, G., Garnier, D., Bernasconi, F., Donatien, Y., 1989. Pulmonary embolism and disseminated intravascular coagulation after being bitten by a *Bothrops lanceolatus* snake. Apropos of a case. Arch. Mal. Coeur. Vaiss 82, 1903–1905.

Fernández, P., Gutiérrez, J.M., 2008. Mortality due to snakebite envenomation in Costa Rica (1993–2006). Toxicon 52, 530–533.

Francischetti, I.M., Gombarovits, M.E., Valenzuela, J.G., Carlini, C.R., Guimaraes, J.A., 2000. Intraspecific variation in the venoms of the South American rattlesnake (*Crotalus durissus terrificus*). Comp. Biochem. Physiol. C. Toxicol. Pharmacol. 127, 23–36.

Friberger, P., 1982. Chromogenic peptide substrates. Their use for the assay of factors in the fibrinolytic and the plasma kallikrein-kinin systems. Scand. J. Clin. Lab. Invest. Suppl. 16, 1–298.

Furtado, M.F., Maruyama, M., Kamiguti, A.S., Antonio, L.C., 1991. Comparative study of nine *Bothrops* snake venoms from adult female snakes and their offspring. Toxicon 29, 219–226.

Furtado, M.F., 2005. Biological and immunological properties of the venom of *Bothrops alcatraz*, an endemic species of pitviper from Brazil. Comp. Biochem. Physiol. C. Toxicol. Pharmacol. 141, 117–123.

Galán, J.A., Sánchez, E.E., Rodríguez-Acosta, A., Pérez, J.C., 2004. Neutralization of venoms from two Southern Pacific Rattlesnakes (*Crotalus helleri*) with commercial antivenoms and endothermic animal sera. Toxicon 43, 791–799.

Girón, M.E., Salazar, A.M., Aguilar, I., Pérez, J.C., Sánchez, E.E., Arocha-Piñango, C.L., Rodríguez-Acosta, A., Guerrero, B., 2008. Hemorrhagic, coagulant and fibrino(geno)lytic activities of crude venom and fractions from mapanare (*Bothrops colombiensis*) snakes. Comp. Biochem. Physiol. C. Toxicol. Pharmacol. 147, 113–121.

Glenn, J.L., Straight, R.C., Wolfe, M.C., Hardy, D.L., 1983. Geographical variation in *Crotalus scutulatus scutulatus* (Mojave rattlesnake) venom properties. Toxicon 21, 119–130.

Glenn, J.L., Straight R., 1978. Mojave rattlesnake *Crotalus scutulatus scutulatus* venom: variation in toxicity with geographical origin. Toxicon 16, 81–84.

Gutiérrez, J.M., Rucavado, A., Escalante, T., Díaz, C., 2005. Hemorrhage induced by snake venom metalloproteinases: biochemical and biophysical mechanisms involved in microvessel damage. Toxicon 45, 997–1011

Kamiguti, A.S., Matsunaga, S., Spir, M., Sano-Martins, I.S., Nahas, L., 1986. Alterations of the blood coagulation system after accidental human inoculation by *Bothrops jararaca* venom. Braz. J. Med. Biol. Res. 19, 199–204.

Kamiguti, A.S., Sano-Martins, I.S., 1995. South American snake venoms affecting haemostasis. J. Toxicol. Toxin Rev. 14, 359–374.

Kornacker, P.M., 1999. Checklist and Key to the Snakes of Venezuela. Lista Sistemática y Clave para las Serpientes de Venezuela. Pako-Verlag, Rheinbach, Germany.

Matsui, T., Fujimura, Y., Titani, K., 2000. Snake venom proteases affecting hemostasis and thrombosis. Biochim. Biophys. Acta 1477, 146–156.

McDiarmid, R.W., Campbell, J.A., Toure, T., 1999. Snake Species of the World: A Toxonomic and Geographic Reference (vol. 1). Herpetologists' League.

Minton, S.A., Weinstein, S.A., 1986. Geographic and ontogenic variation in venom of the western diamondback rattlesnake (*Crotalus atrox*). Toxicon 24, 71–80.

Pućkowska, A., Midura-Nowaczek, K., Bruzgo, I., 2008. Effects of netropsin and pentamidine amino analogues on the amidolytic activity of plasmin, trypsin and urokinase. Acta. Pol. Pharm. 65, 213–215.

Rengifo, C., Rodríguez-Acosta, A., 2005. Serpientes, Venenos y su Tratamiento en Venezuela. Fondo de Publicaciones de la Facultad de Medicina de la Universidad Central de Venezuela, Caracas.

Ribeiro, L.A., Jorge, M.T., 1990. Epidemiology and clinical picture of accidents by adult and young snakes *Bothrops jararaca*. Rev. Inst. Med. Trop. Sao Paulo 32, 436–442.

Rodríguez-Acosta, A., Mondolfi, A., Orihuela, R., 1995. ¿Qué hacer frente a un accidente ofídico? Venediciones, CA, Caracas.

Salazar, A.M., Rodriguez-Acosta, A., Girón, M.E., Aguilar, I., Guerrero, B., 2007. A comparative analysis of the clotting and fibrinolytic activities of the snake venom (Bothrops atrox) from different geographical areas in Venezuela. Thromb. Res. 120, 95–104.

Salazar, A.M., Aguilar, I., Guerrero, B., Girón, M.E., Lucena, S., Sánchez, E.E., Rodríguez-Acosta, A., 2008. Intraspecies differences in hemostatic venom activities of the South American rattlesnakes, *Crotalus durissus cumanensis*, as revealed by a range of protease inhibitors. Blood. Coagul. Fibrinolysis 19, 525–530.

Salazar, A.M., Guerrero, B., Cantu, B., Cantu, E., Rodríguez-Acosta, A., Pérez, J.C., Galán, J.A., Tao, A., Sánchez, E.E., 2009. Venom variation in hemostasis of the southern Pacific rattlesnake (*Crotalus oreganus helleri*): isolation of hellerase. Comp. Biochem. Physiol. C. Toxicol. Pharmacol. 149, 307–316.

Sánchez, E.E., 2004. Aislamiento y Caracterización de Desintegrinas Presentes en el Veneno de Serpientes de los Estados Unidos de América y de Venezuela. Doctoral Thesis, Universidad Central de Venezuela Caracas, Venezuela and Natural Toxins Research Center, Texas A&M University-Kingsville, Kingsville, TX, USA.

Sánchez, E.E., Galán, J.A., Powell, R.L., Reyes, S.R., Soto, J.G., Russell, W.K., Russell, D.H., Pérez, J.C., 2005. Disintegrin, hemorrhagic, and proteolytic activities of Mohave rattlesnake, *Crotalus scutulatus scutulatus* venoms lacking Mojave toxin. Comp. Biochem. Physiol. C. Toxicol. Pharmacol. 141, 124–132.

Sánchez, E.E., Galán, J.A., Russell, W.K., Soto, J.G., Russell, D.H., Pérez, J.C., 2006. Isolation and characterization of two disintegrins inhibiting ADP-induced human platelet aggregation from the venom of *Crotalus scutulatus scutulatus* (Mohave Rattlesnake). Toxicol. Appl. Pharmacol. 212, 59–68.

Sánchez, E.E., Lopez-Johnston, J.C., Rodríguez-Acosta, A., Pérez, J.C., 2008. Neutralization of two North American coral snake venoms with United States and Mexican antivenoms. Toxicon 51, 297–303.

Sano-Martins, I.S., Tomy, S.C., Campolina, D., Dias, M.B., de Castro, S.C, de Sousa-Silva, M.C., Amaral, C.F., Rezende, N.A., Kamiguti, A.S., Warrell, D.A., Theakston, R.D., 2001. Coagulopathy following lethal and non-lethal envenoming of humans by the South American rattlesnake (*Crotalus durissus*) in Brazil. Q. J. Med. 94, 551–559.

Saravia, P., Rojas, E., Arce, V., Guevara, C., López, J.C., Chaves, E., Velásquez, R., Rojas, G., Gutiérrez, J.M., 2002. Geographic and ontogenic variability in the venom of the neotropical rattlesnake *Crotalus durissus*: pathophysiological and therapeutic implications. Rev. Biol. Trop. 50, 337–346.

Sasa, M., 2002. Morphological variation in the lancehead pitviper *Bothrops asper* (Garman) (Serpentes: Viperidae) from Middle America. Rev. Biol. Trop. 50, 259–271.

Swenson, S., Markland, F.S., Jr., 2005. Snake venom fibrin(ogen)olytic enzymes. Toxicon 45, 1021–1039.

Yoshida-Kanashiro, E., Navarrete, L., Rodriguez-Acosta, A., 2003. On the unusual hemorrhagic and necrotic activities caused by the rattlesnake (*Crotalus durissus cumanensis*) in a Venezuelan patient. Rev. Cub. Med. Trop. 55, 36–40.

Young, R.A., Miller, D.M., Ochsner, D.C., 1980. The Grand Canyon rattlesnake (*Crotalus viridis abyssus*): comparison of venom protein profiles with other *viridis* subspecies. Comp. Biochem. Physiol. B. 66, 601–603.

Zelanis, A., de SouzaVentura, J., Chudzinski-Tavassi, A.M., de Fátima Domingues Furtado, M., 2007. Variability in expression of *Bothrops insularis* snake venom proteases: an ontogenetic approach. Comp. Biochem. Physiol. C. Toxicol. Pharmacol. 145, 601–609.

Chapter 8
Tiny Ticks are Vast Sources of Antihaemostatic Factors

Mária Kazimírová, Cho Yeow Koh, and R. Manjunatha Kini

Abstract Ticks are obligate blood-feeding ectoparasites. Damage of the skin of vertebrates leads to hemostatic, inflammatory and immune responses. These would disrupt tick feeding with detrimental consequences. To avoid host reactions, ticks inject saliva, a cocktail of pharmacologically active compounds, such as vasodilators, antiplatelet factors, anticoagulants, anti-inflammatory and immunomodulatory substances. Ticks have evolved powerful tools to prevent or prolong coagulation of the host blood. Majority of the inhibitors identified so far are proteins that display a variety of molecular masses, targets and inhibitory mechanisms. These anticoagulants can be classified as thrombin inhibitors, factor Xa inhibitors, extrinsic and intrinsic tenase complex inhibitors and contact system proteins inhibitors. Based on the diversity of antihemostatic strategies, it has been assumed that the main tick families have adapted to blood feeding independently. The key enzyme of the coagulation cascade – thrombin is often targeted by tick anticoagulants. The most well characterized thrombin inhibitors are the Kunitz-type proteinase inhibitors, i.e., ornithodorin, savignin and monobin from soft ticks, or boophilin, amblin, hemalin from hard ticks. A class of novel thrombin inhibitors is represented by variegin, isolated from the hard tick *Amblyomma variegatum.* This class of inhibitors display structural and functional similarity to hirulog, a peptide designed based on hirudin isolated from leech. TAP, a Kunitz-type FXa inhibitor from the soft tick *Ornithodoros moubata* has been among the best studied tick anticoagulants. Novel tick-derived molecules represent potentially useful therapeutic agents for treatment of hemostatic disorders, cardiovascular diseases and disorders of the immune system.

M. Kazimírová (✉)
Institute of Zoology, Slovak Academy of Sciences, Dúbravská Cesta 9, SK-84506, Bratislava, Slovakia
e-mail: uzaemkaz@savba.sk

R.M. Kini et al. (eds.), *Toxins and Hemostasis*, DOI 10.1007/978-90-481-9295-3_8,

Introduction

Ticks are obligate blood feeding ectoparasites of a large number of terrestrial vertebrates. Adults of the fast feeding soft ticks (Argasidae)[1] can feed repeatedly and rapidly by deep penetration of the host skin, causing considerable damage (Binnington and Kemp, 1980), whereas adults of the slow feeding hard ticks (Ixodidae) feed only once for a prolonged period and penetrate the epidermis either superficially (Metastriate ticks, e.g., *Rhipicephalus* (*Boophilus*) spp., *Dermacentor* spp.), or more deeply (Prostriate ticks, e.g., *Ixodes* spp., *Amblyomma* spp.) (Bowman et al., 1997; Sonenshine, 1991). The volume of the ingested blood as well as the duration of feeding are development stage- and species-specific. Hard ticks may require several days to weeks to complete their blood meal, whereby tick females may ingest more blood than 100-times their initial body weight (Sauer et al., 1995). The insertion of the tick hypostome into the host skin damages the epidermis and ruptures blood vessels. Damage of the skin normally leads to host haemostatic, inflammatory and immune responses. These would disrupt tick feeding and cause rejection of the tick, with detrimental consequences to tick viability and reproduction. However, tick saliva contain compounds such as vasodilators, antiplatelet factors, anticoagulants, anti-inflammatory and immuno-modulatory substances that counteract host defences, enabling ticks to feed for days to weeks at one site (Andrade et al., 2005; Ribeiro and Francischetti, 2003; Steen et al., 2006). Tick saliva composition is complex and in some cases redundant reflecting the complex and redundant host defence responses (Andrade et al., 2005; Brossard and Wikel, 2004; Steen et al., 2006). In addition, many pathogens (viruses, bacteria, protozoa) exploit host vulnerability at the feeding site, transmitting deadly diseases to human and livestock (Bowman et al., 1997; Brossard and Wikel, 2004; Nuttall and Labuda, 2004; Ramamoorthi et al., 2005; Wikel, 1996). Therefore, an understanding of the physiology of tick salivary glands is important for the appreciation of their functions.

Salivary Glands of Ticks

Tick salivary glands are multifunctional complex organs and their anatomy, morphogenesis and physiology have been extensively described (Bowman and Sauer, 2004; Sauer et al., 1995, 2000; Sonenshine, 1991). They consist of an anterior region of acini (generally agranular and primarily involved in osmo-regulation) attached directly to the main duct; the acini are arranged more caudally in lobules connected by intra- and inter-lobular ducts to the main salivary duct. The caudal acini increase greatly in size during feeding and are involved in secretion of salivary bioactive components. The main salivary ducts (one from each gland) pass antecranially into the salivarium which fuses with the pharynx of the tick and forms the oral cavity.

[1] The tick nomenclature used in this chapter is adopted from Barker and Murrell (2004).

During tick attachment, host tissue fluids and tick saliva flow in alternate directions through a common buccal canal. The regulation of salivary gland development, degeneration and fluid secretion are under neuro-hormonal control (Bowman and Sauer, 2004).

Almost all ixodid ticks produce cement proteins that ensure firm attachment of the tick to the host and seal the area around the mouth parts to the wound site. In contrast, argasid ticks do not secrete factors enabling strong attachment to the host. After a tick attaches to a host, expression of a series of new genes and synthesis of proteins is initiated in their salivary glands that reflect the stages of the feeding process. As feeding progresses, the amount of secreted saliva increases and salivary glands undergo a remarkable and rapid structural reorganization. At the peak of the feeding process, the glands can increase 25-fold in size and protein content. Once the tick is engorged and detaches, the glands degenerate through a process of cell apoptosis (Bowman and Sauer, 2004).

Tick Salivary Compounds and Their Effect on Host Haemostasis

Saliva of the same tick species can contain simultaneously many antihaemostatic molecules inhibiting different arms of the haemostatic system, or the same compounds can display multiple functions (Bowman et al., 1997; Francischetti et al., 2009; Mans and Neitz, 2004; Maritz-Olivier et al., 2007; Steen et al., 2006; Valenzuela, 2004). Research on the mechanisms through which ticks inhibit host haemostasis led to the discovery and identification of a variety of compounds with potential to be developed as therapeutic agents (Koh and Kini, 2009; Ledizet et al., 2005). There are several advantages in investigating tick salivary compounds as the starting point of drug discovery process, which are:

1. A same 'library' of molecules can be explored for multiple targets (a comprehensive investigation on a single species, for instance, by high-throughput transcriptomic or proteomic approach, will provide a wide coverage of chemical space).
2. The molecules might be aimed at novel therapeutic targets (identifications of novel targets would otherwise be a lengthy process).
3. Hundreds of millions years of evolution acts as an efficient screening and selection process (molecules which are present should interfere with physiological systems; otherwise they would have been discarded).
4. Similarly, evolution acts as a lead optimization process (by natural selection, mutations that confer greater advantages such as potency and stability, will be selected for).
5. The molecules are likely to be of low toxicity and immunogenicity (it is in the best interest of haematophagous parasites that hosts survive their assaults, thus, exposure to the molecules should not kill the hosts).

In addition to the drug discovery process, such studies can also contribute to our understanding of molecular interactions within the haemostatic system. Moreover, elucidation of the mechanisms of interaction between ectoparasites and their hosts can lead to the discovery of new vaccine targets against ticks and the pathogens ticks transmit (Hovius et al., 2008; Maritz-Olivier et al., 2007). It is important to note that antihaemostatics in ticks differ between species and there is no tick species whose full antihaemostatic capacities have been fully explored and described. Here we describe some of the examples of these exogenous factors.

Vasodilators

Following laceration of blood vessels by tick's mouthparts, arachidonic acid is released by activated platelets and is converted by other platelet enzymes into thromboxane A_2, a platelet-aggregating, platelet-degranulating, and vasoconstricting substance. Activated platelets also release serotonin which, together with thromboxane A_2, is responsible for early vasoconstriction in local inflammation caused by tissue injury. To antagonise vasoconstrictors produced by the host on the site of tissue injury, vasodilators are secreted by ticks to the feeding pool. To date, only non-proteinaceous vasodilatory compounds have been identified in tick saliva. They include lipid derivatives such as prostacyclin and prostaglandins (Bowman et al., 1996; Inokuma et al., 1994; Ribeiro et al., 1985, 1988).

Antiplatelet Factors

Platelet aggregation represents the initial and most immediate stage of haemostasis. Following vascular injury, platelets adhere to the subendothelial tissue, become activated by agonists such as collagen, thrombin, ADP, and thromboxane A_2. Agonists bind to specific receptors on the surface of platelets and initiate a long and highly complex chain of intracellular chemical reactions that lead to platelet aggregation and formation of a platelet thrombus.

Platelet activation and aggregation are inhibited by ticks at several stages. Inhibition can be mediated through disruption of the action of agonists. For example, tick saliva contains the enzyme apyrase, which hydrolyzes the platelet agonist ADP (Valenzuela et al., 2002), moubatin (Keller et al., 1993) and longicornin (Cheng et al., 1999) inhibits collagen-induced aggregation while inhibitors of thrombin also inhibit its platelet activating function (Kazimirova et al., 2002). Molecules inhibiting receptors on the surface of platelets that mediate adhesion and aggregation were also reported. For examples, 'tick adhesion inhibitor' inhibits integrin $\alpha_2\beta_1$-collagen adhesion (Karczewski et al., 1995) while disagregin (Karczewski et al., 1994), savignygrin (Mans et al., 2002b), variabilin (Wang et al., 1996) and ixodegrin (Francischetti et al., 2005) inhibit integrin $\alpha_{IIb}\beta_3$-fibrinogen mediated aggregation. For details, see Chapter 20 by Francischetti, in this volume.

Anticoagulant Factors

In the blood coagulation cascade, circulating zymogens of serine proteinases are sequentially activated by limited proteolysis leading to the formation of a fibrin clot (Davie et al., 1991). Various inhibitory activities on plasma clotting can be observed in tick saliva or salivary gland extracts and other organs such as gut, eggs or haemolymph and whole body extracts. Pioneering works on tick anticoagulants showed inhibition of blood coagulation due to ixodin, a putative enzyme with antithrombin activity detected in whole body extracts from *Ixodes ricinus* (Sabbatini, 1899) or argasin from salivary glands of the fowl tick *Argas persicus* (Nuttall and Strickland, 1908). Studies carried out between the 1950s and early 1990s confirmed the presence of anticoagulants and fibrinolytic factors in salivary glands and other organs of a number of soft and hard tick species, e.g., *Ixodes ricinus, I. holocyclus, Ornithodoros moubata, Dermacentor andersoni* and *Rhipicephalus* (formerly *Boophilus*) *microplus* (Anastopoulos et al., 1991; Hawkins and Hellmann, 1966; Hellmann and Hawkins, 1967; Markwardt and Landmann, 1958, 1961; Willadsen and Riding, 1980). In *O. moubata,* the anticoagulant activities of salivary gland extracts were directed against thrombin and factor IX, but fibrinolytic activity was present in gut, eggs or haemolymph (Hawkins and Hellmann, 1966; Hellmann and Hawkins, 1967). Extracts from eggs and unfed larvae of *R.* (*B.*) *microplus* prolonged activated partial thromboplastin time (APTT) and prothrombin time (PT), but the activity seemed to be important only during the initial phases of feeding (Willadsen and Riding, 1980). Anticoagulant activities have also been demonstrated in embryos, larvae and nymphs of the camel tick *Hyalomma dromedarii* (Ibrahim, 1998; Ibrahim et al., 1998, 2000, 2001a, b) and in salivary glands and extracts of several tick species, but only during the last decade the number of tick anticoagulants that have been isolated and characterized has increased. In addition, with advances in transcriptomic approaches, many sequences that showed similarity to existing anticoagulants are being determined. Majority of the inhibitors identified so far are proteins that display a variety of molecular masses, targets and inhibitory mechanisms. Based on the mechanism of action, these exogenous anticoagulants from ticks can be broadly classified as: (1) thrombin inhibitors; (2) FXa inhibitors; (3) extrinsic and intrinsic tenase complex inhibitors; and (4) contact system proteins inhibitors. Molecules within each group can be sub-divided into many different classes based on their structures (Table 8.1, Fig. 8.1).

Thrombin Inhibitors

Thrombin inhibitors from ticks have many different structures and mechanisms of action, forming at least seven structural classes (Koh and Kini, 2009). Some of the most well characterized thrombin inhibitors from ticks are the Kunitz-type proteinase inhibitors. This class of thrombin inhibitors are present in both soft [e.g. ornithodorin from *O. moubata*) (van de Locht et al., 1996), savignin from *O. savignyi* (Mans et al., 2002a; Nienaber et al., 1999) and monobin from *Argas*

Table 8.1 Anticoagulants from ticks

Class	Inhibitor	Species	MW (kDa)	Stages of feeding	Organ	References
A. Thrombin inhibitors						
Kunitz-type	Ornithodorin	*O. moubata*	12.6	Adult	SG	van de Locht et al. (1996)
	Savignin	*O. savignyi*	12.4	Adult, female	SG	Mans et al. (2002a) and Nienaber et al. (1999)
	Monobin	*A. monolakensis*	13	Adult, female	SG	Mans et al. (2008)
	Boophilin	*R. (B.) microplus*	13.9	Adult	WT	Macedo-Ribeiro et al. (2008)
	Amblin	*A. hebraeum*	17.4	Adult, female	HAE	Lai et al. (2004)
	Hemalin	*H. longicornis*	20	Adult, female, partially engorged	G	Liao et al. (2009)
Variegin-like	Variegin	*A. variegatum*	3.7	Adult, female, partially engorged	SG	Koh et al. (2007, 2009)
Madanin and Chimadanin	Madanin 1	*H. longicornis*	6.8	Adult, engorged	SG	Iwanaga et al. (2003)
	Madanin 2		7.1	Adult, engorged	SG	Iwanaga et al. (2003)
	Chimadanin		7.5	Adult, female, partially engorged & nymph	SG	Nakajima et al. (2006)
NTI	Nymph thrombin inhibitor (NTI)-1	*H. dromedarii*	3.2	Nymph	WT	Ibrahim et al. (2001a)
Microphilin	Microphilin (2 isoforms)	*R. (B.) microplus*	1.7	Adult, female, detached	S	Ciprandi et al. (2006)
BmAP	BmAP	*R. (B.) microplus*	60	Adult, female, detached	S	Horn et al. (2000)

Table 8.1 (continued)

Class	Inhibitor	Species	MW (kDa)	Stages of feeding	Organ	References
BmGTI	BmGTI	*R. (B.) microplus*	26	Adult, female, partially engorged	G	Ricci et al. (2007)
Others	Americanin	*A. americanum*	16	Adult, female, partially engorged	SG	Zhu et al. (1997a)
	Ixin	*I. ricinus*	ND	Adult	SG	Hoffmann et al. (1991)
	Ixodin	*I. ricinus*	ND	Adult	WT	Sabbatini (1899)
	Calcaratin	*R. (B.) calcaratus*	14.5	Adult	SG	Motoyashiki et al. (2003)
	Anticoagulants F1-F6	*H. dromedarii*	ND	Egg, embryo	WT	Ibrahim et al. (1998, 2000)
	Unnamed	*H. inermis*	ND	Adult	SG	Kazimirova et al. (2000)
	Unnamed	*D. reticulatus*	ND	Adult	SG	
B. FXa inhibitors						
Kunitz-type	TAP	*O. moubata*	6.8	Adult	WT	Waxman et al. (1990)
	FXaI	*O. savignyi*	7.2	Adult	SG	Gaspar et al. (1996)
Salp family	Salp14	*I. scapularis*	14	Nymph, engorged	SG	Narasimhan et al. (2002)
	Salp9pac		9	Adult, female, patially engorged	S	
Uncompetitive inhibitors from *Hyalomma*	Unnamed	*H. truncatum*	17	Adult, female, partially engorged	SG	Joubert et al. (1995)
	Unnamed	*H. dromedarii*	15	Nymph, engorged	WT	Ibrahim et al. (2001b)

Table 8.1 (continued)

Class	Inhibitor	Species	MW (kDa)	Stages of feeding	Organ	References
Others	Prothrombinase complex inhibitor	*R. appendiculatus*	65	Adult, partially engorged	SG	Limo et al. (1991)
	Unnamed	*A. americanum*	16	Adult, female, partially engorged	SG	Zhu et al. (1997b)
C. Extrinsic and intrinsic complex inhibitors						
Kunitz-type	Ixolaris	*I. scapularis*	15.7	Adult, female, partially engorged	SG	Francischetti et al. (2002)
	Penthalaris		35	Adult, female, partially engorged	SG	Francischetti et al. (2004)
D. Contact system proteins inhibitors						
Kunitz-type	Haemaphysalin	*H. longicornis*	16	Adult, engorged	SG	Kato et al. (2005)
	BmTI-A	*R. (B.) microplus*	15	Larva	WT	Sasaki et al. (2004); Tanaka et al. (1999)
	BmTI-D		17			
	BmTI-2		8			
	RsTIQ2	*R. sanguineus*	12	Larva	WT	Sant'Anna Azzolini et al. (2003)
	RsTIQ7		8			

SG salivary glands, *S* saliva, *G* gut, *HAE* haemolymph, *WT* whole tick, *ND* not determined

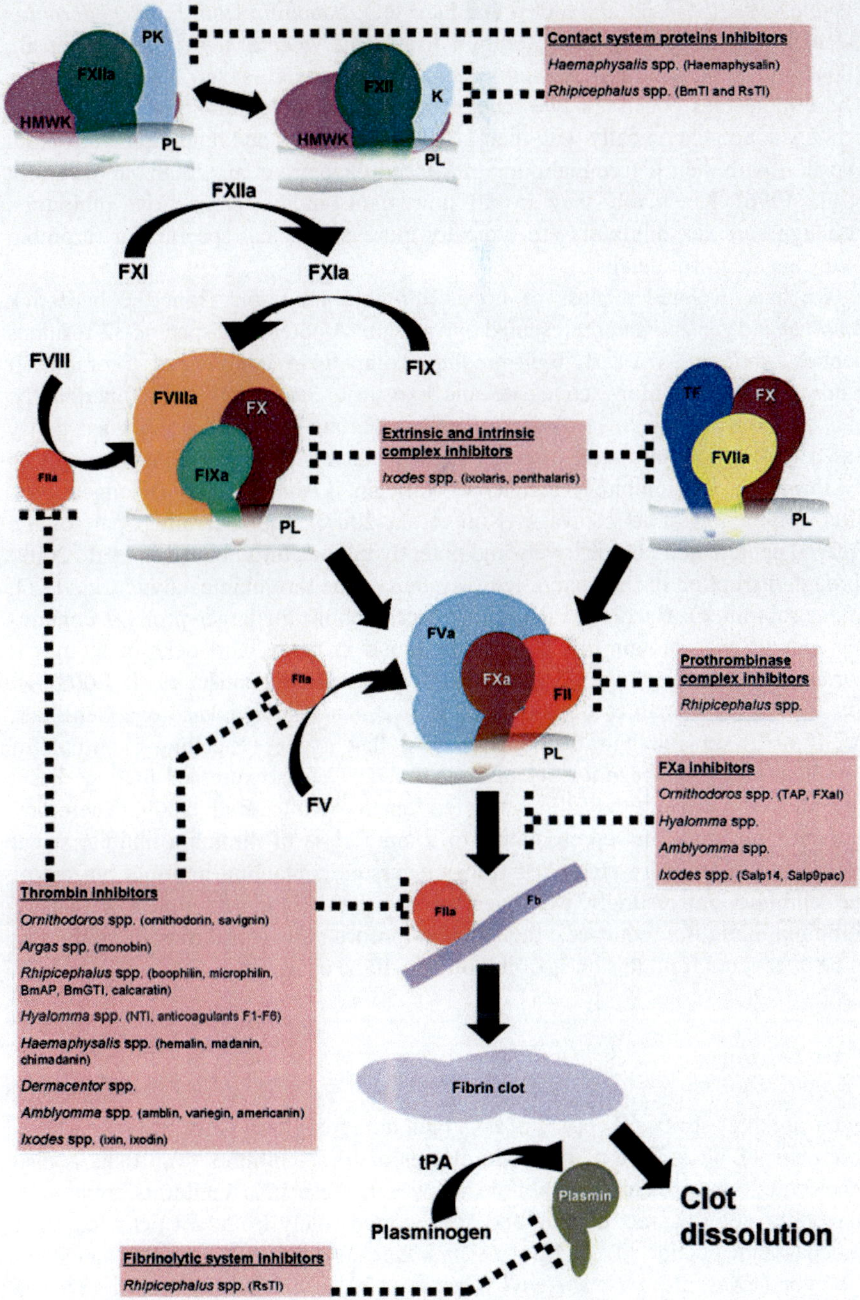

Fig. 8.1 Overview of blood coagulation cascade and fibrinolytic system as well as site of actions of inhibitors identified from ticks. F: factor, HMWK: high molecular weight kininogen, PK: prekallikrein, K, kallikrein, PL: phospholipids, TF: tissue factor, Fb: fibrinogen, tPA: tissue plasminogen activator

monolakensis (Mans et al., 2008)] and hard [e.g. boophilin from *R. (B.) microplus* (Macedo-Ribeiro et al., 2008), amblin from *Amblyomma hebraeum* (Lai et al., 2004) and hemalin from *Haemaphysalis longicornis* (Liao et al., 2009)] ticks. These molecules consist of two tandem Kunitz domains, inhibiting the thrombin active-site non-canonically with their N-terminal domain and binding the thrombin exosite-I with their C-terminal domain (Macedo-Ribeiro et al., 2008; van de Locht et al., 1996). Kinetically they are all slow, tight-binding, competitive inhibitors, although soft tick inhibitors are typically more potent and specific for thrombin (see Chapter 15 for details).

We have isolated a class of novel thrombin inhibitors from the hard tick *Amblyomma variegatum*, represented by variegin. A potent and specific 32-residues peptide, variegin is a fast, tight-binding, competitive inhibitor of thrombin. It binds to both thrombin active site and exosite-I. Structurally and functionally, it is almost identical to hirulog/bivalirudin, a group of peptides designed partly based on hirudin (thrombin inhibitor from medicinal leech). Despite proteolysis by thrombin, the inhibitory activity of variegin is significantly prolonged even after completion of its cleavage (Koh et al., 2007). The C-terminal part of the cleaved peptide non-competitively and potently inhibits thrombin (Koh et al., 2009) through disruption of the charge-relay system of the thrombin catalytic triad (Koh et al., submitted). Recently, cDNA sequences coding for larger proteins containing four to five variegin-like repeats are found in many hard ticks including *A. variegatum* (e.g. GenBank: BAD29729), *A. cajennense* (Batista et al., 2008), *A. americanum* (Aljamali et al., 2009) and *Rhipicephalus sanguineus* (e.g. GenBank: ACX54030). In situ hybridization revealed that a gene encoding a variant of variegin from *A. variegatum* (GenBank: BAD29729) is expressed in large basal cells of salivary acini type III in partly fed females (Roller et al., 2009). Therefore, variegin appears to be representative of a large class of thrombin inhibitors that are post-translationally processed from precursors containing multiple repeats of the inhibitors. Interestingly, with every precursor presenting multiple iterations of thrombin inhibiting sequences, the structural information (for thrombin inhibition) to be uncovered from this class of inhibitors will be enormous.

FXa Inhibitors

There are three structural classes of FXa inhibitors from ticks (Table 8.1). Similar to thrombin inhibitors, one of the main classes of FXa inhibitors from ticks is also the Kunitz-type proteinase inhibitors. However, these FXa inhibitors from ticks have only single Kunitz domain and are reported solely from soft ticks [e.g. tick anticoagulant peptide (TAP) from *O. moubata* (Waxman et al., 1990) and FXa-inhibitor (FXaI) from *O. savignyi* (Gaspar et al., 1996)]. Like the Kunitz-type thrombin inhibitors from soft ticks, the Kunitz-type has a highly distorted reactive-site loop which lacks the basic P1 residue, and thus, inhibits the FXa active-site non-canonically (Wei et al., 1998).

Salp14 (14 kDa) identified in *Ixodes scapularis* was originally described as an immuno-dominant antigen due to its ability to induce early antibody responses in rabbits infested with these ticks (Das et al., 2001). Subsequent activity-based purification of *I. scapularis* saliva uncovered an anticoagulant fraction (9.8 kDa) that resembles N-terminus sequences of salp14. cDNA cloned based on the N-terminal sequence of the anticoagulant fraction translated into a 9.3 kDa protein named salp9pac. Both Salp14 and salp9pac showed high sequence similarity to other salp proteins identified from the same tick but not other proteins. However, recombinant salp14, but not salp9pac, showed anticoagulant activity by inhibiting FXa (Narasimhan et al., 2002). It was later shown that gene silencing of salp14 and salp9pac with RNA interference reduce anti-FXa activities of *I. scupalaris* salivary gland (Narasimhan et al., 2004). Salp14 and its isoforms should represent a new class of FXa inhibitors, hence structural data on these proteins may reveal novel structures that can inhibit FXa.

Extrinsic and Intrinsic Tenase Complex Inhibitors

Inhibitors of extrinsic tenase complex from ticks such as ixolaris (Francischetti et al., 2002) and penthalaris (Francischetti et al., 2004) from *I. scapularis* are Kunitz-type inhibitors. Ixolaris and penthalaris consist of two and five tandem Kunitz domains, respectively. Both proteins bind to FX/FXa on an exosite before binding to FVIIa-TF to form a quaternary complex (see Chapter 14 for details).

So far no specific inhibitor of intrinsic tenase complex from ticks has been characterized in detail. Binding of ixolaris to FX on the heparin binding proexosite prevented the interactions between FX and FVIIIa, attenuating the effect of intrinsic tenase complex but direct binding between ixolaris and FVIIIa are not shown (Monteiro et al., 2008).

Contact System Proteins Inhibitors

All the inhibitors of contact system proteins identified from ticks so far belong to the Kunitz-type proteinase inhibitor family. For example, a two tandem Kunitz domains molecule haemaphysalin identified from the hard tick *H. longicornis* inhibits the association of both FXII and HMWK with activating surfaces in the presence of Zn^{+}, after binding to the molecules (Kato et al., 2005). In addition, amidolytic activity of plasma kallikrein is inhibited by BmTI-A, BmTI-2 and BmTI-D from *R. (B.) microplus* (Sasaki et al., 2004; Tanaka et al., 1999) as well as RsTIQ2 and RsTIQ7 from *R. sanguineus* (Sant'Anna Azzolini et al., 2003).

Inhibitors of Fibrinolytic System

The fibrinolytic system is responsible for dissolving fibrin and the eventual removal of the blood clot. In the presence of fibrin and tissue plasminogen activator, the proenzyme plasminogen is converted to plasmin. Plasmin breaks fibrin into small soluble fractions. Under conditions of physiological haemostasis there is a dynamic equilibrium between fibrin formation by thrombin, its stabilization by factor XIIIa, and the fibrin degrading system. Kunitz-type serine proteinase inhibitors (RsTI) isolated from larvae of the brown dog tick *R. sanguineus* also target plasmin (and a number of other serine proteinases such as neutrophil elastase) in addition to plasma kallikrein as discussed above (Sant'Anna Azzolini et al., 2003) (Fig. 8.1). Fibrinolytic activity (instead of inhibition) in saliva of *I. scapularis* is due to the presence of a metalloproteinase. The role of salivary metalloproteinases in tick feeding appears to be related to their antifibrinogen- and antifibrin-specific activities (Francischetti et al., 2003).

Importance of Tick Biology in the Search for New Antihaemostatics

There is a total of 899 species of ticks (713 species of hard ticks, 185 species of soft ticks and the single species Nutalliellidae) (Barker and Murrell, 2004), presenting enormous opportunities to discover novel inhibitors of haemostatic pathways. Antihaemostatics from ticks are extremely diverse in structures and functions. Although antihaemostatics have been studied and isolated mostly in the saliva or salivary glands of adult tick females, they have also been detected in larvae, nymphs, eggs, hemolymph and gut (Ibrahim, 1998; Ibrahim et al., 1998, 2000, 2001a, b; Iwanaga et al., 2003; Lai et al., 2004; Ricci et al., 2007; Sant'Anna Azzolini et al., 2003; Tanaka et al., 1999; Willadsen and Riding, 1980). Moreover, anticoagulant activities in salivary glands of hard ticks change in the course of prolonged blood meal. For the brown ear tick, *R. appendiculatus* (Limo et al., 1991), the Rocky Mountain wood tick *D. andersoni* (Gordon and Allen, 1991), the lone star tick *A. americanum* (Zhu et al., 1997b) and the tropical bont tick *A. variegatum* (Kazimirova et al, unpublished observations), most potent anticoagulant activities were detected in salivary glands of partially engorged ticks, whereas anticoagulation in unfed and detached ticks was significantly lower.

Adult hard ticks go through extensive changes after attachment to the host, onset of feeding and mating. During the prolonged slow feeding phase female ticks attain a weight that can increase another 2–3 times during the last 12–24 h of the fast feeding phase (Sonenshine, 1991). Feeding is also associated with profound changes in tick salivary gland physiology and structure. Consequently, the spectrum and expression of antihaemostatics also changes during feeding. For example, highly potent antithrombin(s) are present in salivary glands of nymphs and partially fed adult *A. variegatum* ticks, whereas salivary glands derived from unfed female and male ticks strongly inhibit FXa. In addition, feeding adult female ticks have stronger anticoagulant activities than adult male ticks (Kazimirova et al., 2000).

Furthermore, female *A. americanum* show major changes in gene expression in their salivary glands soon after attachment. Mainly expression of genes regulating synthesis of proteins that are involved in the blood feeding process including antihaemostatics, increases soon after onset of feeding. During the slow feeding phase, some of these genes start decreasing in expression, and in the fast-feeding stage, most genes decrease expression to a low level (Aljamali et al., 2009). These changes mostly finish before the ticks detach. Male ticks generally show smaller changes in gene expression during feeding and mating and the pattern of expression differs from that in females – less expression was detected in genes involved in counteracting host defenses and more in those involved in protein synthesis. The biggest change in gene expression, almost entirely upregulation, was detected between the unfed and early fed female ticks. Genes increasing in expression included a number of protease inhibitors (possible anticoagulants). These are again downregulated at the end of the feeding process. Between the fast feeding and detached females, almost no changes were detected (Aljamali et al., 2009).

Investigation of the dynamics of gene expression throughout feeding of *I. ricinus* based on cDNA libraries from four different feeding stages of females – unfed, 24 h, 4 days (partially fed) and 7 days (fully engorged) after attachment also showed that several groups of over-expressed genes were associated with feeding, mainly involved in tick attachment, feeding and evading the host defence systems, including antihaemostatics (Chmelar et al., 2008). Therefore, it would be interesting to investigate the various antihaemostatics that are expressed and secreted in the different organs, life cycles, sexes and phases of feeding, exponentially expanding the possibilities of uncovering novel antihaemostatics from the large number of ticks.

Future Prospects

Ticks have adapted to blood-feeding by counteracting host haemostasis, inflammation, and immune responses and represent a rich source of various pharmacologically active compounds that facilitate blood-feeding and interfere with host defence systems. Novel tick-derived molecules may be useful for development of pharmaceuticals for treatment of a number of haemostatic disorders, cardiovascular diseases and disorders of the immune system. However, the isolation and identification of salivary molecules is often slow and difficult. High-throughput approaches involving molecular biology techniques, proteomics and functional genomics to study salivary components of ticks have been introduced and will enable testing of predicted and novel functions of tick molecules (Valenzuela, 2002, 2004).

Acknowledgments This work was supported by the Slovak Research and Development Agency (contract no. APVV-51-004505) to MK, and Academic Research Grants from National University of Singapore to RMK.

References

Aljamali, M.N., Ramakrishnan, V.G., Weng, H., Tucker, J.S., Sauer, J.R., Essenberg, R.C., 2009. Microarray analysis of gene expression changes in feeding female and male lone star ticks, *Amblyomma americanum* (L). Arch. Insect Biochem. Physiol. 71, 236–253.

Anastopoulos, P., Thurn, M.J., Broady, K.W., 1991. Anticoagulant in the tick *Ixodes holocyclus*. Aust. Vet. J. 68, 366–367.

Andrade, B.B., Teixeira, C.R., Barral, A., Barral-Netto, M., 2005. Haematophagous arthropod saliva and host defense system: a tale of tear and blood. An. Acad. Bras. Cienc. 77, 665–693.

Azzolini, S.S., Sasaki, S.D., Torquato, R.J., Andreotti, R., Andreotti, E., Tanaka, A.S., 2003. *Rhipicephalus sanguineus* trypsin inhibitors present in the tick larvae: isolation, characterization, and partial primary structure determination. Arch. Biochem. Biophys. 417, 176–182.

Barker, S.C., Murrell, A., 2004. Systematics and evolution of ticks with a list of valid genus and species names. Parasitology 129(Suppl), S15–S36.

Batista, I.F., Chudzinski-Tavassi, A.M., Faria, F., Simons, S.M., Barros-Batestti, D.M., Labruna, M.B., Leao, L.I., Ho, P.L., Junqueira-de-Azevedo, I.L., 2008. Expressed sequence tags (ESTs) from the salivary glands of the tick *Amblyomma cajennense* (Acari: Ixodidae). Toxicon 51, 823–834.

Binnington, K.C., Kemp, D.H., 1980. Role of tick salivary glands in feeding and disease transmission. Adv. Parasitol. 18, 315–339.

Bowman, A.S., Coons, L.B., Needham, G.R., Sauer, J.R., 1997. Tick saliva: recent advances and implications for vector competence. Med. Vet. Entomol. 11, 277–285.

Bowman, A.S., Dillwith, J.W., Sauer, J.R., 1996. Tick salivary prostaglandins: presence, origin and significance. Parasitol. Today 12, 388–396.

Bowman, A.S., Sauer, J.R., 2004. Tick salivary glands: function, physiology and future. Parasitology 129(Suppl), S67–S81.

Brossard, M., Wikel, S.K., 2004. Tick immunobiology. Parasitology 129(Suppl), S161–S176.

Cheng, Y., Wu, H., Li, D., 1999. An inhibitor selective for collagen-stimulated platelet aggregation from the salivary glands of hard tick *Haemaphysalis longicornis* and its mechanism of action. Sci. China C. Life Sci. 42, 457–464.

Chmelar, J., Anderson, J.M., Mu, J., Jochim, R.C., Valenzuela, J.G., Kopecky, J., 2008. Insight into the sialome of the castor bean tick, *Ixodes ricinus*. BMC Genomics 9, 233

Ciprandi, A., de Oliveira, S.K., Masuda, A., Horn, F., Termignoni, C., 2006. *Boophilus microplus*: its saliva contains microphilin, a small thrombin inhibitor. Exp. Parasitol. 114, 40–46.

Das, S., Banerjee, G., DePonte, K., Marcantonio, N., Kantor, F.S., Fikrig, E., 2001. Salp25D, an *Ixodes scapularis* antioxidant, is 1 of 14 immunodominant antigens in engorged tick salivary glands. J. Infect. Dis. 184, 1056–1064.

Davie, E.W., Fujikawa, K., Kisiel, W., 1991. The coagulation cascade: initiation, maintenance, and regulation. Biochemistry 30, 10363–10370.

Francischetti, I.M., Mather, T.N., Ribeiro, J.M., 2003. Cloning of a salivary gland metalloprotease and characterization of gelatinase and fibrin(ogen)lytic activities in the saliva of the Lyme disease tick vector *Ixodes scapularis*. Biochem. Biophys. Res. Commun. 305, 869–875.

Francischetti, I.M., Mather, T.N., Ribeiro, J.M., 2004. Penthalaris, a novel recombinant five-Kunitz tissue factor pathway inhibitor (TFPI) from the salivary gland of the tick vector of Lyme disease, *Ixodes scapularis*. Thromb. Haemost. 91, 886–898.

Francischetti, I.M., My, P.V., Mans, B.J., Andersen, J.F., Mather, T.N., Lane, R.S., Ribeiro, J.M., 2005. The transcriptome of the salivary glands of the female western black-legged tick *Ixodes pacificus* (Acari: Ixodidae). Insect Biochem. Mol. Biol. 35, 1142–1161.

Francischetti, I.M., Sa-Nunes, A., Mans, B.J., Santos, I.M., Ribeiro, J.M., 2009. The role of saliva in tick feeding. Front Biosci. 14, 2051–2088.

Francischetti, I.M., Valenzuela, J.G., Andersen, J.F., Mather, T.N., Ribeiro, J.M., 2002. Ixolaris, a novel recombinant tissue factor pathway inhibitor (TFPI) from the salivary gland of the tick, *Ixodes scapularis*: identification of factor X and factor Xa as scaffolds for the inhibition of factor VIIa/tissue factor complex. Blood 99, 3602–3612.

Gaspar, A.R., Joubert, A.M., Crause, J.C., Neitz, A.W., 1996. Isolation and characterization of an anticoagulant from the salivary glands of the tick, *Ornithodoros savignyi* (Acari: Argasidae). Exp. Appl. Acarol. 20, 583–598.

Gordon, J.R., Allen, J.R., 1991. Factors V and VII anticoagulant activities in the salivary glands of feeding *Dermacentor andersoni* ticks. J. Parasitol. 77, 167–170.

Hawkins, R.I., Hellmann, K., 1966. Factors affecting blood clotting from the tick *Ornithodorus moubata*. J. Physiol 185, 70P

Hellmann, K., Hawkins, R.I., 1967. The action of tick extracts on blood coagulation and fibrinolysis. Thromb. Diath. Haemorrh. 18, 617–625.

Hoffmann, A., Walsmann, P., Riesener, G., Paintz, M., Markwardt, F., 1991. Isolation and characterization of a thrombin inhibitor from the tick *Ixodes ricinus*. Pharmazie 46, 209–212.

Horn, F., dos Santos, P.C., Termignoni, C., 2000. *Boophilus microplus* anticoagulant protein: an antithrombin inhibitor isolated from the cattle tick saliva. Arch. Biochem. Biophys. 384, 68–73.

Hovius, J.W., Levi, M., Fikrig, E., 2008. Salivating for knowledge: potential pharmacological agents in tick saliva. PLoS Med. 5, e43.

Ibrahim, M.A., 1998. Purification and properties of protease inhibitor and anticoagulant from the larvae of the camel tick *Hyalomma dromedarii* (Acari: Ixodidae). J. Egypt. Ger. Soc. Zool. A. Comp. Physiol. Biochem. 25, 319–339.

Ibrahim, M.A., Ghazy, A.H., Khalil, M.I., 2000. The embryos of the camel tick *Hyalomma dromedarii* contain a potent inhibitor of both thrombin and FXa. J. Egypt. Ger. Soc. Zool. A. Comp. Physiol. Biochem. 32, 99–114.

Ibrahim, M.A., Ghazy, A.H., Maharem, T., Khalil, M., 2001a. Isolation and properties of two forms of thrombin inhibitor from the nymphs of the camel tick *Hyalomma dromedarii* (Acari: Ixodidae). Exp. Appl. Acarol. 25, 675–698.

Ibrahim, M.A., Ghazy, A.H., Maharem, T.M., Khalil, M.I., 1998. Purification and characterization of peptide anticoagulants from the embryos of the camel tick, *Hyalomma dromedarii* (Acari: Ixodidae). J. Egypt. Ger. Soc. Zool. A. Comp. Physiol. Biochem. 27, 287–305.

Ibrahim, M.A., Ghazy, A.H., Maharem, T.M., Khalil, M.I., 2001b. Factor Xa (FXa) inhibitor from the nymphs of the camel tick *Hyalomma dromedarii*. Comp Biochem. Physiol. B. Biochem. Mol. Biol. 130, 501–512.

Inokuma, H., Kemp, D.H., Willadsen, P., 1994. Comparison of prostaglandin E_2 (PGE_2) in salivary gland of *Boophilus microplus*, *Haemaphysalis longicornis* and *Ixodes holocyclus*, and quantification of PGE_2 in saliva, hemolymph, ovary and gut of *B. microplus*. J. Vet. Med. Sci. 56, 1217–1218.

Iwanaga, S., Okada, M., Isawa, H., Morita, A., Yuda, M., Chinzei, Y., 2003. Identification and characterization of novel salivary thrombin inhibitors from the ixodidae tick, *Haemaphysalis longicornis*. Eur. J. Biochem. 270, 1926–1934.

Joubert, A.M., Crause, J.C., Gaspar, A.R., Clarke, F.C., Spickett, A.M., Neitz, A.W., 1995. Isolation and characterization of an anticoagulant present in the salivary glands of the bont-legged tick, *Hyalomma truncatum*. Exp. Appl. Acarol. 19, 79–92.

Karczewski, J., Endris, R., Connolly, T.M., 1994. Disagregin is a fibrinogen receptor antagonist lacking the Arg-Gly-Asp sequence from the tick, *Ornithodoros moubata*. J. Biol. Chem. 269, 6702–6708.

Karczewski, J., Waxman, L., Endris, R.G., Connolly, T.M., 1995. An inhibitor from the argasid tick *Ornithodoros moubata* of cell adhesion to collagen. Biochem. Biophys. Res. Commun. 208, 532–541.

Kato, N., Iwanaga, S., Okayama, T., Isawa, H., Yuda, M., Chinzei, Y., 2005. Identification and characterization of the plasma kallikrein-kinin system inhibitor, haemaphysalin, from hard tick, *Haemaphysalis longicornis*. Thromb. Haemost. 93, 359–367.

Kazimirova, M., Jancinova, V., Petrikova, M., Takac, P., Labuda, M., Nosal, R., 2002. An inhibitor of thrombin-stimulated blood platelet aggregation from the salivary glands of the hard tick *Amblyomma variegatum* (Acari: Ixodidae). Exp. Appl. Acarol. 28, 97–105.

Kazimirova, M., Silvanova, E., Slovak, M., Balanova, I., Labuda, M., 2000. Anticoagulant activity in salivary glands of ixodid ticks. Proc. 3rd Int. Conf. Tick Tick-Borne Pathogens, 159–164.

Keller, P.M., Waxman, L., Arnold, B.A., Schultz, L.D., Condra, C., Connolly, T.M., 1993. Cloning of the cDNA and expression of moubatin, an inhibitor of platelet aggregation. J. Biol. Chem. 268, 5450–5456.

Koh, C.Y., Kazimirova, M., Nuttall, P.A., Kini, R.M., 2009. Noncompetitive inhibitor of thrombin. ChemBiochem 10, 2155–2158.

Koh, C.Y., Kazimirova, M., Trimnell, A., Takac, P., Labuda, M., Nuttall, P.A., Kini, R.M., 2007. Variegin, a novel fast and tight binding thrombin inhibitor from the tropical bont tick. J. Biol. Chem. 282, 29101–29113.

Koh, C.Y., Kini, R.M., 2009. Molecular diversity of anticoagulants from haematophagous animals. Thromb. Haemost. 102, 437–453.

Lai, R., Takeuchi, H., Jonczy, J., Rees, H.H., Turner, P.C., 2004. A thrombin inhibitor from the ixodid tick, *Amblyomma hebraeum*. Gene 342, 243–249.

Ledizet, M., Harrison, L.M., Koskia, R.A., Cappello, M., 2005. Discovery and pre-clinical development of antithrombotics from hematophagous invertebrates. Curr. Med. Chem. Cardiovasc. Hematol. Agents 3, 1–10.

Liao, M., Zhou, J., Gong, H., Boldbaatar, D., Shirafuji, R., Battur, B., Nishikawa, Y., Fujisaki, K., 2009. Hemalin, a thrombin inhibitor isolated from a midgut cDNA library from the hard tick *Haemaphysalis longicornis*. J. Insect Physiol 55, 164–173.

Limo, M.K., Voigt, W.P., Tumbo-Oeri, A.G., Njogu, R.M., ole-MoiYoi, O.K., 1991. Purification and characterization of an anticoagulant from the salivary glands of the ixodid tick *Rhipicephalus appendiculatus*. Exp. Parasitol. 72, 418–429.

Macedo-Ribeiro, S., Almeida, C., Calisto, B.M., Friedrich, T., Mentele, R., Sturzebecher, J., Fuentes-Prior, P., Pereira, P.J., 2008. Isolation, cloning and structural characterisation of boophilin, a multifunctional Kunitz-type proteinase inhibitor from the cattle tick. PLoS One 3, e1624.

Mans, B.J., Andersen, J.F., Schwan, T.G., Ribeiro, J.M., 2008. Characterization of anti-hemostatic factors in the argasid, *Argas monolakensis*: implications for the evolution of blood-feeding in the soft tick family. Insect Biochem. Mol. Biol. 38, 22–41.

Mans, B.J., Louw, A.I., Neitz, A.W., 2002a. Amino acid sequence and structure modeling of savignin, a thrombin inhibitor from the tick, *Ornithodoros savignyi*. Insect Biochem. Mol. Biol. 32, 821–828.

Mans, B.J., Louw, A.I., Neitz, A.W., 2002b. Savignygrin, a platelet aggregation inhibitor from the soft tick *Ornithodoros savignyi*, presents the RGD integrin recognition motif on the Kunitz-BPTI fold. J. Biol. Chem. 277, 21371–21378.

Mans, B.J., Neitz, A.W., 2004. Adaptation of ticks to a blood-feeding environment: evolution from a functional perspective. Insect Biochem. Mol. Biol. 34, 1–17.

Maritz-Olivier, C., Stutzer, C., Jongejan, F., Neitz, A.W., Gaspar, A.R., 2007. Tick anti-hemostatics: targets for future vaccines and therapeutics. Trends Parasitol. 23, 397–407.

Markwardt, F., Landmann, H., 1958. Untersuchungen uber den blutgerinnungshemmenden wirdstoff der zecke *Ixodes ricinus*. Naturwissenschaften 45, 398–399.

Markwardt, F., Landmann, H., 1961. Ubereinen hemmstoff der thrombokinase aus der lederzecke *Ornithodorous moubata*. Naturwissenschaften 48, 433.

Monteiro, R.Q., Rezaie, A.R., Bae, J.S., Calvo, E., Andersen, J.F., Francischetti, I.M., 2008. Ixolaris binding to factor X reveals a precursor state of factor Xa heparin-binding exosite. Protein Sci. 17, 146–153.

Motoyashiki, T., Tu, A.T., Azimov, D.A., Ibragim, K., 2003. Isolation of anticoagulant from the venom of tick, *Boophilus calcaratus*, from Uzbekistan. Thromb. Res. 110, 235–241.

Nakajima, C., Imamura, S., Konnai, S., Yamada, S., Nishikado, H., Ohashi, K., Onuma, M., 2006. A novel gene encoding a thrombin inhibitory protein in a cDNA library from *Haemaphysalis longicornis* salivary gland. J. Vet. Med. Sci. 68, 447–452.

Narasimhan, S., Koski, R.A., Beaulieu, B., Anderson, J.F., Ramamoorthi, N., Kantor, F., Cappello, M., Fikrig, E., 2002. A novel family of anticoagulants from the saliva of *Ixodes scapularis*. Insect Mol. Biol. 11, 641–650.

Narasimhan, S., Montgomery, R.R., DePonte, K., Tschudi, C., Marcantonio, N., Anderson, J.F., Sauer, J.R., Cappello, M., Kantor, F.S., Fikrig, E., 2004. Disruption of *Ixodes scapularis* anticoagulation by using RNA interference. Proc. Natl. Acad. Sci. U.S.A. 101, 1141–1146.

Nienaber, J., Gaspar, A.R., Neitz, A.W., 1999. Savignin, a potent thrombin inhibitor isolated from the salivary glands of the tick *Ornithodoros savignyi* (Acari: Argasidae). Exp. Parasitol. 93, 82–91.

Nuttall, G.H.F., Strickland, C., 1908. On the presence of an anticoagulin in the salivary glands and intestines of *Argas persicus*. Parasitology 1, 302–310.

Nuttall, P.A., Labuda, M., 2004. Tick-host interactions: saliva-activated transmission. Parasitology 129(Suppl), S177–S189.

Ramamoorthi, N., Narasimhan, S., Pal, U., Bao, F., Yang, X.F., Fish, D., Anguita, J., Norgard, M.V., Kantor, F.S., Anderson, J.F., Koski, R.A., Fikrig, E., 2005. The Lyme disease agent exploits a tick protein to infect the mammalian host. Nature 436, 573–577.

Ribeiro, J.M., Francischetti, I.M., 2003. Role of arthropod saliva in blood feeding: sialome and post-sialome perspectives. Annu. Rev. Entomol. 48, 73–88.

Ribeiro, J.M., Makoul, G.T., Levine, J., Robinson, D.R., Spielman, A., 1985. Antihemostatic, anti-inflammatory, and immunosuppressive properties of the saliva of a tick, *Ixodes dammini*. J. Exp. Med. 161, 332–344.

Ribeiro, J.M., Makoul, G.T., Robinson, D.R., 1988. *Ixodes dammini*: evidence for salivary prostacyclin secretion. J. Parasitol. 74, 1068–1069.

Ricci, C.G., Pinto, A.F., Berger, M., Termignoni, C., 2007. A thrombin inhibitor from the gut of *Boophilus microplus* ticks. Exp. Appl. Acarol. 42, 291–300.

Roller, L., Havlikova, S., Peterkova, K., Kazimirova, M., 2009. Unusual variability of potential thrombin inhibitors in salivary glands of the tropical bont tick *Amblyomma variegatum*, in: Špitalská, E., Kazimírová, M., Kocianová, E., Šustek, Z. (Eds.), Zborník z Konferencie "Labudove dni", Virologický ústav SAV, Bratislava, pp. 55–57.

Sabbatini, L., 1899. Fermento anticoagulante dell *Ixodes ricinus*. Arch. Ital. Biol. 31, 37.

Sasaki, S.D., Azzolini, S.S., Hirata, I.Y., Andreotti, R., Tanaka, A.S., 2004. *Boophilus microplus* tick larvae, a rich source of Kunitz type serine proteinase inhibitors. Biochimie 86, 643–649.

Sauer, J.R., Essenberg, R.C., Bowman, A.S., 2000. Salivary glands in ixodid ticks: control and mechanism of secretion. J. Insect Physiol. 46, 1069–1078.

Sauer, J.R., McSwain, J.L., Bowman, A.S., Essenberg, R.C., 1995. Tick salivary gland physiology. Annu. Rev. Entomol. 40, 245–267.

Sonenshine, D.E., 1991. Biology of Ticks (vol. 1). Oxford University Press, New York, Oxford.

Steen, N.A., Barker, S.C., Alewood, P.F., 2006. Proteins in the saliva of the Ixodida (ticks): pharmacological features and biological significance. Toxicon 47, 1–20.

Tanaka, A.S., Andreotti, R., Gomes, A., Torquato, R.J., Sampaio, M.U., Sampaio, C.A., 1999. A double headed serine proteinase inhibitor – human plasma kallikrein and elastase inhibitor -- from *Boophilus microplus* larvae. Immunopharmacology 45, 171–177.

Valenzuela, J.G., 2002. High-throughput approaches to study salivary proteins and genes from vectors of disease. Insect Biochem. Mol. Biol. 32, 1199–1209.

Valenzuela, J.G., 2004. Exploring tick saliva: from biochemistry to 'sialomes' and functional genomics. Parasitology 129(Suppl), S83–S94.

Valenzuela, J.G., Francischetti, I.M., Pham, V.M., Garfield, M.K., Mather, T.N., Ribeiro, J.M., 2002. Exploring the sialome of the tick *Ixodes scapularis*. J. Exp. Biol. 205, 2843–2864.

van de Locht, A., Stubbs, M.T., Bode, W., Friedrich, T., Bollschweiler, C., Hoffken, W., Huber, R., 1996. The ornithodorin-thrombin crystal structure, a key to the TAP enigma? EMBO J. 15, 6011–6017.

Wang, X., Coons, L.B., Taylor, D.B., Stevens, S.E., Jr., Gartner, T.K., 1996. Variabilin, a novel RGD-containing antagonist of glycoprotein IIb-IIIa and platelet aggregation inhibitor from the hard tick *Dermacentor variabilis*. J. Biol. Chem. 271, 17785–17790.

Waxman, L., Smith, D.E., Arcuri, K.E., Vlasuk, G.P., 1990. Tick anticoagulant peptide (TAP) is a novel inhibitor of blood coagulation factor Xa. Science 248, 593–596.

Wei, A., Alexander, R.S., Duke, J., Ross, H., Rosenfeld, S.A., Chang, C.H., 1998. Unexpected binding mode of tick anticoagulant peptide complexed to bovine factor Xa. J. Mol. Biol. 283, 147–154.

Wikel, S.K., 1996. The Immunology of Host-Ectoparasitic Arthropod Relationships. CAB International, Oxon, UK.

Willadsen, P., Riding, G.A., 1980. On the biological role of a proteolytic-enzyme inhibitor from the ectoparasitic tick *Boophilus microplus*. Biochem. J. 189, 295–303.

Zhu, K., Bowman, A.S., Brigham, D.L., Essenberg, R.C., Dillwith, J.W., Sauer, J.R., 1997a. Isolation and characterization of americanin, a specific inhibitor of thrombin, from the salivary glands of the lone star tick *Amblyomma americanum* (L.). Exp. Parasitol. 87, 30–38.

Zhu, K., Sauer, J.R., Bowman, A.S., Dillwith, J.W., 1997b. Identification and characterization of anticoagulant activities in the saliva of the lone star tick, *Amblyomma americanum* (L.). J. Parasitol. 83, 38–43.

Chapter 9
Sialogenins and Immunomodulators Derived from Blood Feeding Parasites

Anderson Sá-Nunes and Carlo José Freire de Oliveira

Abstract Several molecules and cell types constitute the innate or "natural" and the acquired or "adaptive" arms of the immune response. The innate arm represents the first line of host defense while the acquired arm develops upon recurring contact with the parasites and promotes long-lasting responses and immunological memory. When attempting to intake a blood meal, hematophagous parasites may face the immune system of their vertebrate hosts. In order to counteract the host immune protective barriers and facilitate their feeding, these organisms developed, throughout the evolutionary process, a salivary cocktail with an arsenal of molecules containing several immunomodulatory properties. Advances in biochemical techniques and molecular biology tools made the genome sequencing and the development of sialome catalogs possible, allowing the identification and study of each salivary component. In consequence, an increasing number of immunologists and immunoparasitologists are now focusing their efforts on unveiling the biological role of these molecules and contributing to the creation of a database representing the salivary "immunome" of blood feeding organisms. Thus, the present chapter illustrates the current knowledge of proteins, here termed "sialogenins" (from the Greek *sialo*: saliva; *gen*: origin; *in*: protein), and nonprotein constituents present in the salivary secretion employed by bloodsucking ectoparasites to circumvent such effector mechanisms of host immunity. Also presented is an updated list of complement system and chemokine inhibitors, histamine- and immunoglobulin-binding proteins, and modulators of host cell activation and function described in saliva or salivary gland extract of hematophagous ectoparasites.

A. Sá-Nunes (✉)
Laboratory of Experimental Immunology, Department of Immunology, Institute of Biomedical Sciences, University of São Paulo, São Paulo, SP 05508 900, Brazil
e-mail: sanunes@usp.br

R.M. Kini et al. (eds.), *Toxins and Hemostasis*, DOI 10.1007/978-90-481-9295-3_9,

Introduction

During blood feeding on their hosts, hematophagous parasites face two major challenges. The first is by the host hemostasis barrier represented by platelets, coagulation factors and vascular tonus. In order to counteract these mechanisms, the number and variety of anticlotting, antiplatelet, and vasodilators contained in their salivary secretions is so amazing that these parasites might be considered great invertebrate pharmacologists and the source of many potential therapeutic agents (Mans, 2005; Ribeiro, 1995). A second challenge is posed by the host immune system and begins even before contact with host blood, in the moment that parasite mouthparts are inserted into the skin and mucous membranes, causing the breaking of this physical barrier.

Immune responses are classically divided in innate or "natural" and acquired or "adaptive". The innate arm is the first line of defense and consists of cellular (e.g., neutrophils, macrophages, mast cells, eosinophils, and dendritic cells) and soluble components (e.g., peptides, acute-phase proteins, complement system, cytokines, and chemokines) commonly characterized by acute inflammation (Janeway and Medzhitov, 2002). The adaptive arm, which promotes long-lasting defense and is represented by memory immune response, comprises B lymphocytes, antibody-secreting cells responsible for the so-called humoral adaptive immunity, and T lymphocytes, responsible for helper and cytotoxic functions (Bluestone et al., 2009; McHeyzer-Williams and McHeyzer-Williams, 2005). These arms work together: innate immunity has a critical role in early control of pathogen invasion but also prepares the adaptive immunity, which in turn protects the host through direct action of its products or cells and improving the effector mechanisms of innate immunity (Bluestone et al., 2009; Fazilleau et al., 2007). When everything functions optimally, the parasite is eliminated, and host long-lived specific immunological memory is established.

Resident leukocytes of epidermis and dermis such as mast cells, eosinophils, dendritic cells, and macrophages – as well as keratinocytes and antimicrobial peptides – are the first innate defenses to make contact with parasite mouthparts and parasite-secreted components. Besides their direct effector functions (e.g. phagocytosis and cytotoxicity), these cells release pre-formed mediators in addition to producing chemotactic factors to recruit inflammatory cells such as neutrophils, eosinophils, and macrophages to the bite site. Recurring contact with the same parasite may activate adaptive responses involving T cells and B cells by production of antibodies and sensitization of mast cells and basophils that, in conjunction with eosinophils, are predominant cells in most infestation sites. As a result of a long-time coevolutionary relationship with their hosts, bloodsucking ectoparasites have also developed ways to protect themselves or, at least, minimize the host immune responses when taking their blood meal. The array of mechanisms used to overcome cellular and humoral host responses during the parasitic stage involves a large number of protein (here also termed sialogenins: protein molecules present on saliva secretions; from the Greek *sialo*: saliva; *gen*: origin; *in*: protein) and nonprotein molecules. In general, these molecules are released as soon as the parasite inserts the mouthparts into the skin, and continue to be released during the entire period of parasite infestation. These components have been found to present numerous properties/activities including binding of chemokines, immunoglobulins, and vasoactive amines; inhibition of adhesion molecules and cell-signaling components.

As described above, innate and acquired immune responses elicited by bloodsucking ectoparasites in their hosts are multimediated. The host species and its genetic background will determine which type of immune response will be triggered, as important differences in cell populations and their mediators exist among species. The parasite species and saliva composition is another variable to be considered, as interspecies and intraspecies variations have been extensively found (Anderson et al., 2006; Inokuma et al., 1994; Lanzaro et al., 1999; Ribeiro et al., 2001; Ribeiro and Valenzuela, 2003). The raw result of this complex equation is host susceptibility or resistance to specific parasites. In fact, the pharmacologic activities of the salivary cocktail in conjunction with the elicited host immunity may determine a successful or a deleterious blood meal for these ectoparasites. Although several activities of saliva or salivary gland extracts on host immunity have been described over the years, their molecular identification remains elusive in the majority of cases. This reality is rapidly changing with the publication of the sialomes, sets of mRNA and proteins expressed in the salivary glands of bloodsucking species (Andersen et al., 2007, 2009; Arca et al., 2007; Assumpcao et al., 2008; Calvo et al., 2007; Chmelar et al., 2008; Francischetti et al., 2002, 2008; Ribeiro et al., 2004; Valenzuela et al., 2002a, b). Thus, a huge catalog of secreted salivary molecules is being created, exponentially advancing our knowledge of the immunomodulatory activities of such components. Despite the description of some molecules with related properties in bloodsucking endoparasites secretions, for space limitations this chapter will focus on identified and characterized molecules of ectoparasites only (mosquitoes, sand flies, fleas, bedbugs, leeches, and ticks). Some references on the general activities of whole saliva with no molecular characterization are also provided. The information available on the salivary immunomodulatory activities of these bloodsucking ectoparasites is compiled in the Table 9.1.

Complement System Inhibitors

Complement was first described in 1895 by Jules Bordet as a thermolabile activity in serum that complemented thermostable antibody properties in the killing of bacteria (Bordet, 1895). Now it is well known that the complement system is a proteolytic cascade comprising at least 35 different proteins (including enzymes and their regulators) that can be initiated by four different mechanisms and converge to the assembly of a lytic complex, usually on the surface of a foreign organism (Francischetti et al., 2009; Zipfel and Skerka, 2009). The classical pathway is initiated by the complement component C1q, which binds to specific subclasses of antibodies bound to their antigens (immunocomplexes), or to C-reactive protein; the alternative pathway is initiated when complement component C3 is spontaneously cleaved in the fluid phase and its larger fragment, C3b, binds covalently to

Table 9.1 Summary of the salivary immunomodulatory molecules identified and characterized in bloodsucking ectoparasites

Molecule	Molecular weight	Species	Functions	References
Complement inhibitors				
ISAC	18.14 kDa	*Ixodes scapularis*	Inhibition of alternative pathway	Valenzuela et al. (2000)
SALP20	48.0 kDa	*Ixodes scapularis*	Inhibition of alternative pathway	Tyson et al. (2007, 2008)
IRAC I	18.03 kDa	*Ixodes ricinus*	Inhibition of alternative pathway	Schroeder et al. (2007)
IRAC II	17.46 kDa	*Ixodes ricinus*	Inhibition of alternative pathway	Schroeder et al. (2007)
IxAC-B1	17.58 kDa	*Ixodes ricinus*	Inhibition of alternative pathway	Couvreur et al. (2008)
IxAC-B2	17.69 kDa	*Ixodes ricinus*	Inhibition of alternative pathway	Couvreur et al. (2008)
IxAC-B3	17.85 kDa	*Ixodes ricinus*	Inhibition of alternative pathway	Couvreur et al. (2008)
IxAC-B4	17.88 kDa	*Ixodes ricinus*	Inhibition of alternative pathway	Couvreur et al. (2008)
IxAC-B5	17.66 kDa	*Ixodes ricinus*	Inhibition of alternative pathway	Couvreur et al. (2008)
OmCI	16.77 kDa	*Ornithodoros moubata*	Inhibition of classical and alternative pathways	Nunn et al. (2005)
TSGP2	17.77 kDa	*Ornithodoros savignyi*	Binding of C5 and leukotriene B_4	Mans and Ribeiro (2008)
TSGP3	17.88 kDa	*Ornithodoros savignyi*	Binding of C5 and leukotriene B_4	Mans and Ribeiro (2008)
Histamine-binding proteins				
Ra-HBP1	22.85 kDa	*Rhipicephalus appendiculatus*	Histamine binding	Paesen et al. (1999)
Ra-HBP2	21.37 kDa	*Rhipicephalus appendiculatus*	Histamine binding	Paesen et al. (1999)
Ra-HBP3	21.46 kDa	*Rhipicephalus appendiculatus*	Histamine binding	Paesen et al. (1999)
NP1	20.3 kDa	*Rhodnius prolixus*	Histamine binding	Ribeiro and Walker (1994) and Montfort et al. (2000)
NP2	19.6 kDa	*Rhodnius prolixus*	Histamine binding	Weichsel et al. (1998)
NP3	19.7 kDa	*Rhodnius prolixus*	Histamine binding	Weichsel et al. (1998)
NP4	20.9 kDa	*Rhodnius prolixus*	Histamine binding	Weichsel et al. (1998)
D7r1	16.4 kDa	*Anopheles gambiae*	Histamine binding	Calvo et al. (2006)

Table 9.1 (continued)

Molecule	Molecular weight	Species	Functions	References
D7r2	16.2 kDa	*Anopheles gambiae*	Histamine binding	Calvo et al. (2006)
D7r3	16.4 kDa	*Anopheles gambiae*	Histamine binding	Calvo et al. (2006)
D7r4	16.9 kDa	*Anopheles gambiae*	Histamine binding	Calvo et al. (2006)
D7Long (AeD7)	35.1 kDa	*Aedes aegypti*	Histamine and leukotriene binding	Calvo et al. (2006, 2008)
Chemokine inhibitors				
Evasin-1	10.46 kDa	*Rhipicephalus sanguineus*	Binding of CC chemokines: CCL3, CCL4 and CCL18	Frauenschuh et al. (2007)
Evasin-2	10.46 kDa	*Rhipicephalus sanguineus*	To be determined	Déruaz et al. (2008)
Evasin-3	7.0 kDa	*Rhipicephalus sanguineus*	Binding of CXC chemokines: CXCL1 and CXCL8	Déruaz et al. (2008)
Evasin-4	12,03 kDa	*Rhipicephalus sanguineus*	Binding of CC chemokines: CCL5 and CCL11	Déruaz et al. (2008)
Tick MIF	12.6 kDa	*Amblyomma americanum*	Inhibition of macrophage migration	Jaworski et al. (2001)
HlMIF	12.3 kDa	*Haemaphysalis longicornis*	Inhibition of monocytes/macrophages migration	Umemiya et al. (2007)
Immunoglobulin-binding proteins				
IGBP-MA	29.0 kDa	*Rhipicephalus appendiculatus*	IgG binding	Wang and Nuttall (1998, 1999)
IGBP-MB	25.0 kDa	*Rhipicephalus appendiculatus*	IgG binding	Wang and Nuttall (1998, 1999)
IGBP-MC	21.0 kDa	*Rhipicephalus appendiculatus*	IgG binding Fecundity enhancing	Wang and Nuttall (1998, 1999)
BmPRM	102.0 kDa	*Rhipicephalus (Boophilus) microplus*	IgG binding	Ferreira et al. (2002)

Table 9.1 (continued)

Molecule	Molecular weight	Species	Functions	References
Modulators of cell activation and function				
PGE_2	352.4 Da	*Rhipicephalus (Boophilus) microplus* *Ixodes scapularis* *Ixodes holocyclus* *Haemaphysalis longicornis* *Ambliomma. americanum*	Inhibition of dendritic cell maturation and function; Inhibition of T cell proliferation	Higgs et al. (1976) Ribeiro et al. (1985) Sá-Nunes et al. (2007)
PGI_2	352.4 Da	*Ixodes dammini*	Anti-mast cell degranulating	Ribeiro et al. (1988)
Dopamine	153.2 Da	*Rhipicephalus (Boophilus) microplus* *Amblyomma hebraeum*	Modulation of T cell and dendritic cell function	Binnington and Stone (1977) and Kaufman et al. (1999)
Endocannabinoids: 2-arachidonoylglycerol (2-AG), *N*-acylethanolamines (NAEs: C16:0; C18:0, C18:1)	378.3, 299.5, 327.5, 325.5 Da	*Amblyomma americanum*	Analgesic and anti-inflammatory	Fezza et al. (2003)
Adenosine	267.2 Da	*Phlebotomus argentipes* *Phlebotomus papatasi*	Many immunologic functions on neutrophil, dendritic cell, adhesion, and cell chemotaxis	Ribeiro et al. (1999) and Ribeiro and Modi (2001)
ISL 929, ISL 1373	10.0 kDa	*Ixodes scapularis*	Downregulation of integrin expression and inhibition of O_2^- production	Guo et al. (2009)
Sialostatin L	12.5 kDa	*Ixodes scapularis*	Modulation of neutrophil migration, dendritic cell maturation and T cell function	Kotsyfakis et al. (2006) and Sá-Nunes et al. (2009)
Da-p36	36.0 kDa	*Dermacentor andersoni*	Inhibition of T cell proliferation	Bergman et al. (2000)

Table 9.1 (continued)

Molecule	Molecular weight	Species	Functions	References
SALP15	15.0 kDa	*Ixodes scapularis*	Inhibition of T cell activation	Juncadella et al. (2007) and Ramamoorthi et al. (2008)
BIF	13.0 kDa	*Hyalomma asiaticum asiaticum*	Inhibition of B cell proliferation	Yu et al. (2006)
IRIS	43.0 kDa	*Ixodes ricinus*	Inhibition of IFN-γ production by T cells; inhibition of IL-6 and TNF-α production by macrophages	Leboulle et al. (2002)
Moubatin	17.5 kDa	*Ornithodoros moubata*	Binding of leukotriene B_4	Mans and Ribeiro (2008)
Immunoregulin HA	3.18 kDa	*Hybomitra atriperoides*	Inhibition of IFN-γ and MCP-1 production and increased secretion of IL-10 by splenocytes	Yan et al. (2008)
Immunoregulin TP-1	3.1 kDa	*Tabanus pleskei*	Inhibition of IFN-γ and MCP-1 production and increased secretion of IL-10 by splenocytes	Zhao et al. (2009)

OH-groups on carbohydrates and proteins from pathogen cell membranes (Gasque, 2004). Then, C3b binds factor B, which is then cleaved by factor D, producing the cleavage products Bb and Ba. Bb remains bound to C3b, forming the C3bBb complex (C3 convertase), which cleaves more C3; the lectin pathway is initiated by binding of lectins (such as mannan-binding lectin or MBL) or ficolins to foreign carbohydrate surfaces, leading to activation of MBL-associated serine protease (MASP) (Ip et al., 2009); the fourth pathway, recently discovered, is initiated by activation of C5 by thrombin (Huber-Lang et al., 2006). These different processes converge to form the C5 convertase, generating two C5 convertases, C4b2b3b and C3bBb3b. These convertases can cleave C5 producing one C5b and one smaller C5a anaphylatoxin molecule discussed below. Association of C5b with C6, C7, C8, and C9 ultimately leads to generation of the lytic membrane attack complex (MAC), which may kill the pathogen. Moreover, the smaller cleavage peptides of C3 and C5 – C3a and C5a – are known as the anaphylatoxins, having inflammatory, edema-producing, and chemotactic properties (Bjork et al., 1985; Guo and Ward, 2005).

The alternative pathway of complement activation has been shown to be involved in tick rejection reactions by guinea pigs (Wikel, 1979; Wikel and Allen, 1978), perhaps through the production of inflammatory anaphylatoxins. In fact, the cleavage of host complement protein C5 by component(s) from saliva of the hard tick *Dermacentor variabilis* has been demonstrated, generating chemoctatic fragments for neutrophils (Berenberg et al., 1972). Two decades after these first descriptions, a new molecule displaying regulatory activity on complement activation was characterized: a relatively small (18.5 kDa) *Ixodes scapularis* anticomplement molecule (ISAC) that inhibits the alternative pathway by preventing C3b and factor B binding and by removal of prebound factor B (Valenzuela et al., 2000). Subsequently, several other proteins sharing homology with ISAC were identified, constituting a large family of anticomplement molecules (Couvreur et al., 2008; Ribeiro et al., 2006). Of those, *I. scapularis* Salp20, a 48.0-kDa protein, inhibits the alternative complement pathway directly by binding properdin, a positive regulator that binds and stabilizes C3 convertase (Tyson et al., 2007, 2008); *Ixodes ricinus* orthologs IRAC I and IRAC II, with predicted mature molecular weights of 18.03 and 17.46 kDa, respectively, also inhibits the complement alternative pathway (Schroeder et al., 2007). Recently, five additional sequences – homologous to IRAC (termed IxAC-B1 to -B5) and also able to inhibit the alternative pathway – have been reported in *I. ricinus* (Couvreur et al., 2008). Moreover, the soft tick *Ornithodoros moubata* is able to secrete a protein belonging to the lipocalin superfamily (OmCI) that binds and inhibits the C5 convertase, thus inhibiting both the classical and the alternative pathways (Nunn et al., 2005). In addition, Mans and Ribeiro (2008) have demonstrated that OmCI has high sequence similarities with moubatin, TSGP2, and TSGP3, three other proteins from the lipocalin family found in the tick *Ornithodoros savignyi*. Based on this fact, it was demonstrated by plasmon resonance analysis that TSGP2 and TSGP3 bind to C5 with affinities similar to that observed for OmCI, indicating that these proteins may have the same inhibitory effect on the complement system (Mans and Ribeiro, 2008).

A few uncharacterized complement inhibitors have been described in groups other than ticks. Although the active components have not been identified, excreted/secreted compounds from phlebotomine sand flies and triatomine bugs are also able to inhibit the complement system. For example, saliva from the phlebotomine *Lutzomyia longipalpis* is capable of inhibiting both alternative and classical pathways, while that from *Lutzomyia migonei* acted only on the latter. Moreover, saliva and intestinal contents of the triatomine *Triatoma brasiliensis*, *Triatoma infestans*, and *Rhodnius prolixus* were able to inhibit the classical and alternative pathways (Barros et al., 2009; Cavalcante et al., 2003), while saliva from the bug *Panstrongylus megistus* affects the classical pathway (Cavalcante et al., 2003). It has also been demonstrated that *Aedes aegypti* have no salivary complement inhibitors; nevertheless, soluble contents of its intestine were capable to inhibit C3b deposition by the classical and alternative pathways (Barros et al., 2009). The medicinal leech *Hirudo medicinalis* produces an uncharacterized inhibitor of complement component C1s (Baskova and Zavalova, 2001).

Histamine-Binding Proteins

Release of histamine from basophils and mast cells caused by tissue damage during ectoparasite feeding induces an inflammatory reaction with increased vascular permeability and plasma exudation, causing pruritus and triggering grooming host behavior that might dislodge or even kill such parasites. Thus, these organisms evolved salivary histamine-blocking agents, first described for the *Rhipicephalus sanguineus* tick (Chinery and Ayitey-Smith, 1977) and later identified in the *Rhipicephalus appendiculatus* tick as histamine-binding proteins Ra-HBP1, Ra-HBP2, and Ra-HBP3 (Paesen et al., 1999). These proteins belong to the lipocalin superfamily, extracellular low-molecular weight proteins that transport small hydrophobic ligands that, in ticks, evolved to trap cationic, hydrophilic molecules (Paesen et al., 2000). While histamine-binding activity has been described for lipocalins from virtually all hard and soft ticks studied to date (Beaufays et al., 2008; Mans et al., 2008), this feature independently evolved in one insect at least. Four nitrophorins (NP1-4), nitrosyl-heme proteins with lipocalin folds that bind nitric oxide and histamine, were described in the kissing bug *R. prolixus* (Andersen et al., 2000; Ribeiro and Walker, 1994; Weichsel et al., 1998). Another hemipteran nitrophorin was purified from the salivary glands of *Cimex lectularius*; however, its histamine binding has not yet been tested (Valenzuela and Ribeiro, 1998). Other lipocalins were described in the salivary transcriptome of *T. brasiliensis* (Santos et al., 2007), *T. infestans* (Assumpcao et al., 2008), and *Anopheles gambiae* (Neira Oviedo et al., 2009); however, their putative histamine-binding function remains to be elucidated.

Insects seem to have evolved another efficient scavenger of small molecules, including biogenic amines. The D7 protein family, abundantly expressed in blood-feeding Diptera, is a distant member of the odorant-binding protein superfamily. To date, four short D7 proteins from *An. gambiae* (D7r1, D7r2, D7r3, and D7r4) and one long D7 from *Ae. aegypti* (D7Long) have presented histamine-binding properties (Calvo et al., 2006). Thus, it appears that counteraction of histamine effects has a strong adaptive value for ticks, bugs, and mosquitoes, which co-opted for either members of the lipocalin or the odorant-binding protein families.

Chemokine Inhibitors

Cellular migration to peripheral tissues such as the skin where the ectoparasites take their blood meal is mediated by a superfamily of small chemoattractant cytokines called chemokines. Leukocyte trafficking and positioning are directly regulated by these molecules through a process called haptotaxis, in which chemokine gradients created in the extracellular matrix allow these cells to reach their destination in peripheral tissues (Thelen and Stein, 2008). Production of secreted proteins that bind and neutralize chemokine activity has now been elucidated in bloodsucking arthropods. The presence of anti-CXCL8 (formerly IL-8) activity has been demonstrated in salivary gland extracts from *Dermacentor reticulatus*, *Amblyomma variegatum*, *R. appendiculatus*, *Haemaphysalis inermis*, and *I. ricinus* tick species (Hajnicka et al., 2001; Kocakova et al., 2003). In a later study, it was demonstrated that salivary gland extracts from *D. reticulatus, A. variegatum*, and *I. ricinus* also have antichemokine activity to CCL2 (MCP-1), CCL3 (MIP-1α), CCL5 (RANTES), and CCL11 (eotaxin) (Hajnicka et al., 2005). Evasin-1 was the first protein from bloodsucking ectoparasites with chemokine-binding properties to be cloned and characterized (Frauenschuh et al., 2007). This 94-amino-acid protein from the brown dog tick *R. sanguineus* binds specifically to the CC chemokines CCL3, CCL4 (MIP-1α), and CCL18 (MIP-4). It has been demonstrated that Evasin-1 belongs to a family with three other chemokine-binding proteins presenting stringent selectivity: Evasin-3, with 66 amino acids, binds to CXCL1 (GROα) and CXCL8; Evasin-4, with 110 amino acids, binds to CCL5 and CCL11; and Evasin-2, whose ligands remain to be elucidated (Deruaz et al., 2008).

It has been also demonstrated that saliva of *R. sanguineus* is able to inhibit cell migration by decreasing expression of CCR5 chemokine receptor in dendritic cells (Oliveira et al., 2008). Nevertheless, the molecule(s) responsible for these effects remains to be identified and characterized.

An additional mechanism employed by bloodsucking ectoparasites to inhibit cellular migration is the production of protein homologs. A mammal homolog transcript of the macrophage migration inhibitory factor (MIF) has been identified in a cDNA library from the midgut of fed *Amblyomma americanum* tick, and its expression was later confirmed in salivary glands as well. This was the first human cytokine-like protein identified in an arthropod, and the recombinant protein was confirmed to inhibit migration of macrophages in vitro to the same extent as the human MIF (Jaworski et al., 2001). Another MIF transcript homolog from the *Haemaphysalis longicornis* tick (HlMIF) has been identified in a midgut cDNA library through the same strategy. HlMIF expression has also been confirmed in salivary glands of this tick species, and the recombinant protein displayed similar MIF activity in vitro, (Umemiya et al., 2007).

Immunoglobulin-Binding Proteins

Immunoglobulins are the best characterized group of serum glycoproteins, detected for the first time around 120 years ago (von Behring and Kitasato, 1890). There are five known immunoglobulin classes or isotypes in mammals – IgA, IgD, IgE, IgG, and IgM – all produced and secreted by B lymphocytes. Immunoglobulins act as antibodies in the humoral immune response by binding to specific antigens, but they are also recognized by receptors of a variety of effector cells and systems (Davies and Metzger, 1983). Thus, immunoglobulins can contribute to host immunity in many ways: by neutralizing antigens, promoting opsonization and phagocytosis or antibody-dependent cell-mediated cytotoxicity (ADCC), and/or activating the classical pathway of the complement system (Casadevall and Scharff, 1994; Davies and Metzger, 1983). It has been described that immunoglobulins are ingested during the bloodmeal, invading many parts of the body of blood-feeding arthropods, and can be found on the gut, hemolymph, salivary glands, and ovary (Ackerman et al., 1981; Lackie and Gavin, 1989; Wang and Nuttall, 1994), retaining their immunologic properties (Fujisaki et al., 1984; Sauer et al., 1994). One strategy used by arthropods to counteract the presence of these possibly deleterious immunoglobulins is production of proteins that bind on their structure. Many IgG-binding proteins with molecular weight varying from 14.5 to 54.0 kDa have been isolated from salivary gland extract of the ixodid ticks *R. appendiculatus*, *A. variegatum*, *Ixodes hexagonus*, *I. ricinus* and *I. scapularis* (Packila and Guilfoile, 2002; Wang and Nuttall, 1994, 1995a, b). Moreover, three specific IgG-binding from *R. appendiculatus* ticks have already been identified (GenBank accession numbers IGBP-MA [AF001868]; IGBP-MB [AF001869], and IGBP-MC [AF001870]) (Wang et al., 1998). In vivo importance of IGBP-MC in enhancing fecundity was evidenced by experiments showing that female *R. appendiculatus* ticks co-fed with anti-IGBP-MC serum-injected males, but not nonimmune serum-injected ones, presented decreased body weight (Wang et al., 1998). In addition, a paramyosin from the *Rhipicephalus (Boophilus) microplus* tick (BmPRM), present in all tissues and developmental stages tested, was described; the recombinant protein expressed showed IgG-binding properties. Its biologic role in host evasion, however, remains to be determined (Ferreira et al., 2002).

Modulators of Host Cell Activation and Function

As discussed in the beginning of this chapter, different types of immune cells can mediate host defense against ectoparasites. For example, macrophages, dendritic cells, and neutrophils are key mediators of the early innate immune response by taking up antigens (such as those present in saliva from bloodsucking ectoparasites) but also producing several key cytokines following recognition of pathogen-associated molecular patterns (PAMPs), for example lipopolissacharide (LPS), through sensor molecules known as pattern recognition receptors (PRRs), such as Toll-like receptors (TLRs) (Janeway and Medzhitov, 2002). Mast cells, basophils, and eosinophils contribute to a variety of allergic and other immune responses. In fact, they can be relatively abundant at skin bite sites, where their activation and release of preformed and newly-generated mediators occur as a result of physical damage, presence of ectoparasite mouthparts, and saliva inoculation. T and B lymphocytes are the major components of the cellular and humoral adaptive immunity, and the host long-lasting specific immune response is completely dependent on these cells. Sialogenins and nonprotein modulators of these cells have been described as follows and are summarized in Table 9.1.

Nonprotein Mediators

Regarding nonpeptidic molecules, the presence of prostaglandins (PG) was long demonstrated in tick saliva (Dickinson et al., 1976). PGE_2 was the first lipid mediator to be identified directly from saliva of *R. microplus* (Higgs et al., 1976) and implicated with the antiinflammatory and immunossupressive activities of *I. scapularis* saliva (Ribeiro et al., 1985). Since then, PGE_2 and other prostaglandins such as $PGF_{2\alpha}$ and prostacyclin (PGI_2) have been found in the saliva or salivary glands of several other tick species (Aljamali et al., 2002; Inokuma et al., 1994; Ribeiro et al., 1988). Concerning host immunity, PGE_2 isolated from *I. scapularis* saliva was demonstrated as being the major inhibitor of dendritic cell maturation and diminished antigen-specific T cell proliferation triggered by these cells (Sá-Nunes et al., 2007). PGI_2 is a potent antiplatelet – aggregating that may help tick feeding by preventing host hemostatic and antiinflammatory reactions, by increasing host blood flow at the tick feeding site, or by preventing leukocyte degranulation. In addition, PGI_2 analogs decreased production of proinflammatory cytokines and chemokines, increased production of the antiinflammatory cytokine IL-10 by dendritic cells, and presented an immunossupressive effect on T cells (Zhou et al., 2007a, b). Finally, $PGF_{2\alpha}$ is a vasodilator, but no effect on host cell activation has been described thus far. The presence of prostaglandins and other eicosanoids has been detected in several insect species, as well (Stanley, 2006). Particularly, PGE_2 was found in Malpighian tubules from adult females of *Ae. aegypti* by immunohistochemistry (Petzel et al., 1993) and in several organs of *Anopheles albimanus* by enzyme immunoassay (Garcia Gil de Munoz et al., 2008); however, PGE_2 has not been described in saliva of hematophagous insects to date, and its function in this group seems to be related to physiologic regulation and immune responses in the midgut and fat body rather than host immunomodulation.

Dopamine has been identified in salivary glands of *R. microplus* (Binnington and Stone, 1977; Megaw and Robertson, 1974) and *Amblyomma hebraeum* (Kaufman et al., 1999) ticks and it is likely to be the neurotransmitter at the neuroeffector junction controlling saliva secretion (Sauer et al., 1995). It is also universally found in different tissues of insects, especially in the central nervous system (Osborne, 1996). Although recent studies have demonstrated that dopamine is able to modulate the function of dendritic cells and T cells (Pacheco et al., 2009), the amount of dopamine secreted in saliva of bloodsucking arthropods has not yet been determined, and its influence in host immunity, if any, remains to be evaluated.

The endocannabinoids and/or related congeners, *N*-arachidonoylethanolamine, 2-arachidonoylglycerol, and *N*-palmitoylethanolamine, have been described in salivary glands of the Lone Star tick *A. americanum* (Fezza et al., 2003). The mechanism of action of such molecules suggests their role as analgesic and anti-inflammatory. In fact, the regulatory function of cannabinoids on mast cell function has been established (Samson et al., 2003; Vannacci et al., 2003), suggesting similar activities in tick saliva.

Adenosine and adenosine monophosphate (AMP) have been also found in the salivary glands of *Phlebotomus papatasi* and *Phlebotomus argentipes* (Ribeiro et al., 1999; Ribeiro and Modi, 2001). The critical role of adenosine in modulating the immune response has been widely discussed in recent years. This molecule modulates a variety of immunologic functions such as monocyte, neutrophil, and dendritic cell activation and function, adhesion, and cell chemotaxis (Bours et al., 2006; Hasko and Cronstein, 2004). In fact, salivary gland extract of sand fly *P. papatasi* inhibits in vivo neutrophil migration and production of cytokines, chemokines, and lipid mediators involved in neutrophil recruitment (Carregaro et al., 2008); however, salivary gland extracts from *Phlebotomus duboscqi* and *Lu. longipalpis*, which lack adenosine or AMP activities but rather present an adenosine deaminase, also present similar effects on neutrophil migration (Carregaro et al., 2008; Monteiro et al., 2005). Whether adenosine and/or other molecules are responsible for these activities is still unknown.

Sialogenins

It has been demonstrated that saliva of the *I. scapularis* tick is able to inhibit neutrophil functions (Ribeiro et al., 1990) such as spreading, expression of α2-integrins, and bacterial killing (Montgomery et al., 2004). Two salivary proteins from the ixostatin family, ISL 929 and ISL 1373, were able to reproduce most effects of saliva on neutrophil functions (Guo et al., 2009). Sialostatin L (SialoL), a cysteine protease inhibitor from *I. scapularis* that targets a set of cathepsins, also inhibited in vivo neutrophil migration (Kotsyfakis et al., 2006). In addition, SialoL was demonstrated to down-modulate dendritic cell inflammatory cytokine production, expression of costimulatory molecules, and antigen-dependent T cell proliferation by a cathepsin S-dependent mechanism. Moreover, SialoL administration during the immunization phase of experimental autoimmune encephalomyelitis in mice significantly prevented disease symptoms, which were associated with impaired IFN-γ and IL-17 production and specific T cell proliferation (Sá-Nunes et al., 2009).

Salp15 is another molecule described for *I. scapularis* with multifunctional activities on the host immune system. This protein inhibits activation of T cells through interaction with coreceptor CD4 (Anguita et al., 2002; Garg et al., 2006). It has also been shown to bind DC-SIGN from dendritic cells and, therefore, prevented cytokine secretion by these cells (Hovius et al., 2008). Given these mechanisms of action, Salp15 was successfully employed to prevent development of experimental asthma in mice (Paveglio et al., 2007) and also blocked HIV-1 gp120-CD4 interaction in vitro (Juncadella et al., 2008).

Salivary gland fractions from the tick *Dermacentor andersoni* containing soluble proteins suppressed murine splenocyte proliferation to mitogen concanavalin A in vitro (Bergman et al., 1995). A 36.0-kDa protein termed Da-p36, responsible for this immunossupressant activity, was isolated, characterized, and cloned by the same group (Bergman et al., 1998, 2000).

A protein named B cell inhibitory factor (BIF) has been purified and characterized from the salivary glands of the hard tick, *Hyalomma asiaticum asiaticum*. The cDNA encoding BIF was cloned, and a 13.0-kDa recombinant protein was able to inhibit lipopolysaccharide-induced B cell proliferation (Yu et al., 2006). An 18.0-kDa protein with similar BIF activity was also found in *I. ricinus* saliva, but only a partial purification was performed and its identity remains elusive (Hannier et al., 2004).

Serpins are a superfamily of proteins described in several kingdoms with similar structure but functionally diverse activities including, but not limited to, inhibition of serine proteases (Irving et al., 2000). In vertebrates, serpin activities are related to blood coagulation and fibrinolisis, nevertheless, they can also modulate diverse immunological processes such as complement activation, cell apoptosis, and inflammation (Gettins, 2002). Serpins have been demonstrated in saliva of several mosquito species, initially as modulators of mammalian hemostasis, such as the anticoagulant-factor Xa (AFXa) from *Ae. aegypti* (Stark and James, 1998); more recently, they have been linked to immunity and protection, as well. In fact, *I. ricinus* salivary glands express a immunosuppressive serpin (IRIS) that modulates cell responses through inhibition of interferon-γ production by lymphocytes and inhibition of IL-6 and TNF-α production by human macrophages (Leboulle et al., 2002). In addition to this effect, IRIS was able to inhibit several serine proteases involved in the coagulation cascade and fibrinolysis (Prevot et al., 2006).

Cytokines and lipid mediators play important regulatory roles in cell immune responses. One indirect mechanism employed by hematophagous species to modulate cell migration, activation and function, is the blocking of such activities. A few studies have demonstrated the presence of TNF-α, IL-2, and IL-4 binding proteins on tick saliva (Gillespie et al., 2001; Hajnicka et al., 2005; Konik et al., 2006; Peterkova et al., 2008), but failed to demonstrate the complete identity of such molecules. In addition, lipocalins from the soft ticks *O. moubata* (moubatin) and *O. savignyi* (TSGP2 and TSGP3), originally involved in other immune-related functions, were also described as having leukotriene B_4 binding properties (Mans and Ribeiro, 2008). Similarly, an *Ae. aegypti* D7 protein (AeD7) that presents a histamine-binding property in its C-terminal domain, has also been demonstrated to have leukotriene-binding activity in its N-terminal domain (Calvo et al., 2009). The identification of these molecules will help to clarify the process of successful ectoparasite infestation and may reveal new therapeutic targets to combat inflammatory diseases.

Several peptides have been identified in saliva of hematophagous arthropods, most of them possessing anticoagulant activities (Koh and Kini, 2009), but only a few studies investigated their potential role in host immunity. Currently, at least for horsefly species, some immunoregulatory peptides have been identified and characterized from salivary glands. A 30-amino acid peptide from *Hybomitra atriperoides* salivary glands termed immunoregulin HA was able to inhibit secretion of IFN-γ and monocyte chemoattractant protein-1 and to increase the secretion of IL-10 induced by LPS in rat splenocytes (Yan et al., 2008). Similar activity was found for a 30 residue peptide identified in the salivary glands of *Tabanus pleskei* and named immunoregulin TP1 (Zhao et al., 2009). In addition, leech immunocytes are able to produce mammalian-like proopiomelanocortin-derived peptides, such as the case of adrenocorticotropic hormone (ACTH) and α-melanostimulating hormone (α-MSH) (Salzet et al., 1997). Proopiomelanocortin is a prohormone expressed in pituitary glands and various non-pituitary organs including the skin (Millington, 2006). Both α-MSH and ACTH suppress production of IFN-γ in the skin under cutaneous stress (Brummitt et al., 1988; Slominski et al., 2000). Moreover, α-MSH can function as an antagonist of IL-1 and induces production of many immunosuppressive cytokines while suppressing the expression of accessory molecules CD40 and CD86 on dendritic cells (Becher et al., 1999). These molecules have not been described in leech secretions yet, but are present in their hemolymph, suggesting potential immunoregulatory actions on the feeding site (Salzet et al., 2000). Further demonstrations of immunomodulatory peptides in other hematophagous species is likely in the near future.

Concluding Remarks

In a review almost 30 years ago, Dr. Stephen K. Wikel noted how rapidly knowledge was growing in immunology, especially in the immunoparasitology field. He correctly pointed out that this increase was mainly due to studies focused on protozoa and helminthes and that little attention was paid to ectoparasites (Wikel, 1982). Many years passed before immunologists realized the importance of biological roles of saliva from bloodsucking organisms both on the vertebrate immune system and in host-ectoparasite interactions. Advances in biochemical techniques for protein isolation, purification, and characterization associated with modern molecular biology tools made possible genomes sequencing and development of sialome catalogs. These revolutionary innovations are allowing us to deeply understand and study the effect of individual salivary components, bringing new opportunities of scientific exploration. Indeed, an increasing number of groups are now studying host-ectoparasite interactions from this perspective. This chapter attempted to present one small constellation of this new universe. In the future, we expect that this salivary "immunome" will be helpful as a valuable data source for development of immunomodulators and immunopharmacologic tools, the discovery of molecules that facilitate pathogen infectivity and, perhaps, new vaccine candidates against such parasites and the diseases transmitted by them.

Acknowledgements This work was supported by the Intramural Research Program of the Division of Intramural Research, National Institute of Allergy and Infectious Diseases, National Institutes of Health and by grants from Coordenação de Aperfeiçoamento de Pessoal de Nível Superior (CAPES), Conselho Nacional de Desenvolvimento Científico e Tecnológico (CNPq) and Fundação de Amparo à Pesquisa do Estado de São Paulo (FAPESP). We thank NIAID intramural editor Brenda Rae Marshall for assistance.

References

Ackerman, S., Clare, F.B., McGill, T.W., Sonenshine, D.E., 1981. Passage of host serum components, including antibody, across the digestive tract of *Dermacentor variabilis* (Say). J. Parasitol. 67, 737–740.

Aljamali, M., Bowman, A.S., Dillwith, J.W., Tucker, J.S., Yates, G.W., Essenberg, R.C., Sauer, J.R., 2002. Identity and synthesis of prostaglandins in the lone star tick, *Amblyomma americanum* (L.), as assessed by radio-immunoassay and gas chromatography/mass spectrometry. Insect Biochem. Mol. Biol. 32, 331–341.

Andersen, J.F., Ding, X.D., Balfour, C., Shokhireva, T.K., Champagne, D.E., Walker, F.A., Montfort, W.R., 2000. Kinetics and equilibria in ligand binding by nitrophorins 1–4: evidence for stabilization of a nitric oxide-ferriheme complex through a ligand-induced conformational trap. Biochemistry 39, 10118–10131.

Andersen, J.F., Hinnebusch, B.J., Lucas, D.A., Conrads, T.P., Veenstra, T.D., Pham, V.M., Ribeiro, J.M., 2007. An insight into the sialome of the oriental rat flea, *Xenopsylla cheopis* (Rots). BMC Genomics 8, 102.

Andersen, J.F., Pham, V.M., Meng, Z., Champagne, D.E., Ribeiro, J.M., 2009. Insight into the sialome of the Black Fly, *Simulium vittatum*. J. Proteome Res. 8, 1474–1488.

Anderson, J.M., Oliveira, F., Kamhawi, S., Mans, B.J., Reynoso, D., Seitz, A.E., Lawyer, P., Garfield, M., Pham, M., Valenzuela, J.G., 2006. Comparative salivary gland transcriptomics of sandfly vectors of visceral leishmaniasis. BMC Genomics 7, 52.

Anguita, J., Ramamoorthi, N., Hovius, J.W., Das, S., Thomas, V., Persinski, R., Conze, D., Askenase, P.W., Rincon, M., Kantor, F.S., Fikrig, E., 2002. Salp15, an *Ixodes scapularis* salivary protein, inhibits CD4(+) T cell activation. Immunity 16, 849–859.

Arca, B., Lombardo, F., Francischetti, I.M., Pham, V.M., Mestres-Simon, M., Andersen, J.F., Ribeiro, J.M., 2007. An insight into the sialome of the adult female mosquito *Aedes albopictus*. Insect Biochem. Mol. Biol. 37, 107–127.

Assumpcao, T.C., Francischetti, I.M., Andersen, J.F., Schwarz, A., Santana, J.M., Ribeiro, J.M., 2008. An insight into the sialome of the blood-sucking bug *Triatoma infestans*, a vector of Chagas' disease. Insect Biochem. Mol. Biol. 38, 213–232.

Barros, V.C., Assumpcao, J.G., Cadete, A.M., Santos, V.C., Cavalcante, R.R., Araujo, R.N., Pereira, M.H., Gontijo, N.F., 2009. The role of salivary and intestinal complement system inhibitors in the midgut protection of triatomines and mosquitoes. PLoS One 4, e6047.

Baskova, I.P., Zavalova, L.L., 2001. Proteinase inhibitors from the medicinal leech *Hirudo medicinalis*. Biochemistry (Moscow) 66, 703–714.

Beaufays, J., Adam, B., Decrem, Y., Prevot, P.P., Santini, S., Brasseur, R., Brossard, M., Lins, L., Vanhamme, L., Godfroid, E., 2008. *Ixodes ricinus* tick lipocalins: identification, cloning, phylogenetic analysis and biochemical characterization. PLoS One 3, e3941.

Becher, E., Mahnke, K., Brzoska, T., Kalden, D.H., Grabbe, S., Luger, T.A., 1999. Human peripheral blood-derived dendritic cells express functional melanocortin receptor MC-1R. Ann. N. Y. Acad. Sci. U.S.A. 885, 188–195.

Berenberg, J.L., Ward, P.A., Sonenshine, D.E., 1972. Tick-bite injury: mediation by a complement-derived chemotictic chemotactic factor. J. Immunol. 109, 451–456.

Bergman, D.K., Palmer, M.J., Caimano, M.J., Radolf, J.D., Wikel, S.K., 2000. Isolation and molecular cloning of a secreted immunosuppressant protein from *Dermacentor andersoni* salivary gland. J. Parasitol. 86, 516–525.

Bergman, D.K., Ramachandra, R.N., Wikel, S.K., 1995. *Dermacentor andersoni*: salivary gland proteins suppressing T-lymphocyte responses to concanavalin A in vitro. Exp. Parasitol. 81, 262–271.

Bergman, D.K., Ramachandra, R.N., Wikel, S.K., 1998. Characterization of an immunosuppressant protein from *Dermacentor andersoni* (Acari: Ixodidae) salivary glands. J. Med. Entomol. 35, 505–509.

Binnington, K.C., Stone, B.F., 1977. Distribution of catecholamines in the cattle tick *Boophilus microplus*. Comp. Biochem. Physiol. C. 58, 21–28.

Bjork, J., Hugli, T.E., Smedegard, G., 1985. Microvascular effects of anaphylatoxins C3a and C5a. J. Immunol. 134, 1115–1119.

Bluestone, J.A., Mackay, C.R., O'Shea, J.J., Stockinger, B., 2009. The functional plasticity of T cell subsets. Nat. Rcv. Immunol. 9, 811–816.

Bordet, J., 1895. Les leucocytes et les propriétés actives de serum chez les vaccinés. Ann. Inst. Pasteur 9, 462–506.

Bours, M.J., Swennen, E.L., Di Virgilio, F., Cronstein, B.N., Dagnelie, P.C., 2006. Adenosine 5′-triphosphate and adenosine as endogenous signaling molecules in immunity and inflammation. Pharmacol. Ther. 112, 358–404.

Brummitt, C.F., Sharp, B.M., Gekker, G., Keane, W.F., Peterson, P.K., 1988. Modulatory effects of beta-endorphin on interferon-gamma production by cultured peripheral blood mononuclear cells: heterogeneity among donors and the influence of culture medium. Brain Behav. Immun. 2, 187–197.

Calvo, E., Dao, A., Pham, V.M., Ribeiro, J.M., 2007. An insight into the sialome of *Anopheles funestus* reveals an emerging pattern in anopheline salivary protein families. Insect Biochem. Mol. Biol. 37, 164–175.

Calvo, E., Mans, B.J., Andersen, J.F., Ribeiro, J.M., 2006. Function and evolution of a mosquito salivary protein family. J. Biol. Chem. 281, 1935–1942.

Calvo, E., Mans, B.J., Ribeiro, J.M., Andersen, J.F., 2009. Multifunctionality and mechanism of ligand binding in a mosquito antiinflammatory protein. Proc. Natl. Acad. Sci. U.S.A. 106, 3728–3733.

Carregaro, V., Valenzuela, J.G., Cunha, T.M., Verri, W.A., Jr., Grespan, R., Matsumura, G., Ribeiro, J.M., Elnaiem, D.E., Silva, J.S., Cunha, F.Q., 2008. Phlebotomine salivas inhibit immune inflammation-induced neutrophil migration via an autocrine DC-derived PGE2/IL-10 sequential pathway. J. Leukoc. Biol. 84, 104–114.

Casadevall, A., Scharff, M.D., 1994. Serum therapy revisited: animal models of infection and development of passive antibody therapy. Antimicrob. Agents Chemother. 38, 1695–1702.

Cavalcante, R.R., Pereira, M.H., Gontijo, N.F., 2003. Anti-complement activity in the saliva of phlebotomine sand flies and other haematophagous insects. Parasitology 127, 87–93.

Chinery, W.A., Ayitey-Smith, E., 1977. Histamine blocking agent in the salivary gland homogenate of the tick *Rhipicephalus sanguineus sanguineus*. Nature 265, 366–367.

Chmelar, J., Anderson, J.M., Mu, J., Jochim, R.C., Valenzuela, J.G., Kopecky, J., 2008. Insight into the sialome of the castor bean tick, *Ixodes ricinus*. BMC Genomics 9, 233.

Couvreur, B., Beaufays, J., Charon, C., Lahaye, K., Gensale, F., Denis, V., Charloteaux, B., Decrem, Y., Prevot, P.P., Brossard, M., Vanhamme, L., Godfroid, E., 2008. Variability and action mechanism of a family of anticomplement proteins in *Ixodes ricinus*. PLoS One 3, e1400.

Davies, D.R., Metzger, H., 1983. Structural basis of antibody function. Annu. Rev. Immunol. 1, 87–117.

Deruaz, M., Frauenschuh, A., Alessandri, A.L., Dias, J.M., Coelho, F.M., Russo, R.C., Ferreira, B.R., Graham, G.J., Shaw, J.P., Wells, T.N., Teixeira, M.M., Power, C.A., Proudfoot, A.E., 2008. Ticks produce highly selective chemokine binding proteins with antiinflammatory activity. J. Exp. Med. 205, 2019–2031.

Dickinson, R.G., O'Hagan, J.E., Schotz, M., Binnington, K.C., Hegarty, M.P., 1976. Prostaglandin in the saliva of the cattle tick *Boophilus microplus*. Aust. J. Exp. Biol. Med. Sci. 54, 475–486.

Fazilleau, N., McHeyzer-Williams, L.J., McHeyzer-Williams, M.G., 2007. Local development of effector and memory T helper cells. Curr. Opin. Immunol. 19, 259–267.

Ferreira, C.A., Barbosa, M.C., Silveira, T.C., Valenzuela, J.G., Vaz Ida, S., Jr., Masuda, A., 2002. cDNA cloning, expression and characterization of a *Boophilus microplus* paramyosin. Parasitology 125, 265–274.

Fezza, F., Dillwith, J.W., Bisogno, T., Tucker, J.S., Di Marzo, V., Sauer, J.R., 2003. Endocannabinoids and related fatty acid amides, and their regulation, in the salivary glands of the lone star tick. Biochim. Biophys. Acta 1633, 61–67.

Francischetti, I.M., Mans, B.J., Meng, Z., Gudderra, N., Veenstra, T.D., Pham, V.M., Ribeiro, J.M., 2008. An insight into the sialome of the soft tick, *Ornithodorus parkeri*. Insect Biochem. Mol. Biol. 38, 1–21.

Francischetti, I.M., Sá-Nunes, A., Mans, B.J., Santos, I.M., Ribeiro, J.M., 2009. The role of saliva in tick feeding. Front. Biosci. 14, 2051–2088.

Francischetti, I.M., Valenzuela, J.G., Pham, V.M., Garfield, M.K., Ribeiro, J.M., 2002. Toward a catalog for the transcripts and proteins (sialome) from the salivary gland of the malaria vector *Anopheles gambiae*. J. Exp. Biol. 205, 2429–2451.

Frauenschuh, A., Power, C.A., Deruaz, M., Ferreira, B.R., Silva, J.S., Teixeira, M.M., Dias, J.M., Martin, T., Wells, T.N., Proudfoot, A.E., 2007. Molecular cloning and characterization of a highly selective chemokine-binding protein from the tick *Rhipicephalus sanguineus*. J. Biol. Chem. 282, 27250–27258.

Fujisaki, K., Kamio, T., Kitaoka, S., 1984. Passage of host serum components, including antibodies specific for *Theileria sergenti*, across the digestive tract of argasid and ixodid ticks. Ann. Trop. Med. Parasitol. 78, 449–450.

Garcia Gil de Munoz, F.L., Martinez-Barnetche, J., Lanz-Mendoza, H., Rodriguez, M.H., Hernandez-Hernandez, F.C., 2008. Prostaglandin E2 modulates the expression of antimicrobial peptides in the fat body and midgut of *Anopheles albimanus*. Arch. Insect Biochem. Physiol. 68, 14–25.

Garg, R., Juncadella, I.J., Ramamoorthi, N., Ashish, Ananthanarayanan, S.K., Thomas, V., Rincon, M., Krueger, J.K., Fikrig, E., Yengo, C.M., Anguita, J., 2006. Cutting edge: CD4 is the receptor for the tick saliva immunosuppressor, Salp15. J. Immunol. 177, 6579–6583.

Gasque, P., 2004. Complement: a unique innate immune sensor for danger signals. Mol. Immunol. 41, 1089–1098.

Gettins, P.G., 2002. Serpin structure, mechanism, and function. Chem. Rev. 102, 4751–4804.

Gillespie, R.D., Dolan, M.C., Piesman, J., Titus, R.G., 2001. Identification of an IL-2 binding protein in the saliva of the Lyme disease vector tick, *Ixodes scapularis*. J. Immunol. 166, 4319–4326.

Guo, R.F., Ward, P.A., 2005. Role of C5a in inflammatory responses. Annu. Rev. Immunol. 23, 821–852.

Guo, X., Booth, C.J., Paley, M.A., Wang, X., DePonte, K., Fikrig, E., Narasimhan, S., Montgomery, R.R., 2009. Inhibition of neutrophil function by two tick salivary proteins. Infect. Immun. 77, 2320–2329.

Hajnicka, V., Kocakova, P., Slavikova, M., Slovak, M., Gasperik, J., Fuchsberger, N., Nuttall, P.A., 2001. Anti-interleukin-8 activity of tick salivary gland extracts. Parasite Immunol. 23, 483–489.

Hajnicka, V., Vancova, I., Kocakova, P., Slovak, M., Gasperik, J., Slavikova, M., Hails, R.S., Labuda, M., Nuttall, P.A., 2005. Manipulation of host cytokine network by ticks: a potential gateway for pathogen transmission. Parasitology 130, 333–342.

Hannier, S., Liversidge, J., Sternberg, J.M., Bowman, A.S., 2004. Characterization of the B-cell inhibitory protein factor in Ixodes ricinus tick saliva: a potential role in enhanced *Borrelia burgdoferi* transmission. Immunology 113, 401–408.

Hasko, G., Cronstein, B.N., 2004. Adenosine: an endogenous regulator of innate immunity. Trends Immunol. 25, 33–39.

Higgs, G.A., Vane, J.R., Hart, R.J., Porter, C., Wilson, R.G., 1976. Prostaglandins in saliva of the cattle tick *Boophilus microplus* (Canestrini) (Acarina. Ixodidae). Bull. Entomol. Res. 66, 665–670.

Hovius, J.W., de Jong, M.A., den Dunnen, J., Litjens, M., Fikrig, E., van der Poll, T., Gringhuis, S.I., Geijtenbeek, T.B., 2008. Salp15 binding to DC-SIGN inhibits cytokine expression by impairing both nucleosome remodeling and mRNA stabilization. PLoS Pathog. 4, e31.

Huber-Lang, M., Sarma, J.V., Zetoune, F.S., Rittirsch, D., Neff, T.A., McGuire, S.R., Lambris, J.D., Warner, R.L., Flierl, M.A., Hoesel, L.M., Gebhard, F., Younger, J.G., Drouin, S.M., Wetsel, R.A., Ward, P.A., 2006. Generation of C5a in the absence of C3: a new complement activation pathway. Nat. Med. 12, 682–687.

Inokuma, H., Kemp, D.H., Willadsen, P., 1994. Comparison of prostaglandin E2 (PGE2) in salivary gland of *Boophilus microplus*, *Haemaphysalis longicornis* and *Ixodes holocyclus*, and quantification of PGE2 in saliva, hemolymph, ovary and gut of *B. microplus*. J. Vet. Med. Sci. 56, 1217–1218.

Ip, W.K., Takahashi, K., Ezekowitz, R.A., Stuart, L.M., 2009. Mannose-binding lectin and innate immunity. Immunol. Rev. 230, 9–21.

Irving, J.A., Pike, R.N., Lesk, A.M., Whisstock, J.C., 2000. Phylogeny of the serpin superfamily: implications of patterns of amino acid conservation for structure and function. Genome Res. 10, 1845–1864.

Janeway, C.A., Jr., Medzhitov, R., 2002. Innate immune recognition. Annu. Rev. Immunol. 20, 197–216.

Jaworski, D.C., Jasinskas, A., Metz, C.N., Bucala, R., Barbour, A.G., 2001. Identification and characterization of a homologue of the pro-inflammatory cytokine macrophage migration inhibitory factor in the tick, *Amblyomma americanum*. Insect Mol. Biol. 10, 323–331.

Juncadella, I.J., Garg, R., Bates, T.C., Olivera, E.R., Anguita, J., 2008. The *Ixodes scapularis* salivary protein, salp15, prevents the association of HIV-1 gp120 and CD4. Biochem. Biophys. Res. Commun. 367, 41–46.

Kaufman, W.R., Sloley, B.D., Tatchell, R.J., Zbitnew, G.L., Diefenbach, T.J., Goldberg, J.I., 1999. Quantification and cellular localization of dopamine in the salivary gland of the ixodid tick *Amblyomma hebraeum*. Exp. Appl. Acarol. 23, 251–265.

Kocakova, P., Slavikova, M., Hajnicka, V., Slovak, M., Gasperik, J., Vancova, I., Fuchsberger, N., Nuttall, P.A., 2003. Effect of fast protein liquid chromatography fractionated salivary gland extracts from different ixodid tick species on interleukin-8 binding to its cell receptors. Folia Parasitol. (Praha) 50, 79–84.

Koh, C.Y., Kini, R.M., 2009. Molecular diversity of anticoagulants from haematophagous animals. Thromb. Haemost. 102, 437–453.

Konik, P., Slavikova, V., Salat, J., Reznickova, J., Dvoroznakova, E., Kopecky, J., 2006. Antitumour necrosis factor-alpha activity in *Ixodes ricinus* saliva. Parasite Immunol. 28, 649–656.

Kotsyfakis, M., Sá-Nunes, A., Francischetti, I.M., Mather, T.N., Andersen, J.F., Ribeiro, J.M., 2006. Antiinflammatory and immunosuppressive activity of sialostatin L, a salivary cystatin from the tick *Ixodes scapularis*. J. Biol. Chem. 281, 26298–26307.

Lackie, A.M., Gavin, S., 1989. Uptake and persistence of ingested antibody in the mosquito *Anopheles stephensi*. Med. Vet. Entomol. 3, 225–230.

Lanzaro, G.C., Lopes, A.H., Ribeiro, J.M., Shoemaker, C.B., Warburg, A., Soares, M., Titus, R.G., 1999. Variation in the salivary peptide, maxadilan, from species in the *Lutzomyia longipalpis* complex. Insect Mol. Biol. 8, 267–275.

Leboulle, G., Crippa, M., Decrem, Y., Mejri, N., Brossard, M., Bollen, A., Godfroid, E., 2002. Characterization of a novel salivary immunosuppressive protein from *Ixodes ricinus* ticks. J. Biol. Chem. 277, 10083–10089.

Mans, B.J., 2005. Tick histamine-binding proteins and related lipocalins: potential as therapeutic agents. Curr. Opin. Investig. Drugs 6, 1131–1135.

Mans, B.J., Ribeiro, J.M., 2008. Function, mechanism and evolution of the moubatin-clade of soft tick lipocalins. Insect Biochem. Mol. Biol. 38, 841–852.

Mans, B.J., Ribeiro, J.M., Andersen, J.F., 2008. Structure, function, and evolution of biogenic amine-binding proteins in soft ticks. J. Biol. Chem. 283, 18721–18733.

McHeyzer-Williams, L.J., McHeyzer-Williams, M.G., 2005. Antigen-specific memory B cell development. Annu. Rev. Immunol. 23, 487–513.

Megaw, M.W., Robertson, H.A., 1974. Dopamine and noradrenaline in the salivary glands and brain of the tick, *Boophilus microplus*: effect of reserpine. Experientia 30, 1261–1262.

Millington, G.W., 2006. Proopiomelanocortin (POMC): the cutaneous roles of its melanocortin products and receptors. Clin. Exp. Dermatol. 31, 407–412.

Monteiro, M.C., Nogueira, L.G., Almeida Souza, A.A., Ribeiro, J.M., Silva, J.S., Cunha, F.Q., 2005. Effect of salivary gland extract of Leishmania vector, *Lutzomyia longipalpis*, on leukocyte migration in OVA-induced immune peritonitis. Eur. J. Immunol. 35, 2424–2433.

Montgomery, R.R., Lusitani, D., De Boisfleury Chevance, A., Malawista, S.E., 2004. Tick saliva reduces adherence and area of human neutrophils. Infect. Immun. 72, 2989–2994.

Neira Oviedo, M., Ribeiro, J.M., Heyland, A., VanEkeris, L., Moroz, T., Linser, P.J., 2009. The salivary transcriptome of *Anopheles gambiae* (Diptera: Culicidae) larvae: a microarray-based analysis. Insect Biochem. Mol. Biol. 39, 382–394.

Nunn, M.A., Sharma, A., Paesen, G.C., Adamson, S., Lissina, O., Willis, A.C., Nuttall, P.A., 2005. Complement inhibitor of C5 activation from the soft tick *Ornithodoros moubata*. J. Immunol. 174, 2084–2091.

Oliveira, C.J., Cavassani, K.A., More, D.D., Garlet, G.P., Aliberti, J.C., Silva, J.S., Ferreira, B.R., 2008. Tick saliva inhibits the chemotactic function of MIP-1alpha and selectively impairs chemotaxis of immature dendritic cells by down-regulating cell-surface CCR5. Int. J. Parasitol. 38, 705–716.

Osborne, R.H., 1996. Insect neurotransmission: neurotransmitters and their receptors. Pharmacol. Ther. 69, 117–142.

Pacheco, R., Prado, C.E., Barrientos, M.J., Bernales, S., 2009. Role of dopamine in the physiology of T-cells and dendritic cells. J. Neuroimmunol. 216, 8–19.

Packila, M., Guilfoile, P.G., 2002. Mating, male *Ixodes scapularis* express several genes including those with sequence similarity to immunoglobulin-binding proteins and metalloproteases. Exp. Appl. Acarol. 27, 151–160.

Paesen, G.C., Adams, P.L., Harlos, K., Nuttall, P.A., Stuart, D.I., 1999. Tick histamine-binding proteins: isolation, cloning, and three-dimensional structure. Mol. Cell 3, 661–671.

Paesen, G.C., Adams, P.L., Nuttall, P.A., Stuart, D.L., 2000. Tick histamine-binding proteins: lipocalins with a second binding cavity. Biochim. Biophys. Acta 1482, 92–101.

Paveglio, S.A., Allard, J., Mayette, J., Whittaker, L.A., Juncadella, I., Anguita, J., Poynter, M.E., 2007. The tick salivary protein, Salp15, inhibits the development of experimental asthma. J. Immunol. 178, 7064–7071.

Peterkova, K., Vancova, I., Hajnicka, V., Slovak, M., Simo, L., Nuttall, P.A., 2008. Immunomodulatory arsenal of nymphal ticks. Med. Vet. Entomol. 22, 167–171.

Petzel, D.H., Parrish, A.K., Ogg, C.L., Witters, N.A., Howard, R.W., Stanley-Samuelson, D.W., 1993. Arachidonic acid and prostaglandin E2 in Malpighian tubules of female yellow fever mosquitoes. Insect Biochem. Mol. Biol. 23, 431–437.

Prevot, P.P., Adam, B., Boudjeltia, K.Z., Brossard, M., Lins, L., Cauchie, P., Brasseur, R., Vanhaeverbeek, M., Vanhamme, L., Godfroid, E., 2006. Anti-hemostatic effects of a serpin from the saliva of the tick *Ixodes ricinus*. J. Biol. Chem. 281, 26361–26369.

Ribeiro, J.M., 1995. Blood-feeding arthropods: live syringes or invertebrate pharmacologists? Infect. Agents Dis. 4, 143–152.

Ribeiro, J.M., Alarcon-Chaidez, F., Francischetti, I.M., Mans, B.J., Mather, T.N., Valenzuela, J.G., Wikel, S.K., 2006. An annotated catalog of salivary gland transcripts from *Ixodes scapularis* ticks. Insect Biochem. Mol. Biol. 36, 111–129.

Ribeiro, J.M., Andersen, J., Silva-Neto, M.A., Pham, V.M., Garfield, M.K., Valenzuela, J.G., 2004. Exploring the sialome of the blood-sucking bug *Rhodnius prolixus*. Insect Biochem. Mol. Biol. 34, 61–79.

Ribeiro, J.M., Charlab, R., Valenzuela, J.G., 2001. The salivary adenosine deaminase activity of the mosquitoes *Culex quinquefasciatus* and *Aedes aegypti*. J. Exp. Biol. 204, 2001–2010.

Ribeiro, J.M., Katz, O., Pannell, L.K., Waitumbi, J., Warburg, A., 1999. Salivary glands of the sand fly *Phlebotomus papatasi* contain pharmacologically active amounts of adenosine and 5′-AMP. J. Exp. Biol. 202, 1551–1559.

Ribeiro, J.M., Makoul, G.T., Levine, J., Robinson, D.R., Spielman, A., 1985. Antihemostatic, antiinflammatory, and immunosuppressive properties of the saliva of a tick, *Ixodes dammini*. J. Exp. Med. 161, 332–344.

Ribeiro, J.M., Makoul, G.T., Robinson, D.R., 1988. *Ixodes dammini*: evidence for salivary prostacyclin secretion. J. Parasitol. 74, 1068–1069.

Ribeiro, J.M., Modi, G., 2001. The salivary adenosine/AMP content of *Phlebotomus argentipes* Annandale and Brunetti, the main vector of human kala-azar. J. Parasitol. 87, 915–917.

Ribeiro, J.M., Valenzuela, J.G., 2003. The salivary purine nucleosidase of the mosquito, *Aedes aegypti*. Insect Biochem. Mol. Biol. 33, 13–22.

Ribeiro, J.M., Walker, F.A., 1994. High affinity histamine-binding and antihistaminic activity of the salivary nitric oxide-carrying heme protein (nitrophorin) of *Rhodnius prolixus*. J. Exp. Med. 180, 2251–2257.

Ribeiro, J.M., Weis, J.J., Telford, S.R., 3rd, 1990. Saliva of the tick *Ixodes dammini* inhibits neutrophil function. Exp. Parasitol. 70, 382–388.

Sá-Nunes, A., Bafica, A., Antonelli, L.R., Choi, E.Y., Francischetti, I.M., Andersen, J.F., Shi, G.P., Chavakis, T., Ribeiro, J.M., Kotsyfakis, M., 2009. The immunomodulatory action of sialostatin L on dendritic cells reveals its potential to interfere with autoimmunity. J. Immunol. 182, 7422–7429.

Sá-Nunes, A., Bafica, A., Lucas, D.A., Conrads, T.P., Veenstra, T.D., Andersen, J.F., Mather, T.N., Ribeiro, J.M., Francischetti, I.M., 2007. Prostaglandin E2 is a major inhibitor of dendritic cell maturation and function in *Ixodes scapularis* saliva. J. Immunol. 179, 1497–1505.

Salzet, M., Capron, A., Stefano, G.B., 2000. Molecular crosstalk in host-parasite relationships: schistosome- and leech-host interactions. Parasitol. Today 16, 536–540.

Salzet, M., Salzet-Raveillon, B., Cocquerelle, C., Verger-Bocquet, M., Pryor, S.C., Rialas, C.M., Laurent, V., Stefano, G.B., 1997. Leech immunocytes contain proopiomelanocortin: nitric oxide mediates hemolymph proopiomelanocortin processing. J. Immunol. 159, 5400–5411.

Samson, M.T., Small-Howard, A., Shimoda, L.M., Koblan-Huberson, M., Stokes, A.J., Turner, H., 2003. Differential roles of CB1 and CB2 cannabinoid receptors in mast cells. J. Immunol. 170, 4953–4962.

Santos, A., Ribeiro, J.M., Lehane, M.J., Gontijo, N.F., Veloso, A.B., Sant' Anna, M.R., Nascimento Araujo, R., Grisard, E.C., Pereira, M.H., 2007. The sialotranscriptome of the blood-sucking bug *Triatoma brasiliensis* (Hemiptera, Triatominae). Insect Biochem. Mol. Biol. 37, 702–712.

Sauer, J.R., McSwain, J.L., Bowman, A.S., Essenberg, R.C., 1995. Tick salivary gland physiology. Annu. Rev. Entomol. 40, 245–267.

Sauer, J.R., McSwain, J.L., Essenberg, R.C., 1994. Cell membrane receptors and regulation of cell function in ticks and blood-sucking insects. Int. J. Parasitol. 24, 33–52.

Schroeder, H., Daix, V., Gillet, L., Renauld, J.C., Vanderplasschen, A., 2007. The paralogous salivary anti-complement proteins IRAC I and IRAC II encoded by *Ixodes ricinus* ticks have broad and complementary inhibitory activities against the complement of different host species. Microbes Infect. 9, 247–250.

Slominski, A., Wortsman, J., Luger, T., Paus, R., Solomon, S., 2000. Corticotropin releasing hormone and proopiomelanocortin involvement in the cutaneous response to stress. Physiol. Rev. 80, 979–1020.

Stanley, D., 2006. Prostaglandins and other eicosanoids in insects: biological significance. Annu. Rev. Entomol. 51, 25–44.

Stark, K.R., James, A.A., 1998. Isolation and characterization of the gene encoding a novel factor Xa-directed anticoagulant from the yellow fever mosquito, *Aedes aegypti*. J. Biol. Chem. 273, 20802–20809.

Thelen, M., Stein, J.V., 2008. How chemokines invite leukocytes to dance. Nat. Immunol. 9, 953–959.

Tyson, K., Elkins, C., Patterson, H., Fikrig, E., de Silva, A., 2007. Biochemical and functional characterization of Salp20, an *Ixodes scapularis* tick salivary protein that inhibits the complement pathway. Insect Mol. Biol. 16, 469–479.

Tyson, K.R., Elkins, C., de Silva, A.M., 2008. A novel mechanism of complement inhibition unmasked by a tick salivary protein that binds to properdin. J. Immunol. 180, 3964–3968.

Umemiya, R., Hatta, T., Liao, M., Tanaka, M., Zhou, J., Inoue, N., Fujisaki, K., 2007. *Haemaphysalis longicornis*: molecular characterization of a homologue of the macrophage migration inhibitory factor from the partially fed ticks. Exp. Parasitol. 115, 135–142.

Valenzuela, J.G., Charlab, R., Mather, T.N., Ribeiro, J.M., 2000. Purification, cloning, and expression of a novel salivary anticomplement protein from the tick, *Ixodes scapularis*. J. Biol. Chem. 275, 18717–18723.

Valenzuela, J.G., Francischetti, I.M., Pham, V.M., Garfield, M.K., Mather, T.N., Ribeiro, J.M., 2002a. Exploring the sialome of the tick *Ixodes scapularis*. J. Exp. Biol. 205, 2843–2864.

Valenzuela, J.G., Pham, V.M., Garfield, M.K., Francischetti, I.M., Ribeiro, J.M., 2002b. Toward a description of the sialome of the adult female mosquito *Aedes aegypti*. Insect Biochem. Mol. Biol. 32, 1101–1122.

Valenzuela, J.G., Ribeiro, J.M., 1998. Purification and cloning of the salivary nitrophorin from the hemipteran *Cimex lectularius*. J. Exp. Biol. 201, 2659–2664.

Vannacci, A., Passani, M.B., Pierpaoli, S., Giannini, L., Mannaioni, P.F., Masini, E., 2003. Nitric oxide modulates the inhibitory effect of cannabinoids on the immunological activation of guinea pig mast cells. Inflamm. Res. 52(Suppl. 1), S07–S08.

von Behring, E., Kitasato, S., 1890. Ueber das Zustandekommen der Diphtherie-Immunitat und der Tetanus-Immunitat bei thieren. Deutsch. Med. Wochenscher 16.

Wang, H., Nuttall, P.A., 1994. Excretion of host immunoglobulin in tick saliva and detection of IgG-binding proteins in tick haemolymph and salivary glands. Parasitology 109, 525–530.

Wang, H., Nuttall, P.A., 1995a. Immunoglobulin-G binding proteins in the ixodid ticks, Rhipicephalus appendiculatus, *Amblyomma variegatum* and *Ixodes hexagonus*. Parasitology 111, 161–165.

Wang, H., Nuttall, P.A., 1995b. Immunoglobulin G binding proteins in male *Rhipicephalus appendiculatus* ticks. Parasite Immunol. 17, 517–524.

Wang, H., Paesen, G.C., Nuttall, P.A., Barbour, A.G., 1998. Male ticks help their mates to feed. Nature 391, 753–754.

Weichsel, A., Andersen, J.F., Champagne, D.E., Walker, F.A., Montfort, W.R., 1998. Crystal structures of a nitric oxide transport protein from a blood-sucking insect. Nat. Struct. Biol. 5, 304–309.

Wikel, S.K., 1979. Acquired resistance to ticks: expression of resistance by C4-deficient guinea pigs. Am. J. Trop. Med. Hyg. 28, 586–590.

Wikel, S.K., 1982. Immune responses to arthropods and their products. Annu. Rev. Entomol. 27, 21–48.

Wikel, S.K., Allen, J.R., 1978. Acquired resistance to ticks. III. Cobra venom factor and the resistance response. Immunology 32, 457–465.

Yan, X., Feng, H., Yu, H., Yang, X., Liu, J., Lai, R., 2008. An immunoregulatory peptide from salivary glands of the horsefly, *Hybomitra atriperoides*. Dev. Comp. Immunol. 32, 1242–1247.

Yu, D., Liang, J., Yu, H., Wu, H., Xu, C., Liu, J., Lai, R., 2006. A tick B-cell inhibitory protein from salivary glands of the hard tick, *Hyalomma asiaticum asiaticum*. Biochem. Biophys. Res. Commun. 343, 585–590.

Zhao, R., Yu, X., Yu, H., Han, W., Zhai, L., Han, J., Liu, J., 2009. Immunoregulatory peptides from salivary glands of the horsefly, *Tabanus pleskei*. Comp. Biochem. Physiol. B. Biochem. Mol. Biol. 154, 1–5.

Zhou, W., Blackwell, T.S., Goleniewska, K., O'Neal, J.F., Fitzgerald, G.A., Lucitt, M., Breyer, R.M., Peebles, R.S., Jr., 2007a. Prostaglandin I2 analogs inhibit Th1 and Th2 effector cytokine production by CD4 T cells. J. Leukoc. Biol. 81, 809–817.

Zhou, W., Hashimoto, K., Goleniewska, K., O'Neal, J.F., Ji, S., Blackwell, T.S., Fitzgerald, G.A., Egan, K.M., Geraci, M.W., Peebles, R.S., Jr., 2007b. Prostaglandin I2 analogs inhibit proinflammatory cytokine production and T cell stimulatory function of dendritic cells. J. Immunol. 178, 702–710.

Zipfel, P.F., Skerka, C., 2009. Complement regulators and inhibitory proteins. Nat. Rev. Immunol. 9, 729–740.

Chapter 10
Salivary Protease Inhibitors with Non Anti-Hemostatic Functions

Jindřich Chmelař, Ivo M.B. Francischetti, and Michalis Kotsyfakis

Abstract Arthropod saliva regulates proteolysis in the sites of bite(s). Secreted salivary inhibitors target different proteolytic enzymes (proteases) that have a pivotal role in vertebrate hemostasis, immunity and inflammation. The aim of this chapter is to give an overview of the salivary protease inhibitors that affect the latter two physiological procedures. Functional studies, mainly in ticks, have demonstrated many potent arthropod salivary inhibitors of cysteine proteases (cathepsins) and serine proteases (elastase and tryptase). Emphasis is given to the function of these inhibitors and more specifically to the mechanism by which they facilitate hematophagy.

Introduction

In addition to inhibitors of blood coagulation and platelet aggregation, saliva from hematophagous ticks is a source of protease inhibitors, which target enzymes not associated with hemostasis (Francischetti et al., 2009; Koh and Kini, 2009; Ribeiro and Francischetti, 2003). These molecules, also named sialogenins (from the Greek *sialo*, saliva; *gen*, origin, source; and *ins* for proteins) with anti-protease activity are emerging as important components of the salivary secretion with defined enzyme targets and biologic properties. The aim of this chapter is to describe recent progress in this field with emphasis on inhibitors obtained in recombinant form and how they impact host-vector interactions. These inhibitors are classified as cysteine protease inhibitors or serine protease inhibitors and blockade of protease function may take place in the extracellular space and/or intracellular environment.

M. Kotsyfakis (✉)
Laboratory of Genomics and Proteomics of Disease Vectors, Institute of Parasitology, Biology Centre, ASCR v.v.i., Branisovska 31, 37005, Ceske Budejovice, Czech Republic
e-mail: kotsyfakis@paru.cas.cz

R.M. Kini et al. (eds.), *Toxins and Hemostasis*, DOI 10.1007/978-90-481-9295-3_10,

Cysteine Protease Inhibitors (Cystatins)

Traditionally, cysteine proteases (cathepsins) have been involved in rather non-specific protein degradation in a strongly acidic milieu of lysosomes. In the last fifteen years, a series of studies have identified a number of other functions for cysteine proteases (Turk et al., 2000) including an important role in antigen presentation (Honey and Rudensky, 2003), immune system development (Lombardi et al., 2005), epidermal homeostasis (Reinheckel et al., 2005), neovascularization (Felbor et al., 2000), extracellular matrix degradation and neutrophil chemotaxis during inflammation (Reddy et al., 1995; Serveau-Avesque et al., 2006). The function of these enzymes is highly regulated by inhibitors of cathepsins named cystatins. These are subdivided into three families. The members of family 1 (also known as stefins) are cytosolic molecules with neither disulfide bonds nor carbohydrates. Family 2 contains all of the secreted cystatins that are mainly found in biologic fluids; they have two disulfide bridges and they do not bear sugars. Family 3 cystatins (also known as kininogens) comprise several cystatin modules, and relatively large ml wt (60–120 kDa).

In this chapter we describe family 2 cystatins from tick salivary glands which—through inhibition of cathepsins—interfere with the immune system and modulate vascular and matrix biology. All tick cystatins are efficient inhibitors of cathepsins, while none of them has been demonstrated as efficient inhibitor of any other cysteine protease (legumain, calpain or caspase 3). Even within the cathepsin family, tick cystatins have been shown as targeting only those that have cysteine protease activity, namely cathepsins B, H, C, L, S (Table 10.1), and not those that are aspartic (cathepsin D and E) or serine proteases (cathepsin G) (Kotsyfakis et al., 2006). Two members of family 1 cystatins have been characterized from ticks (Lima et al., 2006; Zhou et al., 2009), but they will not be further discussed herein since they are cytosolic (not secreted) inhibitors. In addition, there is no report on the activity of salivary arthropod kininogens (family 3) against cysteine proteases.

Sialostatins L and L2

Hard ticks feed for a long period of time that can last up to a week. A complex host inflammatory response to injury takes place during this period which is potentially counteracted by the pharmacological properties of saliva. For example, cathepsins can be found in the extracellular environment upon tissue destruction/inflammation (Reddy et al., 1995; Serveau-Avesque et al., 2006) and also present immunomodulatory activities (Zavasnik-Bergant and Turk, 2007). Therefore, targeting cathepsins by tick saliva is an important strategy to attenuate pro-inflammatory tonus at the sites of tick feeding. In this context, salivary gland of the hard tick *Ixodes scapularis* secretes at least two different closely related cystatins named sialostatin L (12.5 kDa) and sialostatin L2 (12.8 kDa). Both inhibitors bind tightly to cathepsin L with a Ki in the low picomolar range (Ki $\sim$ 100 pM) (Kotsyfakis et al., 2006, 2007). Their presence in the tick salivary glands has been demonstrated with both

Table 10.1 Protease inhibitors targeting proteases not related to hemostasis

Name	Species	Protein family	MW (kDa)	Protease(s) targeted	References
Cysteine protease inhibitors					
Sialostatin L	*Ixodes scapularis*	Cystatin	12.5	Cathepsins L, S, C	Kotsyfakis et al. (2006)
Sialostatin L 2	*Ixodes scapularis*	Cystatin	12.8	Cathepsins L, S, C	Kotsyfakis et al. (2007)
HlSC-1	*Haemaphysalis longicornis*	Cystatin	12	Cathepsins L	Yamaji et al. (2009)
Hlcyst-2	*Haemaphysalis longicornis*	Cystatin	12.9	Cathepsin L, B	Zhou et al. (2006)
Om-cystatin 1	*Ornithodoros moubata*	Cystatin	12.4	Cathepsins B, H, C	Grunclova et al. (2006)
Om-cystatin 2	*Ornithodoros moubata*	Cystatin	12.2	Cathepsins B, H, C	Grunclova et al. (2006)
Serine protease inhibitors					
TdPI	*Rhipicephalus appendiculatus*	Kunitz/BPTI	11.1	β-tryptase	Paesen et al. (2007)
Iris	*Ixodes ricinus*	Serpin	43	Elastase	Prevot et al. (2006)
BmSI-6	*Boophilus microplus*	Subtilisin inhibitor-like	7.3	Elastase	Sasaki et al. (2008)
BmSI-7	*Boophilus microplus*	Cysteine-rich/TIL	7.4	Elastase	Sasaki et al. (2008)
BmTI-A	*Boophilus microplus*	Kunitz/BPTI	18	Elastase	Tanaka et al. (1999)
RsTIQ2	*Rhipicephalus sanguineus*	Kunitz/BPTI	12	Elastase	Azzolini et al. (2003)
BmTI-2	*Boophilus microplus*	Kunitz/BPTI	17	Elastase	Sasaki et al. (2004)
BmTI-3	*Boophilus microplus*	Kunitz/BPTI	15	Elastase	Sasaki et al. (2004)
HiTI	*Haematobia irritans*	Kunitz/BPTI	7	Elastase	Azzolini et al. (2004)

biochemical methods and RT-PCR (Kotsyfakis et al., 2006, 2007). Sialostatin L suppresses edema formation and neutrophil migration induced by carragenean injection in the mouse footpad (Kotsyfakis et al., 2006). Sialostatin L, but not sialostatin L2, also suppresses the function of dendritic cells (DCs) in vitro. Accordingly, sialostatin L inhibits LPS-induced maturation of dendritic cells and invariant chain intermediates were accumulated in these cells, resulting in their inability to induce Ag-specific CD4+ T cell proliferation in vitro (Sá-Nunes et al., 2009). Experiments with cathepsin L and cathepsin S knock-out mice further demonstrated that this action is mediated by inhibition of cathepsin S and not of cathepsin L (Sá-Nunes et al., 2009). Given that sialostatin L has at least 500 times higher inhibitory activity than sialostatin L2 against cathepsin S (Kotsyfakis et al., 2007), it is not surprising that only sialostatin L, but not sialostatin L2, had an impact on DC function (Sá-Nunes et al., 2009). Remarkably, the pharmacologic action of sialostatin L on DCs could prevent the onset of experimental autoimmune encephalitis in an animal model commonly used for research on multiple sclerosis pathogenesis (Sá-Nunes et al., 2009). In addition, when sialostatin genes were silenced in the salivary glands of the tick *I. scapularis* by RNAi, the ticks acquired lesser blood from rabbits compared with control ticks; this has been attributed to an increased recognition of tick salivary antigens from the rabbits that resulted increased inflammation in the attachment sites of the sialostatin-silenced ticks (Kotsyfakis et al., 2007). Finally, sialostatin's action can be blocked by vaccinating guinea pigs prior to tick infestation, making them attractive candidate antigens for the development of anti-tick vaccines (Kotsyfakis et al., 2008).

Cystatin from **Amblyomma americanum**

A similar effect on tick feeding success was reported when silencing a gene encoding for a secreted salivary cystatin from the hard tick *A. americanum* (Karim et al., 2005). The target proteases for this cystatin are unknown.

*HlSC-1 (***Haemaphysalis longicornis** *Salivary Cystatin-1) and Hlcyst-2 (***H. longicornis** *Cystatin-2)*

Two secreted cystatins have been characterized from the hard tick *H. longicornis*. HlSC-1 (12 kDa) is highly upregulated in the early phase of tick feeding. Real-time PCR revealed five to ten-fold higher transcript abundance in tick salivary glands compared with that from midguts and a dramatic transcriptional upregulation in the initial phase of blood feeding (24 h) that declined as feeding reaches completion (Yamaji et al., 2009). The protein, immunolocalized in the salivary gland type II acini of an adult tick, inhibits cathepsin L, but not cathepsin B (Yamaji et al., 2009). Hlcyst-2 (12.9 kDa) is expressed in all tick developmental stages. It inhibits cathepsin B and cathepsin L, but its transcripts are mainly localized in the midgut and hemocytes and, to a lesser extent, in the salivary glands (Zhou et al., 2006).

Its expression was up-regulated upon LPS injection in adult ticks and *Babesia gibsoni* infection of larvae ticks, thus revealing an involvement in tick innate immunity (Zhou et al., 2006). When these two cystatins were compared for inhibitory activity against cathepsin L, the salivary one (HlSC-1) was 5–6 times more potent than the one related to tick innate immunity (Hlcyst-2) (Yamaji et al., 2009).

Om-Cystatin 1 and Om-Cystatin 2

Expression of two secreted cystatins (Om-cystatin 1 and Om-cystatin 2) of molecular weight 12.4 and 12.2 kDa, respectively, has been described in the soft tick *Ornithodoros moubata* (Grunclova et al., 2006). Both cystatins are able to inhibit vertebrate cathepsins B, H, and C. While Om-cystatin 1 is expressed strictly in the tick midgut, Om-cystatin 2 is expressed in various tissues including tick salivary glands (Grunclova et al., 2006). Om-cystatins play an additional role in the regulation of endogenous tick cysteine peptidases involved in blood digestion and in heme detoxification (Grunclova et al., 2006). Of note, the salivary cystatin from the soft tick (Om-cystatin 2) inhibits vertebrate cathepsins B and H, but the salivary cystatins from the hard ticks *I. scapularis* and *H. longicornis* do not. As soft ticks imbibe their blood meal much faster than hard ticks, it is tempting to question whether this difference in target specificity (i.e., the inability to inhibit cathepsins B and H) is related to the difference in duration of blood feeding between soft and hard ticks.

Interestingly, cystatins are not found in the two blood-feeding mosquitoes whose genomes are completed—namely *Anopheles gambiae* and *Aedes aegypti*—or in the salivary glands of any other blood-feeding arthropod (sand flies, fleas) other than ticks. An *Ae. aegypti* truncated transcript containing a potential cystatin domain has been reported, but the function of the mature polypeptide and whether it is able to inhibit cysteine proteases remains unknown (Ribeiro et al., 2007). Secreted cystatins have also been reported in *Drosophila*, *Bombyx*, and the flesh fly *Sarcophaga spp.* (Goto and Denlinger, 2002), although it is not known whether these genes are expressed in their salivary glands. Apart from ticks, only lower eukaryotes such as the nematode *Nippostrongylus brasiliensis* utilize cystatin secretion to evade the vertebrate host defense system (Dainichi et al., 2001).

Serine Protease Inhibitors

The catalytic domains of major coagulation factors (VIIa, IXa, Xa, XIa, XIIa and thrombin) belong to the widespread family of trypsin-like serine proteinases. In fact, most of the arthropod salivary serine protease inhibitors studied to date act as anti-hemostatic factors (Corral-Rodriguez et al., 2009). However, other enzymes of the same family, such as-mast cell β-tryptase and leukocyte elastase play an important role in inflammation and tissue response to injury and they are targeted by tick salivary serine protease inhibitors (Table 10.1). Leukocyte elastase cleaves various types of collagens (Gadher et al., 1988) in addition to elastin, fibronectin and laminin (Heck et al., 1990). It also activates procollagenase, prostromelysin and progelatinase (Rice and Banda, 1995; Shamamian et al., 2001) thus contributing to wound healing and tissue repair. Moreover, it plays a role in antimicrobial defense, since patients with Chediak-Higashi syndrome who lack leukocyte elastase in their neutrophils suffer from recurrent bacterial infections (Ganz et al., 1988) and elastase deficient mice are more prone to gram-negative bacterial sepsis (Belaaouaj et al., 1998). β tryptase is another important enzyme involved in different aspects of tissue remodeling. It activates the zymogens of a series of proteins, including prostromelysin (Gruber et al., 1989), pro-urokinase (Stack and Johnson, 1994), PAR-2 (Corvera et al., 1997) and inactivates fibronectin (Kaminska et al., 1999). Recombinant β_1 tryptase has also been reported to play a role in anti-bacterial defenses by enhancing neutrophil recruitment (Huang et al., 2001). In addition, tryptases promote vascular tube formation (Blair et al., 1997), tumor angiogenesis (Coussens et al., 1999) and allergic reactions (Molinari et al., 1995; Molinari et al., 1996). Moreover, β tryptase generates kinins from kininogens; it has been shown that tryptases in concerted action with neutrophil elastase may enhance the production of kinins in inflamed tissues (Kozik et al., 1998). Therefore, these two enzymes are obvious targets for tick saliva since their inhibition may attenuate host response to injury such as tissue repair and angiogenesis (Francischetti et al., 2005). Below we refer to each individual inhibitor that blocks either β-tryptase or elastase and discuss their properties in the context of vector-host interaction.

Tick-Derived Protease Inhibitor (TdPI)

Paesen et al. identified in the hard tick *Rhipicephalus appendiculatus* a protein with molecular weight of 11.1 kDa that inhibits human β_1 tryptase in low nanomolar concentration. Although the protein is related to the Kunitz/BPTI (bovine pancreatic trypsin inhibitor) family, the resolution of its structure in complex with trypsin revealed an "unusual" non-classical Kunitz/BPTI fold (Fig. 10.1a) that makes possible the blockade of three of the four catalytic sites of β_1 tryptase (Paesen et al., 2007). More specifically, the TdPI Kunitz domain adopts an arrow-like structure; this is mainly due to the short length of the loop that connects the β-sheet with the C-terminal α-helix, the altered pattern of disulphide bridges, and the separation of its C-terminus from the N-terminus (Fig. 10.1a). As a result, the highly pointed apex of TdPI gains access to the cramped active sites of $\beta_1$tryptase and its overall triangle-like fold diversifies from the "typical" bullet-like structure of other Kunitz-proteins (Fig. 10.1b, c). Apart from $\beta_1$tryptase, TdPI is also able to inhibit trypsin with a Ki lower than 10 nM and human plasmin with a Ki of 50 nM, but it does not inhibit urokinase, thrombin, factor Xa, factor XIIa, elastase, kallikrein, cathepsin G, granzyme B, chymase, or chymotrypsin, thus having a rather stringent specificity for β_1 tryptase. The inhibitor is active against two different human tryptases, one expressed in the skin (β_1 tryptase) and the other expressed in the lung (β_2 tryptase), while it inhibits the murine tetrameric tryptase MCP-6, as well (Paesen et al., 2007). Interestingly, $\beta_1$tryptase is able to trim TdPI, and this cleavage enhances the ability

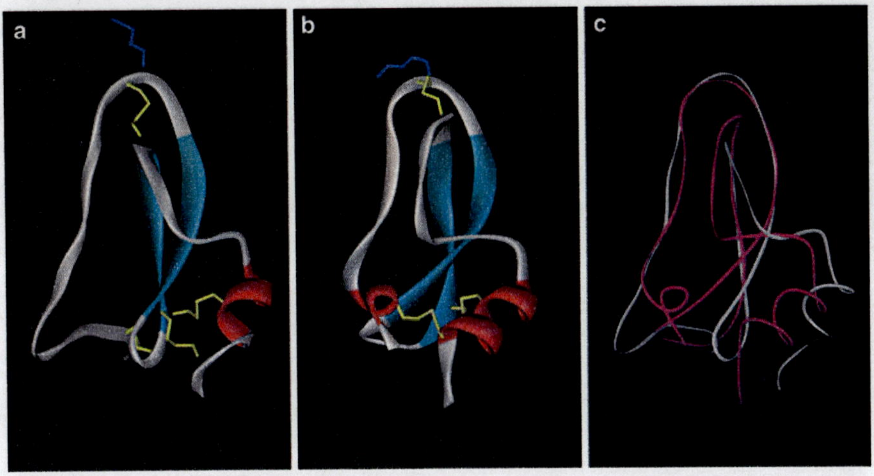

Fig. 10.1 (**a**) The *arrow-like structure* of TdPI (PDB: 2UUX) compared to (**b**) the *bullet-like structure* of BPTI (bovine pancreatic trypsin inhibitor) (PDB: 1BPI). Disulfide bridges are shown in *yellow stick* model and P1 Lys in *blue*. (**c**) The overlaid structures of TdPI (*white*) and BPTI (*pink*)

of TdPI to inhibit β_1 tryptase. Moreover, immunohistochemistry revealed that TdPI is able to penetrate mouse mast cells; it was immunolocalized in the cytosolic granules of dermal mast cells upon TdPI injection into the ears of BALB/c mice (Paesen et al., 2007). Such a result could not be reproduced in mast cell-deficient mice, further verifying the ability of the protein to access mast cell intracellular compartments. Consistent with such an action against mast cells, transcription of TdPI by female *R. appendiculatus* ticks coincides with that of *R. appendiculatus* histamine-binding proteins that sequester histamine, a biogenic amine released upon mast cell stimulation (Paesen et al., 2007). Although this simultaneous transcriptional induction of both effectors can be purely coincidental, it may enhance the overall effect on mast cell activation. The action of TdPI can take place in the extracellular (upon mast cell degranulation) or intracellular environment. Mast cells are able to internalize a range of protease inhibitors (Ide et al., 1999; Kido et al., 1988) including TdPI (Paesen et al., 2007); this internalized TdPI may block the autocatalytic activation of tryptase required for its biologic action (Sakai et al., 1996). Similarly to other synthetic tryptase inhibitors, it may in addition block the release of mast cell mediators (He et al., 2004).

Ixodes ricinus *Immunosuppressor (Iris)*

Iris cDNA was identified among other transcripts to be specifically expressed during the blood meal in the salivary glands of *Ixodes ricinus* female adults; it was chosen to be further characterized because of its similarity to the pig leukocyte inhibitor

(Leboulle et al., 2002b). Immunostaining with polyclonal antibodies against recombinant Iris revealed a protein on the external surface of salivary acini, within the cells, and also in the acini's light in 3-day-fed and 5-day-fed female ticks, but not in unfed ticks (Leboulle et al., 2002a). Iris molecular weight is 43 kDa, it belongs to the serpin supefamily, and it inhibits the active site of most serine proteases tested. The highest Kα was observed for human leukocyte elastase (Prevot et al., 2006). In addition, Iris is able to immunomodulate the vertebrate host independently of its antiprotease activity through the mediation of (an) exosite(s). More specifically, upon peripheral blood mononuclear cell activation by various Toll-like receptor agonists, it disturbs the Th1/Th2 balance by inhibiting interferon-γ production and LPS-induced TNF-α production (Leboulle et al., 2002a; Prevot et al., 2009). Furthermore, Iris protects mice against LPS-induced septic shock (Prevot et al., 2009). The importance of Iris action for tick blood-feeding success was further demonstrated in vaccination experiments, where rabbits were partially protected against tick infestation when vaccinated with Iris (Prevot et al., 2007).

BmSI-6 and BmSI-7 (BmSI: **Boophilus microplus** *Subtilisin Inhibitors)*

Another two inhibitors of elastase have been isolated from the tick *Boophilus microplus*, namely BmSI-6 and BmSI-7, with molecular weight of 7.3 and 7.4 kDa and a Ki for elastase of 0.3 and 0.4 nM, respectively. Both inhibitors also target subtilisin A with a Ki of 1.4 nM. BmSI-7 belongs to the trypsin inhibitor-like cysteine-rich domain family (TIL); it has a clear secretion signal and is transcribed in the salivary glands (Sasaki et al., 2008). There are no data available on the potential pharmacologic action of these inhibitors in the vertebrate host.

Kunitz Inhibitors of Elastase

Activity against elastase has been reported for the *B. microplus* protein BmTI-A (*B. microplus* trypsin inhibitor-A), a member of the Kunitz/BPTI family with molecular weight of 18 kDa (Tanaka et al., 1999), and for the *Rhipicephalus sanguineus* protein RsTIQ2, a 12-kDa member of the same family (Sant'Anna Azzolini et al., 2003). Both inhibitors have high affinity for elastase (Ki of 1.4 and 1.3 nM, respectively), but they were isolated from tick larvae, and therefore their presence in tick saliva requires further investigation. The same occurs for another two Kunitz/BPTI elastase inhibitors isolated from *B. microplus* larvae, namely BmTI-2 and BmTI-3, with estimated molecular weight of approximately 17 and 15 kDa, respectively (Sasaki et al., 2004). A 7-kDa protein belonging also to the Kunitz/BPTI family with similarly high affinity for elastase has been isolated from the head and the thorax of the horn fly *Haematobia irritans*. This protein (HiTI, *Haematobia irritans* trypsin inhibitor)—is not known to be present in saliva (Azzolini et al., 2004).

Concluding Remarks

Saliva from blood-feeding arthropods is a rich cocktail of potent protease inhibitors (Table 10.1). Although inhibition of proteolytic enzymes by tick saliva is often associated with blockade of hemostasis, a picture emerges where blood-sucking arthropods—especially ticks—target specific proteases involved in vertebrate immunity and possibly matrix remodeling, and wound healing. Genomic information available for many arthropod disease vectors is expected to contribute substantially to the field of proteolysis regulation unrelated to hemostasis.

Acknowledgements This work was supported by the Intramural Research Program of the Division of Intramural Research, National Institute of Allergy and Infectious Diseases, National Institutes of Health, by the grant No. IAA600960811 from the Grant Agency of the Academy of Sciences of the Czech Republic and by the Research Center No. LC06009 from the Ministry of Education, Youth and Sports of the Czech Republic. We thank Professor R. Manjunatha Kini for the fruitful discussions and comments on the manuscript, Dr. Koh Cho Yeow for preparing the figure and NIAID intramural editor Brenda Rae Marshall for assistance.

References

Azzolini, S.S., Santos, J.M., Souza, A.F., Torquato, R.J., Hirata, I.Y., Andreotti, R., Tanaka, A.S., 2004. Purification, characterization, and cloning of a serine proteinase inhibitor from the ectoparasite *Haematobia irritans irritans* (Diptera: Muscidae). Exp. Parasitol. 106, 103–109.

Belaaouaj, A., McCarthy, R., Baumann, M., Gao, Z., Ley, T.J., Abraham, S.N., Shapiro, S.D., 1998. Mice lacking neutrophil elastase reveal impaired host defense against gram negative bacterial sepsis. Nat. Med. 4, 615–618.

Blair, R.J., Meng, H., Marchese, M.J., Ren, S., Schwartz, L.B., Tonnesen, M.G., Gruber, B.L., 1997. Human mast cells stimulate vascular tube formation. Tryptase is a novel, potent angiogenic factor. J. Clin. Invest. 99, 2691–2700.

Corral-Rodriguez, M.A., Macedo-Ribeiro, S., Barbosa Pereira, P.J., Fuentes-Prior, P., 2009. Tick-derived Kunitz-type inhibitors as antihemostatic factors. Insect Biochem. Mol. Biol. 39, 579–595.

Corvera, C.U., Dery, O., McConalogue, K., Bohm, S.K., Khitin, L.M., Caughey, G.H., Payan, D.G., Bunnett, N.W., 1997. Mast cell tryptase regulates rat colonic myocytes through proteinase-activated receptor 2. J. Clin. Invest. 100, 1383–1393.

Coussens, L.M., Raymond, W.W., Bergers, G., Laig-Webster, M., Behrendtsen, O., Werb, Z., Caughey, G.H., Hanahan, D., 1999. Inflammatory mast cells up-regulate angiogenesis during squamous epithelial carcinogenesis. Genes Dev. 13, 1382–1397.

Dainichi, T., Maekawa, Y., Ishii, K., Zhang, T., Nashed, B.F., Sakai, T., Takashima, M., Himeno, K., 2001. Nippocystatin, a cysteine protease inhibitor from *Nippostrongylus brasiliensis*, inhibits antigen processing and modulates antigen-specific immune response. Infect. Immun. 69, 7380–7386.

Felbor, U., Dreier, L., Bryant, R.A., Ploegh, H.L., Olsen, B.R., Mothes, W., 2000. Secreted cathepsin L generates endostatin from collagen XVIII. EMBO J. 19, 1187–1194.

Francischetti, I.M., Mather, T.N., Ribeiro, J.M., 2005. Tick saliva is a potent inhibitor of endothelial cell proliferation and angiogenesis. Thromb. Haemost. 94, 167–174.

Francischetti, I.M., Sá-Nunes, A., Mans, B.J., Santos, I.M., Ribeiro, J.M., 2009. The role of saliva in tick feeding. Front. Biosci. 14, 2051–2088.

Gadher, S.J., Eyre, D.R., Duance, V.C., Wotton, S.F., Heck, L.W., Schmid, T.M., Woolley, D.E., 1988. Susceptibility of cartilage collagens type II, IX, X, and XI to human synovial collagenase and neutrophil elastase. Eur. J. Biochem. 175, 1–7.

Ganz, T., Metcalf, J.A., Gallin, J.I., Boxer, L.A., Lehrer, R.I., 1988. Microbicidal/cytotoxic proteins of neutrophils are deficient in two disorders: Chediak-Higashi syndrome and "specific" granule deficiency. J. Clin. Invest. 82, 552–556.

Goto, S.G., Denlinger, D.L., 2002. Genes encoding two cystatins in the flesh fly *Sarcophaga crassipalpis* and their distinct expression patterns in relation to pupal diapause. Gene 292, 121–127.

Gruber, B.L., Marchese, M.J., Suzuki, K., Schwartz, L.B., Okada, Y., Nagase, H., Ramamurthy, N.S., 1989. Synovial procollagenase activation by human mast cell tryptase dependence upon matrix metalloproteinase 3 activation. J. Clin. Invest. 84, 1657–1662.

Grunclova, L., Horn, M., Vancova, M., Sojka, D., Franta, Z., Mares, M., Kopacek, P., 2006. Two secreted cystatins of the soft tick *Ornithodoros moubata*: differential expression pattern and inhibitory specificity. Biol. Chem. 387, 1635–1644.

He, S., Aslam, A., Gaca, M.D., He, Y., Buckley, M.G., Hollenberg, M.D., Walls, A.F., 2004. Inhibitors of tryptase as mast cell-stabilizing agents in the human airways: effects of tryptase and other agonists of proteinase-activated receptor 2 on histamine release. J. Pharmacol. Exp. Ther. 309, 119–126.

Heck, L.W., Blackburn, W.D., Irwin, M.H., Abrahamson, D.R., 1990. Degradation of basement membrane laminin by human neutrophil elastase and cathepsin G. Am. J. Pathol. 136, 1267–1274.

Honey, K., Rudensky, A.Y., 2003. Lysosomal cysteine proteases regulate antigen presentation. Nat. Rev. Immunol. 3, 472–482.

Huang, C., De Sanctis, G.T., O'Brien, P.J., Mizgerd, J.P., Friend, D.S., Drazen, J.M., Brass, L.F., Stevens, R.L., 2001. Evaluation of the substrate specificity of human mast cell tryptase beta I and demonstration of its importance in bacterial infections of the lung. J. Biol. Chem. 276, 26276–26284.

Ide, H., Itoh, H., Yoshida, E., Kobayashi, T., Tomita, M., Maruyama, H., Osada, Y., Nakahata, T., Nawa, Y., 1999. Immunohistochemical demonstration of inter-alpha-trypsin inhibitor light chain (bikunin) in human mast cells. Cell Tissue Res. 297, 149–154.

Kaminska, R., Helisalmi, P., Harvima, R.J., Naukkarinen, A., Horsmanheimo, M., Harvima, I.T., 1999. Focal dermal-epidermal separation and fibronectin cleavage in basement membrane by human mast cell tryptase. J. Invest. Dermatol. 113, 567–573.

Karim, S., Miller, N.J., Valenzuela, J., Sauer, J.R., Mather, T.N., 2005. RNAi-mediated gene silencing to assess the role of synaptobrevin and cystatin in tick blood feeding. Biochem. Biophys. Res. Commun. 334, 1336–1342.

Kido, H., Fukusen, N., Katunuma, N., 1988. Antibodies and inhibitor of chymase are incorporated into mast cell granules and inhibit histamine release. Biol. Chem. Hoppe-Seyler. 369(Suppl), 95–100.

Koh, C.Y., Kini, R.M., 2009. Molecular diversity of anticoagulants from haematophagous animals. Thromb. Haemost. 102, 437–453.

Kotsyfakis, M., Anderson, J.M., Andersen, J.F., Calvo, E., Francischetti, I.M., Mather, T.N., Valenzuela, J.G., Ribeiro, J.M., 2008. Cutting edge: immunity against a "silent" salivary antigen of the Lyme vector *Ixodes scapularis* impairs its ability to feed. J. Immunol. 181, 5209–5212.

Kotsyfakis, M., Karim, S., Andersen, J.F., Mather, T.N., Ribeiro, J.M., 2007. Selective cysteine protease inhibition contributes to blood-feeding success of the tick *Ixodes scapularis*. J. Biol. Chem. 282, 29256–29263.

Kotsyfakis, M., Sá-Nunes, A., Francischetti, I.M., Mather, T.N., Andersen, J.F., Ribeiro, J.M., 2006. Antiinflammatory and immunosuppressive activity of sialostatin L, a salivary cystatin from the tick *Ixodes scapularis*. J. Biol. Chem. 281, 26298–26307.

Kozik, A., Moore, R.B., Potempa, J., Imamura, T., Rapala-Kozik, M., Travis, J., 1998. A novel mechanism for bradykinin production at inflammatory sites. Diverse effects of a mixture of neutrophil elastase and mast cell tryptase versus tissue and plasma kallikreins on native and oxidized kininogens. J. Biol. Chem. 273, 33224–33229.

Leboulle, G., Crippa, M., Decrem, Y., Mejri, N., Brossard, M., Bollen, A., Godfroid, E., 2002a. Characterization of a novel salivary immunosuppressive protein from *Ixodes ricinus* ticks. J. Biol. Chem. 277, 10083–10089.

Leboulle, G., Rochez, C., Louahed, J., Ruti, B., Brossard, M., Bollen, A., Godfroid, E., 2002b. Isolation of *Ixodes ricinus* salivary gland mRNA encoding factors induced during blood feeding. Am. J. Trop. Med. Hyg. 66, 225–233.

Lima, C.A., Sasaki, S.D., Tanaka, A.S., 2006. Bmcystatin, a cysteine proteinase inhibitor characterized from the tick *Boophilus microplus*. Biochem. Biophys. Res. Commun. 347, 44–50.

Lombardi, G., Burzyn, D., Mundinano, J., Berguer, P., Bekinschtein, P., Costa, H., Castillo, L.F., Goldman, A., Meiss, R., Piazzon, I., Nepomnaschy, I., 2005. Cathepsin-L influences the expression of extracellular matrix in lymphoid organs and plays a role in the regulation of thymic output and of peripheral T cell number. J. Immunol. 174, 7022–7032.

Molinari, J.F., Moore, W.R., Clark, J., Tanaka, R., Butterfield, J.H., Abraham, W.M., 1995. Role of tryptase in immediate cutaneous responses in allergic sheep. J. Appl. Physiol. 79, 1966–1970.

Molinari, J.F., Scuri, M., Moore, W.R., Clark, J., Tanaka, R., Abraham, W.M., 1996. Inhaled tryptase causes bronchoconstriction in sheep via histamine release. Am. J. Respir. Crit. Care Med. 154, 649–653.

Paesen, G.C., Siebold, C., Harlos, K., Peacey, M.F., Nuttall, P.A., Stuart, D.I., 2007. A tick protein with a modified Kunitz fold inhibits human tryptase. J. Mol. Biol. 368, 1172–1186.

Prevot, P.P., Adam, B., Boudjeltia, K.Z., Brossard, M., Lins, L., Cauchic, P., Brasseur, R., Vanhaeverbeek, M., Vanhamme, L., Godfroid, E., 2006. Anti-hemostatic effects of a serpin from the saliva of the tick *Ixodes ricinus*. J. Biol. Chem. 281, 26361–26369.

Prevot, P.P., Beschin, A., Lins, L., Beaufays, J., Grosjean, A., Bruys, L., Adam, B., Brossard, M., Brasseur, R., Zouaoui Boudjeltia, K., Vanhamme, L., Godfroid, E., 2009. Exosites mediate the anti-inflammatory effects of a multifunctional serpin from the saliva of the tick *Ixodes ricinus*. FEBS J. 276, 3235–3246.

Prevot, P.P., Couvreur, B., Denis, V., Brossard, M., Vanhamme, L., Godfroid, E., 2007. Protective immunity against *Ixodes ricinus* induced by a salivary serpin. Vaccine 25, 3284–3292.

Reddy, V.Y., Zhang, Q.Y., Weiss, S.J., 1995. Pericellular mobilization of the tissue-destructive cysteine proteinases, cathepsins B, L, and S, by human monocyte-derived macrophages. Proc. Natl. Acad. Sci. U.S.A. 92, 3849–3853.

Reinheckel, T., Hagemann, S., Dollwet-Mack, S., Martinez, E., Lohmuller, T., Zlatkovic, G., Tobin, D.J., Maas-Szabowski, N., Peters, C., 2005. The lysosomal cysteine protease cathepsin L regulates keratinocyte proliferation by control of growth factor recycling. J. Cell Sci. 118, 3387–3395.

Ribeiro, J.M., Arca, B., Lombardo, F., Calvo, E., Phan, V.M., Chandra, P.K., Wikel, S.K., 2007. An annotated catalogue of salivary gland transcripts in the adult female mosquito, *Aedes aegypti*. BMC Genomics 8, 6.

Ribeiro, J.M., Francischetti, I.M., 2003. Role of arthropod saliva in blood feeding: sialome and post-sialome perspectives. Annu. Rev. Entomol. 48, 73–88.

Rice, A., Banda, M.J., 1995. Neutrophil elastase processing of gelatinase A is mediated by extracellular matrix. Biochemistry 34, 9249–9256.

Sá-Nunes, A., Bafica, A., Antonelli, L.R., Choi, E.Y., Francischetti, I.M., Andersen, J.F., Shi, G.P., Chavakis, T., Ribeiro, J.M., Kotsyfakis, M., 2009. The immunomodulatory action of sialostatin L on dendritic cells reveals its potential to interfere with autoimmunity. J. Immunol. 182, 7422–7429.

Sakai, K., Long, S.D., Pettit, D.A., Cabral, G.A., Schwartz, L.B., 1996. Expression and purification of recombinant human tryptase in a baculovirus system. Protein Expr. Purif. 7, 67–73.

Sant'Anna Azzolini, S., Sasaki, S.D., Torquato, R.J., Andreotti, R., Andreotti, E., Tanaka, A.S., 2003. *Rhipicephalus sanguineus* trypsin inhibitors present in the tick larvae: isolation, characterization, and partial primary structure determination. Arch. Biochem. Biophys. 417, 176–182.

Sasaki, S.D., Azzolini, S.S., Hirata, I.Y., Andreotti, R., Tanaka, A.S., 2004. *Boophilus microplus* tick larvae, a rich source of Kunitz type serine proteinase inhibitors. Biochimie 86:643–649.

Sasaki, S.D., de Lima, C.A., Lovato, D.V., Juliano, M.A., Torquato, R.J., Tanaka, A.S., 2008. BmSI-7, a novel subtilisin inhibitor from *Boophilus microplus*, with activity toward Pr1 proteases from the fungus Metarhizium anisopliae. Exp. Parasitol. 118, 214–220.

Serveau-Avesque, C., Martino, M.F., Herve-Grepinet, V., Hazouard, E., Gauthier, F., Diot, E., Lalmanach, G., 2006. Active cathepsins B, H, K, L and S in human inflammatory bronchoalveolar lavage fluids. Biol. Cell 98, 15–22.

Shamamian, P., Schwartz, J.D., Pocock, B.J., Monea, S., Whiting, D., Marcus, S.G., Mignatti, P., 2001. Activation of progelatinase A (MMP-2) by neutrophil elastase, cathepsin G, and proteinase-3: a role for inflammatory cells in tumor invasion and angiogenesis. J. Cell. Physiol. 189, 197–206.

Stack, M.S., Johnson, D.A., 1994. Human mast cell tryptase activates single-chain urinary-type plasminogen activator (pro-urokinase). J. Biol. Chem. 269, 9416–9419.

Tanaka, A.S., Andreotti, R., Gomes, A., Torquato, R.J., Sampaio, M.U., Sampaio, C.A., 1999. A double headed serine proteinase inhibitor--human plasma kallikrein and elastase inhibitor—from *Boophilus microplus* larvae. Immunopharmacology 45, 171–177.

Turk, B., Turk, D., Turk, V., 2000. Lysosomal cysteine proteases: more than scavengers. Biochim. Biophys. Acta 1477, 98–111.

Yamaji, K., Tsuji, N., Miyoshi, T., Islam, M.K., Hatta, T., Alim, M.A., Anisuzzaman, M., Kushibiki, S., Fujisaki, K., 2009. A salivary cystatin, HlSC-1, from the ixodid tick *Haemaphysalis longicornis* play roles in the blood-feeding processes. Parasitol. Res. 106(1), 61–68.

Zavasnik-Bergant, T., Turk, B., 2007. Cysteine proteases: destruction ability versus immunomodulation capacity in immune cells. Biol. Chem. 388, 1141–1149.

Zhou, J., Liao, M., Ueda, M., Gong, H., Xuan, X., Fujisaki, K., 2009. Characterization of an intracellular cystatin homolog from the tick *Haemaphysalis longicornis*. Vet. Parasitol. 160, 180–183.

Zhou, J., Ueda, M., Umemiya, R., Battsetseg, B., Boldbaatar, D., Xuan, X., Fujisaki, K., 2006. A secreted cystatin from the tick Haemaphysalis longicornis and its distinct expression patterns in relation to innate immunity. Insect Biochem. Mol. Biol. 36, 527–535.

Part III
Anticoagulant Proteins

Chapter 11
Blood Coagulation Factor IX/Factor X-Binding Protein

Takashi Morita

Abstract C-type lectin-like proteins of snake have a variety of biological properties, acting for example as an anticoagulant, procoagulant, and agonist/antagonist of platelet activation. Dimerization or oligomer formation of carbohydrate recognition domain (CRD) in C-type lectin by 3D domain swapping generates novel proteins with new functions such as coagulant-, anticoagulant-, and platelet-modulating activities. The structural and functional studies of the first identified C-type lectin-like proteins, IX/X-bp, have been instrumental in defining how new functionally heterodimeric C-type lectin-like proteins are generated from monomeric CRD (carbohydrate recognition domain) in C-type lectin by 3D domain swapping. The crystal structure of IX/X-bp revealed that the two subunits associated by 3 D-domain swapping, and this dimerization resulted in the creation of a concave surface serving as a binding site of Gla domain, the functionally important domain of blood coagulation factors. The strong activities by snaclecs such as IX/X-bp and X-bp are caused by the binding at the Gla domain of factors IX and X. C-type lectin-like proteins of snake venom such as IX/X-bp and its structurally-related proteins recognize various ligands by the higher frequency of mutation in the open reading frames than in the non-coding regions after duplication of a gene.

Introduction

One class of the functional components of snake venom is the C-type lectin-like protein (CLP) family that includes IX/X-bp (factors IX/X-binding protein) (Atoda and Morita, 1989; Atoda et al., 1991; Sekiya et al., 1993), RVV-X (factor X activator of Russell's viper venom) (Takeda et al., 2007; Takeya et al., 1992), alboaggregin-B (platelet agonist) (Peng et al., 1991; Yoshida et al., 1993), and flavocetin-A (platelet

T. Morita (✉)
Department of Biochemistry, Meiji Pharmaceutical University, Noshio, Kiyose, Tokyo 204-0004, Japan; Department of Host Defense and Biological Research, Juntendo University School of Medicine, Tokyo 113-8421, Japan
e-mail: tmorita@my-pharm.ac.jp

R.M. Kini et al. (eds.), *Toxins and Hemostasis*, DOI 10.1007/978-90-481-9295-3_11,

antagonist) (Fukuda et al., 2000; Shin et al., 2000; Taniuchi, et al., 1995). It was recently proposed to call the C-type lectin-like proteins the 'snaclec' protein family (Clemetson et al., 2009a, b). In this review I sometimes use this new name. These snaclec proteins possess a variety of activities, and have proved to be useful tools in elucidating the complex mechanisms involved in clotting and platelet function. These functional and structural studies using snaclecs have led to several important discoveries as follows (Morita, 2004b).

Functional studies of IX/X-bp revealed that Mg^{2+} ions are critical components in the blood coagulation cascade system (Sekiya et al., 1995, 1996b). The crystal structures of Gla domains of coagulation factors X and IX have been elucidated in structural studies of complexes between the Gla domain of factor X and either factor X-binding protein (Mizuno et al., 2001), or factor IX-binding protein (Shikamoto et al., 2003). The discovery of a calcium-dependent prothrombin activator with C-type lectin-like domains, carinactivase (Ca^{2+}-dependent prothrombin activator) (Yamada et al., 1996), has led to its clinical application in measurement of plasma prothrombin levels (Iwahashi et al., 2001; Yamada and Morita, 1999). Many snaclecs are used as critical reagents in the study of platelet activation (Clemetson et al., 1999; Fujimura et al., 1996; Peng et al., 1993). A 3-D domain swapping phenomenon was discovered during the structural study that identified IX/X-bp, the first known C-type lectin like protein.

This review describes the results of studies in which snake venom anticoagulants were used to examine the tertiary structure and function of coagulation factors.

Blood Coagulation Factors-Binding Proteins as Venom Anticoagulants

Representative C-type lectin-like proteins affecting the hemostatic system, including anticoagulants, coagulants in the mammalian blood coagulation cascade (Sekiya et al., 1996b) are shown in Fig. 11.1. Many biologically active C-type lectin-like proteins consist of one or more heterodimers C-type lectin-like subunits. These include the anticoagulant proteins IX/X-bp, IX-bp, and X-bp, which consist of disulfide-linked heterodimers of C-type lectin-like subunits (Atoda and Morita, 1993). Their biological activity is Ca^{2+}-dependent. RVV-X and carinactivase-1 are metalloenzymes containing two C-type lectin-like domains that recognize the Gla domains of factor X and prothrombin, respectively (Morita, 1998; Yamada et al., 1996). The complete amino acid sequence of IX/X-bp, the first C-type lectin like protein to be sequenced, has been reported (Atoda et al., 1991), and its disulfide-bonding pattern has been compared with those of other C-type lectin-like proteins (Atoda and Morita, 1993). The amino acid sequences of the A (α subunit) and B (β subunit) chains of IX/X-bp reveal 15–37% sequence identity with the carbohydrate recognition domain of C-type lectins (Atoda et al., 1991). However, in contrast to the canonical C-type lectins, IX/X-bp has no lectin activity. C-type lectin-like proteins have proved to be useful tools in elucidating the complex mechanisms involved in clotting and platelet formation and the structure-function relationships of clotting factors and glycoproteins of human platelets.

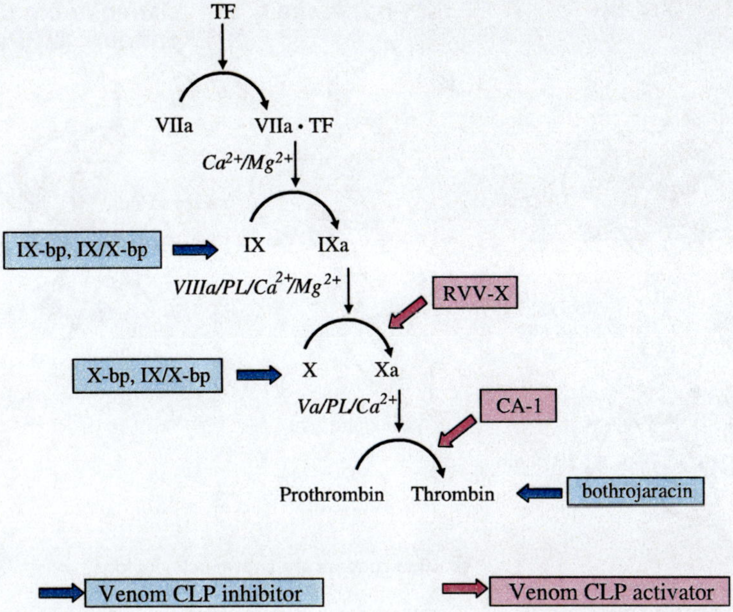

Fig. 11.1 Activators and inhibitors of venom snaclecs, C-type lectin-like proteins on blood coagulation reactions. TF, tissue factor; CA-1, carinactivase-1

The C-type lectin-like protein family includes the anticoagulant proteins habu IX/X-bp (factor IX/factor X-binding protein), habu IX-bp (factor IX-binding protein), and acutus X-bp (factor X-binding protein), which contain disulfide-linked heterodimers of C-type lectin-like subunits (Mizuno et al., 2001; Shikamoto et al., 2003). These C-type lectin-like proteins have strong anticoagulant activities due to their binding to γ-carboxyglutamic acid (Gla)-containing domains of coagulation factors IX and X in the presence of calcium ions (Atoda et al., 1994, 1998). A Gla domain consists of 10–13 Gla residues, and requires Ca^{2+} ions for stabilization of the active conformation for membrane binding.

IX/X-bp, IX-bp and X-bp are heterodimeric proteins that consist of two homologous C-type lectin-like subunits of 14 and 15 kDa. Apparent dissociation constants for the binding of habu IX/X-bp to factor IX and factor X are 0.4 nM and 1.1 nM, respectively, in the presence of Ca^{2+} ions, indicating that these C-type lectin-like proteins have quite high affinities for these coagulation factors (Morita, 2004a, b).

3D-Domain Swapping in the Structure of a Venom Protein IX/X-bp

The crystal structure of IX/X-bp and the structural differences between IX/X-bp and a typical C-type lectin, mannose-binding protein (MBP) are shown in Fig. 11.2a and c (Mizuno et al., 1997). The backbone folds of the subunits are very similar

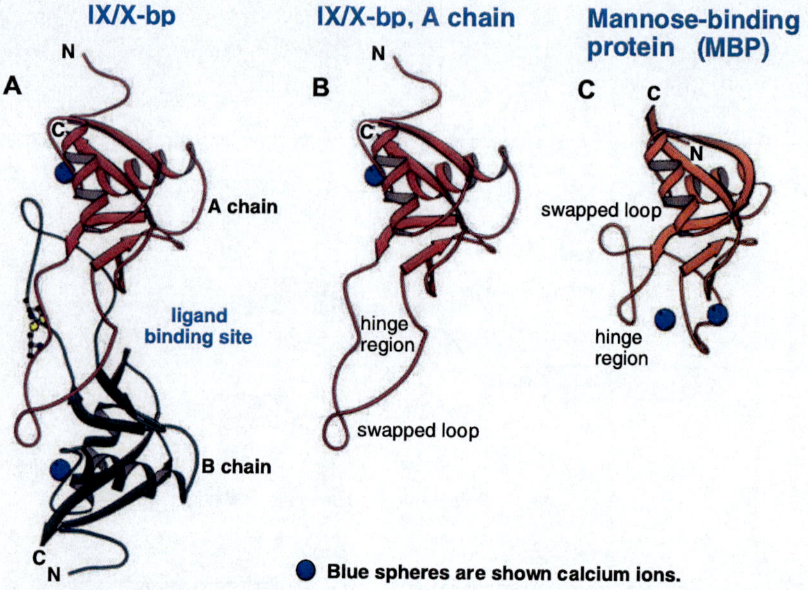

Fig. 11.2 Structural comparison between IX/X-bp and MBP (mannose-binding protein). (*Left*) Structure of IX/X-bp. *Blue spheres* indicate calcium ions. (*Center*) Structure of IX/X-bp A chain. *Spheres* indicate calcium ions. (*Right*) Structure of mannose-binding protein (MBP). *Blue spheres* indicate calcium ions

(Fig. 11.2). The crystal structures of habu IX/X-bp (Mizuno et al., 1997), habu IX-bp (Mizuno et al., 1999) and acutus X-bp (Mizuno et al., 2001) indicate that each subunit of these proteins has a Ca^{2+}-binding site different from that of C-type lectins such as mannose-binding protein (MBP), and that dimerization involves swapping of the central loop of these subunits (Mizuno et al., 1997) (Fig. 11.2). A prominent feature of this structure is a central portion of the polypeptide chain that projects towards the adjoining subunit, forming a tight dimer association (Fig. 11.2a and b). If this feature is excluded, each subunit has a fold similar to the C-type carbohydrate recognition domain (CRD) fold of MBP (compare Fig. 11.2a with c). The polypeptide chain then extends from the surface of the subunit, forming a large open loop, that projects into the adjoining subunit, forming a loop-exchanged dimer, whereas the MBP chain in this region is folded back onto the subunit. The extended loop of IX/X-bp returns to the main body of the subunit. Thus, subunit A generates a structure that is homologous to that of MBP by 'borrowing' the open loop from the adjoining subunit B Subunit B. has similar structural features. Thus, dimerization may be closely related to a function completely different from that of the 'monomer' motif C-type CRD, suggesting economic use of combinations of limited structural domains. This novel structure and function of proteins in snake venom serves as a strategy for survival against animal predators, and represents an evolutionary gain of function for the C-type CRD domain (Mizuno et al., 1997).

Most swapped domains are at either the N or C terminus, and the only exception found to date is IX/X-bp. The two subunits form a heterodimer of IX/X-bp by exchanging a loop in the central part of the molecules (Fig. 11.2). Structural comparison of the two subunits with mannose binding protein (MBP) shows that they adopt the same fold, except that the exchanged loop in the IX/X-bp folds back to the same polypeptide chain in MBP. Thus, IX/X-bp is quasidomain-swapped. IX/X-bp is the only known example of 3D domain swapping taking place in the middle of the molecule, in contrast to other domain-swapped proteins, in which domain swapping takes place at either the N or C terminus. There are two hinge loops in each subunit, see left and center figures of Fig. 11.2. In addition, IX/X-bp is the only known example of a domain-swapped heterodimer. All other domain-swapped proteins are homodimers or homooligomers (Liu and Eisenberg, 2002).

A mechanism of domain swapping for protein dimerization was proposed by Eisenberg and coworkers (Bennett et al., 1995) (Fig. 11.3). In this model, the hinge loop between two domains plays an important role in determining whether domain swapping occurs (Bennett et al., 1995). If a hinge region is shortened by a deletion (Fig. 11.3a), the closed monomer structure is sterically hindered, and the resultant open monomer may be unstable because of the exposure of residues normally buried in the interface. Domain-swapped dimers would then occur. The deletion of the hinge region is very important to the generation of folds in these peptide chains. The deletion in the hinge loop promotes central loop swapping in dimerization to create new functions, because a candidate-binding site is newly formed by a concave surface created by the dimerization of the two subunits. Thus, the process of domain swapping in IX/X-bp would be represented by Fig. 11.3a.

Another mechanism of domain swapping is by amino acid substitution in the hinge region (Fig. 11.3b). Huntington and his coworkers reported the crystallographic structure of a stable serpin (antithrombin mutant) dimer which has domain swapping of more than 50 amino acid residues (Yamasaki et al., 2008). Polymerization of misfolded proteins is the mechanism underlying a multitude of diseases including Alzheimer's, Huntington's and Parkinson's and the prion

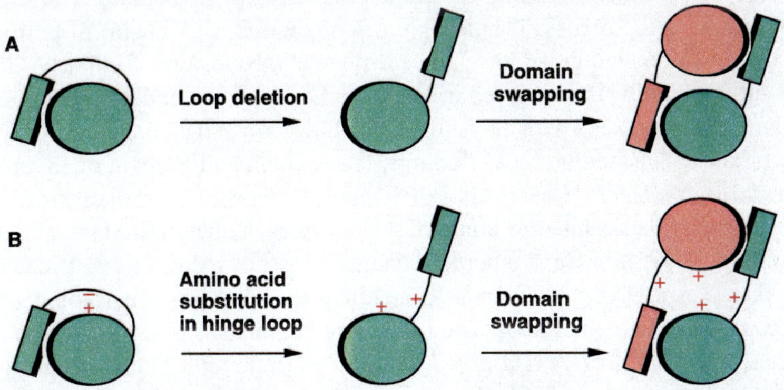

Fig. 11.3 Hinge loop mutations that lead to 3 D domain swapping

encephalopathies. Yamasaki et al also proposed that as the swapping occurs the serpin polymerizes, which explains the molecular basis of the neurodegenerative disorders (Whisstock and Bottomley, 2008; Yamasaki et al., 2008).

The molecular evolution of the myotoxic phospholipase A_2 from snake venom was reported (Ohno et al., 2003). Snake venom contains a variety of venom proteins, such as IX/X-bp and its structurally-related proteins, that recognize various ligands by the higher frequency of mutation in the open reading frames than in the non-coding regions after duplication of a gene (Nakashima et al., 1993; Ogawa et al., 1992; Ohno et al., 2003, Matsuzaki et al., 1996; Tani et al., 2002). Accelerated duplication and mutation of the genes for the A and B chains of IX/X-bp and of other CLPs could have resulted in the present family of heterodimeric CLPs that includes IX/X-bp and various other proteins that are structurally related to but functionally distinct from one another (Matsuzaki et al., 1996; Tani et al., 2002).

The crystal structure of factor X Gla domain peptide complexed with its binding protein X-bp has been determined (Mizuno et al., 1997). Factor X was first cleaved by α-chymotrypsin to isolate the Gla domain peptide residues 1–44. Then, the complex of factor X Gla domain was crystallized with X-bp.

Ca^{2+}-Binding Properties of IX/X-bp and Related Proteins

IX/X-bp, IX-bp, and X-bp have two independent Ca^{2+}-binding sites: a high affinity site with a K_d of 15–25 μM, and a low affinity site with a K_d of 100–200 μM (Atoda et al., 2002). The differences in the Ca^{2+}-binding properties of IX/X-bp and related proteins can be understood by considering their amino acid sequences together with their 3D structures (Atoda et al., 2002). The crystal structures of IX/X-bp, IX-bp, and X-bp (Mizuno et al., 1997, 1999, 2001) demonstrate that these proteins each have two Ca^{2+}-binding sites and the amino acid sequences show that the four amino acid residues forming the Ca^{2+}-binding sites in subunits α (Ser41, Glu43, Glu47, and Glu128 of chain A) and β (Ser41, Gln43, Glu47, and Glu120 of chain B) are conserved. The Ca^{2+}-binding α subunits of IX/X-bp, IX-bp, and X-bp possess Glu128, whereas the non-Ca^{2+}-binding types of subunit α (A chain) of botrocetin, bitiscetin, and alboaggregin-B have a Lys residue at this position. Similarly, Glu120 of subunit β (chain B) is conserved in IX/X-bp, IX-bp, X-bp, and botrocetin, which bind Ca^{2+}, but is replaced with a Lys residue in bitiscetin and alboaggregin-B, which do not bind Ca^{2+}. Based on these findings, the A chain and B chain of IX-bp from *Trimeresurus stejnegeri* (Lee et al., 2003), and B chain of bothrojaracin (Zingali et al., 1993) can be assumed to bind a Ca^{2+} ion in each chain (Atoda et al., 2002). The folding of the non-Ca^{2+}-binding subunits does not require Ca^{2+} because the Lys residue forms hydrogen bonds to stabilize the IX/X-bp-like folding (Atoda et al., 2002; Fukuda et al., 2000). The Lys residue in this position is therefore structurally essential and strictly conserved in non-Ca^{2+}-binding subunits (Batuwangala et al., 2004; Fukuda et al., 2000; Hirotsu et al., 2001; Horii et al., 2003; Jasti et al., 2004).

Significance of RMI Sequence of X-bp and the Related Proteins

The Gla domain of factor Xa is bound to X-bp. The crystal structure of factor X Gla domain peptide complexed with its binding protein X-bp has been determined at 2.3 Å resolution (Mizuno et al., 2001). The peptide chain folding of X-bp is similar to that of IX/X-bp. The Gla domain-binding site is located on the concave face formed by the two subunits. Eight calcium ions are bound to this face; one of these, Ca-1, is involved in the interaction at the X-bp/Gla domain interface.

The Gla domain of factor X has three hydrophobic amino residues, Phe4, Leu5, and Val8 in the N-terminal loop, that contribute to membrane binding. These hydrophobic residues interact with hydrophobic amino acids in the loop of X-bp subunit B. The aromatic ring of Phe4 in the Gla domain stacks over the guanidino group of Arg112 from the X-bp β subunit. Leu5 of the Gla domain participates in a hydrophobic interaction with Ile114 of subunit β. In addition, Val8 of the Gla domain interacts with Met113 and Ile114 of subunit β. Thus, the amino acid sequence Arg112-Met113-Ile114 in the C-terminal region of X-bp subunit β is critical for the formation of a complex between the Gla domain and X-bp. Three other snaclec anticoagulants, IX/X-bp and IX-bp of *T. flavoviridis* (Atoda et al., 1991, 1995) and IX-bp of *Trimeresurus stejnegeri* (Lee et al., 2003), have the homologous anticoagulant sequence RMM (Arg112-Met113-Met114), while all other platelet snaclec modulators found so far lack these sequences. This RMI/RMM sequence in subunit β of anticoagulant snaclecs contributes to their specificity for the Gla domains of factor X or factor IX. The crystal structure further shows that Glu98 and Lys100 of X-bp A chain interact with Arg28 and Gla25 of the factor X Gla domain, respectively. Also, Lys100, Asn103, and Arg107 of X-bp interact with Gla29, Gla32, and Gla29/Gla32 of the Gla domain, respectively. These and other ionic interactions strongly stabilize the formation of the complex through a total of nine salt bridges (Mizuno et al., 2001).

The Gla domain of factor X binds to the concave surface of X-bp. Thus, X-bp acts as an anticoagulant by binding to the Gla domain of factor X, which is essential for Ca^{2+}-dependent binding of factor X to phospholipids membrane.

The crystal structures of the Gla domain from another coagulation factor IX has also recently been determined complexed with another snaclec, IX-bp (Shikamoto et al., 2003). In this case, the Gla domain peptide (amino acid residues 1–46) of factor IX was isolated by limited proteolysis with Asp-N protease (Sekiya et al., 1996a). The crystal structures of the Mg^{2+}/Ca^{2+}-bound and Ca^{2+}-bound (Mg^{2+}-free) factor IX Gla domain (IXGD1-46) in complex with its binding protein (IX-bp) at 1.55 and 1.80 Å resolution, respectively, revealed that three Mg^{2+} and five Ca^{2+} ions were bound in the Mg^{2+}/Ca^{2+}-bound IXGD1-46, and that the Mg^{2+} ions were replaced by Ca^{2+} ions in Mg^{2+}-free IXGD1-46 (Shikamoto et al., 2003). The crystal structure results clearly suggest that Mg^{2+} ions are required to maintain the native conformation and the *in vivo* function of the factor IX Gla domain during blood coagulation. The crystal structure of factor IX Gla domain (residues 1–46) in complex with IX-bp shows that there are three cation-binding sites that are occupied by Mg^{2+} ions when examined within the physiological range of the Mg^{2+} and Ca^{2+}

ion concentrations (Shikamoto et al., 2003). In addition, the three Mg^{2+} ions on the surface of factor IX Gla domain 1–46 were at locations that corresponded to the positions of Ca-1, Ca-7, and Ca-8 in the Gla domain of factor X (residues 1–44) in the crystal structure of the factor X Gla domain (residues 1–46) IX-bp complex. Three Mg^{2+} ion-binding sites have also been identified in the crystal structure of the factor IX Gla domain 1–46 in complex with its venom-binding protein (Shikamoto et al., 2003). A structural study of factor IX Gla domain using snake venom inhibitor has confirmed the importance of Mg^{2+} ions as augmenters of Ca^{2+} ions in factor IX activity (Sekiya et al., 1995, 1996b).

Crystallographic findings indicate that the anticoagulative effect of X-bp is due to the fact that the two patches of the Gla domain essential for membrane binding are buried in the complex formed with X-bp (Mizuno et al., 2001). The binding mode of the interaction of factor Xa with the membrane surface may represent a general feature of the homologous Gla domains of this family of proteins.

Conclusions

Snaclecs, C-type lectin-like proteins of snake, have a variety of biological properties, acting for example as an anticoagulant, procoagulant, and agonist/antagonist of platelet activation. Dimerization or oligomer formation of carbohydrate recognition domain (CRD) in C-type lectin by 3D domain swapping generates novel proteins with new functions such as coagulant-, anticoagulant-, and platelet-modulating activities. The structural and functional studies of the first sequenced snaclec, IX/X-bp, have been instrumental in defining how new heterodimeric C-type lectin-like proteins with different functions are generated from monomeric CRDs (carbohydrate recognition domain) in C-type lectins by 3D domain swapping. The crystal structure of a snaclec, IX/X-bp, revealed that the two subunits associated by 3 D-domain swapping, and this dimerization resulted in the creation of a concave surface serving as a binding site for the Gla domain, the functionally important domain of blood coagulation factors. The strong activity of snaclecs such as IX/X-bp and X-bp are caused by the binding at the Gla domain of factors IX and X. C-type lectin-like proteins from snake venom such as IX/X-bp and its structurally-related proteins evolve to recognize various ligands by a higher frequency of mutations in the open reading frames than in the noncoding regions after duplication of a gene.

References

Atoda, H., Hyuga, M., Morita, T., 1991. The primary structure of coagulation factor IX/factor X-binding protein isolated from the venom of *Trimeresurus flavoviridis*. Homology with asialo-glycoprotein receptors, proteoglycan core protein, tetranectin, and lymphocyte Fcε receptor for immunoglobulin E. J. Biol. Chem. 266, 14903–14911.

Atoda, H., Ishikawa, M., Mizuno, H., Morita, T., 1998. Coagulation factor X-binding protein from *Deinagkistrodon acutus* venom is a Gla domain-binding protein. Biochemistry 37, 17361–17370.

Atoda, H., Ishikawa, M., Yoshihara, E., Sekiya, F., Morita, T., 1995. Blood coagulation factor IX-binding protein from the venom of *Trimeresurus flavoviridis*: purification and characterization. J. Biochem. (Tokyo) 118, 965–973.

Atoda, H., Kaneko, H., Mizuno, H., Morita, T., 2002. Calcium-binding analysis and molecular modeling reveal echis coagulation factor IX/factor X-binding protein has the Ca-binding properties and Ca ion-independent folding of other C-type lectin-like proteins. FEBS Lett. 531, 229–234.

Atoda, H., Morita, T., 1989. A novel blood coagulation factor IX/factor X-binding protein with anticoagulant activity from the venom of *Trimeresurus flavoviridis* (Habu snake): isolation and characterization. J. Biochem. (Tokyo) 106, 808–813.

Atoda, H., Morita, T., 1993. Arrangement of the disulfide bridges in a blood coagulation factor IX/factor X-binding protein from the venom of *Trimeresurus flavoviridis*. J. Biochem. (Tokyo) 113, 159–163.

Atoda, H., Yoshida, N., Ishikawa, M., Morita, T., 1994. Binding properties of the coagulation factor IX/factor X-binding protein isolated from the venom of *Trimeresurus flavoviridis*. Eur. J. Biochem. 224, 703–708.

Batuwangala, T., Leduc, M., Gibbins, J.M., Bon, C., Jones, E.Y., 2004. Structure of the snake-venom toxin convulxin. Acta Crystallogr. D Biol. Crystallogr. 60(Pt 1), 46–53.

Clemetson, J.M., Polgar, J., Magnenat, E., Wells, T.N., Clemetson, K.J., 1999. The platelet collagen receptor glycoprotein VI is a member of the immunoglobulin superfamily closely related to FcalphaR and the natural killer receptors. J. Biol. Chem. 274, 29019–29024.

Bennett, M.J., Schluneggar, M.P., Eisenberg, D., 1995, 3 D domain swapping: mechanism for oligomer assembly. Protein Sci. 4, 2455–2468.

Clemetson, K.J., Morita, T., Kini, R.M., 2009a, Classification and nomenclature of snake venom C-type lectins and related proteins. Toxicon 54, 83.

Clemetson, K.J., Morita, T, Kini, M.R., 2009b. Registry of exogenous hemostatic factors of SSC of the ISTH. Scientific and standardization committee communications: classification and nomenclature of snake venom C-type lectins and related proteins. J. Thromb. Haemost. 7, 360.

Fujimura, Y., Kawasaki, T., Titani, K., 1996. Snake venom proteins modulating the interaction between von Willebrand factor and platelet glycoprotein Ib. Thromb. Haemost. 76, 633–639.

Fukuda, K., Mizuno, H., Atoda, H., Morita, T., 2000. Crystal structure of flavocetin-A, a platelet glycoprotein Ib-binding protein, reveals a novel cyclic tetramer of C-type lectin-like heterodimers. Biochemistry 39, 1915–1923.

Hirotsu, S., Mizuno, H., Fukuda, K., Ma, C.Q., Matsui, T., Hamako, J., Morita, T., Titani, K., 2001. Crystal structure of bitiscetin, a von Willebrand factor-dependent platelet aggregation inducer. Biochemistry 40, 13592–13597.

Horii, K., Okuda, D., Morita, T., Mizuno, H., 2003. Structural characterization of EMS16, an antagonist of collagen receptor (GPIa/IIa) from the venom of *Echis multisquamatus*. Biochemistry 42, 12497–12502.

Iwahashi, H., Kimura, M., Nakajima, K., Yamada, D., Morita, T., 2001. Determination of plasma prothrombin level by Ca^{2+}-dependent prothrombin activator (CA-1) during warfarin anticoagulation. J. Heart Valve Dis. 10, 388–392.

Jasti, J., Paramasivam, M., Srinivasan, A., Singh, T.P., 2004. Crystal structure of echicetin from *Echis carinatus* (Indian saw-scaled viper) at 2.4A resolution. J. Mol. Biol. 335, 167–176.

Lee, W.H., Zhuang, Q.Y., Zhang, Y., 2003. Cloning and characterization of a blood coagulation factor IX-binding protein from the venom of *Trimeresurus stejnegeri*. Toxicon 41, 765–772.

Liu, Y., Eisenberg, D., 2002. 3D domain swapping: as domains continue to swap. Protein Sci. 11, 1285–1299.

Matsuzaki, R., Yoshihara, E., Yamada, M., Shima, K., Atoda, H., Morita, T., 1996. cDNA cloning of IX/X-BP, a heterogeneous two-chain anticoagulant protein from snake venom. Biochem. Biophys. Res. Commun. 220, 382–387.

Mizuno, H., Fujimoto, Z., Atoda, H., Morita, T., 2001. Crystal structure of an anticoagulant protein in complex with the Gla domain of factor X. Proc. Natl. Acad. Sci. U.S.A. 98, 7230–7234.

Mizuno, H., Fujimoto, Z., Koizumi, M., Kano, H., Atoda, H., Morita, T., 1997. Structure of coagulation factors IX/X-binding protein, a heterodimer of C-type lectin domains. Nat. Struct. Biol. 4, 438–441.

Mizuno, H., Fujimoto, Z., Koizumi, M., Kano, H., Atoda, H., Morita, T., 1999. Crystal structure of coagulation factor IX-binding protein from habu snake venom at 2.6 A: implication of central loop swapping based on deletion in the linker region. J. Mol. Biol. 289, 103–112.

Morita, T., 1998. Proteases which activate factor X, in: Bailey, G.S. (Ed.), Snake Venom Enzymes. Alaken, Inc., Colorado, pp. 179–208.

Morita, T., 2004a. C-type lectin-related proteins from snake venoms. Curr. Drug Targets-Cardiovasc. Haematol Disorders 4, 357–373.

Morita, T., 2004b. Use of snake venom inhibitors in studies of the function and tertiary structure of coagulation factors. Int. J. Hematol. 79, 123–129.

Nakashima, K., Ogawa, T., Oda, N., Hattori, M., Sakaki, Y., Kihara, H., Ohno, M., 1993. Accelerated evolution of *Trimeresurus flavoviridis* venom gland phospholipase A_2 isozymes. Proc. Natl. Acad. Sci. U.S.A. 90, 5964–5968.

Ogawa, T., Oda, N., Nakashima, K., Sasaki, H., Hattori, M., Sakaki, Y., Kihara, H., Ohno, M., 1992. Unusually high conservation of untranslated sequences in cDNAs for *Trimeresurus flavoviridis* phospholipase A_2 isozymes. Proc. Natl. Acad. Sci. U.S.A. 89, 8557–8561.

Ohno, M., Chijiwa, T., Oda-Ueda, N., Ogawa, T., Hattori, S., 2003. Molecular evolution of myotoxic phospholipases A_2 from snake venom. Toxicon 42, 841–854.

Peng, M., Lu, W., Beviglia, L., Niewiarowski, S., Kirby, E.P., 1993. Echicetin: a snake venom protein that inhibits binding of von Willebrand factor and alboaggregins to platelet glycoprotein Ib. Blood 81, 2321–2328.

Peng, M., Lu, W., Kirby, E.P., 1991. Alboaggregin-B: a new platelet agonist that binds to platelet membrane glycoprotein Ib. Biochemistry 30, 11529–11536.

Sekiya, F., Atoda, H., Morita, T., 1993. Isolation and characterization of an anticoagulant protein homologous to botrocetin from the venom of *Bothrops jararaca*. Biochemistry 32, 6892–6897.

Sekiya, F., Yamashita, T., Atoda, H., Komiyama, Y., Morita, T., 1995. Regulation of the tertiary structure and function of coagulation factor IX by magnesium (II) ions. J. Biol. Chem. 270, 14325–14331.

Sekiya, F., Yoshida, M., Yamashita, T., Morita, T., 1996a. Localization of the specific binding site for magnesium (II) ions in factor IX. FEBS Lett. 392, 205–208.

Sekiya, F., Yoshida, M., Yamashita, T., Morita, T., 1996b. Magnesium (II) is a crucial constituent of the blood coagulation cascade. Potentiation of coagulant activities of factor IX by Mg^{2+} ions. J. Biol. Chem. 271, 8541–8544.

Shikamoto, Y., Morita, T., Fujimoto, Z., Mizuno, H., 2003. Crystal structure of Mg^{2+}- and Ca^{2+}-bound Gla domain of factor IX complexed with binding protein. J. Biol. Chem. 278, 24090–24094.

Shin, Y., Okuyama, I., Hasegawa, J., Morita, T., 2000. Molecular cloning of glycoprotein Ib-binding protein, flavocetin-A, which inhibits platelet aggregation. Thromb. Res. 99(3), 239–247.

Takeda, S., Igarashi, T., Mori, H., 2007. Crystal structure of RVV-X: an example of evolutionary gain of specificity by ADAM proteinases. FEBS Lett. 581, 5859–5864

Takeya, H., Nishida, S., Miyata, T., Kawada, S., Saisaka, Y., Morita, T., Iwanaga, S., 1992. Coagulation factor X activating enzyme from *Russell's viper* venom (RVV-X). A novel metalloproteinase with disintegrin (platelet aggregation inhibitor)-like and C-type lectin-like domains. J. Biol. Chem. 267, 14109–14117.

Tani, A., Ogawa, T., Nose, T., Nikandrov, N.N., Deshimaru, M., Chijiwa, T., Chang, C.C., Fukumaki, Y., Ohno, M., 2002. Characterization, primary structure and molecular evolution of anticoagulant protein from *Agkistrodon actus* venom. Toxicon 40, 803–813.

Taniuchi, Y., Kawasaki, T., Fujimura, Y., Suzuki, M., Titani, K., Sakai, Y., Kaku, S., Hisamichi, N., Satoh, N., Takenaka, T., et al., 1995. Flavocetin-A and -B, two high molecular mass glycoprotein Ib binding proteins with high affinity purified from *Trimeresurus flavoviridis* venom, inhibit platelet aggregation at high shear stress. Biochim. Biophys. Acta 1244, 331–338.

Whisstock, J.C., Bottomley, S.P., 2008. Serpins' mystery solved. Nature 455, 1189–1190.

Yamada, D., Morita, T., 1999. CA-1 method, a novel assay for quantification of normal prothrombin using a Ca^{2+} -dependent prothrombin activator, carinactivase-1. Thromb. Res. 94, 221–226.

Yamada, D., Sekiya, F., Morita, T., 1996. Isolation and characterization of carinactivase, a novel prothrombin activator in *Echis carinatus* venom with a unique catalytic mechanism. J. Biol. Chem. 271, 5200–5207.

Yamasaki, M., Li, W., Johnson, D.J.D., Huntington, J.A., 2008. Crystal structure of a stable dimer reveals the molecular basis of serpin polymerization. Nature 455, 1255–1259

Yoshida, E., Fujimura, Y., Miura, S., Sugimoto, M., Fukui, H., Narita, N., Usami, Y., Suzuki, M., Titani, K., 1993. Alboaggregin-B and botrocetin, two snake venom proteins with highly homologous amino acid sequences but totally distinct functions on von Willebrand factor binding to platelets. Biochem. Biophys. Res. Commun. 191, 1386–1392.

Zingali, R.B., Jandrot-Perrus, M., Guillin, M.C., Bon, C., 1993. Bothrojaracin, a new thrombin inhibitor isolated from *Bothrops jararaca* venom: characterization and mechanism of thrombin inhibition. Biochemistry 32, 10794–10802.

Chapter 12
Bothrojaracin – A Potent Thrombin Inhibitor

Russolina B. Zingali and Ana Cristina Ferraz Nogueira

Abstract Snake venoms are rich in a large variety of proteins and peptides that can interfere with the hemostatic system. This review focuses on bothrojaracin, which is snaclec molecule, either bind to thrombin, inhibiting its biological activities, such as clotting of fibrinogen, platelet activation or to prothrombin, impairing thrombin formation. Bothrojaracin interacts with both molecules, forming a stable 1:1 complex. The calculated Kd for bothrojaracin was 0.6 nM and 100 nM for thrombin and prothrombin, respectively. Bothrojaracin binds to thrombin exosite I displacing ligands such as fibrin, hirudin, thrombomodulin and factor V and do not block the catalytic site. This protein has helped in our understanding of some molecular aspects of the thrombin and prothrombin structure–function relationship. The knowledge about the mechanism of action and details of structural aspects will certainly result in new medical and pharmacological applications. Furthermore, bothrojaracin offer attractive template for the development of rationally designed therapeutic agents.

Introduction

The great interest of the scientific community in snake venom machinery is due to its ability to affect key elements in almost all animal physiological pathways. The multiple biologically active components, including peptides and proteins, found in snake venom correspond to 90–95% of its dry weight and are responsible for all the biological effects observed following the envenomation of the prey. Additional components in the venom are metallic cations, carbohydrates, nucleosides, biogenic amines and very low levels of free amino acids and lipids (Markland, 1998). Diverse toxins from *Bothrops* venoms can interfere with the hemostatic system and have already been isolated and characterized as procoagulant, anticoagulant or fibrinolytic factors (Assafim et al., 2006; Castro et al., 2004; Marsh, 1994). Several

R.B. Zingali (✉)
Instituto de Bioquímica Medica, Centro de Ciências da Saúde, UFRJ, Rio de Janeiro, Brazil
e-mail: lzingali@gmail.com

R.M. Kini et al. (eds.), *Toxins and Hemostasis*, DOI 10.1007/978-90-481-9295-3_12,

additional components are described in the literature, which alter platelet function by displaying either pro- or antiaggregating properties (Markland, 1998).

Thrombin and the Hemostatic System

Blood clotting involves a large number of proteins that act in response to vascular injury and lead to changes in the vascular endothelial surface with the release of procoagulant factors (Bukys et al., 2006). Hemostasis is a complex process that involves not only platelet activation and clot formation (to ensure the arrest of bleeding when a blood vessel is detached), but also clot dissolution by the fibrinolytic pathway. These two opposite mechanisms are in equilibrium and keep the hemodynamic properties of the blood constant (Braud et al., 2000; Davie et al., 1991). Vascular injury initially promotes recruitment, aggregation and activation of platelets to form the primary plug. Subsequently, the generation of the fibrin plug after sequential activation of serine protease zymogens takes place (Mann, 1999; Braud et al., 2000). Following blood vessel damage, circulating active factor VII (VIIa) forms a complex with its cofactor, cell-bound tissue factor (TF). This "extrinsic tenase" complex, which begins the coagulation process, binds to factor X. Then, the active site of factor VIIa cleaves the peptide bonds necessary to generate activated factor X (Xa). Factor Xa may also be generated in a tissue-factor-independent pathway, which maintains procoagulant activity through the "intrinsic tenase." The central episode of the coagulation process is the generation of thrombin as the product of activation of the zymogen prothrombin (Mann et al., 2003). Thrombin is formed by the cleavage of two peptide bonds in prothrombin by the prothrombinase complex, which requires factor Xa, membranes containing acidic phospholipids, and the protein cofactor FVa (Fig. 12.1). It is also modulated by Ca^{2+}(Kalafatis et al., 1994).

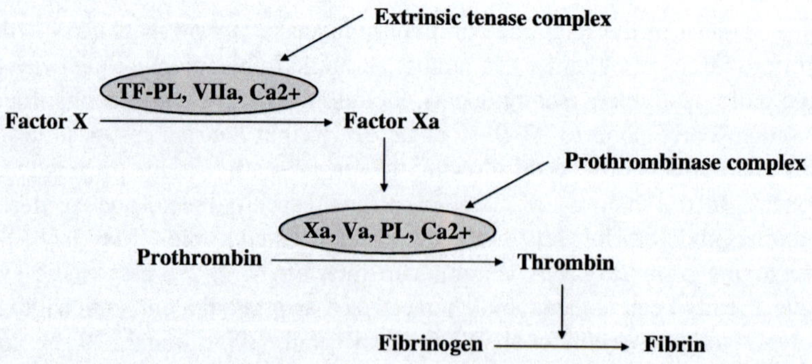

Fig. 12.1 A simplified scheme of fibrin generation. Exposure of tissue factor (TF) upon vascular damage leads to formation of a complex with FVIIa (extrinsic tenase complex) that activates FX. FXa combines with the protein cofactor FVa to form the prothrombinase complex, which in turn activates prothrombin to thrombin. Finally, thrombin cleaves fibrinogen, which leads to fibrin formation. (PL-phospholipids surface from activated platelet)

Thrombin

Thrombin is a two chain enzyme of 36,700 Da comprising an N-terminal "A" chain and a C-terminal "B" chain, which are covalently linked through a single disulfide bond. This multifunctional trypsin-like serine protease (Fig. 12.2) is the last enzyme in the clotting cascade and acts by cleaving fibrinogen to fibrin. It also interacts specifically with other protein substrates, receptors, cofactors, inhibitors, carbohydrates, and modulators (Bode, 2006). Thrombin is a potent cellular agonist, regulates its own production and amplifies the procoagulant process by activating Factors V, VIII and XI. It also activates Factor XIII, which is a glutaminyl transferase that stabilizes the fibrin network (Jenny and Mann, 1998). However, thrombin can also act as an anticoagulant. When thrombin leaves the procoagulant environment, it binds to thrombomodulin, which is a membrane protein present in resting endothelial cells. This complex then activates Protein C, which in turn hydrolyzes Factors Va and VIIIa, and impairs thrombus progression (Kalafatis et al., 1997).

The activity of this multibinding protein is determined not only by the catalytic site and the primary specific pocket, but also by secondary recognition sites called anion-binding exosite-1 (ABE-1) and anion-binding exosite-2 (ABE-2). This is illustrated in the schematic representation of thrombin in Fig. 12.3. ABE-1 has been implicated in the binding of thrombin to fibrinogen (Hofsteenge et al., 1988), thrombomodulin (Hofsteenge et al., 1986), heparin cofactor II (Sheehan et al., 1993), the platelet thrombin receptor PAR-I (Liu et al., 1991) and the leech anticoagulant, hirudin (Stone and Hofsteenge, 1986). On the other hand, ABE-2 is primarily responsible for thrombin binding to heparin (Sheehan et al., 1993), protease nexin

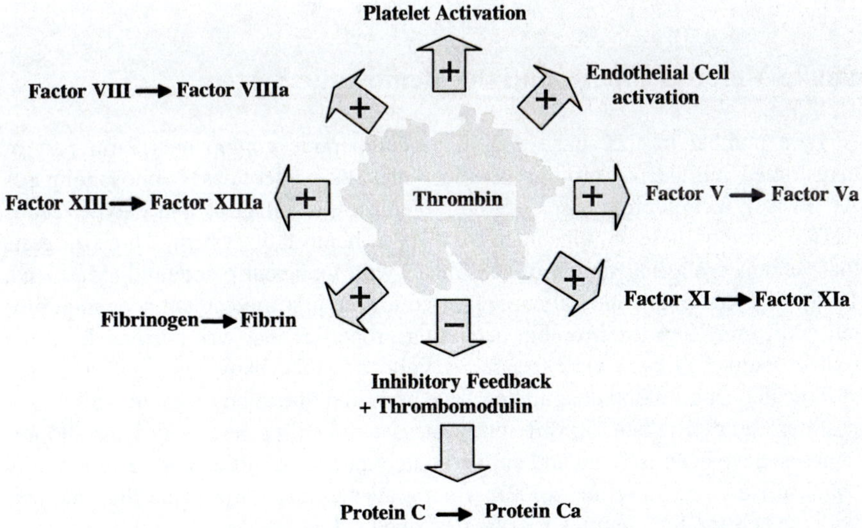

Fig. 12.2 Down (–) and up (+) regulation of blood coagulation pathways by thrombin. Thrombin can function in both procoagulant and anticoagulant pathways

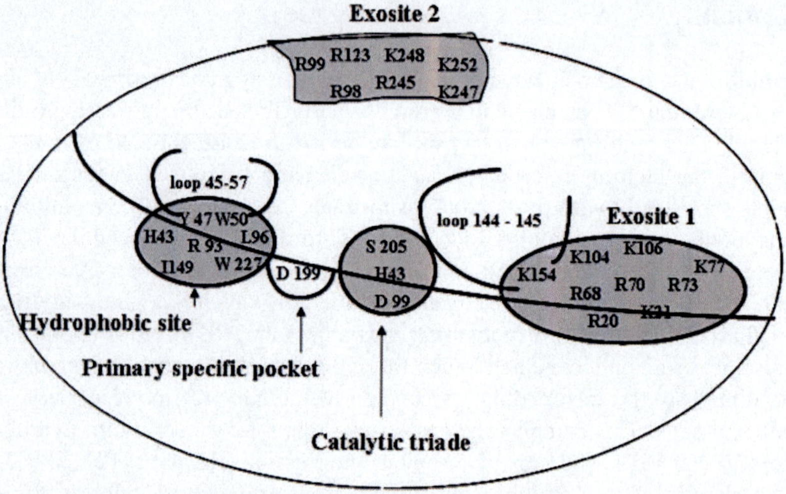

Fig. 12.3 Schematic representation of a thrombin molecule. Thrombin is a globular protein made of two positively charged regions named exosites 1 and 2, a catalytic triad, a primary specific pocket and a hydrophobic site. (Adapted from Guillin et al., 1995)

(Stone et al., 1994) and antithrombin III (Sheehan et al., 1993). Furthermore, both ABEs are involved in the activation of factors V and VIII (Bukys et al., 2006; Esmon and Lollar, 1996; Myles et al., 2001). Due to the central role of this molecule in hemostasis, many new direct and indirect thrombin inhibitors are being developed as antithrombotic drugs (Fareed et al., 1999).

Snake-Venom Proteins and the Hemostatic System

A great number of the snake venom protein/peptide components form part of a restricted number of protein families, such as nucleotidases, phospholipases A_2 (PLA_2), metalloproteinases, serine proteinases, disintegrins and C-type lectins (CTL). Some of these proteins interfere with the hemostatic system, which suggests that venoms are a rich source of substances with therapeutic potential (Markland, 1998). In 1965, Shiau and Ouyang reported for the first time an anticoagulant protein with inhibitory prothrombin activation properties that was purified from the snake venom of *Trimeresurus gramineus*. Following this, many other molecules that specifically act on blood coagulation have been identified and characterized. In the past decade, more than 150 different proteases, including serine- and metalloproteinases, have been isolated and approximately one third of them were structurally characterized (Matsui et al., 2000). Among all the purified proteins that interfere with the hemostatic system, a few are described below.

Serine proteinases from snake venom (SPSV) conserve the trypsin fold. Most of the SPSV that have been studied display activity similar to thrombin towards fibrinogen and are named thrombin-like enzymes, such as batroxobin from *Bothrops atrox moojeni.* Thrombocytin and PA-Bj from *Bothrops atrox* and *B. jararaca* venom, respectively, were reported to activate proteinase-activated receptors 1 and 4 of thrombin (PAR-1 and PAR-4) in human blood platelets (Braud et al., 2000; Kirby et al., 1979; Santos, 2000). On the other hand, the serine proteinase ACC-C purified from *Agkistrodon contortrix* venom activates protein C without requiring thrombomodulin (Murakami and Arni, 2005). Another relevant serine proteinase from *Trimeresurus stejnegeri* venom (TSV-PA) has been described as a specific plasminogen activator (Zhang et al., 1997). Furthermore, there are some reports that RVV-V from Russell's viper venom activates factor V by converting it to Factor Va (Hjort, 1957; Rosing et al., 2001). Some of these enzymes have been used in therapy and for the development of hemostatic diagnostic tests (Funk et al., 1971; Matsuda et al., 1985).

Metalloproteinases are a family of proteins dependent on metal ions for their enzymatic activity, particularly divalent cations like zinc and calcium, which play a critical role in proteolytic and biological activity (Fox and Serrano, 2009). Some metalloproteinases affect coagulation factors and lead to their activation or inactivation (for review see Kamiguti, 2005). For example, ecarin is a metalloproteinase isolated from the venom of *Echis carinatus* that cleaves prothrombin molecules, and it is employed as a diagnostic tool for the detection of abnormal types of prothrombin (Weinger et al., 1980).

Phospholipase A_2 (PLA_2) enzymes inhibit blood coagulation by acting on phospholipids (Boffa et al., 1976). Phospholipases A_2 enzymes, such as superbins I and II or CM-I and CM-II from *Austrelaps supererbus* and *Naja nigicollis,* respectively, have the ability to bind to procoagulant phospholipids. They then compete with coagulation factors and inhibit the extrinsic tenase complex. On the other hand, some PLA_2 can bind directly to coagulation factors like the isoenzyme CM-IV, which binds to factor X ($Kd = 500$ nM) and competes for its binding to factor V during formation of the prothrombinase complex (Kerns et al., 1999; Mounier et al., 2001).

The C-type lectins family can be divided into true C-type lectins, which can bind to saccharides (generally galactose), and C-type lectin-like proteins (CLP) or snaclecs (Clemetson et al., 2009), which have lost the capacity to bind carbohydrates (Guimarães-Gomes et al., 2004). Despite similarities in structure and sequence, proteins of this last class have a wide variety of biological properties and can act as anticoagulants, procoagulants, and as agonists/antagonists of platelet activation (Morita, 2004). Botrocetin was the first snaclec isolated from *Bothrops jararaca* venom. It induces vWF dependent platelet agglutination and is used for the diagnosis of von Willebrand disease and Bernard-Soulier syndrome (Usami et al., 1993). This group also includes echicetin (Peng et al., 1994), agkistin (Yeh et al., 2001) and flavocetin-A (Taniuchi et al., 1995), which strongly inhibits platelet aggregation by specifically binding to platelet GPIb. In addition, alboaggregin B (Peng et al., 1991), TSV-GPIb-BP (Lee and Zhang, 2003) and agglucetin (Wang and Huang, 2001) are snaclecs that stimulate platelet agglutination via GPIb without activating platelets. Proteins that can bind to the Gla domain of Factor IX and/or Factor X are named

FIX/X binding protein (IX/X-bp) and were purified from the venom of *Trimeresurus flavoviridis* (Atoda and Morita, 1989; Atoda et al., 1991, 1995), *B. jararaca* (Sekiya et al., 1993), *Echis carinatus leucogaster* (Chen and Tsai, 1996), *A. halys brevicaudus* (Koo et al., 2002), *T. stejnegri* (Xu et al., 1999), and *A. halys pallas* (Zang et al., 2003) *Deinagkistrodon acutus* (Atoda et al., 1998; Cox, 1993; Tani et al., 2002; Xu et al., 2000). Bothrojaracin, which belongs to this protein family, is the focus of this review.

Thrombin Specific Inhibitors

The literature reports a large number of thrombin specific inhibitors. It is interesting to note that most of the polypeptides that block thrombin activities have been isolated from hematophagous animals, which include jawed leeches (Salzet et al., 2000). One of the most potent anticoagulants known is hirudin. It is a highly specific inhibitor of thrombin *(Ki* = 21 fM) and was first described by Markwardt, who discovered it in the salivary glands of the leech *Hirudo medicinalis* (Maraganore et al., 1990). Over the last few years, recombinant hirudin and hirudin analogues, such as bivalirudin, have come to represent interesting new commercial antithrombotic agents (Stefano et al., 1996; Verstraete and Zoldhelyi, 1995). Other thrombin inhibitors have been purified from different leeches. Theromin, for example, is a potent thrombin inhibitor (*Ki* = 12 fM) from the leech *Theromyzon tessulatum* (Salzet et al., 2000). Haemadin and bufrudin are from jawed leeches (Electricwala et al., 1993; Strube et al., 1993). Other inhibitors have also been purified from blood sucking arthropods (ticks). These include ornithodorin, which was isolated from *Ornithodoros moubata* (*Ki* = 1 pM), and the peptide savignin (12.4 kDa – *Ki* = 4.89 pM), which was isolated from the salivary glands of *Ornithodoros savignyi* (Nienaber et al., 1999; van de Locht et al., 1996). Another relevant peptide, rhodniin (11 kDa) was purified from the blood sucking bug *Rhodnius prolixus* and interacts with thrombin with high affinity to form a stable 1:1 complex (*Ki* = 0.2 pM) (Friedrich et al., 1993). For details see Chapter 15 in this book.

In 1993, Zingali and colleagues described the first thrombin specific inhibitor isolated from snake venom, which was named bothrojaracin. Since then, bothrojaracin-like molecules have been detected by biological and immunoblotting assays from other Brazilian snake venoms. For example, Castro and colleagues purified the protein bothroalternin in 1998 from *Bothrops alternatus* snake venom and in 2008, Oliveira-Carvalho and colleagues also described bothroinsularin molecules. Another snaclec was purified from the snake venom of *Agkistrodon halys brevicaudus* and was named salmorin (Koh et al., 2000). Like bothrojaracin, it has fibrinogen clotting inhibitory activity and because of this, it increases fibrinogen clotting time (induced by thrombin). Nevertheless, as shown below, its mechanism of action is slightly different from that of bothrojaracin. Salmorin forms a complex with prothrombin, but it also inhibits prothrombin activation independently of the presence of factor V (Koh et al., 2000).

Bothrojaracin

Bothrojaracin is a 27,000 Da protein (Fig. 12.4a) comprising two distinct but homologous polypeptides of 13,000 Da and 15,000 Da, which are covalently linked by a disulfide bridge (Fig. 12.4b). These characteristics, as well as the N-terminal sequences of the two chains, establish a structural relationship with the snaclecs (Zingali et al., 1993).

Molecular Cloning and Expression

Bothrops jararaca snake venom contains only very small amounts of bothrojaracin (less than 0.1%). Therefore, attempts have been made to produce a recombinant protein in order to increase the yield of this protein. In 1998, Arocas and colleagues reported the molecular cloning and the expression of bothrojaracin in mammalian COS cells. The strategy used to determine the total sequence of bothrojaracin chains involved the cloning of cDNAs from a library constructed with venom glands extracted from two *B. jararaca* snakes and selected for their ability to produce large quantities of bothrojaracin. The results of this study demonstrated that this combination of experimental strategies used to obtain recombinant bothrojaracin could produce biologically active protein, which can be secreted into the medium

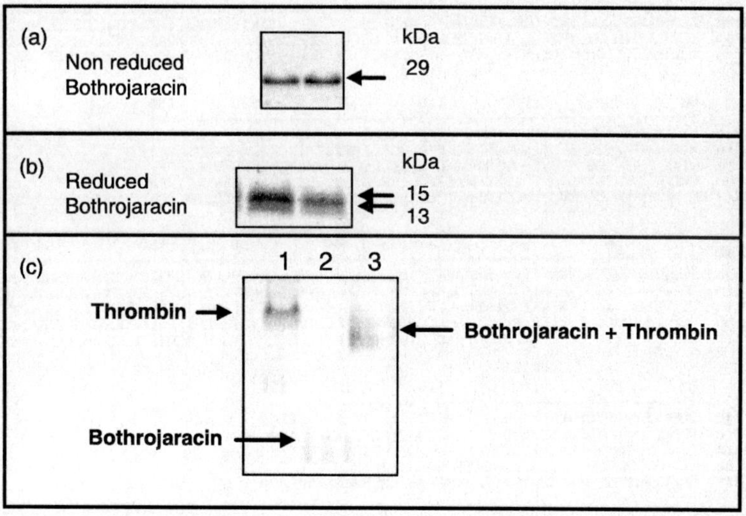

Fig. 12.4 Electrophoresis analyses of bothrojaracin. (**a**) Bothrojaracin whole molecule SDS-PAGE (12%) non reducing conditions; (**b**) alpha and beta bothrojaracin subunits SDS-PAGE (18%) reducing conditions and (**c**) non-denaturing gel electrophoresis of α-thrombin and bothrojaracin molecules. *Lane 1* α-thrombin; *Lane 2* bothrojaracin; *Lane 3* thrombin and bothrojaracin after incubation at 37°C for 10 min

and is able to bind and inhibit thrombin. Notwithstanding, the low level of production of active protein makes this practice not viable. Only 20% of the protein present in the culture media was immunologically reactive with anti-bothrojaracin Ig and was able to inhibit thrombin. A substantial portion of the recombinant bothrojaracin was incorrectly refolded. This result, often observed with recombinant proteins expressed in COS cells, indicates that the association of the two chains of bothrojaracin and their correct folding are required for the inhibitory activity against thrombin. New efforts for expressing this molecule with appropriate structural and biological activity will be necessary to advance structural characterization of the thrombin-bothrojaracin complex.

Structural Features

The entire protein sequence of the two chains (deduced from cDNA) confirmed their great similarity to other snake venom proteins of the snaclec family (Arocas et al., 1997) (Fig. 12.5). Bothrojaracin has only 11 of the 13 amino acid residues that are important for the carbohydrate recognition domain (CRD) and because of this, is not able to bind to carbohydrates (Arocas et al., 1997). In addition, it is clear that

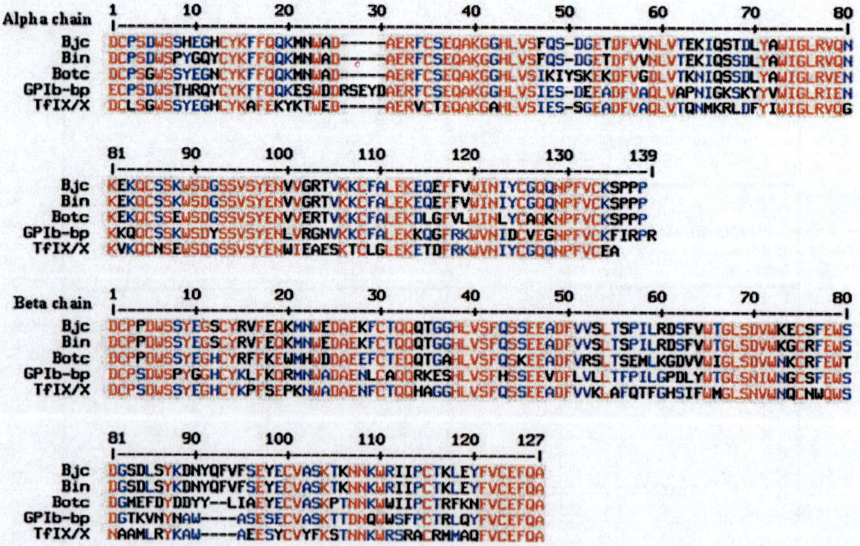

Fig. 12.5 Multiple sequence alignment of the α and β subunits of five snaclecs using MultAlin (http://bioinfo.genotoul.fr/multalin/multalin.html). Abbreviations are bothrojaracin (Bjc) (alpha: gi:62131098 and beta: gi:62131100), bothroinsularin (Bin) (alpha: gi:229621684 and beta: gi:229621685) from *Bothrops insularis* venom, botrocetin (Botc) (alpha: gi:399125 and beta gi:399126) from *Bothrops jararaca* venom, platelet glycoprotein Ib-binding protein (GPIb-bp) (alpha: gi:1839441 and beta: gi:1839442) from *B. jararaca* venom and factor IX/factor X-binding protein (Tf IX/X) (alpha: gi:233489 and beta gi:233490) from *Trimeresurus flavoviridis*

it is a Ca^{2+}-independent interaction, as none of its subunits have all the amino acid residues necessary to bind Ca^{2+} (Atoda et al., 2002).

The two heterologous chains are similar to a large number of molecules isolated from snake venoms, including botrocetin, IX/X-bp and alboaggregin-B, with identities of 80, 57 and 54% for the A chains, and 66, 54 and 56% for the B chains, respectively. More recently, it was demonstrated that bothrojaracin is similar to salmorin molecules (isolated from *Agkistrodonhalys brevicaudus*) with homologies of 56 and 52%, respectively (Kawasaki et al., 1996; Koh et al., 2000; Sekiya et al., 1993; Usami et al., 1993). In 2008, we described bothroinsularin molecules that were isolated from *Bothrops insularis* snake venom as having the highest sequence similarity with bothrojaracin (α 94 and β 98%).

The high degree of sequence identity with other snaclecs, which includes all cysteine residues conserved, allow us to model the arrangement of the intra- and interchain disulfide bonds of bothrojaracin molecules based on the known structures of botrocetin and IX/X-bp. The sequence alignment of bothrojaracin, bothroinsularin and others snaclec family members published in 2008 reaffirmed this striking conservation, which includes the six cysteine residues that form intrachain disulfide bridges (Cys2–Cys13, Cys30–Cys127 and Cys102–Cys119 in α-chain, and Cys2–Cys13, Cys30–Cys123 and Cys100–Cys115 in β-chain). In the model, cysteine 79 in the β chain is covalently linked to cysteine 75 in the α chain. Figure 12.6 also illustrates the six intrachain disulfide bonds proposed for bothrojaracin (Monteiro et al., 2003).

In an effort to construct homology structural models for bothrojaracin to identify important structure features, we compared once again the primary sequences obtained in our laboratory with other snaclec family members, including the new bothrojaracin-like protein named bothroinsularin (BIN). We have used the crystal structures of botrocetin from *Bothrops jararaca* (PDB entry code 1FVU) and of factor IX/factor X-binding protein from *T. flavoviridis* venom (PDB entry code 1IXX) as templates. Molecular modeling showed that each subunit of these proteins has a compact globular unit and an extended long loop, as described for other snaclec family members (Fig. 12.7). Furthermore, BIN and bothrojaracin contain a predominantly negatively-charged region that may favor interaction with the positively-charged regions of thrombin (exosite 1 and 2). They also present some

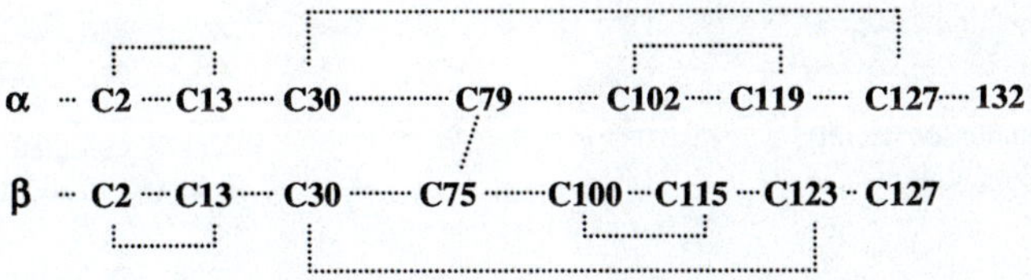

Fig. 12.6 Schematic representation of bothrojaracin showing the locations of interchain and intrachain disulfide bridges (Monteiro et al., 2003). Model constructed by aligning the complete primary sequences deduced from the cloned cDNAs encoding α and β chains of bothrojaracin with sequences reported for IX/X-bp (Atoda et al., 1991) and botrocetin (Usami et al., 1993)

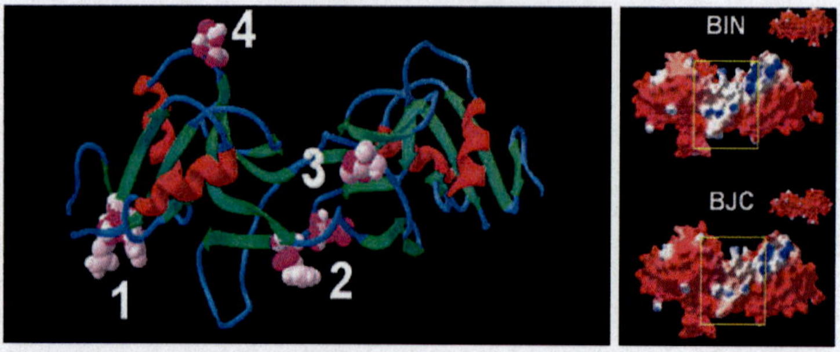

Fig. 12.7 Comparison of bothroinsularin and bothrojaracin models. (**a**) Structural alignment of bothroinsularin (BIN) and bothrojaracin (BJC) homology models in ribbon (RMS=0.08) with the four regions containing the non-conserved positions in solid spheres (BIN=*pink* and BJC=*light pink*) in the full model (1–4). (**b**) Electrostatic potential maps of BIN and BJC homology models (*front view*—larger, *back view*—smaller)

positively charged areas that may help in the orientation of this binding event, as thrombin has negatively-charged regions near the anion-binding exosites (ABEI and ABE-II) (Castro and Rodrigues, 2006). Interestingly, BIN and bothrojaracin homology models pointed to the concave face of these molecules as a feasible region to be directly or indirectly involved in binding thrombin (Oliveira-Carvalho et al., 2008).

In 2003, the first denaturation model was proposed for a snaclec using the thrombin inhibitor bothrojaracin to study subunit dissociation and unfolding (Fig. 12.8). Analyses of the intrinsic fluorescence changes and the circular dichroism (CD) spectra reveal that bothrojaracin can be completely unfolded by the combined action of the denaturant reagent urea at high concentrations (8 M) together with the reducing reagent dithiothreitol (DTT). DTT alone had only a modest effect on the tertiary structure and activity of bothrojaracin. In addition, the effects of urea alone on the tertiary and secondary structures of bothrojaracin are consistent with incomplete

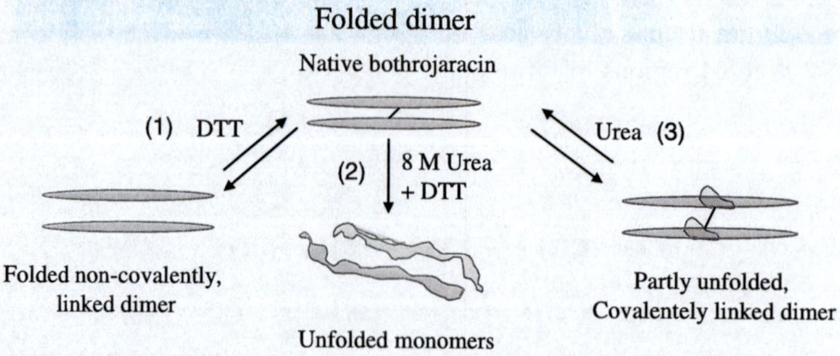

Fig. 12.8 Denaturation model proposed for bothrojaracin. Incubation with DTT has only a modest effect on bothrojaracin tertiary structure and activity. The protein is still linked as a non-covalently folded dimer (1); Treatment of bothrojaracin with urea alone is consistent with incomplete unfolding (3); only the combined action of DTT with high urea concentrations caused complete unfolding of bothrojaracin (2)

unfolding. At a high urea concentration, bothrojaracin loses its tertiary structure while the secondary structure remains mostly unaffected. In fact, unfolding of bothrojaracin produced by the combined action of DTT and urea was irreversible and monomers obtained by denaturation and alkylation events are poorly recognized by polyclonal antibodies. It is clear that both chains are required for full activity (Monteiro et al., 2003).

Distinct Bothrojaracin Isoforms

Variation in snake venom composition has been associated with a number of factors, such as geographical origin (Chippaux et al., 1991; Francischetti et al., 2000; Glenn and Straight, 1978), season (Gubensek et al., 1974), sex (Marsh and Glatston, 1974), age (Jimenez-Porras, 1964), and diet (Magro et al., 2001; Solorzano et al., 1999). In this sense, several molecular isoforms of bothrojaracin have been identified in a pool of venom collected from a large number of animals. Different strategies such as SDS – PAGE electrophoresis, isoelectric focusing and N-terminal sequence using Edman degradation have been used to investigate how these molecules differ from one specimen to another based on amino acid substitution in the primary sequence. In addition to these results, there were clear differences in activity of venom from different pools of animals. Zingali and colleagues showed that pooled venoms obtained from Institute Butantan (Brazil) induced platelet aggregation with an ED_{50} of 50 μg/ml, while venom samples from Miami Serpentarium (U.S.A.) and Latoxan (France) inhibited thrombin-induced platelet aggregation with an IC_{50} of 10 μg/ml. A venom sample purchased from Ventoxin (U.S.A.) showed low levels of both activities.

Subsequently, Monteiro and colleagues (1997) described once again considerable differences in the ability of individual snake venoms to promote coagulation, platelet aggregation, fibrinogenolysis and other biological activities. In addition, differences in peptide sequences were found by microsequencing the amino terminal residues (Monteiro et al., 1998). These involved the replacement of internal amino acids of bothrojaracin, which were collected individually from adult snakes and indicates the existence of more than one isoform of bothrojaracin in the venom of an individual animal. Besides these reported studies, the characterization of new bothrojaracin isoforms using a proteomic approach is in development. In this approach, mass spectrometry experiments revealed new amino acids substitution in both heterodimeric chains thereby corroborating the characteristic heterogeneity of these molecules (data not published). Furthermore, recent studies like surface plasmon resonance, reaffirmed significant differences in kinetic measurements from a pool of different bothrojaracin isoforms (data not published).

Mechanism of Action

Since it was first described in 1993, major advances have been made in the elucidation of mechanisms behind bothrojaracin-thrombin interactions. Studies including

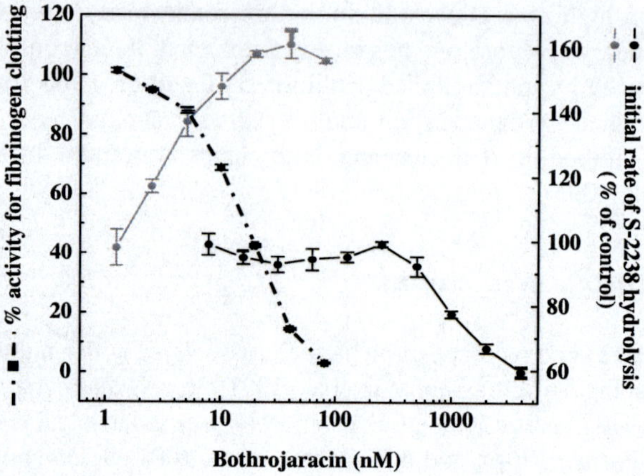

Fig. 12.9 Effects of bothrojaracin on thrombin activity. Hydrolysis of human fibrinogen (2 mg/mL) dashed line and 0.2 mM of S-2238 were assayed in the presence of 2 nM of α-thrombin (*gray*) or γ thrombin (*black*)

in vitro and in vivo experiments demonstrate the potential of this molecule as an antithrombotic agent.

The first report that bothrojaracin interacts specifically with α-thrombin, was made based on its capacity to form a stable complex (Fig. 12.4c) with thrombin and to inhibit most of its enzymatic activity. Competitive assays demonstrated that bothrojaracin does not block the α-thrombin catalytic site, which remains accessible to small chromogenic substrates like Phe-Pip-Arg-pNa (S-2238-Chromogenix). Further studies revealed that bothrojaracin inhibited thrombin-induced platelet aggregation (IC_{50} ranging from 1 to 20 nM), as well as thrombin interactions with fibrinogen (Fig. 12.9), fibrin, thrombomodulin, protein C and hirudin. Bothrojaracin also prolonged the fibrinogen clotting time, inhibited the binding of α-thrombin to thrombomodulin by up to 87%, decreased the rate of protein C activation and antagonized the inhibition of thrombin amidolytic activity by hirudin (Zingali et al., 1993). These findings indicate that the primary site of interaction is the anion binding exosite I.

Double-Site Interaction

The high affinity between bothrojaracin and thrombin molecules is supported by a double-site interaction and results in an efficient inhibition of thrombin in both soluble and clot-bound forms. In 1996, a solid-phase assay was developed in order to determine the dissociation constant (*K*d) of the complex formed. In this study, the high affinity binding of α-thrombin to immobilized bothrojaracin (*K*d 0.6 nM) was inhibited by the C-terminal peptide of hirudin. Bothrojaracin was also found to bind to γ-thrombin, a proteolytic product of α-thrombin that lacks the exosite I. Although it bound with lower affinity (*K*d 0.3 μM), this indicates that bothrojaracin interacts with thrombin mostly through the anion-binding exosite-1. In addition, competition with heparin, antithrombin III-heparin complex or prothrombin fragment 2 for bothrojaracin binding to thrombin confirmed that bothrojaracin binds not only to exosite 1 but also to thrombin exosite II (Arocas et al., 1996).

Allosteric Changes in Thrombin Catalysis

Despite the fact that bothrojaracin interacts with both thrombin anion-binding exosites (I and II), to inhibit macromolecular interactions (but not with its catalytic site), it also induces allosteric changes at the thrombin catalytic site. There is an opposite effect with α-thrombin molecules, which have both exosites, I and II, compared to γ-thrombin, which lacks exosite I. On the one hand, bothrojaracin changed both the kinetic parameters *Km* and *kcat* of α-thrombin towards small synthetic substrates and increased the initial rate of S-2238 hydrolysis, which resulted in an improved efficiency of α-thrombin catalytic activity. On the other hand, bothrojaracin decreased the initial rate of S-2238 hydrolysis of γ-thrombin (2 nM), although this effect was observed only with higher concentrations of bothrojaracin than those used with α-thrombin. As seen in Fig. 12.9 with S-2238, bothrojaracin decreased the rate of hydrolysis by γ-thrombin. This was also the case for other chromogenic substrates such as S-2765, S-2302 and S-2266 (Monteiro et al., 1999).

In agreement with these observations, experiments with thrombin labeled with fluorescein at the active site (fluorescein-Phe-Pro-Arg chloromethylketone)-FPRCK demonstrated conformational changes at the catalytic site in the presence of bothrojaracin molecules. Once again, bothrojaracin had opposite effects on α- and γ-thrombins. The intensity of fluorescence emission of fluorescein-FPRCK-α-thrombin decreases in a dose-dependent manner in the presence of bothrojaracin and reached a maximal value for equimolar concentrations of bothrojaracin. However, bothrojaracin increased the fluorescence emission of fluorescein-FPRCK-γ-thrombin and the maximal fluorescence was obtained at higher concentrations of bothrojaracin (2 μM).

These experiments support the idea that when thrombin interacts with bothrojaracin, two different structural changes are induced in its active site. This depends on whether it interacts exclusively with exosite II, as seen with γ-thrombin, or with exosite I (or both I and II), as observed with α-thrombin.

Recently, real-time binding data from surface plasmon resonance using sensor chips made of α-thrombin with or without FPRCK peptides in its catalytic site, demonstrated differences in binding kinetics with bothrojaracin. These differences include association, dissociation and equilibrium aspects (unpublished observations).

Bothrojaracin and Factor V Activation by Thrombin

The mechanism of action of bothrojaracin is recognized as anticoagulant based on the inhibition of thrombin. In 1998, additional insight into the anticoagulant effects of bothrojaracin revealed that these events occur in the plasma, as illustrated by the prolongation of aPTT (Activated Partial Thromboplastin Time).

The anticoagulant effects of bothrojaracin were not only due to the inhibition of fibrinogen to fibrin conversion, but also to the inhibition of factor V activation by thrombin. The mechanism of this latter inhibition was investigated through the proteolytic cleavage of factor V in parallel with the generation of Va cofactor activity. Bothrojaracin decreased the rate of thrombin-catalyzed proteolysis of factor V and the generation of factor Va cofactor activity was concomitantly measured in a prothrombinase assay. In summary, bothrojaracin blocks the thrombin amplification loop at the level of factor activation. Since factor VIII activation by thrombin also contributes to the amplification of thrombin formation, this is another possible target for bothrojaracin. Due to the similarities between factor V and factor VIII, it appears likely that bothrojaracin might inhibit factor VIII activation, which strengthens the potential antithrombotic effect of bothrojaracin in vivo (Arocas et al., 1998).

Bothrojaracin Interacts with Prothrombin

In addition to its capacity to bind to thrombin molecules, bothrojaracin can bind with high specificity to the thrombin zymogen prothrombin (Arocas et al., 1996, Monteiro and Zingali, 2000, Zingali et al., 2001). Experiments including electrophoresis under non-denaturing conditions demonstrated that this interaction occurs. Further studies showed that prothrombin forms a 1:1 Ca^{2+}-independent, non-covalent complex with immobilized bothrojaracin with a *K*d of 30 nM.

In 2000, additional studies revealed that bothrojaracin is an effective anticoagulant after activation of the intrinsic blood coagulation pathway. These experiments provided evidence that bothrojaracin specifically interacts with prothrombin in human plasma and analysis of prothrombin conversion revealed that bothrojaracin strongly reduced α-thrombin formation. In addition, it was also demonstrated that bothrojaracin inhibits purified prothrombin activation by *Oxyuranus scutellatus* venom, which contains a large quantity of the prothrombin activator scuterin (Monteiro and Zingali, 2000), as shown in Fig. 12.10.

Until 2001, the mechanism by which bothrojaracin binds to prothrombin was not known and the binding site on prothrombin had not been identified. Studies, including competitive binding employing fluorescence probes, revealed that bothrojaracin molecules displace the specific binding of sulfated hirudin$^{54-65}$ ($Hir^{54-65}[SO_3^-]$) to proexosite I of prothrombin. Bothrojaracin similarly disrupts the prothrombin complex with $Hir^{54-65}(SO_3^-)$, which indicates that bothrojaracin and $Hir^{54-65}(SO_3^-)$ bind competitively to proexosite I on human prothrombin. Quantitative characterization reinforces the hypothesis that bothrojaracin binds specifically to the proexosite I of prothrombin with a dissociation constant 76–111 nM and to thrombin with 100-fold higher affinity of 0.7 ± 0.9 nM. Furthermore, isothermal titration calorimetry revealed that bothrojaracin binding to prothrombin is entropically driven with an unfavorable enthalpy (Monteiro et al., 2001).

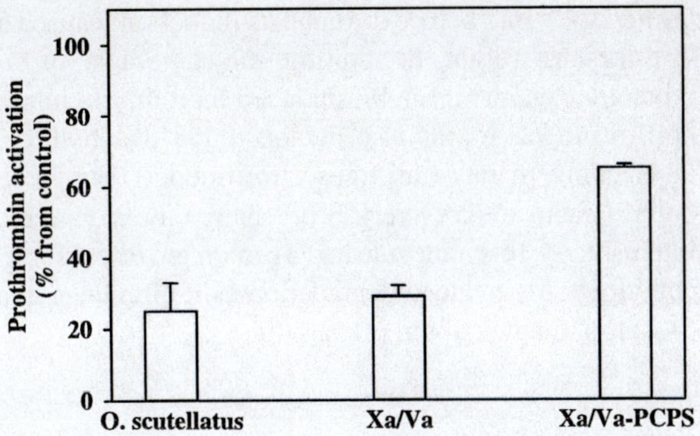

Fig. 12.10 Inhibitory effect of bothrojaracin on prothrombin activation. Experiments were performed in the presence of 1.5 μM bothrojaracin and started by addition of *O. scutellatus* venom, or factor Xa, factor Xa/factor Va, or factor Xa/factor Va/phospholipid vesicles

Since these reports were all based on studies involving bothrojaracin, prothrombin activation and thrombin binding ligands, we can conclude that bothrojaracin exerts its anticoagulant effect by two distinct mechanisms (Fig. 12.11). First, it binds to activated thrombin through exosite 1 to block fibrinogen clotting, platelet activation, factor V activation, and other effects. Alternatively, it interacts with prothrombin to decrease its proteolytic activation, especially that which is dependent on factor V interactions (Monteiro and Zingali, 2000, 2002; Zingali et al., 2001).

Bothrojaracin and In Vivo Anti-Thrombotic Effects

Following the anticoagulant in vitro experiments, we also evaluated in vivo antithrombotic effects of bothrojaracin in a venous thrombosis model in rats, which combines stasis and hypercoagulability. Intravenous administration of bothrojaracin

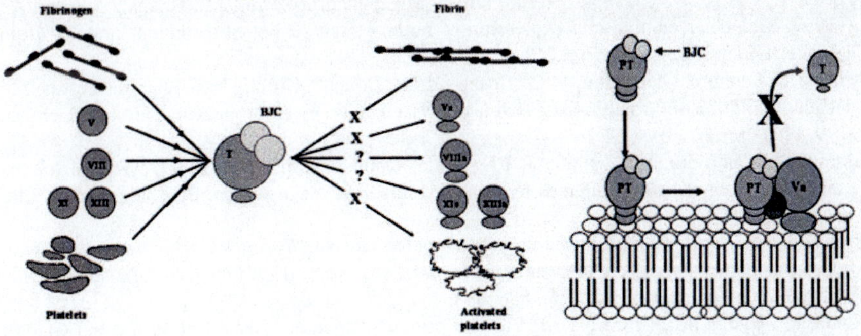

Fig. 12.11 Anticoagulant property of bothrojaracin follows two mechanisms. Mechanism 1 of bothrojaracin inhibition is based on its binding to thrombin exosite I and the subsequent inhibition of its activities. Mechanism 2 is based on the inhibition of prothrombin activation through interaction with prothrombin proexosite I. (BJC, bothrojaracin; PT, prothrombin; Va, activated factor V; Xa, activated factor X; T, thrombin)

(1 mg/kg), 5 min before thrombosis induction, caused a significant decrease of 95% in thrombus weight. In addition, the same dose of bothrojaracin conferred 100% protection against thrombin-induced mortality in mice in a pulmonary thromboembolism model. It was also demonstrated that bothrojaracin produced an increase in bleeding in rats (tail-transection model), but less than heparin. Nevertheless, bothrojaracin induces very little change in ex vivo clotting times (aPTT and PT). Interestingly, this molecule had a prolonged effect for more than 48 h, which is probably due to its reduced degradation as it circulates in a complex with prothrombin (Assafim, manuscript in preparation).

Future Directions

At present, the mechanism of action of bothrojaracin is well characterized. Nevertheless, the structure–function relationship still needs to be better understood in order to use this molecule as a prototype for the development of new antithrombotic drugs. To this end, new bothrojaracin isoforms are being characterized by mass spectrometry and by a "hydroxyl radical – mediated protein footprinting" approach. These methods are being developed to determine the site of interaction between bothrojaracin, thrombin and prothrombin. Also, some aspects of the in vivo effects of this molecule, such as pharmacokinetics, are under investigation. Bothrojaracin has a unique mode of action, and it seems to act in vivo mainly by binding to prothrombin. On the other hand, bothrojaracin is an important tool to evaluate the viability of using prothrombin as a new target for antithrombotic drugs.

References

Arocas, V., Zingali, R.B., Guillin, M.C., Bon, C., Jandrot-Perus, M., 1996. Bothrojaracin: a potent two-site-directed thrombin inhibitor. Biochemistry 35, 9083–9089.

Arocas, V., Castro, H.C., Zingali, R.B., Guillin, M.C., Jandrot-Perrus, M., Bon, C., Wisner, A., 1997. Molecular cloning and expression of bothrojaracin, a potent thrombin inhibitor from snake venom. Eur. J. Biochem. 248, 550–557.

Arocas, V., Lemaire, C., Bouton, M.C., Beazeaud, A., Bon, C., Guillin, M.C., Jandrot-Perrus, M., 1998. Inhibition of thrombin-catalysed factor V activation by bothrojaracin. Thromb. Haemost. 79, 1157–1161.

Assafim, M., Ferreira, M.S., Frattani, F.S., 2006. Counteracting effect of glycyrrhizin on the hemostatic abnormalities induced by *Bothrops jararaca* snake venom. Br. J. Pharmacol. 148, 807–813.

Atoda, H., Morita, T., 1989. A novel blood coagulation factor IX/factor X518 binding protein with anticoagulant activity from the venom of *Trimeresurus flavoviridis* (Habu snake): isolation and characterization. J. Biochem. 106, 808–813.

Atoda, H., Hyuga, M., Morita, T., 1991. The primary structure of coagulation factor IX/factor X-binding protein isolated from the venom of *Trimeresurus flavoviridis*. Homology with asialo-glycoprotein receptors, proteoglycan core protein, tetranectin, and lymphocyte Fc epsilon receptor for immunoglobulin. J. Biol. Chem. 266, 14903–14911.

Atoda, H., Ishikawa, M., Yoshihara, E., Sekiya, F., Morita, T., 1995. Blood coagulation factor IX-binding protein from the venom of *Trimeresurus flavoviridis*: purification and characterization. J. Biochem. 118, 965–973.

Atoda, H., Ishikawa, M., Mizuno, H., Morita, T., 1998. Coagulation factor X binding protein from *Deinagkistrodon acutus* venom is a Gla domain-binding protein. Biochemistry 37, 17361–17370.

Atoda, H., Kaneko, H., Mizuno, H., Morita, T., 2002. Calcium-binding analysis and molecular modeling reveal Echis coagulation factor IX/factor X binding protein has the Ca-binding properties and Ca ion-independent folding of other C-type lectin-like proteins. FEBS Lett. 531, 229–234.

Bode, W., 2006. The structure of thrombin: a janus-headed proteinase. Sem. Throm. Hem. 32, 16–31.

Boffa, G.A., Boffa, M.C., Winchenne, J.J., 1976. A phospholipase A_2 with anticoagulant activity. Isolation from *Vipera berus* venom and properties. Biochem. Biophys. Acta 429, 839–852.

Braud, S., Bon, C., Wisner, A., 2000. Snake venom proteins acting on hemostasis. Biochimie 82, 851–859.

Bukys, M.A., Orban, T., Kim, P.Y. Beck, D.O., Nesheim, M.E., Kalafatis, M., 2006. The structural integrity of anion binding exosite of thrombin is required and sufficient for timely cleavage and activation of factor V and factor VIII. J. Biol. Chem. 281, 18569–18580.

Castro, H.C., Dutra, D.L.S., Oliveira-Carvalho, A.L., Zingali, R.B., 1998. Bothroalternin, a thrombin inhibitor from the venom of *Bothrops alternatus*. Toxicon 36, 1903–1912.

Castro, H.C., Rodrigues, C.R., 2006. Current status of snake venom thrombin-like enzymes. Toxin Rev. 25, 291–318.

Castro, H.C., Zingali, R.B., Albuquerque, M.G., Pujol-luz, M., Rodrigues, C.R., 2004. Snake venom thrombin-like enzymes: from reptilase to now. Cell. Mol. Life Sci. 61, 843–856.

Chen, Y.L., Tsai, I.H., 1996. Functional and sequence characterization of coagulation factor IX/factor X-binding protein from the venom of *Echis carinatus leucogaster*. Biochemistry 35, 5264–5271.

Chippaux, J., Williams, V., White, J., 1991. Snake venom variability: methods of study results and interpretation. Toxicon 29, 1279–1303.

Clemetson, K.J., Morita, T., Kini, R.M., 2009. Scientific and standardization committee communications: classification and nomenclature of snake venom C-type lectins and related proteins. J. Thromb. Haemost. 7, 360.

Cox, A.C., 1993. Coagulation factor X inhibitor from hundred-pace snake (*Deinagkistrodon acutus*) venom. Toxicon 31, 1445–1457.

Davie, E.W., Fujikawa, K., Kisiel, W., 1991. The coagulation cascade: initiation, maintenance, and regulation. Biochemistry 30, 10363–10370.

Electricwala, A., von Sicard, N.A.E., Sawyer, R.T., Atkinson, T., 1993. Biochemical characterisation of a pancreatic elastase inhibitor from the leech *Hirudinaria manillensis*. J. Enzyme Inhib. 6, 293–302.

Esmon, C.T., Lollar, P., 1996. Involvement of thrombin anion-binding exosites 1 and 2 in the activation of factor V and factor VIII. J. Biol. Chem. 271, 13882–13887.

Fareed, J., Lewis, B.E., Callas, D.D., Hoppensteadt, D.A., Walenga, J.M., Bick, R.L., 1999. Antithrombin agents: the new class of anticoagulant and antithrombotic drugs. Clin. Appl. Thromb. Hemost. 5, S45–S55.

Fox, J.W., Serrano, S.M.T., 2009. Timeline of key events in snake venom metalloproteinase research. J. Proteomics 72, 200–209.

Francischetti, I.M., Gombarovits, M.E., Valenzuela, J.G., Carlini, C.R., Guimarães, J.A., 2000. Intraspecific variation in the venoms of the South American rattlesnake (*Crotalus durissus terrificus*). Comp. Biochem. Physiol. C. 127, 23–36.

Friedrich, T., Kroger, B., Bialojan, S., Lemaire, H.G., Hoffken, H.W., Reuschenbach, P., Otte, M., Dodt, J., 1993. A kazal-type inhibitor with thrombin specificity from *Rhodnius prolixus*. J. Biol. Chem. 268, 16216–16222.

Funk, C., Gmur, J., Herold, R., Straub, P.W., 1971. Reptilase-R: a new reagent in blood coagulation, Br. J. Haematol. 21, 43–52.

Glenn, J.L., Straight, R., 1978. Mojave rattle snake *Crotalus scutulatus scutulatus* venom: variation in toxicity with geographical origin. Toxicon 16, 81–84.

Gubensek, F., Sket, D., Turk, V., Lebez, D., 1974. Fractionation of *Vipera ammodytes* venom and seasonal variation of its composition. Toxicon, 12, 167–171.

Guimarães-Gomes, V., Oliveira-Carvalho, A.L., Junqueira-de-Azevedo, I.L., Dutra, D.L.S., Pujol-Luz, M., Castro, H.C., Ho, P.L., Zingali, R.B., 2004. Cloning, characterization and structural analysis of a C-type lectin from *Bothrops insularis* (BiL) venom. Arch. Biochem. Biophys. 432, 1–11.

Hjort, P.F., 1957. Intermediate reactions in the coagulation of blood with tissue thromboplastin. Scand. J. Clin. Lab. Invest. 9, 1–183.

Hofsteenge, J., Taguchi, H., Stone, S.R., 1986. Effect of thrombomodulin on the kinetics of the interaction of thrombin with substrates and inhibitors. Biochem. J. 237, 243–251.

Hofsteenge, J., Kieffer, B., Matthies, R., Hemmings, B.A., Stone, S.R., 1988. Amino acid sequence of the ribonuclease inhnbitor from porcine liver reveals the presence of leucine-rich repeats. Biochemistry 27, 8537–8544.

Jenny, N.S., Mann, K.G., 1998. Coagulation cascade: an overview, in: Loscalzo, J., Schafer, A.I. (Eds.), Thrombosis and Hemorrhage. Williams and Wilkins, Philadelphia, pp. 3–27.

Jimenez-Porras, J.M., 1964. Intraspecific variations in composition of venom of the jumping viper *Bothrops nummifera*. Toxicon 2, 187–196.

Kalafatis, M., Rand, M.D., Mann, K.G., 1994. The mechanism of inactivation of human factor V and human factor Va by activated protein C. J. Biol. Chem. 269, 31869–31880.

Kalafatis, M., Egan, J.O., van't Veer, C., Cawthern, K.M., Mann, K.G., 1997. The regulation of clotting factors. Crit. Rev. Eukaryotic Gene Expression 7, 241–280.

Kamiguti, A.S., 2005. Platelets as targets of snake venom metalloproteinases. Toxicon 45, 1041–1049.

Kawasaki, T., Fujimura, Y., Usami, Y., Suzuki, M., Miura, S., Sakurai, Y., Makita, K., Taniuchi, Y., Hirano, K., Titani, K., 1996. Complete amino acid sequence and identification of the platelet glycoprotein Ib-binding site of jararaca GPIb-BP, a snake venom protein isolated from *Bothrops jararaca*. J. Biol. Chem. 271, 10635–10639.

Kerns, R.T., Kini, R.M., Stefansson, S., Evans, H.J., 1999. Targeting of venom phospholipases: the strong anticoagulant phospholipase A_2 from *Naja nigricollis* venom binds to coagulation factor Xa to inhibit the prothrombinase complex. Arch. Biochem. Biophys. 369, 107–113.

Kirby, E.P., Niewiarowski, S., Stocker, K., Kettner, C., Shaw, E., Brudzynski, T.M. 1979. Thrombocytin, a serine protease from *Bothrops atrox* venom. 1. Purification and characterization of the enzyme. Biochemistry 18, 3564–3570.

Koh, Y., Chung, K., Kim, D., 2000. Purification and cDNA cloning of salmorin that inhibits fibrinogen clotting. Thromb. Res. 99, 389–398.

Koo, B.H., Sohn, Y.D., Hwang, K.C., Jang, Y., Kim, D.S., Chung, K.H., 2002. Characterization and cDNA cloning of halyxin, a heterogeneous three-chain anticoagulant protein from the venom of *Agkistrodon halys brevicaudus*. Toxicon 40, 947–957.

Lee, W.H., Zhang, Y., 2003. Molecular cloning and characterization of a platelet glycoprotein Ib-binding protein from the venom of *Trimeresurus stejnegeri*. Toxicon 41, 885–892.

Liu, L.W., Vu, T.K.H., Esmon, C.T., Coughlin, S.R., 1991. The region of the thrombin receptor resembling hirudin binds to thrombin and alters enzyme specificity. J. Biol. Chem. 266, 16977–16980.

Magro, A.J., Da Silva, R.J., Ramos, P.R.R., Cherubini, M.R., Hatayde, M.R., 2001. Intraspecific variation in the venom electrophoretic profile of recently captured *Crotalus durissus terrificus* (Lurenti, 1768) snakes. J. Venom. Anim. Toxins 7, 276–301.

Mann, K.G., 1999. Biochemistry and physiology of blood coagulation. Thromb. Haemost. 82, 165–174.

Mann, K.G., Butenas, S., Brummel, K., 2003. The dynamics of thrombin formation. Arterioscler. Thromb. Vasc. Biol. 23, 17–25.

Maraganore, J.M., Bourdon, P., Jablonski, J., Ramachandran, K.L., Fenton, J.W., 2nd, 1990. Design and characterization of hirulogs: a novel class of bivalent peptide inhibitors of thrombin. Biochemistry 29, 7095–7101.

Markland, F.S., 1998 Snake venoms and the haemostatic system. Toxicon 36, 1749–1800.

Marsh, N., Glatston, A., 1974. Some observations on the venom of the rhinoceros horned viper *Bitis nasicornis* (Shaw). Toxicon 12, 621–628.

Marsh, N.A., 1994. Snake venoms affecting the haemostatic mechanism – a consideration of their mechanisms, practical applications and biological significance. Blood Coagul. Fibrinol. 5, 399–410.

Matsuda, M., Saeki, E., Kasamatsu, A., Nakamikawa, C., Manabe, S., Samejima, Y., 1985. Fibrinogen Kawaguchi: an abnormal fibrinogencharacterized by defective release of fibrinopeptide A. Thromb. Res. 37, 379–390.

Matsui, T., Fujimura, Y., Titani, K., 2000. Snake venom proteases affecting hemostasis and thrombosis. Biochim. Biophys. Acta 1477, 146–156.

Monteiro, R.Q., Carlini, C.R., Guimarães, J.A., Zingali, R.B., 1997. Distinct bothojaracin isoforms produced by individual jararaca (*Bothrops jararaca*) snakes. Toxicon 35, 649–657.

Monteiro, R.Q., Dutra, D.L.S., Machado, O.L.T., Carlini, C.R., Guimarães, J.A, Bon, C., Zingali, R.B., 1998. *Bothrops jararaca* snakes produce several bothrojaracin isoforms following an individual pattern. Comp. Biochem. Physiol. B. 120, 791–798.

Monteiro, R.Q., Raposo, J.G., Wisner,A., Guimarães, J.A., Bon, C., Zingali, R.B., 1999. Allosteric changes of thrombin catalytic site induced by interaction of bothrojaracin with anion-binding exosities I and II. Biochem. Biophys. Res.Comm. 262, 819–822.

Monteiro, R.Q., Zingali, R.B., 2000. Inhibition of prothrombin activation by bothrojaracin, a C-type lectin from *Bothrops jararaca* venom. Arch. Biochem. Biophys. 382, 123–128.

Monteiro, R.B., Bock, P.E., Bianconi, M.L., Zingali, R.B., 2001. Characterization of bothrojaracin interaction with human prothrombin. Protein Science 10, 1897–1904.

Monteiro, R.Q., Zingali, R.B., 2002. Bothrojaracin, a proexosite I ligand, inhibits factor Va-accelerated prothrombin activation. Thromb. Haemost. 87, 288–293.

Monteiro, R.Q., Foguel, D., Castro, H.C., Zingali, R.B., 2003. Subunit dissociation, unfolding, and inactivation of bothrojaracin, a C-type lectin-like protein from snake venom. Biochemistry 42, 509–515.

Morita, T., 2004. C-type lectin-related proteins from snake venoms. Drug Targets Cardiovasc. Haematol. Disord. 4, 357–373.

Mounier, C.M., Bon, C., Kini, R.M., 2001. Anticoagulant venom and mammalian secreted phospholipase A_2: protein versus phospholipid dependent mechanism of action. Haemostasis 31, 279–287.

Myles, T, Yun, T.H., Hall, S.W., Leung, L.L.K., 2001. An extensive interaction interface between thrombin and factor V is required for factor V activation. J. Biol. Chem. 276, 25143–25149.

Murakami, M.T., Arni, R.K., 2005. Thrombomodulin-independent activation of protein C and specificity of hemostatically active snake venom serine proteinases: crystal structures of native and inhibited *Agkistrodon contortrix contortrix* protein C activator. J. Biol. Chem. 280, 39309–39315.

Nienaber, J., Gaspar, A.R.M., Neitz, A.W., 1999. Savignin, a potent thrombin inhibitor isolated from the salivary glands of the tick *Ornithodoros savignyi* (Acari: Argasidae). Exp. Parasitol. 20, 583–598.

Oliveira-Carvalho, A.L., Guimarães, P.R., Abreu, P.A., Dutra, D.L., Junqueira-de-Azevedo, I.L., Rodrigues, C.R., Ho, P.L., Castro, H.C., Zingali, R.B., 2008. Identification and characterization of a new member of snake venom thrombin inhibitors from *Bothrops insularis* using a proteomic approach. Toxicon 51, 659–671.

Peng, M., Lu, W., Kirby, E.P., 1991. Alboaggregin-B: a new platelet agonist that binds to platelet membrane glycoprotein Ib. Biochemistry 30, 11529–11536.

Peng, M., Holt, J.C., Niewiarowski, S., 1994. Isolation, characterization and amino acid sequence of echicetin b subunit, a specific inhibitor of von Willebrand factor and thrombin interaction with glycoprotein Ib. Biochem. Biophys. Res. Commun. 205, 68–72.

Rosing, J., Govers-Riemslag, J.W.P., Yukelson, L., Tans, G., 2001. Factor V activation and inactivation by venom proteases. Hemostasis 31, 241–246.

Salzet, M., Chopin, V., Baerti, J., Matias, I., Malecha, J., 2000. Theromin, a novel leech thrombin inhibitor. J. Biol. Chem. 275, 30774–30780.

Santos, A.B.F., 2000. Interaction of viper venom serine peptidases with thrombin receptors on human platelets. FEBS Lett. 477, 199–202.

Sekiya, F., Atoda, H., Morita, T., 1993. Isolation and characterization of an anticoagulant protein homologous to botrocetin from the venom of *Bothrops jararaca*. Biochemistry 32, 6892–6897.

Sheehan, J.P., Wu, Q., Tollefsen, D.M., Sadler, J.E., 1993. Mutagenesis selectively modulates inhibition by serpins heparin cofactor II and antithrombin III. Interaction with the anion-binding exosite determines heparin cofactor II specificity. J. Biol. Chem. 268, 3639–3645.

Shiau, S.Y., Ouyang, C., 1965. Isolation of coagulant and anticoagulant principles from the venom of *Trimeresurus gramineus*. Toxicon 69, 213–220.

Solorzano, A., Romero, M., Gutiérrez, J.M., Sasa, M., 1999. Venom composition and diet of the cantil *Agkistrodon bilineatus howardgloydi* (Serpentes: Viperidae). Southwest. Nat. 44, 478–483.

Stefano, G.B., Scharrer, B., Smith, E.M., Hughes, T.K., Magazine, H.I., Bilfinger, T.V., Hartman, A.R., Fricchione, G.L., Liu, Y., Makman, M.H., 1996. Opioid and opiate immunoregulatory processes. Crit. Rev. Immunol. 16, 109–144.

Stone, S.R., Hofsteenge, J., 1986. Kinetics of the inhibition of thrombin by hirudin. Biochemistry 25, 4622–4628.

Stone, S.R., Brown-Luedi, M.L., Rovelli, G., Guidolin, A., McGlynn, E., Monard, D., 1994. Localization of the heparin-binding site of glia-derived nexin/protease nexin 1 by site-directed mutagenesis. Biochemistry 33, 7731–7735.

Strube, K.H., Kroger, B., Biolojan, S., Otte, M., Dodt, J., 1993. Isolation, sequence analysis, and cloning of haemadin. An anticoagulant peptide from the Indian leech. J. Biol. Chem. 268, 8590–8595.

Tani, A., Ogawa, T., Nose, T., Nikandrov, N.N., Deshimaru, M., Chijiwa, T., Chang, C.C., Fukumaki, Y., Ohno, M., 2002. Characterization, primary structure and molecular evolution of anticoagulant protein from *Agkistrodon acutus* venom. Toxicon 40, 803–813.

Taniuchi, Y., Kawasaki, T., Fujimura, Y., Suzuki, M., Titani, K., Sakai, Y., Kaku, S., Hisamichi, N., Satoh, N., Takenaka, T., Handa, M., Sawai, Y., 1995. Flavocetin-A and -B, two high molecular mass glycoprotein Ib binding proteins with high affinity purified from *Trimeresurus flavoviridis* venom, inhibit platelet aggregation at high shear stress. Biochim. Biophys. Acta 1244, 331–338.

Usami, Y., Fujimura, Y., Suzuki, M., Ozeki, Y., Nishio, K., Fukui, H., Titani, K., 1993. Primary structure of two-chain botrocetin, a von Willebrand factor modulator purified from the venom of *Bothrops jararaca*. Proc. Natl. Acad. Sci. U.S.A. 90, 928–932.

Van de Locht, A, Stubbs, M., Bode, W., Friedrich, T., Bol-Ilschweiler, C., Höffken, W, Huber, R., 1996. The ornithodorin-thrombin crystal structure, a key to the TAP enigma? EMBO J. 15, 6011–6017.

Verstraete, M., Zoldhelyi, P., 1995. Novel antithrombotic drugs in development. Drugs 49, 856–884.

Xu, Q., Wu, X.F., Xia, Q.C., Wang, K.Y., 1999. Cloning of a galactose-binding lectin from the venom of *Trimeresurus stejnegeri*. Biochem. J. 341, 733–737.

Xu, X., Liu, Q., Xie, Y., Wu, S.D., 2000. Purification and characterization of anticoagulation factors from the venom of *Agkistrodon acutus*. Toxicon 38, 1517–1528.

Weinger, R.S., Rudy, C., Moake, J.L., Olson, J.D., Cimo, P.L., 1980. Prothrombin Houston: a dysprothrombin identifiable by crossed immunoelectrofocusing and abnormal *Echis carinatus* venom activation. Blood 55, 811–816.

Wang, W.J., Huang, T.F., 2001. A novel tetrameric venom protein, agglucetin, from *Agkistrodon acutus*, acts as a glycoprotein Ib agonist. Thromb. Haemost. 86, 1077–1086.

Yeh, C.H., Peng, H.C., Yang, R.S., Huang, T.F., 2001. Rhodostomin, a snake venom disintegrin, inhibits angiogenesis elicited by basic fibroblast growth factor and suppresses tumor growth by a selective $\alpha_v\beta_3$ blockade of endothelial cells. Mol. Pharmacol. 59, 1333–1342.

Zang, J., Teng, M., Niu, L., 2003. Purification, crystallization and preliminary crystallographic analysis of AHP IX-bp, a zinc ion and pH-dependent coagulation factor IX binding protein from *Agkistrodon halys pallas* venom. Acta Crystallogr. D. Biol. Crystallogr. 59, 730–733.

Zhang, Y., Wisner, A., Maroun, R.C., Choumet, V., Xiong, Y., Bon, C., 1997. *Trimeresurus stejnegeri* snake venom plasminogen activator. Site directed mutagenesis and molecular modeling. J. Biol. Chem. 272, 20531–20537.

Zingali, R.B., Jandrot-Perrus, M., Guillin, M.C., Bon, C., 1993. Bothrojaracin, a new thrombin inhibitor isolated from *Bothrops jararaca* venom: characterization and mechanism of thrombin inhibition. Biochemistry 32, 10794–10802.

Zingali, R.B., Bianconi, M.L., Monteiro, R.Q., 2001. Interaction of bothrojaracin with prothrombin. Haemostasis 31, 273–278.

Chapter 13
Anticoagulant Phospholipases A_2 Which Bind to the Specific Soluble Receptor Coagulation Factor Xa

Grazyna Faure, Haijin Xu, and Frederick Saul

Abstract This review will focus on *Viperidae* venom secreted Phospholipases A_2 ($sPLA_2s$) that exert an anticoagulant effect by a non-enzymatic, phospholipid-independent mechanism through direct binding to human coagulation factor Xa (FXa). We present the potential FXa-binding site of these anticoagulant PLA_2s and the potential PLA_2-binding site of FXa based on molecular docking calculations, site-directed mutagenesis and SPR affinity binding studies. We also discuss the structures of recently crystallized natural isoforms of PLA_2 which interact with FXa with different affinity and which differ in anticoagulant activity. The three-dimensional structure of the PLA_2 isoforms helps to explain the role of natural mutations in the binding mode of PLA_2 with FXa and will be useful in structure-based design of non-competitive FXa inhibitors as potential new anticoagulant drugs.

Introduction

Phospholipases constitute a large superfamily of enzymes that catalyze the hydrolysis of phospholipids (PL) at the lipid-water interface (van Deenen and de Hass, 1963). Based on the cleavage site, five categories of PLA have been described: Phospholipases A_1 (PLA_1: EC 3.1.1.32); Phospholipases A_2 (PLA_2: EC 3.1.1.4); Phospholipases B (PLB: EC 3.1.1.5); Phospholipases C (PLC: EC 3.1.4.3) and Phospholipases D (PLD: EC 3.1.4.4) (Fig. 13.1).

PLA_2 from various organisms comprise extra- and intracellular proteins with different catalytic sites and different functional and structural features. Currently, PLA_2 are classified into 15 structural groups and various subgroups falling into five distinct types of enzymes: the secreted PLA_2s ($sPLA_2$), the cytosolic PLA_2s ($cPLA_2$),

G. Faure (✉)
Institut Pasteur, Unité d'Immunologie Structurale, CNRS, URA 2185, Département de Biologie Structurale et Chimie, F-75015, Paris, France
e-mail: grazyna.faure-kuzminska@pasteur.fr

R.M. Kini et al. (eds.), *Toxins and Hemostasis*, DOI 10.1007/978-90-481-9295-3_13,

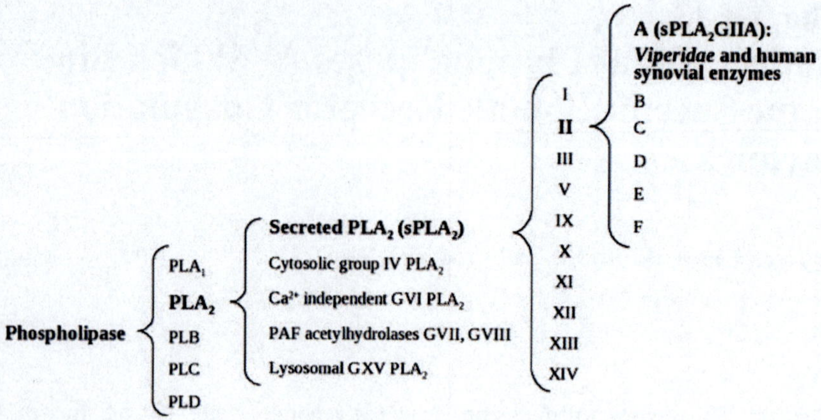

Fig. 13.1 Superfamily of PLA_2. (The figure has been adapted from Schaloske and Dennis, 2006)

the Ca^{2+} independent PLA_2s ($iPLA_2$), the platelet-activating factor acetylhydrolases (PAF-AH), and the lysosomal PLA_2s (Schaloske and Dennis, 2006).

PLA_2 enzymes participate in numerous physiological and pathophysiological processes including digestion of dietary PL, maintenance of cellular PL pools, membrane repair by reacylation/deacylation mechanisms, arachidonic acid release from cellular PL for the biosynthesis of eicosanoids, fertilization, hypersensitization, cell proliferation and muscle contraction as well as cellular functions such as signal transduction, membrane homeostasis and other processes (for reviews, see Dennis, 1997; Kini, 1997; Murakami et al., 1997). PLA_2s play a central role in inflammatory diseases such as rheumatoid arthritis, ARDS, pancreatitis and sepsis (for review, see Nevalainen et al., 2000) and may also contribute to the development of atherosclerosis and cancer (Lambeau and Gelb, 2008; Murakami and Kudo, 2003; Webb, 2005). The antibacterial role of $sPLA_2$ in the killing of Gram-positive and Gram-negative bacteria and in host defense against bacterial infections is well-established (Nevalainen et al., 2008). PLA_2 may also participate in host defense against viruses and parasites (Deregnaucourt and Schrevel, 2000; Fenard et al., 2001; Kim et al., 2007). For reviews concerning regulation and various cellular functions of the five distinct types of PLA_2 see Lambeau and Gelb (2008); Murakami and Kudo (2001); Schaloske and Dennis (2006).

The secreted PLA_2 found both in mammals and in animal venom are small (13–18 kDa) disulfide-rich proteins which bind to cell membranes and catalyze the hydrolysis of glycero-phospholipids at the *sn*–2 position, releasing a fatty acid and a lysophospholipid, thus providing precursors for the biosynthesis of lipid mediators (Dennis, 1994; Dessen, 2000). Interestingly, $sPLA_2$ also selectively interact with various targets such as proteins or heparan sulfate proteoglycans (HSPG) and exert various physiological and patho-physiological effects through protein-protein or protein-sugar interactions (Murakami and Kudo, 2003). The $sPLA_2$ have been classified into 10 Groups (I-(A,B) II–(A,B,C,D,E,F,); III, V, IX, X, XI-(A,B), XII, XIII and XIV) on the basis of their primary structure and number of disulfide bonds (Schaloske and Dennis, 2006) (Fig. 13.1). In the snake venoms, only two groups (IA (cobras and kraits) IB (Australian elapids and king cobra), IIA (rattlesnakes), and IIB (Gabon viper)) of PLA_2 have been identified.

Here, we focus on the secreted phospholipases A_2, group IIA (GIIA) from the *Viperidae* snake family which possess anticoagulant properties and interact with coagulation factor Xa. Two examples are discussed in this short review: Ammodytoxin, a monomeric β-neurotoxin from *Vipera ammodytes ammodytes* and the CB subunit of crotoxin, a heterodimeric β-neurotoxin from *Crotalus durissus terrificus*.

Structure and Properties of Snake Venom GIIA Anticoagulant PLA_2

The $sPLA_2$ from *Viperidae* snake venom are structurally homologous to the inflammatory, non-pancreatic group IIA mammalian $sPLA_2$. They are basic enzymes of 120–125 amino acid residues with a molecular weight of 13–17 kDa containing seven disulfide bonds, of which six are strictly conserved in PLA_2 from *Elapidae* and *Viperidae* snake venom and the seventh is characteristic of each group (for *Viperidae* the seventh disulfide bond occurs at the C-terminal of the polypeptide chain between residues Cys133 and Cys50).

$sPLA_2$ GIIA are Ca^{2+}-dependent interfacial enzymes with a conserved 25-33 Ca^{2+} binding loop (with the consensus sequence motif 25Tyr-Gly-Cys-Tyr-Cys-Gly-X-Gly-Gly33) and the catalytic network His48, Asp49, Asp99 and Tyr52 (Arni and Ward, 1996; Scott et al., 1990a). The PLA_2 possess an Interfacial Binding Site (IBS) formed by a hydrophobic channel (Berg et al., 2001; Snitko et al., 1997) surrounding the entrance to the active site and containing a ring of cationic and hydrophobic residues. The IBS is located in the "front face" of the enzyme, which is suggested to make contact with lipid during association of the PLA_2 with the membrane surface (Chioato and Ward, 2004). Small variations in the topology of the IBS determine the specificity of PL head-group binding (Snitko et al., 1999).

The crystal structures of a number of snake venom PLA_2 have been determined (about 35 are deposited in the Protein Data Bank). The three-dimensional structures of these PLA_2 contain at least 50% α-helical conformation and 10% antiparallel β-pleated sheet structure. Figure 13.2 shows the canonical structural features (protein fold) of group IIA PLA_2s for the case of AtxA from *Vipera ammodytes ammodytes* venom which was solved in our laboratory (Saul et al., 2010). The structure contains an N-terminal α-helix (A), a short helix (B), a Ca^{2+}-binding loop, a long α-helix (C), a loop preceding an anti-parallel two-stranded beta sheet (β-wing), a long α-helix (D) anti-parallel to helix C, and a C-terminal extension. Two anti-parallel disulfide linked α-helices (C and D) form a rigid scaffold to which the

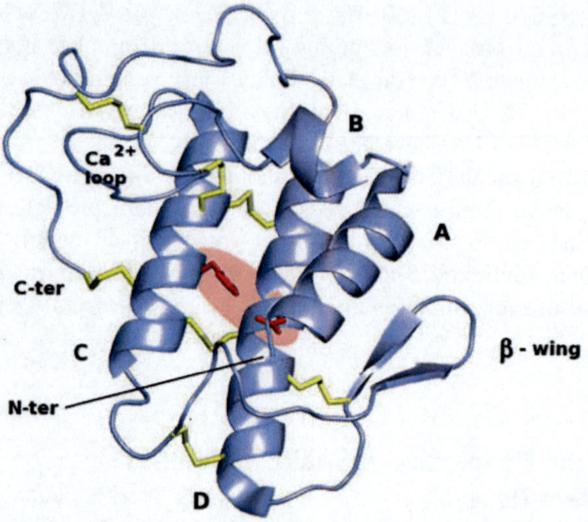

Fig. 13.2 Three-dimensional structure of snake venom anticoagulant PLA_2 (PDB 3G8G). The diad catalytic (His48/Asp99) is shown in *red* and the seven disulfide bridges are in *yellow*

Ca^{2+}-binding loop, the C-terminal loop and the β-wing are covalently linked by disulfide bridges. The highly conserved active site residues are clustered on helices C and D and interact via Asp49 with the Ca^{2+}-binding loop.

Snake venom $sPLA_2$s exhibit different patho-physiological functions, often toxic, but some beneficial (Fenard et al., 1999; Gutierrez and Lomonte, 1995; Kini, 1997, 2003; Nevalainen et al., 2008; Ownby, 1998; Valentin and Lambeau, 2000; Zieler et al., 2001):

I. Toxicity: neurotoxic (pre-synaptic or post-synaptic), myotoxic (local myonecrosis, systemic myotoxicity) and cardiotoxic activity.
II. Effects on haemostasis: anticoagulant effect, initiation and inhibition of platelet aggregation, hemolytic activity, internal hemorrhage, antihemorrhagic activity.
III. Convulsant and hypotensive activity.
IV. Induction of edema, organ necrosis or tissue damage (liver, kidney, lungs, testis, and pituitary damage).
V. Bactericidal, antitumoral, anti-HIV, anti-*Leishmania*, and anti-*Plasmodium* activity.

The presence of pharmacologically distinct sites was first suggested by Kini and there is increasing experimental evidence that PLA_2 enzymes are multifunctional proteins (Chioato and Ward, 2004; Doley and Kini, 2009; Faure, 2002; Faure and Fourier, 2006; Faure and Saul, 2008; Kini and Evans, 1989).

Different Targets and Distinct Pathways to Exert the Patho-Physiological Function of PLA_2

PLA_2 enzymes exhibit multiple biological properties. They can exert physiological and patho-physiological effects through protein-PL, protein-sugar and/or protein-protein interactions. The protein-protein mechanisms, which determine the specific function of $sPLA_2$, are either dependent on or independent of their enzymatic activity. Different protein targets have been characterized for PLA_2:

I. Coagulation factor FXa, the privileged acceptor of certain anticoagulant PLA_2s (Faure et al., 2007; Kerns et al., 1999; Mounier et al., 1998; Prijatelj et al., 2006) (see below)

II. The neuronal or N-type $sPLA_2$ receptors (Lambeau et al., 1989; Lambeau and Lazdunski, 1999) and several extra- and intra-cellular proteins with high binding affinity for snake venom β-neurotoxins: 14-3-3 proteins, calmodulin, protein disulfide isomerase; neuronal pentraxin, crocalbin, TCBP-49, taipoxin-associated calcium binding protein 49 and others (for review see Pungercar and Krizaj, 2007). Also, Crotoxin Acceptor Protein (CAPT-48) was isolated (but not sequenced) from presynaptic membranes of the electric organ of *Torpedo marmorata* (Faure et al., 2003; Krizaj et al., 1997).

III. The M-type $sPLA_2$s receptor located in muscle cell membranes, which display high binding affinity for toxic and non toxic PLA_2s (Lambeau et al., 1990).

IV. Voltage-dependent K^+ channels, the target for β-bungarotoxins (Scott et al., 1990b).

V. Anionic heparan sulfate proteoglycan (HSPG), such us glypican, decorin and biglycan (Murakami and Kudo, 2003; Murakami et al., 1999).

VI. A cytoskeletal protein, vimentin (Boilard et al., 2003).

VII. The natural PLA_2-inhibitors (PLI) present in the blood of snakes. Anti-neurotoxic PLIs inhibit enzymatic activity and neutralize toxicity of neurotoxic PLA_2s by formation of non-covalently linked enzyme-inhibitor soluble complexes (Faure, 2000; Faure and Goyffon, 2008). PLIs were shown to be a 75–180 kDa glycoproteins and have been classified into three types (α, β and γ) based on their structural characteristics. The α-type PLIs (75–120 kDa) display similarities to the mammalian proteins containing Ca^{2+} dependent C-type lectin domain (CTLD), also known as carbohydrate recognition domain (CRD)-like motif; the β-type PLIs (160 kDa) show structural similarity to the proteins containing leucine-rich repeats (LRRs); the γ-type PLIs (90–130 kDa) display structural homology to proteins with three-finger motif. For example, CNF (*Crotalus* Neutralizing Factor)/CICS (Crotoxin Inhibitor from *Crotalus* serum) is a γ-type PLI isolated from the blood of *Crotalus durissus terrificus,* which inhibits PLA_2 activity and toxicity of crotoxin (Faure et al., 2000; Perales et al., 1995).

The $sPLA_2$ also possess other protein targets for which characterization is underway.

Coagulation Factor Xa – The Privileged Protein Target for Some Anticoagulant Group IIA $sPLA_2$

The blood coagulation factor X (FX) is essential for the blood clotting process and hemostasis (Davie, 1995). It is activated to FXa by an extrinsic pathway (the activation peptide is cleaved by factor VIIa) or by an intrinsic pathway (the activation peptide is cleaved by factor IXa). FXa converts prothrombin to thrombin (Fig. 13.3). The prothrombinase complex, which is composed of FXa, negatively charged PL at the cellular surface, calcium ions and cofactor FVa, constitutes an important step of the coagulation cascade. The interaction between FXa and its cofactor FVa results in accelerated conversion of prothrombin to thrombin by selective cleavage of Arg-Thr and Arg-Ile bonds. In the next step of coagulation cascade, thrombin converts fibrinogen into fibrin, which consolidates the primary plug (Fig. 13.3).

Human coagulation factor FX is a vitamin K-dependent glycoprotein synthesized in the liver as a single-chain precursor. In plasma, FX circulates as a two-chain glycoprotein (59 kDa). The light chain (residues 1–139) is linked to the heavy chain (residues 143–448) by a single disulfide bond. The N-terminal region of the light chain (18 kDa) contains the Gla domain (residues 1–45), rich in post-translationally modified γ–carboxyglutamic acid followed by two Epidermal Growth Factor-like domains (EGF-like 1, residues 46–84 and EGF-like 2, residues 85–128). The proteolytic activation of FX consists of removing of the activation peptide (residues 143–194) in a 68-amino-acid external disulfide loop between Cys132 and Cys302. Additional auto-proteolysis of the Arg429–Gly430 peptide bond leads to the removal of a small peptide from the carboxyl terminus of the heavy chain that converts the α-form of FXa to the β-form (Mertens and Bertina, 1980). The heavy chain of FXa (27 kDa), which contains the serine protease

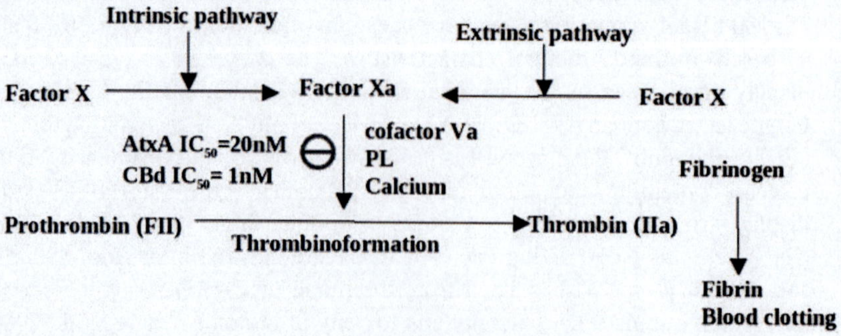

Fig. 13.3 Simplified scheme of blood coagulation reactions

domain consists of two subdomains of antiparallel β-barrel structure each comprising a β-sheet of six strands and four helices. Residues His236, Asp282 and Ser379 form a catalytic triad at the active site cleft between the two subdomains (Fig. 13.4).

Various studies have shown that the anticoagulant activity of $sPLA_2$ could depend on enzymatic activity (PL-dependent mechanisms), since $sPLA_2$ hydrolyse and destroy procoagulant PL (Kini, 2005; Kini and Evans, 1995; Verheij et al., 1980; Zhong et al., 2002) and $sPLA_2$ compete with clotting proteins for binding to the lipid surface (Babu and Gowda, 1994; Prigent-Dachary et al., 1980).

Anticoagulant activity could also be independent of enzymatic activity (PL-independent mechanism). Some snake venom PLA_2 interact directly with coagulation factors thus inhibiting formation of the tenase and/or prothrombinase complexes, using key enzymes of the activation cascade of blood clotting (Inada et al., 1994; Kini, 2005; Kini and Evans, 1995; Mounier et al., 2000).

It has been proposed that $sPLA_2$s inhibit the prothrombinase complex by preventing formation of the FXa/FVa complex (i.e., compete with factor Va for the FXa-binding site) and introducing a lag time in the formation of thrombin (Kini, 2005; Kerns et al., 1999; Mounier et al., 1998, 2000; Prijatelj et al., 2006; Stefansson et al., 1989, 1990). The anticoagulant FXa-binding $sPLA_2$ are therefore promising candidates for treatment of thrombo-embolic disorders since the anticoagulant effects of $sPLA_2$ are progressive, thus provoking less pronounced side effects.

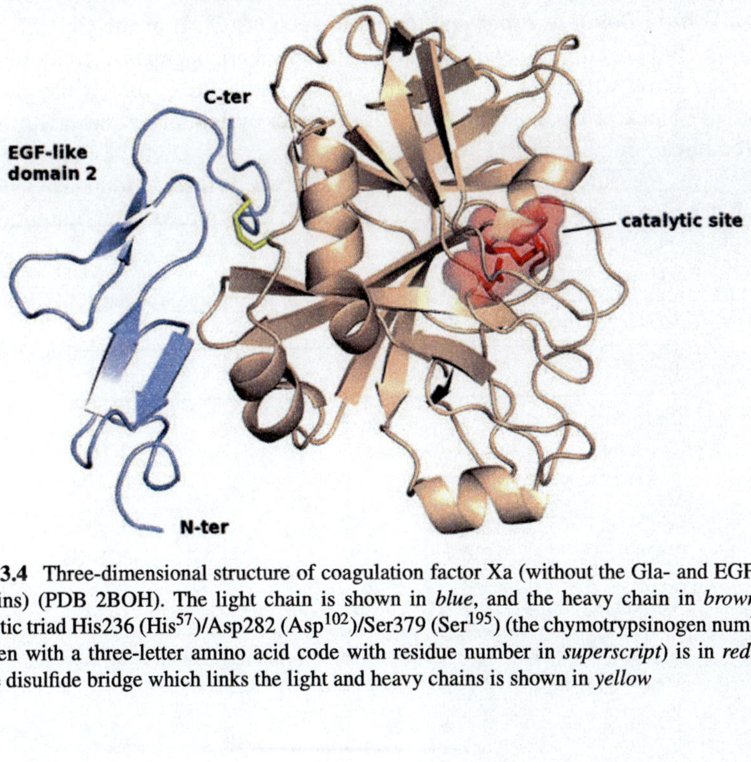

Fig. 13.4 Three-dimensional structure of coagulation factor Xa (without the Gla- and EGF-like1 domains) (PDB 2BOH). The light chain is shown in *blue*, and the heavy chain in *brown*. The catalytic triad His236 (His^{57})/Asp282 (Asp^{102})/Ser379 (Ser^{195}) (the chymotrypsinogen numbering is given with a three-letter amino acid code with residue number in *superscript*) is in *red* and a single disulfide bridge which links the light and heavy chains is shown in *yellow*

Search for the Anticoagulant Site of PLA_2

The Potential Anticoagulant Site Proposed by Site Directed Mutagenesis and SPR Affinity Binding Studies

Kini and Evans first proposed that a pharmacological site, distinct from the catalytic site, could explain the specific biological anticoagulant activity of *Elapidae* group I snake venom PLA_2 (Kini and Evans, 1987). It has been shown that anticoagulant region of group I PLA_2s consists of the basic exposed loop of the polypeptide chain (residues 55–70) immediately preceding the β-wing (Carredano et al., 1998; Singh et al., 2001). More recently it has been proposed that weakly anticoagulant enzymes, which lack the basic anticoagulant region, fail to bind specifically to FXa in the coagulation cascade (Kini, 2005).

For group II PLA_2 previous studies have shown the involvement of basic residues located around the IBS of human group IIA $sPLA_2$ in binding to human FXa (Mounier et al., 2000). Using site-directed mutagenesis and SPR affinity binding studies we demonstrated the importance of the β-wing and C-terminal region for binding of Ammodytoxin A (AtxA) from *Vipera ammodytes ammodytes*, to FXa (Prijatelj et al., 2006). As shown in Table 13.1, $K_D{}^{app}$ of Ammodytoxin A for FXa is 30 nM; in comparison for human PLA_2 group IIA the corresponding value of $K_d{}^{app}$ for FXa is 14 nM (Faure et al., 2007).

Several other $sPLA_2$ from the *Viperidae* family, including the basic CB-subunit of crotoxin from *Crotalus durissus terrificus* (CBc and CBa_2 isoforms), myotoxin II (MtxII) from *Bothrops asper* and the basic subunit (CbII) of the CbICbII complex from *Pseudocerastes fieldi* also display strong anticoagulant activity through direct interaction with FXa (Faure et al., 2007). These PLA_2 inhibit blood coagulation and block procoagulant activity of platelets by inhibiting formation of the prothrombinase complex.

We have measured the binding affinity between these group IIA $sPLA_2$ from *Viperidae* snake venom and FXa, and also determined their inhibition of

Table 13.1 Specific binding affinity to FXa and effect on prothrombinase activity of FXa-binding PLA_2 from *Viperidae family* for which X-ray crystal structures are available

$sPLA_2$	IC_{50} [nM]	$K_d{}^{app}$[nM]	PDB
MtxII	3	1,8	1CLP
CB1(CBc)	0,7	0,6	2QOG
CB2 (CBa_2)	41	52	2QOG
AtxA	25	30	3G8G
AtxC	240	300	3G8H
$DPLA_2$	130	578	1ZWP
bAhp	90	400	1JIA
$hGIIAPLA_2$	9	14	1BBC

IC_{50} corresponds to 50% inhibition of thrombin generation in the absence of phospholipids observed for different PLA_2s prothrombinase activity (Table 13.1). Comparative analysis of these results shows a strong correlation between the dissociation constant K_d^{app} and prothrombinase inhibition, allowing the classification of anticoagulant PLA_2 into three groups displaying strong, medium and weak anticoagulant potency, respectively (Faure et al., 2007) (Table 13.1).

The PLA_2-FXa Binding Interface Identified by Molecular Docking

Using molecular docking simulations between PLA_2 and FXa, taking into account the experimental data obtained by site directed mutagenesis, SPR affinity binding studies and an *in vitro* biological test of inhibition of prothrombinase activity, we have mapped the interaction sites on *Viperidae* snake venom PLA_2s and human FXa (Faure et al., 2007). Interface amino acid residues of the PLA_2 complex with FXa and the potential FXa binding site are shown in Fig. 13.5. The site is composed of two adjacent regions: (i) Region A, (residues 1–19 and 52–77, which include solvent-exposed parts of helices A and B and part of the loop between helix C and the β-wing) and (ii) Region B (residues 23–34 and 118–133, which correspond to part of the Ca^{2+} binding loop and the C-terminal segment (Faure et al., 2007).

Based on models of the complex obtained by molecular docking, the catalytic sites of PLA_2 and FXa are accessible and not involved in the binding interface (Faure et al., 2007). Furthermore, SPR studies have shown that the Gla-domain of FXa is not important for PLA_2-FXa interaction.

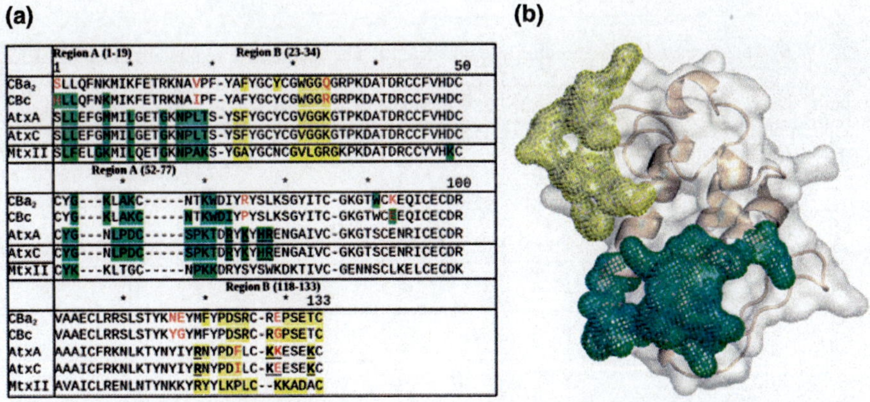

Fig. 13.5 The anticoagulant site. The PLA_2-FXa binding interfaces are mapped by molecular docking, according to Faure et al. (2007). (**a**) The amino acid sequences of two isoforms of CB, Atx, and MtxII showing the identified interface regions A and B critical for binding to FXa and inhibition of prothrombinase activity. (**b**) Possible location of the anticoagulant site on group IIA *Viperidae* PLA_2. Region A (residues 1–19 and 52–77) is shown in *green* and Region B (residues 23–34 and 118–133) is in *yellow*. The alignment reflects the numbering system of Renetseder et al., 1985

The potential PLA_2 binding site on FXa is shown in Fig. 13.6. The site is composed of five regions on the FXa heavy chain and two segments of the EGF-like 2 domain of the light chain. Regions I (yellow), IV (red) and V (orange) of the heavy chain of FXa (Fig. 13.6b) are in contact with Region B of PLA_2 (shown in yellow Fig. 13.5b). Regions II (green) and III (blue) of the heavy chain as well as two segments (cyan and magenta) of the EGF-like 2 domain of the light chain (Fig. 13.6b) are in contact with Region A of PLA_2 (shown in green Fig. 13.5b). A contact map of the PLA_2-FXa of interacting regions is presented in Table 13.2 (Faure et al., 2007).

It has been shown by site directed mutagenesis studies that the basic residues Arg-93, Lys-96, Arg-125, Arg-165, Lys-169, Lys-236 and Arg-240 constitute an exosite in the heavy chain of FXa and that this region can effectively bind heparin (Monteiro, 2005; Rezaie and He, 2000). Since several of these residues are located in regions I, II, III and V identified as the potential PLA_2 binding site on

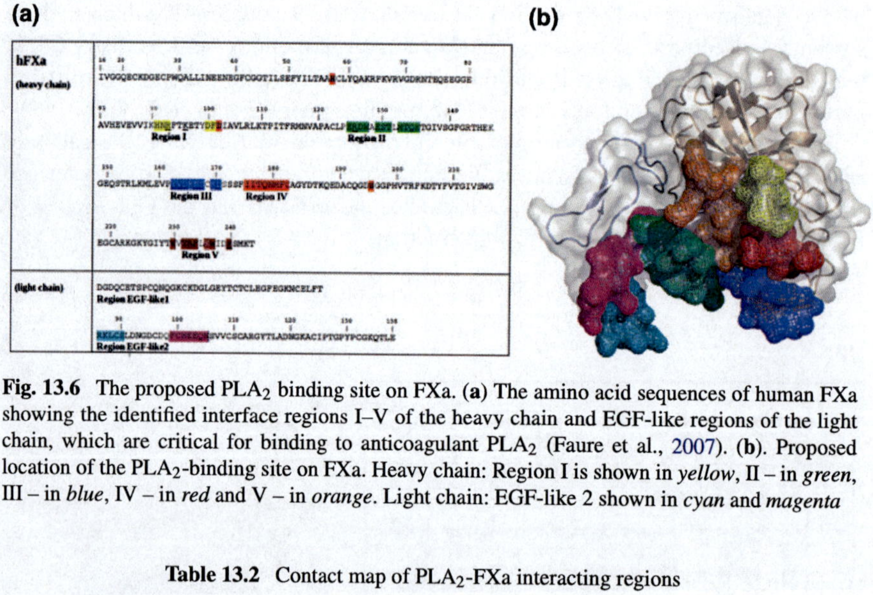

Fig. 13.6 The proposed PLA_2 binding site on FXa. (**a**) The amino acid sequences of human FXa showing the identified interface regions I–V of the heavy chain and EGF-like regions of the light chain, which are critical for binding to anticoagulant PLA_2 (Faure et al., 2007). (**b**). Proposed location of the PLA_2-binding site on FXa. Heavy chain: Region I is shown in *yellow*, II – in *green*, III – in *blue*, IV – in *red* and V – in *orange*. Light chain: EGF-like 2 shown in *cyan* and *magenta*

Table 13.2 Contact map of PLA_2-FXa interacting regions

	FXa						
	HC regions				LC regions		
PLA_2	I	II	III	IV	V	EGF-like1	EGF-like2
Region A (1–19)		+					+
Region A (52–77)			+			+	+
Region B (23–34)	+			+			
Region B (118–133)	+				+		

FXa (see Fig. 13.6a) we suggest that this exosite is also important for interaction with anticoagulant PLA_2.

Structural Basis of the Functional Differences Observed Between Two PLA_2 Isoforms Which Interact with FXa and Differ in Anticoagulant Potency

In order to characterize the structural basis of the functional differences observed between two natural isoforms of ammodytoxin we have determined the crystal structure of AtxA and AtxC (Saul et al., 2010). The two isoforms differ in sequence by only two substitutions (Phe124Ile and Lys128Glu), but display significant differences in toxicity and anticoagulant activity. The crystal structures allow us (after detailed structural comparison of natural and mutated amino acids) to explain the 10-fold decrease in binding affinity of AtxC for FXa (Table 13.1). Briefly, the side chain of Lys128 in AtxA (Fig. 13.7a) (residue identified by mutagenesis as important for interaction with FXa) (Prijatelj et al., 2006) is fully exposed and accessible for binding to FXa, whereas the side chain of Glu128 in AtxC (Fig. 13.7c) makes a stabilizing hydrogen bond with the main chain nitrogen atom of residue Thr35, leading to a displacement in the polypeptide chain backbone at positions 127 and 128 compared with AtxA, and a shift of up to 6 Å in the side chain of the conserved residue Lys127 (Saul et al., 2010). Lys127 has been identified by mutagenesis as important for the interaction of AtxA with FXa (Prijatelj et al., 2006), and has been identified by molecular docking calculations as part of the anticoagulant site (Faure et al., 2007).

This observation is more general, since it can be extended to other FXa-binding PLA_2. In particular, we find similar conformational changes in two isoforms CB1 and CB2 of the anticoagulant CB subunit of crotoxin from *Crotalus durissus terrificus* venom. The pharmacological properties of these individual CB isoforms have been previously reported (Faure et al., 1993, 1994). In a manner similar to AtxA and AtxC, the two monomeric isoforms CB1 and CB2 differ in binding affinity to FXa; CB2 displays a 100-fold lower affinity for FXa and 60-fold lower anticoagulant activity (Faure et al., 2007) and their C-terminal region has been suggested as important for interaction with FXa by docking calculations (Faure et al., 2007). Recently, the crystal structure of a tetrameric complex formed by mixture of two isoforms CB1/CB2 of CB has been reported (Marchi-Salvador et al., 2008) and we compare the individual CB1 and CB2 isoforms (Fig. 13.7b, d) with the crystal structures of AtxA and AtxC (Fig. 13.7a, c). We find that, similarly to the structure of AtxC, residue Glu128 in CB2 induces two events: (i) interaction between Glu128 and the main chain of nitrogen atom of residue 35 (Arg 35 in CB2) which leads to a shift in the side chain of unmutated Arg127 (Lys127 in AtxC) and (ii) a displacement of the main chain at positions 127–128 (Fig. 13.7c, d). Thus, comparison of the crystal structure of the two natural isoforms of CB with the structures of AtxA and AtxC strongly suggests important contributions of Glu128 to the observed decrease in affinity of PLA_2 for FXa.

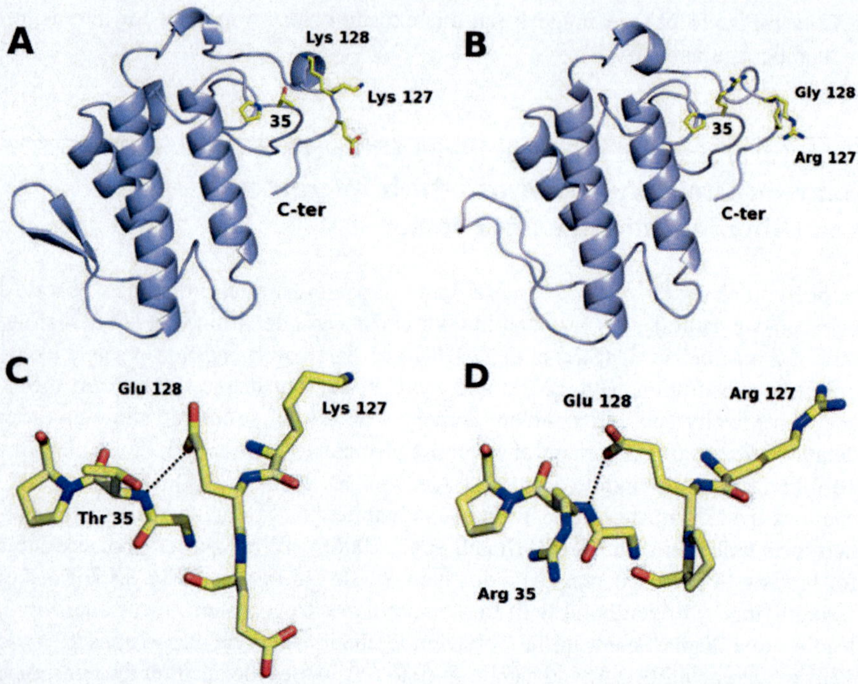

Fig. 13.7 Structural differences between AtxA (PDB 3G8G) and AtxC (PDB 3G8H) and isoforms CB1 and CB2 (PDB 2QOG) which interact with FXa with different binding affinity and differ in anticoagulant potency. Overall view including the C-terminal region of AtxA (**a**) and CB1 (**b**). Detailed view of the C-terminal region of AtxC (**c**) and CB2 (**d**). The stabilizing hydrogen bond formed between the side chain of Glu128 and the main chain of Thr35 in AtxC and Glu 128 and Arg35 in CB2 are shown

Conclusions and Perspectives

Various secreted phospholipases A_2 ($sPLA_2$) obtained from exogenous sources such as snake venoms have been identified as specific blood coagulation inhibitors which interact with human coagulation factor Xa. In this review we describe some *Viperidae* venom $sPLA_2$s that exert an anticoagulant effect by a non-enzymatic, PL-independent mechanism through direct binding to human FXa. The potential FXa-binding site of these anticoagulant PLA_2s and the potential PLA_2-binding site on FXa (located on the exosite in the heavy chain and EGF-like domains in the light chain) were identified based on molecular docking calculations and affinity binding studies. We also discuss the importance of two recently crystallized natural isoforms of ammodytoxin which interact with FXa with different affinity and which differ in anticoagulant potency, to determine the structural basis of functional differences between PLA_2 isoforms. The three-dimensional structure of the two isoforms of ammodytoxins, together with analysis of other FXa-binding PLA_2, helps to explain the role of natural mutations in the binding mode of PLA_2 with FXa.

The key role of coagulation factor FXa in thrombosis is well known but its non-hemostatic activity and potential implications in the pathogenesis of various diseases such as arterial restenosis, venous graft disease, acute inflammation, sepsis and cancer are not yet completely understood (Leadley et al., 2001). It has been proposed that FXa may initiate signaling pathways in various cells by binding to effector cell protease receptor-1 (EPR-1) or by proteolytic activation of protease activated receptor-2 (PAR-2). Moreover, the inter-epidermal growth factor (EGF) sequence ^{83}LeuPheThrArgLysLeu^{88} in FXa was found to be the recognition site for EPR-1 (Ambrosini et al., 1997). The enzymatic activity of FXa is not required for EPR-1 binding. Our PLA_2-FXa docking results strongly suggest that the EGF sequence (^{83}LeuPheThrArgLysLeu^{88}) is also an important part of the recognition site for PLA_2. Does PLA_2 affect FXa binding to EPR-1? The possible implication of PLA_2 in intracellular signaling pathways of FXa will be an interesting topic for further study.

Knowledge of the three-dimensional structure of the anticoagulant snake venom PLA_2 and their sites of interaction with FXa at the level of specific amino acid residues could lead to a detailed understanding of hemostatic and non-hemostatic processes at the molecular level and offer promising perspectives in identifying new drug leads.

References

Ambrosini, G., Plescia, J., Chu, K.C., High, K.A., Altieri, D.C., 1997. Activation-dependent exposure of the inter-EGF sequence Leu^{83}-Leu^{88} in factor Xa mediates ligand binding to effector cell protease receptor-1. J. Biol. Chem. 272, 8340–8345.

Arni, R.K., Ward, R.J., 1996. Phospholipases A_2 – a structural review. Toxicon 34, 827–841.

Babu, A.S., Gowda, T.V., 1994. Dissociation of enzymatic activity from toxic properties of the most basic phospholipase A_2 from *Vipera russelli* snake venom by guanidination of lysine residues. Toxicon, 32, 749–752.

Berg, O.G., Gelb, M.H., Tsai, M.D., Jain, M.K., 2001. Interfacial enzymology: the secreted phospholipases A_2-paradigm. Chem. Rev. 101, 2613–2653.

Boilard, E., Bourgoin, S.G., Bernatchez, C., Surette, M.E., 2003. Identification of an autoantigen on the surface of apoptotic human T cells as a new protein interacting with inflammatory group IIA phospholipase A_2. Blood, 102, 2901–2909.

Carredano, E., Westerlund, B., Persson, B., Saarinen, M., Ramaswamy, S., Eaker, D., Eklund, H., 1998. The three-dimensional structures of two toxins from snake venom throw light on the anticoagulant and neurotoxic sites of phospholipase A_2. Toxicon 36, 75–92.

Chioato, L., Ward, R.J., 2004. Mapping structural determinants of biological activities in snake venom phospholipases A_2 by sequence analysis and site directed mutagenesis. Toxicon, 42, 869–883.

Davie, E.W., 1995. Biochemical and molecular aspects of the coagulation cascade. Thromb. Haemost. s74, 1–6.

Dennis, E.A., 1994. Diversity of group types, regulation, and function of phospholipases A_2. J. Biol. Chem. 269, 13057–13060.

Dennis, E.A., 1997. The growing phospholipase A_2 superfamily of signal transduction enzymes, Trends Biochem. Sci. 22, 1–2.

Deregnaucourt, C., Schrevel, J., 2000. Bee venom phospholipase A_2 induces stage-specific growth arrest of the intraerythrocytic *Plasmodium falciparum* via modifications of human serum components. J. Biol. Chem. 275, 39973–39980.

Dessen, A., 2000. Phospholipase A_2 enzymes: structural diversity in lipid messenger metabolism. Structure, 8, 15–22.

Doley, R., Kini, R.M., 2009. Protein complexes in snake venom. Cell. Mol. Life Sci. 66, 2851–2871.

Faure, G., Harvey, A.L., Thomson, E., Saliou, B., Radvanyi, F., Bon, C. 1993. Comparison of crotoxin isoforms reveals that stability of the complex plays a major role in its pharmacological action. Eur. J. Biochem. 214, 491–496.

Faure, G., Choumet, V., Bouchier, C., Camoin, L., Guillaume, J.L., Monegier, B., Vuilhorgne, M., Bon, C. 1994. The origin of the diversity of crotoxin isoforms in the venom of *Crotalus durissus terrificus*. Eur. J. Biochem. 223, 161–164.

Faure, G., 2000. Natural inhibitors of toxic phospholipases A_2. Biochimie 82, 833–840.

Faure, G., Villela, C., Perales, J., Bon, C. 2000. Interaction of the neurotoxic and non-toxic secretory phospholipases A_2 with the crotoxin inhibitor from *Crotalus* serum. Eur. J. Biochem. 267, 4799–4808.

Faure, G., 2002. Different protein targets for snake venom phospholipases A_2, in: Goudey-Perrière, F., Bon, C., Puiseux-Dao, S., Sauviat, M. (Eds.), Toxines et Recherches Biomédicales. SFET collection, "Rencontres en toxinologie", Elsevier, Paris, France, pp. 305–313.

Faure, G., Copic, A., Le Porrier, S., Gubensek, F., Bon, C., Krizaj, I., 2003. Crotoxin acceptor protein isolated from *Torpedo* electric organ: binding properties to crotoxin by surface plasmon resonance. Toxicon 41, 509–517.

Faure, G., Fourier, A. 2006. SPR study to characterize the snake venom phospholipases A_2 mechanism of action, in: Goudey-Perrière, F., Benoit, E., Goyffon, M., Marchot, P. (Eds.), SFET collection: Toxines et Cancer. Libraire Lavoisier, Cachan, France, pp. 313–318. ISBN 2-7430-0958-6.

Faure, G., Gowda, V.T., Maroun, R., 2007. Characterization of a human coagulation factor Xa-binding site on phospholipases A_2 from Viperidae snake venom by affinity binding studies and molecular bioinformatics. BMC Struc. Biol. 7, 82.

Faure, G., Goyffon, M., 2008. Inhibiteurs naturels de Phospholipases A_2 (PLI) ophidiennes, in: Mion, G., Larréché, S., Goyffon, M. (Eds.), Aspects Cliniques et Thérapeutiques des Envenimations Graves. "Edition spéciale XIV Actualités du Pharo" Edition Xavier Montauban (CD).

Faure, G., Saul, F. 2008. Caractéristiques structurales et fonctionnelles de deux β-neurotoxines: l'ammodytoxine et la crotoxine, in: Benoit, E., Goudey-Perrière, F., Marchot, P., Servent, D. (Eds.), Toxines et Fonctions Cholinergiques Neuronales et Non Neuronales. SFET collection: Rencontres en Toxinologie, E-book RT16, pp. 155–160. ISSN 1760-6004.

Fenard, D., Lambeau, G., Valentin, E., Lefebvre, J.C., Lazdunski, M., Doglio, A., 1999. Secreted phospholipases A_2, a new class of HIV inhibitors that block virus entry into host cells. J. Clin. Invest. 104, 611–618.

Fenard, D., Lambeau, G., Maurin, G., Lefebvre, JC., Doglio, A., 2001. A peptide derived from bee venom-secreted phospholipase A_2 inhibits replication of T-cell tropic HIV-1 strains via interaction with the CXCR4 chemokine receptor. Mol. Pharmacol. 60, 341–347.

Gutierrez, J.M., Lomonte, B., 1995. Phospholipase A_2 myotoxins from *Bothrops* snake venoms. Toxicon 33, 1405–1424.

Inada, M., Crowl, R.M., Bekkers, A.C., Verheij, H., Weiss, J. 1994. Determinants of the inhibitory action of purified 14-kDa phospholipases A_2 on cell-free prothrombinase complex. J. Biol. Chem. 269, 26338–26343.

Kerns, R.T., Kini, R.M., Stefansson, S., Evans, H.J., 1999. Targeting of venom phospholipases: the strongly anticoagulant phospholipase A_2 from *Naja nigricollis* venom binds to coagulation factor Xa to inhibit the prothrombinase complex. Arch. Biochem. Biophys. 369, 107–113.

Kim, J.O., Chakrabarti, B.K., Guha-Niyogi, A., Louder, M.K., Mascola, J.R., 2007. Lysis of human immunodeficiency virus type 1 by a specific secreted human phospholipase A_2. J. Virol. 81, 1444–1450.

Kini, R.M., Evans, H.J., 1987. Structure-function relationships of phospholipases: the anticoagulant region of phospholipases A_2. J. Biol. Chem. 262, 14402–14407.

Kini, R.M., Evans, H.J., 1989. A model to explain the pharmacological effects of snake venom phospholipases A_2. Toxicon 27, 613–635.

Kini, R.M., Evans, H.J., 1995. The role of enzymatic activity in inhibition of the extrinsic tnenase complex by phospholipase A_2 isoenzymes from *Naja nigricollis* venom. Toxicon 33, 1585–1590.

Kini, R.M., 1997. Phospholipase A_2 – a complex multifunctional protein puzzle, in: Kini, R.M. (Ed.), Venom Phospholipases A_2 Enzymes: Structure, Function and Mechanism. John Wiley, Chichester, pp. 1–28.

Kini, R.M., 2003. Excitement ahead: structure, function and mechanism of snake venom phospholipase A_2 enzymes. Toxicon 42, 827–840.

Kini, R.M., 2005. Structure-function relationships and mechanism of anticoagulant phospholipase A_2 enzymes from snake venoms. Toxicon 45, 1147–1161.

Krizaj, I., Faure, G., Gubensek, F., Bon, C., 1997 Neurotoxic phospholipases A_2 ammodytoxin and crotoxin bind to distinct high-affinity protein acceptors in *Torpedo marmorata* electric organ. Biochemistry 36, 2779–2787.

Lambeau, G., Schmid-Alliana, A., Lazdunski, M., Barhanin, J., 1989. Identification and purification of a very high affinity binding protein for toxic phospholipases A_2 in skeletal muscle. J. Biol. Chem. 265, 9526–9532.

Lambeau, G., Schmid-Alliana, A., Lazdunski, M., Barhanin, J., 1990. Identification and purification of a very high affinity binding protein for toxic phospholipases A_2 in skeletal muscle. J. Biol. Chem. 265, 9526–9532.

Lambeau, G., Lazdunski, M., 1999. Receptors for a growing family of secreted phospholipases A_2, Trends Pharmacol. Sci. 20, 162–170.

Lambeau, G., Gelb, M.H., 2008. Biochemistry and physiology of mammalian secreted phospholipases A_2. Annu. Rev. Biochem. 77, 495–520.

Leadley, R.J., Jr., Chi, L., Porcari, A.R., 2001. Non-hemostatic activity of coagulation factor Xa: potential implications for various diseases. Curr. Opin. Pharmacol. 1, 169–175.

Marchi-Salvador, D.P., Corrêa, L.C., Magro, A.J., Oliveira, C.Z., Soares, A.M., Fontes, M.R., 2008. Insights into the role of oligomeric state on the biological activities of crotoxin: crystal structure of a tetrameric phospholipase A_2 formed by two isoforms of crotoxin B from *Crotalus durissus terrificus* venom. Proteins 72, 883–891.

Mertens, K., Bertina, R.M., 1980. Pathways in the activation of human coagulation factor X. Biochem. J. 185, 647–658.

Mounier, C.M., Hackeng, T.M., Schaeffer, F., Faure, G., Bon, C., Griffin, J.H., 1998. Inhibition of prothrombinase by human secretory phospholipase A_2 involves binding to Factor Xa. J. Biol. Chem. 273, 23764–23772.

Mounier, C.M., Luchetta, P., Lecut, C., Koduri, R.S., Faure, G., Lambeau, G., Valentin, E., Singer, A., Ghomashchi, F., Béguin, S., Gelb, M.H., Bon, C., 2000. Basic residues of human group IIA phospholipase A_2 are important for binding to factor Xa and prothrombinase inhibition. Comparison with other mammalian secreted phospholipases A_2. Eur. J. Biochem. 267, 4960–4969.

Murakami, M., Nakatani, Y., Atsumi, G., Inoue, K., Kudo, I., 1997. Regulatory functions of phospholipases A_2. Crit. Rev. Immunol. 17, 225–283.

Murakami, M., Kambe, T., Shimbara, S., Yamamoto, S., Kuwata, H., Kudo, I., 1999. Functional associationof type IIA secretory phospholipase A_2 with the glycosyl phosphatidylinositol-anchoredheparan sulfate proteoglycan in the cyclooxygenase-2-mediated delayedprostanoid biosynthetic pathway. J. Biol. Chem. 20. 162–170.

Murakami, M., Kudo, I., 2001. Diversity and regulatory functions of mammalian secretory phospholipase A_2. Adv. Immunol. 77. 163–194.

Murakami, M., Kudo, I., 2003. Another way for $sPLA_2$ to meet with cells. Blood 102, 2714–2715.

Monteiro, R.Q., 2005. Targeting exosites on blood coagulation proteases. Ann. Braz. Acad. Sci. 77, 275–280.

Nevalainen, T.J., Haapamaki, M.M., Gronroos, J.M., 2000. Roles of secretory phospholipases A_2 in inflammatory diseases and trauma. Biochem. Biophys. Acta 1488, 83–90.

Nevalainen, T.J., Graham, G.G. Scott, K.F., 2008. Antibacterial actions of secreted phospholipases A_2. Biochem. Biophys. Acta 1781, 1–9.

Ownby, C.L., 1998. Structure, function and biophysical aspects of the myotoxins from snake venoms. J.Toxicol.-Toxins Rev. 17, 213–238.

Perales, J., Villela, C., Domont, G., Choumet, V., Saliou, B., Moussatché, H., Bon, C., Faure, G.. 1995. Molecular structure and mechanism of action of the crotoxin inhibitor from *Crotalus durissus terrificus* serum. Eur. J. Biochem. 227, 19–26.

Prijatelj, P., Charnay, M., Ivanovski, G., Jenko, Z., Pungercar, J., Krizaj, I., Faure, G., 2006. The C-terminal and ß-wing regions of ammodytoxin A, a neurotoxic phospholipase A_2 from *Vipera ammodytes ammodytes*, are critical for binding to factor Xa and for anticoagulant effect. Biochimie 88, 69–76.

Prigent-Dachary, J., Boffa, M.C., Boisseau, M.R., Dufourcq, J., 1980. Snake venom phospholipases A_2. A fluorescence study of their binding to phospholipid vesicles correlation with their anticoagulant activities. J. Biol. Chem. 255, 7734–7739.

Pungercar, J., Krizaj, I., 2007. Understanding the molecular mechanism underlying the presynaptic toxicity of secreted phospholipases A_2. Toxicon 50, 871–892.

Renetseder, R., Brunie, S., Dijkstra, B.W., Drenth, J., Sigler, P.B., 1985. A comparison of the crystal structures of phospholipase A_2 from bovine pancreas and *Crotalus atrox* venom. J. Biol. Chem. 260, 11627–11634.

Rezaie, A.R., He, X., 2000. Sodium binding site of factor Xa: role of sodium in the prothrombinase complex. Biochemistry 39, 1817–1825.

Saul, F. A., Prijatelj-Znidarsic, P., Vuillez-le Normand, B., Villette, B., Raynal, B., Pungarcar, J., Krizaj, I., Faure, G., 2010. Comparative structural studies of two natural isoforms of ammodytoxin, phospholipases A_2 from *Vipera ammodytes ammodytes* which differ in neurotoxicity and anticoagulant activity. J. Struct. Biol. 169, 360–369.

Schaloske, R.H., Dennis, E.A., 2006. The phospholipase A_2 superfamily and its group numbering system. Biochem. Biophys. Acta 1761, 1246–1259.

Scott, D.L., White, S.P., Otwinowski, Z., Yuan, W., Gelb, M.H., Sigler, P.B., 1990a. Interfacial catalysis: the mechanism of phospholipase A_2. Science 250, 1541–1546.

Scott, V.E.S., Parcej, D.N., Keen, J.N., Findlay, J.B.C., Dolly, J.O., 1990b. α–Dendrotoxin acceptor from bovine brain is a K^+ channel protein. J. Biol. Chem. 265, 20094–20097.

Singh, G., Gourinath, S., Sharma, S., Paramasivam, M., Srinivasan, A., Singh, T.P., 2001. Sequence and crystal structure determination of a basic phospholipase A_2 from common krait (*Bungarus caeruleus*) at 2.4 A resolution: identification and characterization of its pharmacological sites. J. Mol. Biol. 307, 1049–1059.

Snitko, Y., Koduri, R.S., Han, S.K., Othman, R., Baker, S.F., Molini, B.J., Wilton, D.C., Gelb, M.H., Cho, W., 1997. Mapping the interfacial binding surface of human secertory group IIa phospholipase A_2. Biochemistry 36, 14325–14333.

Snitko, Y., Han, S.K., Lee, B.I., Cho, W., 1999. Differencial interfacial and substrate binding modes of mammalian pancreatic phospholipases A_2: a comparison among human, bovine, and porcine enzymes. Biochemistry 38, 7803–7810.

Stefansson, S., Kini, R.M., Evans, H.J., 1989. The inhibition of clotting complexes of the extrinsic coagulation cascade by the phospholipase A_2 isoenzymes from *Naja nigricollis* venom. Thromb. Res. 55, 481–491.

Stefansson, S., Kini, R.M., Evans, H.J. 1990. The basic phospholipase A_2 from *Naja nigricollis* venom inhibits the prothrombinase complex by a novel nonenzymatic mechanism. Biochemistry 29, 7742–7746.

Valentin, E., Lambeau, G., 2000. What can venom phospholipases A_2 tell us about the functional diversity of mammalian secreted phospholipases A_2? Biochimie 82, 815–831.

van Deenen, L.L.M., de Hass, G.H., 1963. The substrate specificity of phospholipase A_2. Biochem. Biophys. Acta 70, 538–553.

Verheij, H.M., Boffa, M.C., Rothen, C., Bryckaert, M.C., Verger, R., de Haas, G.H., 1980. Correlation of enzymatic activity and anticoagulant properties of phospholipase A_2. Eur. J. Biochem. 112, 25–32.

Webb, N.R., 2005. Secretory phospholipase A_2 enzymes with a potential role in atherogenesis. Curr. Opin. Lipidol. 16, 341–344.

Zhong, X., Jiao, H., Fan, L., Wu, X., Zhou, Y. 2002. Functionally important residues for the anticoagulant activity of a basic phospholipase A_2 from the *Agkistrodon halys pallas*. Protein Pept. Lett, 9, 427–434.

Zieler, H., Keister, D.B., Dvorak, J.A., Ribeiro, J.M., 2001. A snake venom phospholipase A_2 blocks malaria parasite development in the mosquito midgut by inhibiting ookinete association with the midgut surface. J. Exp. Biol. 204, 4157–4167.

Chapter 14
Hematophagy and Inhibition of the Extrinsic and Intrinsic Tenase Complexes

Robson Q. Monteiro, John F. Andersen, and Ivo M.B. Francischetti

Abstract The coagulation cascade involves the regulated sequence of proteolytic activation of a series of zymogens culminating in conversion of fibrinogen to fibrin and clot formation. These reactions are mainly performed by enzymatic complexes comprised of a serine protease, a protein cofactor and membranes containing anionic phospholipids. A number of specific coagulation inhibitors from exogenous sources have been identified from salivary glands of blood-sucking arthropods and herein named sialogenins (from the Greek sialo, saliva; gen, origin, source; and ins for proteins) with anticlotting activity.

Anti-clotting sialogenins target components of the extrinsic (e.g. ixolaris, penthalaris, NAPc2) or intrinsic Xase (e.g. nitrophorin 2, nitrophorin 7) complexes resulting in inhibition of the initiation, propagation or consolidation steps of blood coagulation cascade. In addition, these molecules act in a redundant and synergistic manner in order to keep hemostatic tonus as low as possible so as to facilitate blood-feeding. These molecules may also attenuate inflammatory events associated with vascular injury. Finally, anti-clotting sialogenins have potential therapeutic applications and are valuable tools in pharmacology and cell biology.

Introduction

Coagulation involves the regulated sequence of proteolytic activation of a series of zymogens to achieve appropriate and timely hemostasis in an injured vessel in an environment that overwhelmingly favors an anticoagulant state. In the nonpathologic state, the inciting event involves exposure of circulating factor VII/VIIa to extravascularly expressed tissue factor (TF), which brings into motion a series of steps that result in amplification of the initial stimulus, culminating in conversion

R.Q. Monteiro (✉)
Instituto de Bioquímica Médica, Universidade Federal do Rio de Janeiro, Av. Carlos Chagas Filho 373, Prédio do CCS, Rio de Janeiro, RJ 21941-590, Brazil
e-mail: Robsonqm@bioqmed.ufrj.br

R.M. Kini et al. (eds.), *Toxins and Hemostasis*, DOI 10.1007/978-90-481-9295-3_14,

of fibrinogen to fibrin and clot formation. Traditionally, the coagulation cascade has been separated into three main steps: initiation, propagation, and consolidation. In the initiation phase, TF-dependent activation of coagulation takes place. This step is known as the "extrinsic" pathway, because TF has been reported not to be present in the blood, although more recently is has been found in association with microparticles (Davie et al., 1991; Giesen et al., 1999). TF is abundantly present in membranes of cells surrounding the vascular bed. Once in contact with blood, circulating FVIIa binds to TF. The catalytic active complex formed by FVIIa/TF initiates the coagulation cascade through activation of FIX and FX (Bauer et al., 1990; Broze, 1995; Osterud and Rapaport, 1977; ten Cate et al., 1993). It has been estimated that the FVIIa/TF complex enhances the efficiency of activation of its substrates, FIX and FX, approximately 10^4 fold versus FVIIa alone (Mann et al., 2003). After activation, FIXa and FXa may remain associated with the TF-expressing cell or may bind to the membrane of platelets activated by low amounts of thrombin. Thrombin also activates FV and FVIII, and the generated FVa and FVIIIa serve as important cofactors to FXa and FIXa in the so-called intrinsic "tenase" and "prothrombinase" complexes, respectively (Fig. 14.1). These complexes assemble on the activated platelet membranes, which display negatively charged phospholipids such as phosphatidyl serine (PS) (Mann et al., 2003; Monroe et al., 2002).

The prothrombinase complex activates prothrombin to thrombin in the presence of FVa, phospholipid surface, and Ca^{2+}, while the main function of intrinsic Xase is to convert FX to FXa in the presence of FVIIIa and Ca^{2+}, leading to amplification of the coagulation cascade (Krishnaswamy, 2005; Mann et al., 2003; Monroe et al., 2002). Both complexes convert several thousand substrate molecules per minute. In contrast, free enzymes FXa and FIXa are inefficient, indicating that many blood coagulation reactions are localized and occur on the surface of activated cells (e.g., activated platelets), not during the fluid phase (Roberts et al., 2006). For comparison, the amount of thrombin produced by the prothrombinase complex in one minute would require 6 months if produced by an equivalent concentration of FXa alone acting on 1.4 μM prothrombin solution (Mann et al., 2003). It has been estimated that the fully assembled prothrombinase complex catalyzes thrombin generation at a rate that is at least 300,000 times more efficient than FXa acting alone. Notably, the assembled intrinsic Xase complex on a membrane surface increases the rate of FXa generation approximately 10^9 fold more than FIXa alone (Davie et al., 1991; Mann et al., 2003).

It is important to recognize that FVIIa/TF complex activates limited quantities of FX and FIX (Baugh et al., 1998). With the generation of FXa, the inhibitory effect of TFPI takes place, preventing further production of FXa and FIXa by FVIIa/TF (Baugh et al., 1998; Broze, 1995). Additional FXa can be produced only through the alternative pathway involving intrinsic Xase, which helps to consolidate the coagulation cascade initiated by FVIIa/TF. This is achieved by FXI, which is activated to FXIa by thrombin when bound to the activated platelet surface (Gailani and Broze, 1991; Naito and Fujikawa, 1991; ten Cate et al., 1996). FXIa generates FIXa, which triggers formation of sufficient amounts of FXa to lead to propagation of the coagulation cascade. Accordingly, assembly of enzymes and cofactors on the phospholipid surface is a prerequisite for the coagulation system to operate

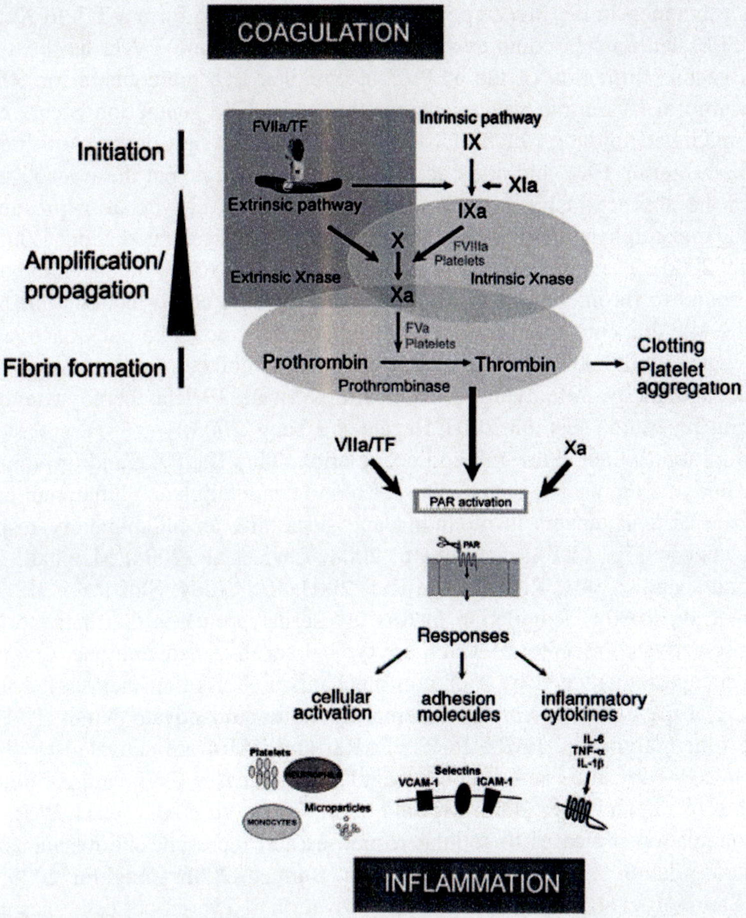

Fig. 14.1 Coagulation cascade and its regulation. Formation of a TF complex with Factor VIIa (FVIIa) leads to activation of FIX and FX. FXa in the presence of phosphatidyl serine and Ca^{2+} (prothrombinase complex) amplifies the coagulation cascade through conversion of prothrombin to thrombin, resulting in platelet aggregation, fibrin formation, and inflammation. Thrombin also activates FXI to FXIa, which activates FIX to FIXa. FIXa in the presence of phosphatidilserine and Ca^{2+} converts FX to FXa, consolidating the coagulation cascade. Activated coagulation factors FVIIa/TF, FXa, and thrombin activate PAR in different cell types including platelets, mononuclear cells and endothelial cells. PAR activation is accompanied by upregulation of molecules (e.g., ICAM-1, VCAM-1, E-selectin, COX-2, NO synthase) and production of proinflammatory cytokines (e.g., TNF-α, IL-1β, IL-6). Modified from (Francischetti, 2008)

properly and also counteracts regulation by anticoagulant mechanisms (Mann et al., 2003; Monroe et al., 2002). Finally, thrombin is critical for formation of the fibrin clot, which is stabilized by thrombin-activated FXIII (FXIIIa), a transglutaminase that catalyses covalent crosslink of fibrinogen (Lorand, 2005).

The coagulation cascade is highly regulated, and there are three major anticoagulant mechanisms in blood: TFPI, ATIII, and PC/thrombomodulin/APC. TFPI is the primary inhibitor of TF/VIIa complex (Sanders et al., 1985) and acts in a two-step manner. In the first step, TFPI inactivates FXa to form a TFPI/FXa complex; TFPI within this complex then inactivates TF-bound FVIIa as the second step. Because formation of the TFPI/FXa complex is a prerequisite for efficient inactivation of FVIIa, the system ensures that some FXa generation occurs before FVIIa-mediated initiation of the coagulation system is shut down. Antithrombin inhibits thrombin, FXa, and other activated clotting factors, but these reactions are slow in the absence of heparin (Sanders et al., 1985). With heparin (presumably with heparan sulphate in vivo), the rate of inhibition is accelerated about 1,000 fold (Rezaie, 2000). Thrombomodulin is a thrombin receptor found on the endothelium. Once bound to thrombomodulin, thrombin undergoes a conformational change at its active site that converts it from a procoagulant enzyme into a potent activator of PC (Rezaie, 2000, 2002). APC, in the presence of its cofactor (protein S), serves as an anticoagulant by proteolytically inactivating FVa and FVIIIa, thereby attenuating thrombin generation (Esmon, 2003; Rezaie and Yang, 2003).

In addition to their roles in blood coagulation, TF/VIIa, FXa, and thrombin are increasingly recognized as the interface of blood coagulation and inflammation and viewed as critical players in mounting and sustaining an inflammatory response as depicted in Fig. 14.1 (Francischetti, 2008; Levi et al., 2004; Marshall, 2001; Opal and Esmon, 2003; Riewald and Ruf, 2003; Ruf, 2004; Slofstra et al., 2003; Taylor et al., 2000). Coagulation factors are serine proteases that activate PARs (protease-activated receptors), which are typical seven-transmembrane, G protein-linked receptors activated by a unique mechanism. Thrombin cleaves the amino terminus of the PAR, allowing the internal ligand to autoactivate (Vu et al., 1991). Of the four mammalian PARs, PAR1, PAR3, and PAR4 are activated by thrombin, while PAR2 can be activated by coagulation proteases FVIIa and FXa but not thrombin (Camerer et al., 2000; Riewald et al., 2001; Vu et al., 1991). PARs connect coagulation proteases to cellular responses and represent one mechanism by which coagulation might affect inflammation (Bunnett, 2006; Coughlin, 2005; Ruf, 2004; Steinhoff et al., 2005). In fact, PAR activation in a variety of cell types in and around blood vessels is associated with production of pro-inflammatory cytokines, expression of adhesion molecules, and cell activation.

Extrinsic Xase Inhibitors

In addition to physiologic blood coagulation inhibitors described above, a number of specific inhibitors from exogenous sources have been identified from salivary glands of blood-sucking arthropods (Francischetti et al., 2009; Koh and Kini, 2009; Ribeiro and Francischetti, 2003) and herein named sialogenins (from the Greek *sialo*, saliva; *gen*, origin, source; and *ins* for proteins) with anticlotting activity. Some of these sialogenins—such as ixolaris, penthalaris, and NAPc2—display important similarities with TFPI and block the extrinsic Xase complex. Other inhibitors, such as nitrophorins-2 and -7, affect the intrinsic Xase complex in different ways, leading to inhibition of FXa generation and blood coagulation as summarized in Table 14.1, and described in detail below.

Table 14.1 Sialogenins that inhibit extrinsic or intrinsic Xase complexes

Targets and inhibitors	Species	Molecular mass	Target	R/S/C	Comments and references
TF/VIIa complex			**Specificity (affinity)**		
Ixolaris	*Ixodes scapularis*	15.5	FVIIa/TF (K_D pM range) FX/FXa (K_D 1 nM)	Y/N/Y	Use FX or FXa as scaffold for blockade of FVIIa/TF (Francischetti et al., 2002) Antithrombotic (Nazareth et al., 2006) and anti-cancer properties in vivo (Carneiro-Lobo et al., 2009)
Penthalaris	*I. scapularis*	35	FVIIa/TF (K_D pM range) FX/FXa (K_D nM range)	Y/N/Y	Use FX or FXa as scaffold for blockade of FVIIa/TF (Francischetti et al., 2004)
NAPc2	*Ancylostoma caninum*	7	FVIIa/TF (K_D 10 pM) FX/FXa (K_D 1 nM)	Y/Y/Y	Use FX or FXa as scaffold for blockade of FVIIa/TF (Stassens et al., 1996) Antithrombotic (Lee et al., 2001) and anti cancer properties in vivo (Zhao et al., 2009)
FIX(a)					
Nitrophorin-2	*Rhodnius prolixus*	20	FIX and FIXa (K_D 10 nM)	Y/Y/Y	Binds to the *gla*-domain of factors IX(a) (Gudderra et al., 2005; Isawa et al., 2000)
Phosphatidylserine					
Nitrophorin-7	*R. prolixus*	20	Phosphatidylserine (K_D 5 nM)	Y/N/Y	Inhibits both prothrombinase and intrinsic Xase complexes Also inhibits platelet aggregation (Andersen et al., 2004)

R/S/C: *R* obtained in recombinant form, *S* structure available, *C* inhibition of coagulation tested with recombinant or purified protein

Ixolaris

Sequencing of a complementary DNA library obtained from the salivary gland of the hard tick *Ixodes scapularis* identified clones with sequence homology to TFPI. One of these cDNAs has been expressed in insect cells and the recombinant protein named ixolaris (Francischetti et al., 2002). Ixolaris is a 15.5-kDa molecule (140 amino acids [aa]) containing ten cysteines and two Kunitz-like domains (Fig. 14.2). Similarly to TFPI, ixolaris inactivates the FVIIa/TF complex very poorly (concentration in the μM range) but becomes a tight inhibitor in the presence of FXa as a scaffold (Francischetti et al., 2002). In contrast to TFPI, however, ixolaris also binds to the zymogen FX with high affinity (K_D ∼1 nM) and forms a quaternary FVIIa/TF/ ixolaris/FX-inhibited complex (Fig. 14.3). Blockade of FVIIa/TF by ixolaris/FX(a) also prevents FIX activation in vitro (Francischetti et al., 2002). Binding of Ixolaris to FX or FXa occurs independently of the *gla* domain, which is a γ-carboxyglutamic domain of vitamin K-dependent coagulation factors that mediate its interaction with phosphatidylserine present on the surface of activated platelets (Furie and Furie, 2003; Kalafatis et al., 1994). The *gla* domain is necessary for FVIIa/TF/ixolaris/FX(a) complex formation.

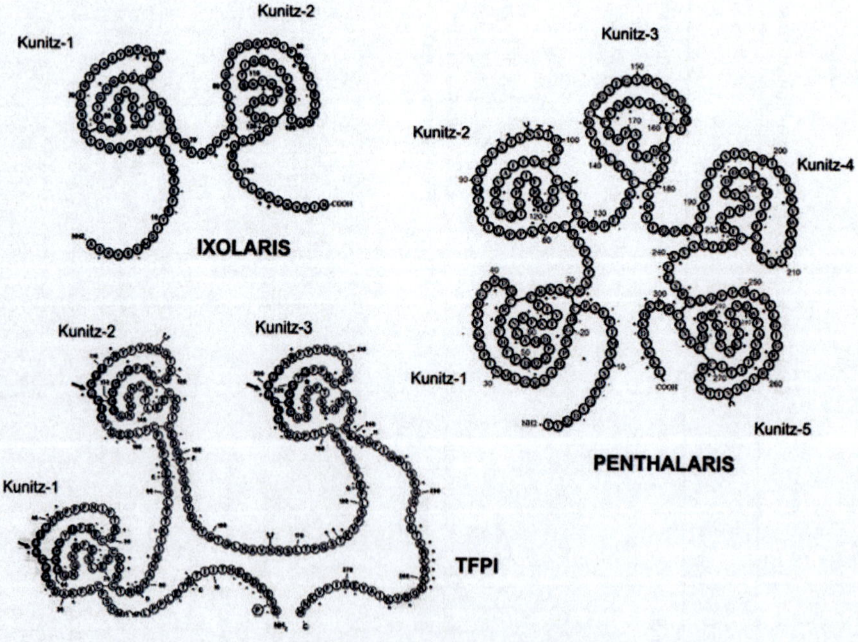

Fig. 14.2 Predicted secondary folding structure for ixolaris, penthalaris and TFPI. Disulfide bonds are assumed on the basis of the crystal structure of bovine pancreatic trypsin inhibitor (Huber et al., 1974). The charges of amino acid side chains are shown. Modified from (Broze, 1995; Francischetti et al., 2002, 2004)

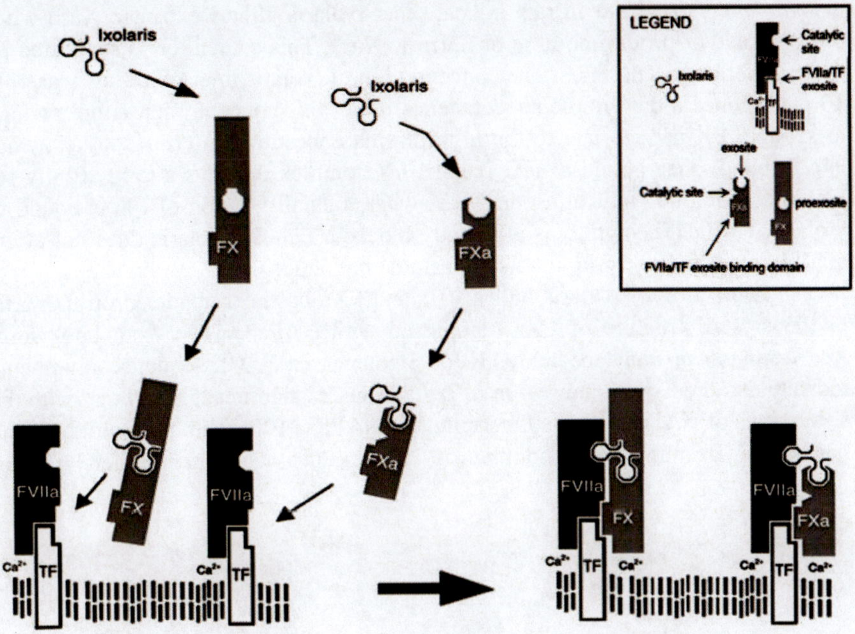

Fig. 14.3 Working hypothesis for the mechanism of action of ixolaris. Ixolaris binds to FXa heparin-binding exosite or FX heparin-binding proexosite leading to the formation of stable ixolaris/FX(a) complexes that assemble into a tightly bound quaternary inhibitory complex composed of FVIIa/TF/Ixolaris/FX(a). Modified from (Francischetti et al., 2002)

Ixolaris does not bind to the active site cleft of FXa, as occurs with TFPI. Instead, studies with a number of mutants of basic residues of the enzyme show that complex formation is mediated by the so-called FXa heparin-binding exosite (FXa HBE) (Monteiro et al., 2005). FXa HBE is an important mediator of the prothrombinase complex assembly, mediating enzyme interaction with its protein cofactor, Factor Va, and/or with its macromolecular substrate, prothrombin (Rezaie, 2000). Functional assays have shown that complex formation between ixolaris and FXa decreases the ability of the enzyme to convert prothrombin into thrombin (Monteiro et al., 2005), explaining why at high concentrations ixolaris may also affect the aPTT in vitro. Further, experiments with FX mutations at basic residues have demonstrated that ixolaris interacts with the zymogen through a precursor state of the heparin-binding exosite or FX proexosite. Complex formation between ixolaris and FX results in decreased recognition of the zymogen by FIXa associated with either FVIIIa or FVIIIa/phospholipids. It has been suggested that ixolaris blocks a FVIIIa-interacting site within FX, as no effect has been observed on FX activation by FIXa alone (Monteiro et al., 2008).

Altogether, ixolaris displays a complex mechanism of action that affects the extrinsic Xase primarily but may also interfere with prothrombinase and intrinsic Xase complex assembly in vitro at higher concentrations; however, in vivo experiments carried out in rats indicate that ixolaris affects extrinsic Xase without blockade of prothrombinase or intrinsic Xase. This contention is supported by three important results. First, the concentration of ixolaris injected in rats was only 10 μg/kg once a day by the subcutaneous route; this concentration is not enough to saturate FX in vivo due to the high plasma concentration (180 nM) of zymogen. Second, once bound to FX, ixolaris/FX complex displays a tight affinity for FVIIa/TF complex, indicating that even when a small fraction of FX is bound to ixolaris, FVIIa/TF complex is efficiently blocked. Finally, ixolaris does not affect aPTT ex vivo at 100 μg/kg in rats (Nazareth et al., 2006).

The interaction of ixolaris with FVIIa and FXa has been modeled on the basis of the crystal structure of FXa complexed with NAPc2, a TF from hookworm *Ancylostoma canynum* (see below) (Rios-Steiner et al., 2007). Sequence alignments indicate that the C-terminal region of ixolaris shares significant homology with the C-terminus of NAPc2 and the C-terminus of NAP5, a FXa inhibitor from the same hookworm. Because the first domain in ixolaris does not contain either lysine or

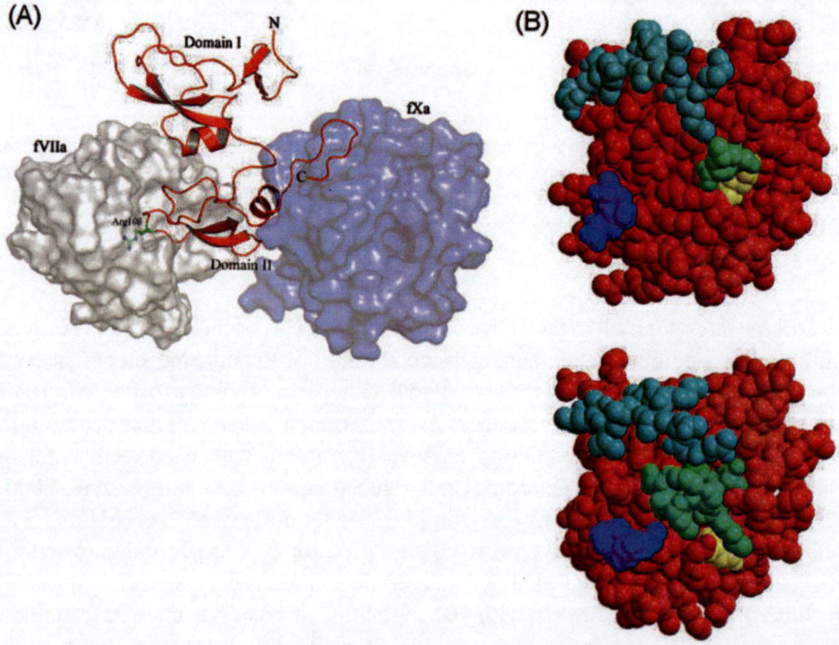

Fig. 14.4 (**a**) Model of ixolaris binding with FVIIa and FXa. This model, based on sequence homology with NAPc2, suggests that the C-terminus of ixolaris binds to FXa heparin-binding exosite and the second Kunitz domain interacts with both FXa and the catalytic site of FVIIa. Figure taken from (Murakami et al., 2007). (**b**) Models of NP2 (*top*) and NP4 (*bottom*) showing the possible interaction interface with coagulation factors IX and IXa. In the structures, the E–F loop is colored *cyan* [•], the B–C loop *blue* [•], and the C-terminus *green* [•]. In the NP2 structure, the shorter C-terminus allows the B–C loop to move away from the β-barrel, exposing the face of the barrel and loop E–F. Mutagenesis of residues in loop E–F strongly affected binding affinity. Figure taken from (Gudderra et al., 2005)

arginine residues at the P1 position, the second domain likely plays a dual role in inhibition of the extrinsic pathway. The P1 position in the second domain is occupied by arginine, which can interact with Asp189 (S1 subsite), whereas the C-terminus interacts with the FX(a) exosite (Fig. 14.4a). These features suggest that the second Kunitz domain and the extended C-terminus of ixolaris likely play roles similar to that of the analogous domains of NAPc2 in inhibiting the extrinsic pathway. It has been suggested that the first domain of ixolaris contains a glutamate residue instead of lysine or arginine at position 36 (numbering is based on TFPI), which is not productive for binding to the active site of FVIIa (Rios-Steiner et al., 2007). Determination of the structure of the ixolaris/FX(a) complex will clarify this issue.

Penthalaris

Further analysis of the salivary gland complementary DNA library from *I. scapularis* identified a clone that codes for a 308-aa protein (35 kDa) with 12 cysteine bridges and 5 tandem Kunitz domains (Fig. 14.2) (Francischetti et al., 2004). This protein, named penthalaris, has been shown to inhibit FVIIa/TF-induced FX activation with an IC_{50} of approximately 100 pM. Similarly to ixolaris, penthalaris tightly binds both zymogen FX and enzyme FXa, independently of the *gla* domain. Because no direct inhibition of FXa catalytic activity on small chromogenic substrates is observed, it is plausible that ixolaris and penthalaris share common binding-sites in FX(a), which resides in the (pro)exosite. While additional Kunitz domains are present in penthalaris, its function(s) remains to be identified.

Nematode Anticoagulant Protein-2 (NAPc2)

Nematode anticoagulant proteins (NAPs) from the hematophagous nematode *Ancylostoma caninum* comprise a family of small (75–84 aa residues) disulfide-linked proteins (Stassens et al., 1996). NAPc2 and isoforms NAPc3 and NAPc4 are potent inhibitors of FVIIa/TF complex upon binding to FX(a), which operate as scaffolds and also allow the molecule to display prolonged elimination half-life (t(1/2)β) with a mean of >50 h (Vlasuk and Rote, 2002). Napc2 displays high affinity (K_D ~1 nM) to FX or the catalytic site-inactivated FXa exosite, which is located in the C-terminus of the molecule (Bergum et al., 2001; Buddai et al., 2002). Inhibition of FVIIa/TF complex by FX(a)/NAPc2 has been calculated as $K_D \sim 10$ pM, which describes a tight interaction. Removal of the γ-carboxyglutamic acid-containing domain from FX did not affect binding to rNAPc2 but abolished the effect of factor Xa as a scaffold for inhibition of FVIIa/TF by NAPc2 (Bergum et al., 2001). Studies in vivo indicate that anticoagulant effects of NAPc2 is primarily due to inhibition of FVIIa/TF complex, and not related to blockade of the intrinsic Xase or prothrombinase complexes (Vlasuk and Rote, 2002).

NMR structures of NAPc2 indicate that Napc2 is particularly flexible in the central acidic loop and C-terminus (Duggan et al., 1999). In addition, crystallographic studies demonstrated that the binding interface of NAPc2-FXa complex consists of an intermolecular antiparallel β-sheet formed by the segment of the polypeptide chain consisting of residues 74–80 of NAPc2 with residues 86–93 of FXa. This interaction is additionally maintained by contacts between the short helical segment (residues 67–73) and a turn (residues 26–29) of NAPc2 with the short C-terminal helix of FXa (residues 233–243) (Murakami et al., 2007). It has been suggested that NAPc2 canonically inhibits the FVIIa active site through its reactive-site loop that contains a P1 Arg residue. This loop is rich in acidic residues and is disordered in the NMR and crystal structure of NAPc2-bound FXa, suggesting that stabilization takes place upon binding to FVIIa (Murakami et al., 2007).

Inhibition of TF in Pathologic States: Experience with Ixolaris and NAPc2

In addition to its essential role in hemostasis, TF may be also implicated in pathologic processes such as cancer, infectious diseases, and thrombosis (Francischetti et al., 2008; Levi et al., 2004; Marshall, 2001; Opal and Esmon, 2003; Riewald and Ruf, 2003; Ruf, 2004; Slofstra et al., 2003; Taylor et al., 2000). In this context, many experimental studies have demonstrated that inhibition of FVIIa/TF procoagulant activity is a powerful inhibitor of in vivo thrombosis and that this approach usually results in a less-pronounced bleeding tendency compared with other, more "classical" antithrombotic interventions, i.e., heparin or vitamin K antagonists such as warfarin. Accordingly, earlier studies of NAPc2 have demonstrated that it safely and effectively inhibits thrombin generation in patients undergoing elective percutaneous coronary interventions (Moons et al., 2003), reduces the incidence of deep venous thrombosis when administered prophylactically in patients undergoing elective total knee arthroplasty (Lee et al., 2001), and reduces ischemia when tested in patients with non-ST-segment elevation acute coronary syndrome (Fluture et al., 2007). Of note, the effects of NAPc2 were obtained at relatively low doses (μg/kg range) and without significantly increasing major and minor bleeding. Results from PT and aPTT indicate that the anticoagulant effect of NAPc2 is primarily due to blockade of TF/VIIa complex, while both prothrombinase and intrinsic Xase complexes are spared despite the binding of the inhibitor to FX and FXa (Vlasuk et al., 2003). NAPc2 has also been experimentally tested with positive results, decreasing progressive tissue injury in burns (Mahajan et al., 2006), abrogating endotoxin-induced coagulation in chimpanzees (Moons et al., 2002) and treating Ebola virus infection in Rhesus monkeys (Geisbert et al., 2003).

Experiments employing animal models have also demonstrated that ixolaris is a potent (μg/kg range) and long-lasting antithrombotic agent (Nazareth et al., 2006). Ixolaris remains active as an antithrombotic agent for up to 24 h after a single intravenous dose, and the elevated half-life has been attributed to its ability to interact with FX in plasma. Remarkably, ixolaris showed minor effects on induced bleeding in rats (Nazareth et al., 2006) and in mice (Carneiro-Lobo et al., 2009), suggesting that it is possibly safe at antithrombotic doses. In addition, repeated injection of ixolaris in mice in the presence of adjuvants used to boost immune response failed to produce antibodies. This indicates that ixolaris is nonimmunogenic, possibly because it shares common epitopes with endogenous TFPI (Francischetti, unpublished observations).

Abnormal elevated TF expression has been well documented in several tumor types and shown to be directly correlated with thromboembolic complications in cancer patients (Rickles and Edwards, 1983). Moreover, studies employing cultured cells as well as patient specimens have demonstrated strong correlation between TF expression and aggressive tumor behaviour. In particular, TF expression correlates with unbalanced production of anti- and/or proangiogenic factors such as vascular endothelial growth factor (VEGF), thus favouring increased tumor vasculature (Rak et al., 2006). Pro-tumoral effects of TF and blood-clotting enzymes (FVIIa, FXa, and thrombin) are intimately related to PARs. In fact, activation of PARs in cancer cells elicits a vast number of cellular responses including migration, invasion, proliferation, metastasis, inhibition of apoptosis, and production of several pro-aggressive factors such as VEGF, interleukin-8 (IL-8), metalloproteases, and others (Belting et al., 2005; Rao and Pendurthi, 2005).

It has been hypothesized that targeting the blood-clotting cascade represents a feasible therapeutic approach for treatment of cancer (Zacharski, 2002). Recently, ixolaris was shown to block in vivo growth of human glioblastoma (U87-MG) cells in a xenograft model (Carneiro-Lobo et al., 2009). This phenomenon is accompanied by a significant decrease in VEGF expression as well as diminished tumor angiogenesis. The antitumor effect of ixolaris has been also observed in a murine melanoma model, in which ixolaris blocks both primary tumor growth and induced metastasis (Monteiro, unpublished observations). Likewise, in vivo studies employing NAPc2 in a model of colon cancer in mice suggest that the antitumor effect of TF inhibitors rely on TF expression by tumor cells (Zhao et al., 2009). Moreover, it seems that inhibition of primary tumor growth is strongly dependent on blockade of FVIIa/TF-mediated signaling through PAR-2 on the tumor cell rather than inhibition of FVIIa/TF procoagulant reactions. In this context, NAPc2 but not NAP5 decreases primary tumor growth in vivo, according to the B16 melanoma model (Hembrough et al., 2003). We have proposed that ixolaris blocks primary tumor growth through inactivation of FVIIa/TF complex on the tumor cell surface, with subsequent interruption of PAR-2 signaling. On the other hand, inactivation of FVIIa/TF complex on the tumor cell surface also blocks downstream coagulation reactions, thus reducing the procoagulant ability of the tumor cell and decreasing its metastatic potential (Fig. 14.5). Thus, ixolaris may attenuate the procoagulant state of cancer patients on one hand and prevent angiogenesis on the other, thus interfering with two important components that contribute to tumor growth and metastasis in vivo.

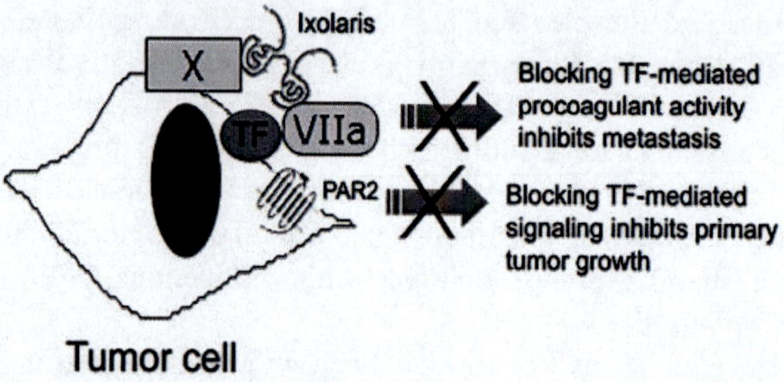

Fig. 14.5 Proposed antitumor mechanism of action of ixolaris. Ixolaris targets FVIIa/TF complex on the tumor cell surface. Blocking of TF-mediated formation of thrombin through activation of FX and FIX leads to decreased metastatic potential. Inhibition of FVIIa/TF-mediated signaling through proteolytic cleavage of PAR-2 leads to reduction in primary tumor growth

Intrinsic Xase Inhibitors

Nitrophorin-2, a Factor IX(a) Inhibitor

Nitrophorins (NPs) are a group of lipocalins from the saliva of *Rhodnius prolixus* that have been most thoroughly studied as nitric oxide (NO) transporters. Four distinct forms (NP1–4) were originally isolated, and several more were identified in the salivary transcriptome (Champagne et al., 1995; Ribeiro et al., 1993). In the saliva, a single molecule of NO is bound per molecule of NP as a ferroheme complex. The heme itself is tethered to the protein via coordination of the imidazole group of a histidine residue (His 57 in NP1) (Weichsel et al., 1998). NO binding is pH-dependent, with higher-affinity binding occurring at lower pH. When the protein is injected into the host, the increase in pH and dilution lead to release of NO, which causes vasodilation and inhibits aggregation of platelets (Ribeiro et al., 1993).

In addition to its role as a NO carrier, one NP form, NP2, is an inhibitor of the intrinsic FXase complex responsible for conversion of FX to FXa (Ribeiro et al., 1995; Sun et al., 1996; Yuda et al., 1997). The protein is effective in preventing the massive accumulation of thrombin normally occurring at the site of a wound. Kinetic studies showed that the inhibition mechanism was hyperbolic mixed type, as indicated by decreases in both V_{max} and K_m of FX activation (Zhang et al., 1998). NP2 binds both the zymogen FIX and mature FIXa in a calcium-dependent manner as measured by surface plasmon resonance (Isawa et al., 2000) but does not inhibit the amidolytic activity FIXa when a small-molecule substrate is used (Sun et al., 1996). Binding studies with "*gla*-less" coagulation factor indicated that an intact γ-carboxyglutamic acid containing the Gla domain is essential for interaction of FIX/IXa with NP2 (Isawa et al., 2000). The calcium ion dependency of the binding reaction suggests that NP2 interacts directly with the calcium-bound form of the Gla domain and that conformational effects may be important in the interaction.

The activity of FIXa is reduced by NP2 binding in the presence of phospholipids alone or FVIIIa alone, indicating that the inhibitor affects interaction with both of these components (Zhang et al., 1998). Activation of FX by FIXa in solution is not inhibited by NP2, further indicating that assembly of the complex is the important factor (Zhang et al., 1998).

Binding of NP2 occurs through a surface interaction that is not related to the heme moiety or to the presence of NO (Ribeiro et al., 1995). NP1 and NP4 are about 45% identical to NP2 at the aa sequence level and show no inhibitory activity and no binding with FIX/IXa in surface plasmon resonance studies (Andersen and Montfort, 2000; Gudderra et al., 2005). NP3 is approximately 80% identical to NP2 and shows weak inhibitory activity and correspondingly weak binding (Andersen and Montfort, 2000; Gudderra et al., 2005). Comparison of the NP4 and NP2 crystal structures demonstrated that the surfaces of the two proteins are significantly different in shape, particularly where the extended C-terminal region of NP4 is packed against the lipocalin β-barrel (Fig. 14.4b) (Andersen and Montfort, 2000). NP2 and NP3 differ at 35 of 179 aa positions. The non-identities are dispersed throughout the sequence, but a segment corresponding to the loop connecting β-strands E and F of the eight-stranded lipocalin β-barrel is particularly rich in substitutions (Andersen and Montfort, 2000; Gudderra et al., 2005). Alanine-scanning mutagenesis of the NP2 sequence at positions differing between NP2 and NP3 that are also located on the surface of the protein was performed to identify points of contact between NP2 and FIX/IXa (Gudderra et al., 2005). As expected, mutagenesis of residues in the E–F loop showed the greatest effect (Gudderra et al., 2005); however, only a quadruple (K92A, A93K, V94A, E97A) mutant reduced the activity to a level near that of NP3. It appears that exposure of a face of the β-barrel as a result of the shortened C-terminal in NP2 and NP3, combined with specific side chain features in the E–F loop of NP2, act to produce a specific, high-affinity interaction surface for the binding of FIX/IXa.

Nitrophorin-7: Phosphatidylserine-Binding Protein

Another NP form, NP7, also has relevance as a coagulation inhibitor (Andersen et al., 2004). This protein shows the same heme- and NO-binding features of the other NPs but differs in displaying a positively charged surface corresponding to a helical region lying outside the β-barrel. This basic patch on the surface of NP7 enables it to bind to anionic phospholipid membranes such as phosphatydilserine (PS) with high affinity (Andersen et al., 2004, Knipp et al., 2007a, b). By targeting PS, NP7 has been shown to block assembly of the prothrombinase complex on vesicles and activated platelets and to also block intrinsic Xase assembly (Andersen, unpublished observations). Membrane binding may also improve its ability to inhibit platelet aggregation at site of hemostasis activation by targeting it to the anionic surfaces of activated platelets and increasing its effective concentration.

Concluding Remarks

Anti-clotting sialogenins target components of the extrinsic or intrinsic Xase complexes resulting in inhibition of the initiation, propagation or consolidation steps of blood coagulation cascade. In addition, these molecules act in a redundant and synergistic manner in order to keep hemostatic tonus as low as possible so as to facilitate blood-feeding as presented in Fig. 14.6. These molecules may also attenuate inflammatory events associated with vascular injury. Finally, anti-clotting sialogenins have potential therapeutic applications and are valuable tools in pharmacology and cell biology.

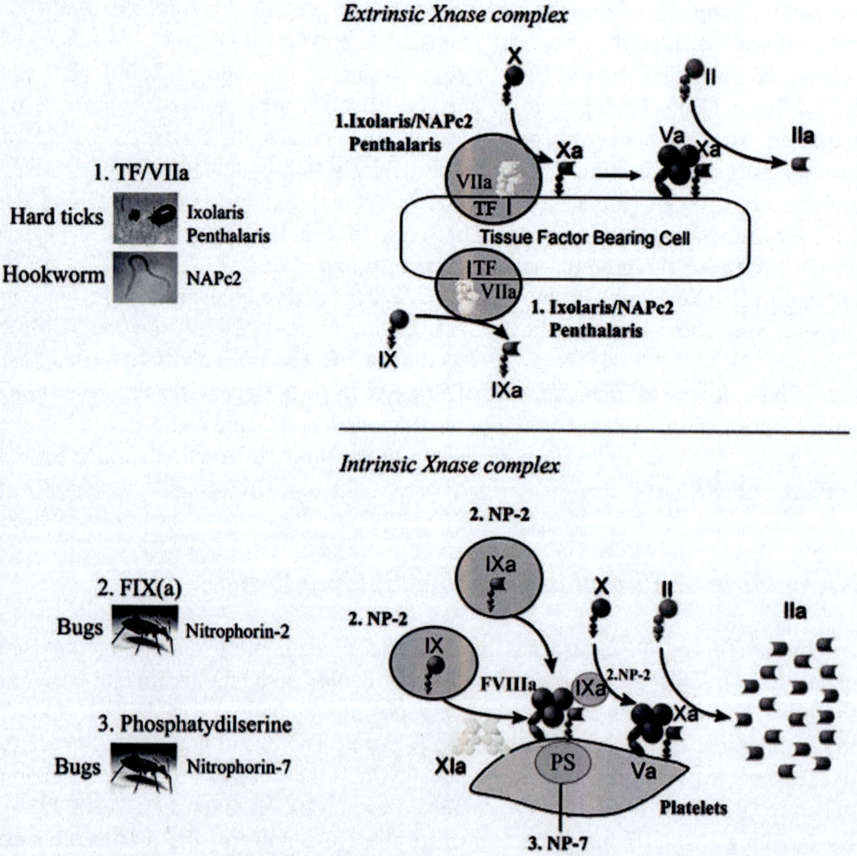

Fig. 14.6 Hematophagy and inhibition of extrinsic or intrinsic Xase complexes. Coagulation cascade is initiatied by FVIIa/TF complex, amplified by the prothrombinase complex and consolidated by the intrinsic Xase complex Modified from (Monroe et al., 2002). The target for each inhibitor is indicated by a number close to a *circle with pink background*. The name of each inhibitor is shown beside each picture exemplifying a different hematophagous animal found in different genus and/or species. PS, phosphatydilserine

Acknowledgements This work was supported by the Intramural Research Program of the Division of Intramural Research, National Institute of Allergy and Infectious Diseases, National Institutes of Health. We thank NIAID intramural editor Brenda Rae Marshall for assistance.

References

Andersen, J.F., Gudderra, N.P., Francischetti, I.M., Valenzuela, J.G., Ribeiro, J.M., 2004. Recognition of anionic phospholipid membranes by an antihemostatic protein from a blood-feeding insect. Biochemistry 43, 6987–6994.

Andersen, J.F., Montfort, W.R., 2000. The crystal structure of nitrophorin 2. A trifunctional antihemostatic protein from the saliva of *Rhodnius prolixus*. J. Biol. Chem. 275, 30496–30503.

Bauer, K.A., Kass, B.L., ten Cate, H., Hawiger, J.J., Rosenberg, R.D., 1990. Factor IX is activated in vivo by the tissue factor mechanism. Blood 76, 731–736.

Baugh, R.J., Broze, G.J., Jr., Krishnaswamy, S., 1998. Regulation of extrinsic pathway factor Xa formation by tissue factor pathway inhibitor. J. Biol. Chem. 273, 4378–4386.

Belting, M., Ahamed, J., Ruf, W., 2005. Signaling of the tissue factor coagulation pathway in angiogenesis cancer. Arterioscler. Thromb. Vasc. Biol. 25, 1545–1550.

Bergum, P.W., Cruikshank, A., Maki, S.L., Kelly, C.R., Ruf, W., Vlasuk, G.P., 2001. Role of zymogen and activated factor X as scaffolds for the inhibition of the blood coagulation factor VIIa-tissue factor complex by recombinant nematode anticoagulant protein c2. J. Biol. Chem. 276, 10063–10071.

Broze, G.J., Jr., 1995. Tissue factor pathway inhibitor and the revised theory of coagulation. Annu. Rev. Med. 46, 103–112.

Buddai, S.K., Toulokhonova, L., Bergum, P.W., Vlasuk, G.P., Krishnaswamy, S., 2002. Nematode anticoagulant protein c2 reveals a site on factor Xa that is important for macromolecular substrate binding to human prothrombinase. J. Biol. Chem. 277, 26689–26698.

Bunnett, N.W., 2006. Protease-activated receptors: how proteases signal to cells to cause inflammation and pain. Semin. Thromb. Hemost. 32(Suppl 1), 39–48.

Camerer, E., Huang, W., Coughlin, S.R., 2000. Tissue factor- and factor X-dependent activation of protease-activated receptor 2 by factor VIIa. Proc. Natl. Acad. Sci. U.S.A. 97, 5255–5260.

Carneiro-Lobo, T.C., Konig, S., Machado, D.E., Nasciutti, L.E., Forni, M.F., Francischetti, I.M., Sogayar, M.C., Monteiro, R.Q., 2009. Ixolaris, a tissue factor inhibitor, blocks primary tumor growth and angiogenesis in a glioblastoma model. J. Thromb. Haemost. 7, 1855–1864.

Champagne, D.E., Nussenzveig, R.H., Ribeiro, J.M., 1995. Purification, partial characterization, and cloning of nitric oxide-carrying heme proteins (nitrophorins) from salivary glands of the blood-sucking insect *Rhodnius prolixus*. J. Biol. Chem. 270, 8691–8695.

Coughlin, S.R., 2005. Protease-activated receptors in hemostasis, thrombosis and vascular biology. J. Thromb. Haemost. 3, 1800–1814.

Davie, E.W., Fujikawa, K., Kisiel, W., 1991. The coagulation cascade: initiation, maintenance, and regulation. Biochemistry 30, 10363–10370.

Duggan, B.M., Dyson, H.J., Wright, P.E., 1999. Inherent flexibility in a potent inhibitor of blood coagulation, recombinant nematode anticoagulant protein c2. Eur. J. Biochem. 265, 539–548.

Esmon, C.T., 2003. The protein C pathway. Chest 124, 26S–32S.

Fluture, A., Giugliano, G.R., Giugliano, R.P., 2007. Recombinant nematode anticoagulant protein c2 in non-ST segment elevation acute coronary syndrome and beyond. Future Cardiol. 3, 365–375.

Francischetti, I.M., 2008. Does activation of the blood coagulation cascade have a role in malaria pathogenesis? Trends Parasitol. 24, 258–263.

Francischetti, I.M., Mather, T.N., Ribeiro, J.M., 2004. Penthalaris, a novel recombinant five-Kunitz tissue factor pathway inhibitor (TFPI) from the salivary gland of the tick vector of Lyme disease, *Ixodes scapularis*. Thromb. Haemost. 91, 886–898.

Francischetti, I.M., Sa-Nunes, A., Mans, B.J., Santos, I.M., Ribeiro, J.M., 2009. The role of saliva in tick feeding. Front. Biosci. 14, 2051–2088.

Francischetti, I.M., Seydel, K.B., Monteiro, R.Q., 2008. Blood coagulation, inflammation, and malaria. Microcirculation 15, 81–107.

Francischetti, I.M., Valenzuela, J.G., Andersen, J.F., Mather, T.N., Ribeiro, J.M., 2002. Ixolaris, a novel recombinant tissue factor pathway inhibitor (TFPI) from the salivary gland of the tick, *Ixodes scapularis*: identification of factor X and factor Xa as scaffolds for the inhibition of factor VIIa/tissue factor complex. Blood 99, 3602–3612.

Furie, B., Furie, B.C., 2003. Real time in vivo imaging of tissue factor-induced thrombus formation. Pathophysiol. Haemost. Thromb. 33(Suppl 1), 26–27.

Gailani, D., Broze, G.J., Jr., 1991. Factor XI activation in a revised model of blood coagulation. Science 253, 909–912.

Geisbert, T.W., Hensley, L.E., Jahrling, P.B., Larsen, T., Geisbert, J.B., Paragas, J., Young, H.A., Fredeking, T.M., Rote, W.E., Vlasuk, G.P., 2003. Treatment of Ebola virus infection with a recombinant inhibitor of factor VIIa/tissue factor: a study in rhesus monkeys. Lancet 362, 1953–1958.

Giesen, P.L., Rauch, U., Bohrmann, B., Kling, D., Roque, M., Fallon, J.T., Badimon, J.J., Himber, J., Riederer, M.A., Nemerson, Y., 1999. Blood-borne tissue factor: another view of thrombosis. Proc. Natl. Acad. Sci. U.S.A. 96, 2311–2315.

Gudderra, N.P., Ribeiro, J.M., Andersen, J.F., 2005. Structural determinants of factor IX(a) binding in nitrophorin 2, a lipocalin inhibitor of the intrinsic coagulation pathway. J. Biol. Chem. 280, 25022–25028.

Hembrough, T.A., Swartz, G.M., Papathanassiu, A., Vlasuk, G.P., Rote, W.E., Green, S.J., Pribluda, V.S., 2003. Tissue factor/factor VIIa inhibitors block angiogenesis and tumor growth through a nonhemostatic mechanism. Cancer Res. 63, 2997–3000.

Isawa, H., Yuda, M., Yoneda, K., Chinzei, Y., 2000. The insect salivary protein, prolixin-S, inhibits factor IXa generation and Xase complex formation in the blood coagulation pathway. J. Biol. Chem. 275, 6636–6641.

Kalafatis, M., Swords, N.A., Rand, M.D., Mann, K.G., 1994. Membrane-dependent reactions in blood coagulation: role of the vitamin K-dependent enzyme complexes. Biochim. Biophys. Acta 1227, 113–129.

Knipp, M., Yang, F., Berry, R.E., Zhang, H., Shokhirev, M.N., Walker, F.A., 2007a. Spectroscopic and functional characterization of nitrophorin 7 from the blood-feeding insect *Rhodnius prolixus* reveals an important role of its isoform-specific N-terminus for proper protein function. Biochemistry 46, 13254–13268.

Knipp, M., Zhang, H., Berry, R.E., Walker, F.A., 2007b. Overexpression in Escherichia coli and functional reconstitution of the liposome binding ferriheme protein nitrophorin 7 from the bloodsucking bug *Rhodnius prolixus*. Protein Expr. Purif. 54, 183–191.

Koh, C.Y., Kini, R.M., 2009. Molecular diversity of anticoagulants from haematophagous animals. Thromb. Haemost. 102, 437–453.

Krishnaswamy, S., 2005. Exosite-driven substrate specificity and function in coagulation. J. Thromb. Haemost. 3, 54–67.

Lee, A., Agnelli, G., Buller, H., Ginsberg, J., Heit, J., Rote, W., Vlasuk, G., Costantini, L., Julian, J., Comp, P., van Der Meer, J., Piovella, F., Raskob, G., Gent, M., 2001. Dose-response study of recombinant factor VIIa/tissue factor inhibitor recombinant nematode anticoagulant protein c2 in prevention of postoperative venous thromboembolism in patients undergoing total knee replacement. Circulation 104, 74–78.

Levi, M., van der Poll, T., Buller, H.R., 2004. Bidirectional relation between inflammation and coagulation. Circulation 109, 2698–2704.

Lorand, L., 2005. Factor XIII and the clotting of fibrinogen: from basic research to medicine. J. Thromb. Haemost. 3, 1337–1348.

Mahajan, A.L., Tenorio, X., Pepper, M.S., Baetens, D., Montandon, D., Schlaudraff, K.U., Pittet, B., 2006. Progressive tissue injury in burns is reduced by rNAPc2. Burns 32, 957–963.

Mann, K.G., Butenas, S., Brummel, K., 2003. The dynamics of thrombin formation. Arterioscler. Thromb. Vasc. Biol. 23, 17–25.

Marshall, J.C., 2001. Inflammation, coagulopathy, and the pathogenesis of multiple organ dysfunction syndrome. Crit. Care Med. 29, S99–S106.

Monroe, D.M., Hoffman, M., Roberts, H.R., 2002. Platelets and thrombin generation. Arterioscler. Thromb. Vasc. Biol. 22, 1381–1389.

Monteiro, R.Q., Rezaie, A.R., Bae, J.S., Calvo, E., Andersen, J.F., Francischetti, I.M., 2008. Ixolaris binding to factor X reveals a precursor state of factor Xa heparin-binding exosite. Protein Sci. 17, 146–153.

Monteiro, R.Q., Rezaie, A.R., Ribeiro, J.M., Francischetti, I.M., 2005. Ixolaris: a factor Xa heparin-binding exosite inhibitor. Biochem. J. 387, 871–877.

Moons, A.H., Peters, R.J., Bijsterveld, N.R., Piek, J.J., Prins, M.H., Vlasuk, G.P., Rote, W.E., Buller, H.R., 2003. Recombinant nematode anticoagulant protein c2, an inhibitor of the tissue factor/factor VIIa complex, in patients undergoing elective coronary angioplasty. J. Am. Coll. Cardiol. 41, 2147–2153.

Moons, A.H., Peters, R.J., Cate, H., Bauer, K.A., Vlasuk, G.P., Buller, H.R., Levi, M., 2002. Recombinant nematode anticoagulant protein c2, a novel inhibitor of tissue factor-factor VIIa activity, abrogates endotoxin-induced coagulation in chimpanzees. Thromb. Haemost. 88, 627–631.

Murakami, M.T., Rios-Steiner, J., Weaver, S.E., Tulinsky, A., Geiger, J.H., Arni, R.K., 2007. Intermolecular interactions and characterization of the novel factor Xa exosite involved in macromolecular recognition and inhibition: crystal structure of human Gla-domainless factor Xa complexed with the anticoagulant protein NAPc2 from the hematophagous nematode *Ancylostoma caninum*. J. Mol. Biol. 366, 602–610.

Naito, K., Fujikawa, K., 1991. Activation of human blood coagulation factor XI independent of factor XII. Factor XI is activated by thrombin and factor XIa in the presence of negatively charged surfaces. J. Biol. Chem. 266, 7353–7358.

Nazareth, R.A., Tomaz, L.S., Ortiz-Costa, S., Atella, G.C., Ribeiro, J.M., Francischetti, I.M., Monteiro, R.Q., 2006. Antithrombotic properties of Ixolaris, a potent inhibitor of the extrinsic pathway of the coagulation cascade. Thromb. Haemost. 96, 7–13.

Opal, S.M., Esmon, C.T., 2003. Bench-to-bedside review: functional relationships between coagulation and the innate immune response and their respective roles in the pathogenesis of sepsis. Crit. Care 7, 23–38.

Osterud, B., Rapaport, S.I., 1977. Activation of factor IX by the reaction product of tissue factor and factor VII: additional pathway for initiating blood coagulation. Proc. Natl. Acad. Sci. U.S.A. 74, 5260–5264.

Rak, J., Milsom, C., May, L., Klement, P., Yu, J., 2006. Tissue factor in cancer and angiogenesis: the molecular link between genetic tumor progression, tumor neovascularization, and cancer coagulopathy. Semin. Thromb. Hemost. 32, 54–70.

Rao, L.V., Pendurthi, U.R., 2005. Tissue factor-factor VIIa signaling. Arterioscler. Thromb. Vasc. Biol. 25, 47–56.

Rezaie, A.R., 2000. Heparin-binding exosite of factor Xa. Trends Cardiovasc. Med. 10, 333–338.

Rezaie, A.R., 2002. Insight into the molecular basis of coagulation proteinase specificity by mutagenesis of the serpin antithrombin. Biochemistry 41, 12179–12185.

Rezaie, A.R., Yang, L., 2003. Thrombomodulin allosterically modulates the activity of the anticoagulant thrombin. Proc. Natl. Acad. Sci. U.S.A. 100, 12051–12056.

Ribeiro, J.M., Francischetti, I.M., 2003. Role of arthropod saliva in blood feeding: sialome and post-sialome perspectives. Annu. Rev. Entomol. 48, 73–88.

Ribeiro, J.M., Hazzard, J.M., Nussenzveig, R.H., Champagne, D.E., Walker, F.A., 1993. Reversible binding of nitric oxide by a salivary heme protein from a bloodsucking insect. Science 260, 539–541.

Ribeiro, J.M., Schneider, M., Guimaraes, J.A., 1995. Purification and characterization of prolixin S (nitrophorin 2), the salivary anticoagulant of the blood-sucking bug *Rhodnius prolixus*. Biochem. J. 308(Pt 1), 243–249.

Rickles, F.R., Edwards, R.L., 1983. Activation of blood coagulation in cancer: Trousseau's syndrome revisited. Blood 62, 14–31.

Riewald, M., Kravchenko, V.V., Petrovan, R.J., O'Brien, P.J., Brass, L.F., Ulevitch, R.J., Ruf, W., 2001. Gene induction by coagulation factor Xa is mediated by activation of protease-activated receptor 1. Blood 97, 3109–3116.

Riewald, M., Ruf, W., 2003. Science review: role of coagulation protease cascades in sepsis. Crit. Care 7, 123–129.

Rios-Steiner, J.L., Murakami, M.T., Tulinsky, A., Arni, R.K., 2007. Active and exo-site inhibition of human factor Xa: structure of des-Gla factor Xa inhibited by NAP5, a potent nematode anticoagulant protein from *Ancylostoma caninum*. J. Mol. Biol. 371, 774–786.

Roberts, H.R., Hoffman, M., Monroe, D.M., 2006. A cell-based model of thrombin generation. Semin. Thromb. Hemost. 32 Suppl 1, 32–38.

Ruf, W., 2004. Protease-activated receptor signaling in the regulation of inflammation. Crit. Care Med. 32, S287–292.

Sanders, N.L., Bajaj, S.P., Zivelin, A., Rapaport, S.I., 1985. Inhibition of tissue factor/factor VIIa activity in plasma requires factor X an additional plasma component. Blood 66, 204–212.

Slofstra, S.H., Spek, C.A., ten Cate, H., 2003. Disseminated intravascular coagulation. Hematol. J. 4, 295–302.

Stassens, P., Bergum, P.W., Gansemans, Y., Jespers, L., Laroche, Y., Huang, S., Maki, S., Messens, J., Lauwereys, M., Cappello, M., Hotez, P.J., Lasters, I., Vlasuk, G.P., 1996. Anticoagulant repertoire of the hookworm *Ancylostoma caninum*. Proc. Natl. Acad. Sci. U.S.A. 93, 2149–2154.

Steinhoff, M., Buddenkotte, J., Shpacovitch, V., Rattenholl, A., Moormann, C., Vergnolle, N., Luger, T.A., Hollenberg, M.D., 2005. Proteinase-activated receptors: transducers of proteinase-mediated signaling in inflammation and immune response. Endocr. Rev. 26, 1–43.

Sun, J., Yamaguchi, M., Yuda, M., Miura, K., Takeya, H., Hirai, M., Matsuoka, H., Ando, K., Watanabe, T., Suzuki, K., Chinzei, Y., 1996. Purification, characterization and cDNA cloning of a novel anticoagulant of the intrinsic pathway, (prolixin-S) from salivary glands of the blood sucking bug, *Rhodnius prolixus*. Thromb. Haemost. 75, 573–577.

Taylor, F.B., Jr., Wada, H., Kinasewitz, G., 2000. Description of compensated and uncompensated disseminated intravascular coagulation (DIC) responses (non-overt and overt DIC) in baboon models of intravenous and intraperitoneal *Escherichia coli* sepsis and in the human model of endotoxemia: toward a better definition of DIC. Crit. Care Med. 28, S12–19.

ten Cate, H., Bauer, K.A., Levi, M., Edgington, T.S., Sublett, R.D., Barzegar, S., Kass, B.L., Rosenberg, R.D., 1993. The activation of factor X and prothrombin by recombinant factor VIIa in vivo is mediated by tissue factor. J. Clin. Invest. 92, 1207–1212.

ten Cate, H., Biemond, B.J., Levi, M., Wuillemin, W.A., Bauer, K.A., Barzegar, S., Buller, H.R., Hack, C.E., ten Cate, J.W., Rosenberg, R.D., 1996. Factor XIa induced activation of the intrinsic cascade in vivo. Thromb. Haemost. 75, 445–449.

Vlasuk, G.P., Bradbury, A., Lopez-Kinninger, L., Colon, S., Bergum, P.W., Maki, S., Rote, W.E., 2003. Pharmacokinetics and anticoagulant properties of the factor VIIa-tissue factor inhibitor recombinant nematode anticoagulant protein c2 following subcutaneous administration in man. Dependence on the stoichiometric binding to circulating factor X. Thromb. Haemost. 90, 803–812.

Vlasuk, G.P., Rote, W.E., 2002. Inhibition of factor VIIa/tissue factor with nematode anticoagulant protein c2: from unique mechanism to a promising new clinical anticoagulant. Trends Cardiovasc. Med. 12, 325–331.

Vu, T.K., Hung, D.T., Wheaton, V.I., Coughlin, S.R., 1991. Molecular cloning of a functional thrombin receptor reveals a novel proteolytic mechanism of receptor activation. Cell 64, 1057 1068.

Weichsel, A., Andersen, J.F., Champagne, D.E., Walker, F.A., Montfort, W.R., 1998. Crystal structures of a nitric oxide transport protein from a blood-sucking insect. Nat. Struct. Biol. 5, 304–309.

Yuda, M., Higuchi, K., Sun, J., Kureishi, Y., Ito, M., Chinzei, Y., 1997. Expression, reconstitution and characterization of prolixin-S as a vasodilator – a salivary gland nitric-oxide-binding hemoprotein of *Rhodnius prolixus*. Eur. J. Biochem. 249, 337–342.

Zacharski, L.R., 2002. Anticoagulants in cancer treatment: malignancy as a solid phase coagulopathy. Cancer Lett. 186, 1–9.

Zhang, Y., Ribeiro, J.M., Guimaraes, J.A., Walsh, P.N., 1998. Nitrophorin-2: a novel mixed-type reversible specific inhibitor of the intrinsic factor-X activating complex. Biochemistry 37, 10681–10690.

Zhao, J., Aguilar, G., Palencia, S., Newton, E., Abo, A., 2009. rNAPc2 inhibits colorectal cancer in mice through tissue factor. Clin. Cancer Res. 15, 208–216.

Chapter 15
Thrombin Inhibitors from Haematophagous Animals

Cho Yeow Koh and R. Manjunatha Kini

Abstract To facilitate the feeding of blood, hematophagous animals use pharmacologically active molecules to ensure free-flowing of blood at both the feeding site and within the digestion system. A number of these exogenous anticoagulants have been identified and characterized in recent years. Among them, inhibitors of thrombin form the largest group both in terms of number and diversity. To date, more than 30 different exogenous thrombin inhibitors from hematophagous animals, grouped into at least 15 unique structural classes, have been reported in the literature. In this chapter, we discuss five representative classes in detail with insights from their high resolution structures in complex with thrombin.

Introduction

Haematophagous animals feed on the blood of their host deriving nutrients needed for survival. A large number of these parasites are arthropods in the order of Ixodida (hard and soft ticks), Diptera (mosquitoes, biting midges, horseflies, tsetse flies, blackflies and sandflies), Hemiptera (kissing bugs), Phthiraptera (sucking lice) and Siphonaptera (fleas). In addition, some annelids in the subclass of Hirudinea (leeches) and parasitic nematodes such as hookworms and even mammals (vampire bats) feed on blood. During feeding, it is essential for haematophagous animals to overcome the common defenses of their hosts, including the skin, vessel walls, the haemostatic, inflammatory and immunological responses to acquire blood (Ribeiro, 1995; Ribeiro and Francischetti, 2003).

Haematophagy has been estimated to evolve since the Jurassic or the Cretaceous periods (200 to 65 million years ago) (Balashov, 1984) against a backdrop of

R.M. Kini (✉)
Protein Science Laboratory, Department of Biological Sciences, National University of Singapore, 117543, Singapore, Singapore
e-mail: dbskinim@nus.edu.sg

R.M. Kini et al. (eds.), *Toxins and Hemostasis*, DOI 10.1007/978-90-481-9295-3_15,

some 430 million years of evolution for vertebrate coagulation network (Davidson et al., 2003). Successful haematophagous parasites evolved potent pharmacological agents that are capable of attenuating the haemostatic response of their hosts (Champagne, 2004). Some of these molecules target important components of the coagulation cascade, halting the process of blood coagulation. The targets of these exogenous anticoagulants include individual coagulation proteinases such as factor Xa (FXa) and thrombin, or complexes such as intrinsic and extrinsic tenase complexes (ITC and ETC) (Koh and Kini, 2008). Owing to the multiple times of independent evolution of hematophagy in more than 15,000 species of blood-feeding parasites (Ribeiro, 1995), a wealth of molecular diversity can be observed in these exogenous anticoagulants (Koh and Kini, 2009). Scientists exploit this rich source of drug candidates for the development of antithrombotic therapeutics. The most notable examples are FDA approved thrombin inhibitors hirudin which was isolated from the medicinal leech *Hirudo medicinalis* (Greinacher and Warkentin, 2008) and bivalirudin which was designed and developed based partly on hirudin structure (Warkentin et al., 2008). Others, such as the ETC inhibitor NAPc2, are in various stages of development (Ledizet et al., 2005).

Thrombin as a Target of Anticoagulants from Haematophagous Animals

Thrombin is one of the most targeted proteinases for exogenous inhibitors from haematophagous animals. This is most likely because thrombin (on its own or through interaction with other zymogens and cofactors) plays unique and important roles in maintaining the integrity of haemostasis (Di Cera, 2003; Huntington, 2005). During blood coagulation, the first traces of thrombin is generated in the initiation phase of blood coagulation and it activates factor V (FV), factor VIII (FVIII) and factor XI (FXI) to provide positive feedback leading to the thrombin burst in the amplification phase. Thrombin cleaves fibrinogen to fibrin, forming insoluble clots. Fibrin polymers are further strengthened and stabilized through covalent cross-linking driven by thrombin activated factor XIII, and eventually a stabilized fibrin thrombus is formed. Thrombin also contributes to the generation of a platelet thrombus, possibly through two mechanisms: (a) it activates platelets by interacting with protease-activated receptors (PARs) and glycoprotein V; and (b) it prevents destabilization of the platelet thrombus (Davie et al., 1991; Di Cera, 2003; Huntington, 2005; Lane et al., 2005).

As a strategy to prevent blood from clotting, inhibition of thrombin appears to be one of the best approaches. It prevents the generation of both fibrin and platelet thrombi, effectively shutting down mechanisms that contribute to the arrest of bleeding. It is thus not surprising that many exogenous anticoagulants from haematophagous animals are thrombin inhibitors. We have recently published a comprehensive catalogue of these thrombin inhibitors. From more than 30 different thrombin inhibitors that have been reported in the literature, we have estimated at least 15 unique structural classes of these inhibitors from haematophagous animals (Koh and Kini, 2009). In this chapter, we will focus on the structure-function relationships of selected classes of thrombin inhibitors using detailed structural information that is available through the crystal structures of the inhibitors in complex with thrombin.

Structural Features of Thrombin

The active site of thrombin contains the catalytic triad – ^{T}His57, ^{T}Asp102 and ^{T}Ser195.[1] Compared to other blood coagulation serine proteinases, thrombin has a prominent, deep and narrow active site cleft (Bode et al., 1992). Two insertion loops (60- and autolysis loops) form the wall of the cleft. The 60-loop (^{T}Leu59–^{T}Asn62) is rigid and hydrophobic while autolysis loop (^{T}Leu144–^{T}Gly150) is flexible and hydrophilic (Fig. 15.1a) (Huntington, 2005). Thrombin active site surfaces that interact with substrate residues N-terminal to the scissile bond are described as 'non-prime subsites' (S subsites). Similarly, the surfaces of active site in contact with substrate residues C-terminal to the scissile bond are typically described as 'prime subsites' (S′ subsites) (Fig. 15.1a) (Page et al., 2005). The active site of thrombin has a negatively charged S1 subsite, hence it preferably cleaves substrates with a positively charged side chain at the P1 position (Bode et al., 1992; Huntington, 2005).

Exosite-I is a surface near the prime subsites of thrombin. The bottom of exosite-I is the deep, canyon-like cleft extending from the prime subsites. The walls of the cleft are formed by two surface loops ^{T}Phe34–^{T}Leu39 and ^{T}Lys70–^{T}Glu80 (described here as the 34-loop and 70-loop, respectively) (Fig. 15.1a) (Rydel et al., 1991). The surface of exosite-I is dominated by several positively-charged residues (Skrzypczak-Jankun et al., 1991). Exosite-I interacts with substrates fibrinogen, FV, FVIII, FXI, ADAMTS13 and PAR-1; with cofactors fibrin and thrombomodulin; and with inhibitor heparin cofactor II (Huntington, 2005). Exosite-II is an even more positively charged surface near the 'non-prime subsites' of the active site (Fig. 15.1a), interacting with substrates FV and FVIII; and with cofactors GpIbα, heparin and chondroitin sulphate moiety on thrombomodulin (Huntington, 2005). Occupancy of either exosites can induce allosteric changes to the active site to enhance catalysis. Na^+ binding loop (^{T}Cys220–^{T}Trp225) (Fig. 15.1a) also provides allosteric control to thrombin function. Na^+-bound 'fast' form favors the procoagulant functions, whereas Na^+-free 'slow' form favors the anticoagulant functions (Di Cera, 2003; Huntington, 2008). By making use of competition for exosites as well as the differences in the local distribution of various substrates and cofactors in the microenvironment, thrombin is able to function as the central control to the blood coagulation system (Lane et al., 2005).

---

[1] Superscripted prefixes 'T', 'H', 'M', 'B', 'O', 'R', 'A' and 'V' were used to indicate residues of thrombin, hirudin, haemadin, boophilin, ornithodorin, rhodniin, triabin and variegin, respectively.

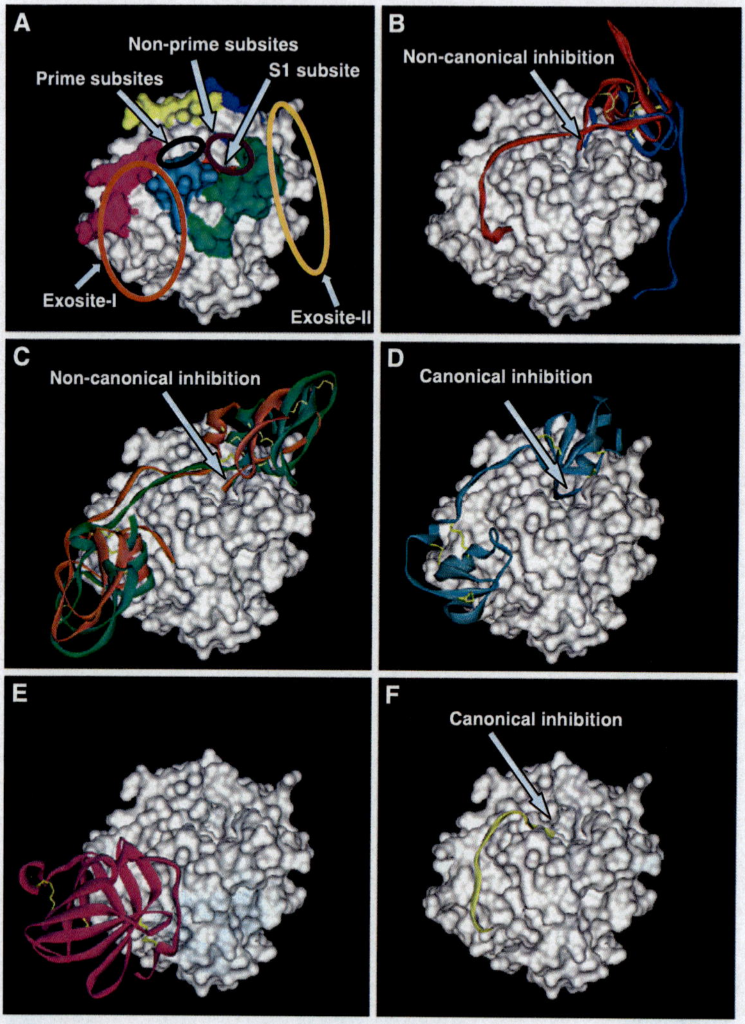

Fig. 15.1 Interactions between thrombin and inhibitors. (**a**) Surface representation of thrombin. Catalytic residues are coloured *red*. Insertion loops are coloured as followed: 60-loop – *green*, autolysis loop – *yellow*, 34-loop – *cyan*, 70-loop – *pink*, Na^{+}-binding loop – *blue*. Exosite-I is circled in *brown* and exosite–II are circled in *gold*. The non-prime and prime subsites of the active site are circled in *violet* and *black*, respectively. S1 subsite is located within non-prime subsites. (**b**) Hirudin (PDB: 2PW8; *red*) and haemadin (PDB: 1E0F; *blue*) showed the same non-canonical inhibition of the thrombin active site but distinct targeting of exosites. (**c**) Boophilin (PDB: 2ODY; *green*) and ornithodorin (PDB: 1TOC; *orange*) showed the same non-canonical inhibition of the thrombin active site with their N-terminal Kunitz domain and binding to the exosite-I with their C-terminal Kunitz domain. (**d**) Rhodniin (PDB: 1TBR; *cyan*) canonically inhibits the thrombin active site with the N-terminal Kazal domain and binds to the exosite-I with the C-terminal Kazal domain. (**e**) Triabin (PDB: 1AVG; *pink*) only binds to the exosite-I. (**f**) Variegin (PDB: 3IWS; *yellow*) remains tightly bound to the active site, prime subsites and the exosite-I after cleavage by thrombin

Hirudin-Like Inhibitors

Hirudin

Hirudin, a 65-residue polypeptide including six cysteines, was first isolated from the peripharyngeal glands of the medicinal leech *Hirudo medicinalis* (Markwardt, 1994). Following its discovery more than 50 years ago, many other isoforms of hirudin were subsequently reported, with notable differences in the sequences of the N-terminus (^{H}Val–^{H}Val or ^{H}Ile–^{H}Thr) (Dodt et al., 1984; Mao et al., 1987; Markwardt, 1994; Scharf et al., 1989). Other molecules with similar primary structures such as bufrudin (Scacheri et al., 1993), hirullin (Steiner et al., 1992) and haemadin (Strube et al., 1993) were also isolated from other species of leeches, forming a large and unique class of thrombin inhibitors.

Hirudin, being the first-in-class molecule is the most studied among them. It is a highly specific and potent inhibitor of thrombin, binding with a K_i value of 22 fM (Stone and Hofsteenge, 1986). Characterized as a fast, tight-binding competitive inhibitor, hirudin behaves as a slow binding inhibitor at high ionic strength solutions (>0.2 M). The presence of multiple negatively charged residues at the C-terminus is thus postulated to be important in mediating electrostatic interactions with thrombin (Myles et al., 2001; Stone and Hofsteenge, 1986). Hirudin is sulfated at ^{H}Tyr64 near the C-terminus (Dodt et al., 1984; Mao et al., 1987; Markwardt, 1994; Scharf et al., 1989). Desulfated hirudin binds to thrombin 10 times weaker, with a K_i of 207 fM (Stone and Hofsteenge, 1986).

Three-dimensional structures of hirudin (Clore et al., 1987; Haruyama and Wuthrich, 1989) and its complex with thrombin (Grutter et al., 1990; Liu et al., 2007; Rydel et al., 1990, 1991; Vitali et al., 1992) has long been a subject of interest due to its unique sequence and function. Residues 1–48 of hirudin forms a globular core unit stabilized by three disulfide bridges (^{H}Cys6–^{H}Cys14, ^{H}Cys16–^{H}Cys28 and ^{H}Cys22–^{H}Cys39) and four short β-strands, whereas residues 49–65 assumes a long and extended conformation. This structural class is only found in thrombin inhibitors from leeches. Two different variants of hirudin, I (N-terminal ^{H}Val-^{H}Val) and II (N-terminal ^{H}Ile–^{H}Thr), were crystallized in complex with thrombin (Grutter et al., 1990; Rydel et al., 1990). Both structures showed extensive interactions between the first three N-terminal hirudin residues with the non-prime subsites, and the long C-terminal tail with the exosite-I of thrombin (Fig. 15.1b). The binding of hirudin to thrombin active site is in the non-canonical form (i.e., in the opposite direction of natural substrates such as fibrinogen), forming a short parallel β-pleated sheet with residues ^{T}Ser214–^{T}Gly216 of thrombin. Interactions between hirudin N-terminal residues with thrombin non-prime subsites are mainly hydrophobic, in agreement to mutagenesis results that showed drastic increase of K_i value when mutated to polar or negatively charged residues (especially ^{H}Glu) (Betz et al., 1992; Wallace et al., 1989). The side chains of hirudin ^{H}Val1/^{H}Ile1 and ^{H}Tyr3 especially, occupy apolar subsites that are lined by the 60-loop (^{T}Tyr60A and ^{T}Trp60D) and other hydrophobic thrombin residues (^{T}Leu99, ^{T}Ile174 and ^{T}Trp215). These apolar subsites are typical targets where other synthetic thrombin inhibitors bind (e.g. DPhe-Pro-Arg chloromethylketone) (Bode et al., 1992), demonstrating the importance of hydrophobic interactions in targeting non-prime subsites of thrombin.

The primary specificity pocket S1 of thrombin, which is typically occupied by a P1 Arg from substrates, is free. Hence, specificity of hirudin for thrombin is derived from the extensive interactions elsewhere in the complex, especially with the exosite-I of thrombin. The N-terminal nitrogen of hirudin forms hydrogen bond with different catalytic residues (^{T}His57 and ^{T}Ser195, respectively) in the respective structures of the two hirudin variants (I and II), although no displacement of catalytic residues have been reported (Grutter et al., 1990; Rydel et al., 1990). More importantly, both structures reported the N-terminal nitrogen to form hydrogen bond with the backbone O of ^{T}Ser214 (Grutter et al., 1990; Rydel et al., 1990). Mutagenesis studies showed the importance of the free amino group at the N-terminus (Wallace et al., 1989), consistent with the structural observations. The rest of the hirudin N-terminal globular core makes limited direct contacts with thrombin. Instead, the main roles of the core appear to be for the positioning of the N-terminal tripeptide in non-prime subsites and providing steric hindrance by closing off the access to the active site pocket. The intermolecular hydrogen bond between ^{H}Thr4 and ^{H}Lys47 and the intramolecular salt bridge between ^{H}Asp5 and ^{T}Arg221A are especially important for fitting the N-terminal of hirudin into thrombin active site (Rydel et al., 1991).

The C-terminal segment of hirudin (^{H}Asp55–^{H}Gln65), rich in negatively charged residues, is inserted into the thrombin anion binding exosite-I groove. The thrombin 34- and 70-loops that flank the exosite-I groove are rich in positively charged residues. As a result, three salt bridges are observed (^{H}Asp55:^{T}Arg73, ^{H}Glu58:^{T}Arg77A and ^{H}Gln65 C-terminus:^{T}Lys36) (Rydel et al., 1991). In addition to direct ionic contacts, hydrophobic contacts also contribute significantly to the interactions. These interactions include the π–π stacking between ^{H}Phe56 and ^{T}Phe34 and the fitting of ^{H}Ile59, ^{H}Pro60, ^{H}Tyr63 and ^{H}Leu64 side chains into a hydrophobic pocket formed by ^{T}Leu64, ^{T}Leu65, ^{T}Tyr76 and ^{T}Ile82 (Betz et al., 1991; Rydel et al., 1991). The structures of hirudin C-terminal peptides with thrombin usually showed lesser number of salt-bridges. For example, only one ion pair is present in thrombin-hirulogs/hirugen structures (Qiu et al., 1992; Skrzypczak-Jankun et al., 1991). Instead, all charged residues help in creating the complementary electrostatic field that is important for the initial association with thrombin (steering effect) before the specific ionic contacts occur (tethering effect) (Myles et al., 2001). The ionic tethering is the rate-limiting step (Jackman et al., 1992) and explained the observation that hirudin binds thrombin slowly in solutions with high ionic strength (>0.2 M) (Stone and Hofsteenge, 1986).

In the recently published structure of sulfo-hirudin-thrombin complex, the sulfate moiety of ^{H}Tyr63 forms a salt bridge with ^{T}Lys81 and a hydrogen bond with ^{T}Tyr76 (Liu et al., 2007). This explains the 10-fold decrease in affinity of desulfated hirudin (Braun et al., 1988; Dodt et al., 1988; Stone and Hofsteenge, 1986). However, no salt bridge was observed for this sulfate moiety in hirugen (synthetic C-terminal peptide of hirudin). It is instead hydrogen bonded to ^{T}Tyr76 and ^{T}Ile82 (Skrzypczak-Jankun et al., 1991). Such disparity might be due to the slightly flexible nature of the hirudin (or hirugen) C-terminus, and sulfation does not appear to restrict a single conformation. For example, the C-terminus is disordered in the hirulog-1 structure (non-sulfated) (Skrzypczak-Jankun et al., 1991), forms a 3_{10} helix turn in hirudin (non-sulfated) (Rydel et al., 1991), hirulog-3 (non-sulfated) (Qiu et al., 1992) and hirugen (sulfated) (Skrzypczak-Jankun et al., 1991), and forms a full α-helical turn in sulfo-hirudin (sulfated) (Liu et al., 2007).

Haemadin

Isolated from the Indian leech *Haemadipsa sylvestrisi*, haemadin is a 57-residue thrombin inhibitor (Strube et al., 1993). There are intermediate sequence similarity between haemadin and hirudin, but more importantly both are of similar length, contain six cysteines and have a cluster of negatively charged residues in the C-terminal tail. It is a slow and tight-binding inhibitor of thrombin, with K_i = 210 fM (Strube et al., 1993). Crystal structure of thrombin-haemadin complex established the structural similarity between haemadin and hirudin (Fig. 15.1b). With the same 1–2, 3–5, 4–6 cysteine pairings, haemadin N-terminal segment (^{M}Ile1–^{M}Gly39) folded into a globular core stabilized by five short β-strands with its C-terminal segment (^{M}Asp40–^{M}Lys57) adopting an extended conformation, similar to the three-dimensional fold of hirudin (Fig. 15.1b) (Richardson et al., 2000).

Although belonging to the same structural class as hirudin, functionally haemadin is distinct from hirudin. Despite inhibiting thrombin active site in non-canonical form with its N-terminal residues (similar to hirudin), the C-terminal segment of haemadin binds to the thrombin exosite-II (instead of exosite-I in the case of hirudin) (Richardson et al., 2000, 2002). The first three N-terminal residues bind to the non-prime subsites of thrombin, forming a parallel β-sheet with thrombin residues ^{T}Ser214–^{T}Gly216. In addition to the hydrophobic contacts mediated through ^{M}Ile1 and ^{M}Phe3 with thrombin, ^{M}Arg2 is inserted into the S1 specificity pocket of thrombin. The N-terminal globular core, again similar to hirudin, is important for the positioning of the tripeptide and for blocking the access to the active site (Richardson et al., 2000).

Structurally the C-terminal tail of haemadin is also similar to hirudin's especially between the stretch of ^{M}Glu46-^{M}Glu51 and ^{H}Asp55-^{H}Pro60 (Richardson et al., 2000). Therefore, the binding of haemadin C-terminal tail to thrombin exosite-II is surprising. The binding of haemadin to exosite-II is ascertained by binding studies (Richardson et al., 2000) and mutagenesis studies (Richardson et al., 2002). Exosite-II is a highly basic surface (more so than exosite-I), situated on the opposite side of exosite-I. Exosite-II is also the heparin-binding site of thrombin (Huntington, 2005). Although haemadin C-terminal tail is rich in negatively charged residues, only two salt bridges are observed in the structure (^{M}Glu49:^{T}Arg93 and ^{M}Glu49:^{T}Arg101). Unlike hirudin-thrombin exosite-I interactions, no significant hydrophobic contacts are observed in the structure (Richardson et al., 2000). Studies of binding energy using modified thrombin and haemadin indicate that interactions between haemadin N-terminal and thrombin active site is more important for the strength of binding (Richardson et al., 2002).

It is intriguing that two structurally similar molecules (hirudin and haemadin) bind to two distinct surfaces (exosite-I and exosite-II) of the same enzyme (thrombin). Significantly, the interactions of hirudin and haemadin with thrombin at the two exosites are also distinct in nature. While hirudin-exosite-I binding is driven by both electrostatic and hydrophobic contacts, haemadin-exosite-II binding is only through electrostastic interactions. A correlation between sequence and exosites targeting was suggested by J.A. Huntington (Huntington, 2005). Haemadin has a higher ratio of negatively charged to hydrophobic residues in the C-terminal tail (2.7:1) hence targets the more positively charged exosite-II. In contrast, the same ratio for hirudin is lower (1.2:1), hence targets the less positively charged exosite-I (Huntington, 2005).

Kunitz-Type Proteinase Inhibitors

Unlike hirudin-like inhibitors that are found only from leeches, some of the other thrombin inhibitors from haematophagous animals belongs to common classes of proteinase inhibitors. For example, many inhibitors from ticks are from the structural class of Kunitz-type serine proteinase inhibitors. The Kunitz-type inhibitors is one of the largest and most studied class of serine proteinase inhibitor, found in many metazoa (Laskowski and Kato, 1980). A common structural feature of a Kunitz-type domain is an exposed reactive-site loop in a three-disulfides-stabilized, alpha/beta core structure. The reactive-site loop usually binds to the active site serine proteinases like a substrate (canonical), thus determinants of potency and specificity are usually restricted within the loop (Grutter, 1994). However, Kunitz-type inhibitors from ticks inhibit thrombin non-canonically without the involvement of the reactive-site loop in active site inhibition (Macedo-Ribeiro et al., 2008; van de Locht et al., 1996).

Boophilin and Ornithodorin

All Kunitz-type thrombin inhibitors from ticks have two tandem Kunitz domains, connected by polypeptide linker. Examples of the molecules include boophilin from the hard tick *Boophilus microplus* (Macedo-Ribeiro et al., 2008) and ornithodorin from the soft tick *Ornithodoros moubata* (van de Locht et al., 1996). Interestingly, primary structure of inhibitors from hard ticks is highly similar to other typical Kunitz-type inhibitors (BPTI) while inhibitors from soft ticks are not. Instead, soft ticks derived Kunitz-type thrombin inhibitors have multiple insertion and deletion throughout, especially in the reactive-site loop (Mans et al., 2002). Hard tick inhibitors typically have a lower affinity for thrombin [e.g. amblin $K_i = 20$ nM (Lai et al., 2004), boophilin $K_i = 1.8$ nM (Macedo-Ribeiro et al., 2008)] compared to the soft tick inhibitors [e.g. ornithodorin $K_i = 1$ pM (van de Locht et al., 1996), savignin $K_i = 4.89$ pM (Nienaber et al., 1999) and monobin $K_i = 7$ pM (Mans et al., 2008)]. In addition, soft tick inhibitors are highly specific for thrombin (Nienaber et al., 1999), but hard tick inhibitors are slightly less so (Lai et al., 2004; Macedo-Ribeiro et al., 2008).

Crystal structures showed that of both ornithodorin (van de Locht et al., 1996) and boophilin (Macedo-Ribeiro et al., 2008) inhibited thrombin active site with their N-terminal Kunitz domain and also bind to thrombin exosite-I with their interdomain linker and C-terminal Kunitz domain (Fig. 15.1c). Inhibition of the thrombin active site is through the non-canonical mechanism where N-terminal residues occupy the thrombin non-prime subsites, similar to both hirudin and haemadin. In both structures, the reactive-site loops are solvent exposed (Macedo-Ribeiro et al., 2008; van de Locht et al., 1996).

In ornithodorin structure, the first four residues form the parallel β-sheet with ^{T}Ser214-^{T}Gly216, while ^{O}Leu1, ^{O}Val3 and ^{O}Asn6 make important hydrophobic contacts within thrombin non-prime subsites (similar to hirudin). Additional contacts are also observed between ^{O}Arg24–^{O}Thr27 and thrombin 60-loop and between ^{O}Tyr40, ^{O}Glu45, ^{O}Gln48 and thrombin autolysis loop. The nature of interactions between ornithodorin and thrombin exosite-I are also similar to hirudin, which is a combination of electrostatic (three salt bridges – ^{O}Glu60:^{T}Arg67, ^{O}Glu100:^{T}Arg77A and ^{O}Glu104:^{T}Lys81) and hydrophobic (involving ^{O}Phe103, ^{O}Val107, ^{O}Val111, ^{O}Ala112, ^{O}Ile117, ^{T}Phe34, ^{T}Leu65, ^{T}Tyr76, ^{T}Met84 and ^{T}Lys36) contacts (van de Locht et al., 1996).

In boophilin structure, the N-terminal residues also forms contacts with thrombin non-prime subsites, although fitting to the hydrophobic pockets are not as close as observed in other non-canonical inhibitors. However, the side chain of the second N-terminal residue Arg is inserted into the S1 pocket (like haemadin but not hirudin and ornithodorin). The interdomain linker and C-terminal Kunitz domain also interact with exosite-I through both electrostatic and hydrophobic contacts. Despite a similar interface area compared to thrombin-ornithodorin structure, most of the interactions between boophilin and thrombin exosite-I is mediated through water molecules. Although most of the residues involved in direct contacts observed for ornthodorin and thrombin are conserved or conservatively replaced in boophilin, no similar interactions are observed in the boophilin structure (Macedo-Ribeiro et al., 2008).

These structures also suggested reasons for the higher affinity and specificity of soft tick inhibitors compared to hard tick inhibitors. The use of tandem domains, with one of them targeting the thrombin exosite-I, confers both molecules thrombin specificity. The narrow and deep active site cleft of thrombin precludes the insertion of Kunitz reactive-site loop, thus a non-canonical mechanism of binding is selected for. However, due to its hydrophobic nature, ornithodorin N-terminal residues fit closer to the similarly hydrophobic non-prime subsites of thrombin. Moreover, the extensive mutations, deletions and insertions of ornithodorin, especially in the reactive-site loop, resulted in a distorted Kunitz scaffold, preventing the molecule from inhibiting other proteinases. In contrast for boophilin, the conserved reactive-site loop and hence the conserved Kunitz scaffold, is available for the inhibition of enzymes other than thrombin, making it a less specific inhibitor.

Kazal-Type Proteinase Inhibitors

Another common class of proteinase inhibitors, the Kazal-type inhibitor, is also found as specific thrombin inhibitors. Structurally, an exposed reactive-loop is being stabilized by a core structure of three disulfide bridges, one or two short α-helices and a three-stranded antiparallel β-sheet. Similar to Kunitz-type inhibitor, the reactive-loop contains the P1 residues that targets S1 subsite of proteinases for canonical inhibition. Other residues in the loop also contribute to the potency and specificity of inhibition (Empie and Laskowski, 1982). Kazal-type thrombin inhibitors are so far isolated only from kissing bugs such as rhodniin from *Rhodnius prolixus* (Friedrich et al., 1993; van de Locht et al., 1995), dipetalogastin from *Dipetalogaster maximus* (Mende et al., 1999) and infestin from *Triatoma infestans* (Campos et al., 2002). Typically these inhibitors have multiple domains and are characterized as slow, tight-binding, competitive inhibitors.

Rhodniin

Rhodniin from *Rhodnius prolixus* has two tandem non-classical Kazal domains and specifically inhibits thrombin with a K_i of 0.2 pM. The first two cysteines of non-classical Kazal domains are separated by one or two residues, unlike the seven or eight spacer residues found in classical Kazal domains (Friedrich et al., 1993). Crystal structure of rhodniin in complex with thrombin showed that the N-terminal Kazal domain inhibit thrombin active site canonically, while the C-terminal Kazal domain binds to exosite-I (Fig. 15.1d) (van de Locht et al., 1995). This tandem domain binding arrangement is similar to ornithodorin and boophilin yet the active site inhibition is through canonical mechanism. Indeed, the reactive-site loop of Kazal-type domain was predicted to be most suited for canonical thrombin inhibition as it experiences least steric hindrance from the 60-loop for access of the narrow active site cleft (Bode et al., 1992).

The reactive-site loop of rhodniin (^{R}Cys6–^{R}Arg14) inserts into thrombin active site like a normal substrate. Residues ^{R}Cys6–^{R}Pro9 bind to thrombin non-prime subsites, forming antiparallel sheet with residues ^{T}Ser214–^{T}Gly216. The side chain of P1 residue ^{R}His10 fits into the S1 pocket, hydrogen bonded directly to ^{T}Glu192 and ^{T}Glu189 via water molecule. The P′ residues ^{R}Ala11 and ^{R}Leu12 run antiparallel to ^{T}Leu40 and ^{T}Leu41 but subsequent residues trace away from the thrombin surface, although hydrogen bonds are possible for side chains of P3′ ^{R}His13 and P4′ ^{R}Arg14 with ^{T}Lys60F and ^{T}Glu39, respectively (van de Locht et al., 1995).

The peptide linking the two Kazal domains has minimum contact with thrombin. The C-terminal Kazal domain binds to thrombin exosite-I though both electrostatic and hydrophobic interactions. Complementary charges appear to be important (a cluster of negative charges formed by ^{R}Asp55, ^{R}Glu59, ^{R}Asp61, ^{R}Asp63 and ^{R}Glu64 are in close proximity to positive charges of thrombin residues ^{T}Arg67, ^{T}Arg75, ^{T}Arg77A and ^{T}Lys87). However, only two salt bridges (^{R}Glu59:^{T}Arg75 and ^{R}Asp63:^{T}Arg77A) are observed between the multiple charged residues (van de Locht et al., 1995).

Lipocalin-Like Family of Inhibitors

Instead of utilizing common proteinase inhibitor scaffolds, other widespread scaffold structures usually without proteinase inhibiting functions were also recruited as thrombin inhibitors in haematophagous animals. For example, triabin isolated from the kissing bug *Triatoma pallidipennis* (Noeske-Jungblut et al., 1995) shared structural similarity with lipocalins, a large group of extracellular protein that usually binds to hydrophobic small molecules, cell surface receptors or other proteins. The lipocalin family of proteins have diverse sequences yet a conserved three-dimensional structure of a single, eight-stranded (A-B-C-D-E-F-G-H), antiparallel β-barrel (Flower, 1996).

Triabin

So far, triabin is the only example of lipocalin-like thrombin inhibitor, although another lipocalin-like anticoagulant, nitrophorin-2 is also being reported as inhibitor of intrinsic tenase complex (Isawa et al., 2000; Zhang et al., 1998). Triabin, with 142-residues, specifically inhibits thrombin with a K_i of 3 pM. Most significantly, unlike all the other thrombin inhibitors discussed so far, it binds only to thrombin exosite-I (Fig. 15.1e). The active site function of thrombin is largely retained but exosite-I related functions are affected (Noeske-Jungblut et al., 1995). The structure of triabin is deviated from a typical lipocalin topology as its B and C strands are inverted in direction. As a result, the first four strands of the β-barrel have an up-up-down-down topology (instead of up-down-up-down). The large and slightly flattened β-barrel surface of triabin precluded the insertion into the exosite-I deep cleft extending from the active site. Instead, the contacts are mainly on a relatively flat surface of the flanking insertion loops (34- and 70-loops) and residues near the end of the cleft. The main interactions are hydrophobic in nature, involving residues of ^{A}Ile99, ^{A}Phe106, ^{A}Leu108, ^{A}Val-126, ^{T}Leu65, ^{T}Tyr76, ^{T}Arg77A and ^{T}Met84. Salt bridges are possible between ^{A}Glu128/^{A}Asp135 with ^{T}Arg77A while polar interactions that are mostly effected through water molecules are observed at the edge of the interface.

Variegin

Recently, we have reported a novel class of thrombin inhibitor, variegin, from the salivary gland extract (SGE) from partially fed female *Amblyomma variegatum* (Koh et al., 2007). Variegin is a novel, fast and tight-binding competitive inhibitor of thrombin. The peptide, 32 residues, is one of the smallest thrombin inhibitors found in nature and it forms a new class of thrombin inhibitors. Despite its small size and flexible structure, variegin binds thrombin with strong affinity ($K_i = 10.4$ pM) and high specificity (Koh et al., 2007). The primary structure of variegin is unique among other thrombin substrates/cofactors/inhibitors. Through chemically synthesized variegin (s-variegin) and two truncation variants we have outlined key structure-function information of variegin: residues 8–14 bind to thrombin active site; residues 15–32 binds to thrombin exosite-I; and the first seven N-terminal residues of variegin are not in direct contact with thrombin but are needed for its fast binding characteristics (Koh et al., 2007).

We showed that variegin binds thrombin with a substrate-like mechanism, and hence is cleaved at the scissile bond between ^{V}Lys10 and ^{V}Met11 by the proteinase. The sequence in the active site binding segment (^{8}EPKMHKT^{14}) is novel and in fact contradicted some of the common knowledge about thrombin substrate preference. However, this unique sequence, especially those residues located C-terminal to the scissile bond (^{11}MHKTA^{15}), was found to be in close contact with thrombin prime subsites. This prime subsites binding segment is mainly responsible for the ability of variegin cleavage product to non-competitively inhibit thrombin. This unique property of variegin to bind thrombin strongly before and after cleavage, results in prolonged inhibition of the enzyme (Koh et al., 2009).

Although we attempted to crystallize the full-length variegin in complex with thrombin, only the electron density for variegin cleaved fragment can be traced. Variegin (cleavage fragment) fitted tightly to thrombin in active site, prime subsites and exosite-I (Fig. 15.1f). The structure describes a novel mechanism of thrombin inhibition and explains the classical non-competitive inhibition observed with variegin cleavage product. The P2′ ^{V}His12 of variegin is in proximity to the catalytic residue ^{T}His57 of thrombin and interacts with the other catalytic residue ^{T}Ser195 by hydrogen bonds. The Oγ of Ser195 is displaced from its normal position, thus breaking the catalytic charge relay system. Overall, the catalytic capability of thrombin is inhibited (Koh et al., manuscript submitted). Based on the information derived from the structure and prior understanding of thrombin interactions, we have subsequently designed and developed multiple variants of variegin, covering a diverse spectrum in terms of potency, kinetics and mechanism of inhibition. The peptides have 'tunable' potential, with affinities ranging from nanomolar to picomolar values, displaying competitive and non-competitive inhibition through fast and slow, tight-binding mechanism (Koh et al., manuscript submitted). We have also demonstrated that the in vivo antithrombotic effects of variegin variants correlate well with their in vitro affinities for thrombin with the exception of slow binding variants. In addition, the thrombin inhibitory activities of the peptides can be reversed by protamine sulfate (Koh et al., manuscript submitted). Therefore, variegin has shown immense potential to be developed into new antithrombotic agent.

Conclusions

Facing the haemostatic system of their hosts, haematophagous animals successfully developed an array of thrombin inhibitors to facilitate their blood-feeding habits. From the molecular and structural point of views, the diversities observed in these molecules are rich and intriguing. While only molecules with detailed three-dimensional information available are discussed here, there are a large number of other interesting and unique inhibitors that should be explored. For example, anophelin, madanin, chimadanin, tick thrombin inhibitor (TTI) and others all have different structural features than inhibitors that are covered here (Koh and Kini, 2009). Moreover, there are many more species of haematophagous animals yet to be studied for their thrombin inhibitors. Therefore, haematophagous animal remains a rich, important and valuable source for both the understanding of blood coagulation molecules and the development of antithrombotic agents.

Acknowledgments This work is supported by academic research grants from National University of Singapore, Singapore.

References

Balashov, Y., 1984. Interaction between blood-sucking arthropods and their hosts, and its influence on vector potential. Annu. Rev. Entomol. 29, 137–156.

Betz, A., Hofsteenge, J., Stone, S.R., 1991. Role of interactions involving C-terminal nonpolar residues of hirudin in the formation of the thrombin-hirudin complex. Biochemistry 30, 9848–9853.

Betz, A., Hofsteenge, J., Stone, S.R., 1992. Interaction of the N-terminal region of hirudin with the active-site cleft of thrombin. Biochemistry 31, 4557–4562.

Bode, W., Turk, D., Karshikov, A., 1992. The refined 1.9-A X-ray crystal structure of D-Phe-Pro-Arg chloromethylketone-inhibited human α-thrombin: structure analysis, overall structure, electrostatic properties, detailed active-site geometry, and structure-function relationships. Protein Sci. 1, 426–471.

Braun, P.J., Dennis, S., Hofsteenge, J., Stone, S.R., 1988. Use of site-directed mutagenesis to investigate the basis for the specificity of hirudin. Biochemistry 27, 6517–6522.

Campos, I.T., Amino, R., Sampaio, C.A., Auerswald, E.A., Friedrich, T., Lemaire, H.G., Schenkman, S., Tanaka, A.S., 2002. Infestin, a thrombin inhibitor presents in *Triatoma infestans* midgut, a Chagas' disease vector: gene cloning, expression and characterization of the inhibitor. Insect Biochem. Mol. Biol. 32, 991–997.

Champagne, D.E., 2004. Antihemostatic strategies of blood-feeding arthropods. Curr. Drug Targets Cardiovasc. Haematol. Disord. 4, 375–396.

Clore, G.M., Sukumaran, D.K., Nilges, M., Zarbock, J., Gronenborn, A.M., 1987. The conformations of hirudin in solution: a study using nuclear magnetic resonance, distance geometry and restrained molecular dynamics. EMBO J. 6, 529–537.

Davidson, C.J., Hirt, R.P., Lal, K., Snell, P., Elgar, G., Tuddenham, E.G., McVey, J.H., 2003. Molecular evolution of the vertebrate blood coagulation network. Thromb. Haemost. 89, 420–428.

Davie, E.W., Fujikawa, K., Kisiel, W., 1991. The coagulation cascade: initiation, maintenance, and regulation. Biochemistry 30, 10363–10370.

Di Cera, E., 2003. Thrombin interactions. Chest 124, 11S–17S.

Dodt, J., Kohler, S., Baici, A., 1988. Interaction of site specific hirudin variants with α-thrombin. FEBS Lett. 229, 87–90.

Dodt, J., Muller, H.P., Seemuller, U., Chang, J.Y., 1984. The complete amino acid sequence of hirudin, a thrombin specific inhibitor: application of colour carboxymethylation. FEBS Lett. 165, 180–184.

Empie, M.W., Laskowski, M., Jr., 1982. Thermodynamics and kinetics of single residue replacements in avian ovomucoid third domains: effect on inhibitor interactions with serine proteinases. Biochemistry 21, 2274–2284.

Flower, D.R., 1996. The lipocalin protein family: structure and function. Biochem. J. 318(Pt 1), 1–14.

Friedrich, T., Kroger, B., Bialojan, S., Lemaire, H.G., Hoffken, H.W., Reuschenbach, P., Otte, M., Dodt, J., 1993. A Kazal-type inhibitor with thrombin specificity from *Rhodnius prolixus*. J. Biol. Chem. 268, 16216–16222.

Greinacher, A., Warkentin, T.E., 2008. The direct thrombin inhibitor hirudin. Thromb. Haemost. 99, 819–829.

Grutter, M.G., 1994. Proteinase inhibitors: another new fold. Structure 2, 575–576.

Grutter, M.G., Priestle, J.P., Rahuel, J., Grossenbacher, H., Bode, W., Hofsteenge, J., Stone, S.R., 1990. Crystal structure of the thrombin-hirudin complex: a novel mode of serine protease inhibition. EMBO J. 9, 2361–2365.

Haruyama, H., Wuthrich, K., 1989. Conformation of recombinant desulfatohirudin in aqueous solution determined by nuclear magnetic resonance. Biochemistry 28, 4301–4312.

Huntington, J.A., 2005. Molecular recognition mechanisms of thrombin. J. Thromb. Haemost. 3, 1861–1872.

Huntington, J.A., 2008. How Na^+ activates thrombin – a review of the functional and structural data. Biol. Chem. 389, 1025–1035.

Isawa, H., Yuda, M., Yoneda, K., Chinzei, Y., 2000. The insect salivary protein, prolixin-S, inhibits factor IXa generation and Xase complex formation in the blood coagulation pathway. J. Biol. Chem. 275, 6636–6641.

Jackman, M.P., Parry, M.A., Hofsteenge, J., Stone, S.R., 1992. Intrinsic fluorescence changes and rapid kinetics of the reaction of thrombin with hirudin. J. Biol. Chem. 267, 15375–15383.

Koh, C.Y., Kazimirova, M., Nuttall, P.A., Kini, R.M., 2009. Noncompetitive inhibitor of thrombin. Chembiochem 10, 2155–2158.

Koh, C.Y., Kazimirova, M., Trimnell, A., Takac, P., Labuda, M., Nuttall, P.A., Kini, R.M., 2007. Variegin, a novel fast and tight binding thrombin inhibitor from the tropical bont tick. J. Biol. Chem. 282, 29101–29113.

Koh, C.Y., Kini, R.M., 2008. Anticoagulants from hematophagous animals. Expert Rev. Hematol. 1, 135–139.

Koh, C.Y., Kini, R.M., 2009. Molecular diversity of anticoagulants from haematophagous animals. Thromb. Haemost. 102, 437–453.

Lai, R., Takeuchi, H., Jonczy, J., Rees, H.H., Turner, P.C., 2004. A thrombin inhibitor from the ixodid tick, *Amblyomma hebraeum*. Gene 342, 243–249.

Lane, D.A., Philippou, H., Huntington, J.A., 2005. Directing thrombin. Blood 106, 2605–2612.

Laskowski, M., Jr., Kato, I., 1980. Protein inhibitors of proteinases. Annu. Rev. Biochem. 49, 593–626.

Ledizet, M., Harrison, L.M., Koskia, R.A., Cappello, M., 2005. Discovery and pre-clinical development of antithrombotics from hematophagous invertebrates. Curr. Med. Chem. Cardiovasc. Hematol. Agents 3, 1–10.

Liu, C.C., Brustad, E., Liu, W., Schultz, P.G., 2007. Crystal structure of a biosynthetic sulfo-hirudin complexed to thrombin. J. Am. Chem. Soc. 129, 10648–10649.

Macedo-Ribeiro, S., Almeida, C., Calisto, B.M., Friedrich, T., Mentele, R., Sturzebecher, J., Fuentes-Prior, P., Pereira, P.J., 2008. Isolation, cloning and structural characterisation of boophilin, a multifunctional Kunitz-type proteinase inhibitor from the cattle tick. PLoS One 3, e1624

Mans, B.J., Andersen, J.F., Schwan, T.G., Ribeiro, J.M., 2008. Characterization of anti-hemostatic factors in the argasid, *Argas monolakensis*: implications for the evolution of blood-feeding in the soft tick family. Insect Biochem. Mol. Biol. 38, 22–41.

Mans, B.J., Louw, A.I., Neitz, A.W., 2002. Evolution of hematophagy in ticks: common origins for blood coagulation and platelet aggregation inhibitors from soft ticks of the genus *Ornithodoros*. Mol. Biol. Evol. 19, 1695–1705.

Mao, S.J., Yates, M.T., Blankenship, D.T., Cardin, A.D., Krstenansky, J.L., Lovenberg, W., Jackson, R.L., 1987. Rapid purification and revised amino-terminal sequence of hirudin: a specific thrombin inhibitor of the bloodsucking leech. Anal. Biochem. 161, 514–518.

Markwardt, F., 1994. The development of hirudin as an antithrombotic drug. Thromb. Res. 74, 1–23.

Mende, K., Petoukhova, O., Koulitchkova, V., Schaub, G.A., Lange, U., Kaufmann, R., Nowak, G., 1999. Dipetalogastin, a potent thrombin inhibitor from the blood-sucking insect. *Dipetalogaster maximus* cDNA cloning, expression and characterization. Eur. J. Biochem. 266, 583–590.

Myles, T., Le Bonniec, B.F., Betz, A., Stone, S.R., 2001. Electrostatic steering and ionic tethering in the formation of thrombin-hirudin complexes: the role of the thrombin anion-binding exosite-I. Biochemistry 40, 4972–4979.

Nienaber, J., Gaspar, A.R., Neitz, A.W., 1999. Savignin, a potent thrombin inhibitor isolated from the salivary glands of the tick *Ornithodoros savignyi* (Acari: Argasidae). Exp. Parasitol. 93, 82–91.

Noeske-Jungblut, C., Haendler, B., Donner, P., Alagon, A., Possani, L., Schleuning, W.D., 1995. Triabin, a highly potent exosite inhibitor of thrombin. J. Biol. Chem. 270, 28629–28634.

Page, M.J., Macgillivray, R.T., Di Cera, E., 2005. Determinants of specificity in coagulation proteases. J. Thromb. Haemost. 3, 2401–2408.

Qiu, X., Padmanabhan, K.P., Carperos, V.E., Tulinsky, A., Kline, T., Maraganore, J.M., Fenton, J.W., 1992. Structure of the hirulog 3-thrombin complex and nature of the S' subsites of substrates and inhibitors. Biochemistry 31, 11689–11697.

Ribeiro, J.M., 1995. Blood-feeding arthropods: live syringes or invertebrate pharmacologists? Infect. Agents Dis. 4, 143–152.

Ribeiro, J.M., Francischetti, I.M., 2003. Role of arthropod saliva in blood feeding: sialome and post-sialome perspectives. Annu. Rev. Entomol. 48, 73–88.

Richardson, J.L., Fuentes-Prior, P., Sadler, J.E., Huber, R., Bode, W., 2002. Characterization of the residues involved in the human α-thrombin-haemadin complex: an exosite II-binding inhibitor. Biochemistry 41, 2535–2542.

Richardson, J.L., Kroger, B., Hoeffken, W., Sadler, J.E., Pereira, P., Huber, R., Bode, W., Fuentes-Prior, P., 2000. Crystal structure of the human α-thrombin-haemadin complex: an exosite II-binding inhibitor. EMBO J. 19, 5650–5660.

Rydel, T.J., Ravichandran, K.G., Tulinsky, A., Bode, W., Huber, R., Roitsch, C., Fenton, J.W., 1990. The structure of a complex of recombinant hirudin and human α-thrombin. Science 249, 277–280.

Rydel, T.J., Tulinsky, A., Bode, W., Huber, R., 1991. Refined structure of the hirudin-thrombin complex. J. Mol. Biol. 221, 583–601.

Scacheri, E., Nitti, G., Valsasina, B., Orsini, G., Visco, C., Ferrera, M., Sawyer, R.T., Sarmientos, P., 1993. Novel hirudin variants from the leech *Hirudinaria manillensis*. Amino acid sequence, cDNA cloning and genomic organization. Eur. J. Biochem. 214, 295–304.

Scharf, M., Engels, J., Tripier, D., 1989. Primary structures of new 'iso-hirudins'. FEBS Lett. 255, 105–110.

Skrzypczak-Jankun, E., Carperos, V.E., Ravichandran, K.G., Tulinsky, A., Westbrook, M., Maraganore, J.M., 1991. Structure of the hirugen and hirulog 1 complexes of α-thrombin. J. Mol. Biol. 221, 1379–1393.

Steiner, V., Knecht, R., Bornsen, K.O., Gassmann, E., Stone, S.R., Raschdorf, F., Schlaeppi, J.M., Maschler, R., 1992. Primary structure and function of novel O-glycosylated hirudins from the leech *Hirudinaria manillensis*. Biochemistry 31, 2294–2298.

Stone, S.R., Hofsteenge, J., 1986. Kinetics of the inhibition of thrombin by hirudin. Biochemistry 25, 4622–4628.

Strube, K.H., Kroger, B., Bialojan, S., Otte, M., Dodt, J., 1993. Isolation, sequence analysis, and cloning of haemadin. An anticoagulant peptide from the Indian leech. J. Biol. Chem. 268, 8590–8595.

van de Locht, A., Lamba, D., Bauer, M., Huber, R., Friedrich, T., Kroger, B., Hoffken, W., Bode, W., 1995. Two heads are better than one: crystal structure of the insect derived double domain Kazal inhibitor rhodniin in complex with thrombin. EMBO J. 14, 5149–5157.

van de Locht, A., Stubbs, M.T., Bode, W., Friedrich, T., Bollschweiler, C., Hoffken, W., Huber, R., 1996. The ornithodorin-thrombin crystal structure, a key to the TAP enigma? EMBO J. 15, 6011–6017.

Vitali, J., Martin, P.D., Malkowski, M.G., Robertson, W.D., Lazar, J.B., Winant, R.C., Johnson, P.H., Edwards, B.F., 1992. The structure of a complex of bovine α-thrombin and recombinant hirudin at 2.8-Å resolution. J. Biol. Chem. 267, 17670–17678.

Wallace, A., Dennis, S., Hofsteenge, J., Stone, S.R., 1989. Contribution of the N-terminal region of hirudin to its interaction with thrombin. Biochemistry 28, 10079–10084.

Warkentin, T.E., Greinacher, A., Koster, A., 2008. Bivalirudin. Thromb. Haemost. 99, 830–839.

Zhang, Y., Ribeiro, J.M., Guimaraes, J.A., Walsh, P.N., 1998. Nitrophorin-2: a novel mixed-type reversible specific inhibitor of the intrinsic factor-X activating complex. Biochemistry 37, 10681–10690.

Chapter 16
Anticoagulants from Scorpion Venoms

N.A. Valdez-Cruz, R. Conde, F.Z. Zamudio, and L.D. Possani

Abstract Venom from at least six distinct families of scorpion: Buthidae, Caraboctonidae, Iuridae, Ischnuridae, Scorpionidae and Vejovidae are known to contain many pharmacologically active peptides and some proteins involved in coagulation. The list of such compounds is increasing with time and newly described cases. Earlier reports performed with venom from *Buthus tamulus* and *Palamneus gravimanus* have shown the presence of coagulopathy in experimental animals. Other scorpion species such as: *Hottentotta judaicus*, *Heterometrus spinnifer*, *Heterometrus fulvipes*, *Parabuthus transvaalicus*, *Androctonus australis*, *Scorpio maurus palmatus* and *Leiurus quinquestriatus habraeus* were reported to contain components implicated in coagulation. From *Buthus marthensi* Karsch a venom active polypeptide (SVAP) that causes platelet aggregation *in vivo* and *in vitro* was described. A gene cloned from the same scorpion coding for a putative peptide named BmKAPi was reported. Its deduced amino acid sequence shows similarities to anticoagulant peptides and proteinase inhibitors from other species of animals. One of the best characterized components was Imperatoxin inhibitor (IpTxi) from the African scorpion *Pandinus imperator* Hector, described as a heterodimeric protein in which the major subunit has an amino acid sequence similar to known phospholipases (PLA2) from pancreas and snake venoms. The venom from the Mexican scorpion *Anuroctonus phaiodactylus* was reported to cause delay of clotting time of human platelet rich plasma (PRP) and poor platelet plasma (PPP). Several heterodimeric phospholipases and genes coding for similar ones were described from the venomous glands of this species. Recently, the venom from the Venezuelan scorpion *Tityus discrepans* was shown to induce inflammation, partial thromboplastin time (PTT) and prothrombin time (PT). Also recently venom from *Hadrurus gertschi and Opisthacanthus cayaporum* were reported to contain phospholipase enzymes. Thus far none of the well documented cases of snake venom

---

L.D. Possani (✉)
Departamento de Medicina Molecular y Bioprocesos, Instituto de Biotecnología, Universidad Nacional Autónoma de México, Avenida Universidad 2001, Col. Chamilpa, Cuernavaca Morelos, 62210, México
e-mail: possani@ibt.unam.mx

R.M. Kini et al. (eds.), *Toxins and Hemostasis*, DOI 10.1007/978-90-481-9295-3_16,

components that interfere with blood coagulation such as: serine proteases, metalloproteinases, C- type lectins, disintegrins have been identified in scorpion venom. This communication revises the state of art in this subject.

Introduction

During more than 350 million years scorpions have developed a variety of proteins in their venom glands to defend themselves from predators or to capture their preys. The venoms have different composition of toxins and components with diverse functions, selectivity and affinities over different types of organisms, such as crustaceans, insects or mammals. The scorpion venom contains bioactive compounds among which are enzymes, short amino acid peptides with toxic action, carbohydrates, free amines, nucleotides, lipids and many other small components of unknown function. Most of the peptides described thus far, recognize K^+, Na^+, Cl^- and Ca^{2+} ion-channels, which are blocked by the peptides or have their gating mechanism (open and closing probability) modified (Possani et al., 1999; Tytgat et al., 1999). Two types of toxic peptides are best described thus far: the "short-chain", specific for K^+-channels, which contain from 23 to 43 amino acid residues and the "long-chain" peptides, specific for Na^+-channels, with a primary structure that contains from 58 to 76 amino acid residues (Martin-Eauclaire and Couraud, 1995; Possani et al., 2000; reviews by Rodríguez de la Vega and Possani, 2004, 2005). The primary structure is closely packed by either three or four disulfide bridges. The three-dimensional structure of several of these peptides was solved (Fontecilla-Camps et al., 1980; see also review by Possani et al., 2006). Practically all of them have a structural arrangement known as the Cysteine αβ motif with a minimum core of αββ topology, although most have one segment of α-helix and three strands of antiparalel β-sheets, maintained by two of the disulfides bridges present in these toxic peptides. Apart from the classical Na^+ and K^+ specific scorpions toxins there are at least four important groups of components: (i) Enzymes, such as phospholipases of 14–19 kDa (Conde et al., 1999; Valdez-Cruz et al., 2004; Zamudio et al., 1997), hyaluronidases of 24–52 kDa (Possani et al., 1977; Tan and Ponnudurai, 1992) and proteases of ~80 kDa (Almeida et al., 2002; Ma et al., 2009; Tan and Ponnudurai, 1992); (ii) Peptides rich in cysteine residues non-toxic to mammals (Ali et al., 1998; Zhu and Li, 2002); (iii) Peptides without cysteine residues but with antimicrobial activity (Dai et al., 2002; Torres-Larios et al., 2000) and (iv) many other components of unknown function (orphan peptides), (Bringans et al., 2008; Diego-García et al., 2007; Elgar et al., 2006).

The main clinical symptoms following a scorpion sting are: excitability, respiratory distress, extensive secretion of nose and eyes, circulatory and cardio toxicity, flaccid paralysis, tissue injury in the sting site, systemic myolysis, pancreatitis, renal damage, and for some venoms also induction of coagulation disorders (coagulopathy and haemorrhage) (Brazón et al., 2001; Berg and Tarantino, 1991; D'Suze et al., 2003; Possani et al., 1999; Sofer et al., 1996). Scorpion envenomation varies on severity probably due to the amount of injected venoms, body mass of the victim and volume of distribution (Devi et al., 1970; Gibly et al., 1999; LoVecchio et al., 1999; Reddy et al., 1972). In humans approximately 8 h after envenomation, the elimination of some venom components occurs in urine with a concentration of 9.0 ng/mL of venom (Chase et al., 2009), but in many cases the venom t1/2 in human body have duration of more than 24 h.

Coagulation Disorders Caused by Scorpion Venoms

Animal venoms are known for causing coagulopathy disorders (Brazón et al., 2008; Han et al., 2008; Marino et al., 2009; Oliveira-Carvalho et al., 2008; Valdez-Cruz et al., 2004). The best anticoagulants or hemorrhagic components studied are those from venomous snakes (Kini, 2005; Kini and Evans, 1987). Several of these specific components are well characterized and grouped into families of proteins such as proteinases and metalloproteinases, C-type lectins, desintegrins and phospholipases. They usually affect: procoagulation, fibrinogen clotting, fibrinolysis, platelet-activation, anticoagulation, and might cause thrombotic or haemorrhagic manifestations (Kini and Evans, 1989). Some scorpion venoms also cause coagulopathy; however the number and type of components thus far characterized is small.

There are reports indicating that the venom from the scorpions *Hottentotta judaicus*, *Heterometrus spinnifer*, *Parabuthus transvaalicus*, *Androctonus australis*, *Scorpio maurus palmatus*, *Leiurus quinquestriatus habraeus* and *Pandinus imperator* species, delay the clotting time of plasma. In particular the venom of *P. imperator* and *P. transvaalicus* venoms delay the clotting time by 2.5 and 2.3 fold respectively; whereas the other venoms delay the clotting around 0.8–2 times (Tan and Ponnudurai, 1992). Venom from *P. imperator* and *P. transvaalicus* compared with snake venoms are 3–4 times less anticoagulant.

The venom from the scorpion *Buthus tamulus*, in general causes coagulopathy, and it has been shown to cause disseminated intravascular coagulopathy (DIC), (Radhakrishna Murthy et al., 1988). Injection of this scorpion venom via intravenously (i.v.) at sub lethal dose into dogs and rabbits causes alteration in the coagulation mechanism (Gajalakshmi, 1982). Human accidents due to sting by the scorpion *Palamneus gravimanus* activates coagulation, probably mediated by a venom component acting on Factor X; it also has an inhibitory effect over thrombin (Hamilton et al., 1974).

Recently, it has been demonstrated that *Tityus discrepans* induces inflammation because of neutrophil infiltration occurs in different tissues, causing vasculitis, arteritis and fibrin deposition. These clinical manifestations can be treated in mice with benzydamine (D'Suze et al., 2007). Also it has been demonstrated that high concentration of *T. discrepans* venom in human plasma fraction, increases the severity of the envenoming symptoms, by modification of the partial thromboplastin time (PTT) and prothrombin time (PT), rising the cytokines levels and increasing amylasemia and glycemia (D'Suze et al., 2003). Brazon et al. (2008) reported separation of *T. discrepans* venom into six sub-fractions by gel filtration on "Protein-Pack 125" and assayed the effects on PTT and PT. Fraction 1 contains high molecular weight proteins, which decreases the PTT time in a dose dependent manner. Fractions 2 and 6 contain peptides that prolong the PTT and PT time. The latter are probably peptides like serpin that inhibit serine-protease activities.

Phospholipase A2 (PLA2) enzymes are also present in scorpion venoms and are responsible for alterations on the coagulation (see below).

Scorpion PLA2s with Anticoagulant Properties

PLA2s (EC 3.1.1.4) catalyze the hydrolysis of fatty acid ester bonds at position 2 of 1,2-diacyl-*sn*-3-phosphoglycerides in presence of Ca^{2+} ions (Dennis, 1994). In addition to the digestion of preys, PLA2 are described as responsible for a wide variety of pharmacological effects, such as: neurotoxicity, cardiotoxicity, myotoxicity, necrosis, anticoagulation, hypotension, hemolysis, haemorrhage and oedema (Kini, and Evans, 1987). However, the correlation between structural and functional relationship of PLA2 have not been well analyzed and understood. For instance, it is not clear why a particular pharmacological functional site overlaps with the catalytic domain (Heinrikson, 1991; Kini and Evans, 1987; Scout and Sigler, 1994; Singh et al., 2001; Zhao et al., 2000).

The anticoagulant PLA2s have been classified into three groups with respect to their efficiency in inhibiting blood coagulation (Verheij et al., 1980): group A includes strongly anticoagulant PLA2s (those from snake venoms); group B contains PLA2s, which exert an anticoagulant effect at high concentrations only; and group C are the PLA2s with marginal or no anticoagulant activity at all (e.g. the pancreatic enzymes). Scorpion PLA2s are classified in between groups B and C; most of them are rather poor anticoagulants.

Anticoagulant PLA2s hydrolyse monomolecular films of phospholipids without any lag time while non-anticoagulant PLA2s present considerable induction times indicative of a low penetrating power (Kini and Evans, 1987; Verheij et al., 1980). When the ester linkages in the procoagulant lipids were replaced by the non-hydrolysable ether bonds, the mixture retained its clotting ability even in the presence of PLA2s, suggesting that anticoagulant PLA2s prevent clot formation by hydrolysing the phospholipids (Verheij et al., 1980). Phospholipids play a crucial role in the process of blood clotting by acting as structures to which some of the coagulation protein factors adhere and subsequently become activated. The process of clot formation in vitro is accelerated by the addition of external phospholipids to platelet-poor plasma. Particularly, the mixture of phosphatidylcholine and phosphatidylserine shows the strongest pro-coagulant properties (Papahadjopoulos et al., 1962).

There are two characteristics of these enzymes that can provide predictive information on the efficiency of the anticoagulant activity. The first is the phospholipase activity itself, i.e. the efficiency by which the enzyme can hydrolyze the ester bond of the phospholipids or can intercalate into the lipid bilayer for rupturing the membrane, and second is the presence of basic amino acid residues located in between residues 55–77 of certain phospholipases, which probably shows affinity towards other proteins of the coagulation cascade, hence deactivating the system, as suggested by Kini and Evans (1987); see also (Cirino et al., 1993; Kramer et al., 1989; Zhao et al., 2000).

For the case of scorpion PLA2s it is not clear to which of these two groups they belong to.

Figure 16.1 shows selected amino acid sequences of scorpion phospholipases in comparison with representative phospholipases from snake, bee and mammalian pancreas. This figure shows the sequence segment that was defined as "Anticoagulant site" (Condrea et al., 1981, 1982; Kini and Evans, 1988). As isoelectric point (pI) of this segment could indicate whether a given phospholipase could bind to proteins implicated in the coagulation cascade, we have included the pI for the entire β-wing. When the pI is basic the possibility that the enzyme is anticoagulant is greater. For example the pI of the highly anticoagulant phospholipases of *Naja nigricollis* and Notexin are over 9.0. The non anticoagulant pancreatic phospholipase has a pI of around 4.4. The pI of imperatoxin, a scorpion phospholipase with moderate anticoagulant activity is 6.0.

So far, at least 12 different PLA2 enzymes are described from scorpion venoms: five from *Anuroctonus phaiodactylus* (Valdez-Cruz et al., 2004), two from *Pandinus imperator* (Conde et al., 1999; Zamudio et al., 1997), two from *Opisthacanthus cayaporum* (Schwartz et al., 2008), and one each of *Hadrurus gertschi* (Schwartz et al., 2009), *Heterometrus fulvipes* (Hariprasad et al., 2007) and *Mesobuthus tamulus* (Hariprasad et al., 2009). Among these enzymes IpTxi of *Pandinus imperator* and Phaiodactylipin of *Anuroctonus pahiodactylus* were shown to delay the clotting time of human platelet poor and rich plasma (Valdez-Cruz et al., 2004).

Imperatoxin I is a heterodimeric protein with a molecular mass of 14,314 Da. The IpTxi dimer consists of a large subunit (104-amino acid residues) with PLA2 activity, covalently linked by a disulfide bridge to a smaller subunit (27 amino acid residues). Both subunits do not seem to be structurally related, although they are coded in a single messenger RNA. The native heterodimer has hydrolytic activity against phospholipids, but it was initially described as an inhibitor of [^{3}H]ryanodine binding to cardiac and skeletal sarcoplasmic reticulum by blocking ryanodine receptor channels (Zamudio et al., 1997). The same IpTxi was shown to delay the anticoagulant activity (Valdez-Cruz et al., 2004).

Phaiodactylipin is a glycosilated heterodimeric PLA2 and has a molecular mass of 19,172 Da. The mature protein is composed of two subunits, the large one consisting of 108 amino acid residues, whereas the small subunit has only 18 residues, and the structure is stabilized by five disulfide bridges. Phaiodactylipin is lethal to arthropods (insects and isopods), but not toxic to mice (up to 20 μg/20 g mouse weight). It causes muscular inflammation, without rupture of the basal membrane of cells in mice. It has a direct haemolytic effect in human erythrocytes and retards the coagulation time of blood. It is an unusual PLA2, with only 36 and 50% amino

```
             1         10        20        30        40        50        60        70        80
             |         |         |         |         |         |         |         |         |
CAX51436     ------------------------PGNIATDESDLGTL-EADRCCRTHDHC-DNIPAGETKYGITNDSFFTKLNCKCEESFDRCLN
PHOSPHOLIPIN -------------FLMWECTKWCGPGNNAKCESDLGPL-EADKCCRTHDHC-DYIASGETKYGITNYAFFTKLNCKCEEAFDRCLT
AAS98377     -------------FLIVSGTKWCGNNNIAANYSDLGFL-EADKCCRDHDHC-DHIASGETKYGLENKGLFTILNCDCDEAFDHCLK
AAR16429     ---------------TMWGTKWCGSGNEAINYTDLGYFSNLDSCCRTHDHC-DSIPAGETKYGLTNEGKYTMMNCKCESAFEKCLR
Q3YAU5       ---------------TMWGTKWCGSGNKAINYTDLGYFSNLDSCCRTHDHC-DNIAAGETKYGLTNEGKYTMMNCKCEATFQQCLR
P59888       ---------------TMWGTKWCGSGNEATDISELGYWSNLDSCCRTHDHC-DNIPSGQTKYGLTNEGKYTMMNCKCETAFEQCLR
P0C8L9       ---------------TVLGTKWCGAGNEAANYSDLGYFNNVDRCCREHDHC-DNIPAGETKYGLKNEGTYTMMNCKCEKAFDKCLS
P00630       -IRDRIGDNELEERIIYPGTLWCGHGNKSSGPNELGRFKHTDACCRTHDMCPDVMSAGESKHGLTNTASHTRLSCDCDDKFYDCLK
1POC                                             hhhhhhhhhhh bbbbbbbbbbbbbbb bbbbbhhhhhhhhhhh

P00608       NLVQFSYLIQCANHGKRPTWHYMDYGCYCGAGGSGTPVDELDRCCKIHDDCYDEAGKK-GCFPKMS-----AYDYYCGENGPYCRN
Q9DF52       NLIQFKNMIQCA--GTRPWTAYVNYGCYCGKGGSGTPVDELDRCCYTHDNCYNEAEKIPGCNPNIK-----TYSYTCTEPNLTCTD
P00605       NLYQFKNMIHCT-VPSRPWWHFADYGCYCGRGGKGTPVDDLDRCCQVHDNCYEKAGKM-GCWPYLT-----LYKYKCSQGKLTCSG
P00603       NLYQFKNMIHCT-VPSRPWWHFADYGCYCGRGGKGTAVDDLDRCCQVHDNCYGEAEKL-GCWPYLT-----LYKYECSQGKLTCSG
1P2P_A1      ALWQFRSMIKCAIPGSHPLMDFNNYGCYCGLGGSGTPVDELDRCCETHDNCYRDAKNLDSCKFLVDNPYTESYSYSCSNTEITCNS
1FE5          hhhhhhhhhhh                              hhhhhhhhhhhhhhhhh                   bbbbb bbbbbb
                          Ca2+ binding site              catalytic site                β-wing
                                                                                 "Anticoagulant site"

               90        100       110       120       130       140       150
               |         |         |         |         |         |         |          PI  β-wing
CAX51436     QLMRMQRIQPR-HRP--KLAKSLLSLYSQTATIAM-------------------------------------       4.56
PHOSPHOLIPIN EAYNKEEKESA-KSSTKRLQNFYFGTYSPECYVVTCNSKR-SGRDAGCENGVATWKKSYKD----------       6.0
AAS98377     EISNNVTTDIRQKGGAENVWRFYFQWYNANCYRLYCKDEK-SARDEACTNQYAVVKKNFTVQ---------       5.45
AAR16429     DVRG-----ILEGKAAAAVRKTYFDLYGNGCFNVKCPSGARSARSEECTNGMATYTGETGYGAWAINKLNG       4.68
Q3YAU5       DVHG-----PLEGKAAFTIRKLYFGLYGNGCFNVQCPS---------------------------------       4.68
P59888       NVTG-----GMEGPAAGFVRKTYFDLYGNGCYNVQCPSQRRLARSEECPDGVATYTGEAGYGAWAINKLNG       6.0
P0C8L9       DISG-----YFTRKAVSAVKFTYFTLYGNGCYNVKCENGR--SPSNECPNGVAEYTGETGLGAKVINFGK-       4.6
P00630       NSAD--------TISSYFVGKMYFNLIDTKCYKLEHPVTGCGERTEGRCLHYTVDKSKPKVYQWFDLRKY-       6.9
1POC         h           hhhhhhhhh         bbbbbbbbbbbbbb     bbbbbb   bbbbbbbb

P00608       IKKK-------CLRFVCDCDVEAAFCFAKAPYNNANWNIDTKKRCQ-------------------------       9.1
Q9DF52       TADT-------CARFLCDCDRTAAICFASAPYNSNNVMISSSTNCQ-------------------------       4.78
P00605       GNSK-------CGAAVCNCDLVAANCFAGARYIDANYNINFKKRCQ-------------------------       9.26
P00603       GNNK-------CAAAVCNCDLVAANCFAGARYIDANYNINLKERCQ-------------------------       8.6
1P2P_A1      KNNA-------CEAFICNCDRNAAICFSKAPYNKEHKNLDTKKYC--------------------------       4.36
1EF5            h       hhhhhhhhhhhhhhhhhh
```

Fig. 16.1 (continued)

acid sequence identities to the closest known PLAs, IpTxi and Phospholipin purified from the venom of the scorpion *Pandinus imperator*.

Phaiodactylipin substrate preference was higher for phosphatidylcholine and phosphatidylethanolamine than phosphatidylserine, especially when the *sn*–2 position is occupied by arachidonic acid. Phaiodactylipin delays the anticoagulant activity on human blood samples, similar to the effect described earlier for IpTxi. Phaiodactylipin exhibits anticoagulant activity by increasing the clotting time in both human PPP and PRP blood samples. IpTxi is more efficient than phaiodactylipin; at concentrations higher than 10 μg IpTxi makes the blood non-coagulable for 30 min. As with other PLA2s (Condrea et al., 1981), the clotting times are longer for the PPP blood samples, as expected.

Novel phospholipases were identified in the venom gland of the scorpions *Hadrurus getschi* (Charaboctonidae) (Schwartz et al., 2007) and *Opisthacanthus cayaporum* (Ichnuridae) (Schwartz et al., 2008). Transcriptome analysis of *Scorpiops jendeki* (Euscorpiidae) indicates the presence of several novel sequences corresponding to various enzymes, but not phospholipases (Ma et al., 2009). Venom proteomic analysis of the scorpion *Heterometrus longimanus* reported the presence of sequences similar to other known proteins that suppress platelet aggregation or inflammation (Bringans et al., 2008).

Finally, some scorpion PLA2s were cloned and the structure of their genes were studied and compared to those of snake PLA2s genes (Valdez-Cruz et al., 2007). PLA2s from both groups appear to follow the same arrangement of the genomic structure. They have introns structurally conserved and exons with high rate of variation (Deshimaru et al., 1996; John et al., 1994, 1996; Nobuhisa et al., 1996; Ogawa et al., 1995, 1996; Valdez-Cruz et al., 2007). The "accelerated evolution" occurring with snake venom phospholipases is discussed in (John et al., 1996; Kini and Evans, 1987; Ogawa et al., 1996). Phaiodactylipin is toxic to insects and crustaceans, but also causes hemolysis and muscular inflammation in mice. Imperatoxin-I causes inhibition of clotting and was reported to affect the ryanodine receptor Ca^{2+} channels. Some of these characteristics are clear evidence that these enzymes have followed a special trend of evolution.

Fig. 16.1 (continued) Alignment of amino acid sequences of PLA2s. PLA2s are classified as strong (P00608: *Notechis scutatus scutatus* "Notexin"; Q9DF52: *Bungarus caeruleus* "KPA2"; P00605: *Naja nigricollis*), weak (P00603: *Naja mossambica* "CM-II"), and non-anticoagulants (1P2P_A1: pancreatic pig). Group III PLA2s included are from *Apis mellifera* (P00630) and from scorpions CAX51436: *Opisthocanthus cayaporum*; Phospholipin: *Pandinus imperator*; P59888: Pandinus imperator IpTxi; AAS98377: *Anuroctonus phaiodactylus* "Phai"; AAR16429: *Mesobuthus tamulus*; Q3YAU5: *Heterometrus fulvipes*; P0C8L9: *Hadrurus gertschi*. 1POC and 1EF5 represent the secondary structures of *Apis mellifera* PLA2 and a basic PLA2 from *Bungarus caeruleus* respectively, as determined by crystallography data. The alignment was performed using Clustal X (Thompson et al., 1997). The isoelectric point from β-wing loop was calculated on line using the Gasteiger et al. (2005) tools on ExPASY. Each sequence starts with its GenBank accession number. Cysteine residues are in **bold**, whereas the β-wing acidic amino acids are marked in *grey* and basic amino acids are *underlined*. The letter "h" represents the α-helix and "b" the β-strands

Peptides with Anticoagulant Properties

Scorpion venom apart from the high molecular weight proteins (over 30 kDa) and the PLA2s of medium size molecular weights (around 14–16 KDa) also contain many small molecular mass peptides that affect coagulation. Among the peptides well characterized are two extracted from the venom of the Chinese scorpion *Buthus martensii* Karsch. The first is named "scorpion venom active polypeptide" (SVAP), which causes platelet aggregation in vivo and in vitro, using rabbit blood (Song et al., 2005). The SVAP also generates the thrombus and change plasma level of Thromboxane B2 and 6-keto-protaglandin F1a. Actually, this peptide SVAP is the most abundant component of this venom (Song et al., 2005).

The other peptide was obtained by cloning and it was called *Buthus martensii* Karsh peptide APi (BmKAPi). The encoded precursor of BmKAPi consists of 89 amino acid residues including a signal peptide of 24 residues, a putative mature peptide of 64 residues and an extra basic residue at the C-terminus which might be removed in the post-translational processing. BmKAPi is stabilized by five disulfide bridges, whereas all other disulfide-bridged scorpion toxins described are cross-linked by three or four disulfide bridges. It suggests that the three-dimensional scaffold of BmKAPi might be different from other scorpion toxins. The amino acid sequence of BmKAPi showed no sequence similarity to any other scorpion venom peptide, but shares a lower degree of similarity with some anticoagulant peptides and proteinase inhibitors isolated from hookworm, honeybee or European frog, respectively (Zeng et al., 2002).

Recently, Brazón et al. (2009) described a peptide called discreplasminin, isolated from the scorpion *Tityus discrepans*, which is a plasmin inhibitor. It has a molecular weight around 6,000 Da and has anti-fibrinolytic activity similar to aprotinin, supposed to interact with the active site of plasmin and tissue-plasminogen-activator.

Indeed, some small size peptides in scorpion venoms display a potential effect in the coagulation cascade, but our knowledge is rather limited.

Conclusion

Research on the subject of scorpion venom components is of great importance, both in terms of scientific interest and practical application. Pharmaceutical companies are searching for leading compounds with biotechnological applications on clinical disorders associated to the blood coagulation system. It is clear that more work must be conducted, first for the isolation and characterization of their structures, as done thus far for the case of PLA2s and a couple of short length peptides, and second for finding out exactly their functions and possible applications. It can be foreseen that scorpion proteins or peptides might became important anticoagulant agents in the treatment of thrombotic accidents. Also the knowledge of the molecular mechanism of the anticoagulant activity of some PLA2s that belongs to the group III and their particular targets is an important research area. Comparison with what is known from the anticoagulants of snake venoms and other organisms related to blood coagulation is important.

Acknowledgements This work was partially supported by grants from *Consejo Nacional de Ciencia y Tencnologia, Fondo-Salud* number 014109 (Mexican Government) and *Instituto Bioclón S.A. de C.V. (Mexico).*

References

Ali, S.A., Stoeva, S., Schütz, J., Kayed, R., Abassi, A., Zaidi, Z.H., Voelter, W., 1998. Purification and primary structure of low molecular mass peptides from scorpion (*Buthus sindicus*) venom. Comp. Biochem. Physiol. A. 121, 323–332.

Almeida, F.M., Pimenta, A.M., De Figueiredo, S.G., Santoro, M.M., Martin-Eauclaire, M.F., Diniz, C.R., De Lima, M.E., 2002. Enzymes with gelatinolytic activity can be found in *Tityus bahiensis* and *Tityus ser*rulatus venoms. Toxicon 40, 1041–1045.

Berg, R.A., Tarantino, M.D., 1991. Envenomation by the scorpion *Centruroides exilicauda sculpturatus*. Severe and unusual manifestations. Pediatrics 87, 930–933.

Brazón, J., D'Suze, G., Guerrero, B., Sevcik, C., Arocha-Piñango, C.L., 2001. The complex effect of *Tityus discrepans* scorpion venom on hemostatic mechanism involves: degradation of fibrinogen a-chain, the presence of plasmin like fractions and/or plasminogen activators and inhibitors of plasmin and/or plasminogen activators. Proceedings of Seventh Symposium: Pan-American Section of the International Society on Toxinology on Animal, Plant and Microbial Toxins, p. 56.

Brazón, J., Guerrero, B., Arocha-Piñango, C.L., Sevcik, C., D'Suze, G., 2008. Effect of *Tityus discrepans* scorpion venom on global coagulation test. Preliminary studies. Invest. Clin. 49, 49–58.

Brazón, J., D'Suze, G., D'Errico, M.L., Arocha-Piñango, C.L., Guerrero, B., 2009. Discreplasminin, a plasmin inhibitor isoltated from *Tityus discrepans* scorpion venom. Arch. Toxicol. 83, 669–678.

Bringans, S., Eriksen, S., Kendrick, T., Gopalakrishnakone, P., Livk, A., Lock, R., Lipscombe, R., 2008. Proteomic analysis of the venom of *Heterometrus longimanus* (Asian black scorpion). Proteomics 8, 1081–1096.

Chase, P., Boyer-hassen, L., Mcnally, J., Vazquez, H.L., Theodorou, A.A., Gardner Walter, F., Alagon, A., 2009. Serum levels and urine detection of *Centruroides sculpturatus* venom in significantly envenomated patients. Clin. Toxicol. 47, 24–28.

Cirino, G., Cicala, C., Sorrentino, L., Browning, J.L., 1993. Human recombinant non pancreatic secreted platelet phospholipase A_2 has anticoagulant activity in vitro on human plasma. Thromb. Res. 70, 337–342.

Conde, R., Zamudio, F., Becerril, B., Possani, L.D., 1999. Phospholipin, a novel heterodimeric phospholipase A_2 from *Pandinus imperator* scorpion venom. FEBS Lett. 460, 447–450.

Condrea, E., Yang, C.C., Rosenberg, P., 1981. Lack of correlation between anticoagulant activity and phospholipid hydrolysis by snake venom phospholipases A_2. Thromb. Haemost. 45, 82–85.

Condrea, E., Rapuano, B.E., Fletcher, J.E., Yang, C.C., Rosenberg, P., 1982. Ethoxyformylation and guanidination of snake venom phospholipases A_2: effects on enzymatic activity, lethality and some pharmacological properties. Toxicon 21, 209–218.

Dai, L., Corzo, G., Naoki, H., Andriantsiferana, M., Nakajima, T., 2002. Purification, structure-function analysis, and molecular characterization of novel linear peptides from scorpion *Opisthacanthus madagascariensis*. Biochem. Biophys. Res. Commun. 293, 1514–1522.

Dennis, E.A., 1994. Diversity of group types, regulation, and function of phospholipase A_2. J. Biol. Chem. 269, 13057–13060.

Deshimaru, M., Ogawa, T., Nakashima, K., Nobuhisa, I., Chijiwa, T., Shimohigashi, Y., Fukumaki, Y., Niwa, M., Yamashina, I., Hattori, S., Ohno, M., 1996. Accelerated evolution of crotalinae snake venom gland serine proteases. FEBS Lett. 397,83–88.

Devi, C.S., Reddy, C.N., Deve, S., Subramaniam Ban, V., Suvarnakumari, G., Murthy, D.P., Reddy, C.R.R.M., 1970. Defibrination syndrome due to scorpion poisoning. Br. Med. J. 1, 345–347.

Diego-García, E., Schwartz, E.F., D'Suze, G., González, S.A.R., Batista, C.V.F., García, B.I., Rodríguez de la Vega, R.C., Possani, L.D., 2007. Wide phylogenetic distribution of Scorpine and long-chain beta-KTx-like peptides in scorpion venoms: identification of "orphan" components. Peptides 28, 31–37.

D'Suze, G., Moncada, S., González, C., Sevcik, C., Aguilar, V., Alagón, A., 2003. Relationship between plasmatic levels of various cytokines, tumour necrosis factor, enzymes, glucose and venom concentration following *Tityus* scorpion sting. Toxicon 41, 367–375.

D'Suze, G., Díaz, P., Salazar, V., Sevcik, C., Brazón, J., 2007. Effect of leukocyte inhibitors benzydamine and cyclophosphamide, on lung injury caused by *Tityus discrepans* scorpion venom. Toxicon 50, 1116–1125.

Elgar, D., Verdonck, F., Grobler, A., Fourie, C., du Plessis, J., 2006. Ion selectivity of scorpion toxin-induced pores in cardiac myocytes. Peptides 27, 55–61.

Fontecilla-Camps, J.C., Almassy, R.J., Suddath, F.L., Watt, D.D., Bugg, C.E., 1980. Three-dimensional structure of a protein from scorpion venom: a new structural class of neurotoxins. Proc. Natl. Acad. Sci. U.S.A. 77, 6496–6550.

Gajalakshmi, B.E., 1982. Coagulant studies following scorpion venom injection in animals. Indian J. Med. Res. 76, 337–341.

Gibly, R., Williams, M., Walter, F.G., McNally, J., Conroy, C., Berg, R.A., 1999. Continuous intravenous midazolam infusion for *Centruroides exilicauda* scorpion envenomation. Ann. Emerg. Med. 34, 620–625.

Gasteiger, E., Hoogland, C., Gattiker, A., Duvaud, S., Wilkins, M.R., Appel, R.D., Bairoch, A., 2005. Protein identification and analysis tools on the ExPASy server, in: Walker, J.M. (Ed.), The Proteomics Protocols Handbook. Humana Press, Totowa, NJ, pp. 571–607.

Hamilton, P.L., Ogston, D., Douglas, A., 1974. Coagulant activity of the scorpion venoms *Palamneus gravimanus* and *Leiurus guinquestriatus*. Toxicon 12, 291–296.

Han, T.S., Teichert, R.W., Olivera, B.M., Bulaj, G., 2008. Conus venoms – a rich source of peptide-based therapeutics. Curr. Pharm. Des. 14, 2462–2479.

Hariprasad, G., Singh, B.S. Das, U., Ethayathulla, A.S., Kaur, P., Singh, T.P., Srinivasan, A., 2007. Cloning, sequence analysis and homology modeling of a novel phospholipase a2 from *Heterometrus fulvipes* (Indian black scorpion). DNA Sequence 18, 242–246.

Hariprasad, G., Saravanan, K., Singh, S.B., Das, U., Sharma, S., Kaur, P., Singh, T.P., Srinivasan, A., 2009. Group III PLA_2 from the scorpion, *Mesobuthus tamulus*: cloning and recombinant expression in *E. coli*. Electronic J. Biotech. 12(3), 1–7.

Heinrikson, R.L., 1991. Dissection and sequence analysis of phospholipases A_2. Meth. Enzymol. 197, 201–214.

John, T.R., Smith, L.A., Kaiser, I.I., 1994. Genomic sequences encoding the acidic and basic subunits of Mojave toxin: unusually high sequence identity of non-coding regions. Gene 139, 229–234.

John, T.R., Smith, J.J., Kaiser, I.I., 1996. A phospholipase A2-like pseudogene retaining the highly conserved introns of Mojave toxin and other snake venom group II PLA2s, but having different exons. DNA Cell Biol. 15, 661–668.

Kini, R.M., Evans, H.J., 1987. Structure-function relationships of phospholipases. The anticoagulant region of phospholipases A_2. J. Biol. Chem. 262, 14402–14407.

Kini, R.M., Evans, H.J., 1988. Correlation between the enzymatic activity, anticoagulant and antiplatelet effects of phospholipase A_2 isoenzymes from *Naja nigricollis* venom. Thromb. Haemost. 60, 170–173.

Kini, R.M., Evans, H.J., 1989. A model to explain the pharmacological effects of snake venom phospholipases A_2. Toxicon 27, 613–635.

Kini, R.M., 2005. Structure-function relationships and mechanism of anticoagulant phospholipase A_2 enzymes from snake venoms. Toxicon 45, 1147–1161.

Kramer, R.M., Hession, C., Johansen, B., Hayes, G., McGray, P., Chow, E.P., Tizard, R., Pepinsky, R.B., 1989. Structure and properties of a human non-pancreatic phospholipase A_2. J. Biol. Chem. 264, 5768–5775.

LoVecchio, F., Welch, S., Klemens, J., Curry, S.C., Thomas, R., 1999. Incidence of immediate and delayed hypersensitivity to *Centruroides* antivenom. Ann, Emerg. Med. 34, 615–619.

Ma, Y., Zhao, R., He, Y., Li, S., Liu, J., Wu, Y., Cao, Z., Li, W., 2009. Transcriptome analysis of the venom gland of the scorpion *Scorpiops jendeki*: implication for the evolution of the scorpion venom arsenal. BMC Genomics 10, 290.

Marino, A., Morabito, R., La Spada, G., 2009. Factors altering the haemolytic power of crude venom from *Aiptasia mutabilis* (Anthozoa) nematocysts. Comp. Biochem. Physiol. A. Mol. Integr. Physiol. 152, 418–422.

Martin-Eauclaire, M.-F., Couraud, F., 1995. Scorpion neurotoxins: effects and mechanisms, in: Chang, L.W., Dyer, R.S. (Eds.), Handbook of Neurotoxicology. Marcel Dekker, New York, pp. 683–716.

Radhakrishna Murthy, K., Zolphaghrian, H., Medh, J.D., Kudalkar, J.A., Yeolekar, M.E., Pandit, S.P., Khopkar, M., Dave, KN., Billimoria, F.R., 1988. Disseminated intravascular coagulation and disturbances in carbohydrate and fat metabolism in acute myocarditis produced by scorpion (*Buthus tamulus*) venom. Indian J. Med. Red., 87, 318–325.

Nobuhisa, I., Nakashima, K., Deshimaru, M., Ogawa, T., Shimohigashi, Y., Fukumaki, Y., Sakaki, Y., Hattori, S., Kihara, H., Ohno, M., 1996. Accelerated evolution of *Trimeresurus okinavensis* venom gland phospholipase A_2 isozyme-encoding genes. Gene 172, 267–272.

Ogawa, T., Kitajima, M., Nakashima, K., Sakaki, Y., Ohno, M., 1995. Molecular evolution of group II phospholipases A_2. J. Mol. Evol. 41, 867–877.

Ogawa, T., Nakashima, K., Nobuhisa, I., Deshimaru, M., Shimohigashi, Y., Fukumaki, Y., Sakaki, Y., Hattori, S., Ohno, M., 1996. Accelerated evolution of snake venom phospholipase A_2 isozymes for acquisition of diverse physiological functions. Toxicon 34, 1229–1236.

Oliveira-Carvalho, A.L., Guimarães, P.R., Abreu, P.A., Dutra, D.L., Junqueira-de-Azevedo, I.L., Rodrigues, C.R., Ho, P.L., Castro, H.C., Zingali, R.B., 2008. Identification and characterization of a new member of snake venom thrombin inhibitors from *Bothrops insularis* using a proteomic approach. Toxicon 51, 659–671.

Papahadjopoulos, D., Hougie, C., Hanahan, D.J., 1962. Influence of surface charge of phospholipids on their clot-promoting. Proc. Soc. Exp. Biol. Med. 111, 412–416.

Possani, L.D., Alagón, A.C., Fletcher, P.L., Jr., Erickson, B.W., 1977. Purification and properties of mammalian toxins from the venom of Brazilian Scorpion *Tityus serrulatus* Lutz and Mello. Arch. Biochem. Biophys. 180, 394–403.

Possani, L.D., Becerril, B., Delepierre, M., Tytgat, J., 1999. Scorpion toxins specific for Na^+-channels. Eur. J. Biochem. 264, 287–300.

Possani, L.D., Merino, E., Corona, M., Bolivar, F., Becerril, B., 2000. Peptides and genes coding for scorpion toxins that affect ion-channels. Biochimie 82, 861–868.

Possani, L.D., Rodríguez de la Vega, R.C., 2006. Scorpion venom peptides, in: Kastin A.J. (Ed.), Handbook of Biologically Active Peptides. Elsevier, Amsterdam, pp. 339–354.

Reddy, C.R., Suvarnakumari, G., Devi, C.S., Reddy, C.N., 1972. Pathology of scorpion venom poisoning. J. Trop. Med. Hyg. 75, 98–100.

Rodríguez de la Vega, R.C., Possani, L.D., 2004. Current views on scorpion toxins specific for K^+-channels. Toxicon 43, 865–875.

Rodríguez de la Vega, R.C., Possani, L.D., 2005. Overview of scorpion toxins specific for Na^+ channels and related peptides: biodiversity, structure-function relationships and evolution. Toxicon 46, 831–44.

Scout, D.L., Sigler, P.B., 1994. The structural and functional roles of calcium ion in secretory phospholipases A_2. Adv. Inorg. Biochem. 10, 139–155.

Schwartz, E.F., Diego-García, E., Rodríguez de la Vega, R.C., Possani, L.D., 2007. Transcriptome analysis of the venom gland of the Mexican scorpion *Hadrurus gertschi* (Arachnida: Scorpiones). BMC Genomics 8, 119.

Schwartz, E.F., Camargos, T.S., Zamudio, F.Z., Silva, L.P., Bloch, C., Caixeta, F., Schwartz, C.A., Possani, L.D., 2008. Mass spectrometry analysis, amino acid sequence and biological activity of venom components from the Brazilian scorpion *Opisthacanthus cayaporum*. Toxicon 51, 1499–1508.

Schwartz, E.F., Capes, E.M., Diego-García, E., Zamudio, F.Z., Fuentes, O., Possani, L.D., Valdivia, H.H., 2009. Characterization of hadrucalcin, a peptide from *Hadrurus gertschi* scorpion venom with pharmacological activity on ryanodine receptors. Br. J. Pharmacol. 157, 392–403.

Singh, G., Gourinath, S., Sharma, S., Paramasivam, M., Srinivasan, A., Singh, T.P., 2001. Sequence and crystal structure determination of a basic phospholipase A_2 from common krait (*Bungarus caeruleus*) at 2.4 Å resolution: identification and characterization of its pharmacological sites. J. Mol. Biol. 307, 1049–1059.

Sofer, S., Gueron, M., White, R., Lifshitz, M., Apte, R., 1996. Interleukin-6 release following scorpion sting in children. Toxicon 34, 389–392.

Song, Y.M., Tang, X.X., Chen, X.G., Gao, B.B., Gao, E., Bai, L., Lv, X.R., 2005. Effects of scorpion venom bioactive polypolypeptides on platelet aggregation and thrombosis and plasma 6-keto-PG F1alpha and TXB2 in rabbits and rats. Toxicon 46, 230–235.

Tan, N.H., Ponnudurai, G., 1992. Comparative study of the enzymatic, hemorrhagic, procoagulant and anticoagulant activities of some animal venoms. Comp. Biochem. Physiol. 103, 299–302.

Thompson, J.D., Gibson, T.J., Plewniak, F., Jeanmougin, F. Higgins, D.G., 1997. The ClustalX windows interface: flexible strategies for multiple sequence alignment aided by quality analysis tools. Nucl. Acids Res. 24, 4876–4882.

Torres-Larios, A., Gurrola, G., Zamudio, F., Possani, L.D., 2000. Hadrurin, a new antimicrobial peptide from the venom of the scorpion *Hadrurus aztecus*. Eur. J. Biochem. 267, 5023–5031.

Tytgat, J., Chandy, K.G., Garcia, M.L., Gutman, G.A., Martin- Eauclaire, M.-F., Van der Walt, J.J., Possani, L.D., 1999. A unified nomenclature for short-chain peptides isolated from scorpion venoms: alpha-KTx molecular subfamilies. Trends Pharmacol. Sci. 20, 444–447.

Valdez-Cruz, N.A., Batista, C.V., Zamudio, F.Z., Bosmans, F., Tytgat, J., Possani, L.D., 2004. Phaiodotoxin, a novel structural class of insect-toxin isolated from the venom of the Mexican scorpion *Anuroctonus phaiodactylus*. Eur. J. Biochem. 271, 4753–4761.

Valdez-Cruz, N.A., Segovia, L., Corona, M., Possani, L.D., 2007. Sequence analysis and phylogenetic relationship of genes encoding heterodimeric phospholipases A_2 from the venom of the scorpion *Anuroctonus phaiodactylus*. Gene 396, 149–158.

Verheij, H.M., Boffa, M.C., Rothen, C., Bryckaert, M.C., Verger, R., de Haas, G.H., 1980. Correlation of enzymatic activity and anticoagulant properties of phospholipase A_2. Eur. J. Biochem. 112, 25–32.

Zamudio, F.Z., Conde, R., Arévalo, C., Becerril, B., Martin, B.M., Valdivia, H.H., Possani, L.D., 1997. The mechanism of inhibition of ryanodine receptor channels by imperatoxin I, a heterodimeric protein from the scorpion *Pandinus imperator*. J. Biol. Chem. 272, 11886–11894.

Zeng, X.C., Wang, S.X., Li, W.X., 2002. Identification of BmKAPi, a novel type of scorpion venom peptide with peculiar disulfide bridge pattern from *Buthus martensii* Karsch. Toxicon 40, 1719–1722.

Zhao, K., Zhou, Y., Lin, Z., 2000. Structure of basic phospholipase A_2 from *Agkistrodon halys* Pallas: implications for its association, hemolytic and anticoagulant activities. Toxicon 38, 901–916.

Zhu, S., Li, W., 2002. Precursors of three unique cysteine-rich peptides from the scorpion *Buthus martensii* Karsch. Comp. Biochem. Physiol. B. Biochem. Mol. Biol. 131, 749–756.

Part IV
Antiplatelet Proteins

Chapter 17
The Discovery of Disintegrins

Tur-Fu Huang

Abstract Disintegrins represent a class of low molecular weight, Arg-Gly-Asp(RGD)/Lys-Gly-Asp(KGD)-containing, cysteine-rich polypeptides derived from venoms of various viper snakes. They bind to various integrins (e.g. αIIbβ3, αvβ3, α5β1 and others) expressed on cell membrane surface, with various degrees of affinity and specificity. Disintegrins were originally discovered as antiplatelet agents by acting as platelet membrane αIIbβ3 antagonists. However, they also have been found to bind αvβ3, α5β1 or α4β1 expressed on endothelial cells, fibroblasts, phagocytes, and tumor cells, thus affecting cell-matrix and cell-cell interaction. The homologous molecular structure among disintegrin, snake venom metalloproteinase (SVMP), and ADAM (a disintegrin and metalloproteinase) reveals their evolutionary relationship. Based on the structure-activity relationship of these molecules and integrins, the potential applications of these disintegrins and their derivatives are briefly discussed in field of arterial thrombosis, cell adhesion, cell migration, angiogenesis, inflammation, and tumor metastasis.

Introduction

Snake venoms contain a mixture of proteins which possess many biological activities. Venoms of the *Elapidae* and *Hydrophidae* families are highly toxic, causing death by blocking neuromuscular transmission, whereas those of *Viperidae* family cause shock, intravascular clotting, systemic and local hemorrhage, edema and necrosis (Teng and Huang, 1991). Some snake venom components are unique in terms of their mode of action and specificity, e.g. neurotoxin α-bungarotoxin blocking neuromuscular transmission by acting on the nicotinic acetylcholine receptor of skeletal muscles, and disintegrins inhibiting platelet aggregation via binding to platelet membrane glycoprotein (GP) IIb/IIIa complex (e.g. αIIbβ3, fibrinogen receptor). Many investigators attempted to isolate the active principles, and asked

T.-F. Huang (✉)
Institute of Pharmacology, College of Medicine, National Taiwan University, Taipei, Taiwan
e-mail: turfu@ntu.edu.tw

R.M. Kini et al. (eds.), *Toxins and Hemostasis*, DOI 10.1007/978-90-481-9295-3_17,

what and how they interact with coagulation cascade, with platelets, or with both. Ouyang et al. (1992) classified them into those which accelerate the process of hemostasis and those which inhibit it. Briefly, the former includes factor X activator, prothrombin activator, thrombin-like enzymes, and platelet aggregation inducers, whereas the latter includes fibrinogenolytic enzymes, prothrombin activating inhibitors, factor X activating inhibitors, and platelet aggregation inhibitors.

Regarding platelet aggregation inhibitors, many components including ADPase, α-fibrinogenase (i.e. snake venom metalloproteinase, SVMP), phospholipase A_2, and disintegrin have been reported to suppress platelet aggregation in vitro. However, these classes of snake venom components affect in vivo hemostasis and cell adhesion, namely GPIb/IX binding proteins (also known as snaclec), hemorrhagins (SVMPs), and disintegrins. GPIb/IX binding proteins can either induce platelet aggregation through vWF modulation, GPIb, α2β1, Clec-2, and GPVI binding; or inhibit platelet aggregation via GPIb blockade (Lu et al., 2005). Hemorrhagins represent SVMP and can be divided into P-I∼P-IV classes according to their possessing a metalloproteinase/disintegrin-like/cysteine-rich and, lectin domain, respectively (Jia et al., 1997; Kini and Evans, 1992). The biological activities of GPIb/IX binding proteins and SVMPs are discussed in detail in Chapters 21, 35 and 39. In this chapter, we will focus on how the disintegrins were discovered and their potential applications in field of arterial thrombosis, inflammation, angiogenesis, tumor metastasis, and other integrin-related diseases.

Role of Integrins in Thrombosis and Hemostasis

Integrins are a family of non-covalently associated α/β heterodimeric cell adhesion receptor. So far at least 19α- and 8β-subunits have been recognized. Integrins play vital roles in platelet aggregation, inflammatory reactions, tissue remodeling, cell adhesion, migration, angiogenesis, and other biological processes (Hynes, 1992).

Circulating platelets quickly respond to vascular injury. The coverage of the exposed endothelium by activated platelets depends on the recognition of adhesive proteins (i.e. vWF, collagen, fibronectin, laminin and fibrinogen) by the respective specific platelet membrane glycoproteins (i.e. GPIb, α2β1/GPVI, α5β1, α6β1, or αIIbβ3). Among these integrins, α2β1 (i.e. GPIa/IIa) /GPVI are known to mediate collagen adhesion/ activation of platelets, whereas αIIbβ3 mediates the adhesion of fibrinogen, and the subsequent platelet-platelet interaction and aggregation, a common final step of platelet aggregation shared by all stimulating agonist (Ginsberg et al., 1988; Watson et al., 2005).

Generally the activation of platelets consists of several signal transduction pathways including PLA_2/PLC activation and the rise of cytosolic free calcium after the ligation of agonist with membrane receptor, leading to shape change, release reaction of ADP and serotonin as well as thromboxane A_2 formation, enhancement of procoagulation activity, and finally the exposure of activated αIIbβ3 (i.e. fibrinogen receptor). Subsequently, the binding of plasma fibrinogen to the activated αIIbβ3 on platelet membrane would bridge adjacent platelets, leading to platelet aggregation, which is reinforced by the subsequent fibrin formation at the injury site, forming a hemostatic plug. However, abnormal thrombi may be formed under pathological conditions, such as chronic endothelial injury due to atherosclerotic lesion (Coller, 1990). On the other hand, patients afflicted with a deficiency or dysfunction of platelet membrane integrin α2β1 or αIIbβ3 show significant bleeding disorder, indicating that these integrins play essential roles in mediating platelet aggregation (Nieuwenhuis et al., 1985; George et al., 1990).

The Discovery of Disintegrin as an Antiplatelet Agent

The frontier studies of Taiwanese investigators showed that the venom of *Trimeresurus gramineus*, *Echis carinatus*, *Agkistrodon halys*, and *Agkistrodon rhodostoma* contain polypeptides which are potent inhibitors of platelet aggregation (Huang and Ouyang, 1984; Huang et al., 1987a; Ouyang and Huang, 1983; Ouyang et al., 1983). These components inhibit platelet aggregation elicited by a variety of aggregation agonists, including ADP, epinephrine, sodium arachidonate, collagen and Ca^{2+} ionophore A-23187, with a similar IC_{50}. However, they neither affect the shape change nor the cAMP level. We proposed that these inhibitors may interfere with a final common step of platelet aggregation. Further studies resulted in the purification and sequencing of trigramin and echistatin from *Trimeresurus gramineus* (Huang et al., 1987b, 1989) and *Echis carinatus* venom (Gan et al., 1988). Trigramin potently inhibited fibrinogen binding to ADP-stimulated platelets, and bound to membrane GPIIb/IIIa with a Kd value of 1.2×10^{-8} M. ^{125}I-trigramin bound to ADP-stimulated platelets from normal people in a saturated manner with around 20,000 bind sites. ^{125}I-trigramin showed only 2.7∼5.4% binding to ADP-stimulated platelets in patients with Glanzmann's thrombasthenia, suggesting that GPIIb/IIIa is the target of trigramin. In addition, mAbs raised against GPIIb/IIIa (e.g. 7E3, A2A6), and RGDS showed inhibitory effect on ^{125}I-trigramin binding to platelets, suggesting that its binding target is GPIIb/IIIa and RGD tripeptide sequence may be essential for its binding activity (Huang et al., 1987b). The subsequent sequencing of trigramin showed that it is RGD-containing single polypeptide chain of 72 amino acid residues and six disulfide bonds (Fig. 17.1, Huang et al., 1989). Trigramin also was shown to inhibit human vWF binding to GPIIb/IIIa complex in thrombin-activated platelets. Reduced and alkylated trigramin lost activity in inhibiting platelet aggregation and binding capacity toward platelets, indicating that biological activity of trigramin depends upon the presence of RGD sequence, and its secondary structure (Huang et al., 1987b).

The In Vivo Antithrombotic Activity of Disintegrin

We first demonstrated the in vivo antithrombotic activity of trigramin in a hamster model (Cook et al., 1989). Intravenous infusion of trigramin significantly prolonged the bleeding time of severed mesentery arteries. The bleeding time immediately

```
Accutin                                            GA QCTAGPCCWP CFKLKEGTIC RRARGDDLDD YCNGISADCP RNPYY
Echistatin                                            ECESGPCCRN CKFLKEGTIC KRARGDDLDD YCNGKTCDCP RNPHKGPAT
Eristicophin                                    QRQEE PCATGPCCRR CKFKRAGKVC RVARGDWNND YCTGKSCDCP RNPWNG
Eristostatin                                      QEE PCATGPCCRR CKFKRAGKVC RVARGDWNDD YCTGKSCDCP KNPWNG

Albolabrin               EAGEDCDCGS PAN...PCCD AATCKLLPGA QCGEGLCCDQ CSFMKKGTIC RRARGDDLDD YCNGISAGCP RNPLHA
Barbourin                EAGEECDCGS PEN...PCCD AATCKLRPGA QCADGLCCDQ CRFMKKGTVC RVAKGDWNDD TCTGQSADCP RNGLYG
Halysin                  EAGEECDCGS PGN...PCCD AATCKLRQGA QCAEGLCCDQ CRFMKKGTVC RIARGDDMDD YCNGISAGCP RNPF
Rhodostomin                GKECDCSS PEN...PCCD AATCKLRPGA QCGEGLCCEQ CKFSRAGKIC RIPRGDMPDD RCTGQSADCP RYH
Triflavin                  GEECDCGS PSN...PCCD AATCKLRPGA QCADGLCCDQ CRFKKKRTIC RIARGDFPDD RCTGQSADCP RWNGL
Trigramin                EAGKDCDCGS PAN...PCCD AATCKLLPGA QCGEGPCCDQ CSFMKKGTIC RRARGDDLDD YCNGRSAGCP RNPFHA

Bitistatin    SPPVCGNKIL EQGEDCDCGS PANCQDRCCN AATCKLTPGS QCNYGECCDQ CRFKKAGTVC RIARGDWNDD YCTGKSSDCP WNH
                       10         20         30         40         50         60         70         80
```

Fig. 17.1 Amino acid sequences of disintegrins. Spaces are inserted into amino acid sequences of medium disintegrins for better alignment with bitistatin (arietin)

returned to normal after cessation of trigramin infusion, indicating that its action is reversible. In the meantime, the disappearance of ^{125}I-trigramin from the circulation fits a two compartment model with $t^{1/2}$, 2 min and 30∼105 min, respectively. Therefore, trigramin may prevent the ability of platelets to form thromboemboli in vivo. In a canine model of coronary thrombosis, a disintegrin kistrin was shown to prevent the reocclusion of coronary artery as in combination with heparin, and to accelerate the recombinant tissue-type plasminogen activator-induced thrombosis (Yasuda et al., 1991). Applagin prevented experimental carotid artery thrombosis in dogs (Rote et al., 1993). The inhibitory effects of many disintegrins on platelets adhesion to surfaces of the extracorporeal circuits was reported (Musial et al., 1990). Therefore, these disintegrins were thought as potential candidates as novel antithrombotic agents. Since then, many pharmaceutical companies paid great interest in developing this class of GPIIb/IIIa antagonists pattered on their steric structure by mean of computer modeling, especially the active motif RGD and their surrounding amino acid sequence.

The Characteristics of Disintegrins

Disintegrins were originally known as low molecular weight (47–84 amino acid residues), cysteine-rich venom proteins that contain an RGD/or KGD loop maintained by specific disulfide bridges and as potent inhibitors of platelet aggregation (Gould et al., 1990; Huang, 1998). The tripeptide, RGD, within the molecule is essential in mediating the binding of αIIbβ3, and the disulfide bridges is also critical for the expression of its biological activities (Huang et al., 1987b, 1989). After the first discovery of trigramin, more than 50 similar polypeptides have been characterized. They bind to αIIbβ3 of both resting and activated platelets with a high binding affinity (Kd, 10^{-7}∼10^{-8} M), about 50∼100 times higher than the binding affinity of the natural ligand fibrinogen. Among these disintegrins, barbourin, a KGD containing peptide, shows a rather specificity toward platelet αIIbβ3 than to integrin αvβ3 (Scarborough et al., 1991). Thus, a cyclic KGD-peptide Integrilin has been successfully developed as antithrombotic agent, and used clinically for prevention of restenosis after percutaneous transluminal coronary angioplasty (PTCA) (Scarborough et al., 1993a, b).

Based on the length and disulfide patterning, disintegrins may be classified into 3 types: the short-sized disintegrin (e.g. echistatin, accutin with 47–51 residues and 4 disulfide bonds), the medium-sized disintegrin (e.g. trigramin, kistrin/rhodostomin, barbourin, with about 70 amino acids and six disulfide bonds), and the long-sized disintegrin (e.g. bitistatin, with 84 amino acids linked by seven disulfide bonds) (Fig. 17.1). In addition, dimeric disintegrins such as contortrostatin, EC3 and EMF-10 were found as another class (Calvete et al., 2003). NMR studies of several disintegrins revealed that the active tripeptide RGD is located at the apex of a mobile loop protruding 14–17 Å from the protein core (Adler et al., 1991; Beer et al., 1992). These RGD-containing disintegrins showed different binding affinity and specificity towards integrins (i.e. $\alpha IIb\beta 3$, $\alpha v\beta 3$ and $\alpha 5\beta 1$). KGD-containing barbourin inhibits the $\alpha IIb\beta 3$ integrin with high degree of selectivity (Scarborough et al., 1991). Dimeric disintegrins show high level of sequence diversity in their integrin binding motif. EC3, a heterodimeric disintegrin from *Echis carinatus* venom, is a selective antagonist of binding of $\alpha 4\beta 1$ and $\alpha 4\beta 7$ integrin toward immobilized VCAM-1 and MAdCAM-1, respectively (Marcinkiewicz et al., 1999). The inhibitory activity of EC3 is associated with MLD sequence (against $\alpha 4$ integrin), and VGD associated with $\alpha 5\beta 1$ binding.

A disintegrin-like domain is also found in a number of hemorrhagic toxins isolated from viper venom. HR1B, a hemorrhagic protein, consists of an N-terminal metalloproteinase domain, a central region homologous with disintegrins and a C-terminal cysteine-rich domain (Takeya et al., 1990). Later on, Hite et al. (1992) and Paine et al. (1992) identified the similar disintegrin-like domain in atrolysin and jarahagin, respectively. However, the conserved sequence of RGD in disintegrin is replaced with ECD sequence. Au et al. showed that the disintegrin rhodostomin (i.e. kistrin) may share a common precursor with a putative hemorrhagic protein, and suggested that disintegrin and hemorrhagic protein may share a common gene sequence (Au et al., 1991, 1993). Although disintegrins intravenously administered prolong bleeding time, they are nontoxic. However, the synergistic effect on bleeding would be expected if hemorrhagins are co-administered with disintegrin.

Disintegrin Interaction with Platelets

Disintegrins are potent inhibitors of platelet aggregation caused by ADP, collagen, thrombin and other agonists. They are active both in washed platelet suspension and platelet-rich plasma. Their potency is at least 1,000 times greater than RGDS. Their inhibitory effect results from the inhibition of fibrinogen binding to receptors associated with GPIIb/IIIa complex (Huang et al., 1987b, 1991a). In general, disintegrins have no effect on platelet shape change, ADP or 5-HT release, thromboxane B_2formation, and calcium mobilization triggered by aggregation agonist (i.e. thrombin) (Gan et al., 1988; Huang et al., 1987b, 1991a, b, c). Disintegrin has no effect on platelet cyclic AMP metabolism.

It is recognized that disintegrins bind to both resting and activated platelets. They do not bind to thrombasthenia platelets, a hereditary defect of GPIIb/IIIa. However, disintegrins can be subclassified to two groups: the first group of disintegrins binds with similar affinity to resting and activated platelets, (e.g. rhodostomin, triflavin), while the other group of disintegrins binds to ADP-activated platelets with a higher binding affinity by about 5- to 10-fold (e.g. halysin, echistatin, bitistatin) (Huang et al., 1991a; Liu et al., 1996). A conformational change of GPIIb/IIIa is observed by preincubating platelets with disintegrins evidenced by a strong binding of anti-L1BS (Ligand induced binding site) antibody (Frelinger et al., 1991). Huang et al. showed that disintegrins including rhodostomin, trigramin, halysin and arietin (bitistatin) compete for the same sites on platelets (Huang et al., 1991a, b, c). Sheu et al. (1992c) first reported cross-linking of ^{125}I-triflavin (flavoridin) to GPIIIa. The monoclonal antibody 7E3 also shares the same binding sites with disintegrins (e.g. trigramin, halysin, triflavin and arietin) (Huang et al., 1987b, 1991a, b, c).

Disintegrin Interaction with Cells Other Than Platelets

Since the disintegrins are RGD-peptides, many researchers attempted to investigate their interaction with RGD-dependent integrins other than αIIbβ3. Knudsen et al. (1988) showed that trigramin inhibited adhesion and spreading of human melanoma cells with both fibronectin and fibrinogen. Then disintegrins were found to inhibit adhesion of human umbilical vein endothelial cells (HUVECs) to vitronectin and fibrin through integrin αvβ3 (Savage et al., 1990). Similarly, disintegrins blocked the adhesion of mouse melanoma cells to fibronectin and vitronectin (Rucinski et al., 1990). Triflavin has been demonstrated to inhibit lung colonization of B16F10 melanoma cells in experimental metastasis model (Sheu et al., 1992b). Furthermore, the adhesion between tumor cells (e.g. hepatoma, cervical carcinoma, and colon carcinoma) and extracellular matrices was blocked by disintegrins (Chiang et al., 1994a, 1996; Sheu et al., 1994a, b, 1996). Scarborough et al. (1993b) also found that echistatin inhibited fibronectin binding to immobilized α5β1, and all disintegrins tested except barbourin blocked vitronectin binding to immobolized αvβ3. Further investigations indicated that the amino acid sequence immediately adjacent to the RGD site can create an extended locus, may be involved in determining the binding affinity and selectivity toward different integrins. The detailed interaction between integrin-ligand complex has been derived from solution X-ray scattering and site-directed mutagenesis (Mould et al., 2003).

The relative selectivity of disintegrins versus various integrins may be relevant in some vitro experiments. Platelets may enhance adhesion of tumor cells to disrupted endothelium (Menter et al., 1987). Disintegrins have been shown to inhibit platelet aggregation induced by B16F10 melanoma cells, cervical carcinoma, hepatoma J5, human prostate (PC-3), breast, and colon (SW-480) carcinoma (Chiang et al., 1994b; Sheu et al., 1992a, b, 1994a, b; Swaim et al., 1996). These tumor carcinoma cells triggered platelet aggregation through different pathways, i.e. ADP release (e.g. B16F10) and tissue factor expression (e.g. MCF-7, PC-3, SW-480, Hela). Triflavin also dose-dependently inhibited B16F10 melanoma cell-induced lung colonization in C57BL/6 mice as tumor cells were intravenously administered. Beviglia et al. (1995) also showed both albolabrin and eristostatin inhibited cell metastasis of B16F10 in mice. It is thought that inhibitory effect of disintegrins on tumor cell adhesion to ECM and tumor cell-induced platelet aggregation may be partially responsible for their in vivo anti-metastasis activity in experimental metastasis model.

Echistatin has been used to isolate pure culture of mammalian osteoclasts (Wesolowski et al., 1995). It appears to be a potent inhibitor of bone resorption in vitro and in vivo. Sato et al. (1990) and Fisher et al. (1993) showed that echistatin efficiently inhibited the resorption of bone by osteoclasts through the blockade of the interaction of integrin $\alpha v\beta 3$ expressed on osteoclasts with bone extracellular matrix. Rhodostomin inhibited the migration and invasion of breast (MDA-MB-231) and prostate (PC-3) carcinoma cells, and specifically inhibited the binding of 7E3, a mAb recognizing $\alpha v\beta 3$, to tumor cells. As breast carcinoma cells were locally injected into tibia in nude mice, histological examination revealed that most of the cancellous bone had been replaced by cancer cells after 28 days' inoculation. Co-administration of trigramin with cancer cells markedly inhibited tumor growth and bone destruction. Therefore disintegrins may be developed as alternate therapy for bone metastasis of cancer cells (Yang et al., 2005).

$\alpha v\beta 3$ integrin expressed on vascular smooth muscle plays a role in mediating cell migration triggered by PDGF or insulin growth factor, and kistrin, echistatin modulated this process (Tsai et al., 1995). Triflavin was reported to inhibit aggregating platelets-induced vasoconstriction in de-endothelialized rat aorta (Sheu et al., 1997).

In addition, triflavin also blocked neuronal sprouting and induction of hyperalgesia induced by nerve injury (Fu et al., 2004). Rhodostomin but not rhodostomin mutant (RGE-rhodostomin) caused the detachment of primary cultured preadipocyte. Rhodostomin inhibited focal adhesion of preadipocyte and cell viability. In addition, rhodostomin also affected the maturation process of preadipocyte into adipocyte in the presence of insulin. Thus disintegrin inhibits processes of adipogenesis and may be developed to treat obesity (Lin et al., 2005).

Disintegrins on Neutrophil and Phagocytes

Rhodostomin blocked the adhesion of activated neutrophil to fibrinogen and attenuated superoxide production, relevant to anti inflammation (Tseng et al., 2004). We also found that rhodostomin decreased the production of tumor necrosis factor-α (TNFα), interleukin-6 and improved cardiovascular dysfunction and thrombocytopenia in vivo in LPS-treated mice (Hsu and Huang, 2009). Rhodostomin inhibited cytokine release and mitogen-activated protein kinase (MAPK) activation of THP-1 induced by LPS. Flow cytometric analysis revealed that rhodostomin bound to LPS-activated THP-1 (monocytic cell) and specifically blocked anti-αvβ3 binding to THP-1, but not anti-β1 or anti-β2 mAb binding. Moreover, rhodostomin blocked the enhanced expression and the procoagulant activity of tissue factor of THP-1 cells stimulated by LPS. These data suggest that disintegrin rhodostomin may interact with monocytes and interfere with the activation and function of monocytes triggered by LPS. Thus rhodostomin may improve endotoxemia syndrome through interaction with αvβ3 integrin of monocytes and inhibition of platelet aggregation via interaction with αIIbβ3 integrin of platelets (Hsu and Huang, 2009).

Disintegrins and Angiogenesis

Integrin αvβ3 expressed on endothelial cells, smooth muscle cells, fibroblasts and transformed cells, modulates cell migration, proliferation, and has a great impact on angiogenesis, restenosis, tumor cell migration and atherosclerosis (Jin and Varner, 2004). The αvβ3 ligands include fibrin(ogen), fibronectin, vWF, osteopontin, and vitronectin (Suehiro et al., 1996). Angiogenesis plays an important role in normal physiological processes, i.e., embryonic development, tissue repair, and luteal formation (Folkman and Shing, 1992). On the other hand, angiogenesis is intimately involved in some pathological conditions, i.e., promoting tumor growth and eliciting diabetic retinopathy and inflammatory disease (Folkman, 1995). The involvement of αvβ3 in angiogenesis was first demonstrated by Brooks et al (1994a, b). In the chick chorioallantoic membrane (CAM) model, LM-609, anti-αvβ3 mAb, inhibited neovascularization induced by implanting a bFGF-containing pellet or an αvβ3-negative melanoma on CAMs of 10-day-old embryos. The cyclic RGD derivative inhibited tumor cell-induced angiogenesis and hypoxia-induced neovascularization in murine retina (Hammes et al., 1996).

As mentioned above, disintegrins inhibit adhesion of tumor cells, and endothelial cells to ECM through blockade of integrin αvβ3 and α5β1. We previously reported that rhodostomin inhibited the morphological change of endothelial cells caused by ancrod (a thrombin-like enzyme protein) generated fibrin through αvβ3 blockade (Chang et al., 1995a, b). Disintegrins (e.g. accutin, rhodostomin, and triflavin) dose-dependently display inhibitory activity on endothelial cell adhesion to ECM, cell proliferation, matrigel-induced capillary tube formation, and neovascularization of CAM model, mainly through the blockade of αvβ3 (Yeh et al., 1998, 2001).

In addition, rhodostomin was found to inhibit neovascularization induced by either B16F10 tumor cells or b-FGF, but not by VEGF, through an αvβ3-dependent mechanism (Yeh et al., 2001). Among them, accutin is the first disintegrin reported to induce apoptosis of HUVECs (Yeh et al., 1998). Usually, cell detachment was observed at the early stage after rhodostomin treatment. Rhodostomin caused a higher percentage of cells at G2/M phase, the cleavage of β-catenin, and poly (ADP-ribose) polymerase during apoptosis (Wu et al., 2003). The activation of caspase-3 was detected, and caspase inhibitor can reverse the above apoptotic events. FAK phosphorylation and actin cytoskeleton were affected upon rhodostomin treatment.

Regarding the anti-angiogenic activity of disintegrins, Markland et al. will discuss in detail in Chapter 19, dealing with therapeutic approaches towards the treatment of tumor angiogenesis, using a novel dimeric disintegrin, contortrostatin in studies of endothelial cells and models of breast, ovarian and prostate cancer (Swenson et al., 2007). Therefore, while current therapies have been promising in this field, disintegrins may become an attractive alterative to conventional antiangiogenic agents.

Translational Medicine Derived from Disintegrins

Over the last two decades, extensive research on RGD/KGD-containing disintegrins focused on their interaction with platelet integrin $\alpha IIb\beta 3$ (e.g. fibrinogen receptor) led to the successful development of novel antithrombotic drugs pattered on the steric structure of these molecules (Tcheng et al., 1995; Weller et al., 1994). Disintegrins were once considered to be highly potential antiplatelet drugs. However, thrombocytopenia was observed as a side effect with echistatin (Frank et al., 1992). In addition, possible antigenicity and a brief half-life (Cook et al., 1989) in circulation limit the development of intact disintegrins as therapeutic agents. However, the above considerations may be avoided if we can modify them with approaches such as PEGylation or conjugation with human serum albumin in minimizing the antigenecity and prolonging their half-life.

Design of Antithrombotic Drugs

Abnormal platelet hyperactivity contributes to formation of arterial thrombi in coronary atherosclerotic diseases and embolic stroke in brain (Coller, 1990). Thus reduction of platelet hyperactivity is an important therapeutic approach. The pivotal role of RGD motif in integrin-mediated cell adhesion with ECM and plasma proteins has inspired many investigators to design RGD-mimetics as antithrombotic agents. Although small linear peptides containing RGD sequence have been shown to inhibit fibrinogen binding to activated-platelets, they usually exhibit low affinity toward $\alpha IIb\beta 3$, with weak antithrombotic efficacy (Gartner and Bennett, 1985; Weller et al., 1994). In contrast, the naturally occurring disintegrins are about 1,000 times more potent than the linear RGD-mimetics in inhibiting platelet aggregation. Based on the specific steric structure of the RGD loop of disintegrins, a series of RGD/KGD-mimetics including cyclic derivatives have been developed by computer modeling. To date, two well-known commercially available products, Tirofiban (Aggrastat, Merck) and Eptifibatide (Integrilin, COR Therapeutics) were developed successfully in this field. In contrast to the first antiadhesive agent, the chimeric

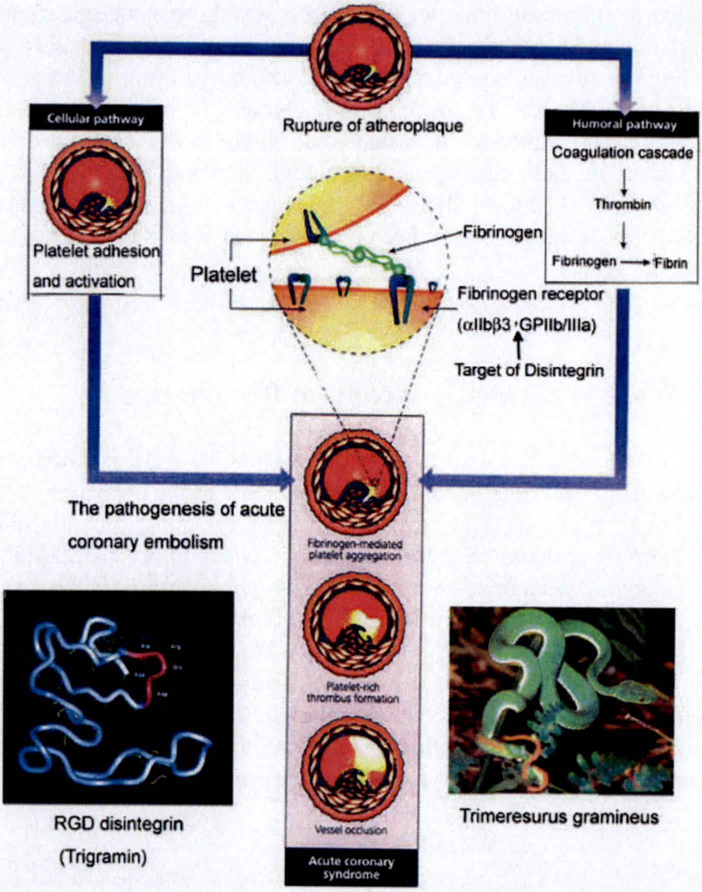

Fig. 17.2 The pathogenesis of acute coronary embolism and the target of disintegrin in blocking platelet aggregation. The acute coronary embolism can be caused by the rupture of atherosclerotic plaque, triggering both the activation of blood coagulation cascade and platelet activation, adhesion and subsequently leading to fibrinogen binding to activated platelets and platelet-rich thrombus formation and vessel occlusion. Disintegrin blocks the final platelet aggregation pathway, i.e. fibrinogen binding to activated GPIIb/IIIa (i.e. αIIbβ3 or fibrinogen receptor) by the preferential binding to αIIbβ3 with very high affinity (Kd, $10^{-7}\sim10^{-8}$M)

7E3 Fab (Abciximab, Repro) mAb raised against αIIbβ3 (The EPIC investigations, 1994), they are small molecule αIIbβ3 antagonists used for preventing restenosis of coronary vessels after PTCA (Fig. 17.2).

Disintegrin as Probe for Platelet αIIbβ3

αIIbβ3 is an abundantly expressed integrin with about 50,000 copies per platelet. It undergoes conformational changes when platelets are activated, and subsequently associates with plasma fibrinogen, leading to platelet aggregation. Unlike fibrinogen, disintegrins bind to resting and activated platelets (Liu et al., 1994, 1996). Activation of platelets by ADP markedly enhances the binding affinity of some disintegrin (e.g., trigramin, halysin, and arietin/bitistatin) to platelet $\alpha IIb\beta 3$ (Huang et al., 1991a, b, c). Therefore, these disintegrins may serve as probes either for measuring $\alpha IIb\beta 3$ content or for distinguishing the status of platelets in resting or activated form.

FITC-conjugated disintegrin can be used as a probe in flow cytometry assay for detecting surface $\alpha IIb\beta 3$ on platelets. Using this technology, we determined the platelet $\alpha IIb\beta 3$ levels of three Glanzmann's thrombasthenia patients and found that less than 5% of normal platelets were detected. Based on our previous observation that the binding sites of disintegrin appear to overlap with those of 7E3 and 7E3Fab, we showed a good correlation between % inhibition on platelet aggregation and blockade of FITC-disintegrin binding to platelets by 7E3 (Liu et al., 1996). This method provides a feasible monitoring method using 5 μl of blood sample coupling with flow cytometry. The methology allow us to find an optimal dosage of Abciximab in order to avoid the overdose of Abciximab in clinic practice.

Perspectives

The discovery of the naturally occurring disintegrins has inspired many researches in studying the molecular interaction of RGD/KGD-containing disintegrin or matrix-derived proteins with integrin $\alpha IIb\beta 3$, $\alpha v\beta 3$, and $\alpha 5\beta 1$, leading to the fruitful discovery of the potential therapeutic agents in field of arterial thrombosis, angiogenesis, tumor metastasis, inflammation, and other integrin-related diseases. The cloning and expression of disintegrins reveal the mysterious evolutionary connection among the common gene of disintegrin, snake venom metalloproteinases, and a disintegrin and metalloproteinase (ADAMs). With the aids of the advanced molecular biology techniques, and importantly the elucidation of the physiological and pathological roles of integrins (Takeda, 2009), we can harvest these disintegrins and their mutants targeting the specific integrin such as $\alpha v\beta 3$ or $\alpha 5\beta 1$ or dual targets ($\alpha v\beta 3$ plus $\alpha 5\beta 1$) for the further evaluation of their potential efficacy on combating angiogenesis and tumor metastasis in animal models. On the other hand, the elucidation of the detailed structural biology of integrins and ligand-integrin complex aided by X-ray crystallography and bioinformatics will accelerate the discovery of therapeutic agents derived from disintegrins in the near future.

Acknowledgements I would like to thank my Ph.D. thesis advisor, Dr. C. Ouyang, who encouraged me to stick on the topic entitled "Platelet inhibitory components in *Trimeresurus gramineus* venom" from which the first disintegrin Trigramin was discovered. Next, I also thank my partners Drs. S. Niewiarowski and Dr. J.C. Holt, Thrombosis Research Center, Temple University School of Medicine, who helped me to elucidate the primary sequence of the first naturally-occurring RGD-containing disintegrin, its primary target of platelet $\alpha IIb\beta 3$, and its in vivo antithrombotic activity. Thanks to my graduate students and assistants for the devoted pursuit of the mysterious biology of snake venom. Finally, I appreciate the long-term financial support from National Science Council of Taiwan.

References

Adler, M., Lazarus, R.A., Dennis, M.S., Wagner, G.., 1991. Solution structure of kistrin, a potent platelet aggregation inhibitor and GPIIb/IIIa antagonist. Science 253, 445–448.

Au, L.C., Chou, J.S., Chang, K.J., Teh, G.W., Lin, S.B., 1993. Nucleotide sequence of a full-length cDNA encoding a common precursor of platelet aggregation inhibitor and hemorrhagic protein from *Calloselasma rhodostoma* venom. Biochim. Biophys. Acta 1173, 243–245.

Au, L.C., Huang, Y.B., Huang, T.F., The, G.W., Lin, H.H., Choo, K.B., 1991. A common precursor for a putative hemorrhagic protein and rhodostomin, a platelet aggregation inhibitor of the venom of *Calloselasma rhodostoma*: molecular cloning and sequence analysis. Biochem. Biophys. Res. Commun. 181, 585–593.

Beer, J.H., Springer, K.T., Coller, B.S., 1992. Immobilized Arg-Gly-Asp (RGD) peptides of varying lengths as structural probes of the platelet glycoprotein IIb/IIIa receptor. Blood 79, 117–128.

Brooks, P.C., Clark, R.A.F., Cheresh, D.A., 1994a. Requirement of vascular integrin alpha-v/beta-3 for angiogenesis. Science 264, 569–571.

Brooks, P.C., Montgomery, A.M., Rosenfeld, M., Reisfeld, R.A., Hu, T., Klier, G., 1994b. Integrin alpha-v/beta-3 antagonists promote tumor regression by inducing apoptosis of angiogenic blood vessels. Cell 79, 1157–1164.

Beviglia, L., Stewart, G.J., Niewiarowski, S., 1995. Effect of four disintegrins on the adhesive and metastatic properties of B16F10 melanoma cells in a murine model. Oncol. Res. 7, 7–20.

Calvete, J.J., Moreno-Murciano, M.P., Theakston, R.D., Kisiel, D.G., Marcinkiewicz, C., 2003. Snake venom disintegrins: novel dimeric disintegrins and structural diversification by disulphide bond engineering. Biochem. J. 372, 725–734.

Chang, M.C., Jeng, J.H., Cheong, T.C., Huang, T.F., 1995b. The morphologic change of endothelial cells by ancrod-generated fibrin is triggered by alpha v beta 3 integrin binding and the subsequent activation of a G-protein coupled phospholipase C. Biochim. Biophys. Acta 1269, 115–121.

Chang, M.C., Wang, B.R., Huang, T.F., 1995a. Characterization of endothelial cell differential attachment to fibrin and fibrinogen and its inhibition by Arg-Gly-Asp-containing peptides. Thromb. Haemost. 74, 764–769.

Chiang, H.S., Peng, H.C., Huang, T.F., 1994a. Characterization of integrin expression and regulation on SW-480 human colon adenocarcinoma cells and the effect of rhodostomin on basal and upregulated tumor cell adhesion. Biochim. Biophys. Acta 1224, 506–516.

Chiang, H.S., Swain, M.W., Huang, T.F., 1994b. Characterization of platelet aggregation induced by human colon adenocarcinoma cells and its inhibition by snake venom peptides, trigramin and rhodostomin. Br. J. Haematol. 87, 325–331.

Chiang, H.S., Yang, R.S., Huang, T.F., 1996. Thrombin enhances the adhesion and migration of human colon adenocarcinoma cells via increased beta 3-integrin expression on the tumour cell surface and their inhibition by the snake venom peptide, rhodostomin. Br. J. Cancer 73, 902–908.

Coller, B.S., 1990. Platelets and thrombolytic therapy. N. Engl. J. Med., 322, 33–42

Cook, J.J., Huang, T.F., Rucinski, B., Strzyzewski, M., Tuma, R.F., Williams, J.A., Niewiarowski, S., 1989. Inhibition of platelet hemostatic plug formation by trigramin, a novel RGD-peptide. Am. J. Physiol. 256, H1038–H1043.

Fisher, J.E., Caulfield, M.P., Sato, M., Quartuccio, H.A., Gould, R.J., Garsky, V.M., Rodan, G.A., Rosenblatt, M., 1993. Inhibition of osteoclastic bone resorption in vivo by echistatin, an "arginyl-glycyl-aspartyl" (RGD)-containing protein. Endocrinology 132, 1411–1413.

Folkman, J., 1995. Angiogenesis in cancer, vascular, rheumatoid and other disease. Nat. Med. 1, 27–31

Folkman, J., Shing, Y., 1992. Angiogenesis. J. Biol. Chem. 267, 10931–10934

Frank, J.D., Gould, R.J., Schaffer, L.W., Davidson, J.T., Gibson, R.E., Patrick, D.H., Vonderfecht, S.L., Cartwright, M.E., 1992. Immunocytochemical localization of platelets in baboon hepatic sinusoids using monoclonal mouse anti-human platelet glycoprotein IIIa following induction of thrombocytopenia. Histochemistry 97, 355–360.

Frelinger, A.L., Du, X.P., Plow, E.F., 1991. Monoclonal antibodies to ligand-occupied conformers of integrin alpha IIb beta 3 (glycoprotein IIb-IIIa) alter receptor affinity, specificity, and function. J. Biol. Chem. 266, 17106–17111.

Fu, W.M., Chang, T.K., Sun, W.Z., Ling, Q.D., Peng, H.C., Liou, H.C., Lu, D.Y., Huang, T.F., 2004. Inhibition of neuropathic pain by a potent disintegrin – triflavin. Neurosci. Lett. 368, 263–268.

Gan, Z.R., Gould, R.J., Jacobs, J.W., Friedman, P.A., Polokoff, M.A., 1988. Echistatin: a potent platelet aggregation inhibitor from the venom of the viper, *Echis carinatus*. J. Biol. Chem. 263, 19827–19832.

Gartner, T.K., Bennett, J.S., 1985. The tetrapeptide analogue of cell attachment site of fibrinogen inhibits platelet aggregation and fibrinogen binding to activated platelets. J. Biol. Chem. 269, 11891–11894.

George, J.N., Caen, J.P., Nurden, A.T., 1990. Glanzmann's thrombasthenia: the spectrum of clinical disease. Blood 75, 1383–1395.

Ginsberg, M.H., Loftus, J.C., Plow, E.F., 1988. Cytoadhsion integrins and platelets. Thromb. Haemost. 59, 1–6.

Gould, R.J., Polokoff, M.A., Friedman, P.A., Huang, T.F., Holt, J.C., Cook, J.J., Niewiarowski, S., 1990. Disintegrins: a family of integrin inhibitory proteins from viper venoms. Proc. Soc. Exp. Biol. Med. 195, 168–171.

Hammes, H.P., Brownlee, M., Jonczyk, A., Sutter, A., Preissner, K.T., 1996. Subcutaneous injection of a cyclic peptide antagonist of vitronectin receptor-type integrins inhibits retinal neovascularization. Nat. Med. 2, 529–533.

Hite, L.A., Shannon, J.D., Bjarnason, J.B., Fox, J.W., 1992. Sequence of a cDNA clone encoding the zinc metalloproteinase hemorrhagic toxin e from *Crotalus atrox*: evidence for signal, zymogen, and disintegrin-like structures. Biochemistry 31, 6203–6211.

Hsu, C.C., Huang, T.F., 2009. A disintegrin, rhodostomin, inhibits activity of LPS-treated monocytes via $\alpha v\beta 3$ integrin. J. Thromb. Haemost. 7, S2, Abstract PP-MO-879.

Huang, T.F., 1998. What have snakes taught us about integrins? Cell. Mol. Life Sci. 54, 527–540.

Huang, T.F., Holt, J.C., Kirby, E.P., Niewiarowski, S., 1989. Trigramin: primary structure and its inhibition of von Willebrand factor binding to glycoprotein IIb/IIIa complex on human platelets. Biochemistry 28, 661–666.

Huang, T.F., Holt, J.C., Lukasiewicz, H., Niewiarowski, S., 1987b. Trigramin. A low molecular weight peptide inhibiting fibrinogen interaction with platelet receptors expressed on glycoprotein IIb-IIIa complex. J. Biol. Chem. 262, 16157–16163.

Huang, T.F., Liu, C.Z., Ouyang, C., Teng, C.M., 1991a. Halysin, an antiplatelet Arg-Gly-Asp-containing snake venom peptides, as fibrinogen receptor antagonist. Biochem. Pharmacol. 42, 1209–1219.

Huang, T.F., Ouyang, C., 1984. Action mechanism of the potent platelet aggregation inhibitor from *Trimeresurus gramineus* snake venom. Thromb. Res. 33, 125–138.

Huang, T.F., Sheu, J.R., Teng, C.M., Chen, S.W., Liu, C.S., 1991b. Triflavin, an antiplatelet Arg-Gly-Asp-containing peptide, is a specific antagonist of platelet membrane glycoprotein IIb-IIIa complex. J. Biochem. (Tokyo) 109, 328–334.

Huang, T.F., Wang, W.J., Teng, C.M., Ouyang, C., 1991c. Mechanism of action of the antiplatelet peptide, arietin, from *Bitis arietans* venom. Biochim. Biophys. Acta. 1074, 144–150.

Huang, T.F., Wu, Y.J., Ouyang, C., 1987a. Characterization of a potent platelet aggregation inhibitor from *Agkistrodon rhodostoma* snake venom. Biochim. Biophys. Acta 925, 248–257.

Hynes, R.O., 1992. Integrins: versatility, modulation and signaling in cell adhesion. Cell 69, 11–25.

Jia, L.G., Wang, X.M., Shannon, J.D., Bjarnason, J.B., Fox, J.W., 1997. Function of disintegrin-like/cysteine-rich domains of atrolysin A. Inhibition of platelet aggregation by recombinant protein and peptide antagonists. J. Biol. Chem. 272, 13094–13102.

Jin, H., Varner, J., 2004. Integrins: roles in cancer development and as treatment targets. Br. J. Cancer 90, 561–565.

Kini, R.M., Evans, H.J., 1992. Structural domains in venom proteins: evidence that metalloproteinases and nonenzymatic platelet aggregation inhibitors (disintegrins) from snake venoms are derived by proteolysis from a common precursor. Toxicon 30, 265–293.

Knudsen, K.A., Tuszynski, G.P., Huang, T.F., Niewiarowski, S., 1988. Trigramin, an RGD-containing peptide from snake venom, inhibits cell-substratum adhesion of human melanoma cells. Exp. Cell Res. 179, 42–49.

Lin, Y.T., Tang, C.H., Chuang, W.J., Wang, S.M., Huang, T.F., Fu, W.M., 2005. Inhibition of adipogenesis by RGD-dependent disintegrin. Biochem. Pharmacol. 70, 1469–1478.

Liu, C.Z., Hur, B.T., Huang, T.F., 1996. Measurement of glycoprotein IIb/IIIa blockade by flow cytometry with fluoresein isothiocyanate-conjugated crotavirin, a member of disintegrins. Thromb. Haemostas. 76, 585–591.

Liu, C.Z., Wang, Y.W., Shen, M.C., Huang, T.F., 1994. Analysis of human platelet glycoprotein IIb/IIIa by fluorescein isothiocyanate-conjugated disintegrins with flow cytometry. Thromb. Haemostas. 72, 919–925.

Lu, Q., Clemetson, J.M., Clemetson, K.J., 2005. Snake venoms and hemostasis. J. Thromb. Haemost. 3, 1791–1799.

Marcinkiewicz, C., Calvete, J.J., Marcinkiewicz, M.M., Raida, M., Vijay-Kumar, S., Huang, Z., Lobb, R.R., Niewiarowski, S., 1999. EC3, a novel heterodimeric disintegrin from Echis carinatus venom, inhibits alpha4 and alpha5 integrins in an RGD-independent manner. J. Biol. Chem. 274, 12468–12473.

Menter, D.G., Steinert, B.W., Sloane, B.F., Gundlach, N., O'Gara, C.Y., Marnett, L.J., Diglio, C., Walz, D., Taylor, J.D., Honn, K.V., 1987. Role of platelet membrane in enhancement of tumor cell adhesion to endothelial cell extracellular matrix. Cancer Res. 47, 6751–6762.

Mould, A.P., Symonds, E.J., Buckley, P.A., Grossmann, J.G., McEwan, P.A., Barton, S.J., Askari, J.A., Craig, S.E., Bella, J., Humphries, M.J., 2003. Structure of an integrin-ligand complex deduced from solution x-ray scattering and site-directed mutagenesis. J. Biol. Chem. 278, 39993–39999.

Musial, J., Niewiarowski, S., Rucinski, B., Stewart, G.J., Cook, J.J., Williams, J.A., Edmunds, L.H., Jr., 1990. Inhibition of platelet adhesion to surfaces of extracorporeal circuits by disintegrins. RGD-containing peptides from viper venoms. Circulation 82, 261–273.

Nieuwenhuis, H.K., Akkerman, J.W., Houdijk, W.P., Sixma, J.J., 1985. Human blood platelets showing no response to collagen fail to express surface glycoprotein Ia. Nature 318, 470–472.

Ouyang, C., Huang, T.F., 1983. Potent platelet aggregation inhibitor from *Trimeresurus gramineus* snake venom. Biochim. Biophys. Acta 757, 332–341.

Ouyang, C., Teng, C.M., Huang, T.F., 1992. Characterization of snake venom components acting on blood coagulation and platelet function. Toxicon 30, 945–966.

Ouyang, C., Yeh, H.I., Huang, T.F., 1983. A potent platelet aggregation inhibitor purified from *Agkistrodon halys* (mamushi) snake venom. Toxicon 21, 797–804.

Paine, M.J., Desmond, H.P., Theakston, R.D., Crampton, J.M., 1992. Purification, cloning, and molecular characterization of a high molecular weight hemorrhagic metalloprotease, jararhagin, from *Bothrops jararaca* venom. Insights into the disintegrin gene family. J. Biol. Chem. 267, 22869–22876.

Rote, W.E., Mu, D.X., Roncinske, R.A., Frelinger, A.L., 3rd, Lucchesi, B.R., 1993. Prevention of experimental carotid artery thrombosis by applaggin. J. Pharmacol. Exp. Ther. 267, 809–814.

Rucinski, B., Niewiarowski, S., Holt, J.C., Soszka, T., Knudsen, K.A., 1990. Batroxostatin, an Arg-Gly-Asp containing peptide from *Bothrops atrox*, a potent inhibitor of platelet aggregation and cell interaction with fibronectin. Biochem. Biophys. Acta 1054, 257–262.

Sato, M., Sardana, M.K., Grasser, W.A., Garsky, V.M., Murray, J.M., Gould, R.J., 1990. Echistatin is a potent inhibitor of bone resorption in culture. J. Cell Biol. 111, 1713–1723.

Savage, B., Marzec, U.M., Chao, B.H., Harker, L.A., Maragnone, J.M., Ruggeri, Z.M., 1990. Binding of the snake venom-derived proteins applagin and echistatin to the arginine-glycine-aspartic acid recognition site(s) on platelet glycoprotein IIb/IIIa complex inhibits receptor function. J. Biol. Chem. 265, 11766–11772.

Scarborough, R.M., Naughton, M.A., Teng, W., Rose, J.W., Phillips, D.R., Nannizzi, L., Arfsten, A., Campbell, A.M., Charo, I.F., 1993a. Design of potent and specific integrin antagonists. Peptide antagonists with high specificity for glycoprotein IIb-IIIa. J. Biol. Chem. 268, 1066–1073.

Scarborough, R.M., Rose, J.W., Hsu, M.A., Phillips, D.R., Fried, V.A., Campbell, A.M. Nannizzi, L., Charo, I.F., 1991. Barbourin. A GPIIb-IIIa-specific integrin antagonists from the venom of *Sistrurus m. barbouri*. J. Biol. Chem. 266, 9359–9362.

Scarborough, R.M., Rose, J.W., Naughton, M.A., Philips, D.R., Nannizzi, L., Arfsten, A., Campbell, A.M., Charo, I.F., 1993b. Characterization of the integrin specificities of disintegrins isolated from American pit viper venoms. J. Biol. Chem. 268, 1058–1065.

Sheu, J.R., Lin, C.H., Chuang, J.L., Teng, C.M., Huang, T.F., 1992a. Triflavin, an Arg-Gly-Asp-containing snake venom peptide, inhibits aggregation of human platelets induced by human hepatoma cell line. Thromb. Res. 66, 679–691.

Sheu, J.R., Lin, C.H., Chung, J.L., Teng, C.M., Huang, T.F., 1992b. Triflavin, an Arg-Gly-Asp-containing antiplatelet peptide inhibits cell-substratum adhesion and melanoma cell-induced lung colonization. Jpn. J. Cancer Res. 83, 885–893.

Sheu, J.R., Lin, C.H., Huang, T.F., 1994a. Triflavin, an antiplatelet peptide inhibits tumor cell-extracellular matrix adhesion via an RGD-dependent mechanism. J. Lab. Clin. Med. 123, 256–263.

Sheu, J.R., Lin, C.H., Peng, H.C., Huang, T.F., 1994b. Triflavin, an Arg-Gly-Asp-containing peptide, inhibits human cervical carcinoma (HeLa) cell-substratum adhesion through an RGD-dependent mechanism. Peptides 15, 1391–1398.

Sheu, J.R., Lin, C.H., Peng, H.C., Huang, T.F., 1996. Triflavin, an Arg-Gly-Asp-containing peptide, inhibits the adhesion of tumor cells to matrix proteins via binding to multiple integrin receptors expressed on human hepatoma cells. Proc. Soc. Exp. Biol. Med. 213, 71–79.

Sheu, J.R., Teng, C.M., Huang, T.F., 1992c. Triflavin, an RGD-containing antiplatelet peptide, binds to GpIIIa of ADP-stimulated platelets. Biochem. Biophys. Res. Commun. 189, 1236–1342.

Sheu, J.R., Yen, M.H., Hung, W.C., Lee, Y.M., Su, C.H., Huang, T.F., 1997. Triflavin inhibits platelet-induced vasoconstriction in de-endothelialized aorta. Arterioscler. Thromb. Vasc. Biol. 17, 3461–3468.

Suehiro, K, Smith, J.W., Plow, E.F., 1996. The ligand recognition specificity of ß3 integrins. J. Biol. Chem. 271, 10365–10371

Swaim, M.W., Chiang, H.S., Huang, T.F., 1996. Characterisation of platelet aggregation induced by PC-3 human prostate adenocarcinoma cells and inhibited by venom peptides, trigramin and rhodostomin. Eur. J. Cancer 32A, 715–721.

Swenson, S., Ramu, S., Markland, F.S., 2007. Anti-angiogenesis and RGD-containing snake venom disintegrins. Curr. Pharm. Des. 13, 2860–3871.

Takeda, S., 2009. Three-dimensional domain architecture of the ADAM family proteinases. Semin. Cell Dev. Biol. 20, 146–152.

Takeya, H., Oda, K., Miyata, J., Omori-Satoh, T., Iwanaga, S., 1990. The complete amino acid sequence of high molecular mass haemorrhagic protein HR1B isolated from the venom *Trimeresurus flavoviridis*. J. Biol. Chem. 265, 16068–16073.

Tcheng, J.E., Harrington, R.A., Kottke-Marchant, K., Kleiman, N.S., Ellis, S.G., Kereiakes, D.J., Mick, M.J., Navetta, F.I., Smith, J.E., Worley, S.J., Miller, J.A., Joseph, D.M., Sigmon, K.N., Kitt, M.M., du Mée, C.P., Califf, R.M., Topol, E.J., 1995. Multicenter, randomized, double-blind, placebo-controlled trial of the platelet integrin glycoprotein IIb/IIIa blocker Integrelin in elective coronary intervention. IMPACT Investigators. Circulation 91, 2151–2157.

Teng, C.M., Huang, T.F., 1991. Snake venom constituents that affect platelet function. Platelets 2, 77–87.

The EPIC investigations. 1994. Use of a monoclonal antibody directed against the platelet glycoprotein IIb/IIIa receptor in high-risk coronary angioplasty. N. Engl. J. Med. 330, 956–961.

Tsai, T.J., Sheu, J.R., Chen, Y.M., Yen, C.J., Chen, C.F., Huang, T.F., 1995. Disintegrin modulates rat glomerular mesangial cell behavior. Nephron 70, 83–90.

Tseng, Y.L., Peng, H.C., Huang, T.F., 2004. Rhodostomin, a disintegrin, inhibits adhesion of neutrophils to fibrinogen and attenuates superoxide production. J. Biomed. Sci. 11, 683–691.

Watson, S.P, Auger, J.M., McCarty, O.J., Pearce, A.C., 2005. GPVI and integrin alphaIIb beta3 signaling in platelets. J. Thromb. Haemost. 3, 1752–1762.

Weller, T., Alig, L., Muller, M.H., Kouns, W.C., Steiner, B., 1994. Fibrinogen receptor antagonists - a novel class of promising antithrombotics. Drugs Future 19, 461–476.

Wesolowski, G., Duong, L.T., Lakkakorpi, P.T., Nagy, R.M., Tezuka, K., Tanaka, H., Rodan, G.A., Rodan, S.B., 1995. Isolation and characterization of highly enriched, prefusion mouse osteoclastic cells. Exp. Cell Res. 219, 679–686.

Wu, W.B., Peng, H.C., Huang, T.F., 2003. Disintegrin causes proteolysis of beta-catenin and apoptosis of endothelial cells. Involvement of cell-cell and cell-ECM interactions in regulating cell viability. Exp. Cell Res. 286, 115–127.

Yang, R.S., Tang, C.H., Chuang, W.J., Huang, T.H., Peng, H.C., Huang, T.F., Fu, W.M., 2005. Inhibition of tumor formation by snake venom disintegrin. Toxicon 45, 661–669.

Yasuda, T., Gold, H.K., Leinbach, R.C., Yaoita, H., Fallon, J.T., Guerrero, L., Napier, M.A., Bunting, S., Collen, D., 1991. Kistrin, a polypeptide platelet GPIIb/IIIa receptor antagonist, enhances and sustains coronary arterial thrombolysis with recombinant tissue-type plasminogen activator in a canine preparation. Circulation 83, 1038–1047.

Yeh, C.H., Peng, H.C., Huang, T.F., 1998. Accutin, a new disintegrin, inhibits angiogenesis in vitro and in vivo by acting as integrin alphavbeta3 antagonist and inducing apoptosis. Blood 92, 3268–3276.

Yeh, C.H., Peng, H.C., Yang, R.S., Huang, T.F., 2001. Rhodostomin, a snake venom disintegrin, inhibits angiogenesis elicited by basic fibroblast growth factor and suppresses tumor growth by a selective $\alpha v\beta 3$ blockade of endothelial cells. Mol. Pharmacol. 59, 1333–1342.

Chapter 18
Brief History and Molecular Determinants of Snake Venom Disintegrin Evolution

Juan J. Calvete

Abstract Disintegrins represent a family of polypeptides released in the venoms of *Viperidae* and *Crotalidae* snakes (vipers and rattlesnakes) by the proteolytic processing of PII Zn^{2+}-metalloproteinases or synthesized from short-coding mRNAs. Disintegrins selectively block the function of β_1 and β_3 integrin receptors. This review summarizes our current view and hypotheses on the emergence and on the structural and functional diversification of disintegrins.

Divergence of Snake Venom Zn^{2+}-Metalloproteinases in Viperidae Led to the Emergence of Disintegrins

Crotalid and viperid venoms contain a number of hemorrhagic proteins. Hemorrhage is primarily the result of the synergistic action of Zn^{2+}-dependent metalloproteinases which degrade the extracellular matrix surrounding blood vessels, and proteins that interfere with hemostasis (reviewed in De Lima et al., 2009; Fox and Serrano, 2005; Mackessy, 2009). Snake venom hemorrhagic metalloproteinases (SVMP) cluster with mammalian matrix-degrading metalloproteinases and proteins of the ADAM (*A Disintegrin And Metalloproteinase*) or reprolysin subfamily of Zn^{2+}-metalloproteinases (PFAM family PF01421; http://pfam.sanger.ac.uk/family) in a monophyletic evolutionary tree (Moura da Silva et al., 1996) (Fig. 18.1a). The monophyletic distribution of mammalian and snake venom proteins indicate that SVMPs have evolved relatively late from a common ancestor gene both by speciation (after mammals and reptiles diverged) and by gene duplication, followed by divergence of the copies through positive Darwinian selection (Glassey and Civetta,

J.J. Calvete (✉)
Laboratorio de Proteinómica Estructural, Instituto de Biomedicina de Valencia, CSIC, Jaime Roig 11, 46010 Valencia, Spain
e-mail: jcalvete@ibv.csic.es

R.M. Kini et al. (eds.), *Toxins and Hemostasis*, DOI 10.1007/978-90-481-9295-3_18,

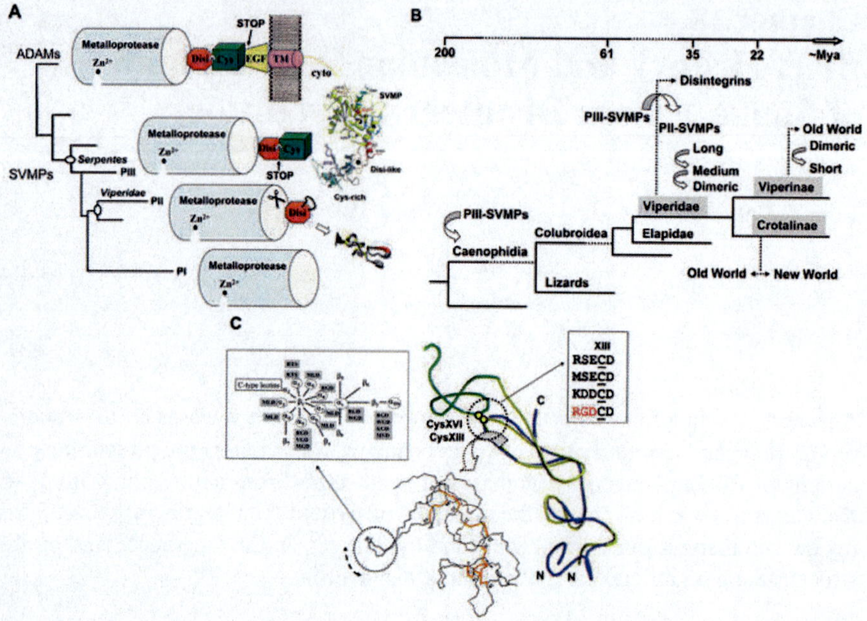

Fig. 18.1 Emergence of disintegrins from a SVMP precursor. (**a**) Cartoon of the proposed emergence of PIII-SVMPs by duplication of a gene coding for a multidomain ADAM integral membrane protein ancestor. A key event in this transition was the deletion from the duplicated ancestor of the regions coding for the domains C-terminal of the disintegrin-like domain, rendering a soluble protein. It is hypothesized that the duplicated gene was recruited by, and selectively expressed in, the venom gland after the divergence of lizards and Caenophidia snakes and represent, therefore, Serpentes-specific toxins. (Panel **b**) displays a proposed evolutionary timing for the emergence and structural diversification of PII-disintegrins and the biogeographic distribution of the disintegrin subfamilies. The PIII-SVMP model depicted in Panel **a** corresponds to the 2.5 Å resolution crystal structure of one subunit of the homodimeric PIII-metalloproteinase vascular apoptosis-inducing protein-1 (VAP1) from *Crotalus atrox* venom reported by Takeda and colleagues (2006) (see also Takeda, 2009). The figure also schematized our current view of the emergence of PII-SVMPs by deletion of the C-terminal cysteine-rich domain and subsequent removal of the structural constraint imposed by the CysXIII-CysXVI linkage in the disintegrin-like domain of PII-SVMP precursors (Panel **c**), enabling thereby the release of the disintegrin domain into the venom (scissors) and the emergence and evolution of the integrin inhibitory motifs at the tip of a mobile loop (highlighted here in a model of the long disintegrin bitistatin (Calvete et al., 1997) by a *discontinuous circle*). The *arrow points* to a diagram of the integrin family and the different disintegrin tripeptide motifs that block specific integrin-ligand interactions. Integrin heterodimers antagonized by snake venom disintegrins are encircled. $\alpha_1\beta_1$ is a receptor for collagen IV (CollIV); $\alpha_2\beta_1$ binds collagen I; $\alpha_4\beta_1$ interacts with fibronectin and VCAM-1; $\alpha_4\beta_7$ bind the same ligands as $\alpha_4\beta_1$ and in addition is a receptor for MdCAM; $\alpha_5\beta_1$ represents the major fibronectin (FN) receptor; integrins $\alpha_3\beta_1$, $\alpha_6\beta_1$, and $\alpha_7\beta_1$ represent major laminin (LM) receptors; $\alpha_8\beta_1$ and $\alpha_9\beta_1$ bind tenascin (TN); $\alpha_v\beta_1$ and $\alpha_v\beta_3$ are major vitronectin (VN) receptors; and $\alpha_{IIb}\beta_3$ is the platelet fibrinogen (FB) receptor involved in platelet aggregation

2004). Studies on the evolution of the venom system in lizards and snakes indicate that SVMPs have been recruited after the divergence of lizards and Caenophidia (some 200 Mya) (Vidal and Hedges, 2002) (Fig. 18.1b) and represent, therefore, Serpentes-specific toxins (Fry et al., 2006). SVMPs have been classified according to their domain structure (Fox and Serrano, 2009). Hemorrhagins of the PIII class, the closest homologues of cellular ADAMs, are large toxins (60–100 kDa) with the most potent activity, and comprise multidomain enzymes built up by an N-terminal metalloproteinase domain and C-terminal disintegrin-like and cysteine-rich domains (Fig. 18.1a). Comparison of full-length cDNA sequences of ADAMs and PIII SVMPs show that the genes coding for the latter molecules possess 3′untranslated regions, which include STOP codons after the cysteine-rich domain. Thus, PIII SVMPs are not simply derived by proteolysis of ADAM molecules but have rather evolved from a common ancestor after having lost the genetic information coding for protein regions downstream of the cysteine-rich domain (EGF-like, transmembrane and cytoplasmic domains) (Fig. 18.1a). The presence of PIII-SVMPs in the venoms of Viperidae, Elapidae, Colubridae and Atractaspididae supports the view that an early recruitment event of an ADAM gene predated the radiation of the advanced (Colubroidea) snakes (Fry et al., 2009) (Fig. 18.1b). On the other hand, metalloproteinases of class PII (30–60 kDa proteins containing a disintegrin domain at the carboxyl terminus of the metalloproteinase domain) occur only in viperid venoms and may thus represent a derivation from ancestral P-III-SVMP genes subsequently to the split of Viperidae (Fig. 18.1a and b), an event which has been dated around the Cretaceous-Tertiary boundary, approximately 61 Mya (Wüster et al., 2008).

The emergence of PII-SVMP precursors involved the deletion of the C-terminal cysteine-rich domain as a result of a mutation causing the appearance of a STOP codon (Calvete et al., 2003; Juárez et al., 2006a). The disintegrin-like domains of PII-SVMP precursors, i.e. BA-5A, amplified from *B. arietans* venom gland (Juárez et al., 2006a), which contain the 16 cysteine residues that are conserved in all known disintegrin-like regions of PIII-SVMPs but lack the cysteine-rich domain. However molecules containing this domain structure have never been found in snake venoms, and may represent pseudogenes, a relic of evolution. On the other hand, disintegrin domains are released into the venom by limited proteolysis of PII-SVMPs (Jia et al., 1996; Kini and Evans, 1992) and comprise a widely distributed venom protein family among Viperinae and Crotalinae (Calvete et al., 2003, 2009; Juárez et al., 2008; McLane et al., 1998). PII disintegrins act autonomously specifically blocking the function of integrin receptors. With the exception of the $\alpha_2\beta_1$ integrin, which is targeted by a number of C-type lectin-like proteins (Ogawa et al., 2005), inhibitory motifs towards β_1 and β_3 integrins have evolved in different members of the disintegrin family (Sanz et al., 2006) (Fig. 18.1c). The integrin inhibitory activity of disintegrins critically depends on the appropriate pairing of cysteines, which determines the conformation of the inhibitory loop that harbors an active tripeptide. Removal of the structural constraint imposed by the CysXIII-CysXVI linkage in the disintegrin-like domain of PII-SVMP precursors has been postulated (Calvete et al., 2003) to represent the key event that paved the way for the emergence and subsequent evolution of the integrin inhibitory activity at the apex of a mobile loop protruding 14–17 Å from the protein core (Monleón et al., 2003, 2005) (Fig. 18.1c).

Functional Diversification of the Integrin Inhibitory Motif

Disintegrins have evolved a restricted panel of integrin blocking sequences, which adapted to the ligand-binding architecture of their target integrin receptors (Sanz et al., 2006). Phylogenetic inference and maximum likelihood-based codon substitution approaches have been used to analyze the evolution of the integrin-binding loop (Juárez et al., 2008). RGD was inferred to represent the ancestral integrin-recognition motif, which emerged from the subgroup of PIII-SVMPs bearing the ^{66}RDECD^{70} sequence after deletion of the PIII-lineage-specific Cys69 (CysXIII residue). Conversion of ^{66}RDE^{68} into ^{66}RGD^{68} can be accomplished with a minimum of two mutations: (GAT or GAC) for (GGT or GGC) and (GAA or GAG) for (GAT or GAC). The most parsimonious nucleotide substitution model required for the emergence of all known disintegrin's integrin inhibitory motifs from an ancestral RGD sequence involves a minimum of three mutations (Fig. 18.2). Among the integrin inhibitory motifs, RGD blocks integrins $\alpha_8\beta_1$, $\alpha_5\beta_1$, $\alpha_v\beta_1$, $\alpha_v\beta_3$, and $\alpha_{IIb}\beta_3$; MLD targets the $\alpha_4\beta_1$, $\alpha_4\beta_7$, $\alpha_3\beta_1$, $\alpha_6\beta_1$, $\alpha_7\beta_1$ and $\alpha_9\beta_1$ integrins; VGD and MGD

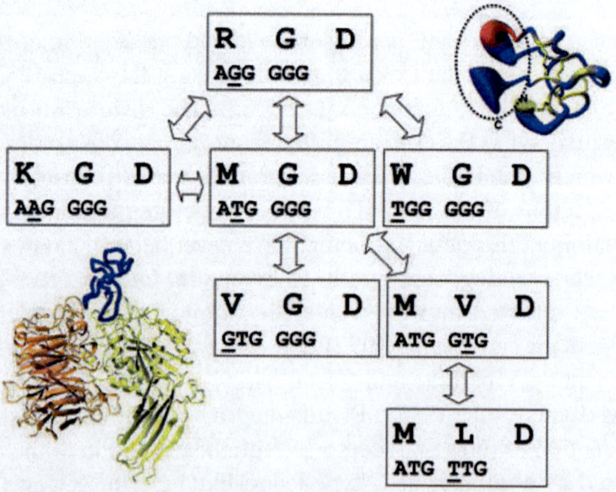

Fig. 18.2 Evolution of the integrin-binding site. Most parsimonious nucleotide substitution events required for the emergence of all known disintegrin's integrin recognition motifs from an ancestral RGD sequence. *Arrows* indicate a single mutational transition at the underlined site. *Upper right*, a "sausage" model of the NMR solution structure of obtustatin (PDB code 1MPZ; http://www.rcsb.org/pdb) highlighting the topology of the active tri-peptide and the C-terminal region, which form the conformational integrin-binding epitope. *Lower left*, ribbon representation of a putative model of the short disintegrin echistatin (*blue*) docked into the RGD-binding crevice of the globular head of integrin αvβ3 (PDB code 1L5G). The αV subunit β-propeller domain has been coloured *orange* and the β3 subunit A- and thigh domains are in *yellow*

impair the function of the $\alpha_5\beta_1$ integrin; KGD inhibits the $\alpha_{IIb}\beta_3$ integrin with a high degree of selectivity; WGD has been reported to be a potent inhibitor of the RGD-dependent integrins $\alpha_5\beta_1$, $\alpha_v\beta_3$, and $\alpha_{IIb}\beta_3$; the adhesive function of the latter integrin is also blocked by MVD; and KTS and RTS represent selective $\alpha_1\beta_1$ inhibitors (Calvete et al., 2009; Sanz et al., 2006). Evolution of disintegrins is also strongly influenced by positive Darwinian selection causing accelerated rate of substitution in a substantial proportion of surface-exposed disintegrin residues (Juárez et al., 2008). However, the adaptive advantage of the emergence of motifs targeting β_1 integrins, and the role of positively selected sites located within nonfunctional disintegrin regions are difficult to rationalize in the context of a predator-prey arms race. Perhaps the neofunctionalization potential of the disintegrin domain is a consequence of the capability of this protein scaffold to accommodate the multiple mutations generated through a process of accelerated evolution, and this feature may have underlayed the recruitment of the disintegrin fold into the venom proteome followed by its successful transformation into a toxin.

The crystal structure of the extracellular segment of integrin $\alpha_v\beta_3$ in complex with an RGD ligand (Xiong et al., 2002) showed that the peptide fits into a crevice between the αV propeller and the β3 A-domain. The Arg side-chain is held in place by interactions with αV carboxylates, the Gly residue makes several hydrophobic interactions with αV, and the Asp ligand interacts primarily with βA residues. Thus, the conserved aspartate residue might be responsible for the binding of disintegrins to integrin receptors which share a β subunit, while the two other residues of the integrin-binding motif (RG, MG, WG, ML, MV, VG) may dictate the primary integrin-recognition specificity. However, it is the composition (in particular the residues flanking the active tripeptide), the conformation, and the concerted dynamics of the integrin inhibitory loop and the C-terminal tail, which define the discontinuous primary functional epitope (Fig. 18.2) and finely tune the potency and integrin receptor selectivity of disintegrins (reviewed by Calvete, 2005; Calvete et al., 2005).

The integrin inhibitory activity of disintegrins depends on the appropriate pairing of cysteines, which determine the conformation of the inhibitory loop (Niewiarowski et al., 1994). As expected, invariant cysteine residues among all disintegrin subfamilies, along with a panel of buried residues, appear to be under strong selection due to their role in maintaining the active conformation of disintegrins (Juárez et al., 2008).

Structural Diversification of Disintegrins – Protein and Gene Structure Minimization Independent of the Evolution of the Integrin-Binding Loop

From a structural standpoint the disintegrin family can be conveniently divided into four groups according to their length (40–100 residues) and the number (4–8) of disulfide bonds (Calvete et al., 2003). The first group includes long disintegrins with ~84-residue polypeptide cross-linked by 7 disulfide bonds. The second group is formed by the medium-sized disintegrins which contain about 70 amino acids and 6 disulfide bonds. The third group is composed of homo- and heterodimers. Dimeric disintegrins contain subunits of about 67 residues with 10 cysteines involved in the formation of 4 intra-chain disulfide bonds and 2 interchain cystine linkages (Bilgrami et al., 2004, 2005; Calvete et al., 2000). Bilitoxin-1 represents another homodimeric disintegrin comprising disulfide-bonded polypeptides, each containing 15 cysteinyl residues (Nikai et al., 2000). The fourth group include the short disintegrins composed of 41–51 residues and 4 disulfide bonds. Figure 18.3 displays a dendrogram for the multiple sequence analysis of the consensus sequences of disintegrins. The most prominent characteristic of this tree is that the members of the different subfamilies are almost perfectly clustered separately with their homologues, suggesting an evolutionary relationship between the different disintegrin subfamilies. This phylogenetic history is compatible with the hypothesis that the PII-dimeric and PII-short represent the latest diverging lineages of disintegrins (Juárez et al., 2008) (see also Fig. 18.1b). Furthermore, whereas functional diversification among disintegrins appears to be due mainly to amino acid substitutions within the active regions, structural diversification was driven through both reduction of the polypeptide chain and a disulfide bond engineering mechanism involving the selective loss of pairs of cysteine residues engaged in the formation of disulfide bonds (Calvete et al., 2003). The cartoon shown in Fig. 18.4a summarizes the postulated evolutionary scenario of the disintegrin family, and Fig. 18.4b displays a scheme of relevant features along the diversification pathways of PII disintegrins. In particular, mutations in the codon of CysI along with a deletion of nine nucleotides coding for the tripeptide CQ(D/N) from a long disintegrin ancestor resulted in removal of the disulfide bond between cysteines I and IV leading to the emergence of medium-sized disintegrins. Further mutations involving the codons of the first two cysteine residues of medium-size disintegrins (CysII and CysIII) yielded polypeptides with 10 cysteines. Formation of two interchain disulfide linkages

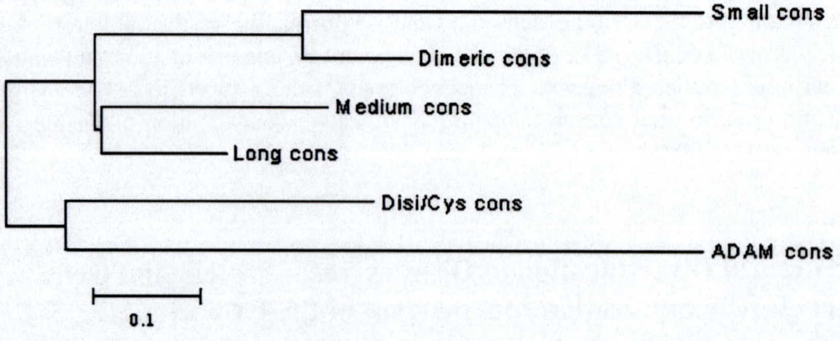

Fig. 18.3 Dendrogram for the multiple consensus sequence analysis of disintegrin domains. The *tree* represents the minimum evolutionary distance estimated through neighbour joining using maximum likelihood distances. Maximum parsimony produced a similar topology. The length of the horizontal scale bar represents 10% divergence. For rooting the tree, the disintegrin-like domains of human ADAM 7 and 10 were used as the outgroup

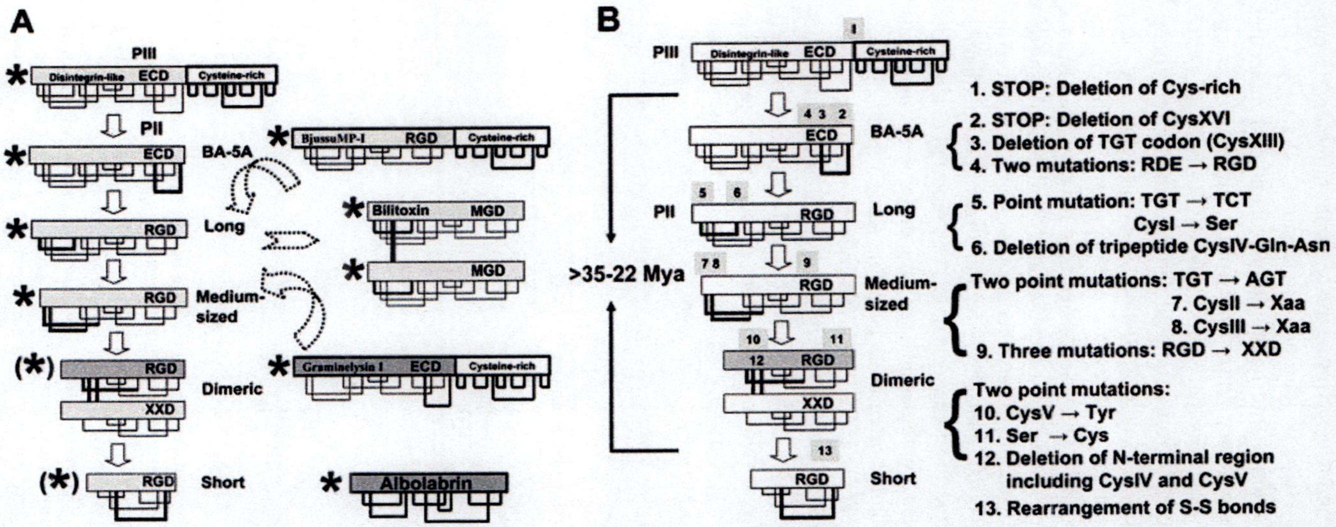

Fig. 18.4 Minimization of structure along the evolutionary diversification pathway of disintegrins. (**a**) Scheme of the domain organization, disulfide bond patterns, and proposed evolutionary pathway from the PIII Disintegrin/Cysteine-rich proteins to short disintegrins. Structural features (the cysteine-rich domain of PIII disintegrin-like molecules, and class-specific disulfides) lost along the canonical disintegrin diversification model depicted at the left are highlighted with *thick lines*. An *asterisk* indicates that the precursor protein is synthesized with an N-terminal metalloproteinase domain. (*), Some, but not all, precursors of dimeric disintegrin subunits and short disintegrins contain the metalloprotease domain; the majority of proteins from these groups are synthesized from short-coding mRNAs coding for subunits lacking the metalloproteinase domain (Juárez et al., 2006b; Okuda et al., 2002). Although most disintegrins may have followed the canonical diversification pathway schematized at the left, the evolutionary scenario might be more complicated: Graminelysin (Wu et al., 2001) represents a PIII metalloprotease from *Trimeresurus gramineus* venom lacking the CysI-CysIV disulphide and might thus represent an alternative intermediate in the evolution of a medium-sized disintegrin; BjussuMP-I is a PII-metalloprotease bearing a C-terminal cysteine-rich domain isolated from the venom of *Bothrops jararacussu* (Mazzi et al., 2007); and Bilitoxin-1, a long disintegrin from *Agkistrodon bilineatus*, possesses an extra cysteine residue between CysIII and CysIV involved in the formation of a disulfide-linked homodimer (Nikai et al., 2000). Furthermore, albolabrin is a medium-sized disintegrins displaying a non-canonical disulfide bond arreangement. (**b**) Timing and relevant features along the canonical diversification route of disintegrins

involving CysVI and CysVII, which in the medium-sized disintegrin precursors are disulfide-bonded to CysIII and CysII, respectively, with homologous cysteine residues from another 10-cysteine-containing disintegrin (CysVIA-CysVIIB and CysVIIA-CysVIB) generates homo- and heterodimeric disintegrin molecules (Bilgrami et al., 2004, 2005; Calvete et al., 2000). Noteworthy, subunits of dimeric disintegrins are synthesized from two distinct gene classes (Okuda et al., 2002) (Fig. 18.5a): β-subunits are synthesized from canonical genes coding for prepropeptide, metalloprotease, and a 10-Cys disintegrin domain, whereas genes encoding α-subunits consist of a signal peptide, a 30-residue pre-peptide domain and the disintegrin domain. Hence, the α-chain precursor genes may have evolved from a duplicated canonical medium-sized disintegrin precursor gene by deletion of a continuous ~1100 bp ORF that in the β-subunit genes encodes the C-terminal part of the pre-peptide, the metalloprotease domain, and the N-terminal region of the disintegrin domain. Homo- and heterodimeric disintegrins have been reported in a number of *Viperidae* snakes and restricted combinations of both dimeric arrangements (which in many cases share a subunit) often co-exist in the same venom (Calvete et al., 2002, 2003; Marcinkiewicz et al., 1999, 2000). The occurrence of precursors of the α and the β subunits of dimeric disintegrins differing in their domain structure suggests a mechanism for regulation of dimerization (Fig. 18.5b). Homodimers may result from disulfide bonding of two α-type subunit precursors, whereas the association of two β-type subunit precursors might be sterically hindered by the

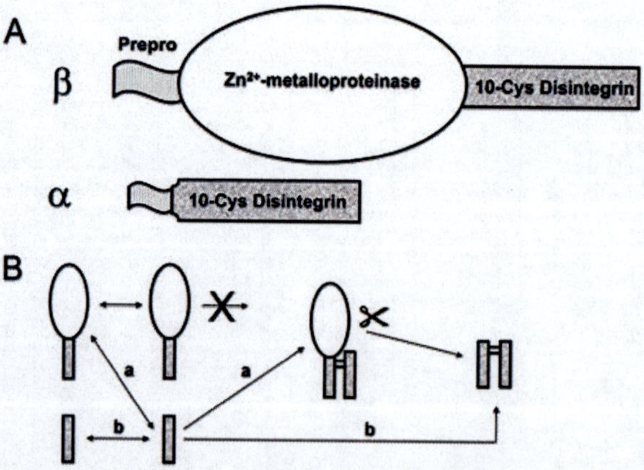

Fig. 18.5 Proposed mechanism for the generation of dimeric disintegrins. (**a**) α-type and the β-type subunits of dimeric disintegrins are synthesized from genes differing in their domain structure. β-subunits transcripts encode signal peptide (SP), prepropeptide, metalloproteinase, and 10-Cys disintegrin domains. α-subunit precursors consist of a SP, a 30-residue prepeptide domain and the 10-Cys disintegrin domain. (**b**) Homodimers may result from disulfide bonding of two α-type subunit precursors, whereas the association of two β-type subunit precursors might be sterically hindered by the large N-terminal multidomain structure. On the other hand, heterodimers may be generated by α-type/α-type and α-type/β-type subunit associations

large N-terminal multidomain structure. On the other hand, heterodimers may be generated by α-type/α-type and α-type/β-type subunit associations.

Disintegrins have been reported in the venoms of a number of genera from the subfamilies Crotalinae and Viperinae of Viperidae (Fig. 18.6). Dimeric disintegrins exhibit the largest sequence diversity in their integrin binding motifs, an indication of fast-evolving lineage with a large number of structural changes accumulating on them. Dimeric disintegrins are widely distributed in *Echis* and *Vipera* venoms, and probably also in the venoms of many other species of Old World *Crotalidae* and *Viperidae*, which are, in addition, rich sources of monomeric disintegrins. The co-existence in many snake species of disintegrins with conserved RGD-motif and disintegrins with variable non-RGD sequences support the hypothesis that, following gene duplication, one copy of the gene (i.e. that coding for an RGD disintegrin)

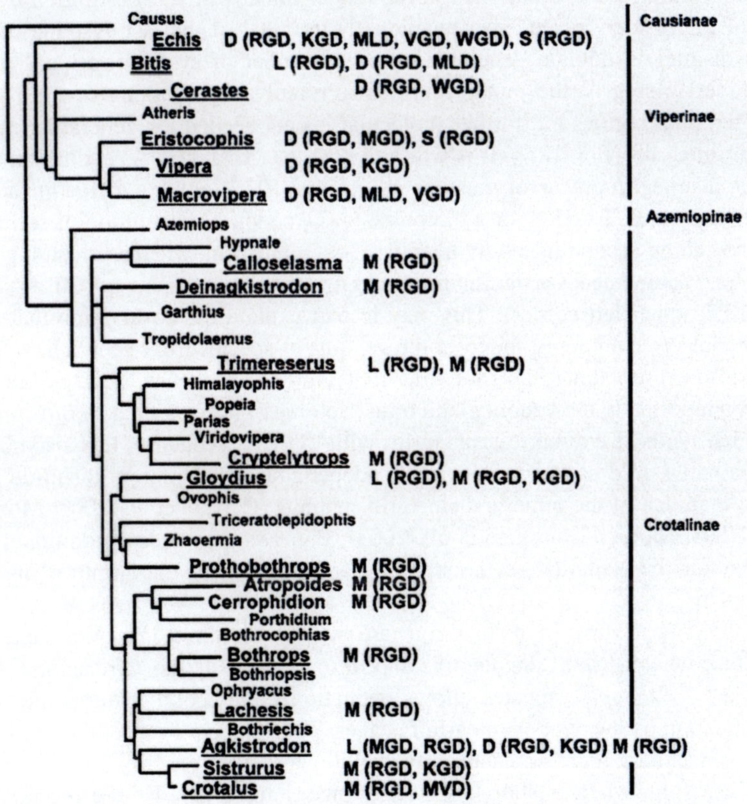

Fig. 18.6 Maximum-parsimony phylogram for genera of the Viperidae family with *Causus* serving as outgroup. Adapted from Lenk et al. (2001) and Castoe and Parkinson (2006). Genera in which PII disintegrins have been reported are highlighted in *bigger letters* and *italicized.* Disintegrin subfamilies are abbreviated as follows: *L*ong, *M*edium-sized, *D*imeric, *S*hort, and the integrin recognition motifs of the disintegrins are given in *parentheses*

will divergently evolve under pressure dictated by the ancestral function (blocking of platelet aggregation), while the duplicate gene(s) (non-RGD disintegrin), unencumbered by a functional role, is(are) free to search for new physiological roles such as inhibition of non-RGD-dependent integrin receptors. In addition, the occurrence in the same venom of short RGD-bearing and dimeric RGD- and/or non-RGD disintegrins, suggests the evolutionary emergence of short disintegrins from a RGD dimeric disintegrin precursor by a process requiring only two nucleotide mutation: CysVI/Tyr and Ser55/Cys (Juárez et al., 2006b) (Fig. 18.7). Crystallographic studies have shown that residue CysVI of each subunit of two *E. carinatus sochureki* dimeric disintegrins are engaged in the two interchain cystine linkages (CysVI$_{A}$-CysVII$_{B}$ and CysVII$_{A}$-CysVI$_{B}$) with cysteineVII from the other subunit of the dimers (Bilgrami et al., 2004, 2005) (Fig. 18.7). The single amino acid substitution CysVI→Tyr, involves a single nucleotide mutation TGT→ TAT or TGC→TAC and is the simplest molecular solution to hinder dimerization. Substitution by cysteine of the residue located two positions N-terminal of the last cysteine residue of the disintegrin domain generated an even number of cysteine residues allowing proper folding of the mutated protein. Proteolytic processing of the N- and the C-terminal regions are further molecular events needed for generating mature short-disintegrins (Fig. 18.7). It is worth to note that in all cDNAs coding for short dimeric disintegrin precursor messengers, residue 102 is either a serine, invariably coded for by a TCT codon, or a threonine (ACT codon). Substitution of serine for cysteine can be accomplished by a single C→G mutation, whereas changing threonine for cysteine needs a minimum of two mutations (i.e. ACT → TCT → TGT) (Fig. 18.7, upper left corner). This may in part explain the observation that short disintegrins are commonly found in the venoms of species from genera like *Echis*, which also express dimeric disintegrins containing the sequence "**S**XDC" but have not been reported in the venom gland transcriptome of species such as *Bitis gabonica* which synthesize dimeric disintegrins with "**T**PDC" sequence. It is also notable that the native fold of short disintegrins adopts a slightly different disulfide bond pattern than that of the dimeric disintegrin subunits (Calvete et al., 2003; Monleón et al., 2005; Moreno-Murciano et al., 2003) (Fig. 18.6), providing additional possibilities for the evolution of the structure and function of this group of integrin antagonists.

The fact that dimeric disintegrins are widely distributed in Viperinae and Crotalinae whereas short disintegrins appear to be restricted to African and Asian *Echis* and *Eristicophis* species allows deduction of temporal features along the diversification pathway of disintegrin lineages. Thus, the emergence of dimeric disintegrins may have represented an early evolutionary event predating the Viperinae-Crotalinae split, whereas short disintegrins may have evolved more recently, i.e. after the radiation of Viperinae. According to Lenk and coworkers (2001), viperines originated in the Oligocene (35–22 Mya) in Africa and successively underwent a first radiation leading to the five basal groups, *Bitis*, *Cerastes*, *Echis*, the *Atherini*, and the Eurasian viperines. Pitvipers (Crotalinae) represent the only viperids found in the Americas. They appear to have dispersed into the New World as a single lineage from Asia during the late Oligocene or the early Miocene, between 22 and

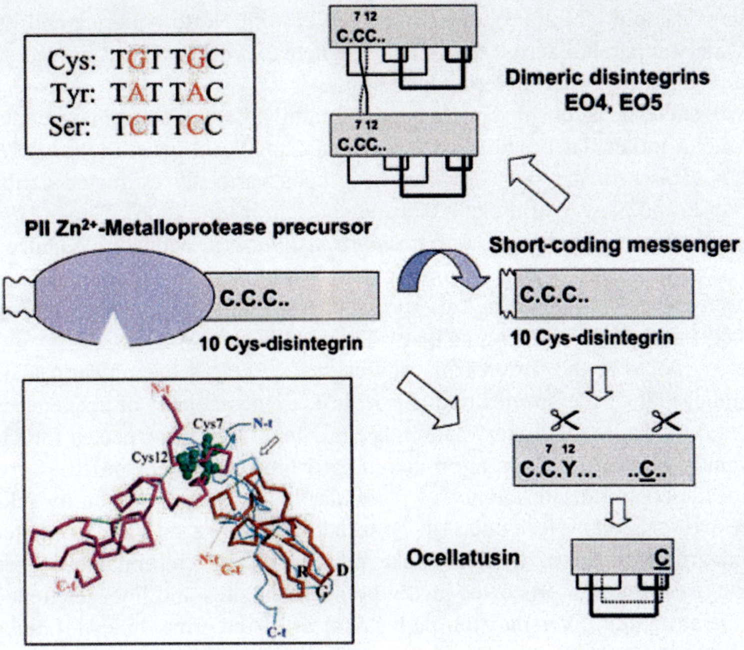

Fig. 18.7 Proposed mechanism for the emergence of the short disintegrin ocellatusin from a dimeric disintegrin subunit precursor (Juárez et al., 2006b). Key events in the emergence of ocellatusin (from *Echis ocellatus*) were (i) the substitution of the second N-terminal cysteine residue (CysVI in the dimeric disintegrin subunit precursor) by tyrosine, impairing thereby dimerization through homologous $CysVI_A$-$CysVII_B$ (Cys7A-Cys12B in the protein secuence numbering) and $CysVII_A$-$CysVI_B$ (Cys12A-Cys7B) linkages, (ii) the appearance of a novel cysteine residue between the 9th and the 10th cysteine of the precursor (**c**) enabling the short-disintegrin-specific disulfide bond depicted by broken line, and (iii) the proteolytic processing of the N-and the C-terminal regions of the expressed protein (scissors). The mechanism responsible for this proteolytic processing (autocatalytic?, specific venom proteinase?, amino- and carboxypeptidases?) remains elusive. However, the fact that disintegrins released into certain venoms (i.e. neotropical Crotalus species (Calvete et al., 2010)) exhibit ragged ends would suggest that the disintegrin-processing protease may have a relaxed cleavage specificity in those venoms. Fox and Serrano (2008) have also speculated about the contribution to venom complexity of the synthesis, folding, disulfide bonding, and post-translational processing of SVMPs. Notice that the two key amino acid changes needed to convert a dimeric disintegrin subunit into a short disintegrin can be accomplished by two nucleotide mutations. The *insert at the left down corner* shows the superposition of the structures of the short disintegrin ocellatusin (modelled based on echistatin structure (1RO3), Monleón et al., 2005) (*red*) and schistain (*magenta and cyan*) (Bilgrami et al., 2004). N-t, N-terminus; C-t, C-terminus. Intrachain disulfide bonds are shown in one subunit of schistatin as green broken lines, and the two interchain cystine linkages of the homodimer are depicted in the space-filling model in *green*. Disulfide bonds of ocellatusin are shown as *broken orange lines*, and the RGD motif of the short disintegrin is labeled. The *arrow* indicates the position of the N-terminal proteolytic cleavage which yields mature ocellatusin

24 Mya (Castoe et al., 2009), at a time when eastern North America and Eurasia were widely separated across the Atlantic, whereas northeastern Asia and Alaska remained connected via the Bering land bridge.

Phylogenetic relationships within the subfamily Crotalinae remain controversial. Among molecular based hypotheses, four Old World genera (*Protobothrops*, *Ovophis*, *Trimeresurus*, and *Gloydius*) have been variously estimated as the sister group to the New World clade (Castoe and Parkinson, 2006). Genus *Gloydius* groups the Asiatic members of *Agkistrodon sensu lato* (previously recognized as a trans-continental genus). To date, among New World pitvipers, dimeric disintegrins have been exclusively found in *Agkistrodon* (Calvete, 2010), pointing to this genus as a likely candidate to the lineage that colonized the New World.

The topology of the disintegrin "phylogenetic tree" is paraphyletic and does not parallel that of the species tree (Fig. 18.6). Clades consist of sequences from different taxa. This evolutionary pattern is consistent with that expected for a multigene family whose members are undergoing a birth-and-death process in which the appearance and disappearance of particular loci are being driven by selection (Hughes, 1994, 2000; Nei et al., 1997). In addition, since different combinations of disintegrins co-exist in different snake venoms, it seems reasonable to postulate that adaptive evolution processes involving different sites and lineages took place along the structural diversification pathway(s) of disintegrins. Indeed, branch-site tests of positive selection indicated that each disintegrin group underwent a process of independent adaptive evolution after divergence. Furthermore, in each group different sites appear to be involved in the process of adaptive evolution driven by positive selection (Juárez et al., 2008), indicating that positive selection in each lineage has followed different evolutionary routes in response to different selective conditions. Lineage-specific adaptations which appeared after structural diversifications splits may represent more recent events of positive selection which allowed the different members of the disintegrin subfamilies to refine their interactions with the corresponding integrin target.

Concluding Remarks and Perspectives

The evolutionary structural diversification pathway of disintegrins comprises the minimization of the protein structures, including the stepwise loss of cystine linkages. Strickingly, and still not understood, protein structure minimization ocurred along with minimization of the gene organization (i.e. loss of introns and coding regions) (Bazaa et al., 2007; Okuda et al., 2002). Functionally, disintegrins have evolved by accelerated evolution and adaptation of a conformational epitope harboring the integrin recognition motif and the C-terminal tail to the active site of the targeted integrin receptor. PII disintegrins block the function of integrin receptors with a high degree of selectivity.

On relatively rare occasions, toxins represent potential therapeutic agents. Selective blockade of integrins is a desirable goal for the therapy of a number of pathological conditions, including acute coronary ischaemia and thrombosis

($\alpha_{IIb}\beta_3$), tumor metastasis, osteoporosis, restenosis, rheumatoid arthritis ($\alpha_v\beta_3$), bacterial infection, vascular diseases ($\alpha_5\beta_1$), inflammation, autoimmune diseases ($\alpha_4\beta_1$, $\alpha_7\beta_1$, $\alpha_9\beta_1$), and tumor angiogenesis ($\alpha_1\beta_1$, $\alpha_v\beta_3$). The relevant integrin receptors involved in the above-listed pathologies are among the targets of many disintegrins (Fig. 18.1b). The potential clinical relevance of blocking specific integrin receptors using snake venom disintegrins has been reviewed (Marcinkiewicz, 2005). In addition, chemically modified synthetic RGD peptides and RGD-disintegrins with nucleids emitting radiation γ (^{99m}Tc, ^{125}I), β particles (^{64}Cu), positrons (^{18}F) or infrared radiation are being used as tools for visualizing integrin-dependent tumor angiogenesis in vivo (reviewed in Calvete et al., 2007). Understanding how toxins evolve in an accelerated fashion may hold the key to understanding speciation at a molecular level but also to learn how to use deadly toxins as therapeutic agents.

Acknowledgments Research on disintegrins has been continuously financed by grants from the Ministerio de Ciencia y Tecnología and Ministerio de Educación y Ciencia, Madrid, Spain.

References

Bazaa, A., Juárez, P., Marrakchi, N., Lasfer, Z.B., El Ayeb, M., Harrison, R.A., Calvete, J.J., Sanz, L., 2007. Loss of introns along the evolutionary diversification pathway of snake venom disintegrins evidenced by sequence analysis of genomic DNA from *Macrovipera lebetina transmediterranea* and *Echis ocellatus*. J. Mol. Evol. 64, 261–271.

Bilgrami, S., Tomar, S. Yadav, S., Kaur, P., Kumar, J., Jabeen, T., Sharma, S., Singh, T.P., 2004. Crystal structure of schistatin, a disintegrin homodimer from saw-scaled viper (*Echis carinatus*) at 2.5 Å resolution. J. Mol. Biol. 341, 829–837.

Bilgrami, S., Yadav, S. Sharma, S., Perbandt, M., Betzel, C., Singh, T.P., 2005. Crystal structure of the disintegrin heterodimer from saw-scaled viper (*Echis carinatus*) at 1.9 Å resolution. Biochemistry 44, 11058–11066.

Calvete, J.J., Schrader, M., Raida, M., McLane, M.A., Romero, A., Niewiarowski, S., 1997. The disulphide bond pattern of bitistatin, a disintegrin isolated from the venom of the viper *Bitis arietans*. FEBS Lett. 416, 197–202.

Calvete, J.J., Jürgens, M., Marcinkiewicz, C., Romero, A., Schrader, M., Niewiarowski, S., 2000. Disulfide bond pattern and molecular modelling of the dimeric disintegrin EMF-10, a potent and selective integrin $\alpha_5\beta_1$ antagonist from *Eristocophis macmahoni* venom. Biochem. J. 345, 573–581.

Calvete, J.J., Fox, J.W., Agelan, A., Niewiarowski, S., Marcinkiewicz, C., 2002. The presence of the WGD motif in CC8 heterodimeric disintegrin increases its inhibitory effect on $\alpha_{IIb}\beta 3$, $\alpha_v\beta_3$, and $\alpha_5\beta_1$ integrins. Biochemistry 41, 2014–2021.

Calvete, J.J., Moreno-Murciano, M.P., Theakston, R.D.G., Kisiel, D.G., Marcinkiewicz, C., 2003. Snake venom disintegrins: novel dimeric disintegrins and structural diversification by disulfide bond engineering. Biochem. J. 372, 725–734.

Calvete, J.J., 2005. Structure-function correlations of snake venom disintegrins. Curr. Pharm. Des. 11, 829–835.

Calvete, J.J., Marcinkiewicz, C., Monleón, D., Esteve, V., Celda, B., Juárez, P., Sanz, L., 2005. Snake venom disintegrins: evolution of structure and function. Toxicon 45, 1063–1074.

Calvete, J.J., Marcinkiewicz, C., Sanz, L., 2007. KTS- and RTS-disintegrins: anti-angiogenic viper venom peptides specifically targeting the $\alpha_1\beta_1$ integrin. Curr. Pharm. Des. 13, 2853–2859.

Calvete, J.J., Juárez, P., Sanz, L., 2009. Snake venomics and disintegrins. Portrait and evolution of a family of snake venom integrin antagonists, in: Mackessy, S.P. (Ed.), Handbook of Venoms and Toxins of Reptiles. CRC Press, Taylor & Francis, Boca Raton, 337–357.

Calvete, J.J., 2010. Snake venomics, antivenomics, and venom phenotyping: the *ménage à trois* of proteomic tools aimed at understanding the biodiversity of venoms, in: Kini, R.M., Clemetson, K.J., Markland, F.S., McLane, M.A., Morita, T. (Eds.), Toxins and Hemostasis: From Bench to Bedside. Springer, Dordrecht, The Netherlands.

Calvete, J.J., Sanz, L., Cid, P., De La Torre, P., Flores-Díaz, M., Dos Santos, M.C., Borges, A., Bremo, A., Angulo, Y., Lomonte, B., Alape-Girón, A., Gutiérrez, J.M., 2010. Snake venomics of the Central American rattlesnake *Crotalus simus* and the South American *Crotalus durissus* complex points to neurotoxicity as an adaptive paedomorphic trend along *Crotalus* dispersal in South America. J. Proteome Res. 9(1), 528–544.

Castoe, T.A., Parkinson, C.L., 2006. Bayesian mixed models and the phylogeny of pitvipers (Viperidae: Serpentes). Mol. Phylogenet. Evol. 39, 91–110.

Castoe, T.A., Daza, J.M., Smith, E.N., Sasa, M., Kuch, U., Campbell, J.A., Chippindale, P.T., Parkinson, C.L., 2009. Comparative phylogeography of pitvipers suggests a consensus of ancient Middle American highland biogeography. J. Biogeogr. 36, 88–103.

De Lima, M.E., Pimenta, A.M.C., Martin-Euclaire, M.F., Zingali, R.B., 2009. Animal Toxins: State of the Art. Perspectives in Health and Biotechnology. Editora UFMG, Belo Horizonte.

Fox, J.W., Serrano, S.M., 2005. Snake toxins and hemostasis. Toxicon 45, 951–1181.

Fox, J.W., Serrano, S.M.T., 2008. Insights into and speculations about snake venom metalloproteinase (SVMP) synthesis, folding and disulfide bond formation and their contribution to venom complexity. FEBS J. 275, 3016–3030.

Fox, J.W., Serrano, S.M., 2009. Timeline of key events in snake venom metalloproteinase research. J. Proteomics 72, 200–209.

Fry, B.G., Vidal, N., Norman, J.A., Vonk, F.J., Scheib, H., Ramjan, S.F., Kuruppu, S., Fung, K., Hedges, S.B., Richardson, M.K., Hodgson, W.C., Ignjatovic, V., Summerhayes, R., Kochva, E., 2006. Early evolution of the venom system in lizards and snakes. Nature 439, 584–588.

Fry, B.G., Vidal, N., van der Weerd, L., Kochva, E., Renjifo, C., 2009. Evolution and diversification of the Toxicofera reptile venom system. J. Proteomics 72, 127–136.

Glassey, B, Civetta, A., 2004. Positive selection at reproductive ADAM genes with potential intercellular binding activity. Mol. Biol. Evol. 21:851–859.

Hughes, A.L., 1994. The evolution of functionally novel proteins after gene duplication. Proc. Roy. Soc. London Scr. B. 256, 119–124.

Hughes, A.L., 2000. Adaptive Evolution of Genes and Genomes. Oxford University Press, Oxford.

Jia, L-G., Shimokawa, K-I., Bjarnason, J.B., Fox, J.W., 1996. Snake venom metalloproteinases: structure, function and relationship to the ADAMs family of proteins. Toxicon 34, 1269–1276.

Juárez, P., Wagstaff, S.C., Oliver, J., Sanz, L., Harrison, R.A., Calvete, J.J., 2006a. Molecular cloning of disintegrin-like transcript BA-5A from *Bitis arietans* venom gland cDNA library: a putative intermediate in the evolution of the long chain disintegrin bitistatin. J. Mol. Evol. 63, 142–152.

Juárez, P., Wagstaff, S.C., Sanz, L., Harrison, R.A., Calvete, J.J., 2006b. Molecular cloning of *Echis ocellatus* disintegrins reveals non-venom-secreted proteins and a pathway for the evolution of ocellatusin. J. Mol. Evol. 63, 183–193.

Juárez, P., Comas, I., González-Candelas, F., Calvete, J.J., 2008. Evolution of snake venom disintegrins by positive Darwinian selection. Mol. Biol. Evol. 25, 2391–2407.

Kini, R.M., Evans, H.J., 1992. Structural domains in venom proteins: evidence that metalloproteinases and nonenzymatic platelet aggregation inhibitors (disintegrins) from snake venoms are derived by proteolysis from a common precursor. Toxicon 30, 265–293.

Lenk, P., Kalyabina, S., Wink, M., Joger, U., 2001. Evolutionary relationships among the true vipers (Reptilia: Viperidae) inferred from mitochondrial dna sequences. Mol. Phylogenet. Evol. 19, 94–104.

Mackessy, S.P., 2009. Handbook of Venoms and Toxins of Reptiles. CRC Press, Taylor & Francis, Boca Ratón.

Mazzi, M.V., Magro, A.J., Amui, S.F., Oliveira, C.Z., Ticli, F.K., Stábeli, R.G., Fuly, A.L., Rosa, J.C., Braz, A.S.K., Fontes, M.R.M., Sampai, S.V., Soares, A.M., 2007. Molecular characterization and phylogenetic analysis of BjussuMP-I: a RGD-P-III class hemorrhagic metalloprotease from *Bothrops jararacussu* snake venom. J. Mol. Graph. Model. 26, 69–85.

McLane, M.A., Marcinkiewicz, C., Vijay-Kumar, S., Wierbzbicka-Patynowski, I., Niewiarowski, S., 1998. Viper venom disintegrins and related molecules. Proc. Soc. Exp. Biol. Med. 219, 109–119.

Marcinkiewicz, C., Calvete, J.J., Marcinkiewicz, M.M., Raida, M., Vijay-Kumar, S., Huang, Z., Lobb, R.R., Niewiarowski, S., 1999. EC3, a novel heterodimeric disintegrin from *Echis carinatus* venom, inhibits α4 and α5 integrins in an RGD-independent manner. J. Biol. Chem. 274, 12468–12473.

Marcinkiewicz, C., Taooka, Y., Yokosaki, Y., Calvete, J.J., Marcinkiewicz, M.M., Lobb, R.R., Niewiarowski, S., Sheppard, D., 2000. Inhibitory effects of MLDG-containing heterodimeric disintegrins reveal distinct structural requirements for interaction of the integrin $\alpha_9\beta_1$ with VCAM-1, tenascin-C, and osteopontin. J. Biol. Chem. 275, 31930–31937.

Marcinkiewicz, C., 2005. Functional characteristics of snake venom disintegrins: potential therapeutic implications. Curr. Pharm. Des. 11, 815–827.

Monleón, D., Moreno-Murciano, M.P., Kovacs, H., Marcinkiewicz, C., Calvete, J.J., Celda, B., 2003. Concerted motions of the integrin-binding loop and the C-terminal tail of the non-RGD disintegrin obtustatin. J. Biol. Chem. 278, 45570–45576.

Monleón, D., Esteve, V., Kovacs, H., Calvete, J.J., Celda, B., 2005. Conformation and concerted dynamics of the integrin-binding site and the C-terminal region of echistatin revealed by homonuclear NMR. Biochem. J. 387, 57–66.

Moreno-Murciano, M.P., Monleón, D., Marcinkiewicz, C., Calvete, J.J., Celda, B., 2003. NMR solution structure of the non-RGD disintegrin obtustatin. J. Mol. Biol. 329, 135–145.

Moura da Silva, A.M., Theakston, R.D.G., Crampton, J.M., 1996. Evolution of disintegrin cysteine-rich and mammalian matrix-degrading metalloproteinases: gene duplication and divergence of a common ancestor rather than convergent evolution. J. Mol. Evol. 43, 263–269.

Nei, M., Gu, X., Sitnikova, T., 1997. Evolution by the birth-and-death process in multigene families of the vertebrate immune system. Proc. Natl. Acad. Sci. U.S.A. 94, 7799–7806.

Niewiarowski, S., McLane, M.A., Kloczewiak, M., Stewart, G.J., 1994. Disintegrins and other naturally occurring antagonists of platelet fibrinogen receptors. Semin. Hematol. 31, 289–300.

Nikai, T., Taniguchi, K., Komori, Y., Masuda, K., Fox, J.W., Sugihara, S., 2000. Primary structure and functional characterization of bilitoxin-1, a novel dimeric P-II snake venom metalloproteinase from *Agkistrodon bilineatus* venom. Arch. Biochem. Biophys. 378, 6–15.

Ogawa, T., Chijiwa, T., Oda-Ueda, N., Ohno, M., 2005. Molecular diversity and accelerated evolution of C-type lectin-like proteins from snake venom. Toxicon 45, 1–14.

Okuda, D., Koike, H., Morita, T., 2002. A new gene structure of the disintegrin family: a subunit of dimeric disintegrin has a short coding region. Biochemistry 41, 14248–14254.

Sanz, L., Bazaa, A., Marrakchi, N., Pérez, A., Chenik, M., Bel Lasfer, Z., El Ayeb, M., Calvete, J.J., 2006. Molecular cloning of disintegrins from *Cerastes vipera* and *Macrovipera lebetina transmediterranea* venom gland cDNA libraries. Insight into the evolution of the snake venom's integrin inhibition system. Biochem. J. 395, 385–392.

Takeda, S., Igarashi, T., Mori, H., Araki, S., 2006. Crystal structures of VAP1 reveal ADAMs' MDC domain architecture and its unique C-shaped scaffold. EMBO J. 25, 2388–2396.

Takeda, S., 2009. Three-dimensional domain architecture of the ADAM family proteinases. Semin. Cell. Dev. Biol., 20, 146–152.

Vidal, N., Hedges, S.B., 2002. Higher-level relationships of snakes inferred from four nuclear and mitochondrial genes. C. R. Biol. 325, 977–985.

Wu, W.B., Chang, S.C., Liau, M.Y., Huang, T.F., 2001. Purification, molecular cloning and mechanism of action of graminelysin I, a snake-venom-derived metalloproteinase that induces apoptosis of human endothelial cells. Biochem. J. 357, 719–728.

Wüster, W., Peppin, L., Pook, C.E., Walker, D.E., 2008. A nesting of vipers: phylogeny and historical biogeography of the Viperidae (Squamata: Serpentes). Mol. Phylogenet. Evol. 49, 445–459.

Xiong, J-P., Stehle, T., Zhang, R., Joachimiak, A., Frech, M., Goodman, S.L., Arnaout, M.A., 2002. The crystal structure of the extracellular segment of integrin $\alpha v\beta 3$ in complex with an Arg-Gly-Asp ligand. Science 296, 151–155.

Chapter 19
Anti-Angiogenesis and Disintegrins

Stephen Swenson, Radu Minea, Samuel Zidovetzki, Corey Helchowski, Fritz Costa, and Francis S. Markland

Abstract Angiogenesis is a critical process in tumor and disease progression. A number of features are central to both tumor growth and development, and the recruitment and invasion of a growing vascular network supplying the tumor with nutrients and a mechanism of escape to allow meastatic growth. One class of molecules important to both tumor growth and blood vessel recruitment are a family of cell surface receptors identified as integrins. Integrins are α/β heterodimeric glycoproteins in which the different α subunits combine with distinct β subunits resulting in a range of specificities toward various extracellular matrix (ECM) proteins. The RGD sequence found in a number of ECM proteins is recognized by several classes of integrins, allowing for linkage of cytoskeletal proteins associated with the integrins to the ECM which leads to involvement in bi-directional signaling that displays profound effects on cellular functions. Among these integrin mediated interactions are the adhesion of both endothelial cells and cancer cells to ECM proteins, an interaction that is integral to metastasis, tumor growth and angiogensis. Antibodies targeted to integrins have been shown to retard tumor growth and subsequent tumor induced angiogenesis. One concern with this approach is that the antibody targets a single integrin, which may allow the tumor to utilize other non-targeted integrins to circumvent this type of blockage. A more broad spectrum agent is available that binds to and blocks the function of several different integrins at a time, this agents is identified as a disintegrin. Originally purified from the venom of *Viperidae* family of snakes, a disintegrins role in nature is presumably to block platelet aggregation following envenomation based on interaction of an integrin on the activated platelet surface with an RGD sequence in the disintegrin. It has been observed that integrins overexpressed on some tumor types and angiogenic vasculature have similar affinity for RGD motifs found in ECM proteins. Based on

S. Swenson (✉)
USC/Norris Comprehensive Cancer Center, University of Southern California, Keck School of Medicine, Cancer Research Laboratory #106, 1303 N. Mission Road, Los Angeles, CA 90033, USA
e-mail: sswenson@usc.edu

R.M. Kini et al. (eds.), *Toxins and Hemostasis*, DOI 10.1007/978-90-481-9295-3_19,

disintegrin structure we have developed a recombinant form of a snake venom disintegrin, which we call vicrostatin (VCN). VCN is a potent anti-angiogenic/anti-tumor agent in in vitro and in vivo laboratory studies. Further development of the recombinant venom derived disintegrin along with new technology looking at additional disintegrin-like proteins may offer a novel therapeutic approach in targeting tumor induced angiogenesis.

Introduction

Angiogenesis, the ingrowth of new blood vessels into a growing tumor, is critical to cancer progression and spread. A number of cancer types such as breast and prostate cancer affect up to 10% of the US population at some point in their lifetime (Jemal et al., 2009). In the description of the development of a novel form of anti-cancer therapy, prostate cancer will be used as the model. With the exception of skin cancer, prostate cancer (PC) is the most prevalent cancer in American men and the second leading cause of cancer death after lung cancer. The American Cancer Society estimates there will be about 192,280 men diagnosed with PC and 27,360 deaths from PC in the United States in 2009 (Jemal et al., 2009). As of 2004 the five year survival rate for men in whom PC has spread to distant sites at time of diagnosis was 32% (Jemal et al., 2009). In advanced PC, different treatment combinations in the past have failed because as prostate cancer progresses it becomes resistant to treatment (Di Lorenzo and De Placido, 2006; Gleave et al., 2005; Harris et al., 2009; Sharifi et al., 2008). It is critical, therefore, that new treatment options be made available for patients with metastatic PC. While cytotoxic therapeutics form the backbone of PC treatment, there are other approaches that may allow for even more effective treatment of PC. For example, the inhibition of blood vessel growth into the tumor, anti-angiogenic therapy, is one of the more promising strategies for enhanced long-term inhibition of PC progression (Aragon-Ching and Dahut, 2008; Choy and Rafii, 2001; Madan and Dahut, 2009; van Moorselaar and Voest, 2002). Abnormal expression of angiogenic factors, such as vascular endothelial growth factor (VEGF), and their receptors have been associated with PC (Aragon-Ching and Dahut, 2009; Fox et al., 2002). Both PC and vascular endothelial cells produce growth factors, including VEGF and basic fibroblast growth factor (bFGF), that promote neovascularization (Beekman and Hussain, 2006; Kwabi-Addo et al., 2004; Pallares et al., 2006; Polnaszek et al., 2003; Strohmeyer et al., 2004; Sun et al., 2004). Angiogenesis is a critical prerequisite for progression to advanced disease and plays a pivotal role for growth and metastasis of PC (Cox et al., 2005; Choy and Rafii, 2001; Madan and Dahut, 2009). Tumor vascularity has been shown to be a prognostic factor in PC, with highly vascular PC having a poor prognosis (Concato et al., 2009; Erbersdobler et al., 2009; Mucci et al., 2009; Weidner et al., 1993). Further, there is a specific correlation between blood vessel density in PC and the metastatic potential (Weidner et al., 1993). PC is identified as an angiogenesis-dependent cancer and an excellent target for anti-angiogenic therapy (Aragon-Ching and Dahut, 2008; Cox et al., 2005; Jimenez et al., 2006; Madan and Dahut, 2009). Anti-angiogenic therapy can produce prolonged tumor dormancy and is cytostatic, which will lead to reduced side effects commonly associated with chemotherapeutic agents (Boehm et al., 1997; Wu and Moses, 2000). However, anti-angiogenic agents when used as monotherapy are not as effective as in combination, suggesting that anti-angiogenic strategies in combination with cytotoxic agents, should be an effective therapeutic approach in PC.

Integrins: Targets for Anti-Angiogenic Therapies

Cell–cell and cell–matrix interactions are of critical importance to newly established cancer cell colonies. These interactions are extremely complex and involve surface interactions between tumor cells and surrounding tissues (Liotta and Kohn, 2001). The integrin class of molecules is particularly important to these interactions. All integrins are α/β heterodimeric glycoproteins (Cheresh, 1992). The 18 α subunits combine with 8 β subunits to create 24 unique αβ heterodimers resulting in a range of specificities toward various extracellular matrix (ECM) proteins (Hynes, 1992; Pignatelli et al., 1992). Several classes of integrins recognize the RGD sequence present in ECM proteins (Ruoslahti, 1991), allowing integrins to link cytoskeletal proteins with the ECM and to be involved in bi-directional signaling that alters cellular functions. Among these interactions are the adhesion of both endothelial cells and cancer cells to ECM proteins (Ruoslahti, 1991), interactions that are integral to tumor growth, metastasis and angiogenesis; integrins also mediate endothelial cell proliferation and migration (Beekman et al., 2006). Importance of the vitronectin receptors, integrins αvβ3 and αvβ5, in angiogenesis is well known. A monoclonal antibody to αvβ3, as well as a cyclic RGD-containing peptide (Brooks et al., 1994), perturbed angiogenesis and produced regression of a human cancer growing on a chick embryo chorioallantoic membrane (CAM). Administration of antagonists of αvβ3 cause apoptosis of vascular endothelial cells responsible for angiogenesis, following selective activation of p53 and increased expression of the p53-inducible cell cycle inhibitor p21WAF1/CIP1 in vivo (Strömblad et al., 1996). In vivo survival of vascular endothelial cells is dependent on attachment and spreading on the ECM. αvβ5 has also been found to play a role in angiogenesis. An anti-αvβ3 antibody blocked angiogenesis induced by bFGF, whereas an anti-αvβ5 antibody blocked VEGF induced angiogenesis (Friedlander et al., 1995). This suggests that down-stream signal transduction pathways of the two growth factors are distinct; there appears to be two angiogenic pathways mediated by αv integrins. α5β1 antagonists inhibit tumor angiogenesis, and block metastases in an animal tumor model (Stoeltzing et al., 2003). Both α5β1 and αvβ3 have been shown to participate in important angiogenic pathways (Bayless et al., 2000); inhibition of endothelial cell lumen formation was only achieved by blocking both αvβ3 and α5β1. Yet integrins are not only central to angiogenesis. In PC, integrins are known to be involved in metastases and exhibit differential expression on PC cells (Demirgoz et al., 2008; Goel et al., 2008; Romanov and Goligorsky, 1999). Zheng et al. (1999) and Sun et al. (2007) observed that the highly invasive human PC cell line, PC-3, expresses αvβ3, which mediated cell adhesion and migration on vitronectin. In contrast, noninvasive LNCaP cells did not express αvβ3, nor adhere to and migrate on vitronectin, although these cells expressed α5β1 (Demirgoz et al., 2008). Exogenous expression of αvβ3 in LNCaP induced the cells to adhere and migrate on vitronectin. Also, primary human PC cells isolated from 16 surgical specimens expressed αvβ3, whereas normal prostate epithelial cells did not (Zheng et al., 1999). This report pointed to αvβ3 as a potential target in PC cell migration and metastasis. Metastasis to remote sites, primarily to the bone, is a major cause of death in prostate cancer. A significant component of metastasis involves adhesion of cancer cells within the vasculature. Adhesion depends on integrins and is dependent on integrin activation. Integrin αvβ3 supports prostate cancer cell attachment under blood flow conditions in an activation-dependent manner (Nemeth et al., 2003; Zheng et al., 1999), and evidence suggests that transition from a locally invasive phenotype to metastatic behavior may be dependent on increased expression of αvβ3 (Sun et al., 2007). Other findings suggest that integrin αvβ3 expression in PC cells accelerates the development of bone metastasis (Cooper et al., 2002), presumably through increased invasion of and adhesion to bone (Kumar, 2003; Nemeth et al., 2003). Other integrins appear to be involved in cancer invasion as well (Cress et al., 1995; Witkowski et al., 1993). Since a single antagonist, targeting a broad range of integrins, can be envisioned to disrupt many of these tumor cell–tissue interactions, integrins present a very attractive target for anticancer drug development. Disintegrins (Gould et al., 1990) avidly attach to a range of human integrins, disrupting normal cellular functions. These findings suggest that metastasis and angiogenesis inhibition should be a feasible treatment strategy in PC. Thus, agents such as the disintegrin vicrostatin (VCN) an antagonist of integrins αvβ3/α5β1/αvβ5 (Minea et al., 2005), as described below should be an effective therapeutic agent for inhibition of metastasis and angiogenesis in PC if targeted to the tumor, and should be highly effective when administered in combination with cytotoxic agents.

Disintegrins as Molecular Weapons Against Cancer

Disintegrins are small, disulfide-rich, RGD-containing peptides that bind to integrins on the surface of normal and malignant cells (Dennis et al., 1990; Huang et al., 1989; McLane et al., 2008; Scarborough et al., 1993). Disintegrins have been characterized from many snake venoms and were originally categorized as platelet aggregation inhibitors (McLane et al., 1998; Niewiarowski et al., 1994; Phillips et al., 1991; Trikha et al., 1994b). Nuclear magnetic resonance has been used to determine the structures of several disintegrins (Adler et al., 1991; Saudek et al., 1991); they display little secondary structure, and are characterized by an RGD or other tripeptide sequence at the tip of a flexible loop protruding from the main body of the peptide chain and anchored by disulfide bonds at the base (Niewiarowski et al., 1994). The Markland lab originally isolated contortrostatin (CN), a homodimeric disintegrin, from the venom of the southern copperhead snake (Trikha et al., 1994b). CN is a homodimer with a molecular mass (Mr) of 13,500; each chain has an RGD sequence and a Mr of 6,750 (Trikha et al., 1994a, b). CN binds competitively to integrins of the β1, β3 and β5 subclasses, including receptors for fibronectin (α5β1), vitronectin (αvβ3, αvβ5), and fibrinogen (αIIbβ3) (Trikha et al., 1994a, b; Zhou et al., 2000a). Human umbilical vein endothelial cells (HUVEC) were exposed to CN for 0–3 days and then tested for processes critical to angiogenesis; migration, invasion, and tube formation (Golubkov et al., 2003). All three activities were inhibited by up to 90%. Further, immunochemical staining showed both actin and VE-cadherin organization to be disrupted by CN in HUVEC (Golubkov et al., 2003). In addition, CN disrupts angiogenesis induced by both bFGF and VEGF, consistent with the observation that cyclic RGD-peptides have no selectivity among αv family members (Friedlander et al., 1995). The lack of integrin specificity by CN is believed to be an advantage in controlling cancer dissemination. PC cell lines, as well as vascular endothelial cells, have been shown to display integrins αvβ3, αvβ5, and α5β1, and the antitumor and anti-angiogenic activity of CN is based on its high affinity interaction with these integrins. CN blocks the function of integrins on both tumor and endothelial cells and has a distinct advantage over many anti-angiogenic agents, which act by blocking a single angiogenic pathway. In animal models of cancer, CN demonstrates anti-angiogenic and antimetastatic activities (Markland et al., 2001; Pyrko et al., 2005; Trikha et al., 1994a; Zhou et al., 2000b). In vivo studies carried out by the Markland lab and other groups have shown that disintegrins are well tolerated and can be infused without detrimental effect on blood pressure, body temperature, or other physiological parameters (Cousins et al., 1995; Shebuski et al., 1989; Yasuda et al., 1991). Acute disintegrin toxicity was studied by the Markland lab in the canine in collaboration with Dr. Benedict Lucchesi, University of Michigan Medical School; disintegrin treated animals did not exhibit changes in heart rate, EKG, or blood coagulation parameters, and there was no evidence of toxicity at the doses tested (Cousins et al., 1995). Although the pharmacological properties of CN have made it an intriguing molecule for potential anticancer therapeutic strategies, the fact that only limited quantities are available from snake venom have highlighted the need to produce a recombinant form of CN. Dr. Minea in Markland's lab designed and engineered a novel recombinant disintegrin, VCN, whose structure is based on the sequence of CN. Interestingly, we found that VCN when expressed in bacteria is a monomer unlike the homodimer CN. Due to the monomeric structure of VCN and the fact that it retains full biological activity, the recombinant system has been used to provide VCN in large quantities. VCN has proven to be a potent anti-angiogenic agent with impressive anticancer activity, as shown in murine models of human breast and prostate cancer (Minea et al., 2005; Pinski et al., 2003). VCN is as active as native CN both in vitro and in vivo and has an additional benefit in that the design of VCN includes a slight carboxyl-terminal sequence alteration that produces much higher affinity for integrin α5β1, which is known to be over expressed on angiogenic endothelial cells and prostate cancer cell lines (Witkowski et al., 1993). In addition, agents that induce endothelial cell apoptosis by antagonizing integrin binding are considered potential therapeutic agents for PC via their ability to inhibit tumor vascularization (Garrison and Kyprianou, 2004). Recent studies by the Markland lab indicate that VCN induces apoptosis of

HUVEC, further supporting its role as an effective anti-angiogenic agent (Minea et al., In Press). Importantly, VCN also induced disruption of the actin cytoskeleton in HUVEC and mediated the dissociation of talin from the cytoplasmic domain of the β1 integrin (unpublished results). These findings provide insight into the molecular mechanism of anti-angiogenic action of VCN, separate from its ability to disrupt integrin adhesion to ECM proteins. In chronic studies in nude mice, a liposomal formulation of VCN (LVCN) was administered by twice weekly intravenous (i.v.) injection over a four week period. There were no visible side effects or signs of internal bleeding, indicating that mice tolerate chronic administration of LVCN very well (Minea et al., 2005). Since there was no evidence of an immune response to a liposomal formulation of CN (LCN), we expect similar results with LVCN. With the availability of the recombinant method for production of VCN, the development of this disintegrin into a commercially viable product is feasible.

Advantages of VCN as an Antitumor/Anti-Angiogenic Agent

The antitumor activity of VCN is based not only on blocking integrin adhesion but also by disrupting the actin cytoskeleton and dissociating the complex between talin and the β1 integrin cytoplasmic tail. Importantly, the binding of talin to integrin β tails represents the final common step in integrin activation (Calderwood, 2004). The effect on integrins mediated by VCN leads to dramatic depression of invasive ability of tumor cells, and of the angiogenic vasculature. Advantages of the disintegrin approach for cancer therapy include: an exclusive and unique recombinant production method that is robust, low cost and easily scalable; the disintegrin sequence can be modified to alter affinity and specificity for integrins; the recombinant expression system can produce venom-derived disintegrins in sufficient quantity to advance this technology to the clinic. Metastasis is a major killer of cancer victims and there are no good anti-metastatic drugs available. We have shown that VCN inhibits processes involved in tumor metastasis (adhesion and invasion) and in a breast cancer model we found that CN inhibited >65% of spontaneous lung metastasis (Zhou et al., 1999), and we are confident VCN will as well.

Importantly, we have shown in rats, dog and mice that there are few if any side effects after chronic exposure to CN or VCN. This indicates that normal, non-migrating cells do not display activated integrins and they are, therefore, not accessible to disintegrin binding. This is particularly true in mature vasculature, which does not display αvβ3. By contrast, αvβ3 is prominently displayed by neovasculature, which serves as the basis for integrin-mediated anti-angiogenic therapy (Gasparini et al., 1998; Gutheil et al., 2000; Hood et al., 2002). This was further emphasized by studies carried out with platelets in vitro wherein we showed that CN only binds to platelets after their exposure to ADP, which induces intracellular signals switching integrins into a ligand competent state (activates integrins by inducing a conformational change in their extracellular domain) (Banno and Ginsberg, 2008; Humphries, 1996; Sun et al., 2007; Xiong et al., 2003), thereby making them accessible to CN.

Avastin (bevacizumab) is a monoclonal antibody that neutralizes VEGF and has shown preclinical evidence of anti-angiogenic efficacy for prostate cancer (Ranieri et al., 2006), it is approved for use in combination therapy in several types of cancer (Courtney and Choueiri, 2009; Grothey and Galanis, 2009). We have shown in head to head comparison of VCN with avastin that there was equivalent efficacy and anti-angiogenic activity of the two agents in a human breast cancer xenograft model. However, whereas avastin has one target (VEGF), VCN targets three pathways based on integrins αvβ3, αvβ5, and α5β1 displayed by tumor cells and angiogenic blood vessels, which could be important when tumor cells develop alternative pathways to bypass VEGF. We expect to see this difference translate to an increased effectiveness of VCN against metastatic disease when compared to avastin. Further, the peptide VCN should be considerably less expensive to manufacture than the antibody avastin. The tyrosine kinase inhibitors (TKIs) sunitinib and sorafenib have shown good efficacy in clinical use, particularly in metastatic renal cell carcinoma, a highly vascular disease (Courtney and Choueiri, 2009; Hiles and Kolesar, 2008; Merseburger et al., 2009; Zhu et al., 2009). Nonetheless, patients treated by all three of these anti-angiogenic agents (avastin and the TKIs) have experienced significant rates of adverse events (AE) (Choueiri et al., 2009). Although many of these AE are minor and clinically manageable, some are severe and life-threatening (Chen and Cleck, 2009). However, two recent reports throw an even more serious spotlight on potential problems with the VEGF/VEGFR targeting agents, including Avastin and the TKIs. In the first report by Kerbel's group (Ebos et al., 2009), sunitinib was shown to accelerate metastatic tumor growth and decrease overall survival in mice receiving short-term therapy in several metastasis assays. Their results suggested possible metastatic conditioning in multiple organs. Similar findings with other TKIs suggest a class-specific effect. Another study (Paez-Ribes et al., 2009) confirmed these findings and provided results suggesting that VEGF-targeting angiogenesis inhibitors initially elicit an antitumor response, but then the tumors appear to adapt and progress to a state of greater malignancy with increased invasiveness and in some cases increased distant metastases. This process is referred to as "evasive resistance". This newly found connection could explain why most current anti-angiogenics have only shown a modest increase in survival in cancer patients (Montagnani et al., 2009). It is important to note that the therapeutic strategy employing VCN is not based solely on the VEGF/VEGFR axis and involves a mechanism that dramatically inhibits the invasive capacity of endothelial and tumor cells, as well as inhibiting growth in the bone in a PC metastatic model (as described below). Thus, the use of disintegrins represents a truly novel therapeutic concept. Finally, cilengitide, cyclo(L-arginyl-L-glycyl-L-aspartyl-D-phenylalanyl-N-methyl-L-valyl), is a potent selective αvβ3-αvβ5 integrin antagonist that has been shown to inhibit integrin-mediated cell adhesion and block in vitro endothelial cell migration. Cilengitide inhibits cytokine-induced bFGF- and VEGF-mediated angiogenesis in a dose-dependent manner, and inhibits tumor growth in several in vivo systems (Beekman et al., 2006). In glioblastoma multiforme (GBM) cilengitide monotherapy is well tolerated with minimal toxicity and exhibits antitumor activity against recurrent disease. It was concluded that integrating cilengitide into combination regimens for GBM is warranted (Reardon et al., 2008). However, when comparing CN and VCN to the parental cyclic RGD peptide cilengitide was derived from, cRGDfV, we found that this cyclic peptide neither induced the profound actin cytoskeleton disruption nor the signaling activity exhibited by CN and VCN. In experiments designed to investigate the mechanism of the anti-invasive activity of CN, we found CN induced dissociation of talin from β1 integrin cytoplasmic tail leading to massive disruption of the actin cytoskeleton in HUVEC plated on Matrigel, whereas cRGDfV had no such activity. Similarly, confocal images of HUVEC plated on Matrigel show a clear discrepancy between VCN and cRGDfV treatment in their ability to collapse the actin cytosleleton of these cells. In further support of these findings, the design of cyclic RGD peptides and RGD mimetics is based on the exposed RGD tripeptide motif in ECM proteins that plays an important role in ligand recognition by integrins, particularly αvβ3 and αvβ5. However, our findings indicate that the presence of the RGD motif alone, although it contributes significantly to the ligand–receptor interaction, may not be sufficient to efficiently and fully modulate the downstream signaling of integrins. Unlike cyclic RGD peptides and peptidomimetics, disintegrins have additional structural elements which enable them to modulate integrin signaling in an efficient and unique manner (Brown et al., 2009; Calvete et al., 2005; Fujii et al., 2003; Marcinkiewicz et al., 1997; Sanz et al., 2005; Yahalom et al., 2002). Thus, unlike cilengitide, the RGD-containing disintegrin loop (11 amino acids) in CN and VCN has additional flanking residues, a structural feature which enables it to make more extensive contacts with the receptor. Furthermore, NMR and crystallographic studies have revealed that the C-terminal tail in disintegrins folds with the RGD-containing disintegrin loop, such that these two structural elements are linked together and form an extended conformational epitope in the three-dimensional structure of disintegrins. This strongly indicates that these two functional regions are engaged in extensive interactions with the target integrin receptor (Brown et al., 2009; Sanz et al., 2005). These observations led us to conclude that, unlike cyclic RGD peptides and peptidomimetics that act only as receptor antagonists, integrin modulation in the case of disintegrins is far more complex. Therefore, by using structural and functional regions in addition to the RGD motif, a tripeptide amino acid sequence that serves as the sole basis for the design of cyclic RGD peptides and non-peptide RGD mimetics, disintegrins are expected to exhibit novel antitumor activities as compared to cyclic RGD peptides and peptidomimetics.

Necessity for Liposomal Delivery of VCN

Due to the short circulatory half-life of peptides such as VCN a delivery system other than direct i.v. administration was needed. Liposomes are submicroscopic nanospheres composed of thin but durable membranes made primarily of phospholipids and cholesterol. The composition, number of lipid layers, size, charge and permeability of the membrane can be altered to enhance delivery of a variety of therapeutic agents encapsulated inside the nanoparticle (Fujii, 1996; Woodle, 1993).

With respect to disintegrin delivery to the tumor bed, we hypothesize that our liposomes are degraded by phospholipases or other enzymes generated in the tumor microenvironment. Due to the incorporation of phospholipids such as phosphatidylcholine or phosphatidylglycerol in the membrane of these unilamellar liposomes, action of these enzymes leads to disruption of the membrane structure with release of the disintegrin payload and subsequent binding of the disintegrin to integrins at the surface of the endothelial and tumor cells. Other advantages associated with our lipid encapsulation process include: (1) enhanced drug delivery to the desired site, (2) prolonged drug half-life and thus reduced dosing frequency, and (3) reduced drug and carrier related toxicities. In the following experimental evidence section, a description of the success is presented for the development of a stable expression system for VCN, a scalable liposomal encapsulation method to prepare LVCN and in vitro and in vivo biological efficacy studies using cells in culture and two PC animal cancer models.

Experimental Evidence for Efficacy of VCN as an Anti-Tumor/Anti-Angiogenic Agent

Recombinant Expression of a Venom Derived Disintegrin

For a number of years the Markland laboratory has worked with contortrostatin (CN), a disintegrin isolated from *Agkistrodon contortrix contortrix* venom. A major obstacle in the pathway to clinical development of CN was the supply of the protein, it is only present as a very small fraction of the total venom protein (~0.01%). Further, for recombinant production, its peculiar structure stabilized by numerous disulfide bonds makes its expression in commonly-employed recombinant systems a very difficult task. Nonetheless, we have successfully employed a recombinant expression system for which we developed a proprietary production method capable of generating approximately 200 mg of purified active recombinant disintegrin per liter of bacterial culture in small-scale laboratory conditions. To generate recombinant disintegrins, we have successfully adapted a commercially-available *E. coli* expression system consisting of the Origami B (DE3) expression host in combination with the pET32a vector (Novagen) for our production needs. A sequence-engineered form of CN, called vicrostatin (VCN), has been directionally cloned into pET32a expression vector incorporating a unique tobacco etch virus (TEV) protease cleavage site, which facilitates the removal of the thioredoxin fusion partner from the expressed VCN.

Following establishment of primary cultures, large scale cultures were inoculated and grown to an OD of 0.6–1.0 and induced with IPTG. Following a 4–5 h induction period the cultures are centrifuged and bacterial pellets lysed by a scalable homogenization method. The insoluble cellular debris was then removed by centrifugation and the soluble cell lysate collected and further analyzed by SDS-PAGE for recombinant protein expression. The expressed fusion protein, thioredoxinA-VCN (Trx-VCN), was proteolysed by adding recombinant TEV protease to the soluble cell lysates. TEV protease treatment efficiently cleaved VCN from its TrxA fusion partner; the status of proteolytic cleavage was monitored by SDS-PAGE. When proteolysis was complete, as assessed by SDS-PAGE, the proteolyzed lysate was filtered, diluted in H_2O and ultrafiltrated through a 50 kDa molecular weight cut-off cartridge that removed most of the higher molecular weight bacterial proteins. The resulting ultrafiltrate was then re-concentrated against a 5 kDa molecular weight cut-off cartridge. VCN was further purified by reverse phase HPLC. The recombinant disintegrin we have produced through this system is recognized by polyclonal antisera raised against native CN and inhibits ADP induced platelet aggregation in a dose dependent manner with an IC50 identical to native CN (~60 nM). Moreover, VCN inhibits tumor cell adhesion, inhibits endothelial cell and tumor cell invasion and inhibits endothelial cell tube formation in a manner indistinguishable from native venom derived contortrostatin. However, VCN is a monomer with a molecular weight of 6,750 Da, whereas CN is a homodimer with a molecular weight of 13,500 Da (Minea et al., 2005)

Liposomal Encapsulation of VCN Using a Homogenization Method

To prepare liposomal vicrostatin (LVCN), stock solutions of phospholipids and cholesterol were prepared by dissolving the lipids in a chloroform/methanol solvent mixture. Thin lipid films were created by pipetting aliquots of the lipid solutions into round bottom glass tubes followed by solvent evaporation. The dried lipids and cholesterol were further dried under vacuum. This process yields lipid powder mixtures that were used to prepare LVCN. For liposome preparation by homogenization, VCN was dissolved in a hydration buffer and added to the dried lipids and incubated briefly at 50°C. LVCN was formed by passing the material through a microfluidizer. The material was processed between 10,000 and 18,000 psi while maintaining an elevated temperature (45–65°C). Samples of the liposome batch were removed during the process and the size distribution of LVCN was determined with an Ultrafine Particle Analyzer. After processing, unencapsulated VCN was removed by 100,000 MWCO ultrafiltration and LVCN sterilized by filtration through a 0.2 μM PVDF filter. We are encouraged that LVCN can be scaled to volumes necessary for commercialization by the homogenization technique.

In Vitro Evaluation of VCN in Limiting Tumor Progression and Angiogenesis

Binding Affinities of VCN to Integrins

To assess the binding affinities of VCN with soluble integrins, fluorescence polarization (FP) was utilized (Jameson and Seifried, 1999). In this method, differing concentrations of functional integrin were incubated with a constant amount of FITC labeled VCN. As VCN is a small molecule it rapidly depolarizes the excitation light. Upon binding to the large integrin, the fluorescent tag on VCN tumbles

Table 19.1 Dissociation constants for interactions of VCN and CN with soluble integrins

	Integrin Kd		
Disintegrin	αvβ3	α5β1	αvβ5
CN	6.6 nM	191.3 nM	19.5 nM
VCN	7.4 nM	15.2 nM	41.2 nM

Values calculated through fluorescence polarization measurements following steady state binding

in solution at a slower rate resulting in increased levels of polarization. The measured FP value is a weighted average of FP values of the bound and free fluorescent VCN and is therefore a direct measure of the fraction bound. Data generated in these experiments can be analyzed like standard radioligand binding, and kinetics of binding can be determined as with Scatchard analysis using a non-linear curve fit. From this set of experiments we determined the dissociation constants for VCN and CN with integrins αvβ3, α5β1 and αvβ5 (Table 19.1). Recombinant VCN was purposely designed with a carboxy terminal extension, which was expected to enhance affinity for α5β1. This was confirmed as CN and VCN exhibit nearly identical affinities for αvβ3 and similar affinity for αvβ5, but there is an order of magnitude difference in the Kd values for binding to α5β1 when comparing VCN (higher affinity binding) to CN.

Inhibition of Cellular Processes Critical to Tumor Progression

To assess the ability of VCN to block processes critical to tumor survival and progression (adhesion, migration and invasion) we have measured the inhibitory effect on different tumor cell lines, as well as endothelial cells. Inhibition of adhesion is evaluated through the ability of VCN to block cell attachment to a number of different extracellular matrix proteins. We observed a dose-dependence in the inhibition of adhesion of PC-3 cells as well as HUVEC, to both vitronectin and fibronectin, ECM proteins that are ligands for integrins targeted by VCN (data not shown). Cellular migration is also inhibited by VCN in a dose dependent manner. To evaluate tumor and endothelial cell migration a phagokinetic tracking assay is employed. In this assay cells are plated on a collagen coated cover-slip with an overlay of colloidal gold. As the cells move they displace or ingest the colloidal gold leaving tracks on the surface of the cover-slip. Then, using dark-field microscopy the tracks can be visualized and photographed. Using image analysis software the area of the tracks in a photographed field can be determined and a "migration index" can be calculated as a percentage of the field lacking gold. Following treatment by increasing concentrations of VCN the migration of both tumor and endothelial cells is significantly limited. Finally, the ability of cells to invade through the ECM was evaluated using modified Boyden chambers. These chambers contain a Matrigel coated porous membrane (pore size 8 μm). A chemoattractant is placed in the lower chamber and untreated cells invade through the membrane toward the attractant. Both CN and

VCN block the invasion of endothelial (HUVEC) and PC-3 cells in a dose dependent manner with IC_{50} at low nM concentrations. These results show that VCN has essentially identical activity to CN in inhibiting invasion of endothelial and prostate cancer (PC-3) cells. The results also convincingly demonstrate one of the important attributes of VCN, that it inhibits endothelial cell as well as tumor cell invasion in the low nM range.

Inhibition of HUVEC Tube Formation

To assess the ability of VCN to interfere with tube formation (an in vitro assay of anti-angiogenic activity), HUVEC were maintained in EGM-2 complete media and grown to confluency. HUVEC cells were harvested and resuspended in basal media. After being maintained in suspension for 15–30 min, cells were seeded onto Endothelial Cell Tube Formation plates, an in vitro endothelial tubulogenesis system, at a concentration of 25,000 cells per well and immediately treated with various concentrations of VCN, CN (positive control), or Suramin (positive control) and incubated for 18 h at 37°C. At the end of the incubation period, cells were washed and then stained with Calcein AM in PBS at 37°C. After 30 min the cells were washed again and imaged (Fig. 19.1) using confocal microscopy at 2.5X and 10X magnifications. On the captured images, the total length of tubes was

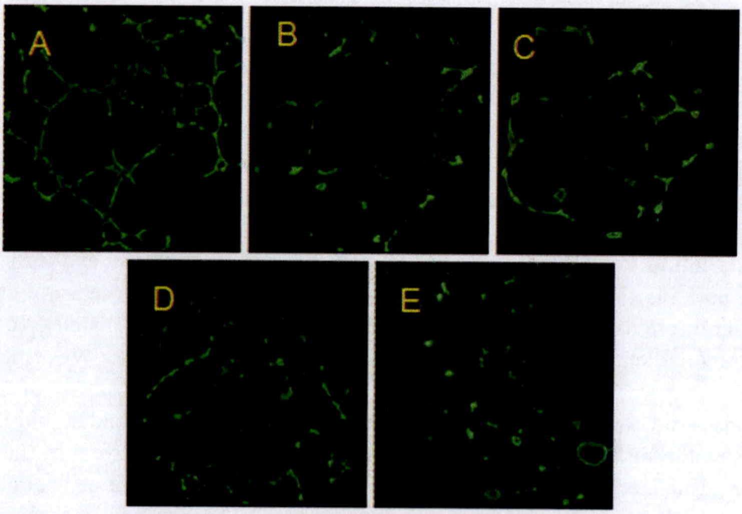

Fig. 19.1 Inhibition of HUVEC tube formation by VCN (representative images from multiple experiments). HUVEC cells were plated on "Endothelial Cell Tube Formation" plates (BD Biosciences) in the presence of various concentrations of VCN (0–1,000 nM), or a known tube formation inhibitor Suramin (used as a positive control). Representative figures from independent experiments were shown above: *panel a* – untreated control; *panel b* – 100 μM suramin; *panel c* – 1 nM VCN; *panel d* –10 nm VCN; and *panel e* – 1,000 nm VCN. Cells were stained with Calcein AM and imaged using confocal microscopy

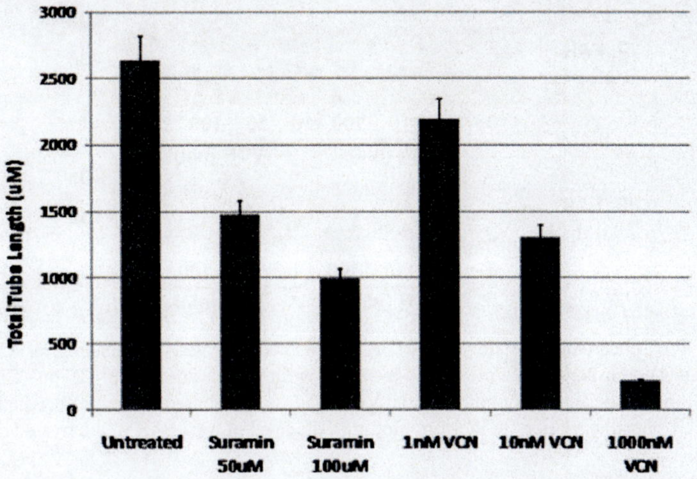

Fig. 19.2 Quantitation of tube formation inhibition by varying concentrations of VCN. The tubes formed by HUVECs were quantitated in multiple fields collected from three repeated experiments by computing the total tube length with Zeiss LSM image software and averaged to form each data point. The data shown above was assembled from multiple independent experiments

quantitated with Zeiss LSM image software and data plotted against the total length of tubes (in μm) generated by untreated cells. Representative tubes from 6 different wells were measured by 3 individuals and averaged to form each data point (Fig. 19.2). VCN exhibits potent dose-dependent inhibition of tube formation at low nM concentrations

Alterations of FAK Phosphorylation Induced by VCN

It has been proposed that the mechanism of action of VCN (delivered as LVCN) involves alterations in integrin mediated signal transduction pathways. Changing the phosphorylation status of FAK has previously been identified as a target of disintegrin action on both tumor and endothelial cells. To carry out these experiments using MDA-MB-435 cancer cells, the cells were grown to confluency and harvested. Cells were then maintained in suspension for 1 h in 2% bovine serum albumin/serum free medium. Various concentrations of VCN or CN (positive control) were used to treat the cells for 10–30 min while still in suspension. Following incubation with disintegrins and washing to remove unbound CN or VCN, cells were lysed with modified radioimmunoprecipitation assay (RIPA) buffer. For further analysis the total amount of protein in lysates was standardized before immunoprecipitation. Rabbit anti-FAK polyclonal antibodies were incubated with the whole cell lysates for 9–12 h at 4°C, and immune complexes were recovered by incubation with protein G immobilized on agarose beads overnight at 4°C. The precipitated protein complexes were centrifuged and the pellets washed, resuspended in SDS-PAGE sample buffer, and

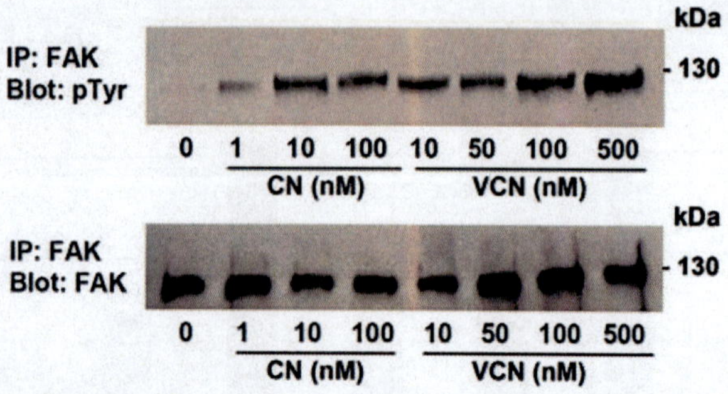

Fig. 19.3 FAK phosphorylation levels in MDA-MB-435 cells treated with disintegrins. Cells were kept in suspension in serum-free media and incubated with various amounts of CN or VCN for 10–30 min. The agonistic activity of both CN and VCN was assessed based on their effect on the global level of FAK phosphorylation by Western blotting. The level of total FAK in the treated cells does not change

boiled. For immunoblotting, proteins were resolved on 10% SDS gels and transferred to nitrocellulose membranes. Blots were incubated with either total anti-FAK monoclonal antibodies or p-TYR antibodies for 2 h at RT. Bound primary antibodies were detected with horseradish peroxidase-conjugated anti-mouse secondary antibodies, followed by enhanced chemiluminescence (Fig. 19.3). Experiments were repeated at least three times to verify results. These studies indicate that the level of phospho-FAK increases with increased concentration of treatment agent, VCN or CN; the difference between CN and VCN is almost indistinguishable, again indicating that the recombinant disintegrin (VCN) has identical bioactivity to the natural protein CN.

Actin Cytoskeleton and Focal Adhesion Disruption

In view of the finding that CN and VCN alter FAK phosphorylation, the effect of disintegrin treatment on focal adhesions and the actin cytoskeleton was evaluated. This was accomplished by visualization of the actin cytoskeleton and talin-based focal adhesions in HUVEC which were actively forming focal adhesions in a polymerized growth factor reduced (GFR) matrigel layer after treatment with either CN or a cyclic RGD peptide (cRGDfV). Eight well chamber slides were coated with GFR matrigel and left to polymerize. Residual media was aspirated out of the chamber slides and HUVECs (2.5×10^4) were seeded in serum free media onto the polymerized matrigel layer. Cells were incubated for 1-h at 37°C before being treated with either 100 nM CN or 10 μM cRGDfV. After overnight incubation, cells were washed with PBS, fixed in ice-cold acetone for 10 min at 4°C, and allowed to air dry for 2 min at 20°C. Each chamber was then blocked with PBS containing 5% BSA for 1 h. Talin was stained with a mouse monoclonal Ab, still in the presence of 5% BSA for 2 h. After washing with PBS, each chamber was counter stained with anti-mouse FITC-conjugated secondary Ab. The nucleus was stained with Hoechst stain and the actin cytoskeleton was stained with rhodamine phalloidin. Following washing in PBS the chambers were separated and mounted with fluorescent mounting media for confocal microscopy. The immunofluorescent staining of talin clearly shows focal adhesions forming when HUVEC cells are plated on polymerized GRF matrigel. For the untreated group, focal adhesions are well-defined, small, cylindrical structures that are easily visualized under the zoomed panel (Fig. 19.4). Every cell also has numerous focal contacts in the leading edge of their lamellipodia. In addition, the actin cytoskeleton of the untreated group is highly organized. In every aspect, from the number of focal contacts made by talin to the organization of the actin cytoskeleton, the staining for cells treated with 10 μM cRGDfV is identical to the untreated group (Fig. 19.4). However, the CN treatment group reveled less

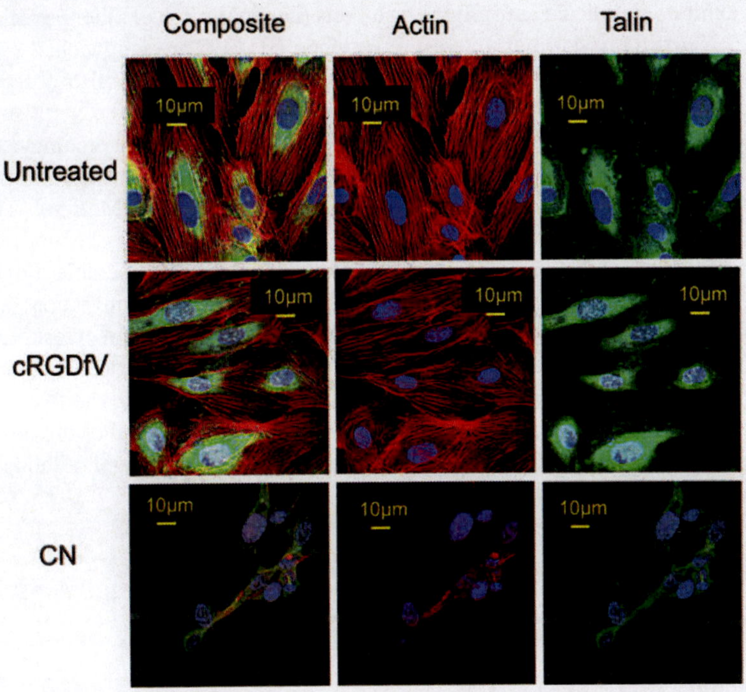

Fig. 19.4 Immunofluorescence of actin cytoskeleton and focal adhesions. Chamber slides coated with growth factor reduced matrigel were seeded with HUVEC (2.5×10^4) in serum free media. Cells were incubated for 1 h at 37°C before either no treatment (*top row*), treatment with 100 nM CN (*bottom row*), or 10 μM cRGDfV (*middle row*). After overnight incubation, cells were washed with PBS, fixed in acetone for 10 min at 4°C, and allowed to air dry for 2 min at 20°C. Each chamber was then blocked with PBS containing 5% BSA for 1 h at 20°C. Talin was stained with a mouse monoclonal Ab (clone TA205) for 2 h at 20°C. Followed by counter staining with anti-mouse FITC-conjugated secondary Ab (488 nM), Hoechst stain (460 nM), and rhodamine phalloidin (580 nM). After three washes in PBS the chambers were separated and mounted with fluorescent mounting media for confocal microscopy (63x). Composite column: superimposed images shown after counter-staining by all three agents; Actin column: rhodamine phalloidin staining; Talin column: talin staining. A 10 μm scale bar is included in each image

defined talin staining of broader sized structures that were hard to classify as focal adhesions (Fig. 19.4). Also, the number of focal contacts each lamellipodia has is greatly reduced. Moreover, the actin cytoskeleton for the CN treatment group is highly disorganized and looks to be under extreme stress conditions. This data indicates that cRGDfV treatment has virtually no effect on cytoskeletal organization and focal adhesions based on their associations with talin, while CN treatment drastically disrupts both of these structures. We assume that VCN will show identical activity based on its similarity to CN.

VCN Induction of Apoptosis in Tubulogenic HUVEC

VCN does not affect HUVEC viability if adherent cells are plated on Matrigel, but VCN exhibits significant anti-migratory effects (i.e., inhibition of tube formation) on HUVEC grown on Matrigel. Interestingly, when HUVEC are sandwiched between two layers of complete Matrigel, a significant apoptotic effect is also observed in the presence of VCN, but not with other integrin ligands (Fig. 19.5). Surprisingly, a similar effect was also seen with Avastin in this setting, though less pronounced than with VCN. It is noteworthy that HUVEC sandwiched between two Matrigel layers migrate and form tubes much faster than when plated on top of Matrigel. Thus, it appears that the rapidly migrating cells sandwiched in Matrigel may be more dependent (than those plated on top of Matrigel) on their ability to assemble a dynamic actin cytoskeleton and this seems to be a requirement not only for migration, but also for survival. Thus, the ability of VCN to efficiently disrupt the actin cytoskeleton of rapidly migrating cells may explain the discrepancy seen in cell survival between the two experimental settings. It is also important to note that, unlike the two integrin binding antibodies, 7E3 and LM609, the c(RGDfV) peptide did alter the morphology of the tubes formed by HUVEC when sandwiched in Matrigel, although to a lesser extent than VCN and with no impact on cell viability.

In Vivo Evaluation of the Anti-Angiogenic Anti-Tumor Activity of VCN

Circulatory Half-Life of LVCN

Previously we had determined the circulatory half-life of CN and liposomal CN (Swenson et al., 2004). We repeated these studies for VCN and LVCN. Blood samples were taken 0.5, 1, 3, 6, 18, 24, 48 and 72 h following i.v. administration of ^{125}I-VCN or L-^{125}I-VCN. Gamma counting of collected blood samples revealed that there was a rapid decrease to <0.1% of the administered counts in the blood 6 h after i.v. injection of ^{125}I-VCN. However, in animals given L-^{125}I-VCN, the percentage of total injected counts in the blood drops to a level of 63% of the injected counts 6 h post-injection and gradually decreases over the following 66 h. By plotting the decrease in radioactivity in blood over time following i.v. administration in tumor-free mice, we observed a circulatory half-life of 0.4 h for ^{125}I-VCN and

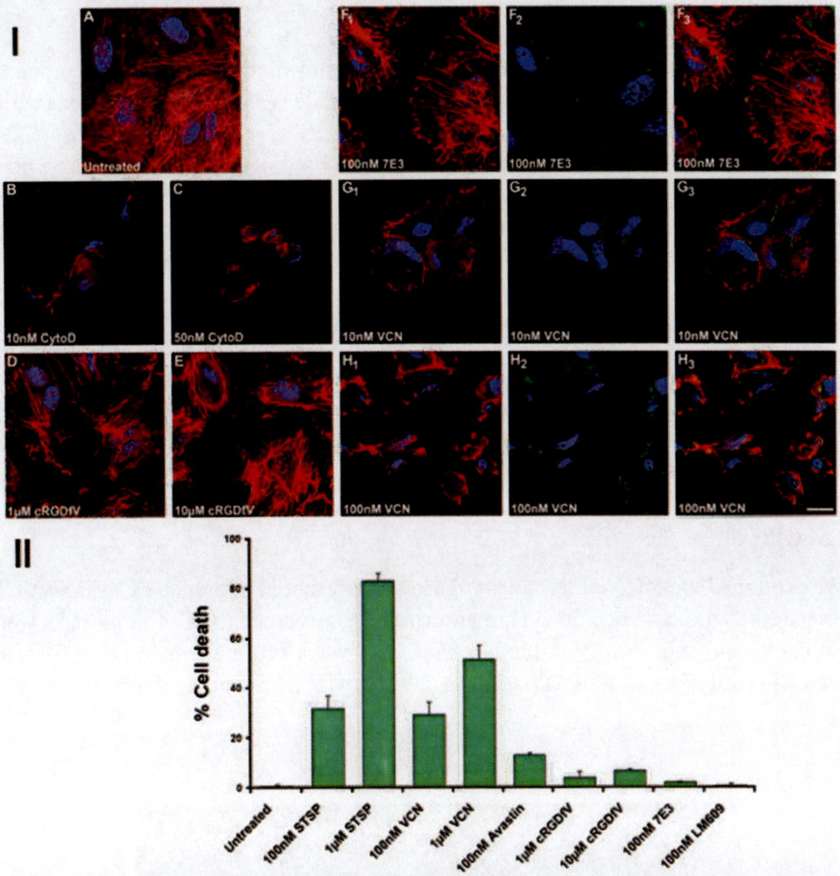

Fig. 19.5 VCN induces massive actin cytoskeleton reorganization in HUVEC seeded on Matrigel. HUVEC were seeded in serum-free media in multiwell chamber slides on complete Matrigel, allowed to adhere, and then treated for 3 h with various concentrations of cRGDfV peptide (1 and 10 μM) or FITC-VCN (10 and 100 nM) or the αvβ3-integrin binding antibody fragment 7E3 (100 nM) followed by a FITC-conjugated secondary antibody. The actin modifier Cytochalasin D was used as a positive control at two different concentrations (10 and 50 nM). At the end of the incubation period, the cells from all conditions were fixed in 4% formaldehyde, permeabilized in 0.1% Triton X-100, stained with Rhodamine-Phalloidin and Hoechst 33342, and imaged by confocal microscopy. The cells in panels F-H (treated with FITC-labeled 7E3 or VCN) are triple stained. The images shown above are Rhodamine-Hoechst (panels A–E and F1–H1), FITC-Hoechst (panels F2–H2) or all three fluorophores (panels F3–H3). Unlike the cRGDfV peptide or the integrin-binding 7E3 antibody fragment (or LM609, not shown), VCN collapses the actin cytoskeleton in HUVEC plated on Matrigel. Representative confocal images from multiple experiments taken at the same magnification (x630) are shown above (scale bar, 20 μm)

20.4 h for L-^{125}I-VCN similar to our previous findings with CN and LCN. Thus, encapsulation of VCN in liposomes not only protects the protein but also maintains it in the circulation for a much longer period of time than unencapsulated VCN, enabling more effective access to the tumor.

Preferential Tumor Binding of VCN

In order to determine if there was preferential tumor binding of VCN as compared to a cyclic peptide cyclo(-RGDfV-), similar to Cilengitide, which is currently in clinical trials for glioma therapy, we evaluated VCN as a PET imaging agent. This experiment was designed to show tumor specific binding of VCN in existing bone metastases, using the androgen dependent PC CWR22. CWR22 cells were injected into the tibia of nude mice and were allowed to grow untreated for ~5 weeks until the tumors were 10–14 mm in diameter. Animals were then injected with ^{64}Cu labeled VCN or cyclo(-RGDfV-), and imaged using a Concorde Systems micro-PET imaging system. As can be seen in Fig. 19.6, injected VCN localizes to the tumor with much higher affinity than the cyclic RGD peptide. This indicates that VCN binds to the tumor and could serve as an effective imaging agent, as well as a therapeutic agent.

In Vivo Biological Efficacy Assay on Both Tumor and Angiogenic Vessel Growth

We examined the effect of treatment of a xenograft model of human PC using LVCN prepared by homogenization. Human prostate cancer cells (PC-3, 2 × 10^6) were implanted subcutaneously in the hip flank of 5-week old male nude mice. Tumors were allowed to grow until palpable (14 days) at which time drug administration was

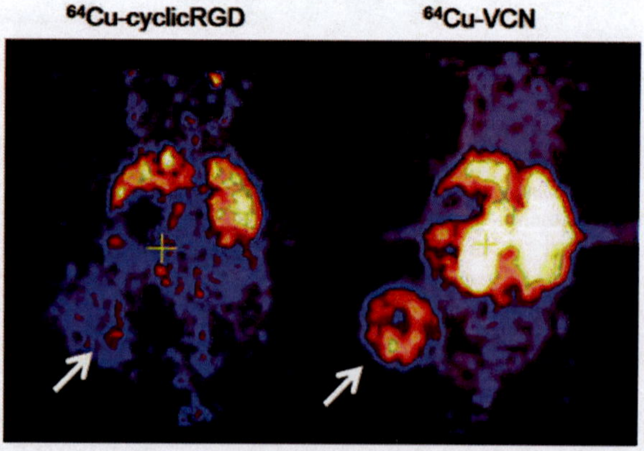

Fig. 19.6 VCN binds to prostate tumor. Comparison of binding efficacy of ^{64}Cu labeled VCN and a cyclic RGD peptide to an existing prostate tumor implanted in the tibia of a nude mouse (*white arrow* points to tumor. As can be seen the VCN binds with higher affinity than the RGD peptide to the existing PC tumor. With VCN a significant amount of the injected radioactivity is associated with the tumor but there also is a high level of non-specific binding in the kidneys liver and intestines. The level of non-specific binding with the RGD peptide appears lower but this can be due to the much shorter circulatory half-life of the RGD as compared to VCN. This indicated that VCN can potentially be used as both a therapeutic or PET imaging agent in prostate cancer. The image registration cross hatch is seen in the RGD panel

initiated. A preparation of LVCN as well as unencapsulated VCN were administered twice weekly via intravenous injection (100 μg per dose for 5 weeks); PBS and empty liposome controls were also included. Tumors were measured weekly via caliper in a blind fashion. At the end of the treatment period there is a significant inhibitory effect on tumor growth (~75% inhibition) when comparing the LVCN treated animals with the PBS treated and empty liposome controls (Fig. 19.7). In addition unencapsulated VCN displayed essentially no anti-tumor activity, which may in part be explained by its relatively short circulatory half-life. In additional PC-3 therapeutic efficacy studies we evaluated the efficacy of LVCN in combination with docetaxel. In these studies we used therapeutic doses of both docetaxel and LVCN and observed no additional effect on inhibition of tumor growth over that of LVCN alone. These studies will be repeated using a reduced LVCN dose and several different doses of docetaxel.

In addition to the inhibitory effect on tumor growth we also evaluated the anti-angiogenic effect of LVCN in the PC-3 model utilizing immunohistochemistry (IHC). Briefly, tumors were embedded in OCT using standard techniques. The frozen blocks were sectioned on a cryostat. The sections were taken from three regions of the tumor, the first, second and third 20 μm segment of the tumor. The thickness per section was 5 μm. This method was used to make sure that the

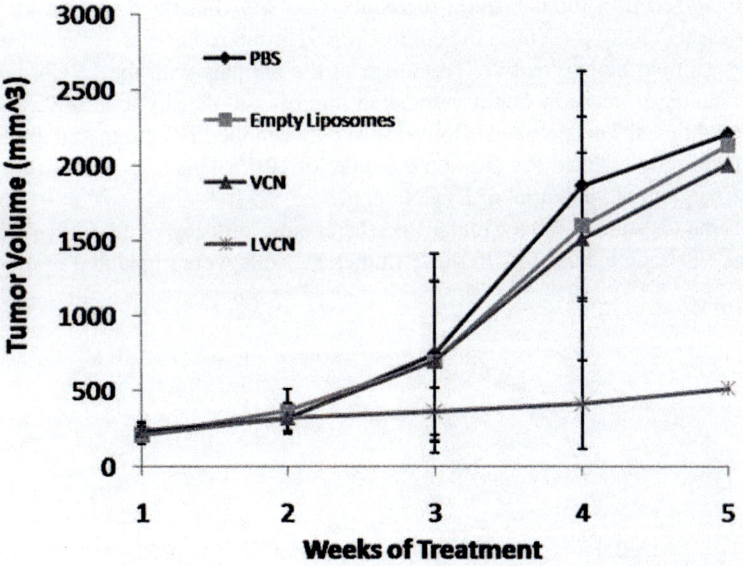

Fig. 19.7 PC-3 xenografts treated with LVCN. Tumor volume measurements in athymic nude mice bearing subcutaneous xenografts of the prostate cancer cell line PC-3 during treatment with LVCN or VCN administrated twice weekly via i.v. injection (100 μg/dose) Control animals were injected with PBS or empty liposomes. Treatment lasted for 5 weeks. Vertical bars indicate SEM. There is statistical significance, with p values less than 0.05 between the LVCN treatment group and the control

staining was representative of the entire tumor, and not due to the specific portion of the tumor tested. Following sectioning, tumor sections were fixed in acetone followed by blocking with 5% goat serum. The primary antibody utilized was a polyclonal rat anti-mouse CD31/PECAM. Following primary antibody incubation, the slides were washed and probed with the secondary antibody, biotinylated goat anti-rat antibody. Slides were then washed in PBS, and incubated with Avidin Binding Complex (ABC), and then stained with 3-amino-9-ethylcarbazole (AEC) chromogen solution. Slides were counterstained with Mayers Hemotoxylin to visualize nuclei. To quantitate the blood vessels the stained sections were subject to "random field" analysis. Four randomly assigned fields were chosen and digitally captured. To insure objectivity, the pictures were taken "blindly" without prior knowledge of treatment group. Four images per section and three sections per tumor (1 from 0–20 μm, 1 from 20–40 μm and 1 from 40–60 μm) were typically analyzed, twelve images per each tumor. Between 2 and 4 tumors per animal group of 5 animals were analyzed to obtain the microvessel density average in each of the experimental groups. Once the images were captured, and loaded onto the computer, they were quantitated using Simple PCI software. Blood vessels were defined as any endothelial cell or endothelial cell cluster, which was distinctly separated from tumor cells or connective tissue. The simple PCI imaging software allows the observer to pick a range of pixel colors denoting blood vessels. Those pixels are then quantified as both a raw number and % pixels per field of view. The images were analyzed in a blind fashion to reduce bias. As illustrated in Fig. 19.8, the microvessel density present in the control (PBS) group represents ~7% of the total pixels in a field being positive. Treatment of the animals with the LVCN resulted in a dramatic decrease in tumor-associated microvessel density to 1.5–2.5% of the visualized fields. The statistical significance between the PBS group and the LVCN treatment group reached P values on the order of 10^{-18}. This data further illustrates the anti-angiogenic potential of LVCN.

In separate studies we evaluated the therapeutic efficacy of LVCN in treating existing PC bone metastases. In these studies PC cells were implanted in the study

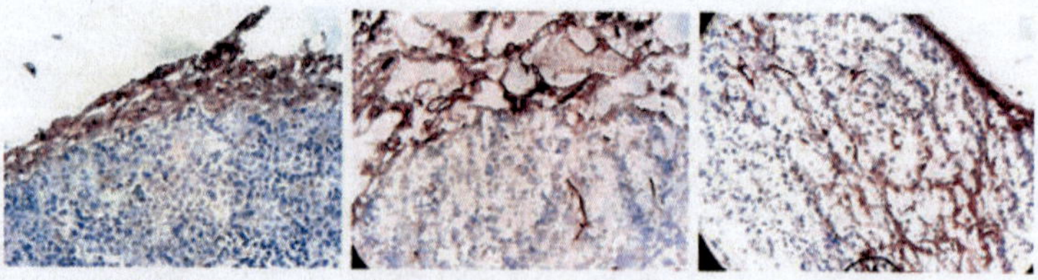

Fig. 19.8 Immunohistochemical Staining of Microvessels with CD31. Tumors were embedded in OCT and frozen prior to sectioning. Blocks were sectioned at 5 μm and probed with an anti-CD31 antibody. Following washing and after initial detection with a secondary antibody the chromagen AEC was used to visualize CD31 mAb. In both the PBS and empty liposome samples numerous vessels are observed while in the LVCN sample considerably fewer vessels are observed. LVCN is an effective antiangiogenic agent in prostate cancer

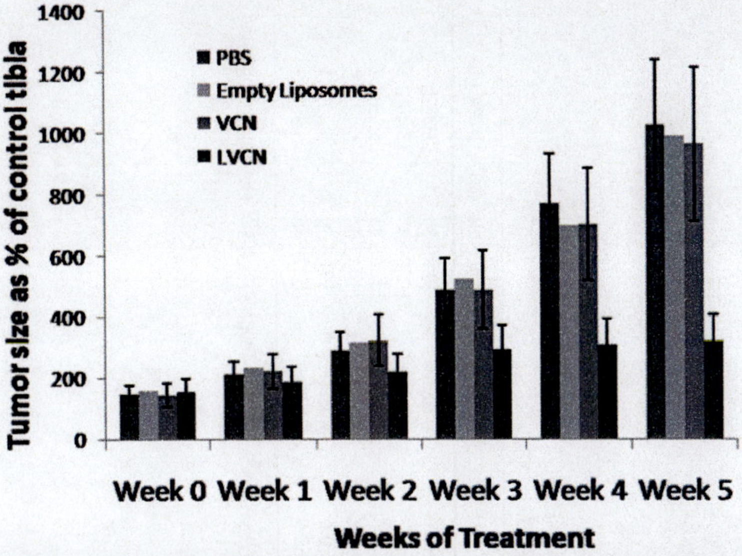

Fig. 19.9 CWR22 model of prostate cancer bone metastasis treated with LVCN. Shown are the tumor sizes as a percentage of the control tibia in athymic nude mice bearing experimental CWR22 PC bone metastasis treated with LVCN or VCN administrated twice weekly via i.v. injection (100 μg/dose). Control animals were injected with PBS or empty liposomes. Treatment lasted for 5 weeks. Vertical bars indicate SEM. There is statistical significance, with p values less than 0.05 between the LVCN treatment group and the control

animals by intra-tibial injection. Briefly, CWR22 prostate cancer cells were suspended in a solution of matrigel in modified Eagles media (MEM) at 2.5×10^6 cells/ ml and 25 μl of this suspension was directly injected into the proximal left tibia of immunodeficient mice. Following injection the diameter of both the left and right (control) tibias were measured by caliper twice weekly. Since the CWR22 cell line is androgen dependent, animals received daily i.p. injections of testosterone suspended in DMSO (1 mg/kg, daily). The tumors grew slowly over the initial 20 days, but at this point tumors were evident and treatment began. As in the subcutaneous model studies, LVCN as well as unencapsulated VCN were administered twice weekly via i.v. injection (100 μg per dose for 5 weeks). LVCN displayed a significant inhibition of tumor growth as compared to VCN, PBS and empty liposomes (Fig. 19.9). Studies to evaluate bone loss are ongoing.

Preliminary Evaluation of LVCN Acute Toxicity

The toxicity of a single intravenous dose of VCN or LVCN prepared by homogenization was evaluated using a rat model. Wistar rats (120–130 g) were assigned to 12 groups of 3 animals each, including 2 control groups (PBS and empty liposome control), and 10 experimental groups that received VCN or LVCN preparations (3, 10, 25, and 75 mg/kg for VCN and 1, 3, 10, 25, 50, and 75 mg/kg for LVCN).

Table 19.2 Average hematologic values for VCN LVCN toxicity test

Agent	Dose	Weight (g)	WBC ($\times 10^3/\mu l$)	RBC ($\times 10^6/\mu l$)	HGB (g/dl)	MCV (10^{-15} l)	MCHC (g/dl)	PLT ($\times 10^3/\mu l$)	MPV (10^{-15} l)
Saline		186±7	5.7±0.6	5.9±0.5	12.2±0.4	55.9±0.6	34.7±1.0	988±39	5.4±0.6
Liposomes		179±9	7.3±1.3	7.1±0.6	13.9±0.8	58.9±0.9	35.0±0.9	1028±44	5.4±0.5
VCN	3 mg/kg	183±7	4.7±0.8	6.8±0.4	13.1±0.9	55.8±0.5	34.6±1.1	1160±83	5.1±0.7
VCN	75 mg/kg	176±9	4.4±0.7	6.5±0.6	12.7±0.6	56.2±0.8	35.2±1.2	1193±64	5.2±0.5
LVCN	3 mg/kg	182±6	5.7±0.6	6.3±0.7	13.0±0.7	55.5±0.6	35.0±1.1	976±59	5.4±0.4
LVCN	75 mg/kg	188±7	5.4±0.9	6.6±0.4	13.6±0.5	57.3±0.9	35.1±0.9	1022±46	5.3±0.3

Weight = animal weight in grams (g); WBC = white blood cell count; RBC = Red Blood Cell count, HGB = Hemoglobin; MCV = Mean corpuscular volume, MCHC = Mean Cell Hemoglobin Concentration, PLT = Platelet Count; MPV = Mean platelet volume

Animals were given a single administration of the test agents and evaluated for signs of physical toxicity or stress over 14 days, and then sacrificed at day 14. Signs of toxicity were monitored via physical status, activity level and total body weight; following sacrifice, gross and microscopic pathology was performed and hematological properties were analyzed. There were no adverse effects observed in the any of the treated animals. Animals in all treatment groups thrived and gained weight indistinguishable from the control groups. There were no observed changes in behavior immediately following agent administration, nor throughout the 14 day study. Gross examination following sacrifice revealed no changes in tissue or organ histopathology between control and treated animals. There were no significant differences in hematological parameters between even the highest dose (75 mg/kg) and control (Table 19.2). Microscopic examination of major body organs revealed that there was no observable inflammation, no significant cellular alterations and no visible hemorrhagic changes in the microscopic sections.

Conclusions and Future Directions of Disintegrins as Anti-Angiogenics

Anti-angiogenic agents currently in clinical trials and clinical use including monoclonal antibodies, such as bevacizumab, which specifically inhibits the VEGF-A signaling cascade, or small-molecule inhibitors of signaling pathways, are limited in that they target a specific single pathway within the tumor or angiogenic vasculature. While such specific targeting strategies may be initially effective, tumors can develop resistance to anti-angiogenic therapies, reviewed by Kerbel (2001a, 2001b). For example, VEGF-A can be replaced by other redundant signaling mechanisms, including other VEGF family members or pathways unrelated to VEGF, during tumor progression. This would enable a tumor to escape the effects of the single target VEGF-A blockade. In addition, targeting a single pathway applies a positive selective pressure upon tumor cells such that resistant variants may be generated through mutation. Tumors can also evade anti-angiogenic agents through the emergence of hypoxia-resistant mutants that are less dependent on angiogenesis. Changes in the structure of tumor vasculature towards a more mature phenotype might also promote resistance. Therefore, the development of drug resistance is a significant concern in the design of novel anti-angiogenic therapies.

As previously discussed, integrins play a central role in a number of key pathways in tumor progression, including cell migration, invasion, adhesion, metastasis and angiogenesis. Integrin antagonists can therefore be expected to impact all of these pathways and have a more broad-spectrum effect upon tumor cells. Such an approach alleviates the problem of drug resistance since multiple pathways are being targeted simultaneously. Since the disintegrin class of integrin antagonists are small peptides, which do not appear to be toxic nor do they activate the immune system, and thus potential side effects are minimized. Utilizing liposome technology, a clinically relevant method for the administration and delivery of the disintegrins has been developed and has proven to be effective in a number of mouse models without significant toxicity. Employing a fully functional recombinant variant of CN, VCN, expressed in a highly modified *E. coli* expression system, which has been described elsewhere (Minea et al., 2005) provides a clear path for the clinical translation of this molecule . This advance in the production and validation of recombinant VCN solves an important supply problem faced by the natural venom purified disintegrin and will allow further development as a clinical agent. Therefore, while current therapies have been promising, disintegrins represent an attractive alternative to conventional anti-angiogenic agents.

Acknowledgements The authors would like to thank Lesley Rakowski, Kyle Brodmann and Barbra Rubino for technical assistance. In addition the authors would like to acknowledge grant support from; National Institutes of Health (FM, 1R41 CA126001-01A1 and 1R41 CA121452-01A1), California Breast Cancer Research Program (SS, 12IB-0153) and Komen for the Cure (FM, BCTR0707423).

References

Adler, M., Lazarus, R.A., Dennis, M.S., Wagner, G., 1991. Solution structure of kistrin, a potent platelet aggregation inhibitor and GP IIb-IIIa antagonist. Science 253, 445–448.

Aragon-Ching, J.B., Dahut, W.L., 2008. The role of angiogenesis inhibitors in prostate cancer. Cancer. J. 14, 20–25.

Aragon-Ching, J.B., Dahut, W.L., 2009. VEGF inhibitors and prostate cancer therapy. Curr. Mol. Pharmacol. 2, 161–168.

Banno, A., Ginsberg, M.H., 2008. Integrin activation. Biochem. Soc. Trans. 36, 229–34.

Bayless, K.J., Salazar, R., Davis, G.E., 2000. RGD-dependent vacuolation and lumen formation observed during endothelial cell morphogenesis in three-dimensional fibrin matrices involves the αvβ3 and α5β1 integrins. Am. J. Pathol. 156, 1673–1683.

Beekman, K.W., Colevas, A.D., Cooney, K., Dipaola, R., Dunn, R.L., Gross, M., Keller, E.T., Pienta, K.J., Ryan, C.J., Smith, D., Hussain, M., 2006. Phase II evaluations of cilengitide in asymptomatic patients with androgen-independent prostate cancer: scientific rationale and study design. Clin. Genitourin. Cancer 4, 299–302.

Beekman, K.W., Hussain, M., 2006. Targeted approaches for the management of metastatic prostate cancer. Curr. Oncol. Rep. 8, 206–212.

Boehm, T., Folkman, J., Browder, T., O'Reilly, M.S., 1997. Antiangiogenic therapy of experimental cancer does not induce acquired drug resistance. Nature 390, 404–407.

Brooks, P.C., Montgomery, A.M., Rosenfeld, M., Reisfeld, R.A., Hu, T., Klier, G., Cheresh, D.A., 1994. Integrin αvβ3 antagonists promote tumor regression by inducing apoptosis of angiogenic blood vessels. Cell 79, 1157–1164.

Brown, M.C., Eble, J.A., Calvete, J.J., Marcinkiewicz, C., 2009. Structural requirements of KTS-disintegrins for inhibition of $\alpha_1\beta_1$ integrin. Biochem. J. 417, 95–101.

Calderwood, D.A., 2004. Talin controls integrin activation. Biochem. Soc. Trans. 32, 434–437.

Calvete, J.J., Marcinkiewicz, C., Monleon, D., Esteve, V., Celda, B., Juarez, P., Sanz, L., 2005. Snake venom disintegrins: evolution of structure and function. Toxicon 45, 1063–1074.

Chen, H.X., Cleck, J.N., 2009. Adverse effects of anticancer agents that target the VEGF pathway. Nat. Rev. Clin. Oncol. 6, 465–477.

Cheresh, D.A., 1992. Structural and biologic properties of integrin-mediated cell adhesion. Clin. Lab. Med. 12, 217–236.

Choueiri, T.K., Duh, M.S., Clement, J., Brick, A.J., Rogers, M.J., Kwabi, C., Shah, K., Percy, A. G., Antras, L., Jayawant, S.S., Chen, K., Wang, S.T., Luka, A., Neary, M.P., McDermott, D., Oh, W.K., 2009. Angiogenesis inhibitor therapies for metastatic renal cell carcinoma: effectiveness, safety and treatment patterns in clinical practice-based on medical chart review. BJU Int. 105(9), 1247–1254.

Choy, M., Rafii, S., 2001. Role of angiogenesis in the progression and treatment of prostate cancer. Cancer Invest. 19, 181–91.

Concato, J., Jain, D., Uchio, E., Risch, H., Li, W.W., Wells, C.K., 2009. Molecular markers and death from prostate cancer. Ann. Intern. Med. 150, 595–603.

Cooper, C.R., Chay, C.H., Pienta, K.J., 2002. The role of $\alpha_v\beta_3$ in prostate cancer progression. Neoplasia 4, 191–194.

Courtney, K.D., Choueiri, T.K., 2009. Optimizing recent advances in metastatic renal cell carcinoma. Curr. Oncol. Rep. 11, 218–226.

Cousins, G.R., Sudo, Y., Friedrichs, G.R., Markland, F.S., Lucchesi, B.R., 1995. Contortrostatin prevents reocclusion after thrombolytic therapy in a canine model of carotid artery thrombosis. FASEB J. 9, A938.

Cox, M.C., Permenter, M., Figg, W.D., 2005. Angiogenesis and prostate cancer: important laboratory and clinical findings. Curr. Oncol. Rep. 7, 215–219.

Cress, A.E., Rabinovitz, I., Zhu, W., Nagle, R.B., 1995. The alpha 6 beta 1 and alpha 6 beta 4 integrins in human prostate cancer progression. Cancer Metastasis Rev. 14, 219–228.

Demirgoz, D., Garg, A., Kokkoli, E., 2008. PR_b-targeted PEGylated liposomes for prostate cancer therapy. Langmuir 24, 13518–13524.

Dennis, M.S., Henzel, W.J., Pitti, R.M., Lipari, M.T., Napier, M.A., Deisher, T.A., Bunting, S., Lazarus, R.A., 1990. Platelet glycoprotein IIb/IIIa protein antagonists from snake venoms: evidence for a family of platelet-aggregation inhibitors. Proc. Natl. Acad. Sci. U.S.A. 87, 2471–2475.

Di Lorenzo, G., De Placido, S., 2006. Hormone refractory prostate cancer (HRPC): present and future approaches of therapy. Int. J. Immunopathol. Pharmacol. 19, 11–34.

Ebos, J.M., Lee, C.R., Cruz-Munoz, W., Bjarnason, G.A., Christensen, J.G., Kerbel, R.S., 2009. Accelerated metastasis after short-term treatment with a potent inhibitor of tumor angiogenesis. Cancer Cell 15, 232–239.

Erbersdobler, A., Isbarn, H., Dix, K., Steiner, I., Schlomm, T., Mirlacher, M., Sauter, G., Haese, A., 2009. Prognostic value of microvessel density in prostate cancer: a tissue microarray study. World J. Urol. In Press

Fox, W.D., Higgins, B., Maiese, K.M., Drobnjak, M., Cordon-Cardo, C., Scher, H.I., Agus, D.B., 2002. Antibody to vascular endothelial growth factor slows growth of an androgen-independent xenograft model of prostate cancer. Clin. Cancer Res. 8, 3226–3231.

Friedlander, M., Brooks, P.C., Shaffer, R.W., Kincaid, C.M., Varner, J.A., Cheresh, D.A., 1995. Definition of two angiogenic pathways by distinct αv integrins. Science 270, 1500–1502.

Fujii, G., 1996. Liposomal amphotericin B(AmBisome): realization of the drug delivery concept, in: Rosoff, M. (Ed.), Vesicles. Marcel Dekker, New York, pp. 491–526.

Fujii, Y., Okuda, D., Fujimoto, Z., Horii, K., Morita, T., Mizuno, H., 2003. Crystal structure of trimestatin, a disintegrin containing a cell adhesion recognition motif RGD. J. Mol. Biol. 332, 1115–1122.

Garrison, J.B., Kyprianou, N., 2004. Novel targeting of apoptosis pathways for prostate cancer therapy. Curr. Cancer Drug Targets 4, 85–95.

Gasparini, G., Brooks, P.C., Biganzoli, E., Vermeulen, P.B., Bonoldi, E., Dirix, L.Y., Ranieri, G., Miceli, R., Cheresh, D.A., 1998. Vascular integrin $\alpha_v\beta_3$: a new prognostic indicator in breast cancer. Clin. Cancer. Res. 4, 2625–2634.

Gleave, M., Miyake, H., Chi, K., 2005. Beyond simple castration: targeting the molecular basis of treatment resistance in advanced prostate cancer. Cancer Chemother. Pharmacol. 56 Suppl 1, 47–57.

Goel, H.L., Li, J., Kogan, S., Languino, L.R., 2008. Integrins in prostate cancer progression. Endocr. Relat. Cancer 15, 657–664.

Golubkov, V., Hawes, D., Markland, F.S., 2003. Anti-angiogenic activity of contortrostatin, a disintegrin from *Agkistrodon contortrix contortrix* snake venom. Angiogenesis 6, 213–224.

Gould, R.J., Polokoff, M.A., Friedman, P.A., Huang, T.F., Holt, J.C., Cook, J.J., Niewiarowski, S., 1990. Disintegrins: a family of integrin inhibitory proteins from viper venoms. Proc. Soc. Exptl. Biol. Med. 195, 168–171.

Grothey, A., Galanis, E., 2009. Targeting angiogenesis: progress with anti-VEGF treatment with large molecules. Nat. Rev. Clin. Oncol. 6, 507–518.

Gutheil, J.C., Campbel, T.N., Pierce, P.R., Watkins, J.D., Huse, W.D., Bodkin, D.J., Cheresh, D.A., 2000. Targeted antiangiogenic therapy for cancer using vitaxin: a humanized monoclonal antibody to the integrin $\alpha v\beta 3$. Clin. Cancer Res. 6, 3056–3061.

Harris, W.P., Mostaghel, E.A., Nelson, P.S., Montgomery, B., 2009. Androgen deprivation therapy: progress in understanding mechanisms of resistance and optimizing androgen depletion. Nat. Clin. Pract. Urol. 6, 76–85.

Hiles, J.J., Kolesar, J.M., 2008. Role of sunitinib and sorafenib in the treatment of metastatic renal cell carcinoma. Am. J. Health Syst. Pharm. 65, 123–131.

Hood, J.D., Bednarski, M., Frausto, R., Guccione, S., Reisfeld, R.A., Xiang, R., Cheresh, D.A., 2002. Tumor regression by targeted gene delivery to the neovasculature. Science 296, 2404–2407.

Huang, T.F., Holt, J.C., Kirby, E.R., Niewiarowski, S., 1989. Trigramin: primary structure and its inhibition of von willebrand factor binding to glycoprotein IIb-IIIa complex on human platelets. Biochemistry 28, 661–666.

Humphries, M.J., 1996. Integrin activation: the link between ligand binding and signal transduction. Curr. Opin. Cell Biol. 8, 632–640.

Hynes, R.O., 1992. Integrins: versatility, modulation, and signaling in cell adhesion. Cell 69, 11–25.

Jameson, D.M., Seifried, S.E., 1999. Quantification of protein–protein interactions using fluorescence polarization. Methods 19, 222–233.

Jemal, A., Siegel, R., Ward, E., Hao, Y., Xu, J., Thun, M.J., 2009. Cancer statistics, 2009. CA. Cancer J. Clin. 59, 225–249.

Jimenez, J.A., Kao, C., Raikwar, S., Gardner, T.A., 2006. Current status of anti-angiogenesis therapy for prostate cancer. Urol. Oncol. 24, 260–268.

Kerbel, R.S., 2001a. Clinical trials of antiangiogenic drugs: opportunities, problems, and assessment of initial results. J. Clin. Oncol. 19, 45S–51S.

Kerbel, R. S., 2001b. Molecular and physiologic mechanisms of drug resistance in cancer: an overview. Cancer Metastasis Rev. 20, 1–2.

Kumar, C.C., 2003. Integrin alpha v beta 3 as a therapeutic target for blocking tumor-induced angiogenesis. Curr. Drug. Targets. 4, 123–311.

Kwabi-Addo, B., Ozen, M., Ittmann, M., 2004. The role of fibroblast growth factors and their receptors in prostate cancer. Endocr. Relat. Cancer 11, 709–724.

Liotta, L.A., Kohn, E.C., 2001. The microenvironment of the tumour-host interface. Nature 411, 375–379.

Madan, R.A., Dahut, W.L., 2009. Angiogenesis inhibition in the treatment of prostate cancer. Anticancer Agents Med. Chem. 27, 5627–5633.

Marcinkiewicz, C., Vijay-Kumar, S., McLane, M.A., Niewiarowski, S., 1997. Significance of RGD loop and C-terminal domain of echistatin for recognition of $\alpha_{IIb}\beta_3$ and $\alpha_v\beta_3$ integrins and expression of ligand-induced binding site. Blood 90, 1565–1575.

Markland, F.S., Shieh, K., Zhou, Q., Golubkov, V., Sherwin, R.P., Richters, V., Sposto, R., 2001. A novel snake venom disintegrin that inhibits human ovarian cancer dissemination and angiogenesis in an orthotopic nude mouse model. Haemostasis 31, 183–191.

McLane, M.A., Joerger, T., Mahmoud, A., 2008. Disintegrins in health and disease. Front. Biosci. 13, 6617–6637.

McLane, M.A., Marcinkiewicz, C., Vijay-Kumar, S., Wierzbicka-Patynowski, I., Niewiarowski, S., 1998. Viper venom disintegrins and related molecules. Proc. Soc. Exp. Biol. Med. 219, 109–119.

Merseburger, A.S., Simon, A., Waalkes, S., Kuczyk, M.A., 2009. Sorafenib reveals efficacy in sequential treatment of metastatic renal cell cancer. Expert Rev. Anticancer Ther. 9, 1429–1434.

Minea, R., Swenson, S., Costa, F., Chen, T.C., Markland, F.S., 2005. Development of a novel recombinant disintegrin, contortrostatin, as an effective anti-tumor and anti-angiogenic agent. Pathophysiol. Haemost. Thromb. 34, 177–183.

Montagnani, F., Migali, C., Fiorentini, G., 2009. Progression-free survival in bevacizumab-based first-line treatment for patients with metastatic colorectal cancer: is it a really good end point? J. Clin. Oncol. 27, e132–e133; author reply e134–e135.

Mucci, L.A., Powolny, A., Giovannucci, E., Liao, Z., Kenfield, S.A., Shen, R., Stampfer, M.J., Clinton, S.K., 2009. Prospective study of prostate tumor angiogenesis and cancer-specific mortality in the health professionals follow-up study. J. Clin. Oncol. 9, 1070–1078.

Nemeth, J.A., Cher, M.L., Zhou, Z., Mullins, C., Bhagat, S., Trikha, M., 2003. Inhibition of $\alpha_v\beta_3$ integrin reduces angiogenesis, bone turnover, and tumor cell proliferation in experimental prostate cancer bone metastases. Clin. Exp. Metastasis 20, 413–420.

Niewiarowski, S., McLane, M.A., Kloczewiak, M., Stewart, G.J., 1994. Disintegrins and other naturally occurring antagonists of platelet fibrinogen receptors. Semin. Hematol. 31, 289–300.

Paez-Ribes, M., Allen, E., Hudock, J., Takeda, T., Okuyama, H., Vinals, F., Inoue, M., Bergers, G., Hanahan, D., Casanovas, O., 2009. Antiangiogenic therapy elicits malignant progression of tumors to increased local invasion and distant metastasis. Cancer Cell 15, 220–231.

Pallares, J., Rojo, F., Iriarte, J., Morote, J., Armadans, L.I., de Torres, I., 2006. Study of microvessel density and the expression of the angiogenic factors VEGF, bFGF and the receptors Flt-1 and FLK-1 in benign, premalignant and malignant prostate tissues. Histol. Histopathol. 21, 857–865.

Phillips, D.R., Charo, I.F., Scarborough, R.M., 1991. GP IIb-IIIa: the responsive integrin. Cell 65, 359–362.

Pignatelli, M., Cardillo, M.R., Hanby, A., Stamp, G.W., 1992. Integrins and their accessory adhesion molecules in mammary carcinomas: loss of polarization in poorly differentiated tumors. Human Pathol. 23, 1159–1166.

Pinski, J., Markland, F., Wang, Q., Horiatis, D., Swenson, S., Costa, F. (2003). A novel therapy for prostate cancer based on the disintegrin contortrostatin. ASCO Annual Meeting. Chicago, IL.

Polnaszek, N., Kwabi-Addo, B., Peterson, L.E., Ozen, M., Greenberg, N.M., Ortega, S., Basilico, C., Ittmann, M., 2003. Fibroblast growth factor 2 promotes tumor progression in an autochthonous mouse model of prostate cancer. Cancer Res. 63, 5754–5760.

Pyrko, P., Wang, W., Markland, F.S., Swenson, S.D., Schmitmeier, S., Schonthal, A.H., Chen, T.C., 2005. The role of contortrostatin, a snake venom disintegrin, in the inhibition of tumor progression and prolongation of survival in a rodent glioma model. J. Neurosurg. 103, 526–537.

Ranieri, G., Patruno, R., Ruggieri, E., Montemurro, S., Valerio, P., Ribatti, D., 2006. Vascular endothelial growth factor (VEGF) as a target of bevacizumab in cancer: from the biology to the clinic. Curr. Med. Chem. 13, 1845–1857.

Reardon, D.A., Fink, K.L., Mikkelsen, T., Cloughesy, T.F., O'Neill, A., Plotkin, S., Glantz, M., Ravin, P., Raizer, J.J., Rich, K.M., Schiff, D., Shapiro, W.R., Burdette-Radoux, S., Dropcho, E.J., Wittemer, S.M., Nippgen, J., Picard, M., Nabors, L.B., 2008. Randomized phase II study of cilengitide, an integrin-targeting arginine-glycine-aspartic acid peptide, in recurrent glioblastoma multiforme. J. Clin. Oncol. 26, 5610–5617.

Romanov, V.I., Goligorsky, M.S., 1999. RGD-recognizing integrins mediate interactions of human prostate carcinoma cells with endothelial cells in vitro. Prostate 39, 108–118.

Ruoslahti, E., 1991. Integrins. J. Clin. Invest. 87, 1–5.

Sanz, L., Chen, R.Q., Perez, A., Hilario, R., Juarez, P., Marcinkiewicz, C., Monleon, D., Celda, B., Xiong, Y.L., Perez-Paya, E., Calvete, J.J., 2005. cDNA cloning and functional expression of jerdostatin, a novel RTS-disintegrin from *Trimeresurus jerdonii* and a specific antagonist of the $\alpha 1\beta 1$ integrin. J. Biol. Chem. 280, 40714–40722.

Saudek, V., Atkinson, R.A., Lepage, P., Pelton, J.T., 1991. The secondary structure of echistatin from ^{1}H-NMR. Eur. J. Biochem. 202, 329–338.

Scarborough, R.M., Rose, J.W., Naughton, M.A., Phillips, D.R., Nannizzi, L., Arfsten, A., Campbell, A.M., Charo, I.F., 1993. Characterization of the integrin specificities of disintegrins isolated from American pit viper venoms. J. Biol. Chem. 268, 1058–1065.

Sharifi, N., Dahut, W.L., Figg, W.D., 2008. The genetics of castration-resistant prostate cancer: what can the germline tell us? Clin. Cancer Res. 14, 4691–4693.

Shebuski, R.J., Ramjit, D.R., Bencen, G.H., Polokoff, M.A., 1989. Characterization and platelet inhibitor activity of bitistatin, a potent arginine-glycine-aspartic acid-containing peptide from the venom of viper *Bitis arietans*. J. Biol. Chem. 264, 21550–21556.

Stoeltzing, O., Liu, W., Reinmuth, N., Fan, F., Parry, G.C., Parikh, A.A., McCarty, M.F., Bucana, C.D., Mazar, A.P., Ellis, L.M., 2003. Inhibition of integrin α5β1 function with a small peptide (ATN-161) plus continuous 5-FU infusion reduces colorectal liver metastases and improves survival in mice. Int. J. Cancer 104, 496–503.

Strohmeyer, D., Strauss, F., Rossing, C., Roberts, C., Kaufmann, O., Bartsch, G., Effert, P., 2004. Expression of bFGF, VEGF and c-met and their correlation with microvessel density and progression in prostate carcinoma. Anticancer Res. 24, 1797–1804.

Strömblad, S., Becker, J.C., Yebra, M., Brooks, P.C., Cheresh, D. A., 1996. Suppression of p53 activity and p21WAF1/CIP1 expression by vascular cell integrin αvβ3 during angiogenesis. J. Clin. Invest. 98, 426–433.

Sun, X.T., Ding, Y.T., Yan, X.G., Wu, L.Y., Li, Q., Cheng, N., Qiu, Y.D., Zhang, M.Y., 2004. Angiogenic synergistic effect of basic fibroblast growth factor and vascular endothelial growth factor in an in vitro quantitative microcarrier-based three-dimensional fibrin angiogenesis system. World J. Gastroenterol. 10, 2524–2528.

Sun, Y.X., Fang, M., Wang, J., Cooper, C.R., Pienta, K.J., Taichman, R.S., 2007. Expression and activation of αvβ3 integrins by SDF-1/CXC12 increases the aggressiveness of prostate cancer cells. Prostate 67, 61–73.

Swenson, S., Costa, F., Minea, R., Sherwin, R.P., Ernst, W., Fujii, G., Yang, D., Markland, F.S., Jr., 2004. Intravenous liposomal delivery of the snake venom disintegrin contortrostatin limits breast cancer progression. Mol. Cancer. Ther. 3, 499–511.

Trikha, M., De Clerck, Y.A., Markland, F.S., 1994a. Contortrostatin, a snake venom disintegrin, inhibits β1 integrin-mediated human metastatic melanoma cell adhesion and blocks experimental metastasis. Cancer Res. 54, 4993–4998.

Trikha, M., Rote, W.E., Manley, P.J., Lucchesi, B.R., Markland, F.S., 1994b. Purification and characterization of platelet aggregation inhibitors from snake venoms. Thromb. Res. 73, 39–52.

van Moorselaar, R.J., Voest, E.E., 2002. Angiogenesis in prostate cancer: its role in disease progression and possible therapeutic approaches. Mol. Cell. Endocrinol. 197, 239–250.

Weidner, N., Carroll, P.R., Flax, J., Blumenfeld, W., Folkman, J., 1993. Tumor angiogenesis correlates with metastasis in invasive prostate carcinoma. Am. J. Pathol. 143, 401–409.

Witkowski, C.M., Rabinovitz, I., Nagle, R.B., Affinito, K.S., Cress, A.E., 1993. Characterization of integrin subunits, cellular adhesion and tumorgenicity of four human prostate cell lines. J. Cancer. Res. Clin. Oncol. 119, 637–644.

Woodle, M.C., 1993. Surface-modified liposomes: assessment and characterization for increased stability and prolonged blood circulation. Chem. Phys. Lipids 64, 249–262.

Wu, I., Moses, M.A., 2000. Angiogenic molecules and mechanisms in breast cancer. Curr. Oncol. Rep. 2, 566–571.

Xiong, J.P., Stehle, T., Goodman, S.L., Arnaout, M.A., 2003. New insights into the structural basis of integrin activation. Blood 102, 1155–1159.

Yahalom, D., Wittelsberger, A., Mierke, D.F., Rosenblatt, M., Alexander, J.M., Chorev, M., 2002. Identification of the principal binding site for RGD-containing ligands in the $\alpha_V\beta_3$ integrin: a photoaffinity cross-linking study. Biochemistry 41, 8321–8331.

Yasuda, T., Gold, H.K., Leinbach, R.C., Yaoita, H., Fallon, J.T., Guerrero, L., Napier, M.A., Bunting, S., Collen, D., 1991. Kistrin, a polypeptide platelet GPIIb/IIIa receptor antagonist, enhances and sustains coronary arterial thrombolysis with recombinant tissue-type plasminogen activator in a canine preparation. Circulation 83, 1038–1047.

Zheng, D.Q., Woodard, A.S., Fornaro, M., Tallini, G., Languino, L.R., 1999. Prostatic carcinoma cell migration via $\alpha_V\beta_3$ integrin is modulated by a focal adhesion kinase pathway. Cancer Res. 59, 1655–1664.

Zhou, Q., Nakada, M.T., Arnold, C., Markland, F.S., 1999. Contortrostatin, a dimeric disintegrin from *Agkistrodon contortrix contortrix*, inhibits angiogenesis. Angiogenesis 3, 259–269.

Zhou, Q., Nakada, M.T., Brooks, P.C., Swenson, S.D., Ritter, M.R., Argounova, S., Arnold, C., Markland, F.S., 2000a. Contortrostatin, a homodimeric disintegrin, binds to integrin $\alpha_v\beta_5$. Biochem. Biophys. Res. Commun. 267, 350–355.

Zhou, Q., Sherwin, R.P., Parrish, C., Richters, V., Groshen, S.G., Tsao-Wei, D., Markland, F.S., 2000b. Contortrostatin, a dimeric disintegrin from *Agkistrodon contortrix contortrix*, inhibits breast cancer progression. Breast Cancer Res. Treat. 61, 249–260.

Zhu, A.X., Duda, D.G., Sahani, D.V., Jain, R.K., 2009. Development of sunitinib in hepatocellular carcinoma: rationale, early clinical experience, and correlative studies. Cancer J. 15, 263–268.

Chapter 20
Hematophagy and Inhibition of Platelet Aggregation

Ivo M.B. Francischetti

Abstract Salivary glands from blood-sucking animals (e.g., mosquitoes, bugs, sandflies, fleas, ticks, leeches, hookworms, bats) are a rich source of bioactive molecules that counteract hemostasis in a redundant and synergistic manner. This review discusses recent progress in the identification of salivary inhibitors of platelet aggregation, their molecular characterization, and detailed mechanism of action. Diversity of inhibitors is remarkable, with distinct families of proteins characterized as apyrases that enzymatically degrade ADP or as collagen-binding proteins that prevent its interaction with vWF, or platelet integrin a2b1 or GPVI. Molecules that bind ADP, TXA_2, epinephrine, or serotonin with high affinity have also been cloned, expressed, and their structure determined. In addition, a repertoire of antithrombins and an increasingly number of RGD and non-RGD disintegrins targeting platelet aIIbb3 have been reported. Moreover, metalloproteases with fibrinogen (olytic) activity and PAF phosphorylcholine hydrolase are enzymes that have been recruited to the salivary gland to block platelet aggregation. Platelet inhibitory prostaglandins, lysophosphatydilcholine, adenosine, and nitric oxide (NO)-carrying proteins are other notable examples of molecules from hematophagous salivary secretions (herein named sialogenins) with antihemostatic properties. Sialogenins have been employed as tools in biochemistry and cell biology and also display potential therapeutic applications.

Hematophagy and Platelet Activation

Hematophagous animals have evolved various strategies to counteract their host hemostatic system and to successfully feed on blood. In evolution, it has been suggested that insects may have been initially nest parasites or predators on other insects. They may also have fed on secretions from vertebrate eyes or other mucosal

I.M.B. Francischetti (✉)
Section of Vector Biology, Laboratory of Malaria and Vector Research, National Institute of Allergy and Infectious Diseases, National Institutes of Health, Rockville, MD 20852-8132, USA
e-mail: ifrancischetti@niaid.nih.gov

R.M. Kini et al. (eds.), *Toxins and Hemostasis*, DOI 10.1007/978-90-481-9295-3_20,

membranes where access to a meal is easier and does not require piercing of the skin. Perhaps this was the first adaptation toward acquiring the ability to penetrate the skin in search of blood (Francischetti et al., 2009; Ribeiro and Francischetti, 2003). Actually, blood-feeding may occur through two different pathways: from hemorrhagic pools that accumulate in tissues following skin lacerations, or through direct cannulation of venules or arterioles. For example, mosquitoes and bugs cannulate the arterioles and venules deep in the skin, sometimes reaching several millimeters. In contrast, sand fly mouthparts penetrate up to 0.5 mm into their host skin and therefore can feed only from superficial hematomas caused by laceration of capillaries. Tabanids, which have cutting scissors, can also lacerate several vessels simultaneously, potentially forming a large hematoma. Further, ticks introduce their mouthparts into the host and generate a feeding cavity from which they feed on blood (Francischetti et al., 2009; Ribeiro and Francischetti, 2003). These animals express molecules from hematophagous salivary secretions, herein named sialogenins (from the Greek *sialo*, saliva; *gen*, origin, source; and *ins* for proteins). Sialogenins target the host response to injury, modulate immune response, prevent pain, and other mechanisms associated with blood feeding. While several functions of sialogenins have been reviewed before (Francischetti et al., 2009; Koh and Kini, 2009; Maritz-Olivier et al., 2007; Ribeiro and Francischetti, 2003; Steen et al., 2006), this review focuses on those that target platelet aggregation.

Platelets have a central role in hemostasis. The first step in the hemostatic cascade is platelet interaction with the exposed extracellular matrix (ECM), which contains a large number of adhesive macromolecules such as collagens and fibronectin. Under conditions of high shear, such as in small arteries and arterioles, the initial tethering of platelets to the ECM is mediated by interaction between the platelet receptor glycoprotein (GP)Ib and vWF bound to collagen (Furie and Furie, 2005; Jackson and Schoenwaelder, 2003; Ruggeri, 2002; Watson et al., 2005). GPIb to vWF binding dissociates rapidly and is insufficient to mediate stable adhesion but rather maintains the platelet in close contact with the exposed surface. This interaction allows the collagen receptor GPVI (Jandrot-Perrus et al., 2000) to bind to collagen. This triggers the conformational change of integrins (e.g., $\alpha2\beta1$) to a high-affinity state, thereby enabling them to mediate firm adhesion to collagen and also promotes the release of TXA_2 and ADP (Furie and Furie, 2005; Jackson and Schoenwaelder, 2003; Ruggeri, 2002; Watson et al., 2005).

ADP binds to the G_q-protein-linked $P2Y_1$ receptor on platelets, which causes a change in cell shape, mobilization of calcium, and initiation of reversible aggregation via stimulation of PLC. It also binds the G_i-linked $P2Y_{12}$ receptor to amplify aggregation via inhibition of adenylyl-cyclase-mediated cyclic AMP production. The resulting platelet activation leads to PI-3-kinase activation, granule secretion, and "inside-out" activation of integrin $\alpha IIb\beta3$, which increases its affinity for fibrinogen and vWF. These ligands then bind to the receptors to form bridges between adjacent platelets, which results in "outside-in" signalling and aggregation (Kahner et al., 2006; Varga-Szabo et al., 2008). Accordingly, sustained ADP-induced platelet aggregation requires activation of both $P2Y_1$ and $P2Y_{12}$ receptors. TXA2 activates platelets through the TP receptors that are coupled to Gq and $G_{12/13}$. While it promotes shape change, activation of this pathway is unable to induce sustained platelet aggregation in the absence of a Gi-coupled receptor agonist such as ADP, explaining the sensitivity of TXA2 mimetic (U46619)-induced platelet aggregation to apyrases (Kahner et al., 2006; Varga-Szabo et al., 2008).

Granule secretion also releases the weak agonists serotonin (5-HT) and epinephrine. While serotonin is not a pro-aggregatory molecule, it binds to G_q-coupled 5-HT_{2A} receptor and activates PLC, leading to shape change. On the other hand, epinephrine inhibits adenylyl cyclase via a G_i-coupled α_{2A} adrenergic receptor without noticeable effects on platelet aggregation; however, when added together, both weak agonists induce platelet aggregation comparable to ADP. Further, epinephrine and serotonin potentiate platelet aggregation by a number of platelet inducers tested at low concentrations, decreasing their threshold for aggregation; this property appears to be particularly relevant in vivo, where aggregation occurs as a result of several different agonists acting synergetically and in a redundant manner (Kahner et al., 2006; Varga-Szabo et al., 2008).

Activated platelets also promote the generation of thrombin—the most potent of all physiological platelet agonists. Thrombin acts predominantly via protease-activated receptors 1 and 4 (PAR1 and PAR4) expressed on platelets. Thrombin cleaves a portion of the N-terminus of these receptors, unmasking the sequence that serves as its ligand; this modification activates the receptor and triggers multiple signal transduction pathways that modulate thrombosis, coagulation, and inflammation (Coughlin, 2005). In addition, activated platelets secrete a number of proinflammatory molecules and may contribute to local inflammatory processes by a number of mechanisms (Gawaz et al., 2005). Finally, physiologic inhibitors of platelet aggregation have been described including endothelium CD39 apyrase, which degrades ADP; nitric oxide (NO), which inhibits platelet aggregation through

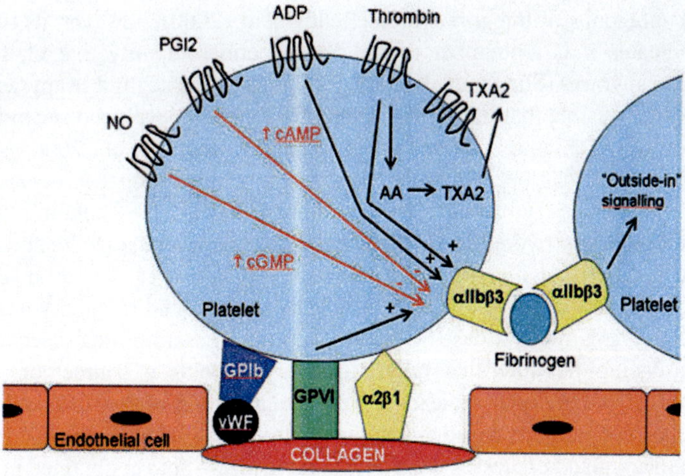

Fig. 20.1 Platelet activation. Pro-aggregatory molecules ADP, thrombin, and collagen activate platelets through specific receptors, leading to PLA2 and PLC activation, granule release, production of TXA2 and activation of integrin αIIbβ3. Inhibitory signals include endothelial apyrase (not shown), prostacyclin, and NO (for details see text)

cGMP; and PGI2, which shuts down different pathways of platelet activation by increasing intracellular cAMP through PGI_2 receptors (Ruggeri, 2002; Varga-Szabo et al., 2008). Figure 20.1 displays the basic mechanisms of platelet aggregation and inhibition. Table 20.1 summarizes the targets and properties for antiplatelet sialogenins.

Sialogenins That Inhibit Platelet Aggregation

Apyrases

The saliva of many hematophagous animals including mosquitoes, bugs, sand flies, fleas, triatomines, and ticks contains large amounts of apyrase activity (ATP-diphosphohydrolase EC 3.6.1.5), which hydrolyses ATP and ADP into AMP and P_i, thus inhibiting platelet aggregation (Ribeiro and Francischetti, 2003). Apyrases are ubiquitous, conceivably because ADP is an important platelet aggregation inducer released by damaged cells at the site of injury and from activated platelets (Watson et al., 2005).

Three classes of apyrases have been characterized at the molecular level in different blood-sucking arthropods. One is the 5′-nucleotidase family, which have been cloned from the salivary gland of mosquitoes *Anopheles gambiae* (Lombardo et al., 2000) and *Aedes aegypti* (Champagne et al., 1995), bugs such as *Triatoma infestans* (Faudry et al., 2004), soft tick *Ornithodorus savignyi* (Stutzer et al., 2009) and from the hard tick *Rhipicephalus (Boophilus) microplus* (Liyou et al., 1999). The platelet aggregation inhibitor chrysoptin has also been cloned from the salivary gland of the tabanid *Chrysops sp*. It has been suggested that it inhibits platelet aggregation as a non-RGD antagonist of integrin αIIbβ3 (Reddy et al., 2000); however, its sequence clearly indicates it is a member of the 5′-nucleotidase family that clades with *An. gambiae* apyrases (Stutzer et al., 2009). Although it has negligible apyrase activity, it remains possible that inhibition of integrin function developed secondarily to degradation by ADP, as it activates integrin αIIbβ3 (Reddy et al., 2000). Another family of apyrases has been initially identified in the bed bug *Cimex lectularius* and confirmed to be enzymatically active (Valenzuela et al., 1998); members of this family were later discovered in sand fly *Phlebotomus papatasi* (Valenzuela et al., 2001) and *Lutzomyia longipalpis* salivary glands (Charlab et al., 1999). Of note, *Cimex*-type apyrase (Valenzuela et al., 1998) has been found in eukaryotes and to be fully active as a nucleotidase (Dai et al., 2004). Therefore, this discovery validates and exemplifies how the study of salivary proteins as sources of (known) biologic activities may provide leads for the identification of the (unknown) function of human counterparts based on sequence similarity. Finally, the flea *Xenopsylla cheopis* salivary gland expresses a third type of apyrase—not yet obtained in recombinant form—whose sequence resembles that of the CD39 family of nucleotidases (Andersen et al., 2007) found in the endothelium. Therefore, degradation of ADP through recruitment of different families of apyrases has been a consistent strategy

Table 20.1 Sialogenins that inhibit platelet aggregation

Targets and inhibitors	Species	Molecular mass (mature protein)	Target	R/S/P	Comments and references
ADP (hydrolysis)			**Specificity/affinity**		
Apyrase	*A. gambiae*	61	ADP	N/N/N	5′-nucleotidase family (Lombardo et al., 2000).
Apyrase	*A. aegypti*	60	ADP	N/N/N	5′-nucleotidase family (Champagne et al., 1995).
Apyrase	*T. infestans*	60	ADP	Y/N/Y	5′-nucleotidase family (Faudry et al., 2004).
Apyrase	*O. savignyi*	63	ADP	Y/N/N	5′-nucleotidase family (Stutzer et al., 2009).
Apyrase	*Chrysops sp.*	59	ADP	Y/N/Y	5′-nucleotidase family (Reddy et al., 2000). Chrysoptin.
Apyrase	*B. microplus*	62	ADP	Y/N/N	5′-nucleotidase family (Liyou et al., 1999).
Apyrase	*C. lectularius*	38	ADP	Y/N/N	Cimex family (Valenzuela et al., 1998).
Apyrase	*P. papatasi*	36	ADP	N/N/N	Cimex family (Valenzuela et al., 2001).
Apyrase	*L. longipalpis*	35	ADP	N/N/N	Cimex family (Charlab et al., 1999).
Apyrase	*X. cheopis*	47	ADP	N/N/N	CD39 family (Andersen et al., 2007).
ADP (binding)			**Specificity/affinity**		
RPAI-1	*R. prolixus*	19	ADP, 50 nM	Y/N/Y	Binds to other adenine nucleotides (Francischetti et al., 2000, 2002). No apyrase activity.
TXA (binding)			**Specificity/affinity**		
Moubatin	*O. moubata*	17	cTXA2, 24 nM	Y/N/Y	Also binds AA, LTB4 (Keller et al., 1993, Mans and Ribeiro, 2008).
TSGP3	*O. savignyi*	16	cTXA2, 5 nM	Y/N/Y	Also binds AA, LTB4, and C5 (Mans and Ribeiro, 2008).

Table 20.1 (continued)

Targets and inhibitors	Species	Molecular mass (mature protein)	Target	R/S/P	Comments and references
Putative ADP or TXA2 (binding)		**Specificity**			
Pallidipin	*T. pallidipennis*	18	ADP?; TXA2?	Y/N/Y	Possible ADP or TXA2 binder (Noeske-Jungblut et al., 1994).
Triplatin	*T. infestans*	19	ADP?; TXA2?	Y/N/Y	Possible ADP or TXA2 binder. GPVI antagonist unclear. (Morita et al., 2006).
Longicornin	*H. longicornis*	16	ADP?; TXA2?	N/N/Y	Possible ADP or TXA2 binder (Cheng et al., 1999).
EPI or 5-HT (binding)			**Specificity/affinity**		
ABP	*R. prolixus*	19	5-HT, 102 nM EPI, 345 nM	Y/N/Y	Blocks potentiation of platelet aggregation induced by serotonin and EPI (Andersen et al., 2003). Binds NE (KD, 24 nM).
D7r1–4 (short)	*A. gambiae*	15	5-HT, 1 nM	Y/Y/N	Displays low affinity for histamine and EPI. D7r2,-3 binds NE (KD 3nM) (Calvo et al., 2006).
AeD7L (long)	*A. Aegypti*	33	5-HT, 0.5 nM	Y/Y/N	High affinity for cysteinyl leukotrienes. Low affinity for histamine and EPI. Binds NE (KD 0.1 nM) (Calvo et al., 2009).
Monotonin	*A. monolakensis*	20	5-HT, < 2 nM	Y/Y/N	Presents a BAB motif (Mans et al., 2008).
SHBP	*D. reticularis*	22	5-HT, 0.6 nM	Y/N/N	Binds histamine with high affinity (Sangamnatdej et al., 2002).
Collagen			**Affinity for collagen/IC_{50} for platelet aggregation**		
Aegyptin	*A. aegypti*	30	1 nM, 50 nM	Y/N/Y	Blocks aggregation and adhesion to collagen at static condition or at high shear rates (Calvo et al., 2007). Antithrombotic in vivo.

Table 20.1 (continued)

Targets and inhibitors	Species	Molecular mass (mature protein)	Target	R/S/P	Comments and references
AAPP	*A. stephensi*	30	?, 25 nM	Y/N/Y	Blocks aggregation and adhesion to collagen at static condition. Antiplatelet ex vivo (Yoshida et al., 2008).
Calin	*H. medicinalis*	65	?, 100 nM	N/N/Y	Molecular identity(ies) unknown. Blocks aggregation and adhesion to collagen at static condition or at high shear rates (Harsfalvi et al., 1995). Antithrombotic in vivo (Deckmyn et al., 1995).
LAPP	*H. officialis*	13	?, 100 nM	Y/Y/Y	Blocks aggregation and adhesion to collagen at static condition or at high shear rates. Antithrombotic in vivo (Keller et al., 1992).
Saratin	*H. medicinalis*	12	50 nM, 20 μM	Y/Y/Y	Preferentially blocks platelet adhesion to collagen at high shear rates (vWF-dependent) (Barnes et al., 2001). Antithrombotic in vivo (Cruz et al., 2001).
Integrin α2β1			**IC_{50} for platelet aggregation**		
Tick adhesion inhibitor (TAI)	*O. moubatta*	5	8 nM	N/N/Y	Inhibits platelet adhesion to soluble collagen but does not affect platelet aggregation. (Karczewski et al., 1995).
Hookworm platelet inhibitor (HPI)	*A. caninum*	20	NA	Y/N/Y	Inhibits α2β1 and αIIbβ3. Recombinant form not active (Del Valle et al., 2003).

Table 20.1 (continued)

Targets and inhibitors	Species	Molecular mass (mature protein)	Target	R/S/P	Comments and references
Nitric oxide (NO)			**IC_{50} for platelet aggregation**		
Nitrophorin (NP7)	*R. prolixus*	19	~10 nM	Y/N/Y	Inhibits ADP-induced platelet aggregation and causes disaggregation (Andersen et al., 2004).
Cimex nitrophorin	*C. lectularius*	32	NA	N/N/N	Putative inhibitor of platelet aggregation (Valenzuela et al., 1995).
Platelet cAMP			**IC_{50} for platelet aggregation**		
PGI2	*I. scapularis*	352	~1 nM	na/na/N	Increases intraplatelet cAMP (Ribeiro et al., 1988).
LysoPC	*R. prolixus*	518	~10 μg/ml	na/na/Y	Increases intraplatelet cAMP (Golodne et al., 2003).
Adenosine	*P. papatasi*	267	~1 μM	na/na/N	Increases intraplatelet cAMP (Ribeiro et al., 1999).
Thrombin			**Affinity and mechanism**		
More than 20 inhibitors	See (∗)	See(∗)	See(∗)	See(∗)	(∗) Usually low mol mass proteins, with or without cysteines. Usually bind to exosite, catalytic site or both. See (Koh and Kini, 2009; Maritz-Olivier et al., 2007; Ribeiro and Francischetti, 2003; Steen et al., 2006).
Integrin αIIbβ3			**IC_{50} for platelet aggregation**		
Disagregin	*O. moubata*	6	104 nM	N/N/Y	Without RGD. Kunitz-containing protein (Karczewski et al., 1994).

Table 20.1 (continued)

Targets and inhibitors	Species	Molecular mass (mature protein)	Target	R/S/P	Comments and references
Decorsin	*M. decora*	4	500 nM	Y/Y/Y	With RGD. Synthetic form active. Protect mice against death from pulmonary thromboembolism (McLane et al., 1995).
Ornatin	*P. ornata*	5	130 nM	N/N/Y	With RGD (Mazur et al., 1991).
Savignygrin	*O. savignyi*	7	130 nM	N/N/Y	With RGD. Kunitz-containing protein (Mans et al., 2002).
Integrin αIIbβ3 (continued)			**IC $_{50}$ for platelet aggregation**		
Variabilin	*D. variabilis*	5	157 nM	N/N/Y	With RGD (Wang et al., 1996).
Tabinhibitin	*T. yao*	25	<40 nM	N/N/Y	With RGD (Ma et al., 2009).
Ixodegrin	*Ixodes* sp	7	NA	N/N/N	With RGD (Francischetti et al., 2005).
Savignygrin-like	*O. coriaceus*	7	NA	N/N/N	With RGD. Kunitz-containing protein (Francischetti et al., 2008).
Monogrin	A. *monolakensis*	10	150 nM	Y/N/Y	With RGD. Kunitz-containing protein (Mans et al., 2008).
Fibrinolytic enzymes			**Specificity/affinity**		
Metalloprotease	*I. scapularis*	37	Fibrin(ogen), ?	N/N/N	Degrades Aα-chain of fibrinogen and fibrin. Degrades fibronectin but not collagen (Francischetti et al., 2003).
Metalloprotease (Tablysin)	*T. yao*	35	Fibrin(ogen), ?	N/N/N	Degrades Aα or Bβ-chains of fibrinogen (Ma et al., 2009).
Metalloprotease (Hementin)	*H. ghilianii*	~80	Fibrin(ogen)/*Km* 1 μM	N/N/N	Degrades Aα, Bβ, and γ chains of fibrinogen and fibrin (Budzynski, 1991).
Metalloprotease (Hementerin)	*H. depressa*	80	Fibrin(ogen)/?	N/N/N	Degrades Aα, Bβ, and γ chains of fibrinogen and fibrin (Chudzinski-Tavassi et al., 1998).

R obtained in recombinant form, *S* structure available, *P* inhibition of platelet aggregation tested with recombinant or purified protein, *na* not applicable, *NA* not available, *5-HT* serotonin, *EPI* epinephrine, *NE* norepinephrine employed by blood-sucking animals to inhibit hemostasis in general and platelet activation in particular.

ADP-Binding Proteins

RPAI-1

Rhodnius prolixus aggregation inhibitor 1 (RPAI-1) is a 19-kDa lipocalin isolated from the salivary gland of *R. prolixus* (Francischetti et al., 2000). RPAI-1 was the first sialogenin whose function was identified as ADP-binding protein (K_D ~50 nM). RPAI-1 also displays high-affinity binding to ATP, AMP, adenosine (Ado), AP4A, and α,β Met ADP but does not bind to inosine, guanosine, uridine, or cytidine, implying that Ado structure is necessary for binding (Francischetti et al., 2002). Despite its high affinity, it only inhibits platelet aggregation triggered by low doses of ADP (<0.5 μM) (Fig. 20.2a) or appropriate concentrations of collagen (Fig. 20.2b), arachidonic acid, U46619, TRAP, PAF, and A23187 (Francischetti et al., 2000, 2002), which is ADP dependent (Kahner et al., 2006; Watson et al., 2005). In fact, high doses of ADP or other pro-aggregatory substances will consistently not (or only marginally) be affected by this family of inhibitors, as the amount of ADP released rapidly saturates the mopping up ability of the inhibitor that occurs at a 1:1 stoichiometry (Francischetti et al., 2000). This is important to keep in mind to avoid a given inhibitor being erroneously classified as a specific inhibitor of collagen, an agonist that—at the usual in vitro concentration of ~1 μg/ml—is particularly sensitive to ADP scavengers. RPAI-1 also inhibits platelet aggregation under high shear observed in the microcirculation as demonstrated by platelet function analyzer (PFA-100) or clot signature analyzer (Francischetti et al., 2002).

Two questions are often raised about the true biological function of lipocalins such as RPAI-1 in vivo. The first is whether the concentration of the inhibitor found in saliva is compatible with the concentrations of pro-aggregatory components produced upon injury at the microcapillary level. The second is whether a high-affinity ADP-binding protein is needed in the saliva, because apyrases effectively inhibit platelet aggregation. Accordingly, computer simulation experiments were performed where the rate of scavenging or hydrolysis of ADP by RPAI-1 or *R. prolixus* (RP) apyrase (a 5′-nucleotidase)—which have distinct K_m and K_{cat} for ADP—were compared. The simulation results indicate that RPAI-1 (high affinity for ADP, no hydrolysis) is functionally more effective at low ADP concentrations (< 1 μM). In contrast, effective degradation of ADP by the RP apyrase (low affinity for ADP and enzymatic degradation) occurs when the ADP concentration is at least 1,000- to 5,000-fold higher (50 μM) than that readily removed by RPAI-1. Of note, ADP physiological plasma concentration is ~0.1 μM, and it can reach 1–2 μM at the site of injury provoked by a Simplate device; 0.2–0.4 μM ADP is known to be biologically active (Francischetti et al., 2000, 2002). Accordingly, RPAI-1 and RP apyrase coexisting in the same secretion have precise and complementary biologic functions in preventing platelet aggregation in vivo.

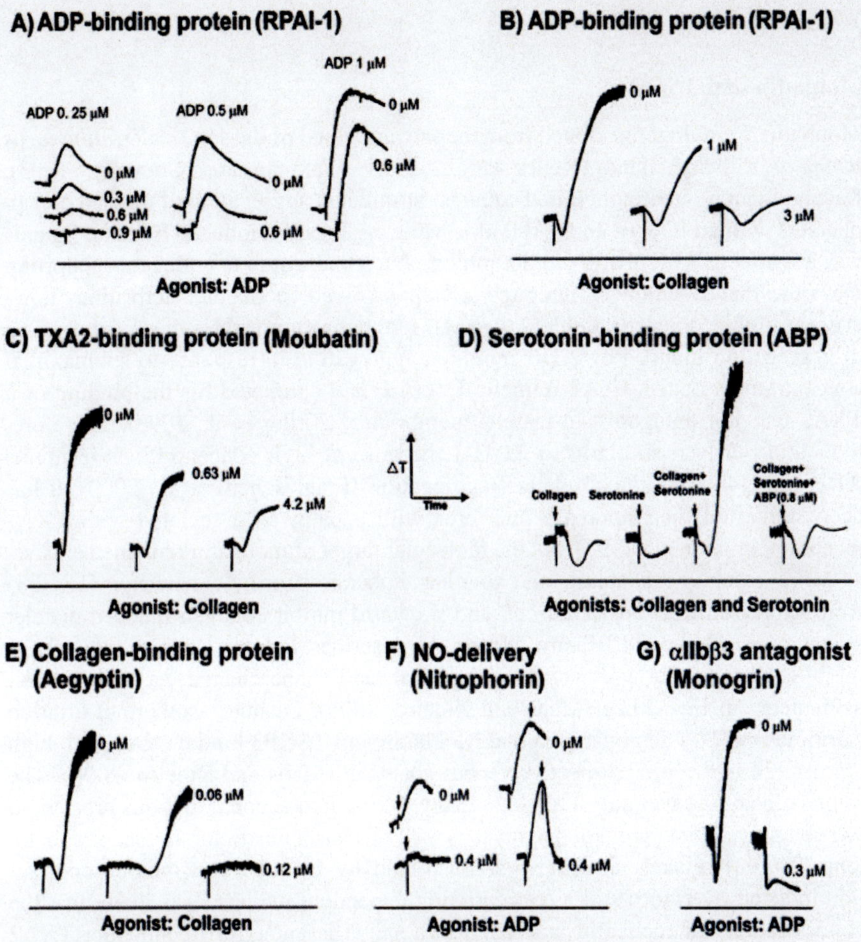

Fig. 20.2 Platelet inhibition by sialogenins. (**a**) ADP-binding proteins, e.g., RPAI-1. Low concentrations of ADP (<0.5 µM)-induced platelet aggregation including shape change is dose-dependently inhibited by RPAI-1 which has only marginal effects at higher concentrations of ADP (>1 µM) (Francischetti et al., 2000). (**b**) Collagen (1 µg/ml)-induced platelet aggregation is attenuated by RPAI-1 without abolishing shape change (Francischetti et al., 2002). (**c**) TXA2-binding protein, e.g., moubatin. Collagen-induced platelet aggregation is attenuated by moubatin without abolishing shape change (Mans and Ribeiro, 2008). (**d**) Serotonin-binding protein, e.g., ABP. Serotonin (0.5 µM)-induced potentiation of aggregation by low doses of collagen (0.75 µg/ml) is blocked by ABP (0.8 µM) (Andersen et al., 2003). (**e**) Collagen-binding protein, e.g., aegyptin. Aggregation induced by collagen (2 µg/ml) is accompanied by delay in onset time for shape change at 60 nM aegyptin; no shape change or aggregation is observable at higher concentrations of the inhibitor (Calvo et al., 2007). (**f**) NO-delivery, e.g., nitrophorins. ADP (3 µM)-induced platelet aggregation and shape change is attenuated by NP2 (0.4 µM) when added before stimulus. NP2 also promotes disaggregation when added (*arrow*) after addition of ADP (Andersen et al., 2004). (**g**) Disintegrin targeting αIIbβ3; e.g., monogrins. ADP (20 µM)-induced platelet aggregation is prevented by monogrins (0.3 µM), while initiation of shape change is not affected (Mans et al., 2008). Concentrations of inhibitors are given next to each tracing of platelet aggregation. ΔT, transmittance

TXA2-Binding Proteins

Moubatin and TSGP3

Moubatin is a sialogenin cloned from the salivary gland of the soft tick *Ornithodorus moubata* salivary gland (Keller et al., 1993; Waxman and Connolly, 1993). Recombinant moubatin inhibited collagen-stimulated aggregation of washed human platelets with an IC_{50} of about 100 nM, while aggregation induced by other ligands was not affected. Moubatin did not inhibit platelet adhesion to collagen, supporting the view that it uniquely inhibited a step involved in platelet activation; however, at higher concentrations (2–6 μM), moubatin diminished the second phase of aggregation induced by ADP, inhibited aggregation in response to submaximal concentrations of the TXA2 mimetic U46619, and competed for the binding of a TXA2 receptor antagonist to platelet membranes. (Keller et al., 1993). Therefore, moubatin behaved similarly to RPAI-1, because at high concentrations it affects ADP and U46619-induced platelet aggregation (Francischetti et al., 2000). It has been suggested that moubatin interferes with a pathway associated with TXA_2 receptor (Keller et al., 1993), but the molecular target of moubatin remained elusive.

More recently, moubatin and another lipocalin from *O. savignyi* (TSGP3) were expressed in *Escherichia coli* and shown to inhibit collagen-induced platelet aggregation (Mans and Ribeiro, 2008). As described before (Keller et al., 1993), inhibition of platelet aggregation was partial, and shape change was still present as depicted in Fig. 20.2c (Mans and Ribeiro, 2008). Of note, isothermal titration calorimetry (ITC) demonstrated that Moubatin and TSGP3 bind cTXA2 with high affinity (24 and 5 nM, respectively), but not ADP (Mans and Ribeiro, 2008). The implications of scavenging TXA_2 are many, taking into account that this prostanoid is a potent platelet aggregation agonist as well as a vasoconstrictor. In fact, moubatin (and TSGP3) relaxes rat aorta pre-constricted by U46619 and inhibits contraction induced by U46619 in a concentration-dependent manner. It is likely that the concentrations of moubatin and TSGP3 in saliva is enough to counteract TXA2 produced in vivo, inhibiting vasoconstriction and platelet aggregation at the site of feeding (Mans and Ribeiro, 2008). Accordingly, it was previously shown to exist as ~5% of the total protein present in salivary gland extracts of *O. savignyi*. It is reasonable to suggest that other lipocalins including moubatin are expressed at a similar level. If it is assumed that only 50% of the salivary gland protein is secreted during feeding and that the feeding site is between 10–50 μl in volume, then moubatin would be present at ~3–18 μM. This is well above the concentrations needed by TXA_2 to induce vasoconstriction or platelet aggregation and given the high affinities for TXA2 (Mans and Ribeiro, 2008).

Putative ADP or TXA2-Binding Proteins

Pallidipin

This is a 19-kDa sialogenin from the salivary gland of *Triatoma pallidipennis*. Pallidipin reportedly blocks collagen-mediated aggregation of platelets

(Noeske-Jungblut et al., 1994). While it does not attenuate platelet aggregation stimulated by other agonists (including ADP, thrombin, U46619, PMA), it is unclear whether it inhibits low doses of ADP-induced platelet aggregation. As described above for RPAI-1, the concentration of platelet agonist tested in vitro is critical when one is investigating the function of a potential ADP-binding lipocalin (Francischetti et al., 2000). It is thus possible that it operates as an ADP-binding protein or a TXA2 scavenger. This contention is based on high sequence similarity to RPAI-1, the fact that it does not block collagen-induced shape change (a GPVI–mediated event), and the finding that relatively high concentrations of pallidipin are needed to inhibit platelet aggregation (Noeske-Jungblut et al., 1994). Isothermal titration calorimetry experiments using recombinant pallidipin may clarify this issue. As reported for moubatin and RPAI-1, pallidipin had no effect on platelet adhesion to immobilized soluble collagen, indicating that it does not affect integrin $\alpha2\beta1$ (Noeske-Jungblut et al., 1994).

Triplatin

Triplatin has been cloned from the salivary gland of *T. infestans* and shown to inhibit platelet aggregation induced by collagen but not ADP, AA, U46619, or thrombin. Triplatin inhibits Fc receptor γ-chain phosphorylation induced by collagen, which is mediated by GPVI. Triplatin also inhibits platelet aggregation induced by collagen-related peptide (CRP), a GPVI agonist. While it has been suggested that triplatin targets GPVI, a direct demonstration has not been provided (Morita et al., 2006). In fact, CRP-induced platelet aggregation is inhibitable by ADP receptor antagonists and cycloxygenase blockers (Watson et al., 2005). It is thus unclear whether the results observed were due to another activity such as ADP- or TXA2-binding properties. Additionally, triplatin displays high sequence similarity to pallidipin-1, works only at high concentrations, and at $> 1\ \mu M$ does not eliminate completely CRP- or collagen-induced shape change (Morita et al., 2006). It is prudent to say that triplatin has not been confirmed as a GPVI antagonist.

Longicornin

Longicornin is a 16-kDa sialogenin from the hard tick *Haemaphysalis longicornis*. Although the N-terminus of longicornin has been identified, it has not been cloned. It inhibits collagen-induced platelet aggregation without affecting other inducers or interfering with platelet adhesion (Cheng et al., 1999); it is not known, however, whether low doses of other agonists are inhibitable by longicornin. It cannot be excluded that longicornin binds either ADP or TXA2.

Finally, a number of lipocalins reportedly bind to AA and to leukotrienes. For example, moubatin, TSGP2, and TSGP3 bind to AA with K_D 34–69 nM, suggesting that it may under certain circumstances mop up AA (Mans and Ribeiro, 2008); however, these three proteins also interact with LTB4 with higher affinity (~20 nM), while moubatin and TSGP3 are TXA2 ligands. While the relative inhibitory contribution of moubatin and related lipocalins to inhibition of AA, TXA2, or LTB4 in vivo is not known, it is plausible that the prostanoid and not AA is the most relevant ligand for this family of lipocalins, due to the higher affinity and intense pharmacologic activity at sites of vascular injury and inflammation.

Epinephrine and Serotonin-Binding Proteins

Amine-Binding Protein (ABP)

A biogenic ABP that belongs to the nitrophorin group of lipocalins has been discovered in *R. prolixus* salivary gland. Using isothermal titration calorimetry and the Hummel-Dreyer method of equilibrium gel filtration, K_D values of 102, 24, and 345 nM were found for serotonin, norepinephrine, and epinephrine, respectively (Andersen et al., 2003). While ABP does not block platelet aggregation by moderate concentrations of collagen or ADP, it inhibits platelet aggregation induced by a combination of low concentrations of ADP and either serotonin or epinephrine. Potentiation of aggregation induced by low concentrations of collagen along with serotonin (or epinephrine) is also inhibited as shown in Fig. 20.2d (Andersen et al., 2003). This occurs because serotonin activates the phospholipase C pathway via the 5-HT_{2A} receptor but not the G_i-dependent pathway and consequently induces only shape change when administered alone. Epinephrine binds to G_i-coupled adrenergic receptors, inhibiting adenylyl cyclase and eliciting no detectable aggregation response by itself (Andersen et al., 2003).

The presence of ABP in the circulation would therefore cause a localized increase in the agonist threshold concentration for platelet aggregation. In *R. prolixus* salivary gland, this protein would act in concert with RPAI-1, NP7, and salivary apyrase, reducing the concentration of weak agonists in the vicinity of the feeding site and thereby attenuating the overall stimulus for platelet aggregation (Andersen et al., 2003).

D7

The D7 protein family is one of the most abundantly expressed sialogenins of mosquitoes and one of the first to be cloned from the salivary glands of insects. Recombinant short D7 of *An. gambiae* were shown to bind serotonin and norepinephrine with high affinity, and it also binds histamine and epinephrine with lower affinity. D7 members from *Aedes* sp. have been identified as high-affinity serotonin-binding proteins with specificity similar to *An. gambiae* short D7 (Calvo et al., 2006). Elucidation of the crystal structure and their binding pockets revealed that the amino domain of the long D7 protein of *Aedes* sp. could bind a hydrophobic compound, which was identified as cysteinyl leukotrienes (Calvo et al., 2009). While the function of short D7 is reportedly to modulate tonus of vessels, it is plausible that scavenging serotonin may block potentiation of platelet activation by serotonin, as described above for ABP from *R. prolixus*.

Monotonin, TSGP-1 and SHBP

Monotonin has been isolated from the salivary gland of the soft tick *Argas monolakensis* and shown to bind to serotonin with high affinity ($K_D < 2$ nM) (Mans et al., 2008b); it may also block platelet aggregation as reported for ABP from *R. prolixus* (Andersen et al., 2003). It has been calculated that the concentration of monotonin in the saliva is compatible with antihemostatic activity at the site of injection (Mans et al., 2008). The crystal structure of monotonin has been determined and indicates that the protein has a single binding site for serotonin. From the conserved features of these proteins, a tick lipocalin biogenic amine-binding motif could be derived that was used to predict biogenic amine-binding (BAB motif) function in other tick lipocalins. Another serotonin-binding protein, TSGP1, has been characterized from *O. savignyi*, and its sequence also contains the BAB motif. It binds 5-HT (K_D ~6 nM) with a stoichiometry of ~1 (Mans et al., 2008). Finally, SHBP from *Dermacentor reticularis* is another lipocalin that binds serotonin (and histamine) with high affinity and may share antiplatelet properties with monotonin and TSGP1 (Sangamnatdej et al., 2002).

Collagen-Binding Proteins

Aegyptin and Anopheline Antiplatelet Protein (AAPP)

Aegyptin is a 30-kDa salivary sialogenin from *Ae. aegypti* that displays a unique sequence characterized by glycine, glutamic acid, and aspartic acid repeats. It specifically blocks collagen-induced human platelet aggregation and granule secretion (IC_{50} ~50 nM) without affecting aggregation induced by other agonists. Figure 20.2e shows tracings of collagen-induced platelet aggregation where, in the presence of 60 nM aegyptin, the onset time for shape change is increased, while at higher doses (120 nM), shape change was abolished. Surface plasmon resonance (SPR) experiments demonstrate that aegyptin binds to collagen (K_D ~1 nM) but does not interact with vitronectin, fibronectin, laminin, fibrinogen, or vWF (Calvo et al., 2007). Aegyptin inhibits collagen-induced platelet aggregation (IC_{50} 100 nM) and blocks vWf interaction with collagen type III under static conditions and platelet adhesion to collagen under flow conditions at high shear rates with similar IC_{50} (~300 nM) (Calvo et al., 2007). More recently, the sequence RGQOGVMGFO, which mediates collagen interaction with vWF, has been identified as a high-affinity binding site for aegyptin. Because it also interacts with the linear peptide RGQPGVMGFP and heat-denatured collagen, triple-helix and hydroxyproline residues are not a prerequisite for binding. Aegyptin also recognizes $(GPO)_{10}$ and GFOGER peptides with low-affinity, which represent GPVI and integrin α2β1 binding sites in collagen, respectively (Calvo et al., 2010[1]).

[1]Reproduced from Toxicon.

Consistent with these results, aegyptin interferes with platelet interaction with soluble collagen and prevents collagen binding to recombinant GPVI. In vivo experiments show that aegyptin prevents laser-induced carotid thrombus formation in the presence of Rose Bengal, a model where collagen reportedly plays an important role. Inhibition of thrombus formation is observable at 50 μg/kg, while occlusion of the carotid took more than 80 min at doses of 100 μg/kg. Rats treated at antithrombotic concentrations did not bleed significantly, suggesting that aegyptin is a suitable molecule to inhibit platelet-collagen interactions in vivo (Calvo et al., 2010).

Anopheline antiplatelet protein (AAPP) from *Anopheles stephensi* saliva is a 30-kDa protein that displays strong sequence similarity to aegyptin. Recombinant AAPP directly binds to immobilized collagen and specifically inhibits collagen-induced platelet aggregation (IC_{50} ~25 nM) and intracellular Ca^{2+} increase. It blocks both platelet integrin α2β1-dependent and GPVI-expressing Jurkat cell-mediated adhesion to collagen. Intravenous injection of AAPP in rats inhibited collagen-induced platelet aggregation ex vivo (Yoshida et al., 2008).

Leech Antiplatelet Protein (LAPP)

LAPP was identified in the salivary glands of the leech *Haementeria officinalis*. Recombinant LAPP inhibits collagen-mediated platelet aggregation under test-tube stirring conditions (IC_{50} ~60–100 nM) without affecting other agonists. It also blocks platelet adhesion to soluble collagen under static conditions (IC_{50} ~80 nM), a step mediated by integrin α2β1 (Connolly et al., 1992; Keller et al., 1992). Consistent with these results, rLAPP prevents integrin α-I domain binding to collagen with IC_{50} ~2 μg/ml (125 nM) (Depraetere et al., 1999). rLAPP also inhibits platelet adhesion to collagen type I (IC_{50} 70 nM) at high shear rate (1,600 s^{-1}) and prevents binding of vWF to collagen type III (van Zanten et al., 1995). While rLAPP inhibits platelet deposition to cross sections of human atherosclerotic coronary arteries (van Zanten et al., 1995) it did not block collagen graft thrombosis in baboons, suggesting that inhibition of collagen alone is not enough to prevent thrombosis (Schaffer et al., 1993), possibly because tissue factor exposure plays an important role in graft models. The structure of LAPP has been determined and consists of a compactly folded C-terminal domain and an N-terminal region that is disordered in the crystal. The C-terminal domain folds similarly to the N-domain of hepatocyte growth factor and has been classified as the so-called PAN domain (Huizinga et al., 2001). A pattern of four conserved cysteines forming two disulfide bonds and of five residues with a conserved hydrophobic character may be characteristic of this type of fold.

Saratin

Saratin, isolated from the saliva of the leech *Hirudo medicinalis*, binds to collagen with high affinity (K_D ~50 nM; SPR experiments). It is a potent inhibitor of vWF binding to collagen (IC_{50} ~1 μg/ml; ~100 nM) according to adhesion experiments performed at high shear rates (Barnes et al., 2001). Platelet adhesion to collagen at 300 s^{-1} is not inhibited at 10 μg/ml (1 μM) saratin, and only very high concentrations of saratin (200 μg/ml; ~20 μM) inhibit platelet aggregation by collagen. Therefore, it has been suggested that saratin, in contrast to aegyptin, LAPP, and calin, is specific for the vWF-collagen interaction. More recently, however, saratin was shown to inhibit binding of the α2 integrin subunit I domain to collagen and prevents platelet adhesion in the presence of ADP and TXA2 inhibitors, suggesting that it interferes with integrin α2β1 binding to collagen in addition to inhibiting vWF-collagen binding (White et al., 2007).

Structural determination of saratin-collagen complex in solution has been revealed by NMR spectroscopy. Saratin has high structural homology to LAPP, which is a distant member of the PAN domain superfamily. In fact, when comparing the secondary structure elements of the solution structure of saratin with the structure of LAPP, both proteins show a very similar arrangement. The structural homology in core regions of the two proteins suggests that both bind to collagen in a similar way (Gronwald et al., 2008).

Because of its therapeutic potential, saratin has been tested in vivo and found to significantly decrease platelet adhesion, intimal hyperplasia, luminal stenosis, and thrombosis after carotid endarterectomy in rats without increasing bleeding time. It also decreases venous anastomotic intimal hyperplasia in a canine dialysis model. Further, under stenotic shear conditions of 800 s^{-1}, saratin reduced (40–60%) platelet deposition triggered by human denuded vessel walls, fatty streaks, severely damaged vessels, and atherosclerotic plaque (Cruz et al., 2001, Vilahur et al., 2004). These results support the view that collagen-binding proteins may be beneficial as antithrombotic agents under certain pathologic conditions.

Calin

Calin is a semipurified substance (~65 kDa.) from *H. medicinalis* that has not yet been molecularly characterized. It inhibits platelet aggregation by collagen (IC_{50} ~6.5 to 13 μg/ml; 100–200 nM) and adhesion to immobilized collagen (IC_{50} ~22 μg/ml; 400 nM) (Deckmyn et al., 1995). Notably, it prevents binding of vWF to coated collagen under static conditions (IC_{50} 10 μg/ml) and high shear rates (1,300 s^{-1}; ~80 μg/ml) (Depraetere et al., 1999; Harsfalvi et al., 1995). Therefore, calin inhibits direct platelet-collagen interactions and vWF-binding to collagen. This dual effect may contribute to prevention of thrombus formation observed under flow (Harsfalvi et al., 1995). In vivo activity of calin has been tested in a thrombosis model in hamsters based on vessel clamp damage to the femoral vein. Results show that intravenous calin inhibited thrombus formation in this model with an ED_{50} of 0.07 mg/kg and complete inhibition with 0.2 mg/kg. No effects were seen on coagulation tests or bleeding times, whereas ex vivo aggregation induced by collagen was inhibited dose dependently (Deckmyn et al., 1995). Because the molecular identity of calin is not known, the precise mechanism of action remains unclear.

Integrin α2β1 Antagonists

TAI is a 15-Kda sialogenin purified from the *O. moubata* salivary gland that has not been molecularly cloned. It inhibits platelet adhesion to soluble collagen under static conditions (IC_{50} 8 nM) without affecting the onset or maximum aggregation triggered by collagen or other platelet agonists. TAI also affects endothelial cell adhesion to collagen and has partial inhibitory activity for fibronectin-mediated platelet adhesion. Further, it outcompetes anti-α2β1 monoclonal antibody Gi9 binding to platelets, suggesting it is an integrin α2β1 antagonist (Karczewski et al., 1995).

HPI has been purified from *Ancylostoma canynum* hookworms and shown to inhibit platelet aggregation by epinephrine, thrombin, and ADP. It also blocks platelet adhesion to fibrinogen and collagen, suggesting that it targets integrin α2β1 and integrin αIIbβ3; however, HPI expressed in *E. coli* is devoid of platelet inhibitory properties. It remains to be proven that is HPI an integrin α2β1 antagonist (Del Valle et al., 2003).

Delivery of NO

Nitrophorins

In *R. prolixus*, NO is stored in the lumen of the salivary glands as a stable nitrophorin(NP)-NO complex. Binding is tight at the pH of the salivary gland (~5), but affinity is lower at the pH of the host (~7.5), a property that facilitates release of NO upon saliva injection (Montfort et al., 2000). Of interest, the ability to transport NO has evolved independently in the bedbug *C. lectularius*, and is again heme-based but apparently does not involve the lipocalin fold (Valenzuela et al., 1995). The structure of the NP4-NO complex has been determined. Binding of NO induces a conformational change in the spacious distal pocket, and the NO moiety becomes completely buried and surrounded by hydrophobic groups. By releasing NO at the site of an arthropod bite, NP conceivably inhibits platelet aggregation. This assumption has been confirmed in vitro in experiments where one NP family member (NP7) loaded with NO reportedly inhibits ADP-induced platelet aggregation and promotes disaggregation of platelets (Fig. 20.2f) in addition to being an anticoagulant through binding to phosphatidylserine (Andersen et al., 2004).

Increase in Intraplatelet cAMP

R. prolixus salivary glands accumulate phospholipids, mainly phosphatidylcholine and lysoPC (75 μg/ml in the lumen), and in the saliva (Golodne et al., 2003). Of note, Lyso-PC (25–150 μg/ml) inhibits platelet aggregation and induces a progressive increase in the cytosolic concentration of cAMP. In addition, salivary lysoPC incubated with porcine arterial endothelial cells for 24 h increases NO production. Accordingly, salivary lysoPC may affect platelets through an increase of intraplatelet cAMP and endothelium-derived NO. The calculated concentration of lysoPC in the feeding site is ~15 μg/ml, which is close to the range capable of full inhibition of platelet aggregation (Golodne et al., 2003), particularly when other salivary inhibitors are delivered simultaneously.

Prostaglandin I_2 (prostacyclin) has been discovered in the saliva of *Ixodes scapularis* at high concentrations (~1.5 μM) (Ribeiro et al., 1988). As an agonist of PGI_2-receptors, it induces cAMP increase—a potent intracellular platelet aggregation inhibitor. PGF2α has also been identified in *Amblyomma americanum* saliva (Aljamali et al., 2002) and at appropriate concentrations may contribute to inhibition of platelet aggregation through a small increase in intraplatelet cAMP or TXA2 receptor antagonism (Jones et al., 2009). Finally, adenosine was identified in sand flies *P. papatasi* and *Phlebotomus perniciosus* at concentrations that may reach ~0.8 mM, at the site of feeding, a concentration above the adenosine IC_{50} (~1 μM) for inhibition of ADP-induced platelet aggregation, via an adenosine receptor and cAMP-dependent pathway (Ribeiro et al., 1999).

Thrombin Inhibitors

A number of sialogenins target thrombin and have been expressed or synthesized in active form. These inhibitors are generally low-molecular-mass molecules, sometimes without cysteines, which allow easy chemical synthesis. They have distinct mechanisms of action targeting exosite I, the catalytic site, or both. Inhibition is either competitive or noncompetitive. The reader is referred to other reviews that discuss in detail the molecular and functional aspects of anticlotting sialogenins (Francischetti et al., 2009; Koh and Kini, 2009; Maritz-Olivier et al., 2007; Ribeiro and Francischetti, 2003; Steen et al., 2006).

Integrin αIIbβ3 Antagonists (Disintegrins)

In the presence of RGD peptides, platelets do not aggregate to any agonist, because fibrinogen cannot crosslink them. Several tick proteins have been characterized that inhibit the activation of platelets through the integrin αIIbβ3. Variabilin (from *Dermacentor variabilis*) is a 4-cysteine, 5 Da RGD-containing disintegrin that blocks ADP-induced platelet aggregation (IC_{50} ~150 nM), and prevents integrin αIIbβ3 binding to immobilized fibrinogen (IC_{50} 9 nM). It also attenuates osteosarcoma cell adhesion to vitronectin (IC_{50} ~100 nM), suggesting that it targets αvβ3 (Wang et al., 1996). Of interest, ixodegrins found in *Ixodes pacificus* and *I. scapularis* transcriptomes (Francischetti et al., 2005) display sequence similarity to variabilin, including RGD position, but have two additional cysteines as shown by the Clustal alignment depicted in Fig. 20.3a. Ixodegrin remains to be produced in recombinant form to confirm its disintegrin activity.

Another tick disintegrin, named disagregin (from *O. moubatta*) lacks RGD motifs and the cysteine stabilized loop that is critical to present RGD motif to integrins. It

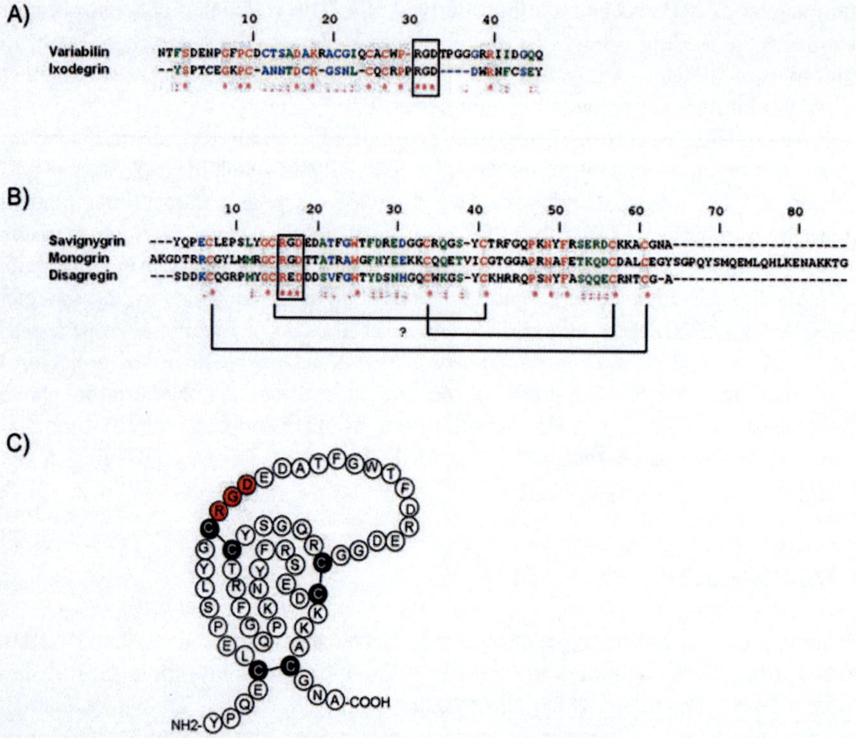

Fig. 20.3 Disintegrin from blood-sucking arthropods. (**a**) Clustal alignment for variabilin (Wang et al., 1996) and ixodegrin (gi 51011476) (Francischetti et al., 2005, 2009) (**b**) Clustal alignment and putative disulfide pattern for savignygrin (gi 21435983) (Mans et al., 2002), disagregin (gi 545738) (Karczewski et al., 1994), and monogrin (gi 114152966) (Mans et al., 2008). (**c**) The predicted secondary folding for savignygrin is shown. The RGD sequence is indicated in *red*

potently blocks ADP-induced platelet aggregation (IC_{50} 150 nM) and platelet adhesion to fibrinogen. Disagregin binds integrin αIIbβ3 in activated platelets with K_D ~40 nM but does not affect endothelial cell adhesion to fibrinogen or vitronectin, confirming its specificity (Karczewski et al., 1994). Likewise, savignygrin (from *O. savigni*) contains a RGD motif and inhibits platelet aggregation by a number of agonists including ADP (IC_{50} 130 nM). It blocks binding of α-CD41 to platelets as well as binding of purified αIIbβ3 to fibrinogen (Mans et al., 2002). Notably, cysteine arrangement of savignygrin is similar to that of the bovine pancreatic trypsin inhibitor (BPTI) family of serine protease inhibitors, suggesting that it presents the RGD motif using the Kunitz-BPTI protein fold (Mans et al., 2002). Savignygrin-like molecules have also been cloned from the soft tick *O. coriaceus*. (Francischetti et al., 2008). More recently, monogrin was purified from the salivary gland of soft tick *A. monolakensis*. It is a RGD-containing 10-kDa protein with sequence homology to savignygrin (and disagregin); it also presents the RGD integrin-recognition motif on the BPTI loop. Both recombinant and purified monogrins block ADP-induced platelet aggregation (IC_{50} ~ 150 nM) but not initiation of shape change (Fig. 20.2g) and were shown by surface plasmon resonance to interact with integrin αIIbβ3. It has been calculated that monogrin concentration in the feeding cavity is between 0.8–4 μM, which is well within the range of its inhibitory properties (Mans et al., 2008). Figure 20.3b shows the Clustal alignment and the putative cysteine pairs for savignygrin, monogrin and disagregin. Figure 20.3c shows the predicted secondary structure for savignygrin which is based on Kunitz protein fold.

In insects, disintegrins have been identified solely in tabanids, where members of the tabinhibitin family acquired such a domain. This 25-kDa RGD-containing protein blocks platelet aggregation by different agonists with concentrations <40 nM (Ma et al., 2009). RGD disintegrin have not been found in fleas, sand flies, or bugs, although recently one was discovered by transcriptome analysis of a anopheline mosquito (*Anopheles darlingi*) and found to belong to the aegyptin-like family.

Leeches are another source of disintegrins. Decorsin from *Macrobdella decora* has been obtained in recombinant and synthetic active forms. It contains an RGD sequence and reportedly inhibits fibrinogen binding to immobilized integrin αIIbβ3 (IC_{50} 1.5 nM); however, it attenuates platelet aggregation at considerable higher concentrations (IC_{50} 500 nM) (McLane et al., 1995; Seymour et al., 1990). At low doses (1.5–3 μg/mouse), it protects mice against death from pulmonary thromboembolism (McLane et al., 1995). The decorsin structure has been determined by NMR and shown to be similar to hirudin despite their distinct primary sequences (Krezel et al., 1994). Ornatin is the name for a family of six isoforms of RGD-containing disintegrins (~5 kDa) purified from whole leech (*Placobdella ornata*) homogenate. It potently inhibits fibrinogen binding to integrin αIIbβ3 (IC_{50} ~5 nM) but, like decorsin, inhibits platelet aggregation at higher concentrations (IC_{50} ~300 nM). A recombinant form of ornatin has been expressed in *E. coli* with a 60-fold decrease in activity, indicating that appropriate folding is critical (Mazur et al., 1991).

Fibrinolytic Metalloproteases and PAF-Acether Hydrolase

Besides apyrases, two other sialogenins with enzymatic activity potentially inhibit platelet aggregation. Metalloprotease activity targeting the Aα chain of fibrinogen and fibrin only has been found in *I. scapularis* saliva, and a number of cDNAs coding for metalloproteases have been sequenced (Francischetti et al., 2003, 2009). Another family of enzymes (tablysin) targeting both Aα and Bβ chains of fibrin(ogen) has been cloned from *Tabanus yao Macquart* (Tabanidae) salivary glands (Ma et al., 2009). These enzymes belong to the reprolysin family of metalloproteases commonly found in Viperidae venom (Francischetti et al., 2003). Hementin and hementerin from leeches *Haementeria ghilianii* and *Haementeria depressa*, respectively, degrade α, β, and γ chains of fibrinogen and fibrin (Budzynski, 1991; Chudzinski-Tavassi et al., 1998). Through cleavage of fibrinogen, these enzymes may inhibit platelet aggregation and clot formation.

A PAF phosphorylcholine hydrolase was found in the salivary glands and saliva of the human-feeding mosquito *Culex quinquefasciatus* and shown to inhibit PAF-induced platelet aggregation. While this enzyme has not yet been cloned, it may prevent platelet activation through neutrophil-derived PAF (Ribeiro and Francischetti, 2001).

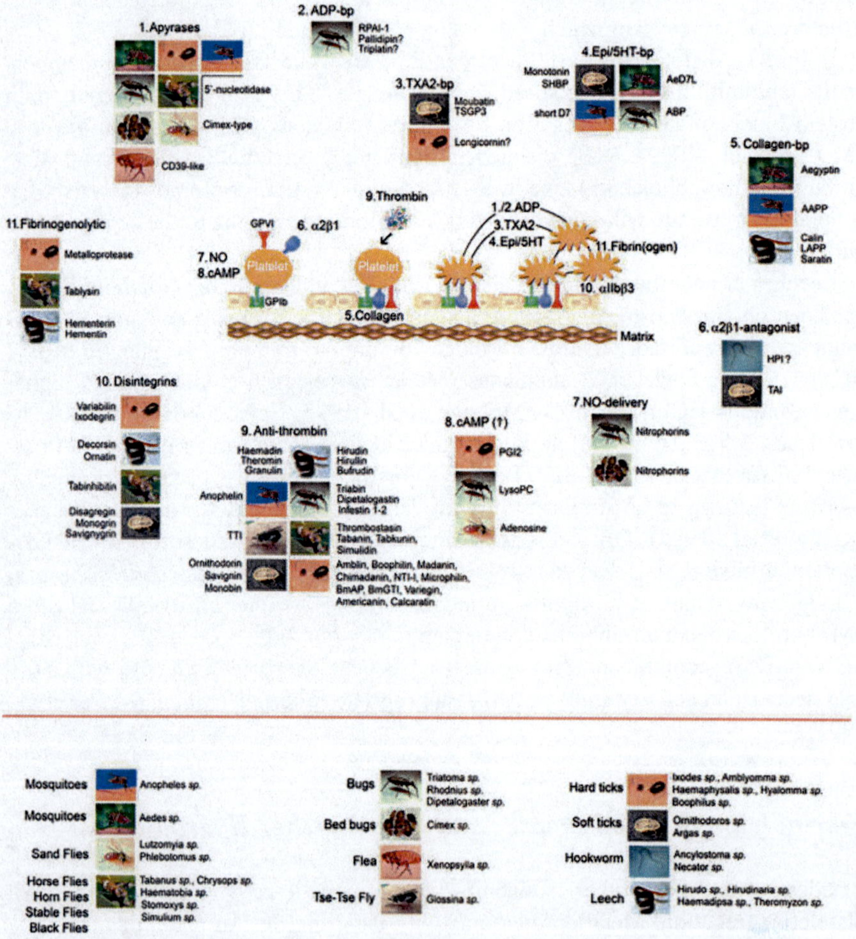

Fig. 20.4 Hematophagy and inhibition of platelet aggregation. Platelets adhere to collagen at sites of vascular injury through vWF, GPIb and specific receptors collagen receptors GPVI and integrin α2β1. This event is accompanied by release of secondary mediators (e.g., ADP, epinephrine and serotonin) and generation of TXA2. During this process, thrombin may be produced through exposure of tissue factor. These agonists activate the signaling pathway, culminating with activation of integrin αIIbβ3, which binds fibrinogen (Gross and Weitz, 2009; Watson et al., 2005). The target for each family of inhibitor is indicated by a number. The target for PAF phosphorylcholine hydrolase is not shown. The name of each inhibitor is shown beside each picture exemplifying a different hematophagous animal found in different genus and/or species. Epi, epinephrine; 5HT, serotonin. Reproduced with permission by Elsevier: Francischetti IM, Platelet aggregation inhibitors from hematophagous animals. Toxicon 2010

Concluding Remarks

Molecules involved in platelet activation are important targets of sialogenins from different hematophagous animals. Notably, only inhibitors targeting molecules directly involved with activation of hemostasis have been discovered. Accordingly, agonists that are only produced upon platelet activation or endothelial denudation—such as thrombin, collagen, ADP, TXA2, and serotonin—are obvious targets for sialogenins with antiplatelet activity. The same is true for disintegrins targeting $\alpha IIb\beta 3$, which only becomes functionally available to bind fibrinogen after platelet activation. In fact, no salivary antagonists have been described so far targeting constitutively expressed receptors such as GPIb, GPVI, PARs or inhibitors of plasma proteins that occurs at high concentrations such as vWF, prothrombin, cofactors V and VIII, *etc.* despite their critical role in hemostasis. It appears that these molecules are likely not ideal targets for sialogenins because they are not easily neutralized due to high plasma concentration, or high receptor density in non-activated cells. They may actually remove sialogenins (usually present in minute amounts) from the site where they have been injected, decreasing their effective concentration. This is very much in contrast to snake venom toxins (usually found in large amounts), which have been reported to bind to GPVI, vWF, prothrombin, and cofactors (Lu et al., 2005). Of interest, some sialogenins have structural features (e.g., positively charged residues) that direct them to the membrane of activated platelets through interaction with phosphatidylserine (Andersen et al., 2004). Conceivably, this targets the sialogenins to activated cells, increasing the effective concentration of the inhibitor at sites of inflammation; this may also avoid diffusion into flowing blood. In conclusion, antiplatelet sialogenins from salivary glands are remarkably functionally diverse. They likely work in a synergistic and redundant manner to keep hemostatic tonus as low as possible so as to facilitate blood-feeding, as depicted in Fig. 20.4. Finally, antiplatelet sialogenins may be useful tools in cell biology and may also have potential for therapeutic applications.

Acknowledgements This work was supported by the Intramural Research Program of the Division of Intramural Research, National Institute of Allergy and Infectious Diseases, National Institutes of Health. We thank NIAID intramural editor Brenda Rae Marshall for assistance.

References

Aljamali, M., Bowman, A.S., Dillwith, J.W., Tucker, J.S., Yates, G.W., Essenberg, R.C., Sauer, J.R., 2002. Identity and synthesis of prostaglandins in the lone star tick, *Amblyomma americanum* (L.), as assessed by radio-immunoassay and gas chromatography/mass spectrometry. Insect Biochem. Mol. Biol. 32, 331–341.

Andersen, J.F., Francischetti, I.M., Valenzuela, J.G., Schuck, P., Ribeiro, J.M., 2003. Inhibition of hemostasis by a high affinity biogenic amine-binding protein from the saliva of a blood-feeding insect. J. Biol. Chem. 278, 4611–4617.

Andersen, J.F., Gudderra, N.P., Francischetti, I.M., Valenzuela, J.G., Ribeiro, J.M., 2004. Recognition of anionic phospholipid membranes by an antihemostatic protein from a blood-feeding insect. Biochemistry 43, 6987–6994.

Andersen, J.F., Hinnebusch, B.J., Lucas, D.A., Conrads, T.P., Veenstra, T.D., Pham, V.M., Ribeiro, J.M., 2007. An insight into the sialome of the oriental rat flea, *Xenopsylla cheopis* (Rots). BMC Genomics 8, 102.

Barnes, C.S., Krafft, B., Frech, M., Hofmann, U.R., Papendieck, A., Dahlems, U., Gellissen, G., Hoylaerts, M.F., 2001. Production and characterization of saratin, an inhibitor of von Willebrand factor-dependent platelet adhesion to collagen. Semin. Thromb. Hemost. 27, 337–348.

Budzynski, A.Z., 1991. Interaction of hementin with fibrinogen and fibrin. Blood Coagul. Fibrinolysis 2, 149–152.

Calvo, E., Mans, B.J., Andersen, J.F., Ribeiro, J.M., 2006. Function and evolution of a mosquito salivary protein family. J. Biol. Chem. 281, 1935–1942.

Calvo, E., Mans, B.J., Ribeiro, J.M., Andersen, J.F., 2009. Multifunctionality and mechanism of ligand binding in a mosquito antiinflammatory protein. Proc. Natl. Acad. Sci. U.S.A. 106, 3728–3733.

Calvo, E., Tokumasu, F., Marinotti, O., Villeval, J.L., Ribeiro, J.M., Francischetti, I.M., 2007. Aegyptin, a novel mosquito salivary gland protein, specifically binds to collagen and prevents its interaction with platelet glycoprotein VI, integrin $\alpha 2\beta 1$, and von Willebrand factor. J. Biol. Chem. 282, 26928–26938.

Calvo, E., Tokumasu, F., Mizurini, D.M., McPhie, P., Narum, D.L., Ribeiro, J.M., Monteiro, R.Q., Francischetti, I.M., 2010. Aegyptin displays high-affinity for the von Willebrand factor binding site (RGQOGVMGF) in collagen and inhibits carotid thrombus formation in vivo. FEBS J. 277(2), 413–427.

Champagne, D.E., Smartt, C.T., Ribeiro, J.M., James, A.A., 1995. The salivary gland-specific apyrase of the mosquito *Aedes aegypti* is a member of the 5′-nucleotidase family. Proc. Natl. Acad. Sci. U.S.A. 92, 694–698.

Charlab, R., Valenzuela, J.G., Rowton, E.D., Ribeiro, J.M., 1999. Toward an understanding of the biochemical and pharmacological complexity of the saliva of a hematophagous sand fly *Lutzomyia longipalpis*. Proc. Natl. Acad. Sci. U.S.A. 96, 15155–15160.

Cheng, Y., Wu, H., Li, D., 1999. An inhibitor selective for collagen-stimulated platelet aggregation from the salivary glands of hard tick *Haemaphysalis longicornis* and its mechanism of action. Sci. China C. Life Sci. 42, 457–464.

Chudzinski-Tavassi, A.M., Kelen, E.M., de Paula Rosa, A.P., Loyau, S., Sampaio, C.A., Bon, C., Angles-Cano, E., 1998. Fibrino(geno)lytic properties of purified hementerin, a metalloproteinase from the leech *Haementeria depressa*. Thromb. Haemost. 80, 155–160.

Connolly, T.M., Jacobs, J.W., Condra, C., 1992. An inhibitor of collagen-stimulated platelet activation from the salivary glands of the *Haementeria officinalis* leech. I. Identification, isolation, and characterization. J. Biol. Chem. 267, 6893–6898.

Coughlin, S.R., 2005. Protease-activated receptors in hemostasis, thrombosis and vascular biology. J. Thromb. Haemost. 3, 1800–1814.

Cruz, C.P., Eidt, J., Drouilhet, J., Brown, A.T., Wang, Y., Barnes, C.S., Moursi, M.M., 2001. Saratin, an inhibitor of von Willebrand factor-dependent platelet adhesion, decreases platelet aggregation and intimal hyperplasia in a rat carotid endarterectomy model. J. Vasc. Surg. 34, 724–729.

Dai, J., Liu, J., Deng, Y., Smith, T.M., Lu, M., 2004. Structure and protein design of a human platelet function inhibitor. Cell 116, 649–659.

Deckmyn, H., Stassen, J.M., Vreys, I., Van Houtte, E., Sawyer, R.T., Vermylen, J., 1995. Calin from *Hirudo medicinalis*, an inhibitor of platelet adhesion to collagen, prevents platelet-rich thrombosis in hamsters. Blood 85, 712–719.

Del Valle, A., Jones, B.F., Harrison, L.M., Chadderdon, R.C., Cappello, M., 2003. Isolation and molecular cloning of a secreted hookworm platelet inhibitor from adult *Ancylostoma caninum*. Mol. Biochem. Parasitol. 129, 167–177.

Depraetere, H., Kerekes, A., Deckmyn, H., 1999. The collagen-binding leech products rLAPP and calin prevent both von Willebrand factor and α2β1(GPIa/IIa)-I-domain binding to collagen in a different manner. Thromb. Haemost. 82, 1160–1163.

Faudry, E., Lozzi, S.P., Santana, J.M., D'Souza-Ault, M., Kieffer, S., Felix, C.R., Ricart, C.A., Sousa, M.V., Vernet, T., Teixeira, A.R., 2004. *Triatoma infestans* apyrases belong to the 5′-nucleotidase family. J. Biol. Chem. 279, 19607–19613.

Francischetti, I.M., Andersen, J.F., Ribeiro, J.M., 2002. Biochemical and functional characterization of recombinant *Rhodnius prolixus* platelet aggregation inhibitor 1 as a novel lipocalin with high affinity for adenosine diphosphate and other adenine nucleotides. Biochemistry 41, 3810–3818.

Francischetti, I.M., Mather, T.N., Ribeiro, J.M., 2003. Cloning of a salivary gland metalloprotease and characterization of gelatinase and fibrin(ogen)lytic activities in the saliva of the Lyme disease tick vector *Ixodes scapularis*. Biochem. Biophys. Res. Commun. 305, 869–875.

Francischetti, I.M., Meng, Z., Mans, B.J., Gudderra, N., Hall, M., Veenstra, T.D., Pham, V.M., Kotsyfakis, M., Ribeiro, J.M., 2008. An insight into the salivary transcriptome and proteome of the soft tick and vector of epizootic bovine abortion, *Ornithodoros coriaceus*. J. Proteomics 71, 493–512.

Francischetti, I.M., My Pham, V., Mans, B.J., Andersen, J.F., Mather, T.N., Lane, R.S., Ribeiro, J.M., 2005. The transcriptome of the salivary glands of the female western black-legged tick *Ixodes pacificus* (Acari: Ixodidae). Insect Biochem. Mol. Biol. 35, 1142–1161.

Francischetti, I.M., Ribeiro, J.M., Champagne, D., Andersen, J., 2000. Purification, cloning, expression, and mechanism of action of a novel platelet aggregation inhibitor from the salivary gland of the blood-sucking bug, *Rhodnius prolixus*. J. Biol. Chem. 275, 12639–12650.

Francischetti, I.M., Sa-Nunes, A., Mans, B.J., Santos, I.M., Ribeiro, J.M., 2009. The role of saliva in tick feeding. Front. Biosci. 14, 2051–2088.

Furie, B., Furie, B.C., 2005. Thrombus formation in vivo. J. Clin. Invest. 115, 3355–3362.

Gawaz, M., Langer, H., May, A.E., 2005. Platelets in inflammation and atherogenesis. J. Clin. Invest. 115, 3378–3384.

Golodne, D.M., Monteiro, R.Q., Graca-Souza, A.V., Silva-Neto, M.A., Atella, G.C., 2003. Lysophosphatidylcholine acts as an anti-hemostatic molecule in the saliva of the blood-sucking bug *Rhodnius prolixus*. J. Biol. Chem. 278, 27766–27771.

Gronwald, W., Bomke, J., Maurer, T., Domogalla, B., Huber, F., Schumann, F., Kremer, W., Fink, F., Rysiok, T., Frech, M., Kalbitzer, H.R., 2008. Structure of the leech protein saratin and characterization of its binding to collagen. J. Mol. Biol. 381, 913–927.

Gross, P.L., Weitz, J.I., 2009. New antithrombotic drugs. Clin. Pharmacol. Ther. 86, 139–146.

Harsfalvi, J., Stassen, J.M., Hoylaerts, M.F., Van Houtte, E., Sawyer, R.T., Vermylen, J., Deckmyn, H., 1995. Calin from *Hirudo medicinalis*, an inhibitor of von Willebrand factor binding to collagen under static and flow conditions. Blood 85, 705–711.

Huizinga, E.G., Schouten, A., Connolly, T.M., Kroon, J., Sixma, J.J., Gros, P., 2001. The structure of leech anti-platelet protein, an inhibitor of haemostasis. Acta Crystallogr. D. Biol. Crystallogr. 57, 1071–1078.

Jackson, S.P., Schoenwaelder, S.M., 2003. Antiplatelet therapy: in search of the 'magic bullet'. Nat Rev. Drug Discov. 2, 775–789.

Jandrot-Perrus, M., Busfield, S., Lagrue, A.H., Xiong, X., Debili, N., Chickering, T., Le Couedic, J.P., Goodearl, A., Dussault, B., Fraser, C., Vainchenker, W., Villeval, J.L., 2000. Cloning, characterization, and functional studies of human and mouse glycoprotein VI: a platelet-specific collagen receptor from the immunoglobulin superfamily. Blood 96, 1798–1807.

Jones, R.L., Giembycz, M.A., Woodward, D.F., 2009. Prostanoid receptor antagonists: development strategies and therapeutic applications. Br. J. Pharmacol. 158, 104–145.

Kahner, B.N., Shankar, H., Murugappan, S., Prasad, G.L., Kunapuli, S.P., 2006. Nucleotide receptor signaling in platelets. J. Thromb. Haemost. 4, 2317–2326.

Karczewski, J., Endris, R., Connolly, T.M., 1994. Disagregin is a fibrinogen receptor antagonist lacking the Arg-Gly-Asp sequence from the tick, *Ornithodoros moubata*. J. Biol. Chem. 269, 6702–6708.

Karczewski, J., Waxman, L., Endris, R.G., Connolly, T.M., 1995. An inhibitor from the argasid tick *Ornithodoros moubata* of cell adhesion to collagen. Biochem. Biophys. Res. Commun. 208, 532–541.

Keller, P.M., Schultz, L.D., Condra, C., Karczewski, J., Connolly, T.M., 1992. An inhibitor of collagen-stimulated platelet activation from the salivary glands of the *Haementeria officinalis* leech. II. Cloning of the cDNA and expression. J. Biol. Chem. 267, 6899–6904.

Keller, P.M., Waxman, L., Arnold, B.A., Schultz, L.D., Condra, C., Connolly, T.M., 1993. Cloning of the cDNA and expression of moubatin, an inhibitor of platelet aggregation. J. Biol. Chem. 268, 5450–5456.

Koh, C.Y., Kini, R.M., 2009. Molecular diversity of anticoagulants from haematophagous animals. Thromb. Haemost. 102, 437–453.

Krezel, A.M., Wagner, G., Seymour-Ulmer, J., Lazarus, R.A., 1994. Structure of the RGD protein decorsin: conserved motif and distinct function in leech proteins that affect blood clotting. Science 264, 1944–1947.

Liyou, N., Hamilton, S., Elvin, C., Willadsen, P., 1999. Cloning and expression of ecto 5-nucleotidase from the cattle tick *Boophilus microplus*. Insect Mol. Biol. 8, 257–266.

Lombardo, F., Di Cristina, M., Spanos, L., Louis, C., Coluzzi, M., Arca, B., 2000. Promoter sequences of the putative *Anopheles gambiae* apyrase confer salivary gland expression in *Drosophila melanogaster*. J. Biol. Chem. 275, 23861–23868.

Lu, Q., Clemetson, J.M., Clemetson, K.J., 2005. Snake venoms and hemostasis. J. Thromb. Haemost. 3, 1791–1799.

Ma, D., Wang, Y., Yang, H., Wu, J., An, S., Gao, L., Xu, X., Lai, R., 2009. Anti-thrombosis repertoire of blood-feeding horsefly salivary glands. Mol. Cell. Proteomics 8, 2071–2079.

Mans, B.J., Andersen, J.F., Schwan, T.G., Ribeiro, J.M., 2008a. Characterization of anti-hemostatic factors in the argasid, *Argas monolakensis*: implications for the evolution of blood-feeding in the soft tick family. Insect Biochem. Mol. Biol. 38, 22–41.

Mans, B.J., Louw, A.I., Neitz, A.W., 2002. Savignygrin, a platelet aggregation inhibitor from the soft tick Ornithodoros savignyi, presents the RGD integrin recognition motif on the Kunitz-BPTI fold. J. Biol. Chem. 277, 21371–21378.

Mans, B.J., Ribeiro, J.M., 2008. Function, mechanism and evolution of the moubatin-clade of soft tick lipocalins. Insect Biochem. Mol. Biol. 38, 841–852.

Mans, B.J., Ribeiro, J.M., Andersen, J.F., 2008b. Structure, function, and evolution of biogenic amine-binding proteins in soft ticks. J. Biol. Chem. 283, 18721–18733.

Maritz-Olivier, C., Stutzer, C., Jongejan, F., Neitz, A.W., Gaspar, A.R., 2007. Tick anti-hemostatics: targets for future vaccines and therapeutics. Trends Parasitol. 23, 397–407.

Mazur, P., Henzel, W.J., Seymour, J.L., Lazarus, R.A., 1991. Ornatins: potent glycoprotein IIb-IIIa antagonists and platelet aggregation inhibitors from the leech *Placobdella ornata*. Eur. J. Biochem. 202, 1073–1082.

McLane, M.A., Gabbeta, J., Rao, A.K., Beviglia, L., Lazarus, R.A., Niewiarowski, S., 1995. A comparison of the effect of decorsin and two disintegrins, albolabrin and eristostatin, on platelet function. Thromb. Haemost. 74, 1316–1322.

Montfort, W.R., Weichsel, A., Andersen, J.F., 2000. Nitrophorins and related antihemostatic lipocalins from *Rhodnius prolixus* and other blood-sucking arthropods. Biochim. Biophys. Acta 1482, 110–118.

Morita, A., Isawa, H., Orito, Y., Iwanaga, S., Chinzei, Y., Yuda, M., 2006. Identification and characterization of a collagen-induced platelet aggregation inhibitor, triplatin, from salivary glands of the assassin bug, *Triatoma infestans*. FEBS J. 273, 2955–2962.

Noeske-Jungblut, C., Kratzschmar, J., Haendler, B., Alagon, A., Possani, L., Verhallen, P., Donner, P., Schleuning, W.D., 1994. An inhibitor of collagen-induced platelet aggregation from the saliva of *Triatoma pallidipennis*. J. Biol. Chem. 269, 5050–5053.

Reddy, V.B., Kounga, K., Mariano, F., Lerner, E.A., 2000. Chrysoptin is a potent glycoprotein IIb/IIIa fibrinogen receptor antagonist present in salivary gland extracts of the deerfly. J. Biol. Chem. 275, 15861–15867.

Ribeiro, J.M., Francischetti, I.M., 2001. Platelet-activating-factor-hydrolyzing phospholipase C in the salivary glands and saliva of the mosquito *Culex quinquefasciatus*. J. Exp. Biol. 204, 3887–3894.

Ribeiro, J.M., Francischetti, I.M., 2003. Role of arthropod saliva in blood feeding: sialome and post-sialome perspectives. Annu. Rev. Entomol. 48, 73–88.

Ribeiro, J.M., Katz, O., Pannell, L.K., Waitumbi, J., Warburg, A., 1999. Salivary glands of the sand fly *Phlebotomus papatasi* contain pharmacologically active amounts of adenosine and 5′-AMP. J. Exp. Biol. 202, 1551–1559.

Ribeiro, J.M., Makoul, G.T., Robinson, D.R., 1988. *Ixodes dammini*: evidence for salivary prostacyclin secretion. J. Parasitol. 74, 1068–1069.

Ruggeri, Z.M., 2002. Platelets in atherothrombosis. Nat. Med. 8, 1227–1234.

Sangamnatdej, S., Paesen, G.C., Slovak, M., Nuttall, P.A., 2002. A high affinity serotonin- and histamine-binding lipocalin from tick saliva. Insect Mol. Biol. 11, 79–86.

Schaffer, L.W., Davidson, J.T., Siegl, P.K., Gould, R.J., Nutt, R.F., Brady, S.F., Connolly, T.M., 1993. Recombinant leech antiplatelet protein prevents collagen-mediated platelet aggregation but not collagen graft thrombosis in baboons. Arterioscler. Thromb. 13, 1593–1601.

Seymour, J.L., Henzel, W.J., Nevins, B., Stults, J.T., Lazarus, R.A., 1990. Decorsin. A potent glycoprotein IIb-IIIa antagonist and platelet aggregation inhibitor from the leech *Macrobdella decora*. J. Biol. Chem. 265, 10143–10147.

Steen, N.A., Barker, S.C., Alewood, P.F., 2006. Proteins in the saliva of the Ixodida (ticks): pharmacological features and biological significance. Toxicon 47, 1–20.

Stutzer, C., Mans, B.J., Gaspar, A.R., Neitz, A.W., Maritz-Olivier, C., 2009. *Ornithodoros savignyi*: soft tick apyrase belongs to the 5′-nucleotidase family. Exp. Parasitol. 122, 318–327.

Valenzuela, J.G., Belkaid, Y., Rowton, E., Ribeiro, J.M., 2001. The salivary apyrase of the blood-sucking sand fly *Phlebotomus papatasi* belongs to the novel Cimex family of apyrases. J. Exp. Biol. 204, 229–237.

Valenzuela, J.G., Charlab, R., Galperin, M.Y., Ribeiro, J.M., 1998. Purification, cloning, and expression of an apyrase from the bed bug *Cimex lectularius*. A new type of nucleotide-binding enzyme. J. Biol. Chem. 273, 30583–30590.

Valenzuela, J.G., Walker, F.A., Ribeiro, J.M., 1995. A salivary nitrophorin (nitric-oxide-carrying hemoprotein) in the bedbug *Cimex lectularius*. J. Exp. Biol. 198, 1519–1526.

van Zanten, G.H., Connolly, T.M., Schiphorst, M.E., de Graaf, S., Slootweg, P.J., Sixma, J.J., 1995. Recombinant leech antiplatelet protein specifically blocks platelet deposition on collagen surfaces under flow conditions. Arterioscler. Thromb. Vasc. Biol. 15, 1424–1431.

Varga-Szabo, D., Pleines, I., Nieswandt, B., 2008. Cell adhesion mechanisms in platelets. Arterioscler. Thromb. Vasc. Biol. 28, 403–412.

Vilahur, G., Duran, X., Juan-Babot, O., Casani, L., Badimon, L., 2004. Antithrombotic effects of saratin on human atherosclerotic plaques. Thromb. Haemost. 92, 191–200.

Wang, X., Coons, L.B., Taylor, D.B., Stevens, S.E., Jr., Gartner, T.K., 1996. Variabilin, a novel RGD-containing antagonist of glycoprotein IIb-IIIa and platelet aggregation inhibitor from the hard tick *Dermacentor variabilis*. J. Biol. Chem. 271, 17785–17790.

Watson, S.P., Auger, J.M., McCarty, O.J., Pearce, A.C., 2005. GPVI and integrin $\alpha IIb\beta 3$ signaling in platelets. J. Thromb. Haemost. 3, 1752–1762.

Waxman, L., Connolly, T.M., 1993. Isolation of an inhibitor selective for collagen-stimulated platelet aggregation from the soft tick *Ornithodoros moubata*. J. Biol. Chem. 268, 5445–5449.

White, T.C., Berny, M.A., Robinson, D.K., Yin, H., DeGrado, W.F., Hanson, S.R., McCarty, O.J., 2007. The leech product saratin is a potent inhibitor of platelet integrin $\alpha 2\beta 1$ and von Willebrand factor binding to collagen. FEBS J. 274, 1481–1491.

Yoshida, S., Sudo, T., Niimi, M., Tao, L., Sun, B., Kambayashi, J., Watanabe, H., Luo, E., Matsuoka, H., 2008. Inhibition of collagen-induced platelet aggregation by anopheline antiplatelet protein, a saliva protein from a malaria vector mosquito. Blood 111, 2007–2014.

Chapter 21
Snaclecs (Snake C-Type Lectins) that Inhibit Platelet Receptors

Kenneth J. Clemetson

Abstract One of the major targets for snake venom proteins is haemostasis. This weakens the prey and helps with swallowing and digestion. The venom proteins act either on coagulation factors or on platelets. Snake venom proteins often adapt physiological mechanisms to inhibit or activate platelets. Inhibitory mechanisms are mainly directed against integrins, which is also true of many of the snaclecs described in this chapter. Their main physiological ligands are von Willebrand factor and collagen. GPIb is also a target of snaclecs described as inhibitory but that may be due to the methods of testing as such snaclecs induce thrombocytopenia when tested in vivo. In the case of those targeting integrins they may indeed be inhibitory but platelets may not be their primary target since $\alpha 2\beta 1$ is also present on endothelial and smooth muscle cells and inhibitory molecules are anti-angiostatic.

Introduction

More than 300 known species of venomous snakes are classified into five families, Hydrophidae, Elapidae, Viperidae, Crotalidae and Colubidae. Venom proteins have also been demonstrated recently in other snakes and some lizards (Fry et al., 2006). Venom components that affect hemostasis are most generally found in Viperidae, and Crotalidae snakes but the others often contain some as well. Snake venom proteins mostly adapt physiological mechanisms to inhibit or activate platelets. Of these, so far only the mechanisms directed against integrins (or their ligands), in particular $\alpha IIb\beta 3$, have really been demonstrated to be inhibitory. This is because it is important for the snake to prevent fibrinogen binding and clot retraction. This is achieved in two ways – blocking of $\alpha IIb\beta 3$ by disintegrins or removal of fibrinogen by haemorrhagic metalloproteases that cleave mainly fibrinogen α or β chains and prevent fibrin formation. Although, some snaclecs that bind $\alpha 2\beta 1$ integrins

---

K.J. Clemetson (✉)
Department of Haematology, University of Berne, Inselspital, CH-3010 Berne, Switzerland
e-mail: kenneth.clemetson@tki.unibe.ch

R.M. Kini et al. (eds.), *Toxins and Hemostasis*, DOI 10.1007/978-90-481-9295-3_21,

inhibit platelet function their main targets in vivo are likely to be endothelial and smooth muscle cells. Recently, there is increasing evidence that other snaclec anti-β1-integrins inhibit angiogenesis and leukocyte function as well, though the benefits for the snake of the effects on immunology remain obscure. It should be noted that snake venoms are used extensively in homeopathy and traditional Chinese and Indian medicine so there may be a practical basis (Koh et al., 2006). This review concentrates on structural and functional properties of venom components of the C-type lectins class (now named snaclecs) that inhibit platelet functions.

C-Type Lectin Canonical/Compact Structure

The C-type type lectins are a widely distributed protein-fold class with various functions. Typically, these proteins bind calcium and sugar residues (hence "C" calcium and lectin). Closely related proteins with Ca^{2+} and sugar (generally galactose) binding properties are also found in snake venoms, mainly as homodimers. However, the main class of these proteins found in snake venoms is the snaclecs. These do not contain the classic calcium/sugar-binding loop and have evolved to bind a wide range of physiologically important proteins and receptors.

Snaclecs

Because of confusion with classic C-type lectins and since names such as C-type lectin-like or –related proteins are almost inevitably abbreviated to CTL or CLP and provide no information about the heterodimeric structure, loop-swapping or higher order multimerization, this group have been named *snaclecs* (*Sna*ke venom *C*-type *lec*tin*s*) in a recent nomenclature proposal by the Exogenous Factor Committee of the International Thrombosis and Haemostasis Society (Clemetson et al., 2009).

A typical structure of a snaclec αβ heterodimer is shown in Fig. 21.1a. This is the basic unit from which the different structures of the snaclec family are assembled. Snaclecs bind to a wide range of coagulation factors, other proteins critical in haemostasis, and membrane receptors on platelets and other cells. However, many have more than one binding site and/or may interact with more than one protein/receptor. Snaclecs have been described that interact with coagulation factors X/XI and X (Mizuno et al., 1999, 2001). Crystallography studies indicate that the binding site on the coagulation factor is the Gla domain whereas the site on the snaclec is the concave surface lying between the two subunits (Morita et al., 1996) (shown in Fig. 21.1b). Binding is largely via electrostatic interactions but shape-fitting also has a role.

Snaclecs have a basic heterodimeric structure with two subunits, α and β, nearly always linked covalently, via a disulphide bond. The heterodimers are often further multimerized either non-covalently, or covalently via additional disulphide bonds, to form larger structures. The interpretation of the effects of snaclecs on platelets presents a problem. Many simple, heterodimeric members of this family have been

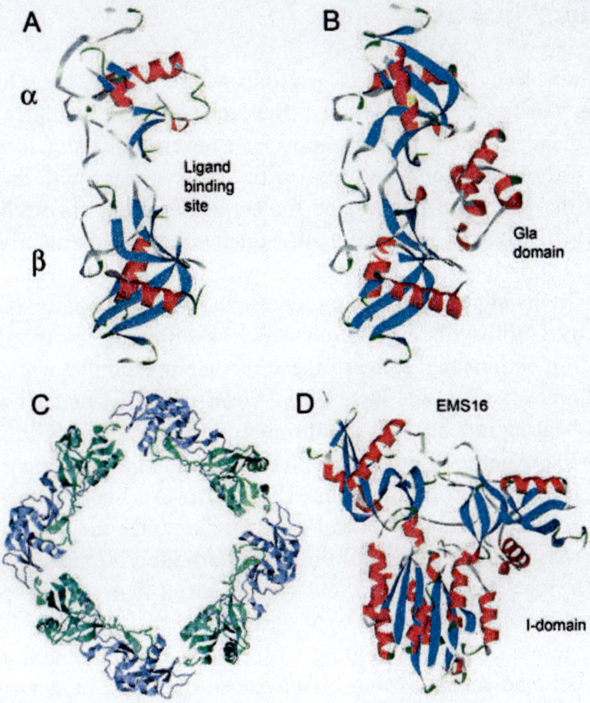

Fig. 21.1 Structure of (**a**) Heterodimeric snaclec (echicetin) with swapped loops. (**b**) Factor IX binding protein (snaclec) in complex with Gla domain of Factor IX. (**c**) Tetrameric heterodimeric snaclec of the flavocetin family. (**d**) Complex of heterodimeric snaclec EMS16 with the I-domain of α2-integrin subunit. The structures were prepared from primary sequences using Swiss-Model and Accelys DS Viewer 5.0

reported to inhibit GPIb-related platelet functions whereas other, more complex, ones activate platelets via the same receptor (see Chapter 35). Most of the studies to characterize these proteins were carried out under conditions far from physiological, with washed platelets, lacking plasma and not in the vasculature, the normal platelet environment. On the other hand, studies in vivo in small animals with "GPIb-inhibiting" snaclecs, seem always to induce thrombocytopenia whenever platelet count was measured, implying that more things were happening than simply inhibition of GPIb.

Although there are quite a few snake venom proteins that inhibit platelet function only a few of these in any given venom are snaclecs. In particular, there is no evidence so far that αIIbβ3 is a target of any of these. This may be because the disintegrins, which are generally smaller molecules, are efficient inhibitors of integrins such as αIIbβ3 by conformationally controlled small peptide loops containing the integrin recognition sequence. Thus, it would have been inefficient to use the larger snaclecs to target these receptors. They are therefore restricted to receptors (and other molecules) that are difficult to target otherwise and need relatively large binding surfaces.

GPIb-Binding Snaclecs

GPIb-binding snaclecs constitute a large group within this class and have been isolated from many different snake venoms. New members are regularly being found. They have sequences similar to one another but nevertheless also differences. The similarities mostly resemble those found with other snaclecs with different targets and represent the structural part of the molecule necessary for folding, which is highly conserved, whereas the divergent sequence represents mainly the binding regions that are individually optimized. As with botrocetin (Fukuda et al., 2002) and bitiscetin (Maita et al., 2003), this suggests that they are optimised for different binding sites on GPIb, an interpretation which is supported by the differences in effects on GPIb functions that some of these show. For example, while most inhibit GPIb-VWF binding, only some have been shown to inhibit platelet activation via thrombin, implicating blockage of the thrombin-binding site on GPIb. Since more ligands for GPIb are being discovered with different binding sites on that molecule, snaclecs may be useful tools to localise these. However, not all these molecules are specific for GPIb alone and several have binding sites for GPIb together with other platelet receptors. Several multimeric (heterodimeric) snaclecs, with specificity for GPIb, have been isolated and characterised that activate platelets with various degrees of efficacity.

An early example of a GPIb-binding snaclec is echicetin, which was shown to block von Willebrand factor as well as low dose thrombin induced platelet activation (Peng et al., 1994). At that point it was therefore classed as inhibitory. However, early reports also noted that it induced thrombocytopenia in small animals. The inhibitory studies had been done with washed platelets. However, when echicetin was tested with platelet rich plasma it induced platelet aggregation (Navdaev et al., 2001) (see Chapter 35). Other GPIb-inhibitory snaclecs do not agglutinate/aggregate platelets in plasma, suggesting that the thrombocytopenia they induce must involve interaction with components/receptors present on other cells such as the endothelium.

Several multimeric C-type lectins with GPIb binding properties, such as the prototype of this class, flavocetin, cause platelet agglutination with some low level activation (Taniuchi et al., 1995). It is difficult to see in such cases how these proteins function for the snake! If they are purely inhibitory the multimeric form might help by increasing the binding constant but biosynthesis of the same end mass of protein would still be necessary to ensure adequate inhibition of the ~80% of GPIb required to produce a visible effect on platelet function.

While activation of platelets is generally the mechanism used by venom components, for the simple reason that this is the most effective strategy for removing platelets from the circulation, using the minimum amount of protein, there are nevertheless a number of venom components, which have been described as inhibitory. These fall predominantly into the disintegrin and snaclec families and, among the snaclecs have either GPIb or integrins as targets. The inhibitory status of these molecules is open to interpretation and can be explained in various ways. In the case of those targeting integrins it can be argued that they are indeed inhibitory but platelets are not their primary target e.g. α2β1 is also present on endothelial and smooth muscle cells and inhibitory molecules are anti-angiogenic.

In the case of GPIb, inhibitory snaclecs are very widespread and very many snake venoms contain at least one. Their ability to bind to GPIb and to inhibit von Willebrand factor binding/activation as well as (in some cases but rarely tested for) responses to low thrombin concentrations is well documented in numerous papers. Several of these GPIb-binding snaclecs were also tested by injection into small animals such as mice or guinea pigs and in most cases caused a profound thrombocytopenia lasting a few hours before the platelet count gradually returned to normal. How should we interpret these results?

GPIb has a fairly high variability in sequence between species, especially in critical binding regions compared to many platelet receptors (e.g. integrins). This is confirmed by the restricted reactivity of monoclonal antibodies prepared against GPIbα. For example, monoclonal antibodies made against human GPIbα react poorly with GPIbα from other species and in most cases cross-react only with a few related primates such as baboons. Thus, it is has so far proved difficult to prepare monoclonals to human GPIb, which cross-react with and can be tested in small animal models. This made research into anti-GPIb antibodies as anti-thrombotic agents much more difficult than with other platelet receptors.

While anti-GPIb snaclecs work across a wide range of species the affect of a given GPIb-binding snaclec on platelets of a given species may vary considerably. The main reason for this broader spectrum of binding compared to monoclonal antibodies is that snaclecs interact with GPIb via a larger protein-protein surface. Thus, even in the case of an imperfect match enough molecular interactions still occur to permit a level of binding adequate for inhibition of von Willebrand factor (or thrombin) binding. While we do not yet have crystal structures of these snaclec/GPIb complexes, we do have quite a lot of information from competition studies with various anti-GPIb monoclonal antibodies that indicate that each GPIb-binding snaclec has individually optimised its binding domain on GPIbα. Although we have no evidence yet, the binding domain might vary somewhat between snaclecs from different species binding to GPIb because of their varying amino acid sequences. One consequence of this variability is that the binding strength of a given snaclec to GPIb varies from one (platelet) species to another. An explanation of this phenomenon may be that investigators have mainly tested these snaclecs against washed human platelets and in a few cases against mouse or rabbit platelets. Although snake venoms have a broad reactivity against mammals, birds and reptiles they each contain 200–300 components so that the possibility that one or more of these is effective in a given species remains fairly high. Thus, snaclecs targeting GPIb will be most effective in certain prey species but there will also be a fairly broad spectrum of prey species where they can assist in immobilising/killing the prey. There may even be species (e.g. human) where they have little or no effect. Although there are only a few snaclecs in the venom of a given species the snake can extend the effective target range by sequence variation (which in humans would be called polymorphism) and also by different α/β subunit combinations. If a given species has a wide range of environments the venom composition may vary considerably across the range. Examples are *Echis carinatus* which has a range from the west coast of North Africa to the deserts of Central Asia. There is some evidence that snakes can adapt their venom protein expression to provide a better match to the prey in their range (Valentin and Lambeau, 2000).

Snaclecs that have been described as inhibiting GPIb receptor function are listed in Table 21.1. They were mainly described in the 1980s and 1990s with only a few in the 2000s. This is probably not because fewer were discovered but rather that it has become progressively more difficult to publish papers dealing with venom proteins that lack a "novel" function. This is a pity because despite their apparent similarity these proteins do all differ in how they interact with GPIb even if only marginally (but mainly we do not know) and have a lot to tell us about how that receptor complex works. GPIb is a receptor for an ever increasing number of ligands and snaclecs are useful tools to help explore these interactions and determine which domains are critical. Compared to monoclonal antibodies snaclecs have the ability to interact with GPIb from a wide range of species although they may vary in their effects. They are therefore also useful tools for comparing inhibition of GPIb between species or for detecting GPIb on platelets from a range of species including those where no monoclonal antibodies have been prepared yet, for example, by flow cytometry. Many studies on snaclecs show sequence comparisons and state that there is a high degree of similarity, even between those with quite different target molecules. While this is certainly true it is not a very useful statement since much of the similarity is due to those parts of the molecule that maintain the structure and not those that have evolved to participate in the binding interaction. It might therefore be more useful to compare only those sequences involved in the latter. Mainly but not exclusively, this would be those sequences making up the concave surface formed between the two subunits. Even if this is considered, two different snaclecs binding to the same target molecule do not necessarily have similar sequences in these domains because they probably have evolved to bind differently to the same target. Botrocetin and bitiscetin are classic examples of this (Fukuda et al., 2005; Maita et al., 2003). Two snaclecs, from different snakes, that bind in different ways to the same two molecules, GPIb and von Willebrand factor but with essentially the same function. This is also why it is so difficult to predict snaclec function (target molecules etc.) based on primary sequence obtained from cloning. Expression of both subunits together and correct refolding is essential to be able to test these snaclecs against real targets and determine their effects. This is quite a difficult operation and has no guarantee of success. Since much snake venom research is now done at the transcriptional level rather than the protein level this means that, compared to many other classes of venom components, snaclec research progresses fairly slowly.

One of the more recent snaclecs, described as a GPIb inhibitor, is dabocetin isolated from *Daboia russellii siamensis* venom (Zhong et al., 2006). Like echicetin, it inhibits ristocetin-induced platelet aggregation but it was also shown by flow cytometry to inhibit monoclonal antibody SZ2 binding to platelets. This implies that it does not bind to the same site on GPIbα as echicetin supporting some of the arguments above.

Table 21.1 Snaclecs inhibiting platelet function

Protein name	Species	Target	Sequence	Structure	References
Agkicetin C	*Deinagkistrodon acutus*	GPIb	+	–	Chen and Tsai (1995); Chen et al. (2000)
Akitonin	*Deinagkistrodon acutus*	GPIb	+	–	Zha et al. (2004)
CHH-A and B	*Crotalus horridus horridus*	GPIb	+	–	Andrews et al. (1996)
Dabocetin	*Daboia russellii siamensis*	GPIb	+	–	Zhong et al. (2006)
Echicetin	*Echis carinatus*	GPIb, IgMk	+	1OZ7	Polgar et al. (1997); Navdaev et al. (2001); Jasti et al. (2004)
GPIB-BP	*Bothrops jararaca*	GPIb	+	–	Kawasaki et al. (1996); Fujimura et al. (1995)
Lebecetin	*Macrovipera lebetina*	GPIb	+	–	Sarray et al. (2003)
TSV-GPIB-BP	*Trimeresurus stejnegeri*	GPIb	+	–	Lee and Zhang (2003)
Tokaracetin	*Trimeresurus tokarensis*	GPIb	–	–	Kawasaki et al. (1995)
Flavocetin-A	*Trimeresurus flavoviridis*	GPIb	+	1C3A	Shin et al. (2000); Fukuda et al. (2000); Taniuchi et al. (1995)
EMS16	*Echis multisquamatus*	α2β1	+	1UKM	Marcinkiewicz et al. (2000); Horii et al. (2003)
Rhodocetin	*Calloselasma rhodostoma*	α2β1	+	3GPR	Wang et al. (1999); Eble and Tuckwell (2003); Eble et al. (2009)
VP12	*Vipera paleastinae*	α2β1	–	–	Staniszewska et al. (2009)

Flavocetin is a tetrameric heterodimer (Fig. 21.1c) capable of forming small agglutinates in washed platelets by binding to GPIb (Taniuchi et al., 1995). However, it appears not to activate platelets based on the usual criteria of αIIbβ3 activation and granule release (P-selectin exposure). The problems of platelet agglutination versus aggregation induced by GPIb binding snaclecs is dealt with in Chapter 35 but, in brief, seems to involve the binding strength, which in turn is related to the GPIb species. While flavocetin binds to GPIb but does not activate platelets it inhibits platelet activation via classic GPIb pathways such as von Willebrand factor presumably by blocking binding sites. Whether this would be the case in all species is an open question and has not been adequately explored.

α2β1-Binding Snaclecs

Several snaclecs inhibit platelet function via α2β1, including rhodocetin from *Calloselasma rhodostoma*, unusual in that the α- and β-subunits of the heterodimer are non-covalently associated (Paaventhan et al., 2005; Wang et al., 1999). It behaves as a monomer under the conditions of testing used so far and as such inhibits platelet responses. Perhaps its main target in vivo is endothelial or smooth muscle cells rather than platelets. Its main platelet binding site lies on the I-domain of the α2-subunit of α2β1 next to the collagen binding site (Eble and Tuckwell, 2003). Rhodocetin has been shown to be α2β1 specific and its crystal structure has been determined. It forms a dimeric heterodimer structure with the concave faces pointing outwards and at right angles to one another (Eble et al., 2009). As with flavocetin the question arises whether a dimeric snaclec is only inhibitory. Does rhodocetin behave as a dimer under physiological conditions? Is its dimeric structure necessary to increase its binding constant? Why does it not cross-link α2β1 on the platelet surface or between platelets?

Another α2β1 specific snaclec was isolated from *Echis multisquamatus* in 2000 and called EMS16 (Horii et al., 2003; Marcinkiewicz et al., 2000). The crystal structure of EMS16 in complex with the I-domain of the human integrin α2-subunit has been determined (Horii et al., 2004) and shows that the concave domain of EMS16 blocks the access of collagen to its key binding site on α2β1 (Fig. 21.1d). VP12 from *Vipera palaestinae* venom (Staniszewska et al., 2009) is a newly described snaclec, directed against α2β1 integrin, shown to inhibit pro-metastatic activities of melanoma cells lines, suggesting that collagen receptors may be an interesting target for development of new anti-metastatic therapies and that snaclecs provide a good method for investigating their function.

As mentioned above snaclecs are not a major route for snake venom to block platelet function. Despite this caveat it remains possible that further snaclecs will be discovered that do this, perhaps acting on receptors other than GPIb or α2β1. Other β1 integrins, in particular remain distinct possibilities. Whether their primary target would be platelets remains dubious.

References

Andrews, R.K., Kroll, M.H., Ward, C.M., Rose, J.W., Scarborough, R.M., Smith, A.I., Lopez, J.A., Berndt, M.C., 1996. Binding of a novel 50-kilodalton alboaggregin from *Trimeresurus albolabris* and related viper venom proteins to the platelet membrane glycoprotein Ib-IX-V complex. Effect on platelet aggregation and glycoprotein Ib-mediated platelet activation. Biochemistry 35, 12629–12639.

Chen, Y.L., Tsai, I.H., 1995. Functional and sequence characterization of agkicetin, a new glycoprotein Ib antagonist isolated from *Agkistrodon acutus* venom. Biochem. Biophys. Res. Commun. 210, 472–477.

Chen, Y.L., Tsai, K.W., Chang, T., Hong, T.M., Tsai, I.H., 2000. Glycoprotein Ib-binding protein from the venom of *Deinagkistrodon acutus* – cDNA sequence, functional characterization, and three-dimensional modeling. Thromb. Haemost. 83, 119–126.

Clemetson, K.J., Morita, T., Kini, R.M., 2009. Scientific and standardization committee communications: classification and nomenclature of snake venom C-type lectins and related proteins. J. Thromb. Haemost. 7, 360.

Eble, J.A., Niland, S., Bracht, T., Mormann, M., Peter-Katalinic, J., Pohlentz, G., Stetefeld, J., 2009. The $\alpha 2\beta 1$ integrin-specific antagonist rhodocetin is a cruciform, heterotetrameric molecule. FASEB J. 23, 2917–2927.

Eble, J.A., Tuckwell, D.S., 2003. The $\alpha 2\beta 1$ integrin inhibitor rhodocetin binds to the A-domain of the integrin $\alpha 2$ subunit proximal to the collagen-binding site. Biochem. J. 376, 77–85.

Fry, B.G., Vidal, N., Norman, J.A., Vonk, F.J., Scheib, H., Ramjan, S.F., Kuruppu, S., Fung, K., Hedges, S.B., Richardson, M.K., Hodgson, W.C., Ignjatovic, V., Summerhayes, R., Kochva, E., 2006. Early evolution of the venom system in lizards and snakes. Nature 439, 584–588.

Fujimura, Y., Ikeda, Y., Miura, S., Yoshida, E., Shima, H., Nishida, S., Suzuki, M., Titani, K., Taniuchi, Y., Kawasaki, T., 1995. Isolation and characterization of jararaca GPIb-BP, a snake venom antagonist specific to platelet glycoprotein Ib. Thromb. Haemost. 74, 743–750.

Fukuda, K., Doggett, T.A., Bankston, L.A., Cruz, M.A., Diacovo, T.G., Liddington, R.C., 2002. Structural basis of von Willebrand factor activation by the snake toxin botrocetin. Structure 10, 943–950.

Fukuda, K., Doggett, T., Laurenzi, I.J., Liddington, R.C., Diacovo, T.G. 2005. The snake venom protein botrocetin acts as a biological brace to promote dysfunctional platelet aggregation. Nat. Struct. Mol. Biol. 12, 152–159.

Fukuda, K., Mizuno, H., Atoda, H., Morita, T., 2000. Crystal structure of flavocetin-A, a platelet glycoprotein Ib-binding protein, reveals a novel cyclic tetramer of C-type lectin-like heterodimers. Biochemistry 39, 1915–1923.

Horii, K., Okuda, D., Morita, T., Mizuno, H., 2003. Structural characterization of EMS16, an antagonist of collagen receptor (GPIa/IIa) from the venom of *Echis multisquamatus*. Biochemistry 42, 12497–12502.

Horii, K., Okuda, D., Morita, T., Mizuno, H., 2004. Crystal structure of EMS16 in complex with the integrin $\alpha 2$-I domain. J. Mol. Biol. 341, 519–527.

Jasti, J., Paramasivam, M., Srinivasan, A., Singh, T.P., 2004. Crystal structure of echicetin from *Echis carinatus* (Indian saw-scaled viper) at 2.4A resolution. J. Mol. Biol. 335, 167–176.

Kawasaki, T., Fujimura, Y., Usami, Y., Suzuki, M., Miura, S., Sakurai, Y., Makita, K., Taniuchi, Y., Hirano, K., Titani, K., 1996. Complete amino acid sequence and identification of the platelet glycoprotein Ib-binding site of jararaca GPIb-BP, a snake venom protein isolated from *Bothrops jararaca*. J. Biol. Chem. 271, 10635–10639.

Kawasaki, T., Taniuchi, Y., Hisamichi, N., Fujimura, Y., Suzuki, M., Titani, K., Sakai, Y., Kaku, S., Satoh, N., Takenaka, T., Handa, M., Sawai, Y., 1995. Tokaracetin, a new platelet antagonist that binds to platelet glycoprotein ib and inhibits von Willebrand factor-dependent shear- induced platelet aggregation. Biochem. J. 308, 947–953.

Koh, D.C., Armugam, A., Jeyaseelan, K., 2006. Snake venom components and their applications in biomedicine. Cell. Mol. Life Sci. 63, 3030–3041.

Lee, W.H., Zhang, Y., 2003. Molecular cloning and characterization of a platelet glycoprotein Ib-binding protein from the venom of *Trimeresurus stejnegeri*. Toxicon 41, 885–892.

Maita, N., Nishio, K., Nishimoto, E., Matsui, T., Shikamoto, Y., Morita, T., Sadler, J.E., Mizuno, H., 2003. Crystal structure of von Willebrand factor A1 domain complexed with snake venom, bitiscetin: insight into glycoprotein Ibα binding mechanism induced by snake venom proteins. J. Biol. Chem. 278, 37777–37781.

Marcinkiewicz, C., Lobb, R.R., Marcinkiewicz, M.M., Daniel, J.L., Smith, J.B., Dangelmaier, C., Weinreb, P.H., Beacham, D.A., Niewiarowski, S., 2000. Isolation and characterization of EMS16, a C-lectin type protein from *Echis multisquamatus* venom, a potent and selective inhibitor of the α2β1 integrin. Biochemistry 39, 9859–9867.

Mizuno, H., Fujimoto, Z., Atoda, H., Morita, T., 2001. Crystal structure of an anticoagulant protein in complex with the Gla domain of factor X. Proc. Natl. Acad. Sci. U.S.A. 98, 7230–7234.

Mizuno, H., Fujimoto, Z., Koizumi, M., Kano, H., Atoda, H., Morita, T., 1999. Crystal structure of coagulation factor IX-binding protein from habu snake venom at 2.6 Å: implication of central loop swapping based on deletion in the linker region. J. Mol. Biol. 289, 103–112.

Morita, T., Atoda, H., Sekiya, F., 1996. Structure and functions of coagulation factor IX/factor X-binding protein isolated from the venom of *Trimeresurus flavoviridis*. Adv. Exp. Med. Biol. 391, 187–196.

Navdaev, A., Dormann, D., Clemetson, J.M., Clemetson, K.J., 2001. Echicetin, a GPIb-binding snake C-type lectin from *Echis carinatus*, also contains a binding site for IgMκ responsible for platelet agglutination in plasma and inducing signal transduction. Blood 97, 2333–2341.

Paaventhan, P., Kong, C., Joseph, J.S., Chung, M.C., Kolatkar, P.R., 2005. Structure of rhodocetin reveals noncovalently bound heterodimer interface. Protein Sci. 14, 169–175.

Peng, M., Holt, J.C., Niewiarowski, S., 1994. Isolation, characterization and amino acid sequence of echicetin β-subunit, a specific inhibitor of von Willebrand factor and thrombin interaction with glycoprotein Ib. Biochem. Biophys. Res. Commun. 205, 68–72.

Polgar, J., Magnenat, E.M., Peitsch, M.C., Wells, T.N., Saqi, M.S., Clemetson, K.J., 1997. Amino acid sequence of the alpha subunit and computer modelling of the alpha and beta subunits of echicetin from the venom of *Echis carinatus* (saw-scaled viper). Biochem. J. 323, 533–537.

Sarray, S., Srairi, N., Hatmi, M., Luis, J., Louzir, H., Regaya, I., Slema, H., Marvaldi, J., El Ayeb, M., Maffakchi, N., 2003. Lebecetin, a potent antiplatelet C-type lectin from *Macrovipera lebetina* venom. Biochim. Biophys. Acta 1651, 30–40.

Shin, Y., Okuyama, I., Hasegawa, J., Morita, T., 2000. Molecular cloning of glycoprotein Ib-binding protein, flavocetin-A, which inhibits platelet aggregation. Thromb. Res. 99, 239–247.

Staniszewska, I., Walsh, E.M., Rothman, V.L., Gaathon, A., Tuszynski, G.P., Calvete, J.J., Lazarovici, P., Marcinkiewicz, C., 2009. Effect of VP12 and viperistatin on inhibition of collagen receptors-dependent melanoma metastasis. Cancer Biol. Therapy 8, 1507–1516.

Taniuchi, Y., Kawasaki, T., Fujimura, Y., Suzuki, M., Titani, K., Sakai, Y., Kaku, S., Hisamichi, N., Satoh, N., Takenaka, T., Handa, M., Sawai, Y., 1995. Flavocetin-A and -B, two high molecular mass glycoprotein Ib binding proteins with high affinity purified from *Trimeresurus flavoviridis* venom, inhibit platelet aggregation at high shear stress. Biochim. Biophys. Acta 1244, 331–338.

Valentin, E., Lambeau, G., 2000. What can venom phospholipases A_2 tell us about the functional diversity of mammalian secreted phospholipases A_2? Biochimie 82, 815–831.

Wang, R., Kini, R.M., Chung, M.C., 1999. Rhodocetin, a novel platelet aggregation inhibitor from the venom of *Calloselasma rhodostoma* (Malayan pit viper): synergistic and noncovalent interaction between its subunits. Biochemistry 38, 7584–7593.

Zha, X.D., Liu, J., Xu, K.S., 2004. cDNA cloning, sequence analysis, and recombinant expression of akitonin β, a C-type lectin-like protein from *Agkistrodon acutus*. Acta Pharmacol. Sin. 25, 372–377.

Zhong, S.R., Jin, Y., Wu, J.B., Chen, R.Q., Jia, Y.H., Wang, W.Y., Xiong, Y.L., Zhang, Y., 2006. Characterization and molecular cloning of dabocetin, a potent antiplatelet C-type lectin-like protein from *Daboia russellii siamensis* venom. Toxicon 47, 104–112.

Part V
Fibrinolytic Proteins

Chapter 22
Plasminogen Activators from Snake Venoms

Bernard F. Le Bonniec and Julie Libraire

Abstract Normal fibrinolysis results from activation of plasminogen to plasmin, the effective blood clot dissolution enzyme. Plasminogen is a remarkably stable proenzyme circulating at a high 2 μM in blood. Plasmin is an efficient trypsin-like protease (clan SA) with broad specificity. Plasmin is highly destructive, but its affinity for fibrin normally localizes its action in the vicinity of blood clot and two powerful inhibitors prevent its uncontrolled dissemination. Five types of plasminogen activators had been described to date. Four are typical serine proteases sharing several distinctive features. The fifth kind includes streptokinase and staphylokinase of bacterial origin. Tissue-type (t-PA) and urokinase-type (u-PA) plasminogen activators are physiologic activators sharing similar vulnerability to inhibitors but differing for activation mechanism and localization of action. An additional activator had been characterized in the saliva of the vampire bat *Desmodus rotundus* (DS-PA) which deliberately lacks regulatory mechanisms of inhibition while retaining a strict dependence to fibrin for activity. Maintaining blood flow in its prey is the goal of bloodsucking animals. During the past 25 years, a number of plasminogen activators had been isolated from various snake venoms: Chinese green tree viper (TSV-PA from *Trimeresurus stejnegeri*), Bushmaster (LV-PA from *Lachesis muta muta*), Korean salmosa snake (Haly-PA from *Agkistrodon halys brevicaudus*) and Ussuri mamushi (ABUSV-PA from *Agkistrodon blomhoffii Ussuriensis*). Most have been characterized, purified, and sequenced; the structure of TSV-PA had been solved by X-ray diffraction. These are uncontrolled proteases triggering systemic plasminogen activation with catastrophic consequence for the prey.

B.F. Le Bonniec (✉)
INSERM U765, Université Paris Descartes, 4 Av. de l'Observatoire, 75006 Paris Cedex, France
e-mail: bernard.le-bonniec@parisdescartes.fr

This chapter is dedicated to Pr. Cassian Bon (1944–2008).

R.M. Kini et al. (eds.), *Toxins and Hemostasis*, DOI 10.1007/978-90-481-9295-3_22,

Introduction

Plasminogen is the zymogen of a serine protease (plasmin) with relatively broad specificity even if it is less efficient than trypsin. Conversion to plasmin requires a single proteolytic event. Plasminogen is peculiar in that activating cleavage occurs within a short disulfide loop. This imposes considerable restrictions on the activation complex because the classical partially extended conformation cannot be achieved (Lamba et al., 1996). Activation by cleavage within a short disulfide loop is unique to plasminogen amongst the thousand precursors of the SA family (Rawlings, 1998). It results that plasminogen is a remarkably stable proenzyme. For instance, plasmin does not retro-activate plasminogen in spite of its broad specificity. Native plasminogen that circulates at a high 2 μM in blood is also called Glu-plasminogen in reference to its N-terminal glutamate. Following activation, the resulting Glu-plasmin is only a transient species because it rapidly retro-cleaves into what could be called Lys-plasmin but is simply called plasmin (Fig. 22.1). Plasmin also retro-cleaves Glu-plasminogen into Lys-plasminogen. This is not a retro-activation because Lys-plasminogen is still an inactive zymogene starting at Lys^{61} or Lys^{76} compared to Glu^{1} of native Glu-plasminogen (Horrevoets et al., 1995). Retro-cleavage constitutes nevertheless an amplification loop because Lys-plasminogen is activated more rapidly to plasmin (Miles et al., 2003). Higher rate of activation seems to originate from a more open conformation of the shorter zymogen rather than an allosteric adaptation (such as induce fit) of the PA. Thus higher sensitivity to proteolysis must concern all PA, including snake venoms PA.

Mammalian fibrinolytic system is mainly involved in the dissolution of blood clot, but it also has a role in processes requiring cell migration (wound healing, tissue remodeling, angiogenesis, and embryogenesis). Plasmin is a powerful serine protease potentially highly destructive. Uncontrolled plasmin dissemination exhausts fibrinogen and a number of plasma proteins with the consequence that haemostasis is no longer efficient. Plasmin also damages proteins of the extracellular matrix, fibronectin in particular. Furthermore, plasmin acts indirectly by activation of a metalloprotease cascade. Ultimately dissemination of uncontrolled plasmin may trigger bleeding. Plasminogen activation and plasmin activity are normally confined to the vicinity of fibrin clot and/or of endothelial cells. Controls occur at the level of plasminogen activation as well as at the level of plasmin localization and diffusion. The first lock is that plasminogen having high affinity for fibrin it is incorporated within the clot. A similar mechanism localizes the main physiologic PA since tissue-type plasminogen activation (t-PA) also has high affinity for fibrin (Lijnen and Collen, 1998). Fibrin being the cofactor for plasminogen activation by t-PA, plasmin formation mostly occurs within the clot. Furthermore plasmin too has high affinity for fibrin thus it is confined to the vicinity of the clot. Plasmin initially creates nicks in the fibrin network but ultimately digests it completely sawing off the branch it sits on because it is no longer protected by fibrin when it starts to disseminate and therefore is rapidly neutralized by α2-antiplasmin and α2-macroglobulin. These inhibitors are inefficient within the clot because fibrin protects plasmin. Even if they function differently these are suicide substrate inhibitors that are consumed during

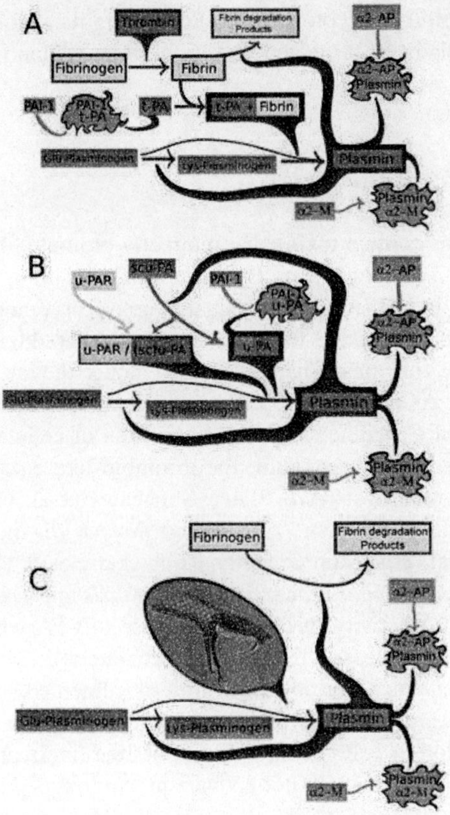

Fig. 22.1 Plasminogen activation pathways. Unless exhausted, α2-antiplasmin (α2-AP) and α2-macroglobulin (α2-M) prevent plasmin dissemination in the three pathways. PAs activate about 10-fold more efficiently Lys-plasminogen than Glu-plasminogen. Plasmin hydrolysis Glu-plasminogen into Lys-plasminogen thus retroclivage constitutes an amplification loop. (**a**) Activation through the t-PA pathway. Fibrin is the key cofactor and dramatically enhances the rate of plasminogen activation. Fibrin also protects t-PA from inhibition by PAI-1 which neutralizes very efficiently t-PA in solution. (**b**) Activation through the u-PA pathway. Binding of u-PA and its zymogen (scu-PA) to their cellular receptor (u-PAR) localizes plasminogen activation in the vicinity of endothelial cells where plasmin degrades the extracellular matrix. Plasmin efficiently activates scu-PA into u-PA retroclivage also constituting an amplification loop. (**c**) Activation by snake PA. Pathway is characterized by the lack of control. Massive production of plasmin exhausts α2-AP and α2-M resulting in fibrinogen consumption

inactivation of their targets. Thus dissemination of plasmin is severely restricted unless inhibitors had been exhausted. Fibrin also protects t-PA from neutralization by its main inhibitor, namely plasminogen activator inhibitor-1 (PAI-1). Thus dissemination of t-PA is also severely restricted unless PAI-1 had been exhausted. Without fibrin as a cofactor, t-PA activates plasminogen slowly and it is neutralized faster by PAI-1 than it produces plasmin. Conversely a similar lock occurs with the second endogeneous PA, namely urokinase-type plasminogen activator (u-PA) which binds to its cellular receptor (u-PAR) confining its action to the cell surface where plasmin is relatively protected from inhibition (Alfano et al., 2005; Miles et al., 2003).

Toxins Promoting Indirect PA

Several snake venoms contain toxins that indirectly promote fibrinolysis by inducing the release of host endogeneous PA. In normal situation, the release of t-PA by endothelial cells is triggered by physical exercise or venous occlusion which increases its plasma level about 3 fold within minutes. Paradoxically, thrombin and several thrombin-like enzymes (Chapter 30) also induce the release of t-PA or u-PA by endothelial cells. As early as 1985, Budzynski published an abstract mentioning that the venom of Crotalinae stimulates secretion of cellular PA and Klöcking et al. (1987) reported that batroxobin, the thrombin-like enzyme from *Bothrops atrox*, induces t-PA release in vivo. Later, Sunagawa et al. (1996) reported that the thrombin-like Habutobin from *Trimeresurus flavoviridis* induces the release of u-PA from endothelial cells. More recently, Rojnuckarin et al. (1999) observed that venoms of the green pit vipers *Trimeresurus albolabris* and *Trimeresurus macrops* induce elevation of PA activity through the release of t-PA which is not counterbalanced by concurrent release of PAI-1. Unfortunately the exact mechanism by which these toxins enhance fibrinolysis is little documented. A number of mechanisms are possible and few had been considered. For instance, independently from direct stimulation of endothelial cells, a possible (indirect) mechanism could be that thrombin-like enzymes do not activate procarboxypeptidase U, also called TAFI for thrombin activatable fibrinolysis inhibitor (Bajzar et al., 1995; Schneider and Nesheim, 2004). In non pathological situation when t-PA is released from endothelial cells, thrombin simultaneously activates TAFI. Thus it is conceivable that uncontrolled fibrinolysis occurs if TAFI remains intact (Mao et al., 2005; Wu et al., 2003). Similarly, a mechanism exhausting PAI-1, the main plasma inhibitor of t-PA and u-PA would mechanically promote fibrinolysis (Gladson et al., 1989). In addition published reports on indirect plasminogen activation do not mention if concurrent direct PA activity was also investigated. A distinctive mechanism of indirect PA seems to be used by Jararafibrase I isolated from the venom of *Bothrops jararaca*. Toxin, which is a metalloproteinase and not a serine protease, cleaves the single chain (inactive) precursor of u-PA (scu-PA) into a lower molecular weight intermediate still inactive but more susceptible to activation (Sugiki et al., 1998).

Direct PA

Beside t-PA and u-PA, a number of endogeneous PA candidates are suspected including plasma kallikrein and matrix metalloproteinases (Lund et al., 2006). Yet, to date, only six kinds of in vivo plasminogen activators have been unquestionably characterized, all being highly specialized molecules. The first two are obviously the mammalian activators t-PA and u-PA with comparable efficiencies as well as vulnerability to inhibitors albeit different mechanism of activation and localization. A third kind (Bergum and Gardell, 1992; Gardell et al., 1989) is represented by *Desmodus* saliva plasminogen activator (DS-PA) that had been characterized in the saliva of the vampire bat *Desmodus rotundus* (Chapter 20). The fourth kind had been extracted from various snake venoms and constitutes the main topic of this chapter. The last two kinds (staphylokinase and streptokinase) are described in detail in Chapter 25. These are of bacterial origin and totally unrelated with the other PA. They act by forming a stoichiometric complex with plasminogen for one, plasmin for the other, and this is the resulting complex which is capable of activating plasminogen.

Only a few direct snake venom PA had been characterized. The first was isolated in 1995 from the Chinese green tree viper *Trimeresurus stejnegeri* and had been named TSV-PA (Zhang et al., 1995). Soon after a second direct PA was characterized in the Korean salmosa snake *Agkistrodon halys brevicaudus* and named Haly-PA (Park et al., 1998). The third had been discovered in the South American bushmaster *Lachesis muta muta* and named LV-PA (Sanchez et al., 2000). This is probably the beginning of a long list as direct PA activity had been investigated in only a limited number of snake venoms. True snake venom PA should not be confused with direct fibrino(geno)lytic enzymes (Markland, 1998a, b; Swenson and Markland, 2005) such as fibrolase from *Agkistrodon contortrix contortrix* and Alfimeprase its recombinant truncated form (Chapter 24). For instance, in addition to the PA, the venom of the Chinese green tree viper *Trimeresurus stejnegeri* contains several fibrino(geno)lytic enzymes named stejnefibrases (Gao et al., 1998). In fact the majority of direct fibrino(geno)lytic enzymes are metalloproteinase, whereas all PA characterized to date are serine proteases belonging to clan SA. In addition, true snake venom PA should not be confused with the toxins which trigger the release of endogenous PA from host. Finally, snake venom PA described herein potentially act in vivo following envenomation, as opposed to toxins routinely used for research or diagnosis purpose which have a specific activity in vitro not occurring in vivo. For instance, Protac from the venom of *Agkistrodon contortrix contortrix* belonging to clan SA specifically activates the anticoagulant protein C but only its calcium-free form (Kisiel et al., 1987; Klein and Walker, 1986; McMullen et al., 1989; Murakami and Arni, 2005; Stocker et al. 1987). In blood, protein C is saturated by calcium ions rendering questionable its susceptibility as a substrate of Protac in case of envenomation. Therefore, characterization as a true PA requires relatively complex techniques to distinguish from other fibrin(ogen)olytic activities.

Although it has not been certified as a true PA, DAV-PA (Wang et al., 2001) is another potential snake venom PA identified through general cloning of the serine proteases expressed within the venomous gland of the hundred-pace snake *Deinagkistrodon acutus*. Independently, two closely related molecules (AaV-SP-I and AaV-SP-II with an identical amino acid sequence as DAV-PA) had been isolated from the venom of the Chinese moccasin snake *Deinagkistrodon acutus* (formerly, *Agkistrodon acutus*). The structures of AaV-SP-I and AaV-SP-II have been solved by X-ray diffraction but, as with DAV-PA, neither had been formally characterized as a true PA (Zhu et al., 2005). Presence of PA activity in the venom of *Agkistrodon halys halys* had been mentioned, but no sequence is available (Karbovs'kyĭ et al, 2006). Another PA candidate had been isolated and partially characterized from the Chinese pit viper venom *Agkistrodon blomhoffii ussuriensis* and named ABUSV-PA (Liu et al., 2006) but its sequence is not yet available. Finally, complementary DNA sequencing from the venomous gland of the western cottonmouth pit viper *Agkistrodon piscivorus leucostoma* allowed prediction of two more PA but neither had been formally characterized (Jia et al., 2008).

Purification of Snake PA

Three of the identified snake PAs were characterized following purification from the venom itself, namely TSV-PA, LV-PA, and ABUSV-PA. TSV-PA had been further characterized following expression in *E. coli* and a report mentions its expression in a baculovirus system (Cao et al., 2001). Haly-PA had not been purified from the venom of *Agkistrodon halys brevicaudus*. Instead, it was identified in a clone isolated from a venom gland cDNA library. The formal characterization of Haly-PA relied on a recombinant form that was expressed in a baculovirus system.

Purification of PAs from crude venom was achieved by a combination of anion-exchange and size-exclusion chromatographies usually performed at 4°C. For example, purification of TSV-PA from the venom of *Trimeresurus stejnegeri* (Zhang et al., 1995) started with a size-exclusion chromatography on Sephadex G-75 resulting in the separation of seven peaks. The plasminogen activation activity separate from the procoagulant and phospholipase A activities was predominantly associated with the fourth peak. Following dialysis against 20 mM Tris-HCl, pH 7.8, TSV-PA was recovered by a double anion-exchange chromatography on a Mono-Q HR 16/10 column using a linear NaCl gradient. SDS-PAGE analysis displayed that final pool contained a major polypeptide of *Mr* 30,000±3,000 in both reducing and nonreducing conditions. PAGE carried out in native conditions revealed, however, the presence of small contaminants. Protocol allowed a 200-fold purification of TSV-PA (about 0.2% of the total venom proteins). TSV-PA was not further purified and most functional studies were performed with a recombinant form expressed in *E. coli* (Braud et al., 2000, 2002; Zhang et al., 1997). Purification of recombinant TSV-PA included a single anion exchange chromatography but required a preliminary refolding step. Inclusion bodies were recovered by centrifugation after cell lysis and washed three times in 50 mM Tris-HCl pH 8.0, containing 2.5 mM EDTA. The protein was fully denatured in 8 M urea containing 50 mM β-mercaptoethanol. Refolding was initiated by a 50-fold sudden dilution in the Tris/EDTA buffer. Progress of the refolding was monitored through the amidolytic activity which reached a plateau in about 48 h. Recombinant TSV-PA was concentrated and dialyzed against 20 mM Tris-HCl, pH 8.8 and purified on a Mono-Q HR16/10 column. The presence of TSV-PA was monitored all along the purifications by A_{280}, amidolytic activity using H-D-Phe-Pip-Arg-pNa (S2238), and plasminogen activation activity. The catalytic activity of recombinant TSV-PA was found comparable to that of TSV-PA purified from the venom (Zhang et al., 1997).

Purification of LV-PA from the venom of *Lachesis muta muta* (Sanchez et al., 2000) also started by a size-exclusion chromatography. Lyophilized crude venom was dissolved in 50 mM ammonium acetate pH 7.3 containing 0.3 M NaCl and centrifuged at 6,000 g to remove insoluble material. Chromatography on Sephadex G-100 yielded 8 peaks, of which the fourth peak contained the PA activity. This material was dialyzed against distilled water, lyophilized, and applied to a DEAE–Sepharose CL-6B column equilibrated in 50 mM Hepes pH 8.0. Fraction was developed through a linear NaCl gradient yielding nine peaks. Fractions containing PA activity were pooled, lyophilized, and rechromatographed on the same column under identical conditions. Active fraction was further purified by size exclusion chromatography on a Sephadex G-75 column leading to a symmetrical peak with constant specific amidolytic activity across the collected fractions. The final product displayed a single band on SDS–PAGE. Apparent *Mr* of LV-PA was 33,000 and 29,000 under reducing and nonreducing conditions, respectively. Purified LV-PA represented about 0.8% of the total venom proteins. All along purification, the presence of LV-PA was monitored by A_{280} and amidolytic activity using H-D-Val-Leu-Lys-pNA (S2251), Tos-Gly-Pro-Lys-pNA, and N-Benzoyl-D-L-Arg-pNA (BAPNA).

In contrast, purification of ABUSV-PA from the venom of *Agkistrodon blomhoffii ussurensis* (Liu et al., 2006) started by an anion-exchange chromatography. Crude snake venom was loaded on a DEAE Sephadex A-50 column equilibrated in 50 mM Tris-HCl pH 7.5 which was developed through a linear NaCl gradient. The active enzyme fractions were desalted, concentrated, and further purified through a size exclusion Sephadex G-75 column equilibrated in 50 mM Tris HCl pH 7.5 containing 0.15 M NaCl. The active enzyme fractions were desalted and loaded on a second DEAE Sephadex A-50 column also developed through a linear NaCl gradient. The final step of purification was a rechromatography on a smaller DEAE Sephadex A50 column in the same condition. Final product was homogeneous yielding a single band by SDS–PAGE analysis with an apparent *Mr* of 31,500. Protocol allowed a 180-fold purification of ABUSV-PA (less than 0.1 % of total venom proteins). The progress of ABUSV-PA purification was monitored all along through A_{280} and esterase activity using N_{α}-tosyl-Arg-methyl-ester (TAME).

Recombinant Haly-PA (Park et al., 1998) was purified from the medium of H5 *Trichoplusia ni* infected cells and also included a combination of anion-exchange and size exclusion chromatographies. Harvested medium was concentrated by ultrafiltration and dialyzed against 20 mM Tris-HCl, pH 7.8. Concentrate was applied to a Q-Sepharose column equilibrated in the Tris-HCl buffer which was developed through a linear gradient of NaCl. Active fractions were concentrated by lyophilization and then dialyzed against 20 mM sodium phosphate buffer pH 7.8. Sample was then applied on a Superose-12 column equilibrated in the same buffer. Final product was homogeneous yielding a single band with an apparent *Mr* of 32,000 by SDS–PAGE analysis. All along purification, Haly-PA was traced through its amidolytic activity using S2238.

Characterization of Snake PA

Conclusive characterization as a true PA relies on the absence of fibrino(geno)lytic activity on its own, conjugated with a strong fibrino(geno)lytic activity in the presence of plasminogen. Direct fibrinolytic activity can be dismissed by deposit on a plasminogen-free fibrin plates, whereas a clear hollow must appear when the PA candidate is co-incubated with plasminogen in the same conditions. This method had been used for the characterization of recombinant Haly-PA (Park et al., 1998). In another approach the PA activity was confirmed through indirect increase of the fibrino(geno)lytic activity, by comparing the supernatant of fibrin clot formed in the presence of the toxin with that formed following further addition of plasminogen. Approach allowed eliminating direct, intrinsic, fibrino(geno)lytic activity with TSV-PA (Zhang et al., 1995). Fibrino(geno)lytic activity had also been evaluated using SDS-PAGE analysis of the progress of fibrinogen degradation resulting from incubation with the PA candidate alone or mixed with plasminogen. Incubation of Haly-PA in the presence of plasminogen resulted in the degradation of the Aα, Bβ and γ chains of fibrinogen in this order, a pattern that is classically found with u-PA in the same conditions. Progress of direct plasminogen activation can also be visualized through SDS-PAGE analysis. This approach allows verifying that the PA candidate hydrolyzes a single peptide bond in plasminogen, as u-PA or t-PA (Zhang et al., 1995). Finally, PA activity is determined by monitoring the increase of the amidolytic activity resulting from the plasmin formed through plasminogen activation, most often through plasmin titration of discrete sampling of the reaction mixture (Park et al., 1998; Sanchez et al., 2000; Zhang et al., 1995). In addition, the uniqueness of LV-PA was demonstrated by raising antibodies against the purified protein which did not cross-react with a number of toxins from various snake venoms (Felicori et al., 2005).

Catalytic Activity and Biological Function

Intriguing evidence emerging from studies of serine protease's specificity is a poor correlation between the catalytic groove preferences and the known biological functions. Overall, hexapeptides spanning the cleavage sites that are derived from sequences rapidly cleaved in macromolecule substrates are only reasonably well hydrolyzed by the protease. Conversely, the most favorable peptidyl substrate rarely corresponds to a naturally cleaved sequence within a macromolecule (Bianchini et al., 2002; Le Bonniec et al., 1993). PAs illustrate this concept well since they have plasminogen as common, if not unique, macromolecule substrate despite diverse peptidyl substrate preferences. The most sensitive chromogenic substrate identified for TSV-PA (S2238) is cleaved with a k_{cat}/K_M value (almost 10^6 M^{-1} s^{-1}) about 10-fold higher than for t-PA. The k_{cat}/K_M value with TSV-PA is comparable to that with trypsin but still much less than the high 8.6×10^7 M^{-1} s^{-1} efficiency of thrombin. The opposite is true with other peptidyl substrates: for instance t-PA cleaves Z-D-Arg-Gly-Arg-*p*NA (S2765) with a k_{cat}/K_M value nearly 20-fold higher than TSV-PA. Other substrates such as pyroGlu-Gly-Arg-*p*NA (S2244) are cleaved with comparable efficiency by TSV-PA and t-PA. S2238 was also the substrate most rapidly cleaved by Haly-PA (Park et al., 1998). Comparison of LV-PA preferences (Sanchez et al., 2000, 2006) with those of TSV-PA reveals however quite dissimilar preferences. In addition, published values suggest that LV-PA is much less efficient. For instance, the k_{cat}/K_M value for H-D-Val-Leu-Arg-*p*NA (S2266) hydrolysis barely reaches 10^3 M^{-1} s^{-1} with LV-PA, this is 16-fold less than with TSV-PA (1.6×10^4 M^{-1} s^{-1}). The difference is even greater for H-D-Pro-Phe-Arg-pNA (S2302) hydrolysis with k_{cat}/K_M values of 1.7×10^2 versus 3×10^4 M^{-1} s^{-1} as well as for S2251 hydrolysis with values of 6.4×10^2 versus 3.3×10^4 M^{-1} s^{-1}. Clearly, the catalytic groove preferences and the efficiency of catalysis for peptidyl substrates vary widely between snakes and mammalian PA.

With respect to plasminogen activation, snake venom PAs have a relatively low efficiency compared with their mammalian counterpart. TSV-PA activates Lys-plasminogen with a k_{cat}/K_M value (6.7×10^4 M^{-1} s^{-1}) just higher than that of t-PA in the absence of fibrin (2.3×10^4 M^{-1} s^{-1}) yet lower than that of u-PA in the absence of cofactor (3×10^5 M^{-1} s^{-1}). The k_{cat}/K_M value reaches 2.2×10^6 M^{-1} s^{-1} with u-PA on cell surface exposing its cellular receptor (Ellis et al., 1991) and plasminogen activation by t-PA is up to 10^7 M^{-1} s^{-1} with optimal fibrin stimulation (Hoylaerts et al., 1982, Madison et al., 1995; Thelwell and Longstaff, 2007). Thus t-PA alone is a relatively poor enzyme but fibrin strikingly enhances (about a hundred fold) the activation rate of plasminogen. Fibrin essentially increases the local plasminogen concentration by creating an additional interaction between t-PA and its substrate (Collen and Lijnen, 2004). Kinetic data support a mechanism where fibrin provides a surface to which t-PA and plasminogen adsorb yielding a cyclic ternary complex. Due to the lack of lysine-binding site in its kringle module, u-PA does not bind to fibrin. Its activity is enhanced about 10-fold by classical cellular receptor binding. Interestingly u-PA's cellular receptor is a relatively good substrate of u-PA with cleavage leading to receptor inactivation thus possibly constituting a regulatory mechanism (Høyer-Hansen et al., 1992). A considerable amount of the work on u-PA had been focused on the hypothesis that degradation of extracellular matrix components is important for tissue remodeling and repair and for invasive cell migration, and thus u-PA has been implicated in many physiological and pathological processes including cancer invasion and metastasis (Danø et al., 2005). The generation of mice with a targeted disruption ('knockout') of the u-PA gene allowed direct demonstration of its biological functions, including physiological fibrinolysis, response of the arterial wall to injury, susceptibility to infection, and ovulation (Carmeliet and Collen, 1996). Snake venom PAs activate plasminogen independently of fibrin: the rate of activation is not affected by CNBr fibrinogen fragments. It is also very unlikely (albeit not proven) that snake venom PA activity could be influenced by u-PA's cellular receptor. Therefore, the rate of Lys-plasminogen activation by TSV-PA is 30 and 100-fold lower than by u-PA and t-PA in optimal conditions. Interestingly, the low k_{cat}/K_M of TSV-PA originates from a low turn over rather than a high K_M (Zhang et al., 1997). In fact, the K_M value of plasminogen for TSV-PA (53 nM) is 35-fold lower than for u-PA. Considering the plasma concentration of plasminogen (2.1 μM) it results that, whereas free u-PA is never saturated, TSV-PA is constantly activating plasminogen at maximum (albeit modest) rate. From this point of view, plasminogen would be a sticky substrate of TSV-PA (Stone et al., 1991). The k_{cat}/K_M for Glu-plasminogen activation by TSV-PA had not been precisely determined, but with u-PA and t-PA value is about 10-fold lower than for Lys-plasminogen (2.9×10^4 and about 10^3 M^{-1} s^{-1} without cofactor, respectively). Rate of plasminogen activation by 30 nM Haly-PA was found comparable to that obtained with 1 nM u-PA suggesting that the activity of Haly-PA is comparable to that of TSV-PA (Park et al., 1998). Detailed kinetics data are not available for plasminogen activation by LV-PA. While kinetic parameters of plasminogen activation by ABUSV-PA have not been determined, a remarkable series of biochemical characterization of its recombinant form has been achieved which had not been performed with any other PA (Liu et al., 2006). ABUSV-PA is for instance a zinc-binding serine protease and metal ion seems important for proper folding rather than for catalysis (Liu et al., 2006). Unfortunately, the capability of ABUSV-PA to activate plasminogen has not been confirmed thus it should still be considered as a putative PA. Sequences identified from mRNAs analysis of the venom gland of *Agkistrodon piscivorus leucostoma* (Ev854880, Fig. 22.2) can also be classified as putative PA candidates. Analysis of the published sequence strongly supports a potential true PA up to Cys^{220}, but sequence thereafter completely diverges from all other PA known (as well as from all trypsin-like enzymes) suggesting a possible cloning artifact.

Inhibition

Snake venom PA are inhibited by general active site probes of trypsin-like enzymes such as phenylmethylsulphonyl fluoride (PMSF), diisopropylfluorophosphate (DFP), benzamidine, and p-aminobenzamidine as well as titrants such as p-nitrophenyl-p'-guanidinobenzoate (NPGB). The active site of snake venom PA is not affected by EDTA or thiol group reagents confirming that they are neither metalloprotease nor cysteine protease. The main physiological inhibitor of t-PA and u-PA is PAI-1. This is a 379 amino acids serine protease inhibitor (serpin) forming a quasi irreversible, inactive, stoichiometric complex with its targets. The second-order rate constant of t-PA inhibition is impressive (over 10^7 M^{-1} s^{-1}). Many serpins are of exquisite specificity exploiting exosites of the protease for interaction (Gettins and Olson, 2009). Precisely, it is well established that to select its targets PAI-1 utilizes the 37-loop of the protease, adjacent to the catalytic groove. In PAI-1 targets, the 37-loop is longer and includes four positively charged amino acids (KHRR in t-PA, RRHR in u-PA). A number of mutagenesis experiments amply confirm the direct relationship linking the nature of the 37-loop of the protease and its susceptibility to PAI-1. For instance substitution in thrombin of the 37-loop by the corresponding region in t-PA increases by three order of magnitude the inhibition rate of the mutant by PAI-1 (Dekker et al., 1999). Conversely, charge reversal or

```
Residue     16      26         36       41       50       * 60
U-PA        -IIGGEFTTIENQPWFAAIY-RRHRGGSVTYVCGGSLISPCWVISATHCFIDY
T-PA        -IKGGLFADIASHPWQAAIFAKHRRSPGERFLCGGILISSCWILSAAHCFQER
DS-PA       HSTGGLFTDITSHPWQAAIFAQNRRSSGERFLCGGILISSCWVLTAAHCFQES
TSV-PA      -VFGGDECNINEHRSLVVLFN------SNGFLCGGTLINQDWVVTAAHCDS--
LV-PA       -VFGGDECNINEHRSLVVLFN------SSGFLCAGTLINKEWVLTAAHCDS--
Haly-PA     -VVGGDECNINEHRSLVVLFN------SSGLICSGTLINQEWVLTAAHCDS--
DAV-PA      -VIGGNECDINEHRFLVAFFN------TTGFFCGGTLINPEWVVTAAHCDS--
AAV-SP-1    -VIGGNECDINEHRFLVAFFN------TTGFFCGGTLINPEWVVTAAHCDS--
Ev854880    -VVGGDECNINEHRSLVVFFN------SSGFLCGGTLINQEWVLTAAHCDS--

Residue      61       71        81       91       100   *
U-PA        PKKEDYIVYLGRSRLNSNTQGEMKFEVENLILHKDYS-ADTLAHHNDIALLKI
T-PA        FPPHHLTVILGRTYRVVPGEEEQKFEVEKYIVHKEFD-DDTYD--NDIALLQL
DS-PA       YLPDQLKVVLGRTYRVKPGEEEQTFKVKKYIVHKEFD-DDTYN--NDIALLQL
TSV-PA      ---NNFQLLFGVHSKKILNEDEQTRDPKEKFFCPNRKKDDEVD--KDIMLIKL
LV-PA       ---ENFQMQLGVHSKKVPNKDEETRDPKEKFICPNRKKNDEKD--KDIMLIRL
Haly-PA     ---KNFQMLFGVHSKKILNEDEQTRDPKEKFICPNKKKDDEKD--KDIMLIRL
DAV-PA      ---TNFQMQLGVHSKKVLNEDEQTRNPKEKFICPNKNNNEVLD--KDIMLIKL
AAV-SP-1    ---TNFQMQLGVHSKKVLNEDEQTRNPKEKFICPNKNNNEVLD--KDIMLIKL
Ev854880    ---KNFQMLFGVHSKKILNEDEQTRNPKEKFICPNRKKDDERD--KDIMLIRL

Residue    110          120      130       140      150
U-PA        RSKEGRCAQPSRTIQTICLPSMYNDPQFGTSCEITGFGKENSTDYLYPEQLKM
T-PA        KSDSSRCAQESSVVRTVCLPPADLQLPDWTECELSGYGKHEALSPFYSERLKE
DS-PA       KSDSPQCAQESDSVRAICLPEANLQLPDWTECELSGYGKHKSSSPFYSEQLKE
TSV-PA      DS----SVSNSEHIAPLSLPS-SPP-SVGSVCRIMGWGKTIPTKEIYPDVPHC
LV-PA       NR----PVSNSEHIALLSLPS-SPP-SVGSVCRIMGWGTISPTKEIYPDVPHC
Haly-PA     DS----PVSNSEHIAPLSLPS-SSP-TVDSVCRIMGWGTIKPADETYPDVPHC
DAV-PA      DK----PISNSKHIAPLSLPS-SPP-SVGSVCRIMGWGSITPVKETFPDVPYC
AAV-SP-1    DK----PISNSKHIAPLSLPS-SPP-SVGSVCRIMGWGSITPVKETFPDVPYC
Ev854880    DS----PVSNSEHIAPLSLPS-SPP-SVGSVCRIMGWGTISPTKVTFPDVPHC

Residue     160        170      180              190      *  200
U-PA        TVVKLISHRECQQPHYYGSEVTTKMLCAAD-----PQ-WKTDSCQGDSGGPLV
T-PA        AHVRLYPSSRCTSQHLLNRTVTDNMLCAGDTRSGGPQANLHDACQGDSGGPLV
DS-PA       GHVRLYPSSRCAPKFLFNKTVTNNMLCAGDTRSGEIYPNVHDACQGDSGGPLV
TSV-PA      ANINILDHAVCR-TAYSWRQVANTTLCAGI-----LQ-GGRDTCHFDSGGPLI
LV-PA       ADINILDHAVCR-AAYSGWLATSTTLCAGI-----LE-GGKDSCHGDSGGPLI
Haly-PA     ANINILDHTVCR-AAYPVLLAGSSTLCAGT-----QQ-GGKDTCVGDSGGPLI
DAV-PA      ANINLLDHAVCQ-AGYPELLAEYRTLCAGI-----VQ-GGKDTCGGDSGGPLI
AAV-SP-1    ANINLLDHAVCQ-AGYPELLAEYRTLCAGI-----VQ-GGKDTCGGDSGGPLI
Ev854880    ANINILDHAVCR-AAYPTLLAESSTVCAGI-----QQ-GGKDTCGGDSGGPLI

Residue           210       220       230       240       245g
U-PA        CSLQGRMTLTGIVSWGR-GCALKDKPGVYTRVSHFLPWIRSHTKEENGLAL
T-PA        CLNDGRMTLVGIISWGL-GCGQKDVPGVYTKVTNYLDWIRDNMRP
DS-PA       CMNDNHMTLLGIISWGV-GCGEKDVPGVYTKVTNYLGWIRDNMHL
TSV-PA      CNGI----FQGIVSWGGHPCGQPGEPGVYTKVFDYLDWIKSIIAGNKDATCPP
LV-PA       CNGQ----FQGIVSLGRHPCGHPDEPGVYTKVFDYTDWIQSIIAGNTDAACPP
Haly-PA     CNGQ----IQGIVSWGAHPCGQGSKPGVYTKVFDHLDWIKSIIAGNTAVTCPP
DAV-PA      CNGQ----FQGIVSYGAHPCGQGPKPGIYTNVFDYTDWIQRNIAGNTDATCPP
AAV-SP-1    CNGQ----FQGIVSYGAHPCGQGPKPGIYTNVFDYTDWIQRNIAGNTDATCPP
Ev854880    CNDK----SRGIVSWGRILVAQVLILMSTPRSSIILTGSRVLFQKYRCDLS
```

Fig. 22.2 (continued)

deletion of the KHRR motif in t-PA dramatically decreases its rate of inhibition by PAI-1 (Madison et al., 1989, 1990). A general characteristic of snake venom proteases is their lack of susceptibility to plasma serpins including PAI-1, antithrombin (with and without heparin), α2-antiplasmin, and α1-antitrypsin. The second order rate constant of TSV-PA inhibition by PAI-1 is for instance less than 10^2 M^{-1} s^{-1} whereas inhibition by the other plasma serpins is undetectable. LV-PA also shows remarkable resistance to inhibition by plasma serpins. TSV-PA, Haly-PA, and LV-PA all lack positive stretch within their 37-loop and it had been postulated that they resist PAI-1 for this reason. Hypothesis had been confirmed in part by mutagenesis of TSV-PA (see below). The bovine pancreatic trypsin inhibitor (BPTI) is the archetype of the Kunits-type 1 inhibitors. There is normally no BPTI in plasma, but because its mechanism of action is so well established just knowing whether BPTI neutralizes a given protease provides precious information. For instance, plasmin is a target of BPTI, whereas thrombin is not (van de Locht et al., 1997). Precisely, TSV-PA shows a remarkable resistance to inhibition by BPTI (Zhang et al., 1995), whereas LV-PA is neutralized (Sanchez et al., 2000). This observation implies that the catalytic groove of TSV-PA is more occluded than that of LV-PA. Indeed, X-ray diffraction and mutagenesis studies of TSV-PA largely confirmed this hypothesis (see below). Finally, α2-macroglobulin is a non specific, general, protease inhibitor (Sottrup-Jensen, 1989). Inhibitor entraps target proteases following its cleavage in a bait region covering most specificities as well as catalytic mechanisms. Catalytic activity of the protease is not irreversibly inhibited and peptidyl substrates are still hydrolysed by the entrapped protease whereas steric hindrance prevents cleavage of macromolecule substrates. Escaping such mechanism seems difficult even for a snake venom protease. The vulnerability of TSV-PA to α2-macroglobulin has not been studied but it had been shown that LV-PA is slowly neutralized by α2-macroglobulin (Hermogenes et al., 2006). At first, the potential of α2-macroglobulin (3-5 μM in plasma) would seem sufficient to neutralize snake venom proteases in case of envenomation. Hence plasma α2-macroglobulin neutralizes a number of toxins in vitro but evidently fails to neutralize snake venom toxins in vivo (Kamiguti et al., 1994). A possible explanation relies on the hypothesis that α2-macroglobulin neutralizes plasmin and other endogenous proteases much more rapidly than the snake venom proteases. Paradoxically, endogenous proteases would exhausts α2-macroglobulin before inhibition of the snake venom toxin could occur. It should be kept in mind that α2-macroglobulin is a suicide substrate inhibitor which is consumed during neutralization of its targets. It would be important to further study the interaction of α2-macroglobulin with snake venoms PA. In particular

Fig. 22.2 (continued) Amino acid sequence alignment of the PA serine protease domains. The numbering is based on the topological equivalence to chymotrypsinogen for structure solved by X-ray diffraction (u-PA, t-PA, DS-PA and TSV-PA). The numbering used for LV-PA, Haly-PA, DAV-PA, AAV-SP-1, and ORF Ev854880 of *Agkistrodon piscivorus leucostoma* is based on sequence similarities. Past the SWG motif preceding Cys^{220}, the sequence of Ev854880 (*in italics characters*) totally diverges from all other PA known to determine whether α2-macroglobulin indeed fails to prevent systemic plasmin generation following envenomation or if it could constitute a valuable (perhaps universal) therapy.

Structural Features

Serine proteases of clan SA have a common framework and share at least 20% identity. Topologies of the catalytic grooves are similar and the major variations arise on surface loops exposed to solvent. Considering a broad specificity protease such as trypsin, only proline immediately following the arginine (or lysine) of the scissile bond prohibits cleavage of isolated peptides (Bianchini et al., 2002). Inversely, a number of macromolecule substrates that are cleaved within analogous sequences are exclusive target of distinctive and specialized proteases. Typically a true PA cleaves only one macromolecular substrate (plasminogen) and is the target of none (snake PA) to two macromolecular inhibitors (t-PA and u-PA). A great diversity is undoubtedly achievable through multiple variations of the protease surface loops. The number of combinations is nevertheless not infinite rising the question as to how specificity of serine proteases can extend from unrestrained to highly specialized. Structurally, snake venom proteases resemble trypsin and therefore would be expected to cleave efficiently many substrates with little specificity. The paradox is that they are of exquisite specificity albeit low efficacy. Given their relative simplicity associated to a surprising selectivity snake venom proteases constitute an exceptional model. Somehow snake venom proteases should permit understanding unknown molecular mechanisms governing their specificity. Identification would not only solve the paradox of snake venom specificity, it would advance our knowledge on proteases in general because it is likely that similar mechanisms are in use elsewhere. In addition to be an attractive topic, studying the molecular basis of snake proteases specificity could facilitate drug design in the future.

Apart from the catalytic groove itself the specificity of serine proteases is achieved to a large extent by edging as well as remote surface loops. Neighboring surface loops often associate to form an exosite. The contribution of these exosites to specificity is nowadays considered as important as that of the catalytic groove itself. Importance is exemplified by thrombin selecting and sorting its numerous substrates through combination between two exosites surrounding a restricted catalytic groove (Bode, 2005; Bode et al., 1989). Double docking permits protease to constrain its macromolecule substrate in a conformation allowing optimum presentation of the scissile bond with secondary binding sites securing geometry. Pending the requirement for arginine (or lysine) prior to the scissile bond and the universal proline prohibition, the amino acids surrounding may or not match the intrinsic preferences of the catalytic groove (Le Bonniec et al., 1996). Conversely a non optimal geometry deters most substrates to bind to the catalytic groove if they are unable to take advantage of a secondary site. Surface loops boarding the catalytic groove are in addition of variable length, longer loop insuring judicious steric hindrance preventing the protease to accommodate a number of potential substrates. Most specialized (thus highly specific) proteases utilize one or both of these mechanisms to insure selectivity and efficiency of catalysis. On the contrary broad specificity serine proteases (i.e., trypsin) expose a widely open catalytic groove to accommodate many peptides or macromolecular substrates. There, catalysis is efficient but specificity is low. It is intriguing to realize that the apparent simplicity of snake venoms proteases is associated with sharp specificity.

An obvious difference between snake venom and the other PA is that the former proteases only function with a catalytic serine protease domain and are devoid of light chain. Mammalian and vampire bat PA are far more complex mosaic proteins. It can be concluded that a light chain is not strictly required for plasminogen activation. Indeed the light chains of t-PA and DS-PA regulate the dependence to fibrin rather than plasminogen activation. Light chain of the 527 amino acids t-PA includes an N-terminal region homologous with the finger domains mediating the fibrin affinity of fibronectin, an epidermal growth factor-like domain, and two kringle domains having a high degree of homology with those of plasminogen. DS-PA has the same domain organization except that it lacks the second kringle domain. Conversely, u-PA is a 411 amino acid protein with a modular structure. Mosaic consists of an epidermal growth factor-like, a single kringle, and a serine protease domain. The kringle and the serine protease domains of u-PA are separated by an unusually long (16 residue) linker region. The epidermal growth factor-like domain of uPA is a critical determinant for binding to its cellular receptor (Appella et al., 1987). Thus lack of regulatory chain in TSV-PA explains that fibrin fragments do not influence the activity and that it activates plasminogen indiscriminately in plasma, without control. An amino acid sequence alignment of the serine protease domains known (or suspected) to activate plasminogen is presented in Fig. 22.2. The numbering is based on the topological equivalence to chymotrypsinogen for the structures solved by X-ray diffraction. The numbering used otherwise is a predicted such chymotrypsinogen numbering, based on sequence similarities. TSV-PA shares 60% similarity with other snake venom serine proteases such as Ancrod the thrombin-like α-fibrinogenase from the Malayan pit viper *Calloselasma rhodostoma* but only 20–23% with the protease domain of t-PA or u-PA.

TSV-PA was the first three-dimensional structure of a snake venom serine protease solved and exhibits the typical fold of serine proteases (Parry et al., 1998). Superposition of TSV-PA (Protein Data Bank code 1BQY) and t-PA (Protein Data Bank code 1BDA) results in 130 topologically equivalent Cα atoms with root mean square deviations of 0.87 angstrom. Amongst the distinctive features of TSV-PA is the negatively charged 99-loop which includes Asp^{97}, as in t-PA, u-PA, and DS-PA. While the folding of the autolysis loop (residues 143–149) differs between TSV-PA and t-PA, both result in partial obstruction of subsite 2′ (which accommodates the second residue following the scissile bond). In t-PA, u-PA, and DS-PA access is restricted by the side chain of Tyr^{151}. In TSV-PA autolysis loop is shifted in a different orientation such that Tyr^{151} is away from subsite 2′ (Fig. 22.3). However, the side chain of the peculiar Phe^{193} of TSV-PA occupies and therefore occludes in part subsite 2′. Partial occlusion of subsite 2′ as well as that of subsite 2 (which accommodate the residue prior to the arginine preceding the scissile bond) is a

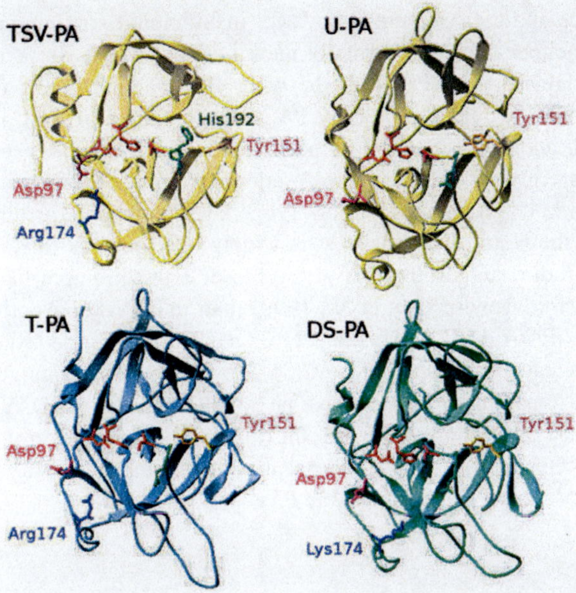

Fig. 22.3 Three-dimensionnal structures of TSV-PA, u-PA, t-PA, and DS-PA. In each structure, the side chains of the charge stabilizing system (His57, Asp102, and Ser195 in chymotrypsinogen numbering system) are in red (not labeled). The folding of the autolysis loop orients Tyr^{151} differently in TSV-PA than in the three other PA. Subsite 2′ is occluded by Tyr^{151} in u-PA, t-PA and DS-PA, by Phe^{193} in TSV-PA (in *green* and preceding His^{192}). Another difference between TSV-PA and the other PA is that Asp^{97} (in *pink*) is engaged in a salt bridge with Arg^{174} in TSV-PA whereas it protrudes towards the catalytic groove in the other PA

characteristic of PA that should cleave within the short disulfide loop of plasminogen. The three-dimensional structure of Haly-PA and LV-PA is not known but they exhibit over 80% sequence similarity with TSV-PA. Neither Haly-PA nor LV-PA carry the peculiar Phe^{193} of TSV-PA, thus it would be interesting to know if Tyr^{51} occludes in part subsite 2′ in Haly-PA and LV-PA. The 37-loop is another distinctive feature in mammalian PA. In t-PA and u-PA this loop is crucial since it constitutes the secondary binding site for inhibition by PAI-1. TSV-PA lacks the sequence responsible for the interaction of t-PA and u-PA with PAI-1. Region 35-39 of TSV-PA is actually 5 and 6 residues shorter than in t-PA and u-PA, respectively (Fig. 22.2). Finally, still in the vicinity of the catalytic groove, the 60-loop is five amino acids shorter in TSV-PA than in the mammalian PA. In thrombin, the 60-loop insertion wraps over subsite 2 and plays a crucial role in its distinctive specificity (Bode et al., 1989; Le Bonniec et al., 1993).

DAV-PA/AaV-SP-I share only 72% sequence identity with TSV-PA (Fig. 22.2). The root mean square deviation of the Cα between TSV-PA (Protein Data Bank code 1BQY) and AaV-SP-1 (Protein Data Bank code 1OP0) is nevertheless 0.81 angstrom (Zhu et al., 2005). The hypothesis that AaV-SP-I (Zhu et al., 2005) is a true PA is based on its identity with DAV-PA (Wang et al., 2001). Knowledge of the sequence and the structure is however insufficient to assign with certainty a function and neither had been formally identified as true PA. In fact, notable differences distinguish AaV-SP-I from the other PA. One surprising feature is that DAV-PA/AaV-SP-I would be the only PA with glutamate instead of aspartate at position 97. In addition, subsite 2′ is not occluded in AaV-SP-I because residues 192 and 193 are glycine whereas Phe^{151} adopts a similar conformation as Tyr^{151} of TSV-PA. Side chains of Phe^{151} and Tyr^{151} overlap when structures are superimposed with the result that subsite 2′ is widely open in AaV-SP-I. Furthermore, phylogenetic studies suggest that DAV-PA is closer to the thrombin-like Acutin from *Agkistrodon acutus* venom (Pan et al., 1999) than to TSV-PA. Another distinctive feature of DAV-PA/AaV-SP-I is residue 215. Over 50% of serine proteases belonging to clan SA have tryptophan at position 215 (in chymotrypsinogen numbering system). Most remaining proteases have phenylalanine, few (as DAV-PA/AaV-SP-I) have tyrosine. Interestingly, LV-PA constitutes a rare exception with leucine at position 215. Overall, it would be essential to characterize AaV-SP-I (and/or DAV-PA) and ascertain that they are true PAs.

Site Directed Mutagenesis

Snake venom proteases are quite different from their mammalian counterpart suggesting that share function results from convergent evolution. It is risky to assign a function to a motif shared by proteases having a common ancestor. For instance, the vast majority of SA proteases carry a motif GWG in 140–141. Mammalian PA and DS-PA have either GFG or GYG, and this is clearly an exception. Snake venom PAs however have the almost universally conserved GWG (as trypsin) ruling out that GFG could be a key motif. On the contrary, if two proteases resulting from convergent evolution have a common function and share a rare motif, then it becomes likely that such particular motif is involved in the function. Sequence alignment is at first deceptive because snake PAs are more alike to other snake proteases than to the mammalian PAs and vice versa. Looking at closer, it is noticeable that all formally characterized PAs carry aspartate at position 97. This position being highly variable in the SA family it appears meaningful that PAs systematically have Asp^{97}. Nevertheless, a number of unrelated proteases also carry Asp^{97}. Thus having Asp^{97} does not make a PA, but it may be required to be a PA. When the structure of TSV-PA was solved by X-ray diffraction it came as a surprise that Asp^{97} forms a salt bridge with Arg^{174} rather than protruding toward the catalytic center (Fig. 22.3). This was surprising because salt bridge is absent in the structure of t-PA which also carries Arg^{174}. Furthermore, the other true PAs characterized do not have Arg^{174} rendering the function of Asp^{97} puzzling. Site directed mutagenesis confirmed the crucial role of Asp^{97} in TSV-PA whereas $Arg^{174}$was found to be dispensable. Clearly, replacing the acidic edge Asp-Asp-Glu of TSV-PA by the corresponding region Asn-Val-Ile of batroxobin abolishes plasminogen activation activity (Zhang et al., 1997). Therefore, while Asp^{97} and Arg^{174} are engaged in a salt bridge in resting TSV-PA, it is likely that following an induced fit mechanism the side chain of Asp^{97} changes conformation during plasminogen binding and/or activation (Parry et al., 1998). Assuming that Arg^{174} is dispensable would rationalize that residue 174 is a tryptophan in LV-PA and a leucine in AaV-SP-I.

In trypsin-like enzymes the most frequent amino acid at position 192 is by far glutamine. Amongst the noticeable exceptions is thrombin with Glu^{192}, which indeed plays a crucial role in its specificity (Le Bonniec and Esmon, 1991; Le Bonniec et al., 1992). All t-PA and DS-PA as well as human and bovine u-PA have the highly conserved glutamine, but mouse and chicken u-PA have Lys^{192}, whereas TSV-PA and LV-PA have His^{192} and Haly-PA has Val^{192}. Remarkably, this position in snake venom serine proteases is quite variable. Even more appealing is residue 193 because this is one of the most conserved glycine within the SA family. TSV-PA with Phe^{193} is one of the very few exceptions and other true PAs carry the classical glycine. A number of mutations have been tested in TSV-PA (H192Q, H192G, F193G, F193R and their combinations). None was deleterious for the catalytic machinery, as attested by studies with chromogenic substrates. On the contrary, mutations slightly improved the k_{cat}/K_M value (3- to 10-fold). Unexpectedly, TSV-PA variant F193G turned out to be 8.5-fold more potent for activating plasminogen. Variant became in addition vulnerable to BPTI. Above all, the F193G mutant was inhibited over a 100-fold more rapidly by PAI-1 and α2-antiplasmin (Braud et al., 2000, 2002). This observation is reminiscent of what had been observed with the E192Q mutation in thrombin. Glu^{192} of thrombin restricts specificity rather than enhancing a particular function. Conversely Phe^{193} of TSV-PA does not drive specificity towards plasminogen activation it rather impedes its inhibition. X-ray diffraction data provides a satisfactory explanation for the presence of Phe^{193} in TSV-PA; it participates in partial occlusion of subsite 2′. In t-PA, u-PA, and DS-PA side chain of Tyr^{151} occupies subsite 2′, in TSV-PA this is the side chain of Phe^{193} which occludes the subsite. Molecular modeling predicted that BPTI collides with the aromatic ring of Phe^{193} and that F193G mutation would permit inhibition by subsite 2′ opening. Mutagenesis confirmed that indeed replacement F193G was sufficient to permit inhibitor docking (Braud et al., 2000). Things are not that simple however, because while the steric hindrance hypothesis is consistent with a better access to inhibitors, it is at odd with the assumption that subsite 2′ should be occluded for docking of the restricted disulfide loop of plasminogen. In fact, F193G TSV-PA activates plasminogen 8.5-fold more efficiently that the wild-type enzyme. An interesting possibility that would merit investigation is that the side chain of Tyr^{151} replaces in the F193G mutant the phenol ring of Phe^{193} in wild-type TSV-PA. Less expected was the little effect of the H192Q and H192G mutations considering the major outcome that corresponding mutation has in thrombin (Le Bonniec and Esmon, 1991; Le Bonniec et al., 1992). Plasminogen activation rate was only slightly decreased (up to 3-fold) and mutant still resisted to BPTI. Mutation H192Q however increased 10-fold the k_{cat}/K_M of S2238 hydrolysis by TSV-PA. Understanding this effect is perhaps more straightforward considering the electrostatic potential that histidine generates at the entrance of the primary binding pocket in wild-type TSV-PA.

Mutagenesis experiments within the 37-loop of TSV-PA confirmed that absence of a positive stretch precludes inhibition by PAI-1. Wild-type TSV-PA inhibition by PAI-1 is hardy detectable with a mere 10^2 M^{-1} s^{-1}. Introducing a polyalanine stretch in the 37-loop of TSV-PA to make it as long as in t-PA or u-PA has little effect on the rate of inhibition by PAI-1. Introducing the 37-loop of t-PA or u-PA increases the value of the second order rate constant 20- or 100-fold, respectively (Braud et al; 2002). These values are still much lower than that of t-PA inhibition by PAI-1 suggesting that at least another structural determinant allows TSV-PA to escape PAI-1. Indeed, when TSV-PA mutants carrying the t-PA or u-PA 37-loop were further mutated to F193G, the second order rate constant reached values that compared well with typical serpin targeting (2.5×10^6 and 6.0×10^6 M^{-1} s^{-1} for the double TSV-PA variants carrying the t-PA and u-PA loop, respectively). The other snake PA also lack positive stretch in region 35-39 and it is likely that this allows them to escape PAI-1.

Conclusion

Only three true snake PA have been conclusively characterized, a surprisingly small figure in comparison with the fibrino(geno)lytic enzymes inventoried. Whether this results from too little studies or reflects an infrequent strategy used by snakes to neutralize preys remains to be determined. A number of relatively simple investigations would greatly improve our knowledge on snake venom PA and of plasminogen activation in general. It would be important to verify whether DAV-PA/AAV-SP-I and sequence Ev854880 are true PA because they diverge appreciably from those of the formally characterized PAs. Knowing if they are indeed true PA could challenge our current understanding of the specific motifs involved in plasminogen activation thus leading to the discovery of new mechanisms. Another important area that would need further investigation is the inhibition of snake venom proteases. Different mechanisms allowing snake venoms to resist serpin have been unveiled, but the potential of α2-macroglobulin needs more examination. Finally, while sequence inventory is obviously a precious tool we should keep in mind that it does not always permit prediction of the function of a protease. Proteases act as a whole and are not simple juxtaposition of conserved and variable regions: even a single amino acid change may have an unexpected consequence. A particular residue may be critical in one protease (Asp^{97} in PA, Phe^{193} in TSV-PA, Glu^{192} in thrombin) while being dispensable in others.

References

Alfano, D., Franco, P., Vocca, I., Gambi, N., Pisa, V., Mancini, A., Caputi, M., Carriero, M.V., Iaccarino, I., Stoppelli, M.P., 2005. The urokinase plasminogen activator and its receptor: role in cell growth and apoptosis. Thromb. Haemost. 93, 205–211.

Appella, E., Robinson, E.A., Ullrich, S.J., Stoppelli, M.P., Corti, A., Cassani, G., Blasi, F., 1987. The receptor-binding sequence of urokinase. A biological function for the growth-factor module of proteases. J. Biol. Chem. 262, 4437–4440.

Bajzar, L., Manuel, R., Nesheim, M.E., 1995. Purification and characterization of TAFI, a thrombin-activable fibrinolysis inhibitor. J. Biol. Chem. 270, 14477–14484.

Bergum, P.W., Gardell, S.J., 1992. Vampire bat salivary plasminogen activator exhibits a strict and fastidious requirement for polymeric fibrin as its cofactor, unlike human tissue-type plasminogen activator. A kinetic analysis. J. Biol. Chem. 267, 17726–17731.

Bianchini, E.P., Louvain, V.B., Marque, P.E., Juliano, M.A., Juliano, L., Le Bonniec, B.F., 2002. Mapping of the catalytic groove preferences of factor Xa reveals an inadequate selectivity for its macromolecule substrates. J. Biol. Chem. 277, 20527–20534.

Bode, W., 2005. The structure of thrombin, a chameleon-like proteinase. J. Thromb. Haemost. 3, 2379–2388.

Bode, W., Mayr, I., Baumann. U., Huber, R., Stone, S.R., Hofsteenge, J., 1989. The refined 1.9 A crystal structure of human alpha-thrombin: interaction with D-Phe-Pro-Arg chloromethylketone and significance of the Tyr-Pro-Pro-Trp insertion segment. EMBO J. 8, 3467–3475.

Braud, S., Le Bonniec, B.F., Bon, C., Wisner, A., 2002. The stratagem utilized by the plasminogen activator from the snake *Trimeresurus stejnegeri* to escape serpins. Biochemistry 41, 8478–8484.

Braud, S., Parry, M.A., Maroun, R., Bon, C., Wisner, A., 2000. The contribution of residues 192 and 193 to the specificity of snake venom serine proteinases. J. Biol. Chem. 275, 1823–1828.

Budzynski, A. Z., 1985. Secretion of cellular plasminogen activators upon stimulation by Crotalinae snake venoms and their components. Thromb. Haemost. 54, 314 (Abstract).

Cao, Y., Yu, G.S., Yang, S.L., Gong, Y., 2001. Expression of a snake venom plasminogen activator TSV-PA in insect cells. Sheng Wu Gong Cheng Xue Baò. 17, 506–509.

Carmeliet, P., Collen, D., 1996. Gene manipulation and transfer of the plasminogen and coagulation system in mice. Semin. Thromb. Hemost. 22, 525–542.

Collen, D., Lijnen, H.R., 2004. Tissue-type plasminogen activator: a historical perspective and personal account. J. Thromb. Haemost. 2, 541–546.

Danø, K., Behrendt, N., Høyer-Hansen, G., Johnsen, M., Lund, L.R., Ploug, M., Rømer, J., 2005. Plasminogen activation and cancer. Thromb. Haemost. 93, 676–681.

Dekker, R.J., Eichinger, A., Stoop, A.A., Bode, W., Pannekoek, H., Horrevoets, A.J., 1999. The variable region-1 from tissue-type plasminogen activator confers specificity for plasminogen activator inhibitor-1 to thrombin by facilitating catalysis: release of a kinetic block by a heterologous protein surface loop. J. Mol. Biol. 293, 613–627.

Ellis, V., Behrendt, N., Danø, K., 1991. Plasminogen activation by receptor-bound urokinase. A kinetic study with both cell-associated and isolated receptor. J. Biol. Chem. 266, 12752–12758.

Felicori, L.F., Chávez-Olórtegui, C., Sánchez, E.F., 2005. Specific identification of *Lachesis muta muta* snake venom using antibodies against the plasminogen activator enzyme, LV-PA. Toxicon 45, 803–806.

Gao, R., Zhang, Y., Meng, Q.X., Lee, W.H., Li, D.S., Xiong, Y.L., Wang, W.Y., 1998. Characterization of three fibrinogenolytic enzymes from Chinese green tree viper (*Trimeresurus stejnegeri*) venom. Toxicon 36, 457–467.

Gardell, S.J., Duong, L.T., Diehl, R.E., York, J.D., Hare, T.R., Register, R.B., Jacobs, J.W., Dixon, R.A., Friedman, P.A., 1989. Isolation, characterization, and cDNA cloning of a vampire bat salivary plasminogen activator. J. Biol. Chem. 264, 17947–17952.

Gettins, P.G., Olson, S.T., 2009. Exosite determinants of serpin specificity. J. Biol. Chem. 284, 20441–20445.

Gladson, C.L., Schleef, R.R., Binder, B.R., Loskutoff, D.J., Griffin, J.H., 1989. A comparison between activated protein C and des-1–41-light chain-activated protein C in reactions with type 1 plasminogen activator inhibitor. Blood 74, 173–181.

Hermogenes, A.L., Richardson, M., Magalhaes, A., Yarleque, A., Rodriguez, E., Sanchez, E.F., 2006. Interaction of a plasminogen activator proteinase, LV-PA with human α2-macroglobulin. Toxicon 47, 490–494.

Horrevoets, A.J., Smilde, A.E., Fredenburgh, J.C., Pannekoek, H., Nesheim, M.E., 1995. The activation-resistant conformation of recombinant human plasminogen is stabilized by basic residues in the amino-terminal hinge region. J. Biol. Chem. 270, 15770–15776.

Høyer-Hansen, G., Rønne, E., Solberg, H., Behrendt, N., Ploug, M., Lund, L.R., Ellis, V., Danø, K., 1992. Urokinase plasminogen activator cleaves its cell surface receptor releasing the ligand-binding domain. J. Biol. Chem. 267, 18224–18229.

Hoylaerts, M., Rijken, D.C., Lijnen, H.R., Collen, D., 1982. Kinetics of the activation of plasminogen by human tissue plasminogen activator. Role of fibrin. J. Biol. Chem. 257, 2912–1919.

Jia, Y., Cantu, B.A., Sánchez, E.E., Pérez, J.C., 2008. Complementary DNA sequencing and identification of mRNAs from the venomous gland of *Agkistrodon piscivorus leucostoma*. Toxicon 51, 1457–1466.

Kamiguti, A.S., Desmond, H.P., Theakston, R.D., Hay, C.R., Zuzel, M., 1994. Ineffectiveness of the inhibition of the main haemorrhagic metalloproteinase from *Bothrops jararaca* venom by its only plasma inhibitor, alpha 2-macroglobulin. Biochim. Biophys. Acta 1200, 307–314.

Karbovs'kyl, V.L., Levkiv, M.I.U., Savchuk, O.M., Hornyts'ka, O.V., Volkov, H.L., Bukhan, T.S., 2006. Plasminogen activator from *Agkistrodon halys halys* venom Ukr. Biokhim. Zh. 78, 32–37.

Kisiel, W., Choi, E., Kondo, S., 1987. Isolation of a protein C activator from southern copperhead venom. Biochem. Biophys. Res. Commun. 143, 917–922.

Klein, J.D., Walker, F.J., 1986. Purification of a protein C activator from the venom of the southern copperhead snake (*Agkistrodon contortrix contortrix*). Biochemistry 25, 4175–4179.

Klöcking, H.P., Hoffmann, A., Markwardt, F., 1987. Release of plasminogen activator by batroxobin. Haemostasis 17, 235–237.

Lamba, D., Bauer, M., Huber, R., Fischer, S., Rudolph, R., Kohnert, U., Bode, W., 1996. The 2.3 A crystal structure of the catalytic domain of recombinant two-chain human tissue-type plasminogen activator. J. Mol. Biol. 258, 117–35.

Le Bonniec, B.F., Esmon, C.T., 1991. Glu-192Gln substitution in thrombin mimics the catalytic switch induced by thrombomodulin. Proc. Natl. Acad. Sci. U.S.A. 88, 7371–7375.

Le Bonniec, B.F., Guinto, E.R., Esmon, C.T., 1992. The role of calcium ions in factor X activation by thrombin E192Q. J. Biol. Chem. 267, 6970–6976.

Le Bonniec, B.F., Guinto, E.R., MacGillivray, R.T., Stone, S.R., Esmon, C.T., 1993. The role of thrombin's Tyr-Pro-Pro-Trp motif in the interaction with fibrinogen, thrombomodulin, protein C, antithrombin III, and the Kunitz inhibitors. J. Biol. Chem. 268, 19055–19061.

Le Bonniec, B.F., Myles, T., Johnson, T., Knight, C.G., Tapparelli, C., Stone, S.R., 1996. Characterization of the P2′ and P3′ specificities of thrombin using fluorescence-quenched substrates and mapping of the subsites by mutagenesis. Biochemistry 35, 7114–7122.

Lijnen, H.R., Collen, D., 1998. t-Plasminogen activator, in: Barrett, A.J., Rawlings, N.D., Woessner, J.F. (Eds.), Handbook of Proteolytic Enzymes. Academic Press, London, UK, pp. 184–190.

Liu, S., Sun, M.Z., Greenaway, F.T., 2006. A novel plasminogen activator from *Agkistrodon blomhoffii ussurensis* venom (ABUSV-PA): purification and characterization. Biochem. Biophys. Res. Commun. 348, 1279–1287.

Lund, L.R., Green, K.A., Stoop, A.A., Ploug, M., Almholt, K., Lilla, J., Nielsen, B.S., Christensen, I.J., Craik, C.S., Werb, Z., Danø, K., Rømer, J., 2006. Plasminogen activation independent of uPA and tPA maintains wound healing in gene-deficient mice. EMBO J. 25, 2686–2697.

Madison, E.L., Coombs, G.S., Corey, D.R., 1995. Substrate specificity of tissue type plasminogen activator. Characterization of the fibrin independent specificity of t-PA for plasminogen. J. Biol. Chem. 270, 7558–7562.

Madison, E.L., Goldsmith, E.J., Gerard, R.D., Gething, M.J., Sambrook, J.F., 1989. Serpin-resistant mutants of human tissue-type plasminogen activator. Nature 339, 721–724.

Madison, E.L., Goldsmith, E.J., Gerard, R.D., Gething, M.J., Sambrook, J.F., Bassel-Duby, R.S., 1990. Amino acid residues that affect interaction of tissue-type plasminogen activator with plasminogen activator inhibitor 1. Proc. Natl. Acad. Sci. U.S.A. 87, 3530–3533.

Mao, S.S., Holahan, M.A., Bailey, C., Wu, G., Colussi, D., Carroll, S.S., Cook, J.J., 2005. Demonstration of enhanced endogenous fibrinolysis in thrombin activatable fibrinolysis inhibitor-deficient mice. Blood Coagul. Fibrinolysis 16, 407–415.

Markland, F.S., 1998a. Snake venoms and the hemostatic system. Toxicon 36, 1749–1800.

Markland, F.S., Jr., 1998b. Snake venom fibrinogenolytic and fibrinolytic enzymes: an updated inventory. Registry of exogenous hemostatic factors of the scientific and standardization committee of the international society on thrombosis and haemostasis. Thromb. Haemost. 79, 668–674.

McMullen, B.A., Fujikawa, K., Kisiel, W., 1989. Primary structure of a protein C activator from *Agkistrodon contortrix contortrix* venom. Biochemistry. 28, 674–679.

Miles, L.A., Castellino, F.J., Gong, Y., 2003. Critical role for conversion of glu-plasminogen to Lys-plasminogen for optimal stimulation of plasminogen activation on cell surfaces. Trends Cardiovasc. Med. 13, 21–30.

Murakami, M.T., Arni, R.K., 2005. Crystallization and preliminary X-ray crystallographic studies of Protac, a commercial protein C activator isolated from *Agkistrodon contortrix contortrix* venom. Biochim. Biophys. Acta 1752, 202–204.

Pan, H., Du, X., Yang, G., Zhou, Y., Wu, X., 1999. cDNA cloning and expression of acutin. Biochem. Biophys. Res. Commun. 255, 412–415.

Park, D., Kim, H., Chung, K., Kim, D.S., Yun, Y., 1998. Expression and characterization of a novel plasminogen activator from *Agkistrodon halys* venom. Toxicon 36, 1807–1819.

Parry, M.A., Jacob, U., Huber, R., Wisner, A., Bon, C., Bode, W., 1998. The crystal structure of the novel snake venom plasminogen activator TSV-PA: a prototype structure for snake venom serine proteinases. Structure 6, 1195–1206.

Rawlings, N.D., 1998. Family S1 of trypsin, in: Barrett, A.J., Rawlings, N.D., Woessner, J.F. (Eds.), Handbook of Proteolytic Enzymes. Academic Press, London, UK, pp. 3–17.

Rojnuckarin, P., Intragumtornchai, T., Sattapiboon, R., Muanpasitporn, C., Pakmanee, N., Khow, O., Swasdikul, D., 1999. The effects of green pit viper (*Trimeresurus albolabris* and *Trimeresurus macrops*) venom on the fibrinolytic system in human. Toxicon 37, 743–755.

Sanchez, E.F., Felicori, L.F., Chavez-Olortegui, C., Magalhaes, H.B., Hermogenes, A.L., Diniz, M.V., Junqueira-de-Azevedo, I.L., Magalhaes, A., Richardson, M., 2006. Biochemical characterization and molecular cloning of a plasminogen activator proteinase (LV-PA) from bushmaster snake venom. Biochim. Biophys. Acta 1760, 1762–1771.

Sanchez, E.F., Santos, C.I., Magalhaes, A., Diniz, C.R., Figueiredo, S., Gilroy, J., Richardson, M., 2000. Isolation of a proteinase with plasminogen-activating activity from *Lachesis muta muta* (Bushmaster) snake venom. Arch. Biochem. Biophys. 378, 131–141.

Schneider, M., Nesheim, M., 2004. A study of the protection of plasmin from antiplasmin inhibition within an intact fibrin clot during the course of clot lysis. J. Biol. Chem. 279, 13333–13339.

Sottrup-Jensen, L., 1989. α-macroglobulins: structure, shape, and mechanism of proteinase complex formation. J. Biol. Chem. 264, 11539–11542.

Stocker, K., Fischer, H., Meier, J., Brogli, M., Svendsen, L., 1987. Characterization of the protein C activator Protac from the venom of the southern copperhead (*Agkistrodon contortrix*) snake. Toxicon 25, 239–252.

Stone, S.R., Betz, A., Hofsteenge, J., 1991. Mechanistic studies on thrombin catalysis. Biochemistry 30, 9841–9848.

Sugiki, M., Yoshida, E., Anai, K., Maruyama, M., 1998. Activation of single-chain urokinase-type plasminogen activator by a hemorrhagic metalloproteinase, jararafibrase I, in *Bothrops jararaca* venom. Toxicon 36, 993–1000.

Sunagawa, M., Hanashiro, K., Nakamura, M., Kosugi, T., 1996. Habutobin releases plasminogen activator (U-PA) from bovine pulmonary artery endothelial cells. Toxicon 34, 691–699.

Swenson, S., Markland, F.S. Jr., 2005. Snake venom fibrin(ogen)olytic enzymes. Toxicon 45, 1021–1039.

Thelwell, C., Longstaff, C., 2007. The regulation by fibrinogen and fibrin of tissue plasminogen activator kinetics and inhibition by plasminogen activator inhibitor 1. J. Thromb. Haemost. 5, 804–811.

van de Locht, A., Bode, W., Huber, R., Le Bonniec, B.F., Stone, S.R., Esmon, C.T., Stubbs, M.T., 1997. The thrombin E192Q-BPTI complex reveals gross structural rearrangements: implications for the interaction with antithrombin and thrombomodulin. EMBO J. 16, 2977–84.

Wang, Y.M., Wang, S.R., Tsai, I.H., 2001. Serine protease isoforms of *Deinagkistrodon acutus* venom: cloning, sequencing and phylogenetic analysis. Biochem. J. 354, 161–168.

Wu, C., Dong, N., da Cunha, V., Martin-McNulty, B., Tran, K., Nagashima, M., Wu, Q., Morser, J., Wang, Y.X., 2003. Activated thrombin-activatable fibrinolysis inhibitor attenuates spontaneous fibrinolysis of batroxobin-induced fibrin deposition in rat lungs. Thromb. Haemost. 90, 414–421.

Zhang, Y., Wisner, A., Maroun, R.C., Choumet, C., Xiong, Y., Bon, C., 1997. *Trimeresurus stejnegeri* snake venom plasminogen activator site-directed mutagenesis and molecular modeling. J. Biol. Chem. 272, 20531–20537.

Zhang, Y., Wisner, A., Xiong, Y., Bon, C., 1995. A novel plasminogen activator from snake venom. Purification, characterization, and molecular cloning. J. Biol. Chem. 270, 10246–10255.

Zhu, Z., Liang, Z., Zhang, T., Zhu, Z., Xu, W., Teng, M., Niu, L., 2005. Crystal structures and amidolytic activities of two glycosylated snake venom serine proteinases. J. Biol. Chem. 280, 10524–10529

Chapter 23
Bat Plasminogen Activator: Desmoteplase – From Bat to Bench to Bedside of Stroke Victims

Wolfgang Söhngen, Karl-Uwe Petersen, and Mariola Söhngen

Abstract Desmoteplase (DSPA) was identified in the salivary venom of *Desmodus rotundus* (a blood-feeding vampire bat common in Latin America) triggered by observations as early as 1964. The initial interest in DSPA as a therapeutic modality was raised by the success of recombinant tissue plasminogen activator (rt-PA) in acute myocardial infarction (AMI) and the evolving paradigm of fibrin-specificity as the key to safe and effective thrombolysis. The early research on DSPA confirmed an extremely high fibrin specificity and a potential for a lower bleeding propensity, demonstrated in a variety of preclinical studies. Obviously, the unique task in the Vampire Bat – curbing clot formation without disintegrating the other salivary proteins – has led to a protease which serves no other known function than activating plasminogen in the presence of fibrin. This fibrin and substrate specificity distinguishes DSPA from rt-PA, which, apart from clot lysis, has a number of additional physiological roles.

The first clinical trial in AMI confirmed the absence of fibrinogen depletion even at doses of 750 μg/kg (90 μg/kg is effective in acute ischaemic stroke, AIS). DSPA was abandoned by Schering AG for strategic reasons and, in 2001, its further development was redirected by PAION to acute ischemic stroke (AIS), based on the assumption that a more fibrin-specific thrombolytic should pose a lower bleeding risk and allow a longer post-stroke treatment window. Also in 2001, the discovery of the ability of rt-PA to enhance NMDA-induced neurotoxicity (which models the glutamate neurotoxicity known in vivo) gave rise to the speculation that DSPA, by virtue of its high specialization, might be different. The subsequent multi-level research indeed confirmed that DSPA is devoid of any neurotoxic properties. The main reason seems to lie in a structural difference: The deleterious augmentation of NMDA neurotoxicity by rt-PA (and its mutants, shown in vitro) requires the kringle 2 (K2) domain, a moiety lacking in DSPA. This distinction may prove advantageous

W. Söhngen (✉)
PAION Deutschland GmbH, Martinstrasse 10–12, 52062 Aachen, Germany
e-mail: w.soehngen@paion.com

R.M. Kini et al. (eds.), *Toxins and Hemostasis*, DOI 10.1007/978-90-481-9295-3_23,

also in other areas, as, more generally, the kringle 2 seems to capacitate (r)t-PA for numerous mechanisms it is known to operate, sometimes to a harmful end. An important example is its K2-dependent ability to activate platelet-derived growth factor (PDGF), which is instrumental in (r)t-PA induced weakening of the blood brain barrier.

DSPA is now in a global phase III program in AIS, sponsored by PAION's partner Lundbeck, treating patients up to 9 h after stroke onset, a wide extension of the current window of 3–4.5 h.

Introduction

The quest for the ideal thrombolytic in ischemic stroke is still ongoing. Most agents under investigation are derived from tissue plasminogen activator (t-PA; rt-PA for the recombinant drug; alteplase), and were designed to cure shortcomings of that protease, for example its fast elimination and limited fibrin specificity (Petersen, 2007). Recent research has corroborated the notion that the physiological role of tissue plasminogen activator is much more than just that of a safeguard of vessel patency. To name a few, physiological functions seem to include facilitation of synaptic remodelling and axon elongation by degrading extracellular matrix, with implications for memory and motor learning, and also regulation of blood brain barrier permeability. Hence it is not surprising that the multifaceted abilities of tissue plasminogen activator can lead to disastrous effects at several sites once the tight physiological control of its function is lost (Benchenane et al., 2004; Tsirka, 2002; Yepes et al., 2009).

The multifold capabilities of t-PA are in stark contrast to the single-purpose design of desmoteplase (DSPAα1; short: DSPA), a plasminogen activator isolatable from the saliva of the vampire bat, *Desmodus rotundus*. This protein is part of a mixture of anticoagulant, antiplatelet and thrombolytic molecules that solely serve the feeding habit of the bat: to bite and lick the blood of its prey, usually cattle in rural Mexico, sometimes for 30 min on end. Without this biochemical arsenal, the blood would clot within a few minutes and the bat would have to bite numerous times to get satiated (Belwood and Morton, 1991). An fibrin dependence of DSPA ensures that endogenous saliva proteins retain their integrity (Renatus et al., 1997). Thus, desmoteplase seems to have been optimized by natural selection just for this feeding act (Schleuning, 2001). It comes as no surprise, then, that preclinical studies support the contention that the actions of this molecule are largely restricted to the desired site, i.e. obstructive blood clots.

Molecular Structure of Desmoteplase

As reviewed by Schleuning (2001), vampire bat DNA contains independent genes coding for four forms of salivary plasminogen activator. To different extents, they share with t-PA typical protease domains, with the notable exception that Kringle 2

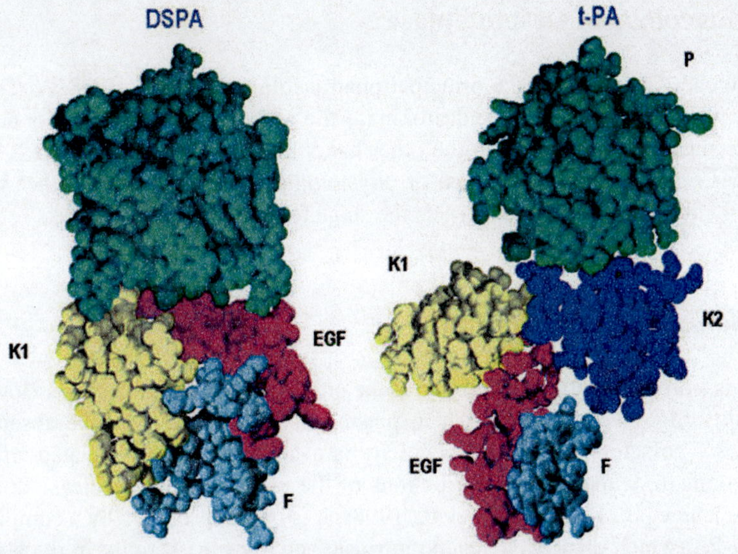

Fig. 23.1 Structure of t-PA and DSPA. EGF, EGF-like domain; F, finger domain; K, kringle domain; P, protease domain. As a notable difference, only t-PA features a kringle 2 domain

is absent in all four variants (Fig. 23.1). Thus, DSPAα features a serine protease, a finger, an epidermal growth factor, and a kringle domain; DSPAß is the same minus the finger domain, while DSPAγ lacks both the finger and the epidermal growth factor domains. DSPAα comes in two closely related forms, DSPAα1 and DSPAα2. Despite a 89% identity in amino acid sequence between the two (Krätzschmar et al., 1991), DSPAα1 shows a slightly lower intrinsic activity, but a larger stimulation by fibrin (factor 10^5 vs. 5×10^4). Mainly for this feature DSPAα1 (named DSPA in the following) was selected for further development. Even so, also DSPAα2 has attracted limited investigative attention under the name of "bat-PA" (e.g., Bergum and Gardell, 1992).

DSPA is a glycoprotein with an amino acid sequence identical to that of its natural counterpart in the saliva of the vampire bat. It is produced by means of Chinese hamster ovary cells endowed with a recombinant plasmid carrying the desmoteplase gene from *Desmodus rotundus*. There is a homology with t-PA of about 72%, the highest grades of conservation being found in the finger, kringle, and epidermal growth factor domains (Gulba et al., 1995). An important difference lies in the Arg275–Ile276 cleavage site. In rt-PA, this is the site at which plasmin hydrolyses the peptide bond to convert the single-chain molecule into a more active two-chain moiety. However, DSPA acts as a single chain enzyme. Even so, DSPA and two-chain rt-PA are very similar in structure and show similar protease activities when fibrin is available (Renatus et al., 1997).

Pharmacology of Desmoteplase

Plasminogen activation is the principal pharmacological action of DSPA. Thus, it shares with drugs like rt-PA and urokinase the ability to lyse fibrin clots and, by this, acts as a thrombolytic. On the other hand, unlike the drugs mentioned, it has no other relevant pharmacological or physiological effects. This, together with a uniquely high fibrin dependence, sets the stage for a high level of safety.

Mechanism of Action

DSPA is a protease with plasminogen as the only known protein substrate. However, its ability to convert plasminogen to plasmin is virtually zero in the absence of fibrin as a cofactor. This fibrin specificity is to a large extent associated with the finger domain, with certain contributions of the kringle and the protease domains (Bringmann et al., 1995; Toschi et al., 1998). It is best appreciated by a comparison with rt-PA, which displays a sizable intrinsic activity (i.e., activity in the absence of cofactors). Addition of molecules like prion protein, ß amyloid, fibrinogen and the complex of D-dimer non-covalently linked to fragment E considerably stimulates plasminogen activation by rt-PA, but hardly by DSPA (Bringmann et al., 1995; Epple et al., 2004; Kruithof and Schleuning, 2004; Stewart et al., 1998). Often misunderstood is the fact that fibrin enhances DSPA activity ($\sim 10^5$-fold) to a much higher extent than that of rt-PA (factor of less than 10^3) (Bringmann et al., 1995): this leads to comparable activities in vitro, as stimulation of rt-PA starts from a much higher level of activity.

DSPA and rt-PA bind to fibrin with similar affinities (Bringmann et al., 1995). In the case of DSPA, this is primarily or solely mediated by the finger domain, whereas rt-PA shows additional low-affinity binding via kringle2 (Stewart et al., 1998).

The anchorage to fibrin seems to turn DSPA into an efficient plasminogen activator by at least two effects: DSPA and its substrate plasminogen are assembled on the fibrin template in a way optimal for interaction and, second, fibrin binding stabilizes DSPA's protease domain in its active conformation (Renatus et al., 1997).

Primary Pharmacodynamics In Vitro

In vitro studies attest to the ability of DSPA to lyse blood clots with a high level of clot specificity. DSPA and rt-PA were found to be approximately equipotent. However, there was no effect of DSPA, studied at concentrations of up to 100 nmol/l over a period of 180 min, on plasma fibrinogen levels. In contrast, exposure to rt-PA resulted in significant fibrinogenolysis at concentrations of 30 nmol/l and above. Similar findings were made when clot-free plasma was incubated with DSPA or rt-PA (Fig. 23.2). Comparable results were obtained upon prolonged exposure (6 h) of human whole blood clots (Schleuning et al., 1992).

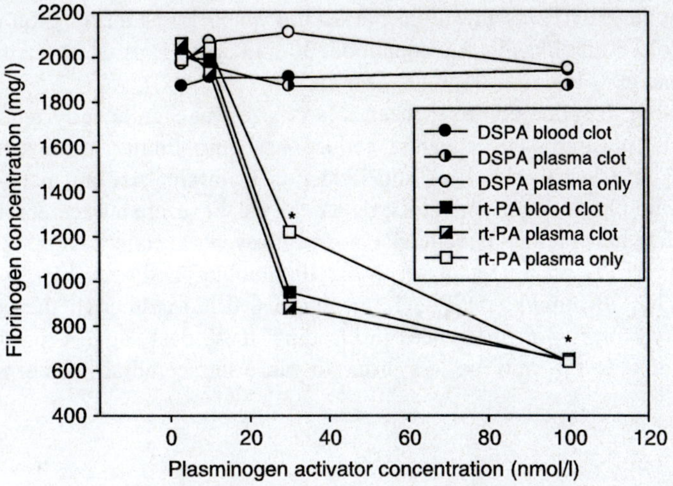

Fig. 23.2 In vitro effects of DSPA and rt-PA on fibrinogen concentration in human plasma. Plasma or whole blood clots or plasma only were incubated with plasminogen activator for 180 min (*n* = 9–13). rt-PA produced significant reductions at concentrations of 30 nmol/l and beyond (Research Report A315, 1992)

Primary Pharmacodynamics *In Vivo*

Peripheral Models

The thrombolytic efficacy and clot specificity of desmoteplase were assessed in numerous animal studies testing arterial and venous thrombosis in vivo including rat, mouse and rabbit stroke models, mostly in head-to-head comparisons with rt-PA (Muschick et al., 1993; Witt et al., 1992; PAION Deutschland GmbH, data on file). Throughout DSPA showed a comparable or greater efficacy compared to rt-PA. Any differences in rate of thrombolysis and maintenance of vascular patency were in favour of DSPA. As predictable from the in-vitro data, rt-PA, but not DSPA produced significant decreases in fibrinogen and plasminogen concentrations in a number of these models.

Where determined, DSPA was eliminated with a substantially longer half life than rt-PA. For example, terminal half lives of rt-PA and DSPA (bolus application) amounted to 5–8 min and 189–199 min, respectively, in a coronary thrombolysis study in dogs (Witt et al., 1994). Thus, the advantage of DSPA over rt-PA in rate and subsistence of success probably relates to a prolonged availability of DSPA.

The efficacies of desmoteplase, reteplase, and rt-PA were compared in a rat model of embolic occlusion of the middle cerebral artery (Research Report PN01 PCD 00002/01, 2001). An autologous blood clot was introduced at the base of the right middle cerebral artery by means of a catheter advanced into the right internal carotid artery. Saline or drugs were administered 1 h post occlusion, DSPA (10, 30, or 90 nmol/kg) by bolus injection, rt-PA (154 nmol/kg) as an infusion over 1 h, reteplase (25.6 nmol/kg) in two equal boli, 30 min apart. Rats were sacrificed 48 h post occlusion.

Compared to saline controls, mean infarct size was significantly reduced with each of the plasminogen activators, reductions being similar in all active treatment groups. Clearly, clot lysis and reduction in infarct size are not surprising findings with plasminogen activators; rather, the risk of severe intracerebral haemorrhage will define potential benefit. Haemorrhagic events accounted for 0–2 fatalities per group (rt-PA, 2/11; reteplase, 1/12; 10 nmol/kg DSPA, 0/13; 30 nmol/kg DSPA, 0/11; 90 nmol/kg DSPA, 1/15; controls, 0/12). Although the study was not powered to test for differences in bleeding, these data suggest that DSPA, at effective dose levels, may be less prone to cause intracerebral haemorrhage (see below).

Lack of Kringle 2 as the Major Discriminant Between DSPA and t-PA Derivatives

There are important functional differences between DSPA on the one side and rt-PA (and its derivatives) on the other side, all of them likely to bear on their clinical utility. To a great extent, these differences can be attributed to DSPA's lack of a kringle 2 domain.

Fibrin Specificity

Kringle 2 mediates rt-PA binding to fibrin, fibrinogen, human prion protein, and the D-dimer, the principal product of plasmin degradation of cross-linked fibrin (Bringmann et al., 1995; Epple et al., 2004; Stewart et al., 1998) and hence is thought to be the major reason for the lower fibrin specificity of rt-PA.

Neurotoxic Potential

Kringle 2 is essential to the ability of plasminogen activators such as rt-PA, reteplase, and tenecteplase to bind and cleave the NR1 subunit of the NMDA receptor (Fernández-Monreal et al., 2004; López-Atalya et al., 2008). Inasmuch as such cleavage may explain the neurotoxic effects of fibrinolytics, the absence of a kringle 2 moiety can account for the fact that DSPA, in head-to-head comparison with rt-PA, did not aggravate NMDA-induced neurotoxicity in any of the preclinical models studied so far, either in vitro (neuronal cultures) or in vivo (intrastriatal and intravenous administration in rat and mouse models) (Liberatore et al., 2003; López-Atalaya et al., 2007, 2008; Reddrop et al., 2005).

That neurotoxicity via the NMDA receptor is highly relevant to the therapeutic profile of rt-PA has been strikingly demonstrated in a recent study using a murine model of ischemic stroke (Benchenane et al., 2007). Mice were immunized against the principal rt-PA binding site at the NMDA receptor (the amino-terminal domain of the NR1 subunit). Such mice showed less brain damage when subjected to cerebral ischemia, consistent with a role for endogenous t-PA in such damage. Likewise, potentiation of NMDA-induced brain lesions by intravenous injection of rt-PA (doubled lesion size) was not observed in immunized mice, indicating that exogenous rt-PA acts by way of the same mechanism.

Potential to Disturb the Blood Brain Barrier

As derived from animal and in vitro studies and supported by evolving evidence in humans, ischaemia will increase permeability of the blood-brain barrier (BBB) in a time-dependent process, the worst outcome being symptomatic intracerebral haemorrhage (Su et al., 2009). This process is promoted by release of t-PA from endothelial and from brain cells (Yepes et al., 2003, 2009), which also helps to understand why administration of rt-PA, especially in the late time window, is associated with enhanced risk of haemorrhagic transformation.

From animal models, it has long been known that thromboembolic cerebral ischaemia will cause BBB disturbances that may be visualized as extravasation of large marker molecules and blood cells. rt-PA, administered at an early time post-occlusion, is able to restore perfusion and reduce infarct size, but, at the same time, enhances BBB injury even in distant non-ischemic brain regions (Busch et al., 1997). Similar rt-PA effects on the BBB have been identified in humans (Kassner et al., 2009; Kidwell et al., 2008).

There is compelling evidence indicating that endogenous t-PA is basically involved in ischemic BBB damage. Its direct downstream target in this process is latent platelet-derived growth factor-CC (PDGF-CC), which has been identified as a specific substrate of t-PA. PDGF-CC contains an N-terminal CUB domain that renders the dimer inactive. The CUB domain (first found in *C*omplement subcomponents Clr/Cls, *U*rinary epidermal growth factor, and *B*one morphogenetic protein), is a structural motif of approximately 110 amino acids that is involved in a wide diversity of regulatory and signalling functions. Removal of the CUB domain from PDGF-CC by t-PA generates active PDGF-CC, which can then activate the PDGF α receptor in the neurovascular unit. The ultimate result is the loss of BBB integrity (Fredriksson et al., 2005; Su et al., 2008).

It is intriguing to note that cleavage and activation of PDGF-CC by t-PA occur at an arginine residue and require a kringle 2 domain (Fredriksson et al., 2005), in striking similarity with NMDA-receptor mediated neurotoxicity (see above). Thus, both unfavourable effects of t-PA, neurotoxicity and blood brain barrier damage, involve proteolytic activity and kringle 2.

It remains to be shown whether the absence of kringle 2 in DSPA, which is thought to underlie its lack of neurotoxicity, confers a similar advantage with regard to BBB integrity.

Clinical Development

Desmoteplase as a General Thrombolytic

In the eighties and nineties of the past millennium, research in thrombolysis was driven by the unsolved need for treatment modalities for acute myocardial infarction (AMI). One of the central paradigms in the quest for the optimal fibrinolytic was the desideratum of fibrin specificity. As pointed out by Collen (1996), the first available thrombolytic agents streptokinase and urokinase had no specific affinity for fibrin and therefore activated circulating and fibrin-bound plasminogen relatively indiscriminately. After exhaustion of plasmin inhibitors, residual plasmin would degrade relevant plasma proteins (fibrinogen, plasminogen, factor V, factor VIII, etc.) to cause a serious bleeding tendency. This was taken to explain the limited efficiency of streptokinase or urokinase and their association with serious, sometimes life-threatening side effects (Sobel et al., 1984). Thus, the advent of tissue plasminogen activator, with its preferential activation of fibrin-bound plasminogen, was seen as a major improvement and further increases in fibrin specificity, such as the modest 14-fold enhancement reached with tenecteplase, were greeted with some enthusiasm (Keyt et al., 1994). Against this background, the level of fibrin specificity found with desmoteplase (almost 200-fold higher than seen with rt-PA; Bringmann et al., 1995) was a striking progress suggesting that DSPA may become a prime choice for all cardiovascular disorders standing to benefit from clot lysis.

It was a logical choice to study DSPA first in the setting of AMI to provide proof of concept for its utility in thromboembolic disorders. The primary aim of the DEEP trial (Desmoteplase in the Establishment of Early Patency; Schering Report, 2002) was the evaluation of efficacy and safety of desmoteplase (single bolus injection accompanied by intravenous heparin) in patients presenting with AMI and to confirm the favourable pharmacokinetics found in volunteers (elimination half life of more than 2 h). The first patients received 500 μg per kg body weight, later increased to 750 μg per kg. These doses appear – with today's knowledge – to be higher than needed, gauged against the background of the current dose of 90 μg/kg for acute ischemic stroke.

Coronary patency was achieved in 21 of the 26 patients. The high fibrin specificity of DSPA was confirmed by normal plasma concentrations of fibrinogen at both dose levels; the time course observed at the higher dose is illustrated in Fig. 23.3. DSPA was well tolerated; there were 2 early serious bleeding events at the lower dose (500 μg/kg) including one case of cerebral haemorrhage, which were attributed to a high dose of heparin used in the first five study patients. No such events were observed in patients enrolled after a reduction of the heparin dose, neither at 500 μg/kg DSPA nor after the increase to 750 μg/kg. Also the long elimination half life of $\sim$ 4.5 h was consistent with the expectations, confirming the feasibility of bolus administration of DSPA (difference in half life due to higher assay sensitivity).

The utility of DSPA in thromboembolic diseases is further supported by a phase II study in patients with acute pulmonary embolism (Tebbe et al., 2009). DSPA at

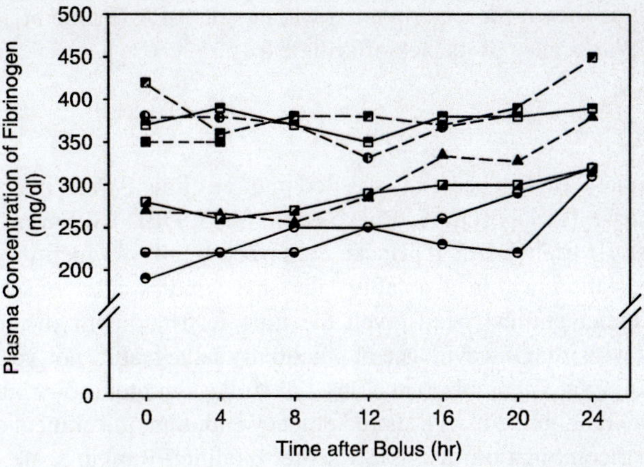

Fig. 23.3 Time course of plasma concentrations of fibrinogen in AMI patients treated with DSPA at 750 μg/kg

doses of 125, 180, and 250 μg/kg body weight was compared with rt-PA (100 mg), using total pulmonary resistance, mean pulmonary artery pressure, and the Miller index as efficacy criteria. The study results suggested that desmoteplase at doses of 180 and 250 μg/kg had similar or greater efficacy compared to rt-PA, with a faster onset of action and comparable safety. The most frequent drug-related serious event was haemorrhage at the catheter and injection site with a drop of haemoglobin (DSPA 180 μg/kg, $n = 2$, 22.2%; alteplase, $n = 1$, 8.3%).

Desmoteplase for Acute Ischemic Stroke

DSPA was abandoned by Schering AG for strategic reasons. PAION took over the drug in 2001, and initiated a change in focus towards acute ischemic stroke, based on the premise that a more fibrin-specific thrombolytic should pose less of a bleeding risk. The attraction of this concept had been enhanced by data from the NINDS trial (1995), which had proven that thrombolysis was an effective option in stroke, but also suggested a particular sensitivity of this condition to intracerebral haemorrhage. This complication had been much less evident from the use of rt-PA in AMI and seemed to increase with the delay at which thrombolysis was implemented. Today, this notion must be modified: Acute ischemic stroke is indeed fraught with a special risk of intracerebral haemorrhage, developing over time and fostered by t-PA released from the brain parenchyma. These effects, however, are supported by, but not primarily due to deficiencies in fibrin specificity. Rather, they are accounted for by the pathomechanisms described above (Su et al., 2008). Thus, the deleterious effects of rt-PA on the BBB are not necessarily a class effect shared by all plasminogen activators. As a point in case, these mechanisms can be dissociated from thrombolysis, as shown for experimental variants of rt-PA (Fanne et al., 2010) and likely for DSPA because of its lack of kringle 2.

Clinical Trials

To date, desmoteplase has been investigated in three clinical trials in acute ischemic stroke: two phase II trials (DIAS, Hacke et al., 2005; DEDAS, Furlan et al., 2006) and one phase III trial (DIAS II, Hacke et al., 2009), all of which investigated the potential application within a notably late time window after onset of stroke – a currently unmet clinical need given the time restriction for the use of rt-PA. Only patients with imaging evidence of potentially salvageable, not-yet irreversibly-infarcted brain tissue (penumbra) and onset of stroke symptoms 3–9 h before DSPA treatment were randomized. As a shared efficacy endpoint, the clinical outcome was evaluated by a combination of NIHSS score, modified Rankin scale, and Barthel Index; the two phase II trials, DIAS and DEDAS, also featured the rate of reperfusion on MRI after 4–8 h as an endpoint. Symptomatic cerebral haemorrhage was the primary safety endpoint.

In the combined analysis of the two phase II trials, the outcome measures of reperfusion after 4–8 h and positive clinical outcome after 90 days (both parameters: 18/30 patients on DSPA 125 μg/kg vs. 8/34 patients on placebo) were strongly suggestive of efficacy. There was an impressive correlation between the improvement in clinical outcome and reperfusion for both the 90 and 125 μg/kg dose groups (Fig. 23.4).

The primary safety endpoint of symptomatic cerebral haemorrhage (1/59 patients treated with either 90 or 125 μg/kg DSPA) was acceptable. After these two favorable studies, doses of 90 and 125 μg/kg were selected for DIAS II, the first phase III trial of a plasminogen activator based on classical dose-range finding studies.

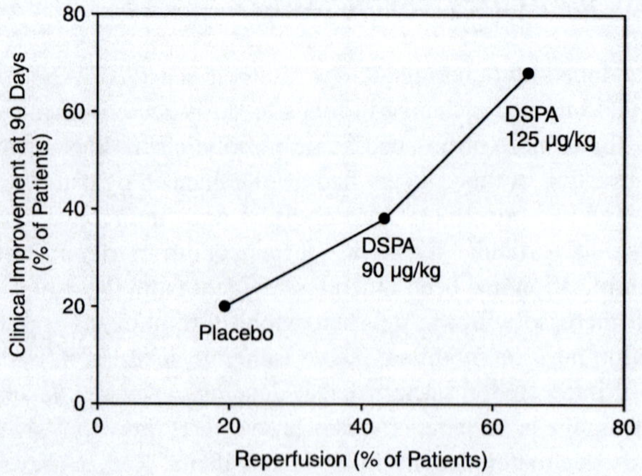

Fig. 23.4 Correlation between clinical improvement after 90 days and restoration of perfusion. Plotted are the percentages of responders identified in a combined analysis of the DIAS and DEDAS studies

DIAS II: Safety

In the phase III trial, DIAS-2, patients with acute ischaemic stroke and tissue at risk were assigned to 90 μg/kg DSPA, 125 μg/kg DSPA, or placebo within 3–9 h of the onset of stroke symptoms. Also this study (Hacke et al., 2009) confirmed the safety of DSPA in the late time window. The rates of symptomatic intracranial haemorrhage were 3.5% (2 of 57) for 90 μg/kg desmoteplase, 4.5% (3 of 66) for 125 μg/kg desmoteplase, and 0% for placebo. As pointed out by the authors, the rates of symptomatic intracranial haemorrhage were comparable to those in the ECASS II study (Hacke et al., 1998) (when the same definition was applied) for the patients who received the higher dose of desmoteplase and significantly lower for those who received the lower dose, which suggests an improved safety profile compared with other studies of thrombolytic treatment, within both the early and late time windows. One of the two cases of symptomatic intracranial haemorrhage (sICH) in the 90 μg/kg desmoteplase group was due to worsening of a baseline haemorrhagic infarction, which constituted a serious protocol violation. 13 major haemorrhagic events were reported by 5% of patients (3 of 57) in the 90 μg/kg desmoteplase group, 8% (5 of 66) in the 125 μg/kg desmoteplase group, and 6% (4 of 63) in the placebo group.

DIAS II: Efficacy

In contrast to the two phase II trials, DIAS II failed to demonstrate efficacy of desmoteplase in acute stroke patients. This failure can be attributed to several interstudy-differences in concert, leading to an unusually high placebo response of 46%. As outlined by the authors (Hacke et al., 2009), baseline strokes were less severe in DIAS II and core lesion volumes smaller, allowing for better spontaneous outcomes, and the occurrence of initial proximal vessel occlusion was unexpectedly low: Overall, only 30% of the patients showed such an occlusion, in stark contrast to the combined rate of 57% observed in DIAS and DEDAS. Stratification of the clinical response by the TIMI score (used to describe vessel occlusion by assessment of residual blood flow) revealed treatment effects vs. placebo as a function of the residual blood flow: no benefit in cases showing no or only partial reduction of flow (TIMI 2–3), but appreciable effects in patients with a complete or sizable obstruction (TIMI 0–1).

Outlook

Stroke is distinguished from other conditions of acute vascular occlusion by a time-dependent disruption of the affected vascular bed, i.e. the BBB. For example, streptokinase, in striking contrast to its relatively safe use of in AMI, was associated with an unacceptably high risk of symptomatic hemorrhagic transformation when tested in stroke (33/156 patients compared to 4/154 patients with placebo; Multicenter Acute Stroke Trial-Europe Study Group, 1996). As another point in case, also rt-PA shows a higher incidence of sICH in stroke than in AMI patients.

Finally, DSPA was associated with a limited rate of sICH in stroke in the face of a virtual absence of such cases at the much higher doses used in AMI and acute pulmonary embolism.

Safety of Thrombolytics in Stroke

Various peculiarities need to be observed for the optimal use of thrombolytics in acute ischemic stroke. For one, lower doses seem to be appropriate in stroke than in other indications; rt-PA, which did not undergo classical dose-range finding for stroke, may be overdosed. Second, the sequence of events initiated by cerebral ischemia seems to impart on the BBB a state of vulnerability that increases over time and will raise the bleeding risk, irrespective of the means used to restore blood flow. There is no doubt about the ability of rt-PA to weaken and finally disrupt the BBB in experimental animals by way of special pathomechanisms (Su et al., 2009). Equivalent effects have been detected in human stroke patients when dedicated imaging techniques were used, even at times within the first 4 h of rt-PA administration (Kassner et al., 2009; Kidwell et al., 2008). The underlying pathomechanism might be shared by other t-PA kringle-2 bearing derivatives, similar to the way such compounds have been found to share neurotoxicity (López-Atalaya et al., 2008).

There is no reasonable doubt that any agent able to lyse a clot should also afford benefit in stroke. The salient point, then, is safety. Apart from its thrombolytic abilities, proven in the blood-feeding bat as well as in numerous in vitro and in vivo studies, DSPA is distinguished by a highly promising safety profile. This profile is consistent with and supported by the growing knowledge of the mechanisms involved in hemorrhagic transformation and neurotoxicity, which are known to be caused by thrombolytics of the rt-PA type in animals and are likely to pertain to humans as well. Currently, DSPA is being studied in two phase III trials, again in the late time window of 3–9 h (DIAS III and DIAS IV) sponsored by Lundbeck. Based on the findings of a thorough re-analysis patient selection is now based on an angiographic verification of a brain artery obstruction. Since the failure of the DIAS II study can largely be explained by a lack of the presence of a clot in the majority of patients, the improved design of the two ongoing studies (i.e. angiography at entry) may allow the demonstration of the benefit expected from restoration of blood flow by a drug that has been adapted by evolution to the single task of plasminogen activation.

References

Belwood, J.J., Morton, P.A., 1991. Vampires: the real story. The truth about the bats people love to hate is even more fascinating than the myths. Bats 9, 11–16.

Benchenane, K., López-Atalaya, J.P., Fernandez-Monreal, M., Touzani, O., Vivien, D., 2004. Equivocal roles of tissue-type plasminogen activator in stroke-induced injury. Trends Neurosci. 27, 155–160.

Benchenane, K., Castel, H., Boulouard, M., Bluthé, R., Fernandez-Monreal, M., Roussel, B.D., López-Atalaya, J.P., Butt-Gueulle, S., Agin, V., Maubert, E., Dantzer, R., Touzani, O.,

Dauphin, F., Vivien, D., Ali, C., 2007. Anti-NR1 N-terminal-domain vaccination unmasks the crucial action of tPA on NMDA-receptor-mediated toxicity and spatial memory. J. Cell. Sci. 120, 578–585.

Bergum, P.W., Gardell, S.J., 1992. Vampire bat salivary plasminogen activator exhibits a strict and fastidious requirement for polymeric fibrin as its cofactor, unlike human tissue-type plasminogen activator. A kinetic analysis. J. Biol. Chem. 267, 17726–17731.

Bringmann, P., Gruber, D., Liese, A., Toschi, L., Krätzchmar, J., Schleuning, W.D., Donner, P., 1995. Structural features mediating fibrin selectivity of vampire bat plasminogen activators. J. Biol. Chem. 270, 25596–25603.

Busch, E., Krüger, K., Fritze, K., Allegrini, P.R., Hoehn-Berlage, M., Hossmann, K.A., 1997. Blood-brain barrier disturbances after rt-PA treatment of thromboembolic stroke in the rat. Acta Neurochir. Suppl. 70, 206–208.

Collen, D., 1996. Fibrin-selective thrombolytic therapy for acute myocardial infarction. Circulation 93, 857–865.

Epple, G., Schleuning, W.D., Kettelgerdes, G., Kottgen, E., Gessner, R., Praus, M., 2004. Prion protein stimulates tissue-type plasminogen activator-mediated plasmin generation via a lysine-binding site on kringle 2. J. Thromb. Haemost. 2, 962–968.

Fanne, R.A., Nassar, T., Yarovoi, S., Rayan, A., Lamensdorf, I., Karakoveski, M., Vadim, P., Fanne, R.A., Jamal, M., Cines, D.B., Al-Roof Higazi, A., 2010. Blood-brain barrier permeability and tPA-mediated neurotoxicity. Neuropharmacology 58, 972–980.

Fernández-Monreal, M., López-Atalaya, J.P., Benchenane, K., Cacquevel, M., Dulin, F., Le Caer, J.P., Rossier, J., Jarrige, A.C., Mackenzie, E.T., Colloc'h, N., Ali, C., Vivien, D., 2004. Arginine 260 of the amino-terminal domain of NR1 subunit is critical for tissue-type plasminogen activator-mediated enhancement of N-methyl-D-aspartate receptor signaling. J. Biol. Chem. 279, 50850–508506.

Fredriksson, L., Ehnman, M., Fieber, C., Eriksson, U., 2005. Structural requirements for activation of latent platelet-derived growth factor CC by tissue plasminogen activator. J. Biol. Chem. 280, 26856–26862.

Furlan, A.J., Eyding, D., Albers, G.W., Al-Rawi, Y., Lees, K.R., Rowley, H.A., Sachara, C., Soehngen, M., Warach, S., Hacke, W., DEDAS Investigators, 2006. Dose escalation of desmoteplase for acute ischemic stroke (DEDAS): evidence of safety and efficacy 3 to 9 hours after stroke onset. Stroke 37, 1227–1231.

Gulba, D.C., Praus, M., Witt, W., 1995. DSPA alpha – properties of the plasminogen activators of the vampire bat *Desmodus rotundus*. Fibrinolysis 9(Suppl.), 91–96.

Hacke, W., Kaste, M., Fieschi, C., von Kummer, R., Davalos, A., Meier, D,. Larrue, V., Bluhmki, E., Davis, S., Donnan, G., Schneider, D., Diez-Tejedor, E., Trouillas, P., 1998. Randomised double-blind placebo-controlled trial of thrombolytic therapy with intravenous alteplase in acute ischaemic stroke (ECASS II). Second European-Australasian Acute Stroke Study Investigators. Lancet, 352: 1245–1251.

Hacke, W., Albers, G., Al-Rawi, Y., Bogousslavsky, J., Davalos, A., Eliasziw, M., Fischer, M., Furlan, A., Kaste, M., Lees, K.R., Soehngen, M., Warach, S., for The DIAS Study Group, 2005. The desmoteplase in acute ischemic stroke trial (DIAS). A phase II MRI-based 9-hour window acute stroke thrombolysis trial with intravenous desmoteplase. Stroke 36, 66–73

Hacke, W., Furlan, A.J., Al-Rawi, Y., Davalos, A., Fiebach, J.B., Gruber, F., Kaste, M., Lipka, L.J., Pedraza, S., Ringleb, P.A., Rowley, H.A., Schneider, D., Schwamm, L.H., Leal, J.S., Söhngen, M., Teal, P.A., Wilhelm-Ogunbiyi, K., Wintermark, M., Warach, S., 2009. Intravenous desmoteplase in patients with acute ischaemic stroke selected by MRI perfusion-diffusion weighted imaging or perfusion CT (DIAS-2): a prospective, randomised, double-blind, placebo-controlled study. Lancet Neurol. 8, 141–50

Kassner, A., Roberts, T.P., Moran, B., Silver, F.L., Mikulis, D.J., 2009. Recombinant tissue plasminogen activator increases blood-brain barrier disruption in acute ischemic stroke: an MR imaging permeability study. Am. J. Neuroradiol. 30, 1864–1869.

Keyt, B.A., Paoni, N.F., Refino, C.J., Berleau, L., Nguyen, H., Chow, A., Lai, J., Peña, L., Pater, C., Ogez, J., Etcheverry, T., Botstzein, D., Bennett, W.F., 1994. A faster-acting and more potent form of tissue plasminogen activator. Proc. Natl. Acad. Sci. U.S.A. 91, 3670–3674.

Kidwell, C.S., Latour, L., Saver, J.L., Alger, J.R., Starkman, S., Duckwiler, G., Jahan, R., Vinuela, F., UCLA Thrombolysis Investigators, Kang, D.W., Warach, S., 2008. Thrombolytic toxicity: blood brain barrier disruption in human ischemic stroke. Cerebrovasc. Dis. 25, 338–343.

Krätzschmar, J., Haendler, B., Langer, G., Boidol, W., Bringmann, P., Alagon, A., Donner, P., Schleuning, W.D., 1991. The plasminogen activator family from the salivary gland of the vampire bat *Desmodus rotundus*: cloning and expression. Gene 105, 229–237.

Kruithof, E.K., Schleuning, W.D., 2004. A comparative study of amyloid-beta (1–42) as a cofactor for plasminogen activation by vampire bat plasminogen activator and recombinant human tissue-type plasminogen activator. Thromb. Haemost. 92, 559–567.

Liberatore, G.T., Samson, A., Bladin, C., Schleuning, W.D., Medcalf, R.L., 2003. Vampire bat salivary plasminogen activator (desmoteplase): a unique fibrinolytic enzyme that does not promote neurodegeneration. Stroke 34, 537–543.

López-Atalaya, J.P., Roussel, B.D., Ali, C., Maubert, E., Petersen, K.-U., Berezowski, V., Cecchelli, R., Orset, C., Vivien, D., 2007. Recombinant *Desmodus rotundus* salivary plasminogen activator (desmoteplase) crosses the blood-brain barrier through a LDL receptor-related protein (LRP)-dependent mechanism without exerting neurotoxic effects. Stroke 38, 1036–1043.

López-Atalaya, J.P., Roussel, B.D., Levrat, D., Parcq, J., Nicole, O., Hommet, Y., Benchenane, K., Castel, H., Leprince, J., van To, D., Bureau, R., Rault, S., Vaudry, H., Petersen, K.-U., Santos, J.S., Ali, C., Vivien, D., 2008. Toward safer thrombolytic agents in stroke: molecular requirements for NMDA receptor-mediated neurotoxicity. J. Cereb. Blood Flow Metab. 28, 1212–1221.

Multicenter Acute Stroke Trial-Europe Study Group, 1996. Thrombolytic therapy with streptokinase in acute ischemic stroke. N. Engl. J. Med. 335, 145–150.

Muschick, P., Zeggert, D., Donner, P., Witt, W., 1993. Thrombolytic properties of *Desmodus* (vampire bat) salivary plasminogen activator DSPA alpha1, alteplase and streptokinase following intravenous bolus injection in a rabbit model of carotid artery thrombosis. Fibrinolysis 7, 284–290

NINDS. The National Institute of Neurological Disorders and Stroke rt-PA Stroke Study Group, 1995. Tissue plasminogen activator for acute ischemic stroke. N. Engl. J. Med. 33, 1581–1587.

Petersen, K.-U., 2007. Thrombolytics – a field in development. Riv. It. Neurobiologia 53, 7–14.

Reddrop, C., Moldrich, R.X., Beart, P.M., Farso, M., Liberatore, G.T., Howells, D.W., Petersen, K.-U., Schleuning, W.D., Medcalf, R.L., 2005. Vampire bat salivary plasminogen activator (desmoteplase) inhibits tissue-type plasminogen activator-induced potentiation of excitotoxic injury. Stroke 36, 1241–1246.

Renatus, M., Stubbs, M.T., Huber, R., Bringmann, P., Donner, P., Schleuning, W.D., Bode, W., 1997. Catalytic domain structure of vampire bat plasminogen activator: a molecular paradigm for proteolysis without activation cleavage. Biochemistry 36, 3483–3493.

Research Report A315, 1992. Lysis of human plasma and whole blood clots in vitro by ZK 152 387 (DSPA alpha1), ZK 153 786 (DSPA alpha2), ZK 155 278 (DSPA beta), ZK 153 858 (alteplase), ZK 130 932 (scu-PA), ZK 113 905 (urokinase), and ZK 156 552 (anistreplase). (PAION, data on file)

Research Report PN01 PCD 00002/01, 2001. Comparing the therapeutic benefit of desmoteplase, reteplase, and alteplase in a rat model of embolic middle cerebral artery (MCA) occlusion. (PAION, data on file)

Schering Report, Gulba, D., 2002. rDSPA alpha1 in the establishment of early patency in myocardial infarction. (PAION, data on file)

Schleuning, W.D., Alagon, A., Boidol, W., Bringmann, P., Petri, T., Krätzschmar, J., Haendler, B., Langer, G., Baldus, B., Witt, W., Donner, P., 1992. Plasminogen activators from the saliva of *Desmodus rotundus* (common vampire bat): unique fibrin specificity. Ann. N. Y. Acad. Sci. 667, 395–403.

Schleuning, W.D., 2001. Vampire bat plasminogen activator DSPA-alpha-1 (desmoteplase): a thrombolytic drug optimized by natural selection. Haemostasis 31, 118–122.

Stewart, R.J., Fredenburgh, J.C., Weitz, J.I., 1998. Characterization of the interactions of plasminogen and tissue and vampire bat plasminogen activators with fibrinogen, fibrin, and the complex of D-dimer noncovalently linked to fragment E. J. Biol. Chem. 273, 18292–18299.

Sobel, B.E., Geltman, E.M., Tiefenbrunn, A.J., Jaffe, A.S., Spadaro, J.J. Jr., Ter-Pogossian, M.M., Collen, D., Ludbrook, P.A., 1984. Improvement of regional myocardial metabolism after coronary thrombolysis induced with tissue-type plasminogen activator or streptokinase. Circulation 69, 983–690.

Su, E.J., Fredriksson, L., Geyer, M., Folestad, E., Cale, J., Andrae, J., Gao, Y., Pietras, K., Mann, K., Yepes, M., Strickland, D.K., Betsholtz, C., Eriksson, U., Lawrence, D.A., 2008. Activation of PDGF-CC by tissue plasminogen activator impairs blood-brain barrier integrity during ischemic stroke. Nat. Med. 14, 731–737.

Su, E.J., Fredriksson, L., Schielke, G.P., Eriksson, U., Lawrence, D.A., 2009. Tissue plasminogen activator-mediated PDGF signaling and neurovascular coupling in stroke. J. Thromb. Haemost. 7(Suppl. 1), 155–158.

Tebbe, U., Bramlage, P., Graf, A., Lechleitner, P., Bode, C., Riess, F.C., Clemens, N., Al-Rawi, Y., Konstantinides, S., Goldhaber, S.Z., 2009. Desmoteplase in acute massive pulmonary thromboembolism. Thromb. Haemost. 101, 557–562.

Toschi, L., Bringmann, P., Petri, T., Donner, P., Schleuning, W.D., 1998. Fibrin selectivity of the isolated protease domains of tissue-type and vampire bat salivary gland plasminogen activators. Eur. J. Biochem. 252, 108–112.

Tsirka, S.E., 2002. Tissue plasminogen activator as a modulator of neuronal survival and function. Biochem. Soc. Trans. 30, 222–225.

Yepes, M., Sandkvist, M., Moore, E.G., Bugge, T.H., Strickland, D.K., Lawrence, D.A., 2003. Tissue-type plasminogen activator induces opening of the blood-brain barrier via the LDL receptor-related protein. J. Clin. Invest. 112, 1533–1540.

Yepes, M., Roussel, B.D., Ali, C., Vivien, D., 2009. Tissue-type plasminogen activator in the ischemic brain: more than a thrombolytic. Trends Neurosci. 32, 48–55.

Witt, W., Balds, B., Bringmann, P., Cashon, L., Donner, P., Schleuning, W.D., 1992. Thrombolytic properties of *Desmodus rotundus* (vampire bat) salivary plasminogen activator in experimental pulmonary embolism in rats. Blood 79, 1213–1217.

Witt, W., Maass, B., Baldus, B., Hildebrand, M., Donner, P., Schleuning, W.D., 1994. Coronary thrombolysis with Desmodus salivary plasminogen activator in dogs. Fast and persistent recanalization by intravenous bolus administration. Circulation 90, 421–426.

Chapter 24
Fibrolase and Its Evolution to Clinical Trials: A Long and Winding Road

Francis S. Markland and Stephen Swenson

Abstract Fibrolase is the fibrinolytic proteinase isolated from *Agkistrodon contortrix contortrix* (southern copperhead snake) venom. The enzyme was purified by a three-step HPLC procedure and was shown to be homogeneous by standard criteria. The purified enzyme is inhibited by EDTA and other chelating agents and is a zinc metalloproteinase containing one mole of zinc per molecule. The enzyme is also rapidly inhibited by alpha2-macroglobulin (α2M). Fibrolase is composed of 203 amino acids with a blocked amino-terminus due to cyclization of the terminal Gln residue. The enzyme is a direct-acting thrombolytic agent and does not rely on plasminogen for clot dissolution. Fibrolase rapidly cleaves the A(α)-chain of fibrinogen and the B(β)-chain at a slower rate; it has no activity on the γ-chain. The enzyme exhibits the same specificity with fibrin. Fibrolase was shown to have very effective thrombolytic activity in a reoccluding carotid arterial thrombosis model in the canine. A recombinant version of the enzyme was made in yeast by Amgen, Inc. (Thousand Oaks, CA, USA) and called alfimeprase. Alfimeprase is identical to fibrolase except for a two amino acid truncation at the amino-terminus and the insertion of a new amino-terminal amino acid in the truncated protein; these changes lead to a more stable enzyme for prolonged storage. Twenty years after it was first purified alfimeprase was taken into clinical trials by Nuvelo, Inc. (San Carlos, CA), which licensed the enzyme from Amgen. Alfimeprase was successful in Phase I and II clinical trials for peripheral arterial occlusion (PAO) and central venous access device (CVAD) occlusion. However, in Phase III trials alfimeprase did not meet the expected end points in either PAO or CVAD occlusion and in a Phase II stroke trial and Nuvelo dropped further development in 2008.

F.S. Markland (✉)
Department of Biochemistry and Molecular Biology, University of Southern California, Keck School of Medicine, Cancer Research Laboratory #106, 1303 N. Mission Road, Los Angeles, CA 90033, USA
e-mail: markland@usc.edu

R.M. Kini et al. (eds.), *Toxins and Hemostasis*, DOI 10.1007/978-90-481-9295-3_24,

Introduction and Background

Kornalik in Czechoslovakia in 1966 was the first to report fibrinolytic activity in *Agkistrodon contortrix contortrix* (southern copperhead snake) venom (Kornalik, 1966), although earlier Didisheim and Lewis had suggested that snake venoms may contain fibrinolytic activity that should be useful for clinical application since it may not be inactivated by inhibitors in mammalian blood (Didisheim and Lewis, 1956). However, it was not until much later that the fibrinolytic enzyme was purified. The Markland laboratory (Bajwa et al., 1982) first identified the enzyme in 1982 and subsequently purified it (Retzios and Markland, 1990). Fibrolase was the name given to this fibrinolytic metalloproteinase.

Fibrolase is a direct acting, 23-kDa fibrinolytic enzyme that cleaves the Aα-chain of fibrinogen (primary cleavage site Lys-Leu bond at residues 413–414). The enzyme also cleaves the Bβ-chain at a slower rate, but it has no effect on the γ-chain (Retzios and Markland, 1988). The enzyme was purified by 3-step HPLC method involving hydrophobic interaction HPLC, hydroxyapatite HPLC and cation exchange HPLC (Retzios and Markland, 1990; Trikha et al., 1994). Fibrolase is a member of family M12, subfamily B (the reprolysins), a grouping of proteolytic enzymes composed of many enzymes of snake venom origin. The active site of the molecule has been identified in the zinc-binding region of fibrolase, spanning amino acids 141–170 (Randolph et al., 1992; Swenson et al., 2004).

Fibrolase differs from the plasminogen activator-based thrombolytics since it acts directly on fibrin; it does not rely on activation of plasminogen (fibrolase neither activates nor degrades plasminogen) (Retzios and Markland, 1988). Fibrolase is a metalloproteinase and not a serine protease (Retzios and Markland, 1988). Therefore, it is not inhibited by the blood serine protease inhibitors, SERPINS (Rau et al., 2007), which are targeted to the blood clotting and fibrinolytic serine proteinases. However, incubation of fibrolase with plasma leads to inhibition of fibrinolytic activity due to the covalent binding of fibrolase by α-2 macroglobulin (α2 M) (Ahmed et al., 1990; Markland et al., 1988). The amino acid sequence of fibrolase was determined in a collaboration between the Markland laboratory and investigators at Chiron Corp. (Emeryville, CA, USA) (Randolph et al., 1992), and the sequence clearly shows that the enzyme is a member of the M12 group of metalloproteinases, the reprolysins. Further, the sequence reveals that there is a complete absence of consensus sites for N-linked glycosylation (Asn-X-Ser/Thr). Separate studies indicated that the enzyme has no thrombin-like activity, no protein C activation activity, no activation nor degradation of plasminogen, no platelet aggregating activity in vitro, no hemolytic activity, and importantly no hemorrhagic activity (Guan et al., 1991).

Evolution of Fibrolase as a Thrombolytic Agent

Based on the direct action of fibrolase on fibrin and the lack of inhibition by blood SERPINS, it appeared that fibrolase should be an effective thrombolytic agent. With this as a background, Dr. Markland and Dr. Benedict Lucchesi, an authority on the use of animal models for evaluation of thrombolytic and anti-platelet agents at the University of Michigan, agreed to collaborate on a study to assess the potential thrombolytic activity of fibrolase in a reoccluding canine carotid arterial thrombosis model. Dr. Markland took fibrolase to the University of Michigan and with Dr. Lucchesi demonstrated that the enzyme isolated from snake venom had excellent lytic activity in this 100% reoccluding arterial thrombosis model (Markland et al., 1994). The collaborative study showed that the enzyme rapidly lysed clots in the carotid artery when administered at a dose of 4 mg/kg in a volume of 3 mL infused over a 5 min period proximal to the site of the thrombus. In this model system, arteries infused with fibrolase in five of five dogs were shown to be cleared of the occluding thrombus within 6 min of initiation of lytic agent infusion. In the contralateral carotid artery that received only vehicle, the vessels remained occluded throughout the course of the experiment. By comparison, animals receiving anisoylated plasminogen streptokinase activator complex (APSAC) lysed the thrombus within 26 min of infusion. Five min after completing fibrolase administration and in one group of the two groups administered APSAC, a fibrinogen receptor antagonist, 7E3 (0.8 mg/kg) was administered intravenously to inhibit platelet aggregation and prevent reocclusion of the thrombolytic-treated arteries. After administration of 7E3, four of five carotid arteries in the dogs treated with fibrolase remained open for the remainder of the experiment, and six of the six arteries treated with APSAC. It is possible that the one dog whose artery did not remain open after fibrolase and 7E3 treatment was treated with saline instead of 7E3, but this was not firmly established. The average time to complete resolution of thrombus for APSAC was 26 min while for recombinant fibrolase the time was 6 min. These studies firmly established that fibrolase is an active lytic agent and that it rapidly lyses a carotid arterial thrombus without any evidence of hemorrhage or compromise of the hemodynamic system. In summary, we concluded that in combination with antiplatelet therapy, fibrolase offers a unique mechanism for clot dissolution that may provide an alternative to plasminogen activator-based thrombolysis and could have significant potential for clinical application. One interesting sidelight to this study was a problem that was experienced with the recombinant version of the enzyme produced by Chiron under a subcontract from Marion Laboratories, Kansas City, MO (later known as Marion Merrell Dow). After trying several different expression methods, Chiron established a yeast expression system and produced the recombinant version of the protein and supplied it to Dr. Lucchesi. Unfortunately, both dogs that were treated with the recombinant fibrolase died shortly after treatment, leading to the suspicion that the recombinant protein was contaminated with endotoxin. Based on these suspicions, Dr. Markland contacted Dr. Pablo Valenzuela at Chiron, who at the time was the Vice President of Research, and obtained more of the recombinant protein that was passed through a detox column to remove any possible endotoxin contamination. Dr. Markland then personally took the purified recombinant enzyme back to the University of Michigan. This time the enzyme worked, so it appeared that the batch of recombinant fibrolase used in the original study was contaminated. The results of this study were published jointly with Dr. Lucchesi's laboratory in 1994

(Markland et al., 1994). But, the damage had already been done. On the basis of the failed test in the two dogs and other Company related issues, Marion Laboratories decided to pull the plug on the project. Funding for the project was halted and the research project lay fallow for several years. However, Dr. Chris Toombs, Research Scientist at Amgen, read the paper describing the thrombolytic activity of recombinant fibrolase (Markland et al., 1994) and contacted Dr. Markland about possible interest by Amgen in the clinical potential of the enzyme. At that point a Material Transfer Agreement (MTA) and then a Research Agreement between Amgen and the University of Southern California (USC) was put into place to develop the clinical potential of fibrolase. This agreement was in place from 1996 until 2001. Table 24.1 summarizes timelines for the evolution of events involved in the long and winding road to clinical trials.

Table 24.1 Chronology of fibrolase-alfimeprase evolution to clinical trials

- 1982 – fibrolase first identified in the Markland laboratory using molecular sieve chromatography of southern copperhead venom
- 1984–1987 – research grant from NIH to Markland lab to study venom fibrinolytic enzymes
- 1986 – US Patent issued to the Markland lab: Fibrinolytic Enzyme from snake venom (direct acting zinc metalloproteinase)
- 1986–1990 – Cortech, Inc., Boulder, CO, and Marion Laboratories, Inc., Kansas City, MO (Marion Merrell Dow, MMD) support research in the Markland laboratory
- 1986 – MMD signs a subcontract with Chiron to produce r-fibrolase in yeast or bacteria (endotoxin contamination)
- 1991 – publication demonstrating action of fibrolase on a- and β-chains of fibrin and inhibition by chelators and a2M
- 1992 – publish amino acid sequence of fibrolase (Chiron and Markland laboratory)
- 1994 – 1st paper published on r-fibrolase thrombolytic activity in vivo; dog carotid arterial thrombosis model collaboration between Drs. Markland at USC and Lucchesi at Univ. Michigan
- 1996 – MTA between USC and Amgen to provide purified venom-derived fibrolase to Amgen.
- 1996–2001 – research agreement between USC and Amgen. Amgen produces an altered recombinant form of fibrolase in yeast and renames it alfimeprase
- 2002 – Nuvelo (then know as Hyseq) obtains rights to alfimeprase from Amgen and initiates clinical trials (20 years after protein is discovered). U.S. FDA grants orphan drug status to alfimeprase for PAO
- 2004 – Amgen licenses alfimeprase to Nuvello with future milestone payments due to Amgen
- 2005 – European Commission grants orphan drug status to alfimeprase for treatment of PAO
- 2006 – FDA grants Nuvelo fast track designation for alfimeprase for the treatment of PAO in NAPA-3 Phase III trial (01/2006)
- 2006 – Bayer Healthcare signs $385 M deal for worldwide (non-USA) rights to alfimeprase with Nuvelo, $50 M upfront based on Phase II PAO results (01/2006)
- 2006 – Alfimeprase fails Phase III clinical trials in PAO and catheter occlusion (CO) (11/2006)
- 2006 – Nuvello initiates Phase II study in stroke prevention: Catheter directed Alfimeprase for Restoration of Neurologic function and Rapid Opening of arteries in Stroke (CARNEROS-2) (12/2007)
- 2007 – Bayer pulls out of agreement with Nuvello; Nuvello reinitiates Phase II study in stroke prevention (06/2007)
- 2008 – Nuvelo abandons development of alfimeprase after it fails to meet endpoints in Phase III PAO and CO trials and low enrollment in the Phase II stroke prevention trial (03/2008)

Fibrolase – Alfimeprase Connection

Dr. Toombs led a group of scientists at Amgen that ultimately produced a recombinant version of fibrolase that was truncated by two amino acids at the amino-terminus and was ultimately called alfimeprase, although an interim terminology of NAT (natural acting thrombolytic) was also used. The truncation was necessitated by the presence of several isoforms of the enzyme encountered by Amgen, and previously experienced by the Markland laboratory with the natural enzyme (Loayza et al., 1994; Markland et al., 1993). Amgen observed the isoforms during expression of fibrolase in an *E. coli* system. Further, in the *E. coli* system a significant percentage of fibrolase was retained in inclusion bodies and was inactive resulting in a very low yield. The truncation of fibrolase into alfimeprase solved the problem of isoforms; however, it did not address the difficulties encountered with recombinant production in *E. coli.* Thus, an alternative expression system was investigated. Since the difficulties involved inactive protein and low yield, a eukaryotic host (a yeast system) was examined for its ability to produce active alfimeprase. The yeast *Pichia pastoris* was chosen as it has been widely used and is an effective host for heterologous expression of recombinant proteins (Swenson et al., 2004). In the *P. pastoris* expression system, the synthetic gene for alfimeprase is incorporated into the genomic DNA of an untransformed yeast strain. A plasmid is used, which encodes alfimeprase DNA and the protein is expressed under the control of the alcohol oxidase 1 promoter, which is tightly regulated and induced by the presence of methanol. Methanol not only induces expression of the target protein, but it also serves as the sole carbon source for this strain of yeast. The yeast cells are cultured in an induction medium that contains methanol; alfimeprase is secreted into the media, where it is purified and formulated into buffer containing Zn^{2+} at physiologic pH.

While identical in enzymatic activity to fibrolase, alfimeprase contains 201 amino acids with an N-terminal sequence of SFPQR- as compared to fibrolase, which contains 203 amino acids with an N-terminal sequence that begins with EQRFPQR-. The CAS Registry Service has assigned registry number 259074-76-5 with the index name 3-203 fibrolase [3-serine] (*Agkistrodon contortrix contortrix*, recombinant). X-ray crystallographic data and modeling of the structure of fibrolase indicated that the amino-terminus of fibrolase was free to move about in three-dimensional space presumably causing instability of the protein (Bolger et al., 2001). The two amino acid truncation and substitution of Ser for Arg at the new amino-terminus led to a more stable protein for long-term storage as well as eliminating the isoform problem, which originated because of sequence variations at the amino-terminus (Loayza et al., 1994). The truncation at the amino-terminus solved this problem and resulted in a recombinant protein that in all other respects was identical to fibrolase, but more stable on storage.

During studies on the lytic activity of fibrolase in a guinea pig model, Dr. Toombs, at Amgen, noted that there was transient hypotension following intra-arterial administration of the enzyme; however, this could be prevented by treatment with a bradykinin antagonist. In order to identify the mechanism involved in the fibrolase-mediated hypotension, the catalytic activity of fibrolase against proteins in

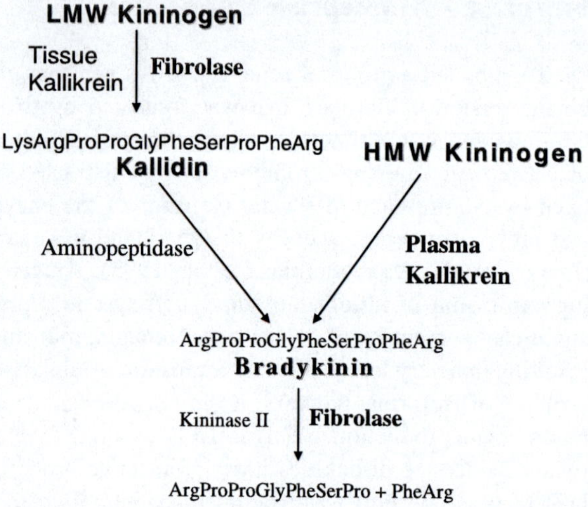

Fig. 24.1 Effect of fibrolase on the bradykinin metabolism pathway. Fibrolase is involved in both the production and degradation of the hypotensive peptide bradykinin. This results in a transient drop in blood pressure following systemic administration of fibrolase at concentrations exceeding that of α2-macroglobulin. LMW, low molecular weight; HMW, high molecular weight

the bradykinin synthetic pathway were assessed. Two biochemical routes contribute to the synthesis of bradykinin (Fig. 24.1) (Burch et al., 1990). Plasma kallikrein directly cleaves high molecular weight kininogen (HMWK) generating bradykinin. By comparison, tissue kallikrein can cleave low molecular weight kininogen (LMWK) to form kallidin a decapeptide. The amino-terminal lysine of kallidin is then removed by a plasma aminopeptidase to form bradykinin. Bradykinin is subsequently cleaved to the inactive heptapeptide Arg-Pro-Pro-Gly-Phe-Ser-Pro by kininase II. Several proteins and peptides of the bradykinin pathway were examined for cleavage by fibrolase. Fibrolase promotes bradykinin formation by cleavage of LMWK with the formation of kallidin. Kallidin is a relatively poor substrate for fibrolase. However, fibrolase does cleave bradykinin between residues Pro7-Phe8 to form the inactive heptapeptide Arg-Pro-Pro-Gly-Phe-Ser-Pro, as confirmed by mass spectrometry analysis. The transient nature of the production and subsequent degradation of bradykinin mimics findings observed with blood pressure alterations in animals treated with fibrolase. Thus, fibrolase promotes bradykinin generation through the LMWK pathway (Fig. 24.1), but it has no effect on HMWK.

Alfimeprase In Vitro Studies

With the recombinant protein available, Dr. Toombs led the preclinical studies of alfimeprase at Amgen. He initiated in vitro studies to characterize the inhibitory activity of α2M on alfimeprase and initiated in vivo studies to characterize thrombolytic activity of the enzyme in a number of different animal models. For studies with α2M, Toombs showed that in the presence of 3-fold molar excess α2M, alfimeprase forms a complex with the inhibitor; the interaction is extremely rapid. Using SDS-PAGE separation and Western blotting with an antibody to alfimeprase, he demonstrated that complex formation begins within 5 sec and is complete by 1.5 min (Deitcher et al., 2006). Dithiothreitol (DTT) was added at different times to completely and instantaneously stop the reaction. As with fibrolase, the interaction between α2M and alfimeprase involves covalent bond cleavage with 1:1 stoichiometry between enzyme and inhibitor. The capacity of serum to bind and neutralize alfimeprase was determined based on the concentration of α2M in serum (in humans ~100–300 mg/dL). The binding capacity of alfimeprase for α2M in human serum was determined and was initially estimated to be 40–50 μg alfimeprase per mL of human serum (Swenson et al., 2004). The effect of alfimeprase on human plasma fibrinogen was examined in a study carried out at Amgen. Plasma was collected from 20 human donors and fibrinogen quantified in all samples prior to the addition of alfimeprase; fibrinogen levels were measured at 30, 60, and 120 min following addition of alfimeprase to the plasma. Plasma fibrinogen was not affected by incubation with alfimeprase at a concentration of 10 μg/mL since the alfimeprase was completely inhibited by α2M in the plasma samples. However, incubation of plasma with alfimeprase at 100 μg/mL resulted in complete degradation of fibrinogen in all plasma samples within 30 min, suggesting that at this alfimeprase concentration the capacity for α2M to bind and inactivate alfimeprase had been exceeded. These studies point out the importance of not over-titrating α2M by alfimeprase during clinical trials since the presence of an active enzyme in blood could potentially cause serious side effects.

Alfimeprase Animal Model Studies

For the in vivo studies, alfimeprase was investigated at Amgen in a number of animal models including: a rat acute carotid artery thrombosis model, two piglet acute carotid artery thrombosis models where the thrombus was aged for 30 min, a dog acute/subacute carotid arterial thrombosis model where the thrombus was aged for 30 min or 24 h, and a baboon model involving acute thrombosis of Dacron grafts in exteriorized arteriovenous shunts (Swenson et al., 2004). In all of these models alfimeprase was delivered locally to avoid the inhibitory effect of α2M, which would have inactivated the enzyme had it been administered intravenously. In the rat model of acute carotid thrombosis, alfimeprase (2 mg total) was compared to urokinase (UK, 250 U/min) and there was 71% incidence of clot lysis (10/14) in 6.3 min in the alfimeprase group vs. 87% lysis (13/15) in 33.5 min for the UK group (Fig. 24.2). Both piglet models involved acute carotid thrombosis with the thrombus aged for 30 min. In one piglet model alfimeprase (5 mg total) was compared to tissue plasminogen activator (t-PA, 2 mg/kg), and there was 100% incidence (11/11) of clot lysis in 4.4 min for alfimeprase as compared to 70% incidence (7/10) of lysis in 17.8 min for t-PA. Importantly, the average blood loss in this study was 1.7 mL in the alfimeprase group, considerably lower than the 17.1 mL average blood loss in the t-PA group. In the other piglet model alfimeprase (3 and 6 mg total) was compared to UK (infused at 500 U/min and 2000 U/Min). There was 83% incidence of clot lysis (10/12) and 91% (10/11) lysis in 7.1 and 10.0 min for the 3 and 6 mg groups, respectively. For UK there was 17% (2/12) and 25% (3/12) incidence of clot lysis in 39.0 and 28.0 min in the 500 and 2,000 U/min groups (Fig. 24.2). In the canine model with the thrombus aged for 30 min alfimeprase (2 mg/kg) was compared to UK (4,500 U/min): alfimeprase resulted in 100% clot lysis (7/7) in

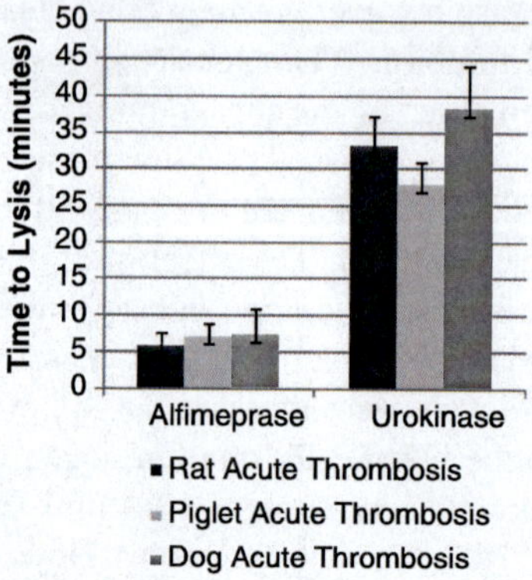

Fig. 24.2 Time to clot lysis for alfimeprase vs. urokinase in animal models of carotid arterial thrombosis. In each study, the administration of alfimeprase produced more rapid clot lysis than a reference dosage of urokinase. Rat study: alfimeprase 2 mg and urokinase 250 U/min × 60 min; piglet study: alfimeprase 3 mg and urokinase 2000 U/min × 60 min; dog study: alfimeprase 2 mg and urokinase 4,500 U/min. Data are expressed as mean ± standard error. Abbreviations: ALF – Alfimeprase; UK – Urokinase. Data reconstructed from (Deitcher et al., 2006)

7.5 min vs. UK with 83% incidence (5/6) in 38.8 min (Fig. 24.2). In the canine model with 24 h aged clots, alfimeprase at 3.8 mg/kg was compared to UK (13,500 U/min). There was 75% clot lysis with alfimeprase (6/8) in 72 min as compared to 67% lysis in 120 min with UK. Finally, in the baboon model of exteriorized arteriovenous shunts, alfimeprase (1 or 10 mg/mL was compared to UK (5,000 or 50,000 U/mL). The agents were infused into the graft segment at 1 μL/min until flow was restored. With 1 mg/mL alfimeprase flow was restored in an average of 15.5 min while with 10 mg/mL flow was restored in 8.7 min. With 5,000 U/mL UK flow was restored in an average of 28.3 min and with 50,000 U/mL flow was restored in 11.7 min. Further, it was noted in animals treated with UK there was a decrease (14–17%) in fibrinogen and an increase (79–176%) in fibrin D-dimer. By comparison there were minimal changes in these parameters (±10% of baseline) in alfimeprase treated animals (Swenson et al., 2004) (Table 24.2).

A porcine model of peripheral arterial occlusion (PAO) was also examined at Amgen; in this model the carotid artery of adult pigs (~20 cm long) is thrombosed along the entire length by balloon injury plus thrombin and stasis. The animals are allowed to recover for 4 days to form a stabilized thrombus. Both the size and age of the thrombus in this model of peripheral arterial occlusion (PAO) are close approximations of the size and duration of ischemic symptoms reported in the largest clinical trial of thrombolysis in PAO, the TOPAS study (Ouriel et al., 1996). In the porcine model a multiple side-hole catheter is advanced for drug delivery into the thrombus. With a dose of 30 mg alfimeprase there is close to complete thrombus resolution and flow restoration within 30 min. By comparison, UK 250,000 U bolus plus 4,000 U/min infusion did not give comparable results despite continuing infusion for up to 4 h.

Table 24.2 Key findings from selected animal pharmacology studies[1,2]

Species	Model	Design	Key findings
Human	In vitro clot lysis	Performed human whole blood clots in tubes or mounted in PTFE graft	Rate of clot lysis is directly proportional to the quantity of alfimeprase added or infused into the graft segment.
Rat	Acute carotid thrombosis	Alfimeprase (2 mg total) vs. UK (250 U/min)	71% incidence (10/14) of clot lysis in 6.3 min in alfimeprase group vs. 87% (13/15) in 33.5 min in UK group.
Piglet	Acute carotid thrombosis; thrombus aged for 30 min	Alfimeprase (5 mg total) vs. tPA (2 mg/kg)	100% incidence (11/11) of clot lysis in 4.4 min with alfimeprase vs. 70% incidence (7/10) in 17.8 min with tPA.
Piglet	Acute carotid thrombosis; thrombus aged for 30 min	Alfimeprase (3 and 6 mg total) vs. UK (500 U/min and 2,000 U/min)	83% (10/12) and 91% (10/11) incidence of clot lysis in 7.1 and 10.0 min for the 3 and 6 mg groups, respectively. 17% (2/12) and 25% (3/12) incidence of clot lysis in 39.0 and 28.0 min 500 U/min and 2,000 U/min groups, respectively.
Dog	Acute/subacute carotid thrombosis thrombus aged for 30 min or 24 h	30 min: Alfimeprase (2 mg/kg) vs. UK (4,500 U/min) 24 h: Alfimeprase (3.8 mg/kg) vs. UK (13,500 U/min)	In 30 min old clots, alfimeprase resulted in 100% incidence (7/7) of clot lysis in 7.5 min vs. UK with 83% incidence (5/6) in 38.8 min. In 24-h clots, 75% success (6/8) in 72 min with alfimeprase vs. 67% success in 120 min using UK.
Baboon	Acute thrombosis of Dacron® grafts in exteriorized arteriovenous shunts	Alfimeprase (1 or 10 mg/mL) vs. UK (5,000 or 50,000 U/mL) Agents are infused into graft segment at 1 μL/min until flow is restored	With 1 mg/mL alfimeprase, flow restored in average of 15.5 min and with 10 mg/mL, flow restored in 8.7 min. With 5,000 U/mL UK, flow restored in average of 28.3 min and with 50,000 U/mL, flow restored in 11.7 min. A decrease (14–17%) in fibrinogen and an increase (79–176%) in D-dimer was noted in the animals treated with UK, whereas negligible changes in these parameters (± 10% of baseline) occurred in alfimeprase treated animals.

[1] *PTFE* polytetrafluoroethylene (Teflon), *UK* urokinase; *tPA* tissue plasminogen activator
[2] Taken from Swenson et al. (2004) with permission from Bentham Science Publishers

Separately, a study was carried out by Dr. Lucchesi and colleagues at the University of Michigan to compare the effect of alfimeprase and rt-PA on extent of myocardial reperfusion injury (Hong et al., 2006). This study was based on the authors' hypothesis that thrombolysis in the absence of plasmin generation could result in improved myocardial salvage. To test the hypothesis the authors examined the effect of thrombolysis by recombinant t-PA (rt-PA) (0.022 mg/kg, 10% delivered as a loading dose and 90% infused over 60 min by intracoronary administration) vs. alfimeprase-induced thrombolysis (0.5 mg/kg over 1 min by intracoronary administration). A canine model of electrolytic injury to the left circumflex coronary artery (LCX) was employed for this comparison. Both agents induced thrombolysis, with alfimeprase being more rapid with onset of reperfusion in 1.5 ± 0.6 vs. 10.1 ± 2.1 min for rt-PA. However, in the absence of adjuvant anti-platelet therapy, time to reocclusion was much shorter with alfimeprase as compared to rt-PA (3.2 ± 0.5 min for alfimeprase vs. 77.5 ± 31.9 min with rt-PA). In a separate group of animals, the presence of the glycoprotein IIb/IIIa platelet receptor antagonist dramatically prolonged time to reocclusion for both agents: for the rt-PA group time to reocclusion was 142.8 ± 34.5 min and in the alfimeprase group it was, 163.3 ± 27.4 min. Separately the effect of each agent on myocardial infarct size was examined after dogs were exposed to 60 min of LCX occlusion followed by intracoronary agent delivery and then, after 4 h of reperfusion, infarct size was measured. The myocardial infarct size was smaller after alfimeprase $19.9 \pm 3.6\%$ of risk region vs. $32.2 \pm 4.0\%$ after rt-PA and $18.5 \pm 3.3\%$ in the saline control (Hong et al., 2006). A significant advantage for local administration of alfimeprase was observed in this model both in the shorter time to onset of reperfusion as well as smaller size of myocardial infarct as compared to rt-PA. It was felt that the generation of plasmin by rt-PA could be associated with secondary proteolytic effects that lead to larger infarct size (this could be a primary proteolytic effect, or secondary, for example, to activation of the complement cascade). Further, the alfimeprase results were obtained with no evidence of altered hemostasis or bleeding at remote sites. α2 M limits alfimeprase lytic action to the site of application. Additionally, other investigators feel that plasminogen activators activate enzyme systems with pro-inflammatory effects, which contribute to the pathogenesis of ischemia-reperfusion injury. Thus, it was concluded that rt-PA is associated with greater risk of myocardial reperfusion injury than the direct-acting thrombolytic agent alfimeprase (Hong et al., 2006).

Further, an acute ischemia-reperfusion stroke model in rats was carried out at the University of Cincinnati (Lu et al., 2009). This study evaluated the activity of alfimeprase vs. rt-PA following reperfusion of the middle cerebral artery after 5-h occlusion. The blinded study compared 10 min infusion immediately after reperfusion using alfimeprase at 0.03, 0.1 and 0.3 mg/kg vs. rt-PA at 1 mg/kg. Analysis of brain sections of treated animals revealed that at the 0.03 mg/kg dose of alfimeprase the hemorrhagic transformation frequency, neurological deficit and mortality rate were much lower than for rt-PA, but at higher doses of alfimeprase there were no differences vs. rt-PA. Infarction and blood-brain barrier permeability were also evaluated in a separate experiment, which compared control, 0.1 mg/kg alfimeprase and 1 mg/kg rt-PA. There were no significant differences between carrier-infused control, alfimeprase and rt-PA for these indices. This study to assess safety of alfimeprase in an ischemic stroke reperfusion model indicated a similar safety profile to rt-PA (Lu et al., 2009).

In conclusion, the in vivo studies revealed that alfimeprase could be used successfully to lyse clots in rabbit, rat, piglet, pig, canine and baboon models of arterial thrombosis (Table 24.2). In all models alfimeprase was administered locally to avoid α2M inhibition and the enzyme was shown to act across species. Alfimeprase did not increase hemorrhage in a rat ischemic stroke model vs. recombinant t-PA (rt-PA), nor lead to other neurological problems (Lu et al., 2009). Thrombi were effectively and quickly cleared, and alfimeprase did not promote re-thrombosis. Alfimeprase is rapidly inactivated in the general circulation by α2M, thus alfimeprase appears to offer promise as a safe, effective and specific agent for thrombolysis when administered locally at the site of the thrombus (Deitcher and Toombs, 2005).

In the United States a number of patents have been issued for alfimeprase and its predecessor fibrolase (these are separate from the original patent 4,610,879, "Fibrinolytic Enzyme from Snake venom" issued to Drs. Francis S. Markland and Nagendranath K. Reddy on September 9, 1986). These patents are: US Patent Nos. 6,261,820 (a fibrinolytically active metalloproteinase protein sequence as a recombinant protein made in yeast, issued July 17, 2001), 6,440,414 (formulation of a fibrinolytic metalloproteinase with a zinc stabilizer, issued August 27, 2002), 6,455,269 (methods for localized administration of a fibrinolytic metalloproteinase for treating vascular thrombi, issued September 24, 2002), 6,617,145 (nucleic acids encoding a fibrinolytically active metalloproteinase and methods for its production, issued September 9, 2003), 7,033,776 (the use of a fibrinolytic metalloproteinase for treating blood clots in indwelling vascular access devices, issued April 25, 2006), 7,138,114 (a method for making a lyophilized fibrinolytic metalloproteinase and reconstituting it, issued November 21, 2006), 7,195,903 (methods for recombinant expression of a fibrinolytic metalloproteinase and its use for local treatment of thrombosis, issued March 27, 2007), 7,244,426 (a method for making a stabilized pharmaceutical composition comprising a therapeutically effective amount of a fibrinolytic metalloproteinase and lyophilizing and reconstituting it, issued July 17, 2007), 7,311,908 (method for making lyophilized fibrinolytic metalloproteinase and a kit for reconstituting it, issued December 25, 2007), 7,314,728 (methods for localized intravascular administration of a fibrinolytic metalloproteinase for treatment of peripheral arterial occlusions, and for restoring patency to an indwelling vascular access device, issued January 1, 2008), and 7,316,911 (method for treatment of an indwelling catheter occlusion using a fibrinolytic metalloproteinase, issued January 8, 2008).

Clinical Trials

The object of all the in vitro and animal model studies on fibrolase and alfimeprase was to advance the protein into clinical trials. In 2002 Nuvelo, San Carlos, CA (then know as Hyseq), obtained the rights to alfimeprase from Amgen and initiated clinical trials. Twenty years had elapsed from the time the protein was first reported by the Markland laboratory (Bajwa et al., 1982) and the initiation of clinical trials (Table 24.1). A Phase I clinical trial was designed to evaluate the safety profile, pharmacokinetics, and thrombolytic activity of alfimeprase in patients with chronic peripheral arterial occlusion (PAO). PAO or "leg attack" is caused by a blood clot that causes blockage of arterial blood flow to a lower limb. There is an underlying peripheral arterial disease and if untreated can lead to nerve and muscle damage, gangrene and in severe cases amputation and death. Treatment involves rapid restoration of arterial patency and blood flow, as well as limb preservation. PAO affects more than 100,000 people per year in the United States and an equal number in Europe.

The Phase I trial was a multicenter, open-label, single-dose, dose-escalation study, involving 20 patients with worsening symptoms of lower extremity ischemia who were treated with alfimeprase in five escalating dose cohorts (0.025, 0.05, 0.1, 0.3, and 0.5 mg/kg) (Deitcher et al., 2006). The United States FDA granted orphan drug status to alfimeprase for the Nuvelo clinical trial for PAO. Orphan drug status is a category created by the FDA for agents used to treat diseases that occur in less than 200,000 cases or where is no hope of recovering the development costs for the Company, so there is little financial incentive to develop these drugs. By being granted orphan drug status, this gives the manufacturer specific financial incentives to provide the drug. An intra-arterial and sometimes intra-thrombic pulsed infusion delivery modality for alfimeprase was used. The primary endpoint was safety assessed by adverse event rates. Additional safety assessments included α2M and anti-alfimeprase antibodies for as long as 3 months after treatment. Pharmacokinetic parameters were evaluated with use of an assay that measures free and α2M-bound (total) alfimeprase. The study was conducted in 7 U.S. hospitals by Nuvelo and was completed in March of 2003. No patients experienced any adverse hemorrhagic events and there were no deaths. The mean plasminogen and fibrinogen concentrations were not altered substantially by treatment. There were three transient, mild treatment-related adverse events; all were reported in the same patient. The pharmacokinetic profile of alfimeprase suggested that the half-life for total alfimeprase ranges from 11 to 54 min (median, 25 min) in patients with PAO. Serum α2M decreased transiently in a dose-dependent manner between 12 and 24 h after treatment initiation and returned to normal $\sim$14 days after alfimeprase exposure. Alfimeprase doses had been selected to be within the alfimeprase-binding capacity of α2M of 1.71 mg/kg. In conclusion this Phase I trial demonstrated that alfimeprase in doses as high as 0.5 mg/kg (within the binding capacity of α2 M) was generally safe in patients with chronic PAO. There were no bleeding complications or systemic thrombolysis noted. The stable fibrinogen concentrations suggest the activity of alfimeprase is limited to the target thrombus. There were no instances of anaphylaxis and no anti-alfimeprase antibodies three months after alfimeprase administration. None of the serious adverse events were attributed to alfimeprase. Thus, this Phase I clinical trial demonstrated that alfimeprase holds the potential to achieve dissolution of PAO with minimal risk of hemorrhage (Moise and Kashyap, 2008; Ouriel et al., 2005).

The Phase II, NAPA-1 (Novel Arterial Perfusion with Alfimeprase) Trial, which started in July 2004, was a multinational, open label, dose escalation trial in 113 acute PAO patients. The primary objective was safety, and a secondary objective was efficacy. Alfimeprase (0.1, 0.3 or 0.6 mg/kg, 2/3 then 1/3 of dose administered 2 h apart via side-hole catheter placed into the thrombus) was administered in 1 ml pulses at 1 pulse/min. Patients were enrolled within 14 days of symptoms and all patients were over 18 years old. The results indicated that alfimeprase lysed thrombi at a rate of up to 76%, restored arterial flow up to 60% within 4 h of dosing and 52–69% of patients avoided surgical intervention. In terms of Adverse Events (AE) there were no intracerebral hemorrhages or deaths, and only one major and three minor bleeds that could possibly be attributed to alfimeprase (patients received aspirin and heparin). There was transient hypotension at high dose but this could be easily managed and α2 M dropped 40–60% but recovered by day 14 (Ouriel et al., 2005).

The Phase 3 Trials, NAPA-2 starting in April 2005, and NAPA-3 starting in April 2006, were two overlapping, randomized, multinational trials, which compared 0.3 mg/kg alfimeprase to placebo. The intention was to enroll ~600 patients worldwide with symptom onset within 14 days. The primary endpoint was avoidance of open vascular surgery within 30 days of alfimeprase treatment; secondary endpoints were rate of arterial flow restoration at 4 h after drug administration and rate of improvement in ankle brachial index at 30 days (Table 24.3). The AE and severe AE (SAE) were major bleeding, intracranial hemorrhage, peripheral embolic events, all cause mortality and pharmacoeconomics such as length of hospital stay. NAPA-2, was a partial-double blind study with 4:3:1 randomization between intra-thrombus alfimeprase, intrathrombus placebo, and perithrombus placebo. Alfimeprase was administered 2/3 of total dose given initially and then 1/3 dose given 2 h later. By

Table 24.3 NAPA-2 and NAPA-3 trials: Primary and secondary endpoint results[1]

	NAPA-2			NAPA-3 (interim)	
	ALF[2] ($n = 149$) (%)	IT Placebo ($n = 113$) (%)	PT Placebo ($n = 38$) (%)	ALF ($n = 51$) (%)	IT Placebo ($n = 51$) (%)
Rate of 30- day open vascular surgery avoidance	34.9	37.2	18.4	29.4	17.6
Rate of restoration of arterial flow at 4 h	46.3	37.2	15.8	35.3	23.5
Rate of 30- day ABI[3] improvement	24.8	23.0	7.0	11.8	7.8

[1]Data taken from (Han et al., 2010)
[2]*ALF* alfimeprase, *ABI* ankle-brachial index
[3]The Ankle-Brachial Index is the ratio of blood pressure in the lower legs to the blood pressure in the arms. Lower blood pressure in the leg is an indication of blocked arteries (PAO)

comparison, NAPA-3 was a double-blind study with 1:1 randomization between intrathrombus alfimeprase and intrathrombus placebo, using the same drug administration schedule, 2/3 of total dose and then 1/3 dose given 2 h later. In January 2006 alfimeprase was granted fast track designation for the NAPA-3 trial for PAO. Fast track can facilitate development and expedited review of new drugs that demonstrate the potential to address an unmet medical need and are intended for treatment of serious or life-threatening conditions. During the NAPA-2 and -3 studies Dr. Fred Weaver, a vascular surgeon at USC and leader of the alfimeprase clinical trial at the Keck School of Medicine site at USC, and his colleagues, noticed that there was a correlation between the length of the thrombus and the success of lysis by alfimeprase: for short occlusion lengths (<10 cm or from 10 cm to <20 cm) the difference in flow restoration at 4 h between alfimeprase and intrathrombus placebo was virtually nonexistent, whereas for an occlusion length of ≥40 cm the 4 h flow restoration rate was 45.5% for alfimeprase and only 16.7% for placebo. This correlation also held when there was no early decline in circulating α2 M level, where the 4-h flow restoration rate for alfimeprase was 70%, whereas for placebo it was 37%. These results suggested to Weaver and colleagues that when alfimeprase is trapped in longer thrombi and is not released into the general circulation, where it interacts with α2 M causing rapid depletion of α2 M, the enzyme is degrading the thrombus leading to more rapid flow restoration in the longer thrombi (Han et al., 2010). These findings suggested that there may have been some flaws in the side hole catheter delivery mechanism and that alternative delivery methodologies with improved residence in the clot should be explored.

In summary, the Phase III NAPA trials did not meet the primary endpoint and showed no significant difference between the intrathrombus alfimeprase and intrathrombus placebo groups (Han et al., 2010). Interestingly, greater efficacy was observed in longer clots and with a smaller drop in α2 M levels, suggesting that improved drug retention in the clot might improve lysis. AE profile in alfimeprase treated patients differed between NAPA-2 and NAPA-3 with the NAPA-3 trials being more favorable for alfimeprase. In both trials the majority of hypotensive episodes occurred within 15 min of alfimeprase administration without apparent clinical sequelae. The risk of hypotension increases with use of anti-hypertensives, particularly renin-angiotensin antagonists. Hypotension may be related to local bradykinin generation induced by alfimeprase as discussed above (see Fig. 24.1). As far as hemorrhage, in NAPA-2 the majority were due to catheter site bleeding (23.0% intrathrombus alfimeprase vs. 10.8% intrathrombus placebo) and surgical bleeding (6.1% intrathrombus alfimeprase vs. 1.7% intrathrombus placebo). Major hemorrhage was higher with intrathrombus alfimeprase (5.4%) vs. intrathrombus placebo (0.9%). In NAPA-3 the catheter site bleeding was confirmed, but no surgical bleeding or major hemorrhage risk was observed. Finally, interim analysis of NAPA-3 showed favorable efficacy trends, but the sponsor decided the delivery method was not optimal and terminated the study. A further evaluation is warranted to improve and optimize the delivery system for alfimeprase to maximize retention of drug in the thrombus and increase lytic activity in PAO.

The second area of clinical application of alfimeprase was central venous access device (CVAD) occlusion. About 5 million catheters are positioned annually in the United States to deliver a variety of chemotherapies, nutrients, antibiotics and blood products, and up to 25% of them become occluded – thus, there is a big market. Due to the direct fibrin degradation activity of alfimeprase, it was hypothesized that in patients with an occluded CVAD there would be rapid clot dissolution following alfimeprase treatment. A Phase II randomized, double-blind, active-control, multi-center, dose-ranging trial was initiated with patients enrolled between May 2003 and August 2004. Safety and efficacy of one or two instillations of 3 intraluminal doses of alfimeprase (0.3, 1.0, and 3.0 mg) or alteplase 2.0 mg were compared (alteplase is a version of recombinant t-PA developed by Genentech) (Cummings-Winfield and Mushani-Kanji, 2008; Lee, 2002). The ability of alfimeprase to re-establish patency to occluded CVADs in 55 adult patients was investigated in this Phase II trial. The trial demonstrated that alfimeprase can restore CVAD function in up to 60% of treated patients in less than 30 min, with the majority of these successes occurring in 15 min or less when initial treatment is with 3 mg of alfimeprase. This rapid activity was associated with an acceptable and favorable safety profile. Treatment with alfimeprase was well tolerated, with no intracranial hemorrhage or any other major hemorrhagic events, or embolic events reported for any study patient and no pattern of adverse events that suggested a safety concern. In conclusion, this Phase II trial demonstrated that all three alfimeprase doses were more successful than alteplase at 5, 15 and 30 min during the first treatment. Alfimeprase at the 3.0-mg dose resulted in 40, 50, and 60% patency restoration rates at 5, 15 and 30 min, compared with 0, 0, and 23% for alteplase. Alfimeprase at 3.0 mg produced the highest patency rate at 120 min after the first (60%) and second (80%) doses (Table 24.4), and no major hemorrhagic or embolic events were reported (Moll et al., 2006; Reddy, 2006).

Table 24.4 Clinical potency of alfimeprase vs. alteplase in patients with CVAD occlusion[1]

Cumulative catheter clearance rate (%)				
	ALF,[2] 0.3 mg (n = 16)	ALF, 1 mg (n = 16)	ALF 3 mg (n = 10)	Alteplase, 2 mg (n = 13)
First dose				
5 min	13	13	40	0
15 min	13	44	50	0
30 min	25	44	60	23
120 min	38	56	60	46
Second dose				
5 min	38	56	60	54
15 min	38	56	60	62
30 min	38	56	60	62
120 min	44	56	80	62

[1]Data reconstructed from (Moll et al., 2006; Reddy, 2006)
[2] *ALF* alfimeprase

A Phase III trial for alfimeprase in CVAD occlusion was then initiated, and it was called SONOMA-2 (Speedy Opening of Non-functional and Occluded catheters with Mini-dose Alfimeprase). The trial was initiated in September 2006 and was a double blind trial comparing efficacy and safety of 3 mg of alfimeprase vs. placebo and was to include 303 patients with occluded central venous catheters. The primary endpoint was the demonstration of restoration of function to occluded catheters in 15 min. Unfortunately, it was announced in December 2006 that the trial did not meet the end point of restoration of function of occluded central venous catheters in 15 min as predicted. Alfimeprase restored catheter function in 15 min but with a p-value of 0.022, it did not meet the more stringent p-value of <0.00125 required for a single pivotal trial. Also, the trial did not meet established secondary endpoints and enrollment in the new SONOMA-3 trial was temporarily suspended in December 2006. However, SONOMA-3 was reinitiated in August 2007 as an open-label, single-arm trial of alfimeprase alone using a 10 mg dose at 5 mg/ml in up to 100 patients. The primary endpoint was safety; however, efficacy was also evaluated. Alfimeprase restored catheter function in ~50% of patients in 15 min and ~60% at 1 h, better than in the SOMOMA-2 trial, but not good enough. By comparison CathfloTM Activase® (alteplase), the recombinant t-PA, cleared ~80% of occluded catheters in patients by 2–4 h. The endpoint for alfimeprase was to restore catheter function with similar efficacy to CathfloTM Activase®, but in a shorter time. As a result of this trial, along with poor results in NAPA-3, Nuvello abandoned development of ALF in March 2008 (http://www.bizjournals.com/sanjose/stories/2008/03/17/daily14.html?t=printable).

Prior to dropping the alfimeprase project, however, Nuvello had initiated a Phase II clinical trial in acute ischemic stroke (http: www.fiercebiotech.com/node/13455/print). Fast track designation for alfimeprase was granted for this proof of concept study. Stroke is the third leading cause of death in the United States, with about 700,000 cases per year, and a significant cause of long-term disability. Stroke is caused when a blood vessel that is carrying nutrients and oxygen to the brain becomes blocked by a blood clot (ischemic stroke) or ruptures due to some causative factor (hemorrhagic stroke). There is an urgent need for agents that can remove the thrombus in ischemic stroke in patients who present at greater than 3 h after the event. The acute ischemic stroke trial with alfimeprase was called CARNEROS-1 (Catheter directed Alfimeprase for Restoration of NEurologic function and Rapid Opening of arteries in Stroke). The study was initiated in June 2007 and the first patient was treated in December 2007. The trial was a multi-center, open label, dose escalation study starting with doses of 1, 5 and 10 mg of alfimeprase in ~100 patients with acute ischemic stroke. Patients were treated within 3–9 h of onset of stroke. The endpoints to be observed were safety and efficacy in stroke patients treated with intra-arterial, catheter directed, bolus alfimeprase. The primary efficacy endpoint was recanalization of main occlusive lesion within 120 min of alfimeprase treatment; safety was assessed by a lack of symptomatic intracerebral hemorrhage within 24 h of drug administration. Due to lack of enrollment in this trial, Nuvelo decided to terminate this study as well as the failed Phase III trials in CVAD and PAO. Thus, in March 2008, Nuvelo discontinued the clinical development of alfimeprase and the program was shut down (No author, 2008).

Summary – the Good, the Bad, the Ugly

The good – the results of the Phase I trials to evaluate the safety profile, pharmacokinetics, and thrombolytic activity of escalating doses of alfimeprase in patients with chronic peripheral arterial occlusion (PAO) were very promising and there was limited toxicity observed (Ouriel et al., 2005). In the Phase II trials (safety and efficacy) there were a low number of major hemorrhagic events in both PAO (Deitcher et al., 2006) and CVAD occlusion (Deitcher et al., 2006; Moll et al., 2006) trials. Importantly, there was a lack of intracerebral hemorrhage in all trials. Alfimeprase facilitates rapid restoration of arterial patency in <4 h in most PAO patients. In the CVAD occlusion trial, alfimeprase restored patency in < 30 min with no evidence of hypotension. In neither PAO nor CVAD occlusion trials was there evidence of detection of anti-alfimeprase antibodies. These trials indicated that alfimeprase has the potential to be a potent direct-acting fibrinolytic agent with an excellent safety profile. As observed for fibrolase, alfimeprase is rapidly inactivated by α2M by forming a 1:1 complex with this inhibitor; a covalent bond forms between α2M and alfimeprase (Deitcher and Toombs, 2005; Toombs, 2001). Inhibition of alfimeprase by α2M in the general circulation limits fibrinolytic action to the site of the clot. Extensive studies with human blood serum from several hundred volunteers were used to accurately estimate the level of α2M in human blood; this established an upper limit for alfimeprase use in humans. This analysis revealed that the alfimeprase binding capacity is directly proportional to the measured α2 M content. When these data were used to calculate a theoretical dose of alfimeprase on a "mg/kg" basis, the mean estimate for the population was 1.7 mg/kg (Swenson et al., 2004). Although some individuals were theoretically capable of tolerating nearly 4 mg/kg of alfimeprase, some in the population might tolerate dosages of only 0.6 mg/kg. In view of these results, clinical dosages in the first human trial of alfimeprase did not exceed 0.5 mg/kg.

The bad – transient bouts of hypotension were experienced at the highest doses in the PAO trial (Deitcher and Toombs, 2005; Deitcher et al., 2006). However, these cases spontaneously resolved or could be easily managed by supportive care including bradykinin receptor antagonist or nitric oxide synthetase inhibitor (Moise and Kashyap, 2008). Bolus delivery of alfimeprase through intrathrombus side-hole catheter in the PAO trial is a mechanical manipulation and this creates channels resulting in clot disruption but with run-off of alfimeprase and its inactivation in the circulation by α2M. The inability to maintain alfimeprase in the thrombus long enough to lyse the full-length of the clot is most likely related to the delivery modality. There also appears to have been a problem related to the dose and dosing schedule in addition to the delivery problem. The use of a catheter that sprays alfimeprase into the thrombus requires that the catheter penetrate into the clot and deliver the lytic agent into the depths of the thrombus mass. The danger is that the disrupted clot could dislodge and move distally, creating an embolization of the clot that would most likely be due to a mechanical effect of the pulsatile delivery.

The ugly – there were failures to meet endpoints in the Phase III CVAD occlusion trial (2008) and the Phase III PAO trial (Han et al., 2010). There was also poor enrollment in the Phase II CARNEROS-1 (stroke) trial (http://clinicaltrials.gov/ct2/show/NCT00499902). In view of these results, in March 2008 Nuvelo discontinued development of alfimeprase (http://www.fiercebiotech.com/node/20663/print). Is there anything that can be done to salvage this technology in the future? Among the areas that might be considered: the use of adjuvant antiplatelet therapy; adjusting the dosage regimen; drug interactions such as with antiplatelet agents, which may be synergistic in nature; the use of a placebo control for the PAO trials based on the report by Han et al. (2010); the use of magnitude of surgical intervention as opposed to avoidance of open vascular surgery as end point for the PAO trial; and finally, further evaluation is warranted to optimize the mode of delivery and dose for alfimeprase. This final issue is critical to the success of alfimeprase and the ability to maximize retention of the drug at the thrombus site and increase lytic activity. Thus, the trials are over and alfimeprase did not pass the test. Will there be another opportunity? Only time will tell! Nonetheless, there are a number of investigators who would like to see a more effective Phase III PAO trial design to assess thrombolytic efficacy of alfimeprase.

References

Ahmed, N.K., Tennant, K.D., Markland, F.S., Lacz, J.P., 1990. Biochemical characteristics of fibrolase, a fibrinolytic protease from snake venom. Haemostasis 20, 147–154.

Bajwa, S.S., Kirakossian, H., Reddy, K.N., Markland, F.S., 1982. Thrombin-like and fibrinolytic enzymes in the venoms from the Gaboon viper (*Bitis gabonica*), eastern cottonmouth moccasin (*Agkistrodon p. piscivorus*) and southern copperhead (*Agkistrodon c. contortrix*) snakes. Toxicon 20, 427–432.

Bolger, M.B., Swenson, S., Markland, F.S., 2001. Three-dimensional structure of fibrolase, the fibrinolytic enzyme from southern copperhead venom, modeled from the X-ray structure of adamalysin II and atrolysin C. AAPS Pharm. Sci. 3, E16.

Burch, R.M., Farmer, S.G., Steranka, L.R., 1990. Bradykinin receptor antagonists. Med. Res. Rev. 10, 237–269.

Cummings-Winfield, C., Mushani-Kanji, T., 2008. Restoring patency to central venous access devices. Clin. J. Oncol. Nurs. 12, 925–934.

Deitcher, S.R., Funk, W.D., Buchanan, J., Liu, S., Levy, M.D., Toombs, C.F., 2006. Alfimeprase: a novel recombinant direct-acting fibrinolytic. Expert Opin. Biol. Ther. 6, 1361–1369.

Deitcher, S.R., Toombs, C.F., 2005. Non-clinical and clinical characterization of a novel acting thrombolytic: alfimeprase. Pathophysiol. Haemost. Thromb. 34, 215–220.

Didisheim, P., Lewis, J.H., 1956. Fibrinolytic and coagulant activities of certain snake venoms and proteases. Proc. Soc. Exp. Biol. Med. 93, 10–13.

Guan, A.L., Retzios, A.D., Henderson, G.N., Markland, F.S., 1991. Purification and characterization of a fibrinolytic enzyme from venom of the southern copperhead snake (*Agkistrodon contortrix contortrix*). Arch. Biochem. Biophys. 289, 197–207.

Han, S.M., Weaver, F.A., Comerota, A.J., Perler, B.A., Joing, M., 2010. Efficacy and safety of alfimeprase in patients with acute peripheral arterial occlusion (PAO). J. Vasc. Surg. 51, 600–609.

Hong, T.T., Huang, J., Lucchesi, B.R., 2006. Effect of thrombolysis on myocardial injury: recombinant tissue plasminogen activator vs. alfimeprase. Am. J. Physiol. Heart Circ. Physiol. 290, H959–H967.

Kornalik, F., 1966. The influence of snake venoms on fibrinogen conversion and fibrinolysis. Mem. Inst. Butantan Simp. Int. 33, 179.

Lee, E.K., 2002. Alteplase use for prevention of central line occlusion in a preterm infant. Ann. Pharmacother. 36, 272–274.

Loayza, S.L., Trikha, M., Markland, F.S., Riquelme, P., Kuo, J., 1994. Resolution of isoforms of natural and recombinant fibrolase, the fibrinolytic enzyme from *Agkistrodon contortrix contortrix* snake venom, and comparison of their EDTA sensitivities. J. Chromatogr. B. Biomed. Appl. 662, 227–243.

Lu, A., Kurosawa, Y., Luskey, K., Pyne-Geithman, G., Caudell, D., Clark, J., 2009. Hemorrhagic profile of the fibrinolytic alfimeprase after ischemia and reperfusion. Neurol. Res. 31, 209–214.

Markland, F.S., Friedrichs, G.S., Pewitt, S.R., Lucchesi, B.R., 1994. Thrombolytic effects of recombinant fibrolase or APSAC in a canine model of carotid artery thrombosis. Circulation 90, 2448–2456.

Markland, F.S., Morris, S., Deschamps, J.R., Ward, K.B., 1993. Resolution of isoforms of natural and recombinant fibrinolytic snake venom enzyme using high performance capillary electrophoresis. J. Liquid Chromatog. 16, 2189–2201.

Markland, F.S., Reddy, K.N.N., Guan, L.F., 1988. Purification and characterization of a direct-acting fibrinolytic enzyme from southern copperhead venom, in: Pirkle, H., Markland, F.S. (Eds.), Hemostasis and Animal venoms (vol. 7). Marcel Dekker, New York, pp. 173–189.

Moise, M.A., Kashyap, V.S., 2008. Alfimeprase for the treatment of acute peripheral arterial occlusion. Expert Opin. Biol. Ther. 8, 683–689.

Moll, S., Kenyon, P., Bertoli, L., De Maio, J., Homesley, H., Deitcher, S.R., 2006. Phase II trial of alfimeprase, a novel-acting fibrin degradation agent, for occluded central venous access devices. J. Clin. Oncol. 24, 3056–3060.

No author, 2008. Alfimeprase. Drugs R D 9, 185–190.

Ouriel, K., Cynamon, J., Weaver, F.A., Dardik, H., Akers, D., Blebea, J., Gruneiro, L., Toombs, C.F., Wang-Clow, F., Mohler, M., Pena, L., Wan, C.Y., Deitcher, S.R., 2005. A phase I trial of alfimeprase for peripheral arterial thrombolysis. J. Vasc. Interv. Radiol. 16, 1075–1083.

Ouriel, K., Veith, F.J., Sasahara, A.A., 1996. Thrombolysis or peripheral arterial surgery: phase I results. TOPAS Investigators. J. Vasc. Surg. 23, 64–73.

Randolph, A., Chamberlain, S.H., Chu, H.L., Retzios, A.D., Markland, F.S., Masiarz, F.R., 1992. Amino acid sequence of fibrolase, a direct-acting fibrinolytic enzyme from *Agkistrodon contortrix contortrix* venom. Protein Sci. 1, 590–600.

Rau, J.C., Beaulieu, L.M., Huntington, J.A., Church, F.C., 2007. Serpins in thrombosis, hemostasis and fibrinolysis. J. Thromb. Haemost. 5(Suppl 1), 102–115.

Reddy, G.K., 2006. Clinical utility of novel agents in the treatment of central venous catheter occlusion. Support Cancer Ther. 3, 135–139.

Retzios, A.D., Markland, F.S., 1990. HPLC-based two-step purification of fibrinolytic enzymes from the venom of *Agkistrodon contortrix contortrix* and *Agkistrodon piscivorus conanti*. Protein Expr. Purif. 1, 33–39.

Retzios, A.D., Markland, F.S., 1988. A direct-acting fibrinolytic enzyme from the venom of *Agkistrodon contortrix contortrix*: effects on various components of the human blood coagulation and fibrinolysis systems. Thromb. Res. 52, 541–52.

Swenson, S., Toombs, C.F., Pena, L., Johansson, J., Markland, F.S., 2004. Alpha-fibrinogenases. Curr. Drug. Targets Cardiovasc. Haematol. Disord. 4, 417–435.

Toombs, C.F., 2001. Alfimeprase: pharmacology of a novel fibrinolytic metalloproteinase for thrombolysis. Haemostasis 31, 141–147.

Trikha, M., Schmitmeier, S., Markland, F.S., 1994. Purification and characterization of fibrolase isoforms from venom of individual southern copperhead (*Agkistrodon contortrix contortrix*) snakes. Toxicon 32, 1521–1531.

Chapter 25
Streptokinase

Paul E. Bock and Pablo Fuentes-Prior

Abstract Streptokinase (SK) is a plasminogen (Pg) activator secreted by a number of streptococcal strains. Kinetic analysis and equilibrium binding studies using active site fluorescently-labeled Pg and plasmin (Pm) has recently allowed dissection of the complicated mechanism of SK-induced Pg activation. Pg binding is aided by interactions between the COOH-terminal lysine residue of SK and one of the five kringle domains of the Pg zymogen. Compelling evidence indicates that the NH_2-terminal residues of the bacterial activator insert into the Pg activation pocket. Interactions mediated by these residues, which mimic the endogenous NH_2-terminus of the Pm catalytic domain, and stabilization of loops surrounding the active site, force SK-bound Pg to adopt a catalytically competent conformation (Pg*). Thus, SK functions according to the "molecular sexuality mechanism" of cofactor-induced zymogen activation. The SK·Pg* complex binds a second Pg molecule as substrate, and proteolytically converts it into Pm. Because SK has a much higher affinity for the product, Pm, than for substrate Pg, SK·Pm complex rapidly becomes the dominant species, which terminates the initial *triggering* catalytic cycle of Pg activation; SK·Pm then converts the remaining Pg molecules into Pm in a second catalytic cycle (the *bullet* cycle). Current investigations focus on areas of Pg*/Pm-bound SK that interact with the Pg substrate; recent results demonstrate that two consecutive Lys residues of the central SK domain play a key role in this regard. Further work on the mechanism of Pm generation would allow developing drugs that selectively target the SK·Pg*/SK·Pm complexes to treat severe streptococcal infections world-wide.

P.E. Bock (✉)
Department of Pathology, Vanderbilt University School of Medicine, Nashville, TN 37232-2561, USA
e-mail: paul.bock@Vanderbilt.Edu

R.M. Kini et al. (eds.), *Toxins and Hemostasis*, DOI 10.1007/978-90-481-9295-3_25,

History

Streptokinase was discovered indirectly in 1933 by Dr. William Tillett in a simple experiment to test the hypothesis that fibrinogen binding to streptococci would prevent its clotting in recalcified human plasma (Sikri and Bardia, 2007). The plasma clotted both in the absence and presence of streptococci, refuting his hypothesis. However, the test tubes were fortuitously neglected and upon reexamination showed liquification only in those containing the bacteria. Dr. Tillett surmised that the bacteria produced a fibrinolytic agent, which he subsequently characterized further and named fibrinolysin. The factor was renamed streptokinase (SK) in 1945 based on the erroneous, but entirely logical idea that it was an enzyme. The use of SK to dissolve venous clots in humans was described by Tillett in 1955, which ultimately led in the mid 80 s to a large enough clinical study, the GISSI trial (GISSI, 1986, 1987), to demonstrate the utility of intravenous SK for treatment of acute myocardial infarction. Today, recombinant tissue-type plasminogen (Pg) activator, tPA, has replaced SK for treatment of heart attack in the United States, because of a small but significant benefit in mortality rates in the GUSTO trial (GUSTO, 1993) (see also Chapter 22). However, streptokinase is still used as a fibrinolytic agent in developing countries, due primarily to its ~10-fold lower cost.

SK in Streptococcal Diseases

The importance of SK as a key virulence factor in streptococcal infections continues to stimulate research efforts to understand its mechanism in this context. Severe streptococcal diseases, including acute rheumatic fever, rheumatic heart disease, post-infection glomerulonephritis, and invasive infections such as endocarditis, toxic shock-like syndrome, and necrotizing fasciitis conservatively account for 500,000 deaths per year globally, with a prevalence of at least 18.1 million cases and 1.8 million new cases each year (Carapetis et al., 2005). It is likely that the frequency of these diseases will increase as more multi-drug resistant bacterial pathogens emerge. This impending major threat to public health is exacerbated by the lack of new antibiotics in pharmaceutical development. The actual enzymatic complexes, SK•Pg* (Pg*, conformationally activated Pg molecule; see below) and SK•plasmin (Pm), are resistant to the natural Pm inhibitor, α_2-antiplasmin (Cederholm-Williams et al., 1979; Wang et al., 1995; Wiman et al., 1979), and there are no known small compound inhibitors that specifically target these complexes.

The host-defense system responds to bacterial infection by tissue factor-initiated blood coagulation, producing thrombin that converts fibrinogen into a localized protective fibrin (Fbn) barrier to prevent bacterial dissemination. The in vivo role of SK from *Streptococcus pyogenes*, which is specific for human Pg, its virulence cofactor, Pg-binding group A streptococcal M-like protein (PAM), and the protective role of fibrin(ogen) were elegantly demonstrated by Sun, Ginsburg, and coworkers in mice either expressing a human Pg transgene (Sun, 2006; Sun et al., 2004) or in

Fbg-deficient mice (Sun et al., 2009). Mortality and systemic spread of the bacteria was increased in mice expressing human Pg, which was dependent on *S. pyogenes* expression of both PAM and SK. Mice lacking Fbg also showed increased mortality and dissemination of the bacteria. PAM, a major Pg receptor expressed by *S. pyogenes*, is a 70% α-helical, coiled-coil protein with a flexible rod-like shape characteristic of M proteins (Berge and Sjobring, 1993). The receptor is covalently linked through an LPXTG motif near its COOH-terminus to the cell wall and covers the surface of the bacteria, from which it extends by about 500 Å (Phillips et al., 1981). Pg bound to PAM is rapidly activated by SK and the resulting Pm remains tightly bound to PAM, forming proteolytic "chain saws" that degrade protective Fbn and extracellular matrix, allowing the bacteria to spread rapidly through soft tissue (Ringdahl and Sjobring, 2000; Ringdahl et al., 1998; Sun, 2006). The role of PAM in dissemination of streptococcal infections is exemplified by the flesh-eating disease, necrotizing fasciitis. Other M-like proteins lack the Pg binding motifs present in PAM, but bind Fbg, allowing the step-wise assembly of M-protein•Fbg•Pg•SK complexes that similarly generate cell surface-bound Pm (Ben Nasr et al., 1994; Boyle and Lottenberg, 1997; Wang et al., 1995). Fbg and Fbn alone act as cofactors in stimulating Pg activation by SK (Cederholm-Williams and Fennell, 1981; Chibber et al., 1985; Fears et al., 1985; Strickland et al., 1982). The molecular mechanisms through which PAM, Fbg/Fbn, SK, and Pg assemble these complexes and facilitate Pm formation remain poorly understood, but there are areas of current research with the ultimate goal of targeting these complexes with mechanism-based therapeutics.

Biochemical Studies of Pg Activation by SK

Three major aspects of the mechanism of SK-mediated Pm generation deserve particular attention, as they connect with the very foundations of enzyme structure and function. Furthermore, thorough understanding of this mechanism might offer targets for therapeutic intervention and/or for improving current fibrinolytic agents. These aspects are: (a) the generation of an active site in SK-bound Pg, which does not involve the usual proteolytic processing step; (b) the progression from SK•Pg* to SK•Pm catalytic complexes; and (c) the ability of these binary complexes to specifically and efficiently recognize and activate other Pg molecules, an activity not possessed by free Pm (Boxrud et al., 2000; Gonzalez-Gronow et al., 1978; Wohl et al., 1980).

A number of ground-breaking early enzymology and protein chemistry studies provided foundational evidence that non-proteolytic activation of the Pg zymogen by SK was involved in the ultimate generation of Pm (McClintock and Bell, 1971; Reddy and Markus, 1972; Schick and Castellino, 1974). The bacterial factor, although devoid of intrinsic enzymatic activity, was shown to generate an active site in bound Pg previous to its conversion to plasmin. In this regard, it is important to recall that trypsin-like serine proteinases (SP) are synthesized as single-chain, inactive zymogens, and require cleavage of a peptide bond with the typical sequence Arg^{15}-Ile^{16} for catalytic site formation. (Residue numbers for SP domains refer to the standard numbering system based on topological equivalence to chymotrypsinogen). Exceptions to this rule are Pg activators such as tPA and the related vampire bat Pg activator, both of which possess notable intrinsic activity even in the uncleaved form (see Chapters 22 and 23).

Pioneering research by Dr. Francis Castellino (Bajaj and Castellino, 1977; Castellino and McCance, 1997; McCance et al., 1994; Menhart et al., 1995; Urano et al., 1987, 1988; Violand and Castellino, 1976; Violand et al., 1978) and many others (Cockell et al., 1998; Mangel et al., 1990; Markus et al., 1978b; Marshall et al., 1994; Ponting et al., 1992a, 1992; Wiman and Wallen, 1975) established the structural and functional properties of the full-length Pg form, [Glu]Pg, a truncated variant obtained upon limited Pm-cleavage, [Lys]Pg, and the active enzyme, [Lys]Pm. These studies and others revealed in particular the roles of the five kringle domains of Pg/Pm in mediating lysine-binding site (LBS) interactions (Lerch et al., 1980; Markus et al., 1978a; Marti et al., 1997; Thewes et al., 1990; Wu et al., 1991) including those subsequently shown to regulate the SK-Pg activation mechanism (Bock et al., 1996; Boxrud and Bock, 2000, 2004; Boxrud et al., 2004; Lin et al. 1996, 2000). We note that the domains of the [Glu]Pg molecule can adopt diverse arrangements, which range from a compact, closed α-conformation at physiological chloride ion concentrations, a partially extended, β-conformation in the presence of benzamidine, to a fully extended γ-conformation in the absence of chloride ions or at saturating concentrations of the lysine analog, 6-aminohexanoic acid (6-AHA). The fact that these forms differ in the accessibility of LBS complicates binding and kinetic data analysis (Bock et al., 1996; Boxrud and Bock, 2000; Boxrud et al., 2004; Cockell et al., 1998; Marshall et al., 1994; Panizzi et al., 2006; Ponting et al., 1992a). Among early studies, Pg activation by SK was proposed to involve the formation of a stable, activated form of Pg, the "virgin enzyme", that was responsible for proteolytic Pm formation (Summaria et al., 1982). By contrast, generation of Pm was also thought to occur by intramolecular proteolytic cleavage of Pg within the activated SK•Pg* complex (Bajaj and Castellino, 1977; Kosow, 1975).

On the other hand, X-ray crystallography studies of trypsin(ogen) and their complexes with active site-directed inhibitors and dipeptides mimicking the NH_2-terminus of the proteinase catalytic domain, coupled with kinetic analysis led Wolfram Bode and Robert Huber to develop the now accepted, standard mechanism of SP activation (Huber and Bode, 1978). Accordingly, zymogens of serine proteinases possess incompletely formed, non-functional active sites. Cleavage of the Arg^{15}-Ile^{16} activation peptide bond allows the newly formed NH_2-terminus of the catalytic domain to insert into a cavity termed the "Ile^{16} pocket", where it engages in a strong, solvent-protected salt bridge with the carboxylate group of the strictly conserved residue, Asp^{194}. This interaction completes formation of the active site machinery, in particular of the oxyanion hole shaped by the main-chain nitrogen atoms of Gly^{193} and the actual catalytic residue, Ser^{195}, and of the substrate-binding cleft (Huber and Bode, 1978).

Realization that the sequence of the SK NH_2-terminal peptide resembles those of SP catalytic domains (Ile-Ala-Gly-Pro/Tyr instead of Ile-Val-Gly-Gly and Val-Val-Gly-Gly in trypsin and Pm, respectively) led Bode and Huber to postulate that generation of a functional active site in SK-bound Pg resulted from the insertion of this peptide into the Ile^{16} pocket of the zymogen, as a proxy of the amino-terminal peptide from the Pm catalytic domain (the "molecular sexuality" hypothesis (Bode and Huber, 1976)). This brilliant hypothesis, although still lacking direct structural proof, has been substantiated by a number of mutagenesis studies (Boxrud et al., 2001; Gladysheva et al., 2002; Wang et al. 1999, 2000), as well as by the recent resolution of the crystal structure of another bacterial activator, staphylocoagulase, bound to the thrombin zymogen ((Friedrich et al., 2003); see also Chapter 32). This structural work, together with the resolution of the crystal structure of SK bound to the catalytic domain of plasmin ((Wang et al., 1998); see also below) allowed further development of this theory. In particular, these investigations showed that the bacterial cofactors cover large surface areas on their cognate serine proteinases, and interact strongly with loops that surround and shape the active site cleft. Thus, non-proteolytic zymogen activation appears to involve a general stabilization of the active site machinery through both Asp^{194}-mediated interactions and contacts with loops surrounding the active site region; this mechanism might therefore be best described as "cofactor-driven zymogen activation".

A major technical barrier to furthering the understanding of the roles of SK-Pg equilibrium binding interactions in the mechanism of Pg activation was the close coupling of formation of the conformationally activated SK•Pg* complex and the ensuing *proteolytic* generation of Pm. This was addressed by the development of fluorescent Pg/Pm derivatives specifically labeled at the active site with extrinsic fluorescence probes. Highly specific, covalent inactivation of SK•Pg* with a thioester tripeptide chloromethyl ketone, followed by generation of a unique free thiol group at the NH_2-terminus of the inhibitor with NH_2OH, permitted various thiol-reactive probes to be incorporated into the catalytic site of the Pg zymogen. Subsequent dissociation of the SK•labeled-Pg* complexes with NaSCN allowed purification of the active site-labeled Pg analogs (Bock, 1988, 1992a, 1992b, 1993; Bock et al., 1996; Boxrud and Bock, 2000, 2004; Boxrud et al., 2000, 2001, 2004). These Pg/Pm analogs enabled quantitative investigation of SK•Pg*/Pm binding interactions under reversible conditions for the first time. The fluorescent Pg/Pm derivatives were used ultimately as competitive binding probes to quantitate the interactions between the native proteins. The results revealed a remarkably higher affinity of SK for Pm (K_D 12 ± 4 pM) than for both [Glu]- and [Lys]Pg (K_D values 130 ± 76 nM and 10 ± 3 nM, respectively); this essential feature was not considered in previous studies of the SK-Pg mechanism. Thus, binding of SK to the substrates, [Glu]Pg and [Lys]Pg, is ~11,000- and ~800-fold weaker than the affinity for the product, [Lys]Pm. This surprising finding accounted for the wide variation in estimates of affinity in studies using experimental approaches that did not strictly exclude generation of Pm (Conejero-Lara et al., 1998; Nihalani et al., 1997; Reed et al., 1995), and the common assumption that the substrates should bind more tightly than the product, as is the case for many enzymes. It should be noted that these affinities represent binding of Pg/Pm in the catalytic mode to SK, while Pg substrate-mode binding to the SK•Pg* and SK•Pm catalytic complexes is distinctly different (see below).

A related development was the establishment of quantitative steady-state kinetic approaches for investigation of the SK-Pg mechanism based on continuous measurement of progress curves in the presence of a chromogenic substrate. These studies resolved the parabolic progress curves into the initial rapid equilibrium step, representing conformational Pg activation and formation of the SK•Pg* complex, from subsequent Pg binding as a substrate and its proteolytic processing to Pm (Boxrud and Bock, 2004; Boxrud et al., 2004). The conformational activation step shows a hyperbolic increase in rate with increasing SK concentration, whereas proteolytic generation of Pm shows an unusual initial increase in rate with SK concentration up to a maximum, and a decrease approaching zero at high SK concentrations. These SK dependencies represent the hyperbolic saturation of SK•Pg* formation, and the binding and cleavage of Pg to Pm, respectively, where the latter process shows an optimum because Pg acts both as the catalyst and the substrate. At high concentrations of SK, its binding to Pg depletes free Pg required to bind as the substrate, thereby inhibiting Pm formation.

Application of the equilibrium binding and steady-state kinetics approaches allowed a unified mechanism for the SK-induced, coupled conformational and proteolytic activation pathway of Pm generation to be postulated (Boxrud and Bock, 2004); Fig. 25.1). In the first catalytic cycle of the model, rapid and reversible

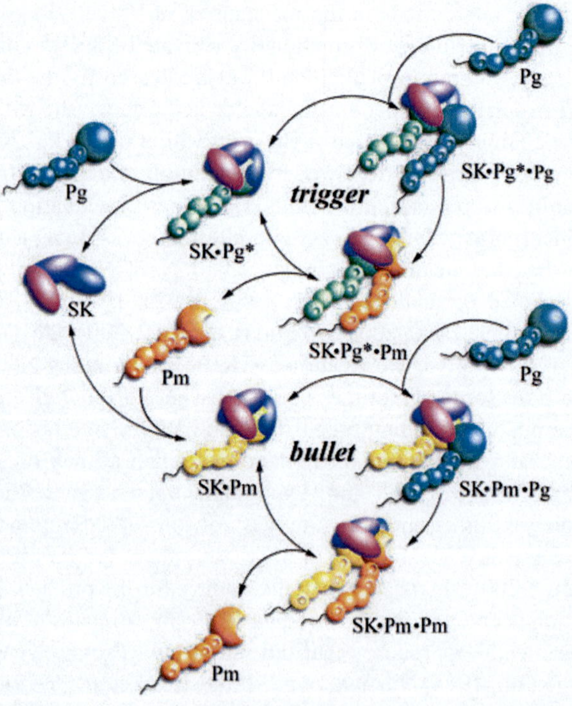

Fig. 25.1 Current view of the mechanism of the coupled Pg conformational and proteolytic activation pathway. SK binding and conformational activation of zymogen Pg in formation of SK•Pg* generates initial amounts of Pm (*trigger*), which is followed by formation of the tighter SK•Pm complex and rapid activation of the remaining Pg (*bullet*; see text for details). Reproduced in modified form Boxrud and Bock (2004). ©2004 The American Society for Biochemistry and Molecular Biology

formation of the activated SK•Pg* complex is followed by binding of a second Pg molecule as a specific substrate, and its intermolecular proteolytic activation to Pm, which is released. Because SK binds Pm with much higher affinity than the zymogen in the catalytic mode (see above), newly generated Pm competes with Pg for SK, forming the second catalytic complex, SK•Pm. This complex catalyzes in turn a second cycle of proteolytic cleavage of Pg to Pm. The first SK•Pg* cycle thus acts as a self-limiting, *triggering* mechanism that generates sufficient Pm to sequester SK in the SK•Pm complex, which shuts off the first catalytic cycle. The tightly bound SK•Pm catalytic complex propagates (*bullet*) the full conversion of the remaining Pg molecules to Pm. The results of the equilibrium binding and kinetic studies embodied in the mechanism are incompatible with previously proposed mechanisms involving slow formation of the SK•Pg* complex, intramolecular cleavage of Pg* within the SK•Pg* complex to Pm, and generation of the so-called virgin enzyme (Bajaj and Castellino, 1977; Kosow, 1975; Summaria et al., 1982).

Structure-Function Studies of SK-Pg Interactions

The advent of the crystal structure of SK bound to the Pm catalytic domain (μPm) spurred a flurry of structure-function studies of SK. SK consists of three β-grasp domains, α, β, and γ from the NH_2-terminus, connected by flexible segments (Wang et al., 1998). In solution, these flexible interdomain linkers allow SK to adopt a variety of conformations (Damaschun et al., 1992). On binding to μPm and presumably, Pg, SK assembles into a structurally defined three-sided, crater-like structure with the Pg/Pm catalytic site at the base of the crater (Fig. 25.2).

Studies with isolated recombinant SK domains, domain fragments, and multiple substitution mutants have yielded divergent, complex, and sometimes conflicting conclusions, in part because the observations were presented before the publication of the crystal structure, and/or the acceptance of the NH_2-terminal insertion mechanism of conformational Pg activation. Among the apparent discrepancies are: (a) conformational activation of Pg by the γ-domain alone; (b) full activity of recombinant SK lacking the NH_2-terminal 16 residues, or in which the NH_2-terminus was blocked; (c) equivalent affinities of the separated domains for Pg; (d) LBS interactions mediated by the α- or β-domains; and (e) mixed functional effects attributed to SK binding, conformational, and/or proteolytic Pg activation. The last problem may be due to experimental designs that do not clearly distinguish between these two events, or are complicated by Pm formation. An unexpected complication was revealed in studies of truncation mutants of the SK α-domain (Bean et al., 2005). Deletion of more than the first 54 residues results in unfolding of the α-domain and normalization of the properties of the further truncated and complete α-domain deletion mutants. Binding studies of the SK mutants and evaluation of the role of LBS interactions from the effect of 6-AHA yielded an unexpected effect on SK mutant binding to fluorescently labeled [Glu]Pg, [Lys]Pg, and Pm. For all Pg/Pm species, the residual affinity of the β-γ-domain fragment is increased by an enhanced contribution to complex stability by LBS-dependent interactions, or free energy

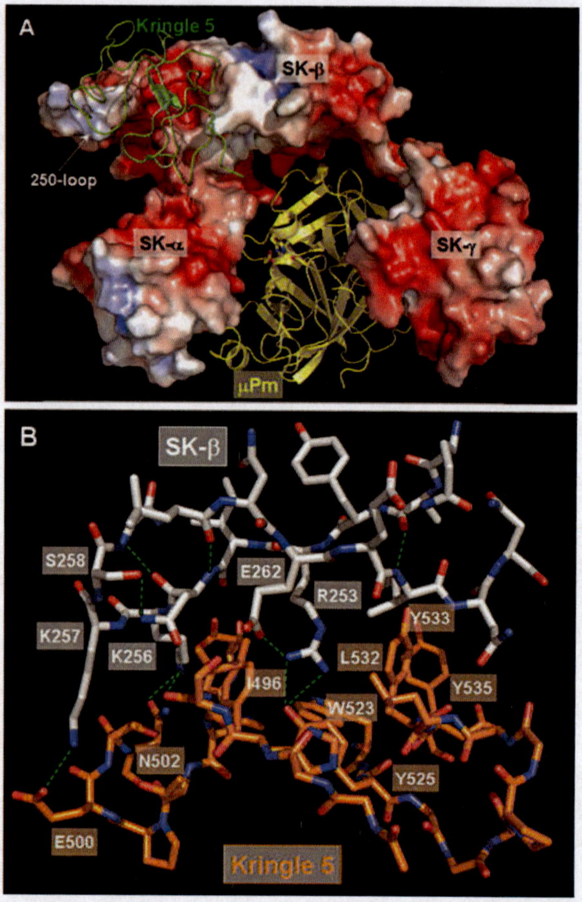

Fig. 25.2 Molecular model of LBS Interactions with kringle 5 in Pg substrate recognition by the SK•Pg*/Pm complexes. (**a**) Crystal structure of SK•μPm binary complex (μPm, *yellow ribbon*; SK, *electrostatic surface potential*), with the predicted location of the K5 domain from an incoming substrate Pg molecule indicated (*green ribbon*). (**b**) Close-up of the putative SK-β:K5 interface, with major side chains from K5 and the SK 250-loop given. Reproduced from Tharp et al. (2009). ©2009 The American Society for Biochemistry and Molecular Biology

coupling between LBS-dependent and –independent interactions. This is thought to reflect the flexibility between SK domains that allows the rearrangement of the SK β-γ-domain fragment into a distinctly different bound conformation accompanying the loss of constraint imposed by interactions of the α-domain. This implies that mutations in SK that affect domain interactions with the Pg/Pm catalytic domain can be manifested as complex functional changes due to the flexibility of SK and the interdependence of its domain interactions.

The fluorescence-based equilibrium binding and steady-state kinetics approaches were applied to identify structural elements of SK responsible for LBS-dependent enhancement of the affinity of SK•Pg*/Pm catalytic complex formation and Pg substrate recognition by these complexes. The SK field neglected to appreciate the potential role of the SK COOH-terminal Lys^{414} residue in LBS interactions, even though it is strictly conserved in all SK sequences, and Pg/Pm kringles 1, 4, and 5 have long been known to interact with COOH-terminal lysine residues, such as those of Pm-degraded Fbn (Bok and Mangel, 1985; Lucas et al., 1983). Kinetics and binding analyses of wild-type SK compared to an SK Lys^{414} deletion mutant (SKΔK414) revealed that binding of Lys^{414} to an unidentified Pg/Pm kringle is responsible for the 14- to 19-fold enhancement of the formation of the SK•[Lys]Pg* and SK•Pm complexes (Panizzi et al., 2006). Interestingly, Pg substrate recognition by these catalytic complexes remains LBS-dependent, indicating that a different structural component of SK is involved.

Substrate Recognition by SK•Pg*/SK•Pm Complexes

A third remarkable element of the mechanism of SK-mediated plasmin generation is the ability of SK•Pg*/Pm complexes to activate free Pg, including zymogen molecules from other species (Collen et al., 1993; Wohl et al., 1978, 1983). In particular, the human SK•Pg*/Pm complexes are capable of activating bovine Pg, which is refractory to activation by SK alone. Unfortunately, information derived from the processing of macromolecular Pm substrates is not helpful to explain this remarkable switch in protease specificity, as the free proteinase does not activate Pg molecules at any physiologically relevant rate (Boxrud et al., 2000; Wohl et al., 1980). It follows that the ability of SK•Pg*/Pm complexes to rapidly convert free Pg into plasmin must be conferred to the SK-bound Pg/Pm moiety by the bacterial cofactor (Fig. 25.2). Two alternative, not mutually exclusive, effects of SK on the Pg*/Pm catalytic domain bound in the enzyme mode could explain their "acquired" ability to recognize and process free plasminogen molecules as substrates: first, allosteric modulation of the immediate active site vicinity to ideally match the Pg activation loop (i.e., the sequence around the $Arg^{15(561)}$–$Val^{16(562)}$ site); and second, generation of an extended binding surface(s) that allows formation of productive SK•Pg*/Pm•$Pg^{substrate}$ complexes. Although some overall stabilization of the Pm active site upon SK binding might be expected, current evidence suggests that direct allosteric modulation of the Pm active site is not involved in proteolytic Pg activation (Boxrud et al., 2000, 2004). Instead, specific interactions between the bacterial cofactor and areas of the Pg substrate molecule distant from the activation site appear to be essential for guiding this scissile peptide bond into the Pg*/Pm active site cleft, as demonstrated directly for SK•Pm (Boxrud et al., 2000). In this regard, SK behaves similarly to intrinsic cofactors of the blood coagulation system, which work primarily by providing exosites for specific substrates (Bock et al., 2007). This hypothesis is also indirectly supported by the observation of a ternary staphylokinase (STA)•μPm^{enzyme}•$\mu Pm^{substrate}$ crystallographic complex ((Parry et al., 1998); PDB entry 1BUI). In this crystal form, in addition to the expected contacts between the bacterial Pg activator, STA, and an enzymatic μPm moiety, a second μPm molecule interacts with STA in a productive manner. In other words, the intact activation loop of a Pg molecule bound to the SK•μPm complex in this conformation would lie close to the active site residues of the μPm moiety. Bode and coworkers therefore hypothesized that this μPm molecule mimics an incoming Pg substrate of the STA•Pm complex. In addition, LBS-mediated interactions with two consecutive Lys residues within an exposed STA loop contribute to Pg recognition and processing by the STA•Pm complex (Rajamohan et al., 2002).

Inspection of the SK•μPm crystal structure immediately suggests that extended regions from all three α, β and γ domains might be potentially involved in interactions with a Pg substrate molecule (Fig. 25.2). A number of investigations have tried to identify those SK elements that could be important for this substrate (Pg)-presentation activity of the bacterial activator at the level of single amino acid residues, and several reports might seem to implicate all SK domains in this regard (Aneja et al., 2009; Chaudhary et al., 1999; Kim et al., 2000; Wakeham et al., 2002; Yadav et al., 2008). These results, however, must be interpreted with caution, as in most cases it cannot be clearly distinguished between SK-Pg interactions in the catalytic and substrate modes. We have recently provided conclusive evidence that basic residues from an exposed loop in SK β-domain, Arg^{253}, Lys^{256} and Lys^{257}, are essential for the LBS-dependence of Pg substrate recognition by SK•Pg*/Pm complexes (Tharp et al., 2009). Indeed, activation of both human and bovine Pg by the SK•human Pm complex was abolished for the double mutant SK(K256A/K257A), indicating that these residues engage in critical interactions with substrate molecules. Furthermore, we identified kringle 5 within the Pg substrate as the domain involved in LBS interactions with the SK loop. These findings, however, do not exclude that other regions of the bacterial cofactor might interact with the Pg catalytic domain in the ternary SK•Pg*/Pm•$Pg^{substrate}$ complexes, as suggested by the overall positioning of α, β, and γ domains relative to the active site cleft.

Summary and Perspectives

Since the discovery of SK 76 years ago, the understanding of the molecular mechanism of SK-initiated fibrinolysis has obviously advanced substantially, but structural studies have generally lagged behind functional studies in solution. Indeed, the only crystal structure containing full-length SK, the SK•μPm complex, was published 11 years ago (Wang et al., 1998) and the corresponding SK•μPg* structure apparently remains elusive. There is no doubt that the SK•μPm structure was one of the most significant advances in understanding of the SK-Pg mechanism, and has driven and enhanced subsequent structure-function studies. A major advance in functional studies of the SK-Pg mechanism was the evidence that conformational activation of Pg by SK obeyed the molecular sexuality, NH_2-terminal insertion mechanism of non-proteolytic zymogen activation predicted by Bode and Huber in 1976 (Bode and Huber, 1976). Although structural proof is still lacking, direct evidence for conformational activation of prothrombin by staphylocoagulase supports further the consensus conclusion that the molecular sexuality mechanism also applies to SK and Pg, despite the profound structural differences between these two conformational activators (see Chapter 32). New crystal structures would certainly advance the SK-Pg research area further.

Related advances in the experimental tools available for equilibrium binding studies by the use of active site-labeled fluorescent Pg/Pm analogs and kinetic analysis of the mechanism of Pg activation by SK enabled a unified mechanism to be proposed. These approaches also proved valuable in structure-function studies defining the features of SK responsible for LBS-dependence of SK•Pg*/Pm catalytic complex formation and subsequent Pg substrate recognition. The fluorescence approaches have been recently extended to the first rapid-reaction kinetic analysis of the SK-Pm binding pathway, and the role of LBS interactions in its regulation (Verhamme and Bock, 2008). These studies define a three-step mechanism of initial rapid equilibrium encounter-complex formation, during which Lys^{414} is engaged by a kringle LBS, followed by two consecutive, favorable conformational changes that increase the affinity of encounter complex formation ~9,000-fold to achieve the tightly bound (K_D 12 pM) final complex. Identifying binding and conformational steps in the SK-Pm binding pathway approaches the ultimate goal of structure-function studies of assigning specific structural changes to individual molecular events. Information from such studies will help to design mechanism-based inhibitors for combating streptococcal infections.

Although SK was the first thrombolytic drug used to treat heart attack in humans, and undoubtedly saved many lives, it has been largely supplanted by tissue-type Pg activator, but is still reducing morbidity and mortality in developing countries. Life-threatening streptococcal diseases are also more prevalent in developing countries (Carapetis et al., 2005) than in the United States, and no mechanism-targeted therapeutic drugs have been developed to be used adjunctively with antibiotics to fight these diseases. The mechanisms by which PAM, Fbg, and Fbn act as cofactors in facilitating SK activation of Pg to Pm and the associated pathogenesis of invasive streptococcal diseases are still poorly understood. New research into these mechanisms may open avenues for the development of novel therapies. The continuing emergence of antibiotic-resistant pathogenic bacteria and the lack of new antibiotics in development will hopefully spur research aimed at addressing the pressing need for new preventative, diagnostic, and therapeutic approaches.

Acknowledgements Research from the authors' laboratories was supported by National Institutes of Health Grant HL056181 from the Heart, Lung, and Blood Institute to P. E. B. and SAF2007-64140 from Spanish Ministerio de Ciencia e Innovación (MICINN) to P. F.-P.

References

Aneja, R., Datt, M., Singh, B., Kumar, S., Sahni, G., 2009. Identification of a new exosite involved in catalytic turnover by the streptokinase-plasmin activator complex during human plasminogen activation. J. Biol. Chem. 284, 32642–32650.

Bajaj, A.P., Castellino, F.J., 1977. Activation of human plasminogen by equimolar levels of streptokinase. J. Biol. Chem. 252, 492–498.

Bean, R.R., Verhamme, I.M., Bock, P.E., 2005. Role of the streptokinase alpha-domain in the interactions of streptokinase with plasminogen and plasmin. J. Biol. Chem. 280, 7504–7510.

Ben Nasr, A., Wistedt, A., Ringdahl, U., Sjobring, U., 1994. Streptokinase activates plasminogen bound to human group C and G streptococci through M-like proteins. Eur. J. Biochem. 222, 267–276.

Berge, A., Sjobring, U., 1993. PAM, a novel plasminogen-binding protein from *Streptococcus pyogenes*. J. Biol. Chem. 268, 25417–25424.

Bock, P.E., 1988. Active site selective labeling of serine proteases with spectroscopic probes using thioester peptide chloromethyl ketones: demonstration of thrombin labeling using N alpha-[(acetylthio)acetyl]-D-Phe-Pro-Arg-CH2Cl. Biochemistry 27, 6633–6639.

Bock, P.E., 1992a. Active-site-selective labeling of blood coagulation proteinases with fluorescence probes by the use of thioester peptide chloromethyl ketones. I. Specificity of thrombin labeling. J. Biol. Chem. 267, 14963–14973.

Bock, P.E., 1992b. Active-site-selective labeling of blood coagulation proteinases with fluorescence probes by the use of thioester peptide chloromethyl ketones. II. Properties of thrombin derivatives as reporters of prothrombin fragment 2 binding and specificity of the labeling approach for other proteinases. J. Biol. Chem. 267, 14974–14981.

Bock, P.E., 1993. Thioester peptide chloromethyl ketones: reagents for active site-selective labeling of serine proteinases with spectroscopic probes. Methods Enzymol. 222, 478–503.

Bock, P.E., Day, D.E., Verhamme, I.M., Bernardo, M.M., Olson, S.T., Shore, J.D., 1996. Analogs of human plasminogen that are labeled with fluorescence probes at the catalytic site of the zymogen. Preparation, characterization, and interaction with streptokinase. J. Biol. Chem. 271, 1072–1080.

Bock, P.E., Panizzi, P., Verhamme, I.M., 2007. Exosites in the substrate specificity of blood coagulation reactions. J. Thromb. Haemost. 5(Suppl 1), 81–94.

Bode, W., Huber, R., 1976. Induction of the bovine trypsinogen-trypsin transition by peptides sequentially similar to the N-terminus of trypsin. FEBS Lett. 68, 231–236.

Bok, R.A., Mangel, W.F., 1985. Quantitative characterization of the binding of plasminogen to intact fibrin clots, lysine-sepharose, and fibrin cleaved by plasmin. Biochemistry 24, 3279–3286.

Boxrud, P.D., Bock, P.E., 2000. Streptokinase binds preferentially to the extended conformation of plasminogen through lysine binding site and catalytic domain interactions. Biochemistry 39, 13974–13981.

Boxrud, P.D., Bock, P.E., 2004. Coupling of conformational and proteolytic activation in the kinetic mechanism of plasminogen activation by streptokinase. J. Biol. Chem. 279, 36642–36649.

Boxrud, P.D., Fay, W.P., Bock, P.E., 2000. Streptokinase binds to human plasmin with high affinity, perturbs the plasmin active site, and induces expression of a substrate recognition exosite for plasminogen. J. Biol. Chem. 275, 14579–14589.

Boxrud, P.D., Verhamme, I.M., Bock, P.E., 2004. Resolution of conformational activation in the kinetic mechanism of plasminogen activation by streptokinase. J. Biol. Chem. 279, 36633–36641.

Boxrud, P.D., Verhamme, I.M., Fay, W.P., Bock, P.E., 2001. Streptokinase triggers conformational activation of plasminogen through specific interactions of the amino-terminal sequence and stabilizes the active zymogen conformation. J. Biol. Chem. 276, 26084–26089.

Boyle, M.D., Lottenberg, R., 1997. Plasminogen activation by invasive human pathogens. Thromb. Haemost. 77, 1–10.

Carapetis, J.R., Steer, A.C., Mulholland, E.K., Weber, M., 2005. The global burden of group A streptococcal diseases. Lancet Infect. Dis. 5, 685–694.

Castellino, F.J., McCance, S.G., 1997. The kringle domains of human plasminogen. Ciba Found. Symp. 212, 46–60.

Cederholm-Williams, S.A., De Cock, F., Lijnen, H.R., Collen, D., 1979. Kinetics of the reactions between streptokinase, plasmin and alpha 2-antiplasmin. Eur. J. Biochem. 100, 125–132.

Cederholm-Williams, S.A., Fennell, S.J., 1981. Binding of plasmin(ogen) to sepharose bound fibrin(ogen) α-chain. Thromb. Res. 21, 503–506.

Chaudhary, A., Vasudha, S., Rajagopal, K., Komath, S.S., Garg, N., Yadav, M., Mande, S.C., Sahni, G., 1999. Function of the central domain of streptokinase in substrate plasminogen docking and processing revealed by site-directed mutagenesis. Protein Sci. 8, 2791–2805.

Chibber, B.A., Morris, J.P., Castellino, F.J., 1985. Effects of human fibrinogen and its cleavage products on activation of human plasminogen by streptokinase. Biochemistry 24, 3429–3434.

Cockell, C.S., Marshall, J.M., Dawson, K.M., Cederholm-Williams, S.A., Ponting, C.P., 1998. Evidence that the conformation of unliganded human plasminogen is maintained via an intramolecular interaction between the lysine-binding site of kringle 5 and the N-terminal peptide. Biochem. J. 333, 99–105.

Collen, D., Van Hoef, B., Schlott, B., Hartmann, M., Guhrs, K.H., Lijnen, H.R., 1993. Mechanisms of activation of mammalian plasma fibrinolytic systems with streptokinase and with recombinant staphylokinase. Eur. J. Biochem. 216, 307–314.

Conejero-Lara, F., Parrado, J., Azuaga, A.I., Dobson, C.M., Ponting, C.P., 1998. Analysis of the interactions between streptokinase domains and human plasminogen. Protein Sci. 7, 2190–2199.

Damaschun, G., Damaschun, H., Gast, K., Gerlach, D., Misselwitz, R., Welfle, H., Zirwer, D., 1992. Streptokinase is a flexible multi-domain protein. Eur. Biophys. J. 20, 355–361.

Fears, R., Hibbs, M.J., Smith, R.A., 1985. Kinetic studies on the interaction of streptokinase and other plasminogen activators with plasminogen and fibrin. Biochem. J. 229, 555–558.

Friedrich, R., Panizzi, P., Fuentes-Prior, P., Richter, K., Verhamme, I., Anderson, P.J., Kawabata, S., Huber, R., Bode, W., Bock, P.E., 2003. Staphylocoagulase is a prototype for the mechanism of cofactor-induced zymogen activation. Nature 425, 535–539.

GISSI, 1986. Effectiveness of intravenous thrombolytic treatment in acute myocardial infarction. Gruppo Italiano per lo Studio della Streptochinasi nell'Infarto Miocardico (GISSI). Lancet 1, 397–402.

GISSI, 1987. Long-term effects of intravenous thrombolysis in acute myocardial infarction: final report of the GISSI study. Gruppo Italiano per lo Studio della Streptochi-nasi nell'Infarto Miocardico (GISSI). Lancet 2, 871–874.

Gladysheva, I.P., Sazonova, I.Y., Chowdhry, S.A., Liu, L., Turner, R.B., Reed, G.L., 2002. Chimerism reveals a role for the streptokinase Beta-domain in nonproteolytic active site formation, substrate, and inhibitor interactions. J. Biol. Chem. 277, 26846–26851.

Gonzalez-Gronow, M., Siefring, G.E., Jr., Castellino, F.J., 1978. Mechanism of activation of human plasminogen by the activator complex, streptokinase-plasmin. J. Biol. Chem. 253, 1090–1094.

GUSTO, 1993. An international randomized trial comparing four thrombolytic strategies for acute myocardial infarction. The GUSTO investigators. N. Engl. J. Med. 329, 673–682.

Huber, R., Bode, W., 1978. Structural basis of the activation and action of trypsin. Acc. Chem. Res. 11, 114–122.

Kim, D.M., Lee, S.J., Kim, I.C., Kim, S.T., Byun, S.M., 2000. Asp41-His48 region of streptokinase is important in binding to a substrate plasminogen. Thromb. Res. 99, 93–98.

Kosow, D.P., 1975. Kinetic mechanism of the activation of human plasminogen by streptokinase. Biochemistry 14, 4459–4465.

Lerch, P.G., Rickli, E.E., Lergier, W., Gillessen, D., 1980. Localization of individual lysine-binding regions in human plasminogen and investigations on their complex-forming properties. Eur. J. Biochem. 107, 7–13.

Lin, L.F., Houng, A., Reed, G.L., 2000. Epsilon amino caproic acid inhibits streptokinase-plasminogen activator complex formation and substrate binding through kringle-dependent mechanisms. Biochemistry 39, 4740–4745.

Lin, L.F., Oeun, S., Houng, A., Reed, G.L., 1996. Mutation of lysines in a plasminogen binding region of streptokinase identifies residues important for generating a functional activator complex. Biochemistry 35, 16879–16885.

Lucas, M.A., Fretto, L.J., McKee, P.A., 1983. The binding of human plasminogen to fibrin and fibrinogen. J. Biol. Chem. 258, 4249–4256.

Mangel, W.F., Lin, B.H., Ramakrishnan, V., 1990. Characterization of an extremely large, ligand-induced conformational change in plasminogen. Science 248, 69–73.

Markus, G., DePasquale, J.L., Wissler, F.C., 1978a. Quantitative determination of the binding of ε-aminocaproic acid to native plasminogen. J. Biol. Chem. 253, 727–732.

Markus, G., Evers, J.L., Hobika, G.H., 1978b. Comparison of some properties of native (Glu) and modified (Lys) human plasminogen. J. Biol. Chem. 253, 733–739.

Marshall, J.M., Brown, A.J., Ponting, C.P., 1994. Conformational studies of human plasminogen and plasminogen fragments: evidence for a novel third conformation of plasminogen. Biochemistry 33, 3599–3606.

Marti, D.N., Hu, C.K., An, S.S., von Haller, P., Schaller, J., Llinas, M., 1997. Ligand preferences of kringle 2 and homologous domains of human plasminogen: canvassing weak, intermediate, and high-affinity binding sites by ^{1}H-NMR. Biochemistry 36, 11591–11604.

McCance, S.G., Menhart, N., Castellino, F.J., 1994. Amino acid residues of the kringle-4 and kringle-5 domains of human plasminogen that stabilize their interactions with omega-amino acid ligands. J. Biol. Chem. 269, 32405–32410.

McClintock, D.K., Bell, P.H., 1971. The mechanism of activation of human plasminogen by streptokinase. Biochem. Biophys. Res. Commun. 43, 694–702.

Menhart, N., Hoover, G.J., McCance, S.G., Castellino, F.J., 1995. Roles of individual kringle domains in the functioning of positive and negative effectors of human plasminogen activation. Biochemistry 34, 1482–1488.

Nihalani, D., Raghava, G.P., Sahni, G., 1997. Mapping of the plasminogen binding site of streptokinase with short synthetic peptides. Protein Sci. 6, 1284–1292.

Panizzi, P., Boxrud, P.D., Verhamme, I.M., Bock, P.E., 2006. Binding of the COOH-terminal lysine residue of streptokinase to plasmin(ogen) kringles enhances formation of the streptokinase.plasmin(ogen) catalytic complexes. J. Biol. Chem. 281, 26774–26778.

Parry, M.A., Fernandez-Catalan, C., Bergner, A., Huber, R., Hopfner, K.P., Schlott, B., Guhrs, K.H., Bode, W., 1998. The ternary microplasmin-staphylokinase-microplasmin complex is a proteinase-cofactor-substrate complex in action. Nat. Struct. Biol. 5, 917–923.

Phillips, G.N., Jr., Flicker, P.F., Cohen, C., Manjula, B.N., Fischetti, V.A., 1981. Streptococcal M protein: alpha-helical coiled-coil structure and arrangement on the cell surface. Proc. Natl. Acad. Sci. U.S.A. 78, 4689–4693.

Ponting, C.P., Holland, S.K., Cederholm-Williams, S.A., Marshall, J.M., Brown, A.J., Spraggon, G., Blake, C.C., 1992a. The compact domain conformation of human Glu-plasminogen in solution. Biochim. Biophys. Acta 1159, 155–161.

Ponting, C.P., Marshall, J.M., Cederholm-Williams, S.A., 1992b. Plasminogen: a structural review. Blood Coagul. Fibrinolysis 3, 605–614.

Rajamohan, G., Dahiya, M., Mande, S.C., Dikshit, K.L., 2002. Function of the 90-loop (Thr90-Glu100) region of staphylokinase in plasminogen activation probed through site-directed mutagenesis and loop deletion. Biochem. J. 365, 379–389.

Reddy, K.N., Markus, G., 1972. Mechanism of activation of human plasminogen by streptokinase. Presence of active center in streptokinase-plasminogen complex. J. Biol. Chem. 247, 1683–1691.

Reed, G.L., Lin, L.F., Parhami-Seren, B., Kussie, P., 1995. Identification of a plasminogen binding region in streptokinase that is necessary for the creation of a functional streptokinase-plasminogen activator complex. Biochemistry 34, 10266–10271.

Ringdahl, U., Sjobring, U., 2000. Analysis of plasminogen-binding M proteins of *Streptococcus pyogenes*. Methods 21, 143–150.

Ringdahl, U., Svensson, M., Wistedt, A.C., Renne, T., Kellner, R., Muller-Esterl, W., Sjobring, U., 1998. Molecular co-operation between protein PAM and streptokinase for plasmin acquisition by *Streptococcus pyogenes*. J. Biol. Chem. 273, 6424–6430.

Schick, L.A., Castellino, F.J., 1974. Direct evidence for the generation of an active site in the plasminogen moiety of the streptokinase-human plasminogen activator complex. Biochem. Biophys. Res. Commun. 57, 47–54.

Sikri, N., Bardia, A., 2007. A history of streptokinase use in acute myocardial infarction. Tex. Heart Inst. J. 34, 318–327.

Strickland, D.K., Morris, J.P., Castellino, F.J., 1982. Enhancement of the streptokinase-catalyzed activation of human plasminogen by human fibrinogen and its plasminolysis products. Biochemistry 21, 721–728.

Summaria, L., Wohl, R.C., Boreisha, I.G., Robbins, K.C., 1982. A virgin enzyme derived from human plasminogen. Specific cleavage of the arginyl-560-valyl peptide bond in the diisopropoxyphosphinyl virgin enzyme by plasminogen activators. Biochemistry 21, 2056–2059.

Sun, H., 2006. The interaction between pathogens and the host coagulation system. Physiology (Bethesda) 21, 281–288.

Sun, H., Ringdahl, U., Homeister, J.W., Fay, W.P., Engleberg, N.C., Yang, A.Y., Rozek, L.S., Wang, X., Sjobring, U., Ginsburg, D., 2004. Plasminogen is a critical host pathogenicity factor for group A streptococcal infection. Science 305, 1283–1286.

Sun, H., Wang, X., Degen, J.L., Ginsburg, D., 2009. Reduced thrombin generation increases host susceptibility to group A streptococcal infection. Blood 113, 1358–1364.

Tharp, A.C., Laha, M., Panizzi, P., Thompson, M.W., Fuentes-Prior, P., Bock, P.E., 2009. Plasminogen substrate recognition by the streptokinase-plasminogen catalytic complex is facilitated by Arg253, Lys256, and Lys257 in the streptokinase beta-domain and kringle 5 of the substrate. J. Biol. Chem. 284, 19511–19521.

Thewes, T., Constantine, K., Byeon, I.J., Llinas, M., 1990. Ligand interactions with the kringle 5 domain of plasminogen. A study by ^{1}H NMR spectroscopy. J. Biol. Chem. 265, 3906–3915.

Urano, T., Sator de Serrano, V., Chibber, B.A., Castellino, F.J., 1987. The control of the urokinase-catalyzed activation of human glutamic acid 1-plasminogen by positive and negative effectors. J. Biol. Chem. 262, 15959–15964.

Urano, T., Sator de Serrano, V., Gaffney, P.J., Castellino, F.J., 1988. Effectors of the activation of human [Glu1]plasminogen by human tissue plasminogen activator. Biochemistry 27, 6522–6528.

Verhamme, I.M., Bock, P.E., 2008. Rapid-reaction kinetic characterization of the pathway of streptokinase.plasmin catalytic complex formation. J. Biol. Chem. 283, 26137–26147.

Violand, B.N., Byrne, R., Castellino, F.J., 1978. The effect of alpha-,omega-amino acids on human plasminogen structure and activation. J. Biol. Chem. 253, 5395–5401.

Violand, B.N., Castellino, F.J., 1976. Mechanism of the urokinase-catalyzed activation of human plasminogen. J. Biol. Chem. 251, 3906–3912.

Wakeham, N., Terzyan, S., Zhai, P., Loy, J.A., Tang, J., Zhang, X.C., 2002. Effects of deletion of streptokinase residues 48–59 on plasminogen activation. Protein Eng. 15, 753–761.

Wang, H., Lottenberg, R., Boyle, M.D., 1995. Analysis of the interaction of group A streptococci with fibrinogen, streptokinase and plasminogen. Microb. Pathog. 18, 153–166.

Wang, S., Reed, G.L., Hedstrom, L., 1999. Deletion of Ile1 changes the mechanism of streptokinase: evidence for the molecular sexuality hypothesis. Biochemistry 38, 5232–5240.

Wang, S., Reed, G.L., Hedstrom, L., 2000. Zymogen activation in the streptokinase-plasminogen complex. Ile1 is required for the formation of a functional active site. Eur. J. Biochem. 267, 3994–4001.

Wang, X., Lin, X., Loy, J.A., Tang, J., Zhang, X.C., 1998. Crystal structure of the catalytic domain of human plasmin complexed with streptokinase. Science 281, 1662–1665.

Wiman, B., Lijnen, H.R., Collen, D., 1979. On the specific interaction between the lysine-binding sites in plasmin and complementary sites in α2-antiplasmin and in fibrinogen. Biochim. Biophys. Acta 579, 142–154.

Wiman, B., Wallen, P., 1975. Structural relationship between "glutamic acid" and "lysine" forms of human plasminogen and their interaction with the NH_2-terminal activation peptide as studied by affinity chromatography. Eur. J. Biochem. 50, 489–494.

Wohl, R.C., Sinio, L., Summaria, L., Robbins, K.C., 1983. Comparative activation kinetics of mammalian plasminogens. Biochim. Biophys. Acta 745, 20–31.

Wohl, R.C., Summaria, L., Arzadon, L., Robbins, K.C., 1978. Steady state kinetics of activation of human and bovine plasminogens by streptokinase and its equimolar complexes with various activated forms of human plasminogen. J. Biol. Chem. 253, 1402–1407.

Wohl, R.C., Summaria, L., Robbins, K.C., 1980. Kinetics of activation of human plasminogen by different activator species at pH 7.4 and 37°C. J. Biol. Chem. 255, 2005–2013.

Wu, T.P., Padmanabhan, K., Tulinsky, A., Mulichak, A.M., 1991. The refined structure of the ε-aminocaproic acid complex of human plasminogen kringle 4. Biochemistry 30, 10589–10594.

Yadav, S., Datt, M., Singh, B., Sahni, G., 2008. Role of the 88–97 loop in plasminogen activation by streptokinase probed through site-specific mutagenesis. Biochim. Biophys. Acta 1784, 1310–1318.

Part VI
Procoagulant Proteins

Chapter 26
Activation of Factor X by Snake Venom Proteases

Jüri Siigur and Ene Siigur

Abstract Human coagulation factor X is a serine protease zymogen, which circulates in blood as a two-chain molecule. A variety of factor X activators have been detected in snake venoms. About 15 activators have been isolated from Viperidae, Crotalidac and Elapidae venoms. Viperidae and Crotalidae venom activators are mainly metalloproteases. Only two factor X activators are characterized from Elapidae venoms, both belonging to serine proteases. Most thoroughly investigated snake venom factor X activators are from *Vipera russellii* (now renamed *Daboia russellii*) - RVV-X, and *Vipera lebetina* (now renamed *Macrovipera lebetina*) - VLFXA. RVV-X is a heterotrimeric metalloproteinase with a mammalian ADAM-like heavy chain and two lectin-like light chains. The crystal structure of RVV-X has recently been determined. VLFXA is the first factor X activator that was cloned and sequenced and its primary structure was deduced from the cDNA sequences. Both activators consist of a heavy chain and two C-type lectin–like light chains which are held together by disulfide bonds. Heavy chains of RVV-X and VLFXA contain metalloprotease, disintegrin-like and cysteine-rich domains. All chains of VLFXA and RVV-X are synthesized from different genes. The primary structures of factor X activating snake venom serine proteases are unknown up to now.

Introduction

Snake venoms, particularly those belonging to the Crotalidae and Viperidae families, contain a large variety of proteases that induce alterations in the blood coagulation cascade. The components of these venoms have been found to have potent effects on coagulation through both pro- and anticoagulant mechanisms. These highly specific proteases which cleave limited bond(s) in the blood coagulation factors are usually divided into two groups: (1) serine proteases

---

J. Siigur (✉)
Laboratory of Bioorganic Chemistry, National Institute of Chemical Physics and Biophysics, Akadeemia tee 23, Tallinn 12618, Estonia
e-mail: juri.siigur@kbfi.ee

R.M. Kini et al. (eds.), *Toxins and Hemostasis*, DOI 10.1007/978-90-481-9295-3_26,

(factor V (FV) activators, protein C activators, plasminogen activators, kinin-releasing and thrombin-like enzymes, α-fibrinogenases, β-fibrinogenases, prothrombin activators); (2) metalloproteases which need Ca^{2+} or Zn^{2+} (or both) for their hydrolytic activity and are inhibited by metal chelating agents (factor X (FX) activators, prothrombin activators, α(β)-fibrinogenases). Various FX activating components are widely distributed in the venoms of Viperidae and Crotalidae. FX activators from different venoms have been characterized (reviews Morita, 1998; Siigur and Siigur, 2006; Tans and Rosing, 2001). This chapter describes, in addition, the recent studies about the FX activators from snake venoms.

Characterization of FX

FX is a vitamin K-dependent multidomain protein that holds a pivotal position in blood coagulation as the only known physiological activator of prothrombin. FX deficiency in blood is characterized by a prolonged prothrombin time. In 1955, Duckert et al. found a coagulation abnormality in a woman named Prower, and in 1957, Hougie et al. found a comparable problem in a man named Stuart. The defect was of plasma protein named Stuart-Prower factor (later FX) (reviewed by Owen, 2001). Congenital (inherited) deficiency is a rare autosomal recessive disorder characterised by considerable phenotypic heterogeneity. Acquired deficiencies have been described in association with amyloid, multiple myeloma, and upper respiratory tract infections (Hertzberg, 1994).

FX (Stuart factor) is synthesized in the liver and secreted into blood. It is synthesized as a precursor molecule with a signal peptide and propeptide which are cleaved during posttranslational processing. It is synthesized as a single-chain polypeptide containing the light (17,500 Da) and heavy (45,000 Da) chains connected by Arg-Lys-Arg tripeptide. The single-chain molecule is then converted to the light and heavy chains by cleavage of two (or more) internal peptide bonds. These two chains are linked together by a disulfide bond (Leytus et al., 1984). All vitamin K-dependent serine proteases are composed of a number of different functional domains. For example, FX contains signal peptide (essential for secretion), propeptide, amino-terminal region (Gla-domain) that contains the γ-carboxyglutamic acid residues necessary for Ca^{2+}-binding; two domains that share homology with epidermal growth factor (EGF); activation peptide, catalytic (protease) domain. Vitamin K is responsible for the generation of γ-carboxyglutamic acid residues within the proteins. FX contains two high affinity Ca^{2+}-binding sites, (1) Gla-domain, (2) besides Gla-domain there is a high affinity Ca^{2+}-binding site in the first EGF domain that includes a β-hydroxyaspartic acid residue (Hertzberg, 1994).

Physiological Activation of FX

FX is a vitamin K-dependent serine protease proenzyme (zymogen), which plays a pivotal role in the coagulation process (Fig. 26.1). This simplified scheme is valid

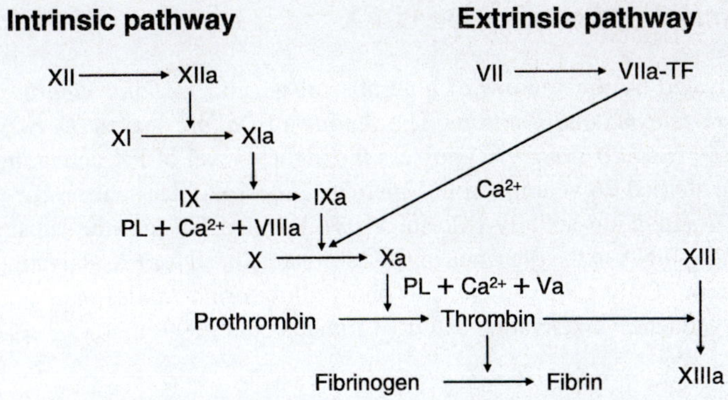

Fig. 26.1 Principal scheme of blood coagulation process. PL-phospholipids; TF-tissue factor

under restricted laboratory conditions and does not accurately represent hemostatic processes that occur in vivo.

Sekiya et al. (1996) proposed a new physiological cascade model. In this model Factor IX is activated by a complex of coagulation initiator Factor VIIa, tissue factor (TF), calcium and magnesium ions. Then, FX and prothrombin are sequentially activated to FXa and thrombin, respectively. The coagulation factors VII/VIIa, IX/IXa, X/Xa, and prothrombin are vitamin K-dependent and contain γ-carboxyglutamic acid (Gla) residues.

In the intrinsic pathway, FX is activated by so-called intrinsic tenase complex comprising factor IXa, factor VIIIa, calcium and phospholipids. In this tenase complex activated factor VIII serves as a cofactor that increases the V_{max} (maximal velocity) of the reaction about 200,000-fold while phospholipid acts to reduce the K_m for the substrate by 5,000-fold (Hertzberg, 1994). In the extrinsic pathway, factor VIIa with TF will activate FX to FXa. Activation results in cleavage of the Arg52-Ile53 bond in the heavy chain of human FX (Arg51-Ile52 in bovine FX) and formation of the active FXaα, that is then converted to FXaβ by autoproteolytic cleavage of Arg-Gly bond in the C-terminal (Fujikawa et al., 1974, 1975). FXaα and FXaβ exhibit similar enzymatic activities, although it has been suggested that they may perform different functions (Pryzdial et al., 1999). The *N*-glycosylation in the activation peptide may be important for the activation while desialylation caused considerable fall in FX activation (Sinha and Wolf, 1993). On the contrary, Baugh and Krishnaswamy (1996) say that the activation does not depend on the propeptide sequence 1–49 and glycosylation. Human and bovine FX have 65% overall sequence identity. The identity is low in the activation peptide (14%) and carboxy-terminal region (5%), though these regions seem to be functionally unimportant (Hertzberg, 1994).

The primary function of FXa is the conversion of prothrombin to thrombin by formation of the prothrombinase complex including FXa, factor Va, phospholipids and Ca^{2+} (on platelet membranes). The binding of FXa to platelets may be by way of factor Va already bound to their membranes.

Nonphysiological Activation of FX

FX is activated by the venoms of a number of genera: *Daboia, Vipera, Cerastes, Echis, Calloselasma and Bothrops*. The venoms of *Daboia* snakes (*D. russellii russellii* and *D. russellii siamensis*) possess the highest level of FX-activating activity among the studied 26 venoms from Viperidae and Crotalidae snakes. Nine venoms had no FX-activating activity (Morita 1998). Purified FX and the substrate Boc-Leu-Gly-Arg–*p*NA in the presence of Ca^{2+} ions were used for FX-activating activity determination. Only *Cerastes cerastes* and *Calloselasma rhodostoma* venoms had Ca^{2+} independent FX activating activity (Yamada et al., 1997).

Daboia russellii (Vipera russellii) and *D. siamensis* Activators

Russell's viper venom (RVV) is the most studied source of FX activating enzymes. Already in 1934, Macfarlane and Barnett found that RVV accelerates blood coagulation, requiring Ca^{2+} ions for maximal activity. Williams and Esnouf (1962) isolated the procoagulant using DEAE-cellulose. They achieved 8-fold purification but the preparation was not homogeneous comprising proteolytic and esterolytic activities. Jackson et al. (1971) used the same method but linear gradient instead of stepwise. They separated esterolytic and clotting activities. Furukawa et al. (1976) purified RVV-X by gel filtration on Sephadex G-50, followed by ion exchange chromatography on DEAE-cellulose and gel filtration on Sephadex G-200. Reduced SDS-PAGE gave four bands: 61,000, 19,800, 16,500 and 13,000 Da. The preparation had weak esterolytic activity. Kisiel et al. (1976) used Sephadex G-150 and QAE-Sephadex A-50 for RVV-X isolation in the presence of 5 and 1 mM benzamidine, respectively. Reduced SDS-PAGE gave one heavy chain band of 59,000 Da and two light chain bands (20,000 and 18,000 Da). Teng et al. (1984) separated *V. russellii* venom into 13 fractions by means of DEAE-Sephadex A-50 column chromatography. Fraction XI containing procoagulant activity was refractionated by gel filtration on Sephacryl S-200. After three gel filtration procedures, a homogeneous preparation was obtained on disc electrophoresis and isoelectric focusing. Two chains (66 and 20 kDa) were separated on reduced SDS-PAGE.

Takeya et al. (1992) isolated RVV-X by modified method of Kisiel et al. (1976) using gel filtration on Sephadex G-75 and ion exchange chromatography on DEAE-Sepharose CL-6B. They sequenced the α- and γ-chains, but not the β-chain of RVV-X. The different names are given for the three chains of RVV-X (Gowda et al., 1994, 1996; Kisiel et al., 1976; Takeya et al., 1992). Here we define the chains as follows: HC-heavy chain (α-chain); LC1- light chain 1 (γ-chain); LC2- light chain 2 (β-chain). The heavy chain (HC, α) consists of 427 amino acid residues including metalloprotease, disintegrin-like and cysteine-rich domains. Light chain 1 (LC1, γ) comprises 123 amino acid residues and shows sequence homology with C-type lectin-like proteins. Takeya et al. (1992) supposed that RVV-X is a two-chain protein consisting of a heavy chain and a heterogeneous light chain. Gowda et al. (1994) disproved this opinion showing that the light chains are two distinct components of the molecule. They also sequenced N-terminals of all chains. The sequences of the heavy chain (α) and LC1 (γ) were identical to those of Takeya et al. (1992); the N-terminal sequence of LC2 (β) was not present within α- or γ-chain. Thus, LC2 must be an individual distinct polypeptide, not a parallel form of LC1 (Gowda et al., 1994).

There were some attempts to purify RVV-X by affinity chromatography. Using a column containing bovine factor X covalently bound to agarose which was equilibrated with Nd(III), Tb(III), Gd(III) or La(III), the RVV-X was purified (Furie and Furie, 1975). Immunoaffinity chromatography with monoclonal antibodies was also used for RVV-X isolation (Durkee et al., 1993; Pukrittayakamee et al., 1983).

Recently Chen et al. (2008) isolated RVV-X from the venom of *Daboia siamensis* by a two-step procedure using Superdex 75 and MonoQ columns with the 3.4% yield. On SDS-PAGE three bands at 62, 21 and 18 kDa under reducing conditions were detected and a single band at 93 kDa under nonreducing conditions, which is in good correlation with the mass obtained by Gowda et al. (1994) on MALDI MS – 92880 Da.

Activation of FX by RVV-X

The activation mechanism of bovine FX by RVV-X does not differ from the physiological mechanism of activation (cleavage of Arg51-Ile52 bond) giving rise to the active serine protease FXaα. Conversion of FXaα to FXaβ is very slow in the case of low concentrations of RVV-X. The activation has an absolute Ca^{2+} requirement. The effect of Ca^{2+} on the activation is mediated by interaction of the cation with the substrate, rather than with the enzyme (Amphlett et al., 1982). Ca^{2+} initiates a conformational change in FX that is a prerequisite for optimal proteolysis by RVV-X. This Ca^{2+}-dependent conformational change requires the Gla domain (Skogen et al., 1983). FX binds Ca^{2+} by at least three so-called "tight" binding sites (K_d 2.2 × 10^{-4} M) associated with the Gla residues, and 25 "weak" binding sites (K_d 2.5 × 10^{-3} M) at other sites in the molecule (Esnouf, 1978). Kosow (1976) showed that Ca^{2+} influenced the V_{max} for the activation reaction without any effect on the K_m. Lindhout et al. (1978) demonstrated that des-γ-carboxy FX is activated by RVV-X only in the presence of Ca^{2+} proving that "weak" Ca^{2+}-binding sites also participate in the activation reaction (although the activation rate is 60 times lower). Skogen et al. (1983) studied the activation of FX and Gla-domainless FX by RVV-X in the presence and absence of Ca^{2+}. Gla-domainless FX was a poor substrate for RVV-X though K_m and k_{cat} could be detected. RVV-X is not able to activate FX molecules bound to phospholipid bilayers (van Dieijen et al., 1981). In terms of its mechanism of action, RVV-X is the simplest activator of FX, requiring only Ca^{2+} as a cofactor and having no feedback requirements (Jesty, 1986).

On the basis of the established three-chain structure Morita (1998) proposed the following mechanism for FX activation: the regulatory subunit (light chains) of RVV-X first recognizes the Gla domain of FX, and then the catalytic subunit (heavy chain) of RVV-X cleaves the peptide bond Arg52-Ile53 of human FX. Lately also the crystal structure of RVV-X has been determined and a FX docking model has been presented (Takeda et al., 2007). The model confirms the catalytic mechanism proposed by Morita (1998).

Role of Glycosylation of RVV-X

RVV-X is a glycoprotein containing over 10% carbohydrate. Furukawa et al. (1976) found 11.1% carbohydrate including 5.1% hexose, 5.0% hexosamine and 1% sialic acid. The preparation characterized by Kisiel et al. (1976) contained 13% carbohydrate: 6% hexose, 5.3% galactoseamine and 1.7% sialic acid.

RVV-X glycosylation was studied by Gowda et al. (1994, 1996) and Chen et al. (2008). RVV-X consists of three disulfide-bound chains – one heavy chain (α-chain) and two light chains: LC1 (γ-chain) and LC2 (β-chain). All chains are glycosylated. Deglycosylation with *N*-glycanase caused decrease in molecular masses on SDS-PAGE under reducing conditions for all three chains showing that all the chains are glycosylated. Deglycosylation lowered the molecular masses of α-chain from 62,000 to 48,000, of β-chain from 21,000 to 17,000, and of γ-chain from 18,000 to 14,000 (reducing SDS-PAGE) (Gowda et al., 1996).

RVV-X contains six *N*-linked oligosaccharides, four in the α-chain and one in each of the β- and γ-chains. The carbohydrate structures resemble rather those in mammalian glycoproteins than in other snake venom glycoproteins. The β-chain contains more galactose and sialic acid than the γ-chain, suggesting that the two chains are glycosylated differentially (Gowda et al., 1994). Carbohydrate compositional and lectin affinity blot analyses indicated that RVV-X contains predominantly complex-type oligosaccharides. Gowda et al. (1994) have shown that the heterogeneity of the light chains is due, at least in part, to differential glycosylation among their subpopulation.

Deglycosylation of the activator essentially affected the activation of FX to FXa, the rate of activation was about 130-fold lower than for the native enzyme (Gowda et al., 1996). Removal of all sialic acid residues, 90% of β-galactosyl residues and 55% of N-acetylglucoseaminyl residues did not affect the activity of RVV-X. The activity was also essentially unchanged (the decrease about 10%) when some of the core mannosyl residues were removed. Circular dichroism spectra of untreated RVV-X and partially deglycosylated RVV-X were almost superimposable. However, complete removal of the carbohydrate moieties significantly changed the CD spectrum of RVV-X (increase in β-sheet content) (Gowda et al., 1996).

The conclusions made from these results suggest that the carbohydrate moieties do not contribute to the binding of FX but are essential for the proteolytic activity of RVV-X. In this case, the binding and catalytic functions are performed by distinct domains. The core sugars seem to be necessary for structural integrity (Gowda et al., 1996).

The *N*-glycosylation profile of *D. siamensis* activator is rather similar to the profile characterized by Gowda et al. (1994) but, as Chen et al. (2008) analysis is based on MALDI MS and advanced MS/MS, they have some important new findings concerning the larger size, multiantennary glycans on the heavy chain. Some of the larger structures were carrying more than one fucose residue; some (or all) of the Fuc residues may be attached to the terminal sequences. From the tryptic peptide analysis, for the heavy chain, four distinct sets of glycopeptides were detected corresponding to peptides carrying the *N*-glycosylated Asn28, Asn69, Asn163 and Asn183 (numbering of mature protein). The glycopeptide corresponding to Asn376 was not found. There are four glycosylated asparagines in the RVV-X heavy chain. For the light chains, a single *N*-glycosylation site in each light chain was identified. Larger glycan structures were found on LC2, which is consistent with previous findings indicating that the mobility of LC2 on SDS-PAGE shifted noticeably after sialidase treatment. The glycoform heterogeneity for LC1 was found to be less complex than that of LC2. Neuraminidase (sialidase) treatment did not remarkably affect the coagulating activity of RVV-X – the decrease was only 5%. The conclusion is that sugars play important role in venom toxicology, not only increasing the solubility and stability of venom glycoproteins but also promoting their target recognition and specific binding in vivo (Chen et al., 2008).

Cloning and Sequencing

RVV-X has been cloned by Chen et al. (2008). The amino acid sequences deduced from the nucleotide sequences were consistent with the published sequence in case of the heavy chain (HC) and the light chain 1 (LC1) (Takeya et al., 1992). The heavy chain contains four additional residues (FSQI) in the C-terminus. The sequence of the LC2 has not been published before. Its predicted mass is 15,983 Da, isoelectric point 5.44, and it has a putative *N*-glycosylation site at Asn59 (Chen et al., 2008). The sequence of mature LC2 is 80.7% identical to the LC2 of VLFXA (Siigur et al., 2004). All three subunits are synthesized from individual genes like in case of VLFXA. The heavy chain of both activators comprises metalloprotease, disintegrin-like and cysteine-rich domains like P-III metalloproteases.

Vipera lebetina Venom FX Activator

FX activator from *Vipera lebetina* venom (VLFXA) was purified by a three-step procedure: gel filtration of crude venom on Sephadex G-100 superfine, ion exchange chromatography on CM-cellulose and TSK DEAE columns. Chromatofocusing of the VLFXA revealed two main peaks with pIs 5.9 and 5.6. (Siigur et al., 2001a). The purified enzyme has a heavy chain of 57.5 kDa and two light chains, 17.4 kDa and 14.5 kDa (reduced SDS-PAGE). Molecular masses of different VLFXA samples detected by MALDI-TOF analyses were in the interval of 85–91 kDa. The sugar content of VLFXA was about 15.5% (8% hexoses, 5% hexoseamines, 2.5% neuraminic acid). VLFXA exists in multiple isoforms (pIs between 5.3 and 6.6); the differences in isoelectric points were due, at least in part, to dissimilarities of the respective sialic acid content of isoforms. No functional differences were discovered between any of isoforms purified from different batches of the venom. The FX activating enzyme is thermolabile. On heating at 70°C for 10 min about 55% of the FX activating activity was lost.

Substrate Specificity

VLFXA hydrolyzed 6–9 amino acid residues containing peptide fragments synthesized according to the physiological cleavage region of human FX and human factor IX. VLFXA cleaved Arg-Ile bond in the nonapeptide NNLTRIVGG (I) and Arg-Val bond in the peptide NDFTRVVGG (II). VLFXA cleaved more effectively FX fragment (I) than factor IX fragment (II). FX nonapeptide fragment was used as substrate for locating the enzyme; the cleavage products were detected by MALDI-TOF mass spectrometry (Siigur et al., 2001b). Hydrolysis of peptide substrates by VLFXA was inhibited with EDTA and *o*-phenanthroline. PMSF had no effect on the activity. These results confirm the metalloprotease nature of VLFXA. The purified VLFXA had no effect on fibrinogen, prothrombin, plasminogen, indicating that it specifically activates FX.

The activator converts the inactive FX to the active form of Xa in the presence of Ca^{2+} ions. The enzyme has optimum activity on FX in the interval of pH 7.5–8.0. VLFXA has no amidolytic activity against FXa substrates S-2337 [Benzoyl-Ile-Glu(Pip)-Gly-Arg-*p*NA-HCl] or S-2222 (Benzoyl-Ile-Glu-Gly-Arg-*p*NA-HCl), while *V. lebetina* venom itself hydrolyzes these substrates (Siigur et al., 2001a).

The catalytic cleavage of human FX by the activators from *V. lebetina* (VLFXA) and *V. russellii* (RVV-X, Sigma) venoms was examined by MALDI-TOF MS. The masses of cleavage products of human FX were 43.1 and 11.6 kDa after treating with VLFXA or RVV-X. On the basis of cleavage of FX nonapeptide fragment (I) and human FX it was concluded that VLFXA hydrolyses human FX at the position Arg52-Ile53. FX can be completely converted to FXa by VLFXA and the specific activity of FXa is the same as that of FXa obtained after activation by RVV-X. It seems that the catalytic mechanism of VLFXA should be similar to that of RVV-X (Morita, 1998).

Structural Features

VLFXA is the first FX activating enzyme whose total structure has been established. The amino acid sequence of RVV-X was published in 1992 without the light chain LC2. The amino acid sequence of VLFXA is deduced from the nucleotide sequence of the cDNAs encoding the three chains of VLFXA. Every chain of VLFXA is encoded by individual cDNA beginning with the 5′-UTR and ending with the 3′-UTR (Siigur et al., 2004). The heavy chain (HC) precursor consists of 2347 basepairs including 5′-UTR (78 bp), signal peptide (54 bp), proenzyme (510 bp), mature metalloprotease (621 bp), spacer (54 bp), disintegrin-like (234 bp) and cysteine-rich (363 bp) domains encoding cDNAs, and 3′-UTR (508 bp). Both light chains – LC1 and LC2 – comprise 5′- and 3′- UTR, propeptide and C-type lectin-like protein encoding domains (Siigur et al., 2004).

The cDNA encoding LC1 comprises 709 bp encoding ORF of 146 amino acids including 23 amino acids of signal peptide and 123 amino acids of LC1. The cDNA encoding LC2 includes 748 bp and encodes an ORF of 158 amino acids including 23 amino acids of propeptide and 135 amino acids of LC2. The parts of sequences are 100% identical to N-terminal and inner sequences determined by LC-MS/MS (Siigur et al., 2004). The nucleotide sequence of the VLFXA LC1 is most similar to the cDNA encoding *D. russellii* factor X activator LC1 (Chen et al., 2008) and *D. r. siamensis* C-type lectin-like protein subunit 2 (DQ060415) – both have 91% identity. The nucleotide sequence of VLFXA light chain 2 is most similar to the cDNA encoding *D. r. siamensis* C-type lectin-like protein subunit 1 (DQ060414) (92% identity) and *D. russellii* factor X activator LC2 (Chen et al., 2008) (91% identity).

The cDNA of the heavy chain encoded a precursor protein of 612 amino acids comprising VLFXA HC of 419 amino acids with calculated molecular mass of 46,881 Da. The VLFXA HC precursor has a "cysteine switch" motif (PKMCGV) in the prodomain. This motif is involved in the activation of snake venom metalloproteases and matrix metalloproteases (Grams et al., 1993). The N-terminal sequence of RVV-X (LVSTSAQFNKIFIELVI. . .) (Takeya et al., 1992) is five amino acids longer than that of VLFXA, and these amino acids exist also in the precursor of VLFXA. It is not clear whether the N-terminal is processed differently or the studied VLFXA has truncated N-terminus. Molecular masses of tryptic fragments of VLFXA HC and the sequences of the internal peptide fragments detected by LC-MS/MS also confirmed the protein sequence deduced from the cDNA. The deduced amino acid sequence of VLFXA HC has a typical Zn^{2+}-chelating sequence H-E-X-X-H-X-X-G-X-X-H-D in the catalytic domain that is conserved in snake venom metalloproteases (Bjarnason and Fox, 1995; Matsui et al., 2000). The disintegrin-like domain contains ECD motif that is characteristic for many P-III type reprolysins (Kishimoto and Takahashi, 2002; Nishida et al., 1995; Silva et al., 2003) and also for ADAM metalloproteases (Cerretti et al., 1999). There are three potential *N*-glycosylation sites in the deduced sequence – Asn259, Asn353 and Asn373 (see Fig. 26.2).

The VLFXA HC precursor showed strong amino acid sequence similarity with *D. russellii* FX activator heavy chain (Chen et al., 2008) – 85% identity and P-III class protease precursors: *Echis ocellatus* metalloprotease (Wagstaff et al., 2006) – 74% and ecarin from *Echis carinatus* (Nishida et al., 1995) – 72% identity. The sequence similarity of mature VLFXA HC was highest with *Vipera russellii* FX activator heavy chain (Takeya et al., 1992) – 83% identity (Fig. 26.2). The sequence analysis demonstrates that VLFXA HC belongs to P-III class of metalloproteases comprising metalloprotease, disintegrin-like and cysteine-rich domains. This is the reason why Fox and Serrano (2008) have transferred RVV-X and VLFXA from PIV class into class P-IIId.

The deduced protein sequence of the VLFXA LC1 has 79% identity with RVV-X LC1 (Chen et al., 2008) (Fig. 26.3). Both LC1 have one putative glycosylation site at

```
Multiple alignment in COBALT

Q7T046.1     1    MMQVLLVTISLAVFPYQGSSIILESGNVNDYEVVYPQKITALPEEAVQQPEQKYEDTMQYEFEVNGEPVVLHLEKNKDLF  80
Q7LZ61.2     1    MMQVLLVTISLAVFPYQGSSIILESGNVNDYEVVYPQKVTALPKGAVQQPEQKYEDTMQYEFEVNGEPVVLHLEKNKILF  80
CAJ01689.1   1    MIQVLLVIICLAVFPYQGCSIILESGNVNDYEVVYPQKVAALPKGAIQQPEQKYEDAIQYEFEVKGEPVVLHLEKNKGLF  80
Q90495.1     1    MIQILLVIICLAVFPYQGCSIILGSGNVNDYEVVYPQKVTALPKGAVQQPEQKYEDAMQYEFEVKGEPVVLHLEKNKELF  80
Q4VM08.1     1    MMQVLLVTISLAVFPYQGSSIILESGNVNDYEVVYPQKVTAMPKGAVKQPEQKYEDAMQYEFKVKGEPVVLLLEKNKDLF  80

Q7T046.1     81   SEDYSETRYSPDGRETTTKPPVQDHCYYHGRIQNDAYSSASISACNGLKGHFKLQGETYLIEPLKIPDSEAHAVYKYENI  160
Q7LZ61.2     81   SEDYSETHYYPDGREITTNPPVEDHCYYHGRIQNDAHSSASISACNGLKGHFKLRGEMYFIEPLKLSNSEAHAVYKYENI  160
CAJ01689.1   81   SEDYSETHYSPDDREITTKPSVEDHCYYHGRVQNDAESTASISACNGLKGHFTLRGETYFIEPLKIPDSEAHAVYKYENI  160
Q90495.1     81   SEDYSETHYSSDDREITTNPSVEDHCYYHGRIQNDAESTASISACNGLKGHFKLRGETYFIEPLKIPDSEAHAVYKYENI  160
Q4VM08.1     81   SEDYSETHYSPDGREITTNPPVEDHCYYHGRIQNDADSSASISACNGLKGHFMLQGETYLIEPLKLPDSEAHAVYKYENV  160

Q7T046.1     161  EKEDEAPKMCGVTQTNWESDEPIKKASQLVATSAKRKFHKT--FIELVIVVDHRVVKKY--DSA--ATNTKIYEIVNTVN  234
Q7LZ61.2     161  EKEDEIPKMCGVTQTNWESDKPIKKASQLVSTSAQ--FNKI--FIELVIIVDHSMAKKC--NST--ATNTKIYEIVNSAN  232
CAJ01689.1   161  EEEDEAPKMCGVKQSNRESDEPIKKASGLIVPSQKRRLDQK--FIELVMVVDHSMVTKYNNDST--AVRTWIYEMVNTVN  236
Q90495.1     161  ENEDEAPKMCGVTQDNWESDEPIKKTLGLIVPPHERKFEKK--FIELVVVVDHSMVTKYNNDST--AIRTWIYEMLNTVN  236
Q4VM08.1     161  EKEDEAPKMCGVTQTNWESDEPIKKASQLNLTPEQRRYLNSPKYIKLVIVADYIMFLKY--GRSLITIRTRIYEIVNILN  238

Q7T046.1     235  EIFIPLNIRLTLIGVEFWCNRDLINVTSSADDTLDSFGEWRGSDLLNRKRHDNAQLFTDMKFDLSTLGITFLDGMCQAYR  314
Q7LZ61.2     233  EIFNPLNIHVTLIGVEFWCDRDLINVTSSADETLNSFGEWRASDLMTRKSHDNALLFTDMRFDLNTLGITFLAGMCQAYR  312
CAJ01689.1   237  EIYLPLNIRVPLVGIVFWSNRDLINVTFTADDTMDSFGEWRASYLLNRKRHDYAQLLTNITLDFDSLGMAFIDGMCKSDR  316
Q90495.1     237  EIYLPFNIRVALVGLEFWCNGDLINVTSTADDTLHSFGEWRASDLLNRKRHDHAQLLTNVTLDHSTLGITFVYGMCKSDR  316
Q4VM08.1     239  VIYRVLNIYIALLGLEIWNNGDKINVLPETKVTLDLFGKWRERDLLNRRKHDNAQLLTDINFNGPTAGLGYVGSMCDPQY  318

Q7T046.1     315  SVGIVQEHGNKNFKTAVIMAHELGHNLGMYHDRKNCICNDSSCIMSAVLSSQPSKLFSNCSNHDYRRYLTTYKPKCILNP  394
Q7LZ61.2     313  SVEIVQEQGNRNFKTAVIMAHELSHNLGMYHDGKNCICNDSSCVMSPVLSDQPSKLFSNCSIHDYQRYLTRYKPKCIFNP  392
CAJ01689.1   317  SVGLIRDDSSTTFRTAVIMAHEMGHSLGMEHDSRSCKCAASPCIMSKALGKQPTKVFSSCSYDDYRMYLAKYKPKCILDP  396
Q90495.1     317  SVELILDYSNITFNMAYIIAHEMGHSLGMLHDTKFCTCGAKPCIMFGKESIPPPKEFSSCSYDQYNKYLLKYNPKCILDP  396
Q4VM08.1     319  SAGIVQDHNKVNFLVALAMAHEMGHNLGMEHDEIHCTCGAKSCIMSGTLSCEASIRFSNCSREEHQKYLINKMPQCILNK  398

Q7T046.1     395  PLRKDIASPPICGNEIWEEGEECDCGSPKDCQNPCCDAATCKLTPGAECGNGLCCEKCKIKTAGTVCRRARDECDVPEHC  474
Q7LZ61.2     393  PLRKDIVSPPVCGNEIWEEGEECDCGSPANCQNPCCDAATCKLKPGAECGNGLCCYQCKIKTAGTVCRRARDECDVPEHC  472
CAJ01689.1   397  PLRKDIASPAVCGNKIWEEGEECDCGSPEDCRNPCCDAETCELFPAAECADGPCCHKCKIRTAGTICRPARDECDVTEHC  476

Q90495.1     397  PLRKDIASPAVCGNEIWEEGEECDCGSPADCRNPCCDAATCKLKPGAECGNGECCDKCKIRKAGTECRPARDDCDVAEHC  476
Q4VM08.1     399  PLKTDIVSPAVCGNYLVELGEDCDCGSPRDCQNPCCNAATCKLTPGSQCADGECCDQCKFRRAGTVCRPANGECDVSDLC  478

Q7T046.1     475  TGQSAECPADGFHANGQPCQNNNGYCYNGDCPIMTKQCISLFGSRATVAEDSCFQENQKGSYYGYCRKE---NGRKIPCA  551
Q7LZ61.2     473  TGQSAECPRDQLQQNGKPCQNNRGYCYNGDCPIMRNQCISLFGSRANVAKDSCFQENLKGSYYGYCRKE---NGRKIPCA  549
CAJ01689.1   477  TGQSAECPRNELQRNGEPCLDKLGYCYNGDCPIMRNQCISLFGSRATVAEDSCFQQNLNGSEHGYCAKE---NGRKIPCA  553
Q90495.1     477  TGQSAECPRNEFQRNGQPCLNNSGYCYNGDCPIMLNQCIALFSPSATVAQDSCFQRNLQGSYYGYCTKEIGYYGKRFPCA  556
Q4VM08.1     479  TGQSAECPTDQFQRNGQPCQNNNGYCYSGTCPIMGKQCISLFGASATVAQDACFQFNSLGNEYGYCRKE---NGRKIPCA  555

                                                                                      Identity
Q7T046.1     552  PQDIKCGRLYCLDNSPGNKNPCKMHYRCRDQHKGMVEPGTKCEDGKVCNNKRQCVDVNTAY---------  612         (100%)
Q7LZ61.2     550  PQDVKCGRLFCLNNSPRNKNPCNMHYSCMDQHKGMVDPGTKCEDGKVCNNKRQCVDVNTAYQSTTGFSQI  619  526/612 (85%)
CAJ01689.1   554  PQDVKCGRLYCLDNSSRKKNPCKMHYLNADQHKGMVEPGTKCEDGKVCIN-RKCVDVKTAYYSTTGFSQI  622  457/614 (74%)
Q90495.1     557  PQDVKCGRLYCLDNSFKKNMRCKNDYSYADENKGIVEPGTKCEDGKVCIN-RKCVDVNTAY---------  616  449/617 (72%)
Q4VM08.1     556  PQDVKCGRLYCFDNLPEHKNPCQIYYTPSDENKGMVDPGTKCGDGKACSSNRQCVDVNTAY---------  616  426/616 (69%)
```

Fig. 26.2 Comparison of the amino acid sequence of the FX activator heavy chain precursor from *Macrovipera lebetina* (VLFXA HC) with those of other snake venom components. The amino acid sequence of VLFXA HC precursor (Accession No. Q7T046) was aligned: Q7LZ61-*Daboia russellii*, RVV-X heavy chain; CAJ01689-*Echis ocellatus*, group III metalloproteinase; Q90495-*E. carinatus*, ecarin precursor; Q4VM08 *Macrovipera lebetina*, VLAIP-A precursor. Gaps are indicated by *dashes*. "Cysteine switch" motif (PKMCGV) is *shaded black*, a typical Zn^{2+}-chelating sequence (H-E-X-X-H-X-X-G-X-X-H-D) is *shaded gray*, ECD motif is *bold*, potential N-glycosylation sites are *underlined*

Asn47. The nearest protein homologs of VLFXA LC2 having 82% identity are FX activator light chain 2 of *D. russellii* (Chen et al., 2008) and *D. russellii siamensis* C-type lectin-like protein subunit 1 (AAY63870) (Fig. 26.4). The main difference between RVV-X and VLFXA LC2 is due to glycosylation. Unlike RVV-X LC2, the LC2 of VLFXA is not glycosylated (Siigur et al., 2004).

The structure of VLFXA closely resembles that of the FX activating enzyme from *V. russellii* venom (RVV-X) having three polypeptide chains bound by S–S-bridges. While LC2 contains 8 cysteines and LC1 7 cysteines, LC1 might be connected with the heavy chain via LC2 (HC-LC2-LC1). The catalytic mechanism of VLFXA should be the same as proposed by Morita for RVV-X (Morita, 1998). The heavy chain should have metalloprotease activity since it contains three domains characteristic to P-III type metalloproteases. LC1 and LC2 resemble

```
AAQ17468  1   MGRFISVSFGLLVVFLSLSGTGADFDCPSDWVSYDQHCYKAFNDLKNWTDAEKFCTEQNKGSHLVSLHSSEEEDFVVNLA  80
AAW69870  1   MGRFISVSFGCLVVFLSLSGTEAVLDCPSGWLSYEQHCYKGFNDLKNWTDAEKFCTEQKKGSHLVSLHSREEEEFVVNLI  80
AAY63871  1   MGRFISVSFGWLVVFLSLSGTEAVLDCPSGWLSYEQHCYKGFNDLKNWTDAEKFCTEQKKGSHLVSLHSREEEKFVVNLI  80
AAY63873  1   MGRFISISFGLLVVFLSLSGTEAAFCCPSGWSAYDQNCYKVFTEEMNWADAEKFCTEQKKGSHLVSLHSREEEKFVVNLI  80
AAY63875  1   MGRFISVSFGLLVVFLSLSGTGAD--CPSEWSSHEGHCYKVFKLLKTWEDAEKFCTEQKKGSHLVSLHSREEEKFVVNLI  78
AAB22478  1   ---------------------VLDCPSGWLSYEQHCYKGFNDLKNWTDAEKFCTEQKKGSHLVSLHSREEEEFVVNLI  57
ABW82673  1   MGRLISVSFGLLVVFLSLSGTGAALNCASGWSGYDQHCYKVFDKPKSWADAEKFCKKQTSGGHLVSFHSSEETDFVVKLV  80
ABW82674  1   MGRFIFVSFGLLVVFLSLSGTGAALNCASGWSGYDQHCYKVFDKPKSWADAEKFCKKQTSGGHLVSFHSSEETDFVVKLV  80

                                                                                        Identity
AAQ17468  81  SQSLQYPVAWIGLGNMWKECRSEWSDGGNVKYKALAEESYCLLINTHKKGWRSMTCNNMAHVICKF--  146  100%
AAW69870  81  SENLEYPATWIGLGNMWKDCRMEWSDRGNVKYKALAEESYCLIMITHEKEWKSMTCNFIAPVVCKF--  146   79%
AAY63871  81  SENLEYPATWIGLGNMWKDCRMEWSDRGNVKYKALAEESYCLIMITHEKVWKSMTCNFIAPVVCKF--  146   79%
AAY63873  81  SENLEYPATWIGLGNMWKDCRMEWSDRGNVKYKALAEESYCLIMITHEKVWKSMTCNFIAPVVCKF--  146   75%
AAY63875  79  SENLEYPATWIGLGNMWKDCRMEWSDRGNVKYKALAEESYCLIMITHEKVWKSMTCNFIAPVVCKF--  144   76%
AAB22478  58  SENLEYPATWIGLGNMWKDCRMEWSDRGNVKYKALAEESYCLIMITHEKEWKSMTCNFIAPVVCKF--  123   77% (95/122)
ABW82673  81  SQTLESQILWMGLSKVWNQCDWGWSNGAKLKYKAWAEESYCVYFSSTKKGWRSRACRLLGHFVCKSPA  148   60% (88/145)
ABW82674  81  SQTLESQILWMGLSKVWNQCDWGWSNGAKLKYKAWAEESYCVYFSSTKKGWRSRACRLLGHFVCKSPA  148   60% (88/145)
```

Fig. 26.3 Comparison of the amino acid sequence of the FX activator light chain 1 precursor from *Macrovipera lebetina* (VLFXA LC1) with those of other snake venom components. The amino acid sequence of VLFXA LC1 (Accession No. AAQ17468) was aligned: AAW69870-*Daboia russellii*, FXA LC1 precursor; AAY63871-*D. russellii siamensis,* C-type lectin like subunit 2 precursor; AAY63873-*D. russellii siamensis,* C-type lectin like subunit 4 precursor; AAY63875-*D. russellii siamensis,* C-type lectin like subunit 6 precursor; AAB22478-*Vipera russellii*, RVV-X LC1 peptide; ABW82673-*M. lebetina,* C-type lectin B2 precursor; ABW82674-*M. lebetina,* C-type lectin B3/B5 precursor. Gaps are indicated by *dashes*, potential N-glycosylation sites are *underlined*

```
AAT91068  1   MGRSISVSFGLLAVFLSLSGTGAGLDCPPDSSPYRYFCYRVFKEQKNWADAERFCAERPNNGHLVSIESMEEAEFVAQLL  80
Q4PRD2    1   MGRFISVSFGLLVVFLSLSGTGAGLDCPPDSSLYRYFCYRVFKEHKTWEAAERFCMEHPNNGHLVSIESMEEAEFVAKLL  80
AAW69869  1   MGRFISVSFGLLVVFLSLSGTGAGLDCPPDSSLYRYFCYRVFKEHKTWEAAERFCMEHPNNGHLVSIESMEEAEFVAKLL  80
2E3X-B    1   ------------------------LDCPPDSSLYRYFCYRVFKEHKTWEAAERFCMEHPNNGHLVSIESMEEAEFVAKLL  56
AAL66391  1   MGRFIFVSFGLLVVFLSLSGTGADFNCPPGWSAYDQYCYQVIKEPKNWDDAERFCTEQADGGHLVSIESKGERDFVAQLV  80
Q8AYA5    1   MGRFIFVSFGLLVVFLSLSGTGADFNCPPGWSAYDQYCYQVIKEPKNWDDAERFCTEQADGGHLVSIESKGERDFVAQLV  80
B4XSY6    1   MGRSISVSFGLLVVFLSLSGTGADFDCPSGWSAYDQHCYQAVDEPKSWADAEKFCTEQANGGHLVSIDSKKEANFVAELV  80
Q8JIV9    1   MGRFIFVSFGLLVVFLSLSGTGADVDCLPGWSAYDQSCYRVFKLLKTWDDAEKFCTERPKGGHLVSIESAGERDFVAQLV  80
B4XSY5    1   MGRSISVSFGLLVVFLSLSGIGADFDCPSGWSAYDQHCYQAVDEPKSWADAEKFCTEQANGGHLVSIDSKKEANFVAELV  80

                                                                                                Identity
AAT91068  81  SKITGKFIT--HFWIGLRIEDKKQQCRSEWSDGSSVSYDNLLKREFRKCFGLEKGTGYRSWFNLNCEEPYPFVCKVPPNC  158  100%
Q4PRD2    81  SNTTGKFIT--HFWIGLMIKDKEQECSSEWSDGSSVSYDKLGKQEFRKCFVLEKESGYRMWFNRNCEERYLFVCKVPPEC  158   82%
AAW69869  81  SNTTGKFIT--HFWIGLMIKDKEQECSSEWSDGSSVSYDKLGKEEFRKCFVLEKESGYRMWFNRNCEERYVFVCKVPPEC  158   82%
2E3X-B    57  SNTTGKFIT--HFWIGLMIKDKEQECSSEWSDGSSVSYDKLGKQEFRKCFVLEKESGYRMWFNRNCEERYLFVCKVPPEC  134   80% (108/134)
AAL66391  81  SQSIESVE--DHVWTGLRVQNKEKQCSTEWSDGSSVSYENLLELYMRKCGALERETGFHKWINLGCIQLNPFVCKFPPQC  158   60%
Q8AYA5    81  SQNIESVE--DHVWTGLRVQNKEKQCSTEWSDGSSVSYENLLELYMRKCGALERETGFHKWINLGCIQLNPFVCKFPPQC  158   60%
B4XSY6    81  SQNIKETRRTDFVWIGLRVEDKRQQCSSEWSDGSSINYQNWIEAESKKCLGLEKQTRYRKWVNLNCGQPYRFTCEI----  156   62% (97/156)
Q8JIV9    81  SENKQT----DNVWLGLKIQSKGQQCSTEWTDGSSVSYENFSEYQSKKCFVLEKNTGFRTWLNLNCGSEYAFVCKSPP--  154   61% (96/156)
B4XSY5    81  SQNIKETRRTDFVWIGLRAEDKRQHCSSEWSDGSSINYQNWIEAESKKCLGLEKQTRYRKWVNLNCGQPYRFTCEI----  156   60% (95/156)
```

Fig. 26.4 Comparison of the amino acid sequence of the FX activator light chain 2 precursor from *Macrovipera lebetina* (VLFXA LC2) with those of other snake venom components. The amino acid sequence of VLFXA LC2 (Accession No.AAT91068) was aligned with: AAW69869-*Daboia russellii*, factor X activator light chain 2 precursor; 2E3X-BChain B, Crystal Structure of Russell's Viper Venom Metalloproteinase; AAL66391-*Deinagkistrodon acutus*, antithrombin 1 A chain; Q8AYA5-Agglucetin subunit alpha-2 precursor; B4XSY6-*M.lebetina*, C-type lectin A11 precursor; Q8JIV9-*D. acutus*, Agglucetin subunit alpha-1 precursor; B4XSY5- *M. lebetina*, C-type lectin A10 precursor. Gaps are indicated by *dashes*, potential N-glycosylation sites are *underlined*

heterodimeric C-type lectin-like proteins and are similar to IX/X binding protein (Matsuzaki et al., 1996). The regulatory subunit (LC1+LC2) binds to the Gla-domain of FX and then the catalytic subunit (HC) cleaves the Arg52-Ile53 bond in the heavy chain of human FX releasing the active FXa. Light chains serve as regulatory regions that determine the specificity of binding of VLFXA to FX.

FX Activator from the Venom of *Vipera ammodytes ammodytes*

FX activator was isolated from the venom of *Vipera ammodytes ammodytes* on Sephacryl S-200 followed by affinity chromatography on Concanavalin A-Sepharose and anion-exchange chromatography on DEAE-Sephacel. An additional separation step on hydroxyapatite column was included followed by gel filtration on Superdex 75 (Leonardi et al., 2008). Two FX activating enzymes – VAFXA-I and VAFXA-II, were isolated with molecular masses of 58 and 70 kDa, respectively, on nonreduced SDS-PAGE; on reducing SDS-PAGE each of them dissociated into three polypeptide chains: VAFXA-I – 42, 15 and 16 kDa; VAFXA-II – 55, 15 and 18 kDa. The partial amino acid sequences of the three chains closely resembled these of RVV-X (Takeya et al., 1992; Chen et al., 2008) and VLFXA (Siigur et al., 2004). Therefore, VAFXA-I and VAFXA-II belong structurally to the P-IIId class of snake venom metalloproteases. VAFXA was partially proteolytically degraded resulting in the appearance of a 30-kDa band on SDS-PAGE whose N-terminal sequence pointed to cleavage of a peptide bond in the heavy chain in the region between metalloprotease and disintegrin-like domains. Analogous effect has been observed in case of VLFXA (Siigur et al., unpublished).

Substrate specificity of VAFXAs was broader than of RVV-X and VLFXA. VAFXA elicited limited hydrolysis of insulin B chain after 24 h and hydrolysed fibrinogen Aα-chain. The parameters of activation of factor X to Xa by VAFXA-I: $K_m = 84$ nM; $V_{\max} = 4.3$ nmol/min. Both activators inhibited collagen-induced platelet aggregation by approximately 30 and 80% at concentrations of 290 and 680 nM. No effect on plasminogen and prothrombin was observed (Leonardi et al., 2008).

Metalloprotease Activators from *Bothrops atrox, Cerastes cerastes, Vipera aspis* and *V. berus* Venoms

Activators resembling RVV-X have been isolated from the venoms of *Bothrops atrox* (Hofmann et al., 1983; Hofmann and Bon, 1987) and *Cerastes cerastes* (Franssen et al., 1983). The activator from *B. atrox* venom was purified by ion exchange chromatography on DEAE-cellulose (in the presence of 5 mM benzamidine), gel filtration on Sephadex G-150, and rechromatography on DEAE-cellulose. Two isoforms of FX activators were isolated with M_r 77,000. SDS-PAGE under reducing conditions gave two bands: heavy chain with M_r 59,000 and one light chain doublet 14,000–15,000. The specific activity was increased 30 times in comparison with the venom but the yield was low (Hofmann et al., 1983). The activator is specific towards FX, being inactive against prothrombin and fibrinogen. The activation is maximal at 2 mM Ca^{2+} and pH 7.5–9.0. Effect on FX was investigated by SDS-PAGE. Both isoforms cleave the heavy chain of FX like RVV-X but, in addition, two other bonds are split giving the polypeptides with M_r of 48,000 and 30,500 (Hofmann and Bon, 1987).

The FX activator from *C. cerastes* was isolated by Sephadex G-150, QAE-Sephadex A-50 and chromatofocusing on PBE94. Native activator gave two close bands on nonreduced SDS-PAGE. After reduction with 2-mercaptoethanol, 3 bands with M_r 58,000, 17,700 and 15,000 were detected. The activation reaction (followed by hydrolysis of a chromogenic substrate) was linear both at high (0.120 μM) and low (0.006 μM) FX concentrations. The kinetic parameters K_m (19.2 nM) and V_{max} (0.11 pmol FXa/min per ng enzyme) (Franssen et al., 1983) differed from RVV-X ($K_m = 710$ nM; $V_{max} = 4.33$ pmol FXa/min per ng enzyme) (van Dieijen et al., 1981). Both FX activators have three polypeptide chains (a heavy chain and two light chains) and most probably the activation mechanism of these venom activators should be similar to that of RVV-X.

Nonspecific metalloprotease FX activators were isolated from *Vipera berus* venom (Samel and Siigur, 1995; Samel et al., 2003) that had weak proteolytic activity. Activator with M_r 38,000 hydrolysed oxidized insulin B chain (Samel and Siigur, 1995), activator with M_r 126 kDa had in addition fibrinolytic, gelatinolytic and caseinolytic activities and cleaved FX fragment NNLTRIVGG at positions Leu-Thr and Arg-Ile. Both activators are glycoproteins, high molecular weight activator is a homodimer (Samel et al., 2003).

Komori et al. (1990) isolated an activator from *V. aspis* venom by gel filtration and anion exchange chromatography. The native enzyme had M_r of 75,000 and pI 4.6. Upon reduction, the activator migrated as two bands with M_r of 16,000 and 14,000 on SDS-PAGE (the ratio in intensities 2:3). The N-terminal sequence of both bands is identical and homologous to C-type lectin-like proteins (e.g. light chains of FX activators RVV-X and VLFXA).

Serine Protease Activators

In 1995, two papers were published concerning FX activation by elapid venoms, *Ophiophagus hannah* (Lee et al., 1995) and *Bungarus fasciatus* (Zhang et al., 1995). These activators differ from the classical RVV-X-type activators by their serine protease character and structural composition representing single-chain glycoproteins with M_r of 62,000 (*O. hannah*) or 70,000 (*B. fasciatus*). They are inhibited by serine protease inhibitors such as PMSF, DFP, benzamidine, but not by EDTA. Both enzymes are stable at neutral pH and temperatures below 55°C. Common with RVV-X-type activators is the absolute Ca^{2+} – dependence (optimal concentrations 4–5 mM). The effect on FX was studied by SDS-PAGE. The disappearance of FX heavy chain was accompanied with the appearance of the 36,000 Da polypeptide. The light chain was uneffected (Zhang et al., 1995). Both activators represent minor components of the venom.

There are some references about serine protease-like activators of FX in the venoms of *Cerastes cerastes* (El-Asmar et al., 1986) and *C. vipera* (Farid et al., 1993). The former isolated a protein with M_r of 14,000 and pI 3.2 from *C. cerastes* venom having proteolytic activity towards albumin and casein and clotting the plasmas deficient in factor VIII, (VIII+IX) or FX, while the latter isolated a component

(12,500 Da, pI 4.4) from *C. vipera* venom that they called FX activator. It had serine protease activity hydrolyzing S-2288 (H-D-Ile-Pro-Arg-*p*NA 2HCl), CBS 33.08 (2 AcOH-H-D-Leu-CHA-Arg-*p*NA) and some small Lys or Glu peptides but had no arginine esterase activity. The enzyme cleaved FX resulting in the formation of a polypeptide with FXa activity. The activation was highly potentiated by phospholipid and Ca^{2+}. The uniqueness of these enzymes is in their extremely low molecular masses. There are no comments concerning the specific activity of the enzymes.

Diagnostic Use of FX Activators

In 1934, Macfarlane and Barnett tested the thromboplastic activity of the venom of the Russell's viper (*Vipera russellii*) in an attempt to treat the severe bleeding of a patient with haemophilia. Their work was based on the earlier observations of Martin, who had noted that certain venoms induced intravascular clotting and that some venoms clotted citrated plasma (reviewed by Owen, 2001). RVV-X has found current application as research and diagnostic reagent of FX assay in human plasma. Usually whole RVV is used for determination of FX but the effect is connected with RVV-X. At defined concentrations other components in the venom do not interfere with the determination. In a clotting assay, FXa, formed by RVV-X in the presence of Ca^{2+} ions, phospholipid and factor V, activates prothrombin to thrombin that causes transformation of fibrinogen to fibrin clot. The clotting time of plasma using RVV-X or RVV is known as the Stypven time (Stocker, 1990, 1998).

A direct assay of FX has been possible after elaboration of the specific method using chromogenic substrate (Aurell et al., 1977) for FXa. The principle of the method is given on Fig. 26.5. The same principle is used for determination of FX activators.

RVV-X has become widely used for FX detection, for distinguishing between FVII and FX deficiency, and lupus anticoagulant assay. The dilute RVV time (dRVVT) is quick, sensitive and inexpensive method for lupus anticoagulant test (Marsh, 2001).

Strachan and Ogden (2000) have used RVV-X in a microsphere coagulation enzyme-linked immunosorbent assay for the detection of *Escherichia coli* O157:H7. They envisage that the method could be useful for the detection of wide range of targets where sensitivity is important (Strachan and Ogden, 2000).

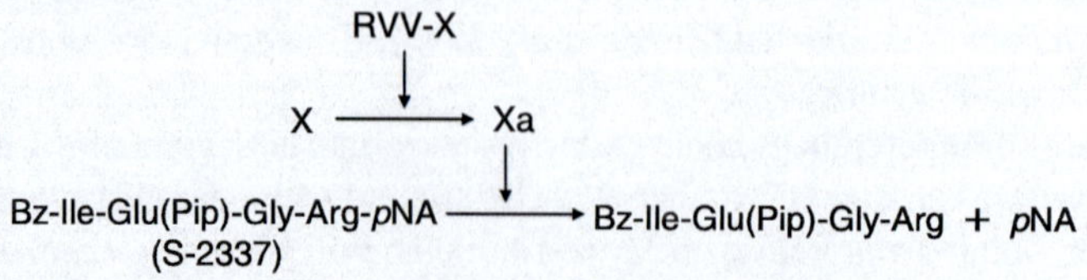

Fig. 26.5 The assay of FX using RVV-X and chromogenic substrate S-2337

Conclusions

Viperidae, Crotalidae and Elapidae snake venoms contain a variety of proteases capable of activating FX. These enzymes are metalloproteases or serine proteases. Most of the metalloprotease activators (e.g. RVV-X, VLFXA) are composed of three chains: a heavy chain with metalloprotease, disintegrin-like and cysteine-rich domains, and two light chains. All these chains are covalently connected with disulfide bonds. The light chains are homologous to C-type lectins. The specific FX activators cleave the Arg52-Ile53 bond in the heavy chain of human FX. The activation of FX by snake venom activators is Ca^{2+} dependent. Whole primary sequences are known for VLFXA and RVV-X. The heavy chain and two light chains of VLFXA and RVV-X are synthesized from different genes. RVV-X is the most potent activator of FX among the snake venom activators and it has become widely used for FX detection.

The serine protease activators of FX are single-chain glycoproteins, unfortunately no structural data are available on any of these proteases.

Acknowledgements The work was financially supported by Estonian Science Foundation grant No. 7251 and by target financing SF0180114Bs08.

References

Amphlett, G.W., Byrne, R., Castellino, F.J., 1982. Cation binding properties of the multiple subforms of RVV-X, the coagulant protein from *Vipera russelli*. Biochemistry 21, 125–132.

Aurell, L., Friberger, P., Karlsson, G., Claeson, G., 1977. A new sensitive and highly specific chromogenic peptide substrate for Factor Xa. Thromb. Res. 11, 595–605.

Baugh, R.J., Krishnaswamy, S., 1996. Role of the activation peptide domain in human factor X activation by the extrinsic Xase complex. J. Biol. Chem. 271, 16126–16134.

Bjarnason, J.B., Fox J.W., 1995. Snake venom metalloendopeptidases: reprolysins. Meth. Enzymol. 248, 345–368.

Cerretti, D.P., DuBose, R.F., Black, R.A., Nelson, N., 1999. Isolation of two novel metalloproteinase-disintegrin (ADAM) cDNAs that show testis–specific gene expression. Biochem. Biophys. Res. Commun. 263, 810–815.

Chen, H.-S., Chen, J.-M., Lin, C.-V., Khoo, K.-H., Tsai, I.-H., 2008. New insights into the functions and N-glycan structures of factor X activator from Russell's viper venom. FEBS J. 275, 3944–3958.

Durkee, K.H., Roh, B.H., Doellgast, G.J., 1993. Immunoaffinity chromatographic purification of Russell's viper venom factor X activator using elution in high concentrations of magnesium chloride. Protein Expr. Purif. 4, 405–411.

El-Asmar, M.F., Shaban, E., Hagag, M., Swelam, N., Tu, A.T., 1986. Coagulant component in *Cerastes cerastes* (Egyptian sand viper) venom. Toxicon 24, 1037–1044.

Esnouf, M.P., 1978. Current perspectives of factor X, in: Recent Progress in Blood Coagulation and Thrombosis Research. Biblthca. Haemat. (vol. 44). Karger, Basel, pp. 75–80.

Farid, T., Nasser, H., Zaki, K., El-Asmar, M.F., 1993. Low molecular weight factor X activator from *Cerastes vipera* (Sahara sand viper) venom. Toxicon 31, 1007–1017.

Fox, J.W., Serrano, S.M.T., 2008. Insights into and speculations about snake venom metalloproteinase (SVMP) synthesis, folding and disulfide bond formation and their contribution to venom complexity. FEBS J. 275, 3016–3030.

Franssen, J.H.L., Janssen-Claessen, T., van Dieijen, G., 1983. Purification and properties of an activating enzyme of blood clotting factor X from the venom of *Cerastes cerastes*. Biochim. Biophys. Acta 747, 186–190.

Fujikawa, K., Coan, M.H., Legaz, M.E., Davie, E.W., 1974. The mechanism of activation of bovine factor X (Stuart factor) by intrinsic and extrinsic pathways. Biochemistry 13, 5290–5299.

Fujikawa, K., Titani, K., Davie, E.W., 1975. Activation of bovine factor X (Stuart factor): conversion of factor $X_{a\alpha}$ to factor $X_{a\beta}$. Proc. Natl. Acad. Sci. U.S.A. 72, 3359–3363.

Furie, B.C., Furie, B., 1975. Interaction of lanthanide ions with bovine factor X and their use in the affinity chromatography of the venom coagulant protein of *Vipera russelli*. J. Biol. Chem. 250, 601–608.

Furukawa, Y., Matsunaga, Y., Hayashi, K., 1976. Purification and characterization of a coagulant protein from the venom of Russell's viper. Biochim. Biophys. Acta 453, 48–61.

Gowda, D.C., Jackson, C.M., Hensley, P., Davidson, E.A., 1994. Factor X-activating glycoprotein of Russell's viper venom. J. Biol. Chem. 269, 10644–10650.

Gowda, D.C., Jackson, C.M., Kurzban, G.P., McPhie, P., Davidson, E.A., 1996. Core sugar residues of the N-linked oligosaccharides of Russell's viper venom factor X-activator maintain functionally active polypeptide structure. Biochemistry 35, 5833–5837.

Grams, F., Huber, R., Kress, L.F., Moroder, L., Bode, W., 1993. Activation of snake venom metalloproteinases by a cysteine switch like mechanism. FEBS Lett. 335, 76–80.

Hertzberg, M., 1994. Biochemistry of factor X. Blood Rev. 8, 56–62.

Hofmann, H., Bon, C., 1987. Blood coagulation induced by the venom of *Bothrops atrox*. 2. Identification, purification and properties of two factor X activators. Biochemistry 26, 780–787.

Hofmann, H., Dumarey, C., Bon, C., 1983. Blood coagulation induced by *Bothrops atrox* venom. Identification and properties of a factor X activator. Biochimie 65, 201–210.

Jackson, C.M., Gordon, J.G., Hanahan, D.J., 1971, Separation of the tosyl arginine esterase activity from the factor X activating enzyme of Russell's viper venom. Biochim. Biophys. Acta 252, 255–261.

Jesty, J., 1986. Analysis of the generation and inhibition of activated coagulation factor X in pure systems and in human plasma. J. Biol. Chem. 261, 8695–8702.

Kishimoto, M., Takahashi, T., 2002. Molecular cloning of HR1a and HR1b, high molecular hemorrhagic factors, from *Trimeresurus flavoviridis* venom. Toxicon 40, 1369–1375.

Kisiel, W., Hermodson, M.A., Davie, E.W., 1976. Factor X activating enzyme from Russell's viper venom: isolation and characterization. Biochemistry 15, 4901–4906.

Komori, Y., Nikai, T., Sugihara, H., 1990. Isolation and characterization of factor X activator from the venom of *Vipera aspis aspis*. Int. J. Biochem. 22, 1053–1060.

Kosow, D.P., 1976. Purification and activation of human factor X: cooperative effect of Ca^{2+} on the activation reaction. Thromb. Res. 9, 565–573.

Lee, W.H., Zhang, Y., Wang, W.-Y., Xiong, Y.-L., Gao, R., 1995. Isolation and properties of a blood coagulation factor X activator from the venom of king cobra (*Ophiophagus hannah*). Toxicon 33, 1263–1276.

Leonardi, A., Fox, J.W., Trampuš-Bakija, A., Križaj, I., 2008. Two coagulation factor X activators from *Vipera a. ammodytes* venom with potential to treat patients with dysfunctional factors IXa or VIIa. Toxicon 52, 628–637.

Leytus, S.P., Chung, D.W., Kisiel, W., Kurachi, K., Davie, E.W., 1984. Characterization of a cDNA coding for human factor X. Proc. Natl. Acad. Sci. U.S.A. 81, 3699–3702.

Lindhout, M.J., Kop-Klaassen, B.H.M., Hemker, H.C., 1978. Activation of decarboxyfactor X by a protein from Russell's viper venom. Purification and partial characterization of activated decarboxyfactor X. Biochim. Biophys. Acta 533, 327–341.

Marsh, N.A., 2001. Diagnostic use of snake venoms. Haemostasis 31, 211–217.

Matsui, T., Fujimura, Y., Titani, K., 2000. Snake venom proteases affecting hemostasis and thrombosis. Biochim. Biophys. Acta 1477, 146–156.

Matsuzaki, R., Yoshiara, E., Yamada, M., Shima, K., Atoda, H., Morita, T., 1996. cDNA cloning of IX/X-BP, a heterogeneous two-chain anticoagulant protein from snake venom. Biochem. Biophys. Res. Commun. 220, 382–387.

Morita, T., 1998. Proteases which activate factor X, in: Bailey, G.S. (Ed.), Enzymes from Snake Venoms. Alaken Inc., Fort Collins, Colorado, pp. 179–208.

Nishida, S., Fujita, T., Kohno, N., Atoda, H., Morita, T., Takeya, H., Kido, I., Paine, M.J., Kawabata, S., Iwanaga, S., 1995. cDNA cloning and deduced amino acid sequence of prothrombin activator (ecarin) from Kenyan *Echis carinatus* venom. Biochemistry 34, 1771–1778.

Owen, C.A., Jr., 2001. Inherited coagulation factor deficiencies, in: Nichols, W.L., Bowie, E.J.W. (Eds.), A History of Blood Coagulation. Mayo Foundation for Medical Education and Research, Rochester, MN, pp. 117–132.

Pryzdial, E.L.G., Lavigne, N., Dupuis, N., Kessler, G.E., 1999. Plasmin converts factor X from coagulation zymogen to fibrinolysis cofactor. J. Biol. Chem. 274, 8500–8505.

Pukrittayakamee, S., Esnouf, M.P., McMichael, A.J., 1983. Purification and inactivation of the factor X activator of Russell's viper venom with monoclonal antibodies. Mol. Biol. Med. 1, 123–135.

Samel, M., Siigur, J., 1995. Medium molecular weight factor X activating enzyme from *Vipera berus berus* venom. Toxicon 33, 41–52.

Samel, M., Vija, H., Subbi, J., Siigur, J., 2003. Metalloproteinase with factor X activating and fibrinolytic activities from *Vipera berus berus* venom. Comp. Biochem. Physiol. B. 135, 575–582.

Sekiya, F., Yoshida, M., Yamashita, T., Morita, T., 1996. Magnesium (II) is a crucial constituent of the blood coagulation cascade. J. Biol. Chem. 271, 8541–8544.

Siigur, E., Aaspõllu, A., Trummal, K., Tõnismägi, K., Tammiste, I., Kalkkinen, N., Siigur, J., 2004. Factor X activator from *Vipera lebetina* venom is synthesized from different genes. Biochim. Biophys. Acta 1702, 41–51.

Siigur, E., Tõnismägi, K., Trummal, K., Samel, M., Vija, H., Subbi, J., Siigur, J., 2001a. Factor X activator from *Vipera lebetina* snake venom, molecular characterization and substrate specificity. Biochim. Biophys. Acta 1568, 90–98.

Siigur, J., Aaspõllu, A., Tõnismägi, K., Trummal, K., Samel, M., Vija, H., Subbi, J., Siigur, E., 2001b. Proteases from *Vipera lebetina* venom affecting coagulation and fibrinolysis. Haemostasis 31, 123–132.

Siigur, J., Siigur, E., 2006. Factor X activating proteases from snake venoms. Toxin Rev. 25, 235–255.

Silva, M.B., Schattner, M., Ramos, C.R.R., Junqueira-De-Azevedo, I.L.M., Guarnieri, M.C., Lazzari, M.A., Sampaio, C.A.M., Pozner, R.G., Ventura, J.S., Ho, P.L., Chudzinski-Tavassi, A.M., 2003. A prothrombin activator from *Bothrops erythromelas* (Jararaca-da-seca) snake venom: characterization and molecular cloning. Biochem. J. 369, 129–139.

Sinha, U., Wolf, D.L., 1993. Carbohydrate residues modulate the activation of coagulation factor X. J. Biol. Chem. 268, 3048–3051.

Skogen, W.F., Bushone, D.S., Johnson, A.E., Cox, A.C., 1983. The role of the Gla domain in the activation of bovine coagulation factor X by the snake venom protein XCP. Biochem. Biophys. Res. Commun. 111, 14–20.

Stocker, K., 1990. Snake venom proteins affecting hemostasis and fibrinolysis, in: Stocker, K. (Ed.), Medical Use of Snake Venom Proteins. CRC Press, Boca Raton, Ann Arbor, Boston, pp. 97–160.

Stocker, K., 1998. Research, diagnostic and medicinal uses of snake venom enzymes, in: Bailey, G.S. (Ed.), Enzymes in Snake Venoms. Alaken Inc., Fort Collins, CO, pp. 705–736.

Strachan, N.J.C., Ogden, I.D., 2000. A sensitive microsphere coagulation ELISA for *Escherichia coli* O157:H7 using Russell's viper venom. FEMS Microbiol. Lett. 186, 79–84.

Takeda, S., Igarashi, T., Mori, H., 2007. Crystal structure of RVV-X: an example of evolutionary gain of specificity by ADAM proteinases. FEBS Lett. 581, 5859–5864.

Takeya, H., Nishida, S., Miyata, T., Kawada, S., Saisaka, Y., Morita, T., Iwanaga, S., 1992. Coagulation factor X activating enzyme from Russell's viper venom (RVV-X). J. Biol. Chem. 267, 14109–14117.

Tans, G., Rosing, J., 2001. Snake venom activators of factor X: an overview. Haemostasis 31, 225–233.

Teng, C.-M., Chen, Y.-H., Ouyang, C., 1984. Purification and properties of the main coagulant and anticoagulant principles of *Vipera russelli* snake venom. Biochim. Biophys. Acta 786, 204–212.

van Dieijen, G., Tans, G., van Rijn, J., Zwaal, R.F.A., Rosing, J., 1981. Simple and rapid method to determine the binding of blood clotting factor X to phospholipid vesicles. Biochemistry 20, 7096–7101.

Wagstaff, S.C., Laing, G.D., Theakston, R.D.G., Pappaspyridis, C., Harrison, R.A., 2006. Bioinformatics and multiepitope DNA immunization to design rational snake antivenom. PloS Med. 3, e184.

Williams, W.J., Esnouf, M.P., 1962. The fractionation of Russell's viper (*Vipera russellii*) venom with special reference to the coagulant protein. Biochem. J. 84, 52–62.

Yamada, D., Sekiya, F., Morita, T., 1997. Prothrombin and factor X activator activities in the venoms of Viperidae snakes. Toxicon 35, 1581–1589.

Zhang, Y., Xiong, Y.-L., Bon, C., 1995. An activator of blood coagulation factor X from the venom of *Bungarus fasciatus*. Toxicon 33, 1277–1288.

Chapter 27
Structural Aspects of the Factor X Activator RVV-X from Russell's Viper Venom

Soichi Takeda

Abstract The venom of Russell's viper *Daboia russelli* contains a potent blood coagulation factor X activator, RVV-X. Because of its high specificity, RVV-X is widely used in laboratories and as a diagnostic tool. RVV-X is a unique heterotrimeric metalloproteinase containing a mammalian a disintegrin and metalloproteinase (ADAM)-like heavy chain and two C-type lectin-like light chains, which are covalently held together by disulfide bonds. The crystal structure of RVV-X indicates that RVV-X adopts a "hook-spanner-wrench"–like structure, in which the metalloproteinase/disintegrin portion constitutes a mobile hook and the lectin-like domains, together with the remainder of the heavy chain, constitute a handle. The lectin-like domains form an intertwined dimer with high structural similarity to anticoagulant factor X-binding proteins. The RVV-X structure displayed a 6.5-nm separation between the catalytic zinc atom and a putative gamma carboxylglutamic acid (Gla) domain–binding exosite, implying molecular mechanism of factor X activation by RVV-X. The three-dimensional structure of RVV-X also provides a typical example of the molecular evolution of protein complexes, giving insight into the molecular basis of substrate recognition and proteolysis by adamalysin/reprolysin/ADAM family proteinases.

Introduction

Numerous snake venoms, particularly those of the Viperidae family, contain procoagulant factors. Such venom procoagulants promote blood coagulation by acting at different steps of the blood coagulation cascade, such as by activating factor V, factor X, or prothrombin, or by directly clotting fibrinogen, thus leading to serious hemostatic disturbances such as disseminated intravascular coagulation (DIC) within prey. Venom coagulant properties vary among species and often differ

S. Takeda (✉)
National Cerebral and Cardiovascular Center Research Institute, Suita, Osaka 565-8565, Japan
e-mail: stakeda@ri.ncvc.go.jp

R.M. Kini et al. (eds.), *Toxins and Hemostasis*, DOI 10.1007/978-90-481-9295-3_27,

considerably from those of mammalian coagulation factors. Factor X–activating proteinases are widely distributed in Viperidae and are also found in some elapid venom (Morita, 1998; Tans and Rosing, 2001). These snake venom factor X activators are divided into metalloproteinase (SVMP) and serine proteinase (SP) categories. In contrast, the proteinases involved in the mammalian coagulation cascade are all SPs.

RVV-X, a factor X activator from Russell's viper *Daboia russelli* venom, is the most potent venom coagulation activator known. Its structural and functional properties and factor X–activation mechanism have been studied in detail (Morita, 1998; Tans and Rosing, 2001). RVV-X specifically activates factor X by cleaving the same Arg194–Ile195 bond in factor X that is cleaved by factors IXa or VIIa on the cell surface during physiological coagulation. RVV-X belongs to the P-IIId class (formerly called P-IV class) of SVMPs (Fox and Serrano, 2005, 2008), and consists of a heavy chain of 57,600 Da and two light chains of 19,400 and 16,400 Da held together by disulfide bonds (Gowda et al., 1994; Kisiel et al., 1976). The 427-residue heavy chain contains distinct metalloproteinase, disintegrin, and cysteine-rich (MDC) domains, which are homologous to those of ADAM/adamalysin/reprolysin family proteins (Fox and Serrano, 2005; Takeya et al., 1992). The ADAM family proteins are associated with numerous disease conditions and comprise an emerging class of mammalian membrane–bound metalloproteinases with potential regulatory roles in cell-cell and cell-matrix adhesion and signaling (Blobel, 2005; Edwards et al., 2009; Seals and Courtneidge, 2003; White, 2003).

The light chains of RVV-X share amino acid sequence homology with C-type (Ca^{2+}-dependent) lectins and the factor X-binding protein (X-bp) of *Deinagkistrodon actus* venoms, as well as with C-type lectin-like proteins (CLPs) isolated from other snake venoms (Morita, 2005; Takeya et al., 1992). The crystal structure of X-bp in complex with the gamma carboxyglutamic acid (Gla) domain of factor X revealed that the two C-type lectin-like domains form a concave surface into which the factor X Gla domain fits. This interaction involves extensive water-mediated hydrogen-bonding and direct contacts between X-bp and the Gla domain (Mizuno et al., 2001). The calcium-dependent prothrombin activator carinactivase-1 in *Echis carinatus* venom has a heterotrimeric P-IIId SVMP structure that is very similar to that of RVV-X. The two isolated C-type lectin-like light chains of carinactivase-1 reportedly recognize the calcium-bound conformation of the prothrombin Gla domain (Yamada et al., 1996). Based on these data, Morita proposed that the two C-type lectin light chains of RVV-X serve as an exosite by which RVV-X recognizes and binds to the factor X Gla domain (Morita, 1998). The recently solved crystal structure of RVV-X revealed that RVV-X exhibits a "hook-spanner-wrench" configuration, in which the catalytic site and a putative Gla domain-binding exosite are located 6.5 nm apart. This suggests that the functional site spatial arrangement substantially contributes to the factor X specificity of recognition by RVV-X (Takeda et al., 2007). The RVV-X structure also provides a good illustration of the evolutionary gain-of-specificity of adamalysin/reprolysin/ADAM family proteinases through multimerization to create an exosite for ligand binding (Takeda, 2009).

This chapter describes the structural aspects and molecular mechanism of factor X activation by RVV-X, after summarizing the current knowledge of the structural and functional properties of SVMPs and related mammalian ADAM family proteinases.

SVMPs, ADAMs and Disintegrins

Snake venom is a complex mixture of bioactive proteins and polypeptides. Most viper venoms are composed of at least 30% SVMPs, suggesting their potentially significant role in envenomation-related pathologies, such as local and systemic hemorrhage, intravascular clotting, edema, inflammation, and necrosis. Based on their domain structures and possible post-translational modifications, SVMPs are classified into P-I to P-III types (Fig. 27.1) (Bjarnason and Fox, 1995; Fox and Serrano, 2005, 2008). P-I SVMPs contain only a metalloproteinase domain, P-II SVMPs contain a metalloproteinase domain followed by a disintegrin domain, and P-III SVMPs contain distinct metalloproteinase, disintegrin-like, and cysteine-rich domains. P-III SVMPs are further divided into subclasses based on their distinct post-translational modifications, such as dimerization (P-IIIc) or proteolytic processing (P-IIIb) (Fox and Serrano, 2008). Formerly called P-IV SVMPs, the heterotrimeric subclass of SVMPs to which RVV-X belongs is classified as P-IIId SVMPs, representing another post-translational modification of the canonical P-III structure (P-IIIa) (Fox and Serrano, 2008; Siigur et al., 2004). SVMPs are phylogenetically most closely related to the mammalian ADAM family of proteins and, together with ADAM and related ADAM with thrombospondin type-1 motif (ADAMTS) family proteinases, constitute the adamalysin/reprolysin subfamily of the M12 family of zinc metalloproteinase (MEROPS classification, http://merops.sanger.ac.uk/).

ADAM family proteins, also called metalloproteinase-disintegrins or MDC proteins, are mammalian membrane-bound glycoproteins that been implicated in cell-cell and cell-matrix adhesion and signaling (Blobel, 2005; Edwards et al., 2009; Seals and Courtneidge, 2003; White et al., 2005). With regard to cellular interactions, fertilin α (ADAM1) and β (ADAM2) have been identified as sperm surface molecules essential for fertilization (Blobel et al., 1992), and meltrins (ADAM9, 12 and 19) have been implicated in myogenesis (Yagami-Hiromasa et al., 1995). The best-characterized in vivo activity of ADAMs is their ectodomain shedding activity. ADAM17, which is also known as tumor necrosis factor-α (TNF-α) converting enzyme (TACE), is the physiological convertase for TNF-α (Black et al., 1997; Moss et al., 1997). ADAMs play key roles in normal development and morphogenesis (Peschon et al., 1998) and are associated with numerous diseases, including rheumatoid arthritis, Alzheimer's disease, heart disease, and cancer (Duffy et al., 2003; Mochizuki and Okada, 2007; Moss and Bartsch, 2004; Murphy, 2008). Excluding pseudogenes, human and mouse ADAMs are encoded by 20 and 37 functional genes, respectively. Although soluble isoforms of certain ADAMs are expressed, canonical ADAMs have an epidermal growth factor (EGF)-like domain, transmembrane and cytoplasmic regions, as well as the MDC domains that are shared by

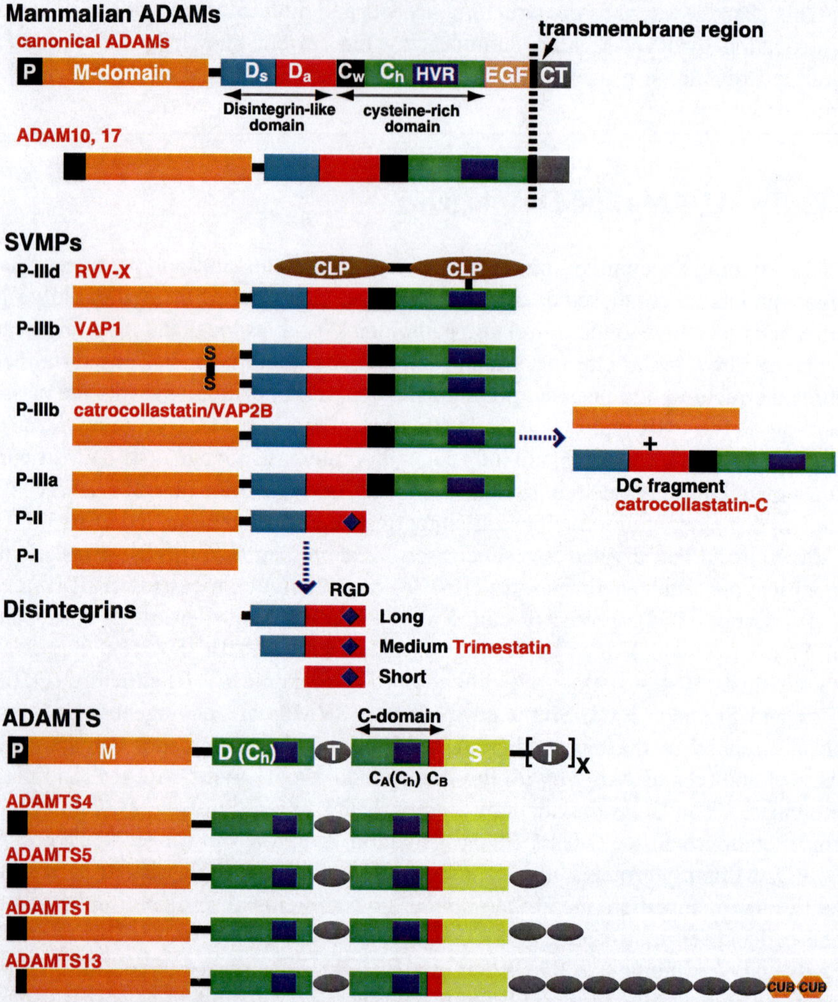

Fig. 27.1 Schematic representation of the domain structure of adamalysin/reprolysin/ADAM family proteins. Each domain or segment is represented by a different color. P, Propeptide; T, thrombospondin type-1 module; EGF, EGF-like module; CT, cytoplasmic region; CLP, C-type lectin-like protein domain. Recent crystallographic studies indicate that the D-domain and N-terminal portion of the C-domain (designated as the C_A domain) of ADAMTS proteinases has a fold similar to the C_h segment of ADAMs and P-III SVMPs (Akiyama et al., 2009; Gerhardt et al., 2007; Mosyak et al., 2008; Takeda, 2009)

all P-III SVMPs (Fig. 27.1) (Takeda, 2009). The ADAMTSs comprise a class of secreted proteinases (19 members in human) that have diverse functions, including procollagen processing, aggrecan degradation, and organogenesis (Apte, 2009; Porter et al., 2005). ADAMTS13 controls platelet thrombus formation through von Willebrand factor (VWF) cleavage. ADAMTS proteinases have varying numbers of carboxyl terminal thrombospondin type-1 repeats and differ from ADAMs in their lack of an EGF-like domain or a transmembrane/cytoplasmic region (Fig. 27.1) (Apte, 2009; Kuno et al., 1997; Porter et al., 2005).

Disintegrins are small proteins (40–90 amino acids) that are generated, albeit not exclusively (Okuda et al., 2002), by the proteolytic processing of larger precursor P-II SVMPs (Fig. 27.1) (Hite et al., 1992; Takeya et al., 1993). Disintegrins typically possess an Arg-Gly-Asp (RGD) recognition sequence on an extended loop (disintegrin-loop) that can inhibit platelet aggregation via integrin binding (Calvete et al., 2005; Huang et al., 1987). ADAMs are unique among cell-surface proteins in that they possess a disintegrin-like sequence, which suggests that integrins are common receptors for ADAMs (Evans, 2001; White, 2003; White et al., 2005). However, the RGD sequence in the disintegrin-loop of ADAMs and P-III SVMPs is usually replaced by XXDC; therefore, the adhesive properties of their D-domains are controversial.

MDC Domains Form a "C Shape"

Adamalysin II, a P-I SVMP from *Crotalus adamantus*, represents the first adamalysin/reprolysin/ADAM proteinase to be solved (Gomis-Ruth et al., 1993, 1994). Since then, the crystal structures of numerous P-I SVMPs and the recombinant M-domains of human ADAMs have been published. The first P-III class SVMP structure was not reported until more than a decade after the first P-I class structure. Homodimeric vascular apoptosis-inducing protein-1 (VAP1) and monomeric catrocollastatin/VAP2B isolated from western diamond back rattle snake *Crotalus atrox* venom were the first and the second P-III SVMP structures, respectively, to be resolved by X-ray crystallography (Igarashi et al., 2007; Takeda, 2008; Takeda et al., 2006). RVV-X was the third of this class of proteinases to be determined (Takeda et al., 2007).

Figure 27.2a depicts the MDC domain architecture of catrocollastatin/VAP2B, the structural prototype of P-III SVMPs and ADAMs (Igarashi et al., 2007). The crystal structures of the P-III SVMPs revealed that the MDC domain folds into a C-shaped configuration in which the distal portion of the C-domain comes close to and faces towards the catalytic site in the M-domain. The M-domain has an oblate ellipsoidal shape with a cleft in its flat side. The catalytic zinc ion is situated at the bottom of the cleft and is tetrahedrally coordinated by the $N\epsilon 2$ atoms of the three histidines in the consensus sequence HEXXHXXGXXHD (Fig. 27.2b). The glutamic acid (Glu334 in catrocollastatin/VAP2) functions as a catalytic base that polarizes a water molecule involved in nucleophilic attack at the sessile peptide bond. The side chain of the conserved methionine (Met357 in catrocollastatin/VAP2) downstream of the catalytic consensus sequence provides a hydrophobic base beneath the three zinc-coordinating histidine side-chains (Fig. 27.2b). All of these structural features are hallmarks of the metzincin clan of metalloendopeptidases (Bode et al., 1993; Gomis-Ruth, 2003).

The globular M-domain is followed by the D-domain, which is further divided into the structurally distinct shoulder (D_S) and arm (D_a) sub-segments (Fig. 27.2a).

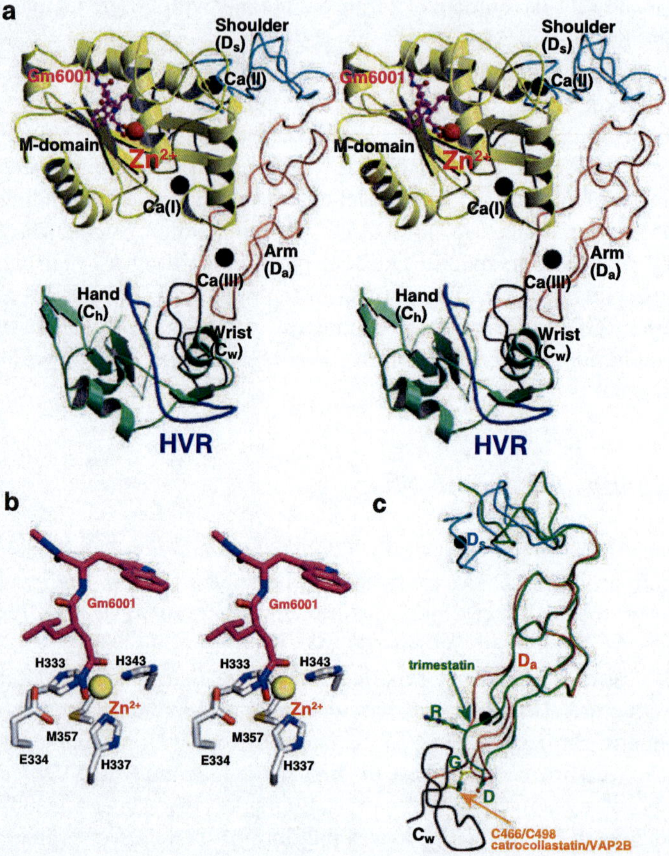

Fig. 27.2 MDC domain architecture. (**a**) Stereo view of the crystal structure of catrocollastatin/VAP2B, a structural prototype of adamalysin/reprolysin/ADAM family proteins (Igarashi et al., 2007). The M-domain, linker, D_s, D_a, C_w and C_h segments, and the HVR are shown in *yellow, grey, cyan, pink, gray, light green*, and *blue*, respectively. Zinc and calcium ions are represented as *red* and *black spheres*, respectively. The hydroxamic inhibitor GM6001 bound to the protein molecule is shown in *ball-and-stick* representation. (**b**) Close-up view of the catalytic site of catrocollastatin/VAP2B in stereo. The side chains of the residues involved in the coordination of the zinc ion (*yellow sphere*), catalytic base (Glue334), and Met-turn (Met357) and bound GM6001 are indicated. (**c**) Superimposition of the D_a-segments of catrocollastatin/VAP2B and trimestatin (Fujii et al., 2003) (shown in *light green*). The Arg-Gly-Asp side chains in trimestatin and the conserved disulfide bond between the D_a and C_w segments in catrocollastatin/VAP2B are shown in ball-and-stick representation. The D_s, D_a, and C_h domains of catrocollastatin/VAP2B are in *cyan, pink*, and *gray*, respectively

These sub-segments are stabilized by numerous disulfide bonds and bound calcium ions. The D_s segment protrudes from the M-domain, and is opposite the catalytic site and close to calcium binding site I. Calcium binding site I is highly conserved among P-III SVMPs and ADAMs (Fig. 27.3) (Takeda, 2008; Takeda et al., 2006). The D- and C-domains together form a C-shaped arm, positioned with its concave surface

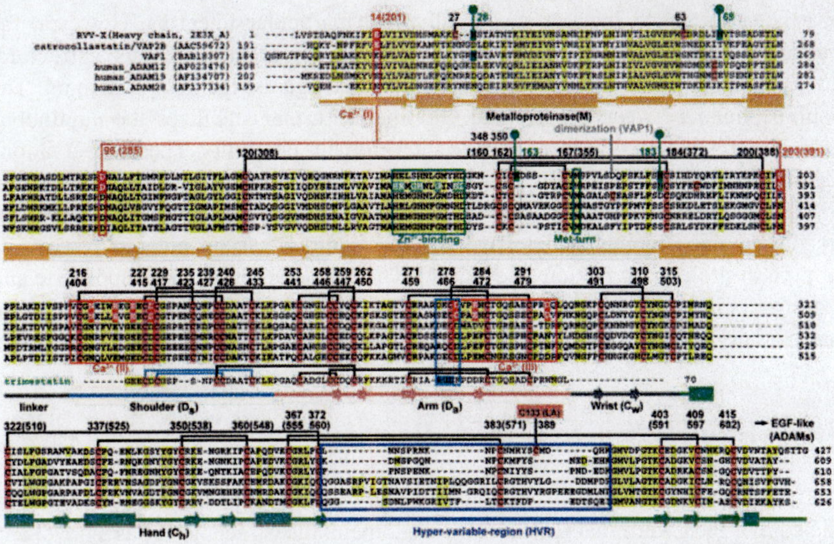

Fig. 27.3 Sequence alignment of RVV-X heavy chain, catrocollastatin/VAP2B, VAP1, trimestatin, and human ADAMs. The cysteinyl and conserved residues are shaded in *pink* and *yellow*, respectively. Disulfide bonds, Asn-linked glycosylations, secondary structures, and domains are drawn schematically. The HVR, Ca^{2+}-binding sites, Zn^{2+}-binding site, and disintegrin-loop are boxed in *blue, red, green*, and *cyan*, respectively. The Ca^{2+}-coordinating residues are shaded in *red*. RVV-X residue numbers are indicated. The residue numbers of catrocollastatin/VAP2B are indicated in *parentheses*

toward the M-domain, resulting in the distal portion of the C-domain coming close to and facing towards the catalytic site in the M-domain (Fig. 27.2a). Both the D_s and D_a segments contain calcium-binding sites II and III in which the calcium ions are situated in pentagonal bipyramidal coordination with the nearby oxygen atoms of the protein (Takeda, 2008; Takeda et al., 2006). The residues that are important for stabilizing the C-shaped MDC domain architecture (e.g., the number and spacing of cysteinyl residues and the residues that form calcium-binding sites II and III) are strictly conserved in the primary structures of all known P-III SVMPs and canonical ADAMs (Fig. 27.3) (Takeda, 2008; Takeda et al., 2006). Recently, the first crystal structure of the entire ectodomain of human ADAM was determined, which revealed that the C-shaped configuration of the MDC domains observed in P-III SVMPs is also conserved in mammalian ADAMs (Liu et al., 2009). Because of the small number of secondary structural elements, bound calcium ions and disulfide bonds are essential for the structural rigidity of this class of proteins.

As predicted by the amino acid sequences, the structure of the D_a segment is very similar to that of the RGD-containing disintegrin trimestatin (Fig. 27.2c) (Fujii et al., 2003) with the exception of the disintegrin-loop and the carboxyl terminus of the D_a segment. These regions in disintegrins are highly mobile candidate sites for integrin-binding (Adler et al., 1991; Saudek et al., 1991; Senn and Klaus, 1993). Using isolated whole or partial D-domains, numerous ADAMs and P-III SVMPs have been shown to interact specifically with particular integrins. However, the crystal structures indicate that the bound calcium ion at site III forms a structural core, which stabilizes the disintegrin-loop packed against the next C-domain. The disulfide bond between the D- and C-domains further stabilizes the continuous structure, suggesting that there is little inter-domain flexibility. These observations suggest that the disintegrin-loop in ADAMs and P-III SVMPs is unavailable for integrin binding due to steric hindrance (Takeda, 2009; Takeda et al., 2006). However, the subsequent C-domain does have surface features that are potentially suitable for protein-protein interaction (see below). The D-domain is located opposite to and apart from the M-domain catalytic site. Thus, it may function primarily as a scaffold that spatially allocates two other functional domains.

Hyper-Variable Region of the C-Domain

The C-domain is structurally subdivided into the wrist (C_w) and hand (C_h) segments. The core of the C-terminal region of the C-domain, the C_h segment, has a novel and unique α/β-fold structure with no structural homology to currently known proteins, except for the corresponding segments of P-III SVMPs (Igarashi et al., 2007; Muniz et al., 2008; Takeda et al., 2006, 2007; Zhu et al., 2009), ADAMs (Janes et al., 2005; Liu et al., 2009), and ADAMTSs (Akiyama et al., 2009; Gerhardt et al., 2007; Mosyak et al., 2008). The C-domain of these adamalysin/reprolysin/ADAM family proteins has been deposited in the Pfam protein domain database as ADAM_CR family (pfam08516). The C_h segment is characterized by a core region that is stabilized by conserved disulfide bonds, and peripheral loops that protrude from the core, providing extended surface areas that are most likely involved in protein binding.

The loop that encompasses residues 561–582 and extends across the central region of the C_h segment of catrocollastatin/VAP2B (blue region in Fig. 27.2a) is the region in which the sequences of P-III SVMPs and ADAMs are most divergent and variable in length (11–55 amino acids) (Fig. 27.3). Therefore, this region has been designated as the hyper-variable region (HVR) (Takeda, 2009; Takeda et al., 2006). Adamalysin/reprolysin/ADAM family proteins have distinct HVR sequences, resulting in molecular surface features that could help determine their specificity for their specific binding counterparts. Interestingly, the HVR is located opposite to the catalytic site within the C-shaped molecule, raising the intriguing possibility that the HVR directly or indirectly creates an exosite for substrate binding (Takeda, 2009; Takeda et al., 2006). Of note, in all the available crystal structures of P-III SVMPs except RVV-X, the HVR interacts with the neighboring molecule to form crystal contacts (Igarashi et al., 2007; Muniz et al., 2008; Takeda et al., 2006; Zhu et al., 2009). This finding suggests its potential function as an interface in protein-protein interactions.

Recent crystallographic and functional studies of the ADAMTS family proteinases have shed new light on the functions of the C_h segment and HVR. The crystal structures of the partial fragments of human ADAMTS1, 4, 5, and

13 revealed that the D-domains of ADAMTSs show no structural homology to the D-domains of ADAMs or P-III SVMPs, but were very similar in structure to the C_h segments of ADAMs and P-III SVMPs, despite relatively low sequence identities (Akiyama et al., 2009; Gerhardt et al., 2007; Mosyak et al., 2008; Takeda, 2009). Therefore, the term D-domain for ADAMTS proteins is a misnomer. Moreover, recent crystallographic determination of a fragment of the disintegrin-like/thrombospondin type-1 motif/cysteine-rich/spacer (DTCS) domains of ADAMTS13 (designated ADAMTS13-DTCS) revealed that the N-terminal portion of the C-domain (residues 440–531, designated the C_A domain) in ADAMTS13 also contains a fold structurally homologous to that of the C_h segment of ADAMs and P-III SVMPs, despite the lack of sequence similarity (Akiyama et al., 2009). Therefore, ADAMTS proteins possess two homologous domains that belong to the ADAM_CR family. In the case of ADAMTS13, the HVR in the D-domain (Akiyama et al., 2009; de Groot et al., 2009) and the variable-loop (V-loop) located adjacent to the HVR in the C-domain (Akiyama et al., 2009) each constitute a VWF-binding exosite.

RVV-X

Overall Structure of RVV-X

GM6001 (*N*-[(2R)-2-(hydroxamidocarbonylmethyl)-4-methylpentanoyl]-*L*-tryptophan methylamide) is a broad spectrum matrix metalloproteinase inhibitor that prevents the degradation and increases the crystallization probability of catrocollastatin/VAP2B (Igarashi et al., 2006). This crystallization inspired the use of GM6001 to crystallize RVV-X, resulting in successful structural determination at a 2.9-Å resolution. In the crystal structure, GM6001 was bound in the catalytic cleft. This result was similar to that with catrocollastatin/VAP2B, where the bulky idol ring of GM6001 was situated at the P2′ position projecting towards the outside of the protein, providing additional contacts with the neighboring molecule in the crystalline lattice (Takeda et al., 2007).

The overall structure of RVV-X resembles a hook-spanner-wrench configuration, in which the $M/D_s/D_a/C_w$ portion of the heavy chain forms a hook and the light chains, together with the C_h segment of the heavy chain, constitute a handle (Fig. 27.4a). The backbone structure of the heavy chain is essentially the same as each monomer of VAP1, catrocollastatin/VAP2B, or other P-III SVMP structures, with the exception of the sub-domain orientations (Fig. 27.4b, c). Structural comparison among these crystal structures revealed a dynamic property of the MDC domains that may be important for the function of adamalysin/reprolysin/ADAM family proteins (Igarashi et al., 2007). Therefore, the C-shaped heavy chain in RVV-X may be flexible in solution and the $M/D_s/D_a/C_w$ portion may behave as a mobile hook in the hook-spanner-wrench structure, while the remaining C_h/LA/LB portion may constitute a relatively solid handle.

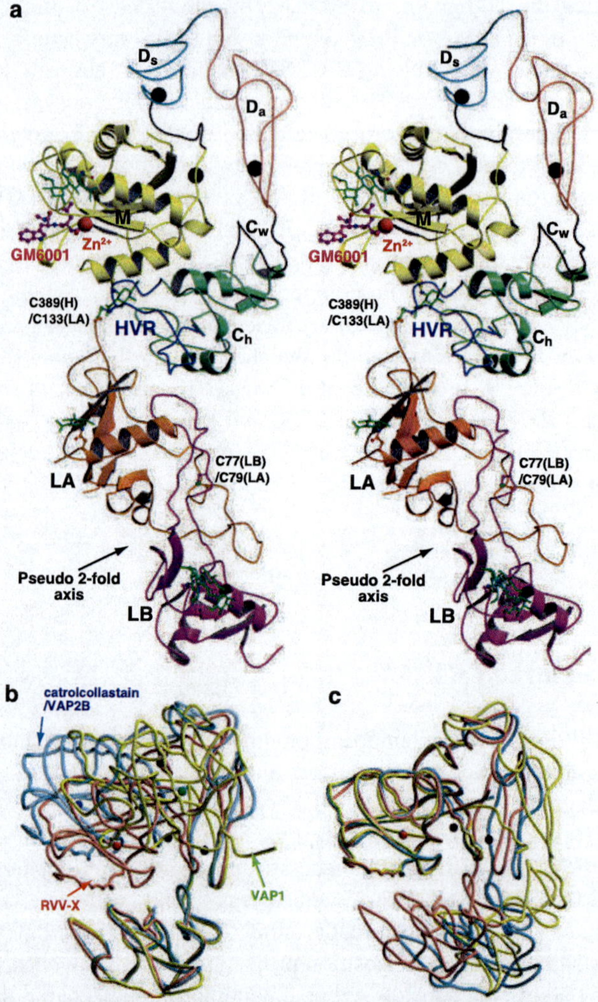

Fig. 27.4 Overall structure of RVV-X. (**a**) Ribbon structure of RVV-X in stereo. Bound calcium and zinc ions are represented by *black* and *red spheres*, respectively. The carbohydrate moieties (*in green*) linked to asparagine residues and GM6001 (*in magenta*) are shown in *ball-and-stick* representations. (**b**) Superimposition of the C_h segment of the RVV-X heavy chain (*in pink*) with that of the VAP1 monomer (chain-A in 2ERO, *in yellow*) and with that of catrocollastatin/VAP2B (chain-A in 2DW0, *in cyan*). The bound zinc and calcium atoms in RVV-X are shown as *red* and *black spheres*, respectively. The zinc atoms in VAP1 and catrocollastain/VAP2B are shown as *green* and *blue spheres*, respectively. (**c**) Superimposition of the M-domain of the RVV-X heavy chain with the M-domains of the VAP1 monomer and catrocollastatin/VAP2B

Hyper-Variable Region – Mediated Protein–Protein Interaction

RVV-X has a unique cysteine residue (Cys389) in the middle of the HVR in the C_h segment (Figs. 27.3 and 27.4a). Cys389 forms a disulfide bond with the C-terminal cysteine residue (Cys133) of light chain A (LA) (Figs. 27.3 and 27.4a). Aside from this inter-chain disulfide bridge, Tyr346, Tyr347, and Met385 in the heavy chain form multiple hydrophobic interactions and hydrogen bonds with Tyr11, Phe12, and Pro131 in LA, further stabilizing the continuous C_h/LA structure (Takeda et al., 2007). The RVV-X structure represents the first example of an HVR-mediated protein-protein interaction. This result supports a model for substrate recognition by adamalysin/reprolysin/ADAM family proteins in which the HVR captures substrates by binding to an associated protein (Takeda, 2009; Takeda et al., 2006).

RVV-X Light Chains Form a Gla Domain-Binding Exosite

The two homologous light chains of RVV-X have a fold similar to the carbohydrate-recognition domain of rat mannose binding protein (Weis et al., 1991). They form an intertwined dimer in which the central portion of each chain is projected towards the adjoining subunit (Fig. 27.4a). This represents an example of three-dimensional domain swapping, which was first identified in the structure of factor IX/X-binding protein (IX/X-bp) (Mizuno et al., 1997; Zelensky and Gready, 2005). The pseudo-symmetrical RVV-X light chain dimer structure is quite similar to the structure of X-bp in complex with the factor X Gla domain (Mizuno et al., 2001) (Fig. 27.5). Factors VII (VIIa), IX (IXa) and X (Xa), protein C (PC) and its active form (PCA), and prothrombin are vitamin K-dependent coagulation factors that contain Gla residues in their N-terminal domains. These residues are essential for their function in plasma (Furie et al., 1999; Mann et al., 1990) and mediate inhibition by snake venom anticoagulants (Morita, 2005). X-bp has strong anticoagulant activities because it binds to the factor X Gla domain, inhibiting its membrane-anchoring function (Atoda et al., 1998). The hydrophobic residues that are critical for this membrane-anchoring function (Phe4, Leu5, and Val8) interact with the hydrophobic patch formed by hydrophobic residues (Met113, Ile114 and Ala115) of the B chain in X-bp (Mizuno et al., 2001). Those residues are conserved in RVV-X (Phe114, Ile115, and Ala116 of LB) (Fig. 27.5b, d). The positively charged patches on X-bp that directly interact with the factor X Gla residues are conserved, but are less prominent in RVV-X because of amino acid substitutions (especially Ile102 and Glu105) (Fig. 27.5b, d).

The similarities in the overall structure and surface features between the RVV-X light chains and X-bp suggest the intriguing possibility that RVV-X recognizes the factor X Gla domain through an exosite formed by its light chains (Fig. 27.5c). When a properly folded Gla domain is absent from factor X, the rate of factor X activation by RVV-X is markedly diminished. In the acarboxy factor X, in which Gla formation has been blocked by a vitamin K antagonist (Lindhout et al., 1978)

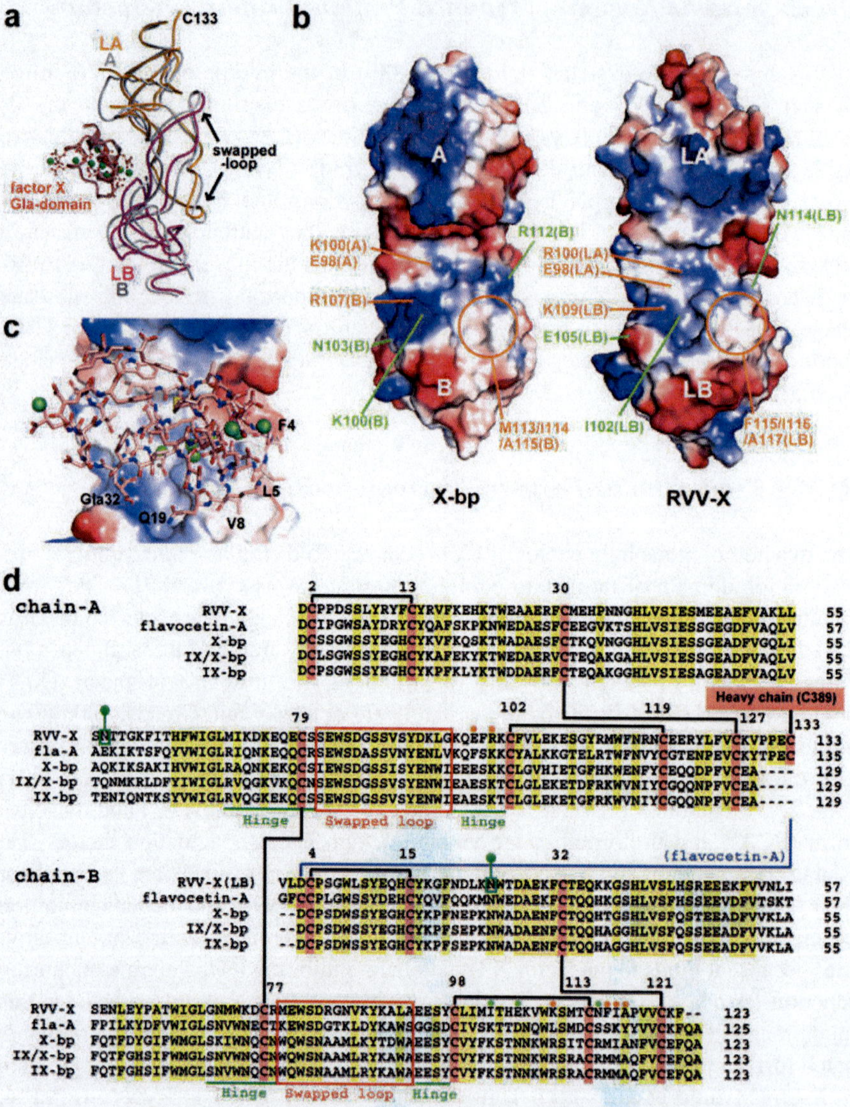

Fig. 27.5 Comparison of the RVV-X light chains and X-bp. (**a**) Superimposition of the RVV-X light chains (LA in *orange* and LB in *magenta*) onto the structure of X-bp (*in gray*) in complex with the factor Xa (1IOD) Gla domain (*in pink*). The Gla residues and Ca^{2+} ions are shown in *ball-and-stick* representation and as *green spheres*, respectively. (**b**) The molecular surfaces of X-bp and the light chains of RVV-X are represented according their electrochemical potentials (*blue* for positive, *red* for negative) and are viewed from the pseudo two-fold axis. Conserved and varied residues are labeled in *orange* and *green*, respectively. (**C**) A model of the RVV-X light chains in complex with the Gla domain, which was positioned based on the X-bp/factor X Gla domain complex structure. Residues in factor X are indicated. (**d**) Amino acid sequence alignment of the two light chains of RVV-X with either the α subunit (A chain) or β subunit (B chain) of flavocetin-A, X-bp, IX/X-bp, and IX-bp. Disulfide bonds and Asn-linked glycosylations are schematically indicated

or the Des [1–44] factor X (Morita and Jackson, 1986), factor X activation occurs at less than 1% of the rate of native factor X. Factor X activation by RVV-X is dramatically enhanced by millimolar Ca^{2+} concentrations, which induce a conformational change in the Gla domain that enhances its binding to RVV-X (Skogen et al., 1983). Moreover, RVV-X–catalyzed factor X activation is inhibited by X-bp (Takeya et al., 1992). The light chains of the structurally homologous prothrombin activator CA-1 serve as a Gla domain-binding exosite (Yamada et al., 1996). Collectively, these observations suggest that the concave cleft created between the two light chains in RVV-X may function as a factor X-binding exosite.

Factor X Docking Model

Factor X consists of a light chain and a heavy chain connected by a single disulfide bond (Mann et al., 1990). The light chain contains an N-terminal Gla domain and two EGF-like domains. The heavy chain contains a trypsin-like SP domain. The crystal structure for factor X zymogen has not been reported; however, several structures are available for fragments of the activated form, factor Xa (Kamata et al., 1998; Mizuno et al., 2001; Padmanabhan et al., 1993; Wang et al., 2003). Therefore, the structures of Des [1–44] factor Xa (PDBID: 1XKA) and the Gla domain (residues 1–44, PDBID: 1IOD) were previously used for a docking study (Takeda et al., 2007) (Fig. 27.6). In the model, the second EGF domain (EGF2) and the SP domain of factor Xa were positioned such that the factor Xa heavy chain N-terminal residue (Ile195) came close to the RVV-X active site and the globular SP domain fitted into the concave surface created by the C_h/LA domains. Because the linker between the two EGF domains was most likely flexible in solution (Kamata et al., 1998), a 50-degree bend was introduced between the two EGF domains (Fig. 27.6a) such that EGF1 fitted into the convex surface of the LA domain (Fig. 27.6b). This displacement successfully positioned the EGF1 domain N-terminus in close proximity to the Gla domain C-terminus, which was positioned according to the X-bp/factor-X Gla domain complex structure.

In the factor Xa structure, Ile195 is buried within the protein (Kamata et al., 1998). However, in factor X zymogen, the intact Arg194–Ile195–containing segment must be situated on the molecular surface, as is the case in the equivalent segments of other SP zymogen structures (Freer et al., 1970; Papagrigoriou et al., 2006; Wang et al., 1985). Therefore, the residues N-terminal to the scissile peptide bond of factor X (the C-terminal portion of AP) may bind to the non-primed cleft region of the RVV-X M-domain. The region of factor X that is C-terminal to the scissile peptide (the pink segment in Fig. 27.6c) is located along the SP domain surface, resulting in its binding to the primed region of RVV-X, similar to the peptidic inhibitor GM6001 in the crystal structure (Fig. 27.6c). In the docking model (Takeda et al., 2007), since each factor Xa segment was positioned as a rigid body without any collision with RVV-X, the active site zinc atom of RVV-X and Ile195 of factor Xa were 16 Å apart. However, the intrinsic hinge motions of the modular $M/D_s/D_a/C_w$ architecture (Igarashi et al., 2007) and conformational changes

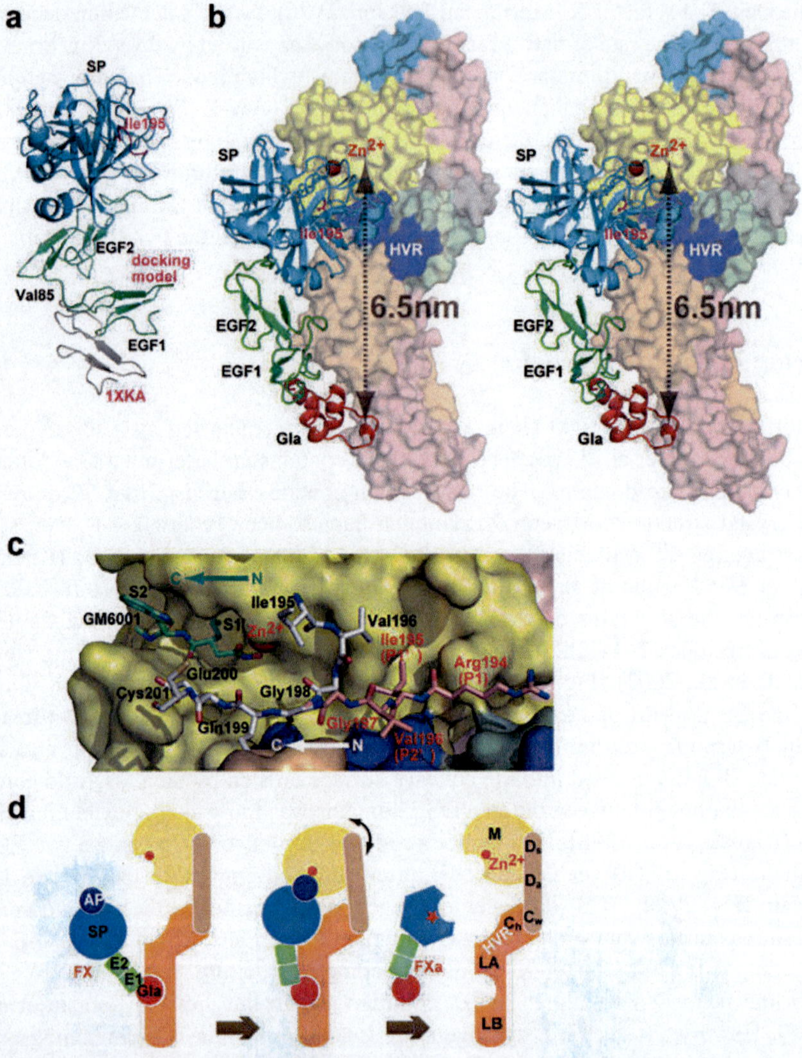

Fig. 27.6 Docking model. (**a**) Structure of factor Xa in the docking model. The EGF1 segment of the original structural model (1XKA) is shown in *gray*. (**b**) Docking model. The surfaces of the RVV-X sub-domains are colored as in Fig. 27.4a. Factor Xa is shown in ribbon representation. Ile195 (*in stick representation*) and the N-terminal region (residues 195–201) of the factor X heavy chain are shown in *magenta*. (**c**) Close-up view of the RVV-X catalytic site of the docking model, viewed from inside the factor Xa molecule. The N-terminal residues of factor Xa are shown in white and those of the model of factor X (zymogen) are shown in *light pink*. Because the factor X structure is currently unavailable, we assumed that this region has an extended structure. (**d**) Schematic model of factor X activation by RVV-X

upon association of RVV-X and factor X zymogen may allow the catalytic site of RVV-X to interact directly with the Arg194-Ile195 bond of factor X when in solution.

Mechanism of Factor X Activation by RVV-X

Based on the crystal structure and docking model, a model of factor X activation by RVV-X has been proposed (Fig. 27.6d) (Takeda et al., 2007). The middle portion of LA/LB forms the Gla domain-binding exosite that may serve as the primary capture site for factor X zymogen in circulation. This interaction also plays a regulatory role in the Ca^{2+}-dependent activation of factor X by RVV-X. Each chain of RVV-X is connected to its adjoining chain by a disulfide bond, and there are extensive surface contact areas between the chains. Thus, the C_h/LA/LB portion of RVV-X may be rigid enough to act as a scaffold to accommodate the elongated tulip-shaped factor X molecule, while separating the Gla domain and the scissile peptide bond in the SP domain. Factor X has a modular structure that is potentially flexible enough to fit along this scaffold. The relatively large separation (~ 65 Å) of the catalytic site and the Gla domain-binding exosite explains the high specificity of RVV-X for factor X. Additionally, the mobile "hook" portion of RVV-X, M/D_s/D_a/C_w, not only exerts a catalytic effect, but may also help regulate the binding affinity between molecules, thus driving the catalytic cycle; the M-domain may associate with the AP region of zymogen (factor X) and release it upon activation. The concave surface created by the boundary of the C_h/LA domains, exposing the HVR region at the middle of its surface, represents another potential exosite for the SP domain of factor X. Conformational changes on the SP surface, associated with the internalization of the NH^{3+} group of Ile195 upon activation (Freer et al., 1970; Huber and Bode, 1978), may also contribute to factor Xa release from RVV-X. The basis of this catalytic mechanism of factor X activation by RVV-X is essentially consistent with that originally proposal by Morita (Morita, 1998).

Factor X is physiologically activated by either the factor IXa/VIIIa complex or the factor IIa/tissue factor (TF) complex in the intrinsic or extrinsic pathway, respectively. Activation of factor X by either of these complexes is selective and involves specific interactions between the substrate (factor X) and enzyme (IXa/VIIIa or TF/VIIa complex). The crystal structure of the complex of active site–inhibited factor VIIa with the subtilisin-treated TF extracellular domain (Banner et al., 1996) revealed that factor VII adopts an extended conformation in the complex. This result suggests that factor X may also adopt a similar extended conformation on the surface of the VIIa/TF complex during its proteolytic activation. The intrinsic Xase complexes are SPs and are therefore structurally unrelated to RVV-X. The intrinsic Xases function on the phospholipid membrane, while RVV-X works in circulation. However, both activators possess an elongated scaffold that might be optimal for specifically targeting the elongated factor X molecule. Structural details of such extended interactions between the factor X and intrinsic Xase complexes are only beginning to be understood. Comparisons between the structures of intrinsic and venom-derived factor X activators may help us to understand how phylogenetically unrelated proteins develop scaffolds that are functionally related to, yet structurally distinct from, one another.

Implications for the Molecular Evolution of Snake Venom Protein Complexes

Protein complexes that exhibit markedly enhanced pharmacological activities compared to the individual components play significant roles in snake venom toxicity (Doley and Kini, 2009). Such complexes include the C-type lectin-like (CLPs) snake venom proteins (snake venom C-type lectins, or Snaclecs) (Clemetson et al., 2009), which are characterized by a unique dimerization mechanism of protein evolution. In this mechanism, two monomers swap a portion of the long loop region, forming a stable functional unit. This mechanism creates a new concave surface for target binding, allowing anticoagulant, procoagulant, platelet-modulating, and other activities (Morita, 2005). Heterodimers can further aggregate with each other to form higher-order cyclic tetramers $(\alpha\beta)_4$ (Batuwangala et al., 2004; Fukuda et al., 2000; Murakami et al., 2003), or dimers $(\alpha\beta)_2$ (Hooley et al., 2008). Alternatively, as is the case for RVV-X, the heterodimers can form covalently linked complexes with a metalloproteinase chain, creating an exosite. The sequences of P-IIId SVMPs and C-type lectin-like light chains, as determined from the cDNA sequences, are increasingly being identified as being provided by separate genes and not by single genes (Fox and Serrano, 2008; Siigur et al., 2004). The RVV-X structure is a good example of evolutionary gain-of-function by snake venom protein complexes for ligand binding specificity.

Concluding Remarks

ADAM proteinases constitute the major membrane-bound sheddases that proteolytically process cell surface proteins for cell-cell communication. As such, they have emerged as potential targets for various diseases. P-III SVMPs are key toxins involved in venom-induced pathogenesis, and thus are important targets for antivenom therapeutics. However, the physiological targets of ADAMs and P-III SVMPs, and the molecular mechanisms of target recognition, remain to be elucidated. Recent crystallographic studies shed new light on the structures and functions of this class of proteinases, and suggest that the HVR constitutes an exosite that captures targets or their associated proteins. The RVV-X structure is consistent with this model and provides insights into the molecular basis of HVR-mediated protein-protein interactions and target recognition for proteolysis by ADAM/adamalysin/reprolysin family proteins.

References

Adler, M., Lazarus, R.A., Dennis, M.S., Wagner, G., 1991. Solution structure of kistrin, a potent platelet aggregation inhibitor and GP IIb-IIIa antagonist. Science 253, 445–448.

Akiyama, M., Takeda, S., Kokame, K., Takagi, J., Miyata, T., 2009. Crystal structures of the non-catalytic domains of ADAMTS13 reveal multiple discontinuous exosites for von Willebrand factor. Proc. Natl. Acad. Sci. U.S.A. 106, 19274–19279.

Apte, S.S., 2009. A disintegrin-like and metalloprotease (reprolysin-type) with thrombospondin type 1 motif (ADAMTS) superfamily-functions and mechanisms. J. Biol. Chem. 284, 31493–31497.

Atoda, H., Ishikawa, M., Mizuno, H., Morita, T., 1998. Coagulation factor X-binding protein from *Deinagkistrodon acutus* venom is a Gla domain-binding protein. Biochemistry 37, 17361–17370.

Banner, D.W., D'Arcy, A., Chene, C., Winkler, F.K., Guha, A., Konigsberg, W.H., Nemerson, Y., Kirchhofer, D., 1996. The crystal structure of the complex of blood coagulation factor VIIa with soluble tissue factor. Nature 380, 41–46.

Batuwangala, T., Leduc, M., Gibbins, J.M., Bon, C., Jones, E.Y., 2004. Structure of the snake-venom toxin convulxin. Acta Crystallogr. D. Biol. Crystallogr. 60, 46–53.

Bjarnason, J.B., Fox, J.W., 1995. Snake venom metalloendopeptidases: reprolysins. Meth. Enzymol. 248, 345–368.

Black, R.A., Rauch, C.T., Kozlosky, C.J., Peschon, J.J., Slack, J.L., Wolfson, M.F., Castner, B.J., Stocking, K.L., Reddy, P., Srinivasan, S., Nelson, N., Boiani, N., Schooley, K.A., Gerhart, M., Davis, R., Fitzner, J.N., Johnson, R.S., Paxton, R.J., March, C.J., Cerretti, D.P., 1997. A metalloproteinase disintegrin that releases tumour-necrosis factor-alpha from cells. Nature 385, 729–733.

Blobel, C.P., 2005. ADAMs: key components in EGFR signalling and development. Nat. Rev. Mol. Cell. Biol. 6, 32–43.

Blobel, C.P., Wolfsberg, T.G., Turck, C.W., Myles, D.G., Primakoff, P., White, J.M., 1992. A potential fusion peptide and an integrin ligand domain in a protein active in sperm-egg fusion. Nature 356, 248–252.

Bode, W., Gomis-Ruth, F.X., Stockler, W., 1993. Astacins, serralysins, snake venom and matrix metalloproteinases exhibit identical zinc-binding environments (HEXXHXXGXXH and Met-turn) and topologies and should be grouped into a common family, the 'metzincins'. FEBS Lett. 331, 134–140.

Calvete, J.J., Marcinkiewicz, C., Monleon, D., Esteve, V., Celda, B., Juarez, P., Sanz, L., 2005. Snake venom disintegrins: evolution of structure and function. Toxicon 45, 1063–1074.

Clemetson, K.J., Morita, T., Kini, R.M., 2009. Scientific and standardization committee communications: classification and nomenclature of snake venom C-type lectins and related proteins. J. Thromb. Haemost. 7, 360.

de Groot, R., Bardhan, A., Ramroop, N., Lane, D.A., Crawley, J.T., 2009. Essential role of the disintegrin-like domain in ADAMTS13 function. Blood 113, 5609–5616.

Doley, R., Kini, R.M., 2009. Protein complexes in snake venom. Cell. Mol. Life Sci. 66, 2851–2871.

Duffy, M.J., Lynn, D.J., Lloyd, A.T., O'Shea, C.M., 2003. The ADAMs family of proteins: from basic studies to potential clinical applications. Thromb. Haemost. 89, 622–631.

Edwards, D.R., Handsley, M.M., Pennington, C.J., 2009. The ADAM metalloproteinases. Mol. Aspects Med. 29, 258–289.

Evans, J.P., 2001. Fertilin beta and other ADAMs as integrin ligands: insights into cell adhesion and fertilization. Bioessays 23, 628–639.

Fox, J.W., Serrano, S.M., 2005. Structural considerations of the snake venom metalloproteinases, key members of the M12 reprolysin family of metalloproteinases. Toxicon 45, 969–985.

Fox, J.W., Serrano, S.M., 2008. Insights into and speculations about snake venom metalloproteinase (SVMP) synthesis, folding and disulfide bond formation and their contribution to venom complexity. FEBS J. 275, 3016–3030.

Freer, S.T., Kraut, J., Robertus, J.D., Wright, H.T., Xuong, N.H., 1970. Chymotrypsinogen: 2.5-Å crystal structure, comparison with α-chymotrypsin, and implications for zymogen activation. Biochemistry 9, 1997–2009.

Fujii, Y., Okuda, D., Fujimoto, Z., Horii, K., Morita, T., Mizuno, H., 2003. Crystal structure of trimestatin, a disintegrin containing a cell adhesion recognition motif RGD. J. Mol. Biol. 332, 1115–1122.

Fukuda, K., Mizuno, H., Atoda, H., Morita, T., 2000. Crystal structure of flavocetin-A, a platelet glycoprotein Ib-binding protein, reveals a novel cyclic tetramer of C-type lectin-like heterodimers. Biochemistry 39, 1915–1923.

Furie, B., Bouchard, B.A., Furie, B.C., 1999. Vitamin K-dependent biosynthesis of gamma-carboxyglutamic acid. Blood 93, 1798–1808.

Gerhardt, S., Hassall, G., Hawtin, P., McCall, E., Flavell, L., Minshull, C., Hargreaves, D., Ting, A., Pauptit, R.A., Parker, A.E., Abbott, W.M., 2007. Crystal structures of human ADAMTS-1 reveal a conserved catalytic domain and a disintegrin-like domain with a fold homologous to cysteine-rich domains. J. Mol. Biol. 373, 891–902.

Gomis-Ruth, F.X., 2003. Structural aspects of the metzincin clan of metalloendopeptidases. Mol. Biotechnol. 24, 157–202.

Gomis-Ruth, F.X., Kress, L.F., Bode, W., 1993. First structure of a snake venom metalloproteinase: a prototype for matrix metalloproteinases/collagenases. EMBO J. 12, 4151–4157.

Gomis-Ruth, F.X., Kress, L.F., Kellermann, J., Mayr, I., Lee, X., Huber, R., Bode, W., 1994. Refined 2.0 Å X-ray crystal structure of the snake venom zinc-endopeptidase adamalysin II. Primary and tertiary structure determination, refinement, molecular structure and comparison with astacin, collagenase and thermolysin. J. Mol. Biol. 239, 513–544.

Gowda, D.C., Jackson, C.M., Hensley, P., Davidson, E.A., 1994. Factor X-activating glycoprotein of Russell's viper venom. Polypeptide composition and characterization of the carbohydrate moieties. J. Biol. Chem. 269, 10644–10650.

Hite, L.A., Shannon, J.D., Bjarnason, J.B., Fox, J.W., 1992. Sequence of a cDNA clone encoding the zinc metalloproteinase hemorrhagic toxin e from *Crotalus atrox*: evidence for signal, zymogen, and disintegrin-like structures. Biochemistry 31, 6203–6211.

Hooley, E., Papagrigoriou, E., Navdaev, A., Pandey, A.V., Clemetson, J.M., Clemetson, K.J., Emsley, J., 2008. The crystal structure of the platelet activator aggretin reveals a novel (αβ)2 dimeric structure. Biochemistry 47, 7831–7837.

Huang, T.F., Holt, J.C., Lukasiewicz, H., Niewiarowski, S., 1987. Trigramin. A low molecular weight peptide inhibiting fibrinogen interaction with platelet receptors expressed on glycoprotein IIb-IIIa complex. J. Biol. Chem. 262, 16157–16163.

Huber, R., Bode, W., 1978. Structural Basis of the activation and action of trypsin. Acc. Chem. Res. 11, 114–122.

Igarashi, T., Araki, S., Mori, H., Takeda, S., 2007. Crystal structures of catrocollastatin/VAP2B reveal a dynamic, modular architecture of ADAM/adamalysin/reprolysin family proteins. FEBS Lett. 581, 2416–2422.

Igarashi, T., Oishi, Y., Araki, S., Mori, H., Takeda, S., 2006. Crystallization and preliminary X-ray crystallographic analysis of two vascular apoptosis-inducing proteins (VAPs) from *Crotalus atrox* venom. Acta Crystallograph. F. Struct. Biol. Cryst. Commun. 62, 688–691.

Janes, P.W., Saha, N., Barton, W.A., Kolev, M.V., Wimmer-Kleikamp, S.H., Nievergall, E., Blobel, C.P., Himanen, J.P., Lackmann, M., Nikolov, D.B., 2005. Adam meets Eph: an ADAM substrate recognition module acts as a molecular switch for ephrin cleavage in trans. Cell 123, 291–304.

Kamata, K., Kawamoto, H., Honma, T., Iwama, T., Kim, S.H., 1998. Structural basis for chemical inhibition of human blood coagulation factor Xa. Proc. Natl. Acad. Sci. U.S.A. 95, 6630–6635.

Kisiel, W., Hermodson, M.A., Davie, E.W., 1976. Factor X activating enzyme from Russell's viper venom: isolation and characterization. Biochemistry 15, 4901–4906.

Kuno, K., Kanada, N., Nakashima, E., Fujiki, F., Ichimura, F., Matsushima, K., 1997. Molecular cloning of a gene encoding a new type of metalloproteinase-disintegrin family protein with thrombospondin motifs as an inflammation associated gene. J. Biol. Chem. 272, 556–562.

Lindhout, M.J., Kop-Klaassen, B.H., Hemker, H.C., 1978. Activation of decarboxyfactor X by a protein from Russell's viper venom. Purification and partial characterization of activated decarboxyfactor X. Biochim. Biophys. Acta, 533, 327–341.

Liu, H., Shim, A.H., He, X., 2009. Structural characterization of the ectodomain of a disintegrin and metalloproteinase-22 (ADAM22), a neural adhesion receptor instead of metalloproteinase: insights on ADAM function. J. Biol. Chem. 284, 29077–29086.

Mann, K.G., Nesheim, M.E., Church, W.R., Haley, P., Krishnaswamy, S., 1990. Surface-dependent reactions of the vitamin K-dependent enzyme complexes. Blood 76, 1–16.

Mizuno, H., Fujimoto, Z., Atoda, H., Morita, T., 2001. Crystal structure of an anticoagulant protein in complex with the Gla domain of factor X. Proc. Natl. Acad. Sci. U.S.A. 98, 7230–7234.

Mizuno, H., Fujimoto, Z., Koizumi, M., Kano, H., Atoda, H., Morita, T., 1997. Structure of coagulation factors IX/X-binding protein, a heterodimer of C-type lectin domains. Nat. Struct. Biol. 4, 438–441.

Mochizuki, S., Okada, Y., 2007. ADAMs in cancer cell proliferation and progression. Cancer Sci. 98, 621–628.

Morita, T., 1998. Proteases which Activate Factor X. Alaken, Colorado.

Morita, T., 2005. Structures and functions of snake venom CLPs (C-type lectin-like proteins) with anticoagulant-, procoagulant-, and platelet-modulating activities. Toxicon 45, 1099–1114.

Morita, T., Jackson, C.M., 1986. Preparation and properties of derivatives of bovine factor X and factor Xa from which the gamma carboxyglutamic acid containing domain has been removed. J. Biol. Chem. 261, 4015–4023.

Moss, M.L., Bartsch, J.W., 2004. Therapeutic benefits from targeting of ADAM family members. Biochemistry 43, 7227–7235.

Moss, M.L., Jin, S.L., Milla, M.E., Bickett, D.M., Burkhart, W., Carter, H.L., Chen, W.J., Clay, W.C., Didsbury, J.R., Hassler, D., Hoffman, C.R., Kost, T.A., Lambert, M.H., Leesnitzer, M.A., McCauley, P., McGeehan, G., Mitchell, J., Moyer, M., Pahel, G., Rocque, W., Overton, L.K., Schoenen, F., Seaton, T., Su, J.L., Warner, J., Willard, D., Becherer, J.D., 1997. Cloning of a disintegrin metalloproteinase that processes precursor tumour-necrosis factor-alpha. Nature 385, 733–736.

Mosyak, L., Georgiadis, K., Shane, T., Svenson, K., Hebert, T., McDonagh, T., Mackie, S., Olland, S., Lin, L., Zhong, X., Kriz, R., Reifenberg, E.L., Collins-Racie, L.A., Corcoran, C., Freeman, B., Zollner, R., Marvell, T., Vera, M., Sum, P.E., Lavallie, E.R., Stahl, M., Somers, W., 2008. Crystal structures of the two major aggrecan degrading enzymes, ADAMTS4 and ADAMTS5. Protein Sci. 17, 16–21.

Muniz, J.R., Ambrosio, A.L., Selistre-de-Araujo, H.S., Cominetti, M.R., Moura-da-Silva, A.M., Oliva, G., Garratt, R.C., Souza, D.H., 2008. The three-dimensional structure of bothropasin, the main hemorrhagic factor from *Bothrops jararaca* venom: insights for a new classification of snake venom metalloprotease subgroups. Toxicon 52, 807–816.

Murakami, M.T., Zela, S.P., Gava, L.M., Michelan-Duarte, S., Cintra, A.C., Arni, R.K., 2003. Crystal structure of the platelet activator convulxin, a disulfide-linked α4β4 cyclic tetramer from the venom of *Crotalus durissus terrificus*. Biochem. Biophys. Res. Commun. 310, 478–482.

Murphy, G., 2008. The ADAMs: signalling scissors in the tumour microenvironment. Nat. Rev. Cancer 8, 929–941.

Okuda, D., Koike, H., Morita, T., 2002. A new gene structure of the disintegrin family: a subunit of dimeric disintegrin has a short coding region. Biochemistry 41, 14248–14254.

Padmanabhan, K., Padmanabhan, K.P., Tulinsky, A., Park, C.H., Bode, W., Huber, R., Blankenship, D.T., Cardin, A.D., Kisiel, W., 1993. Structure of human des(1–45) factor Xa at 2.2 Å resolution. J. Mol. Biol. 232, 947–966.

Papagrigoriou, E., McEwan, P.A., Walsh, P.N., Emsley, J., 2006. Crystal structure of the factor XI zymogen reveals a pathway for transactivation. Nat. Struct. Mol. Biol. 13, 557–558.

Peschon, J.J., Slack, J.L., Reddy, P., Stocking, K.L., Sunnarborg, S.W., Lee, D.C., Russell, W.E., Castner, B.J., Johnson, R.S., Fitzner, J.N., Boyce, R.W., Nelson, N., Kozlosky, C.J., Wolfson, M.F., Rauch, C.T., Cerretti, D.P., Paxton, R.J., March, C.J., Black, R.A., 1998. An essential role for ectodomain shedding in mammalian development. Science 282, 1281–1284.

Porter, S., Clark, I.M., Kevorkian, L., Edwards, D.R., 2005. The ADAMTS metalloproteinases. Biochem. J. 386, 15–27.
Saudek, V., Atkinson, R.A., Pelton, J.T., 1991. Three-dimensional structure of echistatin, the smallest active RGD protein. Biochemistry 30, 7369–7372.
Seals, D.F., Courtneidge, S.A., 2003. The ADAMs family of metalloproteases: multidomain proteins with multiple functions. Genes Dev. 17, 7–30.
Senn, H., Klaus, W., 1993. The nuclear magnetic resonance solution structure of flavoridin, an antagonist of the platelet GP IIb-IIIa receptor. J. Mol. Biol. 232, 907–925.
Siigur, E., Aaspollu, A., Trummal, K., Tonismagi, K., Tammiste, I., Kalkkinen, N., Siigur, J., 2004. Factor X activator from *Vipera lebetina* venom is synthesized from different genes. Biochim. Biophys. Acta 1702, 41–51.
Skogen, W.F., Bushong, D.S., Johnson, A.E., Cox, A.C., 1983. The role of the Gla domain in the activation of bovine coagulation factor X by the snake venom protein XCP. Biochem. Biophys. Res. Commun. 111, 14–20.
Takeda, S., 2008. VAP1: snake venom homolog of mammalian ADAMs, in: Messerschmidt, A. (Ed.), Handbook of Metalloproteins. John Wiley & Sons, Inc., City, pp. 1–15.
Takeda, S., 2009. Three-dimensional domain architecture of the ADAM family proteinases. Semin. Cell Dev. Biol. 20, 146–152.
Takeda, S., Igarashi, T., Mori, H., 2007. Crystal structure of RVV-X: an example of evolutionary gain of specificity by ADAM proteinases. FEBS Lett. 581, 5859–5864.
Takeda, S., Igarashi, T., Mori, H., Araki, S., 2006. Crystal structures of VAP1 reveal ADAMs' MDC domain architecture and its unique C-shaped scaffold. EMBO J. 25, 2388–2396.
Takeya, H., Nishida, S., Miyata, T., Kawada, S., Saisaka, Y., Morita, T., Iwanaga, S., 1992. Coagulation factor X activating enzyme from Russell's viper venom (RVV-X). A novel metalloproteinase with disintegrin (platelet aggregation inhibitor)-like and C-type lectin-like domains. J. Biol. Chem. 267, 14109–14117.
Takeya, H., Nishida, S., Nishino, N., Makinose, Y., Omori-Satoh, T., Nikai, T., Sugihara, H., Iwanaga, S., 1993. Primary structures of platelet aggregation inhibitors (disintegrins) autoproteolytically released from snake venom hemorrhagic metalloproteinases and new fluorogenic peptide substrates for these enzymes. J. Biochem. (Tokyo) 113, 473–483.
Tans, G., Rosing, J., 2001. Snake venom activators of factor X: an overview. Haemostasis 31, 225–233.
Wang, D., Bode, W., Huber, R., 1985. Bovine chymotrypsinogen A X-ray crystal structure analysis and refinement of a new crystal form at 1.8 Å resolution. J. Mol. Biol. 185, 595–624.
Wang, S.X., Hur, E., Sousa, C.A., Brinen, L., Slivka, E.J., Fletterick, R.J., 2003. The extended interactions and Gla domain of blood coagulation factor Xa. Biochemistry 42, 7959–7966.
Weis, W.I., Kahn, R., Fourme, R., Drickamer, K., Hendrickson, W.A., 1991. Structure of the calcium-dependent lectin domain from a rat mannose-binding protein determined by MAD phasing. Science 254, 1608–1615.
White, J.M., 2003. ADAMs: modulators of cell-cell and cell-matrix interactions. Curr. Opin. Cell. Biol. 15, 598–606.
White, J.M., Bidges, L., DeSimone, D.W., Tomczuk, M., Wolfsberg, T.G. 2005. Introduction to the ADAM family, in: Hooper, N.M., Lendeckel, U. (Eds.), The ADAM Family of Proteinases (vol. 4). Springer, Dordrecht, The Netherlands, pp. 1–29.
Yagami-Hiromasa, T., Sato, T., Kurisaki, T., Kamijo, K., Nabeshima, Y., Fujisawa-Sehara, A., 1995. A metalloprotease-disintegrin participating in myoblast fusion. Nature 377, 652–656.
Yamada, D., Sekiya, F., Morita, T., 1996. Isolation and characterization of carinactivase, a novel prothrombin activator in Echis carinatus venom with a unique catalytic mechanism. J. Biol. Chem. 271, 5200–5207.
Zelensky, A.N., Gready, J.E., 2005. The C-type lectin-like domain superfamily. FEBS J. 272, 6179–6217.
Zhu, Z., Gao, Y., Yu, Y., Zhang, X., Zang, J., Teng, M., Niu, L., 2009. Structural basis of the autolysis of AaHIV suggests a novel target recognizing model for ADAM/reprolysin family proteins. Biochem. Biophys. Res. Commun. 386, 159–164.

Chapter 28
Snake Venom Prothrombin Activators – The History

Jan Rosing and Guido Tans

Abstract Snake venom prothrombin activators have been shown to be excellent tools in diagnostic coagulation assays and in studying the mechanism of prothrombin activation. In this review we give an historical overview of the discovery of snake venom prothrombin activators and of their mechanisms of action on prothrombin. Based on the effects of the accessory components of the human prothrombin-activating complex (calcium ions, phospholipids and factor Va), venom prothrombin activators can be divided into four groups comprising activators that are metalloproteinases which either do not require accessory components (group A) or which require calcium ions for expressing activity (group B). The other activators are serine proteases that share homology with blood coagulation factor Xa and which require phospholipids plus calcium ions for optimal activity (Group C) or the activity of which is greatly enhanced by calcium ions, phospholipids and factor Va (group D).

Introduction

Already from the early beginning in blood coagulation research have snake venoms been recognized to be useful in diagnostic assays and as tools to gain insight in the molecular mechanism of blood coagulation factor activation (see e.g. Martin, 1893). Often venom coagulation research has been driven by the need to develop clinical laboratory assays in which snake venoms could be used as a source of (relatively cheap) material with unique properties and which bypassed the need for (partial) purification of clotting factors. The landmark venom reagent is Russell's viper venom (RVV), the factor X-activating principle of which (RVV-X) has found widespread use in the clinical and research laboratory (Marsh and Williams, 2005).

J. Rosing (✉)
Department of Biochemistry, Cardiovascular Research Institute Maastricht (CARIM), Maastricht University, P.O. Box 616, Maastricht, 6200 MD, The Netherlands
e-mail: j.rosing@bioch.unimaas.nl

R.M. Kini et al. (eds.), *Toxins and Hemostasis*, DOI 10.1007/978-90-481-9295-3_28,

But snake venoms are not only used in clinical assays, they also helped to unravel the molecular mechanism of blood coagulation reactions. Prothrombin activation is a prime example of this because studies in which snake venom preparations were used have often played an important role in many of the stages in which significant progress in understanding the biochemistry of prothrombin and its activation has been achieved. As early as in 1937 the venoms from *Notechis scutatus*, *Bothrops atrox*, *Bothrops jararaca* and *Crotalus basilicus* were used to establish that prothrombin is the inactive precursor of thrombin, the enzyme that is ultimately responsible for fibrin formation (Eagle, 1937). Other important contributions of snake venom prothrombin activators came later and three snakes have been of particular significance; they are *Echis carinatus* (saw-scaled viper), *Notechis scutatus scutatus* (Tiger snake) and *Oxyuranus scutellatus scutellatus* (Taipan snake). The venoms of these three snakes are also characteristic representatives of the classes of prothrombin activators to which they are nowadays recognized to belong. As it is our intention to describe the history of understanding how snake venom prothrombin activators act and how they contributed to our current knowledge of prothrombin activation we will focus on these three venoms and we refer the reader to (Kini, 2005; Kini et al., 2001; Rosing and Tans, 1992) for more comprehensive listings of venoms belonging to each group of activators. However, before we turn our attention to the venoms and their history we will first deal briefly with the biochemistry of prothrombin and its activation during normal blood coagulation and define the classes of prothrombin activators currently recognized.

Prothrombin Activation

Human prothrombin is a single chain glycoprotein (MW 72,000) (for reviews, see Davie et al., 1991; Rosing and Tans, 1988) which belongs to the so-called vitamin K-dependent clotting factors, that are proteins that undergo post-ribosomal carboxylation of a limited number of glutamic acid residues at the amino terminal of the molecule to result in γ-carboxyglutamic acid (Gla) residues. This reaction, which occurs in the liver, requires vitamin K. Prothrombin contains 10 Gla residues which are essential for the calcium-dependent binding of prothrombin to negatively charged procoagulant membranes (see below).

The enzyme responsible for in vivo activation of prothrombin is factor Xa but in order for activation to occur efficiently additional components (Ca^{2+}, negatively charged phospholipids and factor Va) are required. Factor Xa cleaves two peptide bonds during the conversion of prothrombin into thrombin (Fig. 28.1). Cleavage between Arg271-Thr272 results in the loss of the amino-terminal half of the molecule, called fragment 1.2. Thrombin is formed from the remaining so-called prethrombin 2 by cleavage between Arg320 and Ile321 upon which the active site is exposed. Since the latter cleavage occurs within a disulfide bond, thrombin is a two-chain enzyme consisting of a heavy (B-chain Mr 32,500) and light chain (A-chain Mr 4,600) and the active site is located in the B-chain originating from the carboxy terminal part of prothrombin. Both peptide bonds in prothrombin are

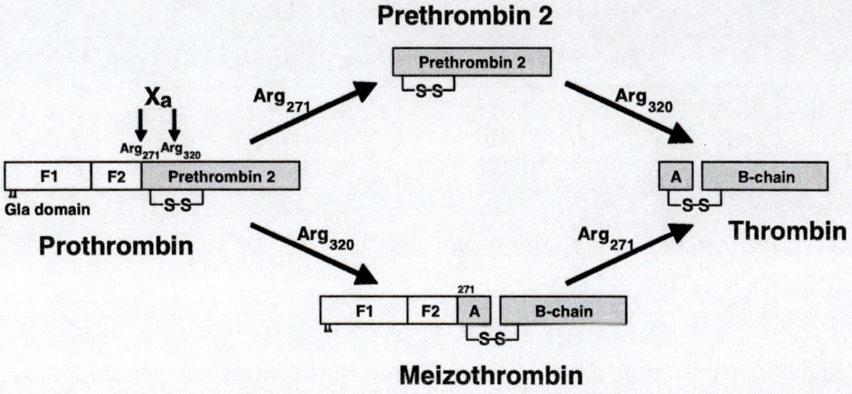

Fig. 28.1 Pathways of prothrombin activation

available for hydrolysis. When cleavage occurs first at Arg320 an active intermediate is formed which is called meizothrombin. Meizothrombin has the same MW as prothrombin but it has a fully functional active site in its B-chain with the rest of the original prothrombin molecule (fragment 1.2-A chain) still attached to it via a disulfide bond.

Meizothrombin and thrombin have equal esterolytic and amidolytic activity (Franza et al., 1975; Kornalik and Blomback, 1975; Morita et al., 1976) but meizothrombin has less than 5% of the activity of thrombin towards fibrinogen, factor V or platelets (Doyle and Mann, 1990; Rosing et al., 1986). Meizothrombin does, however, maintain the capacity to bind to the procoagulant surface via its Gla domain and in the presence of negatively charged phospholipids meizothrombin can in fact efficiently activate factor V (Tans et al., 1994). In addition meizothrombin, like thrombin can activate the anticoagulant protein C at the endothelial cell surface (Hackeng et al., 1996).

Apart from the Arg271–Thr272 and Arg320–Ile321 bonds, additional peptide bonds can be cleaved in prothrombin (or its derivatives) the most important of which is located between Arg155–Ser156. This bond is very efficiently cleaved by (meizo)thrombin (Rosing and Tans, 1988) which results in the removal of the amino terminal Gla-containing fragment 1 and depending on whether cleavage has occurred in prothrombin or meizothrombin either prethrombin 1 or meizothrombin des-fragment 1 is formed.

Since (meizo) thrombin can also cleave an additional peptide bond at Arg284 (Lanchantin et al., 1973; Petrovan et al., 1998; Rabiet et al., 1986) (close to the peptide bond cleaved by factor Xa) it will be clear that the pathway via which prothrombin is converted into thrombin has long been a subject of debate. Snake venom prothrombin activators have played an important role in the elucidation of these pathways and nowadays it is recognized that activation via meizothrombin is the kinetically favored pathway during factor Xa-catalyzed prothrombin activation in plasma (Krishnaswamy et al., 1986; Rabiet et al., 1986; Rosing et al., 1986; Tans et al., 1991).

Table 28.1 Kinetics of prothrombin activation

	Relative rate	K_m (mM)	k_{cat} (min^{-1})
Xa, Ca^{2+}	1	84	0.68
Xa, Ca^{2+}, PL	200	0.11	2.56
Xa, Ca^{2+}, PL, Va	310,000	0.14	4,050

PL phospholipids
Data from (Rosing et al., 1980; van Rijn et al., 1984). The relative rate was calculated assuming a plasma prothrombin concentration of 1.5 μM.

Prothrombin activation by factor Xa shows a particular feature typical for blood coagulation reactions which is that although factor Xa represents the enzymatic unit that catalyses the peptide bond cleavages, efficient activation of prothrombin only occurs when factor Xa is in complex with factor Va at the procoagulant membrane in the so-called prothrombinase complex the formation of which requires the presence of calcium ions. The reason for this is that factor Xa is in fact a very poor activator of prothrombin illustrated by a very high Km for prothrombin and a very low Vmax of thrombin formation (Table 28.1) (Rosing et al., 1980; van Rijn et al., 1984).The presence of the accessory components results in important changes in these kinetic parameters which shows that physiologically relevant thrombin formation will only occur by the complete prothrombinase complex. The procoagulant membrane is thought to greatly facilitate the formation of enzyme-cofactor-substrate complex through coordinate binding of all protein components to the surface. Factor Va not only facilitates the formation of the factor Xa–Va-prothrombin complex (through its affinity for both factor Xa and prothrombin) but it also accelerates one or more forward reactions in the prothrombin activation pathways as illustrated by the approximate 1,000-fold increase in Vmax.

It is this typical feature of additional components required in prothrombin activation which has resulted in a meaningful classification of snake venom prothrombin activators.

Classification of Prothrombin Activators

Denson (1976) first proposed to organize snake venom prothrombin activators with respect to their mechanism of action on prothrombin. He classified them as "complete" or "incomplete" depending on whether or not additional components were needed to obtain maximal prothrombin activating capacity. Later we took this idea a step further and proposed to classify prothrombin activators on the basis of which accessory components of the prothrombinase complex (Ca^{2+}, phospholipids and/or factor Va) were needed (Rosing and Tans, 1992) until finally, Kini (2005) proposed the classification by which prothrombin activators are grouped nowadays (Table 28.2).

Table 28.2 Classes of venom prothrombin activators

Class	Additional cofactor	Example
A	None	*Echis carinatus*
B	Ca^{2+}	*Echis multisquamatus*
C	Ca^{2+}, PL	*Oxyuranus scutellatus*
D	Ca^{2+}, PL, Va	*Notechis scutatus*

PL phospholipids

To be complete an additional group of prothrombin converting proteins from snake venoms can be recognized which concern enzymes that cleave peptide bonds in prothrombin without resulting in the formation of active thrombin. One of these venom proteases, acutin deserves special attention because this activator has been successfully used to specifically remove the amino terminal Gla-domain of prothrombin (Seegers et al., 1981). However, since we will be focusing on venom activators that convert prothrombin into thrombin we will not further discuss these activators here.

To understand how prothrombin activators belonging to each class function and how this knowledge has contributed to our understanding of factor Xa-catalyzed prothrombin activation it is actually best to describe the history of the characteristic examples of each group of activators. In fact, we can go back as far as 1909 where the venoms of *Echis carinatus* and *Notechis scutatus* are already encountered in one of the earliest literature references in this field (Mellanby, 1909). In this paper Mellanby convincingly demonstrated that these venoms must contain principles that were not acting on fibrinogen but instead exerted actions that resulted in the activation of prothrombin. An observation that was extended upon by Eagle in 1937 in the paper already mentioned earlier (Eagle, 1937). However, in these early years the focus in venom coagulation research was not on prothrombin activation but almost all efforts to put snake venom activators to use were directed towards Russell's viper venom as an alternative for thromboplastin (tissue factor/phospholipids) the agent essential for initiating the extrinsic coagulation pathway and determining the prothrombin time (Hjort et al., 1955; Macfarlane, 1948). In fact it was not until the 1960s that the number of papers concerning (specific) prothrombin activators began to rise resulting in substantial contributions to the understanding of the biochemistry of prothrombin activation. The first breakthrough occurred in 1966 when Jobin and Esnouf performed a detailed comparison of the venom prothrombin activator from *Notechis scutatus* with factor Xa (Jobin and Esnouf, 1966). An analysis brought about by the desire to develop a clinical assay in which factor Xa could be replaced by *Notechis* venom but which also yielded the first accurate and systematic quantitation of the actual rate enhancement by which the accessory components Ca^{2+}, factor Va and/or procoagulant phospholipids stimulated factor Xa-catalyzed prothrombin activation. Jobin and Esnouf showed that the *Notechis* venom activator (nowadays known as notecarin) had the same cofactor requirements as and behaved kinetically similar to factor Xa in the activation of prothrombin. An observation that we could very much appreciate later when we purified the activator present in *Notechis scutatus scutatus* venom and found that the enzyme was so much like factor Xa that it was actually very difficult to (kinetically) distinguish between the two (Tans et al., 1985). But in the late sixties early seventies the interest shifted to the prime example of group A activators, *Echis carinatus* which was to play an important role in unraveling the pathways of prothrombin activation.

Group A Activators – Ecarin from *Echis carinatus* (Saw-Scaled Viper) Venom

Kornalik was among the first to attempt purification of the prothrombin activating principle from a snake venom and when he finally succeeded to purify ecarin, the prothrombin activator from *Echis carinatus* venom (Kornalik et al., 1969; Kornalik and Blomback, 1975; Schieck et al., 1972) the whole field of prothrombin activating biochemistry profited. Until Kornalik's studies, in the late 60s and early 70s major efforts had been made to elucidate the pathway of factor Xa-catalyzed prothrombin activation, culminating in a series of excellent papers by the group of Jackson fully describing the prothrombin peptide bond cleavage pattern for all different possible compositions of the prothrombinase complex (Esmon and Jackson, 1974a, b; Esmon et al., 1974a, b, c). These publications led to the consensus that in all probability prothrombin activation had been solved and it was generally accepted that prothrombin activation occurred via prethrombin 2 as intermediate product i.e. that cleavage of prothrombin occurred first at Arg271 (Esmon et al., 1974c). However, the elucidation of prothrombin activation by ecarin showed that the peptide bond cleavage which gives rise to active site exposure (Arg320) was in fact available in the prothrombin molecule and could be efficiently cleaved to result in meizothrombin and/or meizothrombin des-fragment 1 (Briet et al., 1982; Franza et al., 1975; Kornalik and Blomback, 1975; Morita et al., 1976; Morita and Iwanaga, 1978; Rhee et al., 1982). Through these studies characterization of the biochemical properties of meizothrombin became possible which enabled Novoa and Seegers to be the first to suggest that prothrombin activation by factor Xa might actually proceed via meizothrombin as intermediate (Novoa and Seegers, 1980; Seegers et al., 1981). These and our own studies with notecarin (see below) ultimately showed that the pathway in which Arg320 is cleaved first to give rise to meizothrombin is in fact the kinetically favored pathway in factor Xa-catalyzed prothrombin activation.

Ecarin, as well as other group A activators, is a single chain metalloproteinase that efficiently activates prothrombin without requiring additional components. Ecarin was the first snake venom prothrombin activator to be cloned and was shown to consist of 426 amino acids and to share some 64% homology with the heavy chain of RVV-X, the factor X activator from Russell's viper venom (Nishida et al., 1995).Three domains are recognized in ecarin with the metalloprotease domain located at the amino terminal part of the molecule connected via a disintegrin-like domain to a cysteine rich C-terminal domain (Fig. 28.2) (Kini, 2005; Nishida et al., 1995).

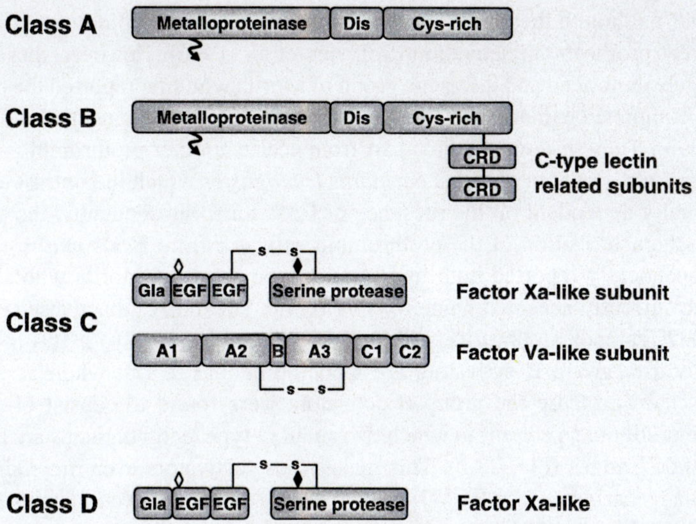

Fig. 28.2 General structure of venom prothrombin activators. Abbreviations used are: Dis = disintegrinlike domain; CRD = carbohydrate recognition domain; Gla = γ-carboxyglutamic acid containing domain; EGF = epidermal growth factor domain. Indicated are glycosilation sites that are not present in mammalian factor Xa; ◇ O-glycosilation; ♦ N-glycosylation site (Rao et al., 2004)

The activators of group A show almost exclusive selectivity for the cleavage at Arg320 although minor cleavages at other peptide bonds have been reported (Briet et al., 1982; Franza et al., 1975; Kornalik and Blomback, 1975). Similar characteristics have been reported for the purified prothrombin activator from *Dispholidus typus* (boomslang) (Bradlow et al., 1980; Guillin et al., 1978), *Bothrops atrox* (fer-de-lance) (Hofmann and Bon, 1987a, b) and *Bothrops neuwiedi* (Wied's lance head) (Govers-Riemslag et al., 1987).

The group A activators are efficient enzymes characterized in general by a low Km for prothrombin and a high kcat of meizothrombin formation. In fact the kcat of 120 s^{-1} calculated for the activator of *Bothrops neuwiedi* even exceeds the kcat reported for the complete prothrombinase complex (Govers-Riemslag et al., 1987)! Moreover, the group A activators are also relatively efficient activators of descarboxy-prothrombin which circulates in the blood of patients treated with vitamin K antagonists such as warfarin or coumarin (Bergstrom and Egberg, 1978; Kirchhof et al., 1978).

Group B Activators

The first report on the characterization of a group B prothrombin activator should be accredited to an Ukrainian group that reported on the prothrombin activator from *Echis multisquamatus* which they called ecamulin (Platonova et al., 1995; Solovjev et al., 1996) although they erroneously reported that ecamulin did not require Ca^{2+} for efficient prothrombin activation (Solovjev et al., 1996). However, these papers went largely unnoticed and it was the group of Morita who first reported the purification and complete characterization of a class B activator i.e. carinactivase (Yamada et al., 1996). They discovered that apart from ecarin another prothrombin activator was hiding in the venom of *Echis carinatus leucogaster* which in contrast to ecarin was critically dependent on the presence of Ca^{2+}-ions. Subsequently, the purification and characterization of the prothrombin activator from *Echis multisquamatus* was independently reported both by us and by the group of Morita who also proposed the currently accepted name for this venom enzyme, multactivase (Petrovan et al., 1997; Yamada and Morita, 1997).

Like ecarin, group B activators are metalloproteinases. But whereas ecarin is a single chain protease the group B activators were found to consist of a heavy chain (homologous to ecarin) to which two small C-type lectin domains are attached via disulfide bridges (Fig. 28.2). This makes these activators even more similar to RVV-X than ecarin because RVV-X also contains two C-type lectin chains (Gowda et al., 1994). Morita (1998) has proposed a model for the molecular structure of carinactivase (and RVV-X) in which these C-type lectin domains are considered to confer the Ca^{2+}-dependence and substrate recognition properties which can explain why group B activators are virtually unable to activate undercarboxylated forms of prothrombin. In fact this very property has made it possible to use this class of prothrombin activators for the development of potential clinically useful assays with which anticoagulant therapy can be monitored in patient plasmas (Petrovan et al., 1997, 1999; Platonova et al., 1995; Yamada and Morita, 1999). In addition to being unable to activate undercarboxylated forms of prothrombin group B activators are also unable to activate prothrombin which is bound to phospholipids a property that we used to develop a quantitative assay for the binding of prothrombin to procoagulant membranes (Govers-Riemslag et al., 1998).

Thus, although the prothrombin activators belonging to this class were only much more recently discovered and characterized and have not contributed as much as ecarin to the understanding of prothrombin activation *per se*, these activators do exhibit interesting features that can potentially be used in laboratory assays.

Group C Activators – Oscutarin from *Oxyuranus s. scutellatus* (Taipan Snake) Venom

With this group of activators we leave the vipers and the metalloproteinases and move to the Australian continent because thus far all activators belonging to group C (and group D) are found in the venom of Australian elapidae. Until the late sixties (individual) variation in factor V plasma levels (an essential accessory component in factor Xa-catalyzed prothrombin activation) presented a problem in plasma prothrombin determinations (Denson, 1969; Denson et al., 1971). Because of this, the venom of *Oxyuranus scutellatus scutellatus* (and to a lesser extent of *Pseudonaja textilis textilis*) became interesting to investigators when it was realized that with these activators prothrombin levels in plasma could be accurately measured independent of the levels of factor V and other clotting factors (Denson, 1969; Denson et al., 1971; Pirkle et al., 1972). This resulted in several studies aimed at investigating the effects of this venom on prothrombin activation with interesting results (Owen and Jackson, 1973; Walker et al., 1980). The activator cleaved prothrombin at the same peptide bonds as factor Xa in a manner independent of factor Va but for optimal prothrombin activation the presence of Ca^{2+} and procoagulant phospholipids was required. The most interesting finding, however, was that it appeared to be a very large multi-subunit protease of estimated 360,000 Mr the prothrombin activating capacity of which could be destroyed by high NaCl without loss of its ability to hydrolyze small peptide substrates (Walker et al., 1980). Following up on these efforts we reported a purification and further characterization of this activator in 1986 and showed it to contain a proteolytic (serine protease) factor Xa-like part and which was comprised of two 30,000 Mr units held together by (a) disulfide bridge(s) and which contained Gla-residues (Speijer et al., 1986). This protease purified in complex with a large protein that consisted of two subunits of 110,000 and 80,000 from which it could be dissociated to result in a virtual loss of prothrombin activation that could, however, be restored by addition of purified factor Va. Thus, it appeared that the Taipan venom activator brings its own factor Va-like cofactor component to ensure efficient prothrombin activation. Similar findings were subsequently reported for the venom activator of the Australian brown snake (*Pseudonaja textilis textilis*) (Masci et al., 1988). The group C venom prothrombin activators are responsible for an astonishingly high percentage of the venom; 10–20% of *Oxyuranus scutellatus scutellatus* venom (Lavin and Masci, 2009)and even 20–40% of *Pseudonaja textilis textilis* venom (Lavin and Masci, 2009; Rao and Kini, 2002) indicating the importance of these activators for the successful capture and subsequent digestion of prey! To this date oscutarin C and pseutarin C represent the most extensively studied group C activators. Until a decade ago the contribution of group C activators to understanding prothrombin activation biochemistry has been relatively minor. But this is about to change drastically, because especially in recent years considerable progress has been achieved with respect to the knowledge on the evolution and biochemistry of these venom prothrombin activators (Filippovich et al., 2005; Kini, 2005; Lavin and Masci, 2009; Minh Le et al., 2005; St. Pierre et al., 2005; Rao and Kini, 2002; Rao et al., 2003b; Reza et al., 2006; Welton and Burnell, 2005). Thus, the cDNA of oscutarin C (St. Pierre et al., 2005; Welton and Burnell, 2005) and pseutarin C (Filippovich et al., 2005; Minh Le et al., 2005) have been cloned and the factor V(a) – like subunit from pseutarin C has recently been subjected to mutagenesis studies (Bos et al., 2009).

Whereas the information obtained from the deduced sequences of the protease subunits more or less confirmed what had been learned from the sequences obtained for group D activators (see below) the information with regard to the factor Va-like subunit has proved to be interesting indeed. Comparison of the sequence of the venom V-like protein with the snakes own blood coagulation factor V and with human factor V has yielded a wealth of information. The venom factor V unit apparently has evolved through gene duplication (Reza et al., 2006) followed by a number of evolutionary events in which almost all of the large B-domain of factor V has been lost rendering the molecule constitutively active (Fig. 28.2) (Bos et al., 2009; Minh Le et al., 2005; Rao et al., 2003b; Reza et al., 2006). Moreover, two of the three cleavage sites by which the activity of blood coagulation factor Va is down regulated by activated protein C (APC) are not present in the venom molecule and one or more additional disulfide bridges are present which will hold the molecule together even after cleavage with an excess amount of APC (Fig. 28.2). Together these evolutionary changes render the prothrombin activating capacity completely resistant to APC (Bos et al., 2009; Minh Le et al., 2005; Rao et al., 2003b; Reza et al., 2006). It is likely that many more interesting lessons concerning the mode of action of factor V(a) in blood coagulation will be learned in the near future from studying these interesting snake venom proteins.

Group D Activators – Notecarin from *Notechis s. scutatus* (Tiger Snake) Venom

Mention of *Notechis s. scutatus* venom appears in one of the earliest literature references concerning procoagulant activities of snake venoms (Mellanby, 1909) and studies on the action of this venom have contributed to nowadays understanding of the biochemistry of prothrombin activation on a regular basis. Thus, as already mentioned earlier this venom was used to show that prothrombin is the inactive precursor of thrombin in plasma (Eagle, 1937) and in their effort to bypass the need for factor Xa purification Jobin and Esnouf carried out a quantitative comparison of prothrombin activation by Tiger snake venom and factor Xa yielding valuable information on factor Xa-catalyzed prothrombin activation in the process (Jobin and Esnouf, 1966). A subsequent big step forward occurred in the 1980s when we purified notecarin, the prothrombin activator from *Notechis scutatus scutatus* venom (Tans et al., 1985). Notecarin turned out to be a serine protease that shared many properties with mammalian blood coagulation factor Xa. Like factor Xa it is a two chain glycoprotein (54,000 Mr) the active site of which is located in the heavy chain of the molecule (Fig. 28.2). The activator is strongly stimulated by Ca^{2+}, factor Va and procoagulant phospholipids. We found the activator to contain Gla-residues which was the first time that the existence of such residues was established in snake venom (Tans et al., 1985). In many functional aspects notecarin proved almost indistinguishable from factor Xa with one noticeable interesting exception. We found that factor Va also stimulated the amidolytic activity of notecarin towards small chromogenic substrates. This is in contrast to human or bovine factor Xa whose amidolytic activity is not stimulated by factor Va. The biggest contribution to prothrombinase biochemistry came, however, from the finding that during the time course of activation of prothrombin activation by notecarin substantial amounts of meizothrombin and meizothrombin-des-fragment 1 could be detected. Since notecarin and factor Xa shared so many properties in common this prompted us to reinvestigate factor Xa-catalyzed prothrombin activation using the tools we developed during our study of notecarin. We showed that meizothrombin occurred as an intermediate in factor Xa-catalyzed prothrombin activation with all possible combinations of the prothrombinase complex and we were able to show that the presence of factor Va caused a major shift in the pathway of prothrombin activation. Thus, in the absence of factor Va large amounts of prethrombin 2 (formed by first cleavage at Arg271) accumulated but with factor Va this intermediate was not detected and we could establish that more than 90% of the initial activated prothrombin that was formed was in fact meizothrombin (Rosing et al., 1986)! Supporting evidence followed shortly thereafter (Krishnaswamy et al., 1986, 1987) and nowadays it is generally accepted that the pathway via meizothrombin is actually the kinetically preferred pathway of factor Xa-catalyzed prothrombin activation (Krishnaswamy et al., 1986; Rabiet et al., 1986; Rosing et al., 1986; Tans et al., 1991). But for the studies with notecarin, the activator of *Notechis scutatus scutatus* venom, we think it might have taken a lot longer to arrive at this conclusion!

Like the group C activators the knowledge concerning group D activators has increased tremendously in the last decade and large progress has been made in unraveling the biochemistry and evolution of group D activators. To date a number of these have been purified and cloned (trocarin D form *Tropidechis carinatus*, notenarin D from *Notechis ater niger* and hopsarin D from *Hoplocephalis stephensii*) (Joseph et al., 1999; Kini, 2005; Kini et al., 2001; Rao et al., 2003a; Reza et al., 2005) and similar to group C activators it has been proposed that these activators most likely have evolved out of a gene duplication event with one of the genes being expressed in the liver, retaining its protective blood coagulation functions, whereas the other gene must have evolved further after being recruited in the venom gland to result in a constitutively active protease (Reza et al., 2005). Similar to the group C activators these activators account for an important part of the total venom protein content (approximately 5%) which on a molar basis is comparable to the amounts of prothrombin activator found in group C venoms (Kini, 2005).

Concluding Remarks

Study of snake venom prothrombin activators has contributed greatly to the current understanding of the biochemistry of prothrombin activation. Especially the diversity of activators has proven crucial for their potential to push knowledge further. The metalloproteinases of group A and B that only cleave one of the two peptide bonds in prothrombin helped solve the puzzle of peptide bond cleavage patterns of factor Xa-catalyzed prothrombin activation. Studies with notecarin paved the road to finally establish that the pathway of prothrombin activation occurs via meizothrombin as intermediate. And the history is not finished! With respect to understanding the molecular basis of the effect of factor V(a) in hemostasis a lot has still to be learned. Human factor V(a) is a large molecule which makes it difficult to establish the function of various domains although a lot has already been learned through mutagenesis and modeling studies. But development has been slow. However, the cloning and characterization of the group C and group D activators show that evolution already has selected successful mutations for us. Most certainly we are at the beginning of exciting discoveries concerning the mode of action of factor V(a) in hemostasis from the studies of these snake venom activators.

References

Bergstrom, K., Egberg, N., 1978. Determination of vitamin K sensitive coagulation factors in plasma: studies on three methods using synthetic chromogenic substrates. Thromb. Res. 12, 531–547.

Bos, M.H., Boltz, M., St. Pierre, L., Masci, P.P., de Jersey, J., Lavin, M.F., Camire, R.M., 2009. Venom factor V from the common brown snake escapes hemostatic regulation through procoagulant adaptations. Blood 114, 686–692.

Bradlow, B.A., Atkinson, P.M., Gomperts, E.D., Gaillard, M.C., 1980. Studies on the coagulant effects of boomslang (*Dispholidus typus*) venom. Clin. Lab. Haemat. 2, 317–331.

Briet, E., Noyens, C.M., Roberts, H.R., Griffith, M.J., 1982. Cleavage and activation of human prothrombin by *Echis carinatus* venom. Thromb. Res. 27, 591–600.

Davie, E.W., Fujikawa, K., Kisiel, W., 1991. The coagulation cascade: initiation, maintenance, and regulation. Biochemistry 30, 10363–10370.

Denson, K.W., 1969. Coagulant and anticoagulant action of snake venoms. Toxicon 7, 5–11.

Denson, K.W., 1976. Clot-inducing substances present in snake venoms with particular reference to *Echis carinatus*. Thromb. Res. 8, 351–360.

Denson, K.W., Borrett, R., Biggs, R., 1971. The specific Assay of prothrombin using the Taipan snake venom. Br. J. Haematol. 21, 219–226.

Doyle, M.F., Mann, K.G., 1990. Multiple active forms of thrombin. IV. Relative activities of meizothrombins. J. Biol. Chem. 265, 10693–10701.

Eagle, H., 1937. The coagulation of blood by snake venoms and its physiologic significance. J. Exp. Med. 65, 613–639.

Esmon, C.T., Jackson, C.M., 1974a. The conversion of prothrombin to thrombin III The factor Xa catalyzed activation of prothrombin. J. Biol. Chem. 249, 7782–7790.

Esmon, C.T., Jackson, C.M. 1974b. Conversion of prothrombin to thrombin. IV. The function of the fragment 2 region during activation in the presence of factor V. J. Biol. Chem. 249, 7791–7797.

Esmon, C.T., Owen, W.G., Jackson, C.M. 1974a. The conversion of prothrombin to thrombin II differentiation between thrombin- and factor Xa-catalyzed proteolysis. J. Biol. Chem. 249, 606–611.

Esmon, C.T., Owen, W.G., Jackson, C.M., 1974b. The conversion of prothrombin to thrombin V The activation of prothrombin by factor Xa in the presence of phospholipid. J. Biol. Chem. 249, 7798–7807.

Esmon, C.T., Owen, W.G., Jackson, C.M., 1974c. A plausible mechanism for prothrombin activation by factor Xa, factor Va, phospholipid and calcium ions. J. Biol. Chem. 249, 8045–8047.

Filippovich, I., Sorokina, N., St. Pierre, L., Flight, S., de Jersey, J., Perry, N., Masci, P.P., Lavin, M.F., 2005. Cloning and functional expression of venom prothrombin activator protease from *Pseudonaja textilis* with whole blood procoagulant activity. Br. J. Haematol. 131, 237–246.

Franza, B.R., Jr., Aronson, D.L., Finlayson, J.S., 1975. Activation of human prothrombin by a procoagulant fraction from the venom of *Echis carinatus*. Identification of a high molecular weight intermediate with thrombin activity. J. Biol. Chem. 250, 7057–7068.

Govers-Riemslag, J.W., Johnsen, L., Petrovan, R.J., Rosing, J., Tans, G., 1998. A kinetic assay to determine prothrombin binding to membranes. Thromb. Res. 92, 239–247.

Govers-Riemslag, J.W., Knapen, M.J., Tans, G., Zwaal, R.F., Rosing, J., 1987. Structural and functional characterization of a prothrombin activator from the venom of *Bothrops neuwiedi*. Biochim. Biophys. Acta 916, 388–401.

Gowda, D.C., Jackson, C.M., Hensley, P., Davidson, E.A., 1994. Factor X-activating glycoprotein of Russell's viper venom. J. Biol. Chem. 269, 10644–10650.

Guillin, M.C., Bezeaud, A., Menache, D., 1978. The mechanism of activation of human prothrombin by an activator isolated from dispholidus types venom. Biochim. Biophys. Acta 537, 160–168.

Hackeng, T.M., Tans, G., Koppelman, S.J., de Groot, P.G., Rosing, J., Bouma, B.N., 1996. Protein C activation on endothelial cells by prothrombin activation products generated in situ: meizothrombin is a better protein C activator than α-thrombin. Biochem. J. 319, 399–405.

Hjort, P., Rapaport, S.I., Owren, P.A., 1955. A simple, specific one-stage prothrombin assay using Russell's viper venom in cephalin suspension. J. Lab. Clin. Med. 46, 89–97.

Hofmann, H., Bon, C., 1987a. Blood coagulation induced by the venom of *Bothrops atrox*. 1. Identification, purification, and properties of a prothrombin activator. Biochemistry 26, 772–780.

Hofmann, H., Bon, C., 1987b. Blood coagulation induced by the venom of *Bothrops atrox*. 2. Identification, purification, and properties of two factor X activators. Biochemistry 26, 780–787.

Jobin, F., Esnouf, M.P., 1966. Coagulant activity of tiger snake (*Notechis scutatus scutatus*) venom. Nature 211, 873–875.

Joseph, J.S., Chung, M.C., Jeyaseelan, K., Kini, R.M., 1999. Amino acid sequence of trocarin, a prothrombin activator from *Tropidechis carinatus* venom: its structural similarity to coagulation factor Xa. Blood 94, 621–631.

Kini, R.M., 2005. The intriguing world of prothrombin activators from snake venom. Toxicon 45, 1133–1145.

Kini, R.M., Rao, V.S., Joseph, J.S., 2001. Procoagulant proteins from snake venoms. Haemostasis 31, 218–224.

Kirchhof, B.R., Vermeer, C., Hemker, H.C., 1978. The determination of prothrombin using synthetic chromogenic substrates; choice of a suitable activator. Thromb. Res. 13, 219–232.

Kornalik, F., Blomback, B., 1975. Prothrombin activation induced by Ecarin – a prothrombin converting enzyme from *Echis carinatus* venom. Thromb. Res. 6, 57–63.

Kornalik, F., Schieck, A., Habermann, E., 1969. Isolation, biochemical and pharmacologic characterization of a prothrombin-activating principle from *Echis carinatus* venom. Naunyn Schmiedebergs Arch. Pharmakol. 264, 259–260.

Krishnaswamy, S., Church, W.R., Nesheim, M.E., Mann, K.G., 1987. Activation of human prothrombin by human prothrombinase. Influence of factor Va on the reaction mechanism. J. Biol. Chem. 262, 3291–3299.

Krishnaswamy, S., Mann, K.G., Nesheim, M.E., 1986. The prothrombinase-catalyzed activation of prothrombin proceeds through the intermediate meizothrombin in an ordered, sequential reaction. J. Biol. Chem. 261, 8977–8984.

Lanchantin, G.F., Friedmann, J.A., Hart, D.W., 1973. Two forms of human thrombin. Isolation and characterization. J. Biol. Chem. 248, 5956–5966.

Lavin, M.F., Masci, P.P., 2009. Prothrombinase complexes with different physiological roles. Thromb. Haemost. 102, 421–423.

Macfarlane, R.G., 1948. Normal and abnormal blood coagulation: a review. J. Clin. Pathol. 1, 113–143.

Marsh, N., Williams, V., 2005. Practical applications of snake venom toxins in haemostasis. Toxicon 45, 1171–1181.

Martin, C.J., 1893. On some effects upon the blood produced by the injection of the venom of the Australian black snake (*Pseudechis porphyriacus*). J. Physiol. 15, 380–400.

Masci, P.P., Whitaker, A.N., de Jersey, J., 1988. Purification and characterization of a prothrombin activator from the venom of the Australian brown snake, *Pseudonaja textilis textilis*. Biochem. Int. 17, 825–835.

Mellanby, J., 1909. The coagulation of blood: Part II. The actions of snake venoms, peptone and leech extract. J. Physiol. 38, 441–503.

Minh Le, T.N., Reza, M.A., Swarup, S., Kini, R.M., 2005. Gene duplication of coagulation factor V and origin of venom prothrombin activator in *Pseudonaja textilis* snake. Thromb. Haemost. 93, 420–429.

Morita, T., 1998. Proteases which activate factor X. in: Bailey, G. (Ed.), Enzymes from Snake Venoms. Alaken Inc., Fort Collins, CO, pp. 179–209.

Morita, T., Iwanaga, S., 1978. Purification and properties of prothrombin activator from the venom of echis carinatus. J. Biochem. 83, 559–570.

Morita, T., Iwanaga, S., Suzuki, T., 1976. The mechanism of activation of bovine prothrombin by an activator isolated from *Echis carinatus* venon and characterization of the new active intermediates. J. Biochem. 79, 1089–1108.

Nishida, S., Fujita, T., Kohno, N., Atoda, H., Morita, T., Takeya, H., Kido, I., Paine, M.J., Kawabata, S., Iwanaga, S., 1995. cDNA cloning and deduced amino acid sequence of prothrombin activator (ecarin) from Kenyan *Echis carinatus* venom. Biochemistry 34, 1771–1778.

Novoa, E., Seegers, W.H., 1980. Mechanisms of α-thrombin and β-thrombin-E formation: use of ecarin for isolation of meizothrombin 1. Thromb. Res. 18, 657–668.

Owen, W.G., Jackson, C.M., 1973. Activation of prothrombin with *Oxyuranus scutellatus scutellatus* (Taipan snake) venom. Thromb. Res. 3, 705–714.

Petrovan, R.J., Govers-Riemslag, J.W., Nowak, G., Hemker, H.C., Rosing, J., Tans, G., 1997. Purification and characterization of multisquamase, the prothrombin activator present in *Echis multisquamatus* venom. Thromb. Res. 88, 309–316.

Petrovan, R.J., Govers-Riemslag, J.W., Nowak, G., Hemker, H.C., Tans, G., Rosing, J., 1998. Autocatalytic peptide bond cleavages in prothrombin and meizothrombin. Biochemistry 37, 1185–1191.

Petrovan, R.J., Rapaport, S.I., Le, D.T., 1999. A novel clotting assay for quantitation of plasma prothrombin (factor II) using *Echis multisquamatus* venom. Am. J. Clin. Pathol. 112, 705–711.

St. Pierre, L., Masci, P.P., Filippovich, I., Sorokina, N., Marsh, N., Miller, D.J., Lavin, M.F., 2005. Comparative analysis of prothrombin activators from the venom of Australian elapids. Mol. Biol. Evol. 22, 1853–1864.

Pirkle, H., McIntosh, M., Theodor, I., Vernon, S., 1972. Activation of prothrombin with taipan snake venom. Thromb. Res. 1, 559–568.

Platonova, T.N., Sushko, E.A., Petrov, A.V., Solov'ev, D.A., 1995. Determination of the general prothrombin level and detection of its functionally inactive forms using the enzyme ecamulin purified from *Echis multisquamatus* venom. Ukr. Biokhim. Zh. 67, 75–80.

Rabiet, M.J., Blashill, A., Furie, B., Furie, B.C., 1986. Prothrombin fragment 1 X 2 X 3, a major product of prothrombin activation in human plasma. J. Biol. Chem. 261, 13210–13215.

Rao, V.S., Joseph, J.S., Kini, R.M., 2003a. Group D prothrombin activators from snake venom are structural homologues of mammalian blood coagulation factor Xa. Biochem. J. 369, 635–642.

Rao, V.S., Kini, R.M., 2002. Pseutarin C, a prothrombin activator from *Pseudonaja textilis* venom: its structural and functional similarity to mammalian coagulation factor Xa-Va complex. Thromb. Haemost. 88, 611–619.

Rao, V.S., Swarup, S., Kini, R.M., 2003b. The nonenzymatic subunit of pseutarin C, a prothrombin activator from eastern brown snake (*Pseudonaja textilis*) venom, shows structural similarity to mammalian coagulation factor V. Blood 102, 1347–1354.

Rao, V.S., Swarup, S., Kini, R.M., 2004. The catalytic subunit of pseutarin C, a group C prothrombin activator from the venom of *Pseudonaja textilis*, is structurally similar to mammalian blood coagulation factor Xa. Thromb. Haemost. 92, 509–521.

Reza, A., Swarup, S., Kini, R.M., 2005. Two parallel prothrombin activator systems in Australian rough-scaled snake, *Tropidechis carinatus*. Structural comparison of venom prothrombin activator with blood coagulation factor X. Thromb. Haemost. 93, 40–47.

Reza, M.A., Minh Le, T.N., Swarup, S., Kini, R.M., 2006. Molecular evolution caught in action: gene duplication and evolution of molecular isoforms of prothrombin activators in *Pseudonaja textilis* (brown snake). J. Thromb. Haemost. 4, 1346–1353.

Rhee, M.J., Morris, S., Kosow, D.P., 1982. Role of meizothrombin and meizothrombin-(des F1) in the conversion of prothrombin to thrombin by the echis carinatus venom coagulant. Biochemistry 21, 3437–3443.

Rosing, J., Tans, G., 1988. Meizothrombin, a major product of factor Xa-catalyzed prothrombin activation. Thromb. Haemost. 60, 355–360.

Rosing, J., Tans, G., 1992. Structural and functional properties of snake venom prothrombin activators. Toxicon 30, 1515–1527.

Rosing, J., Tans, G., Govers-Riemslag, J.W., Zwaal, R.F., Hemker, H.C., 1980. The role of phospholipids and factor Va in the prothrombinase complex. J. Biol. Chem. 255, 274–283.

Rosing, J., Zwaal, R.F., Tans, G., 1986. Formation of meizothrombin as intermediate in factor Xa-catalyzed prothrombin activation. J. Biol. Chem. 261, 4224–4228.

Schieck, A., Habermann, E., Kornalik, F., 1972. The prothrombin-activating principle from *Echis carinatus* venom. II. Coagulation studies in vitro and in vivo. Naunyn Schmiedebergs Arch. Pharmacol. 274, 7–17.

Seegers, W.H., Teng, C.M., Ghosh, A., Novoa, E., 1981. Three aspects of prothrombin activation related to protein M, ecarin, acutin, meizothrombin 1 and prethrombin 2. Ann. N. Y. Acad. Sci. 370, 453–467.

Solovjev, D.A., Platonova, T.N., Ugarova, T.P. 1996. Isolation and characteristics of ekamulin – a prothrombin activator from multiscaled viper (*Echis multisquamatus*) venom. Biokhimiia 61, 1094–1105.

Solovjev, D.A., Platonova, T.N., Ugarova, T.P., 1996. Purification and characterization of ecamulin – a prothrombin activator from the venom of multi-scaled viper (*Echis multisquamatus*). Ukr. Biokhim. Zh. 68, 18–19.

Speijer, H., Govers-Riemslag, J.W., Zwaal, R.F., Rosing, J., 1986. Prothrombin activation by an activator from the venom of *Oxyuranus scutellatus* (Taipan snake). J. Biol. Chem. 261, 13258–13267.

Tans, G., Govers-Riemslag, J.W., van Rijn, J.L., Rosing, J., 1985. Purification and properties of a prothrombin activator from the venom of *Notechis scutatus scutatus*. J. Biol. Chem. 260, 9366–9372.

Tans, G., Janssen-Claessen, T., Hemker, H.C., Zwaal, R.F., Rosing, J., 1991. Meizothrombin formation during factor Xa-catalyzed prothrombin activation. Formation in a purified system and in plasma. J. Biol. Chem. 266, 21864–21873.

Tans, G., Nicolaes, G.A., Thomassen, M.C., Hemker, H.C., van Zonneveld, A.J., Pannekoek, H., Rosing, J., 1994. Activation of human factor V by meizothrombin. J. Biol. Chem. 269, 15969–15972.

van Rijn, J.L., Govers-Riemslag, J.W., Zwaal, R.F., Rosing, J., 1984. Kinetic studies of prothrombin activation: effect of factor Va and phospholipids on the formation of the enzyme-substrate complex. Biochemistry 23, 4557–4564.

Walker, F.J., Owen, W.G., Esmon, C.T., 1980. Characterization of the prothrombin activator from the venom of *Oxyuranus scutellatus scutellatus* (Taipan venom). Biochemistry 19, 1020–1023.

Welton, R.E., Burnell, J.N., 2005. Full length nucleotide sequence of a factor V-like subunit of oscutarin from *Oxyuranus scutellatus scutellatus* (coastal Taipan). Toxicon 46, 328–336.

Yamada, D., Morita, T., 1997. Purification and characterization of a Ca^{2+}-dependent prothrombin activator, multactivase, from the venom of *Echis multisquamatus*. J. Biochem. 122, 991–997.

Yamada, D., Morita, T., 1999. CA-1 method, a novel assay for quantification of normal prothrombin using a Ca^{2+}-dependent prothrombin activator, carinactivase-1. Thromb. Res. 94, 221–226.

Yamada, D., Sekiya, F., Morita, T., 1996. Isolation and characterization of carinactivase, a novel prothrombin activator in *Echis carinatus* venom with a unique catalytic mechanism. J. Biol. Chem. 271, 5200–5207.

Chapter 29
Origin and Evolution of Snake Venom Prothrombin Activators

Md. Abu Reza and R. Manjunatha Kini

Abstract Snake venom is a mixture of various proteins and peptides with distinct pharmacological properties. They have the unique aptitude to attack specific physiological systems of the prey with remarkable specificity and thereby obstruct the natural function culminating in death and debilitation of the victim. The origin of snake venom toxins is one of the most interesting question that intrigued scientists for a long time. In this chapter, we will discuss the origin of a particular group of venom proteins, namely prothrombin activators. The results from our recent studies provide molecular evidence that this class of deadly toxins have evolved by gene duplication and recruitment of blood coagulation factor gene. Thus a new toxin has evolved after recruitment from a simple body protein that is involved in its own lifesaving haemostasis.

Introduction

Snake venom is a mixture of biologically active proteins and peptides. This armory may contain 50 to more than 100 different venomous proteins and peptides. They target a specific physiological system of the prey/victim with pin point accuracy. The intriguing questions that remained unanswered for a long time are from where these toxins came from (origin) and how the snake gathered such a complex arsenal of deadly proteins (recruitment). Since long scientists are trying to predict the origin of these venom proteins using various phylogenetic analyses and alignment tools. Remarkable sequence similarities are evident between some of the venom proteins to normal body proteins. Hence, it was hypothesized that snake venom proteins have originated from the body proteins (Fujimi et al., 2002, 2003; Jeyaseelan et al., 2000; Joseph et al., 1999; Reza et al., 2006, 2007; Tamiya and Fujimi, 2006).

M.A. Reza (✉)
Department of Genetic Engineering and Biotechnology, University of Rajshahi, Rajshahi 6205, Bangladesh; Department of Biological Sciences, National University of Singapore, Singapore 117 543, Singapore
e-mail: rezaru@gmail.com

R.M. Kini et al. (eds.), *Toxins and Hemostasis*, DOI 10.1007/978-90-481-9295-3_29,

However, there was no evidence for such hypothesis until recently when we gathered the molecular evidence for the recruitment of prothrombin activators (PAs) in Australian snake venoms. In this chapter we will discuss how factor X (FX), a liver protein, was probably recruited to be expressed in the venom.

Snake Venoms Targeting Haemostatic System

Neuromuscular and haemostatic systems are the two main targets of snake venom toxins. A wide array of proteins interferes in various aspects of thrombosis and haemostasis includes: procoagulants, anticoagulants, fibrinolytic components, platelet aggregation agonists, platelet antagonists and hemorrhagic proteins (Braud et al., 2000; Chow and Kini, 2001; Hutton and Warrell, 1993; Kini, 2004; Kini and Chow, 2001; Kini and Evans, 1990; Kini et al., 2001b; Markland, 1997, 1998; Morita, 2004). Among procoagulant proteins PAs are one of the most interesting groups. Snake venom PAs (SVPAs) readily clot victim's blood by extrinsically activating prothrombin (Kini et al., 2001b). In the physiological system activated blood coagulation factor X (FXa) forms a complex (prothrombinase complex) with its co-factor FVa, in presence of Ca^{2+} on the phospholipids membrane surface and activates prothrombin to thrombin. This endogenous activation occurs only when there is a vascular injury.

Snake Venom Prothrombin Activators (SVPAs)

A large number of SVPAs have been purified and characterized so far (Gao et al., 2002; Hasson et al., 2003; Joseph and Kini, 2001; Joseph et al., 1999; Kornalik and Blomback, 1975; Morita and Iwanaga, 1978; St. Pierre et al., 2005; Rao and Kini, 2002; Rao et al., 2003a; Rosing and Tans, 1991, 1992; Schieck et al., 1972; Silva et al., 2003; Speijer et al., 1986; Yamada and Morita, 1997; Yamada et al., 1996). They have distinct structures, and cofactor requirements for optimal activity and they produce different end products. Depending on these properties, SVPAs have been classified into four groups (Table 29.1) (Kini et al., 2001a). Group A and B SVPAs are metalloproteinases and are thus structurally and functionally unrelated to blood coagulation factors (Kini and Chow, 2001; Rosing and Tans, 1992). In contrast, group C and D SVPAs are serine proteinases and are found exclusively in Australian elapids. They show significant structural and functional similarities with blood coagulation factors (Joseph and Kini, 2001; Joseph et al., 1999; Lavin and Masci, 2009; Rao and Kini, 2002; Rao et al., 2003a; Reza et al., 2005).

Group C Prothrombin Activator – Pseutarin C

Group C PAs are large protein complexes (~300 kDa) with multiple subunits. To optimally activate prothrombin to thrombin they require only Ca^{2+} ions and phospholipids, but not FVa. Thus, functionally they resemble the 'FVa-FXa' complex.

Table 29.1 Classification of exogenous prothrombin activators from snake venom

Group	Enzyme	Co-factor requirements	End product	Size	Similarity to blood coagulation factors	Examples
A	Metalloproteinase	None	Meizothrombin	~55 kDa	None	Ecarin
B	Metalloproteinase	Ca^{2+}	Meizothrombin	Two subunits of ~25 and ~60 kDa	None	Carinactivase Multactivase
C	Serine proteinase	Ca^{2+} *plus* phospholipids	Thrombin	Two subunits of ~60 and ~220 kDa	'FXa-FVa' association of prothrombinase complex	Oscutarin Pseutarin C
D	Serine proteinase	Ca^{2+} *plus* phospholipids *plus* FVa	Thrombin	~50 kDa	FXa	Trocarin D Hopsarin D Notanarin D

Pseutarin C (PC) is the most extensively studied group C PA so far. It was purified using gel filtration followed by hydroxyapatite chromatography (Rao and Kini, 2002). Under denaturing conditions of reverse-phase high performance liquid chromatography (RP-HPLC), the subunits of pseutarin C are separated (Rao and Kini, 2002). The smaller subunit (~60 kDa) had a catalytic property of serine proteinase, and hence was termed as the pseutarin C catalytic subunit. The larger subunit (~220 kDa) was termed as the pseutarin C non-enzymatic subunit.

The catalytic subunit of PC is a two chain protein (a light chain and a heavy chain), which are linked together by a single disulfide bond (Rao et al., 2004). The N-terminal sequencing of the light and heavy chains of the PC catalytic subunit shows significant similarity (~42%) to mammalian FXa with identical domain arrangement (Rao and Kini, 2002; Rao et al., 2004). The light chain has a Gla domain followed by two EGF-like domains, and the heavy chain contains a serine proteinase domain (Fig. 29.1). The cDNA sequence of PC catalytic subunit shows high level of sequence similarities with that of FXa as well as with group D PAs (Rao et al., 2004). Functionally, PC catalytic subunit has low prothrombin activating

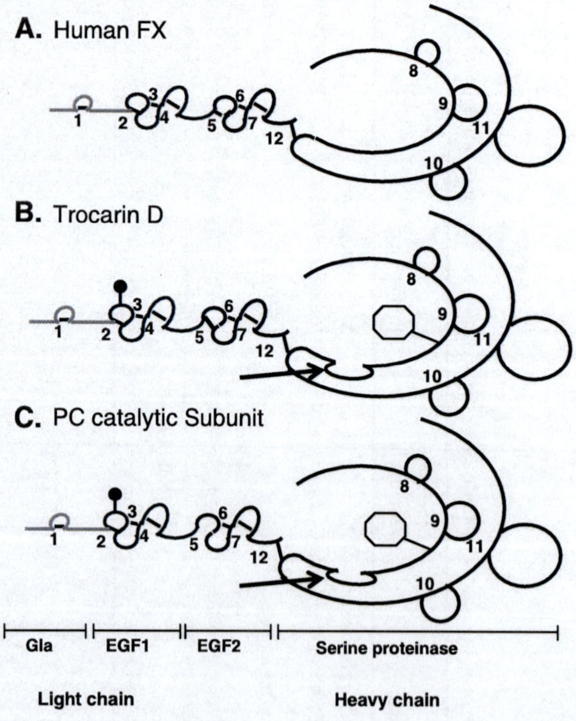

Fig. 29.1 Domain architecture of human blood coagulation FXa (**a**), trocarin D (**b**) and pseutarin C catalytic subunit (**c**). The disulphide bonds are denoted by lines and are numbered. Active FXa does not have any glycosylation. In contrast, trocarin D and the PC catalytic subunit have *O*-glycosylation (solid circle) and *N*-glycosylation (open octagon) in their light and heavy chains, respectively. An insertion of 12 and 13 residues (shown by *arrows*) is present in the heavy chain of trocarin D and PC catalytic subunit respectively

abilities which is enhanced in the presence of FVa similar to mammalian FXa and group D PAs (Rao and Kini, 2002).

The non-enzymatic subunit of PC is devoid of any catalytic properties by itself. However, it was shown that the non-enzymatic subunit of oscutarin C another group C PA isolated from *Oxyuranus scutellatus* venom enhances the prothrombin activation mediated by FXa-like catalytic subunit (Rao and Kini, 2002; Speijer et al., 1986). Thus, PC non-enzymatic subunit is similar to FVa. We determined the complete sequence of PC non-enzymatic subunit using Edman degradation and cDNA sequencing (Rao and Kini, 2002; Rao et al., 2003b). It is ~50% identical to that of mammalian FV with identical domain arrangement of A1-A2-B-A3-C1-C2 (Rao et al., 2003b). Alignment of PC non-enzymatic subunit with FV of other species shows that functionally important A and C domains are highly conserved. On the contrary, they exhibit significant differences in post-translational modifications and the size of B domain. B domain of PC non-enzymatic subunit is much shorter (127 residues) compared to that of fishes (fugu: 530 residues and zebrafish: 756 residues) and mammals (mouse: 843 residues, bovine: 869 residues and human: 882 residues). B domain of FV is poorly conserved even among mammals (~35% identity) and its function is not clear. But it is known that FV lacking the B domain shows markedly increased procoagulant activity (Kane et al., 1990).

PC non-enzymatic subunit has only 11 potential *N*-glycosylation sites in contrast to 29 glycosylation sites in bovine FV (Guinto et al., 1992). Our experimental evidence (Edman degradation) shows that two of these 11 potential sites are not glycosylated (Rao and Kini, 2002). Only three of the remaining nine potential PC non-enzymatic subunit *N*-glycosylation sites are conserved with those of bovine FV. Remaining six are the novel potential *N*-glycosylation sites. Another significant difference of PC non-enzymatic subunit from the mammalian FV is the absence of phosphorylation site at Ser692 position (Kalafatis, 1998). Apart from these, there are six sulfation sites in human FV which are not present in PC non-enzymatic subunit (Rao and Kini, 2002; Rao et al., 2003b). It is speculated that sulfation may play an important role in FV activation by thrombin (Hortin, 1990; Pittman et al., 1994). However, PC non-enzymatic subunit may not be activated by thrombin, as such thrombin or similar proteinase has not been found in *P. textilis* venom (Rao and Kini, 2002). In this circumstance we speculate that either it is activated by some other proteinase or the constant and stable association with PC catalytic subunit, may account for its activation.

Changes in PC Non-Enzymatic Subunit that made Pseutarin C an Efficient Toxin

The snakes use pseutarin C as a toxin. To be evolutionally successful as a toxin, it must be stable and it should not be easily inactivated. In PC non-enzymatic subunit we have identified several changes that confer pseutarin C these properties.

(i) *Lack of APC inactivation sites*: Once the bleeding is stopped, the process of blood coagulation should be terminated to restore the normal blood flow.

In physiological system, FVa is inactivated in a negative feedback loop in which thrombin activates protein C (APC), which in turn cleaves FVa at the heavy chain and inactivates FVa to FVai. This proteolytic inactivation of FVa is critical in the anticoagulation pathway for regulating thrombus formation and the inhibition of acute inflammation triggered by coagulation (Esmon, 2001). Human FVa is inactivated by proteolysis at three cleavage sites on heavy chain: Arg_{306}, Arg_{506} and Arg_{679} (Kalafatis and Mann, 1993; Kalafatis et al., 1994; Mann et al., 1997; Nicolaes et al., 1995). Two out of three APC cleavage sites [Arg_{306} as well as Arg_{316} (an alternative cleavage site for Arg_{306} in mammal FVa) (van der Neut et al., 2004) and Arg_{679}] are mutated and lost in PC non-enzymatic subunit (Rao and Kini, 2002).

(ii) *Absence of phosphorylation*: The phosphorylation at Ser692 at the acidic carboxy terminal tail of the heavy chain by a platelet membrane-associated casein kinase II results in faster inactivation of human FV by APC (Kalafatis, 1998). PC non-enzymatic subunit lacks this phosphorylation site and this may confer the resistance against APC inactivation.

(iii) *PC non-enzymatic subunit exists as a complex*: Pseutarin C exists in the venom as a complex of PC catalytic subunit (FXa paralogue) and PC non-enzymatic subunit (FVa paralogue). Further, unlike the transient nature of the physiological prothrombinase complex, pseutarin C appears to be a stable complex even in the absence of Ca^{2+} and phospholipids (Rao and Kini, 2002). Such complex formation provides pseutarin C an exceptional stability against inactivation by APC (Nesheim et al., 1982; Rao and Kini, 2002).

(iv) *Reduction in the number of activation sites*: During a vascular injury when required FV is activated to FVa by the release of B domain by thrombin or FXa (Foster et al., 1983; Nesheim et al., 1979; Suzuki et al., 1982). This release occurs through proteolytic cleavage at three sites: Arg_{709}, Arg_{1018} and Arg_{1545} (bovine FV numbering). In PC non-enzymatic subunit only Arg_{709} and Arg_{1545} are conserved but not Arg_{1018} (Rao and Kini, 2002). However, the presence of Arg_{709} and Arg_{1545} cleavage sites are adequate for the complete removal of the activation domain B. Besides these three cleavage sites, mammalian FV is also activated by FXa at two other cleavage sites, Arg_{348} and Arg_{1765} (Thorelli et al., 1997). Among these two sites only the first one is conserved in PC non-enzymatic subunit (Rao and Kini, 2002; Thorelli et al., 1997). In addition, the existence of another FXa proteolytic cleavage site at Arg_{346} in PC non-enzymatic subunit was also discovered (Rao and Kini, 2002). However, the overall decrease in the number of sites for thrombin/ FXa to release B domain may have made it more efficient for activation.

Group D Prothrombin Activator – Trocarin D

Group D PAs show remarkable structural and functional similarity to blood coagulation factor FXa. Similar to FXa this group of venom proteins are ~50 kDa proteins with two chains (Joseph and Kini, 2001) joined by a single interchain disulfide bond (Fig. 29.1). Functionally they require Ca^{2+} ion, phospholipids, and FVa for their maximal activity which is identical to the cofactor requirement for FXa during physiological prothrombin activation (Joseph et al., 1999). They readily activate prothrombin to thrombin and thereby induce microclots in prey's plasma, leading to cyanosis and death through disseminated intravascular coagulopathy (DIC) (Joseph et al., 2003). The first member of this group, notecarin, was isolated from the *Notechis scutatus scutatus* venom in 1985 (Tans et al., 1985). However, the most extensively characterized member of group D PA is trocarin D from the *Tropidechis carinatus* venom (Joseph et al., 1999). It was purified using gel filtration followed by anion-exchange chromatography and RP-HPLC (Joseph et al., 1999). Subsequently, we improved the purification technique and were able to purify trocarin D using a single-step RP-HPLC method. This single step protocol was used for the purification of other group D PAs such as notecarin D, notanarin D from *N. ater niger* venom, and hopsarin D from *Hoplocephalus stephensi* venom (Rao et al., 2003a). Group D PAs exhibit potent procoagulant properties (Chester and Crawford, 1982; Jobin and Esnouf, 1966; Joseph et al., 1999; Marshall and Herrmann, 1983; Stocker et al., 1994). Trocarin D shortens the recalcification time of human plasma in a dose-dependent manner and is able to hydrolyze FXa-specific chromogenic peptide substrates such as CBS 31.99, S-2222 and S-2765 (Joseph et al., 1999; Tans et al., 1985). However, kinetic studies reveal that the Vmax for synthetic substrate hydrolysis by trocarin D is two to three orders of magnitude slower than FXa (Joseph et al., 1999; Tans et al., 1985). Interestingly, it cleaves prothrombin faster than FXa in the absence of Ca^{2+} ions (Joseph et al., 1999). Similar to the prothrombin activation, amidolytic activity of trocarin D on chromogenic substrates also increases (~7–14 fold) upon their interaction with FVa (Joseph et al., 1999). In contrast, FXa does not show such increase in its amidolytic activity (Joseph et al., 1999). Trocarin D activates prothrombin in an identical manner like FXa by cleaving the same two peptide bonds (Arg_{274}–Thr_{275} and Arg_{323}–Ile_{324}) (Joseph et al., 1999). These functional data provided the impetus for the structural study of trocarin D.

The complete amino acid sequence of trocarin D was determined via Edman degradation (Joseph et al., 1999). It has identical domain arrangement and high sequence identity (~53–60%) and homology (~62–70%) to mammalian FXa (Joseph et al., 1999) (Fig. 29.2). The light and heavy chains of trocarin D are highly homologous to those of FXa and the positions of cysteine residues are conserved (Joseph et al., 1999). Its light chain contains a Gla domain (with 11 γ-carboxylated

Fig. 29.2 (continued) Alignment of deduced amino acid sequences of venom and plasma PAs from snake with mammalian and fish FX sequences. Alignment was carried out using the software DNAMAN; gaps (–) are introduced for optimal alignment. 100% Conserved residues are shaded in black and 75% conserved ones in grey. The sequences have been taken from the following sources: TrFX (*T. carinatus* FX): AY651849; Tro-D (Trocarin D): DQ017707; HuFX (Human FX): P00742; BoFX (Bovine FX): P00743; MuFX (Murine FX): AAH03877; RtFX (Rat FX): NP 058839; RbFX (Rabbit FX): AF003200; ZbFX (Zebrafish FX): AAM88343; PCCS (Pseutarin C catalytic subunit): AY260939

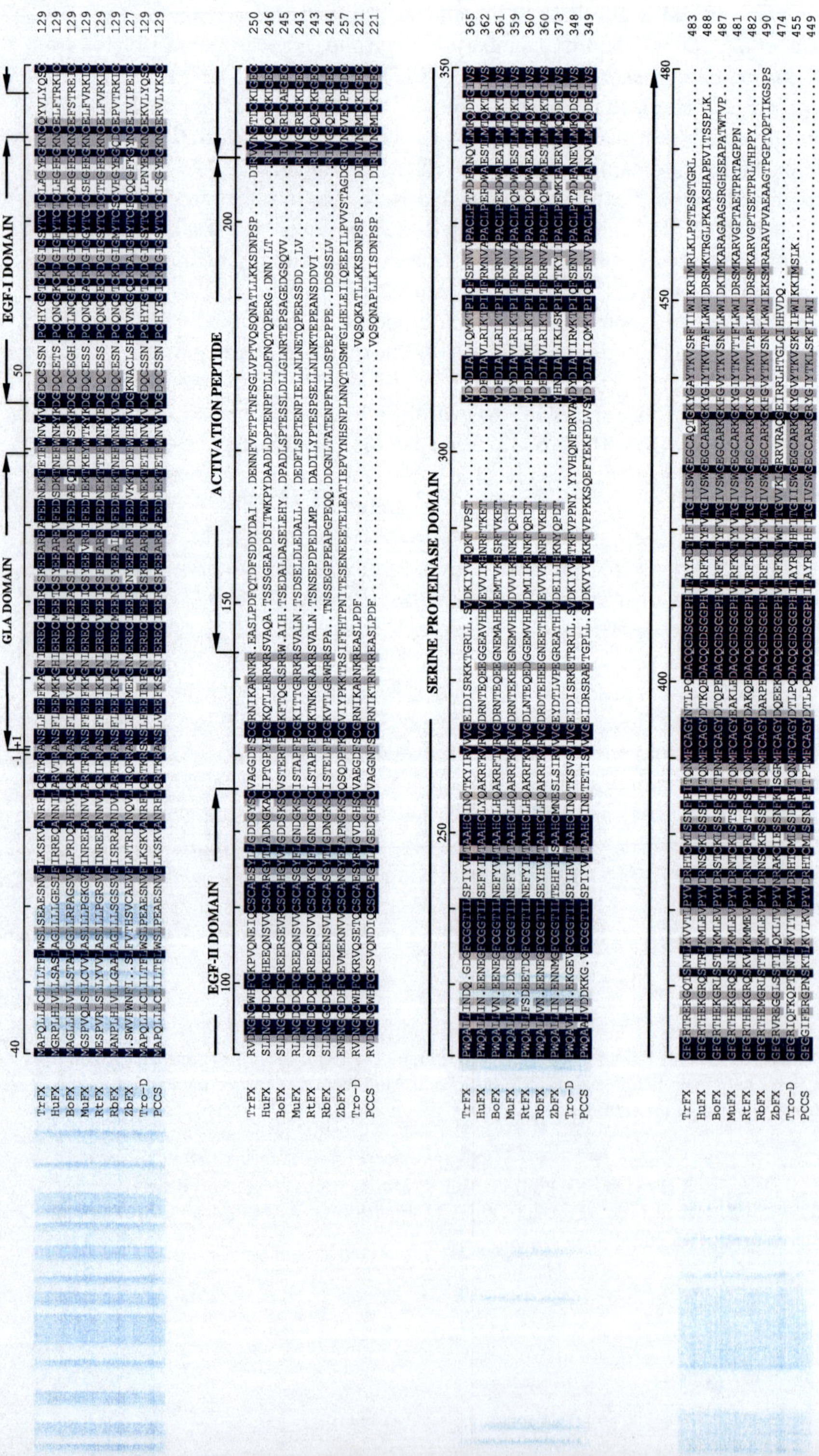

Fig. 29.2 (continued)

Glu residues) followed by EGF-I and EGF-II domains. The heavy chain of trocarin D contains the serine proteinase domain with the characteristic catalytic triad (His_{42}, Asp_{88} and Ser_{185}). These two chains are held together by a single interchain disulfide bond. There are three differences between trocarin D and FXa structure; glycosylation, hydroxylation and an insertion of 12-residue segment. Human FX contains two *N*-glysosylations and two *O*-glycosylations. But, these carbohydrates moieties are present on the activation peptide and are removed when FX is activated. Hence, human FXa is not glycosylated (Inoue and Morita, 1993). In contrast, trocarin D contains *O*-linked carbohydrate at Ser_{52} of the light chain and *N*-linked carbohydrate at Asn_{45} of the heavy chain (Joseph et al., 1999). It is hypothesized that the glycosylation confers thermal stability to proteins and thus probably is more stable as a toxin (Rao et al., 2003a; Wang et al., 1996). In the EGF-I domain of FX Asp_{63} is β-hydroxylated (McMullen et al., 1983; Stenflo et al., 1987), whereas the corresponding residue in trocarin D is not (Joseph et al., 1999). In addition to these posttranslational modifications, there is an insertion of 12-residue segment in the heavy chain. The structural and functional implication of this insertion is not known.

Two Parallel Prothrombin Activator Systems in Australian Elapids

Group C and D PAs act as toxins and induce disseminated intravascular coagulopathy (DIC) and death in the prey/victim. The snakes, being vertebrates, also have their own haemostatic system. As described above, they are structurally similar to blood coagulation factors. This implies that some of the Australian snakes possess two parallel prothrombin activating systems: one in their blood which they use for their own haemostatic purpose and the other in their venom which is used to attack the coagulation system of the prey/victim (Reza et al., 2005). The physiological roles of these two systems are completely different. Therefore, it is interesting to study and compare the molecular aspects of these two parallel systems.

FX, Trocarin D and Pseutarin C Catalytic Subunit and Their Evolution

We completed the cDNA sequence of FX from the liver tissues of *T. carinatus* (TrFX) (Reza et al., 2005). Alignment of the deduced TrFX sequence (482 residues) showed ~50% identity to human and other mammalian FX sequences (Fig. 29.2) and ~80% identity to trocarin D. Thus, TrFX exhibit closer kinship with its own venom PA trocarin D rather than FX from other species. There are some interesting differences between trocarin D and TrFX. For instance, the activation peptide of trocarin D is much shorter, 27 residues compared to 57 residues in TrFX. The size and sequence of activation peptide segment is found to be the most variable region among FX and it reduces during evolution; fish FX have the longest activation peptides (54–66 residues), whereas mammals have the shortest (49–52 residues). TrFX has intermediate size of 57 residues and in its venom counterpart the size seems to have reduced drastically during evolution (Reza et al., 2005).

As PC catalytic subunit is also structurally and functionally similar to mammalian FXa, and group D PAs, we also sequenced FX from the liver of *P. textilis*. Interestingly, there are two molecular isoforms of FX (PFX1 and PFX2) in the liver of *P. textiles* snake (Reza et al., 2006). Both of these isoforms have molecular signatures and domain architecture of factor X. One of the isoform PFX1 is similar to TrFX whereas the other isoform PFX2 is closer to PC catalytic subunit in respect of sequence identity as well as having shorter activation peptide and insertion in the heavy chain (Reza et al., 2006) (Fig. 29.2). The insertion in the heavy chain of PFX2 is shorter compared to those of PC catalytic subunit and group D PAs. Further, in phylogenetic analysis TrFX and PFX1 group together which supports the assumption that PFX1 is the functional FX in *P. textilis*. In contrast, PFX2 groups together with PC catalytic subunit. Quantitative real time expression analysis shows that the expression of PFX2 in the liver ~56,000 times lower than functionally important PFX1 (Reza et al., 2006). Assuming the mRNA levels reflects the protein levels, the concentration of PFX2 would be too low to have any significant physiological role in hemostasis (2.5 pM PFX2 compared to 136 nM PFX1 – calculated based on mammalian plasma FX concentration). Thus, it appears that PFX2 in the liver of *P. textilis* is an evolutionary by-product and possibly a redundant gene and hence one may consider it as an intermediate in the evolution of PC catalytic subunit from the FX gene (Reza et al., 2006). Based on our findings/hypothesis we proposed a model of the probable pathway of evolution and recruitment of FX gene as a toxin in the venom by gene duplication followed by certain changes like insertion and deletion in the sequences (Reza et al., 2006).

FV and PC Non-Enzymatic Subunit and Their Evolution

Because of the similarity of PC non-enzymatic subunit to FV, the cDNA sequence of blood coagulation factor V (PFV) from the liver of *P. textilis* was determined (Minh et al., 2005). The deduced amino acid sequence shows that both of them have a matching domain arrangement. Interestingly, PC non-enzymatic subunit and PFV cDNA shares an unusually high level of sequence identity (96%) to each other (Minh et al., 2005). For proteolytic activation by the release of B domain all of the FXa and thrombin proteolytic cleavage sites are conserved in both PFV and PC non-enzymatic subunit but PFV has an additional FXa proteolytic cleavage (Minh et al., 2005; Rao and Kini, 2002). However, unlike PC non-enzymatic subunit, PFV can be inactivated by APC, as it retains both the Arg_{316} and Lys_{506} which are the proteolytic cleavage sites important for the APC inactivation (discussed above). Apart from these, a careful analysis of the sequence alignment between PFV and PC non-enzymatic subunit reveals that many of the PFV potential post translational modification sites are mutated in PC non-enzymatic subunit. The extremely high level of sequence identity these two genes in the same organism suggest that this venom protein have evolved by the duplication of FV gene. However, on the way of its evolution to a venom protein, FV gene has undergone some small but critical changes by mutating several key residues to become a potent efficient toxin.

Gene Structure of Trocarin D and TrFX

To understand the basis of the differences in the site, mode and level of expression of the plasma and the venom PAs, and to gain insight on their evolutionary relationships, the complete gene structures of trocarin D and TrFX were determined (Reza et al., 2007). Both trocarin D and TrFX genes have analogous gene organization with eight exons interrupted by seven introns and possesses identical exon-intron boundaries. Both genes are nearly identical with the exception of first intron and promoter region (Reza et al., 2007). High identity (92–99%) in the intronic regions indicates the possibility of recent gene duplication event. A comparison of the promoter and intron sequences of these two genes provides us the first molecular evidence of the recruitment (Reza et al., 2007).

Intron 1 of TrFX and Trocarin D

Trocarin D gene intron 1 has two deletions and three unique insertions in contrast to TrFX gene (Fig. 29.3). The two deletions in trocarin D intron 1 are of 255 and 1,406 bp whose respective positions are at 2,610 and 3,770 bp on the TrFX gene. On the other hand, the three insertions in trocarin D intron 1 are at 128, 914 and 3,300 positions (relative to the start codon) of 214, 1,975, and 2,174 bp, respectively. The first insertion is nearly an exact repeat (96.33% identity) of the segment spanning from 3,082 to 3,299 bp. The other two inserts seem to be inverted repeats showing ~71% identities to each other. However, none of the inserts show significant similarity to any known DNA sequences. Interestingly, the third insertion in

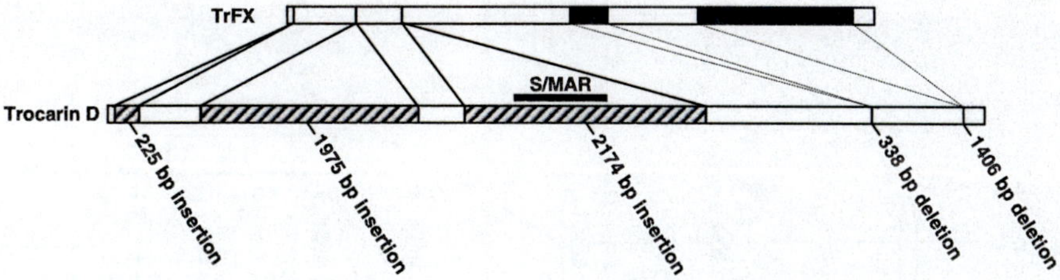

Fig. 29.3 Intron 1 of the trocarin D and TrFX gene. Insertions and deletions are shown as checkered and solid segments, respectively. Identical segments (>95% identity) are shown as *open boxes*. Positions of insertions and deletions are shown by *solid* and *dotted lines*, respectively. The segment showing S/MAR potential is shown

trocarin D intron 1 has a potential Scaffold Matrix Attached Region (S/MAR) which known to have positive regulatory roles in some genes (Allen et al., 1996; Blasquez et al., 1989; Bode et al., 1992; van Drunen et al., 1997; Xu et al., 1989). However, apart from these insertions and deletions, remaining parts of intron 1 (~3.5 kb) are nearly identical (97.14%), similar to other six introns. All these differences in intron 1 strongly suggest that it has some role in the regulation for the differential expression pattern in the venom and plasma PA genes in *T. carinatus* snake.

Promoter of TrFX and Trocarin D Gene

To understand the regulatory mechanism, the putative promoter regions of trocarin D and TrFX gene were determined using genome walking strategy (Reza et al., 2007). Comparison of the promoter of both the genes reveals that they are nearly identical except for an insertion of 264 bp in trocarin D promoter just 29 bp upstream of the start codon (Fig. 29.4). This insertion is unique with significant similarity to any DNA sequence in the database. We hypothesized that this segment may account for switching from constitutive expression of FX in the liver to inducible expression of trocarin D in the venom gland. Hence, this segment was named as *VE*nom *R*ecruitment/*S*witch *E*lement (*VERSE*) (Reza et al., 2007). A similar 271 bp insertion was also found in PC catalytic subunit gene promoter which shares ~95%

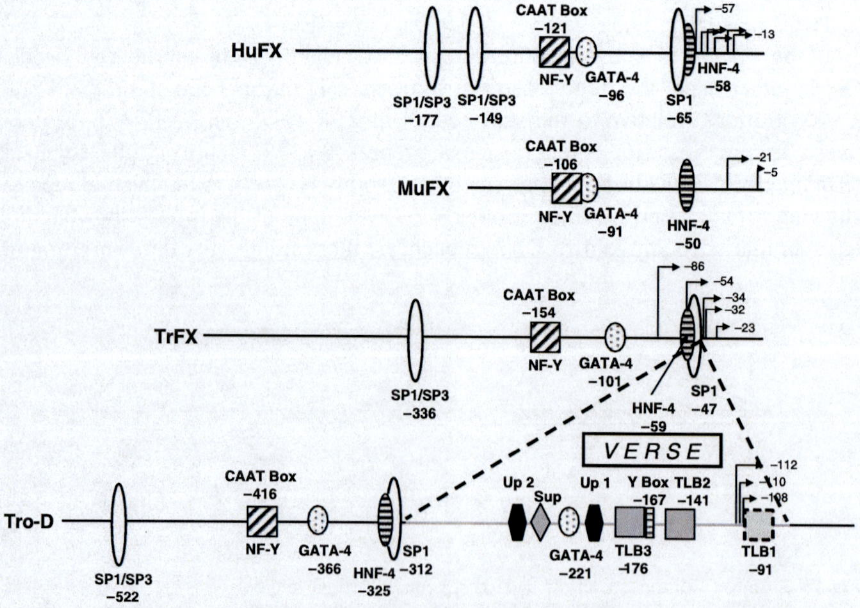

Fig. 29.4 Promoter region of human (HuFX), murine (MuFX), *T. carinatus* FX (TrFX) and trocarin D (Tro-D) gene showing *cis*-elements. The numbering system uses translation start site as +1

identity to trocarin D *VERSE*. Moreover, these insertions are in identical positions CTG-AAA, but 11 bp apart (For details, see Reza et al., 2007).

As with human (Hung and High, 1996; Hung et al., 2001) and murine (Wilberding and Castellino, 2000) FX promoters, TrFX has similar transcription factor binding sites at comparable positions of their promoter (a CAAT box, HNF-4 binding site, GATA-4 binding site and multiple SP1/SP3 binding sites) (Fig. 29.4). Hence, the expression of TrFX is controlled by similar *trans*-regulatory elements (Reza et al., 2007). In the case of trocarin D gene, due to the insertion of *VERSE* at –29 bp position all of these *cis*-regulatory elements are pushed further upstream. Our initial bioinformatics analysis showed that *VERSE* contains three TATA-like boxes (TLB), a Y-box and a GATA-4 binding site (Fig. 29.4) (Reza et al., 2007).

Characterization of TrFX and Trocarin D Promoter

Mammalian cell lines as well as in snake venom gland cells were used to determine the role of *cis*-elements within the promoter of trocarin D *VERSE* segment using luciferase as a reporter (Kwong et al., 2009) (Reza, M.A., 2006. Parallel Prothrombin Activator Systems in Australian Elapids: Their Structure, Gene Organization, Evolution and Regulation of Expression. PhD Dissertation. National University of Singapore, 'Unpublished'). The overall activity of trocarin D promoter was 19, 29, 49 folds higher in comparison with TrFX promoter in HepG2, HEK293T and CHOK1 cell lines, respectively which coincides with the findings of previous real-time data (Kwong et al., 2009; Reza et al., 2007). It was also shown that *VERSE* is the core promoter for trocarin D gene and it is responsible for the elevated level of trocarin D gene expression in the venom gland. Since only *VERSE* region is able to drive expression in HepG2, HEK293T and CHOK1 line, it implies that this segment is not responsible for the tissue-specificity and inducible expression of trocarin D gene. Using the luciferase assays the *cis*-elements were fully characterized. Among the three TLBs only TLB2 is functional; TLB2 plays the role of primary transcription initiator and GATA-4 box and Y box provides the up-regulatory effects. Three other novel *cis*-elements are also present in *VERSE* which was identified by deletion studies and is named as Up1, Up2 and Sup; the former two have up-regulatory effects, while the latter suppresses the expression. Sup works in tandem with Up2 for up-regulation of the expression of trocarin D (For details, see Kwong et al., 2009).

Conclusion

Existence of similar proteins with the same mode of action but divergent functional roles is rare in nature. In general it reflects the outcome of gene duplication and diversification which is the key for the sustainable existence of any population.

Snake venom PAs from Australian Elapids provided an excellent opportunity to study such proteins with similarity in structure and enzymatic properties, but differences in their physiological roles. Our studies showed the origin and recruitment of venom PAs from the blood coagulation factors by gene duplication. We showed the changes that are made in the SVPAs to enhance their function as a successful and potent toxin. The recruitment through insertion of *VERSE* not only contributes to the fundamental process of recruitment, but also provides impetus for the development of new technology for elevated levels expression of recombinant proteins in mammalian cell lines that are useful in biomedical applications.

References

Allen, G.C., Hall, G., Jr., Michalowski, S., Newman, W., Spiker, S., Weissinger, A.K., Thompson, W.F., 1996. High-level transgene expression in plant cells: effects of a strong scaffold attachment region from tobacco. Plant Cell 8, 899–913.

Blasquez, V.C., Xu, M., Moses, S.C., Garrard, W.T., 1989. Immunoglobulin kappa gene expression after stable integration. I. Role of the intronic MAR and enhancer in plasmacytoma cells. J. Biol. Chem. 264, 21183–21189.

Bode, J., Kohwi, Y., Dickinson, L., Joh, T., Klehr, D., Mielke, C., Kohwi-Shigematsu, T., 1992. Biological significance of unwinding capability of nuclear matrix-associating DNAs. Science 255, 195–197.

Braud, S., Bon, C., Wisner, A., 2000. Snake venom proteins acting on hemostasis. Biochimie 82, 851–859.

Chester, A., Crawford, G.P., 1982. *In vitro* coagulant properties of venoms from Australian snakes. Toxicon 20, 501–504.

Chow, G., Kini, R.M., 2001. Exogenous factors from animal sources that induce platelet aggregation. Thromb. Haemost. 85, 177–178.

Esmon, C.T., 2001. Role of coagulation inhibitors in inflammation. Thromb. Haemost. 86, 51–56.

Foster, W.B., Nesheim, M.E., Mann, K.G., 1983. The factor Xa-catalyzed activation of factor V. J. Biol. Chem. 258, 13970–13977.

Fujimi, T.J., Kariya, Y., Tsuchiya, T., Tamiya, T., 2002. Nucleotide sequence of phospholipase A_2 gene expressed in snake pancreas reveals the molecular evolution of toxic phospholipase A_2 genes. Gene 292, 225–231.

Fujimi, T.J., Nakajyo, T., Nishimura, E., Ogura, E., Tsuchiya, T., Tamiya, T., 2003. Molecular evolution and diversification of snake toxin genes, revealed by analysis of intron sequences. Gene 313, 111–118.

Gao, R., Kini, R.M., Gopalakrishnakone, P., 2002. A novel prothrombin activator from the venom of *Micropechis ikaheka*: isolation and characterization. Arch. Biochem. Biophys. 408, 87–92.

Guinto, E.R., Esmon, C.T., Mann, K.G., MacGillivray, R.T., 1992. The complete cDNA sequence of bovine coagulation factor V. J. Biol. Chem. 267, 2971–2978.

Hasson, S.S., Theakston, R.D., Harrison, R.A., 2003. Cloning of a prothrombin activator-like metalloproteinase from the West African saw-scaled viper, *Echis ocellatus*. Toxicon 42, 629–634.

Hortin, G.L., 1990. Sulfation of tyrosine residues in coagulation factor V. Blood 76, 946–952.

Hung, H.L., High, K.A., 1996. Liver-enriched transcription factor HNF-4 and ubiquitous factor NF-Y are critical for expression of blood coagulation factor X. J. Biol. Chem. 271, 2323–2331.

Hung, H.L., Pollak, E.S., Kudaravalli, R.D., Arruda, V., Chu, K., High, K.A., 2001. Regulation of human coagulation factor X gene expression by GATA-4 and the Sp family of transcription factors. Blood 97, 946–951.

Hutton, R.A., Warrell, D.A., 1993. Action of snake venom components on the haemostatic system. Blood Rev. 7, 176–189.

Inoue, K., Morita, T., 1993. Identification of *O*-linked oligosaccharide chains in the activation peptides of blood coagulation factor X. The role of the carbohydrate moieties in the activation of factor X. Eur. J. Biochem. 218, 153–163.

Jeyaseelan, K., Armugam, A., Donghui, M., Tan, N.H., 2000. Structure and phylogeny of the venom group I phospholipase A_2 gene. Mol. Biol. Evol. 17, 1010–1021.

Jobin, F., Esnouf, M.P., 1966. Coagulant activity of tiger snake (*Notechis scutatus scutatus*) venom. Nature 211, 873–875.

Joseph, J.S., Chung, M.C., Jeyaseelan, K., Kini, R.M., 1999. Amino acid sequence of trocarin, a prothrombin activator from *Tropidechis carinatus* venom: its structural similarity to coagulation factor Xa. Blood 94, 621–631.

Joseph, J.S., Kini, R.M., 2001. Snake venom prothrombin activators homologous to blood coagulation factor Xa. Haemostasis 31, 234–240.

Joseph, J.S., Thirumangalathu, S., Tsang, F., Wong, F.W., Kini, R.M., 2003. Trocarin, a blood coagulation factor Xa homologue from snake venom, causes inflammation and mitogenesis. Toxicon 42, 769–776.

Kalafatis, M., 1998. Identification and partial characterization of factor Va heavy chain kinase from human platelets. J. Biol. Chem. 273, 8459–8466.

Kalafatis, M., Mann, K.G., 1993. Role of the membrane in the inactivation of factor Va by activated protein C. J. Biol. Chem. 268, 27246–27257.

Kalafatis, M., Rand, M.D., Mann, K.G., 1994. The mechanism of inactivation of human factor V and human factor Va by activated protein C. J. Biol. Chem. 269, 31869–31880.

Kane, W.H., Devore-Carter, D., Ortel, T.L., 1990. Expression and characterization of recombinant human factor V and a mutant lacking a major portion of the connecting region. Biochemistry 29, 6762–6768.

Kini, R.M., 2004. Platelet aggregation and exogenous factors from animal sources. Curr. Drug Targets Cardiovasc. Haematol. Disord. 4, 301–325.

Kini, R.M., Chow, G., 2001. Exogenous inhibitors of platelet aggregation from animal sources. Thromb. Haemost. 85, 179–181.

Kini, R.M., Evans, H.J., 1990. Effects of snake venom proteins on blood platelets. Toxicon 28, 1387–1422.

Kini, R.M., Morita, T., Rosing, J., 2001a. Classification and nomenclature of prothrombin activators isolated from snake venoms. Thromb. Haemost. 86, 710–711.

Kini, R.M., Rao, V.S., Joseph, J.S., 2001b. Procoagulant proteins from snake venoms. Haemostasis 31, 218–224.

Kornalik, F., Blomback, B., 1975. Prothrombin activation induced by Ecarin – a prothrombin converting enzyme from *Echis carinatus* venom. Thromb. Res. 6, 57–63.

Kwong, S., Woods, A.E., Mirtschin, P.J., Ge, R., Kini, R.M., 2009. The recruitment of blood coagulation factor X into snake venom gland as a toxin: the role of promoter *cis*-elements in its expression. Thromb. Haemost. 102, 469–478.

Lavin, M.F., Masci, P.P., 2009. Prothrombinase complexes with different physiological roles. Thromb. Haemost. 102, 421–423.

Mann, K.G., Hockin, M.F., Begin, K.J., Kalafatis, M., 1997. Activated protein C cleavage of factor Va leads to dissociation of the A2 domain. J. Biol. Chem. 272, 20678–20683.

Markland, F.S., Jr., 1997. Snake venoms. Drugs 54(Suppl3), 1–10.

Markland, F.S., 1998. Snake venoms and the hemostatic system. Toxicon 36, 1749–1800.

Marshall, L.R., Herrmann, R.P., 1983. Coagulant and anticoagulant actions of Australian snake venoms. Thromb. Haemost. 50, 707–711.

McMullen, B.A., Fujikawa, K., Kisiel, W., Sasagawa, T., Howald, W.N., Kwa, E.Y., Weinstein, B., 1983. Complete amino acid sequence of the light chain of human blood coagulation factor X: evidence for identification of residue 63 as beta-hydroxyaspartic acid. Biochemistry 22, 2875–2884.

Minh, L.T., Reza, M.A., Swarup, S., Kini, R.M., 2005. Gene duplication of coagulation factor V and origin of venom prothrombin activator in *Pseudonaja textilis* snake. Thromb. Haemost. 93, 420–429.

Morita, T., 2004. C-type lectin-related proteins from snake venoms. Curr. Drug Targets Cardiovasc. Haematol. Disord. 4, 357–373.

Morita, T., Iwanaga, S., 1978. Purification and properties of prothrombin activator from the venom of *Echis carinatus*. J. Biochem. (Tokyo) 83, 559–570.

Nesheim, M.E., Canfield, W.M., Kisiel, W., Mann, K.G., 1982. Studies of the capacity of factor Xa to protect factor Va from inactivation by activated protein C. J. Biol. Chem. 257, 1443–1447.

Nesheim, M.E., Taswell, J.B., Mann, K.G., 1979. The contribution of bovine factor V and factor Va to the activity of prothrombinase. J. Biol. Chem. 254, 10952–10962.

Nicolaes, G.A., Tans, G., Thomassen, M.C., Hemker, H.C., Pabinger, I., Varadi, K., Schwarz, H.P., Rosing, J., 1995. Peptide bond cleavages and loss of functional activity during inactivation of factor Va and factor VaR506Q by activated protein C. J. Biol. Chem. 270, 21158–21166.

St. Pierre, L., Masci, P.P., Filippovich, I., Sorokina, N., Marsh, N., Miller, D.J., Lavin, M.F., 2005. Comparative analysis of prothrombin activators from the venom of Australian elapids. Mol. Biol. Evol. 22, 1853–1864.

Pittman, D.D., Tomkinson, K.N., Michnick, D., Selighsohn, U., Kaufman, R.J., 1994. Posttranslational sulfation of factor V is required for efficient thrombin cleavage and activation and for full procoagulant activity. Biochemistry 33, 6952–6959.

Rao, V.S., Joseph, J.S., Kini, R.M., 2003a. Group D prothrombin activators from snake venom are structural homologues of mammalian blood coagulation factor Xa. Biochem. J. 369, 635–642.

Rao, V.S., Kini, R.M., 2002. Pseutarin C, a prothrombin activator from *Pseudonaja textilis* venom: its structural and functional similarity to mammalian coagulation factor Xa-Va complex. Thromb. Haemost. 88, 611–619.

Rao, V.S., Swarup, S., Kini, R.M., 2003b. The non-enzymatic subunit of pseutarin C, a prothrombin activator from eastern brown snake (*Pseudonaja textilis*) venom, shows structural similarity to mammalian coagulation factor V. Blood 102, 1347–1354.

Rao, V.S., Swarup, S., Kini, R.M., 2004. The catalytic subunit of pseutarin C, a group C prothrombin activator from the venom of *Pseudonaja textilis*, is structurally similar to mammalian blood coagulation factor Xa. Thromb. Haemost. 92, 509–521.

Reza, M.A., Minh, L.T., Swarup, S., Kini, R.M., 2006. Molecular evolution caught in action: gene duplication and evolution of molecular isoforms of prothrombin activators in Pseudonaja textilis (brown snake). J. Thromb. Haemost. 4, 1346–1353.

Reza, M.A., Swarup, S., Kini, R.M., 2005. Two parallel prothrombin activator systems in Australian rough-scaled snake, *Tropidechis carinatus*. Structural comparison of venom prothrombin activator with blood coagulation factor X. Thromb. Haemost. 93, 40–47.

Reza, M.A., Swarup, S., Kini, R.M., 2007. Structure of two genes encoding parallel prothrombin activators in Tropidechis carinatus snake: gene duplication and recruitment of factor X gene to the venom gland. J. Thromb. Haemost. 5, 117–126.

Rosing, J., Tans, G., 1991. Inventory of exogenous prothrombin activators. For the subcommittee on nomenclature of exogenous hemostatic factors of the scientific and standardization committee of the international society on thrombosis and haemostasis. Thromb. Haemost. 65, 627–630.

Rosing, J., Tans, G., 1992. Structural and functional properties of snake venom prothrombin activators. Toxicon 30, 1515–1527.

Schieck, A., Habermann, E., Kornalik, F., 1972. The prothrombin-activating principle from *Echis carinatus* venom. II. Coagulation studies *in vitro* and *in vivo*. Naunyn Schmiedebergs Arch. Pharmacol. 274, 7–17.

Silva, M.B., Schattner, M., Ramos, C.R., Junqueira-de-Azevedo, I.L., Guarnieri, M.C., Lazzari, M.A., Sampaio, C.A., Pozner, R.G., Ventura, J.S., Ho, P.L., Chudzinski-Tavassi, A.M., 2003. A prothrombin activator from *Bothrops erythromelas* (Jararaca-da-seca) snake venom: characterization and molecular cloning. Biochem. J. 369, 129–139.

Speijer, H., Govers-Riemslag, J.W., Zwaal, R.F., Rosing, J., 1986. Prothrombin activation by an activator from the venom of *Oxyuranus scutellatus* (Taipan snake). J. Biol. Chem. 261, 13258–13267.

Stenflo, J., Lundwall, A., Dahlback, B., 1987. beta-Hydroxyasparagine in domains homologous to the epidermal growth factor precursor in vitamin K-dependent protein S. Proc. Natl. Acad. Sci. U.S.A. 84, 368–372.

Stocker, K., Hauer, H., Muller, C., Triplett, D.A., 1994. Isolation and characterization of Textarin, a prothrombin activator from eastern brown snake (*Pseudonaja textilis*) venom. Toxicon 32, 1227–1236.

Suzuki, K., Dahlback, B., Stenflo, J., 1982. Thrombin-catalyzed activation of human coagulation factor V. J. Biol. Chem. 257, 6556–6564.

Tamiya, T., Fujimi, T.J., 2006. Molecular evolution of toxin genes in Elapidae snakes. Mol. Divers. 10, 529–543.

Tans, G., Govers-Riemslag, J.W., van Rijn, J.L., Rosing, J., 1985. Purification and properties of a prothrombin activator from the venom of *Notechis scutatus scutatus*. J. Biol. Chem. 260, 9366–9372.

Thorelli, E., Kaufman, R.J., Dahlback, B., 1997. Cleavage requirements for activation of factor V by factor Xa. Eur. J. Biochem. 247, 12–20.

van der Neut, K.M., Dirven, R.J., Vos, H.L., Tans, G., Rosing, J., Bertina, R.M., 2004. Factor Va is inactivated by activated protein C in the absence of cleavage sites at Arg-306, Arg-506, and Arg-679. J. Biol. Chem. 279, 6567–6575.

van Drunen, C.M., Oosterling, R.W., Keultjes, G.M., Weisbeek, P.J., van Driel, R., Smeekens, S.C., 1997. Analysis of the chromatin domain organisation around the plastocyanin gene reveals an MAR-specific sequence element in *Arabidopsis thaliana*. Nucl. Acids Res. 25, 3904–3911.

Wang, C., Eufemi, M., Turano, C., Giartosio, A., 1996. Influence of the carbohydrate moiety on the stability of glycoproteins. Biochemistry 35, 7299–7307.

Wilberding, J.A., Castellino, F.J., 2000. Characterization of the murine coagulation factor X promoter. Thromb. Haemost. 84, 1031–1038.

Xu, M., Hammer, R.E., Blasquez, V.C., Jones, S.L., Garrard, W.T., 1989. Immunoglobulin kappa gene expression after stable integration. II. Role of the intronic MAR and enhancer in transgenic mice. J. Biol. Chem. 264, 21190–21195.

Yamada, D., Morita, T., 1997. Purification and characterization of a Ca^{2+}-dependent prothrombin activator, multactivase, from the venom of *Echis multisquamatus*. J. Biochem. (Tokyo) 122, 991–997.

Yamada, D., Sekiya, F., Morita, T., 1996. Isolation and characterization of carinactivase, a novel prothrombin activator in *Echis carinatus* venom with a unique catalytic mechanism. J. Biol. Chem. 271, 5200–5207.

Chapter 30
Thrombin-Like Enzymes in Snake Venoms

Stephen P. Mackessy

Abstract Snake venoms, particularly from vipers, are rich sources of serine proteinases, some of which contain thrombin-like activity. Following human envenomations, these toxins often produce rapid coagulopathies via the depletion of circulating fibrinogen, typically via specific proteolysis of the Aα and Bβ subunits. Hypofibrinogenemia following bites may be prolonged, contributing to hemorrhagic effects of the venom and occasionally leading to life-threatening conditions such as disseminated intravascular coagulation. However, they have also been used as therapeutic drugs for treating a diversity of human disorders, including strokes, deep vein thromboses and cerebral and myocardial infarctions. Many snake venom thrombin-like enzymes (SV-TLEs) have been sequenced, and important structural elements (six disulfides, the catalytic triad) are highly conserved. SV-TLEs are commonly glycosylated, and this modification may confer a high level of stability; unlike trypsin, they are exceptionally stable in aqueous solution. Structurally, they are closely related to other serine proteinases such as trypsin and chymotrypsin, but as a result of gene duplication, accelerated point mutations and ASSET, venom TLEs have evolved a diversity of activities. The relationship between structure and function of the different venom serine proteinases is still unclear, and future studies of substrate specificity of this diverse family of toxins will help resolve this uncertainty.

Introduction

Thrombin is the key element of blood coagulation, and its activation is tightly regulated via a series of factor-dependent reactions of coagulation cascade. Disruption of thrombin's normal activity, via genetic defects, disease or severe injury, can result in serious health concerns, and so it is not surprising that there is an extensive literature

S.P. Mackessy (✉)
School of Biological Sciences, University of Northern Colorado, 501 20th St., CB 92, Greeley, CO 80639-0017, USA
e-mail: stephen.mackessy@unco.edu

R.M. Kini et al. (eds.), *Toxins and Hemostasis*, DOI 10.1007/978-90-481-9295-3_30,

on this critically important serine proteinase. For example, a search of the PubMed citation database (http://www.ncbi.nlm.nih.gov/sites/entrez) in January 2010 using the term "thrombin" returned over 38,500 hits. It is therefore well beyond the scope of this chapter to discuss the many roles of thrombin in hemostasis, but numerous recent reviews are available (e.g., di Cera, 2008), and this chapter will focus on select aspects of thrombin-like enzymes from snake venoms.

A notable result of human envenomations by many species of snakes is the rapid onset of coagulopathies, including the loss of blood coagulability. Though most pronounced following viperid bites, profound (consumptive) coagulopathies are not infrequently observed following elapid bites, particularly by Australian species (Isbister et al., 2009; White, 2005, 2009). Many of the coagulopathies induced by snake venoms result from the activities of snake venom serine proteinases (SVSPs), and among these, the thrombin-like enzymes (TLEs) are among the best characterized and have been the subject of many studies and recent reviews (e.g., Phillips et al., 2009; Serrano and Maroun, 2005; Swenson and Markland, 2005). SVSPs are broadly distributed among venomous reptiles, and like several other toxin families, the diversity of pharmacologies associated with these enzymes has resulted from gene duplication and accelerated evolution (Deshimaru et al., 1996; Kordis and Gubensek, 2000; Nakashima et al., 1993, 1995; Ohno et al., 1998), insertion of new exons (Pawlak and Kini, 2008), extension of intron-exon boundary (Tamiya et al., 1999) and exchange of exon segments (Doley et al., 2008, 2009). Although studies on the additive/synergistic effects of venom components are uncommon, it is likely that TLEs potentiate the effects in vivo of other components, such as the hemorrhagic metalloproteinases. Serine proteinases are most abundant among viperid venoms (e.g., Braud et al., 2000), and members of this protein family may comprise 20% or more of the venom proteome (e.g., Alape-Girón et al., 2008; Calvete et al., 2009; Sanz et al., 2006). Serine proteases which affect other components of hemostasis also have been characterized from a colubrid snake venom (Assakura et al., 1994) and from lizard venoms and saliva (Fry et al., 2006; Utaisincharoen et al., 1993), but at present, none appear to induce life-threatening coagulopathies. However, one colubrid, *Dispholidus typus* (Boomslang), does produce a potent procoagulant glycoprotein of ~55–67 kDa (Hiestand and Hiestand, 1979), but it remains incompletely characterized and may not be a serine protease.

Clotting disorders can occur at many levels in the blood clot cascade, but among venomous reptiles, targeting of fibrinogen patency is most common. Thrombin-like enzymes in venoms are serine proteinases and are typically responsible for the specific cleavage of fibrinogen Aα or/and Bβ chains, and their actions in vivo lead to rapid defibrinogenation (Phillips et al., 2009; Serrano and Maroun, 2005; Swenson and Markland, 2005). Metalloproteinases, abundant in reptile venoms (Fox and Serrano, 2008, 2009), also typically hydrolyze fibrinogen in vitro (e.g., Mukherjee, 2008; Weldon and Mackessy, 2010), and some may also catalyze hydrolysis of the γ chain of fibrinogen and fibrin clots (Mackessy, 1993a, 1996). Many metalloproteinases can be considered to be less specific in their actions toward fibrinogen, as suggested by time course digests which result in production of numerous degradation fragments (Mackessy, 1993a) and their activities toward other structural

Table 30.1 Some properties of thrombin-like snake venom serine proteinases

Species	Name	Activity	UniProtKB accession #	Mass	Inhibitors	References
Agkistrodon bilineatus	Bilineobin	Fibrinogenolytic (Aα and Bβ)	Q9PSN3	57 kDa 26,479 (deglyco-sylated	Heparin + Dithiothreitol + TLCK + Antithrombin III + Leupeptin + Argatroban - Hirudin -	Komori et al. (1993); Nikai et al. (1995)
Agkistrodon contortrix contortrix	Contortrixobin	Fibrinogenolytic (Bβ); Factor V activation; Factor XIII activation	P82981	26 kDa	Benzamidine + DAPI + Antithrombin III -	Amiconi et al. (2000)
Bothrops atrox	Batroxobin	Fibrinogenolytic (Aα)	P04971	41.5 kDa	Benzamidine + α2-macroglobulin + Antithrombin III - Heparin - Hirudin - Aprotinin - SBTI - ε-ACA - Tranexamic acid - Iodoacetamide -	Itoh et al. (1987); Stocker and Barlow (1976); Stocker et al. (1982); Sturzebecher et al. (1986)
Bothrops atrox	Thrombocytin	Fibrinogenolytic (Aα); Factor VIII activation	none	36 kDa	Pro-Phe-ArgCH_2Cl + PRCK + SBTI + Antithrombin III + Heparin + FPRCK + FARCK +	Castro et al. (2004); Kirby et al. (1979); Serrano and Maroun (2005)

Table 30.1 (continued)

Species	Name	Activity	UniProtKB accession #	Mass	Inhibitors	References
Bothrops jararaca	KN-BJ	Fibrinogenolytic (Aα); Kinin release	O13069	38 kDa	Benzamine derivatives +	Serrano et al. (1998)
Bothrops jararaca	Bothrombin	Fibrinogenolytic (Aα); Platelet aggregation; Factor VIII activation	P81661	35 kDa	Platelet Aggregation: Anti-GP IIb/IIIa + Anti-GP Ib +	Nishida et al. (1994)
Bothrops leucurus	Leucurobin	Fibrinogenolytic (Aα); Gyratory	none	35 kDa	Benzamine + β-mercaptoethanol + SBTI - EDTA -	Magalhães et al. (2007)
Calloselasma (Agkistrodon) rhodostoma	Ancrod	Fibrinogenolytic (Aα)	P47797	35.4 kDa	NPGB + Agmatine + α2-macroglobulin + Antithrombin III +	Au et al. (1993); Burkhart et al. (1992); Castro et al. (2004); Nolan et al. (1976)
Cerastes vipera	Cerastobin	Fibrinogenolytic (Aα and Bβ); Platelet aggregation	P18692	38 kDa	Iodoacetamide + Trasylol - SBTI -	Farid et al. (1989, 1990)
Cerastes cerastes	Cerastocytin	Fibrinogenolytic (Aα); Platelet aggregation; Factor X activation	Q7SYF1	38 kDa	SBTI + TLCK + TPCK + Antithrombin III - Hirudin -	Dekhil et al. (2003a); Marrakchi et al. (1995)
Cerastes cerastes	Cerastotin	Fibrinogenolytic (Aα); Platelet aggregation	P81038	40 kDa	TPCK + TPLK + SBTI + Hirudin - Antithrombin III -	Marrakchi et al. (1997)

Table 30.1 (continued)

Species	Name	Activity	UniProtKB accession #	Mass	Inhibitors	References
Crotalus adamanteus	Crotalase	Fibrinogenolytic (Aα); Kinin release	Q9PS55	32.7 kDa	TLCK + Pro-Phe-ArgCH_2Cl + PFRCK + AFRCK + GVRCK + IPRCK + AFKCK + Tetranitromethane + 2-mercaptoethanol + Hirudin - TPCK -	Markland (1976, 1998); Markland et al. (1982); Henschen-Edman et al. (1999)
Crotalus durissus terrificus	Gyroxin-like B2.1	Fibrinogenolytic (Aα); Gyratory	Q58G94	32 kDa (26.6 no glyco)	Dithiothreitol +	Alexander et al. (1988)
Deinagkistrodon acutus	Acutin	Coagulant	Q9YGS1	38 kDa	x	Pan et al. (1999)
Deinagkistrodon acutus	Venom serine proteinase Dav-PA	Fibrinogenolytic; amidolytic	Q9I8X1	28,032	x	Zhu et al. (2005)
Deinagkistrodon acutus	Acutobin	Fibrinogenolytic (Aα)	Q9I8X2	40 kDa (28.8 no glyco)	x	Wang et al. (2001)
Gloydius (Agkistrodon) blomhoffii brevicaudus	Brevinase	Fibrinogenolytic (Aα and Bβ)	Q9PT51	2 Chains: 16.5 and 17 kDa	Pefabloc + Dithiothreitol +	Lee et al. (1999, 2000)

Table 30.1 (continued)

Species	Name	Activity	UniProtKB accession #	Mass	Inhibitors	References
Gloydius (Agkistrodon) halys blomhoffi	Halystase	Fibrinogenolytic (Bβ); bradykinin release	P81176	38 kDa	Leupeptin + Hirudin -	Matsui et al. (1998)
Gloydius ussuriensis (formerly *Agkistrodon caliginosus*)	CPI-enzyme-2	Fibrinogenolytic (Bβ)	O42207	44 kDa	x	Hahn et al. (1998); Shimokawa and Takahashi (1993a, b; 1997)
Gloydius ussuriensis (formerly *Agkistrodon caliginosus*)	Calobin	Fibrinogenolytic (Aα), coagulant	Q91053	34 kDa	Aprotinin + Hirudin +	Hahn et al. (1996)
Lachesis muta	LM-TL	Fibrinogenolytic (Aα); Gyratory	P33589	41–47 kDa	Agmatine + *p*-aminobenzamidine + BPTI - Ecotin - Hirugen - Bothrojaracin - Bothroalternin -	Castro et al. (2001); Magalhaes et al. (1993); Silveira et al. (1989)
Trimeresurus elegans	Elegaxobin II	Fibrinogenolytic (Aα); Kinin release	P84787	35 kDa	*p*-APMSF +	Oyama and Takahashi (2003)
Trimeresurus flavoviridis	Flavoxobin	Fibrinogenolytic (Aα)	P05620	23.5 kDa	x	Shieh et al. (1985, 1988); Yamamoto et al. (2002)

Table 30.1 (continued)

Species	Name	Activity	UniProtKB accession #	Mass	Inhibitors	References
Trimeresurus mucrosquamatus	Mucrosobin	Fibrinogenolytic (Bβ)	U31417	28 kDa	x	Guo et al. (2001)
Viridovipera (formerly *Agkistrodon*) *stejnegeri*	Stejnefibrase-1	Fibrinogenolytic (Aα)	Q8AY80	28.3 kDa	NPGB +, PMSF +	Gao et al. (1998)
Viridovipera (formerly *Agkistrodon*) *stejnegeri*	Stejnobin	Fibrinogenolytic (Aα)	Q8AY81	29.3 kDa	DFP +, PMSF +	Zhang et al. (1998)

+, Inhibition; –, No inhibition. All tested enzymes were also inhibited by DFP and PMSF. AFKCK, Ala-Phe-Lys chloromethyl ketone; AFRCK, Ala-Phe-Arg chloromethyl ketone; BPTI, bovine pancreatic trypsin inhibitor; DAPI, 4',6-diamidino-2-phenylindole; DFP, diisopropyl fluorophosphate; ϵ-ACA, epsilon-aminocaproic acid; FAKCK, Phe-Ala-Lys chloromethyl ketone; FARCK, Phe-Ala-Arg chloromethyl ketone; FPRCK, Phe-Pro-Arg chloromethyl ketone; GVRCK, Gly-Val-Arg chloromethyl ketone; glyco, carbohydrate moiety; IPRCK, Ile-Pro-Arg chloromethyl ketone; NPGB, *p*-nitro-phenyl-*p*-guanidino benzoate HCl; PFRCK, Pro-Phe-Arg chloromethyl ketone; p-APMSF, *p*-amidinophenylmethanesulfonyl fluoride; PMSF, phenylmethylsulfonyl fluoride; PRCK, Pro-Arg chloromethyl ketone; SBTI, Soybean trypsin inhibitor; TLCK *p*-Tosyl-L-lysine chloromethyl ketone hydrochloride; TPCK, *p*-Tosyl-L-Phenylalanine chloromethyl ketone hydrochloride; x, inhibitor not reported. Table largely from Serrano and Maroun (2005) and Phillips et al. (2009)

proteins (Escalante et al., 2006)), and their pro- or anticoagulant activities will not be considered further here. However, it is likely that functionally important interactions occur among these venom components, and hypofibrinogenemia (TLE-catalyzed) accompanied by structural degradation catalyzed by metalloproteinases may produced uncontrollable hemorrhage. For the snake, these actions are important for prey incapacitation and facilitation of digestion; in human envenomations by vipers, these proteinases produce some of the more debilitating and difficult to manage effects (Gutiérrez et al., 2009).

Defining which of the myriad serine proteases often found in a single venom is a TLE-SVSP can be challenging, as many have similar activities toward model substrates such as paranitroaniline-derived peptides as well as toward native protein substrates such as fibrinogen. Further, the term "thrombin-like" is also problematic, because unlike most TLEs, thrombin is a multifunctional enzyme with rather different roles depending on physiological environment (Kini, 2005; Phillips et al., 2009). Additionally, most TLEs have not been assayed with a wide series of substrates (some are defined by sequence homology only), so the true specificity of activity, or lack thereof, is not well defined. Some of the SVSPs labeled as thrombin-like enzymes in the databases are incorrectly assigned to this activity, and some (particularly sequences derived from cDNA libraries) are labeled as TLEs because of sequence homologies but without any activity data. A more limiting definition of venom TLEs is needed, but for the present review, SVSPs which have specific catalytic activity toward fibrinogen Aα or/and Bβ chains will be considered as TLEs. A list of many well-defined venom TLEs is provided in Table 30.1, along with major activities and biochemical attributes.

Structural Features of Snake Venom TLEs

Primary and Secondary Structure

Snake venom thrombin-like enzymes are serine endopeptidases which are members of the trypsin-like serine protease superfamily, the S1 family of peptidases, clan PA and subclan S (Marchler-Bauer et al., 2009; Rawlings et al., 2004a, b). Well over 100 sequences of SVSPs are available in databases such as NCBI (http://www.ncbi.nlm.nih.gov/) and UniProtKB (http://www.uniprot.org/), with 77 entered as thrombin-like enzymes in UniProt (December 2009). These enzymes have a high level of sequence identity (>60%; Kini, 2005) and are structurally constrained by the presence of six highly conserved disulfide bridges (Fig. 30.1), five of which are common to all S1 serine proteinases. The catalytic triad typical of the chymotrypsin/trypsin superfamily (His57, Asp102, Ser195; chymotrypsinogen numbering) is also highly conserved, as are numerous flanking residues, and SVSPs are typically classified as serine proteinases by the common sensitivity to the serine active site inhibitors PMSF and DFP (Serrano and Maroun, 2005).

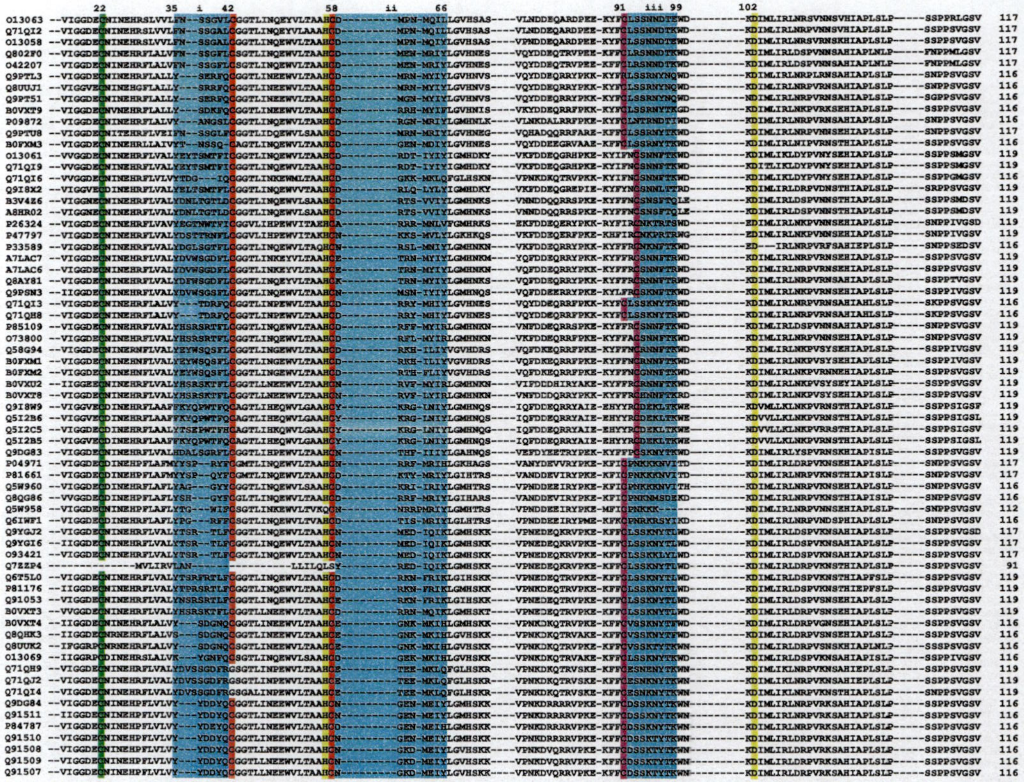

Fig. 30.1 Sequences of snake venom thrombin-like enzymes. TLEs and several additional SVSPs which show greater than 50% sequence identity with batroxobin are shown. Sequences are listed by UniProt Accession numbers, and names and species identities are provided in Appendix. *Shaded residues*: H, D and S – conserved histidine 57, aspartic acid 102 and serine 195 residues of the catalytic triad; residues 35–41, 59–66, 92–99 – segments i, ii and iii undergoing ASSET and likely important surface residues involved in substrate recognition/binding (Doley et al., 2009); C – represent highly conserved cysteine disulfide pairs (residues 22–157, 42–58, 91–245, 136–201, 168–182, 191–220). Numbering above sequences is based on chymotrypsin

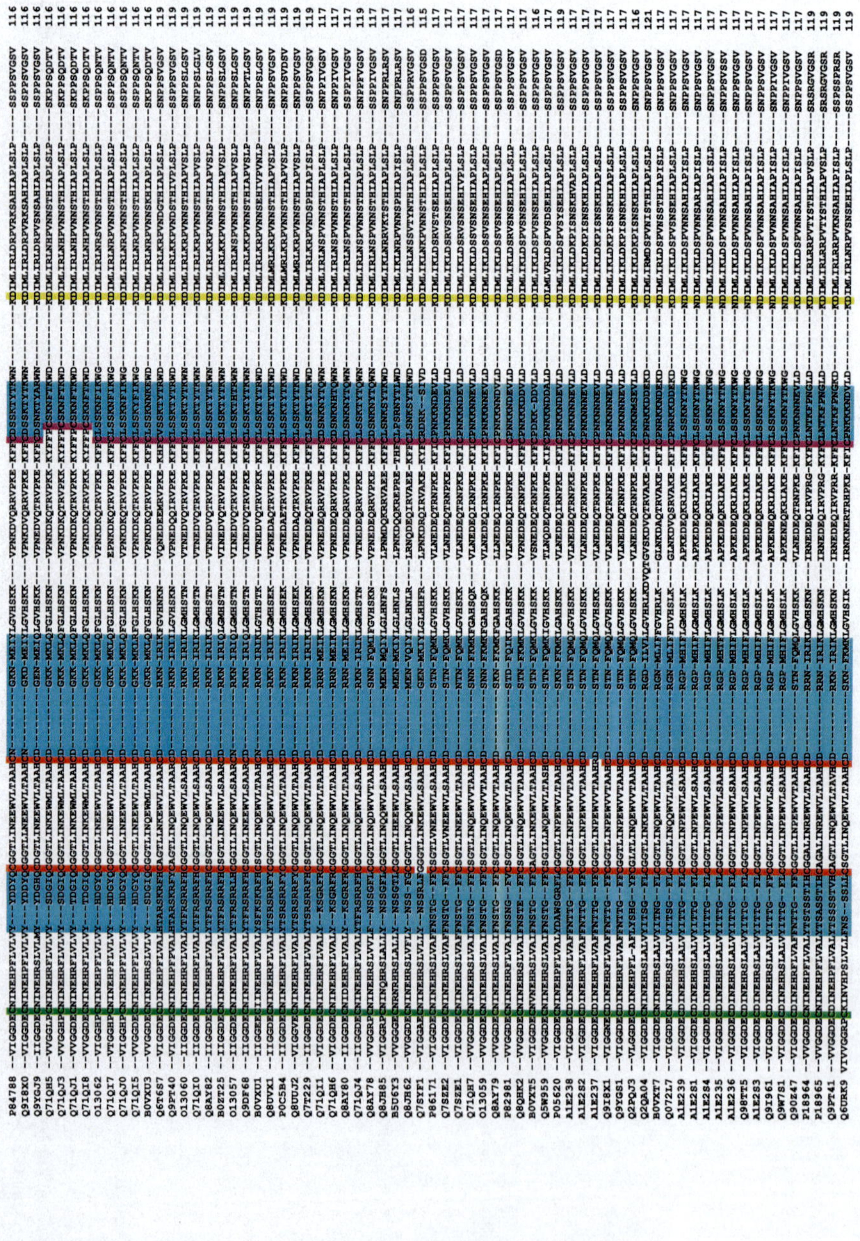

Fig. 30.1 (continued)

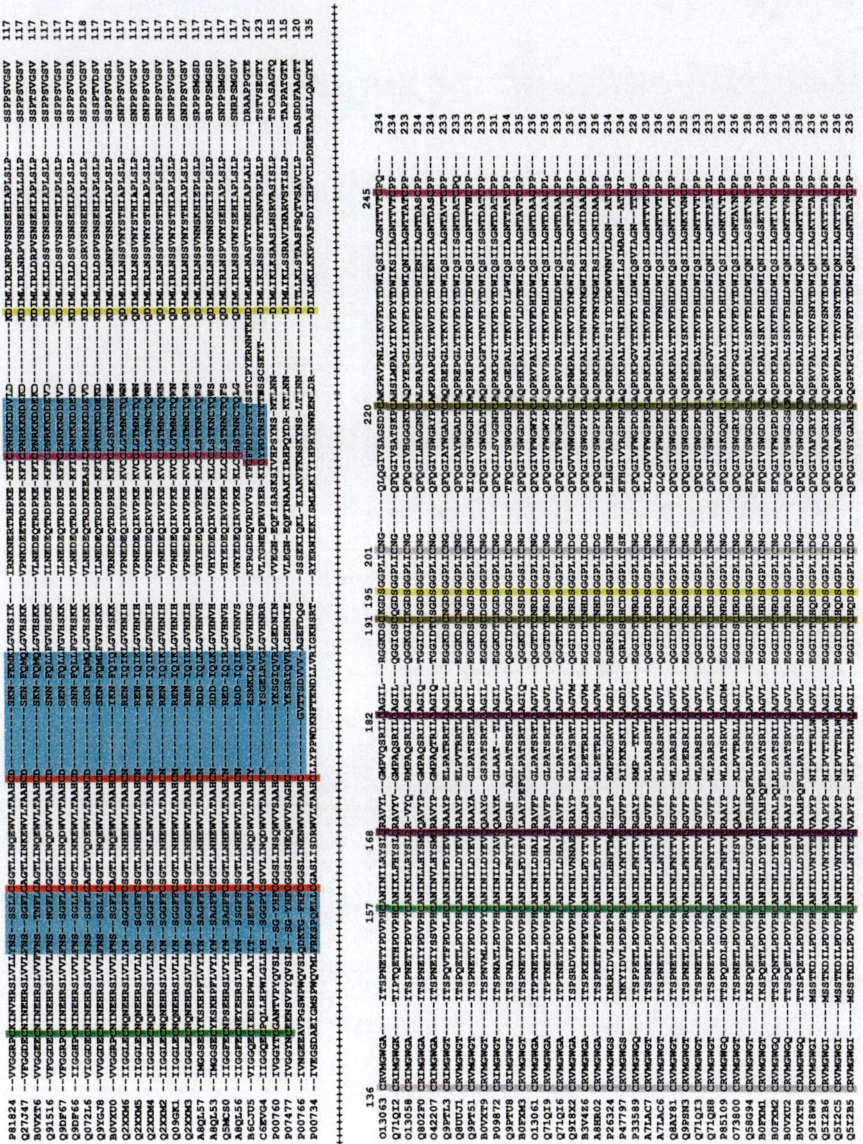

Fig. 30.1 (continued)

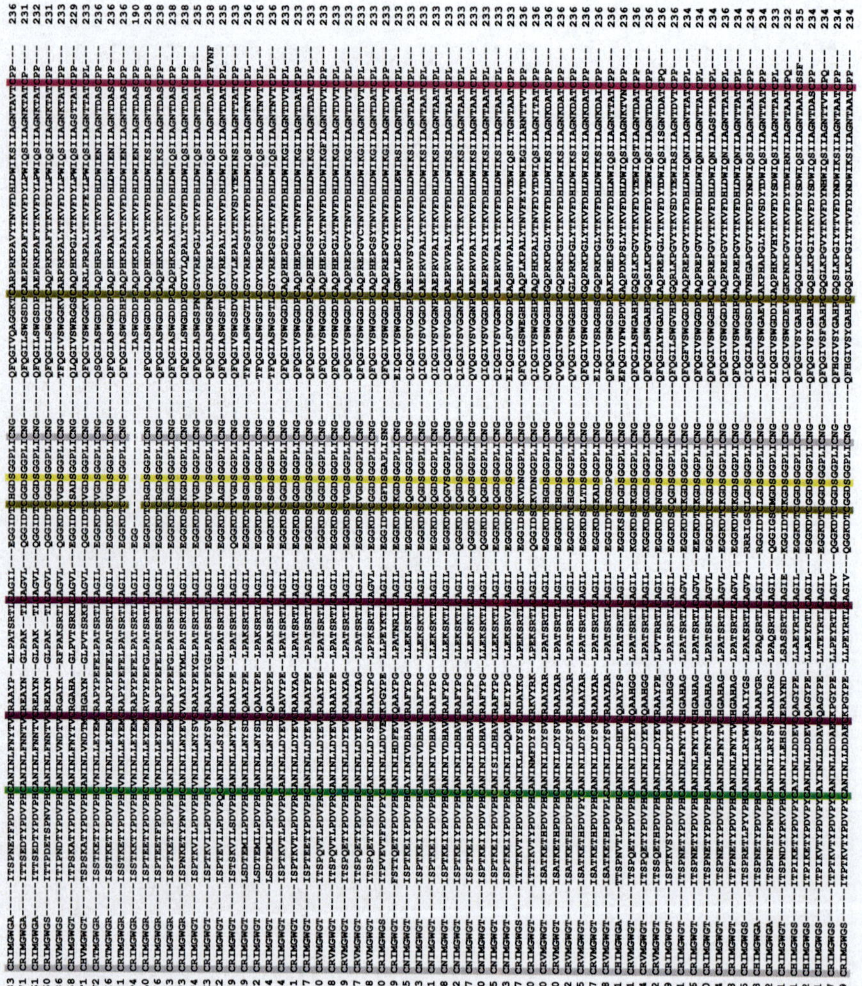

Fig. 30.1 (continued)

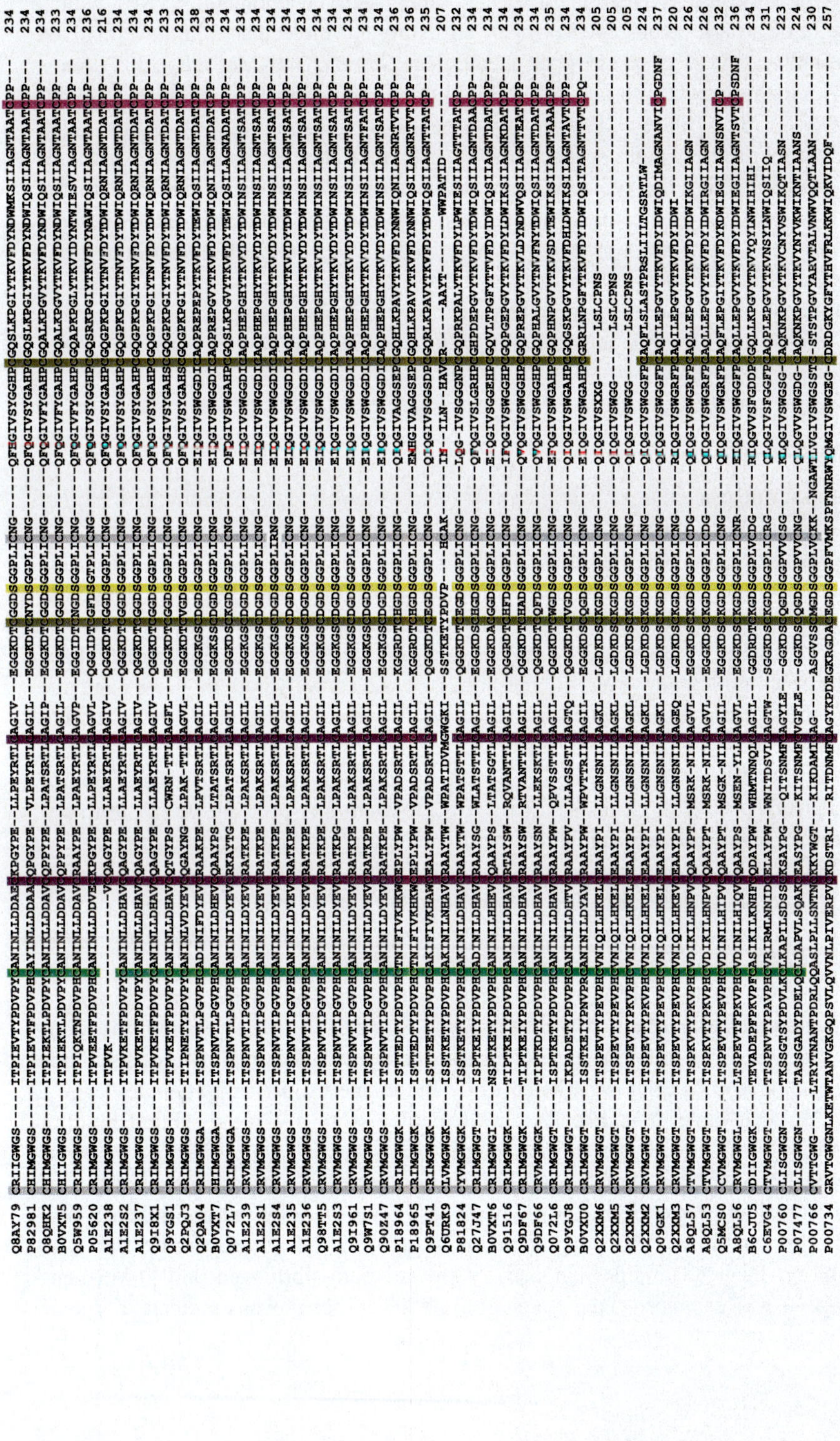

Fig. 30.1 (continued)

Biological solutions to structural needs are often frugal, and biological designs are typically conservative; this truism is notably observed among the SVSPs, which, in spite of a highly conserved primary structure and molecular fold, have evolved into a diversity of different pharmacologies (Serrano and Maroun, 2005), most of which cause dysregulation of the highly ordered coagulation cascade of vertebrates. Thrombin-like SVSPs typically target fibrinogen, but even within this target, TLEs from different sources may catalyze preferential hydrolysis of the Aα and/or Bβ chains. SVSPs are typically very stable molecules, in large part due to the six disulfides found in most. Purified thrombin-like and kallikrein-like SPs from *Crotalus* and *Sistrurus* venoms are unaffected by the relatively harsh conditions of RP-HPLC, and they retain most activity even after long storage (>1 year) in solution at 4°C (Mackessy, unpublished observations). As noted below, part of this exceptional stability may be due to the carbohydrate moiety typical of most SVSPs.

Glycosylation

Venom serine proteinases are commonly glycosylated, and the carbohydrate moiety is commonly Asn-linked (Tanaka et al., 1992). The extent of *N*- or *O*-glycosylation appears to be significant but quite variable, and the functional significance of this variation is incompletely understood (Serrano and Maroun, 2005). Carbohydrate content may account for 5–30% of the total mass, and in a few cases the mass of the carbohydrate moiety (62%) may exceed that of the protein component (Paes Leme et al., 2008). However, the presence of these carbohydrate moieties can have important functional consequences. Glycosylation appears to confer greater thermal stability to BJ-48, a TLE isolated from *B. jararacussu* venom, and deglycosylation also greatly increases susceptibility to trypsin inhibitors such as STI (Silva-Junior et al., 2007). Glycosylation also conferred to *Bothrops* protease A (*B. jararaca* venom) thermal protection and protection against protein inhibitors (soybean trypsin inhibitor {STI}, bovine pancreatic trypsin inhibitor {BPTI}, antithrombin III) but did not protect against inhibition by benzamidine, further indicating a steric protective effect against the larger molecules (Botos and Wlodawer, 2007; Paes Leme et al., 2008). The smaller inhibitor molecule appears to enter/bypass the "protective cage" of the carbohydrate moiety essentially unimpeded. Interestingly, a similar protective mechanism has been demonstrated for glycosylation of the nicotinic acetylcholine receptor of *Naja haje* snake muscle: the small neurotransmitter (acetylcholine) does not experience steric hindrance by the carbohydrate "cage", but the much larger α-neurotoxins are excluded from nAChR α-subunit binding sites, conferring resistance to the snake (Takacs et al., 2001). When glycosylation is eliminated, the receptor showed sensitivity to the neurotoxin similar to that of mammalian preparations. Partial deglycosylation of BPA did not increase susceptibility of the enzyme to protein inhibitors, nor did it significantly increase inhibition by benzamidine, but activity toward both fibrinogen and D-Val-Leu-Arg-pNA was enhanced (Paes Leme et al., 2008). These results indicate a type of functional trade-off between optimal enzyme activity and in vivo stability, suggesting that the protective effect of glycosylation is not without some cost to enzyme efficacy.

Two SVSPs with apparent thrombin-like activity were isolated from *Deinagkistrodon acutus* venom and shown to be *N*-glycosylated at Asn35, as observed from electron density maps of this region of the crystal structures (Zhu et al., 2005). Because this glycosylation site occurs close to the active site, it was interpreted to restrict access of larger molecules like STI and BPTI. Structural analyses using superimposition of the venom SVSPs and trypsin-STI complex demonstrated collision between side-chain residues of STI and the carbohydrate moieties of the SVSPs. Again, it appears that glycosylation of these serine proteases creates steric hindrance of inhibitor binding, thereby protecting the enzyme. Most TLEs also are glycosylated, and it is likely that glycosylation has been selected for as a protective mechanism against endogenous serine protease inhibitors of snake prey species. The effect of glycosylation may therefore be to increase effective half-life in prey tissues and to increase probability of fibrinogen depletion. In human envenomations by rattlesnakes (*Crotalus*), recurrent coagulopathies are commonly encountered following antivenom treatment (Boyer et al., 1999), and it would be of interest to determine if glycosylated TLEs are involved in these persistent and recurrent cases of hypofibrinogenemia and thrombocytopenia.

Crystal Structures and Structural Predictions

At present, no crystal structures of snake venom TLEs have been solved. The structure of a plasminogen activator from *Trimeresurus stejnegeri* venom was resolved at 2.5 Å (Parry et al., 1998), and this protein has a high degree of sequence identity with snake venom TLEs. Several glycosylated serine proteases from *Deinagkistrodon acutus* venom have also been crystallized and their structures solved at 2.1 Å (Zhu et al., 2005); because they have fibrinogenolytic activity, they were considered to be TLEs (Philipps et al., 2009), but activity toward pNA-derived peptide substrates showed that they lacked the specificity often observed for other TLEs, for example from *Crotalus* venoms (e.g., Mackessy, 1993b). A preliminary crystallographic analysis of RVV-V from *Daboia russelli* venom was just published (Nakayama et al., 2009), and the structure should be available soon. Although this enzyme has been considered a TLE, it appears to be a specific activator of Factor V only, although it also showed amidase activity toward the TLE substrate Phe-pipecolyl-Arg-pNA. A theoretical model of this protein has been constructed (Segers et al., 2006), based on the several related SVSP structures which are currently available.

Because of conserved primary structure and disulfide pattern of SVSPs generally, it is probable that venom TLEs have a structure similar to that of other S1

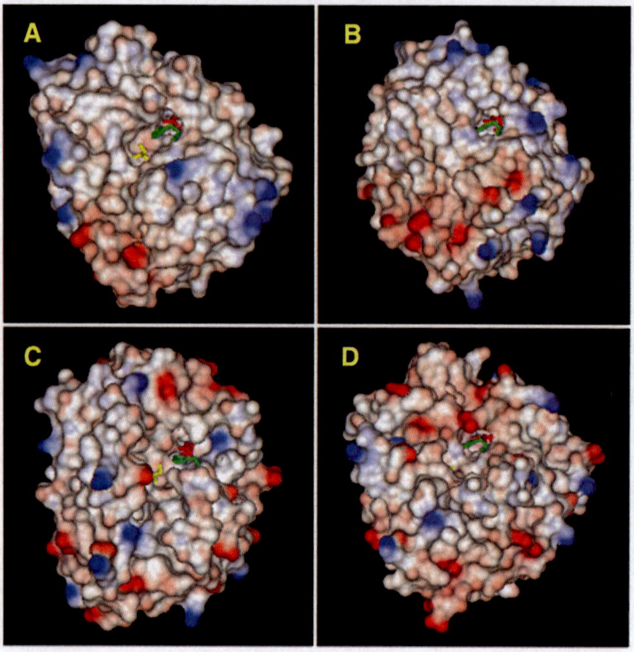

Fig. 30.2 Space-filling models of serine proteases. (**a**) Bovine trypsin (PDB ID: 1AQ7); (**b**) Bovine chymotrypsin B (PDB ID: 1DLK); (**c**) Venom serine proteinase Dav-PA (*Deinagkistrodon acutus* venom; PDB ID: 1OP0). (**d**) Venom plasminogen activator TSV-PA (*Viridovipera* (*Trimeresurus*) *stejnegeri*) venom (PDB ID: 1BQY). Models were drawn with Discovery Studio ViewerPro showing van der Waals radii and partial charges (*red*, negative; *blue*, positive). Active site residues His57 (*green*), Asp102 (*red*) and Ser195 (*yellow*) are shown in stick models. In B and D, Ser195 is barely visible. Note that although overall topology is similar, surface charge density and apparent accessibility to the active site vary between the molecules

serine proteinases, as molecular modeling of LM-TL (a TLE from *Lachesis muta* venom) has indicated (Castro et al., 2001). The overall similarity in molecular fold and active site orientation can be seen from comparisons of structural models of bovine chymotrypsin (MacSweeney et al., 2000) and trypsin and the SVSPs Dav-PA and TSV-PA (Fig. 30.2). Differences in surface charge densities are likely important for conferring observed differences in substrate specificity.

Gene Structures

At present, the gene structure for batroxobin (from *Bothrops moojeni* venom) remains the only published genomic representative of a snake venom TLE (Itoh et al., 1988). The gene is approximately 8 kbp, and the generally shorter exons (exon 1 = 240 bp; exon 2 = 151 bp; exon 3 = 260 bp; exon 4 = 134 bp; exon 5 = 728 bp) are interrupted by introns varying in length from 2.5 kbp (intron 1) to 0.35 kbp (intron 4). In this respect, batroxobin is similar to trypsin, which also has 5 exons and 4 introns (Craik et al., 1984); both differ in organization from chymotrypsin, which has 7 exons and 6 introns (Bell et al., 1984).

Relations to the Charge Relay Catalytic System of Model Serine Proteases: Chymotrypsin/Trypsin

Thrombin-likc serine proteases from snake venoms possess the highly conserved functional residues of the catalytic triad (Ser195, His57, Asp102) characteristic of the well-known serine proteases (e.g., Polgár, 1971) such as chymotrypsin (EC 3.4.21.1). They are members of the S1 serine proteinase family, PA clan of endopeptidases and they share the same basic fold in the catalytic domain with other members of the family, such as trypsin, chymotrypsin and thrombin. Peptide bond hydrolysis occurs via a two step acylation–deacylation reaction, with serine acting as a nucleophile and histidine as a proton donor/acceptor within the catalytic cleft; a hydrogen on the histidine imidazole ring is transferred to the aspartate carboxylate, forming a charge relay system. The mechanism of catalysis has been known for a long time (Hartley and Kilby, 1954; Polgár, 1971; Polgár and Bender, 1969) and has become a classic example in biochemistry courses of enzyme-catalyzed hydrolysis. The importance of the serine residue to snake venom TLEs, as well as unequivocal demonstration of a SVSP, is now routinely confirmed via specific active site inhibitors such as DFP, PMSF and AEBSF, which irreversibly and covalently modify serine in the catalytic cleft.

Distribution and Evolution of Snake Venom TLEs

Occurrence and Relative Abundance in Snake Venoms

Thrombin-like enzymes are found most frequently and as major components in the venoms of viperid snakes (Mackessy, 2009). Uncommonly, TLEs are present in elapid venoms (e.g., Jin et al., 2007), but they appear not to be broadly distributed among species of this family. Although there are reports of serine proteases in venoms from "colubrid" rear-fanged snakes (Assakura et al., 1994; Mackessy, 2002) and several lizards (Fry et al., 2006), TLEs do not appear to occur in venoms and salivas from these species. Serine proteases make up ~2.6% of the venom proteome of *Philodryas olfersii* (Ching et al., 2006), a colubrid snake capable of producing serious human envenomations, but this venom also does not appear to contain TLEs. Venom from *Alsophis portoricensis* (Puerto Rican Racer, family Dipsadidae) was recently shown to contain low activity toward the thrombin substrate Bz-PheValArgpNA (Weldon and Mackessy, 2010), but the levels are an order of magnitude lower than seen in most rattlesnake venoms.

Mechanisms of Evolution of SV-TLEs

Gene Duplication

A common means by which diversity is generated in a specific protein family is via gene duplication. In a critically important component of a complex system, such as thrombin in the blood clot cascade, mutation of specific residues which change activity levels or specificity of substrate recognition could be lethal mutations. However, when the gene is duplicated and one member remains static, the functional product of the original gene remains intact. The other gene copy is freed from selective constraints favoring conservation of original structure/function, and mutations can then lead to production of a novel activity. Repeated gene duplication can result in multiple copies of closely related genes being present, and over evolutionary time, a diverse set of pharmacologies within a structurally conserved protein family may result (Ogawa et al., 1996; Nobuhisa et al., 1996; Kini and Chan, 1999). This appears to be of common occurrence among snake venoms, and venom gene duplication has resulted in a multigene family of SVSPs (Deshimaru et al., 1996), giving rise to venoms with numerous serine proteinases. For example, in the venom of the Desert Massasauga (*Sistrurus catenatus edwardsii*), 24% of the proteome (Sanz et al., 2006) and >37% of the transcriptome consists of serine proteinases (Pahari et al., 2007). The transcriptome analysis showed 12 distinct isoforms of SVSPs, and fractionation of the venom has revealed at least 8 distinct serine proteinases are translated (Mackessy, unpublished observations). At least three of these SVSPs show thrombin-like activity, as assessed by preferential hydrolysis of the substrate Bz-Phe-Val-Arg-pNA over other Arg-terminal substrates (Mackessy, 1993b).

Accelerated Point Mutation

Single nucleotide replacement within an exon can have ramifications on function ranging from neutral to lethal mutation. One of the best documented single point mutations resulting in extreme alteration of function involves the sickle mutant of human hemoglobin which has resulted from the selective pressures of malarial parasites on humans (Huisman, 1993). Among snake venom toxins, point mutations have also occurred commonly and contribute to the primary structure diversity seen among homologous toxins from different species. Many of these point mutations appear to be functionally inconsequential; for example, small variations in sequence of α-neurotoxins from closely related species still results in a three-finger toxin with very high affinity for nicotinic acetylcholine receptors of skeletal muscle (Nirthanan and Gwee, 2004). However, among many toxin families, including SVSPs (Deshimaru et al., 1996), point mutations within protein-coding regions have occurred at a rate greatly accelerated relative to other proteins and even to the highly conserved UTRs within the same protein. Analysis of substitution rates between untranslated and translated regions demonstrated that nucleotide substitutions occurred much more frequently in protein-coding regions, and this unusually high rate of substitution has contributed in part to the diversification of functionality. The ratios of non-synonymous to synonymous substitutions within coding regions were generally greater than 1 (0.67–1.64), whereas these ratios in typical isozyme genes were typically less than 0.2 (Deshimaru et al., 1996), again highlighting the rapid evolution of these toxin genes relative to non-toxins. This accelerated evolution within the SVSP multiple gene family is likely driven by selective pressures favoring multiple toxin activities against snake prey and predators.

ASSET

Recently we have shown that in addition to accelerated point mutations, many snake venom toxins also evolve via accelerated segment switch in exons to alter targeting (ASSET: Doley et al., 2009). The occurrence of ASSET is particularly interesting, as it seems to occur primarily in specific surface segments, while accelerated

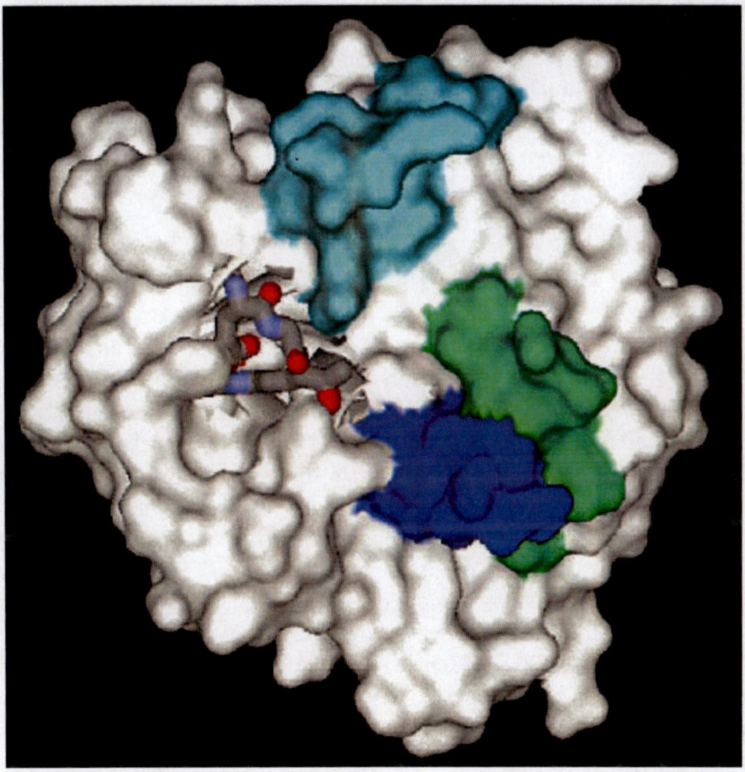

Fig. 30.3 Surface model of venom plasminogen activator (TSV-PA) from *Viridovipera* (*Trimeresurus*) *stejnegeri* venom (PDB ID: 1BQY). The segments that undergo exchange are shown in *green, dark blue* and *turquoise* color; note that the substrate binding area is the region shown in *turquoise*. The side chains of the active site residues are shown as *ball* and *stick*. From Doley et al. (2009)

point mutations occur in the rest of the molecule; among SVSPs, this includes those regions associated with substrate binding. ASSET therefore could produce rapid functional differentiation of gene products which share a highly conserved molecular fold and apparent surface topology (see Fig. 30.2). In viperid venoms in particular, numerous SVSP cDNAs have been sequenced, including many from *Viridovipera stejnegeri* (Tsai and Wang, 2001), *Deinagkistrodon acutus* (Zhang et al., 2006) and *Sistrurus catenatus edwardsii* (Pahari et al., 2007), demonstrating the high level of multiplicity of SVSPs in the venom and gland transcriptome of even a single individual. At least part of this diversification has occurred via ASSET.

ASSET has been hypothesized to be a mechanism of accelerated evolution of venom toxins which can confer new pharmacological functionalities on a conserved molecular fold, as is common among venom proteins. By switching functionally important segments of gene (protein) sequence, such as that important to substrate binding, rapid large scale changes in substrate specificity can occur. Such a mechanism appears to be important in the evolution of SVSPs (Doley et al., 2009), as the regions of exchange include those known to involve substrate binding (Fig. 30.3). ASSET can result in large-scale functional changes, with accelerated point mutations "fine-tuning" substrate fit. This hypothetical scenario may explain the variety of substrate specificities (thrombin-like, kallikrein-like, plasmin-like, arginine esterase, etc) seen among the SVSPs.

Structural and Phylogenetic Relationships

As mentioned above, SVSPs including TLEs show a high level of sequence identity, and the genes are obviously closely related. One might predict that functional classes of the SVSPs (thrombin-like, kallikrein-like, plasmin-like, etc) should cluster following structural cladistic analyses, and this prediction has been borne out by some studies (i.e., Wang et al., 2001). In this study, three functional subtypes clustered into discrete groups (thrombin-like {coagulating}, kallikrein-like {kininogenase} and plasminogen activators). However, a different analysis (Lee and Park, 2000) resulted in the clustering of functionally different SVSPs. A more recent analysis of sequence relationships among TLEs indicated a common ancestry among the SVSPs analyzed but did not demonstrate unequivocal clustering of functional subtypes (Castro et al., 2004). Because of these discrepancies, a phylogenetic analysis of TLEs and other SVSPs was undertaken using ClustalX and bootstrapped neighbor-joining method. One hundred and fifty-one snake venom serine protease sequences were retrieved from the UniProtKB database (http://www.uniprot.org/; January 2010) using the primary sequence of batroxobin in a BLAST search and the criterion of having >50% sequence identity with this target. Five SP sequences from lizards (sequence identity ~40%) were also included. Bovine and human trypsin sequences were used as outgroups, and bovine chymotrypsinogen and human thrombin were included in alignments and subsequent analyses (160 sequences total). Sequences were aligned using ClustalX 1.81(Thompson et al., 1994) and sequence similarities were evaluated using the bootstrapped neighbor-joining algorithm (1,000 bootstrap trials). The resulting NJ tree was drawn using FigTree 1.2.1 (available at http://tree.bio.ed.ac.uk/software/figtree/) and nodes were coded as to

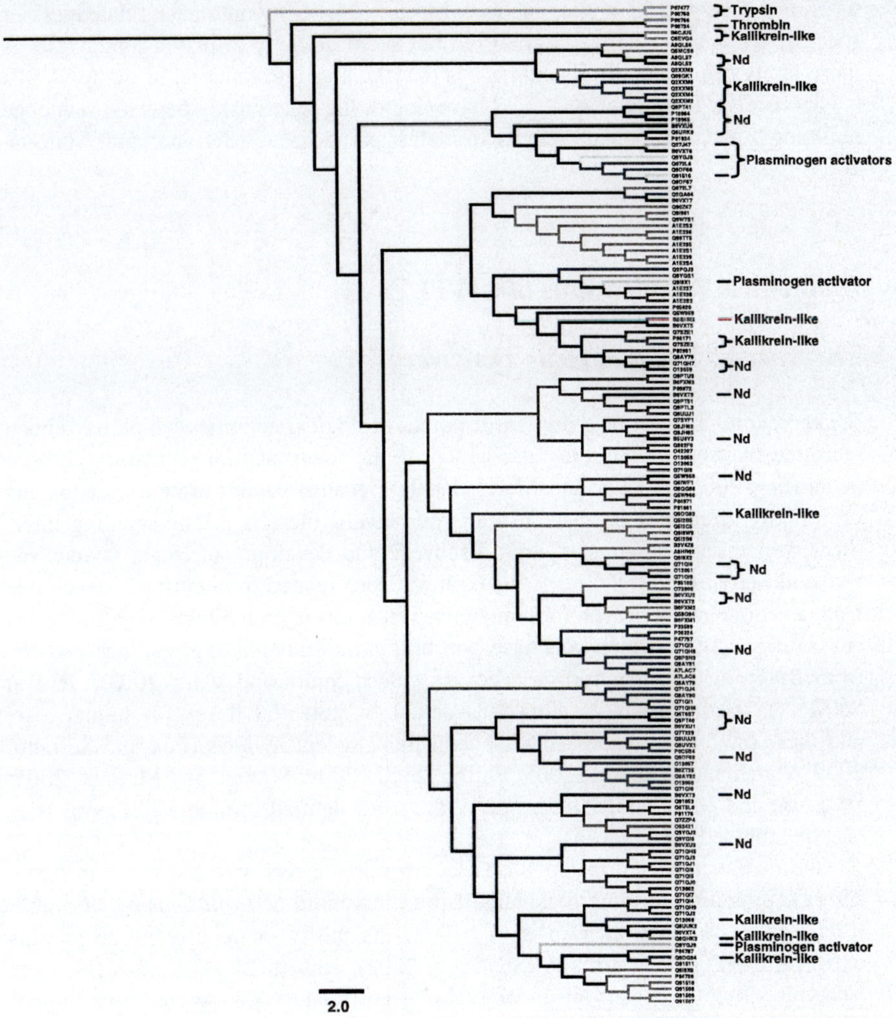

Fig. 30.4 Bootstrapped neighbor-joining tree of relationships among TLE and SVSP. The sequences are shown in Fig. 30.1. Although there is a trend toward clustering of similar activities (such as kallikrein-like activities), kallikrein-like and plasminogen-activating activities are distributed throughout many clades dominated by TLEs, suggesting that similar primary sequence motifs alone are not responsible for differences between activities. Kallikrein and plasminogen activating SVSPs are labeled; unlabeled nodes are all thrombin-like enzymes. Nd, activity not determined

functionality based on database-reported activity. The majority of sequences were identified as thrombin-like enzymes in the UniProt database, with a smaller number of kallikrein-like (14) and plasminogen activator (5) sequences identified; 28 were of unknown/undefined activity. Although there is a tendency for similar function sequences to cluster (Fig. 30.4), there are instances of all three activities occurring within many clades, suggesting that functionality is not dependent on sequence features alone. Effects of phylogenetic constraint (closer evolutionary relatedness of species) were not specifically analyzed but seem unlikely to play a major role. A more likely limitation of this analysis is that the absolute function of many SVSPs is equivocal or unknown. As others have noted, the relationship between sequence homology and biological activity remains paradoxical (Serrano and Maroun, 2005).

Biomedical Applications of SV-TLEs

Therapeutic Use and Drug Discovery

Snake venoms consist of a myriad of potent biological activities which have been recruited from various tissues and which mimic many natural regulatory components (Fry, 2005; Stocker and Meier, 1989). Because venom protein components profoundly affect homeostasis at numerous levels, there is a long-standing interest in venoms as a source of drug discovery and development (for a review, see Fox and Serrano, 2007). One of the earliest drugs related to hemostasis developed from a venom and approved for human use is Captopril, an inhibitor of angiotensin-converting enzyme which was based on bradykinin-potentiating peptides isolated from *Bothrops jararaca* venom (Ferreira, 1965; Smith and Vane, 2003). Several SVSPs have also been developed as actual or potential drugs for human use, and these have been discussed in several more recent reviews (Fox and Serrano, 2007; Marsh and Williams, 2005; Phillips et al., 2009; Serrano and Maroun, 2005; Swenson and Markland, 2005); only those with defined thrombin-like properties will be included here.

One of the most promising SV-TLEs for clinical use was Ancrod (Viprinex), a serine proteinase originally isolated from the venom of *Calloselasma* (formerly *Agkistrodon*) *rhodostoma* (Au et al., 1993). Its utility in treating stroke victims was evaluated some years ago (Sherman, 2002), and initial indications for acute ischemic stroke were very promising. However, in late 2008, Ancrod failed phase 3 clinical trials for acute ischemic stroke, and it is no longer being developed as a drug for human use in strokes (Neurobiological Technologies, http://www.ntii.com). In a recently published study involving 500 patients at several different facilities who received Ancrod within 6 h of symptom onset, Ancrod treatment of acute ischemic stroke victims was halted due to lack of efficacy (Levy et al., 2009). No difference between placebo and Ancrod treatment was seen in positive response of patients or in ninety day mortality levels, and the incidence of symptomatic intracranial bleeding was approximately twice as great in patients receiving Ancrod. This lack of efficacy was particularly disappointing considering that desired changes in fibrinogen levels (rapid initial defibrinogenation and avoidance of prolonged hypofibrinogenemia) were seen in >90% of Ancrod-treated patients (Levy et al., 2009). Ancrod is still used outside of the United States for several coagulopathies, including deep vein thrombosis and coronary artery bypass surgery (Cole, 1998), but the outlook for more extensive clinical use is poor.

Batroxobin (Itoh et al., 1987) is another TLE (from *Bothrops moojeni* venom) which has been investigated extensively for use in a variety of disorders, including cerebral and myocardial infarction, ischemic stroke, angina and for prevention/treatment of surgical bleeding (Phillips et al., 2009; Xu et al., 2007). A subsidiary of the Chinese biopharmaceutical company Sinobiomed was granted a patent for recombinant batroxobin (rBAT, Defibrase) in 2007, and some recent reports indicate that this TLE may be more successful in clinical trials than Ancrod. In a study of a small group of patients with deep vein thrombosis (DVT), a serious and potential fatal disorder of the lower limbs, batroxobin was found to be effective in abolishing symptoms of DVT (pain, swelling/erythema of lower limbs) and achieving limb salvage (Zhang et al., 2009). Patient monitoring for 1–3 month post-treatment indicated no abnormalities or return of symptoms, and fibrinogen levels were lowered for at least 14 days. An interesting effect of batroxobin treatment was the increase in endothelial progenitor cells in peripheral blood, which were greatly elevated from normal levels (statistically significant) at 7 and 14 days post-administration. Because these cells (CD34+, CD31+, VE-cadherin+ cells) are important in neovessel formation in adults and may be an essential component of vasculogenesis following disease/injury (Allegra et al., 2009; Asahara et al., 1999), batroxobin-stimulated mobilization may be an important component of recovery (Zhang et al., 2009). The mechanism of this action is presently unknown, but the positive results obtained support extending clinical trials to include a larger patient group.

Batroxobin has also been recommended for treatment of patients with ischemic stroke and transient ischemic attack who also show hyperfibrinogenemia (Xu et al., 2007). Intravenous batroxobin was administered in clusters immediately following and at 3 and 6 months after ischemic events; fibrinogen levels decreased significantly after administration at 0 and 3 months, but the decrease was not significant at 6 months. One year survivorship was also significantly greater in the batroxobin-treated group, and incidence of intracranial hemorrhage was not different in the two treatment groups. These indications suggest that batroxobin may be a safer and more effective treatment for ischemia than Ancrod. However, a recent review of the effectiveness of batroxobin to control hemorrhage during thoracic surgery concluded that although some differential effects were observed, no clinically relevant benefit was observed following batroxobin use (Zeng et al., 2009). The authors conclude that there is insufficient evidence supporting any benefit of batroxobin for hemorrhage during thoracic surgery. These conflicting reports suggest that while batroxobin may have efficacy for controlling some specific types of coagulopathies, it is not generally indicated for all such conditions.

Several additional TLEs are undergoing evaluation in animal models for antithrombotic use. Acutobin, a TLE derived from *Deinagkistrodon acutus* venom, was reported to be effective in reducing mortality and brain damage following ischemia and reperfusion of the cerebral artery in hyperglycemic rats (Wei et al., 2004). This model mimics a condition particularly at risk of brain tissue injury following ischemia, and acutobin treatment resulted in increased brain tissue perfusion and a reduction in the size of infarct. In a very different application of a snake venom TLE, a dental fibrin adhesive was produced from fibrinogen hydrolyzed by TLE from *Crotalus durissus terrificus* venom (Barbosa et al., 2008). Free gingival grafts were immobilized using either TLE or sutures, and at 7 days post-treatment, inflammatory cell density was lower in the TLE treatment group. By 14 and 45 days, no difference was observed between the two treatments. This study demonstrates that TLEs from snake venom can have utility in production of biological products, with medical applications which are only tangentially related to their in vivo direct actions as fibrinogenolytic serine proteases.

Use in Basic Research and Diagnostics

Thrombin-like enzymes have been developed for use as diagnostic reagents for clinical and research labs, and Reptilase®, derived from *B. jararaca* venom, is commonly used in clinical lab diagnosis of bleeding and other coagulation disorders (Phillips et al., 2009; Stocker, 1998). Like several TLEs, Reptilase® shows high specificity for the Aα chain of fibrinogen and is not affected by fibrin degradation products (FDPs). Clinically, a thrombin time/Reptilase® time (TT/RT) of >1, which is correlated with the presence of FDPs, is indicative of high potential for disseminated intravascular coagulation, a potentially fatal coagulopathy.

Several other TLEs have been used in a variety of other applications, such as in the production of dental adhesives (Barbosa et al., 2008; see above). Gyroxin, purified from *Crotalus durissus terrificus* venom, was labeled with ^{125}I and used to study biodistribution in a mouse model system (Alves da Silva et al., 2006). The authors suggested that distribution and exogenous compound metabolism could be evaluated with these methods, and it was hypothesized that the distribution kinetics/metabolism patterns observed may be correlated with tissue-specific distribution of different protease activated receptors (PARs) for this (and other) serine proteinases. Ancrod, which may be removed from therapeutic use, has been used as a reagent to study polymerized collagen-fibrin matrices (Rowe and Stegemann, 2009). Collagen-fibrin mixed scaffolds showed strength and tensile stiffness greater than collagen alone, and this mixed matrix may have applications in bioengineering of tissues and biomaterials design generally. Potential further applications were indicated for use in cell culture work, in vitro modeling of vascular flow dynamics and in vivo wound healing (Rowe and Stegemann, 2009). These are but a few examples of the creative ways in which venom TLEs are being utilized in both applied clinical and basic research.

Future Potential for SV-TLEs

There is a continuing need for safe and effective drugs to treat coagulation disorders such as venous thromboembolic disease and stroke (Spyropoulos, 2008), and research into thrombin-like enzymes from snake venoms could provide novel lead compounds or enzymes which could be directly useful. There is an obvious diversity of SVSPs from front-fanged snakes which could provide a source of novel compounds. However, though variation in component number can exceed 100 protein/peptide compounds in a single venom, there is a relatively small diversity of protein families so far described from venoms (Juárez et al., 2004), and TLEs described thus far appear to possess conserved functional variation along a common theme. This scenario may change as newer methods allow a deeper probe of the venom proteome and reveal diversity of structure (and likely function) among the much less abundant minor venom components (Bandow, 2010; Calvete et al., 2009; Polaskova et al., 2010), but other sources, such as among venoms from rear-fanged snakes, may prove to contain novel TLEs.

Snake venom serine proteases have proven to be useful in various applications in biotechnology and basic research (Wisner et al., 2001), and their specificities could perhaps be exploited for use in mass spectrometry applications currently dominated by trypsin use, such as peptide fingerprinting, MS/MS sequencing, etc. It may be that the most useful applications of SVSPs like the TLEs may lie in research purposes rather than drug development.

Summary and Conclusion

Snake venom thrombin-like enzymes are important components of most viperid snake venoms and are less broadly occurring among other squamate reptile venoms. As part of the biological weaponry of venomous species, their actions in vivo can cause cataclysmic coagulopathies which may become life-threatening. Purified and characterized, TLEs have many applications in biomedicine as well as basic and applied research. Rapid advances in genomics and proteomics have provided sequences for many venom serine proteinases, including TLEs, and detailed structure/activity data is available for a smaller subset of these. There is a need for rigorous substrate specificity studies to be conducted with the naturally expressed venom serine proteinases, particularly for those species with extensive transcriptome and proteome datasets. Such functional data will help to answer the remaining questions related to the observed diversity of actions of these structurally conservative venom components. Further, there are many species of front-fanged and rear-fanged snakes whose venoms are poorly known, and it is likely that additional interesting variants of this family of proteinases remain to be described.

Appendix: Identification of Serine Proteinases Used in Sequence Alignments (Fig. 30.1) and Sequence Similarity Analysis (Fig. 30.4), Ordered by UniProt Accession Number

Accession	Protein names	Organism	Length
P00760	Trypsin – Bovine cationic	*Bos taurus*	246
P07477	Trypsin-1 (EC 3.4.21.4)	Homo sapiens (Human)	247
P00766	Chymotrypsin – bovine	*Bos taurus*	245
P00734	Thrombin – human	*Homo sapiens* (Human)	259
A1E235	Venom thrombin-like enzyme (Fragment)	*Deinagkistrodon acutus* (Hundred-pace snake)	235
A1E236	Venom thrombin-like enzyme (Fragment)	*Deinagkistrodon acutus* (Hundred-pace snake)	235
A1E237	Venom thrombin-like enzyme (Fragment)	*Deinagkistrodon acutus* (Hundred-pace snake)	235
A1E238	Venom thrombin-like enzyme (Fragment)	*Deinagkistrodon acutus* (Hundred-pace snake)	217
A1E239	Venom thrombin-like enzyme (Fragment)	*Deinagkistrodon acutus* (Hundred-pace snake)	235
A1E2S1	Venom thrombin-like enzyme	*Deinagkistrodon acutus* (Hundred-pace snake)	235
A1E2S2	Venom thrombin-like enzyme	*Deinagkistrodon acutus* (Hundred-pace snake)	235
A1E2S3	Venom thrombin-like enzyme	*Deinagkistrodon acutus* (Hundred-pace snake)	235
A1E2S4	Venom thrombin-like enzyme	*Deinagkistrodon acutus* (Hundred-pace snake)	235
A7LAC6	Thrombin-like serine protease 1	*Trimeresurus albolabris* (White-lipped pit viper)	260
A7LAC7	Thrombin-like serine protease 2	*Trimeresurus albolabris* (White-lipped pit viper)	260
A8HR02	Thrombin-like protein (Fragment)	*Deinagkistrodon acutus* (Hundred-pace snake)	236
A8QL53	Putative serine protease (Fragment)	*Naja atra* (Chinese cobra)	282
A8QL56	Thrombin-like serine protease	*Ophiophagus hannah* (King cobra)	260
A8QL57	Putative serine protease (Fragment)	*Bungarus multicinctus* (Many-banded krait)	282
B0FXM1	Gyroxin-like B1_3 serine protease	*Crotalus durissus terrificus* (South American rattlesnake)	262
B0FXM2	Gyroxin-like B1_4 serine protease	*Crotalus durissus terrificus* (South American rattlesnake)	262
B0FXM3	Gyroxin-like B1_7 serine protease	*Crotalus durissus terrificus* (South American rattlesnake)	259
B0VXT3	Serine proteinase isoform 1	*Sistrurus catenatus edwardsii* (Desert massasauga)	262
B0VXT4	Serine proteinase isoform 2	*Sistrurus catenatus edwardsii* (Desert massasauga)	259

Accession	Protein names	Organism	Length
B0VXT5	Serine proteinase isoform 3	*Sistrurus catenatus edwardsii* (Desert massasauga)	257
B0VXT6	Serine proteinase isoform 4	*Sistrurus catenatus edwardsii* (Desert massasauga)	258
B0VXT7	Serine proteinase isoform 5	*Sistrurus catenatus edwardsii* (Desert massasauga)	258
B0VXT8	Serine proteinase isoform 6 (Fragment)	*Sistrurus catenatus edwardsii* (Desert massasauga)	242
B0VXT9	Serine proteinase isoform 7	*Sistrurus catenatus edwardsii* (Desert massasauga)	273
B0VXU0	Serine proteinase isoform 8	*Sistrurus catenatus edwardsii* (Desert massasauga)	258
B0VXU1	Serine proteinase isoform 9	*Sistrurus catenatus edwardsii* (Desert massasauga)	260
B0VXU2	Serine proteinase isoform 10	*Sistrurus catenatus edwardsii* (Desert massasauga)	260
B0VXU3	Serine proteinase isoform 11	*Sistrurus catenatus edwardsii* (Desert massasauga)	257
B0ZT25	Snake venom serine protease homolog (TjsvSPH)	*Protobothrops* (Trimeresurus) *jerdonii* (Jerdon's pit-viper)	260
B3V4Z6	Thrombin-like protein DAV-WY	*Deinagkistrodon acutus* (Hundred-pace snake)	260
B5U6Y3	Serine proteinase	*Echis ocellatus* (Ocellated saw-scaled viper)	258
B6CJU5	Kallikrein toxin Var13	*Varanus komodoensis* (Komodo dragon)	258
C6EVG4	Kallikrein toxin 1	*Heloderma suspectum cinctum*	255
O13057	Venom serine proteinase 2 (EC 3.4.21)	*Protobothrops* (Trimeresurus) *flavoviridis* (Habu)	260
O13058	Venom serine proteinase 3 (EC 3.4.21)	Protobothrops (Trimeresurus) *flavoviridis* (Habu)	257
O13059	Venom serine proteinase 1 (EC 3.4.21)	*Trimeresurus gramineus* (Indian green tree viper)	258
O13060	Venom serine proteinase 2A (EC 3.4.21)	*Trimeresurus gramineus* (Indian green tree viper)	260
O13061	Venom serine proteinase 2B (EC 3.4.21)	*Trimeresurus gramineus* (Indian green tree viper)	260
O13062	Venom serine proteinase 2C (EC 3.4.21)	*Trimeresurus gramineus* (Indian green tree viper)	257
O13063	Venom serine proteinase 3 (EC 3.4.21)	*Trimeresurus gramineus* (Indian green tree viper)	258
O13069	Kinin-releasing/fibrinogen-clotting serine proteinase 2 (EC 3.4.21) (KN-BJ 2)	*Bothrops jararaca* (Jararaca)	257
O42207	Capillary permeability-increasing enzyme 2 (CPI-enzyme 2) (EC 3.4.21)	*Gloydius ussuriensis* (Ussuri mamushi) (Agkistrodon caliginosus)	258
O73800	Thrombin-like enzyme salmobin (EC 3.4.21)	*Gloydius* (Agkistrodon) *halys pallas* (Chinese water mocassin)	260

Accession	Protein names	Organism	Length
O93421	Thrombin-like enzyme pallase (EC 3.4.21	*Gloydius* (Agkistrodon) *halys pallas* (Chinese water mocassin)	236
P04971	Batroxobin (BX) (EC 3.4.21.74) (Venombin-A) (Defibrase) (Reptilase)	*Bothrops atrox* (Barba amarilla) (Fer-de-lance)	255
P05620	Thrombin-like enzyme flavoxobin (EC 3.4.21) (Venom serine proteinase 1)	*Protobothrops* (Trimeresurus) *flavoviridis* (Habu)	260
P09872	Ancrod (EC 3.4.21.74) (Venombin-A) (Protein C activator) (ACC-C)	*Agkistrodon contortrix contortrix* (Southern copperhead)	231
P0C5B4	Thrombin-like enzyme gloshedobin (EC 3.4.21) (Defibrase)	*Gloydius shedaoensis* (Shedao island pit viper)	260
P18964	Vipera russelli proteinase RVV-V alpha (EC 3.4.21.95) (Factor V-activating)	*Daboia russelli siamensis* (Eastern Russell's viper)	236
P18965	Vipera russelli proteinase RVV-V gamma (EC 3.4.21.95) (Factor V-activating)	*Daboia russelli siamensis* (Eastern Russell's viper)	236
P26324	Ancrod (EC 3.4.21.74) (Venombin-A) (Protein C activator) (ACC-C)	*Calloselasma* (Agkistrodon) *rhodostoma* (Malayan pit viper)	234
P33589	Gyroxin analog (EC 3.4.21.74) (Thrombin-like enzyme) (LM-TL) (Venombin-A)	*Lachesis muta muta* (Bushmaster)	228
P47797	Ancrod (EC 3.4.21.74) (Venombin-A) (Protein C activator) (ACC-C)	*Calloselasma* (Agkistrodon) *rhodostoma* (Malayan pit viper)	258
P81176	Halystase (EC 3.4.21)	*Agkistrodon halys blomhoffi* (Mamushi) (Gloydius blomhoffii)	238
P81661	Thrombin-like enzyme bothrombin (EC 3.4.21.74) (Reptilase)	*Bothrops jararaca* (Jararaca)	232
P81824	Platelet-aggregating proteinase PA-BJ (EC 3.4.21)	*Bothrops jararaca* (Jararaca)	232
P82981	Thrombin-like enzyme contortrixobin (EC 3.4.21)	*Agkistrodon contortrix contortrix* (Southern copperhead)	234
P84787	Thrombin-like enzyme elegaxobin-2 (EC 3.4.21) (Elegaxobin II)	*Protobothrops* (Trimeresurus) *elegans* (Sakishima habu)	233
P84788	Thrombin-like enzyme elegaxobin-1 (EC 3.4.21) (Elegaxobin I)	*Protobothrops* (Trimeresurus) *elegans* (Sakishima habu)	233

Accession	Protein names	Organism	Length
P85109	Thrombin-like enzyme kangshuanmei (EC 3.4.21)	*Gloydius* (Agkistrodon) *halys brevicaudus* (Korean slamosa snake)	236
P86171	Kinin-releasing enzyme KR-E-1 (EC 3.4.21)	*Gloydius ussuriensis* (Ussuri mamushi) (Agkistrodon caliginosus)	235
Q072L6	Venom serine proteinase-like	*Bothrops asper* (Terciopelo)	259
Q072L7	Thrombin-like enzyme (EC 3.4.21)	*Lachesis stenophrys* (Central American bushmaster)	258
Q09GK1	Venom serine protease (EC 3.4.21)	*Philodryas olfersii* (Green snake)	261
Q27J47	Plasminogen-activating proteinase (LV-PA) (EC 3.4.21) (LMUT0402S)	*Lachesis muta muta* (Bushmaster)	258
Q2PQJ3	Venom serine protease 1 (EC 3.4.21) (BjussuSP-I)	*Bothrops jararacussu* (Jararacussu)	232
Q2QA04	Serine proteinase (EC 3.4.21)	*Crotalus durissus durissus* (Central American rattlesnake)	262
Q2XXM2	Kallikrein-Phi5 (Fragment)	*Philodryas olfersii* (Green snake)	248
Q2XXM3	Kallikrein-Phi4 (Fragment)	*Philodryas olfersii* (Green snake)	244
Q2XXM4	Kallikrein-Phi3	*Philodryas olfersii* (Green snake)	229
Q2XXM5	Kallikrein-Phi2	*Philodryas olfersii* (Green snake)	229
Q2XXM6	Kallikrein-Phi1	*Philodryas olfersii* (Green snake)	229
Q58G94	Gyroxin-like B2.1 (EC 3.4.21)	*Crotalus durissus terrificus* (South American rattlesnake)	238
Q5I2B5	Thrombin-like protein 3	*Deinagkistrodon acutus* (Hundred-pace snake)	260
Q5I2B6	Thrombin-like protein 1	*Deinagkistrodon acutus* (Hundred-pace snake)	260
Q5I2C5	Thrombin-like enzyme 2	*Deinagkistrodon acutus* (Hundred-pace snake)	260
Q5MCS0	Serine protease	*Lapemis hardwickii* (Hardwick's sea snake)	265
Q5W958	Venom serine proteinase-like HS120	*Bothrops jararaca* (Jararaca)	253
Q5W959	Venom serine proteinase HS114 (EC 3.4.21)	*Bothrops jararaca* (Jararaca)	258
Q5W960	Venom serine proteinase HS112 (EC 3.4.21)	*Bothrops jararaca* (Jararaca)	255
Q6IWF1	Venom serine protease BthaTL (EC 3.4.21)	*Bothrops alternatus* (Urutu)	233
Q6T5L0	Thrombin-like enzyme shedaoenase (EC 3.4.21)	*Gloydius* (Agkistrodon) *shedaoensis* (Shedao island pit viper)	238
Q6T6S7	Venom serine proteinase-like protein 1	*Bitis gabonica* (Gaboon viper)	260
Q6URK9	Platelet aggregating serine peptidase (Fragment)	*Bothrops jararaca* (Jararaca)	167
Q71QH5	Venom serine protease KN8 (EC 3.4.21)	*Trimeresurus stejnegeri* (Chinese green tree viper)	257
Q71QH6	Venom serine protease KN13 (EC 3.4.21)	*Trimeresurus stejnegeri* (Chinese green tree viper)	258

Accession	Protein names	Organism	Length
Q71QH7	Venom serine protease PA (EC 3.4.21)	*Trimeresurus stejnegeri* (Chinese green tree viper)	258
Q71QH8	Serine protease CL4	*Trimeresurus stejnegeri* (Chinese green tree viper)	257
Q71QH9	Venom serine protease KN14 (EC 3.4.21)	*Trimeresurus stejnegeri* (Chinese green tree viper)	260
Q71QI0	Venom serine protease KN7 homolog	*Trimeresurus stejnegeri* (Chinese green tree viper)	260
Q71QI1	Venom serine protease KN12 (EC 3.4.21)	*Trimeresurus stejnegeri* (Chinese green tree viper)	258
Q71QI2	Venom serine protease CL2 (EC 3.4.21)	*Trimeresurus stejnegeri* (Chinese green tree viper)	258
Q71QI3	Venom serine protease CL5 (EC 3.4.21)	*Trimeresurus stejnegeri* (Chinese green tree viper)	257
Q71QI4	Venom serine protease KN5 (EC 3.4.21)	*Trimeresurus stejnegeri* (Chinese green tree viper)	260
Q71QI5	Venom serine protease KN3 (EC 3.4.21)	*Trimeresurus stejnegeri* (Chinese green tree viper)	257
Q71QI6	Serine protease CL3	*Trimeresurus stejnegeri* (Chinese green tree viper)	257
Q71QI7	Venom serine protease KN11 (EC 3.4.21)	*Trimeresurus stejnegeri* (Chinese green tree viper)	257
Q71QI8	Venom serine protease KN10 (EC 3.4.21)	*Trimeresurus stejnegeri* (Chinese green tree viper)	257
Q71QI9	Serine protease CL1	*Trimeresurus stejnegeri* (Chinese green tree viper)	260
Q71QJ0	Venom serine protease KN2 (EC 3.4.21)	*Trimeresurus stejnegeri* (Chinese green tree viper)	257
Q71QJ1	Venom serine protease KN9 (EC 3.4.21)	*Trimeresurus stejnegeri* (Chinese green tree viper)	257
Q71QJ2	Venom serine protease KN6 (EC 3.4.21)	*Trimeresurus stejnegeri* (Chinese green tree viper)	260
Q71QJ3	Venom serine protease KN1 (EC 3.4.21)	*Trimeresurus stejnegeri* (Chinese green tree viper)	257
Q71QJ4	Venom serine protease KN4 homolog	*Trimeresurus stejnegeri* (Chinese green tree viper)	260
Q7SYF1	Cerastocytin (EC 3.4.21.74) (Proaggregant serine proteinase) (CC-PPP)	*Cerastes cerastes* (Horned desert viper)	256
Q7SZE1	Thrombin-like enzyme defibrase (EC 3.4.21)	*Gloydius* (Agkistrodon) *saxatilis* (Rock mamushi)	258
Q7SZE2	Bradykinin-releasing enzyme KR-E-1 (Thrombin-like defibrase) (EC 3.4.21)	*Gloydius ussuriensis* (Ussuri mamushi) (Agkistrodon caliginosus)	234
Q7T229	Venom serine protease homolog	*Bothrops jararacussu* (Jararacussu)	260
Q7ZZP4	Thrombin-like enzyme PTLE3	*Gloydius* (Agkistrodon) *halys pallas* (Chinese water mocassin)	190

Accession	Protein names	Organism	Length
Q802F0	Thrombin-like enzyme PTLE1 (EC 3.4.21)	*Gloydius* (Agkistrodon) *halys pallas* (Chinese water mocassin)	258
Q8AY78	Venom serine protease 5 (EC 3.4.21)	*Trimeresurus stejnegeri* (Chinese green tree viper)	258
Q8AY79	Thrombin-like enzyme stejnefibrase-2 (EC 3.4.21)	*Trimeresurus stejnegeri* (Chinese green tree viper)	258
Q8AY80	Thrombin-like enzyme stejnefibrase-1 (EC 3.4.21)	*Trimeresurus stejnegeri* (Chinese green tree viper)	258
Q8AY81	Thrombin-like enzyme stejnobin (EC 3.4.21)	*Trimeresurus stejnegeri* (Chinese green tree viper)	260
Q8AY82	Venom serine protease 1 homolog	*Trimeresurus stejnegeri* (Chinese green tree viper)	260
Q8JH62	Serine beta-fibrinogenase (EC 3.4.21) (VLBF)	*Vipera lebetina* (Elephant snake) (Leventine viper)	257
Q8JH85	Serine alpha-fibrinogenase (EC 3.4.21) (VLAF)	*Vipera lebetina* (Elephant snake) (Leventine viper)	258
Q8QG86	Serine proteinase BITS01A (EC 3.4.21)	*Bothrops insularis* (Island jararaca)	257
Q8QHK2	Catroxase-2 (EC 3.4.21) (Catroxase II) (EI)	*Crotalus atrox* (Western diamondback rattlesnake)	258
Q8QHK3	Catroxase-1 (EC 3.4.21) (Catroxase I)	*Crotalus atrox* (Western diamondback rattlesnake)	262
Q8UUJ1	Thrombin-likc cnzymc ussurase (EC 3.4.21)	*Gloydius ussuriensis* (Ussuri mamushi) (Agkistrodon caliginosus)	233
Q8UUJ2	Thrombin-like enzyme ussurin (EC 3.4.21)	*Gloydius ussuriensis* (Ussuri mamushi) (Agkistrodon caliginosus)	236
Q8UUK2	Venom serine proteinase Sp1 (EC 3.4.21)	*Crotalus adamanteus* (Eastern diamondback rattlesnake)	259
Q8UVX1	Thrombin-like enzyme gussurobin (EC 3.4.21)	*Gloydius ussuriensis* (Ussuri mamushi) (Agkistrodon caliginosus)	260
Q90Z47	Venom thrombin-like enzyme	*Deinagkistrodon acutus* (Hundred-pace snake)	235
Q91053	Thrombin-like enzyme calobin-1 (EC 3.4.21) (Calobin I)	*Gloydius ussuriensis* (Ussuri mamushi) (Agkistrodon caliginosus)	262
Q91507	Mucrofibrase-1 (EC 3.4.21)	*Protobothrops* (Trimeresurus) *mucrosquamatus* (Taiwan habu)	257
Q91508	Mucrofibrase-2 (EC 3.4.21) (Trimubin)	*Protobothrops* (Trimeresurus) *mucrosquamatus* (Taiwan habu)	257
Q91509	Mucrofibrase-3 (EC 3.4.21)	*Protobothrops* (Trimeresurus) *mucrosquamatus* (Taiwan habu)	257
Q91510	Mucrofibrase-4 (EC 3.4.21)	*Protobothrops* (Trimeresurus) *mucrosquamatus* (Taiwan habu)	257
Q91511	Mucrofibrase-5 (EC 3.4.21)	*Protobothrops* (Trimeresurus) *mucrosquamatus* (Taiwan habu)	257
Q91516	Venom plasminogen activator (EC 3.4.21) (TSV-PA)	*Trimeresurus stejnegeri* (Chinese green tree viper)	258
Q98TT5	Thrombin-like enzyme	*Deinagkistrodon acutus* (Hundred-pace snake)	258

Accession	Protein names	Organism	Length
Q9DF66	Venom serine proteinase 3 (SP3) (EC 3.4.21)	*Protobothrops* (Trimeresurus) *jerdonii* (Jerdon's pit-viper)	258
Q9DF67	Venom serine proteinase 2 (SP2) (EC 3.4.21)	*Protobothrops* (Trimeresurus) *jerdonii* (Jerdon's pit-viper)	258
Q9DF68	Venom serine proteinase-like protein (SP1)	*Protobothrops* (Trimeresurus) *jerdonii* (Jerdon's pit-viper)	260
Q9DG83	Serpentokallikrein-1 (EC 3.4.21)	*Protobothrops* (Trimeresurus) *mucrosquamatus* (Taiwan habu)	260
Q9DG84	Serpentokallikrein-2 (EC 3.4.21)	*Protobothrops* (Trimeresurus) *mucrosquamatus* (Taiwan habu)	257
Q9I8W9	Venom serine proteinase Dav-X (EC 3.4.21)	*Deinagkistrodon acutus* (Hundred-pace snake)	260
Q9I8X0	Venom serine proteinase Dav-KN (EC 3.4.21)	*Deinagkistrodon acutus* (Hundred-pace snake)	257
Q9I8X1	Venom serine proteinase Dav-PA (AaV-SP-I) (AaV-SP-II) (EC 3.4.21)	*Deinagkistrodon acutus* (Hundred-pace snake)	258
Q9I8X2	Thrombin-like enzyme acutobin (EC 3.4.21) (Acuthrombin) (Acutase)	*Deinagkistrodon acutus* (Hundred-pace snake)	260
Q9I961	Acubin2	*Deinagkistrodon acutus* (Hundred-pace snake)	258
Q9PSN3	Thrombin-like enzyme bilineobin (EC 3.4.21)	*Agkistrodon bilineatus* (Cantil) (Tropical moccasin)	235
Q9PT40	Venom serine proteinase-like protein 2	*Vipera lebetina* (Leventine viper)	260
Q9PT41	Factor V-activating enzyme (FVA) (EC 3.4.21)	*Vipera lebetina* (Leventine viper)	259
Q9PT51	Brevinase (EC 3.4.21) [Brevinase chain A & B]	*Gloydius* (Agkistrodon) *halys blomhoffi* (Mamushi)	233
Q9PTL3	Thrombin-like enzyme salmonase (EC 3.4.21)	*Gloydius* (Agkistrodon) *halys brevicaudus* (Korean slamosa snake)	257
Q9PTU8	Venom serine proteinase A (EC 3.4.21)	*Bothrops jararaca* (Jararaca)	258
Q9W7S1	Acubin	*Deinagkistrodon acutus* (Hundred-pace snake)	258
Q9YGI6	Thrombin-like enzyme pallabin-2 (EC 3.4.21)	*Gloydius* (Agkistrodon) *halys pallas* (Chinese water mocassin)	260
Q9YGJ2	Thrombin-like enzyme pallabin (EC 3.4.21)	*Gloydius* (Agkistrodon) *halys pallas* (Chinese water mocassin)	260
Q9YGJ8	Plasminogen activator Haly-PA (EC 3.4.21)	*Gloydius* (Agkistrodon) *halys pallas* (Chinese water mocassin)	258
Q9YGJ9	Serine protease Haly-2 (EC 3.4.21)	*Gloydius* (Agkistrodon) *halys pallas* (Chinese water mocassin)	257
Q9YGS1	Thrombin-like defibrase 1 (Fragment)	*Deinagkistrodon acutus* (Hundred-pace snake)	234

References

Alape-Girón, A., Sanz, L., Escolano, J., Flores-Díaz, M., Madrigal, M., Sasa, M., Calvete, J.J., 2008. Snake venomics of the lancehead pitviper *Bothrops asper*: geographic, individual, and ontogenetic variations. J. Proteome Res. 7, 3556–3571.

Alexander, G., Grothusen, J., Zepeda, H., Schwartzman, R.J., 1988. Gyroxin, a toxin from the venom of *Crotalus durissus terrificus*, is a thrombin-like enzyme. Toxicon 26, 953–960.

Allegra, A., Coppolino, G., Bolignano, D., Giacobbe, M.S., Alonci, A., D'Angelo, A., Bellomo, G., Teti, D., Loddo, S., Musolino, C., Buemi, M., 2009. Endothelial progenitor cells: pathogenetic role and therapeutic perspectives. J. Nephrol. 22, 463–475.

Alves da Silva, J.A., Muramoto, E., Ribela, M.T.C., Rogero, J.R., Camillo, M.A.P., 2006. Biodistribution of gyroxin using ^{125}I as radiotracer. J. Radioanalyt. Nucl. Chem. 269, 579–583.

Amiconi, G., Amoresano, A., Boumis, G., Brancaccio, A., De Cristofaro, R., De Pascalis, A., Di Girolamo, S., Maras, B., Scaloni, A., 2000. A novel venombin B from *Agkistrodon contortrix contortrix*: evidence for recognition properties in the surface around the primary specificity pocket different from thrombin. Biochemistry 39, 10294–10308.

Asahara, T., Masuda, H., Takahashi, T., Kalka, C., Pastore, C., Silver, M., Kearne, M., Magner, M., Isner, J.M., 1999. Bone marrow origin of endothelial progenitor cells responsible for postnatal vasculogenesis in physiological and pathological neovascularization. Circ. Res. 85, 221–228.

Assakura, M.T., Reichl, A.P., Mandelbaum, F.R., 1994. Isolation and characterization of five fibrin(ogen)olytic enzymes from the venom of *Philodryas olfersii* (green snake). Toxicon 32, 819–831.

Au, L.C., Lin, S.B., Chou, J.S., Teh, G.W., Chang, K.J., Shih, C.M., 1993. Molecular cloning and sequence analysis of the cDNA for ancrod, a thrombin-like enzyme from the venom of *Calloselasma rhodostoma*. Biochem. J. 294, 387–390.

Bandow, J.E., 2010. Comparison of protein enrichment strategies for proteome analysis of plasma. Proteomics 2010 Feb 1. [Epub ahead of print].

Barbosa, M.D.S., Stipp, A.C., Passanezi, E. Greghi, S.L.A., 2008. Fibrin adhesive derived from snake venom in periodontal surgery. Histological analysis. J. Appl. Oral Sci. 16, 310–315.

Bell, G.I., Quinto, C., Quiroga, M., Valenzuela, P., Craik, C.S., Rutter, W.J., 1984. Isolation and sequence of a rat chymotrypsin B gene. J. Biol. Chem. 259, 14265–14270.

Botos, I., Wlodawer, A., 2007. The expanding diversity of serine hydrolases. Curr. Opin. Struct. Biol. 17, 683–690.

Boyer, L.V., Seifert, S.A., Clark, R.F., McNally, J.T., Williams, S.R., Nordt, S.P., Walter, F.G., Dart, R.C., 1999. Recurrent and persistent coagulopathy following pit viper envenomation. Arch Intern Med. 159, 706–710.

Braud, S., Bon, C., Wisner, A., 2000. Snake venom proteins acting on hemostasis. Biochemie 82, 851–859.

Burkhart, W., Smith, G.F., Su, J.L., Parikh, I., LeVine, H., 3rd., 1992. Amino acid sequence determination of ancrod, the thrombin-like alpha-fibrinogenase from the venom of *Agkistrodon rhodostoma*. FEBS Lett. 297, 297–301.

Calvete, J.J., Fasoli, E., Sanz, L., Boschetti, E., Righetti, P.G., 2009. Exploring the venom proteome of the western diamondback rattlesnake, *Crotalus atrox*, via snake venomics and combinatorial peptide ligand library approaches. J. Proteome Res. 8, 3055–67.

Castro, H.C., Silva, D.M., Craik, C., Zingali, R.B., 2001. Structural features of a snake venom thrombin-like enzyme: thrombin and trypsin on a single catalytic platform? Biochim. Biophys. Acta 1547, 183–195.

Castro, H.C., Zingali, R.B., Albuquerque, M.G., Pujol-Luz, M., Rodrigues, C.R., 2004. Snake venom thrombin-like enzymes: from reptilase to now. Cell. Mol. Life Sci. 61, 843–856.

Ching, A.T.C., Rocha, M.M.T., Paes Leme, A.F., Pimenta, D.C., Furtado, M.F.D., Serrano, S.M.T., Ho, P.L., Junqueira-de-Azevedo, I.L.M., 2006. Some aspects of the venom proteome of the

Colubridae snake *Philodryas olfersii* revealed from a Duvernoy's (venom) gland transcriptome. FEBS Lett. 580, 4417–4422.
Cole, C.W., 1998. Controlling acute elevation of plasma fibrinogen with ancrod. Cerebrovasc. Dis. 1, 29–34.
Craik, C.S., Choo, Q.L., Swift, G.H., Quinto, C., MacDonald, R.J., Rutter, W.J., 1984. Structure of two related rat pancreatic trypsin genes. J. Biol. Chem. 259, 14255–14264.
Dekhil, H., Wisner, A., Marrakchi, N., El Ayeb, M., Bon, C., Karoui, H., 2003. Molecular cloning and expression of a functional snake venom serine proteinase, with platelet aggregating activity, from the *Cerastes cerastes* viper. Biochemistry 42, 10609–10618.
Deshimaru, M., Ogawa, T., Nakashima, K., Nobuhisa, I., Chijiwa, T., Shimohigashi, Y., Fukumaki, Y., Niwa, M., Yamashina, I., Hattori, S., Ohno, M., 1996. Accelerated evolution of crotalinae snake venom gland serine proteases. FEBS Lett. 397, 83–88.
Di Cera, E., 2008. Thrombin. Mol. Aspects Med. 29, 203–254.
Doley, R., Mackessy, S.P., Kini, R.M., 2009. Role of accelerated segment switch in exons to alter targeting (ASSET) in the molecular evolution of snake venom proteins. BMC Evol. Biol. 9, 146.
Doley, R., Pahari, S., Mackessy, S.P., Kini, R.M., 2008. Accelerated exchange of exon segments in Viperid three-finger toxin genes (*Sistrurus catenatus edwardsii*; Desert Massasauga). BMC Evol. Biol. 8, 196.
Escalante, T., Shannon, J., Moura-da-Silva, A.M., Gutiérrez, J.M., Fox, J.W., 2006. Novel insights into capillary vessel basement membrane damage by snake venom hemorrhagic metalloproteinases: a biochemical and immunohistochemical study. Arch. Biochem. Biophys. 455, 144–153.
Farid, T.M., Tu, A.T., el-Asmar, M.F., 1989. Characterization of cerastobin, a thrombin-like enzyme from the venom of *Cerastes vipera* (Sahara sand viper). Biochemistry 28, 371–377.
Farid, T.M., Tu, A.T., el-Asmar, M.F., 1990. Effect of cerastobin, a thrombinlike enzyme from *Cerastes vipera* (Egyptian sand snake) venom, on human platelets. Haemostasis 20, 296–304.
Ferreira, S.H., 1965. A bradykinin-potentiating factor (bpf) present in the venom of *Bothrops jararaca*. Br. J. Pharmacol. Chemother. 24, 163–169.
Fox, J.W., Serrano, S.M.T., 2007. Approaching the golden age of natural product pharmaceuticals from venom libraries: an overview of toxins and toxin-derivatives currently involved in therapeutic or diagnostic applications. Curr. Pharm. Des. 13, 2927–2934.
Fox, J.W., Serrano, S.M.T., 2008. Insights into and speculations about snake venom metalloproteinases (SVMP) synthesis, folding and disulfide bond formation and their contribution to venom complexity. FEBS J. 275, 3016–3030.
Fox, J.W., Serrano, S.M.T., 2009. Snake venom metalloproteinases, in: Mackessy, S.P. (Ed.), Handbook of Venoms and Toxins of Reptiles. CRC Press/Taylor & Francis Group, Boca Raton, FL, pp. 95–113.
Fry, B.G., 2005. From genome to "venome": molecular origin and evolution of the snake venom proteome inferred from the phylogenetic analysis of toxin sequences and related body proteins. Genome Res. 15, 403–420.
Fry, B.G., Vidal, N., Norman, J.A., Vonk, F.J., Scheib, H., Ramjan, S.F., Kuruppu, S., Fung, K., Hedges, S.B., Richardson, M.K., Hodgson, W.C., Ignjatovic, V., Summerhayes, R., Kochva, E., 2006. Early evolution of the venom system in lizards and snakes. Nature 439, 584–588.
Gao, R., Zhang, Y., Meng, Q.-X., Lee, W.-H., Li, D.-S., Xiong, Y.-L., Wang, W.-Y., 1998. Characterization of three fibrinogenolytic enzymes from Chinese green tree viper (*Trimeresurus stejnegeri*) venom. Toxicon 36, 457–467.
Guo, Y.W., Chang, T.Y., Lin, K.T., Liu, H.W., Shih, K.C., Cheng, S.H., 2001. Cloning and functional expression of the mucrosobin protein, a beta-fibrinogenase of *Trimeresurus mucrosquamatus* (Taiwan Habu). Protein Exp. Purif. 23, 483–490.
Gutiérrez, J.M., Rucavado, A., Escalante, T. 2009. Snake venom metalloproteinases. Biological roles and participation in the pathology of envenomation, in: Mackessy, S.P. (Ed.), Handbook of Venoms and Toxins of Reptiles. CRC Press/Taylor & Francis Group, Boca Raton, FL, pp. 115–138.

Hahn, B.S., Yang, K.Y., Park, E.M., Chang, I.M., Kim, Y.S., 1996. Purification and molecular cloning of calobin, a thrombin-like enzyme from *Agkistrodon caliginosus* (Korean viper). J. Biochem. (Tokyo) 119, 835–843.

Hahn, B.S., Baek, K., Kim, W.S., Lee, C.S., Chang, I.L., Kim, Y.S., 1998. Molecular cloning of capillary permeability-increasing enzyme-2 from *Agkistrodon caliginosus* (Korean viper). Toxicon 36, 1887–1893.

Hartley, B.S., Kilby, B.A., 1954. The reaction of p-nitrophenyl esters with chymotrypsin and insulin. Biochem. J. 56, 288–297.

Henschen-Edman, A.H., Theodor, I., Edwards, B.F., Pirkle, H., 1999. Crotalase, a fibrinogen-clotting snake venom enzyme: primary structure and evidence for a fibrinogen recognition exosite different from thrombin. Thromb. Haemost. 81, 81–86.

Hiestand, P.C., Hiestand, R.R. 1979. *Dispholidus typus* (Boomslang) snake venom: purification and properties of the coagulant principle. Toxicon 17, 489–498.

Huisman, T.H., 1993. The structure and function of normal and abnormal haemoglobins. Baillieres Clin. Haematol. 6, 1–30.

Isbister, GK, Duffull, SB, Brown, SG, ASP Investigators. 2009. Failure of antivenom to improve recovery in Australian snakebite coagulopathy. QJM 102, 563–568.

Itoh, N., Tanaka, N., Mihashi, S., Yamashina, I., 1987. Molecular cloning and sequence analysis of cDNA for batroxobin, a thrombin-like snake venom enzyme. J. Biol. Chem. 262, 3132–3135.

Itoh, N., Tanaka, N., Funakoshi, I., Kawasaki, T., Mihashi, S., Yamashina, I., 1988. Organization of the gene for batroxobin, a thrombin-like snake venom enzyme. Homology with the trypsin/kallikrein gene family. J. Biol. Chem. 263, 7628–7631.

Jin, Y., Lee, W.H., Zhang, Y., 2007. Molecular cloning of serine proteases from elapid snake venoms. Toxicon 49, 1200–1207.

Juárez, P., Sanz, L., Calvete, J.J., 2004. Snake venomics: characterization of protein families in *Sistrurus barbouri* venom by cysteine mapping, N-terminal sequencing, and tandem mass spectrometry analysis. Proteomics 4, 327–338.

Kini, R.M., 2005. Serine proteases affecting blood coagulation and fibrinolysis from snake venoms. Pathophysiol. Haemost. Thromb. 34, 200–204.

Kini, R.M., Chan, Y.M., 1999. Accelerated evolution and molecular surface of venom phospholipase A_2 enzymes. J. Mol. Evol. 48, 125–132.

Kirby, E.P., Niewiarowski, S., Stocker, K., Kettner, C., Shaw, E., Brudzynski, T.M., 1979. Thrombocytin, a serine protease from Bothrops atrox venom 1. Purification and characterization of the enzyme. Biochemistry 18, 3564–3570.

Komori, Y., Nikai, T., Ohara, A., Yagihashi, S., Sugihara, H., 1993. Effect of bilineobin, a thrombin-like proteinase from the venom of common cantil (*Agkistrodon bilineatus*). Toxicon 31, 257–270.

Kordis, D., Gubensek, F., 2000. Adaptive evolution of animal toxin multigene families. Gene 261, 43–52.

Lee, J.W., Park, W., 2000. cDNA cloning of brevinase, a heterogeneous two-chain fibrinolytic enzyme from *Agkistrodon blomhoffii brevicaudus* snake venom, by serial hybridization polymerase chain reaction. Arch. Biochem. Biophys. 377, 234–240.

Lee, J.W., Seu, J.H., Rhee, I.K., Jin, I., Kawamura, Y., Park, W., 1999. Purification and characterization of brevinase, a heterogeneous two-chain fibrinolytic enzyme from the venom of Korean snake, *Agkistrodon blomhoffii brevicaudus*. Biochem. Biophys. Res. Commun. 260, 665–670.

Levy, D.E., del Zoppo, G.J., Demaerschalk, B.M., Demchuk, A.M., Diener, H.C., Howard, G., Kaste, M., Pancioli, A.M., Ringelstein, E.B., Spatareanu, C., Wasiewski, W.W., 2009. Ancrod in acute ischemic stroke: results of 500 subjects beginning treatment within 6 hours of stroke onset in the ancrod stroke program. Stroke 40, 3796–3803.

Mackessy, S.P., 1993a. Fibrinogenolytic proteases from the venoms of juvenile and adult northern Pacific rattlesnake (*Crotalus viridis oreganus*). Comp. Biochem. Physiol. 106B, 181–189.

Mackessy, S.P., 1993b. Kallikrein-like and thrombin-like proteases from the venom of juvenile northern Pacific rattlesnakes (*Crotalus viridis oreganus*). J. Nat. Toxins 2, 223–239.

Mackessy, S.P., 1996. Characterization of the major metalloprotease isolated from the venom of *Crotalus viridis oreganus*. Toxicon 34, 1277–1285.

Mackessy, S.P. 2002. Biochemistry and pharmacology of colubrid snake venoms. J. Toxicol.-Toxin Rev. 21, 43–83.

Mackessy, S.P., 2009. The field of reptile toxinology: snakes, lizards and their venoms, in: Mackessy, S.P. (Ed.), Handbook of Venoms and Toxins of Reptiles. CRC Press/Taylor & Francis Group, Boca Raton, FL, pp. 3–23.

MacSweeney, A., Birrane, G., Walsh, M.A., O`Connell, T., Malthouse, J.P., Higgins, T.M. 2000. Crystal structure of delta-chymotrypsin bound to a peptidyl chloromethyl ketone inhibitor. Acta Crystallogr., Sect. D 56, 280–286.

Magalhães, A., Da Fonseca, B.C., Diniz, C.R., Gilroy, J., Richardson, M., 1993. The complete amino acid sequence of a thrombin-like enzyme/gyroxin analogue from venom of the bushmaster snake (*Lachesis muta muta*). FEBS Lett. 329:116–20.

Magalhães, A., Magalhães, H.P.B., Richardson, M., Gontijo, S., Ferreira, R.N., Almeida, A.P., Sanchez, E.F., 2007. Purification and properties of a coagulant thrombin-like enzyme from the venom of *Bothrops leucurus*. Comp. Biochem. Physiol. 146A, 565–575.

Marchler-Bauer, A., Anderson, J.B., Chitsaz, F., Derbyshire, M.K., DeWeese-Scott, C., Fong, J.H., Geer, L.Y., Geer, R.C., Gonzales, N.R., Gwadz, M., He, S., Hurwitz, D.I., Jackson, J.D., Ke, Z., Lanczycki, C.J., Liebert, C.A., Liu, C., Lu, F., Lu, S., Marchler, G.H., Mullokandov, M., Song, J.S., Tasneem, A., Thanki, N., Yamashita, R.A., Zhang, D., Zhang, N., Bryant, S.H. 2009. CDD: specific functional annotation with the Conserved Domain Database. Nucleic Acids Res. 37, 205–210.

Markland, F.S., Jr., 1976. Crotalase. Methods Enzymol. 45, 223–236.

Markland, F.S., Jr., 1998. Snake venoms and the hemostatic system. Toxicon 36, 1749–1800.

Markland, F.S., Jr., Kettner, C., Schiffman, S., Shaw, E., Bajwa, S.S., Reddy, K.N., Kirakossian, H., Patkos, G.B., Theodor, I., Pirkle, H., 1982. Kallikrein-like activity of crotalase, a snake venom enzyme that clots fibrinogen. Proc. Natl Acad. Sci. U.S.A. 79, 1688–1692.

Marrakchi, N., Zingali, R.B., Karoui, H., Bon, C., el Ayeb, M., 1995. Cerastocytin, a new thrombin-like platelet activator from the venom of the Tunisian viper *Cerastes cerastes*. Biochim. Biophys. Acta 1244, 147–156.

Marrakchi, N., Barbouche, R., Guermazi, S., Karoui, H., Bon, C., El Ayeb, M., 1997. Cerastotin, a serine protease from *Cerastes cerastes* venom, with platelet-aggregating and agglutinating properties. Eur. J. Biochem. 247, 121–128.

Marsh, N., Williams, V., 2005. Practical applications of snake venom toxins in haemostasis. Toxicon 45, 1171–1181.

Matsui, T., Sakurai, Y., Fujimura, Y., Hayashi, I., Oh-Ishi, S., Suzuki, M., Hamako, J., Yamamoto, Y., Yamazaki, J., Kinoshita, M., Titani, K., 1998. Purification and amino acid sequence of halystase from snake venom of *Agkistrodon halys blomhoffii*, a serine protease that cleaves specifically fibrinogen and kininogen. Eur. J. Biochem. 252, 569–575.

Mukherjee, A.K., 2008. Characterization of a novel pro-coagulant metalloprotease (RVBCMP) possessing α-fibrinogenase and tissue hemorrhagic activity from the venom of *Daboia russelli russelli* (Russell's viper): evidence of distinct coagulant and haemorrhagic sites in RVBCMP. Toxicon 51, 923–933.

Nakashima, K., Nobuhisa, I., Deshimaru, M., Nakai, M., Ogawa, T., Shimohigashi, Y., Fukumaki, Y., Hattori, M., Sakaki, Y., Hattori, S., 1995. Accelerated evolution in the protein-coding regions is universal in crotalinae snake venom gland phospholipase A_2 isozyme genes. Proc. Natl. Acad. Sci. U.S.A. 92, 5605–5609.

Nakashima, K., Ogawa, T., Oda, N., Hattori, M., Sakaki, Y., Kihara, H., Ohno, M., 1993. Accelerated evolution of *Trimeresurus flavoviridis* venom gland phospholipase A_2 isozymes. Proc. Natl. Acad. Sci. U.S.A. 90, 5964–5968.

Nakayama, D., Ben Ammar, Y., Takeda, S., 2009. Crystallization and preliminary X-ray crystallographic analysis of blood coagulation factor V-activating proteinase (RVV-V) from Russell's viper venom. Acta Crystallogr. Sect. F Struct. Biol. Cryst. Commun. 65, 1306–1308.

Nikai, T., Ohara, A., Komori, Y., Fox, J.W., Sugihara, H., 1995. Primary structure of a coagulant enzyme, bilineobin, from *Agkistrodon bilineatus* venom. Arch. Biochem. Biophys. 318, 89–96.

Nishida, S., Fujimura, Y., Miura, S., Ozaki, Y., Usami, Y., Suzuki, M., Titań, K., Yoshida, E., Sugimoto, M., Yoshioka, A., Fukui, H., 1994. Purification and characterization of bothrombin, a fibrinogen-clotting serine protease from the venom of *Bothrops jararaca*. Biochemistry 33, 1843–1849.

Nirthanan, S., Gwee, M.C.E., 2004. Three-finger α-neurotoxins and the nicotinic acetylcholine receptor, forty years on. J. Pharmacol. Sci. 94, 1–17.

Nobuhisa, I., Nakashima, K., Deshimaru, M., Ogawa, T., Shimohigashi, Y., Fukumaki, Y., Sakaki, Y., Hattori, S., Kihara, H., Ohno, M., 1996. Accelerated evolution of *Trimeresurus okinavensis* venom gland phospholipase A_2 isozyme-encoding genes. Gene 172, 267–272.

Nolan, C., Hall, L.S., Barlow, G.H., 1976. Ancrod, the coagulating enzyme from Malayan pit viper (*Agkistrodon rhodostoma*) venom. Methods Enzymol. 45, 205–213.

Ogawa, T., Nakashima, K., Nobuhisa, I., Deshimaru, M., Shimohigashi, Y., Fukumaki, Y., Sakaki, Y., Hattori, S., Ohno, M., 1996. Accelerated evolution of snake venom phospholipase A_2 isozymes for acquisition of diverse physiological functions. Toxicon 34, 1229–1236.

Ohno, M., Menez, R., Ogawa, T., Danse, J.M., Shimohigashi, Y., Fromen, C., Ducancel, F., Zinn-Justin, S., Le Du, M.H., Boulain, J.C., Tamiya, T., Menez, A., 1998. Molecular evolution of snake toxins: is the functional diversity of snake toxins associated with a mechanism of accelerated evolution? Prog. Nucleic Acid Res. Mol. Biol. 59, 307–364.

Oyama, E., Takahashi, H., 2003. Purification and characterization of a thrombin like enzyme, elegaxobin II, with lys-bradykinin releasing activity from the venom of *Trimeresurus elegans* (Sakishima-Habu). Toxicon 41, 559–568.

Paes Leme, A.F., Prezoto, B.C., Yamashiro, E.T., Bertholim, L., Tashima, A.K., Klitzke, C.F., Camargo, A.C., Serrano, S.M., 2008. *Bothrops* protease A, a unique highly glycosylated serine proteinase, is a potent, specific fibrinogenolytic agent. J. Thromb. Haemost. 6, 1363–1372.

Pahari, S., Mackessy, S.P., Kini, M.R., 2007. The venom gland transcriptome of the Desert Massasauga Rattlesnake (*Sistrurus catenatus edwardsii*): towards an understanding of venom composition among advanced snakes (Superfamily Colubroidea). BMC Mol. Biol. 8, 115.

Pan, H., Du, X., Yang, G., Zhou, Y., Wu, X., 1999. cDNA cloning and expression of acutin. Biochem. Biophys. Res. Commun. 255, 412–415.

Parry MA, Jacob U, Huber R, Wisner A, Bon C, Bode, W. 1998. The crystal structure of the novel snake venom plasminogen activator TSV-PA: a prototype structure for snake venom serine proteinases. Structure 6, 1195–1206.

Pawlak, J., Kini, R.M., 2008. Unique gene organization of colubrid three-finger toxins: complete cDNA and gene sequences of denmotoxin, a bird-specific toxin from colubrid snake *Boiga dendrophila* (Mangrove Catsnake). Biochimie 90, 868–877.

Phillips, D.J., Swenson, S.D., Markland, F.S., Jr. 2009. Thrombin-like snake venom serine proteinases, in: Mackessy, S.P. (Ed.), Handbook of Venoms and Toxins of Reptiles. CRC Press/Taylor & Francis Group, Boca Raton, FL, pp. 139–154.

Philipps, J., Thomalla, G., Glahn, J., Schwarze, M., Rother, J., 2009. Treatment of progressive stroke with tirofiban – experience in 35 patients. Cerebrovasc. Dis. 28, 435–438.

Polaskova, V., Kapur, A., Khan, A., Molloy, M.P., Baker, M.S., 2010. High-abundance protein depletion: comparison of methods for human plasma biomarker discovery. Electrophoresis 31, 471–482.

Polgár, L., 1971. On the mechanism of proton transfer in the catalysis by serine proteases. J. Theor. Biol. 31, 165–169.

Polgár, L., Bender, M.L., 1969. The nature of general base-general acid catalysis in serine proteases. Proc. Natl. Acad. Sci. USA 64, 1335–1342.

Rawlings, N.D., Tolle, D.P., Barrett A.J., 2004a. MEROPS: the peptidase database. Nucleic Acids Res. 32, D160–164.

Rawlings, N.D., Tolle, D.P., Barrett A.J., 2004b. Evolutionary families of peptidase inhibitors. Biochem. J. 378, 705–716.

Rowe, S.L., Stegemann, J.P., 2009. Microstructure and mechanics of collagen-fibrin matrices polymerized using ancrod snake venom enzyme. J. Biomech. Eng. 131, 061012. doi: 10.1115/1.3128673.

Sanz, L., Gibbs, H.L., Mackessy, S.P., Calvete, J.J., 2006. Venom proteomes of closely related *Sistrurus* rattlesnakes with divergent diets. J. Proteome Res. 5, 2098–2112.

Segers, K., Rosing, J., Nicolaes, G.A., 2006. Structural models of the snake venom factor V activators from *Daboia russelli* and *Daboia lebetina*. Proteins 64, 968–984.

Serrano, S.M., Hagiwara, Y., Murayama, N., Higuchi, S., Mentele, R., Sampaio, C.A., Camargo, A.C., Fink, E., 1998. Purification and characterization of a kinin-releasing and fibrinogen-clotting serine proteinase (KN-BJ) from the venom of *Bothrops jararaca*, and molecular cloning and sequence analysis of its cDNA. Eur. J. Biochem. 251, 845–853.

Serrano, S.M.T., Maroun, R.C. 2005. Snake venom serine proteinases: sequence homology vs. substrate specificity, a paradox to be solved. Toxicon 45, 1115–1132.

Sherman, D.G., 2002. Ancrod. Curr. Med. Res. Opin. 18, s48–s52.

Shieh, T.-C., Kawabata, S., Kihara, H., Ohno, M., Iwanaga, S., 1988. Amino acid sequence of a coagulant enzyme, flavoxobin, from *Trimeresurus flavoviridis* venom. J. Biochem. 103, 596–605.

Shieh, T.C., Tanaka, S., Kihara, H., Ohno, M., Makisumi, S., 1985. Purification and characterization of a coagulant enzyme from *Trimeresurus flavoviridis* venom. J. Biochem. (Tokyo) 98, 713–721.

Shimokawa, K., Takahashi, H., 1993a. Purification of a capillary permeability-increasing enzyme-2 from the venom of *Agkistrodon caliginosus* (Kankoku-Mamushi). Toxicon 31, 1213–1219.

Shimokawa, K., Takahashi, H., 1993b. Some properties of a capillary permeability-increasing enzyme-2 from the venom of *Agkistrodon caliginosus* (Kankoku-Mamushi). Toxicon 31, 1221–1227.

Silva-Junior, F.P., Guedes, H.L., Garvey, L.C., Aguiar, A.S., Bourguignon, S.C., Di Cera, E., Giovanni-De-Simone, S., 2007. BJ-48, a novel thrombin-like enzyme from the *Bothrops jararacussu* venom with high selectivity for Arg over Lys in P1: role of N-glycosylation in thermostability and active site accessibility. Toxicon 50, 18–31.

Silveira, A.M.V., Magalhães, A., Diniz, C.R., Oliveira, E.B., 1989. Purification and properties of the thrombin-like enzyme from the venom of *Lachesis muta muta*. Int. J. Biochem. 21, 863–868.

Smith, C.G., Vane, J.R., 2003. The discovery of captopril. FASEB J. 17, 788–789.

Spyropoulos, A.C., 2008. Brave new worlds: the current and future use of novel anticoagulants. Thromb. Res. 123, S29-S35.

Stocker, K.F., 1998. Research, diagnostic and medicinal uses of snake venom enzymes, in: Bailey, G.S. (Ed.), Enzymes From Snake Venom. Alaken Press, Fort Collins, CO, pp. 705–736.

Stocker, K., Meier, J., 1989. Snake venom proteins in hemostasis: new results. Folia Haematol. Int. Mag. Klin. Morphol. Blutforsch 116, 935–953.

Stocker, K., Barlow, G.H., 1976. The coagulant enzyme from *Bothrops atrox* venom (batroxobin). Methods Enzymol. 45, 214–223.

Stocker, K., Fischer, H., Meier, J., 1982. Thrombin-like snake venom proteinases. Toxicon 20, 265–273.

Sturzebecher, J., Sturzebecher, U., Markwardt, F., 1986. Inhibition of batroxobin, a serine proteinase from *Bothrops* snake venom, by derivatives of benzamidine. Toxicon 24, 585–595.

Swenson, S., Markland, F.S., Jr., 2005. Snake venom fibrin(ogen)olytic enzymes. Toxicon 45, 1021–1039.

Takacs, Z., Wilhelmsen, K.C., Sorota, S., 2001. Snake α-neurotoxin binding site on the Egyptian Cobra (*Naja haje*) nicotinic acetylcholine receptor is conserved. Mol. Biol. Evol. 18, 1800–1809.

Tamiya, T., Ohno, S., Nishimura, E., Fujimi, T.J., Tsuchiya, T., 1999. Complete nucleotide sequences of cDNAs encoding long chain alpha-neurotoxins from sea krait, *Laticauda semifasciata*. Toxicon 37, 181–185.

Tanaka, N., Nakada, H., Itoh, N., Mizuno, Y., Takanishi, M., Kawasaki, T., Tate, S., Inagaki, F., Yamashina, I., 1992. Novel structure of the N-acetylgalactosamine containing N-glycosidic carbohydrate chain of batroxobin, a thrombin-like snake venom enzyme. J. Biochem. 112, 68–74.

Thompson, J.D., Higgins, D.G., Gibson, T.J., 1994. CLUSTAL W: improving the sensitivity of progressive multiple sequence alignment through sequence weighting, position-specific gap penalties and weight matrix choice. Nucleic Acids Res. 22, 4673–4680.

Tsai, I.-H., Wang, Y.-M., 2001. Identification of geographic variations and cloning of venom proteins of *Trimeresurus stejnegeri*: serine proteases and phospholipases. EMBL/GenBank/DDBJ databases.

Utaisincharoen, P., Mackessy, S.P., Miller, R.A., Tu, A.T. 1993. Complete primary structure and biochemical properties of gilatoxin, a serine protease with kallikrein-like and angiotensin-degrading activities. J. Biol. Chem. 268, 21975–21983.

Wang, Y.M., Wang, S.R., Tsai, I.H., 2001. Serine protease isoforms of *Deinagkistrodon acutus* venom: cloning, sequencing and phylogenetic analysis. Biochem. J. 354, 161–168.

Wei, J.N., Wang, Q.C., Liu, G.F., Ezell, E.L., Quast, M.J., 2004. Reduction of brain injury by antithrombotic agent acutobin after middle cerebral artery ischemia/reperfusion in the hyperglycemic rat. Brain Res. 1022, 234–243.

Weldon, C.L., Mackessy, S.P., 2010. Biological and proteomic analysis of venom from the Puerto Rican Racer (*Alsophis portoricensis*: Dipsadidae). Toxicon 55, 558–569.

White, J., 2005. Snake venoms and coagulopathy. Toxicon 45, 951–967.

White, J., 2009. Envenomation: prevention and treatment in Australia, in: Mackessy, S.P. (Ed.), Handbook of Venoms and Toxins of Reptiles. CRC Press/Taylor & Francis Group, Boca Raton, FL, pp. 423–451.

Wisner, A., Braud, S., Bon, C., 2001. Snake venom proteinases as tools in hemostasis studies: structure-function relationship of a plasminogen activator purified from *Trimeresurus stejnegeri* venom. Haemostasis 31, 133–140.

Yamamoto, C., Tsuru, D., Oda-Ueda, N., Ohno, M., Hattori, S., Kim, S.T., 2002. Flavoxobin, a serine protease from *Trimeresurus flavoviridis* (habu snake) venom, independently cleaves Arg726–Ser727 of human C3 and acts as a novel, heterologous C3 convertase. Immunology 107, 111–117.

Xu, G., Liu, X., Zhu, W., Yin, Q., Zhang, R., Fan, X., 2007. Feasibility of treating hyperfibrinogenemia with intermittently administered batroxobin in patients with ischemic stroke/transient ischemic attack for secondary prevention. Blood Coagul. Fibrinolysis 18, 193–197.

Zeng, Z., Xiao, P., Chen, J., Wei, Y., 2009. Are batroxobin agents effective for perioperative hemorrhage in thoracic surgery? A systematic review of randomized controlled trials. Blood Coagul. Fibrinolysis 20, 101–107.

Zhang, B., Liu, Q., Yin, W., Zhang, X., Huang, Y., Luo, Y., Qiu, P., Su, X., Yu, J., Hu, S., Yan, G., 2006. Transcriptome analysis of *Deinagkistrodon acutus* venomous gland focusing on cellular structure and functional aspects using expressed sequence tags. BMC Genomics, 7, 152.

Zhang, L., Lu, S.H., Li, L., Tao, Y.G., Wan, Y.L., Senga, H., Yang, R., Han, Z.C., 2009. Batroxobin mobilizes circulating endothelial progenitor cells in patients with deep vein thrombosis. Clin. Appl. Thromb. Hemost. 2009 Oct 13. [Epub ahead of print]

Zhang, Y., Gao, R., Lee, W.-H., Zhu, S.-W., Xiong, Y.-L., Wang, W.-Y., 1998. Characterization of a fibrinogen-clotting enzyme from *Trimeresurus stejnegeri* venom, and comparative study with other venom proteases. Toxicon 36, 131–142.

Zhu, Z, Liang, Z, Zhang, T, Zhu, Z, Xu, W, Teng, M, Niu, L., 2005. Crystal structures and amidolytic activities of two glycosylated snake venom serine proteinases. J. Biol. Chem. 280:10524–10529.

Chapter 31
Procoagulant Factors from *Lonomia* Caterpillars

Ana Marisa Chudzinski-Tavassi and Linda Christian Carrijo-Carvalho

Abstract Little has been known about hemostasis of insects and the identity of its clotting factors. In the Lepidoptera Order, few proteins have been described to affect hemolymph. There is the possibility of finding a large diversity of proteins that could play a physiological role in the hemolimph clotting, due to the complexity and variability of their coagulation systems. *Lonomia* has been the most studied genus, due to the large numbers of accidents resulting from human contact with *Lonomia* caterpillars. Those accidents are characterized as a hemorrhagic syndrome resulting from a massive disturbance of the hemostatic system caused by toxins found in the caterpillar bristles. In view of the present knowledge and literature, insects from the Lepidoptera Order, particularly caterpillars, stand as interesting sources for the identification, characterization, cloning and expression of proteins affecting hemostasis, and new candidates for therapeutic and biotechnology uses. Lopap (*L. obliqua* prothrombin activator protease) and Losac (*L. obliqua* Stuart-factor activator protease) toxins isolated from the *L. obliqua* bristles, are among the procoagulant proteins better characterized from Lepitdoptera. Such procoagulant molecules are promising tools for diagnostic and therapeutic uses.

Introduction

The Lepidoptera Order is made up of moths and butterflies, whose studies have focused mainly in their larval stages, due to their importance as agriculture pests, in silk production and in human envenoming. The larvae of lepidopterans are frequently referred as caterpillars. A variety of proteins from these insects has been identified, isolated and characterized, among them proteins capable of displaying procoagulant and anticoagulant effects on human blood, which are discussed here as the exogenous factors affecting hemostasis.

A.M. Chudzinski-Tavassi (✉)
Laboratory of Biochemistry and Biophysics, Instituto Butantan, São Paulo, SP, Brazil
e-mail: amchudzinski@butantan.gov.br

R.M. Kini et al. (eds.), *Toxins and Hemostasis*, DOI 10.1007/978-90-481-9295-3_31,

Two approaches have oriented the research on the identification and characterization of proteins from Lepidoptera insects related to hemostasis. The first is directed on defining how the coagulation system of these insects works and to identify new candidates of clotting factors. Some of these factors have shown domains similar to those found in vertebrate proteins (Dushay, 2009; Patthy, 1985; Theopold et al., 2002). The second is devoted to understanding the role of toxins, mainly found in the caterpillar's bristles, in the pathophysiology of envenoming syndrome in humans caused by accidental contact with some species of lepidopterans.

The Insect Coagulation System

Although the coagulation and fibrinolytic systems are well described in vertebrate animals, little is known about the hemostasis in insects. The studies on hemolymph coagulation in insects have focused mainly on the immunological aspect. In fact, immunological and hemostatic systems are intrinsically associated in these animals. Resembling the largely accepted principle for vertebrates, the hemostatic system in insects represents an initial response to avoid loss of body fluid when a wound happens (Theopold et al., 2002). The coagulation system seems to vary between different insect species and life stages (Dushay, 2009).

In comparison to vertebrates, the main differences in the insect hemostasis are: (1) increased clotting stimuli does not represent thrombotic risk for these animals due to their open circulatory system; (2) they have two systems which are activated in form of a cascade – the protein clotting system and the phenoloxidase activated system; (3) clotting is closely associated to immune response (Dushay, 2009; Theopold et al., 2002). The protein clotting cascade and the prophenoloxidase-activating cascade are supposed to interact with each other during the formation of hemolymph clot (Li et al., 2002), but the mechanism and sequence of reactions are poorly understood.

On the other hand, like vertebrates, the hemostatic system of insects is composed of protein clotting pathways with participation of zymogens and serine proteases linked to a cell-based response. The role of hemocytes in clotting mechanisms is quite similar to the role of vertebrate platelets. This includes the release of granular components, microparticles and the exposure of negatively charged phospholipids (Theopold et al., 2002). In addition, similarly to vertebrates, clotting in insects is likely to rely on calcium-dependent reactions (Li et al., 2002; Schmidt and Theopold, 1997; Theopold et al., 2002). Currently, a generic model of insect clotting system is yet to be proposed due to conflicting experimental data on hemolymph clotting between different species and life stages and the lack of similarity between proteins identified as candidates of clotting factors among these animals (Dushay, 2009). A proposed mechanism was based on the clotting system of the horse-shoe crab, which is a non-insect arthropod whose coagulation has been better characterized and seems to share with insects some similar protein components, particularly the domain structure of factors involved (Theopold et al., 2002). A tentative scheme

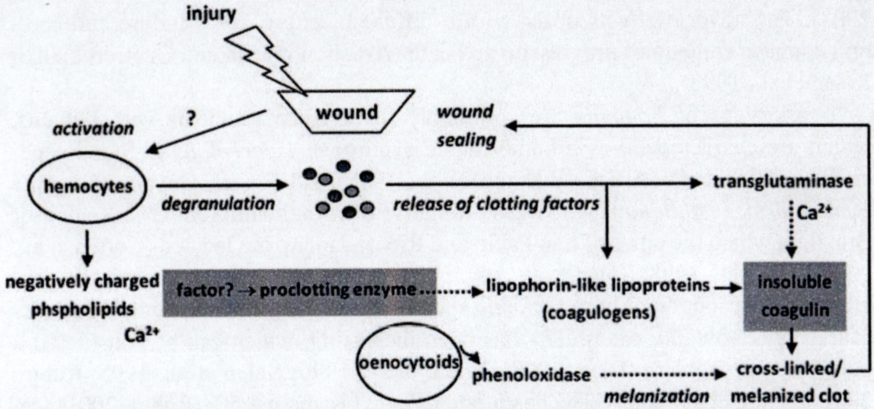

Fig. 31.1 Representative scheme of some steps involved in hemolymph coagulation in insects. Hemocytes are activated, generating granules and microparticles, with exposure of negatively charged phospholipids and release of clotting factors. The sequential activation of clotting factor zymogens in form of a cascade finally results in the precipitation of coagulogen (lipophorin-like lipoproteins), which is substrate of the calcium-dependent enzyme transglutaminase. The soft clot is melanized by the action of phenoloxidase, which is probably released by oenocytoids-type hemocytes, forming a mature hard clot and sealing the wound. Steps corresponding to enzymatic reactions are represented by *dotted arrows*. Indeterminate steps, which are better characterized in the horse-shoe crab, are in *gray boxes*

representing some steps of insect coagulation cascade is shown in Fig. 31.1. After mechanical injury or microorganism invasion, the clotting system is activated in form of a cascade with involvement of hemocytes. The phenoloxidase cascade is also activated, but is not required for initial soft clot formation. However, phenoloxidase is required for melanization and formation of a mature hard clot (Bidla et al., 2005).

The larval stage of *Galleria*, is one of the first and mostly used models from the Lepidoptera Order that has been used for long in the study of hemolymph coagulation in insects. The morphology of the clots, as well as some protein components from this insect that play a role in the hemostasis, were already described (Bidla et al., 2005; Korayem et al., 2007; Li et al., 2002; Rowley and Ratcliffe, 1976).

The Bleeding Syndrome Caused by *Lonomia* Caterpillar

Lonomia is a moth of the Saturniidae family. Species of *Lonomia* are found in South America and their larval stages, commonly referred as caterpillars, represent a risk of poisoning for humans, due to the presence of toxins in the bristles that cover the body of these insects. *Lonomia* toxins affect mainly the hemostatic system of envenomed victims inducing coagulopathy and hemorrhagic syndrome (Arocha-Piñango et al., 1992; Arocha-Piñango and Guerrero, 1999; Carrijo-Carvalho and Chudzinski-Tavassi, 2007; Duarte et al., 1996; Fan et al., 1998; Kelen et al., 1995; Zannin et al.,

2003). The adverse effects observed are diffuse bleeding, renal failure and cerebral damage, sometimes progressing to death (Arocha-Piñango and Guerrero, 2003; Kelen et al., 1995).

Two species of *Lonomia* are commonly involved in accidents with humans, which are characterized as a hemorrhagic syndrome. *L. achelous* is found especially in Venezuela (Arocha-Piñango et al., 1992) and French Guyana (Couppié et al., 1998). *L. obliqua* (Fig. 31.2) is found in the South and Southeast Region of Brazil, in which the cities of Sao Paulo and Rio de Janeiro are located (Corrêa et al., 2004; Fan et al., 1998; Kelen et al., 1995; Rubio, 2001). Figure 31.3 shows the geographical distribution of both *Lonomia* species. Since 1989 the number of accidents caused by *L. obliqua* caterpillars has been increasing, which can be considered a public health problem (Diaz, 2005; Duarte et al., 1990; Kelen et al., 1995; Rubio, 2001; Zannin et al., 2003). The death rate reported is about 2.5% (Rubio, 2001). An antiserum is produced by the Butantan Institute in São Paulo, Brazil. It effectively reverses the coagulation disorders induced by *L. obliqua* venom (Da Silva et al., 1996; Rocha-Campos et al., 2001).

The toxins from *Lonomia* species cause similar clinical effects, but their targets in the hemostatic system may differ. Several reports have demonstrated the procoagulant activity of *L. obliqua* toxins as its major property (Donato et al., 1998; Reis et al., 1999; Zannin et al., 2003) whereas *L. achelous* toxins were described with both procoagulant and anticoagulant activity (Arocha-Piñango and Guerrero, 2001). In Fig. 31.4 a representative scheme of the human coagulation and fibrinolytic system is shown, highlighting the main targets of *L. obliqua* and *L. achelous* toxins (for reviews, see Arocha-Piñango et al., 2000; Arocha-Piñango and Guerrero, 2001, 2003; Carrijo-Carvalho and Chudzinski-Tavassi, 2007; Chudzinski-Tavassi and Carrijo-Carvalho, 2006). Some *L. achelous* toxins identified have not yet been purified to homogeneity for a better characterization. There are three reasonable explanations for the different profiles of toxins from these species: (1) species-specific differences; (2) different larval ages (instars 1–6) and environmental conditions can lead to differential expression of proteins in the caterpillar; (3) biochemical and biological studies of proteins from

Fig. 31.2 *Lonomia obliqua* caterpillar

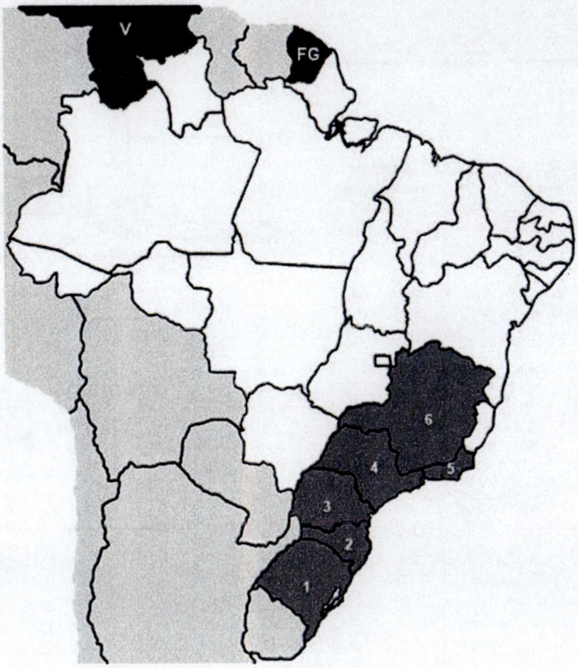

Fig. 31.3 Geographical distribution of *Lonomia* in South America. Accidents with *L. obliqua* caterpillar were reported in the South (States 1. Rio Grande do Sul, 2. Santa Catarina and 3. Paraná) and Southeast Region of Brazil (States 4. São Paulo, 5. Rio de Janeiro and 6. Minas Gerais). Accidents with *L. achelous* caterpillar were reported in Venezuela (V) and French Guiana (FG)

L. obliqua have been conducted mainly with extracts obtained from caterpillar bristles, while the properties of *L. achelous* toxins were investigated mostly in the hemolymph.

In *L. obliqua* bristle extract, two procoagulant toxins have been identified: a factor X activator named Losac (*Lonomia obliqua* Stuart-factor activator) (Alvarez Flores et al., 2006) and a prothrombin activator named Lopap (*Lonomia obliqua* prothrombin activator protease) (Reis et al., 2001a). Lopap seems to play an important role in the hemorrhagic syndrome caused by contact with the *L. obliqua* caterpillar (Reis et al., 1999, 2001b). From *L. achelous,* lonomin II, which has direct fibrinolytic activity, and lonomin V, which degrades coagulation factor XIII, have been suggested as the main active toxins (Arocha-Piñango and Guerrero, 2003; Guerrero et al., 1997a, b).

The coagulation disorders observed in humans after contact with *Lonomia* caterpillars can be reproduced, at least partially, in experimental animals. Various in vivo studies have been carried out to aid understanding of the biological mechanisms triggered by *Lonomia* caterpillar venom (Da Silva et al., 2004; Marval et al., 1999; Prezoto et al., 2002; Rocha-Campos et al., 2001). Administration of the crude extract of *L. obliqua* bristles to rats, rabbits, and mice causes a dose-dependent increase of

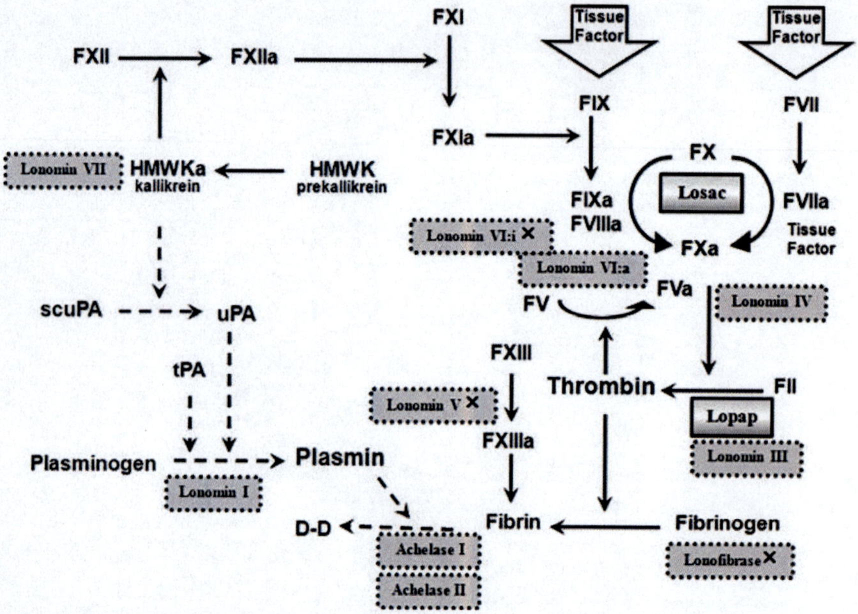

Fig. 31.4 Schematic representation of the human hemostatic system and the targets of *Lonomia* toxins. The action of *L. obliqua* (*solid squares*) and *L. achelous* toxins (*dotted squares*) on components of coagulation (*continuous arrows*) and fibrinolytic (*discontinuous arrows*) pathways are indicated

clotting time (prothrombin time, activated partial thromboplastin time and whole blood clotting time) and bleeding time with fibrinogen depletion (Kelen et al., 1995; Prezoto et al., 2002; Reis et al., 2001b), which characterize a consumption coagulopathy. In addition, *L. obliqua* bristle extract is effective in preventing thrombus formation, but does not exert thrombolytic activity in pre-formed thrombi (Prezoto et al., 2002). The antithrombotic effect of *L. obliqua* bristle extract is most probably due to fibrinogen depletion as a consequence of the massive induction of coagulation by thrombin generation, by the action of *L. obliqua* toxins triggering activation of the coagulation cascade through both prothrombin and factor X activator (Donato et al., 1998).

Studies on *Lonomia* Toxins

Lonomia obliqua

L. obliqua bristle extract contains various protein toxins, as demonstrated by the protein bands recognized by antilonomic serum after electrophoresis (Da Silva et al., 1996; Rocha-Campos et al., 2001). The proteins corresponding to each biological activity have not yet been identified (Chudzinski-Tavassi and Carrijo-Carvalho, 2006). In recent years, proteomic and transcriptomic studies brought new insights

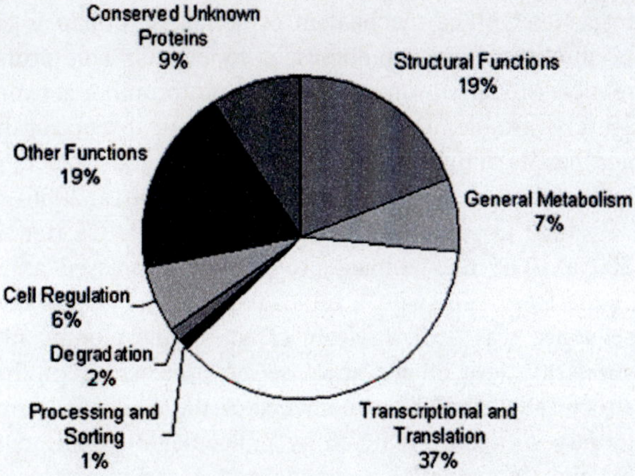

Fig. 31.5 Putative proteins in the cDNA library from *Lonomia obliqua* bristles. The major transcripts in the *L. obliqua* bristles were identified by EST-based strategy. Sequences from 1,270 independent clones were assembled into 702 clusters (509 matching clusters) of distinct genes

for a comprehensive view on the properties of *Lonomia* toxins and the pathophysiology of envenoming (Chudzinski-Tavassi and Alvarez Flores, 2005; Pinto et al., 2008; Ricci-Silva et al., 2008; Veiga et al., 2005). The construction of a cDNA cloning library revealed a variety of proteins expressed in the bristles of *L. obliqua* (Fig. 31.5), among them proteases (serine-like proteases, matrix metalloproteinase, cathepsin B-like cysteine proteinase, cathepsin L-like protease), protease inhibitors (thrombin inhibitor, serpins, Kunitz-type inhibitor), phospholipase A_2, tumor suppressors, and other proteins such as kininogen precursor, hemolin, heat shock-like proteins and antimicrobial peptides (Chudzinski-Tavassi and Alvarez Flores, 2005), and their sequences were deposited in the GenBank at NCBI (accession numbers: CX815710–CX817210, CX820335–CX820336, AY908986). Some of them can have physiological roles for the caterpillar, being part of its hemostatic system. Sequences coding for the prothrombin activator Lopap, also classified as a lipocalin, accounted for the most expressed gene analyzed by the Expressed Sequence Tag (EST) strategy, corresponded to 1.6% of total clones. Likewise, other parallel study (Veiga et al., 2005) reported the sequence coding for Lopap as the major protein expressed in the bristles and tegument of *L. obliqua* caterpillar. Sequences related to trypsin-like enzymes, blood coagulation factors, prophenoloxidase cascade activators, cysteine proteases, phospholipase A_2, serpins, cystatins, antibacterial proteins and other lipocalins were also reported. The abundance of lipocalins in the *L. obliqua* bristle extract was also demonstrated through two-dimensional electrophoresis and mass spectrometry to evaluate the protein profile of *L. obliqua* bristle extract. The majority of proteins with sequence matching to lipocalin were recognized by antilonomic and anti-Lopap serum, thus suggesting the presence of various Lopap isoforms. All these data corroborate previous studies and point out Lopap as one of the most important toxins from *L. obliqua* caterpillar.

Lopap is a prothrombin activator that shows serine protease-like activity (Reis et al., 2001a), triggering the hydrolysis of prothrombin on Arg^{284}–Thr^{285} and Arg^{320}–Ile^{321}, which corresponds to the thrombin and factor Xa cleavage site, respectively. The mechanism of action of Lopap is similar to that of factor Xa in absence of prothrombinase components. This protein does not fit in the current classification for snake venom prothrombin activators (Kini, 2005; Kini et al., 2001). It is able to activate prothrombin in absence of the prothrombinase complex, and has its activity enhanced by calcium ions generating α-thrombin without the intermediate meizothrombin (Reis et al., 2001a, 2006).

Native Lopap was first isolated as a 69-kDa (tetrameric) protein (Reis et al., 2001a). The recombinant protein was expressed as a monomer in the bacteria *Escherichia coli*, with a molecular mass of approximately 21 kDa. Lopap's full sequence was deduced from cDNA of the cloning library and it does not show similarity with other known serine protease or prothrombin activators, but it is structurally similar to members of the lipocalin family (Reis et al., 2006). Its tertiary structure predicted by molecular modeling consists of a β-barrel formed by eight antiparallel β-strands and a serine-protease like catalytic site formed by the amino acid residues His^{168}, Glu^{171} and Ser^{119}, located outside the barrel (Fig. 31.6). Lopap is the first described lipocalin with proteolytic activity (Reis et al., 2006). Lopap sequence has identity with other lepidopteran proteins also classified as lipocalins (accession numbers: BAB85482, BAB84676, NP_001036872,

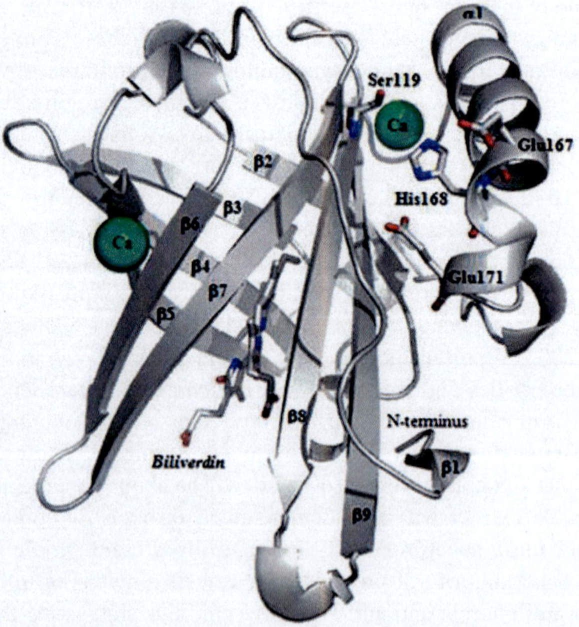

Fig. 31.6 Structural model of Lopap monomer. The structure of the *L. obliqua* prothrombin activator protease (Reis et al., 2006) is characteristic of lipocalin family members. The predicted residues involved in the serine protease-like catalytic activity and the calcium binding sites (Ca) are indicated in the figure. The biliverdin hydrophobic ligand was superposed to indicate the hydrophobic pocket location, inside the β-barrel

AAA85089, BAG12758, P00305, Q00630), such insecticyanin, a blue biliprotein isolated from the hemolymph of the tobacco hornworm *Manduca sexta* (Riley et al., 1984), biliverdin-binding protein, an insecticyanin-type protein from the Eri-silkworm *Samia cynthia ricini* (Saito, 1998), bombyrin from *Bombyx mori,* gallerin from *Galleria mellonella*, and bilin binding protein 1 from *Papilio xuthus*.

Laboratory data on the in vivo effects of Lopap, corroborated the main idea that it plays an important function in the consumptive coagulopathy caused by *L. obliqua*. Injection of Lopap in rats induces a coagulopathy similar to that occurring in human poisoning, triggering a dose-dependent increase of blood clotting time, and fibrinogen depletion (Reis et al., 2001b). Besides its prothrombin activation activity, Lopap can modulate endothelial cell responses, inducing expression of mediators involved in hemostasis, as well as promoting cell survival through antiapoptotic activity (Chudzinski-Tavassi et al., 2001; Fritzen et al., 2005).

Losac is a factor X activator from *L. obliqua,* consisting of a single-chain serine-like protease of approximately 43 kDa (Alvarez Flores et al., 2006; Chudzinski-Tavassi and Alvarez Flores, 2005). This enzyme activates factor X in a dose-dependent manner, in absence of calcium, and the FXa formed is able to integrate the prothrombinase complex. Its mechanism of action differs from the other known FX activators from snake venoms, since they depend on calcium ions as a co-factor for their activity (Tans and Rosing, 2001). The amino acid sequence of Losac showed no similarity with other factor X activators. In addition to its procoagulant activity, Losac also induces the expression of hemostatic mediators, cell proliferation, and inhibition of apoptosis in endothelial cell culture (Alvarez Flores et al., 2006).

Laboratory assays demonstrated that the *L. obliqua* bristle extract is not able to directly activate the fibrinolytic system or degrading cross-linked fibrin (Fritzen et al., 2003; Kelen et al., 1995). These results are in compliance with in vivo studies cited earlier, despite of reports of the presence of fibrinogenolytic toxins in the *L. obliqua* bristle extract (Fritzen et al., 2003; Pinto et al. 2004; Veiga et al., 2003). A 35-kDa fibrinogenolytic enzyme called Lonofibrase was purified from *L. obliqua* hemolymph. This enzyme cleaves preferentially fibrinogen α-chain, and is less effective against the β-chain (Pinto et al. 2004).

Lonomia achelous

Unlike *L. obliqua*, several studies described *L. achelous* toxins that activate both clotting and fibrinolytic pathways. *L. achelous* hemolymph has anticoagulant toxins which can inhibit clot formation and also induce lysis of pre-formed thrombi (Coll-Sangrona and Arocha-Piñango, 1998). The anticoagulant proteins identified in the hemolymph of *L. achelous* are Achelase I and Achelase II (Lonomin II), which have plasmin-like activity (Amarant et al., 1991), and Lonomin V, which degrades fibrin, fibrinogen and human factor XIII, impairing fibrin cross-linking (Guerrero et al., 1997a, b).

Lonomin V has a molecular mass of 16–18 kDa, p*I* > 10, is stable over a wide range of pH values and temperatures and causes a dose-dependent degradation of factor XIII/XIIIa, with the generation of peptide fragments of low molecular weight and concomitant loss of FXIII transglutaminase activity (Arocha-Piñango et al., 2000; Guerrero et al., 1997b). Consequently, Lonomin V impairs fibrin/fibrinogen cross-linking by FXIIIa (Guerrero et al., 1997a). This protein also displayed urokinase-like activity (formerly named Lonomin I) by activating plasminogen (Arocha-Piñango and Pepper, 1981, Arocha-Piñango et al., 2000) and had thrombolytic activity in a vein thrombosis model (Guerrero et al., 2001). Its enzymatic activity is completely inhibited by serine protease inhibitors, and by iodoacetamide, which blocks active site cysteine. This inhibition profile suggests that Lonomin V is a serine protease with a free cysteine residue that is essential for the enzymatic activity (Guerrero et al., 1999). N-terminal sequence analysis of this toxin showed a homology of 75% with plasmin-like toxins from *L. achelous* (Arocha-Piñango and Guerrero, 2003).

The plasmin-like toxins isolated from *L. achelous* venom are called Achelase I and Achelase II, displaying a molecular mass of 22.4 and 22.7 kDa, and a p*I* of 10.5 and 8.5, respectively (Amarant et al., 1991). Although they have a direct fibrinolytic action independently of plasminogen, the fibrinogen degradation products generated by the proteolytic action of these enzymes are different from those generated by the action of plasmin on fibrinogen. The amino acid sequences of Achelase I and II showed high similarity with lepdopteran thrombin-like and trypsin-like serine proteases (Arocha-Piñango et al., 2000). *L. achelous* hemolymph also has a factor V inactivator (Lonomin VI:i), which is thermolabile, has maximal activity at basic pH, and has serine protease-like activity (López et al., 2000).

Two types of prothrombin activators were identified in *L. achelous* hemolymph: a calcium-independent (Lonomin III) and a factor Xa-like (Lonomin IV) (Guerrero and Arocha-Piñango, 1992). In addition to these procoagulant proteins, a factor V activator (Lonomin VI:a) and a kallikrein-like protein (Lonomin VII) were described (Arocha-Piñango and Guerrero, 2001; López et al., 2000). Lonomin VI:a is thermostable, has maximal activity at acid pH, and is inhibited by metalloprotease inhibitors, indicating that it is a metalloproteinase.

Anticoagulant Proteins from Lepidoptera

Besides the anticoagulant toxins from *L. achelous* described above, a small number of anticoagulant proteins from lepidopterans have been characterized. Most of them are serpins, identified mainly through genomic studies, which may have roles in controlling the coagulation stimuli by its inhibitory activities on serine protease components of insect clotting system (Hegedus et al., 2008; Kanost 1999; Tanaka et al., 2008; Tong and Kanost, 2005; Tong et al., 2005; Zou et al., 2009). However, to our knowledge, their effects on vertebrate hemostasis have not been evaluated. Protease inhibitors were also identified in the transcriptomic libraries from bristles and tegument of *L. obliqua* caterpillar (Chudzinski-Tavassi and Alvarez Flores, 2005; Veiga et al., 2005). Endogenous anticoagulant proteins seem to be less necessary for insects, since a strict control of clotting is probably not necessary for them.

Sericin and fibroin are components of the silk produced by the silkworm *Bombyx mori*, which have been used as matters to produce sulfated proteins. These proteins are able to display anticoagulant activity in bovine (Tamada et al., 2004) and human plasma (Sano et al., 2009; Tamada, 2004). The mechanism by which sulfated sericin displays anticoagulant activity seems to be different from heparin and is not dependent of antithrombin III (Sano et al., 2009). These compounds may represent alternative therapeutics to heparin and other anticoagulants commercially available and to development of biomedical material with antithrombogenic property (Tamada et al., 2004).

Procoagulant Proteins from Lepidoptera

Procoagulant proteins from *Lonomia* are well studied and characterized due to their abundance in the caterpillar bristles and hemolymph and its importance in human envenoming outcomes (Carrijo-Carvalho and Chudzinski-Tavassi, 2007). Phenoloxidases are the most studied procoagulant proteins among the important ones for the intrinsic coagulation system of lepidopterans and other invertebrates (Cerenius and Söderhäll, 2004; Jiang et al., 2005). This protein was isolated from the cuticle of the silkworm *Bombyx mori* and its transport from hemolymph to cuticule was demonstrated (Asano and Ashida, 2001). Other proteins have been described, such as scolexin (Finnerty et al., 1999), the lectin M13 (Minnick et al., 1986) and a 22-kDa protein called hemofibrin (Geng and Dunn, 1988) from *Manduca sexta,* and hemocytin from *Bombyx mori*, which is a lectin homologous to the mammalian von Willebrand factor (Kotani et al., 1995). Hexamerin (arylphorin), α-crystallin/small heat shock protein, lipophorin and transglutaminase are other proteins which have been suggested as possible lepidopteran clotting factors (Dushay, 2009; Li et al., 2002). However, little is known about their biochemical properties and possible effects as exogenous factors on vertebrate hemostasis. These effects should be expected, as proteins with high similarity to Lopap can be found among other lepidopterans that do not cause human envenoming. Interestingly, treatment of laboratory animals with extract from the *Galleria mellonella* larvae induces hypocoagulant effect (Rachkov et al., 1994), which can resemble the coagulopathy induced by *Lonomia* caterpillars. Thus, it is reasonable to propose that *Lonomia* toxins could be originated from endogenous clotting factors by gene duplication and differentiation, or these toxins have originally a physiological importance for the insect itself and occasionally acquired roles as toxins.

Future Prospects

Known components involved in hemostasis of insects include members of both the coagulation system and the prophenol-activating cascade, and serpins which negatively regulate clotting pathways. Some proteins in the hemolymph and cuticle of lepidopterans have been identified as candidate clotting factors. Little is known about their roles as procoagulant proteins in vivo or in vitro and its possible effect as exogenous factors has not yet been examined. From the *L. obliqua* transcriptomic library a large number of unknown proteins remain to be identified (Fig. 31.5). An interesting issue is that lepidopteran procoagulant proteins might not be affected by vertebrate inhibitors.

Exogenous procoagulant proteins have a promising application as diagnostic reagents. Several animal toxins have been proved to be useful as reagents in laboratory tests for diagnosis, for example, to detect clotting factor deficiencies and to monitor patients undergoing anticoagulant therapy (Marsh and Williams, 2005; Schoni, 2005). Prothrombin activators, such as Lopap, can be used to detect dysprothrombinemias and disseminated intravascular coagulation. Factor V activators can be useful to monitor patients under anticoagulant therapy, except for vitamin K antagonists, and to detect factor V Leiden mutation genotype (Schoni, 2005). Factor X activators are also used for diagnostic tests, for detection of factor X deficiency (Bezeaud et al., 1995) and lupus anticoagulant (Triplett, 2000).

Exogenous procoagulant proteins can also be used in the therapeutic area, as defibrinogenating agents to prevent thrombosis and decrease blood viscosity (Koh et al., 2006). On the other hand, procoagulant proteins can be envisaged as hemostatic agents to arrest bleeding, for example, during surgical procedures (Patrick et al., 2007), offering advantages on the currently used compounds (Chudzinski-Tavassi et al., 2009).

In view of the value of exogenous factors as therapeutic and biotechnological tools, lepidopterans remain as a potential source for bioprospection of new proteins affecting hemostasis, which can also represent important tools for comparative studies on hemostasis in divergent species and for biochemical characterization. Research in this field can provide new insights on the properties of distinct protein family members and enable the proposal of more suitable classifications of exogenous factors that could incorporate lepidopteran proteins, which are functionally and structurally divergent from snake venom toxins.

References

Alvarez Flores, M.P., Fritzen, M., Reis, C.V., Chudzinski-Tavassi, A.M., 2006. Losac, a factor X activator from *Lonomia obliqua* bristle extract: its role in the pathophysiological mechanisms and cell survival. Biochem. Biophys. Res. Commun. 343, 1216–1223.

Amarant, T., Burkhart, W., LeVine, H., 3rd, Arocha-Pinango, C.L., Parikh, I., 1991. Isolation and complete amino acid sequence of two fibrinolytic proteinases from the toxic Saturnid caterpillar *Lonomia achelous*. Biochim. Biophys. Acta 1079, 214–221.

Arocha-Piñango, C.L., Pepper, D.S., 1981. Studies of a fibrinolytic enzyme from the larvae of *Lonomia achelous* (Cramer) using chromogenic peptide substrates. Thromb. Haemost. 46, 710–713.

Arocha-Piñango, C.L., de Bosch, N.B., Torres, A., Goldstein, C., Nouel, A., Argüello, A., Carvajal, Z., Guerrero, B., Ojeda, A., Rodriguez, A., Rodriguez, S., 1992. Six new cases of a caterpillar-induced bleeding syndrome. Thromb. Haemost. 67, 402–407.

Arocha-Piñango, C.L., Guerrero, B., 1999. *Lonomia obliqua* and haemorrhagic syndrome. Lancet 354, 1304.

Arocha-Piñango, C.L., Marval, E., Guerrero, B., 2000. *Lonomia* genus caterpillar toxins: biochemical aspects. Biochimie 82, 937–942.

Arocha-Piñango, C.L., Guerrero, B., 2001. *Lonomia* genus caterpillar envenomation: clinical and biological aspects. Haemostasis 31, 288–293.

Arocha-Piñango, C.L., Guerrero, B., 2003. Hemorrhagic syndrome induced by caterpillars. Clinical and experimental studies. Invest. Clin. 44, 155–163.

Asano, T., Ashida, M., 2001. Cuticular pro-phenoloxidase of the silkworm, *Bombyx mori*. Purification and demonstration of its transport from hemolymph. J. Biol. Chem. 276, 11100–11112.

Bezeaud, A., Miyata, T., Helley, D., Zeng, Y.Z., Kato, H., Aillaud, M.F., Juhan-Vague, I., Guillin, M.C., 1995. Functional consequences of the Ser334→Pro mutation in a human factor X variant (factor X Marseille). Eur. J. Biochem. 234, 140–147.

Bidla, G., Lindgren, M., Theopold, U., Dushay, M.S., 2005. Hemolymph coagulation and phenoloxidase in *Drosophila* larvae. Dev. Comp. Immunol. 29, 669–679.

Carrijo-Carvalho, L.C., Chudzinski-Tavassi, A.M., 2007. The venom of the *Lonomia* caterpillar: an overview. Toxicon 49, 741–757.

Cerenius, L., Söderhäll, K., 2004. The prophenoloxidase-activating system in invertebrates. Immunol. Rev. 198, 116–126.

Chudzinski-Tavassi, A.M., Schattner, M., Fritzen, M., Pozner, R.G., Reis, C.V., Lourenço, D., Lazzari, M.A., 2001. Effects of lopap on human endothelial cells and platelets. Haemostasis 31, 257–265.

Chudzinski-Tavassi, A.M., Alvarez Flores, M.P., 2005. Exploring new molecules and activities from *Lonomia obliqua* caterpillars. Pathophysiol. Haemost. Thromb. 34, 228–233.

Chudzinski-Tavassi, A.M., Carrijo-Carvalho, L.C., 2006. Biochemical and biological properties of *Lonomia obliqua* bristle extract. J. Venoms Anim. Toxins incl. Trop. Dis. 12, 156–171.

Chudzinski-Tavassi, A.M., Carrijo-Carvalho, L.C., Faria, F., Alvarez Flores, M.P., Simons, S.M., 2009. Exogenous factors affecting hemostasis: therapeutic perspectives and biotechnological approaches, in: de Lima, M.E., Pimenta, A.M.C., Martin-Eauclaire, M.F., Zingali, R.B., Rochat, H. (Eds.), Animal toxins: State of the Art. Perspectives in Health and Biotechnology. UFMG, Belo Horizonte, pp. 495–523.

Coll-Sangrona, E., Arocha-Piñango, C.L., 1998. Fibrinolytic action on fresh human clots of whole body extracts and two semipurified fractions from *Lonomia achelous* caterpillar. Braz. J. Med. Biol. Res. 31, 779–784.

Corrêa, M.S., Siqueira-Batista, R., Gomes, A.P., Franco-Barbosa, A., Verzola, A.C., Oliveira, F.R., Squeff, F.A., Motta-Leal-Filho, J.M., Tavares, R.H., Amorim, D.S., De-Maria-Moreira, N.L., Santos, S.S., 2004. *Lonomia* erucism in Teresópolis, Rio de Janeiro State, Brazil: report of a probable case and review. Rev. Soc. Bras. Med. Trop. 37, 418–421.

Couppié, P., Marty, C., Sainte-Marie, D., Pradinaud, R., 1998. Poisonous caterpillars in French Guyana. 5 cases. Ann. Dermatol. Venereol. 125, 489–491.

Da Silva, W.D., Campos, C.M., Gonçalves, L.R., Sousa-e-Silva, M.C., Higashi, H.G., Yamagushi, I.K., Kelen, E.M., 1996. Development of an antivenom against toxins of *Lonomia obliqua* caterpillars. Toxicon 34, 1045–1049.

Da Silva, G.H., Hyslop, S., da Cruz-Höfling, M.A., 2004. *Lonomia obliqua* caterpillar venom increases permeability of the blood-brain barrier in rats. Toxicon 44, 625–634.

Diaz, J.H., 2005. The evolving global epidemiology, syndromic classification, management, and prevention of caterpillar envenoming. Am. J. Trop. Med. Hyg. 72, 347–357.

Donato, J.L., Moreno, R.A., Hyslop, S., Duarte, A., Antunes, E., Le Bonniec, B.F., Rendu, F., de Nucci, G., 1998. *Lonomia obliqua* caterpillar spicules trigger human blood coagulation via activation of factor X and prothrombin. Thromb. Haemost. 79, 539–542.

Duarte, C.A., Caovilla, J., Lorini, I., Lorini, D., Mantorani, G., Sumida, J., Manfre, P.C., Silveira, R., Moura, S.P., 1990. Insuficiencia renal aguda por acidentes com lagartas. J. Bras. Nefrol. 12, 184–187.

Duarte, A.C., Crusius, P.S., Pires, C.A., Schilling, M.A., Fan, H.W., 1996. Intracerebral haemorrhage after contact with *Lonomia* caterpillars. Lancet 348, 1033.

Dushay, M.S., 2009. Insect hemolymph clotting. Cell. Mol. Life Sci. 66, 2643–2650.

Fan, H.W., Cardoso, J.L., Olmos, R.D., Almeida, F.J., Viana, R.P., Martinez, A.P., 1998. Hemorrhagic syndrome and acute renal failure in a pregnant woman after contact with *Lonomia* caterpillars: a case report. Rev. Inst. Med. Trop. Sao Paulo 40, 119–120.

Finnerty, C.M., Karplus, P.A., Granados, R.R., 1999. The insect immune protein scolexin is a novel serine proteinase homolog. Protein Sci. 8, 242–248.

Fritzen, M., Schattner, M., Ribeiro, A.L., Batista, I.F., Ventura, J., Prezoto, B.C., Chudzinski-Tavassi, A.M., 2003. *Lonomia obliqua* venom action on fibrinolytic system. Thromb. Res. 112, 105–110.

Fritzen, M., Flores, M.P., Reis, C.V., Chudzinski-Tavassi, A.M., 2005. A prothrombin activator (Lopap) modulating inflammation, coagulation and cell survival mechanisms. Biochem. Biophys. Res. Commun. 333, 517–523.

Geng, C., Dunn, P.E., 1988. Hemostasis in larvae of *Manduca sexta*: formation of a fibrous coagulum by hemolymph proteins. Biochem. Biophys. Res. Commun. 155, 1060–1065.

Guerrero, B., Arocha-Piñango, C.L., 1992. Activation of human prothrombin by the venom of *Lonomia achelous* (Cramer) caterpillars. Thromb. Res. 66, 169–177.

Guerrero, B.A., Arocha-Piñango, C.L., San Juan, A.G., 1997a. *Lonomia achelous* caterpillar venom (LACV) selectively inactivates blood clotting factor XIII. Thromb. Res. 87, 83–93.

Guerrero, B.A., Arocha-Piñango, C.L., San Juan, A.G., 1997b. Degradation of human factor XIII by lonomin V, a purified fraction of *Lonomia achelous* caterpillar venom. Thromb. Res. 87, 171–181.

Guerrero, B., Perales, J., Gil, A., Arocha-Piñango, C.L., 1999. Effect on platelet FXIII and partial characterization of Lonomin V, a proteolytic enzyme from *Lonomia achelous* caterpillars. Thromb. Res. 93, 243–252.

Guerrero, B., Arocha-Piñango, C.L., Pinto, M.A., Müller, C.A., San Juan, A.G., Amorim, S., Perales, J., 2001. Thrombolytic effect of lonomin V in a rabbit jugular vein thrombosis model. Blood Coagul. Fibrinolysis 12, 521–529.

Hegedus, D.D., Erlandson, M., Baldwin, D., Hou, X., Chamankhah, M., 2008. Differential expansion and evolution of the exon family encoding the Serpin-1 reactive centre loop has resulted in divergent serpin repertoires among the Lepidoptera. Gene 418, 15–21.

Jiang, H., Wang, Y., Gu, Y., Guo, X., Zou, Z., Scholz, F., Trenczek, T.E., Kanost, M.R., 2005. Molecular identification of a bevy of serine proteinases in *Manduca sexta* hemolymph. Insect Biochem. Mol. Biol. 35, 931–943.

Kanost, M.R., 1999. Serine proteinase inhibitors in arthropod immunity. Dev. Comp. Immunol. 23, 291–301.

Kelen, E.M.A., Picarelli, Z.P., Duarte, A.C., 1995. Hemorrhagic syndrome induced by contact with caterpillars of the genus *Lonomia* (Saturniidae, Hemileucinae). J. Toxicol. Toxin Rev. 14, 283–308.

Kini, R.M., Rao, V.S., Joseph, J.S., 2001. Procoagulant proteins from snake venoms. Haemostasis 31, 218–224.

Kini, R.M., 2005. The intriguing world of prothrombin activators from snake venom. Toxicon 45, 1133–1145.

Koh, D.C., Armugam, A., Jeyaseelan, K., 2006. Snake venom components and their applications in biomedicine. Cell. Mol. Life Sci. 63, 3030–3041.

Korayem, A.M., Hauling, T., Lesch, C., Fabbri, M., Lindgren, M., Loseva, O., Schmidt, O., Dushay, M.S., Theopold, U., 2007. Evidence for an immune function of lepidopteran silk proteins. Biochem. Biophys. Res. Commun. 352, 317–322.

Kotani, E., Yamakawa, M., Iwamoto, S., Tashiro, M., Mori, H., Sumida, M., Matsubara, F., Taniai, K., Kadono-Okuda, K., Kato, Y., Mori, H., 1995. Cloning and expression of the gene of hemocytin, an insect humoral lectin which is homologous with the mammalian von Willebrand factor. Biochim. Biophys. Acta 1260, 245–258.

Li, D., Scherfer, C., Korayem, A.M., Zhao, Z., Schmidt, O., Theopold, U., 2002. Insect hemolymph clotting: evidence for interaction between the coagulation system and the prophenoloxidase activating cascade. Insect Biochem. Mol. Biol. 32, 919–928.

López, M., Gil, A., Arocha-Piñango, C.L., 2000. The action of *Lonomia achelous* caterpillars venom on human factor V. Thromb. Res. 98, 103–110.

Marsh, N., Williams, V., 2005. Practical applications of snake venom toxins in haemostasis. Toxicon 45, 1171–1181.

Marval, E., Guerrero, B., Arocha-Piñango, C.L., 1999. The action of *Lonomia achelous* caterpillar venom on some blood coagulation and fibrinolysis parameters of the rabbit. Toxicon 37, 1491–1504.

Minnick, M.F., Rupp, R.A., Spence, K.D., 1986. A bacterial-induced lectin which triggers hemocyte coagulation in *Manduca sexta*. Biochem. Biophys. Res. Commun. 137, 729–735.

Patrick, A.R., Winkelmayer, W.C., Avorn, J., Fischer, M.A., 2007. Strategies for the management of suspected heparin-induced thrombocytopenia: a cost-effectiveness analysis. Pharmacoeconomics 25, 949–961.

Patthy, L., 1985. Evolution of the proteases of blood coagulation and fibrinolysis by assembly from modules. Cell 41, 657–663.

Pinto, A.F., Dobrovolski, R., Veiga, A.B., Guimarães, J.A., 2004. Lonofibrase, a novel alpha-fibrinogenase from *Lonomia obliqua* caterpillars. Thromb. Res. 113, 147–154.

Pinto, A.F., Dragulev, B., Guimarães, J.A., Fox, J.W., 2008. Novel perspectives on the pathogenesis of *Lonomia obliqua* caterpillar envenomation based on assessment of host response by gene expression analysis. Toxicon 51, 1119–1128.

Prezoto, B.C., Maffei, F.H., Mattar, L., Chudzinski-Tavassi, A.M., Curi, P.R., 2002. Antithrombotic effect of *Lonomia obliqua* caterpillar bristle extract on experimental venous thrombosis. Braz. J. Med. Biol. Res. 35, 703–712.

Rachkov, A.K., Spiridonov, N.A., Kondrashova, M.N., 1994. Adaptogenic and cardioprotective action of *Galleria mellonella* extract in rats and frogs. J. Pharm. Pharmacol. 46, 221–225.

Reis, C.V., Kelen, E.M., Farsky, S.H., Portaro, F.C., Sampaio, C.A., Fernandes, B.L., Camargo, A.C., Chudzinski-Tavassi, A.M., 1999. A Ca^{++} activated serine protease (LOPAP) could be responsible for the haemorrhagic syndrome caused by the caterpillar *Lonomia obliqua. L obliqua* Prothrombin Activator Protease. Lancet 353, 1942.

Reis, C.V., Portaro, F.C., Andrade, S.A., Fritzen, M., Fernandes, B.L., Sampaio, C.A., Camargo, A.C., Chudzinski-Tavassi, A.M., 2001a. A prothrombin activator serine protease from the *Lonomia obliqua* caterpillar venom (Lopap) biochemical characterization. Thromb. Res. 102, 427–436.

Reis, C.V., Farsky, S.H., Fernandes, B.L., Santoro, M.L., Oliva, M.L., Mariano, M., Chudzinski-Tavassi, A.M., 2001b. In vivo characterization of Lopap, a prothrombin activator serine protease from the *Lonomia obliqua* caterpillar venom. Thromb. Res. 102, 437–443.

Reis, C.V., Andrade, S.A., Ramos, O.H., Ramos, C.R., Ho, P.L., Batista, I.F., Chudzinski-Tavassi, A.M., 2006. Lopap, a prothrombin activator from *Lonomia obliqua* belonging to the lipocalin family: recombinant production, biochemical characterization and structure-function insights. Biochem. J. 398, 295–302.

Ricci-Silva, M.E., Valente, R.H., Leon, I.R., Tambourgi, D.V., Ramos, O.H., Perales, J., Chudzinski-Tavassi, A.M., 2008. Immunochemical and proteomic technologies as tools for unravelling toxins involved in envenoming by accidental contact with *Lonomia obliqua* caterpillars. Toxicon 51, 1017–1028.

Riley, C.T., Barbeau, B.K., Keim, P.S., Kézdy, F.J., Heinrikson, R.L., Law, J.H., 1984. The covalent protein structure of insecticyanin, a blue biliprotein from the hemolymph of the tobacco hornworm, *Manduca sexta* L. J. Biol. Chem. 259, 13159–13165.

Rocha-Campos, A.C., Gonçalves, L.R., Higashi, H.G., Yamagushi, I.K., Fernandes, I., Oliveira, J.E., Ribela, M.T., Sousa-E-Silva, M.C., da Silva, W.D., 2001. Specific heterologous F(ab′)2 antibodies revert blood incoagulability resulting from envenoming by *Lonomia obliqua* caterpillars. Am. J. Trop. Med. Hyg. 64, 283–289.

Rowley, A.F., Ratcliffe, N.A., 1976. The granular cells of *Galleria mellonella* during clotting and phagocytic reactions in vitro. Tissue Cell. 8, 437–446.

Rubio, G.B., 2001. Epidemiological surveillance of distribution of the caterpillar *Lonomia obliqua* Walker, 1855, in the State of Paraná, Brazil. Cad. Saude Publica 17, 1036.

Saito, H., 1998. Purification and characterization of two insecticyanin-type proteins from the larval hemolymph of the Eri-silkworm, *Samia cynthia ricini*. Biochim. Biophys. Acta 1380, 141–150.

Sano, M., Tamada, Y., Niwa, K., Morita, T., Yoshino, G., 2009. Sulfated sericin is a novel anticoagulant influencing the blood coagulation cascade. J. Biomater. Sci. Polym. Ed. 20, 773–783.

Schmidt, O., Theopold, U., 1997. *Helix pomatia* lectin and annexin V, two molecular probes for insect microparticles: possible involvement in hemolymph coagulation. J. Insect Physiol. 43, 667–674.

Schoni, R., 2005. The use of snake venom-derived compounds for new functional diagnostic test kits in the field of haemostasis. Pathophysiol. Haemost. Thromb. 34, 234–240.

Tamada, Y., 2004. Sulfation of silk fibroin by chlorosulfonic acid and the anticoagulant activity. Biomaterials 25, 377–383.

Tamada, Y., Sano, M., Niwa, K., Imai, T., Yoshino, G., 2004. Sulfation of silk sericin and anticoagulant activity of sulfated sericin. J. Biomater. Sci. Polym. Ed. 15, 971–980.

Tanaka, H., Ishibashi, J., Fujita, K., Nakajima, Y., Sagisaka, A., Tomimoto, K., Suzuki, N., Yoshiyama, M., Kaneko, Y., Iwasaki, T., Sunagawa, T., Yamaji, K., Asaoka, A., Mita, K., Yamakawa, M., 2008. A genome-wide analysis of genes and gene families involved in innate immunity of *Bombyx mori*. Insect Biochem. Mol. Biol. 38, 1087–1110.

Tans, G., Rosing, J., 2001. Snake venom activators of factor X: an overview. Haemostasis 31, 225–233.

Theopold, U., Li, D., Fabbri, M., Scherfer, C., Schmidt, O., 2002. The coagulation of insect hemolymph. Cell. Mol. Life Sci. 59, 363–372.

Tong, Y., Kanost, M.R., 2005. *Manduca sexta* serpin-4 and serpin-5 inhibit the prophenol oxidase activation pathway: cDNA cloning, protein expression, and characterization. J. Biol. Chem. 280, 14923–14931.

Tong, Y., Jiang, H., Kanost, M.R., 2005. Identification of plasma proteases inhibited by *Manduca sexta* serpin-4 and serpin-5 and their association with components of the prophenol oxidase activation pathway. J. Biol. Chem. 280, 14932–14942.

Triplett, D.A., 2000. Use of the dilute Russell viper venom time (dRVVT): its importance and pitfalls. J. Autoimmun. 15, 173–178.

Veiga, A.B., Pinto, A.F., Guimarães, J.A., 2003. Fibrinogenolytic and procoagulant activities in the hemorrhagic syndrome caused by *Lonomia obliqua* caterpillars. Thromb. Res. 111, 95–101.

Veiga, A.B., Ribeiro, J.M., Guimarães, J.A., Francischetti, I.M., 2005. A catalog for the transcripts from the venomous structures of the caterpillar *Lonomia obliqua*: identification of the proteins potentially involved in the coagulation disorder and hemorrhagic syndrome. Gene. 355, 11–27.

Zannin, M., Lourenço, D.M., Motta, G., Dalla Costa, L.R., Grando, M., Gamborgi, G.P., Noguti, M.A., Chudzinski-Tavassi, A.M., 2003. Blood coagulation and fibrinolytic factors in 105 patients with hemorrhagic syndrome caused by accidental contact with *Lonomia obliqua* caterpillar in Santa Catarina, southern Brazil. Thromb. Haemost. 89, 355–364.

Zou, Z., Picheng, Z., Weng, H., Mita, K., Jiang, H., 2009. A comparative analysis of serpin genes in the silkworm genome. Genomics 93, 367–375.

Chapter 32
Staphylocoagulase

Peter Panizzi, Heather K. Kroh, Pablo Fuentes-Prior, and Paul E. Bock

Abstract Staphylocoagulase (SC) is a protein secreted by most strains of *Staphylococcus aureus*, a potent human pathogenic bacterium. SC causes blood clotting by direct activation of the thrombin precursor, prothrombin (ProT). This occurs through a non-proteolytic, cofactor-induced mechanism of activation called "molecular sexuality". The 2.2 Å crystal structure of a fully active SC fragment in complex with the ProT catalytic domain, prethrombin 2, proved the molecular sexuality mechanism. Recent structural and mechanistic studies have investigated how SC specifically activates ProT to cause fibrin generation through association of active SC·ProT* complexes with fibrinogen. In addition, a predicted structural homolog of SC has been identified from *S. aureus*, called von Willebrand factor-binding protein (VWbp). VWbp also activates human ProT by the molecular sexuality mechanism, but it displays additional regulation of its specificity through a novel hysteretic kinetic mechanism. The mechanisms used by both of these bacterial ProT activators to subvert the human coagulation system offers new insight into the role of secreted exoproteins in staphylococcal infection, as well as alternative pathways for targeted treatment of staphylococcal coagulopathies.

Introduction

Staphylococcus aureus virulence has historically been considered intertwined with the ability of the bacteria to clot blood (Chapman et al., 1934). Clotting of plasma is a result of staphylocoagulase (SC) secreted by *S. aureus*, which circumvents the physiological blood coagulation pathway by binding and non-proteolytically activating the thrombin precursor, prothrombin (ProT), forming an active SC·ProT* complex that converts fibrinogen into insoluble fibrin. This non-proteolytic activation mimics the natural maturation of inactive zymogen precursors to functional

P. Panizzi (✉)
Center for Systems Biology, Massachusetts General Hospital, Boston, MA 02114, USA
e-mail: ppanizzi@mgh.harvard.edu

R.M. Kini et al. (eds.), *Toxins and Hemostasis*, DOI 10.1007/978-90-481-9295-3_32,

proteinases, but also requires SC to create a new fibrinogen substrate-recognition exosite on the SC·ProT* complex to clot blood. Although structural information regarding this new exosite is lacking, the mechanism of SC activation of ProT has been recently established, and further thermodynamic, structural, and kinetic studies have begun to characterize the active SC·ProT* complex. Despite the overwhelming belief that biochemical detection of the presence of SC can reliably identify pathogenic strains of *S. aureus*, to date, there is no study that directly establishes SC as a staphylococcal virulence factor. Herein, we review the broad contributions of pioneering work that initially identified and isolated SC, recent structural and mechanistic studies that illuminate how SC specifically activates ProT to cause fibrin generation, the role of SC in models of human staphylococcal diseases, and future directions required to establish its role as a virulence factor in *S. aureus* infections.

Discovery of SC

The discovery of SC can be separately attributed to the observations that *S. aureus* clots blood (Loeb, 1903) and the isolation of this bacterial zymogen activator, which imparts this clotting potential (Tager, 1948; Tager and Hales, 1948). Cancer and tissue transplantation pathologist, Leo Loeb, while spending a year conducting research at McGill University in 1903, observed that *S. aureus* could clot goose plasma, whereas a number of other bacteria lacked this ability (Loeb, 1903). It is now clear that what Loeb observed was the combined effect of clumping factor A (a fibrinogen-binding protein that will not be discussed in this review) and SC. Three decades later, Chapman et al. proved that the majority of *S. aureus* strains are coagulase-positive (88% of 690 strains tested), and when these coagulase-producing strains where injected into rabbits they caused localized necrosis at the injection site (Shwartzman reaction), and were lethal within 10 days (Chapman et al., 1934). These findings made it apparent that *S. aureus* produced a plasma clotting factor and that this factor may impart virulence to *S. aureus,* potentially useful as a diagnostic marker for identifying the most dangerous staphylococcal strains.

Initial fractionation studies used to purify SC were hampered by poor yields (Tager and Drummond, 1965) and subsequent degradation to NH_2-terminal products ranging from 5 (Murray and Gohdes, 1960) to 44 kDa (Duthie and Haughton, 1958), which later proved important in establishing concepts regarding SC function. Miale determined that SC clotting of plasma was calcium ion-independent and impervious to the thrombin inhibitor, heparin (Miale, 1949a), which was later confirmed by others (Hendrix et al., 1983; Kawabata et al., 1986c; Soulier and Prou-Wartelle, 1967). Tager et al. also determined that the active SC·pro(thrombin)* complex was inactivated by diisopropylfluorophosphate (DFP), but not by soybean trypsin inhibitor (Drummond and Tager, 1962). Generation of anti-SC antibodies was also complicated by the toxicity of injected SC, which caused rapid defibrinogenation of the blood and fibrin clots deposited in the lungs of rabbits (Duthie and Haughton, 1958). Tager noted that when purified SC was injected intravenously into rabbits "...rapid death ensued" (Tager, 1948), reminiscent of *S. aureus* animal studies (Murray and Gohdes, 1960). Similarly, mice injected with lethal doses of SC had pronounced tissue edema and SC deposits in the lungs, as shown by immunostaining (Blobel et al., 1960). These early studies underscored the importance that SC may play in causing disease, and that further investigation was warranted to understand the molecular mechanism by which SC clots plasma.

Pioneering Mechanistic Studies

The pioneering work of Tager showed that partially purified SC could not act on fibrinogen alone, but required another plasma factor, later identified as ProT (Tager, 1956). Hemker et al. demonstrated in titrations of SC with ProT that the concentration of SC was the limiting factor for generation of enzymatic activity, indicating that SC was not acting as an enzyme, and that it most likely formed a stoichiometric complex with ProT (Hemker et al., 1975). Expanding upon these findings, Hendrix, et al. in 1983 confirmed that activation of ProT followed the formation of "... a very stable equimolar complex". It was also determined that the interaction of SC and ProT was of very high affinity, later confirmed by our studies (Panizzi et al., 2006a).

Activation of Serine Proteinase Zymogens

Meanwhile, foundational studies of serine proteinase zymogen activation were being done. Trypsinogen and chymotrypsinogen, the prototypical serine proteinase zymogens, were studied extensively to determine the mechanisms involved in proteolytic conversion to their enzyme forms. Activation requires cleavage of a single peptide bond in the NH_2-terminal activation segment of the zymogen, generating a new NH_2-terminus (Ile^{16}, chymotrypsinogen numbering) that forms a salt bridge with Asp^{194} to induce formation of the active site (reviewed by Khan and James, 1998). Both trypsinogen and chymotrypsinogen exhibit a very low level of activity, indicating that the enzymatic machinery is present in a similar orientation as in the activated enzyme. Trypsinogen was found to have self-activating ability, even in the presence of the specific active-site inhibitors, DFP and soybean trypsin inhibitor (Kay and Kassell, 1971). Further, the active site residue Ser^{195} in bovine trypsinogen and chymotrypsinogen could be alkylphosphorylated by DFP, consistent with a functional conformation of the catalytic triad. However, the reduced rate of inactivation implied misfolding or inaccessibility of the active site to the inhibitor (Morgan et al., 1972). Comparison of the crystal structures of chymotrypsinogen and α-chymotrypsin uncovered the relatively small conformational shifts that were needed to properly form the substrate-specificity cleft and oxyanion hole, supporting a model of activation mediated by slight yet precise movements of specific loops of the zymogen (Freer et al., 1970).

Functional Studies with Proteolytic SC Fragments

Proteolytic fragmentation of SC was performed to isolate the specific regions of SC capable of activating ProT (Kawabata et al., 1986b). From the initial 66-kDa protein, chymotryptic digestion produced two primary fragments with apparent molecular masses (M_r) of 43 and 30 kDa, and a 20-kDa fragment. The 43-kDa fragment (calculated M_r of 38 kDa, corresponding to residues 1–324) displayed clotting activity equivalent to that of intact SC, whereas the 30-kDa and 20-kDa forms, although contained within the longer fragment, had undetectable activity (Kawabata et al., 1986b). These differences, and in particular the ability of the 20-kDa fragment to mediate binding but not activation of ProT, indicated that the functions of SC are localized to discrete regions within the protein (Kawabata et al., 1986b).

Autocatalytic Cleavage of SC·ProT Complexes*

Autocatalytic cleavage of ProT to generate prethrombin 1 (Pre 1) and release fragment 1 was demonstrated in studies of mixtures of SC and ProT. In experiments at low temperature, the clotting activity of SC-ProT mixtures quickly reached a maximum, but amidase activity developed slower and seemed to be associated with production of Pre 1, suggesting that a SC·Pre 1* complex might be the active species (Kawabata et al., 1985b). Pre 1 formation is consistent with thrombin-like activity of the SC·ProT* complex, because removal of fragment 1 can be accomplished by both thrombin and the active ProT intermediate, meizothrombin (Downing et al., 1975; Esmon et al., 1974). The proteolytic pathway for this behavior was further characterized in a more recent study, which monitored proteolytic cleavage of mixtures of SC with both native ProT and active site-inhibited ProT (Panizzi et al., 2006a). ProT was converted first to Pre 1, signifying intermolecular proteolysis of the complexes, but much slower generation of another proteolytic derivative, called prethrombin 2′ (Pre 2′), was also identified. This zymogen species has also been described as an autocatalytic product of meizothrombin (Petrovan et al., 1998). The significance of proteolytic conversion of ProT to Pre 1 and Pre 2′ is unknown, but it may control localization of the SC·ProT* and SC·Pre 1*/Pre 2′* complexes by limiting their binding to phospholipid membranes.

Non-Proteolytic Activation of Zymogens

The concept of non-proteolytic activation of a serine proteinase zymogen was first proposed by Wolfram Bode and Robert Huber in their seminal structural investigations on the activation of trypsinogen. Linkage between the NH_2-terminal Ile^{16} binding pocket and active site formation was illustrated in a number of crystal structures of trypsin(ogen) in both inhibited and free forms (Huber and Bode, 1978). The role of the new NH_2-terminus in coordinating the transition to active proteinase was examined through covalent modification of the active site of trypsinogen with *p*-guanidinobenzoate, which allowed a peptide corresponding to the trypsin NH_2-terminus (Ile-Val) to bind with high affinity to the zymogen. This behavior is consistent with cooperative linkage between the active site and the NH_2-terminal binding pocket. Comparing circular dichroism spectra, the resulting changes in trypsinogen conformation strongly resembled that of trypsin, supporting a highly unfavorable equilibrium between the zymogen and enzyme forms (Bode et al., 1976). This transition was detailed further in the crystal structure of bovine trypsinogen inhibited by pancreatic trypsin inhibitor, which showed a conformation almost exactly like that of trypsin. The presence of an Ile-Val peptide also stabilized the structure of the inhibited proteinase-like zymogen (Bode et al., 1978; Huber and Bode, 1978). The relationship between high-affinity active site binding, NH_2-terminal insertion, and functional rearrangement of the active site indicated that the transition to active proteinase could occur without the strict requirement for proteolytic cleavage of the activation loop.

Based on these observations, a mechanism of zymogen activation, termed "molecular sexuality", was proposed, which relied on conformational changes produced by insertion of the NH_2-terminus of an activator protein into the NH_2-binding cleft of a zymogen. This was originally hypothesized to be the means of plasminogen activation by the secreted streptococcal protein, streptokinase, after realizing that the native Ile-Ala-Gly NH_2-terminus mimics that of plasmin, Val-Val-Gly (Bode and Huber, 1976) (see Chapter 25). Despite the almost strict conservation of Ile-Val (or Val-Val) NH_2-terminal peptides in serine proteinases, the potential role of the homologous Ile-Val NH_2-terminus of SC in ProT activation was never taken into account.

SC Structure, Function, and Mechanism

Substrate Specificity and Inhibitor Profile of SC·ProT Complex*

Characterization of the SC·ProT* complex has demonstrated some common activities with thrombin, such as clotting of fibrinogen, but also several marked differences. Differential coagulation of plasma from various species had been recognized in a number of analyses of the procoagulant activity of staphylococci. Coagulases from human pathogenic *S. aureus* strains showed a strong preference for human and pig plasma, with lesser clotting seen in other species such as cow, rabbit, and mouse (Duthie and Haughton, 1958; Pijoan, 1935). The inability of some common inhibitors of thrombin production or activity, such as sodium oxalate, sodium fluoride, and heparin, to affect SC coagulation gave the first indication that the thrombin-like factor did not behave in the same manner as physiological thrombin (Miale, 1949b). In contrast, small molecule inhibitors (DFP) or substrates (Tosyl-Arg-methyl ester) could access the putative active site on the procoagulant complex (Drummond and Tager, 1962).

A thorough survey of thrombin substrates demonstrated the restricted specificity of the SC·ProT* complex, which was unable to activate coagulation factors V and VIII or platelets, and was not inhibited by antithrombin, α_2-macroglobulin, heparin, or hirudin (Soulier and Prou-Wartelle, 1967). This study also indicated that coagulase could not activate the transglutaminase, factor XIII, which is responsible for covalent cross-linking of fibrin, but this conclusion was contested in later research (Kawabata and Iwanaga, 1994). SC was also shown to protect thrombin but not factor Xa from inhibition by antithrombin (Hendrix et al., 1983). Other coagulation factors (factor X, factor IX, protein C, protein S, and protein Z) could not inhibit formation of the SC·ProT* complex, underscoring the specificity of SC for the thrombin zymogen (Kawabata et al., 1985c). Early enzymatic characterization of the SC complex revealed decreased activity on fibrinogen and small synthetic thrombin substrates (Kawabata et al., 1985a, b), although later investigations found no significant differences, particularly toward fibrinogen (Panizzi et al., 2006b). Together, the altered substrate and inhibitor profile of SC·ProT* and SC·thrombin complexes suggests that binding of the bacterial activator modifies the conformation of the active site to restrict access to certain ligands, and/or sterically blocks interaction of other effectors with the conformationally activated zymogen.

Structural and Functional Evidence in Support of the Molecular Sexuality Mechanism

The molecular mechanism of ProT activation by SC has been conclusively described in recent structural investigations. The fully active NH_2-terminal region of SC (SC(1–325)) was crystallized in complex with α-thrombin and the structure refined at 2.2 Å resolution, allowing visualization of the structure of SC for the first time (Friedrich et al., 2003). In this initial structure and that of the SC(1–325)•bovine thrombin complex refined at lower resolution (Friedrich et al., 2006) the thrombin NH_2-terminus occupies the activation pocket, while the NH_2-terminus of SC is disordered. Intrigued by the proximity of the SC NH_2-terminus to the Ile^{16} pocket, the definitive SC•Pre 2* complex structure was solved. In this structure, the SC NH_2-terminus inserts directly into the NH_2-terminal binding cleft of the zymogen, and SC Ile^1 makes the crucial salt bridge with Asp^{194} that is typically formed by the native NH_2-terminus of active proteinases (Fig. 32.1).

Biochemical studies using SC(1–325) mutants that had Ile^1 or Ile^1-Val^2 deleted, or had an additional Met residue before Ile^1, verified the importance of the native NH_2-terminus. Loss of the first residue decreased the affinity of SC for ProT ~6-fold, while deletion of both Ile^1 and Val^2 resulted in a >98% loss of activity. In contrast, the presence of an initiating Met did not completely block activation, but resulted in ~60-fold lower affinity (Friedrich et al., 2003). These findings indicate a substantial degree of flexibility within the binding cleft, which can accommodate peptides that are one residue longer or shorter than that of the natural activator. The discovery of this thrombin-like structure generated by SC binding to Pre 2, and the

Fig. 32.1 Crystal structure of SC(1–325)·Pre 2 and NH_2-terminal insertion. *Left panel*: Prethrombin 2 (*green/blue*) is in standard thrombin orientation. The NH_2-terminus of SC(1–325) (*orange/red*) can be seen inserting into the *lower left-hand side* of Pre 2. D1 is on the *left* and D2 is on the *right*. *Right panel* shows the completely ordered NH_2-terminus of SC in the Pre 2 NH_2-terminal binding cleft, where Ile^1 forms a salt bridge with Asp^{194} of Pre 2 to trigger conformational rearrangement of the zymogen. *Right panel* reproduced from Friedrich et al. (2003) with permission from *Nature*

requirement for the native NH_2-terminal residues of SC for complete activation of ProT, stands as the first direct evidence for the molecular sexuality mechanism of non-proteolytic zymogen activation.

Previous secondary structure predictions had identified potential α-helical segments in SC, distinct from the predominantly β-sheet organization of the streptococcal protein, streptokinase (Kawabata et al., 1986a). This was confirmed by the SC complex structures, which revealed a unique two-domain fold for the activator. The first NH_2-terminal domain (D1) contains residues 1–142, and is composed of three α-helices in a slightly-twisted bundle, which can participate in formation of a dimeric complex through association with the D1 domain of another SC molecule, as seen in the crystal structure (Friedrich et al., 2003). The D2 domain (residues 150–281) also contains three primary helices, with three smaller helical segments at the distal end.

Together, the two domains are positioned at angle around the catalytic domain of (pre)thrombin (Fig. 32.1), where they contact the thrombin anion-binding exosite I and the 148 loop (autolysis loop). The insertion site for the SC NH_2-terminus is located near to the flexible 148 loop, and this region is significantly distorted by the presence of the SC molecule compared to free thrombin. Residues essential for effector binding are located in exosite I, and D2 of SC interacts with a number of thrombin residues in this area, including Tyr^{76} and several arginines. In addition, domain mutants of SC indicated that binding and activation could be mediated by spatially distinct regions of SC (Friedrich et al., 2003). Domain 1 alone was found to activate ProT, but with considerably lower affinity, while D2 induced no activity in the zymogen, but has ~130-fold higher affinity for ProT than D1, implying a primary function in amplifying binding affinity (Panizzi et al., 2006a).

In accordance with the structural findings, SC was found to compete with a (pro)exosite I-specific ligand (hirudin (54–65)) for binding to either ProT or thrombin, suggesting that the interaction with SC is capable of altering the native ligand recognition capability of the zymogen and proteinase. Exosite I is often referred to as the fibrinogen recognition exosite, as interaction of fibrinogen with a number of residues in this region is required for fibrinopeptide release, such that blockage of exosite I prevents fibrin formation. The potent procoagulant activity of the SC·ProT* complex in spite of an obstructed exosite indicates that a novel fibrinogen-recognition exosite is generated within the complex. The complex binds fibrinogen with ~200–900 fold higher affinity than α-thrombin, with a stoichiometry that indicates binding of two SC·Pre 2* complexes per fibrinogen dimer (Panizzi et al., 2006b). Molecular modeling of a fibrinogen molecule with the putative dimerized SC(1–325)·Pre 2* complex proposed a pentameric structure (Panizzi et al., 2006b), with the two active sites of the complexes straddling the E region of fibrinogen for efficient cleavage and release of fibrinopeptides A and B (Binnie and Lord, 1993; Stubbs et al., 1992). Additional interactions with the COOH-terminal half of SC, a region containing 7 repeats of 27 residues each, may also contribute to presentation of the scissile bonds in fibrinogen, as well as to overall affinity (see below).

Von Willebrand Factor-Binding Protein: Structure, Function, and Mechanism

Determination of the three-dimensional structure of SC led to identification of several other proteins from staphylococcal and streptococcal species that possess primary and secondary structural similarity to the SC D1–D2 domains, as well as putative adhesion protein binding motifs in their COOH-terminal regions. The domain organization reflects the potential bifunctionality of these proteins, and along with SC, they were grouped into a new family called the zymogen activator and adhesion proteins (ZAAPs; Fig. 32.2) (Friedrich et al., 2003; Panizzi et al., 2004). The first additional member of this family to be characterized was von Willebrand factor-binding protein (VWbp), a secreted protein from *Staphylococcus aureus*. VWbp had been first isolated through its specificity for binding von Willebrand factor (VWF), a large oligomeric protein that is responsible for tethering platelets to sites of vascular injury (Bjerketorp et al., 2002). A 26-amino acid sequence in the COOH-terminal half of VWbp (residues 325–351, *S. aureus* strain Mu50) mediates the interaction with VWF, but this threonine- and glutamine-rich region shows no recognizable similarity to any known VWF-binding motifs. Soon after its discovery, VWbp was functionally linked to SC through both its predicted structural homology and ProT-dependent procoagulant activity (Friedrich et al., 2003; Bjerketorp et al., 2004; Kroh et al., 2009; Panizzi et al., 2004).

The reasoning for SC as the structural and functional prototype for the ZAAPs was validated through rigorous biochemical examination of the molecular and kinetic behavior of VWbp in complex with human ProT. Specific activation

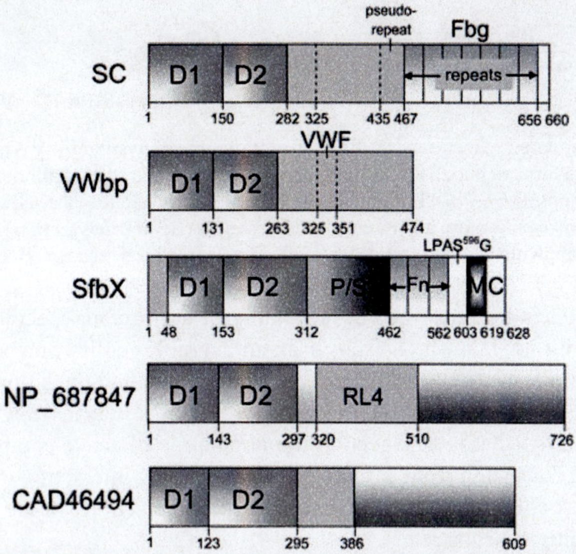

Fig. 32.2 Domain organization of the ZAAPs (zymogen *a*ctivator and *a*dhesion *p*roteins). Proposed members contain structurally homologous D1–D2 domains (NH_2-terminal) and putative adhesion protein binding motifs (COOH-terminal). SC and VWbp are secreted by *Staphylococcus aureus* and recognize fibrinogen (Fbg) and von Willebrand factor (VWF), respectively. SfbX, from *Streptococcus pyogenes*, contains a cell-wall binding motif (LPXTG) and fibronectin (Fn) recognition site. The uncharacterized proteins NP_687847 and CAD46494 are secreted by *Streptococcus agalactiae*, and their adhesion protein specificities have yet to be identified. Reproduced in modified form Panizzi et al. (2004) with permission from *Cellular and Molecular Life Sciences*

of ProT by VWbp was demonstrated, including generation of an active site on the zymogen and thrombin-like enzymatic activity. Like SC, the first two NH_2-terminal residues of VWbp (Val-Val) were necessary for activation of both ProT and Pre 1, and VWbp successfully competed with a fluorescently-labeled hirudin(54–65) peptide for binding to (pro)exosite I (Kroh et al., 2009). Together with the predicted secondary structure, these findings indicate that VWbp operates through the same non-proteolytic, molecular sexuality mechanism of activation as SC.

However, significant differences in the kinetic mechanism of activation were observed in assays with thrombin-specific tripeptide substrates. The initial rates for cleavage of the chromogenic substrate D-Phe-Pip-Arg-*p*NA by SC·ProT* are consistently linear, but with VWbp there is an exponential increase in the rate of substrate cleavage over the time of the assay (minutes). This effect was completely dependent upon the addition of substrate, implicating an uncommon kinetic process called hysteresis in the mechanism of ProT activation by VWbp. Hysteresis is classically defined as a delayed response of an enzyme to a sudden stimulus, resulting in a lag in the observed reaction, an effect that can be produced by processes including slow isomerization or ligand displacement (Frieden, 1970). The

Fig. 32.3 The hysteretic kinetic mechanism of prothrombin activation by VWbp. Prothrombin (ProT) and VWbp form an initial, low affinity complex (K_D), which is in equilibrium with a higher affinity, active complex (k_{C1}/k_{C2}). Binding of a highly-specific substrate to the active complex (K_m) favorably shifts the equilibrium, and increases the observed rate of catalysis (k_{cat}). Reproduced in modified form from Kroh et al. (2009). ©2009 H.K. Kroh, P. Panizzi, and P.E. Bock

particular trigger in this case was tight binding of a thrombin-specific substrate to the VWbp·ProT complex, which exists in an unfavorable equilibrium between active and inactive forms. The slow shifting of the equilibrium and transition to the stabilized, substrate (S)-bound active form, VWbp·S·ProT*, results in a low initial rate of substrate hydrolysis that is enhanced as the population of active complex increases (Fig. 32.3). Such a switch from a relatively weak binding interaction to a very high-affinity complex could serve to regulate and restrict the activity of the complex to a specific physiological substrate.

Substrate Specificity and Inhibitor Profile of VWbp·ProT Complex*

The substrate and inhibitor specificity of VWbp-bound ProT showed a highly restricted profile, similar to that found with SC. VWbp alone or in complex with ProT could not directly activate protein C, factor V, factor IX, factor X, or factor XII, and the enzymatic activity of the VWbp·ProT* complex could not be inhibited by the antithrombin·heparin or heparin cofactor II·heparin complexes, consistent with limited specificity compared to free thrombin. The capacity of VWbp to rapidly produce fibrin clots in human plasma indicates a specific reaction with fibrinogen. Distinct hysteretic behavior in the generation of fibrin was demonstrated in assays with purified proteins, which in combination with VWbp-VWF binding could place temporal and spatial limits on fibrin formation, and emphasizes the potential role for hysteresis and VWF in the overall function of VWbp as a virulence factor.

Animal Models of Human Staphylococcal Diseases

S. aureus infections can lead to acute, severe life-threatening diseases, including most notably, bacteremia, sepsis, endocarditis (Durack, 2001; Korzeniowski and Kaye, 1992), pulmonary infections (Sawai et al., 1997), and mastitis (Haraldsson and Jonsson, 1984; Jonsson et al., 1985). Conflicting results have been derived from animal studies regarding whether SC is a virulence factor for *S. aureus* diseases. Some studies have shown that SC is not involved in the initial attachment stages of endocarditis (Baddour, 1996; Baddour et al., 1994; Moreillon et al., 1995;

Stutzmann Meier et al., 2001), but others have found that SC-deficient mutants had decreased virulence in mouse models of mastitis (Haraldsson and Jonsson, 1984; Jonsson et al., 1985) and in the development of pulmonary infections (Sawai et al., 1997). It can be argued that several of the endocarditis studies focused only on the initial attachment phase of the pathogen to the damaged or denuded endothelium, which is largely mediated by clumping factor A binding to fibrinogen (Stutzmann Meier et al., 2001), and as a result have underestimated the contribution of SC to fibrin-bacterial vegetations that develop during endocarditis (Durack, 1975, 2001; Korzeniowski and Kaye, 1992). This is partly due to the lack of histology confirming or refuting virulence factor expression, and the absence of detailed investigation of the impact that loss of SC has on the morphology of the vegetation and thereby, the pathogenesis of *S. aureus* endocarditis. Contrary to this, Sawai clearly demonstrated in a lung infection model that the bacteria may be responsible for the deposition of a fibrin barrier that appeared to segregate neutrophils and other blood leukocytes to areas distal from the bacterial core abscess. Despite these findings, there is still only a cursory understanding of the in vivo function of SC. The pattern of SC expression in the context of disease models needs to be determined, and it remains to be proven that SC function can be altered in vivo. It is, however, clear that *S. aureus* strains that express little or no SC are significantly less pathogenic (Hasegawa and San Clemente, 1978; Masuda, 1983; Seki et al., 1989). Proteomic profiling of the community-associated methicillin-resistant *S. aureus* secretome indicated that SC was one of eleven secreted proteins that were differentially expressed (Burlak et al., 2007). These differences in the expression of SC and other proteins may result from varied infectivity potential or pathogenicity of certain strains, but the functional significance and cause of the differing expression levels are unknown. Detailed in vivo studies of *S. aureus* diseases are necessary to better understand the role of SC in these diseases, its expression in bacterial abscesses or vegetations, and whether inhibiting SC deposition or the activity of the SC·ProT* fibrin-generating complex is an attractive drug discovery target for treating or preventing staphylococcal diseases.

Summary and Perspectives

Current therapy for *S. aureus* endocarditis relies on intravenous antibiotics, but typically, vegetations are identified only after they have reached a significant size (Cabell and Fowler, 2004). In these cases, valve function would have begun to decline and the risk of vegetation embolism would require immediate surgical intervention. Targeting the SC·(pro)thrombin* complex for mechanism-based inhibitor design is complicated by the remarkable resistance of the activator complexes to the physiologic thrombin inhibitors, heparin, antithrombin, heparin cofactor II (Hendrix et al., 1983; Kawabata and Iwanaga, 1994; Soulier and Prou-Wartelle, 1967; I.M. Verhamme, P. Panizzi, and P.E. Bock, "unpublished results"), and anticoagulant analogs of hirudin. This inhibitor resistance is a result of the fact that SC completely covers exosite I on thrombin, and the precursor form of exosite I, proexosite I (Anderson et al., 2000) on Pre 2 and ProT, which blocks the ability of these inhibitors to regulate thrombin function. The small molecule, active site-directed thrombin inhibitor, argatroban, has been shown to inhibit the SC·ProT* complex (Hijikata-Okunomiya and Kataoka, 2003), but there are no data available concerning the efficacy of this inhibitor in vivo, or whether it can slow vegetation growth without deleterious bleeding. Inhibitory antibodies are also an attractive alternative to small molecules, given their much longer blood half-life and tissue clearance. The use of antibodies to inhibit SC function is not a new idea. Tager in 1948 assayed the plasma of 307 volunteers and observed that 93% were protected from SC-induced clot formation, but the mechanism of protection in these individuals is unclear.

Although the mechanism of ProT activation by SC has been established, there are a number of unanswered mechanistic questions concerning: (a) fibrinogen binding as a substrate to the SC·(pro)thrombin* complex through the expressed fibrinogen-recognition exosite; (b) fibrinogen binding as a potential ligand to localize the SC·(pro)thrombin* complex via the COOH-terminal SC repeat sequences; (c) the variability in the number of repeats among SC from different strains, and their contribution to fibrinogen binding and ProT localization in vivo; and (d) whether new mechanism-based inhibitors of the SC·(pro)thrombin* complex can be developed to decrease the virulence of *S. aureus*. These questions highlight the current gaps in understanding SC function that need to be addressed.

SC can bind fibrinogen through two distinct surfaces, located in the amino terminal 1–325 region and the enigmatic 27 residue repeat sequences at the COOH-terminal end. Historically, the number of repeats has been used to categorize *S. aureus* strains into groups with less genetic variability (Watanabe et al., 2005, 2009), but with no phenotypic understanding of the impact of these repeat variations, little can currently be drawn from these studies. The SC(1–325)·(pro)thrombin* complex can only interact with fibrinogen as a specific substrate to generate fibrin, due to the lack of the COOH-terminal portion of SC that may function independently or interdependently in ProT binding (see Fig. 32.2; Heilmann et al., 2002). Our studies indicate that the SC repeat sequences bind fragment D of fibrinogen (P. Panizzi, A. Maddur, and P.E. Bock, "unpublished results"). Extracellular fibrinogen-binding (Efb) protein, also expressed by pathogenic strains of *S. aureus*, contains two NH_2-terminal 22-residue repeat sequences similar to the 27-residue SC repeats, linked by a 9-residue spacer (Palma et al., 2001). The Efb repeats bind to fibrinogen fragment D, and may compete with the SC repeat sequences (Palma et al., 2001). It is intriguing to speculate that the repeat sequences of SC and Efb may interact similarly to the fibronectin-binding repeats of fibronectin-binding protein A in a β-zipper binding mechanism (Meenan et al., 2007; Schwarz-Linek et al., 2003). Further investigation will be necessary to establish the stoichiometry and affinity of fibrinogen binding to the SC repeats, to identify the binding site(s) on fibrinogen, and to elucidate the structural basis for the interactions.

With the discovery of VWbp and its unique, kinetically-controlled activation mechanism, it is apparent that reassessment of SC and VWbp as dual ProT activators secreted by *S. aureus* in vivo is necessary, especially with regard to their individual contributions to virulence of *S. aureus* in endocarditis. The generation of virulence factor knock-outs has been hampered by poor success rates of homologous recombination of shuttle plasmids (Gotz et al., 1981; Phonimdaeng et al., 1988), when subsequently transduced into the *S. aureus* strain of interest. Creation and in vivo testing of knock-out strains deficient in SC and/or VWbp may be more straightforward with new technology and the advent of the targetron system for rapid knock-out generation (Yao et al., 2006), but this remains to be seen. Assessment of whether these dual ProT activators are expressed in vivo in endocarditis and other staphylococcal diseases, as well as their localization in growing vegetations, will provide new insight into their involvement in the pathobiology of these conditions, in which they may play important roles.

Acknowledgements Research from the authors' laboratories was supported by National Institutes of Health Grant R37 HL071544 from the Heart, Lung, and Blood Institute to P. E. B., F32 HL094010 and K99 HL094533 to P. P., and SAF2007-64140 from Spanish Ministerio de Ciencia e Innovación (MICINN) to P. F.-P. H.K.K. was supported in part by NIH Training Grant T32 HL07751.

References

Anderson, P.J., Nesset, A., Dharmawardana, K.R., Bock, P.E., 2000. Characterization of proexosite I on prothrombin. J. Biol. Chem. 275, 16428–16434.

Baddour, L.M., 1996. A trojan horse mechanism in the pathogenesis of infective endocarditis: a hypothesis. Zentralbl. Bakteriol. 283, 266–270.

Baddour, L.M., Tayidi, M.M., Walker, E., McDevitt, D., Foster, T.J., 1994. Virulence of coagulase-deficient mutants of *Staphylococcus aureus* in experimental endocarditis. J. Med. Microbiol. 41, 259–263.

Binnie, C.G., Lord, S.T., 1993. The fibrinogen sequences that interact with thrombin. Blood 81, 3186–3192.

Bjerketorp, J., Jacobsson, K., Frykberg, L., 2004. The von Willebrand factor-binding protein (vWbp) of *Staphylococcus aureus* is a coagulase. FEMS Microbiol. Lett. 234, 309–314.

Bjerketorp, J., Nilsson, M., Ljungh, A., Flock, J.I., Jacobsson, K., Frykberg, L., 2002. A novel von Willebrand factor binding protein expressed by *Staphylococcus aureus*. Microbiology 148, 2037–2044.

Blobel, H., Berman, D.T., Simon, J., 1960. Purification of staphylococcal coagulase. J. Bacteriol. 79, 807–815.

Bode, W., Fehlhammer, H., Huber, R., 1976. Crystal structure of bovine trypsinogen at 1.8 Å resolution. I. Data collection, application of Patterson search techniques and preliminary structural interpretation. J. Mol. Biol. 106, 325–335.

Bode, W., Huber, R., 1976. Induction of the bovine trypsinogen-trypsin transition by peptides sequentially similar to the N-terminus of trypsin. FEBS Lett. 68, 231–236.

Bode, W., Schwanger, P., Huber, R., 1978. The transition of bovine trypsinogen to a trypsin-like state upon strong ligand binding: the refined crystal structures of the bovine trypsinogen-pancreatic trypsin inhibitor complex and of its ternary complex with Ile-Val at 1.9 Å resolution. J. Mol. Biol. 118, 99–112.

Burlak, C., Hammer, C.H., Robinson, M.A., Whitney, A.R., McGavin, M.J., Kreiswirth, B.N., Deleo, F.R., 2007. Global analysis of community-associated methicillin-resistant *Staphylococcus aureus* exoproteins reveals molecules produced in vitro and during infection. Cell Microbiol. 9, 1172–1190.

Cabell, C.H., Fowler, V.G., Jr., 2004. Repeated echocardiography after the diagnosis of endocarditis: too much of a good thing? Heart 90, 975–976.

Chapman, G.H., Berens, C., Peters, A., Curcio, L., 1934. Coagulase and hemolysin tests as measures of the pathogenicity of *Staphylococci*. J. Bacteriol. 28, 343–363.

Downing, M.R., Butkowski, R.J., Clark, M.M., Mann, K.G., 1975. Human prothrombin activation. J. Biol. Chem. 250, 8897–8906.

Drummond, M.C., Tager, M., 1962. Enzymatic activities associated with clotting of fibrinogen by staphylocoagulase and coagulase-reacting factor and their inhibition by disopropylfluophosphate. J. Bacteriol. 83, 975–980.

Durack, D. 2001. Infective and noninfective endocarditis, in: Hurst, J.W., Sonnenblick, E.H., Wenger, N.K. (Eds.), The Heart: Arteries and Veins (vol. 63). McGraw-Hill, New York, pp. 1230–1255.

Durack, D.T., 1975. Experimental bacterial endocarditis. IV. Structure and evolution of very early lesions. J. Pathol. 115, 81–89.

Duthie, E.S., Haughton, G., 1958. Purification of free staphylococcal coagulase. Biochem. J. 70, 125–134.

Esmon, C.T., Owen, W.G., Jackson, C.M., 1974. The conversion of prothrombin to thrombin II. Differentiation between thrombin- and factor Xa-catalyzed proteolyses. J. Biol. Chem. 249, 606–611.

Freer, S.T., Kraut, J., Robertus, J.D., Wright, H.T., Xuong, N.H., 1970. Chymotrypsinogen: 2.5 Å crystal structure, comparison with α-chymotrypsinogen, and implications for zymogen activation. Biochemistry 9, 1997–2009.

Frieden, C., 1970. Kinetic aspects of regulation of metabolic processes. The hysteretic enzyme concept. J. Biol. Chem. 245, 5788–5799.

Friedrich, R., Panizzi, P., Fuentes-Prior, P., Richter, K., Verhamme, I., Anderson, P.J., Kawabata, S., Huber, R., Bode, W., Bock, P.E., 2003. Staphylocoagulase is a prototype for the mechanism of cofactor-induced zymogen activation. Nature 425, 535–539.

Friedrich, R., Panizzi, P., Kawabata, S., Bode, W., Bock, P.E., Fuentes-Prior, P., 2006. Structural basis for reduced staphylocoagulase-mediated bovine prothrombin activation. J. Biol. Chem. 281, 1188–1195.

Gotz, F., Ahrne, S., Lindberg, M., 1981. Plasmid transfer and genetic recombination by protoplast fusion in staphylococci. J. Bacteriol. 145, 74–81.

Haraldsson, I., Jonsson, P., 1984. Histopathology and pathogenesis of mouse mastitis induced with *Staphylococcus aureus* mutants. J. Comp. Pathol. 94, 183–196.

Hasegawa, N., San Clemente, C.L., 1978. Virulence and immunity of *Staphylococcus aureus* BB and certain deficient mutants. Infect. Immun. 22, 473–479.

Heilmann, C., Herrmann, M., Kehrel, B.E., Peters, G., 2002. Platelet-binding domains in 2 fibrinogen-binding proteins of *Staphylococcus aureus* identified by phage display. J. Infect. Dis. 186, 32–39.

Hemker, H.C., Bas, B.M., Muller, A.D., 1975. Activation of a pro-enzyme by a stoichiometric reaction with another protein: the reaction between prothrombin and staphylocoagulase. Biochim. Biophys. Acta 379, 180–188.

Hendrix, H., Lindhout, T., Mertens, K., Engels, W., Hemker, H.C., 1983. Activation of human prothrombin by stoichiometric levels of staphylocoagulase. J. Biol. Chem. 258, 3637–3644.

Hijikata-Okunomiya, A., Kataoka, N., 2003. Argatroban inhibits staphylothrombin. J. Thromb. Haemost. 1, 2060–2061.

Huber, R., Bode, W., 1978. Structural basis of the activation and action of trypsin. Accts. Chem. Res. 11, 114–122.

Jonsson, P., Lindberg, M., Haraldsson, I., Wadstrom, T., 1985. Virulence of *Staphylococcus aureus* in a mouse mastitis model: studies of alpha hemolysin, coagulase, and protein A as possible virulence determinants with protoplast fusion and gene cloning. Infect. Immun. 49, 765–769.

Kawabata, S.-I., Iwanaga, S., 1994. Structure and function of staphylothrombin. Semin. Thromb. Hemost. 20, 345–350.

Kawabata, S., Miyata, T., Morita, T., Miyata, T., Iwanaga, S., Igarashi, H., 1986a. The amino acid sequence of the procoagulant- and prothrombin-binding domain isolated from staphylocoagulase. J. Biol. Chem. 261, 527–531.

Kawabata, S., Morita, T., Iwanaga, S., Igarashi, H., 1985a. Difference in enzymatic properties between alpha-thrombin-staphylocoagulase complex and free alpha-thrombin. J. Biochem., Tokyo 97, 1073–1078.

Kawabata, S., Morita, T., Iwanaga, S., Igarashi, H., 1985b. Enzymatic properties of staphylothrombin, an active molecular complex formed between staphylocoagulase and human prothrombin. J. Biochem., Tokyo 98, 1603–1614.

Kawabata, S., Morita, T., Iwanaga, S., Igarashi, H., 1985c. Staphylocoagulase-binding region in human prothrombin. J. Biochem. 97, 325–331.

Kawabata, S., Morita, T., Miyata, T., Iwanaga, S., Igarashi, H., 1986b. Isolation and characterization of staphylocoagulase chymotryptic fragment. Localization of the procoagulant- and prothrombin-binding domain of this protein. J. Biol. Chem. 261, 1427–1433.

Kawabata, S., Morita, T., Miyata, T., Kaida, S., Igarashi, H., Iwanaga, S., 1986c. Difference in enzymatic properties between "staphylothrombin" and free alpha-thrombin. Ann. N.Y. Acad. Sci. 485, 27–40.

Kay, J., Kassell, B., 1971. The autoactivation of trypsinogen. J. Biol. Chem. 246, 6661–6665.

Khan, A.R., James, M.N.G., 1998. Molecular mechanisms for the conversion of zymogens to active proteolytic enzymes. Protein Sci. 7, 815–836.

Korzeniowski, O., Kaye, D. 1992. Infective endocarditis, in: Braunwald, E., (Ed.), Heart Disease. A Textbook of Cardiovascular Medicine. W.B. Saunders, Philadelphia, pp. 1078–1105.

Kroh, H.K., Panizzi, P., Bock, P.E., 2009. Von Willebrand factor-binding protein is a hysteretic conformational activator of prothrombin. Proc. Natl. Acad. Sci. U.S.A. 106, 7786–7791.

Loeb, L., 1903. The influence of certain bacteria on the coagulation of the blood. J. Med. Res. 10, 407–419.

Masuda, S., 1983. An efficient method for the isolation of a mutant with an extremely low producibility of coagulase from a *Staphylococcus aureus* strain. Microbiol. Immunol. 27, 801–805.

Meenan, N.A., Visai, L., Valtulina, V., Schwarz-Linek, U., Norris, N.C., Gurusiddappa, S., Hook, M., Speziale, P., Potts, J.R., 2007. The tandem beta-zipper model defines high affinity fibronectin-binding repeats within *Staphylococcus aureus* FnBPA. J. Biol. Chem. 282, 25893–25902.

Miale, J.B., 1949a. The role of staphylocoagulase in blood coagulation; coagulation in the absence of calcium and in the presence of fluorides, heparin, and azo dyes. Blood 4, 1317–1322.

Miale, J.B., 1949b. The role of staphylocoagulase in blood coagulation; the reaction of staphylocoagulase with coagulase-globulin to form coagulase-thrombin. Blood 4, 1039–1048.

Moreillon, P., Entenza, J.M., Francioli, P., McDevitt, D., Foster, T.J., Francois, P., Vaudaux, P., 1995. Role of *Staphylococcus aureus* coagulase and clumping factor in pathogenesis of experimental endocarditis. Infect. Immun. 63, 4738–4743.

Morgan, P.H., Robinson, N.C., Walsh, K.A., Neurath, H., 1972. Inactivation of bovine trypsinogen and chymotrypsinogen by diisopropylphosphorofluoridate. Proc. Natl. Acad. Sci. U.S.A. 69, 3312–3316.

Murray, M., Gohdes, P., 1960. Purification of staphylococcal coagulase. Biochim. Biophys. Acta 40, 518–522.

Palma, M., Shannon, O., Quezada, H.C., Berg, A., Flock, J.I., 2001. Extracellular fibrinogen-binding protein, Efb, from *Staphylococcus aureus* blocks platelet aggregation due to its binding to the α-chain. J. Biol. Chem. 276, 31691–31697.

Panizzi, P., Friedrich, R., Fuentes-Prior, P., Bode, W., Bock, P.E., 2004. The staphylocoagulase family of zymogen activator and adhesion proteins. Cell. Mol. Life Sci. 61, 2793–2798.

Panizzi, P., Friedrich, R., Fuentes-Prior, P., Kroh, H.K., Briggs, J., Tans, G., Bode, W., Bock, P.E., 2006a. Novel fluorescent prothrombin analogs as probes of staphylocoagulase-prothrombin interactions. J. Biol. Chem. 281, 1169–1178.

Panizzi, P., Friedrich, R., Fuentes-Prior, P., Richter, K., Bock, P.E., Bode, W., 2006b. Fibrinogen substrate recognition by staphylocoagulase.(pro)thrombin complexes. J. Biol. Chem. 281, 1179–1187.

Petrovan, R.J., Govers-Riemslag, J.W.P., Nowak, G., Hemker, H.C., Tans, G., Rosing, J., 1998. Autocatalytic peptide bond cleavages in prothrombin and meizothrombin. Biochemistry 37, 1185–1191.

Phonimdaeng, P., O'Reilly, M., O'Toole, P.W., Foster, T.J., 1988. Molecular cloning and expression of the coagulase gene of *Staphylococcus aureus* 8325–4. J. Gen. Microbiol. 134, 75–83.

Pijoan, M., 1935. A study on the blood-coagulating substance produced by staphylococci and its relation to disease. Can. Med. Assoc. J. 32, 476–481.

Sawai, T., Tomono, K., Yanagihara, K., Yamamoto, Y., Kaku, M., Hirakata, Y., Koga, H., Tashiro, T., Kohno, S., 1997. Role of coagulase in a murine model of hematogenous pulmonary infection induced by intravenous injection of *Staphylococcus aureus* enmeshed in agar beads. Infect. Immun. 65, 466–471.

Schwarz-Linek, U., Werner, J.M., Pickford, A.R., Gurusiddappa, S., Kim, J.H., Pilka, E.S., Briggs, J.A., Gough, T.S., Hook, M., Campbell, I.D., Potts, J.R., 2003. Pathogenic bacteria attach to human fibronectin through a tandem β-zipper. Nature 423, 177–181.

Seki, K., Ogasawara, M., Sakurada, J., Murai, M., Masuda, S., 1989. Altered virulence of a pleiotropic *Staphylococcus aureus* mutant with a low producibility of coagulase and other factors in mice. Microbiol. Immunol. 33, 981–990.

Soulier, J.P., Prou-Wartelle, O., 1967. Study of thrombin-coagulase. Thromb. Diath. Haemorrh. 17, 321–334.

Stubbs, M.T., Oschkinat, H., Mayr, I., Huber, R., Angliker, H., Stone, S.R., Bode, W., 1992. The interaction of thrombin with fibrinogen. A structural basis for its specificity. Eur. J. Biochem. 206, 187–195.

Stutzmann Meier, P., Entenza, J.M., Vaudaux, P., Francioli, P., Glauser, M.P., Moreillon, P., 2001. Study of *Staphylococcus aureus* pathogenic genes by transfer and expression in the less virulent organism *Streptococcus gordonii*. Infect. Immun. 69, 657–664.

Tager, M., 1948. Studies on the coagulase-reacting factor; the reaction of staphylocoagulase with the components of human plasma. Yale J. Biol. Med. 20, 369–380.

Tager, M., 1956. Studies on the nature and the purification of the coagulase-reacting factor and its relation to prothrombin. J. Exp. Med. 104, 675–686.

Tager, M., Drummond, M.C., 1965. Staphylocoagulase. Ann. N.Y. Acad. Sci. 128, 92–111.

Tager, M., Hales, H.B., 1948. Properties of coagulase-reacting factor, and relation to blood clotting components. J. Immunol. 60, 1–9.

Watanabe, S., Ito, T., Sasaki, T., Li, S., Uchiyama, I., Kishii, K., Kikuchi, K., Skov, R.L., Hiramatsu, K., 2009. Genetic diversity of staphylocoagulase genes (coa): insight into the evolution of variable chromosomal virulence factors in *Staphylococcus aureus*. PLoS One 4, e5714.

Watanabe, S., Ito, T., Takeuchi, F., Endo, M., Okuno, E., Hiramatsu, K., 2005. Structural comparison of ten serotypes of staphylocoagulases in *Staphylococcus aureus*. J. Bacteriol. 187, 3698–3707.

Yao, J., Zhong, J., Fang, Y., Geisinger, E., Novick, R.P., Lambowitz, A.M., 2006. Use of targetrons to disrupt essential and nonessential genes in *Staphylococcus aureus* reveals temperature sensitivity of Ll.LtrB group II intron splicing. RNA 12, 1271–1281.

Chapter 33
Procoagulant Properties of Plant Latex Proteases

H.V. Shivaprasad, R. Rajesh, M. Yariswamy, and B.S. Vishwanath

Abstract Plant derived molecules have found to be potential intervention with the human physiological events. The foundation laid for the search of pharmacologically active plant molecules is by the traditional knowledge which is passing from generation to generation. Latex of several plants obtained medicinal importance and is exclusively used as a common remedy to stop bleeding on fresh cuts and wound healing. The observed pharmacological activity of plant latices is attributed to the presence of proteolytic enzymes in it. Proteases found in plant latices belong to either cysteine or serine protease family, only one is a member of aspartate protease family. Plant latex proteases exhibit procoagulant action irrespective of the plant species and family. Their mechanism involved in the procoagulant action is yet to be understood clearly. The nature of procoagulant action of plant latices serine proteases is not clear. Where as cysteine proteases from plant latices exhibited specificities upon coagulation factors in inducing plasma coagulation. Ficin derived from *Ficus carica* shown to activate coagulation factor X. Cysteine proteases present in the latices of Asclepiadaceae plants have got thrombin like activity. Thrombin like activity of plant latex cysteine proteases is due to specific cleavage of fibrinogen molecules releasing fibrinopeptides. Apart from blood clot inducing property, both cysteine and serine proteases from plant latices have blood clot dissolving properties (Plasmin like activity). These specific actions of plant latex proteases on the coagulation cascade, could possibly utilized as biological tools in coagulation laboratory and as therapeutic agents.

B.S. Vishwanath (✉)
Department of Studies in Biochemistry, University of Mysore, Manasagangothri, Mysore 570 006, Karnataka, India
e-mail: vishmy@yahoo.co.uk

R.M. Kini et al. (eds.), *Toxins and Hemostasis*, DOI 10.1007/978-90-481-9295-3_33,

Introduction

World Health Organization (WHO) reported that about 80% of the African, 40% of Chinese, 70% of Canadian, 42% of American and 75% of French population meets its primary health needs through traditional medicines utilizing plant source (WHO Traditional Medicine Strategy, 2002–2005). The widespread use of plant based traditional medicine is not only because of lack of modern medical facilities available and the expensive medical care but also because they have proved to be effective in the maintenance and improvement of human health without causing adverse effects. Over centuries researchers are trying to explore the biochemical basis of the remedial effects of plant isolates, plants mixtures or components. Their findings in course of time ultimately gave rise to pharmacologically active molecules. It is not surprising that about 67–70% of modern medicines are derived from natural products (State of the Environment Report, 2001). Still there are significant amount of research is going on to explain the yet-to-be-revealed medicinal properties of plant and plant components.

Plant latices are milky exudates of several Angiosperms and the presence of latex is one of the important characteristic features of Euphorbiaceae, Apocynaceae, Asclepiadaceae and Moraceae families. The latices of several plants are of medicinal importance and are exclusively used as a common remedy to stop bleeding from fresh cuts and assist in wound healing. From past few years we are trying to understand the mechanism of plant components involved in stopping the bleeding and wound healing.

Traditional Application of Plant Latices to Stop Bleeding and Wound Healing

Latices from several plant species such as *Calotropis procera* and *C. gigantea* (Ascelpiadaceae), *Jatropa curcus* and *Euphorbia ligularia* (Euphorbiaceae), *Argemone mexicana* (Papaveraceae), *Abutilon indicum* (Malvaceae), *Litsea glutinosa* (Lauraceae), *Ficus hispida* (Moraceae), *Chromolaena odorata* (Asteraceae), *Cryptolepis buchanani* (Periplocaceae), and *Hyptis suaveolens* (Lamiaceae) are applied over fresh cuts to stop bleeding and subsequently applied over wounds to enhance healing (Ashwani, 1999; Begum and Nath, 2000; Osoniyi and Onajobi, 2003; Prusti and Behera, 2007; Reddy et al., 2003). So far, only a few plant latices used in folk medicine have been well studied and documented. Many developing countries have intensified their efforts in documenting the ethnomedical data on medicinal plants (Samy et al., 2008).

The pharmacological use of plant latices as hemostatic and wound healing agents imply their action on hemostasis and fibrinolysis. Hemostasis and fibrinolysis are two separate events regulated by the action of specific proteolytic enzymes. The observed pharmacological activity of plant latices is attributed to the presence of proteolytic enzymes in it (Rajesh et al., 2005).

Plant Latex Proteases

Winnick et al. (1939) first reported the presence of proteolytic enzymes in plant latex. Over 110 latices of different plant families are known to contain at least one proteolytic enzyme (Domsalla and Melzig, 2008). More than 80 proteases have been purified from latices of different plants belonging to various families and their utilization in industry is well known. Proteases appear to play key roles in the regulation of biological processes in plants; they are involved in growth and development, germination, circadian rhythms as well as in the recognition of pathogens, pests and the induction of effective defense responses (Konno et al., 2004). The precise biological roles of plant latex proteases still remain speculative because of their broad substrate specificity. A systematic study about the occurrence and properties of proteases correlated to plant families, a knowledge about their chemical properties aid for chemotaxonomic studies. Further, such characterization may help in understanding their ability to alter human physiological system.

In contrast, to all the four major classes of proteases found in the mammalian system, most proteases found in plant latices belong to either cysteine or serine protease family (Domsalla and Melzig, 2008). So far, only one is a member of aspartate protease family has been described (Domsalla and Melzig, 2008). A list of proteases isolated from plant latices belonging to Apocynaceae, Asclepiadaceae, Asteraceae, Caricaceae, Convolvulaceae, Euphorbiaceae and Moraceae families are shown in the Table 33.1. The presence of unique type of proteases in latices is characteristic to individual family. In Euphorbiaceae, Asteraceae, and Convolvulaceae only serine proteases are found. In Asclepiadaceae and Caricaceae there are only cysteine proteases and in Apocynaceae both cysteine and serine proteases have been detected. Members of the Moraceae contain serine, cysteine and the only aspartate protease (Domsalla and Melzig, 2008). Thus the determination of the type of latex proteases might aid in chemotaxonomy.

Several commercially available proteases are frequently used in food processing, tenderization of meat, brewing, cheese elaboration, bread manufacture and in leather and textile industries. Besides, some proteases are used as model system for studies on their structure-function relationships, and in the protein folding problem (Dubey and Jagannadham, 2003; Kundu et al., 2000; Patel and Jagannadham, 2003). Plant latices have also received attention in the pharmaceutical industry and biotechnology because of their pharmacological activities.

Isolation and Properties of Plant Latex Proteases

The purification methods of latex proteases often include ammonium sulfate or acetone precipitation, ion-exchange chromatography, gel filtration, affinity chromatography, and hydrophobic interaction chromatography. The molecular weights of plant latex proteases vary from 33 to 117 kDa. Majority of the serine proteases lies between 60 and 80 kDa, whereas cysteine proteases are in the range 21–29 kDa. Latex proteases are stable over broad pH range (3–12) and the temperatures

Table 33.1 List of proteases from plant latices

Family	Plant	Protease(s)	M W (kDa)	References
1. Cysteine proteases				
Apocynaceae	*Ervatamia coronaria* (Jacq.) Stapf.	Ervatamin A, B, C	27.6, 26, 23	Kundu et al. (2000); Nallamsetty et al. (2003); Patel and Jagannadham (2003)
	Ervatamia heyneana (Wall.) T. Cooke	Heynein	23	Patel and Jagannadham (2003).
	Philiberta gilliesii Hook. Et Arn.	Philibertain GI	23.53	Sequeiros et al. (2005)
Asclepiadaceae	*Araujia hortorum* Fourn.	Araujiain HI, HII, HIII	24.03, 23.718, 23.546	Priolo et al. (2000) Obregón et al. (2001)
	Asclepias curassavica L.	Asclepain CI	23.2	Liggieri et al. (2004)
	Asclepias fruticosa L.	Asclepain F	23.652	Trejo et al. (2001)
	Asclepias glaucescens H.B.K.	Asclepain G (10 forms)	Ag_3 22.6, Ag_6 23.5, Ag_7 23, Ag_8 23.5	Tablero et al. (1991)
	Asclepias speciosa Torr.	Asclepain S	–	Winnick et al. (1939)
	Asclepias syrica L.	Asclepains A3, A5	23, 21	Brockbank and Lynn (1979)
	Calotropis gigantea L. Dyrand	Calotropain DI, DII	23.8, 24.2	Sengupta et al. (1984)
	Funastrum clausum (Jacq.) Schlechter	Funastrain CII	23.636	Morcelle et al. (2004)
Asclepiadaceae	*Morrenia brachystephana* Griseb.	Morrenain BI, BII	23.205, 25.5	Cavalli et al. (2001, 2003)
	Morrenia odarata Hook et Arn.	Morrenain OII	25.8	Cavalli et al. (2001)
	Calotropis procera (Aiton) Dryand	Procerain	28.8	Dubey and Jagannadham (2003)
	Pergularia extensa	Pergularain e I	23.356	Shivaprasad et al. (2009)

Table 33.1 (continued)

Family	Plant	Protease(s)	M W (kDa)	References
Caricaceae	*Carica candamarcensis* Hook. f.	Endopeptidases CCI, CCII, CCIII, CCIV, CC28	23–28.6	De Moraes et al. (1994)
	Carica papaya L.	Carician	23.28	Barrett et al. (1998)
		Chymopapain	23.65	
		Glycyl-en-dopeptidase	23.313	
		Papain	23.429	
	Jacartia mexicana A.DC.	Mexicain	23.8	Gavira et al. (2007)
Moraceae	*Ficus carica*	Ficin	23	Devaraj et al. (2008a)
	Ficus carica var. Horaishi	Ficains A, B, C, D	24.0–26.0	Sugiura and Sasaki (1974)
	Ficus glabrata H.B.K.	Ficain	-	Barrett et al. (1998)
	Ficus hispida L. f.	Protease	-	Chetia et al. (1999)
	Ficus pumila L.	Ficains PI	28.6	Perello et al. (2000)
2. Serine proteases				
Apocynaceae	*Wrightia tinctoria* (Roxb.) R. Br.	Wrightin	79.5	Tomar et al. (2008)
Asteraceae	*Parthenium argentatum* A. Gray	Parthenain	63	Lynn (1985)
	Taraxacum officinale Webb s. l.	Taraxalisin	65	Rudenskaya et al. (1998)
Convolvulaceae	*Ipomoea carnea* ssp. *Fistulosa* (Mart.Ex Choisy) D.F. Austin	Carnein	80.23	Patel et al. (2007)
Euphorbiaceae	*Euphorbia cyparissias* L.	Euphorbains Y_1, Y_2, Y_3	67, 33, 67	Lynn and Clevette-Radford (1985a, b, c)
	Euphorbia drupifera (Schum.) Stapf.	Euphorbains D_1, D_2	117, 65	Lynn and Clevette-Radford (1985a, b, c)
	Euphorbia lactea cristata	Euphorbains Lc	70	Lynn and Clevette-Radford (1986a, b, c)
	Euphorbia lactea Haw.	Euphorbains La_1, La_2, La_3	66, 44,33	Lynn and Clevette-Radford (1986a, b, c)

Table 33.1 (continued)

Family	Plant	Protease(s)	M W (kDa)	References
	Euphorbia lathyris L.	Euphorbain L.	43	Lennox and Ellis (1945)
	Euphorbia milii Des Moul.	Milin	51.4	Yadav et al. (2006)
	Euphorbia pseudo-chamaesyce Fisch.	Protease	82	Shimada et al. (2000)
	Euphorbia pulcherrima Willd.	Euphorbains P	74	Lynn and Clevette-Radford (1984a, b)
	Euphorbia supina Raf.	Protease	80	Arima et al. (2000)
	Euphorbia tirucalli L.	Euphorbains T_1, T_2, T_3, T_4	74, 74, 74, 74	Lynn and Clevette-Radford (1985a, b, c)
	Hevea brasiliensis Muell. Arg	Hevains A, B, L	69, 58, 80	Lynn and Clevette-Radford (1986a, b, c, 1984a, b)
Euphorbiaceae	*Euphorbia cyparissias* L.	Euphorbains Y_1, Y_2, Y_3	67, 33, 67	Lynn and Clevette-Radford (1985a, b, c)
	Synandenium grantii Hook 'f'	Three proteases	76 ± 2, 34.4	Menon et al. (2002); Rajesh et al. (2006)
Moraceae	*Artocarpus heterophyllus* Lam.	Artocarpin	79.5	Prasad and Virupaksha (1990)
	Ficus elastica Roxb.	Ficin E	50	Lynn and Clevette-Radford (1986a, b, c)
	Maclura pomifera (Raf.) Schneid.	Macluralisin	65	Rudenskaya et al. (1995)
3. Aspartate proteases				
Moraceae	*Ficus racemosa* L.	Protease	44.5 ± 05	Devaraj et al. (2008b)

(up to 80°C). The serine proteases are inhibited by diisopropyl fluorophosphates (DFP), phenylmethanesulfonyl fluoride (PMSF), p-amidinomethanesulfonyl fluoride (PAMSF), chymostatin and diethyl pyrocarbonte (DEPC), while cysteine proteases are inhibited by iodoacetamide (IAA), p-chloromercury benzoate (PCMB), Sodium tetrathionate, mercuric chloride, transepoxysuccinyl-L-leucylamido-(4-guanidino) butane (E-64).

Plant Latex Proteases and Their Action on Blood Coagulation and Fibrinolysis

Blood coagulation is a specialized event and a response to blood vessel damage (Krem and Di Cera, 2001, 2002). It involves series of inactive precursors of coagulation enzymes and cofactors that are sequentially activated to functional coagulation factors (Davie, 1995). Coagulation cascade comprises of two distinct but closely linked intrinsic and extrinsic pathways. Both intrinsic and extrinsic pathways are initiated differently by different factors. Later both pathways converge by the formation of prothrombinase complex consisting of factor Xa, Va, Ca^{2+} and phospholipids surface (Gailani and Broze, 1991), which proceed further and generates insoluble fibrin clot. Fibrinolysis is the process wherein fibrin clot formed during blood coagulation is broken down to facilitate normal blood flow to tissue repair and is an important event associated with wound healing. The coagulation and fibrinolytic enzymes are generally serine proteases, with few exceptions like, factor VIII and factor V are glycoproteins, and Factor XIII is a transglutaminase.

Snake venom serine and metalloproteases are well studied for their interference in blood coagulation. Some of these proteases exhibit procoagulant or anticoagulant effects by specifically activating or inactivating coagulation factors (Swenson and Markland, 2005). Serine and cysteine proteases from plant latices exhibit procoagulant action irrespective of the plant species and family. But the mechanism involved in pro-coagulant action differs with the type of proteases. Plant latex serine proteases appear to be non-specific (Rajesh et al., 2006) and their mechanism is yet to be understood. They fail to form fibrin from fibrinogen and also fail to induce clot formation in congenital factor X-deficient plasma. They did not alter the prothrombin time but induce clot formation in citrated plasma. These preliminary studies indicated that plant latex serine proteases might act on the intrinsic pathway (Rajesh et al., 2006). In contrast, cysteine proteases exhibit specific effects on coagulation factors. For example, ficin derived from *Ficus carica* shown to activate coagulation factor X (Richter et al., 2002), whereas Cysteine proteases from Asclepiadaceae exhibit thrombin-like activity (Shivaprasad et al., 2009).

Apart from blood clot inducing property, both cysteine and serine proteases from plant latices have blood clot dissolving properties (Plasmin like activity). Latex proteases efficiently hydrolyze the fibrin leading to its dissolution. Thus, latex proteases have dual activities in inducing and dissolving fibrin clot and dissolution of fibrin clot is secondary to fibrin formation. The site of action of proteases from plant latices on blood coagulation is shown in Fig. 33.1.

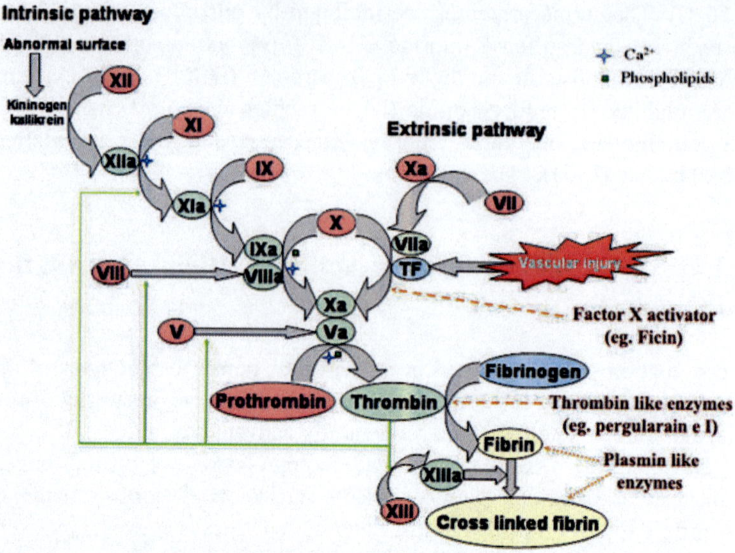

Fig. 33.1 Site of action of plant latex proteases on blood coagulation pathway. (Inspiration from the original source – 316 × 776 – 63 k – jpg – www.new-science-press.com/info/illustration_f

Action of Thrombin on Fibrinogen

Fibrinogen is an elongated molecule has three pairs of non-identical subunits namely Aα, Bβ, and γ which are bridged by disulfide bonds. Fibrinogen is synthesized in liver and found freely in circulating blood at a concentration of 2,500 mg/L

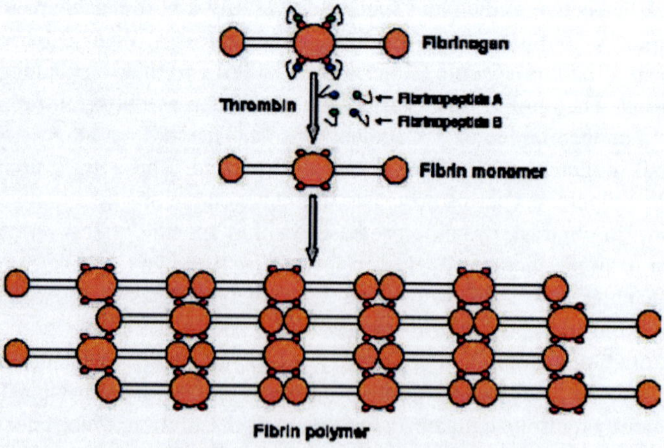

Fig. 33.2 Schematic representation of fibrin formation by thrombin. (Inspiration from the original source – 284 × 400 – 36 k – bmp 2.bp.blogspot.com/.../s400/blood_clotting.bmp

(2–3% of plasma protein). N-terminal sequence of first 16 and 14 amino acid residues of Aα- and Bβ-chains are referred as fibrinopeptides A and B, respectively. Fibrinogen remains free in circulating plasma because fibrinopeptides A and B mask the interaction site of other fibrinogen molecule. In addition, fibrinopeptides region of fibrinogen contain several glutamate and aspartate residues imparting high negative charge which repel other fibrinogen molecules and aid solubility in plasma. Thrombin specifically cleaves between arginine 16 and glycine 17 of Aα chain, and between arginine 14 and glycine 15 of Bβ chain of fibrinogen releasing fibrinopeptides A and B, respectively and thus initiating fibrin polymerization. Fibrin assembly facilitates intermolecular anti-parallel alignment of fibrin monomers which are then covalently cross-linked by the transglutaminase activity of factor XIIIa. The mechanism of thrombin action is shown in Fig. 33.2.

Thrombin-Like Activity of Plant Latex Cysteine Proteases

The term thrombin-like enzymes applied to those proteases which have the capacity to induce fibrin clot and resembles at least in part to that of thrombin hydrolysis of fibrinogen. They specifically hydrolyze the Aα and/or Bβ chains of fibrinogen and their fibrinogen clotting activity is due to the release of fibrinopeptide A and/or fibrinopeptide B (Magalhaes et al., 2007). Several serine proteases which have thrombin-like activity are isolated from snake venoms and are well studied (Matsui et al., 2000; Swenson and Markland, 2005). We described the thrombin-like activity of plant latex cysteine protease pergularin e I from *Pergularia extensa* (Asclepiadaceae) (Shivaprasad et al., 2010). Pergularin e I specifically cleaves Aα and Bβ chains, but not γ-chain of fibrinogen. It selectively releases fibrinopeptide A and B, respectively. The release of fibrinopeptides may be due Arg specific hydrolysis of fibrinogen by pergularin e I. Fibrin clot produced by pergularin e I is soft and friable. Thus this enzyme, similar to snake venom thrombin-like enzymes, do not activate factor XIII. Although several cysteine proteases have been isolated from plant latices, their procoagulant action has not yet been characterized. Plant latex serine protease(s) also hydrolyze Aα and Bβ chains fibrinogen preferentially, but they do not release fibrinopeptides and hence lack thrombin-like activity.

Latex Toxicity

Apart from these pharmacological actions, many reports suggest the toxic nature of plant latex. Some of the plant latex is known to cause contact dermatitis, eye irritation and other deleterious effects (Rietschel and Fowler, 2008; Zuskin et al., 1999). Role of plant latex proteases in inducing toxicity is limited. Plant latex cysteine proteases such as papain, ficin and bromelain showed toxicity against herbivorous insects (Konno et al., 2004). Cysteine proteases from the crude latex extract of *Calotropis gigantea* and other Asclepiadaceae plants were shown to cause skin hemorrhage in experimental animals tested (Rajesh et al., 2005). However, no/low hemorrhagic activity was observed with purified cysteine proteases from the latices of Asclepiadaceae plants. Serine proteases from the plants belonging to Euphorbiaceae and other families do not induce hemorrhage (Rajesh et al., 2007). The major irritants present in plant latices are secondary metabolites such as 12-*O*-tigloyl-4-deoxyphorbol-13-isobutyrate, esters of 12-deoxyphorbol (Kinghorn, 1980; Schmidt and Evans, 1980).

Conclusion

Proteases which interfering with specific action in the coagulation cascade find obvious place in coagulation laboratory for routine assaying of coagulation factors and also proved to have therapeutic value. Several proteases isolated from snake venoms particularly thrombin-like enzymes are being used in the prevention of thrombus formation, in improving the blood circulation in various vascular disorder by reducing blood viscosity. However, the clinical use of thrombin-like enzymes has been limited due to: (a) immunologic reactions in patients; (b) limited availability of the snake venom; and (c) high cost (Warkentin, 1998). Plant latex proteases have marked specific action on the coagulation cascade and potentially can be used as research tools in coagulation laboratory and as therapeutic agents with topical application. Detailed characterization is required to understanding the structure-activity relationships and the diversity of these latex proteases, consequently exploit them for therapeutic use. We hope that this update will facilitate new developments in this field and lead to practical applications of these enzymes.

Acknowledgement H.V. Shivaprasad and R. Rajesh thank the Council of Scientific and Industrial Research, India for financial support.

References

Arima, K., Uchikoba, T., Yonezawa, H., Shimada, M., Kaneda, M., 2000. Cucumisin-like protease from the latex of *Euphorbia supina*. Phytochemistry 53, 639–644.

Ashwani, K., 1999. Ayurvedic medicines: some potential plants for medicine from India. A meeting of the international forum on traditional medicines. Toyama Medical and Pharmaceutical University, Toyama, Japan.

Barrett, A.J., Rawlings, N.D., Woessner, J.F., 1998. Handbook of Proteolytic Enzymes. Academic Press, New York.

Begum, D., Nath, S.C., 2000. Ethnobotanical review of medicinal plants used for skin diseases and related problems in Northeastern India. J. Herbs Spices Med. Plants 7, 55–93.

Brockbank, W.J., Lynn, K.R., 1979. Purification and preliminary characterization of two asclepains from the latex of *Asclepias syriaca* L. (milkweed). Biochim. Biophys. Acta 578, 13–22.

Cavalli, S.V., Arribere, M.C., Cortadi, A., Caffini, N.O., Priolo, N.S., 2003. Morrenain bI, a papain-like endopeptidase from the latex of *Morrenia brachystephana* Griseb. (Asclepiadaceae). J. Protein Chem. 22, 15–22.

Cavalli, S.V., Cortadi, A., Arribére, M.C., Conforti, P., Caffini, N.O., Priolo, N., 2001. Comparison of two cysteine endopeptidases from latices of *Morrenia brachystephana* Griseb. and *Morrenia odorata* (Hook et Arn.) Lindley (Asclepiadaceae). Biol. Chem. 382, 879–883.

Chetia, D., Nath, L.K., Dutta, S.K., 1999. Extraction, purification and physicochemical properties of a proteolytic enzyme from the latex of *Ficus hispida* Linn. Indian J. Pharm. Sci. 61, 29–33.

Davie, E.W., 1995. Biochemical and molecular aspects of the coagulation cascade. Thromb. Haemostat. 74, 1–6.

De Moraes, M.G., Termignoni, C., Salas, C., 1994. Biochemical characterization of new cysteine endopeptidase from *Carica candarmarcensis* L. Plant Sci. 102, 11–18.

Devaraj, K.B., Gowda, L.R., Prakash, V., 2008b. An unusual thermostable aspartic protease from the latex of *Ficus racemosa* (L.). Phytochemistry 69, 647–655.

Devaraj, K.B., Kumar, P.R., Prakash, V., 2008a. Purification, characterization, and solvent-induced thermal stabilization of ficin from *Ficus carica*. J. Agric. Food Chem. 56, 11417–11423.

Domsalla, A., Melzig, M.F., 2008. Occurrence and properties of proteases in plant latices. Planta Med. 74, 699–711.

Dubey, V.K., Jagannadham, M.V., 2003. Procerain, a stable cysteine protease from the latex of *Calotropis procera*. Phytochemistry 62, 1057–1071.

Gailani, D., Broze, G.J., Jr., 1991. Factor XI activation in a revised model of blood coagulation. Science 253, 909–912.

Gavira, J.A., González-Ramírez, L.A., Oliver-Salvador, M.C., Soriano-García, M., García-Ruiz, J.M., 2007. Structure of the mexicain-E-64 complex and comparison with other cysteine proteases of the papain family. Acta Crystallogr. D. Biol. Crystallogr. 63, 555–563.

Kinghorn, A.D., 1980. Major skin-irritant principle from *Synadenium grantii*. J. Pharm. Sci. 69, 1446–1447.

Konno, K., Hirayama, C., Nakamura, M., Tateishi, K., Tamura, Y., Hattori, M., Kohno, K., 2004. Papain protects papaya trees from herbivorous insects: role of cysteine proteases in latex. Plant J. 37, 370–378.

Krem, M.M., Di Cera, E., 2001. Molecular markers of serine protease evolution. EMBO J. 20, 3036–3045.

Krem, M.M., Di Cera, E., 2002. Evolution of enzyme cascades from embryonic development to blood coagulation. Trends Biochem. Sci. 27, 67–74.

Kundu, S., Sundd, M., Jagannadham, M.V., 2000. Purification and characterization of a stable cysteine protease ervatamin B, with two disulfide bridges, from the latex of *Ervatamia coronaria*. J. Agric. Food Chem. 48, 171–179.

Lennox, F.G., Ellis, W.J., 1945. Euphorbain, a protease occurring in the latex of the weed *Euphorbia lathyris*. Biochem. J. 39, 465–470.

Liggieri, C., Arribére, M.C., Trejo, S.A., Canals, F., Avilés, F.X., Priolo, N.S., 2004. Purification and biochemical characterization of asclepain c I from the latex of *Asclepias curassavica* L. Protein J. 23, 403–411.

Lynn, K.R., Clevette-Radford, N.A., 1984a. Euphorbin P, a serine protease from *Euphorbia pulcherrima*. Phytochemistry 23, 682–683.

Lynn, K.R., Clevette-Radford, N.A., 1984b. Purification and characterization of Hevain, a serine protease from the latex of *Hevea brasiliensis*. Phytochemistry 23, 963–964.

Lynn, K.R., Clevette-Radford, N.A., 1985a. Three serine proteases from the latex of *Euphorbia cyparissias*. Phytochemistry 24, 925–928.

Lynn, K.R., Clevette-Radford, N.A., 1985b. Two proteases from the latex of *Eiaeophobia drupifera*. Phytochemistry 24, 2843–2845.

Lynn, K.R., Clevette-Radford, N.A., 1985c. Four latex serine proteases from the latex of *Euphorbia tirucalli*. Can. J. Biochem. Cell Biol. 63, 1093–1096.

Lynn, K.R., Clevette-Radford, N.A., 1986a. Isolation and characterization of proteases from *Euphorbia lacteal* and *Euphorbia lactea cristata*. Phytochemistry 25, 807–810.

Lynn, K.R., Clevette-Radford, N.A., 1986b. Hevains: serine-centred proteases from the latex of *Hevea brasiliensis*. Phytochemistry 25, 2279–2282.

Lynn, K.R., Clevette-Radford, N.A., 1986c. Ficin e, a serine–centred protease from *Ficus elastica*. Phytochemistry 25, 1559–1561.

Magalhaes, A., Magalhaes, H.P., Richardson, M., Gontijo, S., Ferreira, R.N., Almeida, A.P., Sanchez, E.F., 2007. Purification and properties of a coagulant thrombin-like enzyme from the venom of *Bothrops leucurus*. Comp. Biochem. Physiol. A. Comp. Physiol. 146, 565–575.

Matsui, T., Fujimura, Y., Titani, K., 2000. Snake venom proteases affecting hemostasis and thrombosis. Biochim. Biophys. Acta 1477, 146-156.

Menon, M., Vithayathil, P.J., Raju, S.M., Ramadoss, C.S., 2002. Isolation and characterization of proteolytic enzymes from the latex of *Synadenium grantii* Hook, 'f'. Plant Sci. 163, 131–139.

Nallamsetty, S., Kundu, S., Jagannadham, M.V., 2003. Purification and biochemical characterization of a highly active cysteine protease ervatamin A from the latex of *Ervatamia coronaria*. J. Protein Chem. 22, 1–13.

Obregón, W.D., Arribére, M.C., Del Valle, S.M., Liggieri, C., Caffini, N., Priolo, N., 2001. Two new cysteine endopeptidases obtained from the latex of *Araujia hortorum* fruits. J. Protein Chem. 20, 317–325.

Osoniyi, O., Onajobi, F., 2003. Coagulant and anticoagulant activities in *Jatropha curcas* latex. J. Ethnopharmacol. 89, 101–105.

Patel, A.K., Singh, V.K., Jagannadham, M.V., 2007. Carnein, a serine protease from noxious plant weed *Ipomoea carnea* (morning glory). J. Agric. Food Chem. 55, 5809–5818.

Patel, B.K., Jagannadham, M.V., 2003. A high cysteine containing thiol proteinase from the latex of *Ervatamia heyneana*: purification and comparison with ervatamin B and C from *Ervatamia coronaria*. J. Agric. Food Chem. 51, 6326–6334.

Perello, M., Arribere, M.C., Caffini, N.O., Priolo, N.S., 2000. Proteolytic enzymes from the latex of *Ficus pumila* L. (Moraceae). Acta Farm. Bonaerense 19, 257–262.

Prasad, K.M.R., Virupaksha, T.K., 1990. Purification and characterization of a protease from the jackfruit latex. Phytochemistry 29, 1763–1766.

Priolo, N., Morcelle, Del Valle, S., Arribére, M.C., López, L., Caffini, N., 2000. Isolation and characterization of a cysteine protease from the latex of *Araujia hortorum* fruits. J. Protein Chem. 19, 39–49.

Prusti, A.B., Behera, K.K., 2007. Ethnobotanical exploration of Malkangiri district of Orissa, India. Ethnobotanical Leaflets 11, 122–140.

Rajesh, R., Nataraju, A., Raghavendra Gowda, C.D., Frey, B.M., Frey, F.J., Vishwanath, B.S., 2006. Purification and characterization of a 34-kDa, heat stable glycoprotein from *Synadenium grantii* latex: action on human fibrinogen and fibrin clot. Biochimie 88, 1313–1322.

Rajesh, R., Raghavendra Gowda, C.D., Nataraju, A., Dhananjaya, B.L., Kemparaju, K., Vishwanath, B.S., 2005. Procoagulant activity of *Calotropis gigantea* latex associated with fibrin(ogen)olytic activity. Toxicon 46, 84–92.

Rajesh, R., Shivaprasad, H.V., Raghavendra gowda, C.D., Nataraju, A., Dhananjaya, B.L., Vishwanath, B.S., 2007. Comparative study on plant latex proteases and their involvement in hemostasis: a special emphasis on clot inducing and dissolving properties. Planta Med. 73, 1061–1067.

Reddy, P.R., Rao, P.P., Prabhakar, M., 2003. Ethnomedicinal practices amongst Chenchus of Nagarjunasagar Srisailam Tiger Reserve (NSTR), Andhra Pradesh-plant remedies for cuts, wounds and boils. Ethnobotany 15, 67–70.

Richter, G., Hans, P.S., Friedrich, D., Peter, L., 2002. Activation and inactivation of human factor X by proteases derived from *Ficus carica*. Br. J. Haematol. 119, 1042–1051.

Rietschel, R.L., Fowler, J.F., 2008. Allergic sensation to plants, in: Fischer's Contact Dermatitis, 6th edition. BC Decker Inc., Hamilton, ON, Canada, pp. 405–453.

Rudenskaya, G.N., Bogacheva, A.M., Preusser, A., Kuznetsova, A.V., Dunaevsky, YaE., Golovkin, B.N., Stepanov, V.M., 1998. Taraxalisin-a serine proteinase from dandelion *Taraxacum officinale* Webb s.l. FEBS Lett. 437, 237–240.

Rudenskaya, G.N., Bogdanova, E.A., Revina, L.P., Golovkin, B.N., Stepanov, V.M., 1995. Macluralisin – a serine proteinase from fruits of *Maclura pomifera* (Raf.) Schneid. Planta 196, 174–179.

Samy, R.P., Pushparaj, P.N., Gopalakrishnakone, P., 2008. A compilation of bioactive compounds from Ayurveda. Bioinformation 3, 100-110.

Schmidt, R.J., Evans, F.J., 1980. Skin irritants of the sun spurge (*Euphorbia helioscopia* L.). Contact Derm. 6, 204–210.

Sengupta, A., Bhattacharya, D., Pal, G., Sinha, N.K., 1984. Comparative studies on Calotropins DI and DII from the latex of *Calotropis gigantea*. Arch. Biochem. Biophys. 232, 17–25.

Sequeiros, C., Torres, M.J., Trejo, S.A., Esteves, J.L., Natalucci, C.L., López, L.M., 2005. Philibertain g I, the most basic cysteine endopeptidase purified from the latex of *Philibertia gilliesii* Hook. et Arn. (Apocynaceae). Protein J. 24, 445–453.

Shimada, M., Uthikova, T., Arima, K., Kaneda, M., 2000. Isolation and characterization of a Cucumisin-like serine protease from the latex of *Euphorbia seudophamaesyce* Fisch. J. Biochem. Mol. Biol. Biophys. 4, 223–231.

Shivaprasad, H.V., Rajesh, R., Frey, B.M., Frey, F.J., Vishwanath, B.S., 2010. 'Pergularain e I' – a plant cysteine protease with thrombin-like activity from *Pergularia extensa latex*. Thromb. Res. 125, e100–e105.

Shivaprasad, H.V., Riyaz, M., Venkatesh Kumar, R., Dharmappa, K.K., Tarannum, S., Siddesha, J.M., Rajesh, R., Vishwanath, B.S., 2009. Cysteine proteases from the Asclepiadaceae plants latex exhibited thrombin and plasmin like activities. J. Thromb. Thrombolysis 28, 304–308.

State of the Environment Report, 2001. http://www.envfor.nic.in/soer/2001/ind_bio.pdf.

Swenson, S., Markland, F.S., Jr., 2005. Snake venom fibrin(ogen)olytic enzymes. Toxicon 45, 1021–1039.

Tablero, M., Arreguin, B., Soriana, M., Sanchez, R.I., Romera, A.R., Hernandez-Arana, A., 1991. Purification and characterization of multiple forms of Asclepain g from *Asclepias glaucescens* H.B.K. Plant Sci. 74, 7–15.

Tomar, R., Kumar, R., Jagannadham, M.V., 2008. A stable serine protease, wrightin, from the latex of the plant *Wrightia tinctoria* (Roxb.) R. Br.: purification and biochemical properties. J. Agric. Food Chem. 56, 1479–1487.

Trejo, S.A., López, L.M., Cimino, C.V., Caffini, N.O., Natalucci, C.L., 2001. Purification and characterization of a new plant endopeptidase isolated from latex of *Asclepias fruticosa* L. (Asclepiadaceae). J. Protein Chem. 20, 469–477.

Warkentin, T.E., 1998. Limitations of conventional treatment options for heparin-induced thrombocytopenia. Semin Hematol. 35, 17–25.

WHO Traditional Medicine Strategy 2002–2005, http://whqlibdoc.who.int/hq/2002/WHO_EDM_TRM_2002.1.pdf.

Winnick, T., Davis, A.R., Greenberg, D.M., 1939. Physicochemical properties of the proteolytic enzyme from the latex of the milkweed, *Asclepias speciosa* Torr. Some comparisons with other proteases. J. Gen. Physiol. 275, 289–300.

Yadav, S.C., Pande, M., Jagannadham, M.V., 2006. Highly stable glycosylated serine protease from the medicinal plant *Euphorbia milii*. Phytochemistry 67, 1414–1426.

Zuskin, E., Mustajbegović, J., Kanceljak, B., Macan, J., Ivanković, D., Smolej-Narancić, N., 1999. Occurance of nonspecific symptoms of irritation in workers exposed to latex. Lijec Vjesn. 121, 230–233.

Part VII
Platelet Aggregation Inducers

Chapter 34
The Biologic Activity of Aggretin/Rhodocytin, a Snake Venom C-Type Lectin Protein (Snaclec)

Ching-Hu Chung and Tur-Fu Huang

Abstract Aggretin (or rhodocytin), an ab-heterodimeric snake venom C-type lectin (snaclec), was originally found to activate platelets through specific binding to integrin α2β1, leading to activation of phospholipase C, platelet activation and subsequent thrombocytopenia in vivo. Recent reports suggest that the CLEC-2 receptor, involving the Src, Syk and PLCg pathway is critical for mediating platelet activation by aggretin. In this review, we discuss the use of aggretin to explore the role of α2β1 and/or CLEC-2 in different cell types, namely (i) induction of platelet aggregation (α2β1 or CLEC-2), (ii) induction of angiogenesis in endothelial cells (α2β1), (iii) promotion of proliferation, migration of smooth muscle cells and keratinocytes, and induction of cytokine release through CLEC-2 ligation, and also possible signal transduction pathways involving ligation of integrin α2β1 or CLEC-2.

Introduction

Snake venoms affect blood coagulation and platelet function in a complex manner (Ouyang et al., 1992). The most pronounced anti-platelet constituents are Arg-Gly-Asp containing trigramin-like peptides, which have been identified as specific antagonists of the fibrinogen receptor (αIIbβ3) and named disintegrins (Calvete et al., 2003; Gould et al., 1990; Huang et al., 1987). On the other hand, several non-coagulant, non-enzymatic proteins that cause platelet aggregation have been purified from different snake venoms (Huang et al., 1995; Teng and Huang, 1991; Teng et al., 1989). C-type lectin-like proteins (snaclecs) are an important group among the haemorrhagic components in snake venom (Clemetson et al., 2009; Morita, 2005). Classic C-like lectins consist of a carbohydrate-recognition domain (CRD) and most of them are covalently linked homodimers (Hirabayashi et al., 1991). The

T.-F. Huang (✉)
Institute of Pharmacology, College of Medicine, National Taiwan University, Taipei, Taiwan
e-mails: turfu@ntu.edu.tw; turfu@ccms.ntu.edu.tw

R.M. Kini et al. (eds.), *Toxins and Hemostasis*, DOI 10.1007/978-90-481-9295-3_34,

first determination of an X-ray crystal structure of a snaclec (a coagulation factor IX/X binding protein) indicates that the two (hetero) subunits associate covalently to form a concave surface by loop-swapping between the two subunits (Mizuno et al., 1997). Subsequently, many snaclecs have been characterized with specific binding to VWF or platelet receptors such as GPIb, α2β1 and GPVI (Clemetson et al., 2005). Nowadays, it is recognized that integrin α2β1 (also known as GPIa/IIa in platelets) and GPVI are major collagen receptors on platelets in mediating adhesive interactions between platelet and the exposed subendothelial collagen in injured vessels, generating intracellular signals that help to stabilize the thrombus (Herr and Farndale, 2009). The snaclecs aggretin, rhodocetin, bilinexin and EMS16 have all been reported to interact with α2β1.

Aggretin was first identified and reported to bind a collagen receptor with a high affinity ($Kd = 4.0 \pm 1.1$ nM) in a bivalent-cation independent manner, leading to activation of human platelets via endogenous phospholipase C, leading to hydrolysis of phosphatidylinositol 4,5-bisphosphate (PIP2) and liberation of IP3 and intracellular Ca^{2+} mobilization, and subsequent platelet aggregation (Huang et al., 1995). In 1998, Shin and Morita isolated a platelet aggregation inducer designated as rhodocytin from the same snake venom (*Calloselasma rhodostoma*), which bound polyclonal antibodies against habu IX/X-bp (Shin and Morita, 1998). They found that the N-terminal amino acid sequences of rhodocytin α and β chains are homologous to snaclec subunits. As a consequence both aggretin and rhodocytin are used to describe the same protein. In the meantime, based on the N-terminal amino acid sequences of two subunits of aggretin, and the corresponding synthetic mixed-base oligonucleotides, we cloned both chains of aggretin from a cDNA library derived from a Malayan-pit-viper (*Calloselasma rhodostoma*) venom gland (Chung et al., 1999). These results revealed that mature α and β chains consist of 136 and 123 amino acid residues, respectively. Aggretin subunits show high degrees of identity with respective subunits (50–60% for α-chain, 49–58% for β-chain) of snaclecs. The cysteine residues in each chain of aggretin are well conserved and located at positions corresponding to those of snaclecs. Thus, three intracatenary disulfide bridges and an interchain disulfide bond between Cys83(α) and Cys75(β) can be situated. The comparison of the amino acid sequences of aggretin with some other snaclecs is shown in Table 34.1.

The crystal structure of aggretin reveals a non-covalent (αβ)2 dimer, notably different from the covalent αβ multimers of other CLPs (Hooley et al., 2008). Unlike other CLP-like proteins, aggretin has been reported to have multiple binding receptors and diverse functions in many cells other than integrin α2β1. For example, platelets genetically lacking α2β1 with GPIb removed by proteolysis, still showed similar reactivity with aggretin, suggesting that another receptor may also play a critical role in mediating platelet aggregation (Bergmeier et al., 2001). In this chapter, we focus the possible binding sites in mediating its biological activities in a variety of cells, including platelets, phagocytes, vascular endothelial cells, keratinocytes, and smooth muscle cells.

Table 34.1 Comparison of the amino acid sequences deduced from aggretin from cDNAs with those of other snaclecs. The two aggretin subunits are aligned with the corresponding subunits of the following snake venom proteins: convulxin from *Crotalus durissus terrificus* venom (Leduc and Bon, 1998), agglucetin from *Agkistrodon acutus* venom (Wang and Huang, 2001), alboaggregin-A from *Trimeresurus albolabri s* venom (Kowalska et al., 1998). Gaps are inserted to obtain the maximum similarity

α subunit		β subunit	
Aggretin	1 GLEDCDFGWS PYDQHCYQAF	Aggretin	1 DCPSGWSSYE GHCYKPFNEP
Convulxin	GLH-CPSWY YYDQHCYRIF	Convulxin	GFCCPSHWSSYD RYCYKVFKQE
Agglucetin	-DFNCPPGWS AYDQYCYQVI	Agglucetin	DCPSEWSSYE GHCYKAFKQS
Alboaggregin A	-DFHCLPGWS AYDQYCYRVF	Alboaggregin A	GFDCPFGWSSYE FYCYKVYNKK
Aggretin	21 NEQKTWDEAE KFCRAQENGA	Aggretin	21 KNWADAERFC KLQPKHSHLV
Convulxin	NEEMNWEDAE WFCTKQAKGA	Convulxin	MTWADAEKFC TQQHTGSHLV
Agglucetin	KEPKNWDDAE RFCTEQADGG	Agglucetin	KTWADAEKFC TQQHKGSHLA
Alboaggregin A	NEPKNWEDAE RFCAKQADSG	Alboaggregin A	MNWEDAESFC REQHKRSHLV
Aggretin	41 HLASIESNGE ADFVSWLISQ	Aggretin	41 SFQSAEEADF VVKLTRPRLK
Convulxin	HLVSIKSAKE ADFVAWMVTQ	Convulxin	SFHSTEEVDF VVKMTHQSLK
Agglucetin	HLVSIESKGE RDFVAQLVSQ	Agglucetin	SFHSSEEADF VVTLTTPSLK
Alboaggregin A	HLVSIETMGE ADFVAQLISE	Alboaggregin A	SFHSSGEVDF VVSKTFPILR

Table 34.1 (continued)

α subunit		β subunit	
Aggretin	61 KDELADEDYV WIGLRAQNKE	Aggretin	61 ANLVWMGLSN IWHGCNWQWS
Convulxin	NIEESFSH-V SIGLRVQNKE	Convulxin	STFFWIGANN IWNKCNWQWS
Agglucetin	NIESVEDH-V WTGLRVQNKE	Agglucetin	TDLVSIGLKN IWNGCYWKWS
Alboaggregin A	NIQSEKGY-V WIGLKVQNKE	Alboaggregin A	YDFVWMGLSD IWKECTKEWS
Aggretin	81 QQCSSEWSDG SSVSYENLID	Aggretin	81 DGARLNYKDW QEQSECLAFR
Convulxin	KQCSTKWSDG SSVSYDNLLD	Convulxin	DGTKPEYKEW HEEFECLISR
Agglucetin	CSTEWSDGSS VSYENLLELY	Agglucetin	DGTKLDYKDW REQFECLVSR
Alboaggregin A	QQCSSEWSDG SSVTYENLIK	Alboaggregin A	DGARLDYKAW SGKSYCLVSK
Aggretin	101 LHTKKCGALE KLTGFKWVN	Aggretin	101 GVHTEWLNMD CSSTCSFVCK
Convulxin	LYITKCSLLK KETGFRKWFV	Convulxin	TFDNQWLSAP CSDTYSFVCK
Agglucetin	KQMRKCGALE RETGFHKWIN	Agglucetin	TVNNEWLSMD CGTTYSFVCK
Alboaggregin A	LYMRKCGALE QESGFRKWIN	Alboaggregin A	TTNNEWLSMD CSRTRYPVCK
Aggretin	121 YYCEQMHAFV CKLLPY	Aggretin	121 FKA
Convulxin	ASCIGKIPFV CKFPPQC	Convulxin	FEA
Agglucetin	LGCIQLNPFV CKFPPQC	Agglucetin	FQA
Alboaggregin A	LGCIQLNPFV CKFPPQ	Alboaggregin A	FxG

Role of Integrin α2β1 in Platelets

Thrombocytopenia is a common symptom in patients after being bitten by the Malayan pit viper (*Calloselasma rhodostoma*) and its crude venom potently activates washed rabbit platelets (Ouyang et al., 1986; Reid et al., 1963). To identify the possible aggretin receptor on the platelet membrane, we used monoclonal antibodies (6F1 and AP-1) to examine the target of aggretin in mediating platelet aggregation (Huang et al., 1995). At 10 μg/ml, the monoclonal antibody (6F1) raised against integrin α2β1, specifically inhibited both collagen and aggretin-induced platelet shape-change, aggregation and release reaction whereas AP-1, a GPIb mAb, had little inhibitory effect. Another group also tried to investigate whether rhodocytin binds directly or indirectly to integrin α2. They examined the binding of liposomes carrying recombinant fragments of α2β1 (rα2β1 liposomes) to immobilized rhodocytin and suggested that rhodocytin specifically and directly interacts with α2β1 (Suzuki-Inoue et al., 2001). Rhodocytin-coupled Sepharose 4B beads also precipitate α2, but not GPIb or β1. The FcRγ-chain knockout mice was used to investigate if GPVI is involved in the platelet aggregation induced by aggretin/rhodocytin,. Rhodocytin induced aggregation of Fcγ$^{-/-}$ mice platelets, whereas CRP, a GPVI agonist, failed to do so (Suzuki-Inoue et al., 2001), suggesting that rhodocytin induces platelet aggregation independently of GPVI. Li et al. (2004) also found reduced tyrosine phosphorylation of PLCγ2 and Syk in response to rhodocytin in platelets with low integrin α2β1 levels, and suggested that rhodocytin-induced aggregation lag time and signaling are affected by the level of integrin α2β1 expression.

Integrin α2β1 in Endothelial Cells

Integrin α2β1 is a major collagen receptor expressed both on endothelial cells and platelets. In cells expressing integrin α2β1, many signals including the extracellular-regulated kinase (ERK) pathway and matrix remodeling were activated after cell adhesion to collagen (Ivaska et al., 1999; Ravanti et al., 1999). Integrin α2β1 has also been implicated in proliferation, survival, and tube formation of endothelial cells (Davis and Camarillo, 1996; Gamble et al., 1993; Saelman et al., 1995). Therefore, we hypothesized that aggretin, an α2β1 agonist, may promote angiogenesis through ligation of α2β1. We found that aggretin promotes proliferation and migration of HUVECs and these effects are blocked by A2-IIE10, an anti-α2 mAb (Chung et al., 2004). In the ligand-binding study by flow cytometry, we also demonstrated that aggretin specifically interacts with HUVECs via α2β1, whereas 7E3, an anti-αvβ3 mAb, or agkistin, a GPIb antagonist, had little effect. In addition, aggretin stimulates the phosphorylation of PI3K p85α, Akt, and ERK1/2, but not FAK phosphorylation in HUVECs, and the increased VEGF production also induces PI3K, Akt, and ERK1/2 activation in HUVECs. These results have established an association between ligation of integrin α2β1 and VEGF production in angiogenesis. Thus, the effects of aggretin on human endothelial cells may be used as a tool for investigating the involvement of α2β1 in angiogenesis.

Roles of α2β1 in Vascular Smooth Muscles

Since collagen is one of the most abundant ECM proteins and it is known to stimulate the migration and growth of numerous cell types, including vascular smooth muscle cells (VSMCs) (Xiang et al., 2000). We also used aggretin to investigate α2β1 signaling transduction in VSMCs (Chung et al., 2009). We found that the VSMC response to aggretin is mainly mediated by α2, because both its enhancing activities on migration and proliferation of VSMC were blocked by A2IIE10, an anti-α2 mAb. Aggretin also stimulated Src and ERK phosphorylation and, in turn, induced NF-kB activation. NF-kB activation induced by aggretin was abolished by transfection with siRNA for integrin α2, indicating that the NF-kB activation may be primarily through α2β1 ligation. Additionally, an increase in NF-kB nuclear translocation enhanced PDGF-BB expression in aggretin-stimulated VSMCs.

Roles of α2β1 in Keratinocytes

The most abundantly expressed collagen receptor on resting keratinocytes is α2β1, where it is highly expressed in the basal layer of the epidermis. The multiple members of the β1 integrin family help maintain the epithelial barrier, adhesion of the epidermis to the basement membrane, epithelial migration across the dermal extracellular matrix during development and wound healing, and signals that keep keratinocyte stem cells within the cell cycle for normal self-renewal and those that stimulate epithelial differentiation. Mice with keratinocyte-specific deletion of the β1 integrin gene have a mild blistering disorder, partial hair loss with diminished hair follicles, impaired epidermal proliferation, and a mild defect in wound healing (Brakebusch et al., 2000; Grose et al., 2002). Keratinocytes from mice with targeted deletion of the β1 integrin gene demonstrate altered expression of other genes including matrix metalloproteinase 10 (MMP-10; stromelysin-2), MMP-13 (collagenase-3), and the β4 integrin subunit. To evaluate further the role of α2β1 integrin on wound healing, we also tested the function of aggretin in cultured keratinocytes/fibroblasts and in a in vivo wound healing model (Chung et al., 2007). We found that aggretin elicits keratinocyte proliferation and migration, and promotes skin re-epithelialization in vivo and in vitro. Aggretin activates α2β1 in promoting keratinocyte-releasable factors production and this effect was associated with Src, PLC and NF-kB activation. Aggretin may accelerate the closure of wounds by re-epithelization before the completion of dermal reconstruction, thus resulting in the formation of aseptic connective tissue which could prevent life-threatening infections.

The Possible Involvement of GPIb in Aggretin Function

Although α2β1 was considered to be the target receptor of aggretin on platelets, some studies reported that aggretin induces strong platelet activation and aggregation, which coincides with, but is not dependent on α2β1-mediated intracellular signaling events (Chung et al., 2001; Navdaev et al., 2001). On the other hand, Eble et al. (2001) have reported that rhodocytin does not bind to a recombinant α2β1 complex. Because the reversible binding between GPIb-V-IX and von Willebrand factor, associated with collagen is crucial to slow down the platelets and many snaclecs investigated so far affect platelet responses to GPIb, it is important to investigate if GPIb acts as a target receptor for aggretin (Lankhof et al., 1996; Moroi et al., 1997). However, the snake venom protein, agkistin, which inhibits ristocetin-induced human platelet aggregation by acting as a GPIb antagonist, slightly inhibited aggregation induced by aggretin and collagen at a concentration of 20 μg/ml (Chung et al., 2001; Yeh et al., 2000). Nieswandt et al. (2001) reported that hemostasis and thrombosis might involve three steps. First, platelets tether at sites through GPIbα-vWF interactions, which are essential under high shear stress (Savage et al., 1998). GPVI on platelet surface binds to collagen in the second step, and in this step integrins including α2β1 and αIIbβ3 are activated and their affinities are upregulated. The third step is high affinity binding of α2β1 to collagen. Thus, aggretin may interact with platelet GPIb in the first step and participate in third step of platelet aggregation via α2β1. Echicetin, a snake venom GPIb antagonist, prolonged the lag phase of platelet aggregation caused by rhodocytin, which may reflect echicetin blocking the initial binding of rhodocytin, suggesting that GPIb might be involved in the initial step of rhodocytin-induced platelet aggregation (Shin and Morita, 1998). At the same time, Navdaev et al. (2001) also isolated a protein from *C. rhodostoma* venom that is a powerful platelet activator. Based on the molecular mass, sequence of subunits and properties in activating platelets, they identified this protein as aggretin. They also found that many antibodies and snaclecs against GPIb only slightly inhibit aggretin-induced platelet activation, but the MoAb VM16d, directed against the thrombin-binding site in GPIb, inhibits aggretin-induced platelet aggregation completely in a dose-dependent way. Affinity chromatography of platelet lysate prepared in the absence of EDTA gave specific binding of α2β1 and GPIb to biotinylated aggretin/avidin-Sepharose. They also used Fcγ-deficient mice platelets to confirm that GPVI/Fcγ is not involved in platelet activation by aggretin.

A Possible New Target of Aggretin-CLEC-2 in Platelets and Phagocytes

Because the results regarding the target receptor for aggretin in platelet activation were controversial, some groups are still seeking another receptor for aggretin (Bergmeier et al., 2001). A subsequent study, using α2-deficient murine platelets and proteolytic cleavage of the 45-kDa N-terminal domain of GP Ibα had no significant effect on rhodocytin-induced platelet activation. These results demonstrated that rhodocytin induces platelet activation by mechanisms that are fundamentally different from those induced by collagen. Suzuki-Inoue et al. (2006) used rhodocytin affinity chromatography and mass spectrometry to identify further receptors underlying platelet activation by rhodocytin. These approaches identified a novel 32-kDa surface receptor of the C-type lectin class and established CLEC-2 as a receptor that mediates activation by the snake venom toxin rhodocytin in platelets other than integrin α2β1 and GPIb/V/IX.

CLEC-2 was the first member of receptors of the C-type lectin family identified to activate platelets through sequential Src and Syk family tyrosine kinases, thereby initiating a signaling cascade involving other kinase tyrosine phosphorylation and activation of PLCγ2. The signaling cascade of the platelet collagen receptor GPVI activation recruits Syk to a doubly phosphorylated ITAM on FcR γ-chain complex. A unique feature in this signaling cascade, which distinguishes it from that used by other platelet glycoprotein receptors, is that Src kinase–dependent tyrosine phosphorylation of CLEC-2 on a YXXL motif is insufficient to induce binding and activation of Syk and thereby initiate downstream signaling events without either clustering of CLEC-2 or some other receptor being involved. CLEC-2 was first cloned from human bone marrow and its cDNA was originally reported to be expressed selectively in the liver and in some blood cells of myeloid origin, including monocytes, dendritic cells, and granulocytes (Colonna et al., 2000). However, Suzuki-Inoue et al. (2006) also reported CLEC-2 as a megakaryocyte/platelet-specific protein after its identification in 30 serial analysis of gene expression (SAGE) library from mouse megakaryocytes of hematopoietic origin. CLEC-2 potentially could also play an important role in mediating thrombus formation in response to stimulation by an appropriate ligand. Fuller et al. studied the CLEC-2 signaling pathway and its functional motif in platelets and two hematopoietic-derived cell line model systems compared to Dectin-1 and DC-SIGN (Fuller et al., 2007). Their results demonstrate that signaling by CLEC-2 depends completely on the cytoplasmic YXXL motif and requires both SH2 domains of Syk. The signaling pathway activated by CLEC-2 involves Src, Syk, and Tec family kinases and PLCγ, but it is distinct from that of GPVI mediated ITAM signaling. Their studies also demonstrate that some but not all C-type lectin class receptors signal through a single YXXL motif leading to activation of PLCγ.

Although surface expression of CLEC-2 has only been identified on platelets and neutrophils, RT-PCR analysis has shown transcripts in bone marrow cells, monocytes, dendritic cells, and granulocytes (Colonna et al., 2000; Suzuki-Inoue et al., 2006). Many researchers are interested to examine the function of aggretin in inflammatory cells. During the inflammatory process, human peripheral blood monocytes and tissue macrophages play pivotal roles in responses to infection. These host defense responses act to eliminate harmful factors such as tissue damage, pathogen invasion, or diseases, to maintain health. Sustained inflammation may contribute to many pathologies, including asthma, arthritis, multiple sclerosis, and atherosclerosis (Nathan, 2002; Rankin, 2004). When these immune cells are activated, various pro-inflammatory mediators such nitric oxide (NO), cycloxygenase-2, reactive oxygen species, and pro-inflammatory cytokines such as tumor-necrosis factor-α (TNF-α), interleukin-6 (IL-6), and interleukin-1β (IL-1β) are secreted (Chen et al., 2009; Rankin, 2004). Overproduction of these pro-inflammatory cytokines may cause the multi-organ dysfunction in sepsis, leading to shock and death in patients. A wide variety of extracellular stimuli, including lipopolysaccharide (LPS) derived from the outer membrane of Gram-negative bacteria, has been shown to induce the production of these pro-inflammatory cytokines. Macrophages are activated to induce the production of various inflammatory cytokines including TNF-α, IL-6, IL-1β in response to LPS stimulation (Cohen, 2002). Kerrigan et al. (2009) showed that in addition to platelets, CLEC-2 is also expressed on peripheral blood neutrophils as well as on monocytes activated with selected TLR ligands and demonstrate that this receptor can function as an activation receptor on these cells, inducing phagocytosis and proinflammatory cytokine production. Therefore, we treated murine macrophages (RAW 264.7 cells) and human monocytes (THP-1) with different snaclecs, including aggretin, gramicetin, trowaglerix and convulxin, in the absence or presence of lipopolysaccharide for 24 h to examine whether cytokines, such as TNF-α and IL-6 were induced by these agonists (Chang et al., 2009). Among them, only aggretin raised the production of TNF-α and IL-6 in both RAW264.7 and THP-1 cells while the other snaclecs did not. We also found that aggretin induced extracellular signal-regulated kinase 1/2 (ERK1/2) and c-Jun N-terminal kinase (JNK) tyrosine phosphorylation in RAW264.7 cells. Furthermore, the plasma levels of TNF-α and IL-6 were also raised in ICR mice following intra-muscular administration of aggretin. CLEC-2 mAb could displace aggretin bound to RAW264.7 cells. These results indicate that aggretin may induce inflammatory responses by interacting with CLEC-2 on monocytes/macrophages (Chang et al., 2009).

Perspective

According to the above-mentioned studies, we suggest that the integrin $\alpha 2\beta 1$ is important in aggretin-mediated platelet aggregation. Holtkotter et al. observed reduced aggregation to low dose of soluble collagen in $\alpha 2^{-/-}$ mice and Jarvis et al. who used $\alpha 2\beta 1$ blocking antibodies obtained similar results (Holtkotter et al., 2002; Jarvis et al., 2002). These results indicate that integrin $\alpha 2\beta 1$ is not only an aggretin receptor but also is one of the major aggregation receptors in platelets. On the other hand, there are only few studies investigating the physiological and pathological function of another aggretin receptor, CLEC-2. Besides aggretin, HIV-1 has been identified as an exogenous ligand for CLEC-2 and facilitating HIV-1 dissemination in infected patients (Chaipan et al., 2006). However, endogenous ligands for CLEC-2 had remained elusive. Recently, podoplanin was identified as an endogenous ligand for CLEC-2 (Ozaki et al., 2009; Suzuki-Inoue et al., 2007). Podoplanin is a type-I transmembrane sialomucin-like glycoprotein that consists of an extracellular domain, a single transmembrane portion, and a short cytoplasmic tail (Kato et al., 2003). Increased expression of podoplanin/aggrus was observed in various tumor cells and might be associated with tumor cell-induced platelet aggregation in promoting tumor growth and metastases (Morita, 2005). In addition, the podoplanin receptor has been reported to play a role in phagocytic activity of neutrophils (Kerrigan et al., 2009). However, evidence for the importance of CLEC-2 in platelet activation during hemostasis and in the course of thrombotic events is still lacking. May et al. (2009) used anti-CLEC-2 antibody treatment in mice which leads to complete and highly specific loss of CLEC-2 in circulating platelets and studied CLEC-2-deficient platelet function. CLEC-2-deficient platelets displayed normal adhesion under flow but the subsequent aggregate formation was severely deficient in vitro and in vivo. The anti-CLEC-2 antibody also significantly inhibited rhodocytin-induced platelet aggregation while CLEC-2-deficient platelets were resistant to activation with rhodocytin. These results suggest that CLEC-2 serves as an essential receptor in hemostasis, thrombosis and rhodocytin-induced platelet aggregation. Therefore, the discrepancy concerning either α2β1 or CLEC-2 as the critical receptor for activating platelets remains unresolved. The different results may be due to the species difference, i.e. human versus murine platelets. So far, we can not exclude the involvement of α2β1 in mediating platelet activation in human platelets by aggretin. The α2β1 knockout mice experiment may involve some unknown compensatory regulation of CLEC-2 or a related receptor. On the other hand, we propose that α2β1 is involved in mediating the activities of aggretin in endothelial cells, vascular smooth muscle, keratinocytes and CLEC-2 in phagocytes, namely the promotion of angiogenesis, proliferation and migration of smooth muscle cell/keratinocyte, and release of cytokines from phagocyte. We summarize these effects in Table 34.2. Aggretin is unique in targeting receptors in various cell types, and further investigation of its interaction with integrin α2β1 or CLEC-2 at molecular level will shed some light on designing novel drug candidates for α2β1 or CLEC-2 related diseases.

GPIb, α2β1 and GPVI binding proteins show a high degree of sequence similarity although they have different biological activities. Therefore, the detailed study

Table 34.2 Biological activities of aggretin in different cells

Target	Cell preparation	Effects
α2β1 or CLEC-2	Platelets	Platelet activation (+)
α2β1	Endothelial cells	Proliferation (+) Migration (+) Tube formation (+) VEGF production (+)
α2β1	Vascular smooth muscle cells	Proliferation (+) Migration (+) PDGF production (+)
α2β1	Keratinocyte	Proliferation (+) Migration (+)
CLEC-2	Phagocyte (RAW264.7 cell and THP-1)	IL-6 release (+) TNF-α release (+) MAPK activation (+)

of their structure-activity relationship in eliciting cellular responses for aggretin may allow us to determine which polypeptide sequences interact specifically with receptor(s) on the cell surface. Modeling of snaclec structures has been a fairly useful approach for establishing putative binding sites on their surfaces based on the premise that interactions with ligands involve electrostatic as well as shape-matching interactions. It will be worthwhile to explore the involvement of the binding motif of these snaclecs toward specific receptors at a molecular level, namely α2β1, GPIb, GPVI, or CLEC-2 expressed on platelets or other cells, using snaclec fragments and their mutants combined with functional studies, X-ray crystallography and computer modeling. These studies should help to discover the specific small molecular domains targeting these important receptors for eventual therapeutic applications in arterial thrombosis, angiogenesis, wound healing, sepsis and even tumor metastasis.

Acknowledgement We highly appreciate the long-term financial support from grants of National Science Council, Taiwan.

References

Bergmeier, W., Bouvard, D., Eble, J.A., Mokhtari-Nejad, R., Schulte, V., Zirngibl, H., Brakebusch, C., Fassler, R., Nieswandt, B., 2001. Rhodocytin (aggretin) activates platelets lacking $\alpha_2\beta_1$ integrin, glycoprotein VI, and the ligand-binding domain of glycoprotein Ibα. J. Biol. Chem. 276, 25121–25126.

Brakebusch, C., Grose, R., Quondamatteo, F., Ramirez, A., Jorcano, J.L., Pirro, A., Svensson, M., Herken, R., Sasaki, T., Timpl, R., Werner, S., Fassler, R., 2000. Skin and hair follicle integrity is crucially dependent on β1 integrin expression on keratinocytes. EMBO J. 19, 3990–4003.

Calvete, J.J., Moreno-Murciano, M.P., Theakston, R.D., Kisiel, D.G., Marcinkiewicz, C., 2003. Snake venom disintegrins: novel dimeric disintegrins and structural diversification by disulphide bond engineering. Biochem. J. 372, 725–734.

Chaipan, C., Soilleux, E.J., Simpson, P., Hofmann, H., Gramberg, T., Marzi, A., Geier, M., Stewart, E.A., Eisemann, J., Steinkasserer, A., Suzuki-Inoue, K., Fuller, G.L., Pearce, A.C., Watson, S.P., Hoxie, J.A., Baribaud, F., Pohlmann, S., 2006. DC-SIGN and CLEC-2 mediate human immunodeficiency virus type 1 capture by platelets. J. Virol. 80, 8951–8960.

Chang, C.H., Chung, C.H., Hsu, C.C., Huang, T.Y., Huang, T.F., 2009. A novel mechanism of inflammatory response in macrophage induced by a snake venom C-type lectin protein, Aggretin. XXII International Society on Thrombosis and haemostasis, 2009 Poster, PP-MO-878.

Chen, L.Y., Pan, W.W., Chen, M., Li, J.D., Liu, W., Chen, G., Huang, S., Papadimos, T.J., Pan, Z.K., 2009. Synergistic induction of inflammation by bacterial products lipopolysaccharide and fMLP: an important microbial pathogenic mechanism. J. Immunol. 182, 2518–2524.

Chung, C.H., Au, L.C., Huang, T.F., 1999. Molecular cloning and sequence analysis of aggretin, a collagen-like platelet aggregation inducer. Biochem. Biophys. Res. Commun. 263, 723–727.

Chung, C.H., Lin, K.T., Chang, C.H., Peng, H.C., Huang, T.F., 2009. The integrin α2β1 agonist, aggretin, promotes proliferation and migration of VSMC through NF-kB translocation and PDGF production. Br. J. Pharmacol. 156, 846–856.

Chung, C.H., Lin, K.T., Peng, H.C., Huang, T.F., 2007. The interaction of keratinocytes and fibroblasts with integrin α2β1 agonist, aggretin, and its regulation between epithelial/mesenchymal communications. XXI International Society on Thrombosis and Haemostasis, 2009 Poster.

Chung, C.H., Peng, H.C., Huang, T.F., 2001. Aggretin, a C-type lectin protein, induces platelet aggregation via integrin $\alpha_2\beta_1$ and GPIb in a phosphatidylinositol 3-kinase independent pathway. Biochem. Biophys. Res. Commun. 285, 689–695.

Chung, C.H., Wu, W.B., Huang, T.F., 2004. Aggretin, a snake venom-derived endothelial integrin α2β1 agonist, induces angiogenesis via expression of vascular endothelial growth factor. Blood 103, 2105–2113.

Clemetson, K.J., Lu, Q., Clemetson, J.M., 2005. Snake C-type lectin-like proteins and platelet receptors. Pathophysiol. Haemost. Thromb. 34, 150–155.

Clemetson, K.J., Morita, T., Kini, R.M., 2009. Classification and nomenclature of snake venom C-type lectins and related proteins. Toxicon 54, 83.

Cohen, J., 2002. The immunopathogenesis of sepsis. Nature 420, 885–891.

Colonna, M., Samaridis, J., Angman, L., 2000. Molecular characterization of two novel C-type lectin-like receptors, one of which is selectively expressed in human dendritic cells. Eur. J. Immunol. 30, 697–704.

Davis, G.E., Camarillo, C.W., 1996. An α2β1 integrin-dependent pinocytic mechanism involving intracellular vacuole formation and coalescence regulates capillary lumen and tube formation in three-dimensional collagen matrix. Exp. Cell Res. 224, 39–51.

Eble, J.A., Beermann, B., Hinz, H.J., Schmidt-Hederich, A., 2001. α2β1 integrin is not recognized by rhodocytin but is the specific, high affinity target of rhodocetin, an RGD-independent disintegrin and potent inhibitor of cell adhesion to collagen. J. Biol. Chem. 276, 12274–12284.

Fuller, G.L., Williams, J.A., Tomlinson, M.G., Eble, J.A., Hanna, S.L., Pohlmann, S., Suzuki-Inoue, K., Ozaki, Y., Watson, S.P., Pearce, A.C., 2007. The C-type lectin receptors CLEC-2 and Dectin-1, but not DC-SIGN, signal via a novel YXXL-dependent signaling cascade. J. Biol. Chem. 282, 12397–12409.

Gamble, J.R., Matthias, L.J., Meyer, G., Kaur, P., Russ, G., Faull, R., Berndt, M.C., Vadas, M.A., 1993. Regulation of in vitro capillary tube formation by anti-integrin antibodies. J. Cell. Biol. 121, 931–943.

Gould, R.J., Polokoff, M.A., Friedman, P.A., Huang, T.F., Holt, J.C., Cook, J.J., Niewiarowski, S., 1990. Disintegrins: a family of integrin inhibitory proteins from viper venoms. Proc. Soc. Exp. Biol. Med. 195, 168–171.

Grose, R., Hutter, C., Bloch, W., Thorey, I., Watt, F.M., Fassler, R., Brakebusch, C., Werner, S., 2002. A crucial role of beta 1 integrins for keratinocyte migration in vitro and during cutaneous wound repair. Development 129, 2303–2315.

Herr, A.B., Farndale, R.W., 2009. Structural insights into the interactions between platelet receptors and fibrillar collagen. J. Biol. Chem. 284, 19781–19785.

Hirabayashi, J., Kusunoki, T., Kasai, K., 1991. Complete primary structure of a galactose-specific lectin from the venom of the rattlesnake *Crotalus atrox*. Homologies with Ca^{2+}-dependent-type lectins. J. Biol. Chem. 266, 2320–2326.

Holtkotter, O., Nieswandt, B., Smyth, N., Muller, W., Hafner, M., Schulte, V., Krieg, T., Eckes, B., 2002. Integrin α2-deficient mice develop normally, are fertile, but display partially defective platelet interaction with collagen. J. Biol. Chem. 277, 10789–10794.

Hooley, E., Papagrigoriou, E., Navdaev, A., Pandey, A.V., Clemetson, J.M., Clemetson, K.J., Emsley, J., 2008. The crystal structure of the platelet activator aggretin reveals a novel (αβ)2 dimeric structure. Biochemistry 47, 7831–7837.

Huang, T.F., Holt, J.C., Lukasiewicz, H., Niewiarowski, S., 1987. Trigramin. A low molecular weight peptide inhibiting fibrinogen interaction with platelet receptors expressed on glycoprotein IIb-IIIa complex. J. Biol. Chem. 262, 16157–16163.

Huang, T.F., Liu, C.Z., Yang, S.H., 1995. Aggretin, a novel platelet-aggregation inducer from snake (*Calloselasma rhodostoma*) venom, activates phospholipase C by acting as a glycoprotein Ia/IIa agonist. Biochem. J. 309, 1021–1027.

Ivaska, J., Reunanen, H., Westermarck, J., Koivisto, L., Kahari, V.M., Heino, J., 1999. Integrin α2β1 mediates isoform-specific activation of p38 and upregulation of collagen gene transcription by a mechanism involving the α2 cytoplasmic tail. J. Cell Biol. 147, 401–416.

Jarvis, G.E., Atkinson, B.T., Snell, D.C., Watson, S.P., 2002. Distinct roles of GPVI and integrin $\alpha_2\beta_1$ in platelet shape change and aggregation induced by different collagens. Br. J. Pharmacol. 137, 107–117.

Kato, Y., Fujita, N., Kunita, A., Sato, S., Kaneko, M., Osawa, M., Tsuruo, T., 2003. Molecular identification of Aggrus/T1α as a platelet aggregation-inducing factor expressed in colorectal tumors. J. Biol. Chem. 278, 51599–51605.

Kerrigan, A.M., Dennehy, K.M., Mourao-Sa, D., Faro-Trindade, I., Willment, J.A., Taylor, P.R., Eble, J.A., Reis e Sousa, C., Brown, G.D., 2009. CLEC-2 is a phagocytic activation receptor expressed on murine peripheral blood neutrophils. J. Immunol. 182, 4150–4157.

Kowalska, M.A., Tan, L., Holt, J.C., Peng, M., Karczewski, J., Calvete, J.J., Niewiarowski, S., 1998. Alboaggregins A and B. Structure and interaction with human platelets. Thromb. Haemost. 79, 609–613.

Lankhof, H., van Hoeij, M., Schiphorst, M.E., Bracke, M., Wu, Y.P., Ijsseldijk, M.J., Vink, T., de Groot, P.G., Sixma, J.J., 1996. A3 domain is essential for interaction of von Willebrand factor with collagen type III. Thromb. Haemost. 75, 950–958.

Leduc, M., Bon, C., 1998. Cloning of subunits of convulxin, a collagen-like platelet-aggregating protein from *Crotalus durissus terrificus* venom. Biochem. J. 333, 389–393.

Li, T.T., Larrucea, S., Souza, S., Leal, S.M., Lopez, J.A., Rubin, E.M., Nieswandt, B., Bray, P.F., 2004. Genetic variation responsible for mouse strain differences in integrin α2 expression is associated with altered platelet responses to collagen. Blood 103, 3396–3402.

May, F., Hagedorn, I., Pleines, I., Bender, M., Vogtle, T., Eble, J., Elvers, M., Nieswandt, B., 2009. CLEC-2 is an essential platelet activating receptor in hemostasis and thrombosis. Blood 114, 3464–3472.

Mizuno, H., Fujimoto, Z., Koizumi, M., Kano, H., Atoda, H., Morita, T., 1997. Structure of coagulation factors IX/X-binding protein, a heterodimer of C-type lectin domains. Nat. Struc. Biol. 4, 438–441.

Morita, T., 2005. Structures and functions of snake venom CLPs (C-type lectin-like proteins) with anticoagulant-, procoagulant-, and platelet-modulating activities. Toxicon 45, 1099–1114.

Moroi, M., Jung, S.M., Nomura, S., Sekiguchi, S., Ordinas, A., Diaz-Ricart, M., 1997. Analysis of the involvement of the von Willebrand factor-glycoprotein Ib interaction in platelet adhesion to a collagen-coated surface under flow conditions. Blood 90, 4413–4424.

Nathan, C., 2002. Points of control in inflammation. Nature 420, 846–852.

Navdaev, A., Clemetson, J.M., Polgar, J., Kehrel, B.E., Glauner, M., Magnenat, E., Wells, T.N., Clemetson, K.J., 2001. Aggretin, a heterodimeric C-type lectin from *Calloselasma rhodostoma* (Malayan pit viper), stimulates platelets by binding to α2β1 integrin and glycoprotein Ib, activating Syk and phospholipase Cγ2, but does not involve the glycoprotein VI/Fc receptor γ chain collagen receptor. J. Biol. Chem. 276, 20882–20889.

Nieswandt, B., Brakebusch, C., Bergmeier, W., Schulte, V., Bouvard, D., Mokhtari-Nejad, R., Lindhout, T., Heemskerk, J.W., Zirngibl, H., Fassler, R., 2001. Glycoprotein VI but not α2β1 integrin is essential for platelet interaction with collagen. EMBO J. 20, 2120–2130.

Ouyang, C., Teng, C.M., Huang, T.F., 1992. Characterization of snake venom components acting on blood coagulation and platelet function. Toxicon 30, 945–966.

Ouyang, C.H., Yeh, H.I., Huang, T.F., 1986. Purification and characterization of a platelet aggregation inducer from *Calloselasma rhodostoma* (Malayan pit viper) snake venom. Toxicon 24, 633–643.

Ozaki, Y., Suzuki-Inoue, K., Inoue, O., 2009. Novel interactions in platelet biology: CLEC-2/podoplanin and laminin/GPVI. J Thromb. Haemost. 7(Suppl 1), 191–194.

Rankin, J.A., 2004. Biological mediators of acute inflammation. AACN Clin. Issues 15, 3–17.

Ravanti, L., Heino, J., Lopez-Otin, C., Kahari, V.M., 1999. Induction of collagenase-3 (MMP-13) expression in human skin fibroblasts by three-dimensional collagen is mediated by p38 mitogen-activated protein kinase. J. Biol. Chem. 274, 2446–2455.

Reid, H.A., Thean, P.C., Chan, K.E., Baharom, A.R., 1963. Clinical effects of bites by Malayan viper (*Ancistrodon rhodostoma*). Lancet 1, 617–621.

Saelman, E.U., Keely, P.J., Santoro, S.A., 1995. Loss of MDCK cell α2β1 integrin expression results in reduced cyst formation, failure of hepatocyte growth factor/scatter factor-induced branching morphogenesis, and increased apoptosis. J. Cell Sci. 108, 3531–3540.

Savage, B., Almus-Jacobs, F., Ruggeri, Z.M., 1998. Specific synergy of multiple substrate-receptor interactions in platelet thrombus formation under flow. Cell 94, 657–666.

Shin, Y., Morita, T., 1998. Rhodocytin, a functional novel platelet agonist belonging to the heterodimeric C-type lectin family, induces platelet aggregation independently of glycoprotein Ib. Biochem. Biophys. Res. Commun. 245, 741–745.

Suzuki-Inoue, K., Fuller, G.L., Garcia, A., Eble, J.A., Pohlmann, S., Inoue, O., Gartner, T.K., Hughan, S.C., Pearce, A.C., Laing, G.D., Theakston, R.D., Schweighoffer, E., Zitzmann, N., Morita, T., Tybulewicz, V.L., Ozaki, Y., Watson, S.P., 2006. A novel Syk-dependent mechanism of platelet activation by the C-type lectin receptor CLEC-2. Blood 107, 542–549.

Suzuki-Inoue, K., Kato, Y., Inoue, O., Kaneko, M.K., Mishima, K., Yatomi, Y., Yamazaki, Y., Narimatsu, H., Ozaki, Y., 2007. Involvement of the snake toxin receptor CLEC-2, in podoplanin-mediated platelet activation, by cancer cells. J. Biol. Chem. 282, 25993–26001.

Suzuki-Inoue, K., Ozaki, Y., Kainoh, M., Shin, Y., Wu, Y., Yatomi, Y., Ohmori, T., Tanaka, T., Satoh, K., Morita, T., 2001. Rhodocytin induces platelet aggregation by interacting with glycoprotein Ia/IIa (GPIa/IIa, Integrin $\alpha 2\beta 1$). Involvement of GPIa/IIa-associated src and protein tyrosine phosphorylation. J. Biol. Chem. 276, 1643–1652.

Teng, C.M., Huang, T.F., 1991. Inventory of exogenous inhibitors of platelet aggregation. For the Subcommittee on Nomenclature of Exogenous Hemostatic Factors of the Scientific and Standardization Committee of the International Society on Thrombosis and Haemostasis. Thromb. Haemost. 65, 624–626.

Teng, C.M., Hung, M.L., Huang, T.F., Ouyang, C., 1989. Triwaglerin: a potent platelet aggregation inducer purified from *Trimeresurus wagleri* snake venom. Biochim. Biophys. Acta 992, 258–264.

Wang, W.J., Huang, T.F., 2001. A novel tetrameric venom protein, agglucetin from *Agkistrodon acutus*, acts as a glycoprotein Ib agonist. Thromb. Haemost. 86, 1077–1086.

Xiang, D., Huang, D., Gai, L., Liu, H., 2000. Relationship between expression of type III collagen and phenotype of vascular smooth muscle cells in neointimal of stented coronary artery. Chin. Med. J. (Engl) 113, 324–327.

Yeh, C.H., Wang, W.C., Hsieh, T.T., Huang, T.F., 2000. Agkistin, a snake venom-derived glycoprotein Ib antagonist, disrupts von Willebrand factor-endothelial cell interaction and inhibits angiogenesis. J. Biol. Chem. 275, 18615–18618.

Chapter 35
Snaclecs (Snake C-Type Lectins) that Activate Platelets

Kenneth J. Clemetson

Abstract One of the major targets for snake venom proteins is haemostasis. This weakens the prey and helps with swallowing and digestion. The venom proteins act either on coagulation factor or on platelets. Snake venom proteins mainly adapt physiological mechanisms to inhibit or activate platelets. The most efficient way for snake venom to reduce platelet function is not by inhibiting the function of individual or several receptors but rather by activating platelets so that they are removed from the circulation producing thrombocytopenia. Platelets can be activated efficiently by an agonist using low molecule numbers in two main ways already used physiologically. One of these is by proteases acting on proteolytically activated receptors. The other major route is by clustering receptors mimicking physiological ligands such as von Willebrand factor and collagen. The snaclecs described in this chapter fall into this latter category. Their targets are those of the physiological ligands and include GPIb, GPVI, $\alpha 2\beta 1$ and the recently discovered CLEC2.

Introduction

Snake venom proteins that affect hemostasis are most generally found in Viperidae, and Crotalidae snakes but the others often contain some as well.

They mostly adapt physiological mechanisms to inhibit or activate platelets. The most efficient way for snake venom to reduce platelet function is not by inhibiting the function of individual or several receptors but rather by activating platelets so that they are removed from the circulation producing thrombocytopenia. In comparison with the amount of protein which the snake would need to synthesise for efficient inhibition of a major receptor such as GPIb, requiring ~80% inhibition of a receptor present in ~50,000 copies per platelet, only a relatively few molecules of an efficient activating molecule are necessary to activate a platelet and contribute to its

K.J. Clemetson (✉)
Department of Haematology, University of Berne, Inselspital, CH-3010 Berne, Switzerland
e-mail: kenneth.clemetson@tki.unibe.ch

R.M. Kini et al. (eds.), *Toxins and Hemostasis*, DOI 10.1007/978-90-481-9295-3_35,

removal from the circulation. Activating molecules may also act on a range of other cells critical for the snake's aim of immobilising its prey, such as endothelial and smooth muscle cells. There are indeed snake venom components that inhibit platelet functions, such as the disintegrins, but in general they are considerably smaller than snaclecs and thus require less protein synthesis. During evolution the snake with the most efficient poisonous venom has advantages although this is not the only factor playing a role (Fry et al., 2006). Other aspects of venom may come into play as well and have not been widely considered, such as the range of species against which the venom components are effective and whether species change much with the snake's range. Platelets can be activated efficiently by low molecule numbers of an agonist in two main ways which are already used physiologically. One of these is by proteases acting on proteolytically activated receptors. Snake venom often contains thrombin-like proteases that can activate platelets via various PARs. The other major route is by clustering receptors – mainly tyrosine kinase dependent receptors – often by using and adapting normal physiological pathways. This is the main route used by snaclecs though other venom components such as the metalloproteases, which may also contain snaclec domains, may use this mechanism as well. We have no examples of venom components that activate platelets by binding to and changing the conformation of seven transmembrane, G-protein coupled receptors but, superficially at least, there is no obvious reason why they should not exist. They may not yet have been detected. This review concentrates on structural and functional properties of venom components of the C-type lectins class (now named snaclecs) that activate platelets.

Snaclecs

Because of confusion with classic C-type lectins and since names such as C-type lectin-like or – related proteins are almost inevitably abbreviated to CTL or CLP and provide no information about the heterodimeric structure, loop-swapping or higher order multimerization, this group have been named **snaclecs** (*Sna*ke venom *C*-type *lec*tin*s*) in a recent nomenclature proposal by the Exogenous Factor Committee of the International Thrombosis and Haemostasis Society (Clemetson et al., 2009)

A typical structure of a snaclec αβ heterodimer is shown in Fig. 35.1a. This is the basic unit from which the different structures of the snaclec family are assembled. Snaclecs bind to a wide range of coagulation factors, other proteins critical in haemostasis, and membrane receptors on platelets and other cells. However, many have more than one binding site and/or may interact with more than one protein/receptor. Snaclecs have been described that interact with coagulation factors X/XI and X (Mizuno et al., 1999, 2001). Crystallography studies indicate that the binding site on the coagulation factor is the Gla domain whereas the site on the snaclec is the concave surface lying between the two subunits. Binding is largely via electrostatic interactions but shape-fitting also has a role.

Snaclecs have a basic heterodimeric structure with two subunits, α and β, nearly always linked covalently, via a disulphide bond. The heterodimers are often further

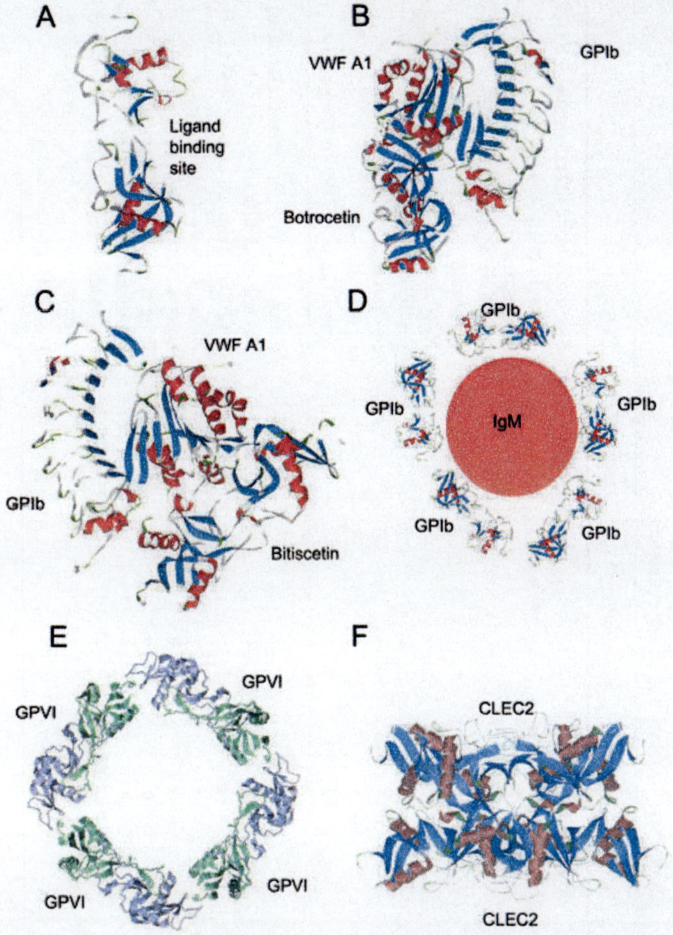

Fig. 35.1 Figure shows the structures of (**a**) Heterodimeric snaclec (echicetin) with swapped loops. (**b**) The botrocetin-von Willebrand factor (VWF) A1 domain-GPIbα complex. (**c**) The bitiscetin-von Willebrand factor (VWF) A1 domain-GPIbα complex. (**d**) The IgMκ-echicetin complex binding five molecules of GPIbα. (**e**) Convulxin, a typical tetrameric heterodimer snaclec. (**f**) Aggretin/rhodocytin, forming a back-to-back dimer, cross-linking CLEC2. The structures were prepared from primary sequences using Swiss-Model or from crystal structures and illustrated using Accelys DS Viewer 5.0

multimerized either non-covalently or covalently, via additional disulphide bonds, to form larger structures. The interpretation of the effects of snaclecs on platelets presents a problem. One of the earliest snaclecs to be described, botrocetin, clearly activates platelets by inducing interactions between GPIb and von Willebrand factor (Brinkhous et al., 1983), but many simple heterodimeric members of this family have been reported to inhibit GPIb-related platelet functions while other, more complex ones, activate platelets via the same receptor. There are many unresolved

Table 35.1 Snaclecs that activate platelets

Protein name	Species	Target	Sequence	Structure	References
Botrocetin	*Bothrops jararaca*	VWF/GPIb	+	1FVU	Fukuda et al. (2002); Sen et al. (2001); Usami et al. (1993);
Bitiscetin	*Bitis arietans*	VWF/GPIb	+	1JWI	Hamako et al. (1996); Hirotsu et al. (2001); Maita et al. (2003)
Aspercetin	*Bothrops asper*	VWF/GPIb	+		Rucavado et al. (2001)
Echicetin	*Echis carinatus*	GPIb, IgMk	+	–	Navdaev et al. (2001); Polgar et al. (1997);
Mamushigin	*Gloydius halys blomhoffii*	GPIb	+	–	Sakurai et al. (1998)
Purpureotin	*Trimeresurus purpureomaculatus*	GPIb	+	–	Li et al. (2004)
Alboaggregin-B	*Cryptelytrops albolabris*	GPIb	+	–	Kowalska et al. (1998); Peng et al. (1991)
Trimecetin	*Trimeresurus mucrosquamatus*	GPIb	–	–	Chiou et al. (1996); Chen et al. (2010)
Flavocetin-A	*Trimeresurus flavoviridis*	GPIb	+	1C3A	Fukuda et al. (2000); Shin et al. (2000); Taniuchi et al. (1995)
Mucrocetin	*Trimeresurus mucrosquamatus*	GPIb	+	1V4L	Huang et al. (2004)
Mucetin	*Trimeresurus mucrosquamatus*	GPIb	+	–	Lu et al. (2004); Wei et al. (2002)
Alboaggregin-C	*Cryptelytrops albolabris*	GPIb	–	–	Peng et al. (1992)
Agglucetin	*Deinagkistrodon acutus*	GPIb	+	–	Wang and Huang (2001)
Convulxin	*Crotalus durissus terrificus*	GPVI, GPIb	+	1UMR	Batuwangala et al. (2004); Leduc and Bon (1998); Murakami et al. (2003); Polgar et al. (1997)

Table 35.1 (continued)

Protein name	Species	Target	Sequence	Structure	References
Alboaggregin-A	*Trimeresurus albolabris*	GPIb, GPVI	+	–	Dormann et al. (2001); Kowalska et al. (1998)
Stejnulxin	*Trimeresurus stejnegeri*	GPVI	+	–	Lee et al. (2003)
Ophioluxin	*Ophiophagus hannah*	GPVI	–	–	Du et al. (2002a)
Alboluxin	*Cryptelytrops albolabris*	GPIb, GPVI	–	–	Du et al. (2002b)
Aggretin/ rhodocytin	*Calloselasma rhodostoma*	CLEC-2 α2β1, GPIb	+	–	Bergmeier et al. (2001); Chung et al. (1999); Navdaev et al. (2001)
Bilinexin	*Agkistrodon bilineatus*	GPIb, α2β1	–	–	Du et al. (2001)
Agkisacutacin	*Deinagkistrodon acutus*	unknown	+	–	Cheng et al. (1999)
Agkaggregin	*Deinagkistrodon acutus*	unknown	–	–	Liu et al. (2002)
Rhodoaggretin	*Calloselasma rhodostoma*	unknown	–	–	Wang et al. (2001)

questions about those snaclecs classified as inhibitory. While those binding to $\alpha 2\beta 1$ probably are, those recognizing GPIb are more problematic and there is a clear case for more investigation being necessary. As discussed in Chapter 21, a major reason why this is not as clear as might be deduced from reading many of the early papers in this area, is that venom components were tested often only against washed human platelets and in best cases only against platelets of one of a narrow range of species; mouse, rat, guinea pig. However, the diet of the snake in question may cover a wide range of species including reptiles and birds and the structure of the target molecule may differ considerably. Although human platelets work for many experiments they may not be ideal for assessing best function. A classic case for reassignment of function is echicetin, which is described below. Snaclecs activate platelets via a number of different clustering strategies, they are either simple heterodimeric structures that link platelet receptors to their physiological ligands under non-physiological conditions or they use "innocent bystander" proteins to help with the cross-linking (Table 35.1). More elaborate strategies involve multimerisation of the snaclec itself. Here, two main techniques have been identified, non-covalent dimerisation of the snaclec, or covalent tetramerisation. It has to be added that under physiological conditions many of these can form non-covalent high multimers that are much more effective in clustering receptors and activating platelets than the simple covalent multimers might suggest. Many aspects of these clustering activation mechanisms remain unclear. Based on these mechanisms and their targets the platelet activating snaclecs can be divided into a number of groups.

GPIb/Von Willebrand Factor Specific Snaclecs

Botrocetin and bitiscetin are two well-known snaclecs that induce binding between von Willebrand factor and GPIb on platelets and agglutinate/aggregate them. Botrocetin was one of the first snaclecs to be described (Brinkhous et al., 1983) and has become a standard reagent for testing von Willebrand factor/platelet interactions and detection of the defects in von Willebrand's disease and in GPIb-related disorders such as Bernard-Soulier syndrome. Another commonly used reagent to induce these interactions is the antibiotic ristocetin (Howard and Firkin, 1971). Differences in the activity of these reagents have puzzled investigators for a number of years. Recently, crystal structures of A1 domains of von Willebrand factor with the 45 kDa domain of GPIbα and botrocetin or bitiscetin have been determined (Fukuda et al., 2005; Maita et al., 2003). The structure of the GPIb/botrocetin/A1 domain complex is shown in Fig. 35.1b. This shows again how the snaclec uses the concave region between the subunits as the main binding site for the A1 domain involving mainly electrostatic binding. It was earlier thought that botrocetin most likely bound to the A1 domain and induced a conformational change in the A1 domain to the GPIb-binding state. However, the crystal structure of the complex shows no major conformational shifts within the A1 domain compared to the free A1 structure, arguing against this explanation. It could also be argued that the recombinant A1 domain used is already in the activated conformation. Be that as it may this structure indicates that, after binding to the VWF A1 domain, botrocetin then slips round to stabilise the "on" form of GPIb/A1 binding by clamping the two molecules together. This probably also involves the anionic peptide region of GPIbα not present in the crystal structure. The β-subunit of botrocetin is very near to the C-terminal region of the 45 kDa domain of GPIbα in the region that would be occupied by the anionic peptide of GPIbα if it were present. In addition, a positively charged patch on the β-subunit formed by lysines 45, 102 and 107 is appropriately placed to be a putative binding site for the 3 sulfated tyrosine residues in the anionic peptide that are critical for this interaction.

The crystal structure of A1-bitiscetin was also published recently and shows that bitisicetin binds the A1 domain at a different site than botrocetin (Maita et al., 2003). A model of the trimeric complex prepared by superimposing the A1-GPIb domain complex, shows that the β-subunit of bitiscetin, like that of botrocetin, also approaches the C-terminal region of the 45 kDa domain of GPIbα, although from the other side of the 45 kDa domain (Fig. 35.1c). Again, three lysines 20, 21, and 120 are positioned to bind to the 3 sulfated tyrosines of the anionic peptide. Thus, both these studies show that botrocetin and bitiscetin most likely facilitate von Willebrand's factor/GPIb binding by clamping both molecules together rather than by forcing confirmational changes (however, the isolated A1 domain has an activated conformation so these results may not reflect completely the situation with native von Willebrand factor). The different binding sites on both the A1 domain and GPIb as well as the different location of the lysine triplets in the two snaclec molecules suggests that they may have arisen by parallel evolution from a snaclec precursor.

Aspercetin, a further member of this family of snaclecs from *Bothrops asper* was described recently (Rucavado et al., 2001) and probably shows similar binding to botrocetin, which it closely resembles. Since this is an efficient way of activating platelets in vivo it is likely that still further snakes will be found that have similar snaclecs in their venom.

GPIb-Binding Snaclecs

The GPIb-binding snaclecs constitute a large group within this class and have been isolated from many different snake venoms. New members are being found regularly. They have sequences similar to one another but nevertheless also differences. The similarities mostly resemble those found with other snaclecs with different targets and represent the structural part of the molecule necessary for folding, which is highly conserved, whereas the divergent sequence represents mainly the binding regions that are individually optimized. As with botrocetin (Fukuda et al., 2002) and bitiscetin (Maita et al., 2003), this suggests that they are optimised for different binding sites on GPIb, an interpretation supported by the differences in effects on GPIb functions that some of these show. For example, among those that inhibit GPIb-VWF binding, only a few have been shown to inhibit platelet activation via thrombin, implicating blockage of the thrombin-binding site on GPIb. However, not all these molecules are specific for GPIb alone and several have binding sites for

GPIb together with other platelet receptors. Several multimeric (heterodimeric) snaclecs, with specificity for GPIb, have been isolated and characterised that activate platelets with various degrees of efficacity.

An early example of a GPIb-binding snaclec is echicetin, which was shown to block von Willebrand factor as well as low dose thrombin induced platelet activation (Peng et al., 1994). At that point it was therefore classed as inhibitory. However, early reports also noted that it induced thrombocytopenia in small animals. The inhibitory studies had been done with washed platelets. However, when echicetin was tested with platelet rich plasma it induced platelet aggregation (Navdaev et al., 2001) (see Chapter 21). When echicetin was tested with platelet rich plasma it was shown to induce platelet aggregation. Further isolation and characterization showed that echicetin binds a specific component of plasma, a subcategory of IgMκ, as well as to GPIb on platelets. The IgMκ fraction can be isolated by affinity chromatography on echicetin. Washed platelets treated with echicetin and purified IgMκ agglutinate and show classic evidence of activation, fibrinogen binding and increased tyrosine phosphorylation of several signalling molecules. Echicetin/IgMκ complexes may activate platelets by clustering GPIb molecules (five per complex) on the platelet surface and also by linking GPIb molecules between platelets (Fig. 35.1d).

Other GPIb-inhibitory snaclecs do not agglutinate/aggregate platelets in plasma, although they induce thrombocytopenia suggesting they interact with as yet unidentified components/receptors present on other cells such as the endothelium. Until we know what is really happening here and what other receptors/proteins are involved it will be difficult to understand how these snaclecs work. In any case their function seems to be to remove platelets from the circulation. After injection of the snaclec and the fall in platelet count the platelets gradually return to the circulation. One mechanism for this might be that GPIb is cleaved from the surface of the platelets together with the snaclec but this has not yet been checked.

There are several multimeric C-type lectins with GPIb binding properties that cause platelet agglutination with varying degrees of activation. One class of these is represented by purpureotin, which is a simple heterodimeric snaclec forming a back-to-back, non-covalent dimer. Presumably it activates platelets by clustering GPIb via this dimeric structure. Another snaclec with a similar structure is aggretin/rhodocytin, dealt with later. Two other similar snaclecs alboaggregin B (Peng et al., 1991) and trimecetin (Chiou et al., 1996) appear to activate platelets via GPIb using similar dimerization mechanisms. It is interesting to compare the induction of aggregation by trimecetin with that by mucrocetin (Huang et al., 2004) implying differences in the overall mechanism. The main group of multimeric snaclecs using GPIb clustering to activate platelets are the tetrameric heterodimers, such as the prototype of this class, flavocetin (Taniuchi et al., 1995). They have been described as having effects on platelets varying from small agglutinates (flavocetin) up to large agglutinates (mucrocetin) (Huang et al., 2004) or to full aggregation (mucetin) (Lu et al., 2004), two other similar molecules from *Trimeresurus mucrosquamatus* venom have been described that have sequences only slightly different from flavocetin and yet show quite important functional differences.

Mucrocetin (Huang et al., 2004) forms large agglutinates with washed platelets and mucetin (Lu et al., 2004), causes washed platelets to aggregate, involving activation of $\alpha_{IIb}\beta_3$ and cross-linking via fibrinogen.

Since they all seem to act via the same receptor there is a lot of bewilderment about these differences and what they mean. We do not see the same level of difference between snaclecs activating platelets via other receptors. The problem therefore seems to lie in the snaclec-GPIb interaction. If we look at the physiological activation of platelets via GPIb we see that there is a requirement for VWF binding to GPIb, most likely for GPIb clustering, and for shear-stress, which translates into pulling on GPIb and the cytoskeleton to which it is attached. This can be simulated under in vitro conditions in a number of ways: using extraneous reagents such as ristocetin or botrocetin to induce VWF–GPIb binding in the absence of high shear; removing sialic acid residues from VWF to produce an "activated" molecule or using mutant forms of VWF of the IIB class that are also preactivated. The use of botrocetin compared to ristocetin gives quite small aggregates, which is also still unexplained. On the other hand, ristocetin-induced platelet aggregation seems to depend on small domains within GPIb and VWF as a recent mutation in VWF demonstrates (Flood et al., 2009) while botrocetin-induced platelet aggregation involves large surface area interactions.

Is it the strength of binding between these snaclecs and GPIb alone that determines the result? One other possibility is that the minor differences in sequence between these snaclecs, which mostly lie outside the concave binding site, affect homointeractions between the snaclecs themselves that may be induced when they bind to GPIb. Thus, mucretin and mucetin may form larger complexes between GPIb on platelets with proportionately more shear stress activation of the platelets in a stirred system.

The sequence of GPIb is fairly variable between species compared to many platelet receptors (e.g. integrins). Thus, there is little cross-reactivity of monoclonal antibodies between species making it difficult to prepare monoclonals to human GPIb, which function i.e. inhibit VWF binding and can be tested in small animal models. The best cases are effective also in other primates such as baboons. Snaclecs work across a wide range of species but the affect of a given GPIb-binding snaclec on platelets of different species may vary considerably. The reason for this broader spectrum of binding compared to monoclonal antibodies is that snaclecs interact with GPIb via a larger protein-protein surface. Thus, even in the case of an imperfect match enough other molecular interactions still occur to permit a level of binding adequate for inhibition of von Willebrand factor (or thrombin) binding. While we do not yet have crystal structures of these snaclec/GPIb complexes, we do have quite a lot of information from competition studies with various anti-GPIb monoclonal antibodies that indicate that each GPIb-binding snaclec has individually optimised its binding domain on GPIbα. Although we have no evidence yet, the binding domain might vary somewhat between snaclecs binding to GPIb from different species because of the varying amino acid sequences. One consequence of this variability is that the binding intensity to GPIb varies from one species to another. Because snake venoms contain 200–300 components, in any given prey species those venom components that are most effective, represent only a small selection of the total. Thus, snaclecs targeting GPIb would be most effective in certain prey species but have a fairly broad spectrum of prey species where they can assist in immobilising/killing the prey. Although there are only a few snaclecs in the venom of a given species the snake can extend the effective target range by sequence variation (which in humans would be called polymorphism) and also by different α/β subunit combinations. If a given species has a wide range of environments the venom composition may vary considerably across the range. Examples are *Echis carinatus* which has a range from the west coast of North Africa to the deserts of Central Asia. There is some evidence that snakes can adapt their venom protein expression to match better the prey in their range (Valentin and Lambeau, 2000).

Some authors think that the different snaclecs affect GPIb differently when they bind, presumably inducing conformational changes which activate the platelet. The mechanism by which GPIb complex "works" is still poorly understood although there are quite a few publications dealing with this. The current model involves binding (any binding may be sufficient) to GPIbα, clustering, and application of tension. This affects the cytoskeleton coupled to the cytoplasmic domain of GPIb complex, within lipid raft domains and causes the sort of shifts in distribution of signaling molecules such as kinases (both src family and tyrosine kinases) and their corresponding phosphatases so as to trigger a signaling cascade leading to Ca^{2+} release in the cytoplasm and eventually activation of αIIbβ3 and release of storage granules. Snaclecs can only partly simulate some of the effects of physiological reagents such as VWF and collagen and so must compensate for this by providing concentrated clustering mechanisms as well as tight binding to "resting" GPIb. If a snaclec can bind to GPIb of a given species but not very tightly – due to an incomplete match of binding site – it may be adequate to induce small scale agglutination. When shear stress is applied – for example, stirring – small agglutinates experience less force/platelet than large agglutinates and the platelets are not activated. Alternatively, weakly binding snaclecs do not cluster GPIb as much as strongly binding ones, limiting the binding force between platelets. Either way a critical cluster size is required for signaling to be induced leading to aggregation. It would be interesting to know whether flavocetin does indeed induce aggregation in platelets/thrombocytes from certain species but it would be quite complicated to check platelets/thrombocytes from a wide range of the presumed prey species of *Protobothrops* (formerly, *Trimeresurus*) *flavoviridis*.

In the long run these differences in snaclec effects on GPIb of different species will probably be exploited to explore GPIb signaling mechanisms.

GPVI Binding Snaclecs

Another important class of snake snaclecs activates platelets via GPVI. GPVI is an important collagen receptor and an ideal target for snake proteins because when clustered it induces massive platelet activation. The first such snaclec to be characterised was convulxin from *Crotalus durissus terrificus* (Leduc and Bon, 1998; Polgar et al., 1997) and it has developed into an important tool for the study of GPVI-specific platelet activation, producing effects similar to those found with another GPVI-specific reagent, collagen related peptide (CRP), made up of cross-linked GPO repeat peptides. The determination of the crystal structure of convulxin by X-ray methods (Batuwangala et al., 2004; Murakami et al., 2003) showed that it too belongs to the tetrameric heterodimer class of snaclecs, like flavocetin etc. and, like these, has an additional cysteine residue near one end of each subunit forming the extra disulfide link between each heterodimer (Fig. 35.1e). Convulxin in solution and in its crystal structure (Horii et al., 2009) is dimeric, i.e. two tetrameric heterodimers complexed non-covalently and this probably provides the main mechanism for platelet-platelet cross-linking. However, it is not clear whether the clustering of four GPVI molecules on a platelet by each convulxin molecule is enough to cause the strong signalling that is observed or whether, like some of the tetrameric GPIb-binding molecules mentioned above there is additional amplification by non-covalent interactions between convulxin molecules. Convulxin also binds weakly to GPIb (Du et al., 2002a; Kanaji et al., 2003) and, although this can contribute to cross-linking platelets, at concentrations normally used to activate platelets has only a very minor role in overall signal transduction. This may be a residual activity in a molecule still actively evolving or may be related to broadening the species specificity of convulxin. Several other snaclecs acting via GPVI similar to convulxin have been described such as ophioluxin from *Ophiophagus hannah* (Du et al., 2002a), stejnulxin from *Trimeresurus stejnegeri* (Lee et al., 2003) and trowaglerix from *Trimeresurus wagleri* (Chang et al., 2008). No residual GPIb-binding activity was detected in any of these. All of these are more active than convulxin on a mole/mole basis varying from 2X for ophioluxin to 10X for trowaglerix. The molecular basis for these differences in activity are unclear but, as in the case of snaclecs of the same tetrameric family activating platelets via GPIb may reflect their binding affinity or their ability to form larger non-covalently associated multimers.

α2β1 Specific Snaclecs

Several snake C-type lectin-like molecules interact with platelets via α2β1, but none of these are completely specific, in all cases reported so far also involving other platelet receptors. These include aggretin (also called rhodocytin) from *Calloselasma rhodostoma* (Bergmeier et al., 2001; Chung et al., 1999; Navdaev et al., 2001), which has a classic heterodimeric structure but forms non-covalent "back-to-back" dimers as shown by X-ray crystallography (Hooley et al., 2008) (Fig. 35.1f). Recent studies found that CLEC-2 was the major platelet receptor for aggretin/rhodocytin but there remains considerable evidence for a role for α2β1 aggretin at least in some cells (see Chapter 34 for more detailed description of this snaclec and its properties).

CLEC-2 Specific Snaclecs

The discovery that aggretin (also called rhodocytin) from *Calloselasma rhodostoma* (Bergmeier et al., 2001; Chung et al., 1999; Navdaev et al., 2001), activates platelets and other cells mainly via CLEC-2 (Suzuki-Inoue et al., 2006) (although there is evidence for a role for α2β1, see above) was a major breakthrough in demonstrating an important role for this C-type lectin membrane receptor that signals via tyrosine phosphorylation. Neutrophils (at least in mice) also express CLEC-2 and are activated by aggretin/rhodocytin (Kerrigan et al., 2009). The platelet studies later lead to the identification of podoplanin as the major physiological CLEC-2 ligand (Kato et al., 2008) and to the discovery, based on CLEC-2 and podoplanin knock-out mice, that CLEC-2-podoplanin interactions are important in haemostasis and in the formation of the blood vessel/lymphatic vessel separation (Uhrin et al., 2010). The podoplanin/CLEC-2 interaction is dependent on the correct glycosylation of podoplanin (Kaneko et al., 2007) since CLEC-2 is a "real" C-type lectin with carbohydrate binding properties. This is a classic example of the utility of studying snake venom components. Without aggretin/rhodocetin the importance of the CLEC-2/popoplanin interaction would certainly have been discovered one day but the snake venom work clearly accelerated this. Podoplanin on some tumour cells interacting with CLEC-2 on platelets was shown to be involved in promotion of pulmonary metastases (Kunita et al., 2007) suggesting that aggretin/rhodocytin might provide a structural basis for development of inhibitors.

It is still unknown whether there are other snake venoms which contain aggretin/rhodocytin-like components acting via CLEC-2 or similar receptors. One of the interesting characteristics marking out aggretin/rhodocytin from other snaclec platelet agonists is the characteristic time lag in the start of aggregation, dependent on agonist concentration, followed by a rapid aggregation. This resembles aggregation induced by collagen and was one of the reasons a role for α2β1 was suspected. There may be other mechanistic causes including the way CLEC-2 is clustered and how downstream signaling is regulated.

Multi-Receptor Snaclecs

Quite a few snake C-type lectins have binding sites for several platelet receptors. In this they mimic natural platelet agonists such as collagen or VWF. The combination of GPIb and GPVI is quite common, e.g. alboaggregin A (Dormann et al., 2001) or alboluxin (Du et al., 2002b), as is GPIb and α2β1 e.g. bilinexin (Du et al., 2001). Strangely, no snaclecs using αIIbβ3 have been reported so far. Other integrins found on other cells and not on platelets have also been detected as binding sites.

Large numbers of new snaclecs from different snake species continue to be reported that activate platelets. Some of these proteins, such as agkisacutacin (Cheng et al., 2000), agkaggregin (Liu et al., 2002) and rhodoaggretin (Wang et al., 2001), use receptors that have not yet been identified. While many of these new snaclecs use receptors and mechanisms similar to those already reported, it is quite likely that other (new?) platelet receptors may also be found to play a role, which offers the possibility of developing new tools to analyze platelet function.

References

Batuwangala, T., Leduc, M., Gibbins, J.M., Bon, C., Jones, E.Y., 2004. Structure of the snake-venom toxin convulxin. Acta Crystallogr. D. Biol. Crystallogr. 60, 46–53.

Bergmeier, W., Bouvard, D., Eble, J.A., Mokhtari-Nejad, R., Schulte, V., Zirngibl, H., Brakebusch, C., Fassler, R., Nieswandt, B., 2001. Rhodocytin (aggretin) activates platelets lacking $\alpha 2\beta 1$ integrin, GPVI, and the ligand binding domain of GPIbα. J. Biol. Chem. 276, 25121–25126.

Brinkhous, K.M., Read, M.S., Fricke, W.A., Wagner, R.H., 1983. Botrocetin (venom coagglutinin): reaction with a broad spectrum of multimeric forms of factor VIII macromolecular complex. Proc. Natl. Acad. Sci. U.S.A. 80, 1463–1466.

Chang, C.H., Chung, C.H., Kuo, H.L., Hsu, C.C., Huang, T.F., 2008. The highly specific platelet glycoprotein (GP) VI agonist trowaglerix impaired collagen-induced platelet aggregation ex vivo through matrix metalloproteinase-dependent GPVI shedding. J. Thromb. Haemost. 6, 669–676.

Chen, Y.S., Huang, C.H., Chiou, S.H., 2010. Characterization and molecular cloning of one novel C-type lectin from the venom of Taiwan habu (*Trimeresurus mucrosquamatus*) Toxicon 55, 762–772.

Cheng, X., Qian, Y., Liu, Q., Li, B.X., Zhang, M., Liu, J., 1999. Purification, characterization, and cDNA cloning of a new fibrinogenlytic venom protein, Agkisacutacin, from Agkistrodon acutus venom. Biochem. Biophys. Res. Commun. 265, 530–535.

Cheng, X., Xu, Z.Y., Liu, Q.D., Li, X.M., Li, X.Y., Liu, J., 2000. Purification and characterization of a platelet agglutinating inhibiting protein (Agkisacutacin) from *Agkistrodon acutus* venom. Sheng Wu Hua Xue Yu Sheng Wu Wu Li Xue Bao (Shanghai) 32, 653–656.

Chiou, S.H., Huang, K.F., Chow, L.P., Tsugita, A., Wu, S.H., 1996. Isolation of a venom factor devoid of proteolytic activity from Taiwan habu (*Trimeresurus mucrosquamatus*): N-terminal sequence homology and no functional similarity to factors IX/X-binding proteins and botrocetin. J. Protein Chem. 15, 667–674.

Chung, C.H., Au, L.C., Huang, T.F., 1999. Molecular cloning and sequence analysis of aggretin, a collagen-like platelet aggregation inducer. Biochem. Biophys. Res. Commun. 263, 723–727.

Clemetson, K.J., Morita, T., Kini, R.M., 2009. Scientific and standardization committee communications: classification and nomenclature of snake venom C-type lectins and related proteins. J. Thromb. Haemost. 7, 360.

Dormann, D., Clemetson, J.M., Navdaev, A., Kehrel, B.E., Clemetson, K.J., 2001. Alboaggregin A activates platelets by a mechanism involving glycoprotein VI as well as glycoprotein Ib. Blood 97, 929–936.

Du, X.Y., Clemetson, J.M., Navdaev, A., Magnenat, E.M., Wells, T N., Clemetson, K.J., 2002a. Ophioluxin, a convulxin-like C-type lectin from *Ophiophagus hannah* (King cobra) is a powerful platelet activator via glycoprotein VI. J. Biol. Chem. 277, 35124–35132.

Du, X.Y., Magnenat, E., Wells, T.N., Clemetson, K.J., 2002b. Alboluxin, a snake C-type lectin from *Trimeresurus albolabris* venom is a potent platelet agonist acting via GPIb and GPVI. Thromb. Haemost. 87, 692–698.

Du, X.Y., Navdaev, A., Clemetson, J.M., Magnenat, E., Wells, T.N., Clemetson, K.J., 2001. Bilinexin, a snake C-type lectin from *Agkistrodon bilineatus* venom agglutinates platelets via GPIb and $\alpha 2\beta 1$. Thromb. Haemost. 86, 1277–1283.

Flood, V.H., Friedman, K.D., Gill, J.C., Morateck, P.A., Wren, J.S., Scott, J.P., Montgomery, R.R., 2009. Limitations of the ristocetin cofactor assay in measurement of von Willebrand factor function. J. Thromb. Haemost. 7, 1832–1839.

Fry, B.G., Vidal, N., Norman, J.A., Vonk, F.J., Scheib, H., Ramjan, S.F., Kuruppu, S., Fung, K., Hedges, S.B., Richardson, M.K., Hodgson, W.C., Ignjatovic, V., Summerhayes, R., Kochva, E., 2006. Early evolution of the venom system in lizards and snakes. Nature 439, 584–588.

Fukuda, K., Doggett, T., Laurenzi, I.J., Liddington, R.C., Diacovo, T.G. 2005. The snake venom protein botrocetin acts as a biological brace to promote dysfunctional platelet aggregation. Nat. Struct. Mol. Biol. 12, 152–159.

Fukuda, K., Doggett, T.A., Bankston, L.A., Cruz, M.A., Diacovo, T.G., Liddington, R.C., 2002. Structural basis of von Willebrand factor activation by the snake toxin botrocetin. Structure 10, 943–950.

Fukuda, K., Mizuno, H., Atoda, H., Morita, T., 2000. Crystal structure of flavocetin-A, a platelet glycoprotein Ib-binding protein, reveals a novel cyclic tetramer of C-type lectin-like heterodimers. Biochemistry 39, 1915–1923.

Hamako, J., Matsui, T., Suzuki, M., Ito, M., Makita, K., Fujimura, Y., Ozeki, Y., Titani, K., 1996. Purification and characterization of bitiscetin, a novel von Willebrand factor modulator protein from *Bitis arietans* snake venom. Biochem. Biophys. Res. Commun. 226, 273–279.

Hirotsu, S., Mizuno, H., Fukuda, K., Qi, M.C., Matsui, T., Hamako, J., Morita, T., Titani, K., 2001. Crystal structure of bitiscetin, a von Willebrand factor-dependent platelet aggregation inducer. Biochemistry 40, 13592–13597.

Hooley, E., Papagrigoriou, E., Navdaev, A., Pandey, A.V., Clemetson, J.M., Clemetson, K.J., Emsley, J., 2008. The crystal structure of the platelet activator aggretin reveals a novel (αβ)2 dimeric structure. Biochemistry 47, 7831–7837.

Horii, K., Brooks, M.T., Herr, A.B., 2009. Convulxin forms a dimer in solution and can bind eight copies of glycoprotein VI: implications for platelet activation. Biochemistry 48, 2907–2914.

Howard, M.A. and Firkin, B.G., 1971. Ristocetin – a new tool in the investigation of platelet aggregation. Thromb. Diath. Haemorrh. 26, 362–369.

Huang, K.F., Ko, T.P., Hung, C.C., Chu, J., Wang, A.H., Chiou, S.H., 2004. Crystal structure of a platelet-agglutinating factor isolated from the venom of Taiwan habu (*Trimeresurus mucrosquamatus*). Biochem. J. 378, 399–407.

Kanaji, S., Kanaji, T., Furihata, K., Kato, K., Ware, J.L., Kunicki, T.J. 2003. Convulxin binds to native, human glycoprotein Ibα (GPIbα). J. Biol. Chem. 278, 39452–39460.

Kaneko, M.K., Kato, Y., Kameyama, A., Ito, H., Kuno, A., Hirabayashi, J., Kubota, T., Amano, K., Chiba, Y., Hasegawa, Y., Sasagawa, I., Mishima, K., Narimatsu, H., 2007. Functional glycosylation of human podoplanin: glycan structure of platelet aggregation-inducing factor. FEBS Lett. 581, 331–336.

Kato, Y., Kaneko, M.K., Kunita, A., Ito, H., Kameyama, A., Ogasawara, S., Matsuura, N., Hasegawa, Y., Suzuki-Inoue, K., Inoue, O., Ozaki, Y., Narimatsu, H., 2008. Molecular analysis of the pathophysiological binding of the platelet aggregation-inducing factor podoplanin to the C-type lectin-like receptor CLEC-2. Cancer Sci. 99, 54–61.

Kerrigan, A.M., Dennehy, K.M., Mourão-Sá, D., Faro-Trindade, I., Willment, J.A., Taylor, P.R., Eble, J.A., Reis e Sousa, C., Brown, G.D., 2009. CLEC-2 is a phagocytic activation receptor expressed on murine peripheral blood neutrophils. J. Immunol. 182, 4150–4157.

Kowalska, M.A., Tan, L., Holt, J.C., Peng, M., Karczewski, J., Calvete, J.J., Niewiarowski, S., 1998. Alboaggregins A and B. Structure and interaction with human platelets. Thromb. Haemost. 79, 609–613.

Kunita, A., Kashima, T.G., Morishita, Y., Fukayama, M., Kato, Y., Tsuruo, T., Fujita, N., 2007. The platelet aggregation-inducing factor aggrus/podoplanin promotes pulmonary metastasis. Am. J. Pathol. 170, 1337–1347.

Leduc, M., Bon, C., 1998. Cloning of subunits of convulxin, a collagen-like platelet-aggregating protein from crotalus durissus terrificus venom. Biochem. J. 333, 389–393.

Lee, W.H., Du, X.Y., Lu, Q.M., Clemetson, K.J., Zhang, Y., 2003. Stejnulxin, a novel snake C-type lectin-like protein from *Trimeresurus stejnegeri* venom is a potent platelet agonist acting specifically via GPVI. Thromb. Haemost. 90, 662–671.

Li, X., Zheng, L., Kong, C., Kolatkar, P.R., Chung, M.C., 2004. Purpureotin: a novel di-dimeric C-type lectin-like protein from *Trimeresurus purpureomaculatus* venom is stabilized by noncovalent interactions. Arch. Biochem. Biophys. 424, 53–62.

Liu, S., Zhu, Z., Sun, J., Huang, Q., Teng, M., Niu, L., 2002. Purification, crystallization and preliminary X-ray crystallographic analysis of agkaggregin, a C-type lectin-like protein from *Agkistrodon acutus* venom. Acta Crystallogr. D. Biol. Crystallogr. 58, 675–678.

Lu, Q., Navdaev, A., Clemetson, J.M., Clemetson, K.J., 2004. GPIb is involved in platelet aggregation induced by mucetin, a snake C-type lectin protein from Chinese habu (*Trimeresurus mucrosquamatus*) venom. Thromb. Haemost. 91, 1168–1176.

Maita, N., Nishio, K., Nishimoto, E., Matsui, T., Shikamoto, Y., Morita, T., Sadler, J.E., Mizuno, H., 2003. Crystal structure of von Willebrand factor A1 domain complexed with snake venom, bitiscetin: insight into glycoprotein Ibα binding mechanism induced by snake venom proteins. J. Biol. Chem. 278, 37777–37781.

Mizuno, H., Fujimoto, Z., Atoda, H., Morita, T., 2001. Crystal structure of an anticoagulant protein in complex with the Gla domain of factor X. Proc. Natl. Acad. Sci. U.S.A. 98, 7230–7234.

Mizuno, H., Fujimoto, Z., Koizumi, M., Kano, H., Atoda, H., Morita, T., 1999. Crystal structure of coagulation factor IX-binding protein from habu snake venom at 2.6 Å: implication of central loop swapping based on deletion in the linker region. J. Mol. Biol. 289, 103–112.

Murakami, M.T., Zela, S.P., Gava, L.M., Michelan-Duarte, S., Cintra, A.C., Arni, R.K., 2003. Crystal structure of the platelet activator convulxin, a disulfide-linked α4β4 cyclic tetramer from the venom of *Crotalus durissus terrificus*. Biochem. Biophys. Res. Commun. 310, 478–482.

Navdaev, A., Clemetson, J.M., Polgar, J., Kehrel, B.E., Glauner, M., Magnenat, E., Wells, T.N., Clemetson, K.J., 2001. Aggretin, a heterodimeric C-type lectin from *Calloselasma rhodostoma* (Malayan pit viper) stimulates platelets by binding to α2β1 integrin and GPIb, activating Syk and PLCγ2, but does not involve the GPVI/Fcγ collagen receptor. J. Biol. Chem. 276, 20882–20889.

Navdaev, A., Dormann, D., Clemetson, J.M., Clemetson, K.J., 2001. Echicetin, a GPIb-binding snake C-type lectin from *Echis carinatus*, also contains a binding site for IgMκ responsible for platelet agglutination in plasma and inducing signal transduction. Blood 97, 2333–2341.

Peng, M., Holt, J.C., Niewiarowski, S., 1994. Isolation, characterization and amino acid sequence of echicetin β-subunit, a specific inhibitor of von Willebrand factor and thrombin interaction with glycoprotein Ib. Biochem. Biophys. Res. Commun. 205, 68–72.

Peng, M., Lu, W., Kirby, E.P., 1991. Alboaggregin-B: A new platelet agonist that binds to platelet membrane glycoprotein Ib. Biochemistry 30, 11529–11536.

Peng, M., Lu, W., Kirby, E.P., 1992. Characterization of three alboaggregins purified from *Trimeresurus albolabris* venom. Thromb. Haemost. 67, 702–707.

Polgar, J., Clemetson, J.M., Kehrel, B.E., Wiedemann, M., Magnenat, E.M., Wells, T.N.C., Clemetson, K.J., 1997. Platelet activation and signal transduction by convulxin, a C-type lectin from *Crotalus durissus terrificus* (tropical rattlesnake) venom via the p62/GPVI collagen receptor. J. Biol. Chem. 272, 13576–13583.

Polgar, J., Magnenat, E.M., Peitsch, M.C., Wells, T.N., Saqi, M.S., Clemetson, K.J., 1997. Amino acid sequence of the alpha subunit and computer modelling of the alpha and beta subunits of echicetin from the venom of *Echis carinatus* (saw-scaled viper). Biochem. J. 323, 533–537.

Rucavado, A., Soto, M., Kamiguti, A.S., Theakston, R.D., Fox, J.W., Escalante, T., Gutierrez, J.M., 2001. Characterization of aspercetin, a platelet aggregating component from the venom of the snake *Bothrops asper* which induces thrombocytopenia and potentiates metalloproteinase-induced hemorrhage. Thromb. Haemost. 85, 710–715.

Sakurai, Y., Fujimura, Y., Kokubo, T., Imamura, K., Kawasaki, T., Handa, M., Suzuki, M., Matsui, T., Titani, K., Yoshioka, A., 1998. The cDNA cloning and molecular characterization of a snake venom platelet glycoprotein Ib-binding protein, mamushigin, from *Agkistrodon halys blomhoffii* venom. Thromb. Haemost. 79, 1199–1207.

Sen, U., Vasudevan, S., Subbarao, G., McClintock, R.A., Celikel, R., Ruggeri, Z.M., Varughese, K.I., 2001. Crystal structure of the von Willebrand factor modulator botrocetin. Biochemistry 40, 345–352.

Shin, Y., Okuyama, I., Hasegawa, J., Morita, T., 2000. Molecular cloning of glycoprotein Ib-binding protein, flavocetin-A, which inhibits platelet aggregation. Thromb. Res. 99, 239–247.
Suzuki-Inoue, K., Fuller, G.L., García, A., Eble, J.A., Pöhlmann, S., Inoue, O., Gartner, T.K., Hughan, S.C., Pearce, A.C., Laing, G.D., Theakston, R.D., Schweighoffer, E., Zitzmann, N., Morita, T., Tybulewicz, V.L., Ozaki, Y., Watson, S.P., 2006. A novel Syk-dependent mechanism of platelet activation by the C-type lectin receptor CLEC-2. Blood 107, 542–549.
Taniuchi, Y., Kawasaki, T., Fujimura, Y., Suzuki, M., Titani, K., Sakai, Y., Kaku, S., Hisamichi, N., Satoh, N., Takenaka, T., Handa, M., Sawai, Y., 1995. Flavocetin-A and -B, two high molecular mass glycoprotein Ib binding proteins with high affinity purified from *Trimeresurus flavoviridis* venom, inhibit platelet aggregation at high shear stress. Biochim. Biophys. Acta 1244, 331–338.
Uhrin, P., Zaujec, J., Breuss, J.M., Olcaydu, D., Chrenek, P., Stockinger, H., Fuertbauer, E., Moser, M., Haiko, P., Fässler, R., Alitalo, K., Binder, B.R., Kerjaschki, D., 2010. Novel function for blood platelets and podoplanin in developmental separation of blood and lymphatic circulation. Blood 115, 3997–4005.
Usami, Y., Fujimura, Y., Suzuki, M., Ozeki, Y., Nishio, K., Fukui, H., Titani, K., 1993. Primary structure of two-chain botrocetin, a von Willebrand factor modulator purified from the venom of *Bothrops jararaca*. Proc. Natl. Acad. Sci. U.S.A. 90, 928–932.
Valentin, E., Lambeau, G., 2000. What can venom phospholipases A_2 tell us about the functional diversity of mammalian secreted phospholipases A_2? Biochimie 82, 815–831.
Wang, R., Kong, C., Kolatkar, P., Chung, M.C., 2001. A novel dimer of a C-type lectin-like heterodimer from the venom of *Calloselasma rhodostoma* (Malayan pit viper). FEBS Lett. 508, 447–453.
Wang, W.J., Huang, T.F., 2001. A novel tetrameric venom protein, agglucetin from *Agkistrodon acutus*, acts as a glycoprotein Ib agonist. Thromb. Haemost. 86, 1077–1086.
Wei, Q., Lu, Q.M., Jin, Y., Li, R., Wei, J.F., Wang, W.Y., Xiong, Y.L., 2002. Purification and cloning of a novel C-type lectin-like protein with platelet aggregation activity from *Trimeresurus mucrosquamatus* venom. Toxicon. 40, 1331–1338.

Chapter 36
The Effect of Bacterial Toxins on Platelet Function

Steven Kerrigan and Dermot Cox

Abstract While the ability of cell wall components of bacteria to interact with platelets has been well established there is also evidence that bacterial toxins have the potential to activate platelets. In particular pore-forming toxins such as pneumolysin, streptolysins and a-toxin can activate platelets probably in a manner similar to the calcium ionophore A23187. Cell wall components such as lipopolysaccharide and lipoteichoic acid can activate platelets via Toll-like receptors although evidence would suggest that this may be indirectly via leucocyte activation. Also, toxins such as Shiga toxin, superantigens, gingipains and M proteins can activate platelets. Ultimately the response of platelets to infection is likely to be due to both direct interaction with bacteria and exposure to secreted bacterial products.

Introduction

It is well established that bacteria can interact with platelets leading to platelet activation and thrombus formation and the best known examples of this are mediated by surface components of the bacteria, usually proteins (Fitzgerald et al., 2006). However, bacteria produce a diverse range of secreted products with multiple effects on host cells (Schmitt et al., 1999). Bacteria and fungi also produce a large number of by-products of metabolism, which may not be virulence factors but have the potential to interact with the host. With this wide variety of molecules it is not surprising that some may potentially interact with platelets. An example is tetrafibricin, a metabolite from *Streptomyces neyagawaensis* that potently inhibits platelet aggregation by binding to GPIIb/IIIa and was identified during a screen of fungal metabolites for anti-platelet activity (Satoh et al., 1993).

While the precise definition of a toxin is unclear, for the purposes of this review we shall define bacterial toxins as secreted or shed bacterial products. A number of

---

D. Cox (✉)
Molecular and Cellular Therapeutics, Royal College of Surgeons in Ireland, Dublin, Ireland
e-mail: dcox@rcsi.ie

R.M. Kini et al. (eds.), *Toxins and Hemostasis*, DOI 10.1007/978-90-481-9295-3_36,

these bacterial toxins are virulence factors and there is evidence that an interaction with platelets is important in their biological activity.

There are a few toxin-producing bacteria that are significant human pathogens. *Streptococcus pneumoniae* is a Gram-positive bacterium that colonises the upper respiratory tract leading to diseases such as pneumonia (van der Poll and Opal, 2009), septicaemia (O'Brien et al., 2009), otitis media (Rodgers et al., 2009) and meningitis (Lynch and Zhanel, 2009). *Streptococcus pyogenes*, a Gram-positive bacterium also known as group A streptococci (GAS) is a ubiquitous microorganism that causes a wide variety of diseases in humans including pharyngitis, impetigo, cellulitis, bacteraemia, necrotizing fasciitis and toxic shock syndrome (extensively reviewed by Carapetis and co-workers (Carapetis et al., 2005)). The Gram-positive bacterium *Staphylococcus aureus* is an opportunistic pathogen that causes diseases such as osteomyelitis (Lew and Waldvogel, 2004), pneumonia (Lowy, 1998), endocarditis (Moreillon and Que, 2004; Mortara and Bayer, 1993), toxic shock syndrome (Davis et al., 1980; Lappin and Ferguson, 2009), and bacteremia (Cockerill et al., 1997; Corey, 2009). *Escherichia coli* is a Gram-negative bacterium that is a normal part of the gastrointestinal flora. However, pathogenic strains can cause serious infections such as haemolytic uremic syndrome (HUS) (Palermo et al., 2009).

Pore-Forming Toxins

Numerous bacteria produce cytotoxic toxins that are important virulence factors. One family of these are pore-forming toxins that insert into the cell membrane creating pores that result in cell death. These include α-toxin and Panton-Valentine leukocidin from *S. aureus* and pneumolysin from *S. pneumoniae*. There is evidence that some of these may also affect platelet function.

Pneumolysin

Pneumolysin is an important *S. pneumoniae* virulence factor in the pathogenesis of pneumococcal disease (Kanclerski and Mollby, 1987). The importance of pneumolysin was first established by Berry and co-workers, who demonstrated that mice infected with a pneumolysin-deficient strain of *S. pneumoniae* showed a reduced inflammatory response compared to mice infected with a pneumolysin-producing strain (Berry et al., 1989, 1992).

Pneumolysin is a 53 kDa thiol-activated, pore-forming cytolysin composed of 471 amino acids (Kanclerski and Mollby, 1987; Paton et al., 1986; Walker et al., 1987) that typically exists as a monomer but can assemble into oligomers (Gilbert et al., 1998). Virtually all clinical strains of *S. pneumoniae* express pneumolysin (Benton et al., 1997; Kanclerski and Mollby, 1987). Its negative charge sequesters the toxin in the cytoplasm of the bacteria. Typically pneumolysin remains within *S. pneumoniae* during the growth phase and can be released by several mechanisms.

Firstly, *S. pneumoniae* undergoes autolysis when the cell wall is degraded by a peptidoglycan hydrolase. *S. pneumoniae* contains a powerful autolytic enzyme that has been characterised as an *N*-acetylmuramoyl-L-alanine amidase commonly called lytA (Lopez et al., 1997). When lytA is activated, the cytoplasmic contents of *S. pneumoniae*, such as pneumolysin are released (Walker et al., 1987). Secondly, pneumolysin can be released from *S. pneumoniae* after destruction by the host immune system where the cell wall of the bacteria is broken down thereby releasing the intracellular contents including pneumolysin (Dessing et al., 2009). Finally some antibiotics (such as the β-lactam class, glycopeptides, fluoroquinolones and aminoglycosides), whose primary function is disruption of bacterial cell wall synthesis can also cause release of pneumolysin into the surrounding environment (Anderson et al., 2007; Charpentier and Tuomanen, 2000; Spreer et al., 2003).

Pneumolysin binds to cholesterol in the cell membrane. This interaction is rapid, pH-dependent but temperature-independent (Johnson et al., 1980). Two models have been proposed to describe pore formation by pneumolysin. In the first, it is suggested that one monomeric pneumolysin inserts into the lipid bilayer and then begins to recruit other monomeric pneumolysins forming a ring or arc shaped oligomeric structure (Morgan et al., 1994). In the second model, several pneumolysin monomers gather on the cell membrane forming a ‘pre-pore’ which then inserts into the lipid bilayer (Tilley et al., 2005). Regardless of the model, pore formation leads to the formation of a channel through which intracellular contents can leak out.

In addition to binding to cholesterol in the plasma membrane of cells, pneumolysin can also activate the classical complement pathway in the absence of antibody (Mitchell et al., 1991; Paton et al., 1984), inhibit polymorphonuclear cell respiratory burst, migration and chemotaxis (Cockeran et al., 2002; Houldsworth et al., 1994; Kadioglu et al., 2004; Paton and Ferrante, 1983), lyse red blood cells (Tilley et al., 2005), induce nitric oxide production by macrophages and stimulate the release of inflammatory mediators such as TNF-α and IL-1 (Malley et al., 2003; Shoma et al., 2008). Together these effects are thought to provide the mechanism through which pneumolysin induces inflammation and membrane damage in pneumococcal disease.

The ability of pneumolysin to induce platelet aggregation is not well characterised. Early studies by Johnson et al., demonstrated that at low concentrations pneumolysin had little effect on platelets, however at higher concentrations platelets were lysed presumably by binding to the sterols present in the platelet plasma membrane and forming multimeric transmembrane pores (Johnson et al., 1981). The ability to form these pores may result in lysis of platelets. Lysis was time-dependent with more than 50% of the platelets lysed. However, more recently a study demonstrated that a concentrated sample of supernatant from an overnight growth of *S. pneumoniae* that contained pneumolysin failed to induce human platelet aggregation; moreover a pneumolysin-negative strain of *S. pneumoniae* also induced platelet aggregation suggesting that pneumolysin is not involved in inducing platelet aggregation.

Streptolysins

There are various toxins produced by GAS that are responsible for host cell damage and the inflammatory response. Streptolysin O (SLO) is amongst the most common. SLO is a thiol-activated cytolysin that is present in almost all clinical strains of group A streptococci (Kehoe et al., 1987). There are at least two forms of SLO, a 50 kDa version and a 70 kDa version. SLO is stable in its native state; however it is rapidly inactivated in the presence of oxygen (Kehoe et al., 1987). SLO binds cholesterol in the lipid membrane of GAS, where it forms multimeric transmembrane pores (Bhakdi et al., 1985, 1996). The ability to form pores in host cells may result in cell lysis or leakage of intracellular contents into the surrounding environment.

Several studies have demonstrated that challenge of animals with GAS leads to platelet thrombi and fibrin clots in capillaries, postcapillary venules and arterioles (Ashbaugh et al., 1998; Bisno and Stevens, 1996; Stevens et al., 1989; Taylor et al., 1999). The mechanism leading to the formation of platelet thrombi has not been successfully characterised, however it is most likely due to leakage of platelet contents which activate adjacent platelets leading to intravascular aggregate formation. More recent studies focused on the formation of platelet/polymorphonuclear leukocyte complexes. Here, the authors demonstrate that addition of SLO to whole blood leads to the expression of p-selectin on platelets, a step essential for initiating the development of platelet/polymorphonuclear leukocyte complex formation. Preincubation of whole blood with an anti-p-selectin antibody prevented the complex formation (Bryant et al., 2005). These results suggest that as infection progresses more toxin is produced by the GAS, which leads to intravascular platelet aggregate formation and platelet/polymorphonuclear leukocyte complex formation. Together these processes accentuate disease progression via occlusion of the surrounding microvascular bed.

α-Toxin

S. aureus is known to secrete several extracellular toxins. α-toxin is a 34 kDa toxin composed of 293 amino acids (Bernheimer, 1965). It is produced by almost all strains of *S. aureus*. It is secreted into the extracellular environment as a monomeric water-soluble protein (Ikigai and Nakae, 1985). The toxin disrupts cell membranes by binding to the lipid bilayer, forming an oligomeric structure that forms a water-filled transmembrane pore (Valeva et al., 1996). Two modes of primary interaction have been defined using rabbit erythrocytes as a model. These include binding specifically to the host target at low concentrations and non-specific adsorption to host target cell membranes at higher concentrations (Hildebrand et al., 1991).

Siegel and Cohen were the first to demonstrate that addition of α-toxin to human platelet-rich plasma induced platelets to undergo shape change and aggregate (Siegel and Cohen, 1964). This study demonstrated that platelets leaked NAD^+, K^+ and ATP but not protein, leading to the conclusion that the platelets were not being lysed. In support of this Bernheimer and Schwartz demonstrated that when platelets were treated with α-toxin it caused them to swell but there were no clear signs of platelet lysis by electron microscopy (Bernheimer and Schwartz, 1965; Manohar et al., 1967). Later studies investigated the molecular mechanism of platelet activation and found that pore formation increased the intracellular calcium concentration (Arvand et al., 1990; Baliakina et al., 1999). Arvand et al. demonstrated that α-toxin causes platelets to undergo the release reaction which secretes large amounts of platelet factor 4 and factor V. Secreted factor V in turn associates with the platelet membrane leading to assembly of the prothrombinase complex (Arvand et al., 1990). This explains the major pathway responsible for the procoagulatory effects of α-toxin. Bayer and colleagues used two models to investigate the role of α-toxin on platelets. In the first and consistent with the above observations, the authors demonstrated that α-toxin caused platelet lysis and release of platelet microbial proteins (PMPs). The release of PMPs from platelets was bactericidal to *S. aureus*. The authors also investigated the role of α-toxin in experimental endocarditis using several different strains of *S. aureus* that differed in the expression of functional versus mutant forms of α-toxin. Under these conditions, the *S. aureus* strains producing either minimal amounts or no α-toxin were less virulent in vivo than wild-type strains (Bayer et al., 1997). Wild-type *S. aureus* strains or indeed an isogenic strain engineered to over-express α-toxin were associated with increased release of PMP from platelets. Together these results suggest that excessive levels of α-toxin released in the vicinity of platelets leads to release of PMPs and therefore protect the host by destroying the α-toxin producing *S. aureus*.

Thus, these pore-forming cytotoxins appear to act by lysing platelets. However, there is some evidence to suggest that low, non-cytotoxic concentrations can activate platelets. This probably occurs in a manner analogous to the calcium ionophore A23187, which induces platelet activation by increasing the intracellular calcium concentration (White et al., 1974).

Lipopolysaccharide (LPS)

Lipopolysaccharides are structural components of Gram-negative bacteria and are also known as endotoxin to reflect the fact that they are not secreted by living bacteria but are released from dead bacteria (David et al., 1998). LPS is a potent stimulator of the innate immune system and mediates its effects through interactions with Toll-like receptors (TLR), primarily TLR-4 on immune cells (Beutler et al., 2003). Shiraki and co-workers were the first to identify TLRs in platelets in, 2004 (Shiraki et al., 2004) and to-date TLR 1, 2, 4, 6 and 9 have been identified on platelets (Andonegui et al., 2005; Aslam et al., 2006; Cognasse et al., 2005; Shiraki et al., 2004; Stahl et al., 2006; Ward et al., 2005). The presence of TLR-4 suggests that platelets may be capable of responding to LPS.

LPS was shown to have no effect on ADP-induced platelet aggregation (Ward et al., 2005) while others have shown inhibition of platelet activation by LPS (Hashimoto et al., 2009) and others have shown some form of platelet activation by LPS (Fabrice et al., 2008; Kuckleburg et al., 2005; Shashkin et al., 2008; Stahl et al.,

2006). Thus, the ex vivo evidence for a TLR 4-mediated response to LPS in platelets is weak. However, there is in vivo evidence to suggest that platelets do respond to LPS. LPS injection in mice leads to thrombocytopenia (Shibazaki et al., 1999) and there was evidence of platelet activation in human volunteers injected with LPS (Wilson et al., 2001). The conflict between the ex vivo and in vivo data may be due to an indirect interaction between LPS and platelets. While TLR 4-dependent platelet activation was seen in mice injected with LPS it was suggested that this was due to megakaryocyte TLR 4 rather than platelet TLR 4 (Jayachandran et al., 2007). Other studies have suggested that LPS binding to leucocyte TLR 4 may mediate the platelet activation. LPS-induced platelet activation was seen in whole blood but not in platelet-rich plasma (Csako et al., 1988; Whitworth et al., 1989) suggesting that other blood cells may play an important role in triggering platelet activation. Pulmonary sequestration of LPS-activated platelets was found to be dependent on a platelet-neutrophil interaction (Andonegui et al., 2005) and platelet activation by LPS in whole blood was found to be dependent on leucocytes (Montrucchio et al., 2003). It was also shown that LPS binding to platelet TLR 4 stimulated binding to neutrophils and the formation of neutrophil extracellular traps (Clark et al., 2007). LPS-activated platelets have also been shown to be internalized by endothelial cells (Kuckleburg et al., 2008a) and to induce apoptosis of endothelial cells (Kuckleburg et al., 2008b).

Thus, while platelets express TLR 4 on their surface the interaction with LPS appears to depend primarily on leucocytes, resulting in the formation of platelet-leucocyte aggregates rather than platelet aggregates.

Lipoteichoic Acid (LTA)

Lipoteichoic acid is a component of Gram-positive bacteria (Morath et al., 2005). Like LPS it is not secreted but shed from the bacteria after lysis or after treatment with β-lactam antibiotics and it also stimulates an immune response through an interaction with Toll-like receptors (Zähringer et al., 2008). The receptor for LTA and also for peptidoglycan is TLR 2 and as detailed in the previous section it is expressed on the platelet surface suggesting a potential for platelets to respond to LTA. The synthetic lipopeptide Pam_3CSK_4 which is a TLR 2 agonist has been shown to induce platelet aggregation (Berg et al., 1994) although others have shown no effect (Ward et al., 2005) while there is also a report that *S. aureus* LTA inhibits platelet activation (Sheu et al., 2000a, 2000b). Thus, the role of TLR 2 in mediating LTA-induced platelet activation is unclear and requires further work to determine if such a role exists.

Shiga Toxin

Shiga toxin (Stx), also known as verotoxin, was first identified in *Shigella dysenteriae* a major cause of dysentery. A Shiga-like toxin (SLT) was subsequently identified in enterohemorrhagic *Escherichia coli* strains. Stx contains an enzymatic domain and 5 binding domains (Ling et al., 1998) which bind to globotriaosylceramide (Gb_3), a glycosphingolipid on the surface of cells (Lindberg et al., 1987) and the complex is subsequently internalized. The internalized Shiga toxin hydrolyses the ribosomes preventing further protein synthesis (Obrig et al., 1987). Shiga toxin binding to platelets has been shown (Cooling et al., 1998; Karpman et al., 2001) while other studies have failed to detect it (Thorpe et al., 1999) although it may be that it only binds to activated platelets (Ghosh et al., 2004). It has been suggested that in blood, Stx binds to leucocytes, which then transfer it on to other cells (te Loo et al., 2000). Infection with *E. coli*-expressing Shiga toxin can lead to haemolytic uremic syndrome (HUS), which is associated with thrombocytopenia, due to extensive endothelial cell activation leading to high-multimer von Willebrand factor secretion.

Shiga toxin has been shown to activate platelets and reduce platelet expression of CD47, an integrin-associated protein. This reduction in CD47 levels was associated with increased phagocytosis of the platelets which may play a role in the thrombocytopenia. However, as a crude bacterial extract was used and as anti-TLR 4 antibodies were found to block this, there is a possibility that this phenomenon may be due to contaminating LPS rather than Shiga toxin (Guo et al., 2009). Rabbits challenged with Shiga toxin have renal damage and thrombosis similar to that seen in HUS (García et al., 2008). Stx was also shown to increase platelet deposition on endothelial cells under high shear (Morigi et al., 2001) and enhanced platelet aggregation under low shear stress although not by direct action on platelets (Yagi et al., 2001). Stx activation of monocytes (Guessous et al., 2005a) and endothelial cells (Guessous et al., 2005b) were shown to lead to platelet activation.

It is likely that in HUS a combination of LPS and Stx from *E. coli* infection lead to increased formation of platelet-leucocyte aggregates and the subsequent release of tissue factor containing microparticles (Stahl et al., 2009).

Superantigens

Gram positive bacteria such as *S. aureus* and *Streptococcus pyogenes* produce a family of toxins known as superantigens. These are known to bind to the major histocompatibility complex (MHC) class II antigens and T-cell receptors resulting in a cytokine storm that causes the toxic shock syndrome. They are known for their potency as they are effective at picogram per litre concentration (Fraser and Proft, 2008). These superantigens are similar to previously known enterotoxins secreted by *S. aureus,* which are associated with fever and severe gastrointestinal disturbance. Their potency and severely debilitating effects has led to interest in their use as a biological weapon (Ler et al., 2006). The best studied member of this family is staphylococcal enterotoxin B (SEB). Another group of related molecules is the staphylococcal superantigen-like (SSL) protein family. These are structurally related to superantigens but have a different mechanism of action (Fraser and Proft, 2008). Together the superantigens and SSL form a large family of extremely potent toxins capable of generating a cytokine storm.

Recently, SSL5, has been shown to induce platelet aggregation by binding to both GPIbα and GPIIb/IIIa (de Haas et al., 2009). However, this only occurred in washed platelets due to albumin binding to SSL5 in plasma. The interaction with GPIbα was abolished by treatment with neuraminidase. The interaction with SSL5 would appear to be specific as SSL11, which also binds to sialic acid residues, did not induce aggregation (de Haas et al., 2009). However, a recent presentation suggests that SSL5 acts by binding to GPIbα and GPVI rather than GPIIb/IIIa (Hu and Peter, 2009). SEB has been shown to inhibit both platelet adhesion and thrombin-induced platelet aggregation (Morganti et al., 2008; Tran et al., 2006). Toxic-shock-syndrome-toxin-1 (TSST-1) another staphylococcal superantigen was also shown to inhibit platelet aggregation (Gareau et al., 1989).

Gingipain

Porphyromonas gingivalis is an oral anaerobe that is associated with periodontal disease. One of the virulence factors that it secretes are a family of three cysteine proteases – gingipains. These proteases play many roles in infection including colonization, generation of nutrients and host defense e.g., complement degradation (Curtis et al., 2001; Fitzpatrick et al., 2009).

Thrombin is a serine protease and a potent activator of platelets. Its ability to activate platelets is dependent on its protease activity as it acts to cleave a portion of its receptor—protease activated receptor (PAR)-1. The action of thrombin on PAR-1 exposes a new terminal peptide (SFLLRN) which subsequently acts as an agonist on the receptor triggering platelet activation (Shah, 2009).

Gingipains have been shown to activate PAR on both endothelial cells and platelets (Lourbakos et al., 2001a, 2001b). It was also found that *P. gingivalis* induces platelet aggregation by shedding vesicles rich in gingipains (Pham et al., 2002). However, it has been suggested that this only happens in washed platelets and that in platelet-rich plasma *P. gingivalis*-induced aggregation is independent of gingipains and is mediated by a direct interaction with platelets (Naito et al., 2006) although it was also proposed that *P. gingivalis* also sensitizes platelets to the actions of adrenaline (Nylander et al., 2008).

Miscellaneous Toxins

S. pyogenes expresses M protein on their surface, a virulence factor with many variants. The M1 protein appears to be associated with the most virulent strains of *S. pyogenes*. It has been shown to be shed from the bacterial surface by enzymatic cleavage and during growth. It has been shown to activate neutrophils and monocytes. It has also been shown to activate platelets by binding to soluble fibrinogen allowing it to bind to GPIIb/IIIa. However, for platelet activation to occur anti-M1 antibodies must also bind to FcγRIIa on the platelet surface (Shannon et al., 2007). This crosslinking of FcγRIIa and GPIIb/IIIa is similar to the process used by many whole bacteria to induce platelet aggregation (Fitzgerald et al., 2006). Streptococcal pyrogenic exotoxin B (SpeB) is an enzyme associated with the M1 virulence factor and its mSpeB2 variant that contains the amino acid sequence RGD has been shown to bind to GPIIb/IIIa (Stockbauer et al., 1999).

When platelets become activated they release cationic anti-microbial peptides (Yeaman and Bayer, 2006) including platelet microbicidal protein (PMP) (Yeaman et al., 1998). Some strains of *S. aureus* secrete an inhibitor of platelet microbicidal proteins (Ivanov et al., 2006).

Conclusion

While some bacterial toxins may induce platelet activation there is a paucity of systematic data to confirm this let alone identify the mechanisms involved. While in some cases the toxins may act directly on platelets, such as with low concentrations of pore-forming toxins, in most cases the evidence would suggest that indirect interactions are more likely to be involved. In many cases leucocytes appear to be the primary target, which, when activated, bind to platelets forming platelet-leucocyte aggregates. This certainly seems to be the case with LPS, Shiga toxin and possibly SLO.

Even if bacterial toxins can activate platelets the clinical relevance for this is unclear. These toxins are likely to be present during infection but the bacteria will also be there and direct bacterial-platelet interactions also occur with many of the toxin-producing bacteria (Fitzgerald et al., 2006). It is not clear if many of the toxins can reach plasma concentrations that could induce platelet aggregation. While some are very potent they have significant levels of binding to plasma proteins, which inactivates them.

Thus, there is a need for further studies to characterize the significance of bacterial toxins in relationship to surface proteins in thrombotic disease. It may be that they act synergistically, with the toxin sensitizing the platelets to activation by the bacteria.

References

Anderson, R., Steel, H.C., Cockeran, R., von Gottberg, A., de Gouveia, L., Klugman, K.P., Mitchell, T.J., Feldman, C., 2007. Comparison of the effects of macrolides, amoxicillin, ceftriaxone, doxycycline, tobramycin and fluoroquinolones, on the production of pneumolysin by *Streptococcus pneumoniae in vitro*. J. Antimicrob. Chemother. 60, 1155–1158.

Andonegui, G., Kerfoot, S.M., McNagny, K., Ebbert, K.V.J., Patel, K.D., Kubes, P., 2005. Platelets express functional Toll-like receptor-4. Blood 106, 2417–2423.

Arvand, M., Bhakdi, S., Dahlback, B., Preissner, K.T., 1990. *Staphylococcus aureus* alpha-toxin attack on human platelets promotes assembly of the prothrombinase complex. J. Biol. Chem. 265, 14377–14381.

Ashbaugh, C.D., Alberti, S., Wessels, M.R., 1998. Molecular analysis of the capsule gene region of group A Streptococcus: the has AB genes are sufficient for capsule expression. J. Bacteriol. 180, 4955–4959.

Aslam, R., Speck, E.R., Kim, M., Crow, A.R., Bang, K.W.A., Nestel, F.P., Ni, H., Lazarus, A.H., Freedman, J., Semple, J.W., 2006. Platelet Toll-like receptor expression modulates lipopolysaccharide-induced thrombocytopenia and tumor necrosis factor-α production *in vivo*. Blood 107, 637–641.

Baliakina, E.V., Gerasimovskaia, E.V., Romanov, Iu.A., Atakhanov, Sh.E., 1999. Role of *Staphylococcus aureus* hemolytic toxin-alpha in pathogenesis of infectious endocarditis: studies in vitro. Ter. Arkh. 71, 28–31.

Bayer, A.S., Ramos, M.D., Menzies, B.E., Yeaman, M.R., Shen, A.J., Cheung, A.L., 1997. Hyperproduction of alpha-toxin by *Staphylococcus aureus* results in paradoxically reduced virulence in experimental endocarditis: a host defense role for platelet microbicidal proteins. Infect. Immun. 65, 4652–4660.

Benton, K.A., Paton, J.C., Briles, D.E., 1997. Differences in virulence for mice among Streptococcus pneumoniae strains of capsular types 2, 3, 4, 5, and 6 are not attributable to differences in pneumolysin production. Infect. Immun. 65, 1237–1244.

Berg, M., Offermanns, S., Seifert, R., Schultz, G., 1994. Synthetic lipopeptide Pam3CysSer(Lys)4 is an effective activator of human platelets. Am. J. Physiol. Cell Physiol. 266, C1684–1691.

Bernheimer, A.W., 1965. Staphylococcal alpha toxin. Ann. N.Y. Acad. Sci. 128, 112–123.

Bernheimer, A.W., Schwartz, L.L., 1965. Effect of staphylococcal and other bacterial toxins on platelets *in vitro*. J. Pathol. Bacteriol. 89, 209–223.

Berry, A.M., Paton, J.C., Hansman, D., 1992. Effect of insertional inactivation of the genes encoding pneumolysin and autolysin on the virulence of *Streptococcus pneumoniae* type 3. Microb. Pathog. 12, 87–93.

Berry, A.M., Yother, J., Briles, D.E., Hansman, D., Paton, J.C., 1989. Reduced virulence of a defined pneumolysin-negative mutant of *Streptococcus pneumoniae*. Infect. Immun. 57, 2037–2042.

Beutler, B., Hoebe, K., Du, X., Ulevitch, R.J., 2003. How we detect microbes and respond to them: the Toll-like receptors and their transducers. J. Leukoc. Biol. 74, 479–485.

Bhakdi, S., Bayley, H., Valeva, A., Walev, I., Walker, B., Kehoe, M., Palmer, M., 1996. Staphylococcal alpha-toxin, streptolysin-O, and *Escherichia coli* hemolysin: prototypes of pore-forming bacterial cytolysins. Arch. Microbiol. 165, 73–79.

Bhakdi, S., Tranum-Jensen, J., Sziegoleit, A., 1985. Mechanism of membrane damage by streptolysin-O. Infect. Immun. 47, 52–60.

Bisno, A.L., Stevens, D.L., 1996. Streptococcal infections of skin and soft tissues. N. Engl. J. Med. 334, 240–245.

Bryant, A.E., Bayer, C.R., Chen, R.Y., Guth, P.H., Wallace, R.J., Stevens, D.L., 2005. Vascular dysfunction and ischemic destruction of tissue in *Streptococcus pyogenes* infection: the role of streptolysin O-induced platelet/neutrophil complexes. J. Infect. Dis. 192, 1014–1022.

Carapetis, J.R., Steer, A.C., Mulholland, E.K., Weber, M., 2005. The global burden of group A streptococcal diseases. Lancet Infect. Dis. 5, 685–694.

Charpentier, E., Tuomanen, E., 2000. Mechanisms of antibiotic resistance and tolerance in *Streptococcus pneumoniae*. Microbes Infect. 2, 1855–1864.

Clark, S.R., Ma, A.C., Tavener, S.A., McDonald, B., Goodarzi, Z., Kelly, M.M., Patel, K.D., Chakrabarti, S., McAvoy, E., Sinclair, G.D., Keys, E.M., Allen-Vercoe, E., Devinney, R., Doig, C.J., Green, F.H., Kubes, P., 2007. Platelet TLR4 activates neutrophil extracellular traps to ensnare bacteria in septic blood. Nat. Med. 13, 463–469.

Cockeran, R., Durandt, C., Feldman, C., Mitchell, T.J., Anderson, R., 2002. Pneumolysin activates the synthesis and release of interleukin-8 by human neutrophils in vitro. J. Infect. Dis. 186, 562–565.

Cockerill, F.R., Hughes, J.G., Vetter, E.A., Mueller, R.A., Weaver, A.L., Ilstrup, D.M., Rosenblatt, J.E., Wilson, W.R., 1997. Analysis of 281,797 consecutive blood cultures performed over an eight-year period: trends in microorganisms isolated and the value of anaerobic culture of blood. Clin. Infect. Dis. 24, 403–418.

Cognasse, F., Hamzeh, H., Chavarin, P., Acquart, S., Genin, C., Garraud, O., 2005. Evidence of Toll-like receptor molecules on human platelets. Immunol. Cell Biol. 83, 196–198.

Cooling, L.L.W., Walker, K.E., Gille, T., Koerner, T.A.W., 1998. Shiga toxin binds human platelets via globotriaosylceramide (Pk Antigen) and a novel platelet glycosphingolipid. Infect. Immun. 66, 4355–4366.

Corey, G.R., 2009. *Staphylococcus aureus* bloodstream infections: definitions and treatment. Clin. Infect. Dis. 48(Suppl 4), S254–S259.

Csako, G., Suba, E., Elin, R., 1988. Endotoxin-induced platelet activation in human whole blood *in vitro*. Thromb. Haemost. 59, 378–382.

Curtis, M.A., Aduse-Opoku, J., Rangarajan, M., 2001. Cysteine Proteases of *Porphyromonas gingivalis*. Crit. Rev. Oral Biol. Med. 12, 192–216.

David, E.H., Jeremy, A.Y., Chris, W., 1998. Molecular basis for structural diversity in the core regions of the lipopolysaccharides of *Escherichia coli* and *Salmonella enterica*. Mol. Micro. 30, 221–232.

Davis, J.P., Chesney, P.J., Wand, P.J., LaVenture, M., 1980. Toxic-shock syndrome: epidemiologic features, recurrence, risk factors, and prevention. N. Engl. J. Med. 303, 1429–1435.

de Haas, C., Weeterings, C., Vughs, M., de Groot, P.G., van Strijp, J., Lisman, T., 2009. Staphylococcal superantigen-like 5 activates platelets and supports platelet adhesion under flow conditions, which involves glycoprotein Ibα and $\alpha_{IIb}\beta_3$. J. Thromb. Haemostas. 7, 1867–1874.

Dessing, M.C., Hirst, R.A., de Vos, A.F., van der Poll, T., 2009. Role of Toll-like receptors 2 and 4 in pulmonary inflammation and injury induced by pneumolysin in mice. PLoS One 4, e7993.

Fabrice, C., Hind, H.-C., Sandrine, L., Olivier, D., Bruno, P., Archie, M., Olivier, G., 2008. Toll-like receptor 4 ligand can differentially modulate the release of cytokines by human platelets. Br. J. Haematol. 141, 84–91.

Fitzgerald, J.R., Foster, T.J., Cox, D., 2006. The interaction of bacterial pathogens with platelets. Nat. Rev. Microbiol. 4, 445–457.

Fitzpatrick, R.E., Wijeyewickrema, L.C., Pike, R.N., 2009. The gingipains: scissors and glue of the periodontal pathogen, *Porphyromonas gingivalis*. Future Microbiol. 4, 471–487.

Fraser, J., Proft, T., 2008. The bacterial superantigen and superantigen-like proteins. Immunol. Rev. 225, 226–243.

García, A., Marini, R.P., Catalfamo, J.L., Knox, K.A., Schauer, D.B., Rogers, A.B., Fox, J.G., 2008. Intravenous Shiga toxin 2 promotes enteritis and renal injury characterized by polymorphonuclear leukocyte infiltration and thrombosis in Dutch Belted rabbits. Microbes Infect. 10, 650–656.

Gareau, R., Gruda, J., Micusan, V., 1989. Effect of toxic shock syndrome toxin-1 on human hemostatic parameters. Thromb. Res. 54, 349–356.

Ghosh, S., Polanowska-Grabowska, R., Fujii, J., Obrig, T., Gear, A., 2004. Shiga toxin binds to activated platelets. J. Thromb. Haemost. 2, 499–506.

Gilbert, R.J., Rossjohn, J., Parker, M.W., Tweten, R.K., Morgan, P.J., Mitchell, T.J., Errington, N., Rowe, A.J., Andrew, P.W., Byron, O., 1998. Self-interaction of pneumolysin, the pore-forming protein toxin of *Streptococcus pneumoniae*. J. Mol. Biol. 284, 1223–1237.

Guessous, F., Marcinkiewicz, M., Polanowska-Grabowska, R., Keepers, T., Obrig, T., Gear, A., 2005a. Shiga toxin 2 and lipopolysaccharide cause monocytic THP-1 cells to release factors which activate platelet function. Thromb. Haemost. 94, 1019–1027.

Guessous, F., Marcinkiewicz, M., Polanowska-Grabowska, R., Kongkhum, S., Heatherly, D., Obrig, T., Gear, A.R.L., 2005b. Shiga Toxin 2 and lipopolysaccharide induce human microvascular endothelial cells to release chemokines and factors that stimulate platelet function. Infect. Immun. 73, 8306–8316.

Guo, Y.-L., Liu, D.-Q., Bian, Z., Zhang, C.-Y., Zen, K., 2009. Down-Regulation of platelet surface CD47 expression in *Escherichia coli* O157:H7 infection-induced thrombocytopenia. PLoS ONE 4, e7131.

Hashimoto, K., Jayachandran, M., Owen, W., Miller, V., 2009. Aggregation and microparticle production through Toll-like receptor 4 activation in platelets from recently menopausal women. J. Cardiovasc. Pharmacol. 54, 57–62.

Hildebrand, A., Pohl, M., Bhakdi, S., 1991. *Staphylococcus aureus* alpha-toxin. Dual mechanism of binding to target cells. J. Biol. Chem. 266, 17195–17200.

Houldsworth, S., Andrew, P.W., Mitchell, T.J., 1994. Pneumolysin stimulates production of tumor necrosis factor alpha and interleukin-1 beta by human mononuclear phagocytes. Infect. Immun. 62, 1501–1503.

Hu, H., Peter, K., 2009. Staphylococcal superantigen-like 5 induces platelet activation and thrombosis via binding to GPIbα and GPVI. Circulation 120, S1080.

Ikigai, H., Nakae, T., 1985. Conformational alteration in alpha-toxin from *Staphylococcus aureus* concomitant with the transformation of the water-soluble monomer to the membrane oligomer. Biochem. Biophys. Res. Commun. 130, 175–181.

Ivanov, I.B., Gritsenko, V.A., Kuzmin, M.D., 2006. Staphylococcal secretory inhibitor of platelet microbicidal protein is associated with prostatitis source. J. Med. Microbiol. 55, 1645–1648.

Jayachandran, M., Brunn, G.J., Karnicki, K., Miller, R.S., Owen, W.G., Miller, V.M., 2007. *In vivo* effects of lipopolysaccharide and TLR4 on platelet production and activity: implications for thrombotic risk. J. Appl. Physiol. 102, 429–433.

Johnson, M.K., Boese-Marrazzo, D., Pierce, W.A., 1981. Effects of pneumolysin on human polymorphonuclear leukocytes and platelets. Infect. Immun. 34, 171–176.

Johnson, M.K., Geoffroy, C., Alouf, J.E., 1980. Binding of cholesterol by sulfhydryl-activated cytolysins. Infect. Immun. 27, 97–101.

Kadioglu, A., Coward, W., Colston, M.J., Hewitt, C.R., Andrew, P.W., 2004. CD4-T-lymphocyte interactions with pneumolysin and pneumococci suggest a crucial protective role in the host response to pneumococcal infection. Infect. Immun. 72, 2689–2697.

Kanclerski, K., Mollby, R., 1987. Production and purification of *Streptococcus pneumoniae* hemolysin (pneumolysin). J. Clin. Microbiol. 25, 222–225.

Karpman, D., Papadopoulou, D., Nilsson, K., Sjogren, A.C., Mikaelsson, C., Lethagen, S., 2001. Platelet activation by Shiga toxin and circulatory factors as a pathogenetic mechanism in the hemolytic uremic syndrome. Blood 97, 3100–3108.

Kehoe, M.A., Miller, L., Walker, J.A., Boulnois, G.J., 1987. Nucleotide sequence of the streptolysin O (SLO) gene: structural homologies between SLO and other membrane-damaging, thiol-activated toxins. Infect. Immun. 55, 3228–3232.

Kuckleburg, C., McClenahan, D., Czuprynski, C., 2008a. Platelet activation by *Histophilus somni* and its lipooligosaccharide induces endothelial cell proinflammatory responses and platelet internalization. Shock 29, 189–196.

Kuckleburg, C., Tiwari, R., Czuprynski, C., 2008b. Endothelial cell apoptosis induced by bacteria-activated platelets requires caspase-8 and -9 and generation of reactive oxygen species. Thromb. Haemost. 99, 363–372.

Kuckleburg, C.J., Sylte, M.J., Inzana, T.J., Corbeil, L.B., Darien, B.J., Czuprynski, C.J., 2005. Bovine platelets activated by *Haemophilus somnus* and its LOS induce apoptosis in bovine endothelial cells. Microb. Pathogen 38, 23–32.

Lappin, E., Ferguson, A.J., 2009. Gram-positive toxic shock syndromes. Lancet Infect. Dis. 9, 281–290.

Ler, S.G., Lee, F.K., Gopalakrishnakone, P., 2006. Trends in detection of warfare agents: detection methods for ricin, staphylococcal enterotoxin B and T-2 toxin. J. Chromatog. A. 1133, 1–12.

Lew, D.P., Waldvogel, F.A., 2004. Osteomyelitis. Lancet 364, 369–379.

Lindberg, A.A., Brown, J.E., Stramberg, N., Westling-Ryd, M., Schultz, J.E., Karlsson, K.A., 1987. Identification of the carbohydrate receptor for Shiga toxin produced by *Shigella dysenteriae* type 1. J. Biol. Chem. 262, 1779–1785.

Ling, H., Boodhoo, A., Hazes, B., Cummings, M.D., Armstrong, G.D., Brunton, J.L., Read, R.J., 1998. Structure of the Shiga-like toxin I B-pentamer complexed with an analogue of its receptor Gb_3. Biochemistry 37, 1777–1788.

Lopez, R., Garcia, E., Garcia, P., Garcia, J.L., 1997. The pneumococcal cell wall degrading enzymes: a modular design to create new lysins? Microb. Drug Resist. 3, 199–211.

Lourbakos, A., Potempa, J., Travis, J., D'Andrea, M.R., Andrade-Gordon, P., Santulli, R., Mackie, E.J., Pike, R.N., 2001a. Arginine-specific protease from *Porphyromonas gingivalis* activates protease-activated receptors on human oral epithelial cells and induces interleukin-6 secretion. Infect. Immun. 69, 5121–5130.

Lourbakos, A., Yuan, Y., Jenkins, A.L., Travis, J., Andrade-Gordon, P., Santulli, R., Potempa, J., Pike, R.N., 2001b. Activation of protease-activated receptors by gingipains from *Porphyromonas gingivalis* leads to platelet aggregation: a new trait in microbial pathogenicity. Blood 97, 3790–3797.

Lowy, F.D., 1998. *Staphylococcus aureus* infections. N. Engl. J. Med. 339, 520–532.

Lynch, J.P., Zhanel, G.G., 2009. *Streptococcus pneumoniae*: epidemiology, risk factors, and strategies for prevention. Semin. Respir. Crit. Care Med. 30, 189–209.

Malley, R., Henneke, P., Morse, S.C., Cieslewicz, M.J., Lipsitch, M., Thompson, C.M., Kurt-Jones, E., Paton, J.C., Wessels, M.R., Golenbock, D.T., 2003. Recognition of pneumolysin by Toll-like receptor 4 confers resistance to pneumococcal infection. Proc. Natl. Acad. Sci. U.S.A. 100, 1966–1971.

Manohar, M., Maheswaran, S.K., Frommes, S.P., Lindorfer, R.K., 1967. Platelet damaging factor, a fifth activity of staphylococcal alpha-toxin. J. Bacteriol. 94, 224–231.

Mitchell, T.J., Andrew, P.W., Saunders, F.K., Smith, A.N., Boulnois, G.J., 1991. Complement activation and antibody binding by pneumolysin via a region of the toxin homologous to a human acute-phase protein. Mol. Microbiol. 5, 1883–1888.

Montrucchio, G., Bosco, O., Del Sorbo, L., Fascio Pecetto, P., Lupia, E., Goffi, A., Omede, P., Emanuelli, G., Camussi, G., 2003. Mechanisms of the priming effect of low doses of lipopolysaccharides on leukocyte-dependent platelet aggregation in whole blood. Thromb. Haemost. 90, 872–881.

Morath, S., von Aulock, S., Hartung, T., 2005. Structure/function relationships of lipoteichoic acids. J. Endotoxin. Res. 11, 348–356.

Moreillon, P., Que, Y.A., 2004. Infective endocarditis. Lancet 363, 139–149.

Morgan, P.J., Hyman, S.C., Byron, O., Andrew, P.W., Mitchell, T.J., Rowe, A.J., 1994. Modeling the bacterial protein toxin, pneumolysin, in its monomeric and oligomeric form. J. Biol. Chem. 269, 25315–25320.

Morganti, R., Marcondes, S., Baldasso, P., Marangoni, S., De Nucci, G., Antunes, E., 2008. Inhibitory effects of staphylococcal enterotoxin type B on human platelet adhesion *in vitro*. Platelets 19, 432–439.

Morigi, M., Galbusera, M., Binda, E., Imberti, B., Gastoldi, S., Remuzzi, A., Zoja, C., Remuzzi, G., 2001. Verotoxin-1-induced up-regulation of adhesive molecules renders microvascular endothelial cells thrombogenic at high shear stress. Blood 98, 1828–1835.

Mortara, L.A., Bayer, A.S., 1993. *Staphylococcus aureus* bacteremia and endocarditis. New diagnostic and therapeutic concepts. Infect. Dis. Clin. North Am. 7, 53–68.

Naito, M., Sakai, E., Shi, Y., Ideguchi, H., Shoji, M., Ohara, N., Yamamoto, K., Nakayama, K., 2006. *Porphyromonas gingivalis*-induced platelet aggregation in plasma depends on Hgp44 adhesin but not Rgp proteinase. Mol. Micro. 59, 152–167.

Nylander, M., Lindahl, T.L., Bengtsson, T., Grenegard, M., 2008. The periodontal pathogen *Porphyromonas gingivalis* sensitises human blood platelets to epinephrine. Platelets 19, 352–358.

O'Brien, K.L., Wolfson, L.J., Watt, J.P., Henkle, E., Deloria-Knoll, M., McCall, N., Lee, E., Mulholland, K., Levine, O.S., Cherian, T., 2009. Burden of disease caused by *Streptococcus pneumoniae* in children younger than 5 years: global estimates. Lancet 374, 893–902.

Obrig, T., Moran, T., Brown, J., 1987. The mode of action of Shiga toxin on peptide elongation of eukaryotic protein synthesis. Biochem. J. 244, 287–294.

Palermo, M.S., Exeni, R.A., Fernandez, G.C., 2009. Hemolytic uremic syndrome: pathogenesis and update of interventions. Expert. Rev. Anti-infective Therapy 7, 697–707.

Paton, J.C., Berry, A.M., Lock, R.A., Hansman, D., Manning, P.A., 1986. Cloning and expression in *Escherichia coli* of the *Streptococcus pneumoniae* gene encoding pneumolysin. Infect. Immun. 54, 50–55.

Paton, J.C., Ferrante, A., 1983. Inhibition of human polymorphonuclear leukocyte respiratory burst, bactericidal activity, and migration by pneumolysin. Infect. Immun. 41, 1212–1216.

Paton, J.C., Rowan-Kelly, B., Ferrante, A., 1984. Activation of human complement by the pneumococcal toxin pneumolysin. Infect. Immun. 43, 1085–1087.

Pham, K., Feik, D., Hammond, B.F., Rams, T.E., Whitaker, E.J., 2002. Aggregation of human platelets by gingipain-R from *Porphyromonas gingivalis* cells and membrane vesicles. Platelets 13, 21–30.

Rodgers, G.L., Arguedas, A., Cohen, R., Dagan, R., 2009. Global serotype distribution among *Streptococcus pneumoniae* isolates causing otitis media in children: potential implications for pneumococcal conjugate vaccines. Vaccine 27, 3802–3810.

Satoh, T., Yamashita, Y., Kamiyama, T., Watanabe, J., Steiner, B., Hadváry, P., Arisawa, M., 1993. Tetrafibricin: a nonpeptidic fibrinogen receptor inhibitor from *Streptomyces neyagawaensis* (I) its GPIIb/IIIa blockage on solid phase binding assay. Thromb. Res. 72, 389–400.

Schmitt, C., Meysick, K., O'Brien, A., 1999. Bacterial toxins: friends or foes? Emerg. Infect. Dis. 5, 224–234.

Shah, R., 2009. Protease-activated receptors in cardiovascular health and diseases. Am. Heart J. 157, 253–262.

Shannon, O., Hertzen, E., Norrby-Teglund, A., Morgelin, M., Sjobring, U., Bjorck, L., 2007. Severe streptococcal infection is associated with M protein-induced platelet activation and thrombus formation. Mol. Microbiol. 65, 1147–1157.

Shashkin, P.N., Brown, G.T., Ghosh, A., Marathe, G.K., McIntyre, T.M., 2008. Lipopolysaccharide is a direct agonist for platelet RNA splicing. J Immunol 181, 3495–3502.

Sheu, J.R., Hsiao, G., Lee, C., Chang, W., Lee, L.W., Su, C.H., Lin, C.H., 2000a. Antiplatelet activity of *Staphylococcus aureus* lipoteichoic acid is mediated through a cyclic AMP pathway. Thromb. Res. 99, 249–258.

Sheu, J.R., Lee, C.R., Lin, C.H., Hsiao, G., Ko, W.C., Chen, Y.C., Yen, M.H., 2000b. Mechanisms involved in the antiplatelet activity of *Staphylococcus aureus* lipoteichoic acid in human platelets. Thromb. Haemost. 83, 777–784.

Shibazaki, M., Kawabata, Y., Yokochi, T., Nishida, A., Takada, H., Endo, Y., 1999. Complement-dependent accumulation and degradation of platelets in the lung and liver induced by injection of lipopolysaccharides. Infect. Immun. 67, 5186–5191.

Shiraki, R., Inoue, N., Kawasaki, S., Takei, A., Kadotani, M., Ohnishi, Y., Ejiri, J., Kobayashi, S., Hirata, K., Kawashima, S., Yokoyama, M., 2004. Expression of Toll-like receptors on human platelets. Thromb. Res. 113, 379–385.

Shoma, S., Tsuchiya, K., Kawamura, I., Nomura, T., Hara, H., Uchiyama, R., Daim, S., Mitsuyama, M., 2008. Critical involvement of pneumolysin in production of interleukin-1α and caspase-1-dependent cytokines in infection with *Streptococcus pneumoniae in vitro*: a novel function of pneumolysin in caspase-1 activation. Infect. Immun. 76, 1547–1557.

Siegel, I., Cohen, S., 1964. Action of staphylococcal toxin on human platelets. J. Infect. Dis. 114, 488–502.

Spreer, A., Kerstan, H., Bottcher, T., Gerber, J., Siemer, A., Zysk, G., Mitchell, T.J., Eiffert, H., Nau, R., 2003. Reduced release of pneumolysin by *Streptococcus pneumoniae* in vitro and in vivo after treatment with nonbacteriolytic antibiotics in comparison to ceftriaxone. Antimicrob. Agents Chemother. 47, 2649–2654.

Stahl, A., Sartz, L., Nelsson, A., Bekassy, Z., Karpman, D., 2009. Shiga toxin and lipopolysaccharide induce platelet-leukocyte aggregates and tissue factor release, a thrombotic mechanism in hemolytic uremic syndrome. PLoS One 4, e6990.

Stahl, A.-l., Svensson, M., Morgelin, M., Svanborg, C., Tarr, P.I., Mooney, J.C., Watkins, S.L., Johnson, R., Karpman, D., 2006. Lipopolysaccharide from enterohemorrhagic *Escherichia coli* binds to platelets through TLR4 and CD62 and is detected on circulating platelets in patients with hemolytic uremic syndrome. Blood 108, 167–176.

Stevens, D.L., Tanner, M.H., Winship, J., Swarts, R., Ries, K.M., Schlievert, P.M., Kaplan, E., 1989. Severe group A streptococcal infections associated with a toxic shock-like syndrome and scarlet fever toxin A. N. Engl. J. Med. 321, 1–7.

Stockbauer, K.E., Magoun, L., Liu, M., Burns, E.H., Gubba, S., Renish, S., Pan, X., Bodary, S.C., Baker, E., Coburn, J., Leong, J.M., Musser, J.M., 1999. A natural variant of the cysteine protease virulence factor of group A Streptococcus with an arginine-glycine-aspartic acid (RGD) motif preferentially binds human integrins $\alpha_V\beta_3$ and $\alpha_{IIb}\beta_3$. Proc. Natl. Acad. Sci. U.S.A. 96, 242–247.

Taylor, F.B., Jr., Bryant, A.E., Blick, K.E., Hack, E., Jansen, P.M., Kosanke, S.D., Stevens, D.L., 1999. Staging of the baboon response to group A streptococci administered intramuscularly: a descriptive study of the clinical symptoms and clinical chemical response patterns. Clin. Infect. Dis. 29, 167–177.

te Loo, D., H., L.A., van der Velden, T., Vermeer, M., Preyers, F., Demacker, P., van den Heuvel, L., van Hinsbergh, V., 2000. Binding and transfer of verocytotoxin by polymorphonuclear leukocytes in hemolytic uremic syndrome. Blood 95, 3396–3402.

Thorpe, C.M., Flaumenhaft, R., Hurley, B., Jacewicz, M., Acheson, D.W.K., Keusch, G.T., 1999. Shiga toxins do not directly stimulate alpha-granule secretion or enhance aggregation of human platelets. Acta Haematologica 102, 51–55.

Tilley, S.J., Orlova, E.V., Gilbert, R.J., Andrew, P.W., Saibil, H.R., 2005. Structural basis of pore formation by the bacterial toxin pneumolysin. Cell 121, 247–256.

Tran, U., Boyle, T., Shupp, J., Hammamieh, R., Jett, M., 2006. Staphylococcal enterotoxin B initiates protein kinase C translocation and eicosanoid metabolism while inhibiting thrombin-induced aggregation in human platelets. Mol. Cell. Biochem. 288, 171–178.

Valeva, A., Weisser, A., Walker, B., Kehoe, M., Bayley, H., Bhakdi, S., Palmer, M., 1996. Molecular architecture of a toxin pore: a 15-residue sequence lines the transmembrane channel of staphylococcal alpha-toxin. EMBO J. 15, 1857–1864.

van der Poll, T., Opal, S.M., 2009. Pathogenesis, treatment, and prevention of pneumococcal pneumonia. Lancet 374, 1543–1556.

Walker, J.A., Allen, R.L., Falmagne, P., Johnson, M.K., Boulnois, G.J., 1987. Molecular cloning, characterization, and complete nucleotide sequence of the gene for pneumolysin, the sulfhydryl-activated toxin of *Streptococcus pneumoniae*. Infect. Immun. 55, 1184–1189.

Ward, J.R., Bingle, L., Judge, H.M., Brown, S.B., Storey, R.F., Whyte, M.K., Dower, S.K., Buttle, D.J., Sabroe, I., 2005. Agonists of toll-like receptor (TLR)2 and TLR4 are unable to modulate platelet activation by adenosine diphosphate and platelet activating factor. Thromb. Haemost. 94, 831–838.

White, J., Rao, G., Gerrard, J., 1974. Effects of the ionophore A23187 on blood platelets I. Influence on aggregation and secretion. Am. J. Pathol. 77, 135–149.

Whitworth, N., Barradas, M., Mikhailidis, D., Dandona, P., 1989. An investigation into the effects of bacterial lipopolysaccharide on human platelets. Eur. J. Haematol. 43, 112–119.

Wilson, M., Blum, R., Dandona, P., Mousa, S., 2001. Effects in humans of intravenously administered endotoxin on soluble cell-adhesion molecule and inflammatory markers: a model of human diseases. Clin. Exper. Pharmacol. Physiol. 28, 376–380.

Yagi, H., Narita, N., Matsumoto, M., Sakurai, Y., Ikari, H., Yoshioka, A., Kita, E., Ikeda, Y., Titani, K., Fujimura, Y., 2001. Enhanced low shear stress induced platelet aggregation by Shiga-like toxin 1 purified from *Escherichia coli* O157. Am. J. Hematol. 66, 105–115.

Yeaman, M., Bayer, A., 2006. Antimicrobial peptides versus invasive infections. Curr. Top. Microbiol. Immunol. 306, 111–152.

Yeaman, M.R., Bayer, A.S., Koo, S.P., Foss, W., Sullam, P.M., 1998. Platelet microbicidal proteins and neutrophil defensin disrupt the *Staphylococcus aureus* cytoplasmic membrane by distinct mechanisms of action. J. Clin. Invest. 101, 178–187.

Zähringer, U., Lindner, B., Inamura, S., Heine, H., Alexander, C., 2008. TLR2 – promiscuous or specific? A critical re-evaluation of a receptor expressing apparent broad specificity. Immunobiology 213, 205–224.

Part VIII
Hypotensive Factors

Chapter 37
Hypertensive and Hypotensive Snake Venom Components

Avner Bdolah

Abstract Most snake venoms of *Viperidae* induce a fall in blood pressure following envenomation. However the mechanism(s) involved in the hypotensive effects of the venoms are largely unknown. VEGF-like peptides, which affect vascular permeability, have been described in *Viperidae* as well as in *Elapidae* venoms. The following chapter reviews some of the cardiovascular toxins that have been isolated and characterized in snake venoms. This includes the sarafotoxins, which are the most potent hypertensive peptides, as well as groups of hypotensive peptides like the natriuretic peptides and calcium channel blockers. The extensive research that has been carried out is now providing us with some new options to treat diseases.

Introduction

Most viperid venoms induce a fall in blood pressure following envenomation. However the mechanism(s) involved in the hypotensive effects of the venoms are largely unknown (see Lee and Lee, 1979). Bicher et al. (1966) reported that intravenous injection of *Vipera palaestinae* venom or an isolated fraction of this venom induced an immediate fall in blood pressure. It was claimed that this effect was due to depression of the central autonomic vasoregulatory mechanism. The "two-component toxin" from the venom of *V. palaestinae* (Simon et al., 1980) was also shown to induce a sudden drop of blood pressure in mice (C. Y. Lee, unpublished observations); however, the mode of action of this toxic fraction is not yet understood. Russell's viper (*Daboia*) induces among other cardiovascular effects a fall in blood pressure (Warrell, 1995). The most conspicuous hydrodynamic effect induced

A. Bdolah (✉)
Department of Zoology, George S. Wise Faculty of Life Sciences, Tel Aviv University, Ramat Aviv, Tel Aviv, 69978 Israel
e-mail: avnerb@post.tau.ac.il

R.M. Kini et al. (eds.), *Toxins and Hemostasis*, DOI 10.1007/978-90-481-9295-3_37,

by rattlesnake venoms is also an immediate fall in systemic blood pressure; similar effects have been described for other crotalid venoms (see Lee and Lee, 1979).

Many snake venoms have been shown to increase vascular permeability. Several proteins possessing this capacity have been isolated and characterized in snake venoms of *Viperidae* (for review, see Joseph et al., 2004). The deduced protein sequence of cDNA of vascular endothelial growth factor (VEGF) from the venom gland of *Bothrops insularis* displayed similarity with vertebrate VEGFs and with the hypotensive factor from *V. aspis*. Western blots with anti-VEGF revealed the presence of this protein in various *Viperidae* venoms and also in *Elapidae* venom (*Micrurus sp.*). These results suggest that VEGF-like peptides, which affect vascular permeability, are widely distributed in snake venoms (Junqueira de Azevedo et al., 2001).

Snake venoms of the genus *Bothrops*, which are responsible for most fatalities in central and South America, comprise a complex mixture of toxins and enzymes. In general, death results from hypotension secondary to hypovolemia (Gutierrez, 1995). Rocha e Silva et al. (1949) first demonstrated that the venom of *B. jararaca* induced the release of bradykinin from plasma globulin resulting in hypotension. Furthermore, it was later demonstrated that the venom contains a so-called "bradykinin potentiating factor", which was found to be an inhibitor of endoproteinases. These studies led to the development of the orally active inhibitor of the angiotensin converting enzyme (Cushman and Ondetti, 1991; Ferreira, 2010).

The following sections review some of the cardiovascular toxins that have been isolated and characterized in snake venoms. These studies, which include the most potent hypertensive peptides as well as several groups of hypotensive peptides, have contributed to our understanding of the physiology and pathology of the processes in which the toxins are involved.

Sarafotoxins

The sarafotoxins (SRTXs) are unique peptides produced by the burrowing asps of the genus *Atractaspis*. About 16 species of this genus have been described. They are widely distributed from South Africa, through the entire breadth of Central Africa, and along the Rift Valley to the north along the Jordan Valley (Gasperetti, 1988). Information on bites by the different species of *Atractaspis* is rather limited. Due to the rapid clinical course in fatal cases of *Atractaspis* bites, and other signs like coma, these venoms have often been described as neurotoxic (e.g., Warrell et al., 1976). Reports on the clinical manifestation following envenomation by the Israeli burrowing asp, *A. engaddensis*, have described both local and systemic symptoms. However, no indications for neurotoxicity of the crude venom were obtained by intra-ventricular injection (unpublished observations) or when tested with nerve-muscle preparations. On the other hand, this venom showed a profound effect on the function of the heart, when tested in anesthetized mice or with isolated heart preparation (Lee et al., 1986; Weiser et al., 1984).

Isolation and Structural Similarity of Sarafotoxins and Endothelins

Initial fractionation of the *A. engaddensis* venom revealed a highly toxic low molecular weight fraction, which constitutes about 30–40% of the venom proteins (Kochva et al., 1982). Further fractionation of this toxic fraction resulted in the isolation of the first three sarafotoxins (SRTXs): SRTX-a, -b, and -c (Takasaki et al., 1988; Wollberg et al., 1988). These are of 21-amino acid peptides with two disulfide bridges (Fig. 37.1). Surprisingly, the SRTXs showed structural similarity with the most potent vasoconstrictor peptide, the endothelin (ET), which was first isolated from vascular endothelial cells and published a few months before the SRTXs (Yanagisawa et al., 1988). Additional ET isopeptides (ET-2 and ET-3) were then isolated (Inoue et al., 1989). The peptides of the two families (the SRTXs from the venom of *A. engaddensis* and the ETs found in different vertebrate tissues) are 21-amino acid peptides with two identical disulfide bridges, Cys 1–15 and Cys 3–11. They reveal a sequence similarity not only within each family (ET, 71–95%; SRTX, 81–95%), but also between the two families (52–67%). All have a hydrophobic carboxy-terminal tail, and the most important differences between the various peptides of the SRTX/ET group reside within the inner loop Cys3-Cys11 (Fig. 37.1).

In an attempt to identify SRTX-like peptides in different species of *Atractaspis*, we used a highly potent rabbit antiserum against SRTX-b. Indeed, immunoreactive-SRTXs, albeit at low levels, were demonstrated in venom samples of several species of this genus: *A. bibroni, A. dahomeyensis, A. microlepidota* (from east Africa), and *A. micropholis* (Bdolah, 2010). Because of the low levels of the SRTX-like peptides in these venoms they could not be isolated by regular gel filtration techniques. A recent report on fractionation of pooled venoms of *Atractaspis* from Saudi Arabia showed a molecular-sieving pattern with a prominent toxic peak of SRTXs, similar to the pattern obtained with *A. engaddensis* (Ismail et al., 2007). Indeed, according to Gasperetti (1988), there are two subspecies of *Atractaspis* in Arabia:

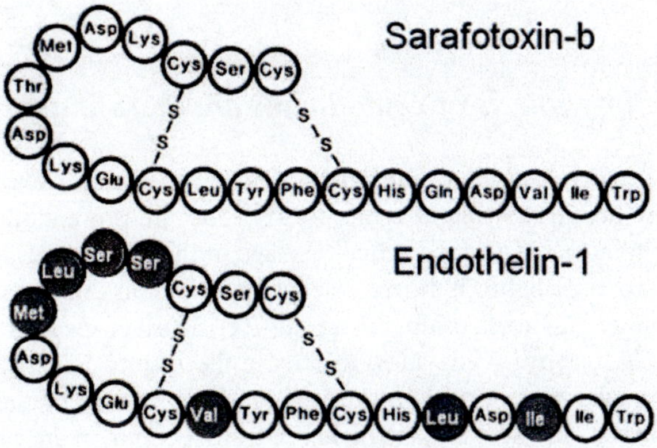

Fig. 37.1 Primary structures of sarafotoxin-b and endothelin-1. The non-identical residues are denoted by a *gray background*

Fig. 37.2 Amino acid sequences of the sarafotoxin isopeptides. Only the most abundant sarafotoxin-like peptides from *A. m. microlepidota* (24-amino acids long) and *A. irregularis* (25-amino acids long) are included in this figure (modified from Ducancel, 2005; Kochva et al., 1993). Residues identical to SRTX-a are *boldfaced*.
[a]BTX=bibrotoxin; [b] SRTX-e is identical to the previously designated SRTX-d (Bdolah, 2010)

1 5 11 15 20	
CSCKDMTDKECLNFCHQDVIW	SRTX-a
CSCKDMTDKECLYFCHQDVIW	SRTX-b
CTCNDMTDEECLNFCHQDVIW	SRTX-c
CSCADMTDKECLYFCHQDVIW	BTX[a]
CTCKDMTDKECLYFCHQGIIW	SRTX-d/e[b]
CSCNDMNDKECMYFCHQDVIWDEP	SRTX-m1
CSCNDINDKECMYFCHQDIIWDEP	SRTX-m2
CSCNDMNDKECVYFCHLDIIWDEP	SRTX-m3
CSCNNMSDKECLNFCNLDIIWENV	SRTX-m4
CSCNDMNDKECVYFCHLDIIWDEP	SRTX-m5
CSCTDMSDLECMNFCHKDVIWINRN	SRTX-i1
CSCADMSDLECMNFCRLDVMWVNRN	SRTX-i2
CSCTDMSDLECMNFCHKDVIWVNRN	SRTX-i3

A. *microlepidota engaddensis* and *A. microlepidota andersonii*. An SRTX-like toxin, bibrotoxin, was isolated from the venom of *A. bibroni* by RP-HPLC (Becker et al., 1993); this peptide differs from SRTX-b in only one residue (Fig. 37.2).

In the last decade a combination of mass spectrometry analysis and molecular cloning has led to the identification of two more families of SRTX-like peptides from the venoms of *A. microlepidota microlepidota* and *A. irregularis* (Ducancel, 2005; Hayashi et al., 2004; Quinton et al., 2005). The SRTX-like peptides of these two species share the conserved cysteine skeleton and several invariant residues, including Trp21, with the "classical" 21-amino acid SRTX/ET peptides. However, these new SRTX isoforms have molecular sizes longer than those isolated from *A. engaddensis*. Most of the SRTX isoforms of *A. m. microlepidota* are 24-amino acid peptides with a C- terminal sequence of Asp-Glu-Pro; they are designated as SRTX-m1–5. A wide range of molecular sizes of SRTX-like peptides was also isolated from *A. irregularis*; the most abundant are 25-amino acids long. They are designated as SRTX-i1–3 (Fig. 37.2).

Biosynthesis of Endothelins and Sarafotoxins

Separate genes, which are located on a separate chromosome, encode each ET isopeptide (Arinami et al., 1991). The pre-pro-endothelins comprise around 200 amino acid residues. After cleavage of the signal peptide, the resulting polypeptide, pro-endothelin, is further cleaved to yield an intermediate form, big-ET, of about 40 amino acid residues. The big-ET is further cleaved by specific endopeptidases, ET-converting enzymes, to release the mature ETs (Masaki, 2000). The study by Landan et al. (1991) on the evolutionary history of the ET/SRTX genes indicated that the SRTXs lineage diverged from a common ancestral ET/SRTX gene prior to the first endothelin gene duplication event. This notion is further supported by the unique organization of the SRTXs genes.

Isolation of the complementary DNA encoding SRTXs was first achieved in the venom gland of *A. engaddensis* (Ducancel et al., 1993). This cDNA comprises 1948 base pairs (bp) including an open reading frame of 1629 bp coding for a long pre-pro-polypeptide of 543 amino acids. The deduced amino acid sequence of this cDNA revealed a unique "rosary-type" organization of the gene encoding the SRTXs in the venom gland of *A. engaddensis*. It consists of 12 successive stretches of 40 residues, each of which begins with a "spacer" of 19 invariant residues, which is followed by a sequence of an SRTX isoform. Six SRTX isoforms were identified within a single precursor molecule; it included SRTX-a, -b, -c, SRTX-a1 and -b1 (which have one residue substitution as compared to SRTX-a, -b), and an additional SRTX-isoform, SRTX-e. Reexamination of the previously isolated SRTX-d (Bdolah et al., 1989a) revealed that the latter was identical to SRTX-e (unpublished observations) and was thus designated as SRTX-d/e (Bdolah, 2010).

An incomplete cDNA encoding SRTX-like precursor was isolated from the venom gland of *A. m. microlepidota* (Hayashi et al., 2004). This fragment of cDNA contains an open reading frame that encodes a 351-amino acid long polypeptide. The structure of this precursor indicates that the SRTX gene of *A. m. microlepidota* has a polycistronic organization, similar to that found in *A. engaddensis*. It comprises seven tandem stretches of 48 residues corresponding to five different SRTX isoforms. Each contains the sequence of 24-amino acid-long SRTX, which shows high homology with the 21-amino acid SRTXs, followed by a highly conserved 24-residue-long spacer. Even though the spacer stretches in the SRTX-precursor of *A. m. microlepidota* are longer than those identified in *A. engaddensis*, the spacer domains in the SRTX-precursors in these two species are highly homologous (Fig. 37.3).

Two cDNAs encoding SRTX-like precursors from the venom gland of *A. irregularis* were cloned (Quinton et al., 2005). The deduced amino acid sequences of the two clones, which show high sequence similarity, include only one stretch of a 25-amino acid long SRTX-like peptide, SRTX-i1 or SRTX-i2, which shows high homology with the 21-amino acid SRTXs (Fig. 37.2). No cDNA clone that encodes SRTX-i3 has yet been identified.

A. engaddensis	**RDT**KQ**AAR**D**PSPQ**RNVE**PL**	
A. microlepidota	VVVSV**RDT**EE**AAR**V**PSPQ**KRSQ**PL**	SRTX-m
	VVVSV**RDT**EE**AAR**V**PSPQ**KRPQ**P**R	SRTX-m1
	VVVSV**RDT**EE**A**T**R**V**PSPQ**KRSQ**PL**	SRTX-m2
	VVVSV**RDT**EE**AAR**V**PSPQ**KRSQ**P**R	SRTX-m3
	VVVSVQ**DT**EE**AAR**V**PSPQ**KRSQ**PL**	SRTX-m4

Fig. 37.3 Amino acid sequences of the spacer domains in the SRTX precursors of *A. engaddensis* and *A. m. microlepidota*. Identical residues are *boldfaced*

To date, SRTX precursors have been identified in only three species of *Atractaspis*. Despite the different overall organization of genes encoding these precursors, they display a high degree of sequence identity in specific domains of the genes. This includes the domain encoding the signal peptide as well as the 61–69 amino acid sequences that precede the stretches that encode the mature SRTXs (Ducancel, 2005).

Pharmacology of Sarafotoxins and Endothelins

The use of radio-labeled SRTX-b made it possible to identify and characterize high affinity binding sites (K_d 3–4 nM, B_{max} ~100 fmol/mg protein) in rat heart and brain preparations. These earlier experiments also showed that the SRTXs are potent activators of phosphoinositide hydrolysis (Fig. 37.4). Competition binding experiments between the labeled SRTX-b and other SRTXs, as well as with ET-1 and ET-3, showed that peptides of the SRTX/ET family compete for similar receptors; however, they also indicated the existence of different SRTX/ET receptor subtypes (Kloog et al., 1988, 1989). Indeed, the first reports of the cloning and characterization of the genes encoding ET receptors appeared two years after the first discovery of the ET. The first ET receptor, designated ET_A, was cloned from a bovine lung cDNA library and expressed in *Xenopus* oocytes; the second receptor subtype, designated ET_B, was cloned from rat lung and then expressed in COS-7 cells (Arai et al., 1990; Sakurai et al., 1990). ET_A showed high selectivity to ET-1 and ET-2, whereas the affinity of ET-3 to this receptor was much lower. ET_B showed similar affinities to ET-1, -2 and -3. SRTX-c usually binds with high affinity to ET_B and is

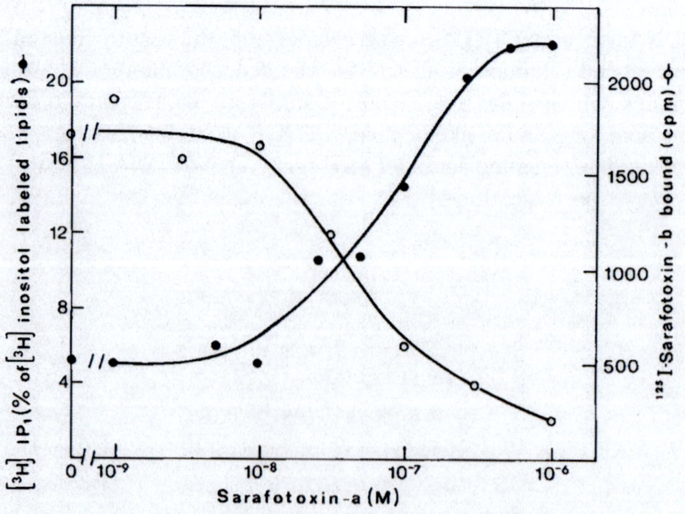

Fig. 37.4 Concentration-dependent inhibition of [^{125}I]SRTX-b and induction of phosphoinositide hydrolysis by SRTX-a, in rat atrial membranes (modified from Kloog et al., 1988)

used as a selective ET_B agonist (Williams et al., 1991). A third subtype of ET receptor, ET_C, which is specific for ET-3, was identified in the dermal melanophores of *Xenopus laevis*; this receptor mediates induction of pigment rearrangement (Karne et al., 1993). All ET receptor subtypes are G-protein coupled receptors (for a review, see Sokolovsky, 1995).

The ET_A receptor appears to be the predominant one in the vascular system. ET_A-selective antagonists, such as BQ-123 (Nambi et al., 1994), potently inhibit the vasoconstriction response ET-1. ET_B receptors are widely distributed throughout the vascular tissues, mediating vasodilation (Kanaide, 1996). ET_B receptors are also present in many non-vascular tissues, including the lung, where they mediate bronchial smooth muscle contraction (e.g., Noguchi et al., 1993). An ET_A subtype receptor was cloned from the South American snake *Bothrps jararaca* (Borgheresi et al., 2006). It was demonstrated that SRTX/ET peptides affect the cardiovascular system in this snake. The presence of ET_A receptor was also demonstrated in the vascular system of a colubrid snake (Mesquita et al., 2008).

Since the discovery of the ET in cultured porcine endothelial cells, ETs have been identified and found to be active in a variety of vertebrate tissues (for review, see Masaki, 2001). The densities of the SRTX/ET receptors and their subtle pharmacological properties may vary between different animals and tissues. For example, binding studies in the mongoose, which is resistant to SRTX-b, revealed SRTX-specific binding in brain and cardiovascular preparations. However, SRTX-b failed to induce contraction in aortal preparations of the mongoose, whereas ET-1 did induce contraction in the same preparation (Bdolah et al., 1997). These results suggest that the SRTX/ET receptors in the mongoose have some structural features that enable them to differentiate between the two peptides. Among other examples of differential effects of SRTX/ET peptides are the effects of SRTX-b and ET-1 on the saphenous vein (Maguire et al., 1996). Binding studies with the two isopeptides, using membrane preparation, revealed similar affinities, with a difference in their maximal binding. This discrepancy was attributed to the existence of two populations of receptors. The above-mentioned examples of pharmacological studies with the different SRTX/ET isopeptides as well as many others raised the possibility that more than the two (or three) conventional endothelin receptor subtypes exist (for a review, see Henry and Goldie, 2001).

The ET/SRTX receptors are able to activate different signal cascades, including activation of phospholipase C, phospholipase D, phospholipase A_2 (Abdel-Latif et al., 1996), as well as to increase production of cyclic nucleotides and other cell signaling systems (for a review, see Simonson, 2001). As mentioned above, activation of ET receptor enhances the inositol lipid turnover, which is one of the major mechanisms for trans-membrane signaling. Inositol 3-phosphate triggers the rapid mobilization of intracellular Ca^{2+}, which affects the contractile mechanism.

The different signal responses have been attributed to differences in the receptor activation, which probably occur at the level of Ligand-Receptor-G-protein coupling (see Sokolovsky and Shraga-Levine, 2001 and references therein). In rat atrial slices, ET-1 stimulated or inhibited cyclic AMP production depending on its concentration. SRTXs showed a different pattern of behavior. ET-1 and SRTX-b also modulated the guanylate cyclase pathway. The stimulation or inhibition of cyclic GMP production was also dependent on peptide concentration. Interestingly, the stimulation of cyclic GMP production via the NO pathway or the CO pathway was found to be different in different tissues, and was activated specifically by the different SRTX/ET peptides that interacted with different receptor subtypes. The study by Shraga-Levine and Sokolovsky (2000) examined the interaction between different SRTX/ET peptides and ET_A or ET_B receptor subtype and their coupling with different G-proteins. It was concluded that upon binding of the peptide agonist with a specific receptor subtype the appropriate G-protein activates a specific intracellular signal pathway.

The major activity of the SRTX/ET peptides involves the cardiovascular system (Kochva et al., 1993 and references therein). The cardiotoxic effects of SRTXs were examined in anesthetized mice, in vitro in Lagendorff's preparations, in rat and human cardiac muscles, and were also followed in human patients bitten by *A. engaddensis*. Typical change in the ECG of anesthetized mice following administration of a lethal dose of SRTX-b included a transient elevation of the ST segment and prolongation of the PR interval, culminating in about 2–3 min in a complete atrioventricular dissociation (Fig. 37.5). ET-1 had similar effects on this system (Bdolah et al., 1989b; Wollberg et al., 1988). Interestingly, administration of ET-1 (1 μg/Kg) into anesthetized rats induced comparable electrocardiographic deteriorations (Yorikane and Koike, 1990), which included a transient elevation of the ST segment that coincided with maximal hypertension. These results as well as those of other systems suggested that the SRTX/ETs have a direct effect on the conducting system and cause coronary vasoconstriction. SRTX-b and ET-1 also showed similar lethal capacity in mice. Altogether, these findings indicate that the SRTXs and ETs activate similar targets. Thus, the difference in the function the two

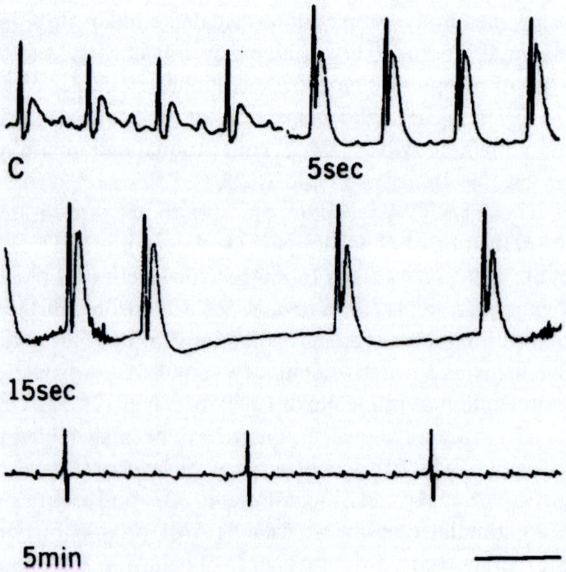

Fig. 37.5 The effect of a lethal dose (25 μg/Kg) of SRTX-b on the ECG of anesthetized mice (modified from Wollberg et al., 1988)

peptide families, as toxic or regulatory agents, derive from their concentration in different tissues and animals: The concentration of the SRTXs in the venom gland of *A. engaddensis* is about five orders of magnitude higher than the concentration of ET-1 in the inner medulla of the rat kidney (9 pg/mg wet tissue), in which the highest concentration of ET-1 was found (Kitamura et al., 1989).

Limited pharmacological studies have been carried out with SRTX-like peptides from the different species of *Atractaspis*: Bibrotoxin (from *A. bibroni*) showed vasoactive effect, but at a quite lower potency as compared to SRTX-b and ET-1 (Becker et al., 1993). SRTX-m (from *A. m. microlepidota*) showed a comparable lethal capacity to SRTX-b. The effect of this peptide on ECG in mice was similar to SRTX-b, but with a longer latency period (Hayashi et al., 2004).

The SRTX/ET peptides are mainly active in smooth muscle systems. In addition to their effects on the vascular system (Fig. 37.6), their effects on several non-vascular systems have been demonstrated (see kochva et al., 1993). These included the effects of SRTX/ET peptides on spontaneous and induced contractions of guinea-pig ileum and of the rat uterus at different stages of the estrus cycle (Wollberg et al., 1991, 1992). Among the ETs VIC (vasoactive intestinal contractor, Saida et al., 1989) showed the highest contraction activity in the ileum; ET-1 and SRTX-b induced similar effects, whereas SRTX-c and ET-3 showed very poor effects. SRTX-d/e had no effect on the guinea-pig ileum; preliminary results suggest that this SRTX iso-peptide, which is practically non-toxic, had even an inhibitory effect on the SRTX-b-induced contraction in the ileum (Wollberg et al., 1998). In the rat uterus preparations the effects of ET-1, ET-3, SRTX-b and SRTX-c on the maximal increase of the peak tension were quite similar; however, the effects on the other parameters of the contractile behavior of the uterus varied between the different SRTX/ET peptides.

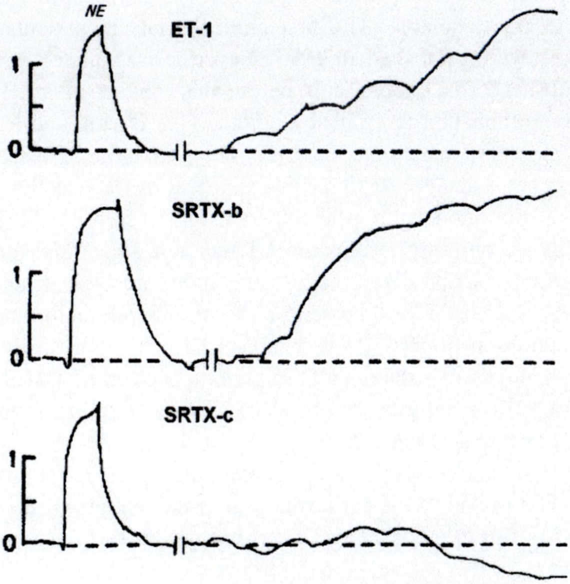

Fig. 37.6 Isometric contractile responses to cumulatively applied ET/SRTX peptides to rabbit aortic rings (modified from Wollberg et al., 1989). Vertical scales, tension force (g). NE, 6 μM norepinephrine

The effects of SRTX/ET peptides have been examined in several non-mammalian systems. Poder et al. (1991) showed that ET-1 induced increase in contraction of vascular preparations of reptile, amphibia, and fish. Profound effects of ET-1 on the cardiovascular system of trout was demonstrated by Olson et al. (1991), using anaesthetized fish, perfused tissues, and isolated vascular rings.

Zigdon-Arad et al. (1992) studied the effects of SRTX-b on isolated heart preparations of fish (*Tilapia* and *Torpedo*), toad, lizard and water snake. The toxin reduced the rate of spontaneous contraction in the torpedo and lizard preparations at a dose of 0.05 μg/ml, whereas much higher concentrations of SRTX-b were required for a similar effect on the *Tilapia* and water snake preparations. ECG deteriorations that led to cardiac arrest were also evident in *Agama* after administration of 15 μg SRTX-b, whereas in toad higher doses of SRTX-b induced only transient disturbances in the ECG.

As mentioned above, ET-1, which was first isolated from porcine endothelial cells (Yanagisawa et al., 1988), was found to be the most potent vasoconstrictor. However, it has been shown by many authors (e.g., Martel et al., 1991), that the response to ET may be biphasic, with a transient period of hypotension apparently caused by vasodilatation, followed by a sustained hypertension. The vasodilation effect is especially evident with the "weak peptides" (SRTX-c, SRTX-d/e and ET-3), which are very poor vasoconstrictors, if at all, and are considerably more effective as vasodilators (Bdolah et al., 1989b; Warner et al., 1989).

Therapeutic Potential of the Endothelin Axis Modulators

Within a short time after cloning of the ET receptors, antagonists for the receptors became available. The first clinical trials in patients with congestive heart failure were published in 1995. Nevertheless, it took several more years and many unsuccessful experiments before the concept of an ET receptors blockade could be established in clinical medicine (see Barton and Yanagisawa, 2008). Bosentan (Tracleer), which is a dual ET_A/ET_B antagonist, became the first orally active ET receptor antagonist to receive approval in US and Europe for treatment of patients with pulmonary arterial hypertension (Rubin et al., 2002). Ongoing clinical trials are currently evaluating ET receptor antagonist for treatment of hypertension, cardiovascular disorders, cancers and more (Aubert and Juillerat-Jeanneret, 2009). Attempts to develop potent inhibitors of endothelin-converting enzymes (ECE) for inhibition of ET-mediated effects have not yet resulted in promising therapeutic potentials. To date, ECE inhibitors have been tested mainly in animal models, in vitro in tumors, and in a few clinical trials in prostate cancer (for review, see Battistini and Jeng, 2001).

In conclusion, the ET receptor antagonists are a promising and innovative drug class to add to the cardiovascular armamentarium. By now, the potential of SRTX isoforms to modify the ET axis has not yet been explored.

Natriuretic Peptides

Sources and Structure

The natriuretic peptides (NPs) are hypotensive hormones; their main actions are implicated in eliciting natriuretic, diuretic and vasorelaxant effects (Pandey, 2005). The atrial natriuretic peptide was first discovered by de Bold (1981). Three mammalian NPs have since been identified and characterized, including the atrial natriuretic peptide (ANP); B-type natriuretic peptide (BNP); and C-type natriuretic peptide (CNP) (Potter et al., 2006). In addition, urodilatin is an NP that has been isolated from human urine and originates from the same precursor as ANP. These four NPs share a structural similarity, which consists of a 17-amino-acid core bridged by an intramolecular disulfide bond (Fig. 37.7). Both ANP and BNP are produced and secreted in the heart, whereas CNP, which was first isolated from the porcine brain, is predominantly localized in the central nervous system and is considered a non-circulatory hormone (Pandey, 2005; Sudoh et al., 1990). Recently, a peptide that exhibits homology with the mammalian CNP has been identified in the platypus (*Ornithorhynchus anatinus*), an egg-laying mammal. This peptide induces hypotension in vivo, relaxes smooth muscle in vitro and elicits cyclic GMP production (de Plater et al., 1998).

Analysis of the evolution of the snake venom proteome has shown that toxins have arisen from recruitment events of genes of different proteins, which include among many others the NPs (Fry, 2005). Indeed many NPs have been identified in snake venom during recent years. The first snake NP was identified in the venom of the green mamba (*Dendraspis angusticeps*) (Schweitz et al., 1992). The *Dendroaspis* natriuretic peptide (DNP) has a 17-amino acid central core similar to the mammalian NPs, but a longer c-terminus (Fig. 37.7). Competition binding experiments with rat aortic myocytes have shown that DNP competes with ANP for the same receptors (see below). Recent experiments with labeled DNP have demonstrated specific NP receptors in human heart and coronary arteries; downregulation of the receptors could be demonstrated in heart failure. Specific binding sites for the DNP have also been revealed in the mammary artery; Like ANP, DNP fully reversed the constrictor response to endothelin-1 in the mammary artery (Singh et al., 2006a,b). The effects of DNP on renal functions and heart failure in dogs have

ANP	SLRRSS**CFG**GRM**DRI**GAQ**SGLGC**NSFRY
BNP	SPKMVQGSG**CFG**RKM**DRI**SSS**SGLGC**KKVLRRH
CNP	GLSKG**CFG**LKL**DRI**GSM**SGLGC**
DNP	EVKYDP**CFG**HKI**DRI**NHV**S**N**LGC**PSLRDPRPNAPSTSA

(Disulfide bond S—S linking the two bold C residues of the core.)

Fig. 37.7 Primary structures of human natriuretic and snake (*Dendroaspis*) natriuretic peptides. The identical amino acid residues in the central core are *bold-faced*

demonstrated its therapeutic potential (Lisy et al., 1999, 2001). Proteomic analysis of several viper venoms revealed the presence of NPs, some of which exhibited anti-platelet activity (Amininasab et al., 2004; Bazza et al., 2005). NPs were also identified in crotalids: N-terminal truncated forms of CNP were isolated from the venom of the habu snake (*Trimeresurus flavoviridis*); one of these peptides exhibited a low potency of vasorelaxation and a weak diuretic effect (Michel et al., 2000); CNP-like sequences have been identified in *Crotalus durissus collilineatus* and in *Bothrops jararaca* (Higuchi et al., 2006; Murayama et al., 1997). Natriuretic-like peptides have also been identified in venoms of Australian elapids (Earl et al., 2006): three NPs were isolated from the venoms of different taipan (*Oxyuranus*) species; only one of the peptides (TPN-c) was equipotent to ANP in hypotensive and vasodilation activity (Fry et al., 2005). Two isoforms of NPs were found in the venom of *Pseudonaja textiles*; one of these exhibited cyclic GMP-stimulating properties (St Pierre et al., 2005).

Receptors and Signal Transduction

Evidence for tissue-specific heterogeneity of NP receptors in different tissues was obtained by ligand-binding experiments and photoaffinity labeling (Pandey et al., 1988). Cloning and expression of cDNAs led to the identification of three distinct types of NP receptors (NPR): NPR-A, NPR-B, and NPR-C (Pandey, 2005). The general topological structures of NPR-A and NPR-B include an extracellular binding domain, a trans-membrane domain, and a guanyl cyclase catalytic domain. NPR-C consists of an extracellular domain, trans-membrane domain and a very short cytoplasmic tail. It was suggested that NPR-C may function as a clearance receptor; however, recent studies have shown that binding of NP to NPR-C resulted in activation of G-proteins (e.g., Zhou and Murthy, 2003). Both ANP and BNP selectively activate NPR-A, whereas, CNP preferentially binds to NPR-B. Both NPR-A and NPR-B generate the 2nd messenger, cyclic GMP.

NPR-A is the dominant receptor in peripheral organs and mediates most actions of ANP. NPR-B is localized mainly in the brain and vascular tissues.

Initial studies indicated that ANP markedly increased cyclic GMP and decreased cyclic AMP. Cyclic GMP is believed to initiate cellular responses by specific interactions with kinases, gated-ion channels and cyclic GMP phosphodiesterase. ATP probably acts allosterically to regulate the catalytic activity of the cyclase. In the kidney the increase in cyclic GMP results in diuretic-sensitive Na^+ and Cl^- transport. It was also shown that ANP has the ability to counteract the renin-angiotensin-aldosterone system (Pandey, 2005). The NP systems also serve as a compensatory mechanism against heart failure and, thus, provide the rational for the development of drugs based on NPs as lead compound (Lee and Burnett, 2007). Recently, a new chimeric natriuretic peptide with a pharmacological profile, which may be advantageous for treatment of acute heart failure, has been synthesized (Lisy et al., 2008).

Calcium Channel Blockers

L-type Ca^{2+} channel blockers are organic molecules, like the dihydropyridines, that are used for treatment of cardiovascular disorders. De Weille et al. (1991) isolated from the venom of the black mamba (*Dendroaspis polylepis*) a 60-amino acid peptide, calciseptine that selectively inhibited L-type Ca channels in the cardiovascular system and in other tissues. FS2, a homologous peptide of calciseptine was isolated from the same venom. The peptides induced vasorelaxation of vasculature preparation and also induced relaxation of pre-contracted trachea and of ileal smooth muscles. They were also shown to induce depressor activity in anaesthetized rats. The potency of the hypotensive activity induced by the mamba toxins, which also had a prolonged effect, was higher as compared to nifedipine (Watanabe et al., 1995).

Calciseptine and FS2 as well as two other mamba peptides, S_4C_8 and $C_{10}S_2C_2$, that were previously isolated from other species of *Dendraspis* (Joubert and Taljaard, 1980a, b) belong to the super-family of three-finger toxins, which have remarkable structural similarity, but show variable pharmacological activities (Kini, 2002). Kini et al. (1998) identified in these peptides a stretch of six amino acid residues (Pro42–Pro47), which interact with the L-type Ca-channel. Indeed, an eight-residue peptide, L-calchin, which was synthesized according to the predicted Ca-channel binding site of the toxins, showed Ca-channel blocking properties. L-calchin also reduced the left ventricular tension in the Lagendorff's system. Another Ca-channel blocker, calcicludine, was identified in the venom of the green mamba. This toxin blocks different Ca-channels (L-, N- and P-type); its three-dimensional structure is similar to that of K+ channel blockers, dendrotoxins (Gilquin et al., 1999).

Taicatoxin, an oligomeric toxin that consists of an α-neurotoxin-like peptide, a phospholipase A_2 and a serine protease inhibitor, was isolated from the Taipan snake venom. The toxin evoked transient but severe arrhythmias in myocyte-enriched cultures and is believed to have a specific blocking activity on voltage-dependent cardiac Ca-channels (Fantini et al., 1996).

Conclusions

Throughout the evolution, venomous animals have engineered sophisticated weapons, known as toxins, which can interfere with any fundamental metabolic function of their prey. Thus, toxins of snakes or of other venomous animals have become an essential tool by which to explore cellular processes. In this chapter several of the snake venom components that affect the cardiovascular system have been discussed in detail. The extensive research that has been carried out is now providing us with some new options to treat diseases. However, additional efforts in the basic and clinical research are, however, needed to explore the therapeutic potential of snake venom.

References

Abdel-Latif, A.A., Ding, K.H., Akhtar, R.A., Yousufzai, S.Y.K., 1996. Effects of endothelin on phospholipases and generation of second messengers in cat iris sphincter and SV-CISM-2 cells. J. Lipid Mediators. Cell Signal. 14, 147–155.

Amininasab, M., Elmi, M.M., Endlich, N., Endlich, K., Parekh, N., Naderi-Manesh, H., Schaller, J., Mostafavi, H., Sattler, M., Sarbolouki, M.N., Muhle-Goll, C., 2004. Functional and structural characterization of a novel member of the natriuretic family of peptides from the venom of *Pseudocerastes persicus*. FEBS Lett. 557, 104–108.

Arai, H., Hori, S., Aramori, I., Ohkubo, H., Nakanishi, S., 1990. Cloning and expression of a cDNA encoding an endothelin receptor. Nature 348, 730–732.

Arinami, T., Ishikawa, M., Inoue, A., Yanagisawa, M., Masaki, T., Yoshida, M.C., Hamaguchi, H., 1991. Chromosomal assignments of the human endothelin family genes – the endothelin-I gene (Edni) to 6P23-P24, the endothelin-2 gene (Edn2) to Ip34, and the endothelin-3 gene (Edn3) to 20Q13.2-Q13.3. Am. J. Human Genet. 48, 990–996.

Aubert, J.D., Juillerat-Jeanneret, L., 2009. Therapeutic potential of endothelin receptor modulators: lessons from human clinical trials. Expert Opin. Ther. Targets 13, 1069–1084.

Bazza, A., Marrakchi, N., El Ayeb, M., Sanz, L., Calvete, J.J., 2005. Snake venomics: comperative analysis of venom proteomes of the Tunisian snakes *Cerastes cerastes, Cerastes vipera* and *Macrovipera lebetina*. Proteomics 4, 4223–4235.

Barton, M., Yanagisawa, M., 2008. Endothelin: 20 years from discovery to therapy. Can. J. Physiol. Pharmacol. 86, 485–498.

Battistini, B., Jeng, A.Y., 2001. Endothelin-converting enzyme inhibitors and their effects, in: Warner, T.D. (Ed.), Endothelin and its Inhibitors. Springer, Berlin, pp. 155–208.

Bdolah, A., 2010. Sarafotoxins, the snake venom homologs of the endothelins, in: Mackessy, S.P. (Ed.), Handbook of Venom and Toxins of Reptiles. CRC Press, Roca Raton, pp. 303–315.

Bdolah, A., Kochva, E., Ovadia, M., Kinamon, S., Wollberg, Z., 1997. Resistance of the Egyptian mongoose to sarafotoxins. Toxicon 35, 1251–1261.

Bdolah, A., Wollberg, Z., Ambar, I., Kloog, Y., Sokolovsky, M., Kochva, E., 1989a. Disturbances in the cardiovascular system caused by endothelin and sarafotoxin. Biochem. Pharmacol. 38, 3145–3146.

Bdolah, A., Wollberg, Z., Fleminger, G., Kochva, E., 1989b. SRTX-d, a new native peptide of the endothelin / sarafotoxin family. FEBS Lett. 256, 1–3.

Becker, A., Dowdle, E.B., Hechler, U., Kauser, K., Donner, P., Schleuning, W.D., 1993. Bibrotoxin, a novel member of the endothelin/sarafotoxin peptide family, from the venom of the burrowing asp *Atractaspis bibroni*. FEBS Lett. 315, 100–103.

Bicher, H.I., Roth, M., Gitter, S., 1966. Neurotoxic activity of *Vipera palestinae* venom – depression of central autonomic vasoregulatory mechanisms. Med. Pharmacol. Exp. 14, 349–359.

Borgheresi, R.A.M.B., Leroy, J.M.G., Yogi, A., DosSantos, R.A., Breno, M.C., Tostes, R.C., 2006. Pharmacologic and molecular characterization of the vascular ETA receptor in the venomous snake *Bothrops jararaca*. Exper. Biol. Med. 231, 729–735.

Cushman, D.W., Ondetti, M.A., 1991. History of the design of captopril and related inhibitors of angiotensin converting enzyme. Hypertension 17, 589–592.

De Bold, A.J., Borenstein, H.B., Veress, A.T., Sonnenberg, H., 1981. A rapid and potent natriuretic response to intravenous injection of atrial myocardial extract in rat. Life Sci. 28, 89–94.

de Plater, G.M., Martin, R.L., Milburn, P.J., 1998. A C-type natriuretic peptide from the venom of the platypus (*Ornithorhynchus anatinus*): structure and pharmacology. Comp. Biochem. Physiol. C. Pharmacol. Toxicol. Endocrinol. 120, 99–110.

De Weille, J.R., Schweitz, H., Maes, P., Tartar, A., Lazdunski, M., 1991. Calciseptine, a peptide isolated from black mamba venom, is a specific blocker of the L-type calcium channel. Proc. Natl. Acad. Sci. U.S.A. 88, 2437–2440.

Ducancel, F., 2005. Endothelin-like peptides. Cell. Mol. Life Sci. 62, 2828–2839.

Ducancel, F., Matre, V., Dupont, C., Lajeunesse, E., Wollberg, Z., Bdolah, A., Kochva, E., Boulain, J.C., Menez, A., 1993. Cloning and sequence analysis of cDNAs encoding precursors of sarafotoxins – evidence for an unusual rosary-type organization. J. Biol. Chem. 268, 3052–3055.

Earl, S.T.H., Birrell, G.W., Wallis, T.P., St. Pierre, L.D., Masci, P.P., de Jersey, J., Gorman, J.J., Lavin, M.F., 2006. Post-translational modification accounts for the presence of varied forms of nerve growth factor in Australian elapid snake venoms. Proteomics 6, 6554–6565.

Fantini, E., Athias, P., Tirosh, R., Pinson, A., 1996. Effect of TaiCatoxin (TCX) on the electrophysiological, mechanical and biochemical characteristics of spontaneously beating ventricular cardiomyocytes. Mol. Cell. Biochem. 161, 61–66.

Ferreira, S.H., 2010. From the *Bothrops jararaca* bradykinin potentiating peptides to angiotensin converting enzyme inhibitors, in: Kini, R.M., K.J. Clemetson, Markland, F.S., McLane, M.A., Morita, T. (Eds.), Toxins and Hemostasis: From Bench to Bedside. Springer, Dordrecht, The Netherlands.

Fry, B.G., 2005. From genome to "venome": molecular origin and evolution of the snake venom proteome inferred from phylogenetic analysis of toxin sequences and related body proteins. Genome Res. 15, 403–420.

Fry, B.G., Wickramaratana, J.C., Lemme, S., Beuve, A., Garbers, D., Hodgson, W.C., Alewood, P., 2005. Novel natriuretic peptides from the venom of the inland taipan (*Oxyuranus microlepidotus*): isolation, chemical and biological characterisation. Biochem. Biophys. Res. Commun. 327, 1011–1015.

Gasperetti, J., 1988. Snake of Arabia, in: Buttiker, W., Krupp, F. (Eds.), Fauna of Saudi Arabia. Karger Libri AG, Basel, pp.169–392.

Gilquin, B., Lecoq, A., Desne, F., Guenneugues, M., Zinn-Justin, S., Menez, A., 1999. Conformational and functional variability supported by the BPTI fold: solution structure of the Ca^{2+} channel blocker calcicludine. Proteins Struc. Func. Gen. 34, 520–532.

Gutierrez, J.M., 1995. Clinical toxicology of snakebite in Central America, in: Meier, J., White, J. (Eds.), Handbook of Clinical Toxicology of Animal Venoms and Poisons. CRC Press, Roca Raton, 645–665.

Hayashi, M.A.F., Ligny-Lemaire, C., Wollberg, Z., Wery, M., Galat, A., Ogawa, T., Muller, B.H., Lamthanh, H., Doljansky, Y., Bdolah, A., Stocklin, R., Ducancel, F., 2004. Long-sarafotoxins: characterization of a new family of endothelin-like peptides. Peptides 25, 1243–1251.

Henry, P.J., Goldie, R.G., 2001. Endothelin receptors, in: Warner, T.D. (Ed.), Endothelin and its Inhibitors. Springer, Berlin, pp. 69–114.

Higuchi, S., Murayama, N., Saguchi, K., Ohi, H., Fujita, Y., da Silva, N.J., de Siqueira, R.J.B., Lahlou, S., Aird, S.D., 2006. A novel peptide from the ACEI/BPP-CNP precursor in the venom of *Crotalus durissus collilineatus*. Comp. Biochem. Phys. C. Toxicol. Pharmacol.144, 107–121.

Inoue, A., Yanagisawa, M., Kimura, S., Kasuya, Y., Miyauchi, T., Goto, K., Masaki, T., 1989. The human endothelin family: three structurally and pharmacologically distinct isopeptides predicted by three separate genes. Proc. Natl. Acad. Sci. U.S.A. 86, 2863–2867.

Ismail, M., Al-Ahaidib, M.S., Abdoon, N., Abd-Elsalam, M.A., 2007. Preparation of a novel antivenom against *Atractaspis* and *Walterinnesia* venoms. Toxicon 49, 8–18.

Joseph, R., Pahri, S., Hodgson, W.C., Kini, R.M., 2004. Hypotensive agents from snake venoms. Curr. Drug Targets – Cardiovasc. Haematol. Disord. 4, 437–459.

Joubert, F.J., Taljaard, N., 1980a. The complete primary structures of 2 reduced and S-carboxymethylated Angusticeps-type toxins from *Dendroaspis angusticeps* (Green Mamba) venom. Biochim. Biophys. Acta 623, 449–456.

Joubert, F.J., Taljaard, N., 1980b. The primary structure of a short neurotoxin homolog (S4C8) from *Dendroaspis jamesoni kaimosae* (Jamesons Mamba) venom. Int. J. Biochem. 12, 567–574.

Junqueira de Azevedo, I.L.M., Farsky, S.H.P., Oliveira, M.L.S., Ho, P.L., 2001. Molecular cloning and expression of a functional snake venom vascular endothelium growth factor (VEGF) from the *Bothrops insularis* pit viper – A new member of the VEGF family of proteins. J. Biol. Chem. 276, 39836–39842.

Kanaide, H., 1996. Endothelin regulation of vascular tonus. Gen. Pharmacol. 27, 559–563.
Karne, S., Jayawickreme, C.K., Lerner, M.R., 1993. Cloning and characterization of an endothelin-3 specific receptor (ET(c) Receptor) from *Xenopus-laevis* dermal melanophores. J. Biol. Chem. 268, 19126–19133.
Kini, R.M., 2002. Molecular moulds with multiple missions: functional sites in three-finger toxins. Clin. Exp. Pharmacol. Physiol. 29, 815–822.
Kini, R.M., Caldwell, R.A., Wu, Q.Y., Baumgarten, C.M., Feher, J.J., Evans, H.J., 1998. Flanking proline residues identify the L-type Ca^{2+} channel binding site of calciseptine and FS2. Biochemistry 37, 9058–9063.
Kitamura, K., Tanaka, T., Kato, J., Ogawa, T., Eto, T., Tanaka, K., 1989. Immunoreactive endothelin in rat kidney inner medulla: marked decrease in spontaneously hypertensive rats. Biochem. Biophys. Res. Commun. 162, 38–44.
Kloog, Y., Ambar, I., Sokolovsky, M., Kochva, E., Wollberg, Z., Bdolah, A., 1988. Sarafotoxin, a novel vasoconstrictor peptide: phosphoinositide hydrolysis in rat heart and brain. Science 242, 268–270.
Kloog, Y., Bousso-Mittler, D., Bdolah, A., Sokolovsky, M., 1989. Three apparent receptor subtypes for the endothelin/sarafotoxin family. FEBS Lett. 253, 199–202.
Kochva, E., Bdolah, A., Wollberg, Z., 1993. Sarafotoxins and endothelins: evolution, structure and function. Toxicon 31, 541–568.
Kochva, E., Viljoen, C.C., Botes, D.P., 1982. A new type of toxin in the venom of snakes of the genus *Atractaspis* (Atractaspidinae). Toxicon 20, 581–592.
Landan, G., Bdolah, A., Wollberg, Z., Kochva, E., Graur, D., 1991. Evolution of the sarafotoxin/endothelin superfamily of proteins. Toxicon 29, 237–244.
Lee, C.Y.W., Burnett, J.C., 2007. Natriuretic peptides and therapeutic applications. Heart Failure Rev. 12, 131–142.
Lee, C.Y., Lee, S.Y., 1979. Cardiovascular effects of snake venoms, in: Lee, C.Y. (Ed.), Snake Venoms, Handbook of Experimental Pharmacology (vol. 52). Spriger-Verlag, Berlin, pp. 547–590.
Lee, S.Y., Lee, C.Y., Chen, Y.M., Kochva, E., 1986. Coronary vasospasm as the primary cause of death due to the venom of the Burrowing Asp, *Atractaspis engaddensis*. Toxicon 24, 285–291.
Lisy, O., Huntley, B.K., McCormick, D.J., Kurlansky, P.A., Burnett, J.C., 2008. Design, synthesis, and actions of a novel chimeric natriuretic peptide: CD-NP. J. Am. Coll. Cardiol. 52, 60–68.
Lisy, O., Jougasaki, M., Heublein, D.M., Schirger, J.A., Chen, H.H., Wennberg, P.W., Burnett, J.C., 1999. Renal actions of synthetic *dendroaspis* natriuretic peptide. Kidney Int. 56, 502–508.
Lisy, O., Lainchbury, J.G., Leskinen, H., Burnett, J.C., 2001. Therapeutic actions of a new synthetic vasoactive and natriuretic peptide, *dendroaspis* natriuretic peptide, in experimental severe congestive heart failure. Hypertension 37, 1089–1094.
Maguire, J.J., Kuc, R.E., Rous, B.A., Davenport, A.P., 1996. Failure of BQ123, a more potent antagonist of sarafotoxin 6b than of endothelin-1, to distinguish between these agonists in binding experiments. Br. J. Pharmacol. 118, 335–342.
Martel, E., Champeroux, P., Brisac, A.-M., Magnier, A., Richard, S., Safar, M., 1991. Pressor responses to endothelin-1 in normotensive and spotaneosly hypertensive rats. Neurochem. Int. 18, 553–557.
Masaki, T., 2000. The endothelin family: an overview. J. Cardiovasc. Pharmacol. 35, S3-S5.
Masaki, T., 2001. Historical prespective and introduction to the endothelin family of peptides, in: Warner, T.D. (Ed.), Endothelin and its Inhibitors. Springer, Berlin, pp.1–10.
Mesquita, L.S.M., Frias, F.T., Carmona, E., Borgheresi, R.A.M.B., 2008. Differences in endothelin receptor types in the vasculature of *Bothrops jararaca* (Viperidae) and *Oxyrhopus guibei* (Colubridae) snakes. Comp. Biochem. Physiol. C. Toxicol. Pharmacol. 148, 61–67.
Michel, G.H., Murayama, N., Sada, T., Nozaki, M., Saguchi, K., Ohi, H., Fujita, Y., Koike, H., Higuchi, S., 2000. Two N-terminally truncated forms of C-type natriuretic peptide from habu snake venom. Peptides 21, 609–615.

Murayama, N., Hayashi, M.A.F., Ohi, H., Ferreira, L.A.F., Hermann, V.V., Saito, H., Fujita, Y., Higuchi, S., Fernandes, B.L., Yamane, T., de Camargo, A.C.M., 1997. Cloning and sequence analysis of a *Bothrops jararaca* cDNA encoding a precursor of seven bradykinin potentiating peptides and a C-type natriuretic peptide. Proc. Natl. Acad. Sci. U.S.A. 94, 1189–1193.

Nambi, P., Elshourbagy, N., Wu, H.L., Pullen, M., Ohlstein, E.H., Brooks, D.P., Lago, M.A., Elliott, J.D., Gleason, J.G., Ruffolo, R.R., Jr., 1994. Nonpeptide endothelin receptor antagonists 1. Effects on binding and signal transduction on human endothelin(A) and endothelin(B) receptors. J. Pharmacol. Exp. Ther. 271, 755–761.

Noguchi, K., Noguchi, Y., Hirose, H., Nishikibe, M., Ihara, M., Ishikawa, K., Yano, M., 1993. Role of endothelin ET(B) receptors in bronchoconstrictor and vasoconstrictor responses in guinea-pigs. Eur. J. Pharmacol. 233, 47–51.

Olson, K.R., Duff, D.W., Farrell, A.P., Keen, J., Kellogg, M.D., Kullman, D., Villa, J., 1991. Cardiovascular effects of endothelin in trout. Am. J. Physiol. 260, H1214–H1223.

Pandey, K.N., 2005. Biology of natriuretic peptides and their receptors. Peptides 26, 901–932.

Pandey, K.N., Pavlou, S.N., Inagami, T., 1988. Identification and characterization of 3 distinct atrial natriuretic factor receptors – evidence for tissue-specific heterogeneity of receptor subtypes in vascular smooth-muscle, kidney tubular epithelium, and leydig tumor-cells by ligand-binding, photoaffinity-labeling, and tryptic proteolysis. J. Biol. Chem. 263, 13406–13413.

St. Pierre, L., Woods, R., Earl, S., Masci, P.P., Lavin, M.F., 2005. Identification and analysis of venom gland-specific genes from the coastal taipan (*Oxyuranus scutellatus*) and related species. Cell. Mol. Life Sci. 62, 2679–2693.

Poder, T.C., Silberberg, S.D., Rampe, D., 1991. Contraction of reptile, amphibian, and fish blood vessels by endothelin-1. Can. J. Physiol. Pharmacol. 69, 215–217.

Potter, L.R., Abbey-Hosch, S., Dickey, D.M., 2006. Natriuretic peptides, their receptors, and cyclic guanosine monophosphate-dependent signaling functions. Endocrine Rev. 27, 47–72.

Quinton, L., Le Caer, J.P., Phan, G., Ligny-Lemaire, C., Bourdais-Jomaron, J., Ducancel, F., Chamot-Rooke, J., 2005. Characterization of toxins within crude venoms by combined use of Fourier transform mass spectrometry and cloning. Anal. Chem. 77, 6630–6639.

Rocha e Silva, M., Beraldo, W.T., Rosenfeld, G., 1949. Bradykinin, a hypotensive and smooth muscle stimulating factor released from plasma globulin by snake venom and by trypsin. Am. J. Physiol. 156, 161–273.

Rubin, L.J., Badesch, D.B., Barst, R.J., Galie, N., Black, C.M., Keogh, A., Pulido, T., Frost, A., Roux, S., Leconte, I., Landzberg, M., Simonneau, G., 2002. Bosentan therapy for pulmonary arterial hypertension. N. Eng. J. Med. 346, 896–903.

Saida, K., Mitsui, Y., Ishida, N., 1989. A novel peptide vasoactive intestinal contractor, of a new (endothelin) peptide family. Molecular cloning, expression, and biological activity. J. Biol. Chem. 264, 14613–14616.

Sakurai, T., Yanagisawa, M., Takuwa, Y., Miyazaki, H., Kimura, S., Goto, K., Masaki, T., 1990. Cloning of a cDNA encoding a non-isopeptide-selective subtype of the endothelin receptor. Nature 348, 732–735.

Schweitz, H., Vigne, P., Moinier, D., Ferlin, C., Lazdunski, M., 1992. A new member of the natriuretic peptide family is present in the venom of the Green Mamba (*Dendroaspis angusticeps*). J. Biol. Chem. 267, 13928–13932.

Shraga-Levine, Z., Sokolovsky, M., 2000. Functional coupling of G proteins to endothelin receptors is ligand and receptor subtype specific. Cell. Mol. Neurobiol. 20, 305–317.

Simon, T., Bdolah, A., Kochva, E., 1980. The two-component toxin of *Vipera palaestinae*: contribution of phospholipase A to its activity. Toxicon 18, 249–259.

Simonson, M.S., 2001. Cell signaling by endothelin peptides, in: Warner, T.D. (Ed.), Endothelin and its Inhibitors. Springer, Berlin, pp. 115–140.

Singh, G., Kuc, R.E., Maguire, J.J., Fidock, M., Davenport, A.P., 2006b. Novel snake venom ligand dendroaspis natriuretic peptide is selective for natriuretic peptide receptor-A in human heart – downregulation of natriuretic peptide receptor-A in heart failure. Circ. Res. 99, 183–190.

Singh, G., Maguire, J.J., Kuc, R.E., Skepper, J.N., Fidock, M., Davenport, A.P., 2006a. Characterization of the snake venom ligand [I-125]-DNP binding to natriuretic peptide receptor-A in human artery and potent DNP mediated vasodilatation. Brit. J. Pharmacol. 149, 838–844.

Sokolovsky, M., 1995. Endothelin receptor subtypes and their role in transmembrane signaling mechanisms. Pharmacol. Ther. 68, 435–471.

Sokolovsky, M., Shraga-Levine, Z., 2001. Sarafotoxins and their relationship to the endothelin family of peptides, in: Warner, T.D. (Ed.), Endothelin and its Inhibitors. Springer, Berlin, pp. 11–34.

Sudoh, T., Minamino, N., Kangawa, K., Matsuo, H., 1990. C-type natriuretic peptide (CNP) – a new member of natriuretic peptide family identified in porcine brain. Biochem. Biophys. Res. Commun. 168, 863–870.

Takasaki, C., Tamiya, N., Bdolah, A., Wollberg, Z., Kochva, E., 1988. Sarafotoxins S6: several isotoxins from *Atractaspis engaddensis* (Burrowing Asp) venom that affect the heart. Toxicon 26, 543–548.

Warner, T.D., De Nucci, G., Vane, J.R., 1989. Rat endothelin is a vasodilator in isolated perfused mesentery of the rat. Eur. J. Pharmacol. 159, 325–326.

Warrell, D.A., 1995. Clinical toxicology of snakebite in Asia, in: Meier, J., White, J. (Eds.), Handbook of Clinical Toxicology of Animal Venoms and Poisons. CRC Press, Roca Raton, pp. 493–594.

Warrell, D.A., Ormerod, L.D., Davidson, N.M., 1976. Bites by the night adder (*Causus maculatus*) and burrowing vipers (genus *Atractaspis*) in Nigeria. Am. J. Trop. Med. Hyg. 25, 517–524.

Watanabe, T.X., Itahara, Y., Kuroda, H., Chen, Y.N., Kimura, T., Sakakibara, S., 1995. Smooth muscle relaxing and hypotensive activities of synthetic calciseptine and homologous snake venom peptide FS_2. Jpn. J. Pharmacol. 68, 305–313.

Weiser, E., Wollberg, Z., Kochva, E., Lee, S.Y., 1984. Cardiotoxic effects of the venom of the burrowing asp, *Atractaspis engaddensis* (Atractaspididae, Ophidia). Toxicon 22, 767–774.

Williams, D.L.J., Jones, K.L., Pettibone, D.J., Lis, E.V., Clineschmidt, B.V., 1991. Sarafotoxin S6c: an agonist which distinguishes between endothelin receptor subtypes. Biochem. Biophys. Res. Commun. 175, 556–561.

Wollberg, Z., Bdolah, A., Galron, R., Sokolovsky, M., Kochva, E., 1991. Contractile effects and binding properties of endothelins/sarafotoxins in the guinea pig ileum. Eur. J. Pharmacol. 198, 31–36.

Wollberg, Z., Bousso-Mittler, D., Bdolah, A., Kloog, Y., Kochva, E., Sokolovsky, M., 1992. Endothelins and sarafotoxins: effects on motility, binding properties and phosphoinositide hydrolysis during the estrus cycle of the rat uterus. J. Basic Clin. Physiol. Pharmacol. 3, 41–58.

Wollberg, Z., Matz, A., Kinamon, S., Bdolah, A. SRTX-d/e inhibits Sarafotoxin-b-induced contraction of guinea pig intestine. Toxicon 36, 1252. 1998. (Abstract)

Wollberg, Z., Shabo-Shina, R., Intrator, N., Bdolah, A., Kochva, E., Shavit, G., Oron, Y., Vidne, B.A., Gitter, S., 1988. A novel cardiotoxic polypeptide from the venom of *Atractaspis engaddensis* (Burrowing Asp): cardiac effects in mice and isolated rat and human heart preparations. Toxicon 26, 525–534.

Yanagisawa, M., Kurihara, H., Kimura, S., Tomobo, Y., Kobayashi, M., Mitsui, Y., Yazaki, Y., Goto, K., Masaki, T., 1988. A novel potent vasoconstrictor peptide produced by vascular endothelial cells. Nature 332, 411–415.

Yorikane, R., Koike, H., 1990. The arrhythmyogenic action of endothelin in rats. Jpn. J. Pharmacol. 53, 259–263.

Zhou, H.P., Murthy, K.S., 2003. Identification of the G protein-activating sequence of the single-transmembrane natriuretic peptide receptor C (NPR-C). Am. J. Physiol. Cell Physiol. 284, C1255-C1261.

Zigdon-Arad, T., Bdolah, A., Kochva, E., Wollberg, Z., 1992. Activity of sarafotoxin/endothelin peptides in the heart and brain of lower vertebrates. Toxicon 30, 439–448.

Chapter 38
Hypotensive Proteins from Hematophagous Animals

P. Takáč, H. Tsujimoto, and D.E. Champagne

Abstract Diverse arthropods exploit vertebrate blood as a food resource, despite the fact that this resource is defended by an array of haemostatic, inflammatory, and immune responses, as well as overt defensive behaviors. These defenses have selected for a broad array of antihaemostatic, anti-inflammatory, and immunomodulatory factors, which are secreted in the saliva of the blood-feeder and injected into the wound during the feeding process. Haemostasis includes three interacting branches: platelet activation and aggregation, vasoconstriction, and coagulation (or clotting). This review deals with salivary factors that counter the vasoconstriction that normally results following vascular injury, but as the branches of haemostasis overlap, and indeed haemostasis interacts with inflammation and immunity, some molecules with vasodilatory (hypotensive) activity may also affect other aspects of the vertebrate response, in which case they are also discussed. Blood-feeding has evolved on numerous independent occasions, leading to a wide diversity of molecules with hypotensive activity amongst extant arthropods. These molecules exploit mechanisms that include direct interaction with vertebrate receptors or signaling pathways leading to vasodilation, sequestration of endogenously generated vasoconstrictors, and enzymatic destruction of vasoconstrictors. Some vasodilators, specifically the nitrophorins from *Rhodnius* and *Cimex*, allow the arthropod to store physiological amounts of nitric oxide and secrete it into the bite site during the course of the blood meal. Finally, several instances are noted where arthropod saliva is known to contain hypotensive activity, but the molecules involved have not been identified.

P. Takáč (✉)
Institute of Zoology, Slovak Academy of Sciences, Bratislava, Slovakia
e-mail: Peter.Takac@savba.sk

R.M. Kini et al. (eds.), *Toxins and Hemostasis*, DOI 10.1007/978-90-481-9295-3_38,

Introduction

Thirty years ago it was proposed (Ribeiro, 1987), based on enzymatic and bioassay data, that saliva of bloodsucking arthropods served mainly an antihemostatic role and, additionally, that hard tick saliva served to evade their hosts' inflammation and immunity. Except for tick salivary prostaglandins (Dickinson et al., 1976; Higgs et al., 1976), no other hematophagous arthropod salivary compound had been molecularly characterized at that time. Thanks to the techniques of molecular biology as well as the miniaturization of high-performance liquid chromatography, mass spectrometry, and Edman degradation techniques, detailed knowledge of such molecules and their role in blood feeding has increased at a fast pace. Accordingly, it was confirmed that almost all bloodsucking arthropods studied (albeit a small sample consisting of only a few medically important species, from over 500 genera and >19,000 known species) (Ribeiro, 1995) have at least one anticoagulant, one vasodilator, and one antiplatelet compound. The vertebrate hemostatic system evolved in the context of the transition from aquatic to terrestrial environments, and so is estimated to be about 300 million years old (Doolittle, 2009; Doolittle and Feng, 1987). As this significantly predates the appearance of the various clades of blood-feeding arthropods in the Mesozoic (Grimaldi and Engel, 2005; Rai and Black, 1999; Ribeiro, 1987, 1995), blood-feeding evolved against the background of a fully functional vertebrate defense. The ability to feed on blood has evolved on multiple occasions, with at least 14 extant arthropod clades having independently developed this specialized feeding strategy. Each of these clades have had to independently solve the problems posed by vertebrate hemostasis, and consequently the molecular diversity of the antihemostatic compounds is very large. Hematophagous insects and ticks have many different and sometimes apparently conflicting salivary strategies, which were not predicted before. In addition, at least one half of the message expressed in the salivary glands of such animals, leading to apparently secreted proteins, has no known function, thus challenging the researchers' knowledge and imagination (Valenzuela, 2002).

Hemostasis

To a bloodsucking animal, paradise is a place where the host blood does not clot, the blood flow at the feeding site is intense, and the host will not bother (or kill) the guest. Real life is different. Vertebrates have three efficient systems that make life potentially difficult for hematophagous animals: hemostasis, inflammation, and immunity (Ribeiro and Francischetti, 2003). These three complex physiological responses interact with each other and, at times, are in opposition.

Hemostasis is the host response that controls the loss of blood following injury to a blood vessel. It consists of platelet aggregation, blood coagulation (clotting), and vasoconstriction. All these phenomena are redundant. Platelets are central to the hemostatic response: the wound generates agonists such as ADP or exposed collagen that interact with receptors on the platelet surface and initiate shape change, aggregation to form a platelet plug, and the release of dense granules containing ADP, thromboxane A_2 (TXA_2), and the biogenic amine serotonin (5-hydroxytryptamine or 5-HT). ADP and especially TXA_2 recruit further platelets to the growing plug, and serotonin and TXA_2 stimulate vasoconstriction, restricting blood flow to the injury site. Injury to the epithelium exposes cells expressing tissue factor, thereby initiating the coagulation cascade, and changes to the platelet surface, especially the movement of negatively charged phosphatidylserine to the outer layer of the cell membrane, provides a favorable surface for formation of the factor Xa-generating complex. Ultimately this results in the activation of thrombin and stabilization of the platelet plug as thrombin generates fibrin, which consolidates the platelet plug into a clot by interacting with receptors and linking platelets together. Hemostasis thus minimizes the host blood loss following injury and places a major barrier to any blood-feeding arthropod (Ribeiro and Francischetti, 2003).

Gaining access to blood necessarily involves creating an injury to the skin. In some cases, such as mosquitoes and triatomines, stylets are inserted as the insect probes, ultimately resulting in cannulation of a blood vessel (Bergman, 1996). Other arthropods such as ixodid ticks, sandflies, blackflies, and tabanids create a hemorrhagic pool in the capillary bed or at the skin surface, from which they feed. In this process, there is a premium placed on speed and efficiency, as contact time with the host is correlated with the probability of detection and interruption of the meal (resulting in fewer eggs being produced) at best, or an untimely demise at worst (Gillett, 1967). Accordingly, the biological fitness of the arthropod is enhanced by the presence of salivary factors that improve the efficiency of blood vessel location and feeding, by countering the normal wounding or hemostatic responses of the host. At the pharmacological level, blood-feeding arthropods counteract these responses through the secretion, into the feeding site, of saliva containing platelet aggregation inhibitors, vasodilators, and anticoagulants (Law et al., 1992). Indeed, vasodilators do double duty for some arthropods, particularly those that must locate arterioles and venules from which to feed. These vessels occupy a limited volume in the skin, and they can be located only by probing repeatedly until a vessel is penetrated. Injection of vasodilators during the probing phase of the meal can enlarge the size of the target vessels, making them easier to locate and decreasing contact time with and irritation of the host, as well as speeding the subsequent ingestion of the meal. In this review we will concentrate on vasodilators, although due to the overlapping and interacting nature of hemostasis we will also discuss some additional effects of vasodilatory molecules in arthropod saliva. We will then specifically discuss one recently described molecule, vasotab, in more detail as an example of current approaches to the anaylsis of vasodilators. For a more comprehensive review of arthropod saliva, readers are referred to Ribeiro and Francischetti (2003) and Champagne (2005), and to Andersen (2009) for a review of structural aspects of interactions between salivary proteins and their ligands. Tick salivary factors and anticoagulant factors from hematophagous animals have been reviewed by Francischetti et al. (2009) and Koh and Kini (2009) respectively.

Vasodilators

In intact vessels, vascular tone is tightly regulated to maintain blood pressure and to regulate heat loss from the skin. The tone is maintained by a balance of endogenous factors that promote vasoconstriction, such as the catecholamine norepinephrine, and factors that promote vasodilation, such as nitric oxide. Before its chemical nature was known, the latter molecule was called "endothelium-dependent relaxation factor", which reflects its synthesis by the vascular endothelium under control of neural factors such as the tachykinin peptide substance P. Other endogenous vasodilators act directly on the smooth muscle of blood vessels. Accordingly, it is informative to discern whether or not a novel vasodilator is endothelium-dependent or endothelium-independent, as that can provide clues as to the mechanism of action. The advantage of using a balance between vasoconstrictors and vasodilators to regulate vascular tone is speed of tissue response: a small shift in one or the other factor can quickly produce a large change in the physiological state of the tissue. This control mechanism suggests that arthropods may cause vasodilation through the secretion of factors that mimic endogenous vasodilators (thereby exploiting existing receptors and signaling pathways), or by removing endogenous vasoconstrictors, either by enzymatic destruction or by sequestration. Each of these strategies is exploited by various arthropods, and indeed some use both strategies simultaneously. The chemical nature of the vasodilators can range from small organic molecules and nitric oxide to peptides and proteins. We will now discuss specific vasodilators, grouped according to their chemical nature and mechanism of action. These molecules are also summarized in Table 38.1.

Small Molecule Vasodilators

Ixodid ticks ("hard ticks") require an extended period to feed, ranging from a few days in the case of larvae to as much as 12–14 days for some adults. During this time they secrete saliva containing the prostaglandins (PGs) PGE_2, $PGF_{2\alpha}$, and prostacyclin, which are potent vasodilators and in addition antagonize platelet activation (Bowman et al., 1996). The concentrations of these molecules can be extremely high. In saliva of *Amblyomma americanum* and *Ixodes scapularis* PGE_2 concentrations (measured at 469 and up to 2,200 ng ml^{-1} respectively) greatly exceed the levels found in inflammatory exudates (2–19 ng ml^{-1}) and the physiological concentration needed to dilate peripheral blood vessels (<10 ng ml^{-1}) (Kemp et al., 1983; Ribeiro et al., 1992). Use of these molecules produces an interesting dilemma for ticks, as they are unstable and must be synthesized and secreted "on demand". However their synthesis requires arachidonic acid, which ticks cannot produce endogenously but must generate using phospholipase to hydrolyze phospholipids from cells in the blood meal (Bowman et al., 1995). As a result prostaglandin levels are low at the beginning of the meal, and increase to high levels (much greater than are present in inflammatory exudates) as the meal progresses. Consequently, although the tick may attach for several days, most of the meal is actually ingested in just a few days at the end of the meal, when inhibition of hemostasis is maximal.

Table 38.1 Vasodilator molecules characterized from arthropd saliva

Molecule	Arthropod	Molecule type	Target/MOA	References[a]
Small molecules(non-peptidic)				
Prostacyclin, PGE_2, PGF_{2a}	Ixodid ticks	Prostaglandins	Increase intracellular cAMP	Bowman et al. (1996)
Adenosine	*Phlebotomus*	Purine nucleoside		Ribeiro et al. (1999)
Lyso-phosphatidylcholine	*Rhodnius*	Phospholipid		Golodne et al. (2003)
Peptide/protein receptor ligands				
Vasotab	Horsefly	Kazal-type peptide	Putative channel modulator	Takac et al. (2006)
SVEP	Blackfly	Novel protein	Putative K+channel modulator	Cupp et al. (1998)
Sialokinin	*Aedes aegypti*	Tachykinin peptide	Substance P receptor	Champagne and Ribeiro (1994)
Maxadilan	*Lutzomyia*	Novel peptide	PACAP type I receptor	Lerner et al. (2007)
Ra-KLP	*Rhipicephalus*	Kunitz-type peptide	Putative channel modulator	Paesen et al. (2009)
Protein carriers of small molecule vasodilators				
Nitrophorin	*Rhodnius*	Lipocalin/heme Protein	NO carrier	Ribeiro et al. (1993) Champagne et al. (1995)
Nitrophorin	*Cimex*	Polyphosphatase/heme Protein	NO carrier	Valenzuela et al. (1995) Valenzuela and Ribeiro (1998)
Enzymatic destruction of endogenous vasodilators				
Catechol	*Anopheles*	Myleoperoxidase	Removes endogenous serotonin,	Ribeiro and Nussenzveig (1993)
Oxidase		Enzyme	Catecholamines	Ribeiro et al. (1994)
Sequestration of endogenous vasodilators				
D7/D7R	Dipterans	Odorant binding Protein family	Sequester biogenic amines (Noradrenalin),leukotrienes	Calvo et al. (2006, 2009)
ABP	*Rhodnius*	Lipocalin	Sequester biogenic amines	Andersen et al. (2003)
Moubatin, TGSP-2	Argasid ticks	Lipocalin	Sequester serotonin, TXA2	Mans and Ribeiro (2008a)

[a]Only the most relevant or a representative reference is given

Additional small molecule vasodilators include the presence of pharmacological amounts (~1 nmole) of adenosine and 5′-AMP in saliva of *Phlebotomus* sandflies (Ribeiro et al., 1999). These purines have both antiplatelet and vasodilatory activities. Saliva of the triatomine bug *Rhodnius prolixus* contains (in addition to other vasodilators to be discussed below) pharmacological amounts of the lipid lysophosphatidylcholine (LPC) (Golodne et al., 2003). LPC, which is produced by hydrolysis of phosphatidylcholine *sn*-2 position by a variety of enzymes including phospholipase A2 (PLA2), platelet activating factor acetylhydrolase, and others, is a major component of low density lipoprotein, with plasma concentrations in the 140–150 μM range. LPC has multiple antihemostatic properties, including antagonizing platelet aggregation and stimulating nitric oxide (NO) secretion from endothelial cells. *Rhodnius* is a vector of *Trypanosoma cruzi*, the causative agent of Chagas Disease, and LPC may enhance the infection process (Mesquita et al., 2008). It is interesting to note that saliva of the ixodid tick *Amblyomma americanum* contains significant levels of PLA2, even at the beginning of the meal (Bowman et al., 1997; Zhu et al., 1998). It seems likely that this enzyme could generate pharmacological amounts of LPC in the feeding lesion, causing an initial vasodilation even before the tick has obtained sufficient blood to generate useful amounts of secreted prostaglandins. This hypothesis is an active area of research.

Peptide and Protein Receptor Ligands

A number of arthropod salivary peptides and small proteins that interact with endogenous receptors, stimulating signaling pathways leading directly or indirectly to smooth muscle relation and vasodilation. The Yellow Fever mosquito *Aedes aegypti* secretes a tachykinin peptide, sialokinin, which stimulates the endothelium to produce NO, resulting in rapid vasodilation (Ribeiro, 1992). Originally the tachykinin nature of the vasodilator was deduced from its pharmacological properties, especially the loss of tissue responsiveness in aorta preparations that had been desensitized by prior challenge with the vertebrate peptide substance P. This experiment demonstrated that the mosquito peptide interacts with the substance P receptor, and so the mosquito achieves its meal by mimicking an endogenous signal (substance P) that is an important regulator of vascular tone. The tachykinin nature of the peptide was confirmed when it was purified and sequenced (Champagne and Ribeiro, 1994). Subsequently the sialokinin gene was sequenced; the lack of sequence similarity outside of the domain that encodes the peptide established that sialokinin is not homologous with vertebrate tachykinins, but instead arose through convergent evolution, with the peptide selected for an efficient interaction with the receptor (Beerntsen et al., 1999).

New World sandflies in the genus *Lutzomyia* secrete a potent vasodilator, the peptide maxadilan (Lerner and Shoemaker, 1992; Lerner et al., 1991; Grevelink et al., 1995). This peptide interacts with the pituitary adenylate cyclase activating peptide (PACAP) receptor-1 (PAC-1) (Moro and Lerner, 1997). Interestingly, a role for PACAP in the regulation of vascular tone in peripheral vessels was not suspected until the discovery of the maxadilan/PAC-1 interaction. Maxadilan is a highly variable peptide; multiple allelic forms, encoded by a multigene family, may be present in a single individual, and at the population level peptide diversity may be very high (Milleron et al., 2004). Despite this variation, all sequences are apparently similarly active as vasodilators (Lanzaro et al., 1999). It is noteworthy that the old world sandfly genus *Phlebotomus* lacks maxadilan, and instead relies on adenosine and 5′-AMP as discussed already. This is only one of several examples in which related insects use different strategies to achieve vasodilation, which cautions against the temptation to assume that related arthropods must rely on homologous molecules to inhibit hemostasis.

The blackfly *Simulium vittatum* also produces a potent vasodilator (Cupp et al., 1994), which has been called *Simulium vittatum* erythema protein or SVEP (Cupp et al., 1998). This 15.4 kDa protein is distinct from both the *Aedes* and the *Lutzomyia* peptides, and in fact has no homology with any known protein family. Although the specific receptor with which it interacts is not known, it is inhibited by glibenclamide and by high concentrations of K^+, which suggests an action on ATP-dependent potassium channels. Recent analysis of the *S. vittatum* sialome (salivary gland transcriptome) (Andersen et al., 2009) revealed a family of SVEP-like proteins, but expression and assay of recombinants suggests that only the original molecule, SVEP-1, is a vasodilator, and the function of the other variants is unknown (JF Andersen, personal communication).

Female horse flies (Diptera, Tabanidae) require substantial amounts of blood (up to 0.5 ml) for egg production. This volume of blood requires multiple feeding episodes; Hollander and Wright (1980) estimated that approximately ten landings on a host were needed to complete one bloodmeal. Tabanids feed predominantly on large mammals, and females of three genera, *Chrysops, Haematopota,* and *Tabanus*, commonly attack humans. Horse flies have chewing and biting mandibulate mouthparts, which restrict their feeding to superficial haematomas that form in the lacerated tissue. This highly procoagulant environment suggests that tabanids rely heavily on the pharmacological properties of their saliva to find blood. Accordingly, saliva of several tabanid genera were shown to have anticoagulant activity (Kazimírová et al., 2001), and salivary gland extracts of the deerfly (genus *Chrysops*) contain a potent inhibitor of platelet aggregation (Grevelink et al., 1993), which was characterized as a novel glycoprotein IIb/IIIa fibrinogen receptor antagonist (Reddy et al., 2000) and subsequently identified as an apyrase paralogue (Ribeiro et al., 2004). Rajská and coworkers (2003) used isolated perfused rat heart to examine the vasoactivity of salivary gland extracts (SGEs) of three horsefly species, *Hybomitra bimaculata* Macquart, *Tabanus bromius* Linnaeus and *Tabanus glaucopis* Meigen. Administration of horsefly SGEs to the heart produced biphasic coronary responses: a decrease and subsequent increase in coronary flow (CF), characterized by initial vasoconstriction followed by prolonged vasodilation of coronary vessels. However, although SGEs of *H. bimaculata* induced a significant decrease in left ventricular pressure (LVP), the effect on changes in CF was not significant except at the highest dose tested. The ability to reduce LVP without significantly lowering CF, or affecting heart rate and rhythm, represents a unique set of properties that have considerable therapeutic potential if they can be reproduced by a single molecule (Rajská et al., 2007). Extracts from *Hybomitra bimaculata* salivary

Fig. 38.1 Head with salivary glands of the Horse fly *Hybomitra bimaculata*. Figure from Takac et al. (2006). Used with permission of The Company of Biologists

glands (Fig. 38.1) also dilates rat femoral artery, and this assay was used to guide isolation of the vasodilatory component from salivary glands (Takac et al., 2006).

Following N-terminal sequencing, degenerate PCR primers were designed to amplify the corresponding cDNA from a cDNA library. The full-length sequence predicted a peptide of 6146.7 Da after accounting for loss of the signal peptide; this matched the observed mass of 6141 Da for the native peptide, after allowing for the loss of 6 protons when 3 disulfide briges are formed. The peptide,which was named Vasotab, has a structure typical of a member of the Kazal family of protease inhibitors, but it is distinguished by a unique insertion of seven amino acids in the loop formed between the third and fourth cysteine residues (Takac et al., 2006) (Fig. 38.2). Orthologous vasodilators are found in other tabanid flies

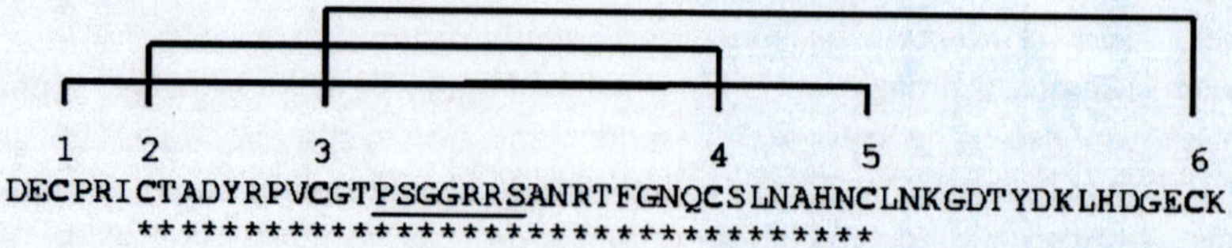

Fig. 38.2 Primary sequence of Vasotab, with positions of the three disulfide bridges typical of Kazal family protease inhibitors. The position of the Pfam consensus domain for the Kazal family is indicated by *asterisks*. The unique seven-residue insert is *underlined*. Figure modified from Takac et al. (2006). Used with permission of The Company of Biologists

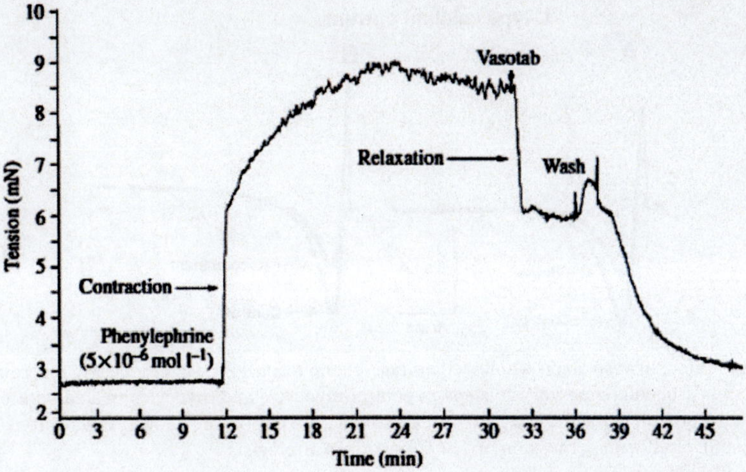

Fig. 38.3 Relaxation of contracted rat femoral artery induced by purified baculovirus-expressed vasotab EV048. The rat femoral artery was contracted with phenylephrine (5×10^{-6}·mol l^{-1}). Figure from Takac et al. (2006). Used with permission of The Company of Biologists

such as *Tabanus yao* (Ma et al., 2009; Xu et al., 2008). Recombinant vasotab, expressed in baculovirus, was shown to vasodilate the rat femoral artery (Fig. 38.3). Voltage-clamp experiments on isolated rat myocytes indicated that both native and recombinant vasotab blocks L-type calcium channels, decreasing the amount of activator Ca^{2+} released from the sarcoplasmic reticulum, and consequently weakening the contractile force.

Kazal-type proteins are a frequent component of insect sialomes, but where functions are known most have anticoagulant activity. The isolation of vasotab as a vasodilator suggests that this common protein scaffold may be modified to provide new functions. Just such adaptation of widespread protein folds to novel functions is a common attribute of venom proteins (Fry et al., 2009), and demonstrates that homology-based deductions about function must be confirmed by experimentation. Whether or not vasotab also retains protease inhibitor activity is not yet known, but comparison with functional Kazal family inhibitors (Lu et al., 2001; Wynn et al., 2001) suggests that it might also retain elastase inhibitory activity (Fig. 38.4).

An additional example of the modification of common protein architecture to provide a novel function is seen in the Kunitz-like protein *Ra*-KLP from the ixodid tick *Rhipicephalus appendiculatus* (Paesen et al., 2009). This protein retains the classical "Kunitz head" domain with two alpha helices joined by a disulphide bridge, but the protease-binding loops have been replaced by β-strands, the disulphide bridge that normally stabilizes the loop is absent, and two new disulphide bridges link the N-terminal region to the Kunitz head domain. Not surprisingly, in light of the extensive modification in structure, *Ra*-KLP lacks protease-inhibitor activity. However, it apparently has acquired the function of interacting with ion channels, as it is able to activate maxiK channels in an *in vitro* system. This suggests

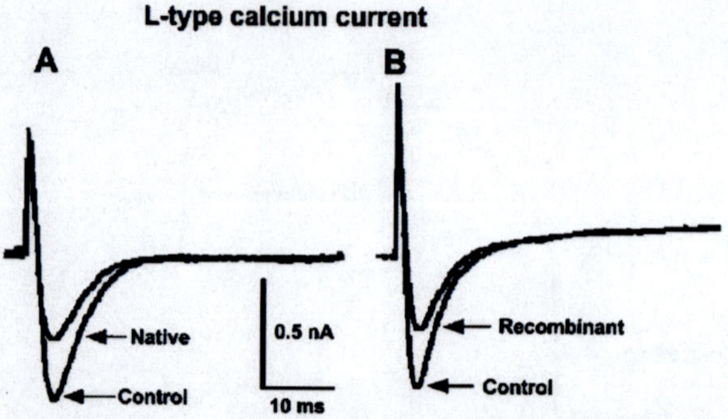

Fig. 38.4 Effect of vasotab on whole-cell voltage-clamp recordings. Representative recordings of I_{Ca} at −5 mV depolarizing voltage steps in control myocytes and myocytes treated with 0.4 μg ml^{-1} native protein (**a**) and 0.4 μg ml^{-1} recombinant protein (**b**) are shown. Figure from Takac et al. (2006). Used with permission of The Company of Biologists

a possible role in enhancing blood flow to the feeding tick, but such a function has not been demonstrated in vivo. Kunitz-like venom components from a number of animals have been found to interact with ion channels, although these mostly tend to inhibit rather than activate these channels (Fry et al., 2009).

Protein Carriers of Small Molecule Vasodilators

Saliva of the triatomine bug *Rhodnius prolixus* has potent vasodilatory activity, and pharmacological assays on rabbit aorta preparations indicated hallmarks of a nitrovasodilator, including potentiation by superoxide dismutase (which prolongs the half-life of nitric oxide [NO] in solution) and inhibition by methylene blue (Ribeiro et al., 1990). This was the first evidence of NO production by an insect. Secretion of NO into the host is an efficient means of countering hemostasis, as NO is both a potent vasodilator and an antagonist of platelet activation. Subsequently four heme proteins were isolated and characterized, and were shown to reversibly bind NO (Champagne et al., 1995; Sun et al., 1996, 1998; Yuda et al., 1997). In recognition of their NO transport function, they were named nitrophorin 1–4 in order of their abundance in saliva of fifth instar nymphs (Ribeiro and Walker, 1994). The relatively weak binding of NO, which allows dissociation as the proteins are injected into the skin of the vertebrate host, was shown to be due to the presence of ferric iron III in the heme group (Ribeiro et al., 1993). This contrasts with ferrous Fe^{2+} hemes, such as are present in hemoglobin, where NO binding is essentially irreversible. Although the affinities of the nitrophorins were not immediately apparent from the sequence, solution of the crystal structure of nitrophorin-1 (NP-1) revealed a typical lipocalin fold, with an antiparallel 8-stranded β-barrel structure (Weichsel et al., 1998). Although typical lipocalins present a hydrophobic pocket which allows them to bind and transport small hydrophobic molecules (such as retinoic acid), the nitrophorins are modified to bind a heme molecule, with the immidazole side chain of His59 occupying the fifth coordination bond to the iron; the unpaired electron of NO interacts with the iron through the sixth coordination bond. This arrangement allows each nitrophorin molecule to ligate a single NO molecule. The NO is produced by a typical nitric oxide synthetase in the secretory cells of the salivary gland epithelium (Ribeiro and Nussenzveig, 1993; Yuda et al., 1996). After depletion of saliva during blood-feeding, the nitrophorins are replenished before the NOS becomes active (Nussenzveig et al., 1995). The epithelium projects as microvillae into the lumen of the gland, greatly increasing the surface area on the lumen-facing surface of the secretory cells. Due to the lipophilic nature of NO, this structure favors the movement of NO towards the gland lumen, where it can be sequestered by the nitrophorins. This architecture facilitates efficient loading of the nitrophorins, and minimizes the risk of toxicity that would result if the NO were to diffuse into the insect's body.

The *Rhodnius* nitrophorins have a number of remarkable properties which allow them to function individually and in concert to affect efficient vasodilation in the host. Firstly, all the nitrophorins ligate and release NO in a pH-dependent manner: the NO is tightly bound at acidic pH, and it is readily released under neutral or basic conditions (Andersen et al., 1997, 2000; Ribeiro et al., 1993). As the pH in the salivary gland is about 5, the nitrophorins are able to efficiently trap and store NO. The environment in the skin/blood of the host, on the other hand, is about pH 7.4, and the reduced affinity allows NO to be released. Elucidation of the crystal structure of NP-2, and NP-4 under acidic and basic conditions revealed the mechanism underlying the shift in NO binding (Andersen et al., 1998, 2000; Andersen and Montfort, 2000; Weichsel et al., 2000). In unligated NP at pH 5, two random coil loops that connect the A-B and the G-H β-sheets interact with three bound water molecules in the distal heme pocket, a configuration that allows ready access to the distal pocket. In this configuration, NO access to the heme is limited only by its diffusion rate. Once NO enters and ligates to the sixth coordination site of the heme iron, though, the three water molecules are expelled from the pocket, and the A-B and G-H loops collapse onto the ligated NO and close off access to the distal pocket, trapping the NO in place. The conformation change likely results from the lipophilic nature of NO; once NO ligates to the heme the distal pocket environment is sufficiently hydrophobic to expel the bound water and drive the conformation change. When the insect feeds and injects saliva into the host, the shift to a more neutral pH results in restored hydrogen bonding to water, such that water reenters the distal pocket and the A-B and G-H loops return to the open conformation allowing egress of NO. In the case of NP-1 the dissociation rate increases 2,500-fold over the pH range from 5 to 8.3 (Andersen et al., 2000). Differential protonation of a specific residue, Asp30 (as numbered in NP-4, or the equivalent reside Asp29 in NP-2), is critical in this conformation change (Maes et al., 2004).

The four *Rhodnius* nitrophorins fall into two groups based on sequence similarity: NP-1 and NP-4 are 90% identical, and NP-2/NP-3 share 79% sequence identity, but overall NP-1/NP-4 has only 39% identity with NP-2/NP-3. It is likely that this diversity is a result of a series of tandem gene duplications. Interestingly, the NP-1/NP-4 releases NO at neutral/basic pH at a faster rate than does the NP-2/NP-3 group (Andersen et al., 2000). This difference is related to differences in the pH-induced conformational change: the change is smaller for NP-2 compared to NP-4, resulting in a less open channel to (and from) the heme and slower egress of NO (Andersen, 2009). Pharmacologically, this difference is likely to result in differential release of NO and subsequent vasodilation as a function of diffusion of saliva from the bite site, which will be related to distance from the bite. NP-1/NP-4 can be expected to release NO in the immediate vicinity of the bite, whereas NP-2/NP-3 may diffuse further before unloading their NO, with the net result being efficient vasodilation along a length of the vessel instead of a localized effect concentrated at the specific site of the bite. *Rhodnius* is a relatively large insect (certainly in comparison to mosquitoes) and the rate of blood ingestion is very high, so vasodilation of a large area may be a significant advantage in feeding efficiency. Indeed, when synthesis of all four NPs is inhibited by dsRNA interference, insects fed from skin at a significantly reduced rate, and had to probe five times more frequently to acquire a suitable blood vessel, compared to untreated insects (Araujo et al., 2009).

The *Rhodnius* nitrophorins contribute additional antihemostatic functions to the saliva, in addition to their vasodilatory properties. Nitrophorin 2 efficiently inhibits the clotting by interfering with the formation of the complex that proteolytically cleaves factor X (fX) to produce enzymatically active fXa (Isawa et al., 2000; Ribeiro et al., 1995; Yuda et al., 1997). The mechanism is independent of the heme moiety, but instead depends on specific molecular interactions with fIXa that prevent interaction between fIXa and fX (Gudderra et al., 2005). Interestingly this activity is absent from the other nitrophorins. However analysis of the *Rhodnius* salivary gland transcriptome (or the "sialome") revealed transcripts of novel nitrophorins in addition to the original four, including nitrophorin 7 that has a distinctive helical domain in the C-terminal region (Andersen et al., 2004). NP-7 is also an anticoagulant, but works by binding to anionic phospholipid membranes, including phosphatidylserine rich membranes. This prevents interaction of coagulation factors with the surface of activated platelets, and as a result the fXa-forming and prothrombinase complexes cannot assemble. Binding to the platelet surface also allows NP-7 to interfere with platelet aggregation stimulated by collagen and ADP. All the nitrophorins efficiently ligate histamine with high affinity, accounting for a pronounced antihistamine activity in saliva (Ribeiro and Walker, 1994). Indeed histamine can accelerate the release of NO from nitrophorin by displacing NO from the iron. The nitrophorins are even able to outcompete the histamine receptor, due to a higher binding affinity. Triatomines are nest feeders that are likely to exploit the same individual repeatedly, a process which will result in seroconversion and a high antibody titer in the host. Antibody (IgE)/antigen complexes will interact with the Fc receptor on mast cells, resulting in release of histamine and consequent inflammation and pain. Removal of histamine from the bite site during the feeding process reduces pain and inflammation, which contributes to the stealthiness of the bug and increased likelihood of surviving the contact with the host. In this context it is interesting to note that first-stage *Rhodnius* nymphs have only NP-2 in their saliva, and other NPs are added with successive developmental stages (Moreira et al., 2003). This may, at least in part, be a strategy to avoid neutralization of the antihemostatic function of saliva that could result from antibody binding to antigenic proteins, by adding additional antigenically novel components to the saliva mix over time. The presence of multiple antihemostatic activities in a single protein also cautions against attempting to understand the structural features of that protein in terms of only one activity, as distinct activities may help to shape different parts of the sequence, or even require compromises in the structure to obtain an optimal balance between the various activities.

After the discovery of the *Rhodnius* nitrophorins, a nitrophorin was discovered in saliva of the bedbug *Cimex lectularius* (Valenzuela et al., 1995). As these are both hemipteran insects it is tempting to assume the nitrophorins will be related, but in fact the *Cimex* nitrophorin is derived from a polyphosphatase, not a lipocalin (Valenzuela and Ribeiro, 1998). This demonstrates that the *Cimex* vasodilator evolved independently from the *Rhodnius* system. The *Cimex* NP is even more efficient than the *Rhodnius* NPs, in that two NO molecules are bound in a cooperative manner (Weichsel et al., 2005). In the unligated protein the fifth coordination bond on the proximal side of the heme is occupied by the side chain of Cys60. Initially one NO will ligate to the distal side of the ferric heme, which breaks the bond to Cys60 and results in oxidation of the heme to the ferrous state, where NO is tightly bound. The side chain of Cys60 is then available to bind a second NO molecule as an S-nitroso (SNO) conjugate. In the host, dissociation of the SNO conjugate partially restores the bond between Cys60 and the heme, reducing the iron to the ferric state and permitting dissociation of the second NO. In this manner the binding and subsequent release of the two NO molecules is cooperative, again favoring storage in the salivary glands and release in the low NO concentration environment in the skin.

Antagonists of Endogenous Vasoconstrictors

As vascular tone is maintained by a balance of endogenously produced vasoconstrictors and vasodilators, relaxation of the blood vessel may be achieved by removing vasoconstrictors from the system. Two strategies have been exploited by arthropods to accomplish this: enzymatic hydrolysis and sequestration of target molecules.

Enzymatic Destruction of Endogenous Vasoconstrictors

In the mosquito *Anopheles albimanus*, vasodilation was shown to be due to a myloperoxidase with catechol oxidase activity (Ribeiro and Valenzuela, 1999). Subsequently, orthologues of this protein have been described from the salivary gland transcriptome of other *Anopheles* species (Arcá et al., 1999; Calvo et al., 2007; Francischetti et al., 2002; Valenzuela et al., 2003). Oxidative inactivation of serotonin and noradrenalin leads to a slow-acting but persistant vasodilation in the target blood vessels (Ribeiro, 1996).

Sequestration of Endogenous Vasoconstrictors

The protein D7 was initially identified as the most abundantly expressed protein in saliva of *Aedes aegypti* (James et al., 1991). Subsequently D7 was shown to be a member of a large protein family found in all blood-feeding nematoceran flies ("lower" Diptera). These proteins exist in long (27–30 kDa) and short (15–20 kDa) (referred to as D7-related or D7R) forms in culicine and anopheline mosquitoes (Calvo et al., 2006; Ribeiro et al., 2004; Valenzuela et al., 2002), in sand flies (Diptera: Psychodidae) (Anderson et al., 2006), and in blackflies (Diptera: Simuliidae) (Andersen et al., 2009). *Culicoides* midges express only D7R-like proteins (Campbell et al., 2005). These proteins were shown to be highly derived members of the odorant-binding protein (OBP) superfamily (Arcá et al., 2002); the long form have two odorant binding domains, and the D7R proteins have a single odorant binding domain that corresponds to the C-terminal domain of the long forms (Calvo et al., 2002, 2009). OBPs function to bind small molecules; the first OBPs to be described transport small hydrophobic odor molecules from pores at the surface of insect antennae across the aqueous receptor lymph, so as to facilitate binding to odor receptors on the surface of antennal neurons. This suggested that salivary D7 proteins might function to bind some small molecule mediator of hemostasis (Arcá et al., 2002; Valenzuela et al., 2002), but their specific ligand (and so their role in facilitating blood-feeding) was not known until Calvo et al. (2006) expressed five short D7s from *An. gambiae* and a long form from *Ae. aegypti*, and showed that the long D7 and four of the short D7s efficiently bound biogenic amines, in particular serotonin, norepinephrine, and histamine, with a stoichiometry of one molecule bound per molecule of D7. The mechanism of ligand binding by D7R was elucidated when Mans et al. (2007) solved the crystal structure of *An. gambiae* D7R4 ligated with various biogenic amines, and in the unligated state. Sensory OBPs have six alpha helices, stabilized by two disulfide bonds, and the odorant binding pocket is predominantly composed of residues from the D, E, and F helices. Although all six helices are also present in D7R4, two novel C-terminal helices (stabilized by a third disulfide bond) make up most of the binding pocket. The aromatic rings of serotonin and norepinephrine are accomodated in a hydrophobic pocket, which positions the phenolic hydroxyl to hydrogen bond to the side chains of Glu-7 and His-35, and the amino group to interact with the side chains of Asp-111 and Glu-114. Sequence alignments suggested that the C-terminal OBP domain of the *Ae. aegypti* long D7 should form a similar structure, but surprisingly in the unligated protein the eighth (H_2) alpha helix is replaced with a disordered loop, and in addition the side chains of Arg-174 and Glu-268 are positioned to create a more open binding pocket than is present in D7R4 (Calvo et al., 2009). However, addition of either serotonin or norepinephrine to the pocket results in a profound structural change, in which Arg-174 and Glu-268 are reoriented and the H_2 helix forms (Fig. 38.5). The result is a binding pocket much like that seen in D7R4. Functionally the difference is that the more open pocket in the unligated protein is better able to accomodate both serotonin and the secondary hydroxyl of catecholamines including norepinephrine; the subsequent structural rearrangement results in trapping the ligand within the pocket.

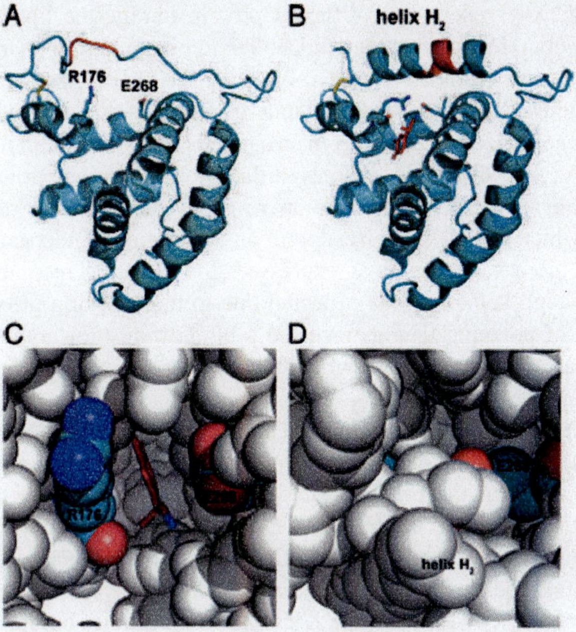

Fig. 38.5 Structure and conformational changes of the C-terminal domain of *Aedes aegypti* long form D7. (**a**) Unliganded structure showing nonhelical C terminus and open conformation of Arg-176 and Glu-268. The poorly ordered portion of the C-terminal coil is shown in red, and the terminal disulfide bond is shown in *yellow*. (**b**) Norepinephrine bound structure showing helical C terminus labeled as helix H2. The poorly ordered region in*A* is shown in red, the ligand is shown as a red stick diagram, and the terminal disulfide bond is shown in *yellow*. The side chains of Glu-268 and Arg-176 are shown in their closed conformations. (**c**) Space-filling diagram of the unliganded D7 structure with a norepinephrine (*red stick diagram*) model placed in the ligand-binding pocket. Arg-176 and Glu-268 (*cyan*, with nitrogen atoms in *blue* and oxygen in *red*) are shown in their open positions. (**d**) Space-filling diagram of the norepinephrine-bound structure with Arg-176 and Glu-268 moved into their closed positions. Helix H2 (*indicated by label*) has ordered to close off the ligand-binding pocket completely burying the ligand. Figure is from Calvo et al. (2009); PNAS 106(10):3728–3733. Used with permission of Proceedings of the National Academy of Science (USA)

In contrast to the C-terminal OBP domain, the N-terminal OBP domain of the long D7 is narrow and hydrophobic (Calvo et al., 2009). This suggested that this domain might bind a fatty acid ligand, such as an eicosanoid. Screening of candidate ligands lead to the discovery that three cysteinyl leukotrienes (LT), LTC4, LTD4, and LTE4, were bound with high and approximately equal affinity (Calvo et al., 2009). As biogenic amines and leukotrienes are highly pro-inflammatory molecules, the multiple forms of D7 and D7R facilitate blood-feeding by inhibiting inflammation and the associated itching response as the insect is biting. In addition, sequestering serotonin released from activated platelets will help abrogate the vasoconstriction arm of the hemostatic response, and so contributes to the vasodilatory properties of saliva.

The role of D7 is taken by a different protein family, the lipocalins, in both triatomine insects (Hemiptera: Rdeuviidae: Triatominae) and in "hard" and "soft" ticks (Argasidae and Ixodidae respectively). Like the OBPs, lipocalins function to bind small ligands. In many cases the ligands are hydrophobic molecules, which the lipocalins help solublize and transport in an aqueous environment. In *R. prolixus*, the nitrophorins (already discussed) belong to this protein family, but analysis of the *R. prolixus* salivary gland transcriptome revealed the presence of two related sequences in which His-59 is replaced with an asparagine (Andersen et al., 2003). When one of these sequences was expressed, the resulting recombinant protein was unable to bind heme (as was expected due to the lack of a proximal ligand), and screening of potential ligands revealed a high affinity for both serotonin and norepinephrine. Accordingly these proteins were named amine-binding proteins or ABPs.

The ixodid tick *Rhipicephalus appendiculatus* also has a salivary amine-binding lipocalin, Ra-HBP (*H*istamine *B*inding *P*rotein), but this differs significantly from the *Rhodnius* ABPs in the presence of two binding sites, one high-affinity (the *H* site) and one with lower affinity (the *L* site), both with a much higher affinity for histamine than for serotonin (Oldham et al., 2003; Paesen et al., 1999, 2000). A second ixodid tick, *Dermacentor reticulatus*, has a homologous protein, SHBP, but in contrast to Ra-HBP the *L* site has high-affinity binding of serotonin (Sangamnatdej et al., 2002). The sialomes of the argasid ticks *Argas monolakensis* and *Ornithodoros savignyi* include several biogenic amine binding lipocalins, which each bind a single ligand molecule (Mans et al., 2008). The *Argas* protein monomine is structurally very similar to the L domain of Ra-HBP, and this protein has high affinity for histamine. On the other hand, the related proteins monotonin and AM-182, also from *Argas*, have high affinity for serotonin (Mans et al., 2008). The conserved motif CD[VIL]X(97,17)EL[WY]X(11,30)C was identified in these tick proteins, and used to identify further biogenic amine binding proteins in the sialome of additional ticks including *O. savignyi* and the ixodid *Ixodes scapularis* (Mans et al., 2008). A related group of single binding domain lipocalins, including moubatin, efficiently sequester both serotonin and TXA_2 (Mans and Ribeiro, 2008a). Completing the parallel with the Dipteran D7/D7R proteins, argasid tick saliva contains yet another set of lipocalins with specificity for binding cysteinyl leukotrienes (Mans and Ribeiro, 2008b).

Redundancy and Cooperation in the Salivary Cocktail

The barriers to a successful blood meal are complex, as inflammation and hemostasis each have multiple facets that are mediated by multiple molecular signals. Accordingly the saliva of blood-feeding arthropods contains a highly complex mixture of anti-inflammatory and antihemostatic molecules, in which cooperative interaction and redundancy are commonly observed. With particular regard to vasodilation, *Anopheles* mosquitoes inactivate serotonin and norepinephrine using a salivary catechol oxidase, while simultaneously sequestering the same molecules with D7 and especially D7R. This cooperative removal of biogenic amines is reminiscent of the removal of the platelet aggregation agonist ADP by *Rhodnius* saliva, where ADP is hydrolysed by apyrase (Sarkis et al., 1986) and also sequestered by a lipocalin, RPAI (Francischetti et al., 2000). In these cases, enzymatic inactivation may serve to reduce high concentrations of a ligand, but that inactivation may become inefficient when substrate concentrations are low and reaction products accumulate to inhibitory levels. Stoichiometric binding may then further reduce the remaining ligand concentration to below the level needed to promote a physiological response, such as vasoconstriction or platelet aggregation. In other instances, a salivary factor with direct vasodilatory activity is accompanied by components that sequester endogenous vasoconstrictors. Examples include the tachykinin peptide sialokinin accompanied by biogenic amine sequestering D7 and D7R in *Aedes* mosquitoes, NO delivered by nitrophorins accompanied by ABPs in *Rhodnius*, and prostacyclin/prostaglandins accompanied by biogenic amine sequestering lipocalins in ixodid ticks. In these cases the salivary components "collaborate", as provision of a vasodilator simultaneously with the removal of vasoconstrictors is likely to provide a more efficient suppression of normal host hemostatic responses. Removal of biogenic amines can also be expected to abrogate the pro-inflammatory effects of these molecules, which illustrates the fact that a single salivary component may effect multiple aspects of the host's defense. An extreme example of this is the *Rhodnius* protein nitrophorin 2, which is a vasodilator by virtue of its NO-transporting activity, and in addition sequesters histamine (a property it shares with the remaining *Rhodnius* nitrophorins) (Ribeiro and Walker, 1994) and uniquely inhibits coagulation by interfering with assembly of the fXa-generating complex (Gudderra et al., 2005). Another aspect of redundancy is the frequent presence of multiple proteins with the same function in the salivary cocktail. Examples include the multiple nitrophorins in *Rhodnius* saliva, the variety of D7/D7R proteins (at least two genes for D7, and five for D7R) in mosquito saliva, and the multiplicity of lipocalins in triatomine and tick saliva. In part this diversity may be due to the need to secrete relatively high concentrations of ligand-binding proteins to efficiently sequester the serotonin (or other target ligand) produced at the bite site. Gene duplication is a well-known mechanism leading to enhanced protein synthesis. Further, salivary proteins injected into a vertebrate host may be expected to be recognized as non-self and elicit an adaptive immune response. Selection to evade this response has lead to the addition of immunomodulatory components to the salivary cocktail (Brossard and Wikel, 2004; Schneider and Higgs, 2008; Titus et al., 2006), but in addition this pressure has likely lead to disruptive selection resulting in the rapid divergence of duplicated genes, outside of the domains encoding the catalytic or binding site, as has been proposed to explain the antigenic diversity of maxadilan peptides in sandflies (Milleron et al., 2004). All these phenomena have contributed to the extraordinary diversity of antihemostatic molecules, including vasodilators, to be found in blood-feeding arthropods.

Uncharacterized Vasodilators

Several arthropods have been shown to have saliva with vasodilatory activity, but the specific components responsible have not been identified. All members of the subfamily Triatominae are blood-feeders, and in an extensive survey Ribeiro et al. (1998) showed that all the tested species (7 *Rhodnius*, 16 *Triatoma*, 2 *Panstrongylus*, and one each for *Dipetalogaster* and *Eratyrus*) expressed vasodilarory activity in their saliva. However, only *Rhodnius* salivary glands had the red color characteristic of nitrophorins. Further, the effect of *Rhodnius* saliva is independent of the presence of an intact endothelium, as would be expected of a nitrovasodilator, but in the majority of the *Triatoma* species the effect is endothelium dependent, and for the remaining genera vasodilation was enhanced in the intact vessel. This suggests that, within the Triatominae, salivary nitrophorins are unique to *Rhodnius*, and a different molecule with a different mechanism of action is expressed in the other genera. Subsequent analysis of the salivary gland transcriptome from *T. infestans* and *T. braziliensis* confirms the absence of nitrophorins, although other lipocalins are highly diverse (Assumpção et al., 2008; Santos et al., 2007). The nature of the vasodilator in these triatomines remains undescribed. Similarly, vasodilatory activity is present in the saliva of the biting midge *Culicoides variipenis* (Perez de Leon et al., 1997). Analysis of the sialome of this midge does not reveal transcripts with homology to known vasodilators (Campbell et al., 2005), and the identity of the molecule involved is unknown. The same situation applies to the West Nile vector *Culex pipiens quinquefasciatus*, which has more vasodilatory activity than *Ae. aegypti* (Ribeiro, 2000), but again the sialome did not contain transcripts suggesting the presence of any known vasodilator. Finally, saliva of the human louse *Pediculus humanus* has an uncharacterized vasodilator (Jones, 1998). In addition, attention to date has understandably focused on disease vectors, particularly those arthropods that vector diseases to humans. This has left the vast majority of blood-feeders unexamined. The search for novel vasodilators will be a fertile field for research for the foreseeable future.

References

Andersen, J.F., 2009. Structure and mechanism in salivary proteins from blood-feeding arthropods. Toxicon [Epub ahead of print]. doi: 10.1016/j.toxicon.2009.11.002

Andersen, J.F., Champagne, D.E., Weichsel, A., Ribeiro, J.M., Balfour, C.A., Dress, V., Montfort, W.R., 1997. Nitric oxide binding and crystallization of recombinant nitrophorin I, a nitric oxide transport protein from the blood-sucking bug *Rhodnius prolixus* . Biochemistry 36, 4423–4428.

Andersen, J.F., Ding, X.D., Balfour, C., Shokhireva, T.K., Champagne, D.E., Walker, F.A., Montfort, W.R., 2000. Kinetics and equilibria in ligand binding by nitrophorins 1–4: evidence for stabilization of a nitric oxide-ferriheme complex through a ligand induced conformational trap. Biochemistry 39, 10118–10131.

Andersen, J.F., Francischetti, I.M., Valenzuela, J.G., Schuck, P., Ribeiro, J.M., 2003. Inhibition of hemostasis by a high affinity biogenic amine-binding protein from the saliva of a blood-feeding insect. J. Biol. Chem. 278, 4611–4617.

Andersen, J.F., Gudderra, N.P., Francischetti, I.M., Valenzuela, J.G., Ribeiro, J.M., 2004. Recognition of anionic phospholipid membranes by an antihemostatic protein from a blood-feeding insect. Biochemistry 43, 6987–6994.

Andersen, J.F., Montfort, W.R., 2000. The crystal structure of nitrophorin 2. A trifunctional antihemostatic protein from the saliva of *Rhodnius prolixus*. J. Biol. Chem. 275, 30496–30503.
Anderson, J.M., Oliveira, F., Kamhawi, S., Mans, B.J., Reynoso, D., Seitz, A.E., Lawyer, P., Garfield, M., Pham, M., Valenzuela, J.G., 2006. Comparative salivary gland transcriptomics of sandfly vectors of visceral leishmaniasis. BMC Genomics 7, 52.
Andersen, J.F., Pham, V.M., Meng, Z., Champagne, D.E., Ribeiro, J.M., 2009. Insight into the sialome of the Black Fly, *Simulium vittatum*. J. Proteome Res. 8, 1474–1488.
Andersen, J.F., Weichsel, A., Balfour, C.A., Champagne, D.E., Montfort, W.R., 1998. The crystal structure of nitrophorin 4 at 1.5 A resolution: transport of nitric oxide by a lipocalin-based heme protein. Structure 6, 1315–1327.
Araujo, R.N., Soares, A.C., Paim, R.M., Gontijo, N.F., Gontijo, A.F., Lehane, M.J., Pereira, M.H., 2009. The role of salivary nitrophorins in the ingestion of blood by the triatomine bug *Rhodnius prolixus* (Reduviidae: Triatominae). Insect Biochem. Mol. Biol .39, 83–89.
Arcá, B., Lombardo, F., de Lara Capurro, M., della Torre, A., Dimopoulos, G., James, A.A., Coluzzi, M., 1999. Trapping cDNAs encoding secreted proteins from the salivary glands of the malaria vector *Anopheles gambiae*. Proc. Natl. Acad. Sci. U.S.A. 96, 1516–1521.
Arcà, B., Lombardo, F., Lanfrancotti, A., Spanos, L., Veneri, M., Louis, C., Coluzzi, M., 2002. A cluster of four D7-related genes is expressed in the salivary glands of the African malaria vector *Anopheles gambiae*. Insect Mol. Biol. 11, 47–55.
Assumpção, T.C., Francischetti, I.M., Andersen, J.F., Schwarz, A., Santana, J.M., Ribeiro, J.M., 2008. An insight into the sialome of the blood-sucking bug *Triatoma infestans*, a vector of Chagas' disease. Insect Biochem. Mol. Biol. 38, 213–232.
Beerntsen, B.T., Champagne, D.E., Coleman, J.L., Campos, Y.A., James, A.A., 1999. Characterization of the Sialokinin I gene encoding the salivary vasodilator of the yellow fever mosquito, *Aedes aegypti*. Insect Mol. Biol. 8, 459–467.
Bergman, D.K., 1996. Mouthparts and feeding mechanisms of haematophagous arthropods, in: Wikel, S.K. (Ed.), The Immunology of Host-Ectoparasitic Arthropod Relationships. CAB International, Wallingford, pp. 30–61.
Bowman, A.S., Dillwith, J.W., Sauer, J.R., 1996. Tick salivary prostaglandins: presence, origin, and significance. Parasitol. Today 12, 388–395.
Bowman, A.S., Gengler, C.L., Surdick, M.R., Zhu, K., Essenberg, R.C., Sauer, J.R., Dillwith, J.W., 1997. A novel phospholipase A2 activity in saliva of the lone star tick, *Amblyomma americanum* (L.). Exp. Parasitol. 87, 121–132.
Bowman, A.S., Sauer, J.R., Neese, P.A., Dillwith, J.W., 1995. Origin of arachidonic acid in the salivary glands of the lone star tick, *Amblyomma americanum*. Insect Biochem. Mol. Biol. 25, 225–233.
Brossard, M., Wikel, S.K., 2004. Tick immunobiology. Parasitology 129(Suppl), S161–S176.
Calvo, E., Dao, A., Pham, V.M., Ribeiro, J.M., 2007. An insight into the sialome of *Anopheles funestus* reveals an emerging pattern in anopheline salivary protein families. Insect Biochem. Mol. Biol. 37, 164–175.
Calvo, E., deBianchi, A.G., James, A.A., Marinotti, O., 2002. The major acid soluble proteins of adult female *Anopheles darlingi* salivary glands include a member of the D7-related family of proteins. Insect Biochem. Mol. Biol. 32, 1419–1427.
Calvo, E., Mans, B.J., Andersen, J.F., Ribeiro, J.M., 2006. Function and evolution of a mosquito salivary protein family. J. Biol. Chem. 281, 1935–1942.
Calvo, E., Mans, B.J., Ribeiro, J.M., Andersen, J.F., 2009. Multifunctionality and mechanism of ligand binding in a mosquito anti-inflammatory protein. Proc. Natl. Acad. Sci. U.S.A. 106, 3728–3733.
Campbell, C.L., Vandyke, K.A., Letchworth, G.J., Drolet, B.S., Hanekamp, T., Wilson, W.C., 2005. Midgut and salivary gland transcriptomes of the arbovirus vector *Culicoides sonorensis* (Diptera: Ceratopogonidae). Insect Mol. Biol. 14, 121–136.
Champagne, D., 2005. Antihemostatic molecules from saliva of blood-feeding arthropods. Pathophysiol. Haemost. Thromb. 34, 221–227
Champagne, D.E., Ribeiro, J.M., 1994. Sialokinin I and II: vasodilatory tachykinins from the yellow fevermosquito *Aedes aegypti*. Proc. Natl. Acad. Sci. U.S.A. 91, 138–142.

Champagne, D.E., Nussenzveig, R.H., Ribeiro, J.M., 1995. Purification, partial characterization, and cloning of nitric oxide-carrying heme proteins (nitrophorins) from salivary glands of the blood-sucking insect *Rhodnius prolixus*. J. Biol. Chem. 270, 8691–8695.

Cupp, M.S., Ribeiro, J.M., Champagne, D.E., Cupp, E.W., 1998. Analyses of cDNA and recombinant protein for a potent vasoactive protein in saliva of a blood-feeding black fly, *Simulium vittatum*. J. Exp. Biol. 201, 1553–1561.

Cupp, M.S., Ribeiro, J.M.C., Cupp, E.W., 1994. Vasodilative activity in black fly salivary glands. Am. J. Trop. Med. Hyg. 50, 241–246.

Dickinson, R.G., O'Hagan, J.E., Shotz, M., Binnington, K.C., Hegarty, M.P., 1976. Prostaglandin in the saliva of the cattle tick *Boophilus microplus*. Aust. J. Exp. Biol. Med. Sci. 54, 475–486.

Doolittle, R.F., 2009. Step-by-step evolution of vertebrate blood coagulation. Cold Spring Harbor Symp. Quant. Biol. [Epub ahead of print]. doi: 10.1101/sqb.2009.74.001

Doolittle, R.F., Feng, D.F., 1987. Reconstructing the evolution of vertebrate blood coagulation from a consideration of the amino acid sequences of clotting proteins. Cold Spring Harbor Symp. Quant. Biol. LII, 869–874.

Francischetti, I.M., Ribeiro, J.M., Champagne, D., Andersen, J., 2000. Purification, cloning, expression, and mechanism of action of a novel platelet aggregation inhibitor from the salivary gland of the bloodsucking bug, *Rhodnius prolixus*. J. Biol. Chem. 275, 12639–12650.

Francischetti, I.M., Sa-Nunes, A., Mans, B.J., Santos, I.M., Ribeiro, J.M., 2009. The role of saliva in tick feeding. Front. Biosci. 14, 2051–2088.

Francischetti, I.M., Valenzuela, J.G., Pham, V.M., Garfield, M.K., Ribeiro ,J.M., 2002. Toward a catalog for the transcripts and proteins (sialome) from the salivary gland of the malaria vector *Anopheles gambiae*. J. Exp. Biol. 205, 2429–2451.

Fry, B.G., Roelants, K., Champagne, D.E., Scheib, H., Tyndall, J.D., King, G.F., Nevalainen, T.J., Norman, J.A., Lewis, R.J., Norton, R.S., Renjifo, C., de la Vega, R.C., 2009. The toxicogenomic multiverse: convergent recruitment of proteins into animal venoms. Annu. Rev. Genomics Hum. Genet. 10, 483–511.

Gillett, J.D., 1967. Natural selection and feeding speed in a blood-sucking insect. Proc. Roy. Soc. Lond. B. Biol. Sci. 167, 316–329.

Golodne, D.M., Monteiro, R.Q., Graca-Souza, A.V., Silva-Neto, M.A., Atella, G.C., 2003. Lysophosphatidylcholine acts as an anti-hemostatic molecule in the saliva of the blood-sucking bug *Rhodnius prolixus*. J. Biol. Chem. 278, 27766–27771.

Grevelink, S.A., Osborne, Loscalzo, J. Lerner, E.A., 1995. Vasorelaxant and second messenger effects of maxadilan. J. Pharmacol. Exp. Ther. 272, 33–37.

Grevelink, S.A., Youssef, D.E., Loscalzo, J., Lerner, E.A., 1993. Salivary-gland extracts from the deerfly contain a potent inhibitor of platelet aggregation. Proc. Natl. Acad. Sci. U.S.A. 90, 9155–9158.

Grimaldi, D., Engel, M. 2005. Evolution of the insects. Cambridge University Press, New York.

Gudderra, N.P., Ribeiro, J.M., Andersen, J.F., 2005. Structural determinants of factor IX(a) binding in nitrophorin 2, a lipocalin inhibitor of the intrinsic coagulation pathway. J. Biol. Chem. 280, 25022–25028.

Higgs, G.A., Vane, J.R., Hart, R.J., Porter, C., Wilson, R.G., 1976. Prostaglandins in the saliva of the cattle tick, *Boophilus microplus* (Canestrini) (Acarina: Ixodidae). Bull. Entomol. Res. 66, 665–670.

Hollander, A.L., Wright, R.E., 1980. Impact of tabanids on cattle: blood meal size and preferred feeding sites. J. Econ. Entomol. 73, 431–433

Isawa, H., Yuda, M., Yoneda, K., Chinzei, Y., 2000. The insect salivary protein, prolixin-S, inhibits factor IXa generation and Xase complex formation in the blood coagulation pathway. J. Biol. Chem. 275, 6636–6641.

James, A.A., Blackmer, K., Marinotti, O., Ghosn, C.R., Racioppi, J.V., 1991. Isolation and characterization of the gene expressing the major salivary gland protein of the female mosquito, *Aedes aegypti*. Mol. Biochem. Parasitol. 44, 245–253.

Jones, D., 1998. The neglected saliva: medically important toxins in the saliva of human lice. Parasitology 116(Suppl), S73–S81.

Kazimírová, M., Šulanová, M., Kozánek, M., Takác, P., Labuda, M., Nuttall, P.A., 2001. Identification of anticoagulant activities in salivary gland extracts of four horsefly species (Diptera: Tabanidae). Haemostasis 31, 294–305.

Kemp, D.H., Hales, J.R., Schleger, A.V. Fawcett, A.A., 1983. Comparison of cutaneous hyperemia in cattle elicited by larvae of *Boophilus microplus* and by prostaglandins and other mediators. Experientia 39, 725–727.

Koh, C.Y., Kini, R.M. 2009. Molecular diversity of anticoagulants from haematophagous animals. Thromb. Haemost. 102, 437–453.

Lanzaro, G.C., Lopes, A.H., Ribeiro, J.M., Shoemaker, C.B., Warburg, A., Soares, M., Titus, R.G., 1999. Variation in the salivary peptide, maxadilan, from species in the *Lutzomyia longipalpis* complex. Insect Mol. Biol. 8, 267–275.

Law, J.H., Ribeiro, J.M., Wells, M.A., 1992. Biochemical insights derived from insect diversity. Annu. Rev. Biochem. 61, 87–111.

Lerner, E.A., Ribeiro, J.M., Nelson, R.J., Lerner, M.R., 1991. Isolation of maxadilan, a potent vasodilatory peptide from the salivary glands of the sand fly *Lutzomyia longipalpis* . J. Biol. Chem. 266, 11234–11236.

Lerner, E.A., Shoemaker, C.B., 1992. Maxadilan: cloning and functional expression of the gene encoding this potent vasodilator peptide. J. Biol. Chem. 267, 1062–1066.

Lu, S.M,. Lu, W., Qasim, M.A., Anderson, S., Apostol, I., Ardelt, W., Bigler, T., Chiang, Y.W., Cook, J., James, M.N., Kato, I., Kelly, C., Kohr, W., Komiyama, T., Lin, T.Y., Ogawa, M., Otlewski, J., Park, S.J., Qasim, S., Ranjbar, M., Tashiro, M., Warne, N., Whatley, H., Wieczorek, A., Wieczorek, M., Wilusz, T., Wynn, R., Zhang, W., Laskowski, M., Jr., 2001. Predicting the reactivity of proteins from their sequence alone: Kazal family of protein inhibitors of serine proteinases. Proc. Natl. Acad. Sci. U.S.A. 98, 1410–1415.

Ma, D., Wang, Y., Yang, H., Wu, J., An, S., Gao, L., Xu, X., Lai, R., 2009. Anti-thrombosisrepertoire of blood-feeding horsefly salivary glands. Mol. Cell Proteomics 8, 2071–2079.

Maes, E.M., Weichsel, A., Andersen, J.F., Shepley, D., Montfort, W.R., 2004. Role of binding site loops in controlling nitric oxide release: structure and kinetics of mutant forms of nitrophorin 4. Biochemistry 43, 6679–6690.

Mans, B.J., Calvo, E., Ribeiro, J.M., Andersen, J.F., 2007. The crystal structure of D7r4, a salivary biogenic amine-binding protein from the malaria mosquito *Anopheles gambiae*. J. Biol. Chem. 282, 36626–36633.

Mans, B.J., Ribeiro, J.M., 2008a. Function, mechanism and evolution of the moubatin-clade of soft tick lipocalins. Insect Biochem. Mol. Biol. 38, 841–852.

Mans, B.J., Ribeiro, J.M., 2008b. A novel clade of cysteinyl leukotriene scavengers in soft ticks. Insect Biochem. Mol. Biol. 38, 862–870.

Mans, B.J., Ribeiro, J.M., Andersen, J.F., 2008. Structure, function, and evolution of biogenic amine-binding proteins in soft ticks. J. Biol. Chem. 283, 18721–18733.

Mesquita, R.D., Carneiro, A.B., Bafica, A., Gazos-Lopes, F., Takiya, C.M., Souto-Padron, T., Vieira, D.P., Ferreira-Pereira, A., Almeida, I.C., Figueiredo, R.T., Porto, B.N., Bozza, M.T., Graça-Souza, AV., Lopes, A.H., Atella, G.C., Silva-Neto, M.A., 2008. *Trypanosoma cruzi* infection is enhanced by vector saliva through immunosuppressant mechanisms mediated by lysophosphatidylcholine. Infect. Immun. 76, 5543–5552.

Milleron, R.S., Mutebi, J.P., Valle, S., Montoya, A., Yin, H., Soong, L., Lanzaro, G.C., 2004. Antigenic diversity in maxadilan, a salivary protein from the sand fly vector of American visceral leishmaniasis. Am. J. Trop. Med. Hyg. 70, 286–293.

Moreira, M.F., Coelho, H.S., Zingali, R.B., Oliveira, P.L., Masuda, H., 2003. Changes in salivary nitrophorin profile during the life cycle of the blood-sucking bug *Rhodnius prolixus*. Insect Biochem. Mol. Biol. 33, 23–28.

Moro, O., Lerner, E.A., 1997. Maxadilan, the vasodilator from sand flies, is a specific pituitary adenylate cyclase activating peptide type I receptor agonist. J. Biol. Chem. 272, 966–970.

Nussenzveig, R.H., Bentley, D.L., Ribeiro, J.M., 1995. Nitric oxide loading of the salivary nitric-oxide-carrying hemoproteins (nitrophorins) in the blood-sucking bug *Rhodnius prolixus*. J. Exp. Biol. 198, 1093–1098.

Oldham, N.J., Lissina, O., Nunn, M.A., Paesen, G.C., 2003. Non-denaturing electrospray ionisation-mass spectrometry reveals ligand selectivity in histamine-binding protein RaHBP2. Org. Biomol. Chem. 1, 3645–3646.

Paesen, G.C., Adams, P.L., Harlos, K., Nuttall, P.A., Stuart, D.I., 1999. Tick histamine-binding proteins: isolation, cloning, and three-dimensional structure. Mol. Cell 3, 661–671.

Paesen, G.C., Adams, P.L., Nuttall, P.A., Stuart, D.L., 2000. Tick histamine-binding proteins: lipocalins with a second binding cavity. Biochim. Biophys. Acta 1482, 92–101.

Paesen, G.C., Siebold, C., Dallas, M.L., Peers, C., Harlos, K., Nuttall, P.A., Nunn, M.A., Stuart, D.I., Esnouf, R.M., 2009. An ion-channel modulator from the saliva of the brown ear tick has a highly modified Kunitz/BPTI structure. J. Mol. Biol. 389, 734–747.

Perez de Leon, A.A., Ribeiro, J.M., Tabachnick, W.J., Valenzuela, J.G., 1997. Identification of a salivary vasodilator in the primary North American vector of bluetongue viruses, *Culicoides variipennis*. Am. J. Trop. Med. Hyg. 57, 375–381.

Rai, K.S., Black, W.C., 4th., 1999. Mosquito genomes: structure, organization, and evolution. Adv. Genet. 41, 1–33.

Rajská, P., Knezl, V., Kazimírová, M., Takáč, P., Roller, L., Vidlička, L., Čiampor, F., Labuda, M., Weston-Davies, W., Nuttall, P.A. 2007. Effects of horsefly (Tabanidae) salivary gland extracts on isolated perfused rat heart. Med. Vet. Entomol. 21, 384–389.

Rajská, P., Pechánová, O., Takác, P., Kazimírová, M., Roller, L., Vidlicka, L., Ciampor, F., Labuda, M., Nuttall, P.A., 2003. Vasodilatory activity in horsefly and deerfly salivary glands. Med. Vet. Entomol. 17, 395–402.

Reddy, V.B., Kounga, K., Mariano, F., Lerner, E.A., 2000. Chrysoptin is a potent glycoprotein IIb/IIIa fibrinogen receptor antagonist present in salivary gland extracts of the deerfly. J. Biol. Chem. 275, 15861–15867.

Ribeiro, J.M.C., 1987. Role of saliva in blood-feeding by arthropods. Annu. Rev. Entomol. 32, 463–478.

Ribeiro, J.M., 1992. Characterization of a vasodilator from the salivary glands of the yellow fever mosquito *Aedes aegypti*. J. Exp. Biol. 165, 61–71.

Ribeiro, J.M.C. 1995. Blood-feeding arthropods: live syringes or invertebrate pharmacologists? Infect Agents Dis. 4, 143–152.

Ribeiro, J.M., 1996. NAD(P)H-dependent production of oxygen reactive species by the salivary glands of the mosquito *Anopheles albimanus*. Insect Biochem. Mol. Biol. 26, 715–720.

Ribeiro, J.M., 2000. Blood-feeding in mosquitoes: probing time and salivary gland anti-haemostatic activities in representatives of three genera (*Aedes*, *Anopheles*, *Culex*). Med. Vet. Entomol. 14, 142–148.

Ribeiro, J.M., Charlab, R., Pham, V.M., Garfield, M., Valenzuela, J.G., 2004. An insight into the salivary transcriptome and proteome of the adult female mosquito *Culex pipiens quinquefasciatus*. Insect Biochem. Mol. Biol. 34, 543–563.

Ribeiro, J.M., Evans, P.M., MacSwain, J.L., Sauer, J.. 1992. *Amblyomma americanum*: characterization of salivary prostaglandins E2 and F2 α by RP-HPLC/bioassay and gas chromatography-mass spectrometry. Exp. Parasitol. 74, 112–116.

Ribeiro, J.M.C., Francischetti, I.M.B., 2003. Role of arthropod saliva in blood feeding: sialome and post-sialome perspectives. Annu. Rev. Entomol. 48, 73–88.

Ribeiro, J.M., Hazzard, J.M., Nussenzveig, R.H., Champagne, D.E., Walker, F.A., 1993. Reversible binding of nitric oxide by a salivary heme protein from a bloodsucking insect. Science 260, 539–541.

Ribeiro, J.M., Katz, O., Pannell, L.K., Waitumbi, J., Warburg, A., 1999. Salivary glands of the sand fly *Phlebotomus papatasi* contain pharmacological amounts of adenosine and 5′-AMP. J. Exp. Biol. 202, 1551–1559.

Ribeiro, J.M., Marinotti, O., Gonzales, R., 1990. A salivary vasodilator in the blood-sucking bug, *Rhodnius prolixus*. Br. J. Pharmacol. 101, 932–936.

Ribeiro, J.M., Nussenzveig, R.H., 1993. Nitric oxide synthase activity from a hematophagous insect salivary gland. FEBS Lett. 330, 165–168.

Ribeiro, J.M., Schneider, M., Guimarães, J.A., 1995. Purification and characterization of prolixin S (nitrophorin 2), the salivary anticoagulant of the blood-sucking bug *Rhodnius prolixus*. Biochem. J. 308, 243–249.

Ribeiro, J.M., Schneider, M., Isaias, T., Jurberg, J., Galvao, C., Guimaraes, J.A., 1998. Role of salivary antihemostatic components in blood feeding by triatomine bugs (Heteroptera). J. Med. Entomol. 35, 599–610.

Ribeiro, J.M., Valenzuela, J.G., 1999. Purification and cloning of the salivary peroxidase/catechol oxidase of the mosquito *Anopheles albimanus*. J. Exp. Biol. 202, 809–816.

Ribeiro, J.M., Walker, F.A., 1994. High affinity histamine- binding and antihistaminic activity of the salivary nitric oxide-carrying heme protein (nitrophorin) of *Rhodnius prolixus*. J. Exp. Med. 180, 2251–2257.

Sangamnatdej, S., Paesen, G.C., Slovak, M., Nuttall, P.A., 2002. A high affinity serotonin- and histamine-binding lipocalin from tick saliva. Insect Mol. Biol. 11, 79–86.

Santos, A., Ribeiro, J.M., Lehane, M.J., Gontijo, N.F., Viols, A.B., Sant'Anna, M.R., Nascimento, Araujo, R., Grisard, E.C., Pereira, M.H., 2007. The sialotranscriptome of the blood-sucking bug *Triatoma brasiliensis* (Hemiptera: Triatominae). Insect Biochem. Mol. Biol. 37, 702–712.

Sarkis, J.J., Guimarães, J.A., Ribeiro, J.M., 1986. Salivary apyrase of *Rhodnius prolixus*: kinetics and purification. Biochem. J. 233, 885–891.

Schneider, B.S., Higgs, S., 2008. The enhancement of arbovirus transmission and disease by mosquito saliva is associated with modulation of the host immune response. Trans. R. Soc. Trop. Med. Hyg. 102, 400–408.

Sun, J., Yamaguchi, M., Yuda, M., Miura, K., Takeya, H., Hirai, M., Matsuoka, H., Ando, K., Watanabe, T., Suzuki, K., Chinzei, Y., 1996. Purification, characterization and cDNA cloning of a novel anticoagulant of the intrinsic pathway, (prolixin-S) from salivary glands of the blood sucking bug, *Rhodnius prolixus*. Thromb. Haemost. 75, 573–577.

Sun, J., Yuda, M., Miura, K., Chinzei, Y., 1998. Characterization and cDNA cloning of a hemoprotein in the salivary glands of the blood-sucking insect, *Rhodnius prolixus*. Insect Biochem. Mol. Biol. 28, 191–200.

Takac, P., Nunn, M.A., Meszaros, J., Pechanova, O., Vrbjar, N., Vlasakova, P., Kozanek, M., Kazimirova, M., Hart, G., Nuttall, P.A., Labuda, M., 2006. Vasotab, a vasoactive peptide from horse fly *Hybomitra bimaculata* (Diptera: Tabanidae) salivary glands. J. Exp. Biol. 209, 343–352

Titus, R.G., Bishop, J.V., Mejia, J.S., 2006. The immunomodulatory factors of arthropod saliva and the potential for these factors to serve as vaccine targets to prevent pathogen transmission. Parasite Immunol. 28, 131–141.

Valenzuela, J.G., 2002. High-throughput approaches to study salivary proteins and genes from vectors of disease. Insect Biochem. Mol. Biol. 32, 1199–1209.

Valenzuela, J.G., Charlab, R., Gonzalez, E.C., de Miranda-Santos, I.K., Marinotti, O., Francischetti, I.M., Ribeiro, J.M., 2002. The D7 family of salivary proteins in blood sucking diptera. Insect Mol. Biol. 11, 149–155.

Valenzuela, J.G., Francischetti, I.M., Pham, V.M., Garfield, M.K., Ribeiro, J.M., 2003. Exploring the salivary gland transcriptome and proteome of the *Anopheles stephensi* mosquito. Insect Biochem. Mol. Biol. 33, 717–732.

Valenzuela, J.G., Ribeiro, J.M., 1998. Purification and cloning of the salivary nitrophorin from the hemipteran *Cimex lectularius*. J. Exp. Biol. 201, 2659–2664.

Valenzuela, J.G., Walker, F.A., Ribeiro, J.M., 1995. A salivary nitrophorin (nitric-oxide-carrying hemoprotein) in the bedbug *Cimex lectularius*. J. Exp. Biol. 198, 1519–1526.

Weichsel, A., Andersen, J.F., Champagne, D.E., Walker, F.A., Montfort, W.R., 1998. Crystal structures of a nitric oxide transport protein from a blood-sucking insect. Nat. Struct. Biol. 5, 304–309.

Weichsel, A., Andersen, J.F., Roberts, S.A., Montfort, W.R., 2000. Nitric oxide binding to nitrophorin 4 induces complete distal pocket burial. Nat. Struct. Biol. 7, 551–554.

Weichsel, A., Maes, E.M., Andersen, J.F., Valenzuela, J.G., Shokhireva, T., Walker, F.A., Montfort, W.R., 2005. Heme-assisted S-nitrosation of a proximal thiolate in a nitric oxide transport protein. Proc. Natl. Acad. Sci. U.S.A. 102, 594–599.

Wynn, R., Zhang, W., Laskowski, M., Jr., 2001. Predicting the reactivity of proteins from their sequence alone: Kazal family of protein inhibitors of serine proteinases. Proc. Natl. Acad. Sci. U.S.A. 98, 1410–1415.

Yuda, M., Higuchi, K., Sun, J., Kureishi, Y., Ito, M., Chinzei, Y., 1997. Expression, reconstitution and characterization of prolixin-S as a vasodilator – a salivary gland nitric-oxide-binding hemoprotein of *Rhodnius prolixus*. Eur. J. Biochem. 249, 337–342.

Yuda, M., Hirai, M., Miura, K., Matsumura, H., Ando, K., Chinzei, Y., 1996. cDNA cloning, expression and characterization of nitric-oxide synthase from the salivary glands of the blood-sucking insect *Rhodnius prolixus*. Eur. J. Biochem. 242, 807–812.

Xu, X., Yang, H., Ma, D., Wu, J., Wang, Y., Song, Y., Wang, X., Lu, Y., Yang, J., Lai, R.. 2008. Toward an understanding of the molecular mechanism for successful blood feeding by coupling proteomics analysis with pharmacological testing of horsefly salivary glands. Mol. Cell. Proteomics 7, 582–590.

Zhu, K., Bowman, A.S., Dillwith, J.W., Sauer, J.R., 1998. Phospholipase A2 activity in salivary glands and saliva of the lone star tick (Acari: Ixodidae) during tick feeding. J. Med. Entomol. 35, 500–504.

Part IX
Hemorrhage and Coagulopathies

Chapter 39
Snakebite-Induced Coagulopathy and Bleeding Disorders

Ponlapat Rojnuckarin

Abstract Snake venoms target mainly neuromuscular and/or hemostatic systems. Each of them is a combination of several toxins. Therefore, coagulopathy is only a part of multi-systemic involvement from envenomation including muscular weakness, rhabdomyolysis, renal failure and hypotension. Kinetics studies reveal that viper venoms comprise long half-life components resulting in a delay onset and prolonged duration of bleeding in a subset of patients. On the other hand, elapid venoms are more rapidly cleared from the circulation showing faster recovery. Remarkably, snake venoms affect almost every component of hemostasis including vascular wall, platelets, coagulation factors, natural anticoagulants and fibrinolysis. They can be stimulatory or inhibitory through enzymatic or binding mechanisms. These effects can contribute to hemorrhagic, as well as thrombotic, manifestations of snakebites. The most prominent clinical syndrome is consumptive coagulopathy from the thrombin-like enzymes and/or coagulation factor activators in the venoms. In addition, anticoagulation syndrome, thromboembolism and thrombotic microangiopathy have been reported in victims of particular snake species. The key treatment of snakebites is antivenom that can promptly reverse coagulopathy in most situations.

Introduction

Venomous snakes concoct their poisons to inflict several physiological systems of their preys. The two main targets are neuromuscular damage causing immobilization and blood clotting activation resulting in circulatory obstruction and rapid death. However, the latter effect displays a different consequence in larger animals, which are not the natural preys. In human, due to greatly larger blood volume, the fibrinolytic system activation causes coagulation factor as well as platelet

P. Rojnuckarin (✉)
Division of Hematology, Department of Medicine, King Chulalongkorn Memorial Hospital and Chulalongkorn University, Rama IV Rd, Patumwan, Bangkok 10330, Thailand
e-mail: rojnuckarinp@gmail.com

R.M. Kini et al. (eds.), *Toxins and Hemostasis*, DOI 10.1007/978-90-481-9295-3_39,

consumption and ensuing bleeding disorders. Furthermore, many proteolytic enzymes, which may aid prey digestion, can disrupt vascular wall, destroy clotting factors and, thus, significantly contribute to the hemorrhagic diathesis.

Snake Envenomation

Snake venom is a mixture of numerous proteins with diverse activities. However, only a subset of them causes clinically apparent effects. For example, coagulation factor activators and anticoagulant activities have been discovered in biochemical studies on king cobra (Lee et al., 1995), cobra (Kini, 2005; Yukelson et al., 1991) and krait (Mitrakul, 1979; Zhang et al., 1995) venoms. Nevertheless, these snakes impose only neuromuscular toxicity and bleeding is not one of the clinical manifestations, although detailed analysis may find subclinical laboratory abnormalities in the coagulation system (Li et al., 2001). This chapter will focus only on the venoms causing clinically significant effects on hemostasis.

From the medical standpoint, coagulopathy is only a part of the multisystem involvements by snake venoms. The protean clinical findings of snake envenomation are classified into local and systemic symptoms and signs. The local tissue injury may yield pain, limb swelling, cutaneous bullae, local muscular necrosis and/or gangrene. Besides hematological disorders, other cardinal features of systemic effects are listed below. The presence of other organ involvement may be helpful in identifying snake species when combined with the data on geographical areas of bites.

1. *Neuromuscular blockade* in the presence of hemorrhagic diathesis (Gold et al., 2004; Warrell, 1989; White, 1998) is caused by selected Australians elapids, which are Tiger snake (*Notechis* spp.), Rough scale snake (*Tropidechis carinatus*), Taipan (*Oxyurenus* spp.), Death adder (*Acanthophis* spp.) and Australian copperhead (*Austrelaps* spp.), and some rattlesnakes, which are South American rattlesnake (*Crotalus durissus*), Mojave rattlesnake (*Crotalus scutulatus*), and Eastern diamondback rattlesnake (*Crotalus adamanteus*). In general, Asian vipers and pit vipers are toxic solely to hemostatic system, except in India and Sri Lanka where Russell's viper (*Daboia russelli*) may also cause muscular weakness.
2. *Systemic rhabdomyolysis* is the prominent feature of sea snakes, although it has been reported in conjunction with hematotoxicity from other snakes. For examples, some Australian elapids, which are Mulga snakes (*Pseudechis* spp.), Tiger snake and Taipan, South American rattlesnake (*Crotalus durissus terrificus*) and lance-head vipers (*Bothrops* spp.), as well as Russell's viper (*Daboia russelli*) particularly in Sri Lanka.
3. *Renal failure* (Kanjanabuch and Sitprija, 2008; Pinho et al., 2008) may be a combinatorial result from direct kidney toxicity, indirect toxicity via cytokines, renal hypoperfusion, rhabdomyolysis (myoglobinuria), hemolysis (hemoglobinuria), disseminated intravascular coagulation, and/or thrombotic microangiopathy (discussed below). It is typically caused by vipers, such as Russell's viper

(*Daboia russelli*), saw scale viper (*Echis carinatus*), South American rattle snake (*Crotalus durissus terrificus*), bushmaster (*Lachesis* spp.) or lance-head vipers (*Bothrops* spp.), as well as a variety of other snakes secondary to rhabdomyolysis.

4. *Hypotension and shock* are the consequences of several factors including vasodilatory venom proteins, fluid leakage, allergic reactions, toxic myocardial damage and/or bleeding.

Toxicokinetics

A proportion of venomous snakebite does not show any symptom because the toxins are not injected. This has been termed 'dry bites'. After venomous bites, toxins are typically absorbed rapidly into circulation within minutes. Hyaluronidases, proteolytic enzymes and inflammatory responses enhance systemic absorption, as well as local tissue damages.

After viper bites, coagulopathy may be detectable as soon as half an hour (Reid et al., 1963). However, in some patients, especially bitten by weakly venomous snakes such as green pit vipers (*Cryptelytrops albolabris*), the maximal toxicity may be delayed up to 3 days after bites (Rojnuckarin et al., 1998). Normal coagulation tests on the first day after bites cannot guarantee patient safety. In general, viper venoms comprise high molecular weight proteins that are slowly cleared from plasma. Due to the long (over 24 h) half-life of venoms in circulation, the toxins can continuously destroy clotting factors and platelets. Consistent with this notion, coagulopathy is not associated with a venom level at a single time point, but correlated with the product of venom level and the period of time after bite or 'venom-hour' that represent the cumulative effects of venom over time (Hutton et al., 1990; Rojnuckarin et al., 1999). Without antivenom treatment, venom antigen may be detectable in blood over 2–3 weeks after the incident (Reid et al., 1963; Visudhiphan et al., 1981) and this persistent phenomenon is correlated with the presence of bleeding tendency (Rojnuckarin et al., 2007). This prolonged effect is also reported in American pit vipers and may cause recurrences of coagulopathy after normalization by a single dose of antivenom. The problem is partly because the low molecular weight Fab antivenom with a short a half-life is used in North America. Therefore, multiple doses of this kind of antivenom are required.

On the contrary, elapid venoms contain smaller molecular weight protein. This may result in very rapid absorption. A mathematical model calculation suggested that the half-life of Australian elapid venom causing coagulopathy was only 1 h (Tanos et al., 2008). This probably resulted in faster clearance and more rapid clinical recovery.

Venoms and Hemostatic System

Snake venom is the most fascinating field for scholars who study hemostasis because it can affect every step and in all aspects of the system, including vascular wall, platelets, coagulation, natural anticoagulation, and fibrinolysis. The toxins can be activating or inhibitory, either via protein-protein interactions or enzymatic proteolysis. To add more complexity, one species of venom contains several toxins, which may both stimulate and inhibit several or even the same molecular or cellular targets. Furthermore, one toxin may possess more than one effect. The detailed discussion of each venom component can be found in other chapters of the book. The brief summary is shown in Fig. 39.1. Novel compounds possessing new activities likely remain to be discovered.

1. *Vascular wall disruption*: Snake venom metalloproteases (SVMPs) digest subendothelial matrix resulting in weakened vessels and susceptible to damages by blood flow. Some also trigger endothelial cell apoptosis. These contribute to both local and systemic bleeding (Gutiérrez and Rucavado, 2000).
2. *Platelet activation*: Several C-type lectin-like proteins (CLPs) activates von Willebrand factor (vWF) receptor (Gp Ib/IX/V), collagen receptors (Integrin $\alpha_2\beta_1$ and Gp VI) and the novel CLEC-2 receptor on platelet surface. These activations may be followed by shedding of GpVI receptor resulting in subsequent inhibition. A thrombin-like enzyme, PA-BJ from *Bothrops jararaca*, stimulates platelet protease-activated receptor. Aggregation of platelets in vivo causes accelerated clearance by phagocytic system and thrombocytopenia.
3. *Platelet inhibition*: Disintegrins block integrin $\alpha_{IIb}\beta_3$ and $\alpha_2\beta_1$ on platelets. In addition, some CLPs are inhibitory by binding and competing with the natural platelet receptor ligands. Furthermore, some SVMPs can degrade platelet receptor or their ligands, such as vWF, contributing to hemorrhage.
4. *Coagulation activation*: This is the principal effect of venom in vivo. Snake venoms mainly target the common pathway of coagulation including factor X, V and prothrombin activators resulting in consumptive coagulopathy. Alternatively, they may directly coagulate fibrinogen through thrombin-like enzymes. In contrast to thrombin, they lack the complete effects and release only one of the fibrinopeptide A (venombin A) or fibrinopeptide B (venombin B). Some may cleave both fibrinopeptides (venombin AB), but incompletely. Therefore, the formed clots are friable and subjected to endogenous and venom-induced fibrinolysis causing defibrination syndrome in humans (Rojnuckarin et al., 1999). Coagulation activators are also implicated in occasional thrombosis after snakebites.
5. *Coagulation inhibition*: Many venom components may inhibit the extrinsic tenase (tissue factor/factor VIIa), intrinsic tenase (factor VIIIa/IXa) or prothrombinase (factor Xa/Va) complexes, as well as prothrombin (CLP, bothrojaracin from *Bothrops jararaca*). A serine protease from *Agkistrodon concortrix*, ACC-C, activates protein C which in turn inactivates factor Va and VIIIa. Furthermore, enzymes directly digesting fibrinogen (fibrinogenolysis) have been described (Muanpasitporn and Rojnuckarin, 2007).
6. *Fibrinolysis*: Fibrin degradation may be a physiologic response to coagulation activation by venoms. Furthermore, toxins may directly digest fibrin clot or indirectly activate plasminogen to plasmin. Moreover, textilinin-1, the antiplasmin from *Pseudonaja textilis*, has been isolated and used as a hemostatic agent (Flight et al., 2009).

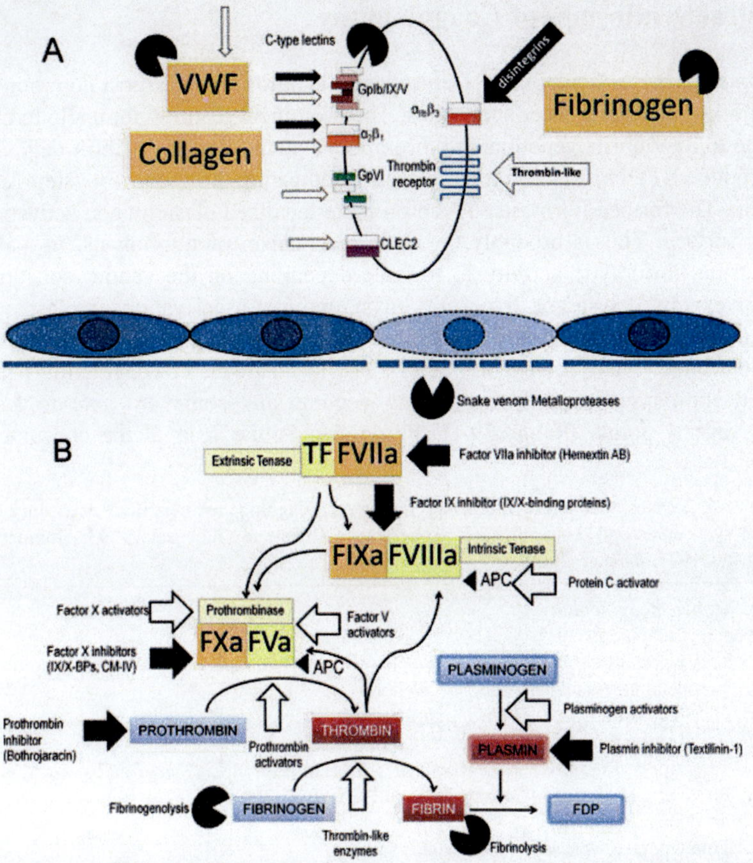

Fig. 39.1 The summary of snake venoms affecting hemostasis. *White arrows* represent activators, while *black arrows* denote inhibitors. The *chipped circles* ( ) indicate digesting proteases. (**a**) Primary hemostasis, a platelet, its receptors and endothelium on basement membrane, is shown. Ligands are displayed in the *boxes* close to their receptors. *Thin arrows* represent a variety of effects of proteins in the C-type lectin-like family. (**b**) Coagulation, anticoagulation and fibrinolytic pathways are shown connected by arrowed lines. The extrinsic tenase complex (Tissue factor/factor VIIa) activates intrinsic tenase complex (Factor VIIIa/IXa on phospholipid surface) and prothrombinase complex (Factor Xa/Va on phospholipid surface) resulting in thrombin generation and, finally, fibrin formation. Fibrinolysis is the effect of plasmin that is activated by endogenous or venom-derived plasminogen activators. Fibrin may be degraded by plasmin or venom fibrinolytic enzymes resulting in fibrin-degradation products (FDP). Venom procoagulants may activate one of the prothrombinase complex components, prothrombin or directly fibrinogen. *Arrow heads* denote the anticoagulant effects of activated protein C (APC) that digests factor Va and VIIIa. APC is stimulated either by thrombin/thrombomodulin complex (The natural activator) or a snake venom protein C activator

Clinical Syndromes of Coagulopathy

1. *Consumptive coagulopathy*: Defibrination syndrome is the most common hemorrhagic syndrome after snakebites. The prominent features are unclotted blood due to hypofibrinogenemia and thrombocytopenia. Levels of fibrin degradation products (FDPs) were highly elevated indicating fibrinolytic system activation. The formed fibrin, itself, can promote localized plasminogen activation on its surface. This is probably the main mechanism of fibrinolysis. In addition, plasma fibrinolytic activity is variable depending on the venom constituents. For example, systemic hyperfibrinolysis defined as elevation of plasma plasminogen activator activity was found in green pit viper, but not in Russell's viper bites (Than-Than et al., 1988). The mechanisms of coagulation pathway activation are also different among venoms and some are able to activate at several points (Table 39.1). Fibrinogen is the sole factor consumed by

Table 39.1 Snakes that cause medically important coagulopathy classified according to the clinical syndromes and their mechanism(s) (Aragon-Ortiz and Gubensek, 1993; Gomperts and Demetriou, 1977; Isbister, 2009)

1. *Consumptive coagulopathy*
   - a. Factor X activation
     - i. True vipers: Russell's viper (*Daboia russelli*), Horned vipers (*Cerastes* spp.), some European vipers (*Vipera* spp.)
     - ii. Pit vipers: Lancehead (*Bothrops* spp.)
   - b. Factor V activation
     - i. True vipers: Russell's viper *Daboia russelli,* European vipers (*Vipera lebetina, V. ursine*)
     - ii. Pit vipers: Lancehead (*Bothrops atrox*)
   - c. Prothrombin activation
     - i. True vipers: Saw scaled vipers (*Echis* spp.)
     - ii. Australian elapids: Brown snakes (*Pseudonaja* spp.), Taipan (*Oxyurenus* spp.), Tiger snakes (*Notechis* spp.), Rough scaled snake (*Tropidechis carinatus*), Broad head snakes (*Hoplocephalus* spp.)
     - iii. Colubrids: Boomslang (*Dyspholidus typhus*), Vine snakes (*Thelotornis* spp.), Red necked keelbacks (*Rhabdophis* spp.)
   - d. Thrombin-like enzymes: Mostly from pit vipers
     - i. Venombin A: Malayan pit viper (*Calloselasma rhodostoma*), Habu (*Protobothrops flavoviridis*), green pit viper (*Cryptelytrops albolabris*), Central and South American pit vipers (*Bothrops* spp.), rattle snakes (*Crotalus adamanteus*)
     - ii. Venombin B: Copperhead (*Agkistrodon contortrix contortrix*), Chinese Habu (*Protobothrops mucrosquamatus*), Halys vipers (*Gloydius halys*)
     - iii. Venombin AB: Cantil (*Agkistrodon bilineatus*), Gaboon viper (*Bitis gabonica*), Bushmaster (*Lachesis muta*)
2. *Anticoagulation syndrome* (Some Australian elapids): Mulga snake (*Pseudechis australis*), Spotted back snake (*Pseudechis guttatus*), Collett's snake (*Pseudechis colletti*), Death adder (*Acanthophis* spp.), Copperhead (*Austrelaps* spp.), New Guinea small-eyed snake (*Micropechis ikaheka*)
3. *Thromboembolism*: Martinique viper (*Bothrops lanceolatus*), Saint Lucia viper (*Bothrops caribbaeus*)
4. *Thrombotic microangiopathy*: Russell's viper (*Daboia russelli*), Horned vipers (*Cerastes cerastes*), Lowland viper (*Proatheris superciliaris*), Australian brown snakes (*Pseudonaja* spp.), Boomslang (*Dispholidus typhus*)

thrombin-like enzymes, while factor X and V levels are depressed by factor X and V activators in Russell's viper envenomation (Mahasandana et al., 1980). On the other hand, prothrombin activators cause low factor V, VIII (thrombin substrates) and prothrombin in human victims (White, 2005). Bleeding symptom severity among snakes depending not only on the strength and quantity of procoagulants, but also the admixed toxins that target several other components of hemostasis as previously mentioned.

Only a subset of patients shows clinical bleeding. The usual sites of hemorrhage were gum, biting sites, venepuncture sites, gastrointestinal and urinary tracts. Multiple sites of bleeding are also common. Furthermore, life-threatening intracranial hemorrhage may occur. The recommended laboratory investigation worldwide is the whole blood clotting time because it is rapid and available everywhere even in remote community hospitals. The only requirements are a clean glass tube and a timer. Furthermore, clotting time is more sensitive to hypofibrinogenemia than the conventional PT or APTT. No clotting after 20-min standing suggests low fibrinogen levels (Sano-Martins et al., 1994) and antivenom is indicated. More standardized tests, e.g. PT, may be used instead of clotting time, but the clinical data are more limited. Thrombocytopenia is often, but not always, correlated with unclotted blood. Blood smear examination is helpful for diagnosis of thrombotic microangiopathy (see below). In some circumstances, clotting factor assays may be helpful to differentiate the species containing thrombin-like effects *vs.* species activating factor X or V or prothrombin.

2. *Anticoagulation syndrome*: Pure anticoagulation without consumptive coagulopathy has been reported in selected Australian elapids. Similar to defibrination syndrome, PT and APTT are prolonged increasing the risk of bleeding. However, fibrinogen and FDP levels are normal. The venom components responsible for this effect are probably phospholipases A_2 (Sharp et al., 1989). However, the exact mechanisms remain to be clearly defined. The clinical significance of the differentiation of this syndrome from the consumption is to aid identifying these snake species (Table 39.1). In addition, bleeding symptoms are usually milder than those of defibrination syndrome (White, 2005).
3. *Thromboembolism*: Thrombosis is rare after snakebites except after bites by two *Bothrops* species on the islands of Martinique (Thomas et al., 1998) and Saint Lucia (Numeric et al., 2002). The patients are presented with deep vein thrombosis with or without pulmonary embolism or ischemic stroke. These complications can effectively be prevented using the specific antivenom (Thomas et al., 1995). In addition, thrombosis has been reported early after other snakebites probably from rapid and strong procoagulant effects of the venoms prior to fibrinolytic activation. This hypothesis is also used to explain sporadic sudden cardiac death early after the brown snake bites (White, 2005).
4. *Thrombotic microangiopathy*: The complication similar to hemolytic uremic syndrome (HUS) that is consisted of microangiopathic hemolytic anemia, thrombocytopenia and renal failure has been reported in Russell's viper (Date et al., 1986), Saharan horned vipers (Schneemann et al., 2004), Lowland viper (Keyler,

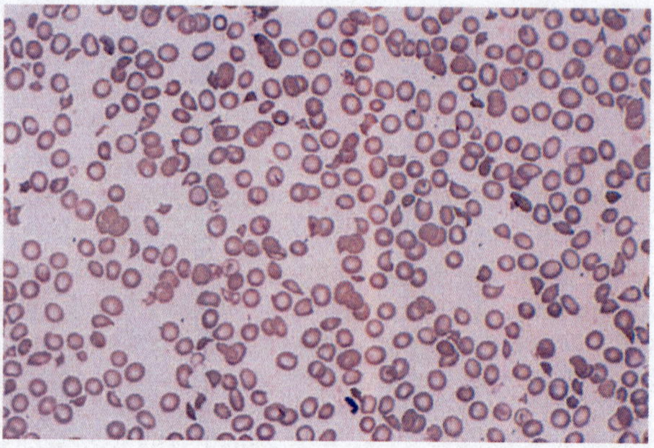

Fig. 39.2 Thrombotic microangiopathy after Russell's viper (*Daboia russelli*) bite showing numerous schistocytes. Polychromasia is not increased due to erythropoietin deficiency from renal failure

2008) and a colubrid, Boomslang (Lakier and Fritz, 1969). Furthermore, it was found in 13% (4/32) of severe brown snake bites (Isbister et al., 2007). This syndrome may be under-recognized because manifestations are very close to the consumptive coagulopathy as described above. The clue is the severe anemia and thrombocytopenia detectable when coagulation time is normal or near normal. Typically, platelet counts, hemoglobin levels and renal functions are markedly declined after fibrinogen levels have been recovered and, sometimes, after antivenom administration. Schistocytes on blood smear (Fig. 39.2), red serum, hemoglobinuria and elevation of lactate dehydrogenase enzyme, as well as unconjugated bilirubin, are seen. Reticulocytes may not be appropriately increased due to impaired kidney functions. Renal biopsy may reveal thrombus in glomeruli (Isbister et al., 2007). The pathogenesis is unknown. Idiopathic thrombotic microangiopathy is caused by vWF cleaving protease deficiency and/or endothelial injury resulting in platelet aggregation by unusually large vWF multimer secreted from endothelium. These two possibilities remain to be explored. The role of plasma exchange in this condition is also unclear.

Management of Venom-Induced Coagulopathy

The current key treatment for hemostatic disorders is antivenom, the polyclonal IgG or parts of IgG purified from immunized horses or sheep. Due to the ethical issues, there has been no placebo-controlled trial to prove it efficacy. However, comparing with the natural history of viper bites, which may result in up to 3 weeks of coagulopathy, antivenom can reverse the defects by 6–12 h (Mitrakul et al., 1991; Rojnuckarin et al., 1998). Antivenom is able to neutralize venom enzyme activity in vitro (Muanpasitporn and Rojnuckarin, 2007) and probably stop the consumptive process. Subsequently, the body produces clotting factors and platelets to restore normal hemostasis. Furthermore, antivenom may enhance toxin clearance. Fresh frozen plasma and/or platelet transfusion is usually ineffective and unnecessary.

On the other hand, antivenom does not appear to be as helpful in coagulopathy induced by Australian elapids. There is no difference in recovery (INR less than 2.0) rates between cases receiving early *vs.* late antivenom. A possible explanation is that the venom is eliminated rapidly before the full action of antivenom. In this model, fresh frozen plasma infusion was correlated with early reversal of coagulopathy (Isbister et al., 2009). A randomized controlled trial will definitely prove the role of plasma infusion after elapid bites.

Antivenom is helpful not only for bleeding problems, but also prevention of thrombotic complications (Thomas et al., 1995). However, the effects on thrombotic microangiopathy are unclear because this condition still occurs after antivenom administration. Nevertheless, it may have been more severe or more frequent without antivenom and most authorities suggest early antivenom in all cases with clinical suggestions of systemic envenomation. Local edema after viper bites showed statistically faster resolution after antivenom compared with placebo (Rojnuckarin et al., 2006), but the effect was too small to recommend antivenom to mild edema. On the other hand, skin necrosis still arises after antivenom (Chotenimitkhun and Rojnuckarin, 2008). These data are consistent with studies in animal models indicating that antivenom is ineffective for local tissue damages by viper venoms (Gutiérrez et al., 1998) and other treatment modalities are required.

The major limitation of antivenom is its early adverse reactions, which are similar to anaphylactic reactions and potentially fatal. It is unpredictable by hypersensitivity skin test and, therefore, close observation during administration is mandatory (Thiansookon and Rojnuckarin, 2008). Advances in antivenom production using caprylic acid to stabilize the IgG molecules and/or cleaving the complement-activating Fc portion of IgG markedly reduced the incidence (Otero-Patiño et al., 1998).

Perspectives

Despite the extreme heterogeneity of snake venoms, the clinical bleeding manifestations are surprising similar. The converging evolution of hematotoxins appears to end up at the defibrination syndrome that can usually be reversed using specific antivenom. Some special exceptions, e.g. anticoagulation by Australian elapids, deserve further studies. In addition, the emerging syndrome of thrombotic microangiopathy requires more clinical recognition and research for pathogenesis and better managements. Nowadays, investigations on venom proteins affecting hemostasis are not only to find the snakebite treatments, but to use them for exploring and dissecting the system. This will give us deeper insights in physiology and may lead to novel diagnostic and therapeutic agents in the future.

References

Aragon-Ortiz, F., Gubensek, F., 1993. A thrombin-like enzyme from bushmaster (*Lachesis muta stenophyrs*) venom. Toxicon 31, 1435–1443.

Chotenimitkhun, R., Rojnuckarin, P., 2008. Systemic antivenom and skin necrosis after green pit viper bites. Clin. Toxicol. 46, 122–125.

Date, A., Pulimood, R., Jacob, C.K., Kirubakaran, M.G., Shastry, J.C., 1986. Haemolytic-uraemic syndrome complicating snake bite. Nephron 42, 89–90.

Flight, S.M., Johnson, L.A., Du, Q.S., Warner, R.L., Trabi, M., Gaffney, P.J., Lavin, M.F., de Jersey, J., Masci, P.P., 2009. Textilinin-1, an alternative anti-bleeding agent to aprotinin: importance of plasmin inhibition in controlling blood loss. Br. J. Haematol. 145, 207–211.

Gold, B.S., Barish, R.A., Dart, R.C., 2004. North American snake envenomation: diagnosis, treatment, and management. Emerg. Med. Clin. North Am. 22, 423–443.

Gomperts, E.D., Demetriou, D., 1977. Laboratory studies and clinical features in a case of boomslang envenomation. S. Afr. Med. J. 51, 173–175.

Gutiérrez, J.M., León, G., Rojas, G., Lomonte, B., Rucavado, A., Chaves, F., 1998. Neutralization of local tissue damage induced by *Bothrops asper* (terciopelo) snake venom. Toxicon 36, 1529–1538.

Gutiérrez, J.M., Rucavado, A., 2000. Snake venom metalloproteinases: their role in the pathogenesis of local tissue damage. Biochimie 82, 841–850.

Hutton, R.A., Looareesuwan, S., Ho, M., Silamut, K., Chanthavanich, P., Karbwang, J., Supanaranond, W., Vejcho, S., Viravan, C., Phillips, R.E., Warrell, D.A., 1990. Arboreal green pit vipers (genus *Trimeresurus*) of South-East Asia: bites by *T. albolabris* and *T. macrops* in Thailand and a review of the literature. Trans. Roy. Soc. Trop. Med. Hyg. 84, 866–874.

Isbister, G.K., Little, M., Cull, G., McCoubrie, D., Lawton, P., Szabo, F., Kennedy, J., Trethewy, C., Luxton, G., Brown, S.G., Currie, B.J., 2007. Thrombotic microangiopathy from Australian brown snake (*Pseudonaja*) envenoming. Intern. Med. J. 37, 523–528.

Isbister, G.K., 2009. Procoagulant snake toxins: laboratory studies, diagnosis, and understanding snakebite coagulopathy. Semin. Thromb. Hemost. 35, 93–103.

Isbister, G.K., Duffull, S.B., Brown, S.G.; ASP Investigators, 2009. Failure of antivenom to improve recovery in Australian snakebite coagulopathy. Q. J. Med. 102, 563–568.

Kanjanabuch, T., Sitprija, V., 2008. Snakebite nephrotoxicity in Asia. Semin. Nephrol. 28, 363–372.

Keyler, D.E., 2008. Envenomation by the lowland viper (*Proatheris superciliaris*): severe case profile documentation. Toxicon 52, 836–841.

Kini, R.M., 2005. Structure-function relationships and mechanism of anticoagulant phospholipase A_2 enzymes from snake venoms. Toxicon 45, 1147–1161.

Lakier, J.B., Fritz, V.U., 1969. Consumptive coagulopathy caused by a boomslang bite. S. Afr. Med. J. 43, 1052–1055.

Lee, W.H., Zhang, Y., Wang, W.Y., Xiong, Y.L., Gao, R., 1995. Isolation and properties of a blood coagulation factor X activator from the venom of king cobra (*Ophiophagus hannah*). Toxicon 33, 1263–1276.

Li, Q.B., Huang, G.W., Kinjoh, K., Nakamura, M., Kosugi, T., 2001. Hematological studies on DIC-like findings observed in patients with snakebite in south China. Toxicon 39, 943–948.

Mahasandana, S., Rungruxsirivorn, Y., Chantarangkul, V., 1980. Clinical manifestations of bleeding following Russell's viper and Green pit viper bites in adults. Southeast Asian J. Trop. Med. Public Health. 11, 285–293.

Mitrakul, C., 1979. Effect of five Thai snake venoms on coagulation, fibrinolysis and platelet aggregation. Southeast Asian J. Trop. Med. Public Health. 10, 266–275.

Mitrakul, C. Juzi, U, Pongrujikorn, W., 1991. Antivenom therapy in Russell's viper bite. Am J Clin. Pathol. 95, 412–417.

Muanpasitporn, C., Rojnuckarin, P., 2007. Expression and characterization of a recombinant fibrinogenolytic serine protease from green pit viper (*Trimeresurus albolabris*) venom. Toxicon 49, 1083–1089.

Numeric, P., Moravie, V., Didier, M., Chatot-Henry, D., Cirille, S., Bucher, B., Thomas, L., 2002. Multiple cerebral infarctions following a snakebite by *Bothrops caribbaeus*. Am. J. Trop. Med. Hyg. 67, 287–288.

Otero-Patiño, R., Cardoso, J.L., Higashi, H.G., Nunez, V., Diaz, A., Toro, M.F., Garcia, M.E., Sierra, A., Garcia, L.F., Moreno, A.M., Medina, M.C., Castañeda, N., Silva-Diaz, J.F., Murcia, M., Cardenas, S.Y., Dias da Silva, W.D., 1998. A randomized, blinded, comparative trial of one pepsin-digested and two whole IgG antivenoms for *Bothrops* snake bites in Uraba, Colombia. The regional group on antivenom therapy research (REGATHER). Am. J. Trop. Med. Hyg. 58, 183–189.

Pinho, F.M., Yu, L., Burdmann, E.A., 2008. Snakebite-induced acute kidney injury in Latin America. Semin. Nephrol. 28, 354–362.

Reid, H.A., Chan, K.E., Thean, P.C., 1963. Prolonged coagulation defect (defibrination syndrome) in Malayan viper bite. Lancet 1, 621–626.

Rojnuckarin, P., Mahasandana, S., Intragumthornchai, T., Sutcharitchan, P., Swasdikul, D., 1998. Prognostic factors of green pit viper bites. Am. J. Trop. Med. Hyg. 58, 22–25.

Rojnuckarin, P., Intragumtornchai, T., Sattapiboon, R., Muanpasitporn, C., Pakmanee, N., Khow, O., Swasdikul, D., 1999. The effects of green pit viper (*Trimeresurus albolabris* and *Trimeresurus macrops*) venom on the fibrinolytic system in human. Toxicon 37, 743–755.

Rojnuckarin, P., Chanthawibun, W., Noiphrom, J., Pakmanee, N., Intragumtornchai, T., 2006. A randomized, double-blind, placebo-controlled trial of antivenom for local effects of green pit viper bites. Trans. Roy. Soc. Trop. Med. Hyg. 100, 879–884.

Rojnuckarin, P., Banjongkit, S., Chantawibun, W., Akkawat, B., Juntiang, J., Noiphrom, J., Pakmanee, N., Intragumtornchai, T., 2007. Green pit viper (*Trimeresurus albolabris* and *T. macrops*) venom antigenaemia and kinetics in humans. Trop. Doct. 37, 207–210.

Sano-Martins, I.S., Fan, H.W., Castro, S.C., Tomy, S.C., Franca, F.O., Jorge, M.T., Kamiguti, A.S., Warrell, D.A., Theakston, R.D., 1994. Reliability of the simple 20 minute whole blood clotting test (WBCT20) as an indicator of low plasma fibrinogen concentration in patients envenomed by *Bothrops* snakes. Butantan Institute Antivenom Study Group. Toxicon 32, 1045–1050.

Schneemann, M., Cathomas, R., Laidlaw, S.T., El Nahas, A.M., Theakston, R.D., Warrell, D.A., 2004. Life-threatening envenoming by the Saharan horned viper (*Cerastes cerastes*) causing micro-angiopathic haemolysis, coagulopathy and acute renal failure: clinical cases and review. Q. J. Med. 97, 717–727.

Sharp, P.J., Berry, S.L., Spence, I., Howden, M.E., 1989. A basic phospholipase A from the venom of the Australian king brown snake (*Pseudechis australis*) showing diverse activities against membranes. Comp. Biochem. Physiol. B. 92, 501–508.

Tanos, P.P., Isbister, G.K., Lalloo, D.G., Kirkpatrick, C.M., Duffull, S.B., 2008. A model for venom-induced consumptive coagulopathy in snake bite. Toxicon 52, 769–780.

Than-Than, Hutton, R.A., Myint-Lwin, Khin-Ei-Han, Soe-Soe, Tin-Nu-Swe, Phillips, R.E., Warrell, D.A., 1988. Haemostatic disturbances in patients bitten by Russell's viper (*Vipera russelli siamensis*) in Burma. Br. J. Haematol. 69, 513–520.

Thiansookon, A., Rojnuckarin, P., 2008. Low incidence of early reactions to horse-derived F(ab')(2) antivenom for snakebites in Thailand. Acta. Trop. 105, 203–205.

Thomas, L., Tyburn, B., Bucher, B., Pecout, F., Ketterle, J., Rieux, D., Smadja, D., Garnier, D., Plumelle, Y., 1995. Prevention of thromboses in human patients with *Bothrops lanceolatus* envenoming in Martinique: failure of anticoagulants and efficacy of a monospecific antivenom. Research group on snake bites in martinique. Am. J. Trop. Med. Hyg. 52, 419–426.

Thomas, L., Tyburn, B., Ketterlé, J., Biao, T., Mehdaoui, H., Moravie, V., Rouvel, C., Plumelle, Y., Bucher, B., Canonge, D., Marie-Nelly, C.A., Lang, J., 1998. Prognostic significance of clinical grading of patients envenomed by *Bothrops lanceolatus* in Martinique. Members of the research group on snake bite in martinique. Trans. Roy. Soc. Trop. Med. Hyg. 92, 542–545.

Visudhiphan, S., Dumavibhat, B., Trishnananda, M., 1981. Prolonged defibrination syndrome after green pit viper bite with persisting venom activity in patient's blood. Am. J. Clin. Pathol. 75, 65–69.

Warrell, D.A., 1989. Snake venoms in science and clinical medicine. 1. Russell's viper: biology, venom and treatment of bites. Trans. Roy. Soc. Trop. Med. Hyg. 83, 732–740.
White, J., 1998. Envenoming and antivenom use in Australia. Toxicon 36, 1483–1492.
White, J., 2005. Snake venoms and coagulopathy. Toxicon 45, 951–967.
Yukelson, L.Y., Tans, G., Thomassen, M.C., Hemker, H.C., Rosing, J., 1991. Procoagulant activities in venoms from central Asian snakes. Toxicon 29, 491–502.
Zhang, Y., Xiong, Y.L., Bon, C., 1995. An activator of blood coagulation factor X from the venom of *Bungarus fasciatus*. Toxicon 33, 1277–1288.

Chapter 40
Antivenoms and Coagulation

Elda E. Sánchez, Alexis Rodríguez-Acosta, Esteban Cantu, and Belsy Guerrero

Abstract The treatment of snake envenomations has been a complicated battle for over a century. Since the development of antivenoms, inevitably fatal venoms have decreased to rarely fatal provided that antivenom is administered early. Snake venoms contain many molecules that act directly at the site of the bite, while others set off cascades of reactions leading to cumulative disruptions. A primary example is the disruption on the coagulation cascade. Venom molecules promote or inhibit molecules involved in the hemostatic system causing major disturbances that can lead to detrimental consequences. Although antivenoms have been present for a long time, the production and availability of these snakebite antidotes have fallen short. In recent years, some pharmaceutical companies have halted their production of antivenoms due to unprofitable circumstances, and the stringent regulations to meet safety standards have deterred others from considering taking on such an endeavor. Physicians, scientists, poison control and government administrators must make a profound effort to come together to bring in new ideas that will aid in resolving those troubled issues surrounding antivenoms. The focus of this chapter is to briefly address venom components acting on the hemostatic pathway and the use and concerns involved with the antivenoms utilized to neutralize them.

Introduction

Snake venoms have been a part of folklore medicine for centuries. Snake venoms have been used to treat leprosy, stop bleeding and identify schizophrenia. In the present day, with current scientific technology, snake venoms have been separated into individual venom components that have potential in the treatment of strokes,

E.E. Sánchez (✉)
Natural Toxins Research Center, College of Arts and Sciences, Texas A&M University-Kingsville, Kingsville, TX, USA
e-mail: elda918@yahoo.com

R.M. Kini et al. (eds.), *Toxins and Hemostasis*, DOI 10.1007/978-90-481-9295-3_40,

heart attacks, cancers, and diabetes among other diseases. One of the most significant uses for snake venom has been the production of antivenom for the treatment of snakebites. In the nineteenth century, the French physician, bacteriologist, and immunologist Léon Charles Albert Calmette developed the first antivenoms for snake envenomations using sera from horses immunized with snake venoms. This task was later taken over by the Brazilian physician, Vital Brazil of the Instituto Buntantan who produced a variety of antivenoms not only for snake venoms, but also for scorpion and spider venoms. Although these antivenoms have saved many lives, even after over a century of existence, there still exist serious concerns associated with antivenoms. Serum sickness and ineffective antivenoms are physiological problems; not to mention the economical problems hindering the production of antivenoms for countries that cannot afford the treatment. Snake venoms contain many molecules that act directly at the site of the bite, while others set off cascades of reactions leading to cumulative disruptions. A prime example is the disruption on the coagulation cascade. Venom molecules promote or inhibit factors involved in the hemostatic system causing major disruptions that can lead to detrimental consequences. The focus of this chapter is to briefly address venom components acting on the hemostatic pathway and the use of antivenom to neutralize them.

Envenomation resulting from snakebites is an important public health problem in many tropical and subtropical countries. A study estimated at least 421,000 envenomations with 20,000 deaths occurring worldwide from snakebites each year; however these figures may be as high as 1,841,000 envenomations with 94,000 deaths (Kasturiratne et al., 2008). Snake venoms contain an arsenal of proteins and peptides that interfere with components of the human hemostatic system, including the vascular wall, platelets, coagulation and fibrinolytic systems (Markland, 1998; Marsh and Williams, 2005) leading to serious injuries and even death.

The coagulation process is a complex set of events involving many proteins that can be altered by snake venom. Teng and Huang (1991) divided the biological activities of snake venoms into two categories, those that accelerate and those that inhibit hemostasis. Venom proteins that accelerate hemostasis include platelet aggregation amplifiers, Factor X, Factor V and prothrombin activators, and thrombin-like enzymes. Venom proteins that inhibit hemostasis include protein C activators, fibrino(geno)lytic enzymes, inhibitors of prothrombin and Factor X activation, thrombin inhibitors, and platelet aggregation inhibitors. The many different molecules in snake venom can alter the clotting cascade but it is not always clear how venom interferes with the clotting process or how antivenom can neutralize the molecules involved in the hemostasis pathway.

Antivenoms

The best and most acceptable treatment of systemically envenomed humans is with antivenom; however, it must be administered as soon as possible since the damage cannot be reversed. After over a century of the existence of antivenom therapy, it is still used empirically. The administration of whole IgG antivenom in envenomated humans poses serious problems due to the severe and sometimes deadly anaphylactic reactions caused by the Fc portion of the IgG molecule. Therefore in recent years, the production of antivenom has focused on the removal of the Fc portion by the proteolytic enzymes papain or pepsin, resulting in F(ab) or $F(ab')_2$ molecules, respectively, to produce safer and more efficient antivenoms. Many antivenoms produced today are in the form of F(ab) (MW: 50 kDa) and $F(ab')_2$ (MW: 100 kDa) fragments instead of the whole IgG (MW: 150 kDa) molecule (Chippaux and Goyffon, 1998). Due to their sizes, their pharmacokinetic profiles are different. The smaller molecules F(ab) can reach extravascular compartments more quickly than the larger molecules; however, the elimination half-life of Fab is shorter (Ismail and Abd-Elsalam, 1998; Riviere et al., 1997) as they are excreted through renal elimination. The premature elimination of the Fab antivenom has caused recurrent coagulopathy in patients who have been treated with this type of antivenom (Boyer et al., 1999). Coagulopathy problems were observed as late as two weeks post envenomation and was proposed that patients with coagulopathies should be closely monitored during the first two weeks of envenomation. In a more recent study, recurrent coagulopathy four days after Fab antivenom administration caused a fatality in a case of envenomation by *Crotalus adamanteus* (Kitchens and Eskin, 2008). In light of these cases, this type of information has not been fully disseminated to many of our health care providers, and snakebite therapy may not be seen as a serious issue. For instance, in a recent personal snakebite encounter, the bitten individual was given four vials of antivenom in an emergency room and sent home that evening, in spite of consultation from a professional. Approximately 24 h later, the individual had to be readmitted and an additional 10 vials of antivenom were given along with two nights in the ICU. In addition, the patient was pressured to have a fasciotomy without prior compartment pressure testing. Because the envenomation was not severe (only one fang had penetrated the tip of the thumb), the patient rejected the fasciotomy and fully recovered.

Although snakebite envenoming has been characterized as a neglected tropical disease (Hotez et al., 2006) especially in low-income countries, the neglect exists even in the state-of-the-art, first-world facilities. Of the six areas that need addressing in the global issue of envenomation (for details, see Gutiérrez et al., 2009), proper training of personnel in charge of treating snakebites remains an area that requires immediate attention in all parts of the world.

The selection of venoms to obtain effective antivenoms is another important issue. It may be taken for granted that if venoms of a certain snake species are used in the immunization protocol, that the antivenom should neutralize the effects of that particular venom regardless of their geographical locations, age and sex. Intraspecies venom variation (Adame et al., 1990; Aguilar et al., 2007; Glenn and Straight, 1978; Glenn et al., 1983; Minton and Weinstein, 1986; Salazar et al., 2008, 2009) poses problems in the treatment with antivenom. Researchers have proposed the thorough use of venom from the same species (Gutiérrez et al.,

2009; Sánchez et al., 2003) suggesting a pool consisting of venoms from wide geographical locations, different seasonal times, age groups and sex.

Are all venom components necessary in the development of effective antivenom? Are there some that merit greater attention? Anai et al. (2002) reported that hemorrhagic metalloproteinases, in addition to causing hemorrhage, also play a key role in spreading toxins into the circulatory system. It is therefore essential for antivenom to be effective in neutralizing these components in addition to other venom molecules that alter the hemostatic pathways at any number of steps along the way.

The advancements of proteomics (Calvete et al., 2007; Gutiérrez et al., 2009) and molecular biology (Jia et al., 2008) in venom will allow the identification of those molecules that exist in high proportions in venoms and can also identify those venom molecules that are not immunodepleted by antivenoms (Gutiérrez et al., 2009). These advancements could identify those molecules that should be used in the immunization process and those which do not elicit immune responses during the immunization process. Immunizations with recombinant proteins, even when activity cannot be achieved, have proven useful in eliciting an immune response, and antibodies have been able to neutralize the native form of the protein (Azofeifa-Cordero et al., 2008). Utilizing a combination of crude venom along with key venom toxins that cause the most problems in envenomation in immunization protocols could render antivenom therapies effective for future generations.

Neutralization of Venom Components Affecting Hemostasis

Alterations in coagulation are the main effects of many snake venoms and are a specific dilemma in the American Viperidae. The most common type is venom stimulated consumption coagulation, which is a result of procoagulant toxins in venoms. Consumption coagulation diverges from disseminated intravascular coagulation. It is believed it may be linked with thrombotic microangiopathy. Anticoagulant coagulopathy is common in the North American vipers, but much less common in South American rattlesnake envenomations, where it is much less severe and responds well to antivenom (Rodríguez-Acosta et al., 1998).

Snakes of the genus *Bothrops* cause approximately 80–90% of the snakebites reported every year in Latin America (Kornacker, 1999; Rengifo and Rodríguez-Acosta, 2005; Ribeiro and Jorge, 1990). Edema and defibrination are the two most significant signs of bothropic envenomation, contributing to local ischemic damage, hypotension and hemorrhage (Otero et al., 2002; Rengifo and Rodríguez-Acosta, 2005). The antivenoms normally used for treatment show a very limited efficacy in the neutralization of edema (Lomonte et al., 1993). Many toxins found in snake venoms causes defibrinating and coagulant effects. Venom components can act as procoagulants causing consumption of coagulation factors, resulting in systemic coagulopathy (Markland, 1998; Rojnuckarin, 2008). Venoms contain thrombin-like proteins that cleave the fibrinopeptide A from fibrinogen, resulting in blood coagulation. However, excessive thrombin-like proteins can also have an ironic effect resulting in excessive bleeding in vivo (Rojnuckarin et al., 1999).

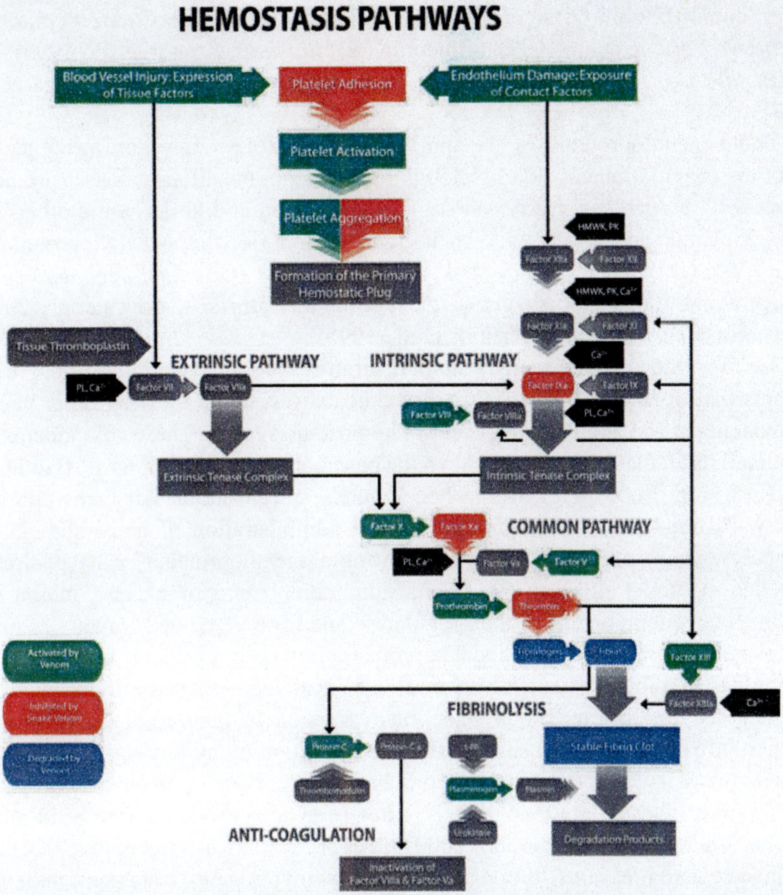

Fig. 40.1 The hemostatic pathways and the sites of action by snake venom components. *Colored areas* indicate their susceptibility to snake venom toxins and areas for antivenom neutralization

Most of the enzymes in the coagulation cascade are serine proteinases (Fig. 40.1). Thrombin plays an important role in hemostasis such that, once activated, it can function in procoagulant, anticoagulant, antifibrinolytic, and anti/pro-inflammatory pathways (Crawley et al., 2007). Thrombin converts fibrinogen to fibrin, resulting in a fibrin clot that is stabilized by FXIIIa, a factor that is also activated by thrombin on the surface of fibrin. Thrombin also activates FV, FVIII and FXI, resulting in a positive feedback loop of coagulation (Fig. 40.1).

The largest group of venom enzymes is the "thrombin-like" enzymes. These enzymes do not act entirely like true thrombin in that they lack the ability to cleave both fibrinopeptide A and B from the Aα and Bβ chains of fibrinogen (Au et al., 1993). Furthermore, they fail to activate FXIII, resulting in a loose fibrin clot.

A few examples of "thrombin-like" enzymes isolated from snake venom are ancrod (Au et al., 1993) (*Calloselasma rhodostoma*), batroxobin (*Bothrops atrox,* Tokunaga et al., 1988), GPV-TLs (*Calloselasma albolabris*, Rojnuckarin et al.,

2006), contortrixobin (Amiconi et al., 2000) and ACC-C (*Agkistrodon contortrix contortrix,* Fortová et al., 1997), bilineobin (*A. bilineatus,* Komori et al., 1993), PA-BJ (*B. jararaca,* Serrano et al., 1995), and RVV-V (*D. russelii,* Tokunaga et al., 1988).

Venom metalloproteinases are another important class of venom proteins that affect the coagulation cascade. Although these venom metalloproteinases are notorious for tissue damage, many can degrade fibrinogen and fibrin, while others can damage platelet membranes and their ligands. Some of these metalloproteinases are RVV-X (*D. russelli*, Kiziel et al., 1976), hellerase (*Crotalus oreganus helleri*, Salazar et al., 2009), alborrhagin (*Calloselasma albolabris*, Andrews et al., 2001), and ecarin (*Echis carinatus*, Nishida et al., 1995).

The first line of defense in venom neutralization is to neutralize those components that open the floodgates of the hemostatic system, allowing other venom components to wreak havoc on the entire hemostatic system. These components are the metalloproteinases that break down the basement membrane of tissue (Gutiérrez and Rucavado, 2000). However, when humans are envenomed it takes approximately 45 min–1 h, at a minimum, for the administration of antivenom in the best-case scenario. By this time, these venom metalloproteinases have already unlocked the doors allowing other hemostatic components to activate, inhibit and bind the many components of the coagulation and fibrinolytic pathways.

It has been reported that local administration of a synthetic matrix metalloproteinase inhibitor effectively neutralized local tissue damage induced, when administered rapidly after venom injection by a venom metalloproteinase (Escalante et al., 2000). These findings suggest that the inhibition of hemorrhagic metalloproteinases may serve to reduce local lesions as well as prevent systemic coagulopathy. Furthermore, phospholipase A_2 (PLA_2) has been identified extensively in snake venoms (Gutiérrez and Rucavado, 2000; Jia et al., 2008; Kashima et al., 2004) and has diverse activities consisting of neurotoxicity, myotoxicity, coagulant, anticoagulant, antibacterial and proinflammatory effects. cDNA libraries have identified the PLA_2 to be a highly expressed gene accounting for 35 and 36% of *Agkistrodon piscivorus lecuostama* and *Bothrops jararacussu* venom glands, respectively (Jia et al., 2008; Kashima et al., 2004). It is for these reasons that antivenoms must be able to not only effectively neutralize the metalloproteinases but also the disintegrins, serine proteinases and PLA_2 that affect the many components of the hemostatic pathways (Fig. 40.1).

Alternative In Vitro Methods for Testing Antivenom Efficacy on Venom Components Affecting Hemostasis

According to the WHO, the most accepted method for determining antivenom efficacy is with an ED_{50} assay done on a mouse model (WHO, 1981). The pre-clinical trial testing of antivenom efficacy on animals is the most acceptable method by authorities as sustaining regulation requirements. However, there are some problems. In vivo antivenom testing requires a large number of animals and it is difficult to obtain permission for antivenom testing, particularly in the United States. The number of mice makes testing antivenom in live animals costly and laborious. Although using animals for testing antivenom efficacy is undesirable, no in vitro assay has yet been developed that will measure the net result pathology as does using live animals. There must be a movement to begin testing in vitro assays that will correlate with those in vivo studies currently in place. Developing other assays to study the neutralization of individual venom molecules is important and will require a battery of in vivo assays (Bogarin, 2000; Gutiérrez et al., 1990, 1996; Sells, 2003).

As previously stated, the clotting cascade is a complex series of chemical reactions that can be altered by snake venoms at any number of steps along the process. A Sonoclot® Coagulation and Platelet Function Analyzer has been used as an alternative in vitro method for testing those venom components affecting coagulation. Measurements are based on the detection of viscoelastic changes of whole blood or plasma (Ganter and Hofer, 2008). The Sonoclot® provides qualitative (Sonoclot® Signature graph) and quantitative (activated clotting time-ACT, clotting rate-CR and platelet function-PF) results on the entire hemostasis process. The ACT is the time in which fibrin formation begins, the CR is the kinetic measurement of fibrin formation and clot development, which is the maximum slope of the Sonoclot® Signature (SS) during initial fibrin polymerization and clot development, and PF is obtained from the timing and quality of the clot retraction. The values for PF range from 0 to 5, where 0 represents no clot retraction and a flat SS as that observed for disintegrins when added to the blood samples such as the signature labeled "V" in Fig. 40.2. A PF higher than 1 represents normal clot retraction and varies from patient to patient. A normal PF contains a sharp peak in the SS after fibrin formation as seen in a normal control patient. The SS can reveal when antivenom neutralizes the venom as indicated by a change in the clotting signature to one that resembled the original human blood control (Fig. 40.2).

Antivenom neutralization can also be assayed using chromogenic substrates, which allows the use of negligible quantities of venom and antivenoms. These chromogenic substrates are synthetic peptides that mimic natural substrates. One example is the chromogenic substrate S-2222, which assays for FXa activity. This

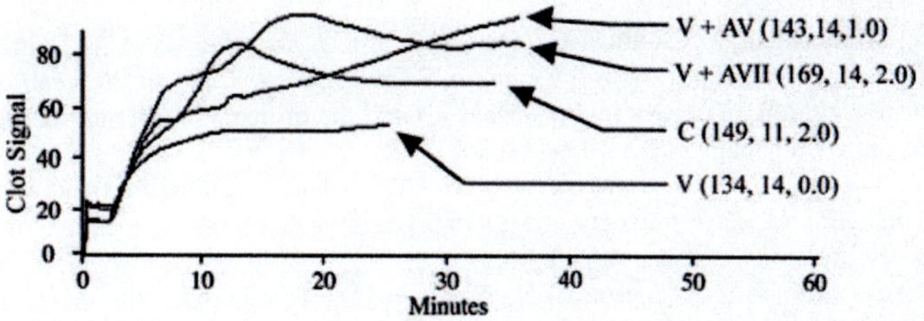

Fig. 40.2 Sonoclot signatures of human blood with venom fractions and the neutralization with antivenoms. AV = Commercial antivenom, AVII = Commercial antivenom II, C = Control human blood, V = Venom component. The values reported in *parentheses* are the activated clotting time, clotting rate and platelet function (clot retraction)

synthetic substrate mimics the molecule prothrombin which FXa will cleave at the C-terminal side of the amino acid arginine. S-2222 is composed of the same amino acids coupled to a chromophore that is hydrolyzed upon cleavage, thus releasing a colored product that can be measured at 405 nm. The change in absorbance (ΔA/min) is directly proportional to the enzyme activity. Salazar et al. (2008) used a battery of these chromogenic substrates to test a series of protease inhibitors on the hemostatic activities of the venom of *Crotalus durissus cumanensis*. This study suggests that antivenoms can also be used to determine the degree of neutralization on the various venom components involved in hemostasis. Diagnostic kits to test snake-venom activities (Schöni, 2005) have proven to have reliable results; however, their full potentials are yet to be explored.

Perspectives

Although the existence of antivenoms has been around for a long time, the production and availability of these snakebite antidotes have fallen short. In recent years, pharmaceutical companies have halted their production of antivenoms due to unprofitable circumstances and the stringent regulations to meet safety standards have deterred others from considering taking on such an endeavor. Snake envenomations causes economic difficulty on poor, rural communities and healthcare systems in many tropical and subtropical countries, a serious issue that remains largely misunderstood and neglected. There should be a strong global effort by physicians, scientists, poison control centers and government administrators to come together to bring about new ideas to resolve these troubled issues involved with antivenoms.

References

Adame, B.L., Soto, J.G., Secraw, D.J., Pérez, J.C., 1990. Regional variation of biochemical characteristics and antigenicity in Great basin rattlesnake (*Crotalus viridis lutosus*) venom. Comp. Biochem. Physiol. B. 97, 95–101.

Aguilar, I., Guerrero, B., Salazar, A.M., Girón, M.E., Pérez, J.C., Sánchez, E.E., Rodríguez-Acosta, A., 2007. Individual venom variability in the South American rattlesnake *Crotalus durissus cumanensis*. Toxicon 50, 214–224.

Amiconi, G., Amoresano, A., Boumis, G., De Cristofaro, R., De Pascali, A., Di Girolamo, S., Maras, B., Scaloni, A., 2000. A novel venombin B from *Agkistrodon contortrix contortrix*: evidence for recognition properties in the surface around the primary specificity pocket different from thrombin. Biochemistry 39, 10294–10308.

Anai, K., Masahiko, S., Yoshida, E., Maruyama, M., 2002. Neutralization of a snake venom hemorrhagic metalloproteinase prevents coagulopathy after subcutaneous injection of *Bothrops jararaca* venom in rats. Toxicon 40, 63–68.

Andrews, R.K., Gardiner, E.E., Asazuma, N., Berlanga, O., Tulasne, D., Nieswandt, B., Smith, A.I., Berndt, M.C., Watson, S.P., 2001. A novel viper venom metalloproteinase, alborhagin, is an agonist at the platelet collagen receptor GPVI. J. Biol. Chem. 276, 28092–28097.

Au, L.C., Lin, S.B., Chou, J.S., Teh, G.W., Chang, K.J., Shih, C.M., 1993. Molecular cloning and sequence analysis of the cDNA for ancrod, a thromin-like enzyme from the venom of *Calloselasma rhodostoma*. Biochem. J. 294, 387–390.

Azofeifa-Cordero, G. Arce-Estrada, V., Flores-Diaz, M., Alape-Giron, A., 2008. Immunization of cDNA of a novel P-III type metalloproteinase from the rattlesnake *Crotalus durissus durissus* elicits antibodies which neutralize 69% of the hemorrhagic induced by the whole venom. Toxicon 52, 302–308.

Bogarin, G., Morais, J.F., Yamaguchi, I.K., Stephano, M.A., Marcelino, J.R., Nishikawa, A.K., Guidolin, R., Rojas, G., Higashi, H.G., Gutierrez, J.M., 2000. Neutralization of crotaline snake venoms from Central and South America by antivenoms produced in Brazil and Costa Rica. Toxicon 38, 1429–1441.

Boyer, L.V., Seifert, M.D., Clark, R.F., McNally, J.T., Williams, S.R., Nordt, S.P., Walter, F.G., Dart, R.C., 1999. Recurrent and persistent coagulopathy following pit viper envenomation. Arch. Intern. Med. 159, 706–710.

Calvete, J.J., Juarez, P., Sanz, L., 2007. Snake venomics. Strategy and applications. J. Mass Spectrom. 42, 1405–1414.

Chippaux, J.P., Goyffon, M., 1998. Venoms, antivenoms, and immunotherapy. Toxicon 36, 823–846.

Crawley, J.T.B., Zanardelli, S., Chion, C.K.N.K., Lane, D.A., 2007. The central role of thrombin in hemostasis. J. Thromb. Haemost. 5(Suppl. 1), 95–101.

Escalante, T., Franceschi, A., Rucavado, A., Gutiérrez, J.M., 2000. Effectiveness of batimastat, a synthetic inhibitor of matrix metalloproteinases, in neutralizing local tissue damage induced by BaP1, a hemorrhagic metalloproteinase from the venom of the snake *Bothrops asper*. Biochem. Pharmacol. 60, 269–274.

Fortová, H., Suttnar, J., Dyr, J.E., Pristach, J., 1997. Simultaneous isolation of protein C activator, fibrin clot promoting enzyme (fibrozyme) and phospholipase A_2 from the venom of the southern copperhead snake. J. Chromatogr/ B. Biomed. Sci. Appl. 694, 49–53.

Ganter, M.T., Hofer, C.K., 2008. Coagulation monitoring: current techniques and clinical use of viscoelastic point-of-care coagulation devices. Int. Anesthesia Res. Soc. 106, 1366–1375.

Glenn, J.L., Straight, R., 1978. Mojave rattlesnake *Crotalus scutulatus scutulatus* venom: variation in toxicity with geographical origin. Toxicon 16, 81–84.

Glenn, J.L., Straight, R.C., Wolfe, M.C., Hardy, D.L., 1983. Geographical variation in *Crotalus scutulatus scutulatus* (Mojave rattlesnake) venom properties. Toxicon 21, 119–130.

Gutiérrez, J.M., Rojas, G., Lomonte, B., Gené, J.A., Chaves, F., Alvarado, J., Rojas, E., 1990. Standardization of assay for testing the neutralizing ability of antivenoms. Toxicon 28, 1127–1129.

Gutiérrez, J.M. Rojas, G., Bogarin, G., Lomonte, B., 1996. Evaluation of the neutralizing ability of antivenoms for the treatment of snake bite envenoming in Central America, in: Bon, C., Goyffon, M. (Eds.), Envenomings and their Treatments. Editions Fondation Marcel Mérieux, Lyon, pp. 223–231.

Gutiérrez, J.M., Rucavado, A., 2000. Snake venom metalloprotineases: their role in the pathogenesis of local tissue damage. Biochimie 82, 841–850.

Gutiérrez, J.M., Lomonte, B., León, G., Alape-Girón, A., Florez-Díaz, M., Sanz, L., Angulo, Y., Calvete, J.J., 2009. Snake venomics and antivenomics: proteomic tools in the design and control of antivenoms for the treatment of snakebite envenoming. J. Proteomics 72, 165–182.

Hotez, P., Ottesen, E., Fenwick, A., Molyneux, D., 2006. The neglected tropical diseases: the ancient afflictions of stigma and poverty and the prospects for their control and elimination. Adv. Exp. Med. Biol. 582, 23–33.

Ismail, M, Abd-Elsalam, M.A. 1998. Pharmacokinetics of ^{125}I-labelled IgG, $F(ab')_2$ and Fab fractions of scorpion and snake antivenins: merits and potential for therapeutic use. Toxicon 36, 1523–1528.

Jia, Y., Cantu, B., Sánchez, E.E., Pérez, J.C., 2008. Complementary DNA sequencing and identification of mRNAs from the venomous gland of *Agkistrodon piscivorus leucostoma*. Toxicon 51, 1457–1566.

Kashima, S., Roberto, P.G. Soares, A.M., Astolfi-Filho, S., Pereira, J.O., Giuliati, S., Faria, M., Jr., Xavier, M.A.S., Fontes, M.R.M., Giglio, J.R., Franca, S.C., 2004. Analysis of *Bothrops jararacussu* venomous gland transcriptome focusing on structural and functional aspects: I-gene expression profile of highly expressed phospholipases A_2. Biochimie 86, 211–219.

Kasturiratne, A., Wickremasinghe, A.R., de Silva, N., Gunawardena, N.K., Pathmeswaran, A., Premaratna, R., Savioli, L., Lalloo, D.G., Silva, H.J., 2008. The global burden of snakebite: a literature analysis and modeling based on regional estimates of envenoming and deaths. PLoS Med. 5, e218.

Kitchens, C.S., Eskin, T.A., 2008. Fatality in a case of envenomation by *Crotalus adamanteus* initially successfully treated with polyvalent ovine antivenom followed by recurrence of defibrinogenation syndrome. J. Med. Toxicol. 4, 180–183.

Kiziel, W., Hermodson, M.A., Davie, E.W., 1976. Factor X-activating enzyme from Russell's viper venom: isolation and characterization. Biochemistry 15, 4901–4906.

Komori, Y., Nikai, T., Ohara, A., Yagihashi, S., Sugihara, H. 1993. Effect of bilineobin, a thrombin-like proteinase from the venom of common cantil (*Agkistrodon bilineatus*). Toxicon 31, 257–270.

Kornacker, P.M., 1999. Checklist and Key to the Snakes of Venezuela. Lista Sistemática y Clave Para Las Serpientes de Venezuela. Pako-Verlag, Rheinbach, Germany.

Lomonte, B., Tarkowski, A., Hanson, H.A., 1993. Host response to *Bothrops asper* snake venom: analysis of edema formation, inflammatory cells, and cytokine release in mouse model. Inflammation 17, 93–105.

Marsh, N., Williams, V., 2005. Practical applications of snake venom toxins in haemostasis. Toxicon 45, 1171–1181.

Markland, F., 1998. Snake venoms and the hemostatic system. Toxicon 36, 1749–1800.

Minton, S.A., Weinstein, S.A., 1986. Geographic and ontogenic variation in venom of the western diamondback rattlesnake (*Crotalus atrox*). Toxicon 24, 71–80.

Nishida, S., Fujita, T., Kohno, N., Atoda, H., Morita, T., Takeya, H., Kido, I., Paine, M.J., Kawabata, S., Iwanaga, S., 1995. cDNA cloning and deduced amino acid sequence of prothrombin activator (ecarin) from Kenyan *Echis carinatus* venom. Biochemistry 34, 1771–1778.

Otero, R. Gutiérrez, J., Mesa, M.B., Duque, E., Rodríguez, O., Luis Arango, J., Gomez, F., Toro, A., Cano, F., Maria Rodríguez, L, Caro, E., Martínez, J., Cornejo, W. Mariano, Gómez, L., Luis Uribe, F., Cárdenas, S. Nuñez, V., Díaz, A., 2002. Complications of *Bothrops*, *Porthidium*, and *Bothriechis* snakebites in Colombia. A clinical and epidemiological study of 39 cases attended in a university hospital. Toxicon 40, 1107–1114.

Rengifo, C., Rodríguez-Acosta, A., 2005. Serpientes, Venenos y su Tratamiento en Venezuela. Fondo de Publicaciones de la Facultad de Medicina de la Universidad Central de Venezuela. Caracas.

Ribeiro, L.A., Jorge, M.T., 1990. Epidemiology and clinical picture of accidents by adult and young snakes *Bothrops jararaca*. Rev. Inst. Med. Trop. Sao Paulo 32, 436–442.

Riviere, G., Choumet, V., Audebert, F., Sabouraud, A., Debray, M., Scherrmann, J-M., Bon, C., 1997. Effect of antivenom on venom pharmacokinetics in experimentally envenomed rabbits: toward and optimization of antivenom therapy. J. Pharmacol. Exp. Ther. 281, 1–8.

Rodríguez-Acosta, A., Aguilar, I., Giron, M.E., Rodríguez-Pulido, V., 1998. Hemorrhagic activity of neotropical rattlesnake (*Crotalus vegrandis* Klauber 1941) venom. Natural Toxins 6, 15–18.

Rojnuckarin, P., Intragumtornchai, T., Sattapiboon, R. Muanpasitporn, C., Pakmanee, N. Khow, O., Swasdikul, D., 1999. The effects of green pit viper (*Trimeresurus albolabris* and *Trimeresurus macrops*) venom on the fibrinolytic system in human. Toxicon 37, 743–755.

Rojnuckarin, P., Maunpasitporn, C., Chanhome, L., Arpijuntarangkoon, J., Intragumtornchai, T., 2006. Molecular cloning of novel serine proteases and phospholipases A_2 from green pit viper (*Trimeresurus albolabris*) venom gland cDNA library. Toxicon 47, 279–287.

Rojnuckarin, P. 2008. Snake venom and haemostasis-an overview. Thrombosis and Haemostasis. Touch Briefings. www.touchbriefings.com.

Salazar, A.M., Aguilar, I., Guerrero, B., Girón, M.E., Lucena, S., Sánchez, E.E., Rodríguez-Acosta, A., 2008. Intraspecies differences in hemostatic venom activities of the South American rattlesnakes, *Crotalus durissus cumanensis*, as revealed by a range of protease inhibitors. Blood. Coagul. Fibrinolysis 19, 525–30.

Salazar, A.M., Guerrero, B., Cantu, B., Cantu, E., Rodríguez-Acosta, A., Pérez, J.C., Galán, J.A., Tao, A., Sánchez, E.E., 2009. Venom variation in hemostasis of the southern Pacific rattlesnake (*Crotalus oreganus helleri*): isolation of hellerase. Comp. Biochem. Physiol. C. Toxicol. Pharmacol. 149, 307–316.

Sánchez, E.E., Galán, J.A., Perez, J.C., Rodríguez-Acosta, A., Chase, P.B., Pérez, J.C., 2003. The efficacy of two antivenoms against the venom of North American snakes. Toxicon 41, 357–365.

Schöni, R. 2005. The use of snake venom-derived compounds for new functional diagnostic test kits in the field of haemostasis. Pathophysiol. Haemost. Thromb. 34, 234–240.

Sells, P.G., 2003. Animal experimentation in snake venom research and in vitro alternatives. Toxicon 42, 115–133.

Serrano, S.M.T., Mentele, R., Sampaio, C.A.M., Fink, E., 1995. Purification, characterization and amino acid sequence of a serine proteinase, PA-BJ, with platelet aggregating activity from the venom of *Bothrops jararaca*. Biochemistry 34, 7186–7193.

Teng, C.M., Huang, T.F., 1991. Inventory of exogenous inhibitors of platelet aggregation. For the subcommittee on nomenclature of exogenous hemostatic factors of the scientific and standardization committee of the international society on thrombosis and haemostasis. Thromb. Haemost. 65, 624–626.

Tokunaga, F., Nagasawa, K., Tamura, S., Miyata, T., Iwanaga, S., Kisiel, W., 1988. The factor V-activating enzyme (RVV-V) from Russell's viper venom. Identification of isoproteins RVV-V alpha, -V beta, and -V gamma and their complete amino acid sequences. J. Biol. Chem. 263, 17471–17481.

World Health Organization, 1981. Progress in the characterization of venoms and standardization of antivenoms. WHO Offset Publication 58, Geneva.

Chapter 41
Biological Activities of Snake Venom Metalloproteinases on Platelets, Neutrophils, Endothelial Cells, and Extracellular Matrices

Chun-Chieh Hsu and Tur-Fu Huang

Abstract Snake venom metalloproteinases (SVMPs) may be classified into P-I~P-IV SVMP. P-I and P-III groups are abundant in viper venoms, and preserve proteolytic activity. These zinc-dependent SVMPs have profound effects on cellular receptors, plasma proteins and extracellular matrices, and thus affecting haemostasis. In this review, we focus on their interaction with platelet glycoprotein (GP) Ib, GP VI, integrin $\alpha 2\beta 1$, neutrophil PSGL-1, endothelial adherens junction, plasma vWF, fibrinogen and other extracellular matrix, e.g. collagen, in causing antiplatelet, antiinflammation, and endothelial apoptosis. In addition, the in vivo antithrombotic and hemorrhagic activities of these SVMPs are also explored. Through these structure-activity relationship studies using SVMPs as tools for elucidating the ligand-receptor interaction, we may devise useful antidotes for thrombosis, inflammation and human victims of snake envenoming.

Introduction

Some snakes store different protein mixture of toxins in the venom gland for killing or weakening the prey efficiently after snakebites. Studies have shown that venoms from these snake families are myotoxic, hemorrhagic and/or neurotoxic (Huang, 1998). After purification and characterization, some venom components are rather specific to their substrates. Based on the molecular structure-relationship of these unique molecules, we can make use of them as tools for academic research, and developing new classes of therapeutics. The cardiovascular system is a delicately balanced system, which is regulated by the integration of the signals and activities from vascular wall, circulating blood cells and plasma proteins. The unbalanced

T.-F. Huang (✉)
Graduate Institute of Pharmacology, College of Medicine, National Taiwan University, Taipei, Taiwan
e-mail: turfu@ntu.edu.tw; turfu@ccms.ntu.edu.tw

R.M. Kini et al. (eds.), *Toxins and Hemostasis*, DOI 10.1007/978-90-481-9295-3_41,

modulations can lead to the disruption of hemostasis. The most well known examples are the bleeding diathesis of Haemophilia A and Bernard-Soulier syndrome patients, who lack coagulation factor VIII (Oldenburg and El-Maarri, 2006) and platelet glycoprotein Ib (GPIb) (Nurden and Nurden, 2008), respectively. Similarly, the hemorrhagic venoms from the Viperidae, Crotalidae and Elapidae families have diverse toxic effects on the cardiovascular system by affecting blood cells, plasma proteins and vessel wall components. These venoms contain different toxins belonging to different protein families, which can trigger biological effect by inhibiting or activating the functions of their targets. Among them, snake C-type lectin, disintegrin/disintegrin-like protein and snake venom metalloproteinase (SVMP) are the most extensively investigated components in affecting haemostasis.

The Structural and Functional Relationship Between Metalloproteinase/Disintegrin and ADAM

Metalloproteinases and disintegrins are important components of most viperid and crotalid venoms (Huang, 1998). Based on the molecular structure, SVMP family is classified into four categories, P-I through P-IV (Fig. 41.1). P-I and P-III groups are abundant in most viper venoms and preserve proteolytic activity. These zinc-dependent SVMPs have profound effects on the interaction of cellular receptors and extracellular matrix and thus affect hemostasis (Fox and Serrano, 2008). P-III metalloproteinases are composed of an N-terminal metalloproteinase domain, a disintegrin-like domain and a cysteine-rich C-terminus (Fig. 41.1). In contrast, disintegrins are small non-enzymatic R/KGD-containing cysteine-rich polypeptides (van Goor et al., 2009). Despite these differences, both components are able to recognize cell surface receptors and thereby to affect cellular response.

Recently, several membrane-anchored adhesion molecules, closely related to soluble venom metalloproteinase or disintegrin, have been described in mammalian cells (Murphy, 2008). This group of membrane-anchored metalloproteinase/disintegrin is ADAM family (A Disintegrin And Metalloproteinase domain), which are involved in recognizing and shedding of many molecules from the cell surface, behaving like venom metalloproteinase and disintegrin (Fig. 41.1).

ADAMs and ADAMTS (ADAM with thrombospondin motifs) play crucial roles in biological processes such as cellular adhesion, cell fusion, shedding of plasma membrane-associated proteins and intracellular signaling (van Goor et al., 2009). Moreover, ADAMs play crucial roles in tumor biology by modulating cell-cell signalling, cellular migration and angiogenesis. The expression of ADAM9, ADAM11, ADAM12, ADAM15, ADAM17, ADAM19 and ADAM28 were found to increase in many human cancers (van Goor et al., 2009). Most widely expressed and best described is ADAM17, the principal sheddase of EGFR ligands, which mediated EGFR transactivation in angiogenesis in colon carcinoma (Blanchot-Jossic et al., 2005). ADAMs also have important roles in chronic CNS disorders and chronic

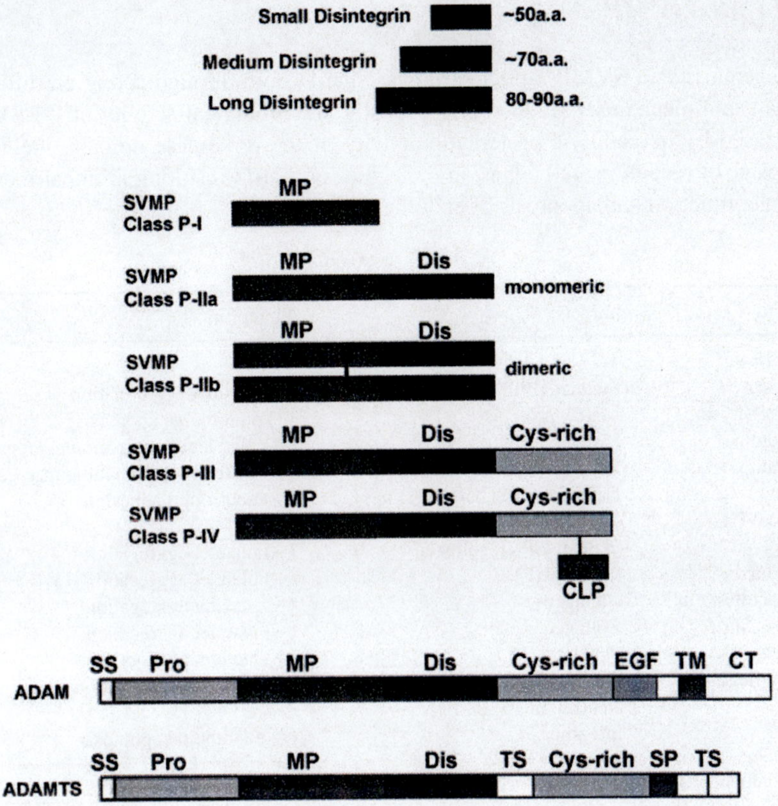

Fig. 41.1 Domain structures of snake venom disintegrins, snake venom metalloproteinases (SVMPs), a disintegrin and metalloproteinase (ADAM) and ADAM with thrombospondin motifs (ADAMTS). The different domain structures of these different proteins are aligned and compared. Snake venom disintegrins fall into one of three size classes: small, medium, or long. SVMPs have four distinct classes (P-I, P-II, P-III, and P-IV) with different domain structures as illustrated: signal sequence (SS), prodomain (Pro), metalloprotease domain (MP), disintegrin domain (Dis), cysteine-rich domain (Cys-rich), and C-type lectin protein (CLP). (Note: For all of these SVMPs, the signal sequence and the prodomain are cleaved from the mature form of the protein.) ADAMs have a domain structure similar to that of class P-III SVMPs, with the addition of an EGF-like repeat (EGF), a transmembrane domain (TM), and a cytoplasmic domain (CT). ADAMTS contain thrombospondin motifs (TS) and spacer domain (SP), but lack EGF-like, transmembrane and cytoplasmic domains

inflammatory diseases (van Goor et al., 2009). However, the exact role of these molecules in the initiation or progression of the disease is generally still poorly understood (Rocks et al., 2008). Based on the structural similarity between ADAM and metalloproteinase/disintegrin families, these snake venom components are useful tools for investigating how the endogenous ADAM regulates the physiologic and pathological functions enabling us to design new strategy for the related pathologic therapy.

P-I Class SVMP

The hemorrhagic SVMPs affect cardiovascular system through acting on different targets and mechanisms (Table 41.1). Mostly, the biological activity of P-I SVMP is caused by its catalytic proteolytic activity of the proteinase domain, including cleavage of receptors, cell adhesion molecules or matrix resulting in impairment of platelet function and apoptosis of endothelial cells.

Table 41.1 Some biologically-active SVMPs, their targets and functions

Protein	Target	Function
SVMP-I		
Crotalin	Platelet GPIb, vWF, fibrinogen	(−) platelet agglutination
Graminelysin	Matrix proteins	(+) endothelial apoptosis
Kistomin	Platelet GPIb, GPVI	(−) platelet agglutination/aggregation
Triflamp	Platelet GPIb	(−) platelet agglutination/aggregation
	Neutrophil PSGL-1	(−) neutrophil adhesion
SVMP-III		
AAV1	Platelet GPVI, collagen	(−) platelet aggregation
Acurhagin	Platelet GPVI	(−) platelet aggregation
Catrocollastatin	Collagen	(−) platelet aggregation
Crovidisin	Collagen	(−) platelet aggregation
Jarahagin	Platelet $\alpha 2\beta 1$	(−) platelet aggregation
Mocarhagin	Platelet GPIb	(−) platelet agglutination
	Neutrophil PSGL-1	
VAP1	Integrins	(+) endothelial apoptosis

(+): induction; (−) inhibition

Antiplatelet Activity

For example, Crotalin, a P-I class SVMP with only a proteinase domain, from venom of *Crotalus atrox*, possesses GP Ib cleaving, vWF binding and cleaving activities in causing antithrombotic activity (Wu et al., 2001a). Kistomin, from *Calloselasma rhodostoma* venom, has been shown to degrade fibrinogen and inhibits ristocetin-induced platelet agglutination, suggesting that it is a GPIb-cleaving protease (Huang et al., 1993). Based on the haemostatic function of vWF/GPIb and collagen/GPVI on platelet and the structural similarity of GPIb/GPVI, we further investigated the cleaving effect of kistomin on platelet GPVI and defined the cleavage sites (Fig. 41.2). We demonstrated that kistomin is capable of binding to platelet GPIbα and cleaves GPIbα and vWF, exhibiting potent antiplatelet and antithrombotic activities both in vitro and in vivo (Hsu et al., 2007). Furthermore, we provided the first evidence demonstrating that kistomin binds and cleaves GPVI, and inhibits collagen-induced platelet aggregation and adhesion at a higher concentration, suggesting that kistomin disrupts these two crucial platelet adhesion receptors, namely GPIb and GPVI, leading to impairment of platelet function (Hsu et al., 2008).

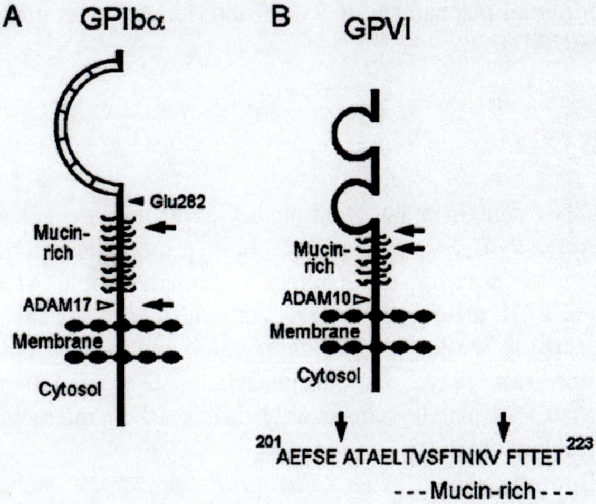

Fig. 41.2 The proposed cleavage sites of kistomin on the glycoprotein VI and Ibα. (**a**) The residues from 1 to 282 of GPIbα are the major ligand-binding globular region, and the C-terminal of this region is followed by mucin-like domain. One possible cleavage site of kistomin on GPIbα is located on this mucin-like domain (*arrow*), and the other site is near the outer membrane of platelet. Endogenous ADAM17 induces ectodomain shedding at Gly464/Val465 (*blank-arrowhead*) on the C-terminal of the mucin-like region. (**b**) Kistomin cleaves GPVI at Glu205/Ala206 and Val218/Phe219 near/on the mucin-like region, however, endogenous ADAM10 induces ectodomain shedding at Arg242/Gln243 (*blank-arrowhead*) on the C-terminal of the mucin-like region. Kistomin cleavage sites are indicated by *arrows* (Hsu et al., 2008)

Effect on Neutrophil-Platelet Interaction

Besides affecting thrombotic system, P-I class SVMP also affects inflammatory system. Triflamp, from venom of *Trimeresurus flavoviridis*, blocks platelet function through GPIb-cleavage (Tseng et al., 2004a), also disrupts P-selectin-mediated adhesion by cleaving PSGL-1 from the neutrophil surface to inhibit heterotypic adhesion between platelets and neutrophils, thus limiting the formation of neutrophil-platelet complexes (Tseng et al., 2004b).

Induction of Apoptosis of Vascular Endothelial Cells

Several SVMPs have been reported to induce apoptosis of human umbilical vein endothelial cells (HUVECs). Graminelysin, purified from venoms of *Trimeresurus gramineus*, inhibited the adhesion of HUVECs to immobilized fibrinogen, but not the adhesion to immobilized collagen, resulting in apoptosis only in the former condition (Wu et al., 2001b). We further showed that graminelysin induced endothelial cells to produce matrix metalloproteinase (MMP)-2, which acts in concert with graminelysin to degrade ECM or to shed VE-cadherin. During apoptosis, adherent junctions, including VE-cadherin, and β- and γ-catenin were cleaved. Graminelysin-induced cleavage in adherent junctions was paralleled with the increased paracellular permeability. The detachment of endothelial cells and the resulting apoptosis could be an additional mechanism of SVMP interference with normal hemostasis (Wu and Huang, 2003).

P-III Class SVMP

P-III class SVMPs contain a metalloprotease, a disintegrin and a cysteine-rich domain. Therefore, P-III SVMP possesses both proteolytic activity of protease domain and binding activity of disintegrin domain (Table 41.1). Acurhagin and AAV1, two P-III metalloproteinase, both purified from *Agkistrodon acutus* venom, selectively inhibit platelet aggregation induced by collagen or convulxin, and suppression of tyrosine phosphorylation of several signaling proteins in convulxin-stimulated platelets are mainly through their interaction with GPVI (Wang, 2007; Wang et al., 2005).

Crovidisin from the venom of *Crotalus viridis* blocks the interaction between platelets and collagen fibers through its binding to collagen fibers, resulting in the blockade of collagen-mediated platelet functions such as adhesion, release reaction, thromboxane formation, and aggregation (Liu and Huang, 1997). A similar P-III SVMP, catrocollastatin, also was reported to block collagen-induced platelet aggregation by binding to collagen via its disintegrin-like domain (Zhou et al., 1996). The other P-III SVMP, jarahagin, from *Bothrops jararaca* viper venom, also inhibits the interaction between collagen and platelets. However, its inhibitory effect was mediated through the binding to platelet integrin α2-subunit via the disintegrin-like domain, followed by proteolysis of the β1-subunit via protease domain (Kamiguti et al., 1996, 1997, 2003). Thus, these P-III SVMPs interrupt cell adhesion between collagen and platelets through three different mechanisms, namely AAV1 and acurhagin interacting with GPVI, crovidisin and catrocollastatin binding to collagen and jarahagin binding to α2- and cleaving β1-subunit.

Mocarhagin, from *Naja mocambique mocambique* venom, is the most intensively investigated P-III SVMP, which cleaves platelet GPIb to generate the fragment His-1-Glu-282 and inhibits the function of GPIb (Ward et al., 1996). Mocarhagin also cleaves a 10-amino acid peptide from the mature N terminus of P-selectin glycoprotein ligand receptor, PSGL-1, and abolished platelet binding to purified P-selectin (De Luca et al., 1995). Mocarhagin affects both thrombotic and inflammatory processes through its protease activity on GPIb and PSGL-1.

VAP1, from *Crotalus atrox*, induces apoptosis in vascular endothelial cells without degrading extracellular matrix or inhibiting adhesion of endothelial cells. However, integrins are involved in VAP1-induced apoptosis by some specific role (Araki et al., 2002; Masuda et al., 1998, 2000).

Functional Roles of SVMPs and ADAM/ADAMTS

Mammalian ADAMs are involved in cell-cell fusion, adhesion and intracellular signaling to regulate biological processes, including fertilization, angiogenesis, and pathological processes, including cancer, inflammation, neurodegeneration and fibrosis (van Goor et al., 2009). Recently, another group of proteins with ADAM domain were investigated, which contain several thrombospondin-like repeats in its C-terminal region and are named ADAMTS. ADAMTS also engage in widely divergent genetic, developmental and pathological aspects, although the functional studies about each domain of ADAMs/ADAMTS are far from obtaining a full understanding (van Goor et al., 2009). However, the structure and function of SVMPs is related to those of ADAMs, in that both possess adhesion (disintegrin-like) and protease (metalloproteinase) domains. P-I class SVMPs, which only contain the metalloprotease domain, could be a tool to investigate the protease domain of ADAMs. P-III class SVMPs are more like ADAMs/ADAMTS structurally in containing a disintegrin, a metalloproteinase and a cysteine-rich domain as ADAMs/ADAMTS (Fig. 41.1). The structure-activity of each domain in SVMP is complex and needs further exploration. Generally, the importance of the catalytic domain in many activities of SVMP such as cleavage of fibrinogen, vWF and fibronectin, and cleavage of GPIb or GPVI are accepted because the use of zinc-chelating agents abolishes both the catalytic and the above-mentioned biological activities including fibrin(ogeno)lysis, platelet-inhibitory and apoptosis of endothelial cells (Hsu et al., 2007, 2008; Tseng et al., 2004a, b; Wu and Huang 2003; Wu et al., 2001a, b). However, the role of the disintegrin-like and cysteine-rich domains in the overall activity of P-III SVMP is still not clear. For example, peptides derived from all three domains of jararhagin can inhibit α2β1 function; therefore, P-III SVMPs might contain several integrin recognition sites (Kamiguti et al., 1996, 1997, 2003).

The binding and cleaving activities of P-III class SVMPs might be good tool in mimicking potential functions of ADAMs and studying the physiological/pathological effects of ADAMs in vitro or in vivo. As some reports indicated, the development of therapeutics targeting ADAMs is a novel strategy in the treatment of many disease states, including inflammation, degeneneration and neoplastic disease (van Goor et al., 2009). SVMPs could be useful for the elucidation of these disorders related to ADAMs/ADAMTS and for drug development.

Concluding Remarks

These SVMPs are versatile components, targeting so many important elements involved in haemostasis, such as basement membrane proteins, blood clotting factors, and various adhesion receptors expressed on platelets, neutrophils and endothelial cells. Their biological activities are closely associated with hemorrhage caused after snakebite in envenomed patients. However, whether these in vitro biological activities would be expressed in vivo system would rely on many factors as these SVMPs were intravenously administered. Human α2-Macroglobulin (α2M) is found in serum and various secretion at high concentration (1.6–4 mg/ml) and is a broad-spectrum proteinase inhibitor and capable of inhibiting the activity of many proteinases including aspartic, metallo, serine and cysteine proteinases (Sottrup-Jensen, 1989). α2M has been shown to neutralize several SVMPs leading to inhibition of fibrinogenolysis and prevention of systemic hemorrhage (Kawano et al., 2002; Saidi et al., 1999; Souza et al., 2001). Triflamp, originally was found to inhibit neutrophil-platelet adhesion via cleavage of platelet GPIb and neutrophil PSGL-1 as tested in vitro (Tseng et al., 2004a, b). However, triflamp failed to affect PSGL-1 on neutrophil and GPIb on platelets in human non-diluted whole blood preparation, because human α2M is primarily responsible for neutralization of its proteolytic activity. In addition, the presence of serum leads mouse PSGL-1 and GPIb to resist being truncated by triflamp (Tseng et al., 2004c). However, there are some exceptions, such as kistomin, crotalin (P-I SVMP), and jararhagin (P-III SVMP); these SVMPs still expressed their cleaving activity on platelet membrane receptors or induction of fibrinolysis, respectively (Chang et al., 1998; Hsu et al., 2007; Kamiguti et al., 1994; Laing and Moura-da-Silva, 2005). Thus, the individual SVMP may exhibit different affinities for α2M escaping from its neutralizing effect. We can only verify this in vivo system, reflecting different animal susceptibility to snakebite bleeding disorder.

In conclusion, we may use these SVMPs as tools for studying these cellular targets such as GPIb, GPVI or α2β1 integrin, and also the structure-activity relationships of SVMPs and ADAMs. Finally, through these studies on the molecular interaction between ligand-receptor, we may devise some clinical useful antidotes for thrombosis, inflammation and even human victims of snake envenoming.

References

Araki, S., Masuda, S., Maeda, H., Ying, M.J., Hayashi, H., 2002. Involvement of specific integrins in apoptosis induced by vascular apoptosis-inducing protein 1. Toxicon 40, 535–542.

Blanchot-Jossic, F., Jarry, A., Masson, D., Bach-Ngohou, K., Paineau, J., Denis, M.G., Laboisse, C.L., Mosnier, J.F., 2005. Up-regulated expression of ADAM17 in human colon carcinoma: co-expression with EGFR in neoplastic and endothelial cells. J. Pathol. 207, 156–163.

Chang, M.C., Lin, H.K., Peng, H.C., Huang, T.F., 1998. Antithrombotic effect of crotalin, a platelet membrane glycoprotein Ib antagonist from venom of *Crotalus atrox*. Blood 91, 1582–1589.

De Luca, M., Dunlop, L.C., Andrews, R.K., Flannery, J.V., Jr., Ettling, R., Cumming, D.A., Veldman, G.M., Berndt, M.C., 1995. A novel cobra venom metalloproteinase, mocarhagin, cleaves a 10-amino acid peptide from the mature N terminus of P-selectin glycoprotein ligand receptor, PSGL-1, and abolishes P-selectin binding. J. Biol. Chem. 270, 26734–26737.

Fox, J.W., Serrano, S.M., 2008. Insights into and speculations about snake venom metalloproteinase (SVMP) synthesis, folding and disulfide bond formation and their contribution to venom complexity. FEBS J. 275, 3016–3030.

Hsu, C.C., Wu, W.B., Chang, Y.H., Kuo, H.L., Huang, T.F., 2007. Antithrombotic effect of a protein-type I class snake venom metalloproteinase, kistomin, is mediated by affecting glycoprotein Ib-von Willebrand factor interaction. Mol. Pharmacol. 72, 984–992.

Hsu, C.C., Wu, W.B., Huang, T.F., 2008. A snake venom metalloproteinase, kistomin, cleaves platelet glycoprotein VI and impairs platelet functions. J. Thromb. Haemost. 6, 1578–1585.

Huang, T.F., Chang, M.C., Teng, C.M., 1993. Antiplatelet protease, kistomin, selectively cleaves human platelet glycoprotein Ib. Biochim. Biophys. Acta 1158, 293–299.

Huang, T.F., 1998. What have snakes taught us about integrins? Cell. Mol. Life Sci. 54, 527–540.

Kamiguti, A.S., Desmond, H.P., Theakston, R.D., Hay, C.R., Zuzel, M., 1994. Ineffectiveness of the inhibition of the main haemorrhagic metalloproteinase from Bothrops jararaca venom by its only plasma inhibitor, alpha 2-macroglobulin. Biochim. Biophys. Acta 1200, 307–314.

Kamiguti, A.S., Hay, C.R., Zuzel, M., 1996. Inhibition of collagen-induced platelet aggregation as the result of cleavage of α2β1-integrin by the snake venom metalloproteinase jararhagin. Biochem. J. 320, 635–641.

Kamiguti, A.S., Markland, F.S., Zhou, Q., Laing, G.D., Theakston, R.D., Zuzel, M., 1997. Proteolytic cleavage of the beta1 subunit of platelet alpha2beta1 integrin by the metalloproteinase jararhagin compromises collagen-stimulated phosphorylation of pp72. J. Biol. Chem. 272, 32599–325605.

Kamiguti, A.S., Gallagher, P., Marcinkiewicz, C., Theakston, R.D., Zuzel, M., Fox, J.W., 2003. Identification of sites in the cysteine-rich domain of the class P-III snake venom metalloproteinases responsible for inhibition of platelet function. FEBS Lett. 549, 129–134.

Kawano, J., Anai, K., Sugiki, M., Yoshida, E., Maruyama, M., 2002. Vascular endothelial cell injury induced by *Bothrops jararaca* venom; non-significance of hemorrhagic metalloproteinase. Toxicon 40, 1553–1562.

Laing, G.D., Moura-da-Silva, A.M., 2005. Jararhagin and its multiple effects on hemostasis. Toxicon 45, 987–996.

Liu, C.Z., Huang, T.F., 1997. Crovidisin, a collagen-binding protein isolated from snake venom of *Crotalus viridis*, prevents platelet-collagen interaction. Arch. Biochem. Biophys. 337, 291–299.

Masuda, S., Hayashi, H., Araki, S., 1998. Two vascular apoptosis-inducing proteins from snake venom are members of the metalloprotease/disintegrin family. Eur. J. Biochem. 253, 36–41.

Masuda, S., Ohta, T., Kaji, K., Fox, J.W., Hayashi, H., Araki, S., 2000. cDNA cloning and characterization of vascular apoptosis-inducing protein 1. Biochem. Biophys. Res. Commun. 278, 197–204.

Murphy, G., 2008. The ADAMs: signalling scissors in the tumour microenvironment. Nat. Rev. Cancer. 8, 929–941.

Nurden, P., Nurden, A.T., 2008. Congenital disorders associated with platelet dysfunctions. Thromb. Haemost. 99, 253–263.

Oldenburg, J., El-Maarri, O., 2006. New insight into the molecular basis of hemophilia A. Int. J. Hematol. 83, 96–102.

Rocks, N., Paulissen, G., El Hour, M., Quesada, F., Crahay, C., Gueders, M., Foidart, J.M., Noel, A., Cataldo, D., 2008. Emerging roles of ADAM and ADAMTS metalloproteinases in cancer. Biochimie 90, 369–379.

Saidi, N., Samel, M., Siigur, J., Jensen, P.E., 1999. Lebetase, an alpha(beta)-fibrin(ogen)olytic metalloproteinase of *Vipera lebetina* snake venom, is inhibited by human alpha-macroglobulins. Biochim. Biophys. Acta 1434, 94–102.

Sottrup-Jensen, L., 1989. Alpha-macroglobulins: structure, shape, and mechanism of proteinase complex formation. J. Biol. Chem. 264, 11539–11542.

Souza, C.T., Moura, M.B., Magalhaes, A., Heneine, L.G., Olortegui, C.C., Diniz, C.R., Sanchez, E.F., 2001. Inhibition of mutalysin II, a metalloproteinase from bushmaster snake venom by human alpha2-macroglobulin and rabbit immunoglobulin. Comp. Biochem. Physiol. B. Biochem. Mol. Biol. 130, 155–168.

Tseng, Y.L., Lee, C.J., Hsu, C.C., Huang, T.F. 2004a. Triflamp, a snake venom metalloproteinase, reduces neutrophil-platelet adhesion through proteolysis of PSGL-1 but not glycoprotein Ib alpha. Thromb. Haemost. 91, 1177–1185.

Tseng, Y.L., Lee, C.J., Huang, T.F., 2004b. Effects of a snake venom metalloproteinase, triflamp, on platelet aggregation, platelet-neutrophil and neutrophil-neutrophil interactions: involvement of platelet GPIbalpha and neutrophil PSGL-1. Thromb. Haemost. 91, 315–324.

Tseng, Y.L., Wu, W.B., Hsu, C.C., Peng, H.C., Huang, T.F., 2004c. Inhibitory effects of human alpha2-macroglobulin and mouse serum on the PSGL-1 and glycoprotein Ib proteolysis by a snake venom metalloproteinase, triflamp. Toxicon 43, 769–777.

van Goor, H., Melenhorst, W.B., Turner, A.J., Holgate, S.T., 2009. Adamalysins in biology and disease. J. Pathol. 219, 277–286.

Wang, W.J., Shih, C.H., Huang, T.F., 2005. Primary structure and antiplatelet mechanism of a snake venom metalloproteinase, acurhagin, from *Agkistrodon acutus* venom. Biochimie 87, 1065–1077.

Wang, W.J., 2007. Purification and functional characterization of AAV1, a novel P-III metalloproteinase, from Formosan *Agkistrodon acutus* venom. Biochimie 89, 105–115.

Ward, C.M., Andrews, R.K., Smith, A.I., Berndt, M.C., 1996. Mocarhagin, a novel cobra venom metalloproteinase, cleaves the platelet von Willebrand factor receptor glycoprotein Ibalpha. Identification of the sulfated tyrosine/anionic sequence Tyr-276-Glu-282 of glycoprotein Ibalpha as a binding site for von Willebrand factor and alpha-thrombin. Biochemistry 35, 4929–4938.

Wu, W.B., Peng, H.C., Huang, T.F., 2001a. Crotalin, a vWF and GP Ib cleaving metalloproteinase from venom of *Crotalus atrox*. Thromb. Haemost. 86, 1501–1511.

Wu, W.B., Chang, S.C., Liau, M.Y., Huang, T.F., 2001b. Purification, molecular cloning and mechanism of action of graminelysin I, a snake-venom-derived metalloproteinase that induces apoptosis of human endothelial cells. Biochem. J. 357, 719–728.

Wu, W.B., Huang, T.F., 2003. Activation of MMP-2, cleavage of matrix proteins, and adherens junctions during a snake venom metalloproteinase-induced endothelial cell apoptosis. Exp. Cell. Res. 288, 143–157.

Zhou, Q., Dangelmaier, C., Smith, J.B., 1996. The hemorrhagin catrocollastatin inhibits collagen-induced platelet aggregation by binding to collagen via its disintegrin-like domain. Biochem. Biophys. Res. Commun. 219, 720–726.

Part X
Applications

Chapter 42
Leeches in Microsurgery – An Evidence-Based Approach

Karsten Knobloch

Abstract Leeches for medical use for blood-letting have a rather long tradition back to 1500 B.C. In microsurgery leeches have been extensively used clinically usually for relief of venous congestion. As such, procedures like scalp replantation, pedicled and free flap surgery, ear replantation as well as finger replantation have been successfully performed with the help of leeches. The chapter reviews the anatomical and physiological basis of medicinal leeches and the clinical applications in an evidence-based medicine approach.

Microsurgery

Microsurgery is a surgical discipline using a microscope and specially designed micro-instruments, such as micro-forceps, micro-scissors and so on. Given the advancement of loupe magnification, the aforementioned term is not necessarily exclusive. From a vessel diameter point of view, anastomosis of vessels with diameters ranging from 0.3 to 3 mm is performed with microsurgical tools. Common microsurgical techniques include: (1) vascular anastomosis with arterial ± venous anastomosis in case of free flap transfer, replantation of digits, scalp, ears; (2) lymphatic anastomosis; (3) infertility repair to reverse vasectomy and fallopian tube surgery; and (4) neural anastomosis for nerve repair in trauma or cancer, nerve grafting or in plexus brachialis repair.

History

The history concerning the medical use of leeches dates back to the ancient times. Notably, "leech" is derived from the Anglo-Saxon word "laece" which, when literally translated, means physician. In 1500 B.C. the Egyptians used leeches for

K. Knobloch (✉)
Plastic, Hand and Reconstructive Surgery, Hannover Medical School, Hannover, Germany
e-mail: knobloch.karsten@mh-hannover.de

R.M. Kini et al. (eds.), *Toxins and Hemostasis*, DOI 10.1007/978-90-481-9295-3_42,

medical indications. In India, leeches were used 3300 years ago. Galen (130–201 B.C.) used leeches for blood-letting in order to correct a humoral dysbalance. The rabbinic responsa literature acknowledges the use of leeching for the prevention and treatment of certain illnesses, and classic Jewish sources including the Bible, Talmud, and Codes of Jewish Law describe leeches and mention their medicinal use. Although the swallowing of a leech is considered dangerous and may lead to abdominal swelling, the Talmud describes an oral concoction containing leeches in wine for patients with enlarged spleens (Rosner, 1999).

Anatomy

To date, there are more than 700 species of leech, all of which are carnivorous and move using suckers. *Hirudo medicinalis* is typically 3–4 cm (Fig. 42.1), but after a sufficient meal might reach up to 12 cm in length. It is composed of 102 segments with 5 annuli in each segment except at each tapered end where there are fewer annuli (Whitaker et al., 2005) (Fig. 42.2). The bite of *Hirudo medicinalis* is created by three jaws, each containing 60–100 pairs of cutting teeth (Fig. 42.3). This leads to a characteristic Y-shaped triradiate conformation 1 mm in diameter and up to 1.5 mm in depth (Sawyer, 1986).

Physiology

The saliva of the leech contains various compounds which might have an effect on the recipient Table 42.1. environment, including hirudin, hyaluronidase and histamine (Table 42.1). Hirudin is one of the most important saliva ingredients. The British physiologist John Berry Haycraft discovered in 1884 that leeches secret a

Fig. 42.1 *Hirudo medicinalis*

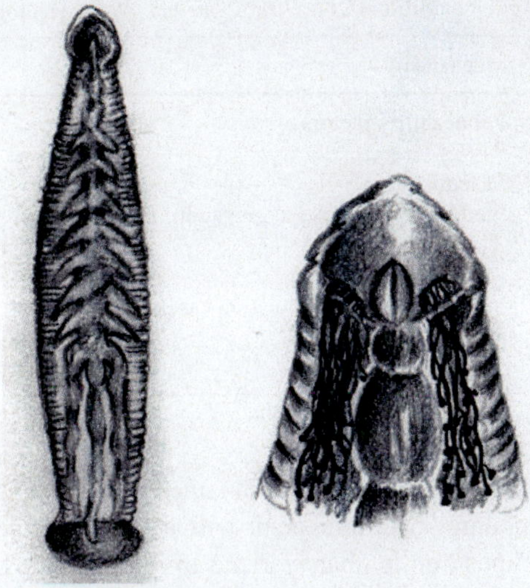

Fig. 42.2 Anatomy of *Hirudo medicinalis*

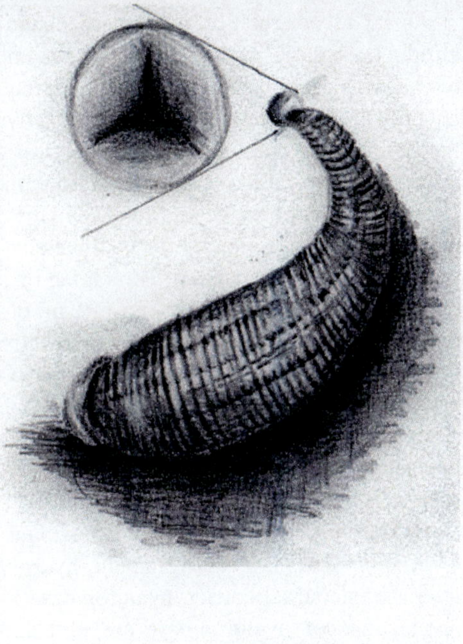

Fig. 42.3 Cutting teeth of *Hirudo medicinalis*

powerful anticoagulant (Haycraft, 1884). Jakobi named the substance hirudin in 1904. Only after the development of protein chemistry and the elucidation of the biochemistry of blood coagulation was the anticoagulant agent, hirudin, isolated and its chemical nature and mode of action clarified. In 1955 Fritz Markwardt was able to extract hirudin from the head of leeches (Markwardt, 1955). Hirudin is extracted from the homogenized heads of medicinal leeches and enriched by precipitation procedures followed by ion-exchange chromatography and gel filtration. The

Table 42.1 *Hirudio medicinalis* physiologically active salivary products

Mechanism	Identified products
Permeability factors	• Collagenase
	• Collagenase
Vasodilatation	• Histamine
Inhibition of platelet aggregation	• Calin
	• Apyrase
	• Saratin
Inhibition of the coagulation cascade	• Hirudin (direct Thrombin inhibitor)
Proteinase inhibitors	• Bdellin
	• Eglin

pure anticoagulant agent obtained is a carboxyhydrate-free single chain miniprotein containing 65 amino acids with a molecular weight of about 7 kDa. The amino acid composition is characterized by a remarkably high content of acidic amino acids at the C-terminus, the absence of arginine, methionine, and tryptophan, and the presence of a sulfated tyrosine residue and three intramolecular disulfide bridges.

As far as coagulation is concerned, the importance of the thrombin inhibitor hirudin becomes especially evident when one takes into account the fact that thrombin has a central position in the coagulation system. The clotting enzyme catalyzes not only the formation of fibrin but also activates clotting factors and blood platelets. Furthermore, thrombin has direct effects on the vascular endothelium and mediates non-hemostatic cellular events. Therefore, the inhibition of thrombin by hirudin represents not only effective interference in the coagulation process, but also modulates the multiple bioregulatory effects of the enzyme. Depending on the hirudin concentration in blood, coagulation is retarded or completely inhibited. Correspondingly, the clotting variables change. From 1 ml of human blood, approximately 100–150 units of thrombin may be generated on activation with thromboplastin. The inhibition of this amount of thrombin requires 100–150 AT-U of hirudin. Therefore, an uncoagulable state in human plasma is reached at concentrations of more than 100 AT-U or 10 pg of hirudid/ml of blood (Markwardt, 1992). Currently, recombinant bivalent direct thrombin inhibitors have been developed and are clinically available, such as Bivalirudin (Angiomax), Lepirudin (Refludan) and Desirudin.

Hyaluronidase facilitates anticoagulant penetration due to degradation of hyaluronic acid. By catalyzing the hydrolysis of hyaluroric acid, which constitutes the interstitial barrier, hyaluronidase increases tissue permeability. Clinically, applications of hyaluronidase are in ophthalmology in combination with local anesthetics. Notably, some bacteria such as *Staphylococcus aureus*, *Streptococcus pyogenes* and *Clostridium perfrigens* produce hyaluronidase to improve mobility. Furthermore, in fertilization hyaluronidase is released by the acrosome of the sperm for conception.

Histamine (Fig. 42.4) maintains vasodilatation acting as a biogenic amine. Histamine is produced by basophile and mast cells in human. Besides vasodilatation, permeability of tissues is increased.

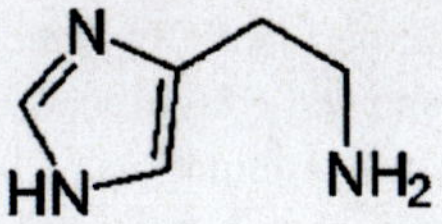

Fig. 42.4 Histamine

Clinical Application in Microsurgery – An Evidence-Based Approach

The medical application of leeches in microsurgery is mainly based on a few cohort studies (evidence level III), and the majority of case series and case reports (Lineaweaver et al., 1991; Wade et al., 1990). To date, no prospective randomized-controlled trials have been published to elucidate the role of leeches for flap survival. Most likely, these data will never be published, since such a trial is rather hard to perform. Thus, we have to accept the lower evidence-based level in this regard today.

Leech therapy has two phases: active bloodletting and passive subsequent bleeding from the wound after detachment of the leech. Notably, the bite of a leech is painless, although the presence of a local anaesthetic in the salivary gland is debated, but not yet discovered. The blood volume varies widely with reports ranging from 1.3 to 50 ml (Conforti et al., 2002). As with active bloodletting, passive bleeding varies greatly. Ninety percent of passive bleeding occurs within 5 h after application with a mean blood loss of 2.5 ml.

The main clinical indication for the use of medical leeches is relief of venous congestion. This is apparent in the transfer of tissues with or without microsurgical anastomosis as in pedicled or free flaps. In addition, replantation of digits, ears or noses are further applications were medical leeches have been applied with reasonable success. Venous congestion might be apparent due to a number of reasons, such as small diameter of the vessels (especially in microsurgery), thrombosis, intimal lesions or a mis-balance between good arterial inflow and limited venous outflow in free flaps. From a surgical point of view, sometimes more than one vein is anastomosed in such instances such as in the case of deep interior epigastric artery perforator flaps for breast reconstruction. Interestingly, the nomenclature of the addition of a vein and/or an additional artery in a microvascular free flap transfer is not incontrovertible. There are two major classifications; the first is "supercharging," and the other "turbocharging" (Semple, 1994). Following the suggestion of Dr. Civelek and co-workers, one should state that any vascular augmentation using a distant source of vessels such as axillary or thoracodorsal vessels should be defined as "supercharging" as opposed to "turbocharging", which should define the vascular augmentation using the vascular sources within the flap territory (Civelek et al., 1998).

Venous thrombosis is more detrimental than arterial occlusion in flap surgery. In experimental models using a porcine latissimus dorsi flap model, venous occlusion led to necrosis of 40% of the latissimus muscle within 3 h. Arterial occlusion, however, did not lead to any muscle necrosis after 3 h (Kerrigan et al., 1994). Venous obstruction leads to microcirculatory thrombosis, trapping of platelets, and stasis.

Experimentally, leech-treated venous-compromised rodent epigastric skin flaps showed a significant increase in flap survival rate (Lee et al., 1992). Interestingly, the combination of hyperbaric oxygenation (HBO) and leeches (67% survival) is superior than leeches (25% survival) or HBO (1%) alone on a venous-congested flap model (Lozano et al., 1999).

Scalp Replantation

The first case report on leeches and microsurgery dates back in 1983, where an avulsion of the scalp treated by microvascular repair in a 28-year-old fitter and turner was successfully replanted with leeches to "decongest the flap when venous drainage appeared inadequate during the first week" (Henderson et al., 1983, evidence level IV, case report). Expanded free scalp flaps were transplanted later between identical twins with postoperative leech therapy (Valauri et al., 1990, evidence level IV, case series). Another case report of a 30-year-old female suffering a scalp avulsion in an industrial accident supports the use of leeches for venous improvement over several days (Rivera et Gross, 1995, evidence level IV, case report).

Pedicled Flap Surgery

Derganc and Zdravid were among the first to report in 1960 the successful use of leeches in pedicled flaps (Derganc et Zdravid, 1960, evidence level III, cohort study). They achieved a 70% salvage rate.

Free Flap Surgery

In free flap microsurgery, Lim reported in 1986 the successful transfer of a free microvascular superficial temporal artery flaps with no obvious venous drainage using leeches for reducing the venous congestion (Lim, 1986, evidence level IV, case report). A cohort study among 74 with extensive skin loss of upper and lower extremity reported 20 patients with venous insufficiency after free tissue transfer within 12 h after transfer (Soucacos et al., 1994a,b, evidence level III, cohort study).

Ear Replantation

Totally amputated ears were successfully microsurgically reattached in a case series with three cases with leeches applied in one case with venous thrombosis (Mutimer et al., 1987, evidence level IV, case series). Another case report from New York encountered venous congestion after total ear replantation, managed by leeches and systemic heparinization (Rapaport et al., 1993, evidence level IV, case report), which was later supported in other case reports (Cho et Ahn, 1999, Funk et al.,

1996, evidence level IV, case report). Using fluorimetry to determine tissue perfusion after ear replantation, leech application restored a normal circulation pattern in a case (Anthony et al., 1989, evidence level IV, case report). An incidence of ear amputation in a child without any venous anastomosis was managed with leeches' therapy only (Concannon and Puckett, 1998, evidence level IV, case report), which was subsequently reported by other groups, as well (Otto et al., 1999; Talbi et al., 2001, evidence level IV, case report). A patient having microvascular ear replantation with arterial anastomosis only and leech therapy survived over 14 days, but ultimately failed due to venous congestion (Akyurek et al., 2001, evidence level IV, case report).

Nasal Tip and Lip Replantation

A case series from Taiwan reported the outcome of seven patients with facial amputations with one scalp, two nasal tips, two ears, one lower lip and one eyebrow amputation (Jeng et al., 1994, evidence level IV, case series). Due to the lack of suitable veins for drainage, leeches were used in addition to arterio-venous fistulas and pin pricks. Four out of seven cases were successful; two had partial loss of the plant and one case failed to due venous congestion. A multicenter retrospective review of a total of 13 lip replantations in 12 institutions found 100% success rate with leeches applied in 11/13 patients (Walton et al., 1998, evidence level III, cohort study). Another case series involving three patients from a single center with lip replantation without any venous anastomosis and leeches over 6 days showed two out of three had full and one partial recovery (Duroure et al., 2004, evidence level IV, case series).

Digit Replantation

Leeches are applied in finger replantation to alleviate venous congestion (Tsai et al., 1989, evidence level IV, case series). Another case report in 1984 states that an upper limb amputee was successfully replanted although it was not possible to restore the venous return. "The venous stasis was successfully overcome by continuous application of leeches during the first 9 postoperative days" (Höltje, 1984, evidence level IV, case report). A larger cohort study in 1991 among 183 injured fingers (80 replantations, 103 revascularizations) stated that "venous drainage should be initially treated medically (2/3 preservation after drainage by leeches), but, when is not effective, surgical revision salvages one third" (Renaud et al., 1991, evidence level III, cohort study). In the same year, another cohort study with 50 patients with finger amputations was published (Baudet, 1991, evidence level III, cohort study). Out of the 50 cases they were able to repair both arteries in 35 patients and only one artery in the other 15. As far as finger veins were concerned, two were anastomosed in 12, only one in 23 and none in 15 because of the severity of the contusion or the very distal site of replantation. Although early arterial thrombosis usually left no solution other than reoperation, venous complications were relieved by ancillary methods such as leeches. They encountered a venous problem in 16 (32%) out of the 50 cases. In severe venous congestion, the authors used leeches over 6 days with four to six leeches per day. They reported that all cases with venous congestion where they applied leeches could be salvaged, and accounted a leech-improved survival rate of 26%. In another cohort study with eleven patients with finger replantation, nine patients were successfully treated with medical leeches for venous congestion (Soucacos et al., 1994a, evidence level III, cohort study). In distal and very distal digital replantations, the use of leeches in the absence of suitable veins for repair does not preclude an acceptable survival rate (68%) in 95 patients (Foucher et Norris, 1992, evidence level III, cohort study).

Nipple Reconstruction

A case series of two females with nipple venous congestion following breast surgery was successfully managed using leeches (Gross et Apesos, 1992, evidence level IV, case series).

Duration of Application

Usually, venous insufficiency is overcome due to venous neovascularization within 5–7 days after the procedure. Thus, the duration of leech application for this instance is likely to be similar. The aforementioned clinical reports usually apply leeches for 5–7 days postoperatively, in limited cases even beyond this time.

Contraindications

In case of arterial malperfusion, leech therapy does not any benefit at all. Furthermore, in heavily congested flaps failing due to surgical errors at the venous anastomosis, leeches are not able to convert a surgical error in a functioning free flap.

Complications

Two major complications are possible during leech therapy: excessive bleeding leading to anemia and infection. A report by White in 1819 highlights the death of a 2-year-old child after a leech bite (Carter, 2001). Massive bleeding occurred in a 65-year-old male with a total of 130 leech bites all over his body (Kose et al., 2008). He had entered a leech pond about 6–8 h prior and remained in the leech pond for about an hour. After getting out of the pond he realized marked hemorrhage on his body, which led to his emergency room appearance. Hemoglobin (Hb) was 9.2 g/dL (RR, 13.6–17.2 g/dL), hematocrit 29%. Despite two units of erythrocyte concentrate, the bleeding did not stop in the 15 h of follow-up. Second day blood count revealed an anemia with Hb 7.3 g/dL, INR 12.46 (RR 0,84–1,2), aPTT 65seconds (RR, 26–37s). A total of eight units of fresh frozen plasma and six units of erythrocyte concentrates were necessary to stabilize this patient. Other reports due to leech bites (Ikizceli et al., 2005), have included gynaecological bleeding (Hernandez and Ramirez Gutierrez, 1998; Saha et al., 2005), rectal bleeding in a young child (Ho and Boyd, 2007) and in adults (Raj et al., 2000).

Another complication during leech therapy is infection with reported incidences varying between 2 and 20%. *Aeromonas hydrophila* lives in the gut of leeches and is essential for leech digestion. However, *A. hydrophila* infection for humans is potentially dangerous. In case reports, *A. hydrophila* infections are published, with *A. septicaemia* reported recently in a patient following replantation of this thumb and leech therapy which was started with ampicillin-sulbactam (Levine et al., 2009, evidence level IV, case report). Sepsis developed with *Aeromonas* blood cultures positive and resistant to ampicillin-sulbactam. The antibiotics were changed to ciprofloxacin and the patient recovered. Another *A. hydrophila* infection was noted following free microvascular osteo-myo-cutaneous flaps from the iliac crest (Kalbermatten et al., 2007). Even delayed *Aeromonas* infections might appear several days after stopping leech therapy in a congested reconstructed breast (Ardehali et al., 2006, evidence level IV, case report).

Notably, elimination of *Aeromonas* from the intestinal tract of the medicinal leech is feasible when fed with ciprofloxacin (100 mg/l, Mumcuoglu et al., 2009). However, to date chinolone antibiotics are recommended clinically during leech therapy to address potential *A. hydrophila* infections (Knobloch et al., 2007).

References

Akyurek, M., Safak, T., Kecik, A., 2001. Microsurgical ear replantation without venous repair: failure of development of venous channels despite patency of arterial anastomosis for 14 days. Ann. Plast. Surg. 46, 439–442.

Anthony, J.P., Lineaweaver, W.C., Davis, J.W. Jr., Buncke, H.J., 1989. Quantitative fluorimetric effects of leeching on a replanted ear. Microsurgery 10, 167–169.

Ardehali, B., Hand, K., Nduka, C., Holmes, A., Wood, S., 2006. Delayed leech-borne infection with *Aeromonas hydrophilia* in escharotic flap wound. J. Plast. Reconstr. Aesthet. Surg. 59, 94–95.

Baudet, J., 1991. The use of leeches in distal digital replantation. Blood Coagul. Fibrinolysis 2, 193–196.

Carter, K.C., 2001. Leechcraft in 19th century British medicine. J. Roy. Soc. Med. 91, 38–42.

Civelek, B., Kargi, E., Aköz, T., Sensöz, O., 1998. Turbocharge or supercharge? Plast. Reconstr. Surg. 102, 1303.

Cho, B.H., Ahn, H.B., 1999. Microsurgical replantation of a partial ear, with leech therapy. Ann. Plast. Surg. 43, 427–429.

Concannon, M.J., Puckett, C.L., 1998. Microsurgical replantation of an ear in a child without venous repair. Plast. Reconstr. Surg. 102, 2088–2093.

Conforti, M.L., Connor, N.P., Heisey, D.M., Hartig, G.K., 2002. Evaluation of performance characteristics of the medicinal leech (*Hirudo medicinalis*) fort he treatment of venous congestion. Plast. Reconstr. Surg. 109, 228–235.

Derganc, M., Zdravid, F., 1960. Venous congestion of flaps treated by application of leeches. Br. J. Plast. Surg. 13, 187–192.

Duroure, F., Simon, E., Fadhul, S., Fyad, J.P., Chassagne, J.F., Stricker, M., 2004. Microsurgical lip replantation: evaluation of functional and aesthetic results of three cases. Microsurgery 24, 265–269.

Foucher, G., Norris, R.W., 1992. Distal and very distal digital replantations. Br. J. Plast. Surg. 45, 199–203.

Funk, G.F., Bauman, N.M., Rinehart, R.J., Mankarious, L.A., 1996. Microvascular replantation of a traumatically amputated ear. Arch. Otolarnygol. Head Neck Surg. 122, 184–186.

Gross, M.P., Apesos, J., 1992. The use of leeches for treatment of venous congestion of the nipple following breast surgery. Aesthetic Plast. Surg. 16, 343–348.

Haycraft, J.B., 1884. Über die Einwirkung eines Sekretes des officinellen Blutegels auf die Gerinnbarkeit des Blutes. Arch. Exp. Pathol. Pharmakol. 18, 209–217.

Henderson, H.P., Matti, B., Laing, A.G., Morelli, S., Sully, L., 1983. Avulsion of the scalp treated by microvascular repair: the use of leeches for post-operative decongestion. Br. J. Plast. Surg. 36, 235–239.

Hernandez, M., Ramirez Gutierrez, R.E., 1998. Internal hirudiniasis: vaginal bleeding resulting from leech bite. Ginecol. Obestet. Mex. 66, 284–286.

Höltje, W.J., 1984. Successful replantation of an amputated upper limb. Plast. Reconstr. Surg. 73, 664–670.

Ho, V., Boyd, P., 2007. Rectal bleeding due to leech bite in a young child. Indian J. Gastroenterol. 26, 255–256.

Ikizceli, I., Avsarogullari, L., Sözüer, E., Yürümez, Y., Akdur, O., 2005. Bleeding due to a medicinal leech bite. Emerg. Med. J. 22, 458–460.

Jeng, S.F., Wie, F.C., Noordhoff, M.S., 1994. Replantation of amputated facial tissues with microvascular anastomosis. Microsurgery 15, 327–333.

Kalbermatten, D.F., Rieger, U.M., Uike, K., Erba, P., Laifer, G., Hintermann, B., Pierer, G., 2007. Infection with *Aeromonas hydrophila* after use of leeches (*Hirudo medicinalis*) in a free microvascular osteo-myo-cutaneous flap-suggestions for successful management. Handchir. Mikrochir. Plast. Chir. 39, 108–111.

Kerrigan, C.L., Wizman, P., Hjortdal, V.E., Sarnplis, J., 1994. Global flap ischaemia: a comparison of arterial versus venous etiology. Plast. Reconstr. Surg. 93, 1485–1495.

Knobloch, K., Gohritz, A., Busch, K., Spies, M., Vogt, P.M., 2007. *Hirudo medicinalis* – leech applications in plastic and reconstructive microsurgery – a literature review. Handchir. Mikrochir. Plast. Chir. 39, 103–107.

Kose, A., Zengin, S., Kose, B., Gunay, N., Yildrim, C., Kilinc, H., Togun, I., 2008. Leech bites: massive bleeding, coagulation disorders, and severe anemia. Am. J. Emerg. Med. 26, 1067.e3-e6.

Lee, C., Mehran, R.J., Lessard, M.L., Kerrigan, C.L., 1992. Leeches: controlled trial in venous compromised rat epigastric flaps. Br. J. Plast. Surg. 45, 235–238.

Levine, S.M., Frangos, S.G., Hanna, B., Colen, K., Levine, J.P., 2009. *Aeromonas* septicemia after medicinal leech use following replantation of severed digits. Am. J. Crit. Care Nov 25 [Epub ahead of print].

Lim, C.L., 1986. Successful transfer of "free" microvascular superficial temporal artery flap with no obvious venous drainage and use of leeches for reducing venous congestion: case report. Microsurgery 7, 87–88.

Lineaweaver, W.C., O'Hara, M., Stridde, B., Valauri, F.A., Buncke, H.J., 1991. Clinical leech use in a microsurgical unit: the San Francisco experience. Blood Coagul. Fibrinolysis 2, 189–192.

Lozano, D.D., Stephenson, L.L., Zamboni, W.A., 1999. Effect of hyperbaric oxygen and medicinal leeching on survival of axial skin flaps subjected to total venous occlusion. Plast. Reconstr. Surg. 104, 1029–1032.

Markwardt, F., 1955. Untersuchungen über Hirudin. Naturwissenschaften 42, 537–538.

Markwardt, F., 1992. Hirudin: the promising antithrombotic. Cardiovasc. Drug Rev. 10, 211–232.

Mumcuoglu, K.Y., Huberman, L., Cohen, R., Temper, V., Adler, A., Galun, R., Block, C., 2009. Elimination of symbiotic *Aeromonas spp.* from the intestinal tract of the medicinal leech, *Hirudo medicinalis*, using ciprofloxacin feeding. Clin. Microbiol. Infect. Jun 11 [Epub ahead of print].

Mutimer, K.L., Banis, J.C., Upton, J., 1987. Microsurgical reattachment of totally amputated ears. Plast. Reconstr. Surg. 79, 535–541.

Otto, A., Schoeller, T., Wechselberger, G., Ohler, K., 1999. Successful ear replantation without venous anastomosis by using leeches. Handchir. Mikrochir. Plast. Chir. 31, 98–101.

Raj, S.M., Radzi, M., Tee, M.H., 2000. Severe rectal bleeding due to leech bite. Am. J. Gastroenterol. 95, 1607.

Rapaport, D.P., Breitbart, A.S., Karp, N.S., Siebert, J.W., 1993. Successful microvascular replantation of a completely amputated ear. Microsurgery 14, 312–314.

Rcnaud, B., Langlais, F., Colmar, M., Thomazeau, H., 1991. Digital reimplantations and revascularizations: factors of preservation. 183 fingers. Ann. Chir. Main Memb. Super. 10, 385–398.

Rivera, M.L., Gross, J.E., 1995. Scalp replantation after traumatic injury. AORN J. 62, 175–180.

Rosner, F., 1999. The medicinal leech in Jewish writings. Isr. Med. Assoc. J. 1, 296–298.

Saha, P.K., Roy, S., Bhattacharya, D., Mukherjee, P., Naskar, T., Bhuiya, A., 2005. Leech bite: a rare gynaecologic emergency. MedGenMed 7, 73.

Sawyer, R.T., 1986. Feeding and Digestive System. Leech Biology and Behaviour (vol. 2). Clarendon Press, Oxford, pp. 467–518.

Semple, J.L., 1994. Retrograde microvascular augmentation (turbocharging) of a single-pedicle TRAM flap through a deep inferior epigastric arterial and venous loop. Plast. Reconstr. Surg. 93, 109–117.

Soucacos, P.N., Beris, A.E., Malizos, K.N., Kabani, C.T., Pakos, S., 1994a. The use of medicinal leeches, *Hirudo medicinalis*, to restore venous circulation in trauma and reconstructive microsurgery. Int. Angiol. 13, 251–258.

Soucacos, P.N., Beris, A.E., Malizos, K.N., Xenakis, T.A., Georgoulis, A., 1994b. Successful treatment of venous congestion in free skin flaps using medical leeches. Microsurgery 15, 496–501.

Talbi, M., Stussi, J.D., Meley, M., 2001. Microsurgical replantation of a totally amputated ear without venous repair. J. Reconstr. Microsurg. 17, 417–420.

Tsai, T.M., McCabe, S.J., Maki, Y., 1989. A technique for replantation of the finger tip. Microsurgery 10, 1–4.

Valauri, F.A., Buncke, H.J., Alpert, B.S., Lineaweaver, W.C., Argenta, L.C., 1990. Microvascular transplantation of expanded free scalp flaps between identical twins. Plast. Reconstr. Surg. 85, 432–436.

Wade, J.W., Brabham, R.F., Allen, R.J., 1990. Medicinal leeches: once again at the forefront of medicine. South Med. J. 83, 1168–1173.

Walton, R.L., Beahm, E.K., Brown, R.E., Upton, J., Reinke, K., Fudem, G., Banis, J., Davidson, J., Dabb, R., Kalimuthu, R., Kitzmiller, W.J., Gottlieb, L.J., Buncke, H.J., 1998. Microsurgical replantation of the lip: a multi-institutional experience. Plast. Reconstr. Surg. 102, 358–368.

Whitaker, I.S., Cheung, C.K., Chahal, C.A.A., Karoo, R.O.S., Gulati, A., Foo. I.T.H., 2005. By what mechanism do leeches help to salvage ischaemic tissues? Br. J. Oral Maxillofac. Surg. 43, 155–160.

Chapter 43
Diagnostic Use of Snake Venom Components in the Coagulation Laboratory

Anna Maria Perchuc and Marianne Wilmer

Abstract For almost every factor involved in haemostasis there exists a snake venom protein that can mimic, activate or deactivate it. Many of these compounds are insensitive to the physiological and therapeutically used coagulation inhibitors and, because of their unique features, are applied as molecular tools for diagnosis of haemostatic disorders. Different snake venom proteins are nowadays widely used in the coagulation laboratory and have facilitated extensively the routine assays of haemostatic parameters and understanding of basic biological mechanisms involved in the clotting and fibrinolysis processes. Some of these routine applications have been adopted as the preferred option of the diagnostic tests. The following parameters can be tested by means of snake venom components: antithrombin III, fibrinogen, its breakdown products and its dysfunctions, prothrombin and dysprothrombinaemias, blood clotting factors V, VII and X, lupus anticoagulants, protein C and its pathway, as well as activated protein C resistance, von Willebrand factor (vWF) and related syndromes. Further, immediately acting anticoagulants, such as heparins and direct thrombin inhibitors, which are among the most frequently applied drugs, can be monitored by an assay using snake venom enzyme.

Ongoing research leads to isolation and characterization of new snake venom components, which are potential tools for future applications in the field of haemostasis, in diagnostic as well as in therapeutic approaches.

Introduction

Venoms and venomous animals have always fascinated man and already very early in human history they were used as medicines. Nowadays, there are many research groups all over the world investigating the potential of snake venoms in order to find novel tools useful in different fields of medicine and diagnostic.

A.M. Perchuc (✉)
Swiss Tropical and Public Health Institute, Medical Parasitology and Infection Biology, Socinstrasse 57, 4051 Basel, Switzerland; August Lenz Foundation at the Institute for Prevention of Cardiovascular Diseases, University of Munich, Munich, Germany
e-mail: annaperchuc@gmail.com

R.M. Kini et al. (eds.), *Toxins and Hemostasis*, DOI 10.1007/978-90-481-9295-3_43,

Interference with the functions of the human haemostatic system is a common mechanism of snake venoms, to a greater or lesser degree depending on the snake family. For almost every factor involved in coagulation or fibrinolysis there is a venom protein that can mimic, activate or deactivate it (Koh et al., 2006; Lu et al., 2005).

Snake venom toxins affecting haemostasis have been classified by virtue of their overall effect into the following groups (Braud et al., 2000; Hutton and Warrell, 1993; Koh et al., 2006; Lu et al., 2005; Markland, 1997, 1998; Marsh, 1994; Marsh and Williams, 2005; Matsui et al., 2000):

- coagulant factors (thrombin-like enzymes, prothrombin activators, and FV and FX activators),
- anticoagulant factors (including FIX/FX binding proteins, protein C activators, thrombin inhibitors, and phospholipases A_2),
- proteins acting as activators or inhibitors of blood platelets (proteinases including metalloproteinases, C-type lectins, disintegrins, phospholipases A_2, and L-amino acid oxidases),
- factors acting on fibrinolysis (fibrinolytic enzymes and plasminogen activator), and
- haemorrhagins (metalloproteinases degrading the blood vessel's extracellular matrix).

Many of these components are insensitive to both, 'in vivo available' and in vitro commonly applied inhibitors of the haemostatic system. Thus, they are used for the treatment of haemostatic disorders and routine assays of various parameters in the coagulation laboratory. The following parameters can be tested by means of snake venom components (Kornalik, 1985; Lu et al., 2005; Marsh, 1994, 2001; Marsh and Fyffe, 1996; Marsh and Williams, 2005; Schöni, 2005; Stocker, 1986, 1998):

1) antithrombin III, fibrinogen and fibrinogen breakdown products in heparin-containing samples, as well as the detection of fibrinogen dysfunction, can be performed by the use of snake venom thrombin like enzymes (SVTLEs), such as batroxobin, Reptilase® or ancrod;
2) prothrombin, studies of dysprothrombinaemias, as well as preparation of meizothrombin and non-enzymatic forms of thrombin, can be performed using different snake venom prothrombin activators in relation to their cofactor dependence (e.g. ecarin, noscarin, oscutarin);
3) blood clotting factors V, VII and X, and, importantly, lupus anticoagulants (LA) can be assayed by means of toxins contained in Russell's viper venom: RVV-V and RVV-X;
4) LA screening, can be performed applying a number of snake venom activators, including RVV-X and prothrombin activators, such as: oscutarin, textarin, and ecarin;

5) protein C and the protein C pathway, as well as activated protein C resistance, the last being concerned as one of the major causes of thrombophilia, can be measured by means of RVV and Protac®;
6) anticoagulants, like indirect (unfractionated heparins (UFH), low molecular weight heparins (LMWH)) or direct FXa inhibitors and direct thrombin inhibitors (DTIs) can be monitored by means of a relatively novel test (Pefakit® PiCT®) using RVV-V;
7) von Willebrand factor (vWF) and related syndromes (Bernard-Soulier disease and Type IIa von Willebrand disease) can be studied with Botrocetin®.

Additionally to the assays mentioned above and commonly performed in the coagulation laboratory, snake venom proteins, such as disintegrins, metalloproteinases, and C-type lectins are used to study the properties of platelet glycoprotein receptors, platelet-platelet and platelet-endothelium interactions. Furthermore, bitistatin, a disintegrin from the *Bitis arietans* venom, was also shown to be promising in imaging both thrombi and emboli (Baidoo et al., 2004; Knight and Romano, 2005; Marsh and Williams, 2005). Nevertheless, up to date there is no diagnostic application of these venom components. Snake venom toxins affecting platelet functions were precisely reviewed elsewhere (Andrews et al., 2001; Clemetson et al., 2001; Kamiguti, 2005; Lu et al., 2005; Wijeyewickrema et al., 2005).

SVTLEs and the Detection of Antithrombin III and Other Haemostatic Variables in Heparin-Containing Samples: Detection of Fibrinogen, Fibrinogen Breakdown Products, and Fibrinogen Dysfunctions

Over 90 SVTLEs have been reported from over 30 snake species (Marsh, 2001; Pirkle, 1998). They are widely distributed in several pit viper genera (*Agkistrodon*, *Bothrops*, *Crotalus*, *Gloydius*, *Lachesis* and *Protobothrops* (formerly *Trimeresurus*)), as well as in some true vipers (*Bitis* and *Cerastes*) and the colubrid, *Dispholidus typus* (Castro et al., 2004; Lu et al., 2005; Pirkle, 1998). The most widely studied and used SVTLEs are those from *Bothrops atrox* (batroxobin, Reptilase®, both from DSM Nutritional Products AG Branch Pentapharm, Basel Switzerland) and *Callosellasma rhodostoma* (ancrod, from Neurobiological Technologies, Inc., Emeryville, California, USA) (Marsh and Williams, 2005).

Sequences of about 30 SVTLEs have been determined and are published in Swiss-Prot and TrEMBL database (http://us.expasy.org/sprot/). Although called thrombin-like enzymes, showing in vitro blood clotting activity and a high level of similarity among each other, SVTLEs share rather a weak sequence identity with thrombin (26–33%) (Castro et al., 2004) and do not possess all its properties. Unlike thrombin, cleaving both fibrinopeptide A and B (FPA and FPB) from fibrinogen and activating FXIII, the SVTLEs usually cleave FPA alone; only a few release

FPB, or FPA and FPB together. Based on this feature, SVTLEs have been classified as FP-A, FP-B or FP-AB groups (Kornalik, 1990). Because SVTLEs, in general, do not activate factor XIII, clots produced in vitro are imperfectly formed, much smaller in size and easily broken down. The lack of cross-linked clots in vivo leads to their enhanced degradation by the fibrinolytic system and effective removal of fibrinogen from the circulation (Koh et al., 2006). Unlike thrombin, SVTLEs neither activate other coagulation factors nor induce platelet aggregation and release reaction. They are neither inhibited by thrombin inhibitors like antithrombin III and hirudin (Stocker, 1990).

Among the SVTLEs, the most commonly used diagnostic tool is the Reptilase® reagent. Pefakit® Reptilase® Time is a plasma based functional assay for the determination of the reptilase induced clotting time (reptilase time). Reptilase time is used for the determination of fibrinogen polymerization disorders and other clinical conditions connected with the last phase of coagulation. A prolongation of the clotting time, while clotting is triggered by reptilase, indicates fibrinogen disorders or an interference with normal fibrin polymerization.

The well established reptilase time is a convenient alternative or a supplementation to the thrombin time (TT). As reptilase is not inhibited by heparin, hirudin or antithrombins it can be used, in combination with the TT, as a rapid fibrinogen assay in samples containing heparin and hirudin. The presence of fibrin degradation products (FDPs), hypofibrinogenaemia, as well as defects in fibrin polymerisation (dysfibrinogenaemia) will lead to a prolongation of both, the TT and the reptilase time. In the presence of FDPs there is an obvious disproportion between the prolongation of reptilase time and the TT. In the latter case the TT is more prolonged, while in patients with inherited or acquired dysfibrinogenemia the opposite is found (Funk et al., 1971; Latallo and Teisseyre, 1971). The laboratory testing of dysfibrinogenaemia including the use of reptilase has been reviewed by Cunningham (Cunningham et al., 2002).

Interestingly, it has also been observed that very high fibrinogen levels may be associated with prolonged reptilase times (Van Cott et al., 2002). Reptilase or ancrod are also applied for plasma preparations in assays measuring antithrombin III (ATIII). Using these enzymes plasma can be prepared free of fibrinogen without the addition of thrombin which would react with ATIII and interfere with the assay (Canoso et al., 1979; Howie et al., 1973).

SVTLEs are also useful tools for detailed investigations, such as (Marsh and Williams, 2005): (i) fibrin function including magnetic birefringence studies of fibrin assembly (Torbet, 1987), (ii) thrombin dependent feedback activation of the clotting system (Kumar et al., 1994), (iii) platelet-force development (Carr et al., 1996), (iv) preparation of desaa-fibrinogen (by means of batroxobin) in place of CNBr-fragments of fibrinogen as a fibrin stimulant in the functional assay of tissue plasminogen activator (Ranby et al., 1982), or (v) desbb-fibrinogen by means of ACTE, the enzyme from *Agkistrodon c. contortrix*, which cleaves only fibrinopeptide B from fibrinogen.

Prothrombin Activators and the Detection of Prothrombin, Monitoring of DTIs, Studies of Dysprothrombinaemias, and Preparation of Meizothrombin

Activation of prothrombin to mature thrombin occurs by the proteolytic action of the prothrombinase complex consisting of FXa, FVa, phospholipids (PLs), and calcium ions. A number of exogenous prothrombin activators are found in snake venoms. According to the recent classification, they are categorized into four classes based on their structures, functional characteristics and cofactor dependency (Kini, 2005; Kini et al., 2001, 2002; Lu et al., 2005; St. Pierre et al., 2005; Yamada et al., 1996).

Ecarin (DSM Nutritional Products AG Branch Pentapharm, Basel Switzerland), a metalloproteinase from the venom of *Echis carinatus*, is the best characterized group A prothrombin inhibitor. It cleaves prothrombin producing meizothrombin, which is then finally converted into a-thrombin by autolysis (Kini, 2005; Kini et al., 2002). Meizothrombin is inhibited by hirudin and its analogues, but not by glycosaminoglycans like heparin or heparinoids.

The ecarin clotting time (ECT), developed in 1991/1992 and published in 1993 (Nowak, 2003/2004; Nowak and Bucha, 1996), is a meizothrombin generation test that allows for precise quantification of DTIs in citrated and heparinised blood. It has demonstrated its usefulness for biochemical and pharmaceutical investigations, as well as in clinical research and routine applications. ECT is neither affected by oral anticoagulants nor by heparin, which is very favourable in clinical practice, especially when patients have to be bridged to an alternative anticoagulant. For instance, in thrombophilic states, the residual hirudin level must often be assayed in the case of concomitant oral anticoagulation. As ecarin also activates PIVKA- (Proteins Induced by Vitamin K Absence/Antagonists) prothrombin to PIVKA-meizothrombin, measurement of hirudin in plasma is possible even with concomitant anticoagulant treatment. This property has also been utilised to assay PIVKA in liver disease (Kornalik and Vorlova, 1988). During many years ECT has been adjusted to clinical requirements and was proven to be a valuable tool especially for drug-monitoring in r-hirudin therapy (Nowak, 2003/2004). Moreover, ECT is a bioanalytical tool for the definition of a new specific inhibitor family, the 'meizothrombin inhibitors'. Meizothrombin plays an important role in the activation of PARs, most of all in cancer cells (Kaufmann et al., 2000). However, further investigations must prove if this substance class has clinical relevance (Nowak, 2003/2004).

Ecarin chromogenic assay (ECA) is another method that uses the advantages of the ECT and compensates for its disadvantages (Lange et al., 2003). In the ECA the cleavage of a chromogenic substrate by meizothrombin is inhibited in a concentration-dependent way by DTIs. This reaction is independent of variations in the concentration of either fibrinogen or prothrombin. Moreover, the ECA can easily be performed both as a point-of-care test on manual coagulation analyzers provided with an option for optical measurement and carried out on automated coagulation analyzers (Lange et al., 2003). It is also suitable for quantitative determination of DTIs other than hirudin like argatroban, a synthetic small molecule that selectively and reversibly inhibits the catalytic site of thrombin (Kathiresan et al., 2002).

Further, the ECT and also the Taipan snake venom time (TSVT) have been used in the following applications (Marsh and Williams, 2005): (i) studying dysprothrombinaemias (Collados et al., 1997, Weinger et al., 1980), (ii) diagnosing disseminated intravascular coagulation (Sakuragawa et al., 1975), and (iii) controlling the PEG–hirudin anticoagulation (Moser et al., 2001).

Carinactivase (CA-1) is another prothrombin activator isolated from the venom of *E. carinatus*. Its activity depends on the presence of calcium ions and, unlike ecarin, it does not activate prothrombin derivatives in which the binding of calcium ions has been perturbed, e.g. decarboxy prothrombin, called PIVKA-II (Yamada et al., 1996, Yamada and Morita, 1999). Thus, based on the use of CA-1 a chromogenic microplate assay, called CA-1 method, has been developed for the quantification of normal prothrombin (Yamada and Morita, 1999). Applying the CA-1 method the levels of normal, fully carboxylated prothrombin can be determined in patients undergoing the warfarin treatment. This is not possible using ecarin which does not discriminate between normal prothrombin and the PIVKA-II. CA-1 method is a rapid and highly sensitive assay suitable for analyzing plasma samples treated with EDTA or heparin (Yamada and Morita, 1999). Further, because of the small sample size (1 μl of patient plasma), its simplicity and rapidness, the CA-1 method has been shown to be suitable for use in neonates (Fukushima et al., 2002). Similar chromogenic method for determination of normal prothrombin levels has been described applying multactivase, a CA-1 analogue isolated from the venom of *E. multisquamatus* (Yamada and Morita, 1997)

Other snake venom prothrombin activating enzymes are also used in preparing meizothrombin (Novoa and Seegers, 1980; Rosing and Tans, 1988), non-enzymic forms of thrombin and meizothrombin (Rosing and Tans, 1992), in studies of recombinant prothrombin cleavage (Stevens et al., 1996), as well as for the detection of lupus anticoagulants and for the improved distinction of FV wild-type and FV Leiden mutation, both applications are precisely described below.

RVV-V and RVV-X – Detection of Blood Clotting Factors V, VII, and X

Detection of Factor V

Several FV activators have been described from *Bothrops atrox*, *Vipera russelli*, *V. labetina*, *V. ursine* and *Naja oxiana* venoms (Gerads et al., 1992; Rosing et al., 2001). RVV-V from *V. russelli* (DSM Nutritional Products AG Branch Pentapharm, Basel Switzerland) activates FV to FVa by cleavage of a single peptide bond at Arg1545-Ser1546, which is one of three thrombin cleavage sites in FV (Lu et al., 2005; Rosing et al., 2001; Tokunaga et al., 1988). Unlike thrombin, RVV-V shows no apparent effects on coagulation FVIII, FXIII, fibrinogen or prothrombin (Tokunaga et al., 1988).

RVV-V can be applied for the preparation of a reagent used in the routine assay of FV in plasma (Kisiel and Canfield, 1981). The utility of RVV-V for FV assay is somewhat limited but the reagent still remains a useful tool in the study of FV activation and structure-function relationship of FVa (Marsh and Williams, 2005; Rosing et al., 2001). RVV-V has been also employed in the following applications: (i) to define the functional importance of cleavage sites in both wild type and recombinant factor V (Keller et al., 1995), (ii) for the improved distinction of FV wild-type and FV Leiden mutation, and recently (iii) in the prothrombinase-induced clotting time (PiCT®), which will be precisely discussed below.

Detection of Factor VII and X

FX activators have been isolated from many Viperide and Crotalide venoms as well as from a few Elapid species and were reviewed by Tans and Rosing (2001). The best known and most extensively characterized FX activator is RVV-X (DSM Nutritional Products AG Branch Pentapharm, Basel Switzerland). The activation of FX to FXa by RVV-X involves the specific cleavage of the peptide bond Arg52–Ile53 in the heavy chain of FX. The same bond is cleaved by FIXa and FVIIa during normal blood coagulation (Gowda et al., 1994; Morita, 1998). However, unlike the vitamin K-dependent clotting factors, RVV-X does not require a negatively charged surface such as PLs, but just the calcium ions for FX activation (Takeya et al., 1992).

RVV-X has been employed in a number of clotting assays (Marsh and Williams, 2005), such as the measurement of factor X itself (Bachmann et al., 1958), for distinguishing between factor VII and factor X deficiency (Quick, 1971), and also in lupus anticoagulant assay (Thiagarajan et al., 1986). The plasma clotting time triggered by RVV-X is known as the Stypven time™ (Denson, 1961) and it measures the clotting time proportional to FX concentration in the following manner: FX is being activated to Xa by RVV-X in the presence of calcium ions, FVa and PLs, forming the classical prothrombinase complex. The prothrombinase complex in turn activates prothrombin to thrombin initiating the clot formation. A normal Stypven time™ determined in conjunction with a prolonged prothrombin time (PT) suggests FVII deficiency, whereas a prolonged Stypven time™ indicates FX deficiency (Quick, 1971). Furthermore, after a RVV-X triggered activation of FX a direct measurement of FXa activity can be performed based on the cleavage of a chromogenic FXa substrate (Marsh and Williams, 2005). Further, RVV-X was also used in the following applications (Marsh and Williams, 2005): (i) identification of FX recognition sites (Chattopadhyay and Fair, 1989), (ii) investigation of deglycosylation effects on FX activation (Inoue and Morita, 1993), and (iii) in platelet factor 3 assay (Hardisty and Hutton, 1966; Jy et al., 1995).

Detection of Lupus Anticoagulants

Lupus anticoagulants (LA) are identified in the laboratory by their interference with one or more phospholipids dependent coagulation tests. Many assays have been developed and a number of snake venom activators are employed in the LA detection. These include the following activators (Marsh and Williams, 2005): RVV-X (LA Screen and LA Confirm, Life Therapeutics, previously Gradipore Pty Ltd, Sydney, Australia), as well as prothrombin activators from taipan snake (*Oxyuranus scutellatus*) venom and the venoms of Australian brown snake (*Pseudonaja textilis*) and saw-scaled viper (*Echis carinatus*). The diluted Russell's viper venom time (dRVVT) has been used as both a screening test and a confirmatory test in the LA diagnosis (Court, 1997). The test is quick, sensitive and inexpensive, as well as adaptable for use in most automated coagulometers. It uses diluted PLs and diluted Russell's viper venom containing RVV-X, thus it evaluates two phospholipids-dependent coagulation steps. Nevertheless, its sensitivity to LA varies with the phospholipid components of the reagent (Court, 1997), as well as it is dependent on the heterogeneity of the venom (Moore and Savidge, 2004). Another test has been developed based on the taipan snake venom time (TSVT) combined with a platelet neutralization procedure (PNP) (Rooney et al., 1994). The taipan snake venom directly activates prothrombin in the presence of PLs and calcium ions. A PNP is recommended due to the ability of platelets to 'neutralize' or 'bypass' the in vitro effect of LA (Court, 1997). The TSVT was found to be influenced by the presence of heparin, but suitable for use in patients receiving oral anticoagulants and patients with mild liver dysfunction (Parmar et al., 2009; Rooney et al., 1994). Recently, it has also been shown that combining the TSVT/ECT screening with the mixing studies of conventional assays increases the detection rates of LA in orally anticoagulated patients (Moore, 2007; Moore et al., 2003). A further snake-venom based assay has been developed with the potential for combined screening and confirmation of LA. This test employs two snake venom prothrombin activators with different mode of action, i.e. ecarin which, as previously mentioned, activates prothrombin independently of any cofactors, and textarin, a protein fraction of *Pseudonaja textilis* venom, which activates prothrombin in the presence of PL, FV and calcium ions (Triplett et al., 1993). Thus, the ECT is unaffected by the presence of LA, whereas the textarin time is typically prolonged. The textarin/ecarin clotting time ratio is a two-stage procedure, delivering the results as a ratio of two clotting times (ratio >1.3 being considered as a positive result). The test was shown to be sensitive, reliable and relatively simple method, unaffected by oral anticoagulants (Court, 1997).

Detection of Protein C and Activated Protein C Resistance (APC-R)

The protein C (PC) pathway is probably the most important anticoagulant mechanism of the blood coagulation. Proteinases converting the zymogen PC into activated

PC (APC) have been found in venoms of different snakes. Most PC activators have been purified from *Agkistrodon* venoms; others come from *Bothrops*, *Trimeresurus*, *Vipera* or *Cerastes* venoms (Gempeler-Messina et al., 2001; Kini, 2006; Lu et al., 2005; Meier and Stocker, 1991; Stocker et al., 1986). Protac® from *A. contortrix contortrix* venom (DSM Nutritional Products AG Branch Pentapharm, Basel Switzerland) is a fast-acting PC activator and unlike the activation product generated by thrombin, Protac®-activated PC remains unaffected by the plasma PC inhibitor. In contrast to thrombin, Protac® itself exerts neither a coagulant nor a chromogenic substrate-splitting activity, thus is has been used for activation of plasma PC in a chromogenic assay. This test is simple and specific, can easily be automated for clinical use and shows an extraordinary accuracy (Gempeler-Messina et al., 2001; Klein and Walker, 1986; Meier, 1998; Stocker et al., 1987). In a PC clotting assay, PC of patient's plasma is activated with Protac® and the inhibitory effect of APC is determined by aPTT. The prolongation of the clotting time is proportional to the activity of patient's PC (Gempeler-Messina et al., 2001). Also a screening assay for the PC pathway has been developed based on the aPTT ratio performed in the presence and absence of Protac® (Robert et al., 1996). This test and its newer version (ProC Global®, Siemens Healthcare Diagnostics, Eschborn, Germany) have been successfully applied for the identification of all patients with defects in the PC anticoagulant pathway such as PC deficiency, PC resistance/FV Leiden and protein S (PS) deficiency. ProC Global® was further used for the detection of FV Leiden and prothrombin mutation 20210 G to A (Sillero et al., 2001), for evaluation of a venous thromboembolism risk in woman undergoing hormonal therapy (Sarig et al., 2006), as well as it was successfully applied in a large multicenter trial of thrombotic patients for the identification of all FV Leiden carriers and those with low levels of APC (Toulon et al., 2000). The use of Protac® enabled also a development of a simplified PS assay, in which PS activity can be measured as a prolongation of the prothrombin-time-based clotting assay. The PS assay is performed in the following manner: diluted patient plasma is mixed with PS-depleted plasma containing Protac®-activated PC, bovine brain tissue thromboplastin and calcium ions. Thus, the measured clotting time prolongation is proportional to the PS content in the patient plasma (Suzuki and Nishioka, 1988).

Further tests applying snake venom enzymes/coagulation activators were designed to screen for FV Leiden mutation in patient's plasma. Such tests include the (i) GradiLeiden V kit (Life Therapeutics, previously Gradipore Pty Ltd, Sydney, Australia) and similar CRYOcheck Clot APCR Assay (Precision BioLogic, Dartmouth, Canada) both based on the dRVVT, (ii) STA-STACLOT APC-R Test (Diagnostica Stago, Asnieres, France) based on a FX activator from the *Crotalus viridis helleri* venom, (iii) textarin time (Hoagland et al., 1996), and (iv) Pefakit® APC-R Factor V Leiden (DSM Nutritional Products AG Branch Pentapharm, Basel Switzerland) using RVV-V and noscarin.

The GradiLeiden V kit method is based on the dRVVT clot procedure. The kit provides RVVT reagent and a reagent used as a source of PC activator, containing the whole diluted *A. contortrix contortrix* venom. The dRVVT assay is performed on patient plasma using the RVVT reagent and saline, and then repeated using the RVVT reagent/PC activator combination. As a result a ratio of both clotting times is calculated (i.e. RVVT[+PCA]/RVVT[+saline]) (Favaloro et al., 2006). The GradiLeiden V can be performed on plasma from patients on stabilized oral anticoagulant or heparin therapy.

In the STA-STACLOT APC-R Test patient plasma sample is mixed with the diluent buffer and FV-deficient plasma. The clotting time is measured following the addition of *Crotalus viridis helleri* venom containing a specific FX activator and APC. This test can also be applied in patents treated with oral anticoagulants (Quehenberger et al., 2000).

Finally, the Pefakit® APC-R Factor V Leiden relies on a FVa inactivation and a FVa detection step. In the FVa inactivation step patient plasma is mixed with standardized FV-depleted plasma and a reagent containing RVV-V and APC is added. Thus the activated FV, in the wild-type situation, is immediately inactivated. In the FVa detection step, a reagent containing noscarin, a FVa-dependend prothrombin activator from *Notechis scutatus scutatus* is added. Due to the FVa dependency of the prothrombin activator the velocity of prothrombin activation is decreased in the non FV Leiden carriers (wild-type) resulting in a prolonged clotting time. If FVa elimination has been incomplete, as it occurs in the carriers of the FV Leiden mutation, the velocity of prothrombin activation is high and the clotting time short. A second determination is performed under identical assay conditions, with the exception that no APC is added in the FVa elimination step. This determination is called the APC(–) measurement, as opposed to APC(+) measurement and a ratio between both measurements is calculated. Neither heparin therapy nor LA interfere with the Pefakit® APC-R Factor V Leiden assay (Schöni, 2005; Wilmer et al., 2004).

Monitoring of Anticoagulants like ATIII Dependent FXa Inhibitors (UFH, LMWH), Direct FXa Inhibitors (e.g. Rivaroxaban) and Direct Thrombin Inhibitors (e.g. Argatroban)

Immediately acting anticoagulants, such as heparins and DTIs are among the most frequently applied drugs in any field of medicine. The common mode of action of these drugs is their inhibition of FXa and/or thrombin (Calatzis et al., 2008). The prothrombinase-induced clotting time (PiCT®, DSM Nutritional Products AG Branch Pentapharm, Basel Switzerland) is designed to monitor all kinds of FXa and FIIa inhibitors. These could be either ATIII dependent FXa inhibitors like heparins (UFH, LMWH), the pentasaccharide Fondaparinux (Arixtra®, GlaxoSmithKline, UK), direct synthetic FXa inhibitors like Rivaroxaban (Xarelto®, Bayer Healthcare AG, Leverkusen, Germany) or DTIs like r-Hirudin, Argatroban (Argatra®, GlaxoSmithKline, UK) and Darbigatran (Pradaxa®, Boehringer Ingelheim, Germany) (Calatzis et al., 2008; Graff et al., 2007; Harder et al., 2008; Harenberg et al., 2007, 2008). Oral anticoagulants acting on vitamin K-dependent coagulation factors are not detected by this test. The PiCT assay is based on the addition of a defined supplementation of FXa, RVV-V and PLs to platelet-poor plasma. Following an incubation time, the mixture is recalcified and the clotting time is determined. The addition of RVV-V assures the complete activation of available FV contained in patient's plasma and makes the PiCT® assay independent of coagulation factors responsible for FV activation. Addition of FXa, PLs and calcium allows for the immediate formation of prothrombinase complex. The formed prothrombinase complex activates thrombin at optimal speed, which guarantees a very short, clear and repeatable onset of the test. In a concentration dependent manner all FXa-inhibitors slow down thrombin generation, thus leading to prolonged clotting times. As well, all FIIa inhibitors will result in increased clotting times depending on their plasma concentrations (Calatzis et al., 2008). Because the PiCT assay focuses specifically on the prothrombinase complex, subsequently thrombin activation and fibrin generation, it is insensitive to influencing or disturbing factors (Schöni, 2005).

Studies of Von Willebrand Factor and Related Syndromes

Botrocetin® (DSM Nutritional Products AG Branch Pentapharm, Basel Switzerland) is a platelet-aggregating protein found in a number of species of snake genus *Bothrops*, although its richest source is the *Bothrops jararaca* venom. Botrocetin modulates vWF to elicit platelet glycoprotein Ib (GPIb)-binding activity, leading to platelet agglutination (Matsui and Hamako, 2005). This property has been used for assaying vWF-dependent platelet aggregation and its abnormalities in both human and animals. The reactivity of botrocetin in platelet agglutination is essentially independent of the multimer size of the vWF molecule and the species source of vWF (Read et al., 1989).

The vWF values in human plasma obtained with assays using either botrocetin or ristocetin were found to be comparable. Nevertheless, botrocetin has a slightly but significantly different effect from that of ristocetin. Botrocetin does not have the limitations of ristocetin, which is highly restrictive in its reaction with plasma vWF. Ristocetin induces little or no platelet aggregation with most mammalian plasmas and is unreactive with most species of platelets except for those of human origin. It is also known to cause platelet aggregation with only the high molecular weight multimers of vWF. If the distribution of the array of multimeric forms of vWF is abnormal, with a reduction or absence of the higher molecular weight forms, the apparent vWF/ristocetin concentration is lower than vWF/botrocetin values (Brinkhous et al., 1988). Thus, plasma from the patients with type IIA von Willebrand disease (vWD), which lacks the highest molecular weight multimers, does not aggregate platelets with ristocetin (Brinkhous et al., 1983). Further, botrocetin partially aggregates platelets from patients suffering from Bernard-Soulier disease. However, ristocetin does not induce platelet aggregation in plasma from these patients since it depends on GPIb.

Botrocetin has been used for assaying plasma vWF in both human and animal plasmas, for characterizing antibody-inhibitors generated in vWD and autoimmune

Table 43.1 Diagnostic applications of snake venom components in the coagulation laboratory

Snake venom components	Diagnostic applications	Examples of diagnostic assays based on the use of snake venom component
Snake venom thrombin-like enzymes (SVTLEs)	• Determination of clotting time in samples containing heparin, hirudin or antithrombins; • Determination of fibrinogen polymerization disorders; • Determination of clinical conditions connected with the last phase of coagulation; • Preparation of plasma for assays measuring AT III; • Preparations of desaa- or desbb-fibrinogen; • Different investigations, including: fibrin function, thrombin dependent feedback activation of the clotting system and platelet-force development.	• Pefakit® Reptilase® Time, using reptilase
Prothrombin activators	• Monitoring of DTIs, like hirudin or argatroban, in citrated or heparanised blood samples, as well as by patients on concomitant anticoagulant therapy; • Quantification of normal prothrombin in comparison to PIVKA-II; • Assessment of PIVKA in liver disease; • Diagnosing of disseminated intravascular coagulation; • Studying of meizothrombin generation and new specific 'meizothrombin inhibitors'; • Studying of dysprothrombinaemias; • Preparations of meizothrombin, non-enzymatic forms of thrombin and meizothrombin; • Studying of recombinant prothrombins; • Detection of lupus anticoagulants; • Application in the novel method for the distinction of FV wild-type and FV Leiden mutation.	• Ecarin Clotting Time (ECT), using ecarin • Ecarin Chromogenic Assay (ECA), using ecarin • CA-1 method, using carinactivase • Taipan Snake Venom Time (TSVT), using prothrombin activators from taipan snake venom • Pefakit® APC-R Factor V Leiden, using RVV-V and noscarin

Table 43.1 (continued)

Snake venom components	Diagnostic applications	Examples of diagnostic assays based on the use of snake venom component
FV activators	• Preparation of reagent used in the routine assay of FV in plasma; • Studying of FV activation and structure-function relationship of FVa; • Studying and determination of FV wild-type and FV Leiden mutation; • Monitoring of anticoagulants.	• Pefakit® APC-R Factor V Leiden, using RVV-V and noscarin • Pefakit® PiCT®, using RVV-V
FX activators	• Measurement of FX (either by a clotting method or chromogenic); • Measurement of clotting time proportional to FX concentration; • Distinguishing between FVII and FX deficiency; • Detection of lupus anticoagulants; • Determination of FV Leiden mutation; • Identification of FX recognition sites; • Investigation of deglycosylation effects on FX; • Application in platelet factor 3 assay.	• Stypven time™, using RVV-X • LA Screen and LA Confirm, using RVV-X • dRVVT, using the whole Russell's viper venom • GradiLeiden V kit and similar CRYOcheck Clot APCR Assay, using dRVVT clotting procedure • STA-STACLOT APC-R, using FX activator from *Crotalus viridis helleri* venom

Table 43.1 (continued)

Snake venom components	Diagnostic applications	Examples of diagnostic assays based on the use of snake venom component
PC activators	• Determination of PC activity (either by a clotting method or chromogenic); • Screening of the PC pathway; • Determination of FV Leiden mutation; • Measurement of PS activity; • Evaluation of the thromboembolism risk.	• ProC Global®, using Protac®
Botrocetin®	• Assaying of vWF-dependent platelet aggregation; • Determination of type IIA vWD and Bernard Soulier disease; • Characterization of antibody-inhibitors generated in vWD and autoimmune von Willebrand syndrome; • Screening of novel antithrombotic agents blocking vWF interactions; • Characterization of human platelet vWF and its mutations.	

von Willebrand syndrome, and for analyzing thrombopathies such as Bernard-Soulier (Read et al., 1989). Also a simple inhibition assay for ristocetin- or botrocetin-induced biotinylated vWF binding to platelet GPIb, based on an enzyme-linked immunosorbent assay (ELISA) using platelet-bound microplates, has been established. This assay was considered to be convenient for the screening of novel antithrombotic agents that block the interaction of vWF with GPIb. (Miura et al., 1996). Further, the ristocetin-botrocetin combination has also been applied in the characterization of human platelet vWF (Williams et al., 1994) and for detection of missence mutations in Type B vWD (Rabinowitz et al., 1992).

Conclusions

Snake venoms with their cocktail of proteins represent a rich source of active compounds many of which have found application as diagnostic tools in the field of haemostasis. These are nowadays widely used in the coagulation laboratory and have facilitated extensively the routine assays of haemostatic parameters and understanding of basic biological mechanisms involved in the clotting and fibrinolysis processes. Diagnostic use of the snake venom components described in this chapter is summarized in Table 43.1. Some of these routine applications have been adopted as the preferred option of the diagnostic tests. Ongoing research leads to isolation and characterization of new snake venoms components, which are potential tools for future applications in the field of haemostasis, in diagnostic as well as therapeutic approaches.

References

Andrews, R.K., Kamiguti, A.S., Berlanga, O., Leduc, M., Theakston, R.D., Watson, S.P., 2001. The use of snake venom toxins as tools to study platelet receptors for collagen and von Willebrand factor. Haemostasis 31, 155–172.

Baidoo, K.E., Knight, L.C., Lin, K.S., Gabriel, J.L., Romano, J.E., 2004. Design and synthesis of a short-chain bitistatin analogue for imaging thrombi and emboli. Bioconjug. Chem. 15, 1068–1075.

Bachmann, F., Duckert, F., Koller, F., 1958. The Stuart–Prower assay and its clinical significance. Thromb. Diath. Haemorrh. 2, 24–38.

Braud, S., Bon, C., Wisner, A., 2000. Snake venom proteins acting on hemostasis. Biochimie 82, 851–859.

Brinkhous, K.M., Read, M.S., Fricke, W.A., Wagner, R.H., 1983. Botrocetin (venom coagglutinin): reaction with a broad spectrum of multimeric forms of factor VIII macromolecular complex. Proc. Natl. Acad. Sci. U.S.A. 80, 1463–1466.

Brinkhous, K.M., Smith, S.V., Read, M.S., 1988. Botrocetin and von Willebrand factor, in: Pirkle, H., Markland, F.S., Jr. (Eds.), Hemostasis and Animal Venoms. Marcel Dekker Inc., New York, pp. 377–398.

Calatzis, A., Peetz, D., Haas, S., Spannagl, M., Rudin, K., Wilmer, M., 2008. Prothrombinase-induced clotting time assay for determination of the anticoagulant effects of unfractionated and low-molecular-weight heparins, fondaparinux, and thrombin inhibitors. Am. J. Clin. Pathol. 130, 446–454.

Canoso, R.T., Hutton, R.A., Deykin, D., 1979. The haemostatic defect of chronic liver disease. Gastroenterology 76, 540–547.

Carr, M.E., Carr, S.L., Greilich, P.E., 1996. Heparin ablates force development during platelet mediated clot retraction. Thromb. Haemost. 75, 674–678.

Castro, H.C., Zingali, R.B., Albuquerque, M.G., Pujol-Luz, M., Rodrigues, C.R., 2004. Snake venom thrombin-like enzymes: from reptilase to now. Cell Mol. Life Sci. 61, 843–856.

Chattopadhyay, A., Fair, D.S., 1989. Molecular recognition in the activation of human blood coagulation factor X. J. Biol. Chem. 264, 11035–11043.

Clemetson, K.J., Navdaev, A., Dörmann, D., Du, X.Y., Clemetson, J.M., 2001. Multifunctional snake C-type lectins affecting platelets. Haemostasis 31, 148–154.

Collados, M.T., Fernandez, J., Paramo, J.A., Montes, R., Borbolla, J.R., Montano, L.F., Rocha, E., 1997. Purification and characterization of a variant of human prothrombin: prothrombin Segovia. Thromb. Res. 85, 465–477.

Court, E.L., 1997. Lupus anticoagulants: pathogenesis and laboratory diagnosis. Br. J. Biomed. Sci. 54, 287–298.

Cunningham, M.T., Brandt, J.T., Laposata, M., Olson, J.D., 2002. Laboratory diagnosis of dysfibrinogenaemia. Arch. Pathol. Lab. Med. 126, 499–505.

Denson, K.W.E., 1961. The specific assay of Prower Stuart factor and factor VII. Acta. Haematol. 25, 105–120.

Favaloro, E.J., Soltani, S., McDonald, J., Grezchnik, E., Easton, L., 2006. Activated protein C resistance: the influence of ABO-blood group, gender and age. Thromb. Res. 117, 665–670.

Fukushima, S., Tanaka, T., Sato, T., Shirakawa, Y., Asayama, K., Shirahata, A., 2002. Prothrombin levels in newborn infants by the carinactivase-1 method. Semin. Thromb. Hemost. 28, 539–544.

Funk, C., Gmur, J., Herold, R., Straub, P.W., 1971. Reptilase-R, a new reagent in blood coagulation. Br. J. Haematol. 21, 43–52.

Gempeler-Messina, Volz, K., Bühler, B., Müller, C., 2001. Protein C activators from snake venoms and their diagnostic use, in: Bon, C., Kini, M., Markland, F.S., Marsh, N.A., Rosing, J. (Eds.), International Conference on Exogenous Factors Affecting Thrombosis and Haemostasis. Haemostasis 31, S. Karger Medical and Scientific Publishers, pp. 266–272.

Gerads, I., Tans, G., Yukelson, L.Ya, Zwaal, R.F., Rosing, J., 1992. Activation of bovine factor V by an activator purified from the venom of *Naja naja oxiana*. Toxicon 30, 1065–1079.

Gowda, D.C., Jackson, C.M., Hensley, P., Davidson, E.A., 1994. Factor X-activating glycoprotein of Russell's viper venom. Polypeptide composition and characterization of the carbohydrate moieties. J. Biol. Chem. 269, 10644–10650.

Graff, J., Picard-Willems, B., Harder, S., 2007. Monitoring effects of direct FXa-inhibitors with a new one-step prothrombinase-induced clotting time (PiCT) assay: comparative in vitro investigation with heparin, enoxaparin, fondaparinux and DX 9065a. Int. J. Clin. Pharmacol. Ther. 45, 237–243.

Harder, S., Parisius, J., Picard-Willems, B., 2008. Monitoring direct FXa-inhibitors and fondaparinux by Prothrombinase-induced Clotting Time (PiCT): relation to FXa-activity and influence of assay modifications. Thromb Res. 123, 396–403.

Hardisty, R.M., Hutton, R.A., 1966. Platelet aggregation and the availability of platelet factor 3. Br. J. Haematol. 12, 764–776.

Harenberg, J., Giese, C., Hagedorn, A., Traeger, I., Fenyvesi, T., 2007. Determination of antithrombin-dependent factor Xa inhibitors by prothrombin-induced clotting time. Semin Thromb Hemost. 33, 503–507.

Harenberg, J., Jörg, I., Vukojevic, Y., Mikus, G., Weiss, C., 2008. Anticoagulant effects of idraparinux after termination of therapy for prevention of recurrent venous thromboembolism: observations from the van Gogh trials. Eur. J. Clin. Pharmacol. 64, 555–563.

Hoagland, L.E., Triplett, D.A., Peng, F., Barna, L., 1996. APC-resistance as measured by a textarin time assay—comparison to the APTT-based method. Thromb. Res. 83, 363–373.

Howie, P.W., Prentice, C.R.M., McNicol, G.P., 1973. A method of antithrombin estimation using plasma defibrinated with ancrod. Br. J. Haematol. 25, 101–110.

Hutton, R.A., Warrell, D.A., 1993. Action of snake venom components on the haemostatic system. Blood Rev. 7, 176–189.

Inoue, K., Morita, T., 1993. Identification of O-linked oligosaccharide chains in the activation peptides of blood coagulation factor X. Eur. J. Biochem. 218, 153–163. (Leipzig) 115, 483–487.

Jy, W.C., Horstman, L.L., Wang, F., Duncan, R.C., Ahn, Y.S., 1995. Platelet factor 3 in plasma fractions—its relation to micro-particle size and thromboses. Thromb. Res. 80, 471–482.

Kamiguti, A.S., 2005. Platelets as targets of snake venom metalloproteinases. Toxicon 45, 1041–1049

Kathiresan, S., Shiomura, J., Jang, I.K., 2002. Argatroban. J. Thromb. Thrombolysis 13, 41–47.

Kaufmann, R., Zieger, M., Tausch, S., Henklein, P., Nowak, G., 2000. Meizothrombin, an intermediate of prothrombin activation, stimulates human glioblastoma cells by interaction with PAR-1-type thrombin receptors. J. Neurosci. Res. 59, 643–648.

Keller, F.G., Ortel, T.L., Quinn-Allen, M.A., Kane, W.H., 1995. Thrombin-catalysed activation of recombinant human factor V. Biochemistry 34, 4118–4124.

Kini, R.M., 2005. The intriguing world of prothrombin activators from snake venom. Toxicon 45, 1133–1145.

Kini, R.M., 2006, Anticoagulant proteins from snake venoms: structure, function and mechanism. Biochem. J. 397, 377–387.

Kini, R.M., Rao, V.S., Joseph, J.S., 2001. Procoagulant proteins from snake venoms, in: Bon, C., Kini, M., Markland, F.S., Marsh, N.A., Rosing, J. (Eds.), International Conference on Exogenous Factors Affecting Thrombosis and Haemostasis. Haemostasis 31, S. Karger Medical and Scientific Publishers, pp. 218–224.

Kini, R.M., Joseph, J.S., Rao, V.S., 2002. Prothrombin activators from snake venoms, in: Menez, A. (Ed.), Perspectives in Molecular Toxinology. John Wiley & Sons Ltd., Chichester, pp. 341–355.

Kisiel, W., Canfield, W.M., 1981. Snake venom proteases that activate blood coagulation factor V, in: Lorand, L. (Ed.), Methods in Enzymology (vol. 80). Academic Press, New York, pp. 275–285.

Klein, J.D., Walker, F.D., 1986. Purification of a Protein C activator from the venom of the Southern copperhead snake (*Agkistrodon contortrix contortrix*). Biochemistry 25, 4175–4178.

Knight, L.C., Romano, J.E., 2005. Functional expression of bitistatin, a disintegrin with potential use in molecular imaging of thromboembolic disease. Protein Expr. Purif. 39, 307–319.

Koh, D.C., Armugam, A., Jeyaseelan, K., 2006. Snake venom components and their applications in biomedicine. Cell Mol. Life Sci. 63, 3030–3041.

Kornalik, F., 1985. The influence of snake venom enzymes on blood coagulation. Pharmacol. Ther. 29, 353–405.

Kornalik, F., 1990. Toxins affecting blood coagulation and fibrinolysis, in: Shier, W.T., Mebs, D. (Eds.), Handbook of Toxicology. Marcel Dekker Inc., New York, pp. 683–760.

Kornalik, F., Vorlova, Z., 1988. Ecarin test in diagnosis of dicoumarol therapy, liver diseases and DIC. Folia Haematol. (Leipzig) 115, 483–487.

Kumar, R., Beguin, S., Hemker, H.C., 1994. The influence of fibrinogen and fibrin on thrombin generation—evidence for feedback activation of the clotting system by clot bound thrombin. Thromb. Haemost. 72, 713–721.

Lange, U., Nowak, G., Bucha, E., 2003. Ecarin chromogenic assay—a new method for quantitative determination of direct thrombin inhibitors like hirudin. Pathophysiol. Haemost. Thromb. 33, 184–191.

Latallo, Z., Teisseyre, E., 1971. Evaluation of reptilase-R and thrombin clotting time in the presence of fibrinogen degradation products and heparin. Scand. J. Haematol. 1971; 261–266.

Lu, Q., Clemetson, J.M., Clemetson, K.J., 2005. Snake venoms and hemostasis. J. Thromb. Haemost. 3, 1791–1799.

Markland, F.S., Jr., 1997. Snake venoms. Drugs 54(Suppl 3), 1–10.

Markland, F.S., 1998. Snake venoms and the hemostatic system. Toxicon 36, 1749–1800.

Marsh, N.A., 1994. Snake venoms affecting the haemostatic mechanism—a consideration of their mechanisms, practical applications and biological significance. Blood Coagul. Fibrinolysis 5, 399–410.

Marsh, N., 2001. Diagnostic uses of snake venom, in: Bon, C., Kini, M., Markland, F.S., Marsh, N.A., Rosing, J. (Eds.), International Conference on Exogenous Factors Affecting Thrombosis and Haemostasis. Haemostasis 31, S. Karger Medical and Scientific Publishers, pp. 211–217.

Marsh, N.A., Fyffe, T.L., 1996. Practical applications of snake venom toxins in haemostasis. Boll. Soc. Ital., Biol. Sper. 72, 263–278.

Marsh, N., Williams, V., 2005. Practical applications of snake venom toxins in haemostasis. Toxicon 45, 1171–1181.

Matsui, T., Fujimura, Y., Titani, K., 2000. Snake venom proteases affecting hemostasis and thrombosis. Biochim. Biophys. Acta 1477, 146–156.

Matsui, T., Hamako, J., 2005. Structure and function of snake venom toxins interacting with human von Willebrand factor. Toxicon 45, 1075–1087.

Meier, J., Stocker, K., 1991. Snake venom protein C activators, in: Tu, A.T. (Ed.), Handbook of Natural Toxins. Marcel Dekker Inc., New York-Basel-Hong Kong, pp. 265–280.

Meier, J., 1998. Proteinases activating protein C, in: Bailey, G.S. (Ed.), Enzymes from Snake Venom. Alaken Inc., Fort Collins, CO, pp. 253–285.

Miura, S., Nishida, S., Makita, K., Sakurai, Y., Shimoyama, T., Sugimoto, M., Yoshioka, A., Ishii, K., Kito, M., Kobayashi, T., Fujimura, Y., 1996. Inhibition assay for the binding of biotinylated von Willebrand factor to platelet-bound microtiter wells in the presence of ristocetin or botrocetin. Anal., Biochem. 236, 215–220.

Moore, G.W., 2007. Combining Taipan snake venom time/Ecarin time screening with the mixing studies of conventional assays increases detection rates of lupus anticoagulants in orally anticoagulated patients. Thromb. J. 5, 12–16

Moore, G.W., Savidge, G.F., 2004. Heterogeneity of Russell's viper venom affects the sensitivity of the dilute Russell's viper venom time to lupus anticoagulants. Blood Coagul. Fibrinolysis 15, 279–82

Moore, G.W., Smith, M.P., Savidge, G.F., 2003. The Ecarin time is an improved confirmatory test for the Taipan snake venom time in warfarinised patients with lupus anticoagulants. Blood Coagul. Fibrinolysis 14, 307–312.

Morita, T., 1998. Proteases which activate factor X, in: Bailey, G.S. (Ed.), Enzymes from Snake Venom. Alaken Inc., Fort Collins, CO, pp. 179–208.

Moser, M., Ruef, J., Peter, K., Kohler, B., Gulba, D.C., Paterna, N., Nordt, T., Kubler, W., Bode, C., 2001. Ecarin clotting time but not aPTT correlates with PEG-hirudin plasma activity. J. Thromb. Thrombolysis 12, 165–169.

Nowak, G., 2003/2004. The ecarin clotting time, a universal method to quantify direct thrombin inhibitors. Pathophysiol. Haemost. Thromb. 33, 173–183.

Nowak, G., Bucha, E., 1996. Quantitative determination of hirudin in blood and body fluids. Semin. Thromb. Hemost. 22, 197–202.

Novoa, E., Seegers, W.H., 1980. Mechanisms of α-thrombin and β- thrombin-E formation: use of ecarin for isolation of meizothrombin 1. Thromb. Res. 18, 657–668.

Parmar, K., Lefkou, E., Doughty, H., Connor, P., Hunt, B.J., 2009. The utility of the Taipan snake venom assay in assessing lupus anticoagulant status in individuals receiving or not receiving an oral vitamin K antagonist. Blood Coagul. Fibrinolysis 20, 271–275.

St. Pierre, L., Masci, P.P., Filippovich, I., Sorokina, N., Marsh, N., Miller, D.J., Lavin, M.F., 2005. Comparative analysis of prothrombin activators from the venom of Australian elapids. Mol. Biol. Evol. 22, 1853–1864.

Pirkle, H., 1998. Thrombin-like enzymes from snake venoms: an updated inventory. Scientific and standardization committee's registry of exogenous hemostatic factors. Thromb. Haemost. 79, 675–683.

Quehenberger, P., Handler, S., Mannhalter, C., Pabinger-Fasching, I., Speiser, W., 2000. Evaluation of a highly specific functional test for the detection of factor V Leiden. Int. J. Clin. Lab. Res. 30, 113–117.

Quick, A.J., 1971. Thromboplastin generation: effect of the Bell-Alton reagent and Russell's viper venom on prothrombin consumption. Am. J. Clin. Pathol. 55, 555–560.

Rabinowitz, I., Tuley, E.A., Mancuso, D.J., Randi, A.M., Firkin, B.G., Howard, M., Sadler, J.E., 1992. Von Willebrand disease type B: a missense mutation selectively abolishes ristocetin-induced von Willebrand factor binding to platelet glycoprotein Ib. Proc. Natl. Acad. Sci. U.S.A. 89, 9846–9849.

Ranby, M., Norrman, B., Wallen, P., 1982. A sensitive assay for tissue plasminogen activator. Thromb. Res. 27, 743–749.

Read, M.S., Smith, S.V., Lamb, M.A., Brinkhous, K.M., 1989. Role of botrocetin in platelet agglutination: formation of an activated complex of botrocetin and von Willebrand factor. Blood 74, 1031–1035.

Robert, A., Eschwege, V., Hameg, H., Drouet, L., Aillaud, M.F., 1996. Anticoagulant response to Agkistrodon contortrix venom (ACVtest): a new global test to screen for defects in the anticoagulant protein C pathway. Thromb. Haemost. 75, 562–566.

Rooney, A.M., McNally, T., Mackie, I.J., Machin, S.J., 1994. The taipan snake venom time: a new test for lupus anticoagulant. J. Clin. Pathol. 47, 497–501.

Rosing, J., Govers-Riemslag, J.W.P., Yukelson, L., Tans, G., 2001. Factor V activation and inactivation by venom proteases, in: Bon, C., Kini, M., Markland, F.S., Marsh, N.A., Rosing, J. (Eds.), International Conference on Exogenous Factors Affecting Thrombosis and Haemostasis. Haemostasis 31, S. Karger Medical and Scientific Publishers, pp. 241–246.

Rosing, J., Tans, G., 1988. Meizothrombin, a major product of factor Xa-catalysed prothrombin activation. Thromb. Haemost. 60, 355–360.

Rosing, J., Tans, G., 1992. Structural and functional properties of snake venom prothrombin activators. Toxicon 30, 1515–1527.

Sakuragawa, N., Takahashi, K., Hoshiyama, M., Jimbo, C., Matsuoka, M., Onishi, Y., 1975. Significance of a prothrombin assay method using Echis carinatus venom for diagnostic information in disseminated intravascular coagulation syndrome. Thromb. Res. 7, 643–653.

Sarig, G., Aberbach, I., Schliamser, L., Blumenfeld, Z., Brenner, B., 2006. Evaluation of ProC Global assay in women with a history of venous thromboembolism on hormonal therapy. Thromb. Haemost. 96, 578–583.

Schöni, R. 2005. The use of snake venom-derived compounds for new functional diagnostic test kits in the field of haemostasis, in: Kini, R.M., Marsh, N., Markland, F.S., Broady, K.W. (Eds.), Proceedings of the Third International Conference, EFATH 2005. Pathophysiol. Haemost. Thromb. 34, S. Karger Medical and Scientific Publishers, pp. 234–240.

Sillero, P.L., Fernández de Velasco, J., Loscertales, J., Espinoza, J., Soto, C., Tomás, J.F., 2001. ProC global: an automated screening test for factor V Leiden and prothrombin mutation 20210 G to A detection. Thromb. Res. 101, 215–216.

Stevens, W.K., Cote, H.C.F., MacGillivray, R.T.A., Nesheim, M.E., 1996. Calcium ion modulation of meizothrombin autolysis at Arg55–Asp56 and catalytic activity. J. Biol. Chem. 271, 8062–8067.

Stocker, K., 1986. Use of snake venom proteins in the diagnosis and therapy of haemostatic disorders, in: Kornalik, K., Mebs, D. (Eds.), 7th European Symposium on Animal, Plant and Microbial Toxins. Prague, pp. 9–40.

Stocker, K., 1998. Research, diagnostic and medicinal uses of snake venom enzymes, in: Bailey, G.S. (Ed.), Enzymes from Snake Venom. Alaken Inc., Fort Collins, CO, pp. 705–736.

Stocker, K., Fischer, H., Meier, J., Brogli, M., Svendsen, L., 1986. Protein C activators in snake venoms. Behring Inst. Mitt. 79, 37–47.

Stocker, K., Fischer, H., Meier, J., Brogli, M., Svendsen, L., 1987. Characterization of the protein C activator Protac from the venom of the southern copperhead (*Agkistrodon contortrix*) snake. Toxicon 25, 239–252.

Stocker, K., 1990. Snake venom proteins affecting hemostasis and fibrinolysis, in: Stocker, K. (Ed.), Medical Use of Snake Venom Proteins. CRC Press, Inc., Boca Raton, pp. 97–160.

Suzuki, K., Nishioka, J., 1988. Plasma protein S activity measured using Protac, a snake venom derived activator of protein C. Thromb. Res. 49, 241–251.

Takeya, H., Nishida, S., Miyata, T., Kawada, S., Saisaka, Y., Morita, T., Iwanaga, S., 1992. Coagulation factor X activating enzyme from Russell's viper venom (RVV-X). A novel metalloproteinase with disintegrin (platelet aggregation inhibitor)-like and C-type lectin-like domains. J. Biol. Chem. 267, 14109–14117.

Tans, G., Rosing, J., 2001 Snake venom activators of factor X: an overview, in: Bon, C., Kini, M., Markland, F.S., Marsh, N.A., Rosing, J. (Eds.), International Conference on Exogenous Factors Affecting Thrombosis and Haemostasis. Haemostasis 31, S. Karger Medical and Scientific Publishers, pp. 225–233.

Thiagarajan, P., Pengo, V., Shapiro, S.S., 1986. The use of dilute Russell viper venom time for the diagnosis of lupus coagulants. Blood 68, 869–874.

Tokunaga, F., Nagasawa, K., Tamura, S., Miyata, T., Iwanaga, S., Kisiel, W., 1988. The factor V-activating enzyme (RVV-V) from Russell's viper venom. Identification of isoproteins RVV-V alpha, -V beta, and -V gamma and their complete amino acid sequences. J. Biol. Chem. 263, 17471–17481.

Torbet, J., 1987. Fibrin assembly after fibrinopeptide A release in model systems and human plasma studied with magnetic birefringence. Biochem. J. 15, 633–637.

Toulon, P., Halbmeyer, W.M., Hafner, G., Schmitt, Y., Randgard, B., Odpadlik, M., van den Eynden, C., Wagner, C., 2000. Screening for abnormalities of the protein C anticoagulant pathway using the ProC Global assay. Results of a European multicenter evaluation. Blood Coagul. Fibrinolysis 11, 447–454.

Triplett, D.A., Stocker, K., Unger, G.A., Barna, L.K., 1993. The textarin/ecarin ratio: a confirmatory test for lupus anticoagulants. Thromb. Haemost. 70, 925–931.

Van Cott, E.M., Smith, E.Y., Galanakis, D.K., 2002. Elevated fibrinogen in an acute phase reaction prolongs the reptilase time but typically not the thrombin time. Am. J. Clin. Path. 118, 263–268.

Weinger, R.S., Rudy, C., Moake, J.L., Olson, J.D., Cimo, P.L.,1980. Prothrombin Houston: a dysprothrombin identifiable by crossed immunoelectrofocussing and abnormal Echis carinatus venom activation. Blood 55, 811–815.

Wijeyewickrema, L., Berndt, M.C., Andrews, R.K., 2005. Snake venom probes of platelet adhesion receptors and their ligands. Toxicon 45, 1051–61.

Williams, S.B., McKeown, L.P., Krutzsch, H., Hansmann, K., Gralnick, H.R., 1994. Purification and characterization of human platelet von Willebrand factor. Br. J. Haematol. 88, 582–591.

Wilmer, M., Stocker, C., Bühler, B., Conell, B., Calatzis, A., 2004. Improved distinction of factor V wild-type and factor V Leiden using a novel prothrombin-based activated protein C resistance assay. Am. J. Clin. Pathol. 122, 836–842.

Yamada, D., Morita, T., 1997. Purification and characterization of a Ca2+-dependent prothrombin activator, multactivase, from the venom of *Echis multisquamatus*. J. Biochem. 122, 991–997.

Yamada, D., Morita, T., 1999. CA-1 method, a novel assay for quantification of normal prothrombin using a Ca^{2+}-dependent prothrombin activator, carinactivase-1. Thromb. Res. 94, 221–226.

Yamada, D., Sekiya, F., Morita, T., 1996. Isolation and characterization of carinactivase, a novel prothrombin activator in *Echis carinatus* venom with a unique catalytic mechanism. J. Biol. Chem. 271, 5200–5207.

Index

Note: The letters 'f' and 't' following locators refer to figures and tables respectively.

R.M. Kini et al. (eds.), *Toxins and Hemostasis,* DOI 10.1007/978-90-481-9295-3,

B

C

D

E

F

G